THE OXFORD BOOK OF INVERTEBRATES

THE OXFORD BOOK
OF INVERTEBRATES

PROTOZOA, SPONGES, COELENTERATES, WORMS, MOLLUSCS,
ECHINODERMS, AND ARTHROPODS (other than Insects)

Illustrations by
DEREK WHITELEY

Text by
DAVID NICHOLS
and
JOHN A. L. COOKE (ARTHROPODS)

OXFORD UNIVERSITY PRESS
1971

Oxford University Press, Ely House, London W.1

GLASGOW NEW YORK TORONTO MELBOURNE WELLINGTON
CAPE TOWN SALISBURY IBADAN NAIROBI DAR ES SALAAM LUSAKA
ADDIS ABABA BOMBAY CALCUTTA MADRAS KARACHI LAHORE
DACCA KUALA LUMPUR SINGAPORE HONG KONG TOKYO

Printed in Great Britain by Jesse Broad & Co. Ltd., Old Trafford, Manchester

Contents

		Page
INTRODUCTION	. . . .	vi
PROTOZOA and other tiny Invertebrates	. . . . . .	viii
SPONGES	. . . .	4
COELENTERATES		
Polyps	. . . .	6
Jellyfish	. . . .	12
Sea Anemones	. . . .	14
PLATYHELMINTHES		
Flatworms	. . . .	20
Flukes	. . . .	22
Tapeworms	. . . .	26
LOWLY WORMS	. . . .	28
MOLLUSCS		
Chitons and Tusk Shells	. . . . . .	32
Marine Limpets and Snails	. . . . .	34
Sea Slugs	. . . .	46
Land Snails	. . . .	50
Land Slugs	. . . .	62
Freshwater Snails	. . . . . .	64
Marine Bivalves	. . . .	72
Freshwater Bivalves	. . . . . .	90
Squids, Cuttlefish and Octopuses	. . . .	92
ANNELID WORMS	.	
Bristleworms	. . . .	94
Earthworms	. . . .	112
Leeches	. . . .	116
Aberrant Annelids	. . . . . .	120
ARTHROPODS		
Seashore	. . . .	122
Freshwater	. . . .	132
Parasitic and Commensal	. . . . .	140
Terrestrial	. . . .	148
HIGHER WORMS AND LAMP SHELLS	. . . .	168
BRYOZOANS	. . . .	172
ARROW WORMS AND LANCELETS	. . . . .	176
ECHINODERMS		
Sea-urchins	. . . .	178
Crinoids and Brittle-stars	. . . . .	180, 184
Starfishes	. . . .	180
Sea-cucumbers	. . . .	186
TUNICATES	. . . .	188
A CLASSIFICATION OF INVERTEBRATES	. . .	192
GLOSSARY	. . . .	208
SOURCES OF FURTHER INFORMATION	. . .	210
INDEX	. . . .	211

INTRODUCTION

The animal world is mostly invertebrate — lacking a backbone. Vertebrates, for all the honour and prominence accorded to them, represent a mere part of one group (phylum) of the animal kingdom, whereas invertebrates occupy about twenty-five phyla of widely different animals. The purpose of this book is to direct the attention of naturalists and students to the profusion of invertebrate wildlife that surrounds them.

Invertebrates exert a profound influence on Man. Their beneficial effects on modern living, such as their place in pollination, in the processes taking place in agricultural soil, in scavenging, in biological control of pests, and of course, as part of the complex food webs of higher animals, are of extreme importance to Man: and their detrimental effects, in parasitism, in disease carrying, as pests of food and so on, are of great consequence to us all.

Life is thought to have begun in the sea about two thousand million years ago. Just how the transition from non-living to living things occurred is not known, but it seems likely that among the earliest true organisms were virus- and bacterium-like forms which fed on organic substances and reproduced themselves by a process of splitting. From forms such as these there came simple plants which used the sun's energy to help build more complex substances from inorganic ones. Then there arose other simple organisms which fed on these plants to obtain nourishment. These were the first animals, with simple bodies composed of single units of living material. Although all subsequent animals have probably evolved from forms like these and have, for the most part, become very much more efficient in the way they tackle vital processes, there remain to this day representatives of these simple animals, the PROTOZOA (pp. viii to 3), which play an important role in the economy of aquatic habitats, in providing the basic food material on which many higher animals ultimately depend for their existence.

One could fill many volumes on the Protozoa alone; in this book a mere handful of the larger forms are included from the two major aquatic habitats, the sea and freshwater. Like all the other animals included in this book, they are forms which can be seen either with the naked eye or with a handlens.

There are many examples of Protozoa which form colonies, such as *Volvox* on p. 2 (really a plant, if judged by its way of feeding); and it is thought that some of these formations might resemble the way in which cells first aggregated together to form higher, many-celled animals. One such excursion into a many-celled condition is seen in the SPONGES (pp. 4 – 7), but this is not altogether a successful line, principally because there is no nervous system integrating the activities of the various parts of the body.

Better attempts are seen in the COELENTERATES (pp. 6 – 19) and FLATWORMS (pp. 20 – 27); both of these have been able to evolve a nervous system which, although primitive, is at least a good deal better than anything the sponges achieved. Also, they both show the beginnings of an alimentary tract or gut for taking in food material, breaking it down to simpler substances and absorbing it for use by the body. But their guts, like their nervous systems, are primitive, being provided with a single opening only, for use both as mouth and as anus. One very important feature of flatworms is that advanced forms have taken to parasitism — living on or in the tissues of other organisms to extract sufficient food for themselves without inflicting too much damage on the host. Many other invertebrate groups have taken to parasitism, or to various other forms of association.

Slightly more advanced worms called ribbon worms, or NEMERTINES (pp. 28 – 29), show two further steps forward: first, a one-way gut, with anus as well as mouth; and, secondly, the beginnings of a blood system for carrying essential substances around the body. These advances really paved the way for the evolution of higher animals, and marked an enormous step forward in the history of invertebrate animals.

One evolutionary line led to the true worms, or ANNELIDS (pp. 94 – 121), in which the body is longitudinally subdivided or segmented, and in which the full potential of the worm-shape is realized for burrowing, boring, swimming, preying and tube-living. These have much more sophisticated nervous and blood systems than the lowlier worms, and have been able to centralize certain organs in one place in the body rather than repeat the organ many times in easy reach of all parts of the body.

Of course, there are many excursions from this central evolutionary theme in the early invertebrates, and the sequence of events giving rise to these (and indeed the interrelationships of almost all invertebrate groups) is far from clear. One branch which went on to produce a major invertebrate group is the MOLLUSCS (pp. 32 – 93), most of them having a heavy shell and slow gait that do not immediately give an impression of success. But the molluscs have infiltrated almost every possible habitat, have assumed a wide variety of body forms and, because of their numbers and their hefty shells, are among the most obvious of the world's invertebrates. This is why they make up nearly a third of this book.

The great group of the annelid worms almost certainly gave rise to another, even greater, group, the ARTHROPODS (pp. 122 – 167), a part of which has already formed a separate volume in this series, the *Oxford Book of Insects*. The other groups of the arthropods, including the crustaceans, myriapods and spiders, are dealt with in this volume. This is a phylum which has adopted a hard exoskeleton for both protection and movement with such success that it has infiltrated not only all major aquatic and terrestrial habitats but aerial ones as well.

At the top of the invertebrates is a strange assortment of animals which do not immediately suggest an advanced mode. The main ones are the ECHINODERMS (pp. 178 – 187) and the PROTOCHORDATES (pp. 176 – 177 and 188 – 191), and these, together with a number of minor groups (pp. 168 – 177), are apparently all quite closely related, and most assume a rather passive mode of life on the sea-floor. But within this group lies a hidden potential: somewhere among these unlikely invertebrates lay the group which eventually gave rise to the chordates, that great phylum which includes all the vertebrates and Man himself. The Protochordates betray their closeness to the chordate line because at some time in their life they possess a rod of turgid cells, the *notochord*, as a support for the body, and this rod acts in the same way as, and is a precursor to, the spinal column of the vertebrates.

So invertebrates range from tiny protozoans to protochordates. At several points in their long evolutionary history they have reached remarkable heights of achievement: the social organisation seen among the arthropods, the extraordinary adaptations of the cephalopod molluscs which so closely parallel those of some vertebrates, and the adoption of a notochord by the protochordates are particularly momentous milestones of invertebrate success. But above all, so many of them display not only a fascinating beauty but also remarkable adaptations in their way of life; the figures in this book serve to demonstrate the one, and the written account tells something of the other.

In the pages that follow, the invertebrates are surveyed phylum-by-phylum; but within this framework representative genera are selected from the main habitats in which each group is found, so that animals illustrated on any one plate have a similar background. Obviously the selection has omitted some important animals, but the aim has been to show examples of all major groups within each phylum that are likely to be encountered in their natural habitats in the United Kingdom, or in zoos and aquaria, or those that have a special 'zoological' reason for being included, such as an unusual body form or a curious habitat.

Latin names have been used throughout, because in so few cases do invertebrates have common English names; where there are common names, they are mentioned too.

Animals drawn on the same plate have been drawn to the same scale except where great differences in size have made this impossible. In such cases the larger animals have been placed in the background so that they appear at a reduced scale compared with those in the foreground; and microscopic creatures, on plates where larger animals also appear, have been shown as insets. On some plates a familiar animal has been drawn to indicate scale. In general, the size or range of size of each animal is indicated in the text.

TINY INVERTEBRATES OF FRESHWATER
VISIBLE WITH A LENS

All the tiny animals at the lower end of the animal kingdom require the nurture of an aquatic environment. The most primitive are the Protozoa (pp. 1–3), in which the body is not divided into component cells. Since the division into plants and animals probably occurred within this phylum, some examples do not readily betray the kingdom to which they belong. Members of the Protozoa move by three main methods: by the beating of a few long processes from the body called *flagella;* or by the integrated beat of many smaller processes called *cilia;* or they may extrude protoplasmic processes into which the rest of the body flows. Some tiny protozoans form large colonies (1), visible to the naked eye; some have a permanent stalk (2), while others make a temporary stem but can swim away if necessary (5); some secrete a skeleton for support or protection (6 and 8), while others build a 'house' out of sand (9). A few of the larger protozoans are illustrated on this page, together with examples of two other phyla of tiny animals (4 and 7).

1 **Volvox.** This is an example of a colony in which several hundred individual plant-like protozoans, each with two whip-like flagella, form the wall of a jelly-filled sphere. The beat of the flagella is synchronized, and there is a definite anterior pole directed forwards, with slightly longer flagella which are thought to test the environment. The colony reproduces by budding off special individuals from its wall which come to lie in the jelly-filled interior; here they multiply by dividing down the middle, and form spherical colonies within the parent. They are later released into the water. Sometimes, even the young within the parent have their own young. This organism conforms to the definition of a plant, in that its individuals have a cellulose cell wall and can undergo photosynthesis.

2 **Vorticella** is usually found in groups of several individuals, all adhering to a piece of weed or an animal by a thin stem which has a single muscle fibril within it; the colony appears to twitch as its individuals contract one after the other. Every so often, an individual breaks free from its stalk and swims off to settle elsewhere and grow a new stalk, divide to form new individuals and found a new colony.

3 **Spirostomum** is covered by tiny vibratile cilia, similar to, but smaller than, flagella (1). The beat of all the body's cilia is co-ordinated so that the organism moves steadily through the water in one direction, but with the ability to reverse. An oddly-shaped oral groove on one side, where food is taken in, makes the animal tend to spiral on its longitudinal axis as it moves.

4 **A rotifer.** These metazoan (many-celled) animals may be confused with protozoans such as *Stentor* (5), but even with a handlens the feeding jaws may be seen grinding the food that has been collected by the ciliary crown round the top. The animal may attach temporarily by means of an adhesive organ on its stalk.

5 **Stentor** is covered with cilia, most conspicuous round the mouth. The oral cilia form a spiral feeding funnel which draws a current of water, carrying food, into the gullet. The animal attaches to something hard by an adhesive holdfast, but every so often it detaches, foreshortens, and swims off to a new site. The nuclear apparatus, where the body's activities are co-ordinated, resembles a sausage-string suspended in the body. *Stentor* is most often found coloured green because of symbiotic algae within its protoplasm where they gain protection from the host and give some food and oxygen in exchange.

6 **Actinosphaerium,** one of the wonders of minute pond life, is a 'sun-animal', or heliozoan, in which the protoplasmic processes are rigid and needle-like.

7 **A gastrotrich.** This is another example of a phylum of tiny multi-cellular animals, not much bigger than many Protozoa, that can sometimes be seen crawling over vegetation.

8 **Arcella.** One of the most common freshwater amoebae, this animal builds a dome-shaped shell of protein with a hole in the middle of the flattened base through which the protoplasmic processes emerge.

9 **Difflugia** builds a protective 'house' out of sand grains carefully selected for size and shape, from the bottom of the pond in which it lives. The grains are stuck together with a secreted matrix, and the protoplasmic processes emerge through an opening in one end.

10 **Amoeba.** This is the much-famed and seemingly simple protozoan which actually has a highly complex organization. Protoplasmic processes flow out from the main mass of the body in one or several directions and the rest of the body flows into them; the protoplasm also flows round the prey, and encloses it in a tiny vacuole (10A). Amoeboid protozoans, either *Amoeba* itself or a relative, can often be seen against a dark background in pond-water.

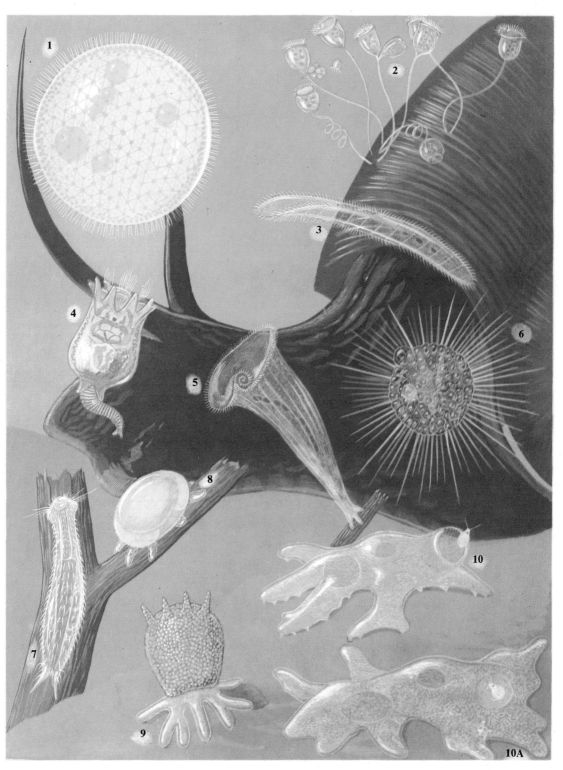

1 Volvox globator, a colonial flagellate   2 Vorticella, a ciliate (on Ramshorn Snail)
3 Spirostomum, a large ciliate
4 Brachionus, a rotifer   5 Stentor, a ciliate   6 Actinosphaerium, a heliozoan
7 Chaetonotus, a gastrotrich   8 Arcella, a shelled rhizopod   10 Amoeba, feeding
9 Difflugia, a rhizopod with sand house   10a Amoeba, after ingesting prey

INVERTEBRATES OF THE SEA
VISIBLE WITH A HANDLENS

By far the majority of marine invertebrates are not the large, conspicuous creatures of the shore, nor the floating animals seen from a ship, nor those in an aquarium, nor again the many forms that live on or in the sea bottom: they are the tiny single-celled creatures, the Protozoa, that float and drift with the plankton in the upper levels of the world's oceans. These minute creatures feed on the even smaller single-celled plants which get their nourishment, as do land plants, with the help of the sun's energy. Most of the protozoans, as in freshwater (p. 1), are too small to be seen except with a microscope, but a few, illustrated here, can be picked out with a handlens.

1-2 The Dinoflagellates. This protozoan group shows very well how difficult it is to decide whether some of these lowly organisms are plant or animal. Some of them, such as *Ceratium* (2), have typical plant photosynthetic pigments and feed in a truly plant-like way, while others, such as *Noctiluca* (1) feed, animal-like, on other organisms. The body of *Noctiluca* is filled with a greenish gelatinous substance across which strands of protoplasm radiate from the central mass containing a nucleus. The jelly apparently makes the animal lighter than water, so it floats passively in the upper layers. Near the central mass is a mouth alongside which is a curious tentacle-like structure, said to be used by the animal to help catch its prey and thrust it into the mouth. As its name suggests, it is markedly luminescent, and large numbers of this organism, when stirred up by a passing boat or as the waves strike an obstacle in the water, can cause a spectacular display as though the very sea is on fire. *Ceratium* is a curiously shaped dinoflagellate that has plant pigments in its protoplasm so that it can make use of the sun's energy for feeding. The processes which project from the body are almost certainly devices to increase its powers of suspension in the water, and it has been observed that the organism can change the length of its spines according to the viscosity of the water in which it finds itself: warm water is less viscous than colder water, and individuals of a species in warm water will have longer spines than individuals of the same species in colder water. There are two flagella for locomotion, one projecting from the body and the second lying in a groove encircling the body.

3 The Radiolaria. This protozoan order contains some of the most beautiful of all sea creatures, yet few people ever see the exquisite architecture of their shells because of their small size. Some may reach a few millimetres, but in the main they are seen only with a handlens, and, in the living state, only by chance from a net haul. They were favourite objects of Victorian microscopists and slide-makers. The body is usually spherical with a sculptured skeleton of silica and a series of radiating silica rods emerging from the soft body.

4-6 The Foraminifera belong to the same class, Rhizopoda, as the amoebae, but here the main part of the body is protected by a shell, usually made of chalk. From holes in this shell (giving the group its name) the protoplasmic processes emerge, and in the outside water they branch and interconnect to form an effective trap for food. Both this group and the radiolarians (3) are very important contributors to the sediments on the ocean floor: vast areas may have an immense thickness of silt composed entirely of the dead shells of protozoans (*see* 5A). Foraminiferans have been an important constituent of past seas too, and in ancient sediments that are now rock these organisms can be used for dating and zoning the rock, an important aspect of exploration for oil and other deposits. Not all foraminiferans are tiny: some reach the size of a tenpenny piece, though our British forms are not so impressive. *Polystomella* (4) forms a flat spiral of chambers, each whorl of which overlaps the previous whorl, and therefore almost hides it. The first whorl, at the centre, occurs in two sizes in the same species; this represents an alternation of generations in the life history, the megaspheric form producing gametes which fuse to produce a zygote which develops into the microspheric form. This then breaks up into many tiny amoebulae, each of which can grow into a new megaspheric form. *Globigerina* (5) is one of the few planktonic Foraminifera. Some foraminiferans live within a single chamber, but more often the animal secretes a series of bigger and bigger chambers as it grows. In this case, the end result is a symmetrical spiral of chambers, each occupied by part of the animal. Oil droplets, which are lighter than water, are thought to aid its buoyancy. The shells of this form (5A) may grow to about the size of a pin's head. In the case of *Nodosaria* (6), which is usually rather bigger, the succession of chambers is arranged in a straight line, the mouths of each chamber opening into the chambers to either side.

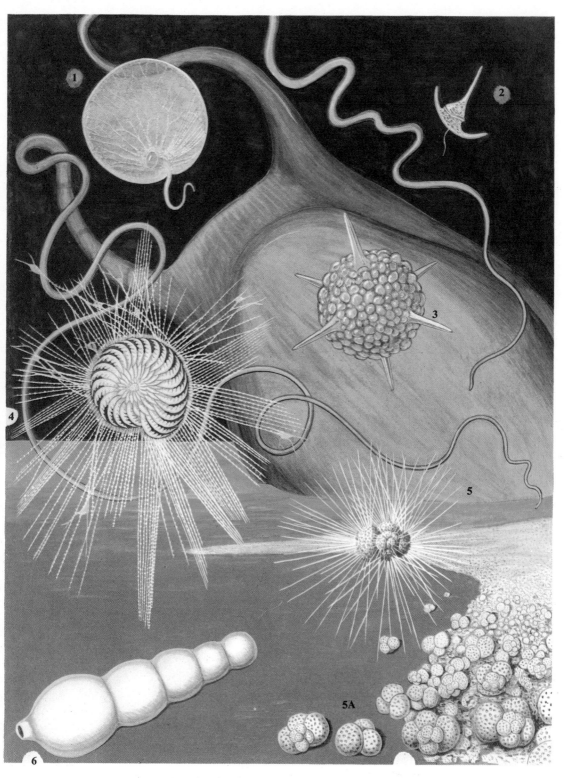

1 Noctiluca, a dinoflagellate 2 Ceratium tripos, a dinoflagellate
3 Hexacontium, a radiolarian
4 Polystomella, a foraminiferan 5 Globigerina, a foraminiferan
6 Nodosaria, a foraminiferan 5a Globigerina, empty shells as ooze

SPONGES OF THE SEA

This group of very primitive metazoan (multicellular) animals is almost exclusively marine, though there is a single family of freshwater sponges (p. 7). The sponges almost certainly represent an evolution of the multicellular condition separate from all other lines, and it is probable that no further groups of animals evolved from them — in other words, that they are a dead-end, from an evolutionary point of view. Basically, a sponge's body consists of a cup-shaped aggregation of various types of cell which all adhere to one another in a set pattern, but between which there is no nervous co-ordination — unlike all other metazoans. The most important cells are those lining the cup, which produce water currents by means of flagella. Their combined activity causes water to flow out of the top of the cup, and displacement is made good by water passing into the cup through pores in its side. This can often be shown in a living sponge in a dish, if some ink is added to the water in the region of the sponge with a pipette. In very simple sponges, like *Leucosolenia* (6) the exhalant pore is at the top of the cup, while the inhalant pores merely pierce the side walls; in higher sponges (1–5, and 7) the wall becomes more complex by folding and multiplying the canal system. Sponges are supported by a skeleton of either calcareous spicules, siliceous spicules, or protein fibres. It is the latter skeletal system from some Mexican, Caribbean or Mediterranean sponges that is used for the familiar bath sponge. The calcareous or siliceous spicules of sponges can easily be isolated by leaving a small fragment of sponge overnight in domestic bleach which dissolves away the soft tissues.

1 **Sycon.** Though this sponge and the next both look superficially like a simple sponge, such as *Leucosolenia* (6), and indeed all three are usually called 'purse sponges', the structure of this and *Grantia* is more complex than that of *Leucosolenia* in that its walls are 'folded' to increase the canal systems, and in consequence the inhalant pores do not pass directly through the walls to the central cavity. In this sponge, the supporting spicules often protrude beyond the soft tissues round the exhalant opening, forming a jagged fringe. It occurs in places on the lower shore where it will not be exposed to too much wave action; it is widely distributed.

2 **Grantia.** Whereas in *Sycon* the complication of the capsule wall can be regarded as a mere folding, in *Grantia* it is as though the folded wall has become covered by an extra skin or cortex, so that the surface is smoother. The shape of the capsule varies, though it is usually borne on a short stalk, and the wider part is usually rather flat. This is one of the commonest of the so-called 'purse-sponges', and is found clinging to rock and weed in crevices and pools.

3 **Halichondria** (Bread-crumb Sponge). Its English name aptly describes the appearance, at least of the white colour variety, though it may occur, like most other sponges, in a range of colours, particularly green. Different colour varieties may be found on the same rock, and may even touch one another. The large round exhalant pores, borne on conical projections, give the appearance of a number of tiny volcanoes. The inhalant pores are often too small to be seen clearly with the naked eye.

4 **Hymeniacidon.** The surface of this sponge has a puckered appearance, and the exhalant pores are small and randomly arranged. Very often it will be found with a small, flat, almost oval polychaete worm, *Spinther*, browsing on it, the worm accurately matching the sponge in colour.

5 **Suberites** may encrust stones, unoccupied shells, or snail shells with hermit crabs within. The surface may be thrown into folds, and it may not be easy to distinguish the exhalant from the inhalant pores, or even to see either of them.

6 **Leucosolenia.** A simple sponge. Though it usually grows in branching colonies, each branch is a mere perforated cup with an exhalant pore at its tip. Several species are common: one encrusts the stems of fucoid sea-weeds; several have anastomosing tubes attached to stones and rocks; and one, *L. complicata* (illustrated), has tall, branching tubes forming a compact colony. The form of these colonies and the way they branch may be considerably influenced by the currents impinging on them while they are growing.

7 **Cliona.** This sponge bores into limestone rocks and shells so that only a small lobule of the sponge is visible on the outside, projecting about 2 mm from the hole it has made. The projecting part bears either a collection of inhalant pores, or a single exhalant one. The boring process is not well understood, but it appears that an acid secreted by the growing sponge may help in the process. It can make quite a mess of the outer layers of a thick shell, such as that of an oyster.

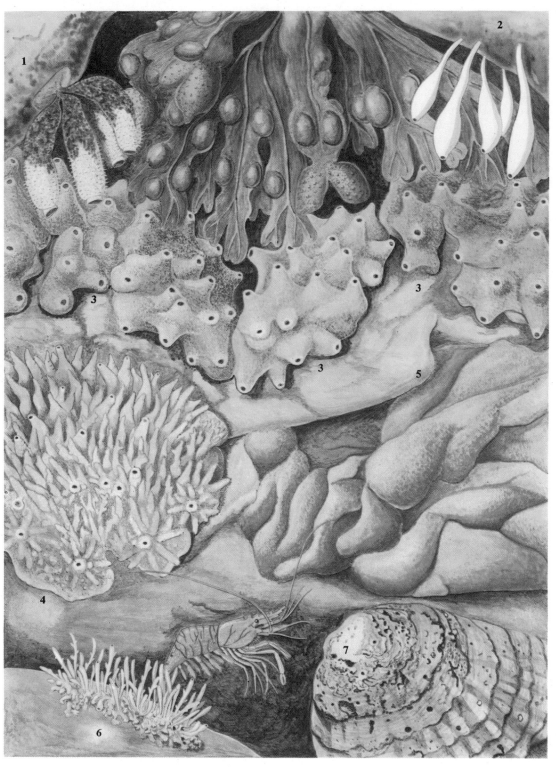

1 SYCON CORONATUM, a purse sponge   2 GRANTIA COMPRESSA, a purse sponge
3 HALICHONDRIA PANICEA, encrusting, various colours
4 HYMENIACIDON, encrusting   5 SUBERITES, encrusting
6 LEUCOSOLENIA, a purse sponge   7 CLIONA, in holes it has bored in oyster shell

FRESHWATER HYDROIDS, MEDUSAE, AND SPONGES

The Coelenterata represent another evolutionary line of muticellular animals which probably arose from the Protozoa. Unlike sponges, the coelenterates have a body which is integrated by a nervous system, so that considerable change in body shape is possible, and more complex activities can be achieved. In common with most of the lower invertebrate phyla, the main radiation of coelenterates has taken place in the sea; but in freshwater lakes, ponds, and ditches one finds a few representatives of one coelenterate group, the Hydrozoa (1–5). The coelenterate polyp consists of a bag-like body, attached to something solid at its bottom end, with a single opening, the mouth, at the top. Surrounding the mouth is a ring of tentacles, heavily charged with stinging cells, to catch the prey (4) and transfer it to the mouth.

Most sponges live in the sea (p. 4), but one family contains members which live in clear freshwater. None of these show the simple sponge grade: all have complex wall and canal structure, and the external appearance is therefore of a large number of small inhalant pores and fewer exhalant ones. Most of them are encrusting forms. They differ from marine forms in having overwintering reproductive stages called *gemmules* (7). In autumn, some of the sponge's cells become aggregated and surrounded by a hard covering, complete with spicules, which is highly resistant. With the onset of winter, the sponge itself disintegrates, but the gemmules carry sufficient component cells through to the spring, to aggregate into the beginnings of a new sponge.

1-2 Craspedacusta and **Microhydra.** Although the great majority of hydrozoan coelenterates that have colonized freshwater are hydroids, one rare limnomedusan has a dominant medusoid phase, called *Craspedacusta* (1), about 10 mm in diameter; its tiny polyp is called *Microhydra* (2), but for many years after its discovery the relationship was not realized. They occur, but rarely, in canals and rivers in the South-west.

3 Hydra. If pond weed is placed in a jar of clear water, one may later find the *Hydra* polyp hanging from the weed, its tentacles moving gently. When a water-flea or other small animal touches a tentacle, the stinging cells are immediately fired into the body of the prey to hold and paralyse it, while the tentacles slowly contract to pull it towards the mouth. It is digested within the gut cavity, and the skin voided through the mouth. *Hydra* may move from place to place by 'cartwheeling' on to its tentacles then back on to its base (3A). In conditions of food storage, the *Hydra* will produce gonads, either male or female or both, as dome-shaped swellings of the body wall, the ovary generally lower down. The germ cells are liberated into the water, where fertilization takes place, and the fertilized egg develops into a ciliated larva which either settles to produce a new *Hydra* immediately, or surrounds itself with a thick protective covering to withstand the unfavourable conditions which caused the *Hydra* to undergo sexual reproduction.

4 Chlorohydra differs from *Hydra* by harbouring in its cells symbiotic algae which are taken into the primitive gut with the food, absorbed in the living state, and retained in the cells to utilize the waste and provide oxygen as a by-product of the photosynthetic process.

In both this and *Hydra*, when conditions are good, reproduction takes place by budding a new individual from its side, as shown in the upper of the three individuals here; sometimes the bud has a bud itself.

5 Cordylophora is colonial. The first individual to settle buds as normal, but the bud fails to separate from the parent, so that after some time a cluster of individuals is formed. But the polyps differ from *Hydra* in two ways: first, the tentacles do not all arise from round the mouth; and, secondly, each polyp secretes a thin tube of hard brown material round itself as an aid to support.

6 Euspongilla (Pond Sponge) forms greenish encrustations on stones, sticks or weed in still water. The green colour is due to symbiotic algae living within some of the sponge's cells, and because the algae are dependent on light, any of the sponge encrustation which is hidden from light is less green. Usually, the Pond Sponge has finger-like projections from the ends of which some of the supporting spicules can be seen protruding.

7 Ephydatia (River Sponge) is generally less green than the Pond Sponge, and seldom has such long projections. It is generally found on the underside of stones. Both this and *Euspongilla* produce masses of gemmules for overwintering, visible in late autumn on stones, etc., as tiny brown spots (shown on right). Most sponges harbour tiny animals in their canal systems. The Spongilla Fly Larva, *Sisyra* (7A) closely matches the River Sponge, its host, in colour. After this larva has left the water, it spins a silk cocoon, then emerges as a lacewing (see *Oxford Book of Insects*, p. 36, 5).

1 CRASPEDACUSTA, a freshwater medusa   3 HYDRA OLIGACTIS
2 'MICROHYDRA' budding its medusoid form *Craspedacusta*
6 EUSPONGILLA LACUSTRIS   5 CORDYLOPHORA LACUSTRIS   3A H. OLIGACTIS, somersaulting
4 CHLOROHYDRA VIRIDISSIMA   7 EPHYDATIA FLUVIATILIS

7A E. FLUVIATILIS, enlarged, with larva of sponge-fly, *Sisyra*

MARINE POLYPS

Most of the radiation of coelenterates has taken place in the sea, where these exquisite polyps are found. They are from the class Hydrozoa, which mostly pass through a polyp and a medusoid phase. The polyp phase is easier to find because it is sedentary and usually shore-dwelling; the medusae, miniature jellyfish, swim away from the parent organism to become sexually mature while floating in the upper levels of the sea. In the polyps the mouth faces upwards, while in the medusae it faces downwards. These compound hydrozoans start their colonies by producing buds like *Hydra* (4, p. 7); but the hydrozoan buds fail to separate, and so a branching system, each with a terminal polyp, is produced. The colony is usually rooted to some hard object by a creeping holdfast. The medusae are budded off in a special region of the colony.

1 **Coryne.** The tentacles have knobbed ends containing stinging cells, and the animal commonly holds them curved inwards. The reproductive capsules are developed irregularly among the tentacles, and when they are ripe, up to 25 eggs may be liberated from a single female capsule in about a minute. It is not a very common animal, but it is usually found in rock pools. The colony might be up to 100 mm high.

2 **Clava.** The club-shaped stems arise from a dense mass of spreading tubes, usually attached to brown seaweeds such as *Ascophyllum* or *Fucus*. The tentacles, paler than the body, are scattered on the body of the polyp, with clusters of sexual capsules below them, forming a collar, though these capsules are too small to be seen in the figure.

3 **Dynamena,** probably the commonest British hydrozoan, is found mainly on fucoid sea-weeds. The stems, up to 40 mm high, are crowded on to creeping stolons, and little transparent cups, containing the polyps, arise from the main stems, opposing in pairs so that the stem appears serrated. The eggs and sperm are released into the sea from tiny 'medusae' which remain attached to the parent colony, and do not have a separate existence.

4 **Sertularia** (Sea Hair) is a pale yellow colony, often found in dense tufts on *Fucus* or *Laminaria* sea-weed. It is more abundant offshore, but plenty is washed in, to be found along the strand-line at ebb-tide. The rim of the cup surrounding each polyp is produced into a little point laterally, with a small tooth-like projection on each side of it. Here too, the medusae remain attached, and are scattered over the stems, surrounded by little rigid cups which bear tiny lids for protection. This is the 'Sea Fern' or 'White Weed' of commerce which is collected from the shores of the North Sea, and dried and stained for use in decorations.
Shown browsing on the polyps of this form is a common predator, the nudibranch *Doto coronata*.

5 **Plumularia.** This is found on the sides of rock pools, on pier piles and other hard objects in shallow water, and on weed. The delicate fronds have alternating branches, with the polyps arranged on one side of each branch. Between adjacent hydranths there may be small finger-like polyps bearing stinging cells. The medusae are produced in special vase-like structures in the axils of some of the branches.

6 **Tubularia,** richly-coloured, with long, drooping polyps, is common locally and found at low tide under rocky overhangs and in rock pools. It has two circlets of thread-like tentacles, a small set round the mouth and a larger, drooping set lower down, between which a circlet of medusa-producing structures arises. The medusae remain attached to the parent and the genital products are shed from them.

7 **Bougainvillea** has a tree-like appearance, with a thick main stem and luxuriant, irregular branches. It produces free-living medusae, one at a time in special regions of the colony (illustrated).

8 **Obelia** is the famous coelenterate of elementary biology courses, since it is fairly common and shows the typical alternation between hydroid and medusoid phases. The stems arise from a branched, creeping holdfast. They are branched alternately on either side, and have hoop-like rings to provide flexibility. The medusae escape through a hole in the medusa-producing polyp. When free, the medusae develop gonads which liberate eggs or sperm, and fertilization takes place in the sea, producing a ciliated larva, the *planula*. After free life for a short time, the planula settles on a suitable substratum and develops into a polyp. Subsequently, the holdfast grows and spreads and the stem of the polyp elongates, budding off new polyps in the typical zig-zag fashion of the mature colony.

9 **Myriothela** is a solitary zooid, with two groups of tentacles. Below the mouth is a region of small tentacles with reddish tips charged with stinging cells; at the base is a region of branching tentacles bearing the sexual capsules. It is found attached by a holdfast to stones at low tide.

1 Coryne muscoides
2 Clava squamata   3 Dynamena pumila
4 Sertularia Operculata, with the sea-slug *Doto coronata*   5 Plumularia setacea
6 Tubularia larynx   7 Bougainvillea ramosa   8 Obelia geniculata   9 Myriothela cocksi

MEN-O'-WAR AND OTHER COLONIAL FORMS

Colonial coelenterates are not necessarily rooted to one spot, as in the hydrozoan polyps (p. 9): on this page a selection of an important group of *pelagic* colonial hydrozoans is illustrated. These include the notorious Portuguese Man-o'-war *Physalia* (1) whose sting can be most unpleasant. The siphonophores exhibit the greatest variety of polyp structure seen in the phylum. The polyp and medusoid forms may be modified to form floats, swimming bells, feeding or reproductive structures, or defensive organs heavily charged with stinging cells. In addition, there may be long, trailing tentacles, some with branches, dangling below the rest of the colony like a fisherman's paternoster, also heavily charged with stings. Besides the surface-dwelling forms like *Physalia*, there are mid-water forms such as *Muggiaea* (3) which have one or more jet-propulsion units at the head of the colony and tentacles trailing below; while siphonophores have both polyps and medusae in their colonies, chondrophores (2) have only modified polyps, though otherwise the two groups are very similar.

When a fish or other prey has been caught and paralysed by a tentacle, it is drawn towards a group of feeding polyps by contraction of the tentacle so that the mouths of the polyps are pressed against the prey. Digestive juices are exuded on to it and the products of digestion, taken into the feeding polyps, are shared with the rest of the colony, since the cavities of all the various forms and tentacles are confluent.

Siphonophore and chondrophore colonies are formed through a process of budding. The original polyp forms a float which develops budding zones from which the various members of the colony arise. In the case of the reproductive polyps of siphonophores, medusoid individuals may form on them, but, like some of the polyps shown on p. 9, these do not separate from the parent colony but develop gonads while still attached.

1 **Physalia** (Portuguese Man-o'-war). Though subtropical, this siphonophore strays into the waters of the Gulf Stream fairly frequently and may be seen in British coastal waters or cast up on the beaches. The float, looking like an inflated plastic bag, may be coloured in delicate shades of pink or blue. From the float hang the polyps and tentacles, sometimes extending to 2 metres below the float. A fish which wanders into the deadly mass of tentacles and polyps is almost immediately stunned by the stinging cells and hauled up towards the feeding polyps. A good-sized *Physalia* can catch a full-grown mackerel, so effective is its trap. The sting can be most unpleasant to bathers, as one of the authors has found; he relates that it 'feels like being stuck with red-hot pins — a very nasty sensation indeed'. It is rarely fatal, but anybody suffering an encounter would be well advised to seek medical help as soon as possible. It has been reported that the animal's float sometimes flops over to one side, then to the other, to keep the float moist. It has also been said that the buoyancy in the float can be altered to such an extent that the animal can sink below the surface from time to time.

2 **Velella** (Jack Sail-by-the-wind). Common in warm seas, this chondrophore is only an occasional visitor to Britain, where shoals may be driven ashore if the wind is in that direction. The colony has a flat, oval float with an erect sail projecting vertically, at an angle to the axis of the body, so that the animal can take best advantage of the wind to drive it along. It is said that the progeny of any one parent shows a polymorphism

in this feature: a proportion will bear the sail from left to right, while the others have it from right to left. This means that the young forms will be driven in different directions away from the parent, and so the risks of being driven ashore are reduced. The same is true of *Physalia* (1). Beneath the disk is a single large feeding polyp, surrounded by many reproductive polyps and a fringe of stinging polyps. In this case the reproductive polyps produce medusae which do break free, to produce gonads and then gametes as an alternate generation in the life cycle.

Sometimes swarms of these animals are encountered with millions of individuals. One expedition reports that it sailed through a swarm 260 kilometres long; and occasionally they are cast up on the beach in mounds half a metre high.

3 **Muggiaea** is a pelagic form with a swimming bell at its head. From the lower part of the bell a stem emerges, and at intervals this has feeding and reproductive polyps and tentacles. It swims by jet-propulsion, and is said to be able to escape by withdrawing its stem towards the swimming bell, turning over and shooting off downwards.

4 **Physophora.** This is a short-stemmed siphonophore with several swimming bells and a single float at the apex. These animals may swim quite rapidly, not always upwards but sometimes horizontally, their tentacles trailing behind like streamers from a coach-outing.

1 Physalia physalis, with captured fish   2 Velella velella
3 Muggiaea atlantica
4 Physophora hydrostatica

JELLYFISH AND SESSILE MEDUSAE

Jellyfish belong to the coelenterate class Scyphozoa, in which the medusoid phase is predominant. In some jellyfish a sessile polyp phase is passed through briefly as a larval stage (1A). The majority live in coastal waters — a notorious nuisance to bathers, because of the stinging cells on their tentacles. Some are very large — *Cyanea* (3) may be 3600 mm diameter, and its 30-metre tentacles may well be the basis for some 'sea-serpent' stories. A typical jellyfish life history is: the adult sheds gametes from gonads near the 'hub' and the gametes fuse with those from another individual. The resulting zygote forms a *planula* larva, a tiny undifferentiated mass of cells with external cilia by which it swims. This settles down on to a hard substratum and produces the polyp phase which may bud at certain times of the year. Generally, the polyp, called the *scyphistoma*, and its buds (if any) start to undergo transverse fission in winter and spring to produce immature medusae called *ephyrae* (1B). The ephyra may either settle to form another scyphistoma or it may develop directly into an adult medusa. Its eight original tentacles multiply by the addition of others in the angles between them; the gut and canal systems (radial canals and a ring canal round the rim of the umbrella) begin to form, and the early stages of the gonads are laid down. The body consists of three main layers of cells: an outer epithelium, an inner epithelium lining the gut and canals, and a thick layer of jelly between. The jelly acts as an antagonist to the body muscles. When these contract, they force water from beneath the bell so that the animal moves upwards, and when they relax the jelly layer restores the original shape. So the jellyfish maintains its position in the water by pulsations of its bell; if it ceases to pulsate, it slowly sinks.

1   **Aurelia** is probably the commonest jellyfish throughout the world. It grows to about 400 mm and has a wavy edge to its bell. It is nearly colourless, except for four violet circles or arcs near its hub, which are the gonads. The mouth is at the centre of a four-armed star, which is itself on the end of a short neck hanging from the bell's centre.
After fusion of eggs and sperm the resulting planula larvae settle in a rocky place to form a polyp-like scyphistoma (1A) which buds transversely to produce miniature eight-armed ephyrae which separate from the column of buds and float free (1B). These very young jellyfish catch tiny fish for their food. Surprisingly, later in life they eat plankton instead, which they catch and transfer to the mouth by means of the stinging cells and cilia in the grooves leading to the mouth, which act like tiny conveyor-belts.

2   **Pelagia** is toadstool-shaped, with a frilled edge to the bell and eight thin tentacles. As in *Aurelia* (1), there are four lobes hanging down from the mouth. This strongly phosphorescent jellyfish glows in the sea at night if disturbed, and if handled leaves a luminous mucus on the hands.

3   **Cyanea.** This form, mainly from the East coast of Britain, has a lobed rim with eight bunches of tentacles. Its sting can raise a nasty weal on the skin. The specimen shown here is a small one; in British seas they grow to about 800 mm maximum diameter, but in some other seas to 1800 mm.

4   **Chrysaora** (Sea Nettle) is milky-white and grows to about 500 mm diameter. The bell, shaped like an inverted soup-plate, has lobed edges and each lobe has a

dark brown spot. From the rim hang 24 tentacles, heavily charged with stinging cells. With these it stuns its prey — fairly large animals of the plankton — then 'licks' it off with the trailing frills round the edge of the mouth and takes it into the body. Some species avoid the nuisance of finding a hard substratum for their polyp phase by retaining the larva in little cysts on the parent.

5   **Rhizostoma,** when full-grown, has a bell as big as a football, grey or pale green with darker margins which have no tentacles. The mouth lobes form heavy, bunched masses with thousands of tiny openings leading into the gut cavity. It feeds by taking in plankton through its many mouths. Seen from below, there are four cavities in the subumbrella surface, in which a tiny crustacean, *Hyperia*, often lurks.

6   **Haliclystus** is a stauromedusan, or sessile jellyfish, with an adhesive disk on the side away from the mouth. Stauromedusans are found in protected coastal waters in colder seas, sometimes on the sea-grass, *Zostera*, and sometimes on fucoid seaweeds. It is usually pale green or pinkish with white marginal anchors lying between the tentacle-clusters; these anchors are used in the looping movement by which the animal sometimes changes its position. The larvae often settle in groups and in this way can feed on animals, such as rotifers, nematodes and copepods, which they would be unable to kill alone.

7   **Lucernariopsis,** another stauromedusan, is very similar to *Haliclystus* in shape but lacks the marginal anchors. It may be found clinging to fronds of weed in quiet rock-pools.

1 Aurelia aurita 2 Pelagia noctiluca
1A A. aurita, scyphistoma stage 1B ephyra stage 3 Cyanea capillata var. lamarcki
6 Haliclystus auricula 4 Chrysaora isosceles 5 Rhizostoma octopus
7 Lucernariopsis campanulata

13

SEA ANEMONES

These belong to the coelenterate class Anthozoa, which, like the Hydrozoa, has polypoid members. In the Anthozoa the polyps are very much bigger, the gut cavity is traversed by at least four radiating partitions, and the mouth is elongated, so that water passes in at one end of the slit and out at the other; a primitive form of continuous passage of water for feeding, respiration, and removal of waste is therefore created, which is an advance on the hydrozoan condition.

1 **Actinia** (Beadlet Anemone) is the commonest of the British shore-living sea anemones, and is found in rock pools and crevices round the whole country. Well able to withstand exposed conditions and buffeting, it will often be found high on the beach. Out of the water, and sometimes in the water too, its tentacles are retracted, so that it appears like a little blob of reddish-brown jelly, sometimes with a rim of bright blue spots. Like many other anemones, the tentacles are in several circlets, 6 in this case, with multiples of 6 tentacles in each, giving a total of 192. The column is usually rust-red, and the tentacles slightly lighter but there may be considerable variation in shape and colour. Particularly common is the 'strawberry' variant (which may be a separate species) in which the body is crimson with green spots. Reproduction is viviparous; that is, the eggs are fertilized in the female's gut and the young expelled from the mouth when they are old enough to fend for themselves. They are wafted or creep to a hard surface some way from the parent and grow.

2 **Anemonia** (Opelet Anemone). This genus prefers crevices to open rock faces, and is more common on South and West coasts than elsewhere. It has about as many tentacles as *Actinia*, but less regularly arranged in circlets. Only rarely are the tentacles withdrawn into the column: they often extend rather than retract under even quite severe mechanical stimulation. It is not as variable in colour as *Actinia:* it is usually light brown, but in one fairly common variety (2A) the tentacles are an iridescent green with purple tips. As well as the normal sexual process of reproduction, this genus multiplies by longitudinal fission; a single individual may break up, in the surprisingly quick time of 1 to 3 hours, into two or three new individuals, each provided with some of the parent's tentacles.

3 **Metridium** (Plumose Anemone). This has a deeply lobed, rather slimy column, and the tentacles are so numerous, slender, and pointed that they give a feathery appearance to the head of the anemone. It feeds mainly by ciliary action of its tentacles, but can take in larger food. It is usually found at extreme low water, in pools or clinging to the underside of overhangs, and it may be locally quite common. It occurs on all British coasts.

4 **Tealia** (Dahlia Anemone) is another generally-distributed and common anemone. It is usually found in shady places, for instance under weed, and it has the habit of gathering stones and shells on to its column, where they adhere to grey, sticky warts. There are 80 tentacles in circlets which are multiples of 10; the tentacles are stocky, conical, sometimes with transverse banding. When they are withdrawn into the body (4A), there is a characteristic lip round the top. If poked when it is withdrawn, it is likely to squirt you in the eye with water from its gut cavity.

5 **Bunodactis** (Gem Anemone) is rather variable in colour, and some specimens are beautifully marked — hence the common name. The column is usually rose-pink, with 6 or 7 vertical lines of warts. The 48 tentacles have a mark like the letter 'B' at their bases. It is commoner towards the South-West of the British Isles, on the middle and lower shore.

6-7 **Sagartia.** In several families of anemones, including the Sagartiidae, there are curious structures called *acontia* which are actually prolongations of the radiating partitions inside the gut cavity; they are thin threads, heavily charged with stinging cells, which are emitted when needed through slits in the body wall of the column or out through the mouth. They may act in defense or offense, and in *Sagartia* they are particularly readily emitted. Like *Tealia*, this anemone tends to gather stones and shells to its column. It may become more elongate at night than during the day, and sometimes one or two of its tentacles may extend further than the others. It occurs on all coasts of the British Isles, generally at middle to low tide, but more frequently at the lower end of the beach.

8 **Aiptasia** has very long tentacles when they are fully extended, tapering to a fairly fine point; they are not readily retractile. This form extrudes acontia, like *Sagartia*. In addition to the normal sexual process of reproduction, it may divide by transverse fission. The oral end so produced attaches to the rock by the side of its column until it has healed and produced a new base; and the other end regenerates a disk and tentacles in about 10 weeks. This genus is found only in the South-West at low water spring tides.

1 ACTINIA EQUINA (three colour forms)   2 ANEMONIA SULCATA (two colour forms)   3 METRIDIUM SENILE
1A A. EQUINA, contracted   2A A. SULCATA   4 TEALIA FELINA
5 BUNODACTIS VERRUCOSA   4A T. FELINA, contracted
8 AIPTASIA COUCHII   6 SAGARTIA TROGLODYTES   7 SAGARTIA ELEGANS

15

CTENOPHORES AND BURROWING ANEMONES

The Ctenophora are related to the medusoid coelenterates (p. 13). The commonest and most typical members are the Sea Gooseberries such as *Pleurobrachia* (2), which are often found stranded on sandy beaches looking like blobs of jelly the size of a plum. The gastric cavity with a single opening forms a system of canals within the body. On the outside are eight meridional rows of transverse plates made up of fused cilia whose beat drives the animal, mouth forwards, through the water. A sheathed tentacle, armed with stinging cells, lies on either side of the body. From this basic ctenophore two other main trends are seen: to the very elongate and flattened forms called Venus's Girdles (1); and to creeping forms flattened in the other plane (3).

The anemones shown on this page are those that habitually burrow in sand or mud, or live in rock crevices. The burrowers arch the body so that the narrow foot can penetrate the surface; then the mouth is closed and the fluid in the gastric cavity forced by muscle action towards the foot. The expanded foot grips the cavity wall, and the muscles of the column contract to pull the body a little way into the hole. The process is repeated for about an hour until the anemone is in its final position with its tentacles just clear of the surface.

1 **Cestus** (Venus's Girdle). In the tropical Atlantic and the Mediterranean, and hardly ever in British waters, this elongated ctenophore may attain 1·5 metres in length. Sometimes skin-divers see it lying in the water as straight as a ruler, almost entirely transparent, and discernible only by the iridescent flicker of the cilia along its top edge. When disturbed, it will undulate and move away.

2 **Pleurobrachia** (Sea Gooseberry or Comb Jelly). If this beautiful animal is watched in water, the intermittent shudder of its iridescent cilia will be seen passing down the comb-rows, and the tentacles emerging from their sheaths. The body, being 99% water, is almost the same density as its surrounding medium and so it can maintain the same level effortlessly. When a small animal is caught by the tentacles, these are 'wiped' past the mouth to transfer the food to the gastric cavity.

3 **Coeloplana.** This remarkable animal is not a member of the British fauna, but it is included because of its superficial resemblance to some of the turbellarian flatworms (pp. 21 and 23). In fact, some zoologists see this animal as indicating a relationship between the platyhelminths and the ctenophores, but this is more generally denied. *Coeloplana* occurs in warmer waters, creeping over encrustations of algae or corals.

4 **Milne-Edwardsia.** This anemone is named after a French zoologist who produced a monumental treatise on polyps in the 1850s. This form normally intrudes its base into a crevice in the rock, so that only the upper part of the column protrudes, and this upper part has a cuticular sheath, giving it an opaque appearance. The gut may show through the body wall as a red streak, and the tentacles are long, delicate, and translucent. It is mainly found on the lower shore.

5 **Cereus** (Daisy Anemone). This vase-shaped animal may live either in a crevice or with its base thrust into the sand and attached to some buried object, such as a stone or shell. When disturbed, the animal will close up and withdraw below the surface.

6 **Peachia** is a very apt name for this anemone, since it is both peach-coloured and named after an 18th-century zoologist called Peach. It burrows at extreme low water and below, in sand, mud, or gravel, and its burrow may be up to 300 mm deep, the anemone occupying the top part until disturbed, when it will retreat to the bottom. The young larvae attach themselves to jellyfish, and when they are big enough to settle they may often devour their host as their first adult meal. The column has 12 fine vertical lines down its sides, and may be swollen towards the base, where the column grips the burrow walls.

7 **Halcampa,** very variable in form, may be thin and worm-like to almost spherical. It is usually found burrowing in sand and below rocks and stones. It normally reproduces sexually; the males usually extrude their sperm first, which induces the females to ovulate as the sperm is taken into their gastric cavity. Fertilization occurs within the cavity, and the zygotes are then covered with a sticky substance before being expelled, which enables them to adhere to the substratum to hatch.

8 **Cerianthus.** This anemone occupies a separate order, the Cerianthidea, from the others illustrated here and on p. 15, differing from them in internal anatomy and mode of growth, particularly of the septa in the gut. It lives in a tube made of discharged nematocysts and sand grains. It has two groups of tentacles, one round the mouth, the other round the edge of the disk.

1 Cestus veneris, a ctenophore   2 Pleurobrachia pileus, a ctenophore
4 Milne-edwardsia carnea, an anemone   3 Coeloplana, a ctenophore
6a Peachia hastata (normally only the crown is visible)
5 Cereus pedunculatus   6 Peachia hastata   7 Halcampa chrysanthemum   8 Cerianthus lloydi

CORALS, COMPOUND ANEMONES
AND ANEMONE ASSOCIATIONS

Some anemones (such as 3, 4, 6, shown here) are always or often associated with mollusc shells occupied by hermit crabs. The crabs benefit by the extra protection of the stinging tentacles of their passengers, and the anemones obtain scraps of food as the crab eats.

The alcyonarians are colonial anthozoans which always have 8 tentacles and 8 radiating septa in the gut cavity. Among them are the Soft Corals (7), so called because the greater part of each polyp body is embedded in a thick fleshy secretion, and the Horny Corals, or Sea Fans (1), where the polyps are borne on fan-shaped supports.

1 **Eunicella** (Sea Fan). Most of the sea-fans are found in warmer waters, but this one occurs in the English Channel, and is familiar in marine aquaria. Sea-fans differ from alcyonarians (7) in the presence of a horny axial skeleton which grows, tree-like, from a holdfast. The polyps arise from the sides of this skeleton (1A), each being borne in a tiny pit. The fan's flat surface is usually presented perpendicular to the prevailing water currents to provide the maximum catchment area for food.

2 **Caryophyllia** (Devonshire Cup Coral) is a true madreporarian coral, secreting a calcareous skeleton, on which the polyp sits. The skeleton is very similar to, but much smaller than, those of reef-building tropical corals. It is a solitary coral, though several individuals may be found on the same patch of rock. The soft tissues, which resemble those of an anemone, sit upon the skeleton and when the soft parts are withdrawn or have died (*see* top polyp, 2A) the skeleton can be seen to have radiating partitions which in life support the septa in the gastric cavity. The knobs on the ends of the tentacles are heavily charged with stinging cells. This is a most delicately beautiful animal, a rare gem for the naturalist to find.

3 **Calliactis.** This anemone can live separately, but is most often found associated with a whelk shell, either empty or occupied by a hermit crab, *Pagurus bernhardus*. Sometimes it is found attached to the claws of crabs. When the shell is empty, the anemone tends to stand erect, but when a crab is within, and hence moving the shell around, the anemone droops over so that its tentacles sweep the sea floor. A complex behaviour pattern has evolved between the anemone and the crab. As the hermit grows, it needs to move to a larger shell; rather than lose the protection the anemone affords, the crab will cause it to detach by striking the side of its column. When part of the base has lifted from the shell, the crab inserts its claw under the base and lifts the anemone on to its new shell.
The thick body-wall of the anemone is pierced for the extrusion of acontia (6, p. 14), which, like the tentacles, are heavily charged with stinging cells — no doubt a defence to the crab as well as the anemone. (*Calliactis* is shown also on the book-jacket.)

4 **Epizoanthus** is one of the Zoanthidea, a group more abundant in tropical seas; but several species of this genus turn up rather rarely in our colder waters. The Zoanthidea habitually grow on other animals, and *Epizoanthus* usually occurs on mollusc shells occupied by hermit crabs. After wrapping itself around the shell, it dissolves the chalky material away, so that it comes to enclose the crab directly.

5 **Corynactis** (Jewel Anemone). This brilliantly coloured, anemone-like animal is placed in a separate group, the Corallimorpha, and considered more closely related to the true corals than to the anemones despite the absence of a chalky skeleton. Its individuals are small and squat. It grows in numbers together and touching, thus looking like a colony. It is usually found on the lower shore, in crevices or overhangs.

6 **Adamsia.** This anemone is always found on a shell with a hermit, *Pagurus prideauxi*, and never alone or on an empty shell. There is generally only one anemone on each shell. It usually wraps itself round the shell so that the disk, very much elongated, lies below the hermit's body and the spread-out base meets along the top of the shell. If the shell is too small, then the anemone secretes a horny 'extension' to adhere to. (In 6, the meeting edges of the base are uppermost; the ends of the elongated opening, with the fringes of tentacles, are just visible round the sides of the shell.) This is an example of a true symbiosis, living together for the good of both partners, because if the hermit-crab leaves the shell permanently, the anemone drops off and dies unless it finds another host soon.

7 **Alcyonium** (Dead Men's Fingers). This soft coral has spongy lobes which in life bear the polyps. But when it is washed up on the beach its lobes look horribly like its gruesome English name. The matrix of the colony is stiffened by chalky spicules, and the greater part of each polyp is embedded in the matrix with only the oral end protruding. When molested, the oral end can withdraw below the surface of the matrix. There is a system of canals connecting the polyps.

1 EUNICELLA VERRUCOSA   2 CARYOPHYLLIA SMITHI   2A C. SMITHI, skeleton
1A E. VERRUCOSA, polyps enlarged   3 CALLIACTIS PARASITICA, on a whelk shell
4 EPIZOANTHUS INCRUSTATUS   5 CORYNACTIS VIRIDIS
6 ADAMSIA PALLIATA, with a hermit crab   7 ALCYONIUM DIGITATA, a soft coral

19

FRESHWATER AND TERRESTRIAL FLATWORMS

Of the many phyla which have worm-like members, the phylum Platyhelminthes is evolutionarily the lowliest with many primitive features. There is no blood system to transport essential substances to the tissues of the body, and this has two consequences: typically, the body is flattened from above, ensuring a large surface area to get respiratory gases to and from the tissues; and the alimentary system is extensively branched (3A) so that the products of digestion can be conveyed in it within diffusion distance of every cell of the body.

These are unsegmented worms, though some (p. 27) have serial repetition for reproductive efficiency. As there is no body cavity between the gut and outside wall, the body can undergo only limited activities and movement. There is no anus; waste products are voided through the mouth. The mouth usually has a sucking pharynx by which the prey is attacked and the soft parts sucked into the gut. This mode of feeding seems very apt for evolving towards parasitism, for within the phylum we see a splendid transition from free-living members, in the class Turbellaria (p. 21), through parasites that feed actively in the classes Monogenea, Cestodaria and Digenea (pp. 22 – 25), to those that absorb digested food from the host, in the class Cestoda (pp. 26 – 27).

1-2 Land planarians. This is a small group of turbellarian flatworms that live in humus and under bark. Neither is very common in Britain. *Bipalium* (1) is a cosmopolitan form that was — like many another unusual animal — first found in Kew Gardens. Its shovel-shaped head has a fringe of tiny eyes and there is an aggregation of nerves where the head meets the body. It normally reproduces asexually. *Rhynchodemus* (2) is mainly nocturnal; it can sense earthworms and other potential food up to 8 mm away in dry soil, and when necessary can make a fast and accurate attack on such swift moving prey as spring-tails and fruit-flies. Both animals move by a combination of the activity of cilia on the under surface and muscular flexures of the body.

3 Dendrocoelum. This milky-white flatworm is a triclad, that is, it has three main branches to its intestine which can be clearly seen through the body wall in a recently-fed specimen (3A). It is a carnivore, feeding on small animals and eggs. The body wall secretes a mucus by which the animal adheres to the substratum or to the surface film of water, and it moves, even under the surface film, by ciliary gliding. In addition, it can swim, rather ineptly, by using its body muscles to undulate. It has a pair of pigment cup eye-spots at the anterior end, and lateral sensory grooves which are chemoreceptive: the animal will move towards a source of food. It has a complex hermaphrodite reproductive system, and a mechanism to ensure cross fertilization, but sometimes it will reproduce by tearing in two and regenerating the missing parts on each half. This, *Polycelis* (6) and *Dugesia* (7) are all freshwater planarians.

4 Mesostoma. This is a rather small rhabdocoel, that is, a flatworm with a simple rod-like gut, which seldom exceeds 2 mm in length. There are tiny projections at both ends: sensory hairs anteriorly, and fixation papillae posteriorly. A pair of eyes and a mass of nervous tissue beneath them can usually be made out with a handlens. The pharynx is very muscular, and the animal can swallow small earthworms and crustaceans. Sexually mature animals are rare, the animal apparently relying on asexual fission (dividing into two) for reproduction.

5 Stenostomum is another freshwater rhabdocoel. Under good conditions it reproduces by transverse fission, but very often the new individuals do not separate at once, so you might find chains of 4 or 5 individuals fastened together (as shown by the animals at lower-left of the group illustrated). When winter approaches, the adults usually lay eggs which can survive until the next spring, and there may be several young in each egg. It feeds by sucking up bacteria, flagellates, and ciliates — and even its own young — by means of cilia in its pharynx wall. It is common in detritus in ponds and ditches.

6 Polycelis, another planarian, is darkly pigmented. The name refers to the many eyes along the sides of the head, though they may be difficult to see. The incredible powers of asexual reproduction and regeneration shown by some turbellarians is particularly well seen in this form, since the regenerated parts of the animal are generally less pigmented than the original parts. This animal occurs in considerable numbers in standing or gently-running water, and you may come across many individuals under one stone or water-lily leaf.

7 Dugesia is the best-known flatworm, and a favourite of elementary biology courses. It has one pair of eyes with a pigmentless area surrounding each eye, giving it a slightly cross-eyed look. Like the others, this flatworm is carnivorous, but if the prey is too big to be taken into the gut, it is first wrapped in a mucous envelope, then the pharynx sucks bits off until the animal is devoured. It lays eggs in a cocoon which can sometimes be seen within the body before it is laid. The right-hand of the two illustrated is shown flexed to display the pharynx protruded from its opening.

1 Bipalium kewense, a land planarian   2 Rhynchodemus terrestris, a land planarian
3 Dendrocoelum lacteum, at the water surface
4 Mesostoma tetragonum   5 Stenostomum sp.   6 Polycelis nigra
7 Dugesia subtentaculata   3A D. lacteum, on the bottom

MARINE FLATWORMS AND FLUKES

Turbellarians are primarily aquatic, and by far the greatest number of them are marine; a few of these sea-dwelling forms are illustrated on this page. Most of them are bottom dwellers that live in sand or mud, or under stones and weed inshore. But many of the marine flukes are parasites of fish and two examples are shown.

1 Convoluta. This tiny worm (not more than 2 mm long) belongs to the order Acoela, a turbellarian order which is entirely marine. Most members of the order lack an alimentary tract, though in some a tiny mouth is present, but closed off. Some species of this worm feed on diatoms, protozoans, and small arthropods, which are captured by little extensions from the cells within the mouth, or simply straddled by the body until the food can be pressed into the mouth region. Others exhibit a remarkable example of symbiosis, or the coexistence of two organisms to their mutual benefit. The young of *Convoluta roscoffensis* will take in tiny flagellates rather like the individual cells of which a colonial *Volvox* (p. 1) is composed; then when there are sufficient in the body, the temporary mouth will 'seal over' and the animal will feed exclusively on the materials produced by the flagellates. Being plants, the flagellates require a supply of carbon dioxide for photosynthesis, and this is a waste product of the animal; conversely, the flagellates produce oxygen as a by-product of the process, which is required by the animal. But one prerequisite of a healthy metabolism for the flagellates is sunlight, and the animal must ensure that it is well placed to provide optimum conditions for this. It could prove rather disastrous for this shore-dwelling animal to remain on the surface of the sand as the tide rises and falls and the waves race up the shore. So there is an inherent rhythm to its activity which takes the animal below the surface as the tide comes in and causes it to move upwards to the surface when the tide has dropped again, to expose its body to the energy-giving rays of the sun at low tide (1A). The interesting thing here is that specimens taken to the laboratory and kept in a constant aquarium still maintain this rhythmic behaviour for some time, synchronously with the tides.

2 Procerodes. This, like the freshwater planarians (p. 21), is a triclad, and the three main branches of the gut can be seen through the body wall if it has been feeding recently. It is up to 5 mm long, with two rounded horns at the head end. It occurs on the upper and middle shore, usually where a freshwater stream trickles over the beach. It lays reddish brown egg capsules 1 mm in diameter in groups on stones.

3 Cycloporus. This animal belongs to the order Polycladida, which is exclusively marine; the name of the order refers to the fact that its members have many branches to the gut. This worm has a papillate surface and two small tentacles anteriorly. There are four groups of numerous tiny eyes. It is normally found on tunicates and sponges, and is shown opposite near patches of red tunicates (p. 189) and the Breadcrumb Sponge, *Halichondria* (p. 5).

4 Leptoplana is another polyclad. It is a delicate oval form, up to 25 mm long, generally pale brown with white patches, and with groups of black eye-spots at the front end, to either side of a concentration of nerve tissue. There are no tentacles. It occurs under stones, or around sponges and weed, though it may swim by sinuous movements of the sides of its body.

5-6 The Flukes. Flukes constitute one of the major groups of metazoan parasites. They may be external or internal parasites on a wide variety of hosts but, though common, they are not often encountered by the casual collector. Older workers united all the flukes into a platyhelminth class Trematoda, but more recently it has become evident that there are such differences between the two orders Monogenea and Digenea (separated on whether the life-cycle involves one or more than one host) that these categories have been raised to classes. On this page one monogenean and one digenean example are included.
Derogenes (5) is a digenean the final host of which is a fish, and the fluke is usually found in its stomach or intestine. A wide variety of fish may be host to it, such as sea bream, gurnard, blenny, cod, whiting, sole, plaice, salmon and trout, and it is the most widely distributed digenean of marine fishes. The life-cycle is not well known, but the intermediate host, as in other digeneans, is a mollusc. *Gyrodactylus* (6), a monogenean, occurs on the gills of the rockling, *Spinachia*, the fish illustrated. As is usual in parasites, the reproductive system is particularly well developed, and betrays a rapid reproductive cycle. Sometimes embryos can be seen within the adult, and a second generation within the embryo, and a third within that, to improve the fecundity of the fluke and therefore the chances of infesting the host. There are adhesive organs at the front and at the back, and these have tiny hooks which irritate the host's skin, so that it produces a defense reaction in the form of a mucous secretion containing cells on which the fluke feeds. Heavy infestation causes the death of the fish, and therefore it cannot be considered a very highly evolved parasite.

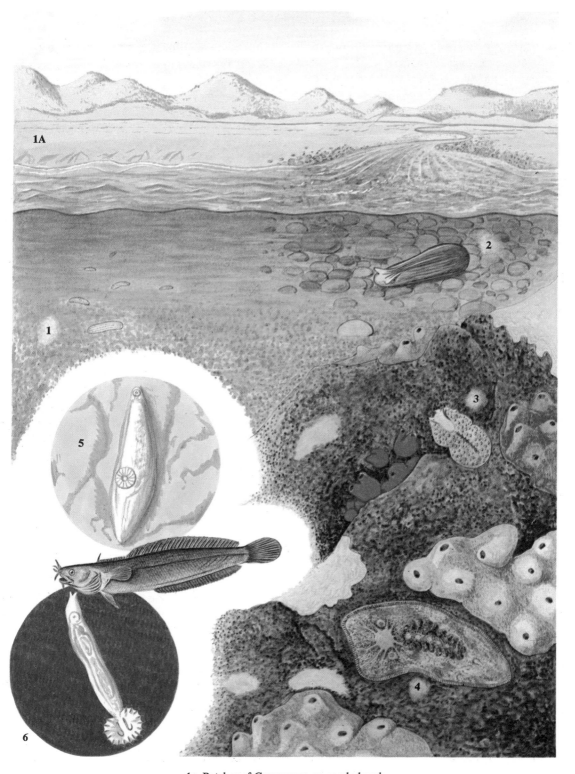

1A Patches of CONVOLUTA on sandy beach
1 CONVOLUTA ROSCOFFENSIS, an acoel 2 PROCERODES ULVAE, a triclad
5 DEROGENES VARICUS, a fluke 3 CYCLOPORUS PAPILLOSUS, a polyclad
6 GYRODACTYLUS sp., a fluke parasite on rockling 4 LEPTOPLANA TREMELLARIS, a polyclad

FLUKE PARASITES OF LAND ANIMALS

There are, as mentioned on p. 22, two groups of flukes which appear to have evolved convergently with each other from different ancestral stocks. Both groups differ from the free-living turbellarians (pp. 20 – 23) in having lost the external covering of cilia in the adult, and in possessing, as one would expect, special organs of adhesion. These usually consist of an anterior sucker surrounding the mouth, which also functions in feeding, and another sucker towards the posterior end. Sense organs are poorly developed in the adult, which is again understandable in animals which are normally permanently attached or insinuated snugly into their hosts' tissue; but sense organs may be highly developed in some of the young stages which have to seek out the host and find the correct place for attaching.

1 **Fasciola** (Liver Fluke). This inhabitant of the bile passages in the livers of cattle often inflicts severe damage on the host, which can result in death. The life cycle is a complex one involving several intermediate larval forms, many of which themselves multiply to increase the fecundity of the animal so that the chances of spreading the organism to other hosts and maintaining the population are increased. The adult may lay about 45,000 eggs (1A), which pass out with the host sheep's faeces on to the pasture. These eggs can survive for a year or so in dry conditions, but when the pasture is damp, the first larva, the *miracidium* (1B), hatches and this swims by its cilia in the film of water on the pasture until it encounters a snail of the genus *Limnaea*, when it burrows through the skin into the body cavity, losing its cilia and its eyes. There it grows to a large sac, the *sporocyst* (1C), in which the next larval stage, the *redia* (1D), develops. This tiny creature has a mouth and gut, etc., and migrates to the snail's digestive gland, where daughter rediae may form within it. The redia then develops into a tiny tailed version of the adult called a *cercaria* (1E) which drills its way out of the snail and swims and crawls up plants where it may either be eaten fairly soon by the primary host, the sheep (1F), or will encyst as a *metacercaria* for up to 6 weeks. If it is swallowed, it hatches out in the host's intestine, then migrates to the body cavity and thence to the liver, where it attaches by its suckers and feeds on the blood in the vessels permeating the liver.

There are many other flukes, including parasites of Man such as the Chinese liver fluke, which might become established after a man has eaten undercooked fish; *Fasciola* is merely an example which is common and economically important in Britain.

2 **Schistosoma** is one example of a blood fluke, most of which occur in the blood vessels of the human intestine. Modern systems of sewage disposal in Britain do not allow the parasite to occur here, but immigrants are occasionally infected when they arrive. These flukes are unisexual and copulation occurs; the males and females are generally found together, because the male has a grooved body in the folds of which the female is held. The fertilized eggs break into the host's

bladder and pass out with the urine on to the soil. Then the miracidium enters a snail and passes through the typical fluke larval stages of sporocyst, redia, and cercaria, the latter swimming in the water which a human may later wade barefooted in (2A), or unwisely drink.

The greatest concentration of these flukes occurs in Japan and China, where the rice fields provide breeding grounds, but other species are fairly common as far north and west as Portugal. It is no exaggeration to say that hundreds of millions of the world's population are infected with blood flukes. Occasionally, the cercariae of a non-human species will burrow into human skin, to give rise to a complaint known as 'swimmer's itch'.

3 **Pneumonoeces** is a digenetic fluke from the lungs of frogs and toads, and very often encountered by students while dissecting. There may be as many as 12 in each lung, though usually there are only 2 or 3. There are two intermediate hosts: a snail and an insect.

4 **Polystoma.** A blood-sucking parasite from the frog's bladder, this fluke has a complex attachment organ or *haptor* at the posterior end consisting of hooks and 6 large suckers. The adults release their eggs as the frogs enter the water to breed, possibly due to stimulation by the sex hormones that are being carried around the host's blood stream at that time. In about 4 weeks, the eggs become ciliated larvae which attach to the gills of the tadpole. Each larva now produces many hundreds of eggs which enter the tissues of the older tadpoles and migrate to the bladder.

5 **Aspidogaster.** The presence of a large and characteristic adhesive organ ventrally and the absence of an anterior sucker has posed a problem to systematists, some of whom think this is so different from other flukes that it ought to be classified in its own class, the Aspidobothrea. It normally inhabits the pericardium (the heart cavity) of molluscs, sometimes 20 or 30 to an individual, but it has also been found in cold-blooded vertebrates such as fish and turtles.

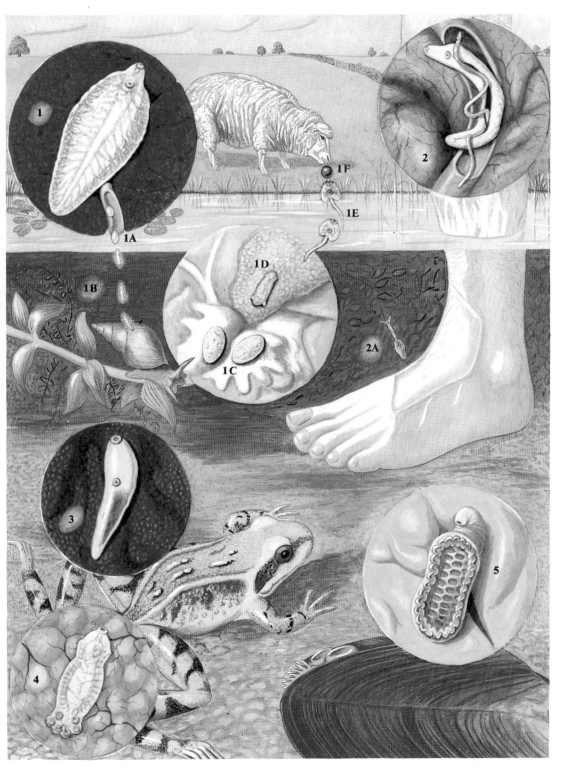

1 Fasciola hepatica, in sheep's liver   2 Schistosoma mansoni, in Man
1a to f F. hepatica, early stages in pond snail   2a S. mansoni, larva entering host
3 Pneumonoeces variegatus, in frog's lung
4 Polystoma integerrimum, in frog's bladder   5 Aspidogaster conchicola, in mussel

TAPEWORMS

These flatworms, possibly the most horrific of all animal parasites, inhabit the alimentary tracts of a wide variety of vertebrate hosts. They have complex life cycles which involve the transfer of eggs *via* intermediate hosts; if the host's faeces are disposed of privily, or if the meat from the intermediate host is properly cooked, then the cycle cannot be completed. The human tapeworm is therefore now rare in Britain, but dogs fed on raw or undercooked meat are still liable to infection. Most effective cures rely on drugs with an antimony base. The tapeworm body usually consists of a tiny head, or *scolex*, provided with hooks and suckers, which attaches to the wall of the host's gut. Just behind the head is a growing region where *proglottids*, looking like segments, are budded off. Each proglottid contains an entire reproductive apparatus, of which the parts of one sex mature before the other. If there is more than one tapeworm in a host, then cross fertilization will occur between individuals; but if only one is present, then the proglottids that are male-ripe will copulate with a proglottid that is female-ripe, thus giving more chance of genetic variability than if eggs and sperm from the same proglottid were to fuse. An enormous number of offspring can be produced from one tapeworm.

1 **Dipylidium** occurs fairly commonly in dogs and cats, and occasionally in man. Each proglottid has a double set of gonads, just to increase the fecundity even more. The intermediate host is the flea, and as dogs rid themselves of fleas by biting, the cycle is not difficult to complete.

2 **Echinococcus.** Sometimes Man is the intermediate host for tapeworms, and this is one example. The adult, which lives in the intestines of dogs, consists of less than 5 proglottids, only the last of which is gravid. The eggs of the worm are often carried on dogs' tongues and can be transferred if the dog licks a human face or hands. Infection may also occur if dogs are allowed to foul a human water supply. The interesting thing about this worm is the relative sizes of the two stages; the adult is no more than 5 mm long, but the *cysticercus* larva may be up to 150 mm in diameter — as big as an orange. It is this cyst, sometimes called a bladder-worm, which causes the trouble in Man, and may lead to epilepsy. The bladder is not specific to Man, but may occur in other vertebrates, chiefly sheep and cattle.

3 **Taenia.** There are many species of this tapeworm, most of them being specific to either their final or their intermediate host. For instance, *Taenia solium*, illustrated here, is a human tapeworm the intermediate host of which is usually a pig; *T. saginata* (now usually called *Taeniarhynchus saginatus*) is another human tapeworm, but the intermediate host is usually a cow.

When the final ripe proglottid breaks off from the worm, it passes out of the man's body with the faeces, but the eggs do not hatch unless eaten by a pig. The pig's digestive juice dissolves away the outer protective shell of the egg, and a tiny embryo, the *onchosphere*, with 6 hooks, breaks free and burrows through the intestine wall to a blood vessel in which it is transported to some tissue of the body. This is usually muscle, though it can be a vital part of the nervous system, with disastrous results. When embedded, the embryo develops into a cysticercus, or bladder-worm, in which the primordium of the adult worm develops. The bladder-worm will remain in the pig until it is eaten by a man, but it will be killed if the pork is properly cooked. Carcass inspection at wholesale butchers usually detects infected 'measly' pork before it is sold. Home reared, killed and cured pork may be a source of danger if the tapeworm is already present in the human population. In countries where meat is cooked in large pieces over an open fire and where sanitation is poor — perhaps under wartime conditions — a large proportion of the population may be affected.

4 **Diphyllobothrium.** This human parasite is notable, not only because it is the largest of the human tapeworms, but also because it is spreading in some parts of the United States and Canada. There are two intermediate hosts: copepods (p. 134) and fish, particularly pike and trout. The habit of eating smoked or raw fish is causing the spread. If human effluent flows untreated into rivers and thence into lakes, thousands of eggs may be liberated into freshwater containing the copepod. Then the embryos are eaten by the copepod, and the copepod by a fish which might then be exported to another part of the country, thus spreading the curse.

5-6 **Cestodarians.** These parasites lack a gut, but the body is a single unit, not forming proglottids. *Amphilina* (5) lives in such fish as the sturgeon, into which it bores like an animated drill, by twisting its body. *Gyrocotyle* (6) has an oval sucker posteriorly, which some authorities think may be equivalent to the tapeworm scolex, and thus the 'head end' of a true tapeworm may really be the posterior end.

1 Dipylidium canium, from dog   2 Echinococcus granulosus, from dog
3 Taenia solium, from Man   4 Diphyllobothrium latum, from Man
5 Amphilina foliacea, from fish   4a eggs of Diphyllobothrium
6 Gyrocotyle fimbriata, from fish

PROBOSCIS WORMS OF THE SEA SHORE

The Nemertina are sometimes known as the Rhynchocoela, and the two alternative names give a clue to the outstanding feature these worms possess: a thread-like proboscis which can be forced out at great speed and accuracy by the contraction of a fluid-filled cavity surrounding the proboscis. (Nemertina means 'unerring thread'; Rhynchocoela means 'proboscis cavity'.) They represent an advance on the platyhelminth worms in two respects: first, the gut has an anus as well as a mouth, so that food can now pass along the tract continuously; and, secondly, for the first time in the invertebrates there is a blood vascular system, to convey essential substances round the body. In some other features, however, such as the serial repetition of certain organs — particularly the reproductive system — down the length of the body, they resemble platyhelminths (pp. 20 – 27), and the two phyla are generally considered quite closely related. Nemertines are almost all carnivorous, feeding mainly on annelid worms, though they will also take dead molluscs and crustaceans. When a living animal is caught, it is taken alive into the gut, and killed by an acid secretion while it lies in the first part of the alimentary tract. As with leeches (p. 116), starvation leads to great reduction in size.

Also included on this page is a representative of another phylum of worms with a large proboscis: the Priapulida. However, the proboscis is very different from that of the Nemertina, and is not shot out rapidly but rolled inside-out and back again for feeding.

1 **Eunemertes (Emplectonema).** This worm has the power of rapid contraction if disturbed, and is normally found as a wriggling coil underneath stones. Its eyes are in two clusters on the head, but are not easy to see. As in all other nemertines, the proboscis, which is so important a feature of these animals to biologists, is not readily visible, though occasionally if the animal has been damaged the proboscis may remain extruded from the front end when the animal is discovered.

2 **Tubulanus.** This worm may grow to 300 mm long, and is one of the most beautiful of the nemertines. It lives under stones and crevices, especially where there is a sandy floor, and it secretes a tube of mucus to live in. It has no eyes.

3 **Cephalothrix.** The body of this worm may be thick or thin, depending on its state of contraction. It is gregarious, and quite a few may be found together, under the same stone or nestling in a cavity in clean sand, among coralline algae, or in shells.

4 **Lineus** (Bootlace Worm). This is probably the longest invertebrate in the country. Specimens of 5 metres are fairly common, but one of the authors found a noble animal 10 metres long which took some time to unravel before it could be measured. When found, it is usually in coils under rocks and in crevices, its velvety, iridescent body almost inextricably knotted. It can take in quite large annelids, including tube-dwelling polychaetes (p. 107) such as *Branchiomma;* since the prey is thicker than the worm itself, there is a tell-tale swelling in the body of the nemertine as digestion proceeds. As anybody who has tried to unknot a specimen will know, the worm fragments easily, but regeneration is comparatively simple, and the worm uses fragmentation as a means of reproduction. But

the normal, sexual, method is also used; the sexes are separate, and the eggs are laid in a gelatinous rope.

5 **Nemertopsis (Tetrastemma).** The distinguishing feature of this nemertine is the 4 eyes arranged in a rectangle. Compared with *Eunemertes* (1), this is a sluggish worm, and is often gregarious. The head is snake-like.

6 **Amphiporus** (Pink Ribbon Worm). This worm, up to 80 mm long, has eyes in four groups on its spatulate head: a double row on either side in front and a pair of clusters further back. Its proboscis is as long as the body, and armed with one or two stylets at the end which grow as they are worn down.

7 **The Priapulids.** This phylum probably does not belong close to the nemertines, but then it is not at all clear where it does belong. It is a very small phylum (only two genera) of highly distinctive animals, with an appearance that justifies the somewhat lubricious name. The cylindrical body has a proboscis and either one or two clusters of respiratory appendages at the posterior end. The proboscis contains a spiny introvert which turns in on itself to force the prey into the alimentary tract. There is a body cavity, but it does not appear to be equivalent to the coelom of higher invertebrates (*see*, for instance, p. 94), because muscle fibres stretch across it, and it is not bounded by a true skin or lining. *Priapulus* is a mud-dwelling worm found in the glutinous mud of estuaries and very slow-moving waters. It normally sits vertically in the substratum, with its mouth at the surface, though it may move through the silt to catch its prey. It feeds on soft-bodied slow-moving invertebrates, particularly polychaetes (p. 103).

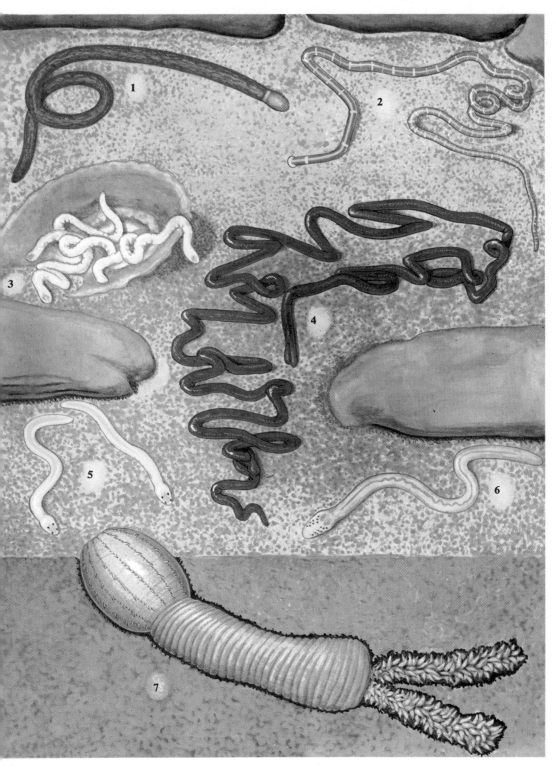

1 Eunemertes (Emplectonema) neesii 2 Tubulanus annulatus
3 Cephalothrix rufifrons 4 Lineus longissimus
5 Nemertopsis (Tetrastemma) flavida 6 Amphiporus lactifloreus
7 Priapulus bicaudatus, a priapulid

LOWLY WORMS

Judged on the criteria of numbers of individuals and numbers of habitats exploited, the phylum Nematoda, the Round Worms, is one of the most successful in the animal kingdom. Not only are there free-living nematodes in every type of freshwater, marine and terrestrial habitat, but many more are parasitic in animals and plants. Humans have their share (these are the 'worms', that are a fairly common ailment in children). Scarcely any animal, including Man, goes through life without at some time supporting a population of nematodes, and most of the time they are so harmless that we, or other animals, do not even know they are there. But sometimes they can grow to an enormous size, or occupy a most inconvenient site, to the great distress of the host. For instance, in Man the Eye-Worm (not illustrated) inhabits the epithelial tissues and migrates around the body, sometimes passing across the cornea of the eye. Again, the adults of the Filaria penetrate the skin and block the lymph channels, producing the dreadful condition known as elephantiasis; the Whipworm lives in the intestine, but its larva burrows through the gut wall to various muscles causing cysts which give great pain and sometimes cause death; and the American Hookworm invades human tissues and causes anaemia and general lack of energy and may result in physical and mental retardation. But in Britain the worst most people see of these parasites is the Pinworm in children, a less spectacular manifestation of a highly successful phylum. The external structure of all nematodes is surprisingly uniform. They have slender, elongated bodies with tapered ends, and except for specialised mouthparts in a few members of the phylum, the only projecting parts are concerned with reproduction. The other phyla of lowly worms illustrated on this page (3 and 4) are both parasitic at some stage in their life-history.

1 **Ascaris** is one of the largest of the parasitic nematodes, which is most often encountered in Britain in pigs. The adult may be from 20 mm to about 300 mm in length, the males slightly smaller than the females and with a curved posterior end. A single female may lay 200,000 eggs in a day, with thick, resistant shells. They pass out to the exterior with the host's faeces. Then, if an egg is swallowed by the right host, its digestive juices dissolve away the shell and liberate the young worm. For some reason, the young worms do not remain in the intestine, but bore their way through the gut tissues into blood vessels and other organs, finally attaching to the walls of the lungs. Then they climb the bronchial tubes to the mouth cavity and are swallowed into the alimentary canal again.

2 **Rhabditis** is a free-living nematode which lives in soil on decaying animal and vegetable matter. It is quite an easy matter to see these animals: every earthworm appears to be infected with the larval stages, the development of which is inhibited by the host. But as soon as the earthworm dies, the larvae develop and can be seen all over the rotting corpse. So if an earthworm is killed in a dish with damp filter paper and left for a few days, the worms can be collected in large numbers.

3 **The Acanthocephala** (Proboscis Roundworms). These worms with a spiny retractile proboscis have two hosts: the juveniles are in arthropods and the adults in the digestive tract of vertebrates such as fish, birds or mammals. There may be many of these worms in the vertebrate host, and copulation occurs between the separate sexes. The eggs are shed with the host's faeces, as is normal with this sort of parasite, and a larva emerges when the egg is eaten by the arthropod secondary host. After boring through the gut, the larva becomes lodged in the body cavity and develops until nearly adult. Development continues only when the intermediate host is eaten by the final host. Once within its gut, the worm attaches by its spiny proboscis. The intermediate and final host may be either terrestrial or aquatic. There are common acanthocephalans from rats and pigs, and it is not unknown for Man to become infected, by dirty eating habits. Only one example is shown here, *Echinorhynchus*, a parasite of freshwater fish.

4 **The Nematomorpha** (Hair Worms). These long, thin worms (up to 800 mm) are free-living as adults in freshwater and damp soil, but the juveniles are parasitic in insects and other arthropods. The females are rather inactive, while the males swim or crawl to copulate. The eggs are deposited in water, and when the juveniles hatch they have a protrusible proboscis armed with spines (4A), by means of which they enter the arthropod host at the water's edge. This host is very often a beetle or a grasshopper. They feed parasitically for a time, then emerge as adults, with a very featureless body. They are usually found in the process of emerging, as a thread protruding from the host's body when it is near water. The adults are said not to feed, and the gut is almost non-existent. The example shown, *Gordius*, is often found almost inextricably knotted. It takes its name from the intricate knot tied by Gordius, a king in Phrygia. Whoever should untie the knot, the oracle proclaimed, should rule Asia; Alexander the Great cut through it with his sword.

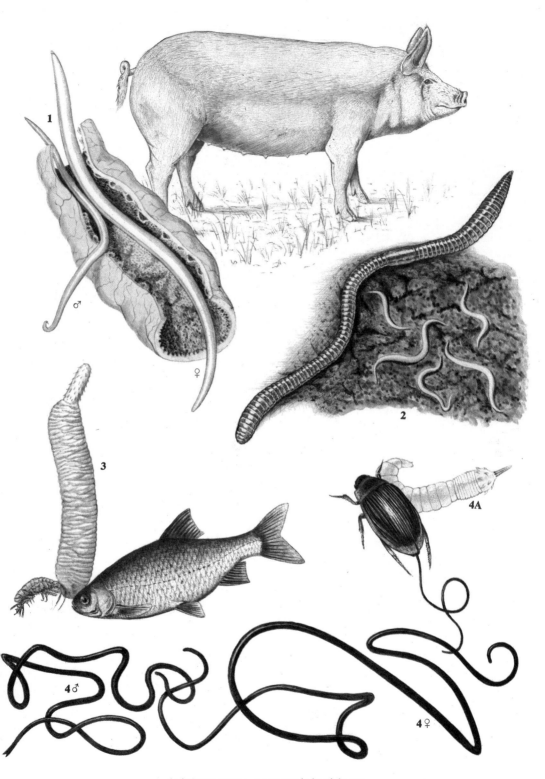

1 Ascaris suilla, a nematode in pig's gut
2 Rhabditis maupasi, from earthworm
3 Echinorhynchus clavula, an acanthocephalan, from fish   4a Gordius, larva
4 Gordius, a nematomorphan, emerging from beetle

CHITONS AND TUSK-SHELLS OF THE SEA SHORE

The phylum Mollusca (pp. 32 – 93) is important for several reasons: first, it is second to the Arthropoda in the number of species so far described (about 107,000 to the Arthropoda 850,000); secondly, its members are among the most conspicuous invertebrate groups to casual collectors, gardeners, and skin divers; and lastly, its members occupy marine, freshwater and terrestrial habitats. Though very diverse in appearance, they all share common features which, though perhaps seeming trivial in themselves, have undoubtedly contributed to the success of the group. For instance, most possess a calcareous *shell* secreted by a soft skin or *mantle* which covers the body like a tarpaulin over a hay-rick; the edges of the mantle form a skirt beneath which is an important space, the *mantle cavity*, which contains gills and other important structures and into which open the various ducts. Most molluscs are slow-movers, relying on muscular ripples down a flat, sticky pad, the *foot*; but in their ultimate evolutionary form, the squids and octopuses (p. 93), they use a kind of jet propulsion, by shooting water from the mantle cavity. There is no denying the success of molluscs in terms of actual numbers of individuals, numbers of species, and the sort of habitats they occupy; yet the general body form of the lower members does not suggest a very promising plan, with its simple nervous system, 'open' blood system (not confined to vessels), and limited locomotory and feeding possibilities. But at the other end of the scale the cephalopods (p. 92) closely parallel vertebrates in some of their features, such as the organization of their sense organs and nervous system.

Among the most primitive molluscs are flattened, slow-moving forms called chitons, or coat-of-mail shells, which possess a series of shelly plates along the mid-line of the back, rather than a single shell. The mantle forms a peripheral flap, like the sides of a tent, which can close down tightly to the substratum. Chitons mostly live on rocky shores, where they are well adapted to withstand the buffeting of the waves. If they are dislodged, they can curl up like a woodlouse and suffer little damage from being thrown about by the currents. One representative of another small class, the Scaphopoda, is included on this page. These offshore sand burrowers are unlike any of the other molluscs in having a tubular shell.

1 **Lepidochitona** has a dull shell and the surrounding mantle flap has 'granules' of irregular size and colour. It is the commonest of the intertidal chitons, and exhibits a variegated colour pattern of reds, greens and browns.

2 **Ischnochiton** has an oval yellowish body with narrow, glossy valves each having a median keel. The mantle flap has small spines round its edges and small globular granules on its surface.

3 **Acanthochitona** is easily distinguished from the others by its tufts of bristles on the mantle flap. The shell valves have unevenly distributed pear-shaped granules and the shell colour is variable. It is widely distributed on the lower shore, but particularly common in the South and West.

4 **Tonicella** is one of the largest chitons, and many grow to 40 mm long. Its shell is more shiny than that of *Lepidochitona* and *Acanthochitona*, and the mantle flap has small granules embedded in it.

5 **Scaphopods** (Tusk-Shells). This, the smallest of the mollusc groups and exclusively marine, has the mantle forming a tube and producing round the body a quill-like shell, open at both ends. The animal lives buried 'head first' in the sand with the narrower end projecting from the surface. It is through this end that respiratory currents enter and leave the mantle cavity. At the head end are tentacle-like structures for collecting food particles. *Dentalium* is a British genus which usually occurs offshore, but there are a few localities in the North where it has been found between tides. In the rest of the country the shells are sometimes cast up on the shore, but very infrequently in recent years, and there are fears that it is becoming increasingly rare.

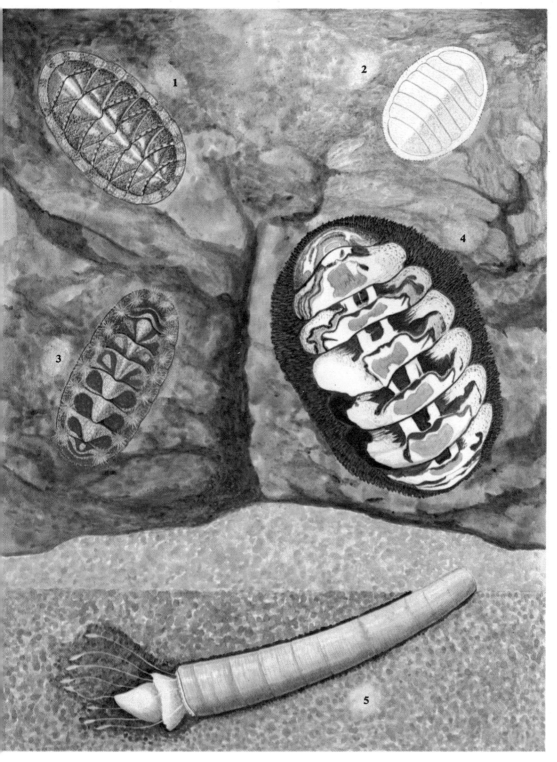

1 Lepidochitona cinerea   2 Ischnochiton albus
3 Acanthochitona crinitus   4 Tonicella marmorea
5 Dentalium entalis, a scaphopod

LIMPETS OF THE ROCKY SHORE

The class Gastropoda (pp. 34 – 71) shows well how many different habitats and ways of life can be exploited by animals of one basic design. The single shell covers the main organs of the body, and a head-and-foot forms a base by which the animal moves and feeds. Compared with the chitons (p. 33), the evolution of gastropods has involved increase in size of the soft parts: in the limpets (p. 35) the result is a tent-shaped animal; while in others (p. 41) the dorsal parts of the body project upwards and are thrown into a coil.

In all members of the subclass Prosobranchia (pp. 34 – 43) the mantle cavity, with its gills and its apertures from the organ systems of the body, twists round from the posterior to the anterior end. This causes the formerly straight gut to become U-shaped and turns the once-parallel nerves into a figure-of-8. What could be the advantage in the extraordinary shift? Probably the currents flowing into the mantle cavity were murky with sediment stirred up by the animal when the cavity was posterior; but in the forward position it receives 'clean' water. On this page are examples of the archaeogastropods, the limpets, and topshells. These are thought to be the most primitive prosobranchs, mostly retaining the paired gills in the mantle cavity.

1 **Diodora** (Keyhole Limpet). One of the problems of bringing the mantle cavity towards the front is that the anus now opens close to the mouth, and prosobranchs show various adaptations for counteracting this rather nasty disadvantage. This one has the anus opening by a hole in the top-centre of the shell. The anus, on a spire-like tube, may protrude through the 'keyhole'. This is a lower shore form, mainly found in the South and West.

2 **Emarginula** (Slit Limpet) has a marginal slit in its white, ribbed shell, which may show the anal spire protruding through its upper part. A widely distributed form generally found at low-water spring tides.

3 **Acmaea** (Tortoiseshell Limpet) has a rather flattened and delicate shell with irregular red-brown markings giving it the tortoiseshell appearance. This limpet occurs particularly among the holdfasts of the beltweed, *Laminaria*.

4-6 **Patella** (Limpet). This mollusc is probably the most characteristic one of the shore; its conical form and hefty shell fit it to live in exposed places. When the water has receded, we see them as static animals, strongly attached by the broad foot to one place on the rock to which their shell accurately fits, awaiting the battering tide. But when covered by water again, they make feeding forays over the rocks, scraping off encrusting growth with their rasp-like radula. Then, as the tide turns, they return by the same path to their 'home' to await exposure at low tide. The three species are not easily distinguished except by examining the radula; but as a rough guide the common limpet, *P. vulgata* (4), usually found higher on the beach, has a tall ribbed, shell; *P intermedia* (6) is flatter with finer ribs and a darker shell margin; *P. aspera* (5) is even flatter and has the apex more eccentric.

7 **Patina** (Blue-Rayed Limpet) occurs in two quite different forms in its life-history, When young, it is smooth and somewhat translucent, with three vivid, shimmering blue rays across the shell, a beautiful jewel of a mollusc which cannot fail to cause its finder a moment of joy. With age (7A) the shell becomes thicker, loses the rays and sometimes becomes flatter. The young specimens frequent the fronds of *Laminaria*, while the older ones are found more among the holdfasts.

8 **Haliotis** (Ormer, from the French 'oreille de mer', or 'Sea-Ear'), has a flattened shell with a series of holes of increasing size towards the edge of the whorls. These are used as exhalant pores as the animal grows, and usually only the last four or five are functional; the others may close over. This mollusc is found in the Channel Isles only. The inside of the shell (8A) is a shimmering mother-of-pearl.

9 **Crepidula** (Slipper Limpet) is a mesogastropod, the characteristics of which are described on p. 38. It is becoming a pest in some parts, especially where it competes with the stocks in oyster farms. It was introduced accidentally at the end of the last century, possibly at Southampton, and is rapidly spreading round the coast; it has even got to the North Sea and Holland. The shell (9A) has an internal ledge like a slipper. It is a static animal, and tends to settle on the back of a slightly older individual, until a chain of about 9 is formed. The sex of the individuals is determined by their position in the chain; the oldest (underneath) are the females, and the youngest (on top) the males, so that there is a change of sex with age.

10 **Calyptraea** (Chinaman's Hat Shell), like *Crepidula*, is a filter feeder. It also has an internal shelf (visible in 10A, an upturned shell), though this is not as obvious as in *Crepidula*.

1 DIODORA APERTURA   4 PATELLA VULGATA
2 EMARGINULA RETICULATA   6 P. INTERMEDIA
3 ACMAEA VIRGINEA   5 P. ASPERA
7 PATINA PELLUCIDA   7A old specimen   8 HALIOTIS TUBERCULATA
9 CREPIDULA FORNICATA, stack   9A shell   10 CALYPTRAEA CHINENSIS   10A shell

TOPSHELLS AND COWRIES

There is scarcely a rock-pool which contains no topshells, and some have gay, variegated colours which makes them conspicuous and popular with collectors. Most of the topshells discharge their eggs into the water, where they are fertilized and pass through a planktonic *trochophore* stage, totally unlike the adult, with rings of vibratile cilia to propel the tiny organism and keep it in the upper levels of the sea; the shell begins to be formed just before the larva sinks to the bottom to change into the adult form. Some topshells, however, such as *Calliostoma* (3) lay gelatinous ribbons of eggs on the rocks, and these hatch out as shelled *veligers*, omitting the planktonic larval stage.

Some archaeogastropods show the primitive feature of paired gills within the mantle cavity (the so-called 'diotocardian' condition), and some of these are illustrated on the previous page. But others, the 'monotocardians', some of which are illustrated here, have lost the second gill. The primitive nature of these molluscs is indicated, however, by the fact that the remaining gill retains the double row of gill filaments.

1 **Gibbula magus.** The genus *Gibbula* is locally very common, with several important species. This one is probably commonest of all. The various species show a certain zonal distribution down the shore, like the limpets (p. 34) and the periwinkles (p. 38): *G. magus* is usually found lowest on the shore. It has a rather flat, solid shell with irregular bumps, particularly along the tops of the larger whorls; it is usually yellowish-white with pink or purple markings, and incorporates sand and gravel particles into the mucus on its foot, possibly as some form of concealment of the soft parts of the body.

2 **Monodonta** (Thick Topshell) is often placed in the genus *Gibbula*, as *G. lineata*. It has a rounded shell of rather dull colours, with zig-zag grey streaks over it, and the apex may be worn away to show the inner layers of the shell. The opening of the shell has a single tooth-like projection. This topshell is locally very common, on the coast between Dorset and N. Wales, and is generally restricted to a narrow zonal band around mid-tide level.

3 **Calliostoma** (Painted Topshell) is a noble inhabitant of pools on the lower shore. It occurs in many colour varieties, two of which are shown. (The specimen 3A was from Ayr.) The shell has almost straight sides and a flat base, so that it looks very much like a bell-tent. In addition to the usual pair of tentacles on the head, there are four additional pairs along the sides of the body above the foot, which can be seen when the animal is emerged from the shell; *Gibbula*, on the other hand, has only three pairs of these so-called epipodial tentacles. The eggs are laid in ribbons on a rock surface and the young emerge and creep away to establish themselves in the same general area as the parents.

4 **Gibbula umbilicalis** (Flat Top or Purple Top) is usually far higher on the shore than either of the other two species shown. It is flatter and has fewer pink markings

than *G. cineraria* (5). This topshell is more likely to be found in the South-West.

5 **Gibbula cineraria** (Grey Top or Silver Tommy) is usually found lower on the shore than *G. umbilicalis* but higher than *G. magus*. It has a rather taller shell than the other two shown here.

6 **Cantharidus** (Grooved Topshell) is smaller and taller than the others. A pinkish ridge forms four or five spirals round the shell, and the apex is red or pink. Most species of this genus are rather local in occurrence, and most are confined to the South and West coasts; they are sometimes found on the eel-grass *Zostera*.

7 **Tricolia** (Pheasant Shell, because of its many colours) is taller than it is broad and is the smallest of the topshells. Its shell is glossy and in spite of its name it occasionally occurs in a uniform chocolate-brown colour. It is a southern form.

8-9 **Cowries** are mesogastropods, the characteristics of which are described on p. 38: they are included here because they often occur in rock pools with the topshells. This is a remarkable group, because apart from their beauty of design and colour, they have a protective mechanism which involves folding the edges of the mantle over the top of the shell; by this means they cannot be attacked by such predators as starfishes. The adult shell has a toothed opening, notched at both ends. Its markings are variable, sometimes with bright spots and contrasting bands. The large handsome cowries of tropical seas do not extend to British waters: here, the group is represented by one genus, *Trivia*, which is about 10 mm long with 20 – 25 ribs. *T. monacha* (9) has three brownish purple spots on the shell. *T. arctica* (8) lacks the spots and is slightly smaller. A ventral view of the shell is shown in 8A.

1 Gibbula magus   2 Monodonta (Gibbula) lineata

3 Calliostoma zizyphinum   3a C. zizyphinum, colour variety   4 Gibbula umbilicalis

5 Gibbula cineraria

6 Cantharidus exasperatus   7 Tricolia pullus   8 Trivia arctica   8a ventral view

9 Trivia monacha

PERIWINKLES AND MARINE PULMONATES

The Mesogastropoda is a large group containing more than half the known species of molluscs. Almost without exception they have a single gill with one row of filaments and hence are regarded as higher in the evolutionary scale than those prosobranchs described on pp. 34 and 36. Typically, mesogastropods have a proboscis which can take water into the mantle cavity; it is particularly useful when the rest of the animal is buried, enabling it to acquire sufficient suitable water when it would be difficult otherwise. The proboscis is a modification of the mantle, and projects forwards or upwards; it acts as a kind of 'movable nostril'. This feature is also seen in the next order, the Neogastropoda, and a particularly good example is illustrated (1 on p. 43). There are a few mesogastropods that have become wholly terrestrial (*see*, for instance, 1 and 4, p. 55). Also included on this page is one of the only shore-dwelling pulmonate (lunged) snails (1).

1 Otina is, strangely, a member of the Pulmonata, the gastropod group mainly composed of land and fresh-water snails, the characteristics of which are described on p. 50. Very few pulmonates occur on the shore, and those that do are small: *Otina* is scarcely more than 3 mm long. It is confined to crevices and crannies, particularly the empty shells of barnacles, and among the lichen *Lichina pygmaea*, and hence it is usually found on the middle to upper shore. The shell is red-brown, thin and somewhat ear-shaped and has two whorls, the second of which contains the animal, which is white with distinct eyes. This mollusc occurs only in the South and South-West of the country.

2-5 Littorina (Periwinkles). These well-known molluscs occur abundantly all over the world, sometimes in clusters of enormous numbers. Of the several hundred known species, four are common in Britain, and they provide a splendid example of zonation: each species occupies a zone on the beach relative to tide marks, which, though they overlap, often give the shore collector a very good idea of his position on the shore relative to the tidal rise and fall. For instance, *L. neritoides* (2), the smallest species, occurs above high water mark, in clefts and slits in the rock which are merely splashed by the waves when the tide is high; *L. saxatilis* (=*rudis*) (3), the Rough Winkle, with a larger and usually sculptured shell, is next and occurs among the barnacles on those parts of the beach that are covered by only the highest tides; then comes *L. obtusata* (=*littoralis*) (4), the Flat Winkle, with a flatter shell than all the others, highly variable in colour, which occurs below neap high tide level, among the fucoid sea weeds; lastly there is *L. littorea* (5) the

Common Periwinkle, the largest of the species, which occurs in rock pools and on weed at low tide.

All the periwinkles are herbivorous, some feeding almost exclusively on lichens, others on fucoids. They have solved the problems of shore-living by adaptations not seen in many other molluscs. For instance, they do not merely liberate eggs and sperm into the sea but rely on a process of copulation for fertilization. In *L. neritoides* the fertilized eggs, in a protective shell, are liberated at spring tides and the young spend some time in the plankton before settling on the upper shore and climbing to the splash zone; in *L. saxatilis* the eggs develop within the female and are born with a shell and resembling the parent; the eggs of *L. obtusata* are laid in gelatinous masses on weed, and the young crawl from these; lastly, *L. littorea* sheds its eggs into the sea in groups of three to each capsule, and the young spend some time in the plankton, like those of *L. neritoides*.

Unlike limpets, the periwinkles cannot adhere closely to the rock for protection. Thus, they tend to nestle in crevices and cracks; but a strong sea may 'winkle' them free from their crannies, and they then roll about at the whim of the sea, closed up within their tough shell by a horny operculum.

6-7 Lacuna (Chink Shells) are so called because the shell opening has a tiny chink on one side of it; these are also littorinids, and they normally occur on weed on the lower shore or below. There are about 5 species in British waters; *L. pallidula* (6) and *L. vincta* (7) are about the same height but *pallidula* is slightly fatter. The shells are yellowish, and *vincta* has reddish bands.

2 LITTORINA NERITOIDES   1 OTINA OTIS
3 LITTORINA SAXATILIS (RUDIS)
4 LITTORINA OBTUSATA   5 LITTORINA LITTOREA
6 LACUNA PALLIDULA   7 LACUNA VINCTA

SPIRE SHELLS AND OTHER SAND GASTROPODS

The animals shown on this page mostly live on sand and other deposits, but the Pelican's Foot (2) and the Turret Shell (3) are generally not found alive on the shore, but only washed up from deeper water. These are all mesogastropods, the characteristics of which are described on p. 38. The Wendletrap or Staircase Shell (1) belongs to the family Epitoniidae, the Pelican's Foot Shell (2) to the Aporrhaidae, the Turret Shell (3) to the Turritellidae, and the Needle Whelk (4) to the Cerithiidae.

Included here is one example of a Necklace or Moon Shell (5) of the family Naticidae, which is a burrowing carnivore of the sand. The large, shiny shells of these gastropods are often found empty on the beach; the living animal, being a burrower, is not so often encountered, except at low tide, by digging.

1 **Clathrus** (Common Wendletrap, from a Dutch word meaning 'spiral staircase') is one of the most beautiful of the British shells. Like most of the others on this page, it really lives below tide levels, and is unlikely to be encountered alive, except at extra low tide levels; but its shell is often cast up, and gladdens the eye of the collector. It lays eggs in a string, held together with mucus (1A), one or both ends of which are inserted into the sand for anchorage.

2 **Aporrhais** (Pelican's Foot Shell) is also a sand burrower from deeper water, the shell of which may be cast up on the beach. The whorls are heavily ornamented, and as the animal reaches maturity it lays down a wide flange to the end of the last whorl which, of course, is not present in the immature shell (2A). The animal itself has a pair of whip-like pointed tentacles and eyes standing out like organ-stops.

3 **Turritella** (Turret Shell) is a sand-dweller from offshore water, the shell of which is sometimes cast up on the beach. It is often very common where it occurs, and its empty shell is a favourite shelter for young hermit crabs. It feeds by creating an inhalant current with its gill cilia.

4 **Bittium** (Needle Whelk) has a small, darkly coloured pointed shell with regular tubercles making the shell rough to the touch. It is found alive only at the lowest spring tides, on the South and West coasts of Ireland, and Wales, in Devon and Cornwall, and occasionally on the West coast of Scotland.

5 **Natica** (Necklace Shell) is so called because its egg mass resembles a wide string of beads or choker. This gastropod is a voracious carnivore that burrows in the sand and attacks bivalve molluscs which also burrow (5A). While the Necklace Shell holds the prey with its very mobile foot, it plays its proboscis against the bivalve's shell, generally just below the umbo (5B). A small gland at the proboscis tip secretes an acid which drills a neat round hole through the shell, so that the bivalve can be killed, its adductor muscle relaxed to allow the shell to open, and the contents scraped out by the predator's radula. In this animal, the shell is submerged in the fleshy lobes of the mantle and foot; in fact, when the animal is about to be attacked by a spinulosan starfish, such as *Asterias* (p. 181) it pulls the front and side flaps of the foot over its shell almost completely, so that the starfish's tube-feet cannot get purchase on the shell to hold it. There is an inhalant siphon (not shown in the picture) which emerges from the head region to be projected above the surface of the sand when the animal is burrowing, so that respiratory water can be brought to the gill.

One interesting point in the life of this animal requires further research. Most molluscs that extend soft parts, such as the foot, use a build-up in pressure of the blood in special spaces within the tissues. *Natica* uses this method too: but some workers report that such an agency is insufficient to account for the observed extensions of so large a structure, and it appears that sea water might be taken in by special pores to other spaces within the tissues to assist the protraction of the foot.

This gastropod, as its name suggests, lays eggs in a flat C-shaped ribbon which encircles the body as it is formed. The eggs are embedded in a jelly-like matrix which tends to get sand grains stuck to it, as one would expect. These ribbons, looking like miniature horse-collars, are often cast up on the beach at the strand-line. Normally they lie out on the sandy bottom, and the eggs hatch into tiny larvae.

1 Clathrus clathrus
1A String of egg cases of Clathrus clathrus
2 Aporrhais pespelicani   3 Turritella communis   4 Bittium reticulatum
2A A. pespelicani, young shell   5 Natica catena, moving towards its prey
5A Prey of Natica   5B bivalve bored by Natica

41

WHELKS, DRILLS, AND SPINDLE SHELLS

The order Neogastropoda appears to represent the culmination of the evolution of prosobranch gastropods, and the effect of torsion (*see* p. 34) on the internal anatomy of the body is seen at its ultimate. For instance, the nervous system is more concentrated and the sensory organs better developed. Its members may be either rock dwellers or dwellers on or in sand; most are carnivorous, feeding on other molluscs particularly bivalves (*see*, for instance, 5), and echinoderms (pp. 178 – 187). For this purpose they are provided with an external proboscis, and in most cases the action of this in boring through the shell is mechanical, in contrast to that of *Natica* (p. 40) which is chemical. The proboscis contains the radula, with its rasp-like teeth, which, as they wear out, are replaced from a sac-like formative region in the mouth cavity. When the animal is not feeding, the proboscis is contained in a sheath in the head, which has a mouth-like opening in the front. To feed, the proboscis is protruded from this opening by pressure of the blood from behind and the radula is exposed at the proboscis tip. Some of these predatory carnivores, such as *Buccinum* (1) and *Neptunea* (2), grip the prey with the foot and use the edge of their shell to wedge open the two valves of a bivalve shell, so that the proboscis can be insinuated to the soft tissue within, while others, such as *Urosalpinx* (3) and *Nucella* (4), smother the victim with the foot to render it easier to attack.

1 **Buccinum** (Whelk) has a large shell which is often cast upon the beach, sometimes heavily encrusted with epizoic organisms and even inhabited by a hermit-crab (*see* book-jacket). Common too are the egg masses (1A), often cast up empty, along the strand-line. There is a prominent siphon and when the animal is on the move, the operculum is carried on the rear of the foot; when the animal withdraws into the shell, the operculum acts as a lid to close off the shell opening.

2 **Neptunea** (Spindle Shell) may be even bigger than *Buccinum*, but the shell is smoother. This gastropod is found only in the North. The picture shows particularly well the groove in the shell opening through which the siphon is protruded.

3 **Urosalpinx** (Oyster Drill) can be a harmful pest to oyster beds. The sole of the foot has a protractible gland which is applied to the area to be drilled; the secretion has not yet been fully elucidated, but is *not* an acid: more likely, it is a chemical which denatures the chalk of the shell. The drilling operation consists in alternately applying the gland and the end of the proboscis to the place to be drilled; the gland softens the area, and the radula on the proboscis scrapes at it, until the shell is perforated, when the proboscis is intruded to the interior of the shell. The radula then works on the soft tissues, which are taken into the gut.

The shell resembles the commoner *Ocenebra* (7) but is smaller, darker, and less ridged. It was apparently introduced from America to British water in the 1920s with American oysters but so far it has not penetrated much beyond Essex and Kent. Unfortunately, wherever oysters are extensively farmed in the waters of these counties, it is a serious pest.

4 **Nucella** (Dog Whelk) is often called *Thais*, and there is uncertainty about which generic name has priority. This is one of the commonest and most widely distributed of the shore gastropods, and one which occurs in many colour varieties, three of which are shown. Scarcely a rock is overturned in the middle zones of a rocky shore which does not yield specimens of this mollusc, and crevices are sometimes stuffed to overflowing with their brightly coloured shells. Another common sight is the egg masses, little vase-like capsules laid sometimes rather sparsely (*see* 4C), and sometimes in vast carpets covering many square millimetres. Each capsule may contain several dozen eggs, and the young whelks emerge by climbing out of the open end of the capsule. The adult shell has a smooth rim to the opening and a notch at the lower end. This is where the siphon lies, through which water is drawn to the mantle cavity when the animal is active. This animal, like so many other neogastropods, is a voracious carnivore, feeding on limpets, topshells, and particularly barnacles, by means of the protrusible proboscis.

5-6 **Nassarius** (Netted Dog Whelk) looks rather like a skein of brown hair retained by a hair net. *N. reticulatus* (5) often occurs on sand, where it preys on bivalves, as shown in the picture. It sometimes wedges open the valves with the edge of its shell and rasps away at the meat with its proboscis. It occurs on the lower shore and below. *N. incrassatus* (6) is half as big as the other common species, and often occurs on stonier ground.

7 **Ocenebra** (Sting Winkle) uses mechanical and chemical means to drill holes in the shells of its prey in a manner very similar to that of *Urosalpinx* (3), except that here the accessory drilling gland is on the edge of the mouth.

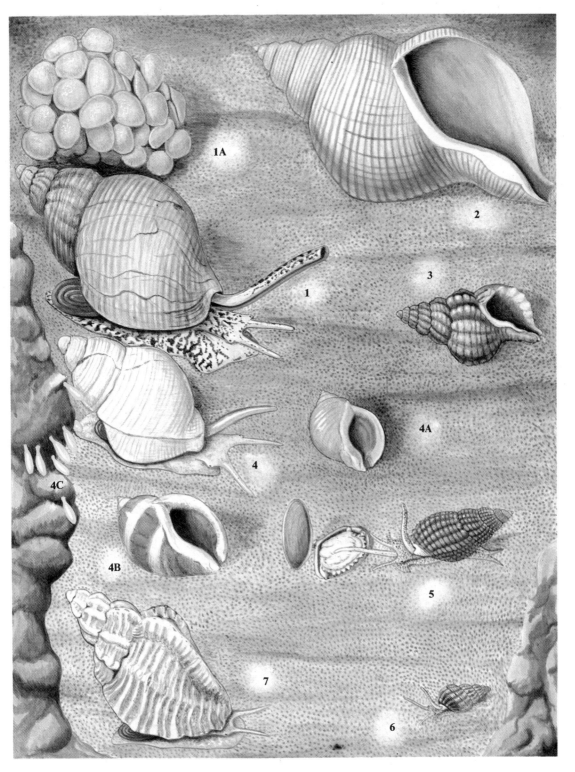

1A Egg-mass of BUCCINUM
1 BUCCINUM UNDATUM
4 NUCELLA LAPILLUS
4C N. LAPILLUS, egg cases
7 OCENEBRA ERINACEA

2 NEPTUNEA ANTIQUA
3 UROSALPINX CINEREA
4A, B N. LAPILLUS, colour varieties
5 NASSARIUS RETICULATUS
6 N. INCRASSATUS

OPISTHOBRANCHS OF SAND AND SHALLOW SEA

The Opisthobranchia is an exclusively marine group in which we see a progressive loss of the shell, with corresponding changes to the soft parts including, remarkably, a loss of torsion (*see* p. 34) and a return to a more symmetrical form to the nervous system and other soft parts. The mantle cavity is also lost in the advanced members, and the gills are often visible externally (*see*, for instance, 3, p. 45, and all the dorids illustrated on p. 47). Why, if the shell of other molluscs affords such protection, can members of this group manage without? The answer almost certainly lies largely in chemical defense. The bodies of these gastropods are made highly distasteful by means of secretions, and any predator taking one nibble will be instantly put off. Some opisthobranchs are coloured to blend with the animal on which they browse, so that they escape detection (*see*, for instance, 2, p. 47, and 5, p. 49). Surprisingly, most of the molluscs illustrated on the opposite page can swim with varying efficiency: only *Actaeon* (6) which is a burrower, appears to be confined to the bottom. The swimming tendency teaches its acme in the pteropods, such as *Clione* (1).

1 Clione (Naked Sea Butterfly) is a member of the order Gymnosomata ('naked body'), in which the shell is lost and the sides of the foot are elongated into parapodial wings which propel the animal through the water. There is no mantle cavity. This deceptively beautiful little carnivore swims quite rapidly after its prey, often the shelled sea butterflies (not shown). These it seizes with its mouth, a heavily armoured structure with teeth, jaws, hooks and papillae by which the prey is crushed and partly masticated.

2 Aplysia (Sea Hare) differs from most other opisthobranchs in being a herbivore; it crops sea-weed by means of a pair of jaws and a broad rasp-like radula. The shell, though not visible externally, except by probing among the folds on the top of the body, is reduced to a mere vestige, thin, transparent, and flexible — and whether such a pathetic structure affords any protection is questionable. The anterior tentacles project forwards like a pair of spatulae, and the posterior tentacles project upwards like thin ears. The animal lives low on the shore or in shallow water offshore, but comes inshore to spawn in the spring, sometimes in vast numbers. The eggs are laid in pink gelatinous strings (2A), coiled around weed or rocks. When handled, the animal tends to secrete a purple mucus from glands in the mantle.

3 Pleurobranchus has a shell of even more miserable size than that of *Aplysia:* it is wholly internal, and covered by the tuberculated mantle. The mantle cavity is non-existent and the single gill is carried on one side of the body, looking as though it has flopped out on that side. There are prominent tentacles on the head. The animal feeds on large simple sea-squirts, such as *Ascidia* (p. 189) and may come inshore in the summer in quite large numbers.

4 Akera. This, like *Aplysia* (2) and *Haminoea* (5), belongs to the order Anaspidea, meaning 'without a shell'; but this is not altogether a valid description, since the shell is there as in *Aplysia*, but is almost hidden beneath the lateral flaps of the mantle (as shown). In spring, at spawning time, the animal uses these flaps to swim in short bursts. There is a long tentaculate projection of the mantle which protrudes from the spire of the shell. The head is elongated. This is a dweller on mud-flats at low-tide. particularly in the South and West.

5 Haminoea, rather similar to *Akera*, also has a thin, coiled shell round which the mantle flaps wrap when it is not swimming. It is a deposit feeder, like *Actaeon*, picking up particles of food from the surface of the sand as it moves across it.

6 Actaeon is a member of the order Bullomorpha, meaning 'bubble-shell', the alternative name of which is Cephalaspidea, 'protected by a shield'. This pretty little mollusc is sometimes found low down on sandy beaches as the tide recedes, trying to burrow into the wet sand or anchoring its gelatinous egg case into the sand by a sticky thread. The edges of the mantle invest the sides of the shell, and two flaps from the head protrude backwards over the front part of the shell. It feeds by sucking in the film of organic matter lying on the surface of the sand.

7 Philine is a bulloid like *Actaeon*, but here the shell is internal and the body flattened. This mollusc has a strong calcified gizzard with which it crushes its prey.

1 Clione limacina
2a A. punctata, eggs 2 Aplysia punctata
3 Pleurobranchus membranaceus 4 Akera bullata
5 Haminoea navicula
6 Actaeon tornatalis 6a empty shell 7 Philine quadripartita

DORID NUDIBRANCHS

The order Nudibranchia contains the Sea Slugs. Here, both the shell and the mantle cavity are finally lost, and the animals, as though relieved of the repressive burden of these typically molluscan characters, assume nearly perfect bilateral symmetry, and take on a slug-like form. The upper surface of the animal becomes important in many of the vital functions — such as sensation, respiration, getting rid of waste, camouflage and other defense mechanisms — and the under surface retains the pad-like foot and usually the mouth. The name of the order means 'naked gill' and refers to the tufts and processes which sprout from the dorsal side. Logically, the name 'nudibranch' should refer only to the members of the suborder Doridacea, shown on this page, since only in these sea slugs are the dorsal structures actually gills; in the suborder Eolidacea, shown on the next page, the dorsal processes serve other functions in addition to respiration. These are most beautiful animals, sometimes bearing striking colours. They often come onshore to spawn, and some lay long ribbons of eggs which develop into larvae sometimes having a small shell and mantle cavity, to betray their ancestry.

1 **Onchidoris** has a single ring of gills surrounding the anus on the upper surface; this ring has a variable number of processes, but usually between 11 and 30, the circlet being incomplete behind. The mantle overhangs the foot, and has tubercles scattered over it. It feeds on barnacles (p. 129) and polyzoans (p. 171).

2 **Rostanga** is not more than 10 mm long; its colour is red with black spots, and with yellow tentacles. There are usually 10 rather squat gill processes. The red pigmentation of this nudibranch almost exactly matches that of the red sponges on which it feeds and it is easily overlooked by the collector. It is common in the South.

3 **Archidoris** (Sea Lemon) is the commonest of the sea slugs. Elliptical in shape, it has a yellowish mantle with tuberculated surface and nine 3-armed gills. In this form and in *Rostanga* (2), both of which belong to the family Dorididae, the gills are retractile, while in the others, of the family Polyceridae, they are not. The one shown in the picture is a yellowish species *A. pseudoargus*, which feeds on the Crumb-of-bread Sponge, *Halichondria* (p. 5), on which it is often found in the spring; if it does feed on red sponges, it generally assumes a redder colour itself.

4 **Polycera** has a translucent white body with several rows of yellow tubercles. In addition to the other processes which protrude from the body, this form has six pointed, orange-tipped processes to the mantle flaps. There are seven gills. This one is often found on the belt-weed, *Laminaria*, feeding on the polyzoans (p. 173) which encrust the fronds of the weed.

5 **Acanthodoris** has a very convex body, and a colour ranging from white to purple-brown. The tentacles curve backwards, and there are nine gills. This one is common on polyzoans such as *Flustrella* and *Alcyonidium* (p. 173).

6 **Goniodoris** has a smooth-surfaced body with a fairly pronounced keel running down its upper mid-line. *G. nodosa*, shown here, has a pink tinge, speckled with white, and the tentacles may be yellow, while *G. castanea* (not illustrated) is reddish brown with white spots. Both species feed on sea squirts (p. 191) and polyzoans (p. 173).

7 **Ancula** has a transparent body with yellow and orange markings. It is superficially similar to *Polycera* (4), but the three large two-process gills are surrounded by a ring of 10 orange-tipped sensory processes.

8 **Adalaria** has a yellow or white elliptical body with large tubercles covering the upper surface, and eleven gills. It feeds on polyzoans (p. 173).

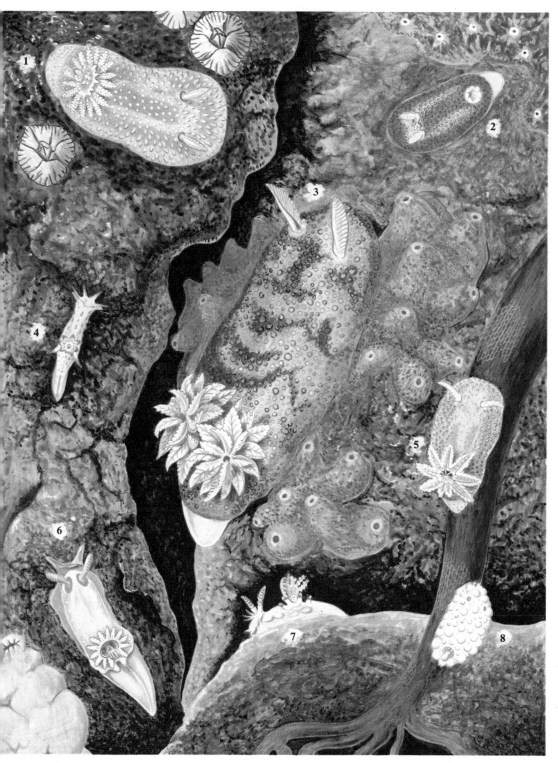

1 ONCHIDORIS OBLONGA   2 ROSTANGA RUFESCENS
3 ARCHIDORIS PSEUDOARGUS
4 POLYCERA QUADRILINEATA   5 ACANTHODORIS PILOSA
6 GONIODORIS NODOSA   7 ANCULA CRISTATA   8 ADALARIA PROXIMA

EOLID NUDIBRANCHS

The suborder Eolidacea has most graceful and highly colourful members. Although they are 'nudibranchs' ('naked gills', *see* p. 46), the processes on the dorsal side of the body called *cerata* (singular: *ceras*) are not really gills, though they do take part in the respiratory process. Each ceras contains elements of the blood-filled body cavity, and it is between these elements and the sea water that respiratory exchange mostly takes place. Each one also has diverticula of the gut within it, and it is in connexion with these blind channels from the gut that we see the most interesting feature of these animals: their ability to store the poison cells called nematocysts from the hydroid coelenterates (p. 9) on which they feed. What apparently happens is that the eolid creeps over the hydroid (as *Tergipes* (6) is shown creeping over a patch of *Sertularia*) and browses on the polyps. As it does so it prevents their discharge by some mechanism not understood, bundles them into convenient packets with mucus, and passes them up into the lumina of the cerata, where they are stored. Each ceras opens by a tiny pore to the outside, and when the eolid is molested slightly, it exudes a few packets of nematocysts which float out into the surrounding water, to be taken in by the attacking animal, thus releasing the poison. Stronger attacks from predators release more and more packets of nematocysts, and ultimately the eolid will sever one or more cerata at their bases (a process called *autotomy*) so that the predator gets a mouthful of poison instead of a tasty meal. Further defense of the eolid may be provided by the vivid colours some of them show; these may act as warning to a persistent predator so that, having once tasted a brightly coloured ceras, it will keep well away next time. On the other hand, in some cases there is little doubt that the bright colours are camouflage, since the animals on which they browse are also highly coloured.

Included also on this page are nudibranch-like molluscs of two other orders, Onchidiacea (1) and Sacoglossa (2, 4 and 5). The onchidiaceans are in some ways like shell-less limpets, though their true affinities are far from clear. The sacoglossans are modified to suck the cell-sap of algae, and so they have very modified feeding structures compared with other opisthobranchs.

1 **Onchidella (Onchydium)** has a slug-like body with warts but no cerata. A veil of the mantle covers the head, giving the whole animal a dome-like appearance. This animal, confined to the South-West, creeps about on rocks on the upper shore, making for crevices when the weather becomes too dry. It resembles land slugs (p. 63) in possessing a lung-like structure posteriorly and was formerly regarded as a pulmonate mollusc; but now it appears to be related to other opisthobranchs, and is placed in an order separate from the rest of the nudibranchs.

2 **Alderia** is a sacoglossan (see above). Its name honours the memory of J. Alder, whose work with A. Hancock on the British nudibranchs is a milestone in their study. The brown body is less than 10 mm long, flattened and with light greenish-brown cerata restricted to the sides. This sea slug is typically found in East Coast estuaries and inlets, where it occurs on the eel-grass *Zostera* and on other marine plants.

3 **Aeolidia** (Grey Sea Slug) is the largest British nudibranch, sometimes exceeding 90 mm in length. The upper surface of the body is almost covered by brownish-green cerata, except for a patch in the front mid-line. This is found under stones on the middle and lower shores; it feeds on sea anemones and is fairly common. It lays eggs in white spiral ribbons on the undersides of rocks and overhangs.

4 **Limapontia** is a tiny sacoglossan, scarcely exceeding 3 mm long, in which the tentacles are reduced to orange or yellow patches at the front end. It often occurs in large numbers on filaments of green algae. *L. capitata* (shown here) is generally on the middle or lower shore, while *L. depressa*, which lacks the tentacular patches, (not illustrated) is commonly found on the green weed of salt marshes.

5 **Acteonia.** This well-camouflaged sacoglossan occurs on green weed in rock-pools where it is often overlooked. It is about 5 mm long, velvety brown with lighter sides and tentacles.

6 **Tergipes** has eight relatively large cerata projecting alternately from each side of the rather thin body. The green colour of the gut and its branches into the cerata show through the transparent body wall. This is a tiny nudibranch, seldom more than 5 mm long, and it is often hard to spot on its prey, the hydroids and polyps (p. 9) which grow on rocks and weeds.

7 **Facelina.** This beautiful creature has a slender body from which clusters of cerata emerge, each with crimson sides and purple tip. The tentacles are ringed and usually yellow-brown.

1 Onchidella (Onchydium) celtica   2 Alderia modesta
3 Aeolidia papillosa   4 Limapontia capitata
5 Acteonia senestra   6 Tergipes despectus
7 Facelina auriculata

GARDEN SNAILS

The Pulmonata (pp. 50 – 71) is the third great order of the gastropod molluscs. The difference between these gastropods and those on the preceding pages is that here the water-breathing gills have been lost and the mantle cavity has become an air-breathing *lung*, with a hole to the outside, the *pneumostome*, by which air is taken in and expelled (shown best in *Helix pomatia*, p. 55, 3, and in the slugs, p. 63). They generally show the typical gastropod spiral shell, borne by an elongated head-and-foot, though in slugs the shell is nearly or completely lost. With the modifications to the mantle cavity, there are modifications to the arrangement of the nerves, too: no longer is there a cross-over of nerves going to the viscera, as there is in the prosobranchs (p. 34), but instead the nerve cords are shortened and come to lie, uncrossed, near the head. Pulmonates are hermaphrodite and undergo copulation to fertilize the eggs. One feature of the process in land snails is particularly interesting: prior to copulation the two snails who will exchange sperm approach each other and drive chalky darts into the body wall of each other. This drastic process, reminiscent of a matador's preparation of a bull, apparently stimulates each partner for the sex act. A snail is born with a tiny shell which becomes bigger as the animal grows.

1-2 **Monacha (Theba) cartusiana** (Chartreuse Snail) and **M. (T.) cantiana** (Kentish Snail) both have a tell-tale orange tinge to the last part of the largest whorl of the shell. *M. cantiana* (2) is the larger of the two and has a slightly higher spire and a more translucent shell. They both inhabit the same general environment, banks and hedges, and are usually most conspicuous in hot dry weather, attached to the stems of fairly tall plants. *M. cantiana* is widespread throughout the country, while *M. cartusiana* is more confined to South-Eastern counties.

3 **Helix aspersa** (Common or Garden Snail — the original anglicised name which gave this phrase to our everyday speech). The saying 'familiarity breeds contempt' was never more apt, for this maligned creature which the gardener drops unceremoniously into a jar of brine carries an exquisite shell and a gentle beauty in the sculpturing of its body wall. There is considerable variation, not only in colour but in shell shape and thickness; for instance, in some non-calcareous districts, such as in the Channel Islands, there is a small, thin-shelled form very different from the usual one. Lift a rockery stone in the winter months, and generally you will find aggregations of this snail hibernating, with a thick mucous seal across the shell opening. The 'home' is used all the year round, in summer as a base from which to make its foraging trips to find leaves and fruit on which to feed, and in winter as a den in which to await the warmer weather. This snail is not common in industrial areas, where the smoke probably poisons the leaves on which it feeds. It has apparently been eaten since Roman times until comparatively recently, and even now unscrupulous delicatessens sell the species instead of the larger and more tender *H. pomatia* (p. 55).

4 **Hygromia (Trichia) striolata** (Strawberry Snail) has regular striations on each whorl of the shell and a lighter colour to the end of the last whorl. There may be a slight keel to the shell, particularly when young, and young specimens are also rather 'hairy'. This snail lives in damp places, particularly under vegetation and in hedges, and is a pest on strawberry beds. It is found throughout England and in Scotland as far north as Perth.

5 **Trichia hispida** (Hairy Snail) appears dull because of the fine covering of 'hairs' on the shell. It is common throughout Britain under mossy stones and logs, in gardens and in the wild. The 'hairs' are tiny curved projections from the horny outer layer of the shell, and probably serve to help anchor the shell in crevices. Sometimes the hairs get rubbed off, and the shell appears shiny in patches.

6 **Oxychilus cellarius** (Cellar Glass Snail) has a flat, semi-transparent shell of yellowish-brown, with white near the hole in the centre of the underside. The animal itself is a light-grey with tiny brown spots on the edge of the mantle and has an unpleasant pungent smell. This snail, common throughout the British Isles, lives in damp places in gardens and woods, and particularly in damp buildings, outhouses and cellars. It feeds on fungi.

7 **Oxychilus alliarius** (Garlic Glass Snail) is smaller than *O. cellarius*, and smells strongly of garlic. The shell is glossier than that of *cellarius* and the brown colour of the underlying body shows through more strongly. This snail lives on walls and banks, in gardens and elsewhere all over Britain.

8 **Pyramidula** (Rock Snail) is always found on stone walls, buildings or rocks, especially limestone. The shell is shaped like a depressed cone, light brown in colour, with four or five whorls. It does not occur in East Anglia or North-East Ireland, and is rather local in Scotland.

1 Monacha (Theba) cartusiana   2 M. (T.) cantiana   3 Helix aspersa

4 Hygromia (Trichia) striolata

5 Trichia hispida   6 Oxychilus cellarius

8 Pyramidula rupestris   7 O. alliarius

SNAILS OF THE HEDGE-BOTTOM

The debris under the canopy of a hedge provides a wonderful habitat for so many snails; the air, the soil and the leaf-litter retain moisture because the sun cannot penetrate, and such conditions are favourable to molluscs, which must retain a moist body surface for locomotion and other vital functions.

1   **Acanthinula aculeata** (Prickly Snail) has a crest of sharp spine-like processes making a rough keel round each whorl of the shell. It has a tiny brown shell, scarcely more than 2 mm high and 2·3 mm broad. It occurs throughout Britain, and is chiefly found among dead leaves under hedges, but also in woodlands.

2   **Ashfordia** (Silky Snail) is sometimes referred to the genus *Monacha* (*see* 1 and 2, p 51). The thin shell has a covering of fine, stiff whitish hairs which are rather thick at the base. The animal itself has a somewhat translucent body, speckled with white. It is usually found under hedges in low-lying places, where the water may lie during winter months; in fact, this is one of the snails that is almost amphibious.

3   **Cepaea hortensis** (White-Lipped Banded Snail) shows considerable variation both in background colour and in the number and thickness of the dark bands on the whorls. The commonest forms are yellow with five bands or yellow with no bands, though varieties with one, two, three or four bands are all known. There is also a form with a dark lip, which makes it difficult to distinguish on shell characteristics alone from its relative *C. nemoralis* (4).

4   **Cepaea nemoralis** (Grove Snail or Dark-Lipped Banded Snail). Here again, there is great variation in background colour and banding pattern: the background colour may be pink (4), yellow (4A) or brown, and any or all of five bands may be present, either showing singly or fused in some way, or there may be no bands at all. Sometimes even the bands are broken up into mere spots. Both species of *Cepaea* are preyed on by the thrush, which cracks open the shell on convenient stones called 'anvils' (4B). It is clear that the banding and colour patterns serve to camouflage the snails when they are in different habitats: a yellow background, for instance, will be preferable in grassy places and a brown in woodlands; a many-banded form will be less visible in vegetation, and a non-banded or fuse-banded form will be at an advantage on a plain background. Since these snails occur in a variety of habitats in addition to hedges, such polymorphism in colour pattern, genetically determined by the parents, is an obvious advantage.

5   **Cochlicopa lubrica** (Slippery Snail or Moss Snail) has a tall spire to its very shiny, semi-transparent shell, and the sutures between the whorls are fairly deep. Interestingly, it has a sexual excitatory organ by which it prods its co-copulant into sexual activity, taking the place of the love-darts of snails like *Helix* (p. 50). One finds this mollusc in moss and dead leaves under the hedge, nestling in rotten wood or in turf.

6   **Helicella (Candidula) gigaxi** (Excentric Snail) is so-called because the canal running from the centre of the underside up the inside of the spire, the *umbilicus*, appears excentrically placed because the last whorl meets the previous one about half way across its thickness. The opening to the shell is sharp, and just inside, as seen in the figure, there is a faint pinkish internal rib. It feeds on dead and decaying plants. Like its relatives *Helicella caperata*, *H. virgata*, and *H. itala* (shown on p. 61), it is tolerant of dry conditions.

7   **Oxychilus helveticus** (Glossy Snail) is another species of the same genus of which two others are shown on p. 51 (6 and 7). This one is best distinguished from the others by the light brown colour of its shell, with the slightly darker band near the opening, and by its lack of a garlic smell. It lives in hedges, woods, and sometimes quarries; it is rare in Scotland and apparently absent from Ireland.

8   **Lauria** (Chrysalis Snail) looks rather like a moth chrysalis. It has a tiny shell which is, according to one authority, 'not above one quarter part of a barleycorn'. The shell is glossy and faintly striated, and the white-rimmed opening is the shape of half an oval, across which projects a single tooth, giving the opening the appearance of a very young baby yawning.

1 Acanthinula aculeata   3 Cepaea hortensis   2 Ashfordia (Monacha) granulata
4 C. nemoralis   4a C. nemoralis, colour variant   3a C. hortensis, colour variant
4b Thrush anvil, with broken Cepaea shells
5 Cochlicopa lubrica   6 Helicella (Candidula) gigaxi
7 Oxychilus helveticus   8 Lauria cylindracea

SNAILS OF THE DOWNLAND

The dry, friable soil of a typical downland supports a particular collection of snails. This is not to say these snails live nowhere else: no rigid ecological conditions can be quoted for most land snails, but downland is the sort of habitat in which these snails are particularly found. Almost all land snails are, of course, members of the gastropod subclass Pulmonata (lung-breathing — see p. 50 for description); but two of them, *Acicula* (*Acme*) and *Pomatias*, both shown on this page, are highly adapted members of the primitive subclass Prosobranchia, whose members are more appropriately marine (*see* p. 34). They both have an operculum for closing off the shell opening when the animal is withdrawn; this is a permanent structure, not like the temporary calcareous seal which *Helix pomatia* (3) secretes before hibernating. In *Acicula* and *Pomatias*, in parallel with pulmonates, the original gill has been lost and the mantle cavity has become a lung for air-breathing.

1 **Acicula lineata** (= **Acme fusca**) (Point Snail). The shell of this mesogastropod is long and narrow, yellowish-brown and shiny. The operculum, a rare feature in land snails, and present here because this is a proso-branch rather than a pulmonate (*see* above), is thin and horny. This snail lives in downland grass tussocks, among moss and dead leaves. It is more common in Ireland than in most parts of England, Wales and Scotland.

2 **Arianta arbustorum** (Copse Snail) may be mistaken, at first sight, for a small specimen of *Helix aspersa*, but it is more globular and has a dark peripheral band, with a white border to the shell opening; its overall colour is a richer brown, sometimes described as 'chestnut'. Although often found on downland and sandhills, it may also occur in damp places such as marshes, river banks and damp woodland.

3 **Helix pomatia** (Roman Snail or Edible Snail) is the largest of the British snails, with a thick, globular shell of cream or pale-yellow colour, rather coarsely striated. The animal itself is pale yellow-grey with a yellowish border to the foot. Interestingly, in places where the chalk or limestone comes to the surface, this snail tends to lose the horny outer layer to the shell, so that it appears almost chalky-white, which may aid its concealment in such localities. Like *H. aspersa* (p. 50), this snail hibernates for about half the British year, and seals its opening with a false operculum, composed of mucus with chalky fragments in it; when it emerges in April or May, this seal is discarded. Even some English people eat this snail (the 'escargot' of French cuisine) and apparently have been doing so since Roman times. In fact, the Romans had nurseries in which the snails were fattened up on bran soaked in wine. Some of our forebears would attribute healing powers to it; one knight of the realm is said to have used it to 'cure his beloved wife of decay', but of what and whether it worked is not clear.

4 **Pomatias elegans** (Round Mouthed Snail or Land Winkle) is the other prosobranch found on land (*see* also *Acicula*, 1). It used to be known as *Cyclostoma* ('round hole') a name which reflected its outstanding character, namely, an almost perfectly circular opening to the shell; the name *Pomatias* comes from the Greek word for 'door', referring to the operculum, a hefty calcareous plate. The shell has strong spiral ridges and transverse striations. Usually, the tip of the shell has a mauve tinge to it while the rest is pinkish-brown. It spends most of its time burrowing in the loose friable soil that is so characteristic of chalky downlands. The snail will not usually be found in those parts of the downland where the soil is basically clay-with-flints, but only where the soil is underlain by limestone. It eats mostly dead leaves. The front end of the foot is forked, and the snail adopts a kind of bipedal gait, shuffling along by moving each half independently of the other.

5 **Carychium minimum** (Herald Snail). This stubby spire-shell has three small teeth within the thick-lipped mouth opening. It is common in grass-roots and leaf mould throughout Britain. There is a close relative, *C. tridentatum*, which is slightly longer and has larger teeth in a more elongated shell-opening, living in drier conditions than *C. minimum*.

6 **Abida secale** (Large Chrysalis Snail) has a more pointed spire than the Chrysalis Snail, *Lauria* (shown on p. 53). The outer lip of the opening is slightly splayed, and there are about 9 teeth projecting inwards on the inside of the opening; these teeth help to anchor the body within the shell. This is another of the snails that live in the rather dry soil and rock fragments of the downland, and the best way to find it is to shake the soil out of a root-tussock of grass on a dry limestone hillside. It occurs in most of England except the Eastern counties, but is absent from Ireland and Scotland.

7 **Caecilioides acicula** (Blind Snail). This little snail, with a semi-transparent white shell, lives underground in friable soil. Eyes are present, though they are un-pigmented. This is another snail of dry grassy places. It is absent from Northern Ireland and Scotland.

1 ACICULA LINEATA (ACME FUSCA)   2 ARIANTA ARBUSTORUM
3 HELIX POMATIA
4 POMATIAS ELEGANS   5 CARYCHIUM MINIMUM   6 ABIDA SECALE
7 CAECILIOIDES ACICULA

MARSHLAND SNAILS

Snails vary markedly in their ability to withstand desiccation. On the previous page we saw a group which managed to cope with conditions that are usually very dry indeed; but on this page we see snails which seek out wet places such as stream banks, marshes, and water-meadows; their tolerance of dry conditions is much lower than that of other pulmonates. Lift a fallen log from squelchy mud, and these are among the snails you are most likely to find beneath it.

1 **Retinella radiatula** (Rayed Snail) has a flat spiral shell, very glossy and rather thin, with strong regular striations across each whorl which appear to radiate from the shell centre. The shell is light brown with a slightly lighter underside. It occurs in damp places generally, particularly in willow-beds and other marshy ground. It is never very common, but occurs all over Britain.

2 **Succinea putris** (Amber Snail) is large enough to be easily noticed in wet places. It has a thin shell of amber colour which becomes thicker and more opaque with age. This animal, like so many other molluscs, is the intermediate host for a fluke (a digenean flatworm — *see* p. 25), in this case *Distomum macrostomum* whose final host is a bird. One of the problems of parasites is reinfecting a new host, and normally the intermediate host is, of course, preyed on by the final host. But in this case, the larval fluke in the mollusc actually advertises its presence by becoming lodged in the tentacles and turning bright orange, so that the bird catches sight of the snail more easily; the bird pecks off the tentacles, thus reinfecting itself, while the snail regenerates new tentacles. The snail is found in water-meadows and other damp places.

3 **Vertigo** (Whorl Snail) has a squat little shell of chestnut-brown with four or five teeth within the shell opening. Besides damp places, this snail is found in grass-roots and under stones and logs. *V. pygmaea* (illustrated here) is the commonest of nine or so species of this genus which occur in Britain.

4 **Columella edentula** (Toothless Chrysalis Snail) resembles *Vertigo* (3), *Lauria* (p. 53) and *Abida* (p. 55), but there are no teeth within the shell opening. It is a typical damp-loving snail, and the Butterbur is said to be a favourite plant it frequents. It occurs throughout Britain.

5 **Euconulus fulvus** (Tawny Glass Snail) has a conical shell with gradually-enlarging whorls and a prominent spire. It lives throughout Britain, anywhere there is a bit of moisture, and it can even survive flooding for fairly long periods.

6 **Vitrea crystallina** (Crystal Snail) is flatter than *Euconulus*. Its shell is thin and greenish-white in colour. It usually lives underground or among the roots of marsh-plants.

7 **Zonitoides nitidus** (Shiny Snail) has a glossy brown shell and the animal itself is very shiny. This snail is so much at home in wet situations that it frequently occurs in association with true aquatic snails such as *Planorbis* (p. 69) and *Sphaerium* (p. 91). It occurs throughout Britain.

1 RETINELLA RADIATULA   2 SUCCINEA PUTRIS   3 VERTIGO PYGMAEA
4 COLUMELLA EDENTULA
5 EUCONULUS FULVUS   6 VITREA CRYSTALLINA
7 ZONITOIDES NITIDUS

WOODLAND SNAILS

A damp environment is demanded by a great many land snails, and the humid crannies and spaces in the litter on a woodland floor, with an abundance of rotting vegetation, fungi, and soft roots, make an ideal habitat for some members of this group. Most molluscs require lime for the healthy construction of their shells, and so a woodland in a limestone district is a particularly good place to search for many land snails. Find what the poet called a 'mossy moot' at the foot of a beech tree, pull away the ivy carpet round its base or pick off the bark from a fallen branch, and most of the snails illustrated here will be found easily.

1 **Azeca goodalli** (Three-Toothed Snail) has seven rather flattened whorls, with very shallow sutures between each. This snail resembles *Cochlicopa lubrica* (5, p. 53), and the best way to distinguish between them is that in *Azeca* the shell is held horizontally, not tipped up at the end. Though there is some variation, most members of this species have three teeth within the mouth. This is a limestone-loving snail, found among the dead leaves and moss in woodland in England and Wales and a few places in Scotland.

2 **Balea perversa** (Tree Snail) has, like *Marpessa* (3) and *Clausilia* (6), a sinistral shell, that is, one in which the spiral twists in the opposite way to most other gastropods, and hence the rather insulting specific name. It has a pear-shaped opening in the shell, with one small tooth sometimes present; the lip round the opening is thin and slightly flared. This is one of the snails that usually lives in crevices of trees, especially beech, elm, ash, willow and apple, and is scarcely ever found on the ground. It is also said to avoid the presence of other animals. It occurs throughout Britain.

3 **Marpessa laminata** (Plaited Door Snail), with sinistral shell (*see* 2), is a noble woodland dweller, the biggest of the chrysalis-shaped snails in this country. The shell is a glossy yellow-brown with delicate striations across each whorl. The opening has a reflected lip which is lighter in colour than the rest of the shell. There are complex folds and processes within the shell opening. This shell is very common on beech-woods, where it is easily mistaken for beech buds, though whether this bestows any selective advantage on the mollusc is not known. It is frequent in England, but less common in Wales, Scotland, and Ireland.

4 **Goniodiscus rotundatus** (Rounded Snail) has a lens-shaped shell with very characteristic brown stripes

curving across each whorl and fairly strong ribs on the upper side. During the day this snail lies concealed under leaves or moss, but makes night-time feeding forays for fungi and decaying matter. It is common throughout England, Wales and Ireland, but less common in Scotland.

5 **Ena obscura** (Lesser Bulin) is often hard to find, as the specific name implies, because the sutures between the whorls collect grit and other foreign matter which tends to camouflage the shell. It is found among dead leaves and moss and sometimes climbs trees and rocks in wet weather. It is found throughout Britain and Ireland.

6 **Clausilia bidentata (C. rugosa)** (Two-Toothed Door Snail) is shaped like a long club. The generic name refers to a notch-like groove on the peristome which accommodates a sliding 'door' to the mantle cavity, the *clausilium*, which is said to prevent entry of tiny beetles, etc, through the pneumostome when air is being exchanged. *Marpessa* (3) also has a clausilium, but it is missing on the other member of the door snails shown here, *Balea perversa*, which may help to account for its specific name.

7 **Retinella nitidula** (Smooth Snail) has a dull waxy light-brown shell with a slightly raised spire; the animal itself is grey. The last whorl of the shell is nearly twice as wide as the previous whorl which makes the shell look oval instead of round. It lives in litter, moss, and vegetation below trees and hedges.

8 **Punctum pygmaeum** (Dwarf Snail) is a tiny creature with a glossy, light-brown shell; the size alone is sufficient to identify it. This is another inhabitant of damp leaf litter on the woodland floor.

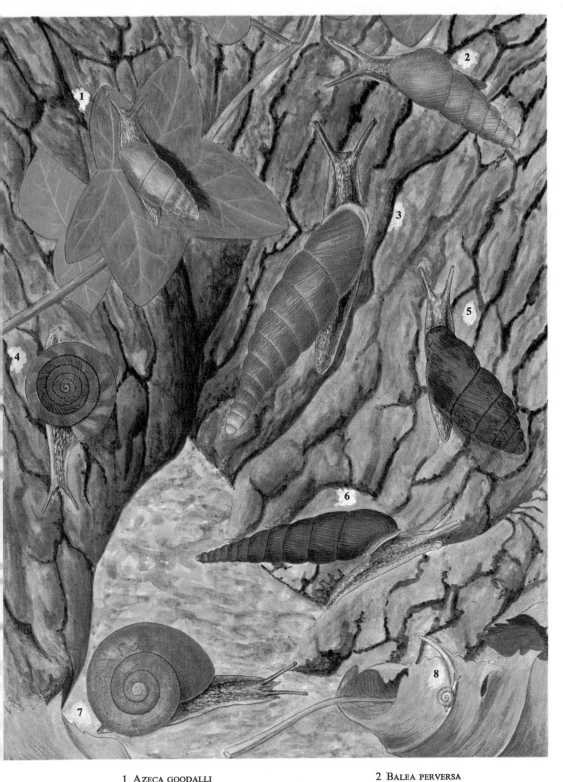

1 AZECA GOODALLI   2 BALEA PERVERSA

3 MARPESSA LAMINATA

4 GONIODISCUS ROTUNDATUS   5 ENA OBSCURA

6 CLAUSILIA BIDENTATA (RUGOSA)

7 RETINELLA NITIDULA   8 PUNCTUM PYGMAEUM

SNAILS OF THE SAND-DUNES

Sandhills and dunelands have a characteristic fauna of snails which appear to take to the dry conditions so characteristic of these places. Actually, some of them burrow or shelter in very dry weather, but after it has rained they swarm out on the grass, and that is the time to observe these lovely shells in their hundreds and collect the most representative samples of species. One genus, *Helicella*, is prominent in dunes, and about six species of it occur in such places, of which three are shown here (2–4); *Vallonia*, too, has three species occurring in much the same habitat, of which only one is shown (6). Of course, these snails are not restricted to dunes; most of the ones illustrated occur in almost any dry place, such as walls and cliffs. Similarly, many of the snails illustrated for other habitats, such as *Cepaea* (3–4, p. 53) and *Helix* (3, p. 51) are also abundant in sand-dunes.

1 Cochlicella acuta (Pointed Snail) has a turret-shaped shell which is usually almost white, streaked with brown; but many other variants exist, such as forms with an almost uniform brown colour and others with one or more bands round the whorls. This snail loves dry grassy places in the South, West and North of Britain, and in Ireland.

2 Helicella caperata (Wrinkled Snail). The genus *Helicella* has about six species, all of which are dwellers on arid open ground, and all of which show variants in the background and banding patterns on the shell; three of the common species are included here. *H. caperata* has prominent transverse striations on the shell, the most usual colour-pattern being dark-brown, somewhat irregular bands on cream background colour. This snail emerges after rain, and feeds on decaying vegetation, especially the leaves of yarrow, but it usually remains under cover and does not shin up the stems of plants as do the other species of the genus. It occurs all over the British Isles.

3 H. virgata (Striped Snail) is slightly bigger than *H. caperata* but has less prominent striations on the shell. The light-cream shell normally has a single broad brown band above the widest part and about seven narrower bands below. This snail usually remains in hibernation until about mid-summer, and then will emerge only in dampish weather, when it will creep over thistle, knapweed, and other composite plants.

4 H. itala (Heath Snail) is usually very slightly bigger than *H. virgata*, and there are normally only three narrower brown bands below the widest part of the shell, in addition to the wide band above it. It inhabits a similar range of plants to *H. virgata* (3). One other member of this genus, *H. gigaxi*, which some authorities place in a different genus, *Candidula*, is shown on p. 53 (6).

5 Euparypha (Theba) pisana (White or Sandhill Snail) has a thick-walled shell with many bands, some complete and some broken, on a very light cream background. Its most usual habitat in this country is close to the sea, where this snail, unlike *Helicella* (2), will remain attached to the stems of plants even in very hot weather, though it feeds only after rain.

6 Vallonia costata (Ribbed Grass Snail) is a tiny snail with a thickened trumpet-shaped opening to the shell. There are two other species, not illustrated here, *V. pulchella* (the 'Beautiful Snail') and *V. excentrica*. But *V. costata* is the commonest, and slightly the biggest of the three, all of which live in the same sort of situation.

7 Pupilla muscorum is very similar in appearance to *Lauria* (8, p. 53), though it is slightly smaller and more cylindrical, and within the shell opening is a prominent white rib, not present in *Lauria*. This snail is usually found among moss and grass throughout the British Isles.

1 Cochlicella acuta

2 Helicella caperata   3 H. virgata

4 H. itala   5 Euparypha (Theba) pisana

6 Vallonia costata   7 Pupilla muscorum

7A P. muscorum, empty shell, showing opening

SLUGS

Slugs do not generally attract avid collectors. The primeval dislike of anything that creeps upon its belly, plus an aversion to slime and a justifiable anger at anything that damages our coveted young plants, helps to put them high in the list of unloved creatures. Yet the sleek contours of their bodies and the special adaptations they show to their particular way of life surely should command something other than a shudder. Mostly, they have tough shell-less bodies which they invest in secreted mucus, and the mantle sits on top of the body, with the 'lung' beneath it, opening to the outside by a prominent pneumostome. There are twenty-three British species of slug, and it is almost certain that the slug-like form of the body has been attained independently several times, so that not all slugs are necessarily all that closely related. Only one group of slugs, the Testacellidae (*see*, for instance, 5) has a shell which is visible externally; the others have internal shells, sometimes so reduced that a mere collection of calcareous grains is all that is left. Some slugs breed for a large part of the year, and they lay their eggs in little caches below the surface as shiny white balls.

1 **Agriolimax reticulatus** (Netted Slug) is our commonest British slug, and is the one that probably does most damage in the garden. It is a pale buff or cream colour with brown blotches on it. Its mucus is milky-white. Contracted (1A), its body is very dome-shaped.

2 **Arion hortensis** (Garden Slug) has a characteristic groove running round the upper edge of the foot and a dark band just above it. The sole of the foot is orange; the shell is reduced to a few chalky fragments under the rear part of the mantle. This slug breeds throughout most of the year, and is another pest of crops such as lettuce, fruit and tubers.

3 **Limax maximus** (Great Grey Slug) may reach 200 mm in length. It has many rows of long elliptical tubercles along its body, and there is often a keel down the last third of the animal's back. Before mating they perform a curious circling 'dance' (if so turgid a movement can be so described), after which the co-copulants climb a shrub or tree, crawl out on to a branch, attach a thread of mucus to the branch and descend by it so that they entwine and copulate in mid-air. This

species lives in gardens and yards, under logs, stones, and vegetation, and feeds on fungi and decaying matter. It is found throughout the British Isles.

4 **Milax sowerbyi** (Keeled Slug) has rather a dry body for a slug, and is rather compressed from side to side. There are furrows on the sides of the body, and the furrows tend to be more highly pigmented than the crests.

As the English name suggests, there is a prominent dorsal keel which is paler than the rest of the body. The slug spends most of its time underground, and is a pest to root crops such as potatoes. In gardens, it is often found in rubbish and compost heaps. It occurs everywhere in these islands except the extreme North.

5 **Testacella haliotidea** (Shield-Shelled Slug). This is the only British genus which has an externally visible shell. The body broadens gradually towards the posterior, where the tiny ear-shaped shell sits on top of the very end. This slug burrows during most of the day, and makes nocturnal forays, to feed carnivorously on earthworms, centipedes, and other slugs. It occurs more commonly in the Southern parts of the country, but is found as far north as about Stirling.

1 AGRIOLIMAX RETICULATUS   1A A. RETICULATUS, contracted   2 ARION HORTENSIS

3 LIMAX MAXIMUS

4 MILAX SOWERBYI

4A M. SOWERBYI, contracted

5 TESTACELLA HALIOTIDEA

POND SNAILS

Most permanent and temporary water-masses have representatives of a very common genus of gastropod, *Limnaea*, the pond snails so treasured by small boys with jam-jars. The animals have brown conical shells with lighter-coloured soft parts, and have non-retractable conical tentacles with eyes at the base. They crawl over weed, or across the surface film of the pond or ditch, but usually emerge into the open only on dull days: when the sun is bright they tend to find their way to a dimly-lit part of the pond. Three of the species illustrated here, *L. truncatula* (2) *L. glabra* (4) and *L. palustris* (5) are virtually amphibious, and are often found out of standing water — which has dire consequences for sheep-farmers (*see* 2).

All these pond-snails used to be referred to the same generic name, *Limnaea*. More recently, taxonomists have attempted to subdivide the genus, in the case of the British forms into about four separate genera. As a compromise, the newer names are generally quoted as sub-genera, that is, given in brackets after the usual generic name, and this practice is followed here.

1 **Limnaea stagnalis** (Great Pond Snail) is the largest species of this common and well-known group, and the commonest of all the freshwater snails. It may reach up to 50 mm in height. As its name suggests, this one prefers rather stagnant water, while the others often live in streams with varying speeds of flow. It has a fairly slender spire, and its last whorl looks proportionally rather bigger than it should. It is found in large ponds and lakes, in almost stagnant ditches and long-standing marshes. It lays its eggs in a gelatinous egg-mass (1A) on water weed or other solid objects in the pond.

2 **L. (Galba) truncatula** (Dwarf Pond Snail) has a much smaller shell with rather ovoid whorls; the spire is usually taller than the shell-opening, and the colour is a horny yellow. This mollusc is infamous in being the intermediate (secondary) host for the common sheep liver fluke, *Distomum* (*Fasciola*) *hepaticum* (p. 25). In the tissues of the snail the fluke passes through several larval stages; then the final larva, called a *cercaria*, bores its way out of the snail and climbs up plants. It is here, in the damp pasture or along the edge of a ditch, that it is eaten by sheep with the grass so that the primary host is reinfected. The snail is amphibious and can resist drought for long periods, so this is why it so often lives in association with sheep.

3 **L. (Radix) pereger** (Wandering Snail) is probably the most ubiquitous of all our pond snails, and occurs in both still and running water all over Britain, Europe, North Africa, and a good deal of Asia: it can be found in ditches, rivers, ponds, lakes, and marshes. The last whorl is very large, and splays out into a trumpet-shaped opening in the adult. This snail is extremely variable in shape and colour, probably due to environmental influences, and some forms of it grow an even larger whorl than normal, and come to look superficially very much like *L.* (*R.*) *auricularia* (6). Some are quoted as having flat shells, looking rather like the Ramshorn Snails (p. 69), and others have a taller-than-usual spire, looking rather like *L. stagnalis* (1). In Ireland it seems that each mountain lake may have its own peculiar form of this snail, due to slight differences in the environment in which the young snails grow.

4 **L. (Leptolimnea) glabra** (Mud Snail) has a slender spire and seven or eight enlarging whorls, of which the last makes up less than half the shell. Like *L.* (*G.*) *truncatula* (2), this snail is amphibious, and tends to be found in those wet places in meadows which are liable to dry up in the summer. In fact, in such temporary ponds it may be the only water-snail found.

5 **L. (Galba) palustris** (Bog or Marsh Snail) is much smaller than *L. stagnalis* but somewhat resembles it in general shape. The shell apex is sometimes abraded so that the lighter colour of the crystalline layers shows through. It is not particularly common, but is usually found on the edges of water-masses or in marshy places.

6 **L. (Radix) auricularia** (Ear Pond Snail) has a large last whorl with a marked ear-like flare to the opening. Like *L.* (*R.*) *pereger* (3) it shows some variety of form, probably due to environmental influences during growth. It is fairly common in rivers, lakes and canals particularly in Southern and Eastern England, but as breeding takes place early in the year, after which the adults die off, specimens are often not found later than about June.

1 Limnaea stagnalis   2 L. (Galba) truncatula

1A L. stagnalis, egg masses

3 L. (Radix) pereger   4 L. (Leptolimnea) glabra

5 L. (Galba) palustris   6 L. (Radix) auricularia

RIVER SNAILS

On this page we return to prosobranch snails that belong taxonomically nearer the periwinkles (pp. 38 – 39) than the pulmonate pond-snails. In fact, one of the forms illustrated opposite, *Hydrobia* (5), appears to have migrated to true freshwater only comparatively recently. Because they are prosobranchs, they respire by means of gills rather than the secondarily-adapted lungs seen in pond-snails; and in some of them, such as *Valvata* (1 – 3), the gills are visible in the live animal, protruding from the shell opening. They all have an operculum, or lid, by which they can close off the shell opening when they have withdrawn. Any piece of debris taken from a river, such as a tin-can or an old boot, will probably have at least one of these snails nestling in it; where the river widens before entering the sea, the shells of some may be found inches deep, having been brought down by the current and deposited where the current slackens.

1 **Valvata cristata** (Flat Valve Snail) has a disk-like shell, very flat and rather thin. It may look superficially like a Ramshorn Snail (p. 69) but is immediately distinguished by the operculum (or valve — hence the common name) carried on the side of the foot. The anterior end of the head is prolonged forwards into a sensory snout, and the front of the foot is splayed out into a wide leading-edge. Just in front of the operculum the gill filament is usually visible. In this genus the gill has two series of filaments, unlike most of the mesogastropods, which have only one (*see* p. 38). This species is absent from Cornwall and parts of the Scottish Highlands.

2 **V. macrostoma** (Large-Mouthed Valve Snail) is about the same diameter as *V. cristata*, but has a higher spire (about 2 mm, as opposed to 1·25 mm). The two are easily distinguished, however, by the shape of the shell opening (compare figures 1A and 2A, which show the dead shells of *cristata* and *macrostoma* lying on the river bed). This species is far less widely distributed than the other two, and apparently occurs only in the Midlands, East Anglia, and parts of Southern England.

3 **V. piscinalis** (Common Valve Snail) is slightly larger than the other two and has a higher spire. A good way to collect samples of the shells of valvatids, where they occur, is to collect the cases of the caddis fly larva, *Limnophilus flavicornis*, which habitually uses small mollusc shells for protection, building them into the wall of its tube-like case. These shells are about the right size for the larva, and very often caddis-cases incorporating them are the first signs that this genus of mollusc is present in a stream or ditch. Unlike most prosobranchs, *Valvata* is hermaphrodite, and to avoid self-fertilization, which is not good for the genetic

'health' of the species, this genus shows consecutive sexuality, that is, the male organs mature first, then the female, and finally the male organs again.

4 **Bithynia**, strangely, does not have a common name, though it occurs widely and commonly. The shell is stubby and glossy, and slightly variegated in colour. The animal has a prominent snout, clearly seen in the figure opposite, which it uses to pick up small particles of food.

5 **Hydrobia (Potamopyrgus) jenkinsi** (Jenkins's Spire Shell) is a remarkable little snail that appears to have entered freshwater only since the middle of the nineteenth century. It first occurred in non-saline waters at Gravesend in 1859, and has steadily invaded rivers and streams throughout the country, becoming extremely common — so common that one estimate puts the number in one square metre as high as 42,000, though normally the highest densities are in the region of 5,000 per square metre. There are several species which show different salinity preferences, from *H. ulvae*, at 10 – 33 per thousand (33‰ is normal seawater) to *H. jenkinsi*, in freshwater.

6 **Viviparus** (River Snail). This is easily the largest of our freshwater operculate snails, and is found in slow-running weedy rivers, canals and ditches throughout England and Wales, and particularly in hard-water districts. The eggs are retained in the oviduct of the parent, from whence the young, already with a shell, escape. The young snail (6A) has a shell with hairs in bands round each whorl, but these are lost in older forms.

1 Valvata cristata   2 V. macrostoma

4 Bithynia tentaculata

5 Hydrobia (Potamopyrgus) jenkinsi   3 V. piscinalis

6a V. viviparus, young   6 Viviparus viviparus

1a V. cristata, shell   2a V. macrostoma, shell   3a V. piscinalis, shell

RAMSHORN SNAILS

Like the limnaeas, these snails are found in most bodies of water, especially where there is plenty of weed; for they are entirely herbivorous, and browse tiny plants from the surfaces of any solid object in the pond, ditch, or lake they inhabit. But unlike the limnaeas, their blood contains the pigment haemoglobin, which is red, and this makes the blood more efficient as an oxygen-carrier so that the ramshorns can possibly live in more stagnant conditions than other water-snails. When the snails are crawling over weed or other objects, they carry the shell with the underside uppermost (*see*, for instance, 3 and 5); the correct orientation for the shell is with its opening to the *left* of the observer, that is, the shells are sinistrally coiled, unlike the limnaeas (p. 65) or the river snails (p. 67). As with the limnaeas, the ramshorns are usually placed in one or at most two, genera: but several new genera have been proposed which are here given in brackets after the older name, where appropriate.

1 **Segmentina nitida** (Shiny Ramshorn) has a small, rather thin-walled shell, convex on one side and flat on the other. The last whorl of the shell, containing the body, makes up most of the width; sometimes the internal strengthening ribs of the shell, in groups of three, show through to the outside as white radiating marks. At first sight, this snail might be confused with *Planorbis albus* (2) or its closer relative *P. (Segmentina) complanatus* (8), but it has a higher shell than either of these in proportion to its breadth. This is not a common snail, and appears to be restricted to Southern, Eastern and Midland counties of Britain, and to a few places in Wales. It is found mainly in ditches and marshy places.

2 **Planorbis (Gyraulus) albus** (White Ramshorn) has a thin, transparent shell with very fine transverse striations and spiral ridges marking the shell. These ridges bear fine hair-like processes which usually get rubbed off as the snail grows older, so that the markings come to look rather like a lattice-work. Strangely, this species is said not to contain the blood-pigment haemoglobin, as do all the others. It is found commonly in ditches all over Britain except the very North-West of Scotland.

3 **P. carinatus** (Keeled Ramshorn) has a sharp keel round the widest part of the shell. The only other two species with any sort of keel are *P. planorbis* (5) and *P. vortex* (6), in both of which the keel is to one side of the shell, and the rare *P. verticulus* (not illustrated here) which is much flatter. *P. carinatus* occurs in slow-running or standing water in most counties of England, Wales and Ireland except Cornwall; it is also found in the lowlands of Scotland.

4 **P. (Planorbarius) corneus** (Great Ramshorn). This is by far the largest of its family, with well-rounded whorls and a strong-walled shell with fine striations. As with the larger species of *Limnaea* (pp. 64 – 65) this noble snail is often used by aquarists to rid their tanks of algae; some dealers sell a rarer albino variety which lacks pigment in its shell so that the red colour of the

haemoglobin-containing blood shows through, and the whole snail appears an attractive pink. The adult lays about 60 eggs at a time in a flat gelatinous egg-mass (4A) on the surface of a plant or some other hard surface. The snail occurs throughout most of England and Wales except Western coastal counties.

5 **P. planorbis** (The Ramshorn). This snail has a keeled shell, rather like that of *P. carinatus* (3) but here the keel is close to the top of the shell rather than in the centre. In the figure opposite this is obscured, as it is in life, because the animal carries its shell upside-down. This species favours shallow weedy water and is found throughout England, except some Western counties, and in Wales, parts of Ireland, and the Scottish lowlands.

6 **P. (Anisus) vortex** (Whirlpool Ramshorn) has a shell which resembles a tiny catherine-wheel because the whorls only gradually increase in size. It is small and thin-walled with a sharp keel slightly to one side of the centre of the outer whorl. As in *P. planorbis*, this species prefers weedy water, and can even be found on weed above the surface. It occurs throughout England, East Wales, and a few places in Scotland.

7 **P. spirorbis** (Button Ramshorn) resembles *P. vortex* in general shape, but is smaller. They very often occur together. In *P. spirorbis*, the shell is quite thick and ornamented with very fine striations. It can survive the drying up of a pond or ditch, by withdrawing into the shell and closing off the apertures with a temporary plug called an epiphragm. It occurs in still and flowing water throughout the British Isles.

8 **P. (Segmentina) complanatus** (Flat Ramshorn) is flatter than its relative *Segmentina nitida* (1), and looks more the shape of an athlete's discus, as can be seen by the empty shell in the figure. It occurs throughout Britain, except in Cornwall and West and North Scotland.

1 Segmentina nitida   2 Planorbis (Gyraulus) albus   3 P. carinatus
4 P. (Planorbarius) corneus
5 P. planorbis   4a P. corneus, egg mass
6 P. (Anisus) vortex   7 P. spirorbis   8 P. (Segmentina) complanatus

SNAILS OF SLOW-RUNNING WATER

On this page we see examples of two orders of gastropods: the prosobranchs are represented by 3 and 4, and the pulmonates by 1, 2, 5 and 6. The main characteristics of the prosobranchs are given on p. 34, and of the pulmonates on p. 50. In the one case a mainly marine group has taken to freshwater, while in the other a mainly terrestrial group, with primitively an air-breathing lung, has become aquatic. The typical places these snails most often inhabit are slow-running canals, and this is why the corner of a bridge pillar is used as a background in the figure opposite.

1 **Aplexa** (Moss Bladder Snail) is a pulmonate that has a thin, sinistrally-coiled shell which is glossy and semi-transparent. The actual animal is almost black, with very slender tentacles. It is often found in very slow-running or still water with plenty of weed, but it is tolerant of drought and may even climb out of the water on occasions and be found on weed on the banks.

2 **Physa** (Bladder Snail) is the other fairly common spire-shaped shell with a sinistral coil, but it is squatter than its relative, *Aplexa* (1), and when active the mantle edges are produced into finger-like processes that protrude over the shell. This snail is commoner than *Aplexa*, and prefers water that is slow-running and has plenty of weed, such as water-cress beds. It is reported that the commonest species of this genus, *P. fontinalis*, has several races in different habitats: those near the edge of a body of water frequently come to the surface to take in a bubble of air to the mantle cavity which they then draw on for their respiratory needs, while those in deeper water have the mantle permanently full of water, so that respiration is entirely aquatic. Whether there are genetical differences between the two is not known, but it shows an interesting adaptability of the pulmonate 'lung'.

3 **Bythinella** (Amnicola) (Taylor's Spire Shell). This is a tiny mesogastropod prosobranch found only in Lancashire, Cheshire and Stirlingshire, particularly in the canals. The shell is chubby and blunt-spired, and the animal itself has yellow markings near the eyes, which shimmer in the light so as to appear almost luminescent. The tentacles can be retracted to no more than tiny spheres. This snail is related to *Hydrobia* (*Potamopyrgus*) (p. 66), and appears to have been present in this country prehistorically, but to have died out until being re-introduced from America around 1900.

4 **Theodoxus** (Neritina) (Freshwater Nerite) is an archaeogastropod, the main characteristics of which are given on p. 34 in connexion with this snail's marine relatives. It possesses the primitive feature of paired gills, and resembles a miniature version of *Crepidula*, the Slipper Limpet (p. 35). The most usual places to find these snails are on the shells of freshwater mussels (p. 91) or on the surfaces of stones. The eggs are laid in capsules containing 50 or 60 together, and the capsule may be stuck to another shell, such as a mussel; when they hatch there is often cannibalism among the off-spring, so that very few — sometimes only one — of the original number survive.

5 **Ancylastrum** (River Limpet). This and its relative *Ancylus* (6) are pulmonates, and they show examples of the way the evolution of the pulmonates has paralleled that of the prosobranchs, in producing a limpet-like body form (compare with some of the proso-branch limpets on p. 35). In both these snails the shell has the shape of a cone with a bent-over top, though the curvature of the River Limpet is more marked than that of the Lake Limpet (compare the shells, shown side-view, in 5 and 6 opposite). Both have thin shells with particularly delicate margins, so that the shell can be pulled down on to the hard surface over which it is moving, and fit snugly on the slight irregularities of the surface. This is as much a protection against sediment, which these limpets cannot tolerate, as against predators.

6 **Ancylus** (Acroloxus) (Lake Limpet) has a more conical shell than its relative *Ancylastrum* (5) and the tip of the shell is twisted slightly to one side. It is also slightly smaller, seldom reaching more than 5 mm long. As its common name implies, it prefers slower-running water, even still water, though it may occur at the edges of rivers, where the current is slight.

1 Aplexa hypnorum   2 Physa fontinalis
3 Bythinella scholtzi (Amnicola taylori)   4 Theodoxus (Neritina) fluviatilis
5 Ancylastrum fluviatilis   6 Ancylus (Acroloxus) lacustris

MUSSELS OF THE SEA-SHORE

The important class of molluscs, Bivalvia, (pp. 72 – 91) includes those with two shells, such as the mussels, clams and oysters. Here, the shell is divided by a dorsal hinge into right and left valves, the mantle cavity is huge and contains sheet-like gills, the foot is laterally compressed and the head is greatly reduced. On this page are illustrated some of the commonest forms of the group, the mussels, which are found attached to a hard surface by thin brown 'guy-ropes' called *byssus threads* (*see* for instance, 8); these threads are placed on to the hard surface by means of the foot, and once they are established they can anchor the mussel against a tremendous battering from the sea. When the tide recedes and we walk across a bed of mussels, we can hear them pulling their shells together tightly to prevent desiccation of their bodies. But when the water covers them again, they open slightly and the cilia covering the gills pump water through the mantle cavity; food is trapped on the gills and carried to the mouth.

1 **Pteria** (Wing Oyster). The shell of this form has wing-like extensions of the valves which increase the length of the hinge-line; the backward-pointing wing is about four times as long as the forward-pointing wing, and there is a small notch in the forward-pointing wing through which the byssus threads emerge, to attach the shell to the gravel or sand on which it lives. The shell surface is scaly and shows irregular concentric growth-lines.

2 **Pinna** (Fan Mussel) is the largest of the British bivalves, and may grow up to 300 mm long, with a brown triangular shell showing growth lines and often showing splits and fractures, particularly towards the wide posterior end. Though this huge mussel may be fairly common in deeper water, it is only rarely taken near the shore, and when it is, it occurs in areas of sheltered water with a bottom of sandy mud. The byssus, as shown opposite, resembles a pony-tail. The shell edge is somewhat flexible in places, particularly at the wide posterior end, so that the shells can be closed together completely. Very often, the shells of the bigger individuals are bored into by other animals, such as sponges (p. 5).

3 **Modiolus modiolus** (Horse Mussel) sometimes occurs in vast populations several miles long and several miles wide, the individuals being attached to one another or to the gravel bottom by their threads and forming a continuous carpet. Such regions must have vast suspensions of food in the water above them to support such numbers; the mussels will filter the water through the gills and retain the food particles. *Modiolus* can be distinguished from *Mytilus* (8) by the position of the apex of each valve: in *Modiolus* it is slightly above the narrow end of the shell, while in *Mytilus* it is right at the end.

4 **M. phaseolinus** (Bean Horse Mussel) is the smallest species in a genus which contains about 5 species in British waters. Shaped like a kidney-bean, it has a slightly crenulated top edge, which distinguishes it from small specimens of *M. modiolus* (3), which otherwise resemble it. The other surface of the shell has short furry projections pointing towards the posterior edge.

5 **M. barbatus** (Bearded Horse Mussel) has even more marked processes from the shell, arising from concentric fringes, but has no crenulations along the top edge. The processes and the fringes tend to collect debris and sand among them, possibly helping in concealment. This species lives from low-tide, where it inhabits *Laminaria* holdfasts or lives under rocks, to about 60 fathoms.

6 **Musculus** is common everywhere round the British coasts, where it lives on the lower half of the shore among coralline algae and other plants. It too is bean shaped, but unlike *Modiolus* the shell is smooth and redder in colour.

7 **Crenella** lives on bottoms of sand or gravel, usually offshore, and will be encountered by the casual collector only when it is washed up on the shore. It has a sculpturing of radiating ribs and concentric growth lines. This is mainly a form from Northern waters, though it has been taken from the Bristol Channel.

8 **Mytilus** (Common Mussel) occurs on many British shores from near high-water mark to below low-water. Like *Modiolus* (3 – 5), it may occur in vast beds when there is a suitable surface for attachment. On the East Coast, particularly around the Wash, and in Wales and Scotland, small specimens are commercially transplanted to sheltered areas in estuaries where food is plentiful, and these provide the supplies for many a sea-food stall. The optimum size for commercial use is about 100 mm long, but some specimens may reach 230 mm.

1 PTERIA HIRUNDO   2 PINNA FRAGILIS
3 MODIOLUS MODIOLUS
5 M. BARBATUS   4 M. PHASEOLINUS
6 MUSCULUS DISCORS   7 CRENELLA DECUSSATA   8 MYTILUS EDULIS

73

SAND-BURROWING BIVALVES OF THE SHORE (1): VENUS CLAMS AND CARPET SHELLS

In the first of a series of plates illustrating a very important group of animals, the sand-burrowing bivalves (pp. 74 – 83), are the venus clams and carpet shells that are so often washed up on the beach and collected by the beachcomber. Most of these habitually live below the lowest point of the tide, but occasionally they survive further up the shore. They burrow only a little way below the surface of the sand or gravel which means that their schnorkel-like siphons are only short; but those that live on the shore usually burrow deeper as the tide recedes. In almost all these shells the concentric growth-lines are clearly visible, representing successive stages in the deposition of the shell by the growing mantle-edge which underlies it.

1 **Venerupis pullastra** (Pullet Carpet Shell). The common name of this shell comes from the apparent similarity of its colouring to the plumage of a domestic hen. Although the presence of teeth to help keep the two valves together is a feature of this group of shell, the genus *Venerupis* often suffers damage to the teeth, so that they become inconspicuous. The shell is somewhat oval in shape, with many fine ribs radiating from the apex; these ribs are sometimes more marked on the posterior side (left side in figure). In life, the animal has the usual two siphons projecting above the sand surface, the intake siphon being shorter and fatter than the exhalant one. The animal is common at and below low tide, where it lives buried an inch or so in hard, stony sand or gravel under rocks with its byssus threads usually attached to something hard under the surface.

2 **V. decussata** (Crosscut Carpet Shell). In this species, which is similar in general appearance to *V. pullastra* (1), the sculpturing is more marked, and the concentric lines are crossed by radiating lines — which is reflected in the common name. Again, the sculpturing becomes more marked on the posterior (left) side of the shell. In life, the animal's siphons have pigmented spots near the apertures. This is mainly a Southern and Western form, living in muddy gravel and stiff clay at and below low-tide.

3 **V. saxatilis** (Banded Carpet Shell) is usually about half the size of the other two species illustrated. The shell is sometimes rather irregular in shape, and this is reflected in the concentric growth lines, as shown in the figure. This species is usually found, not buried beneath sand or gravel as are the other two included here, but inhabiting the abandoned holes made by burrowing bivalves such as *Pholas* and *Hiatella* (p. 87), and adhering to the sides of the cavity with its byssus. This bivalve is locally common on the South coast, but has also been recorded from the West of Scotland.

4 **Venus (Mercenaria) mercenaria** (Quahog or Hard Shell Clam). This noble bivalve, living in mud with stones and shells at or just below low tide, is a native of the Atlantic coast of North America, but it appeared in our waters at about the middle of the last century. Like the Soft Shell Clam *Mya arenaria* (p. 81), this

mollusc is harvested as a splendid source of food by the Americans, who gave it its common names, 'Quahog' or 'Quahang' being an eighteenth-century name bestowed by the Narraganset Indians. In English waters, colonies are found now in the Solent, Southampton Water, and Portsmouth Harbour, possibly started by the kitchen waste thrown from the hatches and portholes of trans-Atlantic liners.

5 **Venus (Clausinella) fasciata** (Banded Venus) has a very solid-looking shell which has strong concentric ridges looking rather like the steps on a semi-circular dais. These ridges, more numerous in the young shell, become fewer but stronger with age. *Venus* is a shallow-burrowing bivalve, preferring to live in coarse gravel just off-shore, but also occasionally turning up under stones at or near low-water. The colour, which is varied, generally has a good measure of reds and browns with lighter-coloured rays. The siphons are united, except at the very tip.

6 **V. (Timoclea) ovata** (Oval Venus). This is a much smaller species than the other two illustrated (4 and 5), and seldom exceeds 15 mm width. It has 40 to 50 radiating ribs so that it resembles a cockle (4, p. 77). The shell is usually yellowish with tinges of pink or orange and brown blotches. As in *V. fasciata*, the siphons are united together, except right at the end. This prefers a slightly finer substratum to burrow in than the previous species.

7 **Dosinia** (Rayed Artemis) has a pronounced hollow in front of the hinge (the anterior side is to the *right* in the picture). The shell is larger than in any species of *Venus* from British waters, and the foot is proportionally bigger than that in the venerids. This bivalve is widely distributed and common.

8 **V. (Chamelea) striatula** (Striped Venus) is the commonest species of this genus found between tide marks. The shell has many concentric ribs crossed by finer radiating ones, but the outstanding feature in most individuals is the three reddish rays radiating from the apex to the shell edge. Although frequently found as a washed-up shell and occasionally living at low tide, this species prefers clean sand offshore.

1 Venerupis pullastra   2 Venerupis decussata

3 Venerupis saxatilis

4 Venus (Mercenaria) mercenaria   5 Venus (Clausinella) fasciata

6 Venus (Timoclea) ovata

7 Dosinia exoleta   8 Venus (Chamelea) striatula

75

SAND-BURROWING BIVALVES OF THE SHORE (2): COCKLES AND TROUGH SHELLS

Most of the shells illustrated on this page are likely to be found among the debris on the shore, and not in the place where the animals live: most of them occur from just below the shore down to a depth of a few fathoms; but when they are displaced or die, their shells are trundled in towards the shore by currents and deposited as the tide recedes. The family Cardiidae comprises the cockles, one species of which, the Edible Cockle (4) does, of course, occur between tide-marks. There are about eleven species in British waters, and though formerly they were placed in the same genus, today they are grouped into various sub-genera within the genus *Cardium*; these sub-generic names are given in brackets below. While cockles almost always have chubby ribbed shells, the Trough Shells, of which two examples, *Mactra* (3) and *Spisula* (6) are shown, are somewhat thinner and smoother. All these forms, however, live just below the surface of the sand, and protrude two short siphons above the surface to take in and eject currents of water for feeding and respiration.

1 **Cardium (Acanthocardia) tuberculata** is a large (up to 90 mm long) cockle with 21 or 22 strong ribs radiating from the apex to the margin and blunt tubercles in rows, one to each rib. This is commonly found along the shore of Southern England, particularly after a good gale when the animals have been churned out of their burrows in the offshore sand and brought by the currents to the strand-line on the beach. Sometimes the. large red foot lolls out from between the shells, like a pointed tongue, and if large numbers of individuals are stranded together they can produce a hideous smell of rotting shellfish.

2 **C. (A.) echinata** (Prickly Cockle) is a smaller species. It has rather pointed spines on its ribs and the spines in each row are joined together at their base, whereas in *C. (A.) tuberculata* they are entirely separate. This is another offshore form that will usually be found as a stranded shell only.

3 **Mactra** (Rayed Trough Shell) has a thin brittle shell which is creamy-white in colour tending to purple near the apex, with brown radiating rays across the valves from apex to margin. This bivalve occurs in clean sand, sometimes at very low tide but more usually off-shore. It is another of the shells that will be found most frequently along the strand-line, or in those interesting collections of offshore shells that get deposited where there is a tidal eddy near the rocky outcrops on a sandy shore. When alive, it lives in much the same situation as the other Trough Shell illustrated here, *Spisula* (6).

4 **Cardium (Cerastoderma) edule** (Edible Cockle) is a famous animal that has caused the smacking of many a holidaymaker's lips. Vast beds of this mollusc occur in such places as the Wash, Thames Estuary, More-cambe Bay, Dee Estuary and the North Gower coast in Wales; acre upon acre, crowded with individuals in some places (maybe 10,000 per square metre) occur in these localities. The local people rake them up when the tide is low and heave sackfuls on to carts drawn by ponies which stand out on the sand banks for many an hour, their patient heads drooping. Cockles live just below the surface, as shown in the picture, and when the water covers them they extend two short papillate siphons just above the surface of the sand or mud and take in water by ciliary action to filter off the suspended food material it contains. Naturally, a great deal of the suspended matter is fine sand and mud particles, and it is essential to keep the cockles in sea-water for several hours before attempting to cook them, so that the gritty particles are ejected by the ciliary currents.

5 **Cardium (Parvicardium) exiguum** (Little Cockle) is a plump species seldom growing to more than 13 mm in length. The ribbed shell may have slight tubercles in young specimens, but these tend to rub off in full-grown individuals, particularly from the posterior side of the shell (*see* left side in the figure opposite). The growth stages can usually be seen on the shell.

6 **Spisula** (Thick Trough Shell) is similar in habit and appearance to its relative, *Mactra* (3) but it lacks the brownish rays on the shell. It is usually an offshore form, and sometimes occurs in vast numbers. Both *Spisula* and *Mactra* are thinner than the cockles and can move more easily in the sand, and burrow deeper.

1 CARDIUM (ACANTHOCARDIA) TUBERCULATA
2 C. (A.) ECHINATA 3 MACTRA CORALLINA
4 C. (CERASTODERMA) EDULE 5 C. (PARVICARDIUM) EXIGUUM 6 SPISULA ELLIPTICA

SAND-BURROWING BIVALVES OF THE SHORE (3): TELLINS

The bivalves shown on the previous page were filter-feeders; they pump water down one limb of the siphon and across the gills, where food particles are trapped. On this page are a group of superficially similar burrowing bivalves, but most of these are deposit-feeders: they mostly have long mobile siphons which mouth over the surface of the sand like vacuum-cleaners, picking up tiny diatoms and organic debris. Most of these bivalves have smooth shells and are very flat from side to side — both features which allow them to burrow easily and deeply into the sand. In consequence, they can protect themselves from desiccation as the tide recedes, and most of them are found further up the shore than the cockles. Another tellin, *Pharus legumen*, is illustrated alongside the razor-shells on p. 81, because it is so elongated as to resemble the true razors rather than its closer relatives.

1 **Tellina tenuis** (Thin Tellin) is a very common shell, up to 20 mm long, with a glossy and often strikingly beautiful shell. It inhabits clean sand from mid-tide down to a few fathoms, sometimes in very large aggregations. A strong ligament keeps the two valves together in life, and when the animal dies the elasticity of the ligament causes the valves to open like the wings of a butterfly.

2 **T. fabula** is very similar in size and appearance to *T. tenuis*, but the right valve is obliquely striated, and the posterior side of the shell (in the picture the upper side, from which the siphons originate) is slightly straighter near the apex and more angular near the siphons. It overlaps *T. tenuis* in its zonation on the shore, but in general is commoner at low tide and down to 20 fathoms.

3 **Macoma** (Baltic Tellin) has a rather dull shell with irregular concentric striations. It can tolerate low salinities, and therefore is frequently found in estuaries; the Baltic Sea has a reduced salinity and it occurs in large numbers there — hence its common name. It usually lives in a substratum which has an admixture of mud. The shell is variable in colour, as the three variants illustrated show (3, 3A, 3B).

4 **Abra (Syndosmya)** has a thin shell, often snowy white and polished. Its light brown siphons are cylindrical and do not have prominent fingers. This is another mud-dweller, from estuaries and creeks, as well as offshore.

5 **Scrobicularia** (Peppery Furrow Shell) is common fairly high up estuaries and in salt marshes. It is the largest of the deposit-feeding bivalves, growing up to 65 mm long, and it is extremely flat. It may burrow to a depth of 200 mm, its long mobile siphons stretching up to the sand or mud surface to maintain vital services to the animal. To see these siphons to advantage, it is rewarding to place a specimen of this mollusc in a dish of sea-water, when the siphons will gradually emerge and extend to their full length.

6 **Gari** (Sunset Shell) has a sharply angulated shell on the side adjacent to the siphons. It may be very common in places, especially further North, in coarse sand or gravel.

7 **Tellina crassa** (Blunt Tellin). The shell is larger than those of the other species illustrated here (1 and 2), with concentric rings; and the shell valves are less flattened. This species inhabits coarse sand and shell gravel. Its siphons are long, the exhalant one being about three times as long as the inhalant.

8 **Donax** (Banded Wedge Shell) is wedge-shaped when viewed end-on, and has concentric bands and five radiating striations on each valve. The siphons, each with a fringe of sensory papillae, are about equal in length. This is a shell that thrives on exposed sandy shores.

1 Tellina tenuis   2 T. fabula   3 Macoma balthica
3A, B M. balthica, colour forms
4 Abra (Syndosmya) alba   5 Scrobicularia plana   6 Gari fervensis
7 Tellina crassa   8 Donax vittatus

SAND-BURROWING BIVALVES OF THE SHORE (4): GAPERS AND RAZOR SHELLS

The razor shells, looking like the old-fashioned cut-throat razors, are among the most specialized of the burrowing bivalves. They can escape from predators by very rapid burrowing, and just how efficient they are at getting to great depths can best be appreciated by trying to dig one up: to be successful, you must either creep up on them and plunge a spike or a fork into the place where the siphons emerged when the tide was in, or dig a very deep hole into the sand. The foot, which is responsible for the efficient burrowing, emerges not from the underside of the shell, but from the anterior end (*compare* 6 and 7) so that, by repeatedly filling with blood and then contracting, the foot can anchor at its base and pull the long, thin shell down towards it. This process can occur again and again, if the sensitive animal picks up vibrations from an attacker. The process can be reversed to bring the shell to the surface when the predator has gone or the tide has returned. Razor shells never quite close the ends of the shell, so that the foot and the siphons can protrude even when the valves are otherwise tightly shut. The same is true of the other group included here, the gapers, but though most of them live deep in the sand they do not have the same burrowing power.

1 **Ensis ensis** has curved valves in which the anterior end (downwards in the figure) is rounded and the posterior end tapers slightly. The foot is usually reddish-brown. It is common around Britain from low on the shore to a depth of a few fathoms.

2 **E. arcuatus** is half as long again as *E. ensis* (1) when fully grown, and differs also in the colour of the foot which here is creamy-white. It extends to deeper water than *E. ensis*.

3 **Pharus legumen** (Pod Shell) is really related to the tellins (p. 79), but parallels the true razor shells in general shape so closely that it is included here for comparison. The position of the hinge between the valves gives the game away, because whereas in razor shells the hinge is near the anterior (lower) end, in *Pharus* it is almost central. There is a sculpturing of fine concentric lines on the shell, and the surface is often slightly iridescent. The siphons are separate. Though the empty shells are extremely common locally around Britain, the intact animal is not all that easy to obtain by digging, since it appears to live at just below low-tide.

4 **Solen** (Grooved Razor) has a deep groove between the hinge and the anterior end (bottom left corner in the figure). It may burrow to a depth of 450 mm, usually in muddy sand. Though it does occur up the West coast as far as the Clyde, it is commoner on the South coast.

5 **Corbula** (Basket Shell) is a very common bivalve belonging to the same family as *Mya* (6), in which the left valve fits snugly into the much larger right valve, somewhat like a basket with a lid. The shell rarely exceeds 10 mm in length.

6 **Mya** (Sand Gaper, Soft Shell Clam, or Brallion). This bivalve, drawn opposite with its magnificent siphons almost retracted, would occupy more than the length of the page if drawn fully extended at the same scale. The animal is the same species (*M. arenaria*) as is farmed in the coastal waters and estuaries of the United States. There, as with its counterpart, the Hard Shell Clam (4, p. 75), they do justice to its delicious meat by baking, frying or steaming clams and serving them with melted butter or Cranberry sauce — and make a sociable occasion of it to boot, with a noisy, neighbourly clam-bake. The English do not generally eat it, though it is very common in some places, particularly in estuaries where there is a proportion of mud in the sand.
The gaper burrows deeper as it grows, until it lives about 400 mm below the surface, its siphons stretching up to the sand surface to maintain the vital connexion. But if it is dug up and placed on the sand, which does not normally happen to it in life, it is unable to burrow down again.

7 **Ensis siliqua** (Pod Razor) is the largest British razor shell, growing to about 200 mm, and it is this one that the shellfish hunters seek at low tide.

8 **Lutraria** (Otter Shell) is superficially like *Mya*, but differs in the character of the projections of the shell near the hinge: in *Lutraria* there is a spoon-shaped projection on each valve, while in *Mya* the projection is present on the left valve only. The shells that are cast up on the beach are usually a dull pink. This bivalve is related to *Mactra* (3, p. 77).

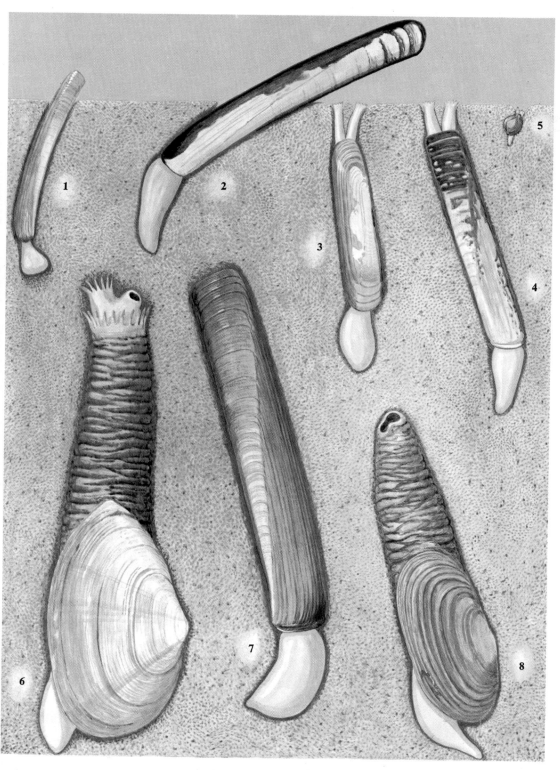

1 ENSIS ENSIS   2 E. ARCUATUS   5 CORBULA GIBBA
3 PHARUS LEGUMEN   4 SOLEN MARGINATUS
6 MYA ARENARIA   7 ENSIS SILIQUA   8 LUTRARIA LUTRARIA

SAND-BURROWING BIVALVES OF DEEPER WATER

Shown on this page are some rather special bivalves which do not readily tie in with other burrowers shown on previous pages. Some of them, such as *Pandora* and *Cuspidaria* (1 and 2) represent animals of groups which are considered highly advanced in the class: others, such as *Glycymeris* and *Nucula* (3 and 4), are thought to be rather primitive. All are inhabitants of offshore deposits of mud or sand, which are often thrown into ripples by the currents over them (as shown in the figure). These are shells which are most likely to be found dead on the shore after being washed in by the currents.

1 **Pandora** (Pandora's Box Shell) is a rather rare bivalve with a somewhat fragile white shell, rather unexciting for the shell collector because it is not brightly coloured, but biologically interesting because it is another bivalve with unequal valves. In this case the left valve (underneath in the picture) is convex, while the right valve is flat; further, the left valve is bigger than the right and overlaps it, so that the whole shell resembles a little round-bottomed box which someone must have thought was like Jupiter's gift to Pandora. The siphons are short, and though rather little appears to be known about its habits, it is said that the shell burrows just below the surface, so that the flat right valve becomes covered by a thin layer of sand, and the siphons just protrude above the surface. It can occasionally be taken at very low tide, but is usually offshore, sometimes in *Zostera* beds, and restricted to South and West Coasts.

2 **Cuspidaria.** Among the general characters of almost all bivalve molluscs is their ability to filter the sea-water for food by means of the gills. In this form, however, the gills as such are replaced by a flexible and muscular membrane which stretches across the mantle cavity to provide a pump by which water and food are taken into the body. The animal is carnivorous and moves over the surface of sand in deeper waters while its siphons, which arise from a special spout at the posterior end of the shell, mouth over the sand to take in dead or dying animals.

This animal is often placed, with a few other genera, in a separate order of the Bivalvia, the Septibranchia, in recognition of the very special nature of the septum-like 'gill' but it is not clear whether they ought to be so separated. Some authorities say the septibranchs are related to the order Laternulacea, which contains *Pandora* (1), and suggest they both should reside in the same subclass Anomalodesmacea, but they are insufficiently well known to confirm or deny this classificatory point.

3 **Glycymeris** (Dog Cockle). This has a large thick shell, up to 70 mm in diameter, with highly convex valves held together by a ligament which shows clearly on the outside (upper left in the picture). It has a yellowish background colour with reddish-brown markings. It will probably not be dug up by the casual collector, but will often occur washed up on the beach. In its native place, it lives just below the surface in sandy or muddy gravel.

4 **Nucula** (Nut Shell). Like *Cuspidaria* (2), its method of feeding provides another exception to the usual gill-feeding habit of bivalves. Here, the gills are fairly small and perform only a respiratory function, while the job of filtering food from the sea-water passing into the mantle cavity is performed by special mucus-secreting projections from near the mouth called *palps*. The ends of these palps are produced into *palp proboscides* which can emerge from the shell to collect food particles from the substratum. The foot of this bivalve is a flattish shape, and for movement the animal thrusts it ahead, fills it with blood so that the end expands, then contracts its longitudinal muscles so that the shell is dragged forwards through the sand. It is a small shell, seldom more than 18 mm long, and the inside margin of the shell (4A) is crenulated.

5 **Astarte** has a solid shell with strong concentric ridges on it. The margin of the shell is thickened and often finely crenulated, if the sculpturing has not been rubbed away by wear. The largest specimens the casual collector may come across washed up on the beach are about 25 mm across. In life, the animal lives below about 3 fathoms, in gravel or mud and is apparently fairly common all round Britain.

1 Pandora albida   2 Cuspidaria cuspidata
3 Glycymeris glycymeris
4 Nucula turgida   5 Astarte sulcata
4A N. turgida, inside of shell

SCALLOPS AND OYSTERS

Among the most familiar of bivalves are the beautifully-coloured scallops, frills, and file-shells. No other group of bivalves shows quite the same variety of colour and pattern to delight the naturalist and the shell-collector. No less remarkable is their way of life, for here we see bivalves that in later life swim actively to escape their sea-bed predators. In the swimming process water is expelled forcibly from the mantle cavity in a sheet-like jet, usually from between the valves at the gape, but occasionally from near the hinge, so that the animal appears to be 'biting' its way through the water. Here too is the oyster, prized as food and in more exotic climes as the manufacturer of pearls.

1 **Pecten** (Great Scallop). This splendid shell-fish has a convex right valve (downwards in life) and a flat left valve which are hinged together along two ear-like flaps at the top; the ligament holding the valves together looks like glue oozing out along the hinge. The edge of the mantle bears projecting finger-like processes which are highly sensory and between them are spherical, opalescent eyes. When a predator such as a starfish approaches, the scallop senses its presence chemically and forcibly slams the two valves together, causing a rapid extrusion of water from between them, so that the whole animal leaps away, hinge first, from the predator. Several successive flaps are possible, so the process can be called 'swimming'. Early in life *Pecten* is attached to the substratum by a byssus; later it becomes free. It is highly prized as food; in the Isle of Man, the Sussex and Devon coasts, and parts of Scotland, it is brought up from the sea bottom by special metal-framed dredges.

2 **Anomia** (Saddle Oyster) looks superficially very like a brachiopod (2, p. 171) because the valves are attached to the substratum by a collection of byssus threads which emerge from the inside of the shell through an embayment in the right valve. The left valve tends to overgrow the right and become convex. It is common on the middle shore and below.

3 **Lima** (Gaping File-Shell) has the beauty that attracts photographers. The finger-like pink and orange projections from its mantle edge project like rakers across the gaping slit between the valves. These processes cannot be withdrawn, so the animal is unmistakable. A file-shell swimming is one of the sights of nature: it lazily flaps its valves while hanging vertical in the water, the mantle processes waving like the hair of some tiny mermaid. Curiously, this bivalve is the only one which constructs a nest from byssus threads, lying safely within it and drawing water through the mantle cavity.

4 **Chlamys opercularis** (Queen Scallop or Frill) is, like *Pecten* (1), sometimes fished for the table. Both valves are convex, but the left (upper) valve is slightly more so. The colour varies enormously; generally, reds, pinks, and oranges form the basis of the many shell patterns.

The ear-like flaps bearing the hinge are not quite so different in size or shape as those of *C. varia* (5). It 'swims' better than *Pecten* for it can move either way—with hinge leading or with hinge trailing. The direction is determined by the position across the mantle cavity of a flap of tissue under muscular control, by means of which water is directed either from the 'corners' of the ear-like projections, or from between the valves, or from one corner only, so that the shell gyrates away from its attacker. With several flaps of the shell it can move quite considerable distances.

5 **C. varia** (Variegated Scallop) has very unequal ear-like projections bearing the hinge, considerable variation in colour and pattern, and the colours are in blotches. Even in the adult, a byssus may still attach the animal to the substratum.

6 **C. distorta** (Hunchback Scallop) also has unequal ear-like projections, but in this case the right (lower) valve becomes cemented to a hard substratum. This is the only scallop which becomes permanently fixed in an oyster-like manner. It is found in *Laminaria* holdfasts on the lower shore, or in rock crevices or old shells, its shape often distorted to fit theirs.

7 **Arca** (Ark Shell) is another bivalve which attaches itself to the substratum by means of a byssus — a massive greenish one which attaches the animal to rocks and old shells at low tide and below; sometimes the whole animal becomes encrusted, and so camouflaged, with weed and tube-dwelling polychaetes (p. 111). It is not common or easy to detach.

8 **Ostrea** (Oyster) may be encountered in natural or cultivated beds. In this case it is the *left* valve that is underneath, cemented to the substratum, with the rough and often worn upper-valve assuming bizarre shapes according to the situation in which it grew. Tiny pearls sometimes occur in English oysters, but very seldom grow to any size. An irritation produced by a foreign object such as a grain of sand can induce concentric layer upon layer of chalk to be deposited round it . . . but what a way to describe so beautiful a treasure as a pearl.

1 Pecten maximus   2 Anomia ephippium

5 Chlamys varia   4 C. opercularis   3 Lima hians

6 C. distorta

7 Arca tetragona   8 Ostrea virgata

ROCK AND WOOD BORING BIVALVES

Another way in which bivalves protect themselves is shown on this page. Some, such as the hiatellids (1) and rock piddocks (4–7) can bore into rocks, leaving a small entrance to the exterior; others, such as the Wood Piddock and the so-called 'shipworm' (2 and 3) can excavate burrows in wood which happens to be in the sea. In all these forms, the excavating is accomplished by using the edges of the shell-valves, though in the shipworm this may be aided by slow digestion of cellulose, a constituent of wood. Naturally, the shell edges wear down during the process, but every so often the shell ceases boring and the mantle edge creeps over the shell edge and deposits a replacement layer of chalk. The bivalves obtain purchase on the walls of the rock or wood by means of the foot, or, if they are well down in a burrow or crevice already, by the sides of the siphons. It might seem odd that the shipworms could have evolved before men took to using wooden objects at sea; but one has only to see the estuary of a great river or an eroding coastline to realize how many logs, which can provide a habitat for these molluscs, find their way into the sea naturally.

1 Hiatella is a rock-borer (if the rock is soft enough) or a rock-nestler (if it is hard). It is very common round our coasts. Its presence is indicated by two red-rimmed holes at the rock surface, which are the ends of the two siphons. If you just touch the surface of the rock near the animal, that is often enough to cause the shell to contract, sending a jet of sea-water into your eye. The shell is about the size and shape of a haricot-bean, fairly solid and irregular in shape and outline. The colour is usually off-white.

2 Xylophaga (Wood Piddock) has a shell that surrounds the whole body except for the siphons, and in this respect it differs from *Teredo* (3). The galleries it bores are shallow compared with those of *Teredo*, and the siphons emerge about 3 mm from the opening.

3 Teredo ('Shipworm'). The common name merely reflects the worm-like appearance of its body; but it is an unfortunate colloquialism which may suggest to some people that this animal belongs to a totally different phylum. The larva settles on to a piece of timber, such as a pier-pile, ship's bottom, or floating tree, and immediately starts to burrow, using the reduced valves of the shell (shown in the lower part of the animal in the figure) as a drill for boring into the wood. The foot projects beyond the shell and grips the side of the burrow as the valves are twisted relative to each other on a fulcrum-like hinge. The animal's viscera are not all contained between the shells but occupy the first region of the limp, worm-like part of the body; the rest of this part contains the siphons which project as separate tubes beyond the end and are protected by two calcareous flaps. The worm-like part of the body also secretes a chalky lining to the excavated tube which may show marks and constrictions caused by uneven deposition (as shown in the figure). Though a good deal of the food is obtained from the water which is taken in by one siphon, this mollusc has the power to digest cellulose from the wood in which it bores, to provide part of its requirements. *Teredo* causes considerable damage to wooden structures in

the sea and has caused the loss of many a good ship in the past.

4 Pholas (Common Piddock) bores into rock or stiff clay. The valves have a permanent gape both in front and behind, the one allowing emergence of the foot, the other of the siphons (not shown in the figure). In addition to the shell-valves there are accessory shields, a pair towards the front (lower in the figure), a small one between and above these, and one almost in the centre. As in other borers, the foot holds the animal in its burrow while the shell-valves are rocked back and forth to abrade the rock; then the foot changes its hold and the shell rotates slightly and starts excavating again. This mollusc is common locally in the South and South-West, and is slightly phosphorescent.

5 Zirphaea (Oval Piddock) bores into the same sort of hard substances as *Pholas*. It has a more compact, almost oval, shell and a furrow with spines down each valve, almost dividing the anterior half of the valves from the posterior. It is probably more widely distributed than *Pholas*, but is in local patches.

6 Barnea (White Piddock) has up to 30 concentric rows of spines on its rather thin shell and its siphons are relatively short. The cavity it excavates tends to be horizontal, if it can establish itself on a vertical rock face. This piddock occurs round the whole of Britain, perhaps slightly more commonly in the South-West, but is local.

7 Petricola (American Piddock) is actually a relative of *Venus* and *Venerupis* (1 – 8, p. 75), but it too has adopted the boring habit, making short and rather irregular galleries in soft limestone and hard sand. This is another example of a relatively recent arrival to our shores; it probably came over with American oysters when they were being established in the East Coast oyster-beds. Although it has been reported as far West as Cornwall and North as Lincolnshire, it is commonest around the Thames estuary.

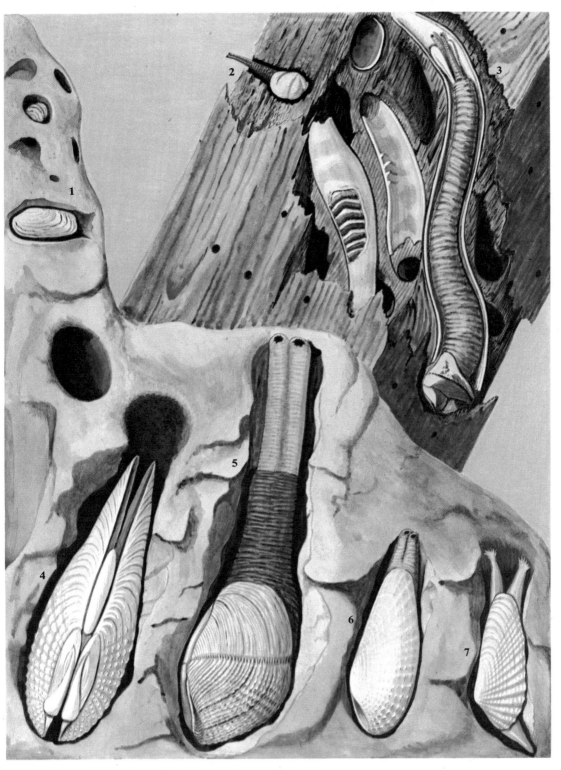

1 Hiatella arctica   2 Xylophaga dorsalis   3 Teredo norvegica
5 Zirphaea crispata
4 Pholas dactylus   6 Barnea candida   7 Petricola pholadiformis

EPIZOIC AND COMMENSAL BIVALVES

A rather unusual way of life adopted by bivalves is one in which they attach to the outer surface of some other organism. This is no mere settling on the nearest hard substratum, but is apparently a host-specific relationship, that is, a sort of parasitism in which one species of bivalve is always associated with a certain kind of host, or at most a narrow range of hosts. But in this case, because the bivalve is merely seeking shelter and the odd crumb of food and is apparently not conferring a noticeable disadvantage on the host, the relationship is called *commensalism* ('sharing a table'). The most usual sort of association is that in which the bivalve clings to the outside of a burrower (3, 4). Almost all burrowing animals need to maintain a circulating current to and from their burrows to bring in vital substance and remove waste, and one might expect that other animals would seize on such a situation and use the protection of the burrow, plus the 'services' provided by the host, to their own advantage. And this is what some of these bivalves do. In some cases, such as *Montacuta* (4), it is said that the presence of the commensal bivalve may actually be of advantage to its host, because it sometimes feeds on waste organic matter that might otherwise foul the burrow, so that the relationship tends towards a *symbiosis* ('living together', to the benefit of both associates). In any case, a commensal will never over-exploit its host, because such behaviour might be detrimental to the host's survival — which would be rather like a tenant killing off his landlord and then having his neglected house collapse around him.

1 **Lasaea,** on the lichen *Lichina pygmaea*. This tiny mollusc, never more than about 1 mm long, often occurs in large numbers among the tufts of the lichen, or in the roots of the coralline alga *Corallina*. It has an oval shell, white with a reddish tinge. It does not appear to be at all host specific, since it occurs also in crevices and in empty shells, but it certainly seems commoner among the lichen tufts than elsewhere. It occurs all round Britain, between tide marks.

2 **Lepton,** in the burrow of the prawn *Eupogebia*. The fast-moving callianassid prawn *Eupogebia* burrows in sand or mud, particularly among the roots of the sea-grass *Zostera*. At the mouth of the burrow lives this bivalve, moving over the sand surface by means of its foot. The mantle edge is sometimes extended into frills of white tentacles which presumably help it to collect food. It is restricted to the South-west coasts of England and Ireland, and is nowhere very common. There are related species which appear to be free-living, but very little is known about this group.

3 **Devonia,** on the burrowing sea-cucumber *Leptosynapta*. The foot, when visible, is extremely flat and forms a sucker by means of which the bivalve is attached to the outside of the holothuroid, about one-third of the way down the body from the anterior end. Since food-collection and exchange of respiratory gases both take place at the anterior end of this particular holothuroid's body, and as the burrow is usually blind, there cannot be too much current passing down the sides of its body for the benefit of the commensal bivalve. But every so often the host is obliged to withdraw into its burrow for protection, and one must assume that in doing so it takes with it fragments of food that get left in the first part of the burrow, where the bivalve can utilize them — and, by thus removing them, benefit the host.

4 **Montacuta,** on the burrowing brittle-star *Acrocnida* or the heart-urchins *Spatangus* or *Echinocardium*, is another tiny bivalve, rarely more than 1·5 mm long. It is found attached to the spines of the heart-urchins or brittle-stars. In the figure opposite it is shown among the spines of *Spatangus purpureus*, near the anus. It attaches to the spines by means of byssus threads, and because the ends of echinoderm spines more often than not lack a covering of skin, the mollusc can attach directly to the crystalline skeletal material. Moreover, the skeletal crystals are always perforated by tiny canals, and so the byssus threads can 'sew' the mollusc to its host.

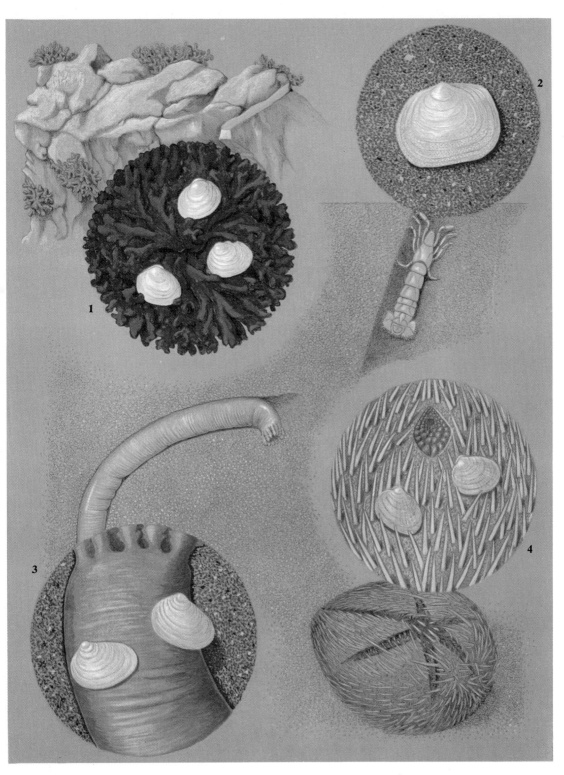

1 Lasaea rubra, on *Lichina pygmaea*   2 Lepton squamosum, on *Eupogebia*
3 Devonia perrieri, on *Leptosynapta*   4 Montacuta substriata, on *Spatangus*

BIVALVES OF RIVERS, STREAMS AND PONDS

Any mass of freshwater will probably have its representative bivalves, particularly rivers and streams, where the sand or gravel at the bottom will act as a substratum for them. The larger mussels are mainly of two genera: *Anodonta* and *Unio*; a third genus is also shown here, the rarer immigrant *Dreissena* (9). Then there are two genera of pea-sized bivalves, *Sphaerium* (1) and *Pisidium* (3), which so resemble the gravel in which they live that they are often overlooked. There is some evidence that the shape of the shell of at least some of these bivalves is influenced during growth by the speed of the current flowing over them, and consequently they may provide a good indication of the average flow of the river in which they live.

1 **Sphaerium rivicola** (River Orb Mussel). This is the largest of the pea-sized bivalves that occur in Britain, growing to about 30 mm in length. While generally living among the gravel in the bed of a river, stream or pond, it can also climb the stems of water weeds, as shown in the figure, using the leverage and stickiness of its foot to do so. This species has a strongly striated shell and is found in hard-water districts as far north as Yorkshire.

2 **S. corneum** (Common Orb Mussel). *S. corneum* is the commonest species of the four represented in Britain. It has a pale brown shell with a long pointed foot that protrudes from the anterior end (on the right in the figure).

3 **Pisidium** (Pea Mussel) is smaller than *Sphaerium*. It has only a single siphon, whereas *Sphaerium* has two. All pea shells and orb shells are hermaphrodite (have both sexes) and produce fully developed young which restock the same area as their parents. Few bodies of freshwater are without their representatives of these tiny bivalves.

4 **Anodonta cygnea** (Swan Mussel) is the largest of this group of animals — 150 mm up to 230 mm long. It uses its foot to pull the ventral part of the shell a little way into the substratum, but the siphons at the posterior end (right in the figure) are always exposed above the surface, the inhalant one (the lower) having a frilled rim, and the exhalant one usually having no frills. The cilia on the extensive gills hanging within the mantle cavity draw water into the cavity, across the gills for filtering out food and exchanging respiratory gases, then expel the water again at about 3 litres per hour. Sometimes when a specimen is collected its gills are peppered with its own young. These young stages, called *glochidia*, will have grown from the egg in the safety of the gills, and escape in due time to the outside through the exhalant siphon; each valve has a pair of enormous teeth at the free margins of the shells (the name 'glochidium' means 'arrowhead'). Successful glochidia will attach themselves to the gills of fish, where they live as parasites on the fish's blood and tissues until able to drop off and assume an independent existence.

5 **A. anatina** (Duck Mussel) is smaller than *A. cygnea*, has more oval and more swollen shells and is darker in colour. There are finer striations on the shell surface too, and the concentric ridges are often wavy, compared with the straight ones of *cygnea*. Also, whereas *A. cygnea* prefers a muddy bottom to inhabit, this species prefers sand.

6 **Unio pictorum** (Painter's Mussel). The genus *Unio* is distinguished from *Anodonta* by the presence of a depression just in front of the umbo, that is, to the left in each of the species illustrated opposite. This shell was sufficiently common in days gone by, and of such a size that it provided a convenient pallet for mixing artist's colours.

7 **U. tumidus** (Swollen River Mussel) has a slightly less pointed posterior end (to the right in the figure) than *U. pictorum*. It often has alternating light and dark radiating bands from umbo to shell margin. This is a hard-water species, living in slow rivers and canals.

8 **U. (Margaritifer) margaritifer** (Pearl Mussel) has an elongated shell which is somewhat kidney-shaped, and the lower margins of the valves have a slight concavity. This species occasionally forms tiny pearls in the mantle, usually as a result of irritation from a parasite; such pearls were treasured in Roman times and a flourishing fishery was established in this country.

9 **Dreissena** (Zebra Mussel) is more like a marine mussel (8, p. 73) in having byssus threads for attachment to a hard substratum. The shell is thick and keeled and there are alternating light and dark zigzag bands over the valves. This mollusc is said to have arrived in this country from its native region, the Baltic, about 1824 and to have spread through the extensive canal system during the subsequent 20 years. Interestingly, this was probably a re-establishment of a mussel that was formerly here, since it has been found sub-fossil in Cheshire.

1 Sᴘʜᴀᴇʀɪᴜᴍ ʀɪᴠɪᴄᴏʟᴀ 2 S. ᴄᴏʀɴᴇᴜᴍ 3 Pɪsɪᴅɪᴜᴍ ᴀᴍɴɪᴄᴜᴍ
4 Aɴᴏᴅᴏɴᴛᴀ ᴄʏɢɴᴇᴀ
5 A. ᴀɴᴀᴛɪɴᴀ 6 Uɴɪᴏ ᴘɪᴄᴛᴏʀᴜᴍ 7 U. ᴛᴜᴍɪᴅᴜs
9 Dʀᴇɪssᴇɴᴀ ᴘᴏʟʏᴍᴏʀᴘʜᴀ 8 U. (Mᴀʀɢᴀʀɪᴛɪғᴇʀ) ᴍᴀʀɢᴀʀɪᴛɪғᴇʀ

SQUIDS, CUTTLEFISH, AND OCTOPUSES

Probably because of the sinuous way they move their suckered tentacles, the squids and octopuses often engender in the human breast the same fears and reactions as those created by snakes. But, studied more deeply, they reveal one of the most fantastic evolutionary success stories in the invertebrates. They belong to the class Cephalopoda which, on many criteria, is the most advanced in the Mollusca. All of them can move by jet propulsion; all show tremendous enlargement of the central parts of the nervous system to form a structure much more akin to the vertebrate brain than to the ganglia of their molluscan forebears; all have vertebrate-like eyes with a lens capable of being focused by muscles; all have a marvellous mechanism for creating a 'smoke-screen' to confuse predators; but above all, they all show a remarkable capacity for learning. The name cephalopod means 'head-foot', and refers to the fact that most of the foot has evolved into a ring of tentacles surrounding the head to provide a formidable array of capturing devices. The mantle has become a muscular bag behind the head which contains the gills and provides the jet-propulsion unit for the escape reaction of the animal. A bag of pigment, suspended in fluid, opens into the mantle cavity so that when the animal retracts backwards by operating the jet, it can also emit a black cloud at the front. Coupled with a rapid change to a lighter body colour itself, this cloud can confuse a predator into taking a bite at it rather than at the cephalopod. Other molluscan classes have prominent shells but in this class the shell is either internal (in the squids and cuttlefish) and used mainly as a buoyancy device, or missing altogether (in the octopuses). Their learning capability enables cephalopods to distinguish between various shapes after experience, which probably allows the mollusc to learn the best way of attacking, say, a crab with dangerous pincers.

1 **Loligo** (Common Squid) has a torpedo-shaped body with a ring of eight tentacles plus a pair of longer, spatula-shaped ones at the front end, and a pair of stabilizing fins at the rear. It may reach a length of 800 mm, but 200 to 300 mm is more usual. It catches small fish, crustaceans, and even other squids with its spatulate tentacles, and because it is so common in offshore waters at certain times of the year, it is considered harmful to some of our fisheries. Its shell is a long, horny 'pen' which is sometimes found on the beach.

2 **Sepia** (Common Cuttlefish) has a broader body than *Loligo* and a lateral fin by undulations of which it swims slowly in either direction. The pattern of black and white stripes over the body is distinctive of the male; during breeding the colour intensifies at the approach of another cuttlefish, and those individuals that do not change colour as they approach are assumed to be females. The internal skeleton is the familiar cuttle 'bone'. The cuttlefish hunts for shrimps. It blows jets of water at the sand surface to uncover them. Even then it can spot its transparent prey only if they move to cover themselves again, when it catches them in the pair of long tentacles (as shown).

3 **Sepiola** (Little Cuttle) is the smallest British cephalopod, never more than 40 mm long, and is the one most likely to be encountered on the beach, where it occasionally gets stranded in a crevice or in a ripple-mark in the sand. The body is rounded, but there are dia-

mond-shaped lateral fins. It often burrows rear first into sand and feeds on passing shrimps and other prey.

4 **Eledone** (Curled or Lesser Octopus), shown in the picture settling down towards its rocky haunt, has tentacles with a single row of suckers down them, whereas its relative *Octopus* (5) has a double row. The body is bag-like and has no shell at all. The intake to the mantle cavity (towards the reader and wide open in the figure) lies at the side of the head, and the funnel, for directing the jet in emergency dashes backwards, is in the centre of the under-surface. This is more common in Northern waters.

5 **Octopus** (Common Octopus) is confined to the southern parts of England, because it is really a native of warmer waters, such as the African and Central American coasts. Octopods, as the name suggests, have only the outer ring of eight tentacles. *Octopus* itself uses these to catch its food, usually crabs and lobsters, which it holds in the tentacles while it bites into the prey with strong jaws just inside the mouth, which look like the beak of a parrot.
Usually not common, *Octopus* may appear off our coasts in huge numbers following a mild winter. It is said that the animal hardly ever breeds on our shores, but new individuals restock our waters from the coast of France. The eggs, in capsules, are laid in large numbers in rock crevices, looking like long bunches of grapes (as seen in the lair behind the animal figured opposite). After hatching, the larva spends several weeks in the plankton before settling down in some rocky place to assume its more characteristic place on the sea-bed.

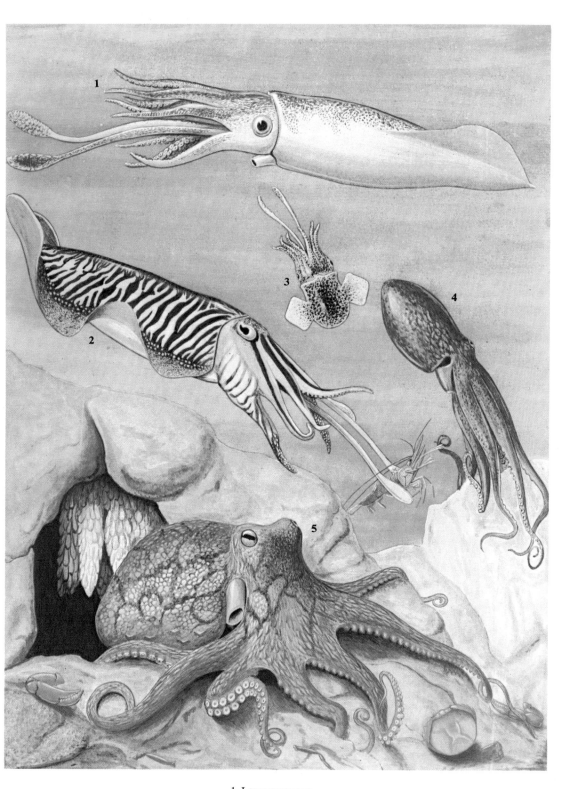

1 Loligo forbesi

2 Sepia officinalis   3 Sepiola atlantica   4 Eledone cirrhosa

5 Octopus vulgaris

SCALEWORMS OF THE SHORE

The important phylum Annelida, the segmented worms, has marine, freshwater, and terrestrial representatives. The three main groups are the polychaetes or bristle worms (pp. 94 – 111), the oligochaetes or earthworms and pondworms (pp. 112 – 115), and the hirudineans or leeches (pp. 116 – 119). Most annelids have soft, cylindrical bodies with a gut running from an anterior mouth to an anus at the posterior end. The features which place the annelids in a higher grade than the nemertines and other lowly worms (pp. 20 – 31) are, first, the division of the body into *segments*, and, secondly, the possession of an extensive body cavity or *coelom* which they put to important use in locomotion and other activities. Both these features are related to movement and, as one would expect, they become modified in less active annelids, such as tube-dwellers and parasites.

The polychaetes are almost entirely marine. They are distinguished by possession of limb-like projections from the body, called *parapodia*, from which emerge bunches of stiff, hair-like *chaetae*. A typical polychaete has well-marked segments, is an active carnivore, and uses both the sinuous movement of its body and its parapodia for locomotion. A number of highly specialized worms may well be aberrant polychaetes, but so unlike the typical forms are they that some are placed in a separate class, the Archiannelida (pp. 120 – 121) and others even in separate phyla (pp. 168 – 169).

Aphroditids and polynoids (sea mice and scale worms) are crawling or *errant* polychaetes that have plate-like extensions of the upper body wall called *elytra*, imbricating like tiles on a roof, as a protective and respiratory device. These plates are visible from above in polynoids, but are hidden beneath a covering of matted chaetae in aphroditids. Many polynoids have formed semi-permanent associations with other animals, usually by inhabiting their burrows or clinging to the outside of their bodies, if they can thus gain protection; some of these are illustrated with their hosts elsewhere (*see* pp. 103, 179).

1 **Halosydna (Alentia)** has 18 pairs of gelatinous elytra which are covered with fine papillae. It is semi-transparent, and its eyes are larger than in other scale-worms. It has long dorsal cirri and two long anal processes. It is usually found under stones on the lower shore.

2 **Lepidonotus** has a short, stiffish body of roughly uniform width. There are 12 pairs of elytra visible from above, since, unlike *Aphrodita*, it does not have a covering 'felt' of matted chaetae. The colour is variable, from yellow to brown with a mozaic pattern on the elytra. It is found quite commonly under stones in the middle to lower shore and among seaweeds.

3 **Eunoe** has 15 pairs of elytra which cover the body completely. The elytra may be grey or brownish-purple, and may bear small projections on the dorsal surface. This worm is both iridescent and phosphorescent; it is found not only in muddy sand but also in worm tubes, and in cracks and crevices in rocks.

4 **Lagisca** also has 15 pairs of elytra, but a naked tail region. The elytra are covered with conical papillae, with larger ones on the posterior border of each elytron. This is probably the commonest polynoid of the South Coast, where it is found under stones and in cracks; it is common elsewhere round our shores too.

5 **Aphrodita (Sea Mouse)**, easily the most massive of the British polychaetes, is hardly worm-like at all. It lives beneath the sand surface with only its hind end exposed and muscular activity of its body causes currents to pass in towards the animal along its underside, up beneath the parapodia and thence beneath the arching dorsal scales where gaseous exchange takes place; the scales are then drawn down to expel water out through the upper side of the hind end. This is rather a lethargic animal which lives at low tide or below; but sometimes, after a storm, vast numbers may be stranded ashore. An outstanding feature of this animal is the iridescence of its chaetae, which give green, yellow, blue and red interference colours as they catch the light.

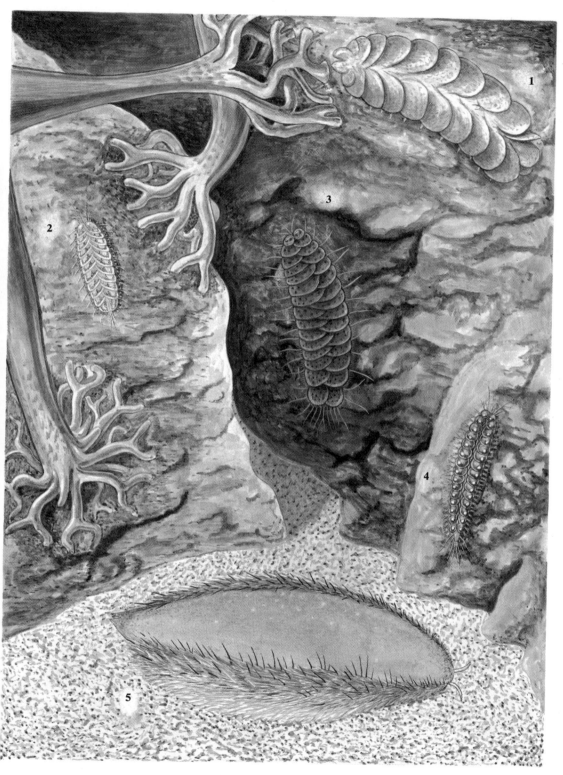

1 Halosydna (Alentia) gelatinosa
2 Lepidonotus squamatus   3 Eunoe nodosa
4 Lagisca extenuata
5 Aphrodita aculeata

95

ERRANT POLYCHAETES OF THE ROCKY SHORE (1)

Three families of active, carnivorous polychaetes are represented on this page. All possess a muscular pharynx which can be extended by a build-up of pressure in the body fluids. The family Phyllodocidae is a large group of polychaetes distinguished by leaf-like extensions of the parapodia. They have slender bodies, by flexure of which they crawl over the surfaces of boulders or soft bottoms. Two examples are shown on this page (2, 3). They look superficially similar to the nereids (p. 101), but, unlike them, they lack jaws and usually have a simple, balloon-like proboscis which is extruded to catch the prey. The family Tomopteridae, of which one example is shown (1), consists of transparent, pelagic annelids that have their parapodia modified into biramous paddles without chaetae, by means of which they swim in the upper levels of the sea, preying on the rich supplies of food there, such as the smaller members of the plankton, fish eggs, and even larval fish. There is a pair of trailing tentacles from the head region.

The family Syllidae are mostly small worms, usually less than 10 mm long, distinguished by possession of three tentacles on the head, a median one and two laterals. Syllids usually cling to sea-vegetation or nestle in empty shells and crevices. They are not immediately evident to the casual collector, and the best way to see them is to leave some weed in a jar of sea-water in the light, when they will emerge and gather along the edge of the jar. The capacity of annelids to regenerate new parts if they are cut into several lengths is well known, and it is not surprising that an extension of this ability is used as a method of reproduction. Nowhere is this better demonstrated than in the syllids. Some produce new individuals by forming new heads at intervals down the length of the body; others bud off new individuals in a clump from one segment, so that a little knot of new offspring is produced. Three examples are shown here (4–6).

1 **Tomopteris** is a frequent find in plankton samples from offshore but it is not often encountered near shore. It is of great importance to sea fisheries because it takes considerable toll of very young herring.

2 **Phyllodoce** (Paddle Worm) can swim as well as crawl, by undulations of its flat body. Some specimens of *P. lamelligera*, which are found under stones on the lower shore, reach a length of 300 mm, and usually the colour is a splendid green. Its smaller relative, *P. maculata*, grows to 100 mm long and is yellow or greenish. This species prefers to nestle beneath stones in sandy areas. Phyllodocids lay eggs in green gelatinous pear-shaped masses on stalks, which they attach to weeds and other hard objects.

3 **Eulalia** is a common animal on rocky shores and the bases of cliffs, where it nestles among barnacles and mussels. It is often seen just after the tide has receded as a thin, dark green strand, weaving its way to a damp crevice in the rock face. It has a very long cylindrical proboscis with a crown of 15 to 30 rounded papillae. It differs from *Phyllodoce* (2) in having a bluntly rounded head, compared with the heart-shaped head of the other.

4 **Autolytus** produces new individuals by budding (as shown) and these break free and swim away from the parent to the surface waters of the sea, where they become sexually mature and spawn. The female carries her eggs in a large mass attached to the ventral side of the body. Such individuals can often be seen at the surface of the sea at night if a light is shone on to the surface; they are attracted to the light and glisten as they swim near it. Interestingly, the male and female forms were not immediately recognised as being a part of an otherwise benthic animal's life history, and were originally called by separate names, *Sacconereis* for the female and *Polybostrichus* for the male.

5 **Myrianida** is usually found among tunicates (pp. 188 – 191) and sponges (p. 5). It too has the capacity to bud off new individuals, but here the budding process appears to be precisely delimited: a new head will form at the 14th segment down the body initially, then further new heads will form at intervals of 3 segments posteriorly. In some, a chain of nearly 30 individuals may be formed in this way.

6 **Syllis** shows well the typical syllid processes arising from the sides of the body. Syllids all show considerable variation in the superficial markings on the body.

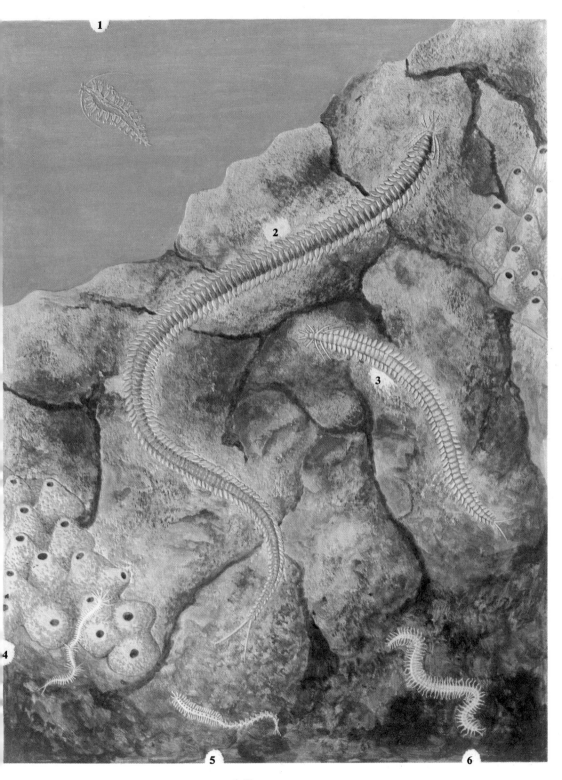

1 Tomopteris sp.

2 Phyllodoce lamelligera   3 Eulalia viridis

4 Autolytus sp.   5 Myrianida pinnigera   6 Syllis hyalina

ERRANT POLYCHAETES OF THE ROCKY SHORE (2)

Though the members of the three families shown on this page generally burrow or live in crevices, they are nonetheless highly active worms. The eunicids (rock worms and palolos) form a large and varied family of rock-dwellers and mud-burrowers, all of which have a unique rasp-like proboscis which can be protracted from the mouth and used to scrape encrustations from the rock face, or to take in particles of food from the mud. The proboscis has a series of replaceable teeth. Two examples of the family Eunicidae are shown here (1 and 2). Some eunicids, such as the arabellids and lumbrinereids (not illustrated) are remarkably earthworm-like in appearance, and some authorities place them in a separate family; these cylindrical worms are usually found among the holdfasts of large algae, a splendid habitat for a great range of animals. Many eunicids have gills, elaborately branched, which sometimes arch over the dorsal surface, and when they live in crevices or burrows they ensure a circulation of sea-water over the gills by gentle movements of the body. A highly modified eunicid, *Histriobdella*, has taken to a parasitic life. It has lost all but the last pair of parapodia, and with these it scrambles over the gills and eggs of lobsters in berry. The family Nephtyidae is composed of beautiful worms, mostly having an iridescent pearly sheen to their rather rectangular bodies. They burrow in cleaner sand and may therefore be encountered on beaches of seaside resorts where the sand is not muddy enough for some other worms. If taken from the sand and dropped into water, they swim well, but as soon as the head touches the sand they immediately stop swimming and commence digging into the sand with the proboscis. The proboscis is also used in feeding: as the worm burrows through the sand, it is protracted to take in food in the form of other small organisms or detritus. Three species of the principal genus, *Nephtys*, are shown here (3–5).

The glycerids, like the nephtyids, average about 300 mm long, but they differ from them in having a circular cross section. They are unpigmented, but the red colour of the blood, taken round the body not in closed blood vessels but in the body cavity (coelom), often shows through the body wall. There is a slender conical projection from the head which bears four antennae at its tip. The group is notable for the large balloon-like proboscis which can be everted from the mouth. One example is shown (6).

1 **Marphysa** (Red Rock Worm) can grow to about a metre in length, though it is seldom found more than 300 mm long. Each segment has a bunch of bright red, feathery gills lying close to the dorsal body wall. *M. sanguinea* occurs in rock crevices on the lower shore, particularly in the English Channel, while its relative *M. bellii*, with comb-like gills, is common among *Zostera*, the eel-grass, or under stones.

2 **Eunice** is shorter and browner than *Marphysa* (1), though it too lives in rock crevices and under stones, where it often makes a membranous tube in which to lie. The head has a median antenna and two pairs of lateral antennae, which are much more prominent than the tiny processes of *Marphysa*.

3 **Nephtys caeca.** The three common species of *Nephtys* (3 – 5) are found in muddy or gravelly shores and in estuaries, as well as in clean sand. *N. caeca* is probably the commonest of the three, and is often dug up in the same place as the lugworms fishermen seek for bait,

that is, rather muddy sand. It usually grows to about 250 mm long — generally the largest of the three — and is pearl-grey with pink or reddish tints.

4 **N. hombergi** (White Worm) is usually bluish-white with iridescent pink shading and there is a wide greyish line down its back. It frequents all sediments but the finest mud.

5 **N. cirrosa** is not common, but appears to be more fussy than the other two, and is usually found only on beaches of cleaner sand. It is usually between 50 and 100 mm in length, though much larger specimens sometimes occur. It has yellow chaetae, some of which are bent.

6 **Glycera** reacts to disturbance by twisting its body into tight coils. This is a common worm of the lower sandy shore.

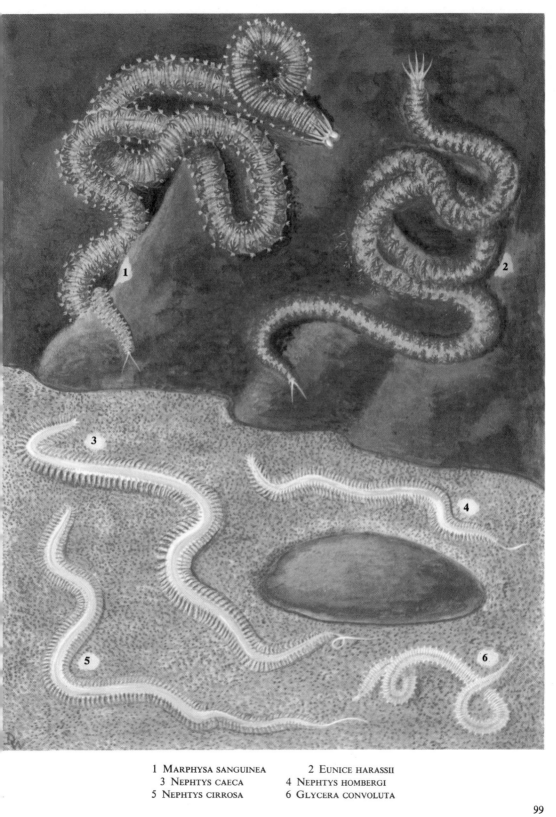

1 Marphysa sanguinea   2 Eunice harassii
3 Nephtys caeca   4 Nephtys hombergi
5 Nephtys cirrosa   6 Glycera convoluta

ERRANT POLYCHAETES OF THE SHORE (3)

On this page are shown members of a single family, the Nereidae, comprising the rag-worm which anglers find so useful as bait. Pulled from the sand, these hang limply from the hand, and the jagged appearance given to the body by the parapodia have earned them their common name, because they resemble a piece of ragged cloth. Though most burrow in sand or mud, some live under stones on the shore; some have adapted to living in the reduced salinity of estuaries and can actively maintain the difference in concentration between their body fluids and the only slightly saline water around them. Those that do not have this facility, if placed in estuarine water, would be unable to prevent water entering their bodies by osmosis thus seriously diluting their body fluids and distending every cell and cavity of the body. The nereid head has a prostomium, or projection in front of the first segment, which bears two short antennae and four eyes. There is a proboscis armed with groups of teeth and strong jaws for feeding and burrowing. Some species undergo a kind of metamorphosis as sexual maturity approaches: the eyes enlarge; the parapodia, sometimes in the rear part of the body only, expand to form paddles; and the new form, either male or female, swims upwards from the bottom towards the sea surface. Originally, this stage in the life cycle was thought to be a separate genus and was given the name *Heteronereis* — a name now used to describe the pelagic phase of any nereid worm. The males and females make a nuptial journey to the sea surface to spawn when the fertilized eggs develop into the typical annelid *trochophore* larva which remains for a short time in the plankton before settling to the bottom again for the benthic phase of the life history.

1 **Nereis (Eunereis) longissima.** Whereas none of the other species have more than about 140 segments, this sand-dwelling form may have 200 or more. The tentacles are short and the processes arising from the parapodia are also smaller than in the other species.

2 **Nereis (Neanthes) virens (King Rag).** This is a noble polychaete, sometimes growing to the thickness of a man's finger. The parapodia develop foliaceous gills over the animal's back. It is usually found in sticky mud beneath stones, where it builds a burrow for itself, lined with mucus; undulations of the body keep a steady current of water through the burrow for respiration. It will emerge from the burrow when necessary to seize a passing animal with its powerful jaws.

3 **Nereis pelagica.** Typically golden-brown with greenish flanks, this species is found on rocky shores among seaweeds and shells and is rather commoner in the North of Britain.

4 **Nereis (Neanthes) fucata** has long parapodia and is usually brick red with a white longitudinal band on each side of the red line of the dorsal blood vessel. In its young stages, this worm is free-living, but older individuals usually inhabit old whelk shells occupied by the fully grown hermit crab *Pagurus* (*see* p. 123 and book jacket).

5 **Nereis (Neanthes) diversicolor** is often found in estuaries and can tolerate surprisingly low salinities, to which its physiology is able to adjust, as mentioned above. Its colour is variable, from green to red, but most forms have a distinct red line down the back. There is no heteronereis stage.

6 **Platynereis** has long tentacles which may reach back to a quarter the length of the worm, that is, 15 mm in a 60 mm long worm. It tends to make tubes for itself in holdfasts of *Laminaria* on the lower shore.

7 **Perinereis** lives in gravelly mud or sand, under stones and in pools. Its proboscis has two groups of teeth behind the jaws, looking like eyebrows. It is quite common.

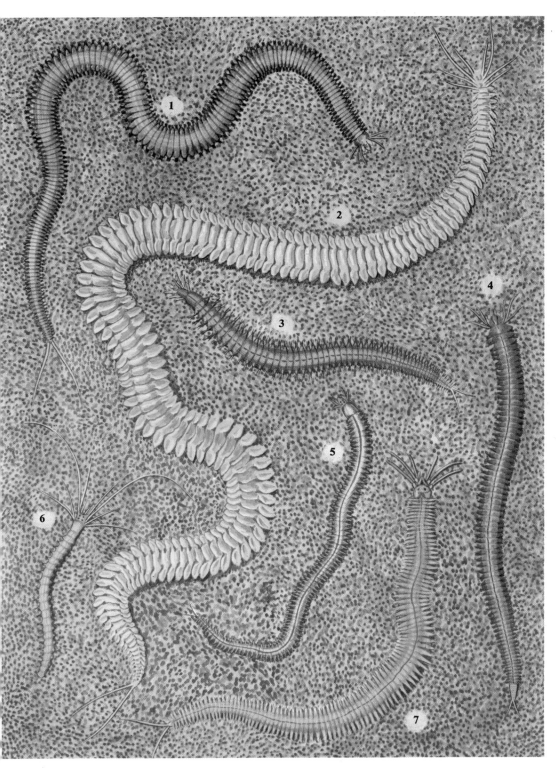

1 NEREIS (EUNEREIS) LONGISSIMA   2 NEREIS (NEANTHES) VIRENS
3 NEREIS PELAGICA   4 N. (NEANTHES) FUCATA
5 N. (NEANTHES) DIVERSICOLOR
6 PLATYNEREIS DUMERILI   7 PERINEREIS CULTRIFERA

POLYCHAETES OF SAND AND MUD

So difficult are the evolutionary relationships of the polychaetes to assess, that one is forced to classify them in a way that may not necessarily reflect the evolution of the group, but which is according to behavioural similarity. In this scheme, the polychaetes are divided into the active or *errant* forms (Errantia, pp. 94 – 101) and the sedentary forms (Sedentaria, pp. 102 – 111). Most sedentary forms build permanent tubes or burrows within which they move to perform their vital activities but which they seldom leave. Most have various modifications for feeding, respiration, and ridding the vicinity of the animal of excretory waste. On this page sedentary forms of various families which build a mucous tube in sand or mud are shown.

1 **Cirratulus** lives just beneath the surface of rather foul mud full of rotting algae, with its thin, pink respiratory tentacles protruding above the surface. Other superficially similar tentacles near the anterior end are used for feeding, which they do by trapping particles of food and conveying them to the mouth. The body tapers at both ends.

2 **Amphitrite.** Some members of the family Terebellidae, like this one, burrow in sandy mud, while others, such as *Terebella* and *Lanice* (p. 105) build sandy tubes for their own occupation; yet others, such as *Polycirrus*, do not live in tubes or burrows but among algal holdfasts and other sheltered places. All terebellids may be recognized by the mass of tangled tentacles at the head end; just behind the feeding tentacles are three pairs of dark red gills. Sharing the burrow of *Amphitrite* there may be another polychaete, the scaleworm *Gattyana* (p. 94). This is an example of commensalism in which one organism gains protection from the activities of another, without conferring any appreciable advantage or disadvantage on its host.

3 **Magelona** has a spatula-shaped prostomium, behind which arise two papillate tentacles, probably more than half as long as the body, which it uses for picking up detritus and small organisms from the sand surface. There are no gills and the chaetae are short and simple. The anterior end of the body, including the tentacles, is usually pink while the posterior end is green.

4 **Notomastus**, also a capitellid, is between 150 and 300 mm long; the first 12 of its 100 or so segments have chaetae and the red colour often does not extend to the posterior end. The proboscis is larger than that of *Capitella* (5).

5 **Capitella** lives in rather foul mud, where it may be locally very abundant. Most capitellids are bright red and superficially like earthworms, having no bulky appendages projecting from the body. They are very fragile. *Capitella* is between 20 and 100 mm long and the first 7 of its segments have chaetae. The red coloration extends to the posterior end.

6 **Nerine** is a spionid, distinguished by the pair of flexible tentacles just behind the head, which are frequently carried folded back along the body and concealed between the gills. Each tentacle has a ciliated gutter down one side along which food particles are carried to the mouth. In *N. cirratulus* (illustrated) the head (6A) is sharply pointed, whereas in the rarer *N. foliosa* the head is blunt.

7 **Arenicola** (Lugworm). This most familiar polychaete is used as bait by fishermen, and its casts betray its presence in many a sandy beach. Though rather inactive, it is highly specialized for a successful existence between tide marks, and can withstand adverse conditions of temperature, oxygen availability, and salinity. It builds for protection an L-shaped burrow and lives with its head thrusting into the blind end to 'work' the sand — so that there is continuous movement of sand from above its head. This activity produces a little pit on the surface of the sand, about 150 mm from the burrow opening, in which detritus and potential prey collect; they then sink as the sand is worked from below, until they reach the vicinity of the worm and can be devoured with some of the sand. After passing through the worm, where the covering of organic matter is digested off, the sand is voided from the vertical limb of the burrow as a cast. This happens every 40 minutes or so. After defaecation, the worm returns to the toe of its burrow to resume feeding, which it performs in bursts every 7 minutes or so. Respiratory water is brought down the vertical limb of the burrow by periodic contractions of the body; it flows over the gills in roughly the middle of the body; then it picks up excretory waste from the ducts which open just in front of the gill region; then it passes over the head, to filter away between the sand particles at the blind end of the burrow, thus keeping the sand semi-fluid and mobile. *Arenicola* is not a good burrower: once out of its tube, it can burrow again only slowly.

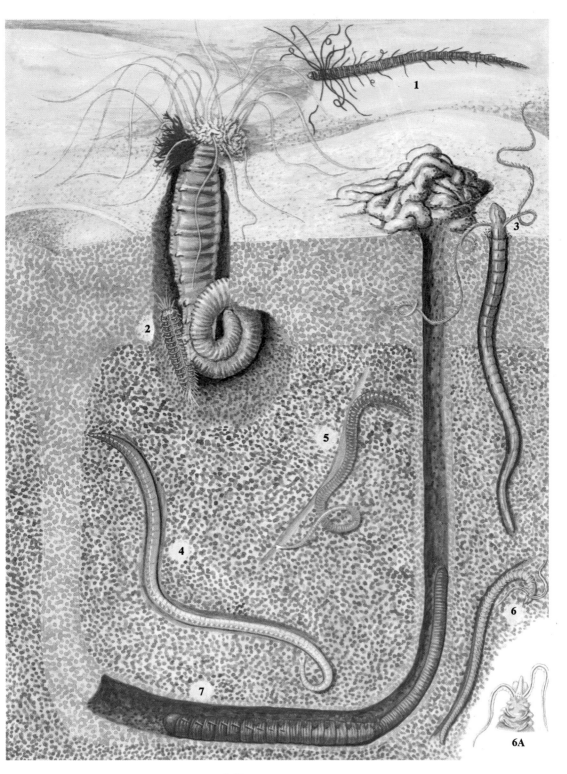

1 Cirratulus cirratus
2 Amphitrite johnstoni   3 Magelona papillicornis
4 Notomastus latericeus   5 Capitella capitata
6 Nerine cirratulus
7 Arenicola marina   6a N. cirratulus, head

POLYCHAETES WITH SANDY TUBES

On the previous page those sedentary polychaetes living in burrows or nestling under rocks are illustrated; some of them manipulate sand particles to form the walls of the burrow. In the ones illustrated on this page this ability is extended to building special tubes of sand, in which the animal is protected. Some of these forms have a marked ability to select the right size or shape of particle, and some carefully orientate each particle as it is manipulated into place.

1 **Sabella** (Peacock Worm) is generally found only at extra low-water, when forests of their protective tubes may be seen, the base of the tube being embedded in the substratum while the open end projects high above the disturbed silt near the sea bed. The anterior end of the animal bears a magnificent funnel-shaped crown of tentacles which fan out from the open end when water covers the worm to filter out food particles and pass them by the action of cilia to the mouth. The rest of the animal grips the inside of the tube with its chaetae, then, when danger threatens, it moves down the tube and in doing so collapses the fan as it is pulled in.

2 **Lanice** (Sand Mason) builds a ragged-ended tube in sandy beaches, and this is the worm whose tubes give the beach at low tide a bristly appearance when viewed towards the sun. This is a terebellid like *Amphitrite* (p. 103), and so has a thick clump of reddish tentacles at the head end. The dorsal blood vessel is usually very wide, looking like a red ribbon embedded in the body wall, and in the living animal its contractions can be clearly seen. *Lanice* incorporates a variety of materials in its tube and this accounts for the ragged end, which always appears to be in a state of collapse even when perfectly healthy.

3 **Terebella.** All terebellids have a characteristic appearance, with a crown of writhing tentacles at the head end for feeding and, just behind this, three pairs of feathery gills. *Terebella*, like *Lanice* (2), makes a tube in the silt or beneath a stone, which it irrigates with a piston-like action of its body. The animal protrudes its head end from the tube and extends the tentacles almost straight out, to attach to the rock surface by a special flat zone near the tip. The very end of the tentacle will move sinuously to explore, and food particles will be captured here and elsewhere along the tentacle length and passed along the ciliated gutter on one side to the mouth. It seems that the cilia in the region of the attachment zone can cause the zone to 'creep' across the rock surface, thus extending and moving the tentacle.

4 **Owenia.** Dig in fairly clean sand at low tide and you are quite likely to come across the long, flexible tubes built by this thin worm. It is seldom more than 2 mm in diameter and 50 mm in length, but it builds its sandy tube in such a way that each grain or fragment of debris overlaps the one below it, like tiles on a roof. The worm moves along the tube to feed at its open end by means of a crown of short, frilly tentacles. The animal can apparently move its tube in the sand.

5 **Megalomma.** This sabellid makes a leathery tube about 300 mm long in gravelly sand, to the lower end of which are attached fragments of shell and pebbles, so that the tube is a heavier affair than that of *Sabella* (1). The tube projects from the mud about 20 mm and closes off when the tide recedes. The tentacles are feathery and darker in colour at the base; at each tip is a dark eye-spot. The colour of the worm itself varies from yellow, through red to brown.

6 **Pectinaria.** This polychaete builds a beautifully regular conical tube, open at both ends, which is often found empty in debris left behind by the receding tide. Sand grains of selected size only are used in its construction. In its living position, the animal orientates the narrow end of the tube to point upwards just projecting above the surface of the sand, and lives head-downwards. Near the head the chaetae form a golden comb-like structure ('pectinate', hence the generic name) which it uses to dig into the sand or to form a cage in which the feeding tentacles work. Feeding as they do, and selecting sand grains for their house, these worms surely exemplify the words of Deuteronomy: 'They shall seek of the abundance of the seas, and of treasures hid in the sand'.

1 Sabella pavonina   2 Lanice conchilega
3 Terebella lapidaria
4 Owenia fusiformis   5 Megalomma vesiculosum
6 Pectinaria belgica

POLYCHAETES WITH MUCOUS AND OTHER TUBES

Some sedentary polychaetes make mucous tubes incorporating material so fine that the tube collapses when disturbed and cannot be collected; others make tubes of hardened mucus which withstand a certain amount of handling; others again lie beneath stones and line their flimsy burrows with mucus. The animals illustrated on this page are not necessarily closely related; neither is there very much difference between the way some of these burrow and the method used by some shown on pages 99 and 103, but it is convenient to consider the forms with mucoid tubes together.

1 **Myxicola** is a sabellid polychaete which builds a tube about 150 to 200 mm long of several layers of gelatinous, translucent mucus. The tube is thick and rather flabby, and the single opening is wider than the worm. The tube is usually completely buried in the substratum, which is often muddy gravel on the lower shore. The crown of tentacles is held so that the ends of the tentacles are just flush with the surface of the substratum. Each tentacle is joined to its neighbour by a membrane, rather like the web of a duck's foot. This arrangement means that water can be taken in only at the tip of the crown, but the membrane will increase the area covered by cilia, and hence the strength of the pump; a medium-sized *Myxicola* can pump nearly 300 ml per hour, which is about the same pumping efficiency as *Arenicola* (p. 103) though the latter uses a totally different method — peristaltic movements of its body. In the case of *Myxicola*, the worm does not irrigate its tube, but relies solely on its crown of tentacles for respiratory exchange.

2 **Hyalinoecia** (Quill Worm). This eunicid from offshore waters constructs a fairly thick hyaline tube (hence the generic name) of parchment-like mucoid material, somewhat resembling the shaft of a feather (hence the common name). The tube has several valves near each end which the worm can push aside as it passes through them, and which spring back to close off the tube when the worm withdraws. The worm is a fairly typical eunicid (compare it, for instance, with 1 and 2 on p. 99), with tentacles on the head and a pair of posterior processes. The worm usually crawls about on the surface by means of its anterior parapodia while keeping the posterior part of its body in its tube.

3 **Flabelligera** is a dweller under stones and in crevices. It builds a transparent mucous case through which only its bristle tips protrude. The body is short, having a mere 30 to 50 segments, compared with 150 in the case of *Myxicola* (1). The front part of the head forms a short, retractile tube into which the sense organs and gills can be withdrawn.

4 **Chaetopterus** builds a U-shaped mucus-lined burrow in sand or mud, both ends of which protrude slightly above the surface. More rarely, it lives beneath rocks which are lying on sand. Instead of collecting food by means of flexible tentacles at the front end of the body, this worm relies on food coming to it in currents which are continually passed through its tube by the muscular action of fan-like parapodia in the middle region of its body. The worm makes for itself a mucous filtering bag by means of parapodia at its anterior end. This bag is held in the tube in such a way that the water has to pass through it, and in doing so the food particles are retained and concentrated into a bolus which is periodically passed along a ciliary gutter on the animal's dorsal side to its mouth, where the much-reduced head tentacles help it into the mouth. In passing across the body, the current of water also fulfils the respiratory needs of the animal. In this case, the currents pass across the body from front to back, which is the reverse of such tube-dwellers as *Arenicola* (p. 103).

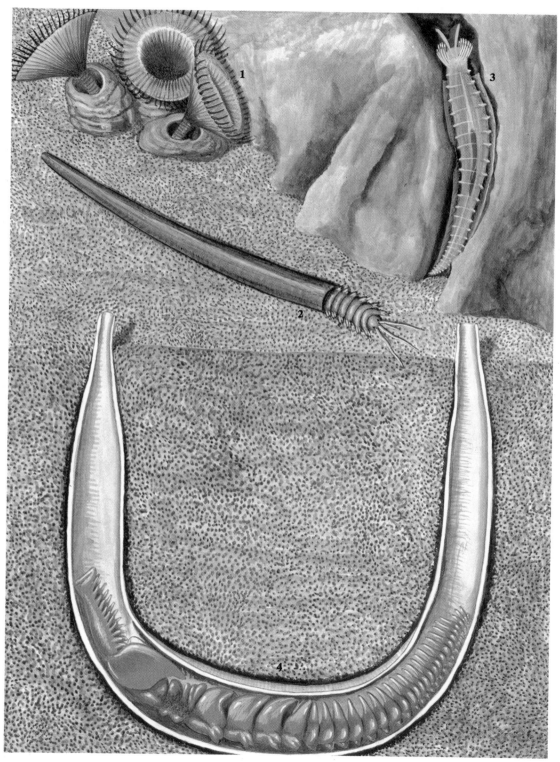

1 Myxicola infundibulum   3 Flabelligera diplochaitus
2 Hyalinoecia tubicola
4 Chaetopterus variopedatus

POLYCHAETES BUILDING REEFS AND BURROWS

Members of several families of sedentary polychaetes are found in rocky places, nestling in available crevices, boring into the rock itself or building massive tubes on the surface of existing rocks, sometimes in such numbers as to give rise to a thick reef. Three such worms, all from different families, are shown on this page. *Sabellaria* (1), the only member of the family Sabellaridae illustrated, builds reefs of sand in rocky places; both *Potamilla* (2), a member of the family Sabellidae (*see also* 1 and 5, p. 104, and 1, p. 106), and *Polydora* (3), of the family Spionidae (*see also* 6, p. 102), build U-shaped burrows in the rock, either excavating the rock themselves or building their tubes in its fissures and cracks.

1 **Sabellaria** is a gregarious rock-dweller that builds tubes of compacted sand and shell fragments on the surface of rocks and boulders. The whole colony looks like an irregular honeycomb of porous sandstone and grows to several inches thick when really prolific, producing quite prominent reefs on existing rock. The worm itself is about 30 mm long, with a heavy anterior end crowned by a mass of flattened chaetae which act as a bung to the tube when the animal retracts. Behind the chaetae are a mass of short tentacles which spread out from the open end of the tube to form a feeding funnel when water covers the reef. Naturally, reefs of this worm can establish themselves only in regions where there is both a hard substratum and a good deal of sandy material with which to build tubes.

2 **Potamilla** is a sabellid (*see* p. 104) which builds a brown, horny tube to which sand adheres, in fissures and crevices, discarded shells, or other worm tubes on the lower shore. When the worm retreats into its tube, the opening is protected by the tube bending over at the end, and sometimes rolling up, to close it off. The worm itself is about 60 to 80 mm long, with up to 100 segments; its anterior end is greenish brown, the posterior end orange. It has about 20 short, translucent tentacles in two groups of 10 each, with long, nearly white branches.

3 **Polydora** is a tiny worm belonging to the family Spionidae (*see also* p. 102). It lives in a U-shaped boring in limestone or other soft rock, the two ends of the burrow protruding slightly from the rock surface a short distance apart. From one tube there emerges a pair of long flexible tentacles which arise from just behind the head, and these flail around in the water in search of food. There is a ciliated gutter down one side of each tentacle, for transferring the food to the mouth. The tubes, which are a mere 2 mm in diameter can often go unnoticed except when the worm is living in calm waters, for then sand particles too big for ingestion are thrust to one side of the tube, and in time a little chimney of rejected particles may stand up from the rock face for a few millimetres.

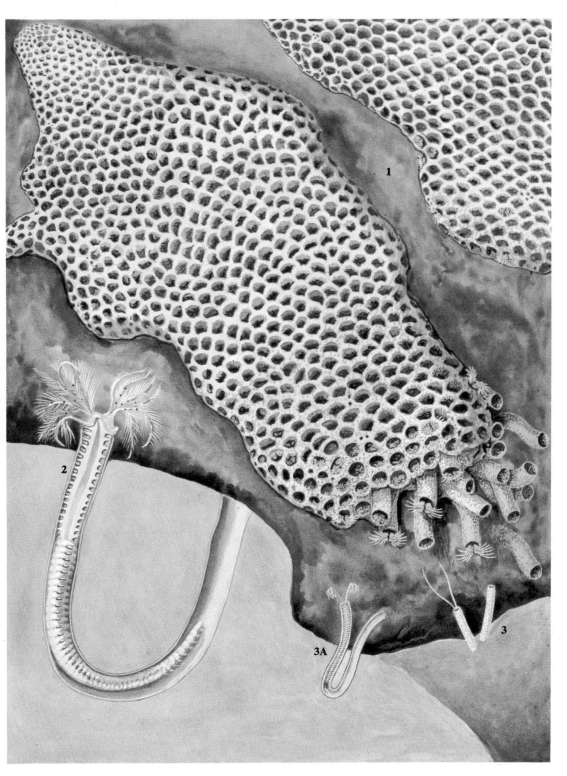

1 Sabellaria alveolata
2 Potamilla reniformis   3 Polydora sp.
3a Polydora, tube in section

POLYCHAETES WITH CALCAREOUS TUBES

In this final selection of polychaete worms are those that deposit calcium carbonate on a groundwork, or foundation, of mucoid material. These are the members of the family Serpulidae, which make up the majority of the worm tubes that encrust the larger shells, such as scallops, whelks and crabs, and that attach to the fronds of algae, and that form aggregated, twisted masses like spaghetti on rocky surfaces. The tubes are secreted by special glands in the first body segment and moulded by the worm's anterior collar. They have characteristic shapes and arrangements and vary in gregarity: thus, *Spirorbis* (1) has a spiral tube and is either solitary or aggregated; *Serpula* (3) has a straight, cylindrical tube and is aggregated; *Pomatoceros* (5) has a keeled tube and is aggregated; and *Filograna* (2) has cylindrical tubes and is semi-colonial, that is, it always appears in fairly dense masses. The worms themselves are very much like sabellids (p. 104), but are much smaller. Most of them have one of their crown tentacles modified as a sort of stopper to block the open end of the tube when the animal has retreated into it. They either shed their eggs into the sea, where fertilization takes place, or the eggs may be fertilized within the tube.

1 Spirorbis is here drawn on the alga *Fucus serratus*, the Serrated Wrack, a common habitat for it. There are several common species which are likely to be encountered: *S. borealis*, usually on *Fucus*, has a fairly smooth tube and a clockwise coil, whereas *S. spirillum* is superficially similar but with an anticlockwise coil; *S. pagenstecheri* also has an anticlockwise coil but has a toothed opening; *S. corallinae* usually settles on coralline algae, whereas *S. tridentatus* usually settles on rock and has three well-marked ridges down its tube.

2 Filograna. Where *Pomatoceros* (5), *Serpula* (3), and *Hydroides* (4) are sometimes single, this serpulid always occurs in fairly dense masses, the tubes aggregated into irregular, twisted lumps up to 120 mm square. It is found on the lower shore or among algal holdfasts, or low down on pier piles, etc. Lumps of the tube masses are often thrown ashore. The operculum of the worm itself is small compared with that of *Serpula*, and the opercular stem has small spine-like projections on it whereas that of *Serpula* is smooth.

3 Serpula is drawn here on a *Buccinum* shell. This worm is a shallow-water form rather than an inshore one, but is often cast up on the beach attached to shells and other objects. The tube is round, becoming trumpet shaped at the end with growth rings at intervals. The operculum, or lid closing the tube when the animal has withdrawn, has a smooth stem, unlike *Filograna* (2), and a funnel-shaped, toothed head. The tube, though fixed to a hard object basally, may be elevated away from the surface at the open end, sometimes standing nearly erect, several tubes intertwining.

4 Hydroides. This is another gregarious form, in which the thin, delicate tubes often entwine to form aggregated masses. The animal lives below tide, but its tubes are often cast up on the beach, attached to shells or stones. *Hydroides*, like some other serpulids, is said to reproduce asexually as well as sexually, the budded young emerging from the side of the parent's tube by dissolving away the chalky material and pushing a hole in it.

5 Pomatoceros is an abundant serpulid, usually on shells, both living and dead. There is one British species, *P. triqueter*. The tube, irregularly bent, is keeled and therefore looks rather triangular in transverse section; at the open end the keel ends in a point, and anybody who has collected on the shore will know how the sharp point can cut the hands when encrusted rocks are turned over. When the tentacles are extended underwater, they appear either orange, brown, or blue, depending on the proportions of several pigments present in their skin. The top surface of the body is generally purplish-brown, but the colour of the undersurface differs in the two sexes; in the males it is generally white, while in the females generally red.

1 SPIRORBIS BOREALIS   2 FILOGRANA IMPLEXA
3 SERPULA VERMICULARIS   4 HYDROIDES NORVEGICA
5 POMATOCEROS TRIQUETER

GARDEN EARTHWORMS

The class Oligochaeta includes the familiar earthworms (p. 113), and some very common fresh-water forms (p. 115). Though there are a few true marine forms, mainly below tide level, they are rarely encountered. Oligochaetes are distinguished from polychaetes, as the name suggests, by the small number and size of the chaetae. There are no parapodia to bear the chaetae, and indeed because they are mainly burrowers any lateral projections would be disadvantageous. The body is annulated and each ring represents a single segment. As in polychaetes, every segment has certain structures in common with every other segment, such as body muscles, chaetae, excretory organs, and nerve branches; but certain organ systems, such as reproductive organs, are restricted to a few segments only. All oligochaetes are hermaphrodite, that is, each worm has a set of male and female organs, but an elaborate mechanism normally exists for ensuring an exchange of sperm and for preventing self-fertilization. It is when garden earthworms are exchanging sperm that they are often encountered: on moist days in late summer the worms emerge from their burrows until they encounter another of the same species also in reproductive state; they come to lie side by side with their anterior ends overlapping. Then the overlapping ends become enveloped in a common mucous tube which holds them tightly together; sperm flows from the testes down a groove in the side of each worm to a small sac in the other. Both worms do this at the same time — a case of give and take. In most oligochaetes there is a *clitellum* or saddle near the reproductive segments that becomes prominent seasonally. This produces a mucous belt when the eggs are ready to be laid, out of which the worm wriggles; as the belt passes the female opening, the eggs are deposited in it and as it passes the male opening, a drop of sperm is exuded to fertilize them. Then the worm wriggles from the mucous belt, the ends close over, and it forms a cocoon for the embryos to develop in. Oligochaetes lack respiratory organs, and gas exchange takes place over the entire body surface; this means that the skin must be kept moist, or the worm may suffocate. The importance of the 'working' of soils by earthworms is well known — worms in one acre of land may bring to the surface 18 tonnes of soil per year. The effect is to mix, aerate, and break up the soil, and germination and growth are thus improved.

1 **Dendrobaena** is flattened from top to bottom and has a red or purple pigmentation with lighter clitellum. It is common in moist places rich in decaying organic matter, such as beneath the bark of fallen trees.

2 **Lumbricus** is the fattest of the English garden earthworms and is easily the best known, being much used to introduce students to invertebrate anatomy. And with good reason, for in this easily obtained animal we have a generalized member of a predominantly important phylum of invertebrates, that illustrates well the way functional problems are overcome in soft-bodied animals. The effects of this worm on lawns and flowerbeds in winter are well known: it plugs up the end of its burrow with sticks and leaves, sometimes making a little clump several inches high. It feeds on the organic matter pulled in like this.

3 **Bimastus** occurs under the bark of old trees and in leaf mould.

4 **Allolobophora** differs from *Lumbricus* not only in diameter but in the nature of the prostomium: in *Lumbricus* it reaches back to the second segment dorsally, thus dividing the first segment, but in *Allolobophora* no such posterior extension occurs. Some species, such as the long, thin *A. longa*, produce casts at the soil surface (shown at top of picture), a nuisance on bowling greens and lawns. *A. longa* is usually muddy-brown, while *A. chlorotica* may be various shades of green, blue, or pink, the greenish forms preferring wetter places than the pink ones. This worm will come to the surface if the soil in which it lives is vibrated.

5 **Enchytraeus** (Pot Worm) is a member of a separate family, the Enchytraeidae, from the other worms illustrated here. It is small and almost white, and extremely common in soil. Unlike the lumbricids, the enchytraeids are unable to burrow but merely squeeze through cracks between the lumps of soil. They occur in soils not prone to drying out.

6 **Octolasium** is a large bluish-grey worm in which the posterior few segments are much lighter in colour. It is usually found under stones or in mossy places.

7 **Eisenia** can be recognized by the pink colour of the anterior few segments and the orange tint of its clitellum. *E. rosea* is common in gardens; *E. foetida* is found in dung heaps and compost piles, especially when fermentation has set in.

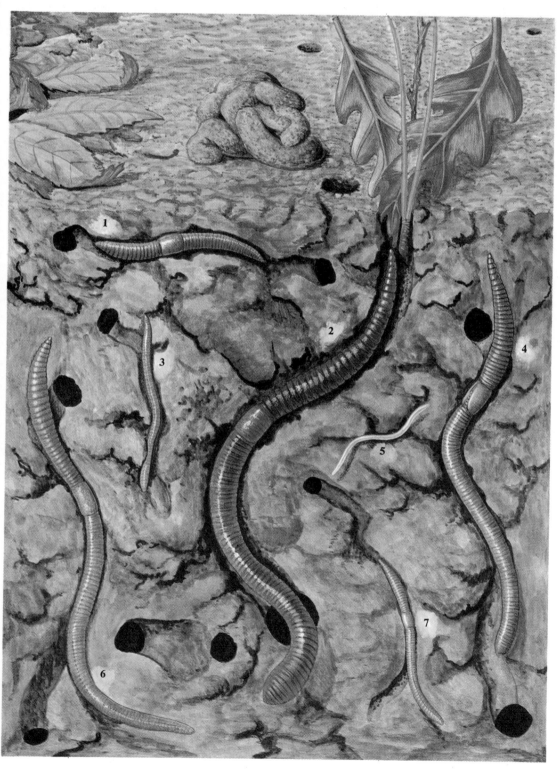

1 Dendrobaena subrubicunda   2 Lumbricus terrestris
3 Bimastus tenuis   4 Allolobophora longa
5 Enchytraeus sp.
6 Octolasium cyaneum   7 Eisenia rosea

FRESHWATER OLIGOCHAETES

Aquatic oligochaetes are most likely in shallow water places, such as ponds and ditches, and the edges of rivers and streams. Tubificids, however, occur also at the bottom of deep lakes, where they may be in concentrations of as much as 8,000 per square metre. Most aquatic oligochaetes burrow in the silt or debris, but some crawl over submerged weed and hard objects, feeding on adhering organic matter as they go. The burrowers usually build tubes to line their burrows, and the ends of the tubes sometimes project above the silt surface. Their relatives, the terrestrial earthworms (p. 113), are able to regenerate a few posterior segments, and some aquatic oligochaetes use an extension of this facility as their normal reproductive method: they multiply by fragmentation, rather as some polychaetes do (*see*, for instance, p. 96), and the sexual method appears to occur only very infrequently.

1 **Chaetogaster** is a member of a large family of aquatic oligochaetes, the Naididae, characterized by having cleft ends to the ventral chaetae. This one is almost transparent with a fat front end. It is found particularly in mollusc shells, or in the tubes of insect larvae, such as chironomids and caddises.

2 **Eiseniella** (Square Tailed Worm). This is a lumbricid, that is, it belongs to the same family as the terrestrial earthworms *Lumbricus*, *Allolobophora*, etc. (p. 113), but it is wholly aquatic. It resembles a small earthworm, with a clitellum seasonally present about a third the way along the body, and tiny hair-like chaetae projecting from its sides. Though terrestrial lumbricids do fall into water and cause confusion to collectors, this one can easily be recognized by the squareness of the hind end of its body. It lives in mud at the edges of lakes, in streams, or in places where vegetation is constantly sprayed with water such as the sides of mill-races and waterfalls.

3 **Stylaria**, another naidid, is distinguished by the presence of a style-like proboscis at the front end. A common habitat for this worm is floating vegetation near the surface of ponds and ditches.

4 **Tubifex** is a member of a large family of aquatic oligochaetes, the Tubificidae. It is more than 30 mm long, is usually red, and coils tightly when disturbed. It usually lives in the bottom of rivers and streams in thick mud. At first sight such mud looks unlikely to support any life at all — but place a thick layer of it in a glass dish with water and let it settle in the light and it is quite probable that tubificids will emerge, tail upwards, to undulate their bodies gently in the water for respiratory exchange. The more the tail projects from the surface of the mud, the less oxygen there is in the surrounding water. Tubificids feed by passing mud through the gut, to digest off the adhering organic matter. This is the worm that is sold in tight knots of several hundred individuals, as food for freshwater aquarium fish.

5 **Lumbriculus** belongs to another family, Lumbriculidae. It may be up to 80 mm long and its body usually appears bright red because the body wall is transparent and the colour of the blood shows through. This is an example of an oligochaete which has never been known to show sexual reproduction: it apparently always reproduces by fragmentation, each piece of the parent worm growing a new head or tail, as necessary. It lives along the edges of ditches, streams, rivers, and lakes, and constructs a flimsy tube in the silt.

6 **Nais** is another naidid, usually less than 20 mm long, often pale pinkish or brown in colour, with a pair of eyespots. It is frequently found in chains of budding individuals, rather like the rhabdocoel *Stenostomum* (p. 21), showing that it reproduces by fragmentation. It builds a flimsy tube in detritus at the bottoms of ponds.

7 **Dero**, also a naidid, has a bunch of ciliated gills at the posterior end which sometimes appear red due to the blood in their thin-walled vessels. This worm is also a tube-builder, living in the mud at the bottom of ponds and ditches.

8 **Aeolosoma** is in a family of its own, the Aeolosomatidae, whose members have chaetae all of which are hair-like, unlike the naidids (3 and 6) and tubificids (1, 3, 6 and 7) which have some chaetae with bifid tips. *Aeolosoma* is a small worm less than 10 mm long. The skin contains numerous oil droplets, looking like pink spheres. These worms feed on tiny single-celled plants which are driven into the mouth by cilia at the anterior end of the body. They can produce resistant cysts in unfavourable conditions, such as drying up or freezing, in which the worm curls up in an outer protective case, from which it breaks free when water returns.

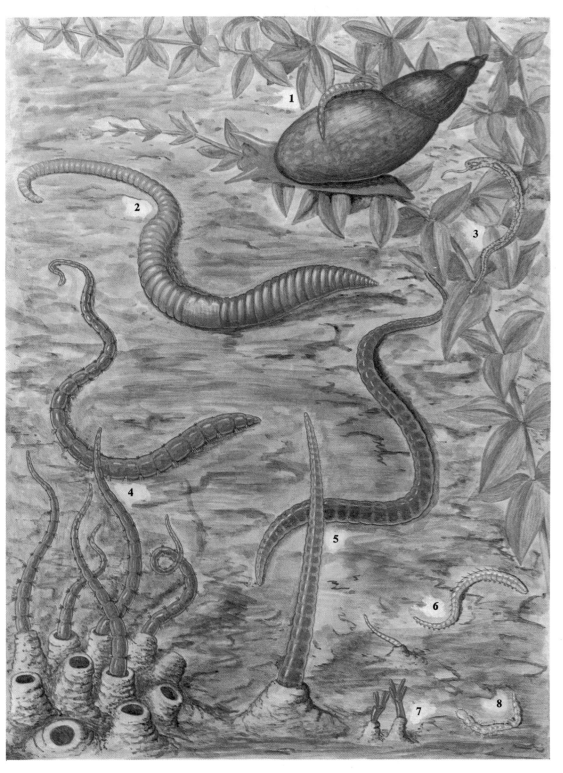

1 Chaetogaster limnaea
2 Eiseniella sp.   3 Stylaria lacustris
4 Tubifex tubifex   5 Lumbriculus rivulorum   6 nais sp.
7 Dero sp.   8 Aeolosoma sp.

FRESHWATER LEECHES

Nearly three-quarters of the world's leeches are found in freshwater or terrestrial habitats. One may regard them either as external parasites or as rather specialized predators, for they attach themselves to their host only long enough to have one meal at the host's expense; they then detach and 'sleep it off' before seeking the next victim. Some feed on the body fluids of invertebrates, others on cold-blooded vertebrates, and yet others, of course, on warm-blooded vertebrates. For the mediaeval practice of 'leeching' patients suffering from certain ailments, the freshwater leech *Hirudo* was cultured. Leeches belong to the annelid class Hirudinea whose members differ from other annelids in having a sucker at each end. In addition, the number of body segments is fixed at 33, though each segment may have several rings to it externally. (For further characteristics of leeches, see p. 118.)

1 **Hemiclepsis.** Normally a fish predator, this leech may also attack tadpoles and some molluscs. It is often mistaken for a colour variant of *Glossiphonia* (2), but the body is much more flask-shaped and the posterior sucker more prominent, and it is found in small ponds or slow running rivers rather than the fairly swift streams *Glossiphonia* prefers.

2 **Glossiphonia** feeds on aquatic snails, into which it plunges its proboscis to suck out the body fluids and usually the soft parts too. It has a small anterior sucker and a large posterior one. On the dorsal side of the body there are three pairs of eyes in two parallel rows and two almost parallel lines of dark pigment interrupted by tiny sensory papillae. This is probably the commonest freshwater leech of streams and brooks, often found under stones, or in the shells of molluscs on which it has been feeding.

3 **Theromyzon.** This leech enters the nostrils of wading and swimming birds and feeds on the mucous membrane. When hungry, it becomes positively phototactic, that is, it moves towards the light, and as it does so it may gradually (over an hour or so) change its external colour by either concentrating or dispersing the pigment in cells called chromatophores, so that it still matches its surroundings while awaiting its prey near the surface of the water.

4 **Hirudo** (Medicinal Leech). This vicious animal, formerly used in medical practice, is a native of Europe. It was believed extinct in Britain until discovered in the wild parts of the New Forest, the Lake District, South Wales, Anglesey and Islay, perhaps surviving from monasteries and other places of healing. These wild populations will find it increasingly difficult to maintain themselves, now that bridges have replaced fords and drinking troughs have replaced ponds. It is a noble animal. Greenish, with longitudinal red stripes, it has a pattern of irregular markings over its body, and is easily the biggest of the British freshwater leeches, at least when contracted. It is a gnathobdellid, that is, a jawed leech, and has strong teeth which make a Y-shaped cut in the skin of its prey, through which it sucks the blood. The anti-coagulant in its digestive juices causes the wound it makes to continue bleeding, sometimes for a considerable time, but eventually its effect wears off.

5 **Piscicola** (Fish Leech) is a freshwater member of a mainly marine family (p. 118). Most freshwater leeches lie in wait in crevices for their prey, but *Piscicola* will remain in open water, ready to attack passing fish. Disturbance in the water caused by a passing fish, for instance. will at once cause it to swim — a behavioural adaptation of obvious advantage. Anglers often find specimens clinging by one or both suckers to fish landed from larger lakes and streams. The unusual feature is the presence down the sides of the body of small hemispherical gills.

6 **Erpobdella** (Worm Leech) has lost the power to penetrate its host's tissues, though it still has a conspicuous pharynx with prominent ridges with which it attacks worms, insect larvae, and other soft invertebrates, which it then swallows whole. It seems to prefer small ponds in soft water districts.

7 **Helobdella.** This leech, too, is a predator rather than a parasite, for it sucks out the body fluids of snails and insect larvae, killing the host. It has one pair of simple eyes, well separated, and a horny scute on the dorsal surface about one-sixth of the way between the anterior and posterior ends. It prefers ponds in hard water districts.

8 **Haemopis** (Horse Leech). Thoroughly misnamed, this leech has nothing to do with horses. Its jaws are weak and its teeth blunt, and it has abandoned the blood-sucking habit in favour of a carnivorous one, spending much of its time out of water feeding on earthworms, insects, molluscs and decaying flesh. It grows to an impressive size and may measure more than 300 mm when extended. Its sinuous movement through water is a beautiful sight.

1 HEMICLEPSIS MARGINATA   2 GLOSSIPHONIA COMPLANATA   3 THEROMYZON TESSELATUM

4 HIRUDO MEDICINALIS

5 PISCICOLA GEOMETRA   6 ERPOBDELLA OCTOCULATA   7 HELOBDELLA STAGNALIS

8 HAEMOPIS SANGUISUGA

MARINE LEECHES

The leeches are advanced annelids with certain very obvious similarities with the oligochaetes. For instance, like the oligochaetes they have a saddle or clitellum at certain seasons for producing a cocoon in which the eggs develop. In fact, in some classificatory schemes the oligochaetes and hirudineans are contained in the same class, the Clitellata, to emphasise this similarity. But the leeches are more active than the earthworms, and have taken up a very specialized predatory habit: sucking the blood and body fluids of other animals. To do so, they are provided with some sort of piercing mechanism and body muscles which can suck out the juices from their prey; to get to the prey they can swim strongly, and to hold on to it they have their strong suckers at front and back.

When a blood-containing animal is injured, the escape of blood is normally checked by a clotting mechanism brought about by substances in the blood itself. To a blood-sucking animal, such a mechanism is a severe disadvantage, and would result in the partial healing over of the wound before the predator is satisfied. To prevent this, and to prevent the blood clotting when it is in the leech's gut, the salivary secretion of leeches contains an anticoagulant. This often means that the wound a leech has made will continue bleeding for some time after the leech has disengaged. By its very nature, leech feeding is fortuitous. The animal must try to attach on the right host when it is hungry and then detach when it is full. In many cases there are complex behavioural patterns which help; many leeches are positively vibrotactic and will 'home' on a passing animal which might provide a source of food. But there will be times when long periods elapse between one meal and the next, and to tide them over such times leeches are able to draw on stores of food occupying what is, in other annelids, the body cavity or coelom. When these stores are used up, they are still able to live on, by drawing on the very materials of their own bodies, so that they decrease in size to a fraction of what they were originally. Most leeches are freshwater (*see* p. 116), but there are a few important members of the group which occur in Atlantic waters. They attack both bony and elasmobranch (cartilaginous) fish, though individual species appear to be highly host-specific. Our knowledge of marine leeches is poor, because of the difficulty in getting specimens for study.

1 **Pontobdella.** This leech is found on rays, skate, and other elasmobranchs. While waiting for its prey it rests on a stone or rock on the sea-bed in a characteristic spiral form, attached by its posterior sucker. But when its host passes, it will swim off by undulatory movements towards the fish and attach by its large front sucker. This is a large leech, with a rounded body studded with tubercles of various sizes. It may attain 200 mm in length. It is said to be an important vector in transmitting blood parasites, such as trypanosomes, from fish to fish.

2 **Branchellion** attacks *Raja*, the ray. In this leech, the body is in two distinct regions: a short, narrow anterior portion which includes the clitellum (in season), and a longer abdomen posteriorly. Each segment of the body has three separate annuli, and in the abdominal region each annulus bears a leaf-like gill, through the lumen of which coelomic fluid flows to carry gases

to and from the tissue of the body. The presence of such large gills is rather unusual: in most leeches exchange of respiratory gases occurs across the body wall and there are either no gills at all or they are mere bumps on the sides of the body, as in *Piscicola* (5, p. 117), a close relative of these marine forms.

3 **Janusion** is usually found on *Cottus scorpius*, the bullhead or sea-scorpion. It is rather small, scarcely ever exceeding 15 mm in length. It has two pairs of eyes, consisting of clusters of black spots.

4 **Calliobdella.** There are several species of this genus, but the commonest appears to be very host-specific to the angler-fish, *Lophius piscatorius*. It has no eyes though the other species of the genus have. It has a very complex copulatory mechanism to ensure cross-fertilization from one animal to another, even though, like other leeches, it is hermaphrodite.

1 PONTOBDELLA MURICATA, resting position
2 BRANCHELLION BOREALE, from ray   3 JANUSION SCORPII, from sea-scorpion
4 CALLIOBDELLA LOPHII, from angler fish

ABERRANT ANNELIDS

The class Archiannelida is a rag-bag assortment of mostly unrelated annelids, almost solely marine and concealing their relationships beneath specializations for particular modes of life. Many live in the interstices of sedimentary particles on the sea bottom or beach. They may lose the parapodia and chaetae and so come to look rather unlike other annelids. The name 'Archiannelida' was suggested by early zoologists who thought them primitive; so some of them might be, but as with all soft-bodied animals the relationships are almost impossible to assess because of the lack of fossils. We can only say that this group represents curious worm-like animals, in many ways resembling polychaete annelids, but with special features. Most are scavengers and have a protrusible tongue for conveying food to the mouth. Two other groups, Myzostomida (6) and Echiurida (7 and 8), regarded by some as classes within the annelids and by others as separate phyla, are included here.

1 Nerilla is the archiannelid which looks most like a typical polychaete, since it has two bundles of chaetae on each side of each segment. It has a ciliated food groove on the ventral side, five head tentacles, and a pair of blunt-ended palps.

2 Dinophilus is more like a larval annelid (p. 177) than an adult worm, since it has locomotory rings of cilia girdling the body. The ventral surface is also ciliated, and on it the worm creeps, flatworm-like, across the substratum. The body, about 3 mm long, shows 5 or 6 segments externally, but the body cavity is not divided into compartments. There is a fairly distinct head and a pointed tail. This bright orange animal is found in empty shells in rock pools, vivid under bright light against the pure white of the shell's nacre. The sexes are separate.

3 Protodrilus is cylindrical, has no chaetae or parapodia and, like *Dinophilus*, has ciliated rings round the body and cilia on the ventral surface. There is a pair of anterior tentacles acting both as feelers and for respiration. *Protodrilus* is hermaphrodite, the first 7 segments producing ova and the others sperm. This and *Polygordius* (5) were the original members of the Archiannelida, though probably not closely related.

4 Saccocirrus, like *Protodrilus* (3), has long tentacles on the head. It also has simple chaetae but in only a single bundle on each side of each segment. The sexes are separate; the males have intromittent organs for copulation, and the females sperm receptacles, in each segment.

5 Polygordius is similar to *Protodrilus* (3), but the segments are delimited by shallow grooves. It lacks the archiannelid 'tongue' and feeds by ciliary activity. At the posterior end the anus is surrounded by glandular papillae by means of which the animal can adhere temporarily to the substratum. The sexes are separate, and external fertilization gives rise to a typical annelid trochophore larva with an equatorial double girdle of cilia, between which lies the mouth. Almost the whole of the original larva becomes the head of the adult, the body being produced by extension of the posterior part of the larva.

6 The Myzostomids. Animals can be expected to exploit any situation where food is plentiful. One such place is the food groove of crinoids (sea-lilies and feather-stars, p. 181), and ophiuroids (brittle-stars, p. 185), and some myzostomids have become specialized to straddle the grooves or sit over the mouth and intercept the stream of food before it reaches the host's mouth. The body is flattened and there are special suckers, claws and tentacles by which the animal adheres to its host's body and a protrusible proboscis by which food is taken in. So unlike other annelids are these worms that some authorities place them in a separate phylum, but their obvious annelid affinities are betrayed by their mode of development.

Myzostoma itself is ectoparasitic on *Antedon*, the European feather-star. It closely resembles its host in colour; in the general confusion of arms and pinnules of a specimen of *Antedon* plucked from the water the worm is often missed, sitting as it normally does flat against the soft upper surface of its host's disk.

7-8 The Echiuroids. These strange bag-like animals have undoubted affinities with polychaete annelids. In some ways resembling sipunculoids (p. 186), they live in burrows in sand, mud, and rock with a food-collecting proboscis protruding. *Bonellia* (7) will probably be found only in warmer waters such as the Mediterranean. This animal shows extreme sexual dimorphism: the female body may be 80 mm long, and the proboscis a further metre in length, but the male may be less than 3 mm, with no proboscis and a very much reduced gut. Further, the male is parasitic on the female, living in her body cavity and thus enabling internal fertilization to take place. *Thalassema* (8) is not a common animal but turns up occasionally in holes in rock, or under stones on the lower shore. The body is up to 60 mm long, with a long, grooved proboscis.

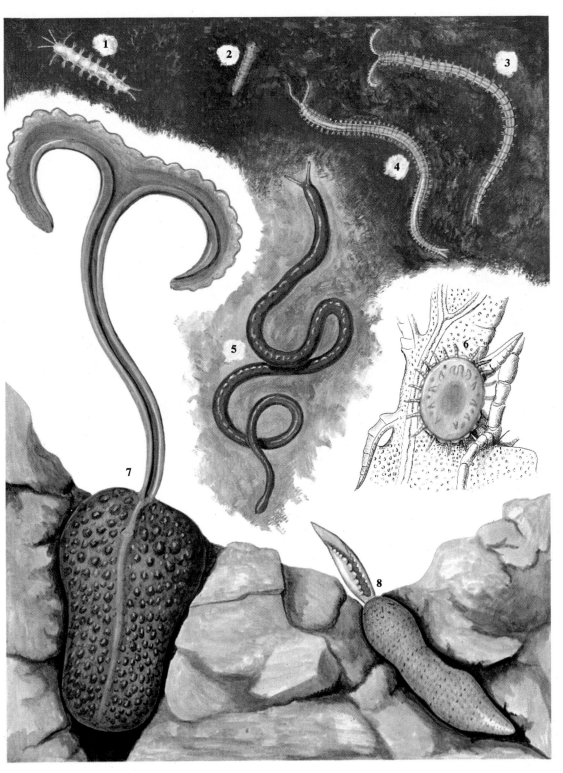

1 NERILLA ANTENNATA   2 DINOPHILUS sp.   3 PROTODRILUS FLAVOCAPITATUS
                     4 SACCOCIRRUS sp.
        5 POLYGORDIUS LACTEUS   6 MYZOSTOMA CIRRIFERUM, on *Antedon*
   7 BONELLIA VIRIDIS   8 THALASSEMA NEPTUNI

SEASHORE CRUSTACEANS

The arthropods (pp. 122 – 167) include more than 850,000 species and form by far the largest phylum in the animal kingdom, exceeding in number all the other phyla combined. The characteristic tough exoskeleton and jointed limbs are superimposed on a segmental body plan that reflects the evolution of arthropods from ancestors of the annelid worms (pp. 94 – 121). The insects (covered in a separate volume in this series), the myriapods, and the arachnids are essentially terrestrial while the crustacea are the dominant group in aquatic habitats.

1 **Palinurus** (Spiny Lobster). Known also as 'rock lobster' and 'crawfish', *Palinurus* is a deep-water creature that only occasionally turns up in rock pools. The protection afforded by the heavily armoured exoskeleton with its numerous sharp spines more than compensates for the absence of lobster-like pincers, and the flailing antennae in particular can cause painful injuries to the bare hands of a would-be captor. In Britain the gastronomic importance of *Palinurus* is sadly overlooked in favour of the less palatable 'Norway Lobster' (*Nephrops*) but the French, who call it 'langouste', have fully exploited its potential. The striking *phyllosoma* larva of *Palinurus* is illustrated on p. 127.

2 **Pagurus.** The largest and commonest of the British hermit crabs is *P. bernhardus*. These animals have soft, unprotected bodies and must seek security in abandoned mollusc shells. Although hermit crabs' shells become encrusted with a wealth of sponges, bryozoans, barnacles, and serpulid worms, this is fortuitous and there is no particular relationship between the crab and these animals. However a highly specific and apparently mutually rewarding relationship does exist with the anemone *Calliactis*, which is deliberately placed on the shell by the crab. Within the shell, co-existing with the crab and apparently keeping it free from encrusting organisms, lives a ragworm, *Nereis fucata* (p. 100), that is found nowhere else. The hermit crabs are voracious feeders, not only scavenging like the true crabs but also filtering suspended particles from the water.

3 **Galathea.** Two common species of 'Squat Lobster' are found on British coasts. During the winter they live in deeper water but as the weather becomes warmer they move inshore to intertidal rock pools and may occasionally be found under stones. Normally the abdomen is kept flexed beneath the body as the animal crawls slowly over the bottom but in emergencies the body is straightened and the broad tail used as a paddle for darting swiftly backwards. However, unlike *Homarus* (4), *Galathea* cannot sustain this rapid backwards swimming for any distance.

4 **Homarus** (Common Lobster). This, the largest British crustacean, is a widespread species that usually lives in the sublittoral zone. However, during the summer it tends to move into shallow water and may occasionally be found lurking in rock pools. Although

normally associated with rocky shores, in some areas it is also found commonly living on sand.
The female lobster breeds only once in two years, producing as many as 160,000 eggs at a time, which are carried adhering to the underside of the abdomen. Development of the eggs takes nearly a year and results in the formation of a planktonic *mysis* larva nearly 10 mm in length resembling the adult in most features. A distinction should be made between *Homarus* and the related *Nephrops*, which is marketed as 'Norway Lobster' or 'Dublin Bay Prawn'. This is a much smaller animal with long slender claws.

5 **Crangon** (Common Shrimp). The shrimps differ from their close relatives the prawns (p. 124) principally in their flatter body and in possessing only a single pair of stout pincers. They are sand-dwelling creatures and during the day bury themselves completely with only two small branches of their antennae protruding. If disturbed, they are able to control their colour to match the background. They normally emerge at night to hunt, feeding on a wide range of diet, either scavenging, or preying on worms, mollusc eggs, young fish, and other crustaceans. *Crangon* is markedly tolerant of changes in temperature and salinity, which enables it to penetrate estuarine waters, particularly in summer.

6 **Cancer** (Edible Crab). During the summer small individuals, seldom large enough for eating, may be found beneath rocks and in crevices around the low tide mark. As the weather becomes colder there is a migration to deep water, where spawning takes place during the winter. The eggs hatch the following summer in shallow coastal waters.

7 **Macropipus** (Velvet Swimming Crab). The swimming crabs are immediately recognizable by the shape of the hind legs, which terminate in a broad, flattened swimming paddle. *Macropipus* is a large and aggressive species, quite common on southwestern shores where it usually lives under stones and seaweed in rock pools. However, in some areas it is found living on sandy bottoms, using the swimming paddles to excavate a burrow in which to hide. When the crab is buried, the respiratory current is drawn in through the teeth of the large pincers, which are held tightly in front of the body and serve as strainers to remove particles of sand.

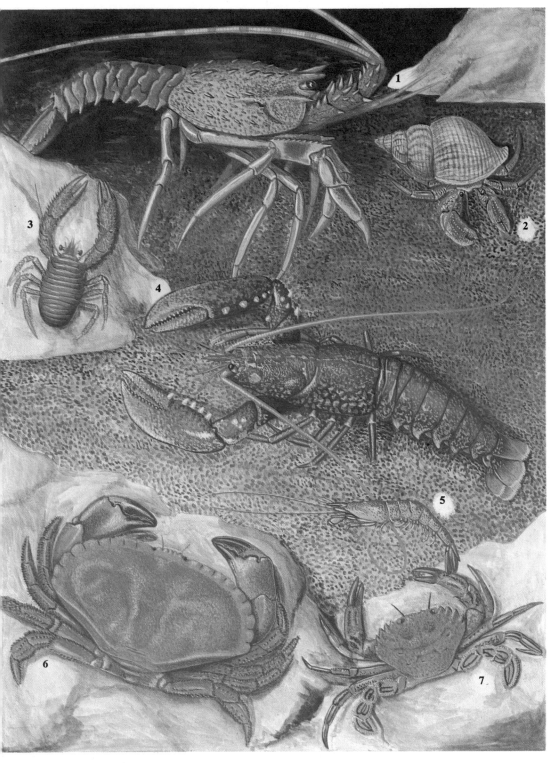

1 Palinurus vulgaris   2 Pagurus (Eupagurus) bernhardus, in whelk shell
3 Galathea squamifera   4 Homarus vulgaris
5 Crangon vulgaris
6 Cancer pagurus   7 Macropipus puber

SEASHORE CRUSTACEANS

1 **Palaemon (Leander)** is the largest of the prawns found in British waters. During the winter they live in sub-littoral waters but as the temperature rises they move into rock pools, where they may be found throughout the summer. Although capable of rapid movement backwards in emergencies, using the tail fan as a paddle, prawns usually move slowly over the bottom, picking up fragments of weed and animal debris with the two pairs of fine pincers.

2 **Bopyrus** is an isopod crustacean belonging to a group that is entirely parasitic on other crustaceans. The eggs hatch into a typical larva, resembling miniature free-living isopods, but with piercing mouthparts and large, hooked claws. The larva swims actively until it finds a copepod (p. 126) on which it settles and feeds. A week later it leaves the copepod and searches for a second host, eventually entering the gill chamber of a young decapod (1). Here it feeds by sucking blood, periodically moulting in synchrony with its host. The first *Bopyrus* to arrive in a gill chamber develops into a female but a subsequent arrival becomes a male — a method of sex determination also shown by the annelid *Bonellia* (p. 121). As with *Sacculina* (7) the parasite destroys the reproductive organs of its host.

3 **Praunus** is the commonest British Opossum-shrimp, closely related to the freshwater *Mysis* (p. 132). It occurs in rock pools and on shallow sandy shores, hovering in swarms just above the bottom. *Praunus* exhibits outstanding colour control and is able to blend accurately against its background. The opossum-shrimps get their name from the way in which eggs and young are carried in a ventral brood pouch.

4-5 **Gammarus** (4) is an amphipod that occurs in large numbers under driftwood and sea-weed in the inter-tidal zone, living somewhat further down the shore than its close relative *Orchestia* (5), the sand-hopper. It is often found in pairs, the male being held within the curve of the female's body.

6 **Carcinus** (Shore Crab). This tough and aggressive crab is one of the most frequently encountered animals of the sea shore. Its resistance to exposure and to changes in salinity enable it to live throughout the intertidal zone where it shelters not only in rock pools but beneath stones and in cracks and crevices often far up the shore. Among young *Carcinus* many colour varieties exist, made up of a combination of red, yellow, green and white patches. However adults tend to be more uniformly blackish-green. *Carcinus* is frequently found to be parasitized by the profoundly modified cirripede *Sacculina* (7).

7 **Sacculina** is a crustacean parasite related to the bar-nacles (Cirripedia). The eggs develop into typical *nauplius* larvae (p. 128) in the usual barnacle manner. The *cypris* larva attaches itself to a young crab and, casting off its limbs, becomes reduced to little more than a simple sac. It then develops a syringe-like dart which injects a small mass of cells into the body of the host. These settle beneath the gut and proliferate, sending roots throughout the body. Eventually the central body penetrates the external skeleton of the crab and appears in its adult form as a lump under the abdomen. The parasite destroys the reproductive organs of its host and also prevents it from moulting again. The whole life cycle of *Sacculina* is completed in about nine months.

8 **Idotea.** This isopod, a marine relative of the common terrestrial woodlice, is found in the lower parts of the intertidal zone. It is a nocturnal animal and is usually found living in sea-weed, where it feeds on almost any-thing that becomes available. It is more slender and graceful than *Ligia*, the Sea-slater, and, living further down the shore, it is also a more active swimmer. There are seven species of *Idotea* in Britain but only two are intertidal residents.

9 **Macropodia** is one of the spider crabs, distinguished by their long slender legs and rather compact triangular bodies. *Macropodia* is common in rock pools but is easily overlooked because the body is usually camou-flaged with living algae and sponges.

10 **Porcellana** (Porcelain Crab). These are not true crabs but related to the squat lobsters and hermit crabs (p. 122) as they possess a distinct tail-fan and are filter-feeders. Two species occur on British coasts, the small reddish *P. longicornis* living amongst stones or the holdfasts of sea-weeds in clear water, and the larger hairy *P. platycheles* found under stones in muddy water. The flattened body and spiny legs allow these crabs to scuttle about safely on exposed rocks without being washed away by heavy seas.

11 **Pinnotheres** (Pea Crab). These very small globular soft-bodied crabs live commensally in the mantle cavity of bivalve molluscs such as mussels, cockles and oysters (p. 73 on). Within its adopted protective shell, *Pinno-theres* filters out particles of food from the water drawn past by the respiratory currents of its host. The pre-sence of Pea Crabs in bivalves was known to the ancients, who believed that the crab gave warning to its host of the approach of predators.

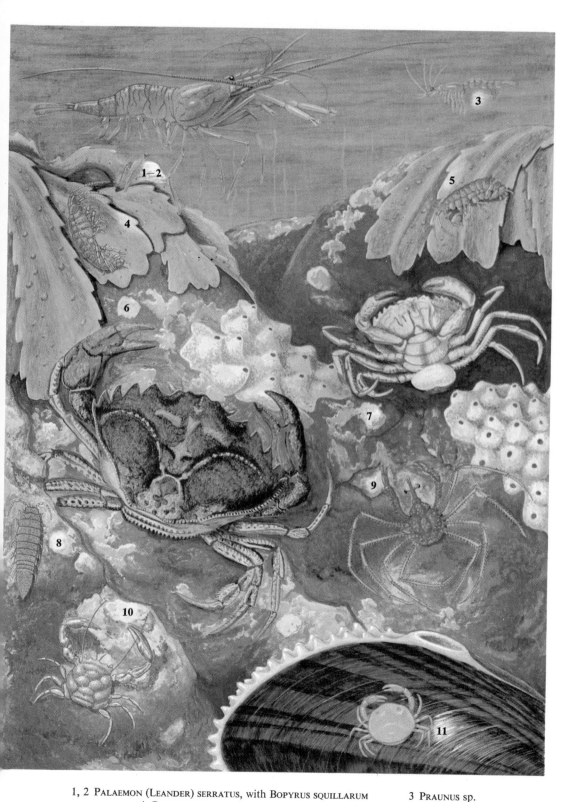

1, 2 Palaemon (Leander) serratus, with Bopyrus squillarum   3 Praunus sp.

4 Gammarus locusta   5 Orchestia gammarella

6 Carcinus maenas   7 Carcinus with sacculina carcini

8 Idotea granulosa   9 Macropodia rostrata

10 Porcellana longicornis   11 Pinnotheres pisum, on mussel

SEASHORE CRUSTACEANS AND SEA SPIDERS

1 Calanus. The marine copepod *Calanus* was discovered and first described by the Bishop of Trondheim in northern Norway in 1767. Since then it has been studied more extensively than any other member of the plankton. It is found in oceans and coastal waters throughout the world, often occurring in enormous numbers. It feeds on a variety of small organisms such as diatoms, protozoans and minute crustaceans, which it filters out by means of maxillary bristles from feeding currents produced by the antennae and limbs. *Calanus* in turn is fed upon extensively by fish, particularly herring, and it is this economic importance that has resulted in much of the research into its biology.

The female *Calanus* is fertilized internally, which is unusual in a marine creature, and the eggs are laid singly, subsequently being carried in a pair of egg-sacs slung beneath the abdomen.

2-4 Decapod larval forms. The great majority of crustaceans hatch from the egg into a larva that is very different in form from the adult. The most typical crustacean larva is the *nauplius* (p. 128) but amongst the decapods there are found a number of highly curious and bizarre forms that were long thought to be distinct species until their life histories were worked out. Characteristic larval forms occur in most groups of marine organisms, particularly amongst those with sessile adults, and their function seems to be primarily dispersal and site selection. For this reason they exhibit various adaptations to enable them to remain for as long as possible in the plankton. Larvae of some phyla make use of bands of beating cilia but cilia are absent in the crustacea. Instead they delay sinking by the development of spines and other protuberances, which sometimes reach extraordinary proportions.

The early stages of the true crabs are known as *zoea* larvae. As an active, and often extremely abundant, member of the plankton, the zoea feeds voraciously on diatoms and protozoa, moulting regularly. The zoea of the common porcelain crab (4) at this time develops its enormous and quite characteristic spines. Further growth transforms the zoea into a rather more crab-like creature with reduced spines known as a *megalopa* (3). This too lives in the plankton for a while, the abdominal limbs developing into swimmerets to help keep it afloat. Eventually the megalopa sinks to the bottom, no longer able to sustain itself in the plankton, the swimmerets disappear, the abdomen curls up beneath the body and the crab assumes its final identity. Perhaps the strangest of the decapod larvae is the paper thin, transparent *phyllosoma* larva of *Palinurus* (2), which reaches half an inch in length.

5 Caprella. Popularly known as 'skeleton shrimps', caprellids are often found clinging to colonial hydroids

(p. 9), such as *Tubularia* and *Coryne*, in the lowest parts of the intertidal zone.

Caprella feeds on small planktonic animals, particularly copepods, which are seized by the large pincers as they swim by. Eggs are laid mainly in the spring and when they hatch the young are removed by the mother from her brood pouch and placed on her limbs. Here they live for several weeks, often forty or more in number, being fed by the mother with fragments of prey she has captured. In time, having undergone several moults, they drop off and search for a suitable hydroid on which to live.

6 Pycnogonid larvae. The mating biology and life histories of sea-spiders are very poorly known; indeed fertilization has never been observed. As the female pours out her eggs, the male gathers them up and glues them on to his ovigers — a small pair of leg-like appendages — using secretions from special glands on his legs. It is presumably whilst gathering the eggs in this way that he fertilizes them. Males may often be found carrying up to 1,000 eggs at a time in two large balls. The eggs hatch into small, highly distinctive larvae possessing only two pairs of legs but with a pair of massive pincers, which are unlike the larvae of any other group of animals. The larvae of many species become parasitic on hydroids such as *Tubularia* and *Obelia*, where they may cause gall-like growths, but others infest bivalve and nudibranch molluscs. In a few species the larvae are suppressed and the eggs hatch directly into small sea-spiders which are then carried by the male until they are nearly half-grown.

7 Nymphon. The 'Sea-spiders' are entirely marine. Although the largest, with a leg-span of over half a metre, occur in deep water, many small species live in rock pools, where they may be found creeping slowly over sponges, hydroids or sea-weed. The body of sea-spiders is greatly reduced, the abdomen being usually little more than a wart between the hind legs. Organs displaced from the body extend into the legs, where eggs may sometimes be seen developing within the ovaries. Although most sea-spiders possess eight legs, some of the oceanic forms have ten or even twelve.

8 Pycnogonum belongs to a group of sea-spiders with comparatively short and massive legs. *Pycnogonum*, like *Nymphon* (7) is quite common in rock pools exposed by the low spring tides, also extending into quite deep water. It is often seen clinging with legs outstretched on the column of the Beadlet Anemone (p. 15) but it is also quite commonly found wandering. The massive proboscis, which might easily be mistaken for the abdomen, carries the mouth at its tip. This is inserted into the tissues of its coelenterate prey and the juices sucked out by means of a strong, muscular pump.

1 Calanus finmarchicus
2 Phyllosoma larva of Palinurus
5 Caprella linearis
3 Megalopa larva of Cancer
6 Pycnogonid 'Protonymphon' larva
4 Zoea larva of Porcellana
7 Nymphon sp.
8 Pycnogonum littorale

BARNACLES

The barnacles form a distinct and very remarkable group of Crustacea known as the cirripedes. Until 1830, when the development of *Balanus* (6) from a *cypris* larva (4) was observed, the crustacean affinities of the barnacles were not realized and they were classified with the molluscs. To understand the organization of barnacles it is necessary to remember that they are attached to the rock by cement glands on the larval antennules — upside-down in fact. The abdomen is reduced to a mere vestige and the feather-like cirri that filter diatoms and other food from the surrounding water with rhythmic sweeps are thoracic limbs. The barnacles alone amongst the Crustacea are hermaphrodite, a condition that is reversed in a few species possessing parasitic dwarf males. Some cirripedes such as *Sacculina* (p. 124) have become endoparasites, thereby concealing their relationships still further.

1 **Lepas** (Goose Barnacle). This is one of the stalked barnacles, which are commonly found attached to bits of wood and other floating objects that have drifted up from the warmer parts of the Atlantic ocean. One species, *L. fascicularis*, is able to secrete its own float of gas-filled bubbles and is consequently found on occasions with other species of *Lepas* hanging from it. Before the days of steam, the bottoms of slow-moving sailing ships used to become seriously fouled by large numbers of *Lepas*, but modern ships move through the water too rapidly for them to settle. The popular name refers to an old belief, probably derived from the feathery appearance of the cirri, that *Lepas* gave rise to geese. The myth was well worth perpetuating in the days of strict religious observance, for the supposed marine origin of geese meant that they could be consumed on fast-days with a clear conscience.

2 **Verruca.** This small, flattened acorn barnacle is particularly abundant in the lowest parts of the intertidal zone, although its range extends down to about 300 fathoms. It is found sticking to any hard surface and is quite common on the roots of kelp as well as on rocks. Like *Balanus* (6) and other cold water barnacles, *Verruca* produces a very large spring brood but the nauplius larvae are common in the plankton throughout the year.

3 **Scalpellum.** This is another stalked barnacle, a relative of *Lepas* (1). *S. scalpellum* is the only representative of the genus occurring in shallow water although it may also be found down to about 200 fathoms, living on colonial hydroids and polyzoans. A remarkable feature of the genus *Scalpellum* is the presence of dwarf males in some species, which may become parasitic on their partner. *S. scalpellum* represents an intermediate condition, in which the large specimens, whilst predominantly female, are still hermaphrodite and retain a small functional male reproductive system. However, attached to the edge are small parasitic individuals with no mouth-parts, blind gut, and reduced cirri that are wholly male. In *S. velutinum* large specimens are entirely female and the dwarf parasites are reduced to little more than a male reproductive system.

4-5 **Barnacle larvae.** Although some crustacean eggs hatch directly into small animals resembling the adult in form, the majority of species produce a characteristic larval stage known as the *nauplius*. Initially a very simple creature with unsegmented body and only three pairs of appendages, the nauplius develops through a series of five moults becoming larger and more complex in structure with extra segments added on behind. The first pair of appendages, the antennules, are primarily sensory but the antennae and mandibles are used for swimming and in later stages for filtering food from the water with the aid of long setae. The nauplii of barnacles may be readily recognized by their possessing a pair of anterior horn-like projections not found in other groups. In the last nauplius stages, some three weeks after hatching, paired compound eyes develop on either side of the single nauplius eye and the antennules begin to show attachment organs.

The nauplius is followed by another larval stage to which the name *cypris* is given due to its superficial resemblance to an ostracod (p. 136). The body is enclosed in a bivalve shell from which only the antennules project. Six pairs of beating limbs within the shell help the antennules propel the cypris, which is a non-feeding stage whose sole function is to locate a suitable site on which the adult barnacle will be able to thrive. Once suitable conditions of substrate and current have been found, the cypris anchors itself by means of the cement glands on the antennules and orients its body across the direction of water movement. Once settled, it undergoes a profound metamorphosis to become a typical sedentary barnacle.

6 **Balanus.** One of the most abundant and widespread British acorn barnacles, *B. balanoides* is essentially an intertidal species although it may sometimes be found surprisingly high up the splash zone. They are most numerous on exposed coasts, but if they do settle in protected places they may grow to an abnormally large size. Fertilization takes place during the winter months and involves the insertion of an intromittent organ by an adjacent hermaphrodite animal — a somewhat surprising occurrence in a marine creature.

1 Lepas anatifera

2 Verruca stroemia

3 Scalpellum sp.

4 Cypris larva of Balanus

5 Nauplius larva of Balanus

6 Balanus balanoides

SEASHORE ARACHNIDS AND MYRIAPODS

1 Hydroschendyla. This is one of the smaller, ribbon-like, geophilomorph centipedes. It is rare in Britain and is restricted to the sea shore, living in crevices or beneath stones lying on mud. It is able to withstand immersion in sea water for several days without any apparent ill effects. Little is known of its biology, but it is reported to prey on polychaete worms, biting them and carrying off autotomized portions.

2 Strigamia. Three species of *Strigamia* occur in Britain but only *S. maritima* (2) is found in coastal situations. It occurs in many localities living beneath stones and in sea-weed and is capable of withstanding prolonged submersion. However it is not confined to the inter-tidal zone like *Hydroschendyla* (1) and is sometimes found well above the high-tide mark.

3 Necrophleophagus. Although commonly found in tide litter, *Necrophleophagus* is really a ubiquitous species that is also found in many other habitats. It is one of many centipedes that can produce a phosphorescent secretion when disturbed, but as these animals are blind, the ability to produce light is almost certainly only a defensive reaction.

4 Parasitus. *P. kempersi* is a characteristic mite of decaying sea-weed on the strand line although it is not confined to this environment. It is a member of a large and diverse genus that is found in many types of habitat. Other species live in forest litter, some are pests of stored food products, and others occur in dung. A great many species of *Parasitus*, whilst not being parasites, are found attached to insects. These are usually nymphal stages taking advantage of the insects, mainly scarabeid beetles and bumble bees, for dispersal.

5 Erigone. It does not take long for a large population of arthropods to build up in the litter deposited by the high spring tides. A major component of this fauna consists of spiders. In addition to the more conspicuous wolf-spiders such as *Lycosa* (p. 150) and *Arctosa* (p. 162), there are myriads of small, black shiny spiders belonging to the family Linyphiidae. These money-spiders, as they are often called, make up two-fifths of the British spider fauna and present serious problems of identification even to experts. *E. arctica* is a typical representative of the group to be found in this habitat.

6 Neobisium. This pseudoscorpion is wholly littoral and lives in rock crevices where it spins silken cells for protection during moulting. It is usual to find quite large numbers of cells together and although a proportion of them will be deserted, a cast 'skin' will often be found within. Cells are also constructed for hibernation and by females for brooding eggs and young larvae. Mating takes place during the summer and autumn, each female producing about a dozen young. The majority overwinter as nymphs, maturing to mate the following summer; but a few adults also overwinter. Little is known of the food of *Neobisium* but presumably they feed on any small arthropods they can catch. They have been observed to capture small springtails. On the East Anglian coastal salt marshes the largest British pseudoscorpion, *N. carpenteri*, has recently been discovered in large numbers living amongst rotting *Spartina* stems.

7 Halacarus. The halacarids may be regarded as the marine counterparts of the Water-mites (p. 138). However, they are not active swimmers and do not possess the fringes of swimming hairs found on the legs of many water-mites. Associated with their aquatic existence, the tracheal system is much reduced and may be lost entirely in some species. Most halacarines live in the intertidal zone crawling over rocks, sand, or sea-weed but a few species are found in deeper water down to about 700 fathoms. Like some of the water-mites, a number of halacarines are cave-dwelling, and show marked adaptations to this type of habitat such as loss of pigment and total reduction of the eye spots. Halacarines feed in a variety of ways, some being predatory, some living exclusively on algae, whilst a few are even parasitic.

8 Halolaelaps. Unlike *Bdella* (9) which is essentially a terrestrial mite that comes down to the shore, *Halolaelaps* is restricted to the lower intertidal zone. It is found beneath stones from about the mid-tide mark down to the upper limit of the *Laminaria* zone. The respiratory systems of this and related species living in the intertidal zone are perfectly normal and show no special adaptations. They probably do not become immersed but retreat into air-filled cracks and crevices as the tide comes in. The only structural adaptations appear to be hair-like lobes on the legs to assist movement over the wet substrate. Little is known of the feeding habits of *Halolaelaps* but they probably prey on bryozoans, hydroids, minute worms, and hala-carine mites.

9 Bdella belongs to the group known as 'Snout Mites'. They are generally associated with damp environments and this species is found not only in rotting drift litter but also living in rock crevices towards the upper end of the intertidal zone. At low tide they emerge, often in large numbers, to hunt over the bare rock. Little is known of their prey but they have been seen to feed on dipterous larvae and small oligochaete worms.

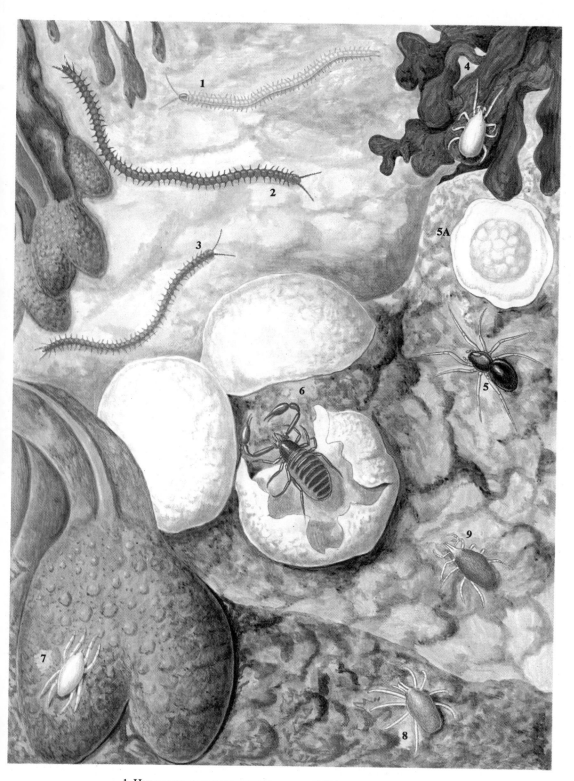

1 Hydroschendyla submarina
2 Strigamia maritima
3 Necrophleophagus longicornis
6 Neobisium maritimum
7 Halacarus bisulcus

4 Parasitus kempersi
5a egg case
5 Erigone arctica
9 Bdella sp.
8 Halolaelaps marinus

131

FRESHWATER SPIDERS AND CRUSTACEANS

1 **Argyroneta** (Water Spider). No other spider in the world has invaded the water as completely or successfully as *Argyroneta*, which is found in ponds, bogs and ditches throughout Britain. Out of water it appears dull velvety-black in colour but when submerged it is covered by a silvery layer of air. The spider's whole life is passed under water, where it spins a dome of silk amongst the vegetation, filled with bubbles of air brought from the surface (1A) — for like all spiders, *Argyroneta* is air-breathing and must have a constant supply of oxygen. Mating can occur at any season, but the eggs are laid in early summer inside the air-filled 'diving bell'. Soon after hatching the baby spiderlings disperse, often taking up residence in empty snail shells, which they fill with air bubbles from the surface. Although the 'diving bell' is often associated with quite an extensive web, this serves only as a system of guy-lines and is not used to catch prey. The main food is the water-louse (6). *Argyroneta* is widespread in Europe but because it lives well in captivity and makes an attractive aquarium animal, it has been taken to America, where specimens have escaped and bred in the wild. It is remarkable that no other species, even amongst the bizarre tropical forms, have managed to conquer the aquatic environment like *Argyroneta*.

2 **Leptodora** is perhaps one of the most transparent animals known. It is the largest of the Cladocera and belongs to the same group as *Polyphemus* (3), having the shell reduced to nothing more than a covering for the brood pouch, which projects from its back like a narrow-stemmed goblet. It is a planktonic animal and occurs only in large bodies of water such as lakes and reservoirs. It is primarily a northern creature, only rarely being found in the South or East of Britain. During the day it lives in deeper water but rises to the surface at night, swimming by means of its large second pair of antennae. Adults occur from May to October. *Leptodora* is unique among cladocerans in that its winter eggs, which are a resistant, resting stage in the life cycle, hatch into a typical crustacean nauplius larva (p. 128). During the summer, successive batches of eggs are produced at regular intervals and these develop, in the typical cladoceran manner, into miniature replicas of the adult.

3 **Polyphemus** belongs to a group of Cladocera that is separated from the Water-fleas (p. 136) by having the shell greatly reduced and serving only as a covering for the brood pouch. It is a transparent animal and as it swims all that is usually visible is the very conspicuous eye. It is a widespread species, often occurring in large numbers around the edges of lakes and clear pools, particularly in late summer. *Polyphemus* is a predatory animal and feeds on rotifers, protozoa, and minute crustaceans.

4 **Bythotrephes** is another colourless, transparent cladoceran belonging to the same group as *Poly-*

phemus (3) and *Leptodora* (2). It swims as an active predator in the surface waters of northern lakes.

5 **Mysis.** This is the only freshwater representative of a group known as the Opossum Shrimps (*see Praunus*, p. 124). In Britain it has been found only in Ennerdale and a number of Irish loughs and rivers but it is widespread in northern Europe and North America. *M. relicta* is believed to be derived from populations of the marine species *M. oculata* that became isolated due to glacial activity during the Ice Ages.

It is a planktonic species, rising to the surface at night to feed. It is primarily a particle-feeder and its main food consists of diatoms and algae, but it also consumes ostracods and other minute crustacea. By day, *Mysis* lives in deep water, rising to the surface at night, but it also migrates to shallower waters in order to breed. Breeding takes place during the winter, when the temperature drops to below 7°C. Between 10 and 40 eggs are laid and these are carried in a brood pouch. After hatching, the young remain in the brood pouch for 2 – 3 months, growing to about a third of their adult size. Maturity is reached in 10 months but growth continues for about a year and a half.

6 **Asellus** (Water-louse). This is a close relative of the terrestrial woodlice, both belonging to the order Isopoda. Two of the three British species of *Asellus* are common and widely distributed, living in well-vegetated ponds and streams. They can withstand low concentrations of oxygen in the water and are often particularly abundant in areas that have been polluted by sewage. They do not swim but crawl over mud and plants scavenging decaying organic debris and occasionally feeding on algae. The eggs, which are laid in the spring, are carried beneath the front part of the body in a chamber formed from plates projecting from the base of the legs. After hatching, the young animals, which resemble their parents, continue to live for some time within the brood chamber before dispersing to live as free individuals.

7 **Astacus** (Crayfish). The only indigenous freshwater decapod in Britain is *A. pallipes*, although the larger, red-clawed *A. fluviatilis* has been introduced from France on several occasions for culinary purposes and may still be found living in a few areas. Crayfish live in flowing water, particularly streams and rivers in chalk or limestone districts, hiding by day beneath stones or in excavated burrows and emerging at night to feed on snails, tadpoles, and insect larvae. They normally crawl slowly over the bottom but can, in an emergency, dart swiftly backwards by flicking their fan-like tail. Eggs are laid in the autumn and adhere to the female abdomen, where they are fertilized by sperm deposited earlier by the male. On hatching in the following spring, the baby crayfish remain clinging to the female by their pincers for some time.

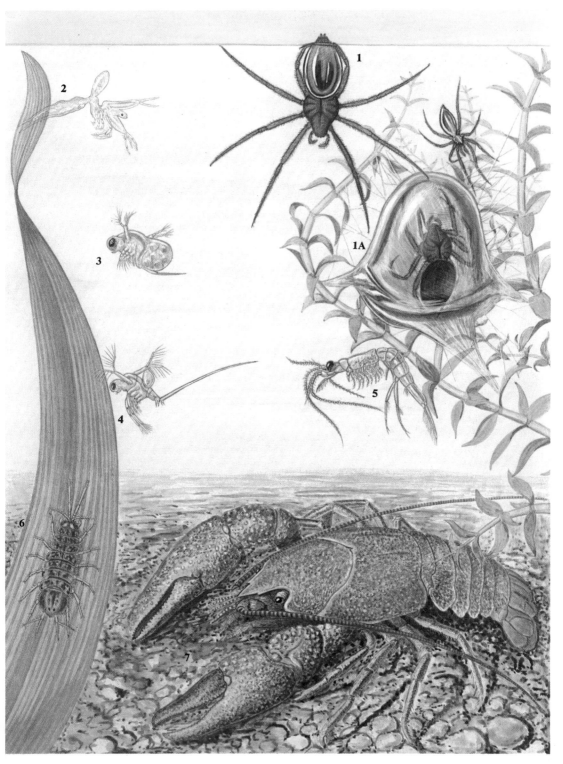

2 Leptodora kindti   1 Argyroneta aquatica, collecting air at surface
3 Polyphemus pediculus   1a A. aquatica inside 'diving bell'
4 Bythotrephes longimanus   5 Mysis relicta
6 Asellus aquaticus   7 Astacus fluviatilis

133

FRESHWATER CRUSTACEANS

1 Triops is a close relative of the Fairy-shrimp, *Cheirocephalus* (2), although quite different in appearance. Much of the body is concealed beneath a large carapace from which only the abdomen projects. The numerous leaf-like limbs are used for respiration and beat continuously. Although widespread in Europe, *Triops* is extremely rare in Britain. It has been found in isolated populations from Scotland to the south coast but most infrequently. It lives in temporary bodies of freshwater, often little more than puddles, and feeds on insect larvae. The males are seldom seen, even on the continent, and most individuals develop from unfertilized eggs.

2 Cheirocephalus (Fairy-shrimp). These beautiful, transparent animals grow to more than an inch in length and may occasionally be found in southern Britain living in rainwater puddles and similar temporary water bodies in considerable numbers. They belong to a group of crustaceans considered primitive because of the large number of body segments and the uniform structure of their limbs, which are used for respiration. They swim continuously on their backs, filtering organic debris, algae, and minute animals out of the water with special hairs on the legs. The food is clearly visible in the gut due to the animal's transparency and can be a variety of colours. The eggs are carried in a brood pouch that is highly resistant to desiccation and they survive for long periods in dried mud, being frequently carried to new localities on the feet of birds. On becoming submerged, the eggs hatch into nauplius larvae which grow rapidly.

3 Diaptomus. The six British species of *Diaptomus* resemble one another closely but may be distinguished from the other free-living freshwater copepods by the fact that their antennae are almost as long as their body. They are filter-feeders, sifting small particles of organic debris, algae, etc., from the water as they swim about on their backs. Unlike *Cyclops* (4), females of *Diaptomus* carry their eggs in only a single sac, which is often reddish in colour. *D. castor* is only found in the winter, and may be abundant in shallow ponds and ditches, even those that dry out completely in summer. Its eggs are highly resistant to desiccation. *D. gracilis*, however, prefers larger open bodies of water and may be found at all seasons although it is most abundant in the spring. It does not appear to produce resistant eggs. *Diaptomus* eggs hatch into nauplius larvae which moult several times before assuming the adult form.

4 Cyclops. There are nearly forty British species of *Cyclops*, all very similar in appearance. They are found, often in very large numbers, in almost all types of still, freshwater habitat. The well-developed antennae, which the males use to grasp the females during mating, may be used for swimming but the main propulsion comes from the body appendages. Female *Cyclops* may be immediately recognized by the pair of trailing egg-sacs on either side of the body. The eggs hatch into nauplius larvae, which moult five times before assuming adult proportions. A further four moults are needed to reach maturity and the whole cycle is completed in under a month. *Cyclops* is an active feeder, seizing particles of food and also consuming the bodies of dead animals.

5 Canthocamptus belongs to a group of very small copepods with short antennae. They do not swim but crawl over vegetation or through mud on the bottom of ponds. There are many British species, all similar in appearance, of which *C. staphylinus* is perhaps the commonest. It is found throughout the country but is less frequent in the north. It lives in shallow ponds and ditches, even those that become dry in summer, being most abundant in the winter months. Individuals maturing in the spring are thought to secrete protective cysts around themselves and aestivate during the summer. Mature females carry a single egg-sac and also a spermatophore in which sperm from the male is stored until required.

6 Gammarus (Freshwater Shrimp). The commonest of the three British species is *G. pulex*. They are, of course, not shrimps at all but amphipods and closely related to the common 'sandhoppers' of the sea shore (p. 124). *Gammarus* is found throughout Britain living in well-oxygenated water, usually streams or shallow lakes. It can swim rapidly if disturbed but normally spends most time hidden beneath stones or leaves. Although *Gammarus* feeds principally as a scavenger on decaying organic debris, it is also predatory and will on occasion attack small animals. In spring and summer, females are often found carrying eggs in a brood-pouch under the thorax. These eggs hatch directly into miniature replicas of the adults, without passing through a larval stage, and may remain within the brood pouch for a week or more.

7 Bathynella. This strange little crustacean has a rather curious history. It was first discovered living in a well in Prague in 1880. It was taken in Britain for the first time in 1927 from a spring at Corsham near Bath and was not found elsewhere until 1959, when a single specimen turned up near Oxford in a cattle trough fed from a spring. However, since 1959 it has been found in at least six counties from Devon to Scotland, sometimes in quite large numbers. The apparent scarcity of *Bathynella* is explained by the fact that it is a member of the interstitial fauna that lives in deep subterranean cracks and crevices, which for many years were not searched for small arthropods. In Yorkshire *Bathynella* has been found to be quite common in certain underground streams and pools, but little is yet known about its biology.

1 TRIOPS CANCRIFORMIS   2 CHEIROCEPHALUS GRUBEI
3 DIAPTOMUS GRACILIS   4 CYCLOPS STRENUUS   5 CANTHOCAMPTUS STAPHYLINUS
6 GAMMARUS PULEX   7 BATHYNELLA NATANS

FRESHWATER CRUSTACEANS AND WATER-BEARS

The Water-fleas are a large and important group of freshwater Crustacea. The body is enclosed within a transparent shell through which can be seen the gut and other organs, particularly the rapidly-beating heart and the brood-pouch with eggs or developing young. They swim jerkily, using their antennae for propulsion, the legs being mainly employed in filtering out the algae and bacteria on which they feed from the respiratory water current drawn through the shell. In summer they multiply rapidly, usually by parthenogenesis. However when conditions become unfavourable, males are produced which fertilize special eggs that are protected within a modified brood-pouch (ephippium) that is cast off when the female moults. This dispersal stage is often spread far and wide by sticking to the feet of birds. Water-fleas are a very important link in the conversion of plant matter into animal food and although many are eaten by *Hydra* (p. 7), flatworms (p. 21) and even *Stentor* (p. 1), their major predators are small fish.

1 **Simocephalus** is an extremely widespread water-flea which occurs in all types of water, living amongst vegetation. As in all water-fleas, the eggs hatch within the brood pouch of the female directly into miniatures of the adult and do not pass through a larval nauplius stage as do most other groups of Crustacea.

2 **Sida.** This very transparent water-flea is widespread in Britain but is not usually found to be particularly abundant in any given spot. It is frequently observed adhering to the vegetation in which it lives, by means of a special gland at the back of the head, and when caught will often fasten itself to the net in the same way.

3 **Daphnia.** Probably the best known of all the British water-fleas is *D. pulex*, which is found in almost all parts of the country. In water containing little dissolved oxygen *Daphnia* becomes quite markedly reddish in colour due to the presence of the red blood pigment, haemoglobin. During the summer each brood consists of some 30 – 40 young, occasionally more, produced from unfertilized eggs at intervals of about three days throughout the warm months.

4 **Holopedium.** This striking little water-flea is only found in a few Northern lakes where the water is clear and soft. It is exclusively planktonic and may be immediately recognized by the fact that it swims on its back within a jelly-like 'house' which it secretes itself. Although this habit is found in a few other animals, for example *Oikopleura*, (p. 188), *Holopedium* is the only crustacean known to build a 'house' in this way. It is to be found during the summer and autumn and may assume a wide range of colours, including both bright red and blue.

5-6 **Ostracods.** The ostracods comprise a distinctive and uniform group of mostly very small crustacea. The body is completely enclosed within a hinged shell from which normally only the antennae and a pair of legs

may be seen protruding. Some species swim actively using the long hairs on the antennae as paddles but most are to be found crawling on plants or over the muddy bottoms of ponds. The majority of crustaceans carry their eggs, but the ostracods lay theirs in small brightly-coloured clumps on water plants. The eggs are extremely resistant to desiccation and are able to survive at least twenty years in dry mud. On hatching, there emerges what is in effect a nauplius larva, but enclosed within a shell like an adult ostracod. Growth takes place by means of frequent moults and the cast shells are often found floating on the water surface or embedded in mud. The best known genus of freshwater ostracods is *Cypris* (5) which occurs in huge numbers in weedy ponds and ditches. It is very variable in colour, being usually brown, green, or yellow according to the background on which it lives.

7 **Macrobiotus** (Water-bear). These are minute segmented animals that until recently have been regarded as aberrant arthropods but they are now placed in a phylum on their own, the Tardigrada. Although they are incredibly widespread and abundant, occurring everywhere in soil, moss, lichens, and in water bodies of all kinds, particularly guttering on roofs, they remain a poorly known group of animals. The clue to their abundance is their striking resistance to desiccation. All development stages are able to withstand prolonged drying, the most famous example being the specimens that emerged from some dried moss after it had been stored for more than one hundred years in a museum. There are some fifty species known from the British Isles, few of them more than 1 mm in length. Most clamber along algal strands, piercing cells and sucking out the juices, but some species are predators, attacking rotifers and nematodes, and at least one species is known to be parasitic on sea-cucumbers. Little is known about their life histories or mating. The eggs, which are laid within the mother's moulted cuticle, hatch directly into miniature replicas of the adults, sometimes with fewer appendages, and growth proceeds by a series of moults.

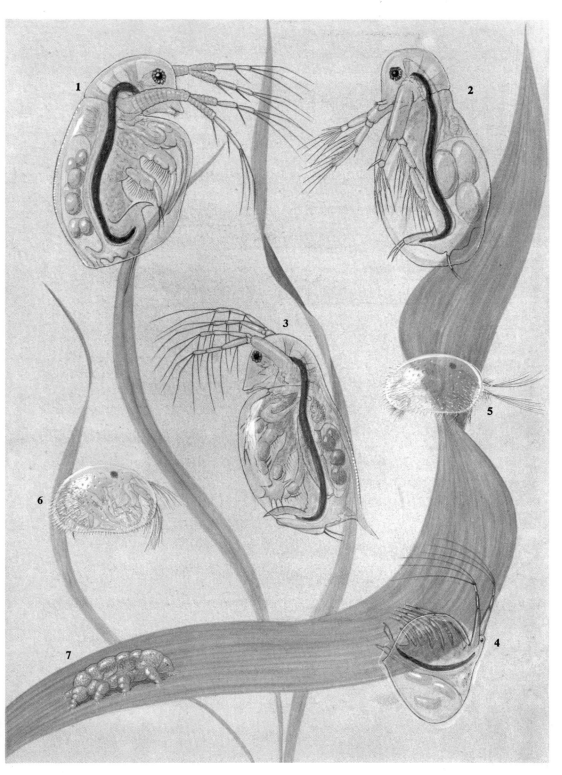

1 Simocephalus vetulus   2 Sida crystallina

3 Daphnia pulex

6 Pionocypris vidua   5 Cypris sp.

7 Macrobiotus sp.   4 Holopedium gibberum

WATER-MITES

The Water-mites form a distinctive group within the diverse order Acari on account of their mode of life, even though they lack any common anatomical distinguishing feature. They show few structural adaptations to their mode of life, although some species have developed swimming hairs on the legs. Some water-mites swim actively in surface waters, but the majority move slowly over the bottom or on aquatic vegetation. Most are active predators, feeding particularly on cladocerans, ostracods, and chironomid midge larvae by piercing the cuticle, digesting the soft parts, and sucking out the resulting juices. The life histories of most water-mites have not been worked out — largely due to the problems posed by the possession of an ecto-parasitic larval stage. The name 'larva' is applied to the six-legged, parasitic stage and 'nymph' to the immature eight-legged stage. Approximately two hundred and twenty species are known in Britain.

1 **Hydrachna.** Some twenty species of *Hydrachna* are found in Britain and all are rotund and bright red in colour. Unlike most other water-mites, *Hydrachna* does not lay its eggs in a gelatinous mass. Instead the female uses her mouthparts to make a small slit in the stems of water plants, and the eggs are laid without a protective covering inside the spongy tissues of the plant. Six weeks later the eggs hatch into a typical six-legged larva which attaches itself to an aquatic insect, often the water-scorpion (*Nepa*), where it feeds as a parasite, degenerating into a formless blob. After feeding on its host for several weeks the larva moults into a free-living nymph.

2 **Hygrobates.** The commonest member of the genus is the soft-bodied, oval water-mite *H. longipalpis*. Although it is sometimes found in rivers and other running water, it is more frequently encountered in ponds and ditches. *Hygrobates*, unlike many other water-mites, is not seasonal and adults may be found fully active even in winter.

3 **Neumania.** Although closely related to *Unionicola* (8), in which the adults are parasitic, *Neumania* conforms to the normal water-mite life history pattern of parasitic larva and free-living adults. On the back is a more or less distinct reddish marking in the shape of a Y. *N. spinipes* is the commonest species and is widely distributed both in Europe and America in slow moving waters.

4 **Mideopsis.** The commoner of the two British representatives of this genus is *M. orbicularis*. The body, which is heavily sclerotised and patterned with pores and sculpturing, is very markedly flattened. This gives the body an almost circular appearance. Although the legs are provided with swimming hairs, the animal is not an actively swimming species but spends its life on the bottom in flowing water, burrowing in mud. It is interesting to notice that water-mites living in fast-flowing water tend to be small with flattened, heavily armoured bodies, and live on the bottom by holding on with strong claws. As a further adaptation, the parasitic larval stage is suppressed and the eggs hatch directly into a 'nymph'. *Mideopsis* does not exhibit

these adaptations to the full as it is not an inhabitant of torrent waters. Thus it possesses a larval stage, parasitic on aquatic insects, that shows the same flattened form as the adult but transparent.

5 **Arrhenurus.** The genus *Arrhenurus* consists of some twenty species, mostly green or brown in colour. They are active swimmers and the legs are well provided with swimming hairs. The most striking feature of the genus *Arrhenurus* is the marked difference in shape between the sexes. Whereas the females are more or less oval in shape, the males bear a remarkable tail-like projection which in some species is drawn out into two distinct lobes.

6 **Limnesia.** These attractively marked water-mites are very widespread. They are powerful, active swimmers and very voracious.

7 **Limnochares** is a very sluggish bright red animal that may be found crawling slowly over the muddy bottom of slow-moving streams and rivers. The eggs, which are also scarlet, are laid in clumps enclosed within a protective gelatinous covering on stones and roots early in summer. The larvae emerge in four to six weeks and swim to the water surface where they attach themselves on to passing insects, particularly pondskaters (*Gerris*). Sometimes pondskaters may be found covered in the small red parasitic larvae, usually from June through to September.
Limnochares is ubiquitous in Britain wherever the right kind of habitat conditions are to be found, and it is also widely distributed throughout Europe. *Limnochares* is reported to be highly resistant to cold and is said to be able to withstand freezing in ice for at least two weeks.

8 **Unionicola** is unusual because it reverses the normal pattern of water-mite life history. Thus the larva, at least for the early part of its existence, is free-living whilst the adult life is spent as a parasite. Adult *Unionicola* may be found at any season of the year, often in considerable numbers, living on the gills and mantle of the freshwater mussels *Anodonta* and *Unio* (p. 90).

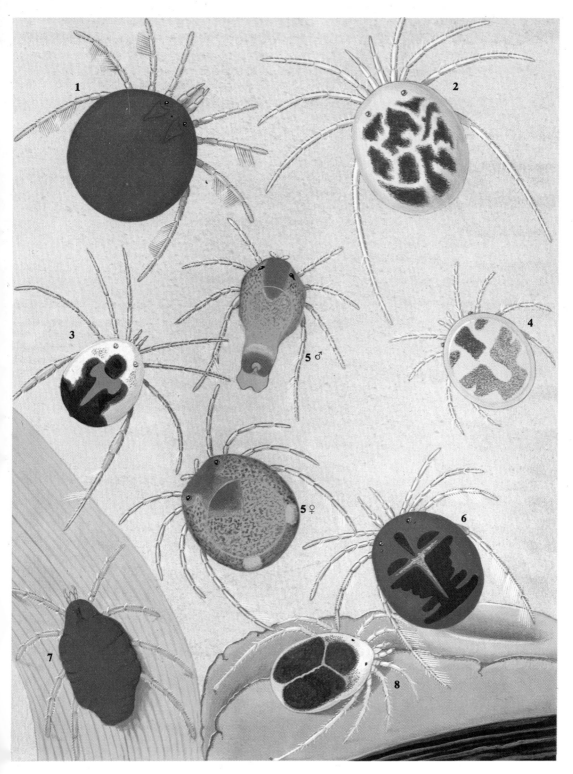

1 HYDRACHNA GLOBOSA   2 HYGROBATES LONGIPALPIS
3 NEUMANIA SPINIPES   4 MIDEOPSIS ORBICULARIS
5 ARRHENURUS CAUDATUS   6 LIMNESIA FULGIDA
7 LIMNOCHARES AQUATICA   8 UNIONICOLA INTERMIDIA, on mussel

PARASITIC CRUSTACEANS

Parasitic representatives occur amongst many orders of Crustacea, particularly the copepods, isopods and cirripedes (barnacles). The relationship can range from the purely symbiotic to some of the most extreme forms of parasitism known. In each group there are parasitic species that differ but little from the typical free-living forms: however, each group also includes species so profoundly modified that their place among the Crustacea may be determined only by studying their larval development. No matter how bizarre the adult, most parasitic crustaceans pass through a typical nauplius larval stage (p. 128). Total reduction of the body is not confined to internal parasites, and such ectoparasites as *Lernaeocera* (7) may also lack all traces of appendages. However, the most striking examples are found with internal parasites, such as *Xenocoeloma*, which in the adult consists of nothing more than a mass of tissues, lacking even integument, embedded in the host's body.

1 **Argulus** (Fish-louse). These flattened, transparent animals belong to a separate subclass, the Branchiura. They are quite frequently found sticking to the fins and bodies of freshwater fishes, often in some numbers. They hold on by means of two large suckers with additional help from numerous hooks and spines, and are able to move about on their host without difficulty. They are also active swimmers and can move from one host to another. Beneath the body they have a poison spine and a long protrusible proboscis that is inserted into the host for extracting blood. Mating can occur either on or off the host and the eggs are laid in a gelatinous mass of one hundred or more, attached to roots or stones. The larvae resemble small adults, but without suckers, and there are no nauplii.

2 **Thersitina.** This is a cyclopoid copepod that is parasitic on sticklebacks, particularly in brackish water. It lives beneath the gill covers and can sometimes be present in large numbers. *Thersitina* is one of the less extreme parasitic copepods and retains an essentially *Cyclops*-like form but with markedly enlarged egg sacs. Only the females are parasitic, both the males and the nauplius larvae being free-living.

3 **Ergasilus** is another cyclopoid copepod, closely resembling *Thersitina* (2). The adult female is a rare parasite living on the gills of the Grey Mullet, but the males and nauplius larvae are free-living.

4 **Salmincola.** This is an ectoparasite of salmon, and is known to fishermen as 'Gill Maggot'. Only the females are known and the species presumably reproduces parthenogenetically. They are really parasites of freshwater habitats, this being where the nauplii live, but the adult parasite can withstand immersion in sea water and thus survive when the salmon migrate.

5 **Lepeophtheirus.** This copepod is sometimes referred to by fishermen as a 'Sea Louse'. It is an ectoparasite of salmon, living mainly in the region of the anus. It is usually found on fish when they return to rivers to spawn but in fact *Lepeophtheirus* is a marine creature and does not survive long in fresh water. Both sexes are parasitic but the males are very scarce and reproduction is probably largely parthenogenetic.

6 **Xenobalanus.** This is a highly modified barnacle that is found associated with porpoises. It is probably not truly parasitic as it derives nothing from its host apart from mobility. *Xenobalanus* is found in groups attached to the ends of the tail flukes and flippers, where it feeds like other barnacles by filtering particles from the water with its modified limbs or cirri. Superficially *Xenobalanus* resembles one of the stalked barnacles such as *Lepas* (p. 128) but it is in fact a true 'acorn' barnacle like *Balanus* (p. 128) in which the soft parts have become elongated and the opercular plates reduced to a circlet that is embedded in the tissues of the host and acts as an anchor.

7 **Lernaeocera** is one of the most remarkable of the parasitic copepods. Not only does the body undergo profound modifications, totally losing its crustacean characters in the adult, but there is also a complex life history involving two separate hosts. The eggs hatch into nauplius larvae, which last for only a day before moulting into typical miniature copepods. After swimming freely the young animal attaches to the gills of a flatfish. On moulting it produces an attachment thread from which it hangs whilst moulting several more times. On reaching maturity, *Lernaeocera* again becomes mobile and swims actively, during which time mating occurs. The males immediately die but the females enter the gill chamber of a cod or whiting and attach themselves. The body grows into a large, worm-like sac, losing all trace of limbs, whilst the head penetrates the host tissues and branches out to provide an anchor. The parasite lives on blood drawn from its host, digesting it so thoroughly that nothing remains, and the anus, having become unnecessary, is lost completely. Eggs develop in large numbers in paired coiled threads that grow out of the parasite's sac-like body and the whole life cycle is completed within about ten weeks.

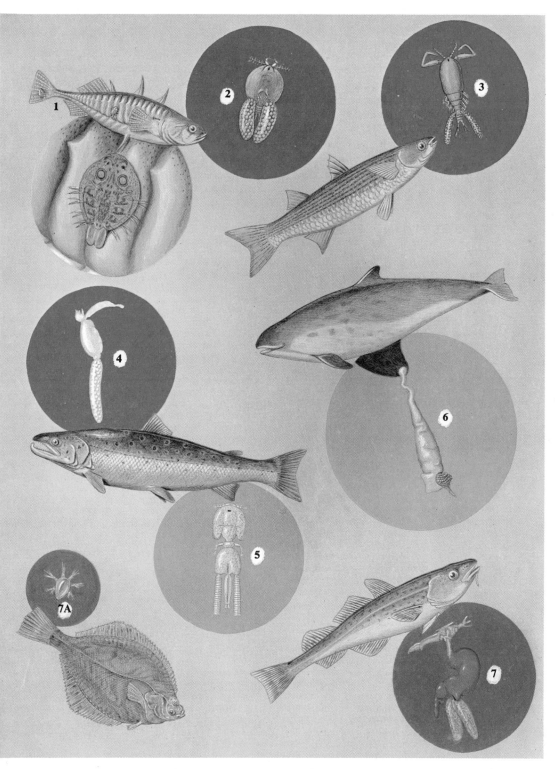

1 Argulus foliaceus, on stickleback   2 Thersitina gasterostei   3 Ergasilus sieboldi
4 Salmincola salmonea   6 Xenobalanus sp.
5 Lepeophtheirus salmonis
7a Nauplius larva of Lernaeocera   7 Lernaeocera branchialis

TONGUE-WORMS AND TICKS

With the exception of *Linguatula* (1) the animals on this page are all 'ticks', ectoparasites belonging to the acarine order Metastigmata. To complete their development ticks must take three meals, each from separate hosts. Their sac-like bodies become enormously distended as they consume large quantities of their hosts' tissue fluids, but most species are capable of withstanding prolonged starvation, though not desiccation, in between such meals. The Metastigmata are divided into the ixodids (hard ticks) and argasids (soft ticks).

1 **Linguatula** belongs to a strange group of endoparasites, the Pentastomida, that live in the respiratory tract of vertebrates, particularly reptiles. Their relationships are obscured by the anatomical adaptations associated with a parasitic mode of life but they appear to be derived from arthropod stock. The only suggestions of limbs are the paired hooks borne on small stumps at the front end of the long, worm-like body. Little is known of their habits, but most species are believed to produce small mobile larvae that will enter a wide range of vertebrate hosts, including man. However, the adults tend to be more exacting and often show specific host preferences. In all cases the adults infect the respiratory tract and live with their front ends deeply embedded in the host tissue feeding on blood, mucus, or lymph.

2 **Dermacentor.** This tick has a wide distribution in Europe but in Britain appears to be restricted to the South-west. In the spring, considerable numbers of adults may be found feeding on the backs of cattle, although quite a number of other large mammals will also serve as hosts, such as horses, dogs and foxes, as well as smaller creatures such as hares and hedgehogs. Mating takes place on the host, a single male fertilizing several females. The eggs are not laid until the female has left the host and is on the ground, when several thousand will be produced. The larvae can survive for several weeks before finding a host, usually a small mammal, on which they feed before dropping to the ground again. After metamorphosing into nymphs, a second host has to be found, again usually a mammal but occasionally a bird. After feeding, the nymph returns to the ground to moult into the adult state and it has been reported that the unfed adult can survive for two years before finding a host. *Dermacentor* is involved in the transmission of several diseases of dogs, cattle and horses.

3 **Haemaphysalis.** In Europe, *Haemaphysalis* is thought to transmit redwater fever to cattle but in Britain, where it has a rather local distribution, it has not yet been found feeding on this host. The larvae and nymphs feed on bird or small-mammal hosts. Only the adults gorge themselves on larger mammals. The life cycle is usually completed in about two years.

4 **Ixodes** (Castor Bean or Sheep Tick). This important parasite is probably the commonest tick in Britain. It is most often found on sheep and cattle but can also occur on most other warm-blooded animals and will occasionally bite man. It is economically important because it transmits a number of serious animal diseases, notably the virus causing the encephalomyelitis of sheep known as 'louping ill' and the protozoan parasite *Babesia* causing redwater fever in cattle. The complex life cycle takes about three years and because of the high mortality at intermediate stages many eggs must be laid. On hatching in the autumn a six-legged larva emerges. The following spring it climbs adjacent vegetation and in time attaches itself on to a passing host animal. After feeding, the swollen larva falls to the ground, where it moults into an eight-legged nymph resembling the adult. The nymphs remain quiescent until spring the year after, when they too climb on to the vegetation and wait for a passing host. After feeding for several days they drop to the ground and in due course moult to maturity. In the spring of the third year the adults in turn climb in search of a host, from which the female consumes large quantities of blood. Mating takes place either on the host or soon after dropping to the ground, after which the males die. The eggs are laid during mid-summer in crevices in the ground, each one being covered in a special waterproofing layer of wax. Once the eggs are laid, the females also die. It is extraordinary that during the three years of its life the animal takes only three meals and spends less than a month on its host.

5 **Argas.** The three British argasids are parasitic on birds and bats. They are able to survive for several years without feeding and people are sometimes bitten near long-deserted pigeon cotes, the wound itching for several weeks afterwards. These ticks are usually found not on their hosts but in their nests or resting places. Although the larvae may remain attached to the host for several days. the nymphs and adults are seldom carried from the nest because they feed very rapidly whilst the host is asleep. Because of this association with nests, argasids have little difficulty in finding hosts and they therefore lay far fewer eggs than other ticks.

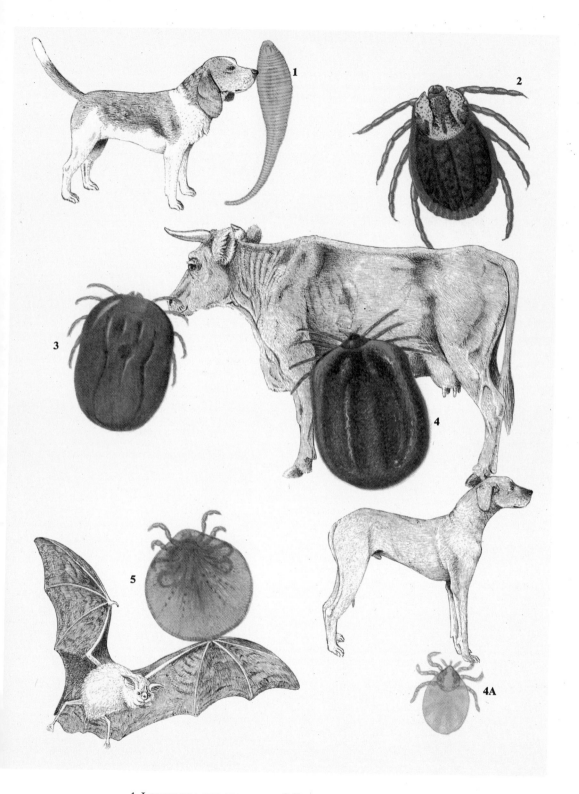

1 Linguatula serrata   2 Dermacentor reticulatus
3 Haemaphysalis punctata   4 Ixodes ricinus
5 Argas vespertilionis   4a I. ricinus, larva

PARASITIC MITES

1 Demodex (Follicle Mite). Mites of the genus *Demodex* are highly specialized animals living in the hair follicles and sebaceous glands of mammals. The body is soft and elongate whilst the legs are reduced to little more than attachment organs. One species (*D. folliculorum*) occurs very commonly on man and although it is not thought to cause any disease, it is frequently found associated with a variety of skin disorders including black-heads, acne, and impetigo. Another species (*D. canis*) occurs on dogs and is associated with the unpleasant and malodorous condition known as follicular or red mange. It is thought that the presence of the mite enlarges the follicles allowing the causative agent, a *Staphylococcus*, to enter. The whole life cycle is spent on the host and is completed in about two weeks. After mating at the mouth of a follicle, the female descends to lay her eggs in the sebaceous gland. The larvae, which soon moult into the first nymphal stage, are carried to the surface by sebaceous secretions. The second nymphal stage leaves the follicle to crawl for some hours in search of a fresh home. Females remain in or near their follicle until copulation but males wander about the skin feeding in different follicles and searching for a mate. Once the adult female has laid her eggs she returns to the top of the follicle and dies, thus blocking the entrance and discouraging further infestation.

2 Halarachne is the only mesostigmatid mite known to be endoparasitic on mammals in Britain. It is only found infecting the Grey Seal, where it lives inside the nasal cavity. As in so many parasitic mites, the body has become a soft, elongated sac and the legs reduced in length with powerful claws for attachment to the host. The life cycle has not yet been fully worked out, but it appears that young may well be produced by parthenogenesis. The larvae are either produced alive or else hatch immediately after the eggs have been laid and metamorphose directly into adults without an intermediate nymphal stage. The infective larvae are spread from one seal to another by the animals rubbing noses on the breeding beaches.

3 Trombicula (Harvest Mite or Bracken Bug). The popular name of this mite is strictly applicable only to the bright red parasitic larval stage, the nymphs and adults being free-living. It is found throughout Britain and is often very common. Fortunately, unlike certain tropical relatives, the British species is not known to transmit any diseases, although the bite can cause considerable discomfort. The inflamed feeding sites are often intensely irritating for several days and some exceptionally sensitive people suffer from an allergic dermatitis after being bitten. The larvae (3A, B) will infest all warm-blooded animals, the rabbit being a very common host. On man they tend to attach themselves

to areas where the skin is thin (ankle, groin and armpit) or where clothing is tight. The blade-like mouthparts are inserted deep into the skin and 'salivary' secretion injected. If left undisturbed the larva will feed for about three days taking in lymph fluid but no blood. After feeding, the mite drops to the ground where it eventually moults into a nymph which lives freely in the soil feeding on small arthropods and their eggs. A final moult produces a free-living adult capable of burrowing to a depth of three feet if surface conditions become extreme.

4 Psoroptes. These parasitic mites belong to a family that infests the skin of domestic animals and are known collectively as 'Scab Mites'. *Psoroptes* causes the extremely contagious and unpleasant condition known as 'sheep scab'. The mites penetrate the skin, usually on the back, and form small lymph-filled vesicles. These grow into large pustules that give rise to hard scabs. As the infected area does not provide an amenable habitat for the mites, they move away causing the infected area to grow. The acute irritation causes the unfortunate host to scratch and bite, so damaging more of its wool coat. The general health of the host is soon affected and a severe infestation can cause death. The same parasite will also attack horses, cattle, and rabbits, causing similar distressing symptoms. The entire life cycle, which is completed within a fortnight, takes place on the host. Whereas the females have a larval stage followed by two nymphal stages, the males have only a single nymphal stage. The males attach themselves by means of anal suckers to the second stage nymphal females and remain there until they moult to maturity when mating occurs.

5 Sarcoptes (Itch Mite). The Sarcoptidae are a family of skin parasites closely related to the Scab Mite *Psoroptes* (4). Sarcoptic mange affects domestic animals, and similar infestations in domestic fowls cause 'scaly leg' and also loss of feathers. However, the best known of sarcoptid infections is that of scabies, occurring in man. The adult female mite, having been fertilized, burrows down into the tough outer layers of the skin, usually selecting the hands or wrists although other regions of the body may also be affected. Burrowing is carried out using the mouthparts and special cutting surfaces on the front legs. Whilst these are being used, the animal anchors itself with suckers on its feet. Eggs are laid in small numbers as the mite burrows and as these hatch the larvae climb out on to the skin and search for hair follicles, where they feed and moult. Although the life cycle is only about a fortnight, individual patients are seldom found to have more than about a dozen mites on them. Even so, this number can cause agonizing itching, especially at night, and severe damage to the skin often results from scratching, particularly by the introduction of infective bacteria.

1 Demodex canis   2 Halarachne halichoeri
3a T. autumnalis larva   3b Fully-fed larva
3 Trombicula autumnalis
4 Psoroptes ovis   5 Sarcoptes scabei

PARASITIC AND PHORETIC FALSE-SCORPIONS AND MITES

1 **Lamprochernes.** This widespread and common pseudoscorpion lives mainly in moss and vegetable debris but is also found inhabiting dung heaps. It is perhaps the most frequently encountered of the British pseudoscorpions because it is regularly phoretic (see 7). In autumn it is quite common to find houseflies, hoverflies, beetles, and even harvestmen with numbers of *Lamprochernes* clinging on to their legs. However, this is only a means of distribution and no food is taken from the 'host'; indeed, no pseudoscorpions are parasitic. Only the females are phoretic, males of this species being extremely rare.

2 **Dermanyssus.** The mites of this genus are all ecto-parasites of birds, and the mouthparts are modified to form sucking stylets. One species, *D. gallinae*, is economically important and is known to poultry keepers as 'Red Fowl Mite'. All the active stages except the larvae are blood feeders and attack the birds whilst they are roosting. Feeding is rapid and the mites quickly return to cracks and crevices in the perches. The membranes joining the plates of the body are highly elastic and the mites can swell enormously as they gorge with blood and lymph. Even moderate infestations can reduce the numbers of eggs laid by the birds whilst serious attacks can kill. In the absence of birds *Dermanyssus* will turn to mammals, and man can receive irritating bites, particularly in disused poultry houses.

3 **Megninia** (Feather Mite). Most of the feather-mites belong to the family Analgesidae. They are mainly scavengers feeding on the scurf and oily exudations of the skin, but some species living on sea birds have been found to feed on diatoms. A few genera live within the quills and may be truly parasitic. Many of the feather-mites possess specialized hooks and claws to enable them to remain attached to the host whilst in flight. It is interesting to note that different species of mite, although seldom restricted to a single host species, are extremely particular about which part of the bird they live on. Thus some occur only on the primary wing feathers whilst others live only on certain parts of a feather. This specialization is apparently based on different temperature preferences. On the whole feather-mites are not of great economic importance although heavy infestations of *Megninia* are some-times a nuisance on poultry.

4 **Spinturnix** belongs to a group of mesostigmatid mites that are all parasitic on bats. The body is markedly flattened and the legs are short and very stout. Each leg terminates in powerful claws that provide the main means of attachment to the body of the host. The females are viviparous, meaning that they do not lay eggs but give birth to nymphs that are miniature replicas of themselves. The five British species of *Spinturnix* have been collected from noctules, pipi-strelles, whiskered bats and both greater and lesser horseshoe bats.

5 **Eutrombidium** (Velvet Mite). These widespread and attractive red mites appear to have no economic importance. The larvae hatch from eggs that have been buried in the ground, and seek a suitable insect host, usually a cricket or grasshopper, on which to feed. They attach themselves to the thinner parts of the cuticle, particularly the intersegmental membranes, where they drive in their knife-like mouthparts and suck blood. Normally the host does not appear to suffer, although a heavy infestation on the hind wings may cause damage to these delicate organs. After gorging for several days and swelling to a consider-able size, they drop to the ground and burrow. Here moulting occurs and the resulting nymphs emerge as free-living individuals. Both nymphs and adults hunt over the soil, feeding on insect eggs, especially those of orthopterans. The life cycle is completed in just under a year and although most velvet-mites overwinter as adults, some are able to complete their parasitic stage by the autumn and pass the winter as nymphs.

6 **Acarapis** is an endoparasitic mite of considerable economic importance being the cause of 'Acarine' or 'Isle of Wight disease' in the common Honey Bee. The mites enter the respiratory system of young bees before the protective hairs guarding the spiracles have hardened and there they feed by piercing the tracheal wall and sucking the bee's blood. Eggs are laid and soon a thriving population builds up which severely impedes the bee's breathing. Eventually a heavily in-fested bee becomes unable to fly because of this. It is interesting to note that like many parasitic mites, *Acarapis* has suppressed the nymphal stage and the larvae develop directly into adults. Not all individuals of *Acarapis* live inside the tracheal system and some may be found hiding in crevices on the body surface. The status of these surface-living individuals is not yet clear. They are most probably mites that arrived after the spiracular guard-hairs had hardened, and are therefore living in what may be called a second-best habitat.

7 **Parasitus.** Despite their name, mites of this genus are not truly parasitic. They are gamasids that are norm-ally free-living predators in a variety of habitats in-cluding vegetable debris and also the nests of small mammals. At some stage in their lives, usually as nymphs but also occasionally as adults they become phoretic — that is, they attach themselves to an insect host, without attacking it in any way, and use it as a means of transport. *Parasitus* may be encountered, often in very large numbers, on a variety of hosts, but the most common are bumble-bees and scarabeid beetles.

2 DERMANYSSUS GALLINAE   1 LAMPROCHERNES NODOSUS   1A, on leg of crane-fly

3 MEGNINIA CUBITALIS   4 SPINTURNIX VESPERTILIONIS

5 EUTROMBIDIUM ROSTRATUS   6 ACARAPIS WOODI   7 PARASITUS COLEOPTRATORUM

5A larvae on grasshopper   7A larvae on scarab beetle

DOMESTIC ARACHNIDS AND MYRIAPODS

1 **Chelifer** is a widespread pseudoscorpion that may occasionally be found hunting actively on the walls of buildings in broad daylight. It should not be confused, however, with the far smaller 'Book Scorpion' (*Cheiridium*), which is a secretive animal. The male performs a vigorous display before the female, waving his pincers and extending his inflatable 'ram's-horn' organ — a striking structure of unknown function that protrudes from the ventral genital opening. The male deposits a small drop of seminal fluid on a stalk stuck to the ground and slowly leads the female on to it, so fertilizing her.

2 **Oonops.** This small pink spider is probably common throughout Britain although there are as yet few records from Scotland. It is a rather secretive animal, emerging only at night from its silken cell, when it hunts minute insects on walls and ceilings. It moves with a very characteristic slow, steady gait, interrupted by sudden rapid darting movements, with the front legs held out in front groping for prey.

3 **Pholcus.** Most people living south of the Thames will be familiar with this curious long-legged spider, that spins flimsy but extensive cobwebs in attics and the corners of rooms. In the Midlands *Pholcus* is confined to cellars, but is largely absent from the North. Females may frequently be seen during the summer holding a ball of eggs or newly hatched young in their jaws. When disturbed, *Pholcus* vibrates rapidly in its web and becomes almost invisible.

4 **Scytodes** (Spitting Spider). At one time *Scytodes* was thought to be very rare but now, although restricted to the South-east, it is being seen rather more often. *Scytodes* usually hides away during the day, emerging at night to hunt over walls and ceilings. Its method of catching prey is most remarkable. Within the high-domed cephalothorax — the front part of the body — are greatly enlarged poison glands. As the spider draws near its victim, these glands empty two streams of poisonous glue which cover the unfortunate insect, firmly sticking it to the ground. *Scytodes* is sometimes found carrying a bundle of eggs loosely wrapped in a few strands of silk beneath its body.

5 **Scutigerella** belongs to the class Symphyla, and should not be confused with *Scutigera* (6) which is a centipede. Symphylans are small colourless myriapods superficially resembling centipedes, the largest being only 8 mm in length. The most striking difference, apart from the absence of poison claws, is that whereas there are twelve pairs of legs, the body is protected by a greater number of dorsal plates, varying from 15 to 24. Symphyla occur throughout the country from the seashore to mountain tops, living mainly in soil but also in moss, organic debris, and under bark; and

some species occur in hot-houses. In soil, populations of as many as 80 million per acre have been recorded and as their main food appears to be young plant roots it is not surprising that they are regarded as serious economic pests. In spite of this, however, they remain a neglected and rather poorly known group of animals, and only fourteen species have been taken in Britain. Their main enemies are probably centipedes and the larger predatory mites. The Symphyla lay small batches of eggs in the soil, which hatch into miniatures of the adults but with fewer legs and shorter antennae. They live for about four years during which time they may moult as many as fifty times.

6 **Scutigera** is the only British representative of the centipede suborder Scutigeromorpha, which is largely tropical. Indeed the inclusion of *Scutigera* as a true member of the British fauna could be disputed, for although it is quite common in the Channel Isles, there are very few records from this country. *Scutigera*, with its very long legs, is extremely active and it is interesting to note that associated with this it has good eyesight and a well-developed respiratory system. Although tending to avoid bright light, *Scutigera* does not hide away in crevices like most other centipedes, but is found hunting prey with great agility in more open situations. Like other centipedes it possesses powerful poison fangs, but its bite is not painful to man.

7 **Tegenaria.** Three species of *Tegenaria* are commonly found in houses and they are probably the best known and best hated of all British spiders. The most imposing species is *T. parietina*, which is confined to the South-eastern part of the country. The other large species is *T. saeva*, which is more widespread but uncommon in the North. It is not confined to houses and may often be found away from human habitation. In the autumn particularly they have the habit of climbing into baths in search of water but because they are essentially web-dwellers their feet lack the adhesive pads that enable free-living spiders to climb smooth surfaces. Hence they remain trapped until discovered by the horrified housewife. Male spiders, whose legs are proportionately longer than those of the female, are the more frequently encountered. This is because they wander widely in their active search for a mate. The smallest species, *T. domestica*, is found in buildings throughout Britain and is very common. *Tegenaria* spins a large sheet web, often more than 12 inches across, with a tubular retreat in one corner. The struggles of an insect that has fallen on to the sheet alert the spider which rushes out and poisons the unfortunate victim. The meal is then consumed at leisure within the retreat. In the days when the ague — better known as malaria — was common in Britain, an efficacious remedy was thought to be a *Tegenaria* 'gently bruised and wrapped up in a raisin or spread upon bread and butter'.

1 Chelifer cancroides   2 Oonops domesticus   3 Pholcus phalangioides, with egg sac
5 Scutigerella sp.   4 Scytodes thoracica, with prey
6 Scutigera coleoptrata   7 Tegenaria saeva, and web

GARDEN SPIDERS

1 Araneus. The most familiar member of this genus is the 'Cross'- or 'Garden-spider' *A. diadematus*, which matures in the autumn. The web of *Araneus* is a familiar object, with its radiating threads and catching spiral covered in drops of gum, and many people have watched the garden spider rush down from its retreat in one corner and swathe an unfortunate flying insect in a shroud of silk before taking it back to consume it. *A. umbraticus*, however, does not respond to insects in the web and indeed has no special 'telegraph wire' to its retreat as do other orb-weavers. Its prey is mainly small midges, often in great numbers, and each night it breaks down the web and eats it, together with its haul, before spinning a fresh snare. In late autumn *Araneus* constructs a cocoon of yellow silk containing some 400 – 800 eggs under a piece of loose bark or similar protection. She lingers on for a few weeks, a shrunken remnant of her former glory, before dying, but her eggs do not hatch until the spring, when the little spiderlings disperse on strands of silk wafted upwards by rising air currents. Maturity is probably not reached until the following summer.

2 Segestria. The commonest member of the genus, *S. senoculata*, constructs a tubular retreat, often under bark or in masonry, that opens on to the surface with about a dozen threads radiating from the rim. Passing insects touching one of these radiating threads are immediately seized by the spider, who rushes out from the depths of the tube with great speed. A vibrating tuning-fork gently touched against one of the threads will also bring the spider out.

3 Meta is a common orb-web spinner, closely related to *Araneus* (1), but rather smaller and less heavily built. The webs are spun on trees, bushes, low plants, and buildings, and are rather open in the centre. Courtship is complex and occurs in both spring and autumn. The male lures the female, after extensive preliminaries, on to a special strand of silk he has attached to her web and there mates with her. Two large, dark species of *Meta* are associated with cellars, caves, culverts and similar deeply shaded locations.

4 Amaurobius. These spiders live beneath bark or stones or in the cracks and crevices of walls. The web, which is made of a bluish flocculent silk, appears as a small patch some 50 mm across with a tunnel leading from it. Special spinning glands opening through a sieve-like plate produce numerous fine strands which are combed into a carefully tangled snare by a row of little hooks on the hind legs. This web relies on its mechanical properties to catch prey or impede would-be predators and parasites, and has no gum like the webs of *Araneus* (1) or *Meta* (3). Thus there is no need for the web to be replaced at regular intervals and it slowly grows as the spider lays down a few more threads each night. In the autumn the adult male courts the female by tapping violently on her web, slowly advancing towards her in the retreat. After being driven away several times, he eventually manages to approach her and mate. He does not live long after mating, dying when the weather becomes colder. The females, however, survive a further year at least, laying their eggs the following summer in a cell within the retreat and guarding them until the spiderlings hatch.

5 Dysdera is a large nocturnal hunter that preys on woodlice. Whilst most spiders are driven off by the secretions of repugnatorial glands in the woodlouse integument, *Dysdera* is unaffected by them. The powerful chelicerae and long fangs appear specially adapted to dealing with woodlice, although *Armadillidium* (p. 164) is more or less immune to attack. *Dysdera* lays some forty eggs during the summer in a silk cell beneath a stone. The female remains in the cell until the young hatch out in the autumn and slowly disperse on foot. Maturity is not reached for eighteen months, and the adult spider may live and breed for a further two or three years, unlike the great majority of British spiders, which complete their life cycle in a single year.

6 Drassodes is a common and widespread species that lives under stones or in clumps of dry grass, where it spins small temporary shelters. These are quite frequently found to contain the remains of a moulted 'skin'. *Drassodes* does not spin a web but emerges at night to hunt its prey, which it does mainly by touch. However, if it encounters another large spider or other formidable adversary, *Drassodes* will endeavour to snare it with a band of silk drawn from the spinnerets, biting fiercely at the same time.

7 Lycosa is one of the wolf-spiders, a family of generally fast-moving forms that hunt by sight, chasing their prey and leaping on it. Males mature in the spring and on sunny days they may be seen performing elaborate courtship dances in front of the females. These visual displays usually involve waving the legs or palps, which in the males are clothed in conspicuous black hairs, in a highly specific pattern that resembles a sort of semaphore signalling. When the eggs are laid soon afterwards, they are bundled into a compact silk cocoon which the female carries round with her fastened to her spinnerets. On hatching, the young emerge from the cocoon and clamber on to their mother's back where they remain for a week or more before dispersing to fend for themselves.

1 ARANEUS DIADEMATUS, in web   2 SEGESTRIA SENOCULATA
3 META SEGMENTATA, with web   4 AMAUROBIUS SIMILIS
5 DYSDERA CROCATA   6 DRASSODES LAPIDOSUS   7 LYCOSA AMENTATA, with egg sac

GARDEN ARACHNIDS, MYRIAPODS, AND WOODLICE

1 **Panonychus** (Fruit Tree Red Spider Mite). This belongs to the family Tetranychidae, which includes the most important of the plant-feeding mites. Since orchards were sprayed with D.D.T. to control cotton moths, *Panonychus* has become the greatest single orchard pest in Britain, attacking apple, plum, damson, and pear in particular. The eggs, which are somewhat onion-shaped, are highly resistant and are covered by both wax and tough protective layers. They are attached to the tree by strands of silk, those laid in autumn overwintering until the following spring. Four or five generations are produced during the summer, resulting in very dense populations on occasion. Like spiders, these mites disperse by dropping on a silk thread, being wafted away by the wind to start a new infection elsewhere.

2 **Cecidophyopsis** belongs to the economically important group of prostigmatid mites known as the Eriophyidae or Gall Mites which are remarkable for possessing only two pairs of legs. They are exclusively plant feeders, a minority inducing the formation of galls but most being vagrants over leaf surfaces or living in galls formed by other species. Many are confined to only a single species of plant. *C. ribis* is best known to gardeners for causing 'Big Bud' on blackcurrants. Immatures and adults enter new buds towards the end of May and slowly burrow towards the centre. Having mated previously, the adults lay a hundred or more eggs a few days after entry. During the warmer months hatching occurs within a week so that by the time the mites leave the buds the following spring there will be several overlapping generations living there, as mating takes place throughout the year. Departure from the buds can often result in massive swarming as the mites wander over the foliage searching for fresh buds to invade.

3 **Steneotarsonemus**. One of the mites belonging to this genus, *S. pallidus*, is a serious pest of strawberries and ornamental hot-house plants, especially cyclamen. They feed on the upper surfaces of developing leaflets and lay their eggs in folds of distorted leaves. A related species *S. laticeps* is a major pest of bulbs, particularly narcissus, living between the fleshy scales and piercing them to suck out the juices.

4 **Phalangium**. This ubiquitous harvestman is found in a wide range of open habitats throughout Britain. It is particularly common in the autumn and is more frequently seen than other species being less markedly nocturnal. The males are readily recognized because the chelicerae (jaws) are drawn upwards into long, forward-curving horns. Like other harvestmen, the female *Phalangium* has a long protrusible ovipositor with which she lays eggs deep into the ground.

5 **Blaniulus** (Spotted Snake Millipede). This attractive animal gets its common name from the fact that the repugnatorial glands lying along the length of its pale body show up as conspicuous bright orange-red spots. *Blaniulus* lives in the soil and although found in woodland and beneath stones, it is particularly frequent on cultivated ground. Indeed it is quite the commonest millipede of arable land and can be something of a pest, attacking sugar beet, potatoes, wheat, oats, strawberries, and other crops. The tough covering of such crops as potatoes is normally resistant to the small jaws of this millipede, but once a hole has been made, for example by a wireworm (a beetle larva), the *Blaniulus* can enter. Over one hundred individuals have been taken from a single potato. Although *Blaniulus* is comparatively resistant to desiccation, it is nevertheless stimulated by drought and will burrow deep into the soil until moisture is reached. Similarly, despite the absence of eyes it possesses a sensitivity to light and will, if illuminated, continue to crawl until it finds itself in darkness.

6 **Polydesmus**. Five species of these characteristically flattened and sculptured millipedes are found in Britain. They are rather more susceptible to desiccation than many other millipedes, and thus tend to be associated with fairly moist environments. They are usually found under stones or bark or in deciduous litter, but in gardens they appear to have a special affinity for lupin roots and strawberries.

7 **Lithobius**. The commonest of the sixteen British species of this genus is *L. forficatus*, which is frequently found in gardens and quite often attracts attention by wandering into houses and outbuildings at night. *Lithobius* is an aggressive carnivore and will feed on almost any animal of suitable size including worms, slugs, and other centipedes.

8 **Oniscus** is one of our largest and commonest woodlice. Like all woodlice, the cuticle of *Oniscus* is not waterproof and the whole life of the animal revolves around the avoidance of desiccation. Thus there is a strong physiologically based tendency to congregate in areas of high humidity. Although normally shunning light, if the daytime retreat starts to dry out this reaction is reversed and the woodlouse emerges to hunt for a safer spot. However, most activity takes place at night when the humidity is high. Females are often encountered with eggs or young, which are carried in a special brood-pouch beneath the body. Growth is slow, with periodic moults, and maturity is not reached for about two years.

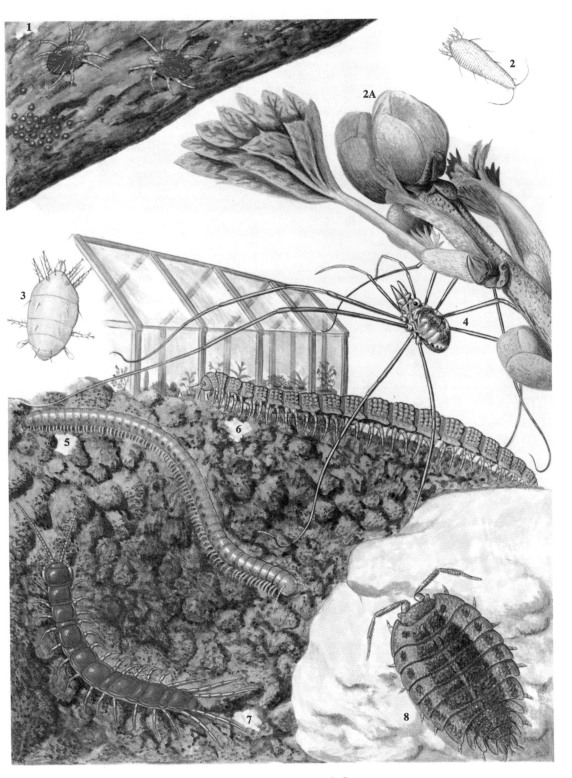

1 PANONYCHUS ULMI   2 CECIDOPHYOPSIS RIBIS

2A 'Big Bud'

3 STENEOTARSONEMUS PALLIDUS   4 PHALANGIUM OPILIO

5 BLANIULUS GUTTULATUS   6 POLYDESMUS COMPLANATUS

7 LITHOBIUS FORFICATUS   8 ONISCUS ASELLUS

WOODLAND FALSE-SCORPIONS, MITES, AND MYRIAPODS

1 **Allochernes.** This false-scorpion is found almost exclusively under the bark of trees. It is not particularly common and appears to be confined to the southern part of the country. It is one of several species that appears to be frequently associated with ants, with which it can live in harmony. Like other pseudo-scorpions, *Allochernes* constructs small silken cells in which to moult and in which to rear young, and these cells may sometimes be found beneath loose bark, deserted but containing the moulted 'skin'. The cell is constructed with silk produced from special glands on the chelicerae (jaws). The animal slowly revolves, laying down a wall of silk threads into which it incorporates any wood fragments it can find. Slowly the wall rises higher until eventually only a small hole remains at the top and is duly sealed. The eggs are laid within the cell but remain fastened to the mother within a secreted sac. After hatching, the larvae stay attached, feeding through a tube from the degenerating ovaries. After swelling enormously, the larvae moult into the first nymphal stage, become separated from the parent, and leave the cell as free individuals. Maturity is reached about a year later.

2 **Bryobia** belongs to the family Tetranychidae, which includes the most important of the plant-feeding mites. It occurs mainly on fruit trees, particularly apple and pear. Feeding takes place on the leaves but there are periodic migrations back to the bark for moulting and egg-laying, which takes place throughout the summer. Indeed there may be four generations before special overwintering eggs are laid in the autumn. Like many mites, *Bryobia* is parthenogenetic, that is to say the females produce fertile eggs without mating. Other species infest gooseberries, ivy, grasses, and herbaceous plants. One of them, *B. praetiosa*, which normally lives on herbaceous plants, quite often becomes an additional pest by swarming into houses. This is particularly associated with new buildings in early summer, when the mites may be seen leaving the herbage in great numbers to moult and to lay their eggs in crevices in the concrete.

3 **Polyxenus** is a somewhat aberrant millipede, placed in a separate subclass on its own. It is very small — only 2 – 3 mm in length — with a rather soft integument lacking the armour plating characteristic of other millipedes. The body is instead covered with strangely shaped tufts of bristles. Although thought to be widely distributed in Britain, the records are rather scattered as it is usually a somewhat cryptic animal. It lives mainly beneath impacted stones and bark and may also be found in organic litter. It is one of several millipedes that is myrmecophilous, being found particularly in the nests of the Wood Ant (*Formica rufa*). There are also records of its being seen running in the open on paths and exposed walls in the South.

4 **Lithobius.** The only centipede found in Britain but nowhere else is *L. variegatus*. It is common in most parts of the country but appears absent in some western districts. It is primarily a woodland species, living in deciduous litter, but it also climbs trees and is the only species to be found higher than about three metres above the ground. It is active at night, emerging to hunt small arthropods, which it grasps in its powerful poison claws. These formidable weapons lie just behind the head beneath the first body segment. Although tropical species can inflict serious damage, it is doubtful whether even a large British centipede can cause more than slight discomfort. Eggs are laid singly, camouflaged with particles of earth before being hidden in cracks and crevices. The newly hatched larva has fewer legs and segments than the adult. Initially fresh legs are added at each moult but once the full complement of fifteen has been acquired the last three moults result only in an increase in size.

5 **Cryptops.** The three British species of *Cryptops* are the only representatives we have of the Scolopendromorpha, an order of centipedes that is widespread and abundant in the tropics. The tropical species include both the largest and the most venomous of the centipedes. However, the British *Cryptops* are not formidable animals, although they possess the characteristic chilopod poison fangs. They are slender, colourless, fast moving creatures that, somewhat surprisingly, are totally blind. They differ from lithobiomorphs by having the dorsal plates equal in size, instead of having alternate long and short ones. The last pair of legs are not used for locomotion but are modified as organs of defence and are also used to grasp prey. *Cryptops* lives in dead or dry wood and has a rather irregular distribution in Britain, being generally rare.

6 **Pauropus.** The class Pauropoda includes the smallest myriapods, no species exceeding 2 mm in length. Their small size, secretive habits, and lack of colour resulted in their remaining undiscovered until 1886. They are immediately distinguished from other myriapods by the fact that their antennae are forked and also because the number of dorsal plates protecting the body (usually six) is less than the number of legs. Pauropods live mainly in damp situations among rotting mouldy leaves and under rotting logs and are probably worldwide in distribution. They are exceedingly sensitive to desiccation and consequently prove difficult to keep in captivity. Although it is believed that some species may be predatory, most are thought to be humus-feeders, probably browsing on fungal mycelia. Eggs are laid singly or in clumps in secluded crevices, and hatch into minute replicas of the adult but with fewer segments and legs. After four or five moults maturity is reached, after which no further moulting occurs. The whole life cycle is completed in about three months.

1 Allochernes panzeri   3 Polyxenus lagurus
2 Bryobia rubrioculus
4 Lithobius variegatus
6 Pauropus sp.   5 Cryptops hortensis

TREE-TRUNK HARVESTMEN AND SPIDERS

1 **Drapetisca** is seldom found away from trees. Its cryptic colouring, which can be very variable, makes it almost invisible on the bark of the Beech or Pine on which it usually lives. In Midland areas with heavy air pollution an increasing number of entirely black (melanic) individuals is being found which are well camouflaged on the darkened tree trunks. Although the spider spins a web, this is almost invisible, and amounts to little more than a small scaffolding platform. This is not used as a snare but as a base from which to hunt. It is also used when the spider moults, and cast 'skins' may sometimes be found suspended in it.

2 **Megabunus.** The abdomen of this harvestman is often brightly marked with a colourful pattern of green, silver and black and the paired eyes mounted on a projecting turret are surrounded by a double circlet of five long spines. *Megabunus* occurs throughout the British Isles and may be found at altitudes up to 2,000 ft. It is primarily an inhabitant of tree trunks and may also be encountered on rocks and boulders but never in ground vegetation. Little is known of its food preferences but it appears to hunt actively during the day for small arthropods and has been seen feeding on chironomid midges. Maturity is reached in May and June when mating occurs and soon afterwards the females descend to lay their eggs in the ground before dying. Hatching takes place in the autumn and immatures may be found through the winter from about November onwards.

3 **Harpactea** is a six-eyed spider which passes the day hidden in a silk cell, often under bark but also in dry matted vegetation, birds' nests, and similar spots. Hunting by night, *Harpactea* relies solely on its sense of touch to locate and identify prey. The legs are held forwards and waved to and fro like feelers as it prowls slowly. Many small invertebrates are attacked and eaten but if anything too formidable is encountered the spider backs away and retreats at high speed. Mating occurs in early summer and soon afterwards the eggs are laid in batches of about twenty. Whereas most spiders spin a cocoon, the female *Harpactea* lays her eggs loose within her retreat, which she then seals. The mother remains with the eggs inside the retreat guarding them until the young have undergone their first moult and are strong enough to leave and fend for themselves.

4 **Marpissa** is one of the largest British Jumping Spiders, but quite uncommon. The flattened body reflects its habit of hiding beneath bark, where it spins a small silk retreat. Two of its eight eyes are conspicuously large and provide acute vision which is used when stalking flies and other prey over the bark in bright sunshine. In April, when the sexes mature, the male performs a courtship dance before the female who watches intently. With his thickened front legs extended in front he zig-zags from side to side, slowly drawing closer to her. After mating the eggs are laid in a silk cocoon incorporated into the retreat.

5 **Leiobunum** is a harvestman distinctive for its small globular body and quite excessively long legs, the last joint of which is subdivided into more than fifty units to help the animal run and climb through vegetation. *Leiobunum* is common throughout Britain and is frequently seen during the day on tree trunks, although it also occurs in other habitats. This is related to its custom of migrating to low vegetation at dusk to feed. Hunting occurs during the night on the undersides of leaves, particularly Bluebells, where a wide range of small arthropods are preyed upon. Towards dawn there is a general return to the tree trunks again. Eggs are laid underground in the autumn, hatching in the spring. As the young grow, they gradually migrate upwards through the vegetation and reappear on tree trunks in about May.

6 **Lepthyphantes.** Despite its name, *L. minutus* is the largest representative of the genus, which has twenty species in Britain. The snare is of the 'hammock web' type. Beneath a loose, tangled superstructure the spider spins a finely meshed sheet, slightly domed, with a few supporting threads below. Passing insects are impeded by the superstructure and drop on to the sheet and the spider rushes up to bite them from underneath before they are dragged through and eaten. *L. minutus* is common all over Britain but often passes unnoticed. By day the spider hides in crevices in the bark but at night a torch will often reveal large numbers each suspended in its web.

7 **Zygiella.** The three British species of *Zygiella* all spin orb-webs that are immediately recognizable because one of the radiating strands from the centre is isolated and not joined to the sticky spiral, which is thus incomplete. The spider waits in a silken retreat, one foot resting on the isolated radial thread. As soon as a flying insect becomes snared in the web the vibrations generated by its struggles are immediately transmitted to the spider, who descends to the central platform, discovers the position of the prey, and rushes out to bite it. Once poisoned and wrapped, the victim is carried to the retreat to be consumed at leisure. Maturity is reached in late summer and autumn, when mating occurs. The adults die off with the onset of cold weather and the young spiderlings emerge from the cocoon the following spring.

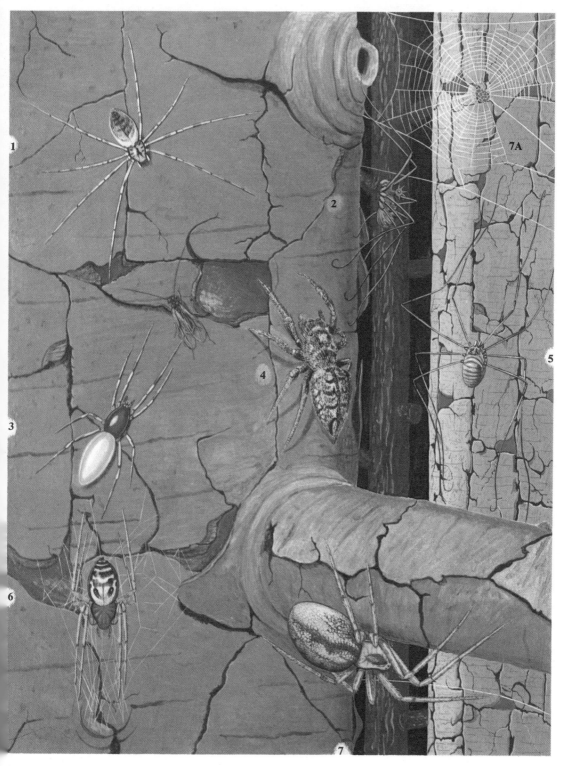

1 Drapetisca socialis   2 Megabunus diadema   7a web

3 Harpactea hombergi   4 Marpissa muscosa   5 Leiobunum rotundum

6 Lepthyphantes minutus   7 Zygiella atrica, holding signal line

WOODLAND SPIDERS

Spiders, in the number of described species, form the largest arachnid order and are represented in Britain by over 600 species. The body is composed of two units, the cephalothorax and abdomen, the former bearing the chelicerae or 'jaws', a pair of leg-like pedipalps which act as feelers, and four pairs of walking legs. The abdomen bears the spinnerets, finger-like projections through which silk is drawn, at the rear end of the body. In male spiders the pedipalps are swollen and modified to transfer sperms to the female during mating. The males, although generally smaller than the females, are very seldom eaten by their mate as is popularly believed, although this does happen regularly in just a few species.

1 **Hyptiotes** is a very rare species, found only in a few areas of southern England, where it spins a characteristic web on conifers, especially Yew. It possesses spinning organs similar to those of *Amaurobius* (p. 150) and is able to produce the same kind of flocculent, bluish silk with extremely good catching powers. However, its triangular snare is very different and resembles a group of three segments from an orb-web with a single strand running from the apex. This strand of silk is not anchored to a twig, as might be expected, but is held by the spider, who thus spans a break in the line. As soon as an insect touches the web, the spider slackens its hold and the victim is entangled in the sagging snare. On reaching it *Hyptiotes* does not bite it, being the only British spider lacking poison glands, but wraps it vigorously in swathes of silk. The capture of prey usually results in the destruction of the web which must subsequently be rebuilt and the spider retires to eat its meal under the shelter of a convenient branch.

2 **Anyphaena** (Buzzing Spider). This distinctive species is related to *Clubiona* (3) but is more arboreal. Its life is spent hunting over the leaves of trees, particularly oaks, for small insects. Found throughout Britain, it is a very common species and often occurs in large numbers. Maturity is reached in May and the male then begins to court the female. This he does by drumming with his abdomen on the leaves, advancing towards her with his long front legs vibrating. The high-pitched buzzing sound produced by the abdomen is quite audible to the human ear. The eggs are laid in a flattened cocoon constructed within a silk-lined cell made from a rolled leaf and here the female waits until the young are ready to disperse.

3 **Clubiona.** The spiders of this genus are mostly found living on low vegetation and tree trunks. Several species protect their cocoons by constructing them in the shelter of a rolled leaf.

4 **Philodromus.** Except for the very cryptic *P. fallax* which lives on coastal sand-hills, the members of this genus are active on foliage. Although they are crab-spiders, related to *Xysticus* (6) and *Thomisus* (p. 162) they have comparatively long legs and slender bodies which reflect their active search for prey. The male of *P. dispar* appears to have a shiny black abdomen. *P. aureolus* is the commonest and most widespread species and may be encountered throughout the warmer months.

5 **Linyphia.** Several species of this genus spin in low vegetation and in woodland. *L. hortensis* is particularly common on Dog's Mercury. It belongs to the same very large family as *Erigone* (p. 130) and *Lepthyphantes* (p. 156) and like the latter constructs a web of the 'hammock' type. It is widely distributed but only abundant locally. The commonest species is *L. triangularis* which may often be found in large numbers on gorse, brambles, and other low herbage in more open habitats.

6 **Xysticus.** Members of this genus are typical crab-spiders. Some species wait motionless on the ground for their prey, with front legs extended, but others are found on low vegetation. *X. lanio* is particularly associated with young (2 – 3 m) oak trees. The courtship of *Xysticus* is extraordinary. The male, which is slightly smaller, clambers all over the quiescent female lacing her to the substrate with numerous fine silk threads. In due course he tilts her abdomen up and clambers underneath to insert a palpal organ. The function of the tying down is not clear for when mating is completed, the female is able to get up and walk off without much difficulty.

7 **Tetragnatha.** These long-legged thin-bodied spiders are particularly associated with damp places and often occur in large numbers. They are found in all parts of Britain, spinning on trees and low vegetation, particularly close to water. *Tetragnatha* spins an orb-web like that of *Araneus* (p. 150) but the silk is finer and the spacing wider. The spiders tend to sit in their webs, which are often inclined or even horizontal, with their legs outstretched looking just like fragments of dead grass. The slightest disturbance will usually cause them to drop immediately into the undergrowth. *Tetragnatha* may also be found stretched out along a grass stem, concealing itself rather like *Tibellus* (p. 166).

1 Hyptiotes paradoxus

2 Anyphaena accentuata   3 Clubioña corticalis   4 Philodromus aureolus

5 Linyphia hortensis   6 Xysticus lanio   7 Tetragnatha extensa

FOREST MITES, MYRIAPODS AND WOODLICE

Mites, together with the ticks (p. 142) comprise the arachnid order Acari. They are a large and very diverse group, often present in enormous numbers. Their biology is complex and amongst the parasitic forms the life cycles are often very involved. The classification and identification of mites is notoriously difficult and very much the province of specialists.

1 **Uropoda.** The Uropodidae form a large and very important group of soil-dwelling mesostigmatid mites. They are heavily armoured and have markedly dorso-ventrally flattened bodies. It has been suggested that this shape of body facilitates progress through the narrow spaces between the compressed leaves that make up the litter layer in forests. In fact, however, they tend to be sedentary and rather inactive animals. They are found in a wide range of damp habitats, particularly in moss, grass, and forest litter, and many species are associated with animal dung. They are mainly feeders on fungal mycelia and spores, with some eating decaying plant material. Because their armour plating makes them practically immune to the smaller arthropod predators, many species are able to live commensally in ants' nests. A characteristic feature of uropodid life histories is the inclusion of a phoretic stage. Phoresy, the scientific name for 'hitch-hiking', is a means of dispersal. The young nymphs attach themselves to some larger animal, not as parasites but simply as passengers. The uropodids fasten themselves to their 'host' by means of a stalk formed from hardened secretions of the anal glands. They are most commonly found on dung-living beetles but are also frequent on ants and woodlice. Some littoral species inhabiting the litter left by the high tide are also found attached to sand hoppers.

2-4 **Litter Mites.** The surface layers of soil, with their overlying cover of decaying vegetable debris and varying stages of humus production beneath, provide a diverse and complex environment for numerous small animals, particularly arthropods. Of all soil animals the mites are the most abundant both in numbers of species and in numbers of individuals. In damp deciduous forests, leaves are broken down by large invertebrates such as woodlice, millipedes and earthworms, and the litter layer tends to be comparatively thin — but even so the mite populations are considerable. In coniferous forests on well drained acid soils, earthworms are absent and a considerable depth of organic debris accumulates and it is here that the greatest mite populations are to be found. Most of the mite population is to be found in the upper layers, where some heavily armoured forms such as *Steganacarus* (3) may be found feeding on decaying leaves and wood. *Steganacarus* belongs to the Oribatidae, popularly known as 'beetle mites'. Also abundant in the upper layers are many species of large predatory gamasid mites such as *Pergamasus* (4). These hunt actively, preying on springtails, smaller mites and particularly nematode worms. The related genus *Rhodacarellus* (2), another gamasid, is one of the few mites to be found living deep in the soil and it is interesting to note that like other deep-dwelling forms they are small, weakly

armoured, and lightly pigmented. Amongst the least well-known soil mites are the small, delicate prostigmatids such as the 'Snout Mites' (p. 130). Little is known of their biology but some prey on small insects and mites and their eggs whilst others are plant feeders. They often occur in very large numbers.

5 **Polyzonium** is a millipede belonging to an order that is largely tropical in distribution. The group is characterized by a tendency for the jaws to become reduced and modified for sucking, a feature seen in *Polyzonium*. In Britain *Polyzonium* is a rather rare animal, being confined to woodland on chalk in the South-east with one record from Yorkshire. Despite its bright colour, it is not easily seen as it closely resembles the beech bud scales in the litter which it inhabits. Like other millipedes, *Polyzonium* lacks a waterproofing wax layer to the cuticle and is hence very susceptible to desiccation. A slight reduction in atmospheric moisture causes the animal to become active and seek a damper environment. In summer, individuals may be found crawling continuously on the undersides of smooth chalk boulders whilst in winter they will be found buried deep in moist humus, tightly curled up and unmoving.

6 **Polymicrodon.** This is the largest and commonest of the silk-producing millipedes in Britain. It is widespread throughout the country but elsewhere occurs only in France. It is predominantly a woodland species living under stones, logs and bark on old tree stumps. It is particularly associated with calcareous soils although not restricted to them. In addition *Polymicrodon* is the commonest cave-living millipede in Britain. They are very active animals, particularly during winter, but even more susceptible to desiccation than most millipedes. Conversely they must also avoid free water too, as this is rapidly absorbed by osmosis. Mating takes place in the autumn and winter and soon afterwards the female constructs a silk cell to protect the eggs, using special silk-producing glands at the rear end of her body. The young grow rapidly, moulting some nine times before maturing in about six or seven months. For each moult the millipede constructs a silken cell, which it afterwards eats.

7 **Trichoniscus** is a small woodlouse that occurs throughout Britain in moist, well-shaded situations where it feeds on decaying vegetable matter. It is particularly common amongst wet, matted leaves on the forest floor and in the litter of ungrazed grassland, although it is easily overlooked on account of its size and colouring. The most conspicuous woodland woodlice are probably *Oniscus* (p. 152) and *Porcellio* (p. 166).

1 Uropoda sp.   2 Rhodacarellus sp.
6A Polymicrodon, egg cell   5 Polyzonium sp.
3 Steganacarus sp.
6 Polymicrodon polydesmoides, eating moulting cocoon   4 Pergamasus   7 Trichoniscus pusillus

HEATHLAND SPIDERS

1 **Thomisus.** This beautiful pink crab-spider is confined to the most southern counties, where it may be found quite commonly in some areas living on heather flowers. Its shape and colour make it extremely difficult to spot even when it is lying in a sweep net. It preys on flying insects coming to feed on the heather flowers, grasping them from below and sucking out their juices through a small hole. The male is very much smaller than the female and more lightly built, but still basically pink in colour despite darkened front legs.

2 **Evarcha** is one of the jumping spiders, a family distinguished by the pair of extra-large eyes on the front of the carapace. Like *Marpissa* (p. 156) they stalk their prey with slow deliberation, finally leaping on it at the last moment. *Evarcha* is confined to the more southern counties and may be found throughout the summer, often in large numbers. Females, which are rather more drab than the males, can sometimes be discovered guarding their eggs, which are laid in a substantial cocoon made of white silk laced between sprigs of heather.

3 **Singa.** These small, attractive spiders, with a shiny, globular abdomen, are related to *Araneus* (p. 150). They are rather uncommon and confined to the southern counties, where they spin small orb-webs amongst heather.

4 **Achaearanea** is a specialist feeder on ants. From an untidy superstructure a series of tensioned threads run vertically to the ground, each bearing drops of gum near the bottom. A passing insect, usually an ant, sticks to the thread, which breaks from the ground and jerks the hapless victim into the air. The waiting spider descends from its domed retreat at the top of the web and hauls in its catch, further ensnaring it with silk and finally biting it. In summer when the temperature in the retreat rises unduly, the spider carries out the several egg-sacs it has produced and hangs them in the cooler parts of the web. As the temperature drops later in the day, the egg-sacs are carefully gathered up and brought back again. When the eggs hatch, the mother carries ants up to the retreat and the baby spiderlings come to join her in a meal. *A. saxatile* is a rather uncommon species found mainly on southern heathlands, where it spins beneath clumps of heather.

5 **Ero.** The three British species of *Ero* prey exclusively on other spiders. They climb onto a web and with great stealth edge their way towards the rightful occupant. Drawing near they extend their long front legs around their victim and with scarcely a movement seize it by a limb, injecting venom of high potency. Without further ado they start to feed, slowly sucking out the juices, finally leaving the empty husk still hanging in its web as though still alive.

6 **Atypus** (Purse Web Spider). Generally regarded as rare, *Atypus* is in fact very abundant locally in parts of the South. The spider lives in a silk tube running about 230 mm into the ground, with the top 50 mm lying on the surface and carefully camouflaged with bits of earth and dried vegetation. An insect or woodlouse touching the tube is impaled through the silk by the spider's huge fangs and paralyzed by its poison. When the victim is still, the spider tears a hole through the silk, using the tip of each fang in turn, and the body is dragged in. Before settling down to feed, the spider returns to the tear and carefully mends it, laying down fresh layers of silk inside. In the autumn male spiders leave their tubes and search for mates. On finding a female's tube, the male drums on it with his pedipalps, tears a hole, and enters. After mating he remains in the tube, usually dying a natural death by the end of the winter. The eggs are laid during the summer and hatch in September but the young spiders do not leave the parental tube until the following spring, when they disperse to build their own burrows.

7 **Arctosa** is a common and widespread wolf-spider that is particularly associated with sandy areas both coastal and inland. It hides for much of the time in a horizontal silk burrow camouflaged with grains of sand, darting out from time to time to seize passing ants, which make up its main diet. *Arctosa* is preyed upon in turn by the hunting wasp *Pompilius*, which may sometimes be found dragging or carrying a paralyzed spider over the sand. The wasp lays its eggs on the spider's abdomen and on hatching the wasp larva is assured of a fresh meat supply. On sand *Arctosa* is a cryptic animal and quite difficult to spot. It is of interest, therefore, that a population has recently been found living on dark-coloured colliery spoil tips in the Midlands in which the spiders are blackened (melanic) genetically to blend with their specialized background environment.

8 **Eresus.** Britain's rarest and most sought-after spider, *Eresus* has not been collected with certainty since 1906, although there is good reason to think a specimen was seen in Cornwall in 1932. The female, which is a uniform velvety-black, spins a tough funnel-shaped web incorporating fragments of past meals. The male, which is most likely to be seen wandering in southern heathery areas near the sea in early spring, is unmistakable, its black cephalothorax contrasting strikingly with the vivid scarlet abdomen.

1 Thomisus onustus   2 Evarcha arcuata   3 Singa pygmaea
4 Achaearanea saxatile   5 Ero furcata
6 Atypus affinis, in burrow   7 Arctosa perita   8 Eresus niger

GRASSLAND ARACHNIDS, WOODLICE, AND MYRIAPODS

1 **Trogulus** belongs to a group of rather primitive harvestmen in which the body is somewhat elongate and the legs are short. Another characteristic feature is the two-pronged hood projecting forward above the mouthparts. *Trogulus* is a rather rare animal in Britain, being confined to a few southern localities where it lives amongst grass and litter at the edge of woods, particularly beech, on chalk. Like its close relative *Analasmocephalus*, with which it is sometimes confused, *Trogulus* camouflages itself with particles of earth and moves rather ponderously. The young are often a rather striking purple colour beneath their dirt. All stages may be found throughout the year and although maturity may be reached in as little as six months, the adults live for several years. The principal food is believed to consist of snails but they have also been seen feeding on julid millipedes.

2 **Misumena** is a common species of crab-spider in the South frequently found sitting in the middle of flowers with its long front legs held out ready to envelop any insect that comes to feed. To escape notice both from potential prey and would-be predators such as birds, *Misumena* is able to alter its colour from white to yellow or even pink to blend with the flowers it is sitting in. The male is far smaller than its mate and could easily be mistaken for a different species. Its front legs are dark and proportionately much longer and the abdomen has black markings.

3 **Walckenaera** is a widely distributed spider, usually being found in shaded situations where it spins minute horizontal platforms of silk in litter and grass tussocks. The heads of male linyphiid spiders sometimes bear extraordinary lumps and protuberances. The male of *Walckenaera*, which is adult during the colder months of the year, is one of the strangest in this respect, with all eight eyes carried on a tall stalk. The reasons for this male extravagance are quite unknown.

4 **Micrommata** is one of the most beautiful British spiders. The female is uniformly green but the male carries yellow and crimson stripes in addition once he reaches maturity. It is a rather local species more commonly found in the South than in the North. In July a cluster of bright green eggs is laid in a silk-lined cell made of leaves. This is guarded by the shrunken female until the little green spiderlings emerge late in August.

5 **Zelotes.** Ten species of the spider genus *Zelotes* occur in Britain, all of very similar appearance. Except for *Z. electus*, which is confined to coastal sandhills, members of this genus are usually found beneath stones, particularly on chalk grassland. They move with great speed when disturbed and seldom wait to be studied, but sometimes a female will be found guarding her egg-sac and hesitant to flee. The flattened egg-sacs, which are securely fastened to the stone, are often pinkish in colour and have a smooth, shiny finish. This is made by the adult female polishing the silk with her mouthparts, using saliva and excreted matter to produce a toughened skin that reduces desiccation of the eggs and protects them against would-be predators and parasites, particularly Hymenoptera.

6 **Armadillidium** (Pill-bug). Like the Pill-millipede *Glomeris* (8) with which it is frequently confused, *Armadillidium* is able to escape from predators by rolling up into an impregnable ball. It is able to tolerate rather drier situations than other woodlice and is consequently found in disturbed areas with short vegetation, such as roadsides and railway cuttings. It shows a marked preference for chalky areas, and is often associated with loose cement in builders' yards. This is probably connected with the heavy impregnation of calcium carbonate in the armoured cuticle. One rather rare species, *A. album*, is confined to tide litter on western beaches. Other abundant woodlice in grassland are *Philoscia* and *Trichoniscus* (p. 160).

7 **Haplophilus** is a centipede belonging to the order Geophilomorpha. These are very elongate animals, sometimes with as many as one hundred pairs of short legs. They are all blind, somewhat sluggish animals, usually found in rather damp situations. *Haplophilus* is the most frequently found geophilomorph in grassland and is also the only species that sometimes eats living plants. It has been known to cause damage to root crops.

8 **Glomeris** (Pill-millipede). This smooth, shiny animal, with its strongly armoured plates constructed from cuticle impregnated with calcium salts, is extremely common in Britain. Although most abundant in woodland, it is also able to live in areas of undisturbed grassland, particularly on calcareous soils, because it can withstand considerably drier conditions than other millipedes. The popular name refers to the animal's habit of rolling up into a tight ball when attacked. In this it resembles the woodlouse *Armadillidium* (6) with which it is frequently confused. Eggs are laid singly, each being given a protective covering of digested earth by the eversible rectum. Nine moults are required to reach maturity, spread over at least three years. By continuing to moult at intervals the adult is able to live for a further three or four years. *Glomeris*, unlike the majority of diplopods, does not construct a special protective chamber when it moults.

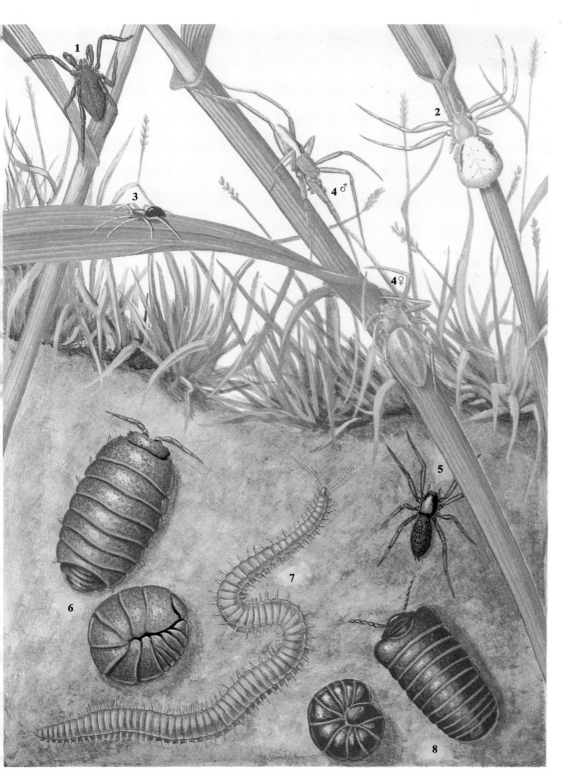

1 Trogulus tricarinatus   2 Misumena vatia
3 Walckenaera accuminata   4 Micrommata virescens
6 Armadillidium sp.   7 Haplophilus subterraneus   5 Zelotes sp.
8 Glomeris sp.

165

SANDHILL WOODLICE, MYRIAPODS, AND ARACHNIDS

1 **Porcellio** is one of the large and very common British woodlice. It normally occurs in a variety of humid habitats, including under loose bark, in heaps of stones, and also quite often in the cellars of houses. However, *P. scaber* is also the species most usually encountered on sand dunes. Only one woodlouse, the rather rare *Armadillidium album*, is confined to sand dunes. It occurs on the West coast, living in litter and under drift wood washed up by the spring tides.

2 **Cylindroiulus** belongs to a large family of rather stout, cylindrical, well armoured millipedes. *C. latestriatus* is the only one of the six British species to be associated with coastal districts. It is often found on sand dunes but may also occur on sandy areas well inland. In Devon it has been taken on the forest floor, a more typical habitat for the genus, and in Europe it has become synanthropic, being found in areas where man has disturbed the natural habitat.

3 **Dactylochelifer** lives exclusively on the sea shore, most commonly on sandhills but occasionally in rock crevices, under stones or beneath the bark of drift-wood. It is one of the largest British false-scorpions, growing to some 3 mm in length, and is found on the East coast from Scotland to Kent. On sandhills *Dactylochelifer* is usually to be found in clumps of marram grass, either living among the sand grains or hiding inside the hollow stems. It is one of the easiest species to keep in captivity needing only sand in which to burrow and some moist cottonwool to maintain the humidity. Feeding presents no difficulties for it will feed avidly on any small insects offered. Prey is grasped in the large pincers while the mouthparts tear the body apart and pour in digestive juices, for like almost all arachnids, false-scorpions can only take in liquid food.

4 **Hyctia.** Like many of the British Jumping-spiders, *Hyctia* is rather uncommon. It is confined to coastal sandhills, mostly along the South and East coasts. Both sexes are quite large, with long, slender bodies, but what makes them distinctive is the size of the front pair of legs. In the male particularly, they are not only large but dark and shiny. In early spring, during the brief season of male maturity, the legs form the focal point of the male's courtship dance before the female. Like all jumping-spiders, *Hyctia* has good eyesight and the male, in order to make clear to the female that he is a potential mate and not just another meal, performs an elaborate courtship dance in front of her. This is similar to the dances performed by the wolf-spiders such as *Lycosa* (p. 150) but usually more vigorous and complex. *Hyctia* lives mostly on the long, swaying stems of marram grass and the scope for such displays is limited, the spider having to remain comparatively stationary. The legs are held out side-ways at 45 degrees and are alternately raised and lowered. At the same time the abdomen is held out so that the dark chevrons are visible to the female. For

half an hour or more he holds the posture, rocking the body from side to side until at last he can draw close enough to climb on top of the female's body and lean over to insert a palpal organ.

5 **Synageles** is another Jumping-spider, like *Hyctia* (4), that is rather rare, although less so than used to be thought. Although it has been taken from reed stems in a fen in Huntingdon, it is otherwise known only from sandhills along the South coast and in South Wales. In certain localities, such as Braunton Burrows on the North coast of Devon, it is very numerous and on a fine sunny day in summer many specimens are active, either running on the sand or climbing the stems of marram grass. In dull weather they may seem at first to be absent, having hidden themselves within the dead hollow marram stems. However, even in fine weather the untrained eye will not readily spot them. Like its relative *Myrmarachne*, *Synageles* is an accomplished ant-mimic, and all too easily the spiders escape notice as one of the ubiquitous ants. The courtship of *Synageles*, like that of *Hyctia*, usually takes place on the marram stems. Both sexes carry conspicuous white bands around the abdomen and as the male approaches the object of his attentions he raises his abdomen at right angles, thereby displaying his white bands. If the female has already mated or for some other reason is not interested in his advances, she too raises her abdomen and displays, thereby warning the male away.

6 **Cheiracanthium.** These are hunting spiders that seek their prey on grass and other low vegetation. They somewhat resemble *Clubiona* (p. 158), to which they are related, but are rather more attractively coloured, and have longer legs and much larger chelicerae. *C. erraticum*, which is a widespread species, has a brilliant crimson stripe along the abdomen but in *C. virescens*, the most usual species on coastal sandhills, the abdomen is unicolorous. In early summer, the females of *Cheiracanthium* construct a cocoon by tying the tops of grass stems together with silk to form a protective dome and these can be quite conspicuous. A third species, *C. pennyi*, has recently been rediscovered after an interval of almost one hundred years.

7 **Tibellus.** Although the relationship is not obvious at first sight, these spiders are in fact relatives of *Xysticus* (p. 158) and *Thomisus* (p. 162). Their elongate bodies make them very inconspicuous among the grass stems on which they live, particularly as they spend most of the day lying stretched out along the stems. Although *Tibellus* is not by any means restricted to sandhills and is often found in grassy areas, few spiders are as common on the marram grass of our coastal dunes. Unlike the typical crab-spiders, which are rather sedentary animals that wait for their prey with outstretched legs, *Tibellus* can move rapidly and with considerable agility to leap upon any small insects that venture within range.

1 PORCELLIO SCABER   2 CYLINDROIULUS sp.
5 SYNAGELES VENATOR   3 DACTYLOCHELIFER LATREILLEI   4 HYCTIA NIVOYI
6 CHEIRACANTHIUM VIRESCENS   7 TIBELLUS MARITIMUS

ACORN WORMS AND OTHER WORM-LIKE ANIMALS OF THE SHORE

The phylum Sipunculoida is a small phylum of rather strange animals. Once, they were included in the Annelida (94 to 121), because their mode of development is identical to that of the annelids; more recently, however, it has become clear that the differences between these and the true annelids warrant placing them in different phyla. In the first place, the gut of these animals is recurved, that is, it is U-shaped so that the anus opens quite near the mouth, an important feature for an animal which burrows in the substratum with only one end protruding. Secondly, there is a special feeding organ surrounding the mouth consisting of a set of ciliated, hydraulically operated tentacles which filter out food particles and pass them to the mouth. A lophophore is also possessed by pterobranch hemichordates (see below), bryozoans and phoronids (170–175), and, in a modified form, by brachiopods (170–171) and echinoderms (178–187).

The phylum Hemichordata was formerly included in the Protochordata (*see* 176 and 188), under the mistaken impression that the adults possess a *notochord*, the stiffening rod that all primitive chordates possess at some stage in their life history. They do, however, have another feature which they share with the lower chordates: gill slits. These are openings from the first part of the gut to the exterior, for the exit of water which has passed over the respiratory surfaces. There are two main classes within the hemichordates: Enteropneusta (the acorn worms) and Pterobranchia. The pterobranchs are rather uncommon colonial tube-dwelling forms from deep water, and as they are unlikely to be encountered, they are omitted.

The body of an enteropneust has three regions: a proboscis, a collar, and a trunk. The proboscis and collar together look like an acorn (hence the common name) and the long slender trunk usually has folds of the body-wall covering the gill slits. The body is very soft and may break when handled. They are rather sluggish animals which burrow by thrusts of the proboscis, making a U-shaped tube in sand or mud.

1 **Aspidosiphon.** This is a crevice-living sipunculoid. The anterior end is often swollen, to block the opening of the crevice it lives in, which is often the discarded burrow of piddocks such as *Pholas* (p. 87).

2 **Phascolion.** This sipunculoid usually inhabits discarded turret shells, such as *Aporrhais* or *Turritella* (p. 41), and because of this the body acquires a spiral twist. The worm lines the mollusc shell with sand or mud and leaves a small hole through which it emerges. It grows up to 80 mm long.

3 **Golfingia.** This pale straw-coloured sipunculoid with brownish ends inhabits the lower shore of sandy beaches. It received its extraordinary name from the great nineteenth-century naturalist, Ray Lankester, whose golf-ball sailed over a bunker at Ainsdale and came to rest on the shore in a small depression in the sand. Sharing the depression was a worm which Lankester realized was new to science, so he named it after the activity he was indulging in when it came to light.

4 **Glossobalanus.** This hemichordate from muddy sand and shell gravel has a conical proboscis, and the liver pouches show through the skin of the trunk region. It is found on the lower shore, mainly on South and West coasts and Northern Ireland.

5 **Protoglossus** differs from *Glossobalanus* in having a spatulate proboscis and no liver pouches visible through the body wall. It occurs on the middle and lower shore, principally in silty sand and gravel, mainly in the Irish Sea.

6 **Saccoglossus.** This salmon-pink hemichordate has an elongate proboscis. When found on the beach, it often emits the pungent smell of iodoform. It differs from the previous two in having direct development, that is, it does not pass through a larval stage before the adult form is assumed. It spawns in May, June, or July, and tidal and climatic factors are said to influence the timing. It occurs on the lower shore, in clean sand and gravel, and is most likely to be found in Wales, Scotland, and Ireland.

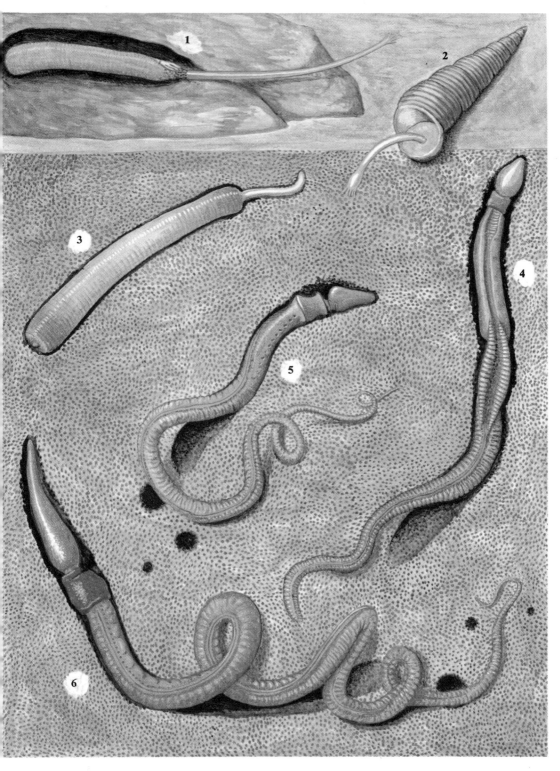

1 ASPIDOSIPHON MULLERI   2 PHASCOLION STROMBI
3 GOLFINGIA ELONGATA   4 GLOSSOBALANUS SARNIENSIS
5 PROTOGLOSSUS KOEHLERI
6 SACCOGLOSSUS CAMBRENSIS

LAMP-SHELLS AND PHORONIDS

Animals of these two phyla, though of startlingly different external appearance, are quite closely related. They are united by possession of a tentaculate *lophophore* (p. 168), for feeding and respiration. They are all protected by secreted cases of some sort, the lamp-shells by a bivalved shell and the phoronids by horny tubes. All are entirely marine, but, it must be admitted, few representatives of these two phyla are likely to be encountered by the casual collector. Nonetheless, they are zoologically very important, as representatives of distinct phyla of higher invertebrates which illustrate the extreme adaptive divergence which has occurred in animals having a lophophore.

Showing striking convergence with bivalve molluscs (pp. 72 – 91), the brachiopods or lamp-shells all have a two-valved shell, but here the valves are anatomically dorsal and ventral, whereas in bivalves the valves are left and right. The two valves are dissimilar, and usually the ventral valve has a hole near the hinge from which a stalk emerges, to fasten the animal to the substratum. The whole shell resembles a Roman lamp upside down (2B) which gives the group its common name; but in some there is no stalk and the ventral valve is cemented to the substratum and the dorsal valve lifts off it like the lid of a powder compact. There are two classes of brachiopods: the Inarticulata, which have only muscles connecting the valves; and the Articulata which have proper hinges. In British waters there are representatives of both classes, and one each is included here. The lophophore in brachiopods is internal and it can be seen protruding in 1 and 2A; it is supported internally in articulate brachiopods (e.g. 2) by a skeletal projection of the ventral valve, but in inarticulates (e.g. 1), the tentacles can extend beyond the limits of the valves. Though some of the phoronids are solitary, most members of this phylum are gregarious and occur in tangles of membranous tubes. From the open end of the tube the tiny animal can protrude its lophophore, which has a double row of tentacles in the shape of a horseshoe. The mouth is within the pattern of tentacles and the anus, at the end of a U-shaped gut, opens a little way away, outside the tentacles, so that faecal waste is not taken in again with the food. As a further precaution, the mouth has a small flap near it on the anal side which deflects any currents from the anus away from it. Phoronids are hermaphrodite, the genital products being formed in gonads that hang in the coelom. The products are shed through a pair of excretory tubules opening one on each side of the anus. The eggs may adhere to the tentacles, where fertilization occurs by sperm from another individual. Each fertilized egg hatches into a free-swimming larva called an *actinotrocha* which joins the plankton for a period to disperse the species.

1 **Crania** is an inarticulate brachiopod, that is, the shell is more horny than calcified, there is no proper hinge, and the gut ends in an anus. *Crania* illustrates a special way of life of brachiopods: adhering by one valve to a hard object. In this case, the ventral valve is flat and adheres by the whole of its under surface to a rock, while the dorsal valve can open. The tentacles of the lophophore, unsupported by a skeleton, can be protruded beyond the shell edges for feeding.

2 **Terebratulina** is an example of a typical articulate brachiopod, both in shape and in way of life. It usually lives in offshore waters, particularly round the West coasts of Britain and Ireland. The stalk adheres to a rocky place, and the shell hangs free in the water. Sometimes there is a clump of several shells arising from the same spot. Usually, the shell hangs in such a way that it gains maximum benefit from the currents around it, so all the shells in a clump may point the same way. As it is an articulate, the shell is calcified, the lophophore is supported by a skeletal frame, and there is no anus. It is not common.

3 **Phoronis** occurs in dense aggregations, like miniature spaghetti, on rocky shores, sometimes embedded in crevices and empty shells; the tubes usually have sand and debris adhering to them. The British species, *Ph. hippocrepia*, seldom exceeds 10 mm in length, and though its *actinotrocha* larva is a fairly common occurrence in plankton samples, the adult is surprisingly rare, being found only locally in Devon, in Wales, and on the North-East coast. The individuals will withdraw into their tubes at the slightest disturbance; even a shadow passing over them will cause retraction.

1 CRANIA ANOMALA   2 TEREBRATULINA CAPUTSERPENTIS
2A T. CAPUTSERPENTIS, showing lophophore
2B T. CAPUTSERPENTIS, side view
3 PHORONIS HIPPOCREPIA, aggregated

171

SEA MATS

These small animals, usually called bryozoans ('moss animals') or sometimes polyzoans ('many animals') or ectoprocts, almost always occur in colonies. They either encrust algae or other solid objects of the shallow sea, or form frond-like colonies somewhat resembling algae. The majority are marine, but there are a few important freshwater members (p. 175). Each colony is made up of individuals, called *zooids*, usually in skeletal capsules into which they retract for protection. Sometimes the individuals are organically connected; others merely touch their neighbours. They are superficially similar to some hydroids (p. 9), but have a quite different anatomical structure. For one thing, the gut has both mouth and anus; for another, the feeding tentacles round the mouth form a *lophophore* (*see* p. 168), that is, they have a coelomic lumen the pressure in which protracts the tentacles and keeps them turgid. Zooid eversion is brought about by increasing the pressure between the zooid walls and the skeletal box, and various devices are employed in the group to do this: some (3) have a flexible top to the box which is pulled down by muscles to decrease the volume of the box; others (5) have flexible sides as well; others have a small compensation sac with a small opening to the outside which fills with sea water as its volume is expanded by muscles. When the zooid is retracted, a lid or operculum may close over the opening for further protection. In some forms (6), there are modified zooids forming pincer-like structures, the *avicularia* (because they resemble a bird's head), to fend off settling larvae of other encrusting organisms.

1 **Alcyonidium** may occur in a variety of forms, according to the water currents in the place in which it is growing: it may be either erect with branches, or encrusting. It is often on algae, and may form colonies up to 300 mm long. There is no calcification, and the individual capsules are set in a gelatinous mass. The aperture through which the zooid is everted is closed by a folded membrane when the lophophore is retracted.

2 **Flustra.** Because this forms fronds rather than encrustations, it is often mistaken for an alga. But each 'frond' consists of two layers of capsules, back-to-back. Though not really a shore organism, it may be cast up by the receding tide, when it dries and becomes brittle. When living, it is yellow or dull white in colour. Each capsule (2A) is horny and flexible with 4 or 5 spines round its rim, and the opening is closed by an operculum.

3 **Membranipora.** This makes flat encrustations on stones and weed, usually *Laminaria* or *Fucus*, with the capsules arranged in fairly regular, alternating lines. Colonies may be large; one colony, about 2 metres long and 200 mm wide on a huge *Laminaria* frond, was estimated to contain 2·3 million individuals. Each capsule (3A) has calcareous sides and a membranous top (hence the generic name). This is one of the bryozoans that pull down this top to decrease the volume within the box and force out the zooid by hydraulic pressure. There is a blunt spine at the corner of each capsule.

4 **Crisia** occurs as colonies forming dense white tufts attached to stones or to algae, usually the red algae. The branches are erect and jointed, attached to the substratum by rooting fibres. The individual capsules (4A) alternate, in two rows, and the walls are calcareous but membranous at the joints in the colony.

5 **Flustrella** forms a thickish brown crust on algae such as *Fucus*, often around the base of the frond. When the zooids are expanded, the mass of tentacles gives the impression of a blue-grey film over the colony, though at intervals there are reddish-brown horny spines. The zooids are expanded by muscle activity on the walls of the capsule, which are membranous, not calcified, so they can flex inwards to decrease the volume within and thus force out the zooid head.

6 **Bugula** consists of bushy tufts of regular-dividing branches which often hang in quantity from rocky overhangs on the middle and lower shore. The colony usually tapers to a point at its apex, and may attain 70 mm in height. Alongside the capsules containing the zooids one can often see with a handlens the modified pincer-like zooids called avicularia, which protect the colony from settlement of other encrusting organisms.

7 **Mucronella** has skeletal capsules with convex tops, through which there are holes which help in zooid eversion. Each capsule has a pair of avicularia near the opening. When on shells, it may grow to about 50 mm across.

8 **Umbonula** forms rose-red encrustations on stones, shells, and weed such as *Laminaria*, particularly on the holdfasts. The capsules are oval, each with a small mound in its centre, from which lines radiate, forming strengthening girders.

1 ALCYONIDIUM HIRSUTUM   2 FLUSTRA FOLIACEA   2A detail of capsules

3A M. MEMBRANIPORA, detail   3 MEMBRANIPORA MEMBRANIPORA

4A CRISIA, detail   4 CRISIA sp.   5 FLUSTRELLA HISPIDA   5A F. HISPIDA, detail

6 BUGULA TURBINATA   7 MUCRONELLA COCCINEA   7A M. COCCINEA, detail

8 UMBONULA VERRUCOSA   8A U. VERRUCOSA, detail

FRESHWATER BRYOZOANS

Most of the bryozoans are marine (p. 172) but a few are freshwater, and as these are among the most beautiful animals found in streams, ditches, and ponds, it is unfortunate that the casual collector may be unaware of their existence. These animals do not construct calcareous capsules as do their marine relatives, but mostly have their zooids embedded in a jelly-like ground substance which forms an encrusting mat over stones and weed. Again unlike the marine forms, the body cavities of all individuals of a colony are confluent. Two classes of bryozoans are recognized: Gymnolaemata are almost exclusively marine, and Phylactolaemata (1 – 3) are all freshwater. Freshwater bryozoans reproduce in three ways: asexually by budding; sexually; and by the production of internal buds with strong walls for overwintering. These overwintering bodies, called *statoblasts*, are lens-shaped envelopes containing a mass of cells. They bud off from the inside of the colony during the summer and float away from the decaying parent in the autumn. They can often be found in the debris round the edges of ponds and ditches, and are a sure sign that adult colonies of bryozoans can be found there at the right time of year. Since they float on the surface of the water and get stranded in the debris, they must be able to withstand both freezing and desiccation, to tide the animal over until the following spring, when 'germination' starts again to produce another colony. The statoblast itself resembles two saucers stuck together, with the vital cell mass between the halves. In the spring, the two halves separate and fall apart so that the cells can start growth again. The statoblasts have another function too: they provide a means of dispersal from one water mass to another either by wind or by the agency of other animals, such as the feet of water birds. Most freshwater bryozoans, like their marine relatives, are static and encrusting, but some, (e.g. 3) move slug-like over the surface of plants and stones. This movement is thought to be brought about by the combined activity of the muscles of each individual zooid.

1 **Plumatella.** This lives typically on reeds, roots, and the underside of water-lily leaves, but it does not have the power to move. It will mainly be found in ditches, and (a favourite place) under canal bridges and landing stages. In still water, part of the colony may hang free and project into the water, but in moving water the whole colony is attached to the substratum. The animal grows very much larger than the other two illustrated here; sometimes, new individuals may grow on old ones, forming a mat nearly 100 mm thick and 300 mm long. It occasionally causes trouble in waterworks by encrusting the interior of pipes and filters, thus restricting the flow. It is notoriously shy, and once the zooids have retracted after being disturbed, it may be up to an hour before they emerge again.

The classification of this genus is far from clear: there may be as many as eight species in British waters, or there may be only two. In any case, the two forms illustrated (1 and 1A) appear to be mere polymorphs, or variants, of a single species, the form which appears depending on the local conditions in which it grew. The zooids of both forms are identical (1B) and bear a horseshoe-shaped lophophore with about 20 tentacles.

2 **Lophopus** colonies are usually less than 10 mm long, that is, less than a quarter the length of *Cristatella* (3), but the individual zooids are larger and each colony consists of about 12 of them. It is often found on the rootlets of duckweed and pondweed, over which the colonies are able to move slightly, like *Cristatella*. The winter cysts are lemon-shaped in plan view, with slightly raised centres. It has been reported that *Lophopus* is hardier than most freshwater bryozoans, and in some cases even the adult colony can overwinter.

3 **Cristatella** is often mistaken for a 'rope' of snails' eggs, particularly if the zooids are all retracted. The colony may be up to 50 mm long and is usually found in clear ponds and lakes, on the underside of water-lily leaves or the upper sides of stones. Of all the freshwater bryozoans, this one will tolerate most disturbance before retracting its zooids, and they will extend again pretty soon after the disturbance has ceased, which contrasts with forms like *Plumatella* (1). A colony can move over weed and stones at a speed of about 100 mm a day, and occasionally a full-grown colony may split into two sub-colonies, each of which will subsequently grow to full size again. Sexual reproduction shows an element of brood-care: fertilization occurs within the ovaries, each of which harbours a single egg. This develops *in situ* until a free-living larva escapes from the colony through a hole formed when another adult individual dies.

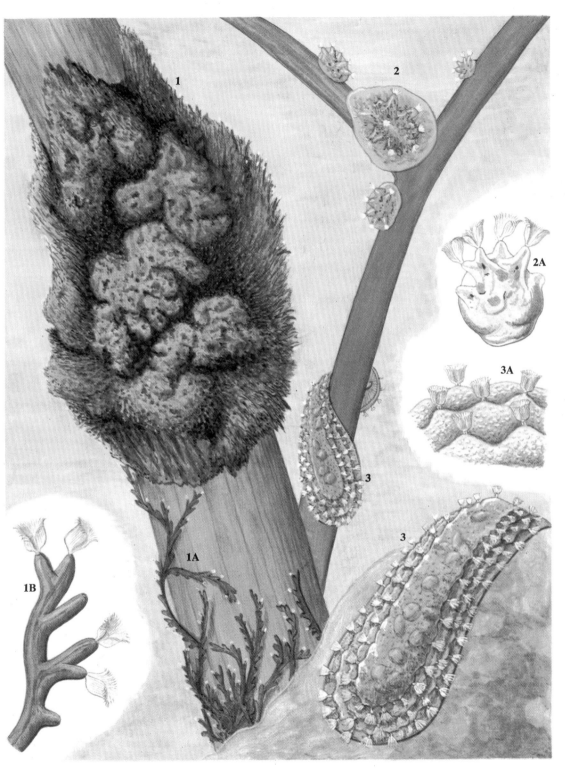

1 PLUMATELLA FUNGOSA, mat form   2 LOPHOPUS CRYSTALLINA

2A L. CRYSTALLINA, small colony enlarged

1A P. FUNGOSA, linear form   3 CRISTATELLA MUCEDO   3A C. MUCEDO, detail

1B P. FUNGOSA, detail

ARROW WORMS AND LANCELETS

The needle-like chaetognaths (1–3) are among the most conspicuous members of the marine plankton and will almost certainly be collected if a small plankton net is attached to a pier-pile or dragged behind a boat. All members are much alike, even though one genus (3) is benthic, and most are specialized for a fast-swimming, carnivorous life at the expense of weaker members of the plankton. Arrow worms alternately swim and float, and when they see their prey (they have a pair of multi-faceted eyes on the head giving all-round vision) they dart to attack by flicks of the tail and hold the prey with rows of horny hooks or teeth. They may dart to escape predators too, and escape is aided by their semi-transparent bodies, They are hermaphrodite and mating has been observed in at least some genera for the exchange of sperm, though information on the process is sparse. Self-fertilization has been reported to occur in *Sagitta* (1 and 2) and indeed it is not easy to see how a planktonic and carnivorous animal could ensure mating with another of its own kind. The eggs are shed into the sea, either singly or in clumps, and hatch into miniature adults.

Arrow worms (1–3) and lancelets (4) are not all that closely related, though current opinion suggests that the arrow worms may lie near the evolutionary line leading via the echinoderms and hemichordates, to the protochordates. The Cephalochordata is one class of protochordates. In it, we have a group of offshore, burrowing animals that seem to provide a primitive model on which the organization of fishes may have been based. Chordate animals, including the great group of the vertebrates, differ from invertebrates in possessing a flexible rod of turgid cells, the *notochord*, which provides not only a structural support for the body but also an elastic protagonist for the muscles used in swimming (*see also* p. 188). In the higher chordates the notochord is augmented by a segmental bony rod called the vertebral column. No animals mentioned in this book have a vertebral column, but a number of groups (pp. 177, 189, 191) have a notochord during at least part of the life history, and these animals mark the ultimate evolutionary attainment of the invertebrates on the line leading towards the vertebrates, which are the subject of other books in this series.

Protochordates have other features which hint at vertebrate organization, such as gill slits and, in some, a hollow, dorsally placed nerve cord. But they also possess features more appropriate to invertebrates, such as the nature of their excretory system and the ciliary-mucous feeding method.

1-2 **Sagitta** is a chaetognath that has anterior and posterior lateral fins for stability, and a spatula-like tail fin for propulsion. It preys on any small, slow members of the plankton (see picture), such as veliger larvae of molluscs, plutei of echinoderms, small crustacea, annelid trochophores and fish larvae; it is preyed on by jellyfish and pelagic annelids. The two principal species of *Sagitta* occupy different qualities of sea-water, and illustrate well the concept of 'indicator species' to distinguish one body of water from another. *S. elegans* (1) is an 'oceanic' form, that is, it occurs only in water of high salinity and purity, with a high winter phosphate content and rich in plankton. *S. setosa* (2) is a 'coastal' form, from water of lower phosphate content and sparser in plankton. On the whole, *S. elegans* is a northern form, and the English Channel marks its southernmost limit. Until 1930 its range extended to the East of Plymouth, but it has since been found only to the West and this points to a change in the Channel water which might account for the decline in its fisheries. The main difference between the two species is in the shape and extent of the lateral fins; another difference is that *S. elegans* becomes more opaque in

formalin than does *S. setosa*, and stiffer when handled with needles in a plankton sample.

3 **Spadella** is the only benthic genus in the phylum. It has adhesive papillae near the tail for attaching temporarily to the sea-bottom, where it waits for its prey before darting to the kill. Apparently unlike other chaetognaths, self-fertilization does not occur and reciprocal exchange of sperm has been observed.

4 **Amphioxus (Branchiostoma)**, a protochordate, is a microphagous feeder usually found in shell gravel or coarse sand. It normally burrows into the sea-bed then works its way round until its head emerges again, so that it can take in a stream of water and food particles by ciliary action of its pharynx. The water passes first through a coarse filter of oral spines to keep out larger particles, then enters the mouth and pharynx, where it passes across the gills for respiratory exchange and to trap the food particles in mucus, and thence to the exterior.

Shown floating in the water are various examples of planktonic organisms and larval forms, the largest of which are the arrow worms (1-3)

1 Sagitta elegans   2 Sagitta setosa
3 Spadella sp.
4 Amphioxus (Branchiostoma) lanceolatus

SEA-URCHINS OF THE SHORE AND SHALLOW SEA

For symmetrical beauty, the phylum Echinodermata is without match. It contains the familiar starfishes, brittle-stars, feather-stars, spherical sea-urchins, and worm-like sea-cucumbers. All members have skeletal plates, but the skeleton is most prominent in the urchins, where it forms a rigid protection for the body and a spiny covering; it is least conspicuous in the sea-cucumbers, where the plates are mostly tiny and embedded within the skin. One of the fascinating aspects of this phylum is the almost universal display of 5-rayed symmetry. Many members have spines (the name means spiny-skinned). All have tiny hydraulic organs, the *tube-feet*, protruding from the body, important soft structures which serve vital functions such as locomotion, feeding, respiration, and mucus-producing.

The class Echinoidea is the only one in which the skeleton forms a rigid capsule. Because the skeleton is formed inside the body, it is not a shell, but a *test*. When a spiny sea-urchin dies, the spines drop off and the ornamented test is revealed in all its beauty. One can see the tubercles which bore the spines, the holes from which the tube-feet protruded, and the limits of the individual plates of which the test is composed.

1 **Psammechinus** (Purple Sea-urchin). This hardy little urchin lives under rocks and overhangs inshore round most of our coasts, but is found only at low tide. In places, the shores and rock pools may be littered with its denuded tests, a delight to those who appreciate its symmetrical beauty. It is often hard to find, because it may cover itself with bits of gravel, shell, algae and other debris, which it holds to itself with its tube-feet. Three tube-feet emerge from each tube-foot-bearing plate.

2 **Paracentrotus.** This small urchin replaces *Psammechinus* in the West of Ireland and in places in Scotland. It is distinguished from *Psammechinus* in that it has 5 or 6 tube-feet to each tube-foot-bearing plate, and the whole of every spine is purple instead of only the tips of the primary spines. The urchin normally excavates in the rock a hole or burrow for itself from which, as it grows, it cannot escape.

3 **Echinus** (Edible Sea-urchin) is the largest of the British echinoids, and is said to provide the tastiest sea-urchin 'caviar' from its five creamy-white gonads within the test. It comes inshore occasionally, and can happily cling to rocks that are pounded by waves, though more usually it lives on a gravelly bottom just offshore. Mostly, it is found below tide level, and fishermen's trawls and skin-divers bring up some which find their way into curio shops.

4 **Echinocyamus** (Pea Urchin). This echinoid, seldom more than 10 mm long, is the only British representative of the order Clypeasteroida, which includes the tropical sand dollars. It is an 'irregular urchin', in that the anus has moved from the primitive position on the dorsal side to the ventral side, not far from the mouth. The tube-feet on the dorsal side form a star-shaped area for respiration; ten round the mouth are chemosensory to test the food; the other hundred or

so on the body have suckered disks, rather like those of the 'regular' urchins, to help the animal move among the shell gravel in which it lives. This urchin usually occurs offshore, but in rare places is found in coarse sand at low tide. It usually turns pea-green when handled — hence the common name.

5 **Spatangus** (Purple Heart-urchin). This is another 'irregular' urchin, but from the order Spatangoida. Its normal habitat is offshore, but occasionally the tests are washed inshore, or turn up in curio shops. Spatangoids show various adaptations for burrowing in sand, mud, or gravel: the mouth lies forward; the tube-feet do not have suckered disks; and the spines are variously modified — for example, on the ventral side, the spines in a shield-shaped area are paddle-shaped for locomotion and the spines, in bands called *fascioles*, are highly ciliated to produce especially strong currents for pumping water round the burrow. In *Spatangus* there is a single fasciole just below the anus on the posterior side of the body.

6 **Brissopsis** (Lyre Urchin). On the dorsal surface of this heart-urchin is a fasciole shaped like the frame of a lyre, but it is not easily seen unless the test is denuded of its spines. In British waters it is not found inshore.

7 **Echinocardium** (Sand Urchin or Sea Potato) is the urchin most likely to be encountered; on a sandy beach at low tide, one may be stepping on the burrows of these creatures. A small, irregular hole is the give-away, and the urchin can be dug out with the fingers. It has three fascioles, clearly visible when the living urchin cleans itself of its sandy covering in clean water. There is great division of labour in the tube-feet, some building the funnel from the burrow to the sand surface, others helping in feeding, respiration, or sensation, and yet others helping to rid the body of waste by building a little soakaway from the burrow.

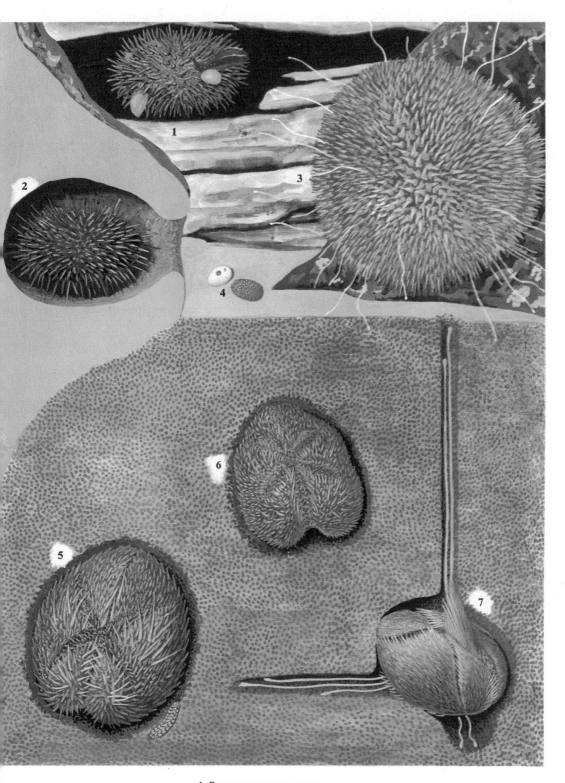

1 Psammechinus miliaris

2 Paracentrotus lividus   3 Echinus esculentus

4 Echinocyamus pusillus, test and with spines

6 Brissopsis lyrifera

5 Spatangus purpureus and commensal Malmgrenia   7 Echinocardium cordatum

CRINOIDS, BRITTLE-STARS AND STARFISHES
OF ROCKY SHORES

The class Crinoidea is the most primitive class of the phylum, and is the only one which has retained an upwardly-directed mouth. The body consists of a tiny central disk which carries the main organs, and a number of long, feathery arms. In the first crinoids, the disk was anchored to the sea-floor by a stem, but during evolution there has been a tendency to lose the stem, and today there are far more of the free living crinoids than stemmed forms; the only British crinoid is free-living. The class Asteroidea contains the starfishes. These familiar objects of the shore either burrow in sand and gravel or live among rocks. The mouth is in the centre of the underside and leading to it are grooves on the underside of each arm, grooves that are avenued by hundreds of tube-feet. On this page are shown the common starfishes likely to be encountered on rocky shores. Another class of star-shaped echinoderms, the Ophiuroidea, are also found among rocks. The characteristics of the class are outlined on p. 184.

1 **Antedon** (Rosy Feather-Star). A beautiful, fragile creature that would be smashed by waves, this sole representative of the crinoids in British waters is usually found below the shore, clinging to wrecks, pier piles, and rocky outcrops by means of its jointed cirri. Its tiny disk houses the main organs of the body, while its ten arms, each with numerous pinnules branching from them, form a food-catching funnel. But the arms can be used for locomotion too: to move its position, the animal will release its cirri, then five of the arms beat downwards while the other five beat upwards, in alternate slow beats. The gonads hang from the pinnules near the arm bases, and release their gametes into the sea, where fertilization takes place. Subsequently, a swimming larva is formed which changes into a stalked, attached stage, and it is from this that the adult breaks free, to find a suitable hard object to cling to. Sometimes a commensal annelid worm, *Myzostoma*, is found attached near the mouth, scooping food from the host's food channels (*see also* 6, p. 121).

2 **Asterias** (Common Starfish). This carnivorous starfish preys on bivalves, gastropods, and crabs. Unlike the more primitive starfishes (pp. 182 – 3), *Asterias* cannot burrow, but instead has evolved suckered tube-feet which not only help it in locomotion, but also help it to obtain purchase on the shell of its prey. There is an astonishing feature to the feeding process: the starfish stomach can be extruded from the mouth, so that it can be inserted into any small hole in the prey's armour and digestion can begin. In the case of bivalves, it is often necessary for the valves to be pulled apart before the gap between them is big enough for insinuation of the stomach; the combined pull of the tube-feet and body muscles of the starfish is astonishingly powerful and a prolonged effort will gradually break down the resistance of the bivalve's adductor muscle so that the valves part.

3 **Asterina** (Gibbous Starlet). This tiny blunt-armed starfish lives under stones in rock pools and in the shelter of overhangs. It may also nestle in the holdfasts of algae, among sea-grass stems and between the 'fingers' of sponge masses. It feeds not only on molluscs but also on worms and ophiuroids. The colour may vary from reddish-brown, through blue to yellow. The eggs when shed do not float free in the water, but are deposited in small groups under stones, to which they remain attached. During egg-laying the female may be partly covered by two or three males, which fertilize the eggs as they are laid, so that a form of copulation may be said to occur.

4 **Marthasterias** (Spiny Starfish) is locally a fairly common starfish, especially in regions where certain of its favourite molluscs, such as *Chlamys* (4 – 6, p. 85), are abundant. It differs from *Asterias* in having fewer and larger spines on the upper surface of the body. There is probably more colour variation than in *Asterias*, varying from yellow, through greens and blues to red. It is confined to the South and West coasts only.

5 **Amphipholis**. This is a true shore-dwelling ophiuroid, being found mainly under stones or among the roots of algae, especially *Corallina*. It is one of the many marine animals which exhibit phosphorescence. This is one of the echinoderms that does not pass through a pelagic larval stage: the adults are hermaphrodite, and viviparous, that is, the young emerge from the parent as tiny brittle stars.

6 **Ophiopsila**. This rather rare ophiuroid is found in rocky crevices, especially holes bored by piddocks in rocks. It is normally reddish-brown finely spotted with white, and has banded arms. The disk may be up to 12 mm in diameter.

7 **Ophiactis**. This tiny brittle star, the disk of which seldom exceeds 5 mm diameter, is a favourite food of the cod, enormous numbers of which may be found in a cod's stomach. It prefers deeper water, but often squeezes its way into shells and worm-tubes which are cast up on the beach. Like most ophiuroids, its colours are variable, but normally it has a disk of red, variegated with white, and banded arms of red-brown and white.

1 Antedon bifida   2 Asterias rubens
5 Amphipholis squamata   3 Asterina gibbosa   6 Ophiopsila aranea
4 Marthasterias glacialis   7 Ophiactis balli

STARFISHES OF THE SHORE AND SHALLOW SEA

On the previous page are shown the starfishes most likely to be encountered in rock pools on the shore. Other starfishes, usually living in deeper water, may be encountered by accident on the shore, or seen while skin-diving, and some of these are shown on this page.

While rock-dwelling starfishes mainly feed carnivorously on big prey, most of the ones shown here feed on microscopic particles of food which settle on and around the animals, or are drawn towards them by ciliary currents. In some, such as *Porania* (1) the upper surface collects quite a bit of food which is wafted in streams of sticky mucus down between the arms to the under-surface, where it is passed in a conveyor-belt of mucus along the arm-grooves to the mouth. The main part of the gut is in the centre of the star, but ciliated canals lead out to each arm, in which there is a highly folded sac in which digestion and absorption take place. The anus opens near the centre of the upper surface, but is generally so small and hidden by spines as to be invisible, except when voiding faeces. At the tips of each arm are special sensory tube-feet and a pigmented light-sensitive spot.

1 **Porania** (Cushion Star). This brightly-coloured starfish has not been seen to attack large prey, and no largish animals have been found in its stomach. Rather, it seems that this is one starfish that relies almost solely on ciliary-mucous feeding. Particles of food that settle on the animal's broad upper surface are transferred via the grooves between the arms to the mouth in moving belts and strings of mucus. This is mainly an offshore form.

2 **Solaster** (Sunstar). This splendid starfish has up to 13 arms; it usually has a reddish-purple disk and rather lighter arms. It lives on coarse sand and gravel, from low tide mark down to about 20 fathoms. It is an Arctic form, of which the English Channel forms its lower limit. Though it takes molluscs like some other asteroids, it will also eat other starfishes, and is said to be able to tackle and devour an *Asterias* almost as big as itself. There are two species: *S. papposus* is probably commoner, and is usually slightly smaller than *S. endeca*, and in *S. endeca* the distance between the centre and an arm tip is $2\frac{1}{2}$ times the distance from the centre to the edge of the disk, whereas the same ratio is only 2 times in *S. papposus*.

3 **Henricia** (Scarlet Starfish). This blood-red starfish may occasionally be found on the shore, under rocks and on gravel. The interesting feature is that the female parent broods the young under her body for three weeks after the eggs are laid, fasting as she does so. She forms a brooding tent with her body, standing on arm-tips, with the disk raised in a dome.

4 **Astropecten.** In contrast to *Henricia*, the marginal plates bordering the arms are clearly visible here. The starfish burrows in sand, with only its arm-tips protruding, though occasionally it may thrust its disk above the surface for defaecation. Some experiments on its feeding habits have shown that it 'prefers' to eat those molluscs, such as *Spisula* (see 6, p. 77), which have a high metabolic rate. The reason is that the molluscs are taken whole into the starfish's stomach, and the digestive process relies on the fact that they have to gape open after some time, so that the digestive juices immediately pour inside and kill the mollusc. If molluscs with a slow metabolic rate, such as *Venus* (see 4 – 7, p. 75), are taken in, the chances are that the valves will remain tightly shut until the mollusc is voided again. Unlike *Asterias* and *Marthasterias*, the tube-feet of *Astropecten* are not suckered, but taper to a point. They are heavily charged with mucous glands and are used to assist in burrowing.

5 **Luidia** is a voracious carnivore, which pursues its prey over the rough ground on which it lives and makes a final leap on top of it. The mouth frame of plates can stretch to admit a medium-sized heart-urchin (p. 174) or a full-grown ophiuroid (pp. 181 and 185). Feeding experiments on this form have shown it to have an order of preference for its food: it will eat *Ophiura* in preference to *Amphiura* or *Echinocardium*; it will not touch those echinoderms known to have highly acid mucus, such as *Ophiocomina*, neither will it take non-echinoderm food. There are two species: *L. sarsi* has five arms, and *L. ciliaris* seven. One rather curious feature of both species is that one seldom comes across a perfect specimen: almost always at least one of the arms shows evidence of having been broken and the end regenerated.

6 **Anseropoda** (Goose Foot Star) is a starfish which is almost pentagonal in shape, and wafer-thin at the margin. It may reach a size of about 200 mm diameter but at this size the outline becomes broken and frayed rather like the ears of an old tom-cat. It prefers gravel and sand, and lies almost buried in it. For food, it takes in crustaceans, molluscs and other echinoderms

1 Porania pulvillus   2 Solaster papposus
3 Henricia oculata   4 Astropecten irregularis
5 Luidia sarsi   6 Anseropoda placenta

BRITTLE-STARS OF SAND, MUD, AND GRAVEL

The class Ophiuroidea contains the brittle-stars — and most aptly named they are, for if handled with anything but the most gentle care they break off their arms or split their disks. This happens in nature, of course, but the powers of regeneration are excellent, as shown by the many individuals one finds which are clearly regenerating parts. The arms have a sensuous movement, mostly in the horizontal plane, and it is by arm flexures that the animals move themselves over the sea-bed, helped by the sticky tube-feet. Within the disk is the blind-ending gut and ten sacs, called *bursae*, into which water is pumped for respiration and which are also used by some brittle-stars for brooding their young. Judged on the criterion of numbers of individuals, this is the most successful class of the echinoderms, as will be pointed out (1). The brittle-stars most often encountered inshore are either small rock-nestling forms, such as those shown on p. 181, or larger burrowing ones, such as *Acrocnida*, *Ophiura* and *Amphiura*, shown on this page. The largest ones, *Ophiocomina* and *Ophiothrix*, also shown here, are usually offshore, though smaller individuals do turn up between tides, or caught up in debris along the strand-line.

1 **Ophiothrix** is easily distinguished from *Ophiocomina* by the presence of a pair of triangular radial plates on the dorsal side of the disk at the origin of each arm. These plates are an insertion for muscles used in respiration; the contractions of these muscles cause sea-water to be taken into and out of the bursae (*see above*), the entrances to which can be seen alongside the origins of each arm on the ventral surface. The bursae are also used in this brittle-star as a brood chamber for the young. Slightly older immature stages may often be found inshore, clinging to algae, sponges, etc., by means of specially hooked spines on the arms. In the adult, the spines, according to Forbes, have 'a lightness and beauty which might serve as a model for the spire of a cathedral'.

In some regions around our coasts there may be vast beds of this ophiuroid, sometimes lying two or three deep on the seafloor. Underwater photography has shown thick carpets of them in the English Channel and Irish Sea, and skin-divers have reported seething masses of them over weed just below tide level. When they occur in such profusion, there are usually fast tidal currents racing over the sea-bottom, and in these conditions the ophiuroids need to cling to the gravelly bottom; but when the tide turns there is a slack period during which each individual raises one or two of its arms, to catch particles of food in the water above it. The food is then passed down the arm groove to the mouth by means of the tube-feet.

2 **Ophiocomina.** This and *Ophiothrix* (1) are often found together, but the nature of the plates of the disk easily separates the two, the disk in this one being granulated, and the radial plates — the two triangular plates at the origin of each arm — hidden below the surface. The colour is usually black, but many other colour variants are found, such as shades of brown, red, and fawn to yellow-orange. Locally, an entire population may be jet black, whereas elsewhere there may be all colours together; sometimes it is spotted. It is able to climb, and sometimes it is found in vast numbers clinging to the fronds of *Laminaria*, held there by the sticky secretion of its tube-feet. More often, however, it is found on gravel or sand and, like

Ophiothrix, it may raise several arms into the current to filter out food particles. A study of its feeding methods has revealed that it can utilize food of many kinds, employing a variety of techniques to get the food to the mouth. One of the main methods is to catch small particles, work them into a tiny pellet with mucus, and pass this from tube-foot to tube-foot down the arm groove to the mouth.

3 **Ophiura.** If you walk along the water's edge in a sandy bay as the tide drops, you may see an irregular pattern in the wet sand, somewhat resembling a gull's foot mark. Gently thrust a finger to one side of the mark and lift, and this sand-burrowing ophiuroid will come to light. When the tide is in, the creature will emerge and forage for food on the sand surface.

4 **Amphiura.** This form likes slimy mud, and though it is generally taken offshore, it may venture between tide marks. Its tube-feet are highly active, and not only help the animal to burrow but help to catch food as well. When the animal is in its burrow, most of each arm is below the surface of the mud, but a centimetre or two projects above the surface, gently waving in the currents. The tube-feet and spines, both producing copious mucus, make an efficient filter to pluck tiny organisms from the water, then the tube-feet nearer the animal's disk transport the food to the mouth. Meanwhile, each arm gently undulates within its tubular burrow to maintain a respiratory current.

5 **Acrocnida.** This long-armed brittle-star is not colourful. Like *Amphiura*, it burrows, but prefers sand rather than mud. The arm tips are exposed above the sand surface, two or three at a time, the remaining complement of arms being kept as 'spares', curled up under the disk in the burrow. Then, as though changing watch, a protruding arm will be withdrawn and its place taken by one previously not being used. The commensal polynoid worm (*see* pp. 94 – 95), *Harmothoe*, when very young will choose the burrows of this brittle-star for its home; then, as it grows, it moves house to the burrow of the sea-cucumber, *Labidoplax* (4, p. 187).

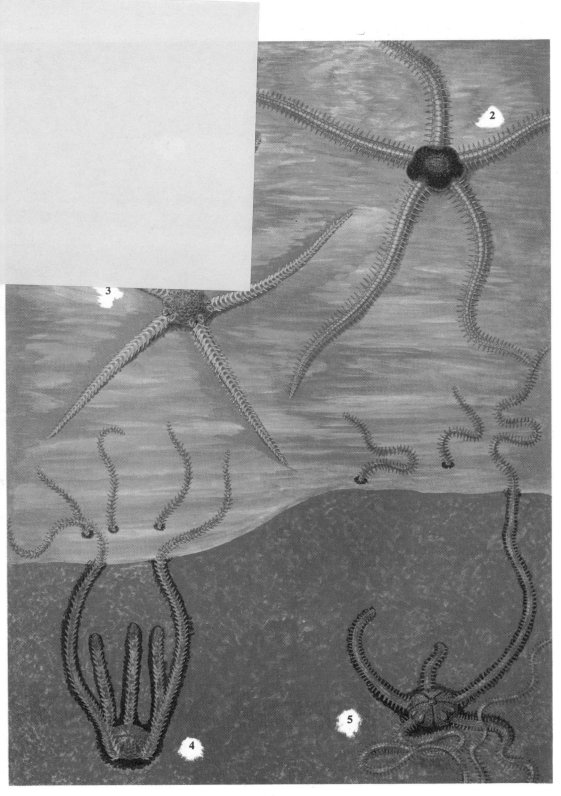

1 OPHIOTHRIX FRAGILIS   2 OPHIOCOMINA NIGRA

3 OPHIURA TEXTURATA

4 AMPHIURA FILIFORMIS   5 ACROCNIDA BRACHIATA

SEA-CUCUMBERS OF THE SHORE AND SHALLOW SEA

The class Holothuroidea has worm-like members with leathery integument and a slimy feel. Most of the skeletal elements are reduced to mere spicules, but there is a ring of larger ossicles round the first part of the gut, which serves as an attachment for body muscles. All have feeding tube-feet round the mouth, usually feathery structures, to pick up detritus from the substratum or pluck food from the surrounding water. Most have tube-feet on the rest of the body too, either suckered, like those of starfishes and sea-urchins or papillate for burrowing. Respiration in most is via the anus, through which water is drawn into special organs within the body called *respiratory trees*. In those which lack such trees the body wall is generally thin enough for gaseous exchange to take place across it.

British sea-cucumbers are not a culinary thrill, unlike some from elsewhere in the world, where dried holothuroids, called 'trepang' or 'bêche de mer', are said to be both tasty and nutritious.

1 Aslia (Cucumaria) is superficially like a small *Holothuria* (2), but is typically a shore-dweller, in holes and crevices. The tube-feet of the body all have tiny suckers at the end, to obtain all-round grip, whereas only those on the ventral side of *Holothuria* are suckered. The brown and white oral tube-feet collect food in suspension, and sweep the 'fore-court' of the animal's rocky home where it nestles so efficiently that it is very difficult to remove .This is all to the good, since it is far from common. The rather similar genus *Pawsonia* has dorsal tube-feet which are papillate rather than suckered.

2 Holothuria (Cotton Spinner). The name 'holothuroid' (Greek, 'forceful expulsion') refers to the habit possessed by many sea-cucumbers of eviscerating themselves with some violence from the terminal cloaca when molested. *Holothuria* has special organs, the *cuvierian organs*, which serve the same function as evisceration in distracting a predator from the main part of the body, but in a way which is less expensive to the animal. The organs are composed of arrow-headed threads endowed with their own musculature. When disturbed, the animal contracts and forces the ends of the threads out through the anus (hence its common name). In the water, the threads rapidly elongate by squeezing on the fluid in the lumen and become sticky. Sometimes found inshore, especially where rocky gullies give shelter, *Holothuria* takes in particles of the substratum together with detritus by means of the foliaceous tube-feet round the mouth.

3 Thyone is usually found on shelly bottoms and tends to cover itself with broken shell. Its skin is soft and thin, usually whitish or delicately pink in colour. The tube-feet emerge from all over the body, and not confined to columns. The oral tube-feet are rather finger-like.

4 Labidoplax is one of several burrowing holothuroids. It usually burrows tail-first into the sand or mud, showing only the fringe of oral tube-feet which it can withdraw into its burrow if danger threatens. This and *Leptosynapta* (6) are apodous holothuroids, that is, having tube-feet only round the mouth. *Labidoplax* is rather less translucent than *Leptosynapta*, with a red or brownish dorsal side and slightly lighter ventral. The gut does have a loop in it, and the animal takes in particles of the substratum and digests off encrusting organic matter as the particles pass through. The commensal polynoid worm *Harmothoe* shares the sea-cucumber's burrow at times.

5 Leptopentacta (Cucumaria) is a burrower, but it is also found curiously out of its element, in rocky crevices. It prefers thick, slimy mud, into which it burrows head first, producing a U-shaped burrow, from one limb of which its head emerges for feeding. Round the mouth are 10 feeding tube-feet, eight long and two Y-shaped short ones. The large tube-feet collect the food from the suspended matter just above the mud surface, then bend round towards the mouth, where the food is wiped off in the fork of one of the short tube-feet, like a child licking jam from its fingers. It backs up the burrow to defaecate with some force so that it does not pollute its burrow with its own waste.

6 Leptosynapta is another apodous holothuroid (*see* 4). Like *Labidoplax* it burrows in muddy sand, particularly among the roots of *Zostera* inshore. It is somewhat transparent, and through the skin the gut can be seen to pass directly from mouth to anus, without a loop. As mentioned in the introduction to this phylum (p. 178) and above, the holothuroid skeleton is reduced to mere spicules in the body wall. In this genus some of the spicules are anchor-shaped and lie just beneath the surface; when the body wall is taut, as it is when handled, the points of the anchors protrude through the skin and give the body wall a rough feel to the finger if stroked in one direction.

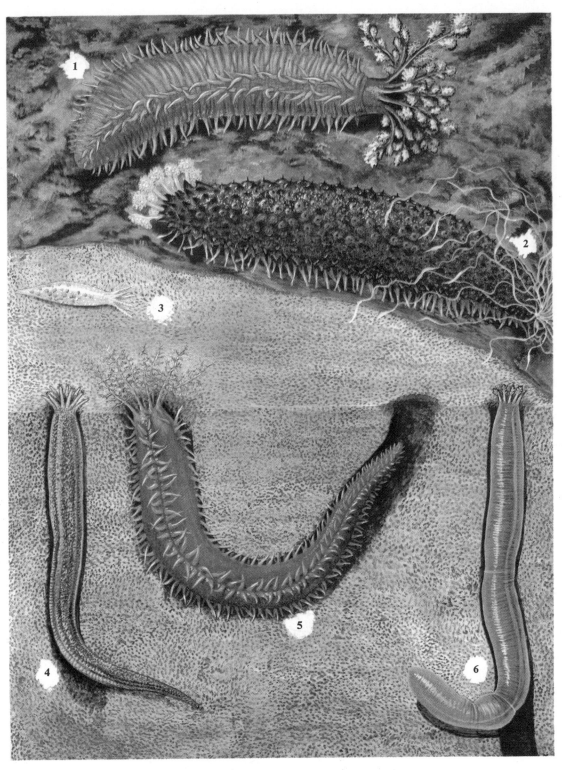

1 Aslia (Cucumaria) lefevrei   2 Holothuria forskali
3 Thyone fusus
5 Leptopentacta (Cucumaria) elongata
4 Labidoplax digitata   6 Leptosynapta inhaerens

SIMPLE SEA-SQUIRTS

These fascinating animals are sometimes called 'tunicates', because a tunic of cellulose-like substance envelopes the body. A generalized tunicate (6) is sessile, with a bag-shaped body having two openings: an inhalant one leading to a sieve-like branchial basket, and an exhalant one to one side. The basket acts both as a filtering mechanism for feeding, and in gaseous exchange. Their 'tadpole' larva, which is the dispersive phase in the life cycle, has within its tail an elastic rod called a notochord (*see* p. 176). Because of this, sea-squirts are thought to occupy a special evolutionary position, since the notochord and branchial basket are mainly chordate structural features. There are three main groups within the Urochordata: pelagic forms with a barrel-shaped body (1 and 3); pelagic forms which closely resemble the tadpole, though they reach maturity (2); and ones with a sessile adult and a larval tadpole (5 – 8).

1 **Salpa** has a barrel-shaped body with openings fore and aft. Water taken in through the front opening passes through a single pair of fine gill slits in the pharynx where minute organisms are filtered out, and is squeezed out through the rear by contractions of the body muscles, which can be seen like struts through the tunic. There is an alternation of generations during the life cycle. The adult is hermaphrodite, but the two sexes develop at different times, so eggs from one individual are fertilized by sperm from another. Two or three eggs at a time develop within the adult's body, which splits to liberate them. When free, they grow into barrel-shaped asexual individuals, each of which grows a trailing 'tail' on which buds form. These buds grow into miniature adults, which may remain united for some time (1). Eventually, they split off and become the sexual salps (1A).

2 **Oikopleura.** This remarkable, tiny tadpole-like creature has a well-developed tail with skeletal notochord, and a pair of gills. It is probably an example of an adult animal which has retained larval shape, a process called *paedomorphosis*. Even more remarkable is this animal's mode of feeding. It builds out of mucus an elaborate plankton-filtering device (2A), in which it sits and through which it creates a current by undulating its tail. The mucous 'house' has two inhalant openings each protected by a screen which allows only the smallest particles to enter; within the house is a pair of filters made of the finest mucous threads, against which minute organisms from the plankton are caught. There is an escape hatch in the wall from which *Oikopleura* can emerge should danger threaten; if it needs to abandon the filter-house, it can build another in a short time.

3 **Doliolum.** Like *Salpa* (1), this pelagic tunicate moves by jet-propulsion, and the water passing through the body is forced through fine gill slits in the pharynx wall, so that minute organisms are filtered out for food. It has a complicated life history. The fertilized egg develops into a tadpole larva and then a barrel-shaped individual. This undergoes budding in one region; the buds migrate to a tail-like projection from the body and there adhere, to be pulled along by the

so-called 'nurse' form as they develop into barrel-shaped individuals themselves. When they break free, the process, slightly modified, is repeated, then their daughter buds break off to form the adult sexual individuals.

4 **Perophora** is a colonial tunicate, notable for its reproduction by budding. The individuals thus formed may stay as colonies, connected by rootlets, or the individuals may break free, to attach elsewhere. Each individual is about 10 mm high.

5 **Ascidia.** This sea-squirt, about 50 mm long, is usually attached to rocks or shells on the lower shore along part of one of its sides. Its tunic is quite tough and usually greenish. The inhalant siphon is at one end, while the other is about two-thirds the way down one side of the body.

6 **Phallusia.** This large, solitary form is always attached to stones, usually in hard clay or mud. Its thick, smooth tunic has dome-like prominences all over it, which distinguishes it from other forms.

7 **Ciona** is larger than *Ascidia* (5) when fully grown — up to about 120 mm long — and is distinguished from it by the softness of the tunic and the contractility of the body. The upper part of the body can be drawn into the lower like the finger of a glove. It has prominent siphons with yellow margins and red spots on the side walls, and five longitudinal muscle bands visible in the body wall. It is found on hard objects, and can cause problems in marine aquaria, by settling within the plumbing pipes and filters, unless they are made from plastics.

8 **Molgula** (Sand Egg). Though a 'solitary' sea-squirt, it is often found in groups, each individual being a round, hard body with short, wide siphons. The tunic is covered with hair-like processes which tend to collect sand, except round the siphons. As they live in sand or mud, or in holes in rocks, generally only the siphons can be seen.

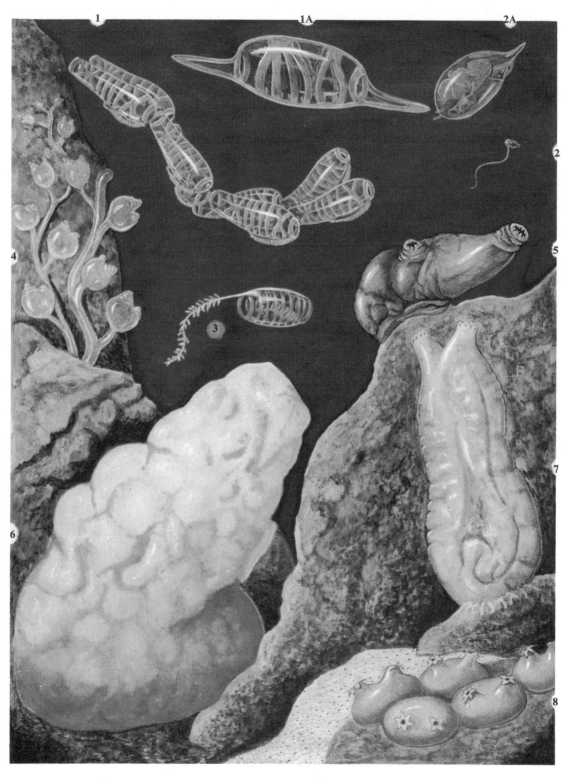

1 Salpa fusiformis   1a aggregate form   2a Oikopleura, house
4 Perophora listeri   2 Oikopleura sp.
3 Doliolum gegenbauri   5 Ascidia mentula
6 Phallusia mammillata   7 Ciona intestinalis
8 Molgula citrina

189

COLONIAL AND COMPOUND SEA-SQUIRTS

We have seen on the previous page that reproduction by budding is commonly seen in tunicates. Sometimes, this results in a whole group of so-called solitary forms aggregating in a mass (8, p. 189, and 5, p. 191); sometimes it results in physical connection by a branching stolon (4, p. 189, and 6, p. 191), or a fleshy mass (7). Also shown on this page are forms in which the process has proceeded even further. These are the Compound Sea-squirts, in which the relation between individuals is much more intimate than is seen in those which are merely connected by rootlets. Here, while the inhalant apertures are separate, the exhalant ones are usually united. The commonest pattern is for the individuals to be arranged in a star shape, with the common exhalant aperture at the centre; but various other arrangements occur.

1 **Botryllus.** This is a true 'compound tunicate', in which the individuals are grouped together round a common exhalant opening to form a star-shaped system of between three and twelve individuals, depending on age and physical conditions. In shallow water it is usually flat and encrusting, while in deeper water the colony is thicker and more fleshy, with individuals on all sides of it. Sometimes, it may form hanging, finger-like masses, as shown in the upper part of the colony illustrated here.

This is a widespread sea-squirt, common in shallow water and on the lower shore. It is normally on and under stones, but is also found on algae, pier-piles, ship's bottoms, and even on other tunicates. There is great variety to the colonial form, but the star-shaped systems are always conspicuous.

2 **Botrylloides.** Here, the individuals are not usually arranged in star systems but in two parallel rows with the inhalant apertures laterally, and the branchial cavity leading into a common exhalant duct in the mid-line. This duct opens to the exterior at intervals. The colonies encrust the surfaces of stones at low tide or in shallow water.

3 **Aplydium.** The form of this colony is very variable, but often it is found wrapped round the stems of algae, such as *Fucus*, and on *Zostera* stems, or it may be attached by a short stalk to a hard object, or it may encrust, as shown here. The colonies are embedded in a fleshy, often fairly bright coloured matrix which may be sand encrusted. Even the arrangement of the zooids may vary on one colony, as shown here.

4 **Sidnyum.** In this form, the individuals may have a long attachment stalk from the base of the body. It may form flattish colonies on the blades of sea plants or more erect structures on hard objects. Each of the little cup-shaped structures is itself a colony, and several may arise from a common holdfast. This is fairly common locally on all our shores except the East Coast.

5 **Dendrodoa (Styelopsis).** This occurs as either a solitary or a colonial form. When solitary, the body is short and its base splays out to form an attachment pad; when colonial, which is generally in sheltered places, individuals are taller and form compact masses. The tadpoles tend to settle near their parents, so the colony often has the appearance of being formed by budding, though this does not occur. It is normally found on hard stony ground, on the lower shore and in deeper water.

6 **Clavelina** is loosely colonial, though there is one species which is solitary. The individuals, about 40 mm high, are elongated and erect, with inhalant and exhalant openings at the apex. The individuals are held together by root-like stolons attached to the substratum.

7 **Distomus.** This is rather similar to the aggregated form of *Dendrodoa*, but in this case, unlike *Dendrodoa*, the colony is formed by budding, not by crowded settlement. It is often found on the upper side of rocks, or around the holdfast of *Laminaria*. This form is viviparous, that is, the eggs are retained within the branchial cavity of the adult until the tadpoles hatch, when they swim out of the exhalant opening to join the plankton. They may swim for several days, but usually within 24 hours they settle and change into the adult form, and become fully functional in about two weeks.

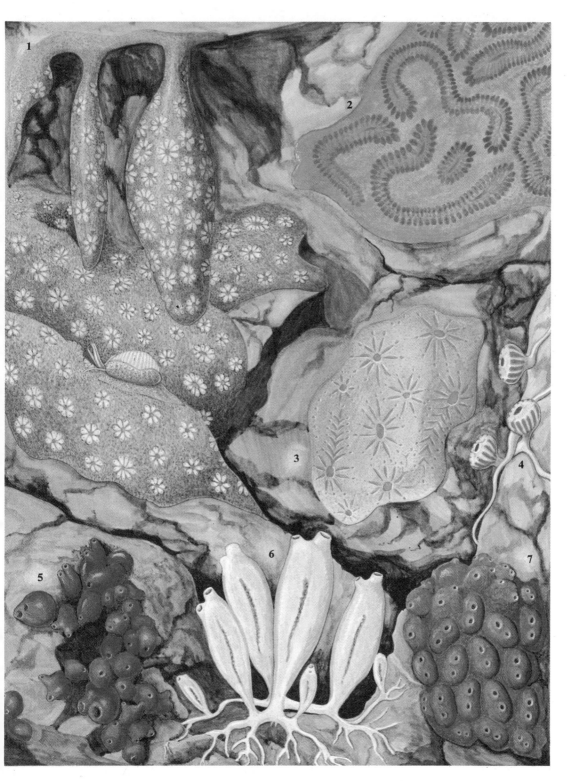

1 Botryllus schlosseri   2 Botrylloides leachi

3 Aplydium proliferum   4 Sidnyum turbinatum

5 Dendrodoa (Styelópsis) grossularia   6 Clavelina lepadiformis   7 Distomus variolosus

A CLASSIFICATION OF INVERTEBRATES

The ideal classification of the animal kingdom has not yet been compiled: expert opinion is still divided upon which is the best system, and probably always will be. Even the major categories into which animals should be placed are still not settled. In the pages that follow one possible classificatory scheme is given, with brief criteria for most of the categories, and we believe that the scheme adopted will prove acceptable to most zoologists.

Animals that have a fundamentally similar pattern are placed in the same PHYLUM. Within each phylum the categories, in descending order, are CLASS, ORDER, FAMILY and GENUS. The lowest level is that of SPECIES, and all animals belonging to a particular species are, broadly speaking, capable of interbreeding.

When the principles of biological classification were laid down in the eighteenth century, the universal language of science was Latin. Consequently the scientific names of all animals are in a latinised form which consists of the generic name first, followed by the specific name. Thus the water-flea, *Daphnia pulex*, is just one of many closely-similar species united in the genus *Daphnia*.

In the following synopsis of invertebrate animals all phyla and classes of living forms are listed, but many extinct groups are omitted. However, only a few representative orders and families are included, chiefly in the more important phyla. All genera to which reference is made in the text are listed in *italics*. Groups in square brackets are not illustrated or described in the text.

Phylum PROTOZOA

Organisms of minute size in which the body is not divided into cells.

Class **MASTIGOPHORA**
Protozoans with one or more flagella in the main part of their life cycle.

Subclass PHYTOMASTIGOPHORA Organelles containing photosynthetic pigments present and therefore more like plants than animals.

Ceratium	*Noctiluca*	*Volvox*

[*Subclass* ZOOMASTIGOPHORA] Organelles with photosynthetic pigment not present; free-living or parasitic.

Class **SARCODINA**
Amoeboid in the main phase of their life-cycle.

Subclass RHIZOPODA Without internal skeleton.

Order **Amoebida** Cell processes lobose; no sexual reproduction.

Amoeba

Order **Testacida** Cell processes lobose; single chambered 'house' or rigid external membrane.

Arcella	*Difflugia*

Order **Foraminiferida** Cell processes usually form a network; single- or many-chambered test.

Polystomella	*Globigerina*	*Nodosaria*

[*Order* **Mycetozoida**] The slime fungi, more the province of botanists, but included here because most have an amoeboid phase.

Subclass ACTINOPODA Body has inner and outer zones and often an internal skeleton; cell processes usually radial and stiff.

Order **Heliozoida** Mainly freshwater; siliceous skeleton usually present.

Actinosphaerium

Order **Radiolarida** Entirely marine; skeleton of silica or strontium sulphate always present.

Hexacontium

[*Class* **SPOROZOA**]
Parasitic protozoans, usually within the cells of another organism; complex life-cycle.

[*Class* **CNIDOSPORA**]
Amoeboid internal parasites producing spores.

Class **CILIATA**
        Possess cilia in at least part of the life-cycle.
    [*Subclass* HOLOTRICHA]   Usually with uniform ciliary covering.
    Subclass PERITRICHA   Cilia usually only round mouth; often attached by contractile stalk.
                                          Vorticella
    Subclass SPIROTRICHA   Cilia near mouth often fused or specialized in other ways for feeding.
                        Stentor   *Spirostomum*

[*Phylum* MESOZOA]

A group of minute enigmatic cellular endoparasites.

Phylum PORIFERA

Sponges. Multicellular animals lacking nervous integration; body with a single cavity or interconnected canals.

Class **CALCAREA**
        With calcareous spicules.
    Order **Homocoela**   Inner layer of flagellated cells continuous.
                                    Leucosolenia
    Order **Heterocoela**   Flagellated cells in chambers.
                      Sycon   *Grantia*

[*Class* **HEXACTINELLIDA**]
        Glass sponges. With only six-rayed spicules of silica. Mainly deep-water; no British representatives.

Class **DEMOSPONGIAE**
        Skeleton, if present, of horny spongin or silica or both; silica spicules simple or four-rayed.
    [*Subclass* TETRACTINELLIDA]   Skeleton, if present, only four-rayed spicules of silica.
    Subclass MONAXONIDA   Skeleton of two-rayed spicules.
              Suberites   *Cliona*   *Hymeniacidon*
              Halichondria   *Ephydatia*   *Euspongilla*
    [*Subclass* KERATOSA]   No spicules; skeleton of spongin only.

Phylum COELENTERATA

Polyps, medusae and sea-anemones. Multicellular animals mainly with radial symmetry; simple gut cavity with only one opening; body wall of two main layers containing nerve network; often possessing stinging cells.

Class **HYDROZOA**
        With polyp and (usually) medusoid phases; gut cavity has no internal septa; mostly marine.
    Order **Trachylina**   Medusae usually well developed; polyp small or absent.
                         Craspedacusta (and *Microhydra*)
    Order **Hydroida**   Polyp phase prominent; medusa may be absent.
      Suborder **Limnomedusae**   Polyp with no theca.
             Hydra   *Chlorohydra*   *Cordylophora*
      Suborder **Anthomedusae**   Theca surrounds only stem and holdfast; polyp itself naked; medusoid phase bell-shaped.
             Coryne   *Clava*   *Dynamena*
             Sertularia   *Plumularia*   *Tubularia*
                         Bougainvillea   *Myriothela*
      Suborder **Leptomedusae**   Polyp surrounded by cup-like theca; medusoid phase saucer-shaped.
                                          Obelia
      Suborder **Chondrophora**   Pelagic colonies of modified polyps.
                                          Velella

Order **Siphonophora**   Complex floating colonies of polyps and medusae, both highly modified.

Physalia *Muggiaea* *Physophora*

Class **SCYPHOZOA**

True jellyfish. Medusa the main phase; entirely marine.

Order **Stauromedusae**   Sessile jellyfish.

Haliclystus *Lucernariopsis*

Order **Semaeostomae**   Scalloped margin to bell; sometimes with tentacles.

Aurelia *Chrysaora* *Pelagia*
Cyanea

Order **Rhizostomae**   Bell margin lacks tentacles; 'mouth' openings subdivided.

Rhizostoma

Class **ANTHOZOA**

Sea anemones and corals. Exclusively polypoid; medusae absent. Gut cavity contains radiating septa; entirely marine.

Subclass ALCYONARIA   Mostly colonial; polyp has eight tentacles.

Alcyonium *Eunicella*

Subclass ZOANTHARIA   Solitary or colonial; polyp has more than eight tentacles.

Order **Actiniaria**   Sea anemones. Solitary; without skeleton.

Actinia	*Anemonia*	*Metridium*
Tealia	*Bunodactis*	*Sagartia*
Aiptasia	*Milne-Edwardsia*	*Cereus*
Peachia	*Halcampa*	*Calliactis*
		Adamsia

Order **Madreporaria**   Corals. With calcareous skeleton.

Caryophyllia

Order **Zoanthidea**   Anemone-like but with simpler mouth.

Epizoanthus

Order **Ceriantharia**   Anemone-like but with many gut septa.

Cerianthus

Order **Corallimorpha**   Coral-like but without skeleton.

Corynactis

[Order **Antipatharia**]   Black corals. Branching colonies.

Phylum CTENOPHORA

Sea gooseberries. Body plan somewhat similar to coelenterates (and sometimes included in the same phylum) but mostly with extrusible tentacles and groups of locomotory cilia; lacking stinging cells.

Class **TENTACULATA**

Possessing tentacles.

Cestus *Pleurobrachia* *Coeloplana*

[Class **NUDA**]

Without tentacles.

Phylum PLATYHELMINTHES

The flatworms. Ribbon-like, simple worms without coelom or circulatory system and with only one opening to the gut.

Class **TURBELLARIA**

Free-living, with simple life-cycle.

Order **Acoela**   No intestine or excretory organs; entirely marine.

Convoluta

Order **Rhabdocoela**   Small, with rod-like or sac-like gut; marine, freshwater and terrestrial forms.

Mesostoma *Stenostomum*

Order **Tricladida**   Fairly large forms, with intestine having three branches; marine, freshwater and terrestrial forms.

Dendrocoelum	*Polycelis*	*Dugesia*
Procerodes	*Bipalium*	*Rhynchodemus*

Order **Polycladida**   Fairly large forms, with intestine having many branches; entirely marine.

Cycloporus	*Leptoplana*

[*Order* **Alloeocoela**]   A rather artificial group of specialized marine and freshwater forms.

Class **MONOGENEA**

Flukes parasitic on a single host only.

Subclass MONOPISTHOCOTYLEA   Attachment organs simple; few testes.

Gyrodactylus

Subclass POLYOPISTHOCOTYLEA   Attachment organs complex; many testes.

Polystoma

Subclass ASPIDOBOTHREA   With large, compartmentalized ventral sucker.

Aspidogaster

Class **CESTODARIA**

Fluke-like parasites, but lacking a gut.

Amphilina	*Gyrocotyle*

Class **CESTODA**

The tapeworms. Body subdivided into compartments for reproductive efficiency; without gut.

Order **Pseudophyllidea**   Parasites of teleost fish and fish-eating vertebrates. Yolk glands scattered.

Diphyllobothrium

Order **Cyclophyllidea**   Usually parasites of birds and mammals. Yolk glands compact.

Taenia	*Echinococcus*	*Dipylidium*

Class **DIGENEA**

Flukes which parasitize more than one host during the life-cycle.

Derogenes	*Fasciola*	*Schistosoma*
		Pneumonoeces

Phylum NEMERTINA

The ribbon worms. Similar to platyhelminths, but with protrusible proboscis above alimentary tract; with anus; primitive circulatory system.

Class **ANOPLA**

With mouth behind the brain.

Tubulanus	*Cephalothrix*	*Lineus*

Class **ENOPLA**

With mouth in front of the brain.

Nemertopsis	*Amphiporus*	*Eunemertes*
(*Tetrastemma*)		(*Emplectonema*)

Phylum PRIAPULIDA

The proboscis worms. Mouth on protrusible introvert.

Priapulus

Phylum ACANTHOCEPHALA

The spiny-headed worms. Endoparasites without mouth or gut.

Echinorhynchus

[*Phylum* KINORHYNCHA]

Minute interstitial marine animals; body covered in spines.

Phylum GASTROTRICHA

Minute freshwater animals; body with spines, scales or bristles.

Chaetonotus

Phylum ROTIFERA

The wheel animalcules. Minute aquatic forms; well developed internal jaws; characteristic crown of cilia anteriorly.

Branchionus

Phylum NEMATODA

The roundworms. Ubiquitous and frequently parasitic; with tough outer cuticle.

Ascaris   *Rhabditis*

Phylum NEMATOMORPHA

The threadworms. Extremely long and thin; juveniles parasitic in arthropods, but adults free-living.

Gordius

[Phylum ENTOPROCTA or KAMPTOZOA]

Aquatic animals on stalks; with circlet of tentacles surrounding both mouth and anus.

Phylum MOLLUSCA

Body divided into a combined 'head/foot' and a visceral hump, between which is a mantle cavity; hump often secretes a calcareous shell.

Class **POLYPLACOPHORA**

The chitons. Body covered by a series of calcareous plates; foot ringed by gills.

Lepidochitona   *Acanthochitona*   *Ischnochiton*
Tonicella

[*Class* **APLACOPHORA**]

Worm-like; no covering of plates; foot reduced or absent.
Note: Sometimes the Polyplacophora and Aplacophora are united in a class AMPHINEURA, when they themselves become subclasses.

[*Class* **MONOPLACOPHORA**]

Superficially segmented but with single shell; rare deep-water forms.

Class **GASTROPODA**

Snails and slugs. Shell single or absent; possess tentacles.

Subclass PROSOBRANCHIA Mantle cavity opening at the front of the body; aquatic.

Order **Archaeogastropoda** Shell usually cone-shaped; mostly with paired gills, each gill having two rows of respiratory filaments.

Superfamily **Zeugobranchia** Paired gills, depressed spiral shell.

Emarginula   *Haliotis*   *Diodora*

Superfamily **Patellacea** Shell not spiral.

Patella   *Acmaea*   *Patina*

Superfamily **Trochacea** Right hand gill lost.

Calliostoma   *Cantharidus*   *Gibbula*
Monodonta   *Tricolia*

Superfamily **Neritacea** Freshwater and terrestrial; the only archaeogastropods with special genital duct.

Theodoxus   *Neritina*

Order **Mesogastropoda** Usually with single gill having one row of filaments.

Superfamily **Archaeotaenioglossa** (=**Cyclophoracea**) River snails.

Viviparus

Superfamily **Valvatacea** Valve snails.

Valvata

Superfamily **Littorinacea** Periwinkles and round-mouthed snails.

Littorina	*Lacuna*	*Pomatias*
	Acme	*Acicula*

Superfamily **Rissoacea** Spire shells.

Hydrobia	*Potamopyrgus*	*Bithynia*
	Amnicola	*Bythinella*

Superfamily **Cerithiacea** Turret shells.

Turritella	*Bittium*

Superfamily **Ptenoglossa** Wendletraps and violet snails.

Clathrus

Superfamily **Calyptraeacea** Slipper limpets.

Calyptraea	*Crepidula*

Superfamily **Cypraeacea** Cowries.

Trivia

Superfamily **Naticacea** Necklace shells.

Natica

Superfamily **Strombacea** Pelican's foot shells.

Aporrhais

Order **Neogastropoda** (=**Stenoglossa**) The most advanced prosobranchs; carnivorous, with protractible proboscis emerging through spout-like canal in shell.

Superfamily **Buccinacea** Whelks.

Buccinum	*Neptunea*	*Nassarius*

Superfamily **Muricacea** Dog whelks.

Nucella (Thais)	*Ocenebra*	*Urosalpinx*

Subclass **Opisthobranchia** Mantle cavity opens at side or rear of body; mostly lacking shell.

Order **Bullomorpha** (=**Cephalaspidea**) Shell still present.

Actaeon	*Philine*

Order **Anaspidea** (=**Aplysiomorpha**) Shell retained but reduced and internal; small mantle cavity at right side.

Aplysia	*Haminoea*	*Akera*

[*Order* **Thecosomata**] Planktonic 'sea-butterflies' with shell.

Order **Gymnosomata** Planktonic 'sea-butterflies' without shell or mantle cavity.

Clione

Order **Notaspidea** (=**Pleurobranchomorpha**) Shell retained but reduced and internal; no mantle cavity.

Pleurobranchus

Order **Nudibranchia** Sea slugs. Shell and mantle cavity entirely lost: gills exposed.

Suborder **Doridacea** Circlet of gills at rear of body.

Onchidoris	*Rostanga*	*Archidoris*
Polycera	*Acanthodoris*	*Goniodoris*
Ancula	*Adalaria*	

Suborder **Eolidacea** Body covered with finger-like cerata.

Aeolidia	*Tergipes*	*Facelina*

Suborder **Dendronotacea** Anus opens on right side; dorsal surface usually has foliaceous processes.

Doto	*Dendronotus*

Order **Onchidiacea** Slug-like body with warty skin.

Onchidella (Onchydium)

Order **Sacoglossa** Possess sap-sucking mouthparts.

Alderia	*Limapontia*	*Acteonia*

Subclass **Pulmonata** Snails and slugs. Lining of mantle cavity forms a lung, usually for air-breathing.

Order **Basommatophora** One pair of non-retractile tentacles; eyes at base of tentacles.

Family **Ellobiidae** Hollow-shelled snails.

Carychium

Family **Physidae** Bladder snails.

Aplexa	*Physa*

Family **Limnaeidae** Pond snails.

Limnaea	*Galba*	*Radix*
		Leptolimnaea

Family **Planorbidae** Ramshorn snails.

Planorbis	*Segmentina*	*Anisus*
Planorbarius	*Gyraulus*	*Armiger*

Family **Ancylidae** River limpets.

Acroloxus	*Ancylus*	*Ancylastrum*

Order **Stylommatophora** Two pairs of retractile tentacles; eyes at tip of hinder pair of tentacles. Mainly terrestrial.

Superfamily **Succineacea** Amber Snails.

Succinea	*Pyramidula*

Superfamily **Vertiginacea** Toothed snails and hairy snails.

Cochlicopa	*Vertigo*	*Ena*
Vallonia	*Acanthinula*	*Clausilia*
Balea	*Marpessa*	*Azeca*
Abida	*Lauria*	*Columella*
		Pupilla

Superfamily **Achatinacea** Spire snails.

Caecilioides

Superfamily **Oleacinacea** Shelled slugs.

Testacella

Superfamily **Endodontacea** Discus snails and one group of slugs.

Arion	*Punctum*	*Goniodiscus*

Superfamily **Zonitacea** Glass snails and several groups of slugs.

Zonitoides	*Limax*	*Milax*
Agriolimax	*Vitrina*	*Retinella*
Oxychilus	*Vitrea*	*Euconulus*

Superfamily **Helicacea** Typical snails.

Helix	*Helicella*	*Cochlicella*
Cepaea	*Candidula*	*Hygromia*
Monacha	*Theba*	*Ashfordia*
Euparypha	*Arianta*	*Trichia*

Class SCAPHOPODA

The tusk shells. Burrowing forms with long, narrow shell open at both ends.

Dentalium

Class BIVALVIA (=PELECYPODA)

The bivalves. Two shells hinged on one side; complex folded gills used for both feeding and respiration

Subclass **Protobranchia** Simple, leaf-like gills; feed by means of tentacles from labial palps.

Superfamily **Nuculacea** The nut shells.

Nucula

Subclass **Lamellibranchia** Complex gills with fused filaments.

Order **Taxodonta** Row of uniform teeth near hinge.

Superfamily **Arcacea** Ark shells.

Arca	*Glycymeris*

Order **Anisomyaria** Shell usually lies on its right side; valves unequal.

Superfamily **Mytilacea** Mussels.

Mytilus	*Musculus*	*Modiolus*
		Crenella

Superfamily **Pteriacea** Wing oysters and fan mussels.

Pteria	*Pinna*

Superfamily **Pectinacea** Scallops and file shells.
                Pecten   *Chlamys*   *Lima*

Superfamily **Anomiacea** Saddle oysters.
                                                   Anomia

Superfamily **Ostreacea** Oysters.
                                                   Ostrea

Order **Schizodonta** Divided teeth; mainly freshwater.
  Superfamily **Unionacea** River mussels.
                Unio   *Anodonta*   *Margaritifer*

Order **Heterodonta** Teeth not uniform; usually with well-developed siphons.
  Superfamily **Astartacea** Astartes.
                                                   Astarte

  Superfamily **Sphaeriacea** Pea mussels.
                             Sphaerium   *Pisidium*

  Superfamily **Dreissenacea** Zebra mussels.
                                                   Dreissena

  Superfamily **Erycinacea** Coin shells.
                Lepton   *Lasaea*   *Montacuta*
                                                   Devonia

  Superfamily **Cardiacea** Cockles.
                Cardium   *Parvicardium*   *Acanthocardia*
                                                   Cerastoderma

  Superfamily **Veneracea** Venus clams and carpet shells.
                Venus   *Venerupis*   *Clausinella*
                Timoclea   *Mercenaria*   *Chamelea*
                                 Dosinia   *Petricola*

  Superfamily **Mactracea** Trough shells.
                Mactra   *Lutraria*   *Spisula*

  Superfamily **Tellinacea** Tellins and wedge shells.
                Donax   *Tellina*   *Scrobicularia*
                Abra   *Macoma*   *Syndosmya*
                                 Gari   *Pharus*

Order **Adapedonta** Teeth reduced; mantle extensively fused; shells gape at each end.
  Superfamily **Solenacea** Razor shells.
                                       Solen   *Ensis*

  Superfamily **Myacea** Gapers and basket shells.
                                   Mya   *Corbula*

  Superfamily **Saxicavacea (=Hiatellacea)** Rock borers.
                                                   Hiatella

  Superfamily **Adesmacea (=Pholadacea)** Piddocks and 'ship worms'.
                Pholas   *Teredo*   *Xylophaga*
                Zirphaea   *Barnea*

Order **Anomalodesmacea** Outer part of gill reduced; specialized forms.
  Superfamily **Pandoracea** Lantern shells and Pandora's boxes.
                                                   Pandora

Order **Septibranchia** Gill a muscular septum acting as a pump; carnivorous.
  Superfamily **Poromyacea** Cusp shells.
                                                   Cuspidaria

Class **CEPHALOPODA**

The squids and octopuses. Head well-developed, with large eyes and bearing tentacles derived from the modified foot; shell either internal, external or absent.

[*Subclass* **NAUTILOIDEA**] Mainly extinct; represented in today's seas by the Pearly Nautilus.

[*Subclass*   AMMONOIDEA] Entirely extinct.

  Subclass COLEOIDEA Shell internal or absent.

   Order **Decapoda** Ten tentacles.

     Superfamily **Sepiacea** Cuttlefish.

Sepia	*Sepiola*

     Superfamily **Loliginacea** Squids.

	Loligo

   Order **Octopoda** Eight tentacles.

     Superfamily **Octopodacea** Octopuses.

Eledone	*Octopus*

Phylum ANNELIDA

The true worms. Body segmented and with separate coelom.

Class **POLYCHAETA**

  Bristle worms. Many chaetae borne on parapodia; mainly marine.

   Order **Errantia** Actively moving.

     Family **Aphroditidae** Body has dorsal scales.

Aphrodita	*Lepidonotus*	*Halosydna* (*Alentia*),
Eunoe	*Lagisca*	*Gattyana*
Harmothoe		*Malmgrenia*

     Family **Phyllodocidae** Simple proboscis; leaf-like cirri borne on parapodia.

Phyllodoce	*Eulalia*

     Family **Tomopteridae** Pelagic; parapodia modified as paddles.

Tomopteris

     Family **Syllidae** Small; three tentacles on head.

Syllis	*Autolytus*	*Myrianida*

     Family **Eunicidae** Rockworms and palolos. Rasp-like proboscis, which emerges ventrally.

Eunice	*Marphysa*	*Hyalonoecia*

     Family **Nephtyidae** Body rectangular in cross-section; proboscis papillate with two jaws.

Nephtys

     Family **Glyceridae** Body circular in cross-section; proboscis has four jaws.

Glycera

     Family **Nereidae** Ragworms. Proboscis has two toothed jaws and small teeth.

Nereis	*Neanthes*	*Eunereis*
	Perinereis	*Platynereis*

   Order **Sedentaria** Inactive; mostly burrowers and tube-dwellers.

     Family **Capitellidae** Slender worms lacking appendages.

Capitella	*Notomastus*

     Family **Arenicolidae** Lugworms. Each segment has several annuli externally; gills in middle region of body.

Arenicola

     Family **Magelonidae** Prostomium spatulate; head bears one pair of long feeding tentacles.

Magelona

     Family **Cirratulidae** Body usually tapers at both ends; often with respiratory cirri emerging all along body.

Cirratulus

     Family **Terebellidae** Prostomium bears large number of feeding tentacles; several gills behind tentacles.

Terebella	*Amphitrite*	*Lanice*
		Polycirrus

     Family **Spionidae** Single pair of feeding palps.

Nerine	*Polydora*

Family **Sabellidae** Peacock worms. Prostomium expanded to form feathery crown; build tube of sand, mud or shell fragments.

Sabella	*Megalomma*	*Myxicola*
		Potamilla

Family **Oweniidae** Prostomium bears ciliated crown for feeding; build tube of sand.

Owenia

Family **Pectinariidae** (=**Amphictenidae**) Build conical houses of sand; anterior chaetae form prominent rakes for burrowing.

Pectinaria

Family **Sabellariidae** Usually build sandy tubes, often reef-forming; front end of body forms an operculum and mass of feeding tentacles.

Sabellaria

Family **Chaetopteridae** Mostly build mucous tubes in sand; segments differentiated for various functions.

Chaetopterus

Family **Flabelligeridae** (=**Chloraemidae**) Head extended into tube for protection of gills.

Flabelligera

Family **Serpulidae** Mostly build calcareous tubes; prostomium expanded into feathery crown.

Serpula	*Spirorbis*	*Pomatoceros*
	Filograna	*Hydroides*

Class OLIGOCHAETA

The earthworms. Few chaetae, and not on parapodia; clitellum seasonally; mainly terrestrial and freshwater.

Order **Plesiopora Plesiothecata** Male duct opens in segment behind that containing the testes; spermathecae adjacent to testes.

Family **Aeolosomatidae** Sexual reproduction rare; clitellum restricted to ventral side; chaetae hairlike.

Aelosoma

Family **Naididae** Chaetae have forked ends; spermathecae in fifth segment.

Nais	*Stylaria*	*Chaetogaster*
		Dero

Family **Tubificidae** Spermathecae in tenth segment.

Tubifex

Order **Plesiopora Prosothecata** Spermathecae well anterior to testes.

Family **Enchytraeidae** The pot worms.

Enchytraeus

Order **Prosopora** Male ducts open in same segment as that containing testes.

Family **Lumbriculidae** Two pairs of spermathecae.

Lumbriculus

Order **Opisthopora** Male ducts open in segments well posterior to those containing testes.

Family **Lumbricidae** The earthworms.

Lumbricus	*Dendrobaena*	*Bimastus*
Eisenia	*Eiseniella*	*Octolasium*
		Allolobophora

Class HIRUDINEA

The leeches. Ectoparasites, with suckers; almost all lack chaetae; clitellum present seasonally.

[*Order* **Acanthobdellae**] Coelom compartmentalized; four pairs of chaetae.

Order **Rhynchobdellae** With eversible proboscis.

Family **Glossiphoniidae** Flattened; anterior sucker poorly developed.

Hemiclepsis	*Glossiphonia*	*Theromyzon*
		Helobdella

Family **Piscicolidae** (=**Ichthyobdellidae**) Fish leeches. Body usually cylindrical; well-marked suckers.

Piscicola	*Branchellion*	*Calliobdella*
	Pontobdella	*Janusion*

Order **Gnathobdellae** With pharyngeal teeth.

Family **Hirudidae** Medicinal and horse leeches.

Haemopis	*Hirudo*

Order **Pharyngobdellae** With sucking pharynx.

Family **Erpobdellidae**

Erpobdella

Class **ARCHIANELLIDA**

Small, simplified marine worms, usually without parapodia and chaetae.

Dinophilus	*Protodrilus*	*Polygordius*
	Nerilla	*Saccocirrus*

Class **MYZOSTOMARIA**

Highly modified, flattened worms with false parapodia but lacking visible segmentation; ectoparasites on sea-lilies and brittle-stars.

Myzostoma

Class **ECHIUROIDEA**

Unsegmented; body divided into bladder-like trunk and non-eversible proboscis; male often parasitic on female.

Thalassema	*Bonellia*

Phylum TARDIGRADA

Water-bears. Minute, stubby-legged animals with thin cuticle; aquatic, particularly associated with small temporary water bodies.

Macrobiotus

[*Phylum* ONYCHOPHORA]

Elongate terrestrial animals with soft cuticle and stumpy legs; sharing annelid and arthropod characters; no British representatives.

Peripatus

Phylum ARTHROPODA

Segmented animals with jointed limbs and chitinous exoskeleton.

Class **CRUSTACEA**

Mainly aquatic, with two pairs of antennae; limbs often branched.

Subclass **BRANCHIOPODA** Free-living; predominantly freshwater; with flattened limbs used for swimming, respiration and filter-feeding.

Order **Anostraca** Primitive forms lacking carapace; eyes on stalks.

Cheirocephalus

Order **Notostraca** Body largely covered by carapace; eyes not stalked.

Triops

Order **Cladocera** Mostly small freshwater forms with branched antennae bearing long swimming hairs; includes water-fleas.

Sida	*Holopedium*	*Simocephalus*
Daphnia	*Polyphemus*	*Bythotrephes*
		Leptodora

Subclass **OSTRACODA** Small; free-living; body enclosed within bivalved carapace.

Cypris	*Pionocypris*

Subclass **COPEPODA** Includes both free-living and parasitic forms; lacking compound eyes and carapace.

Calanus	*Cyclops*	*Canthocamptus*
Diaptomus	*Lernaeocera*	*Salmincola*
Ergasilus	*Thersitina*	*Lepeophtheirus*

[*Subclass* MYSTACOCARIDA] Small littoral interstitial animals allied to copepods.

Subclass CIRRIPEDIA Barnacles. Sessile, free-living or parasitic.

Balanus	*Verruca*	*Xenobalanus*
Lepas	*Scalpellum*	*Sacculina*

Subclass BRANCHIURA Temporary fish parasites bearing suckers.

Argulus

Subclass MALACOSTRACA A diverse group containing most of the larger crustacea; possessing compound eyes, usually on stalks.

Order **Mysidacea** Opossum shrimps.

Mysis   *Praunus*

Order **Isopoda** Mainly dorso-ventrally flattened; includes woodlice.

Idotea	*Bopyrus*	*Asellus*
Oniscus	*Porcellio*	*Armadillidium*
		Trichoniscus

Order **Amphipoda** Usually laterally flattened; includes sand-hoppers.

Gammarus   *Orchestia*   *Caprella*

Order **Bathynella** Small, freshwater interstitial forms; lacking carapace and eyes.

Bathynella

Order **Decapoda** The major group of malacostracans; includes shrimps, crabs and lobsters.

Astacus	*Crangon*	*Cancer*
Palaemon	*Carcinus*	*Pinnotheres*
Macropodia	*Porcellana*	*Macropipus*
Palinurus	*Homarus*	*Galathea*
		Pagurus

Subphylum MANDIBULATA

Arthropoda with true jaws and a single pair of antennae.

Class **INSECTA**

See *Oxford Book of Insects*.

'Myriapoda'

An assemblage of four classes of terrestrial mandibulates; possessing distinct head and with more than eight pairs of legs.

Class **PAUROPODA**

Small cryptozoic myriapods with forked antennae.

Pauropus

Class **DIPLOPODA**

Millipedes. Body segments cylindrical, mostly bearing two pairs of legs; cuticle calcareous.

Glomeris	*Polyxenus*	*Polydesmus*
Polyzonium	*Polymicrodon*	*Blaniulus*
		Cylindroiulus

Class **CHILOPODA**

Centipedes. Flattened; carnivorous; with poison claws on first body segment.

Lithobius	*Cryptops*	*Scutigera*
Geophilus	*Haplophilus*	*Hydroschendyla*
		Necrophlaeophagus

Class **SYMPHYLA**

Small cryptozoic animals resembling centipedes; possessing silk glands but lacking eyes.

Scutigerella

Subphylum CHELICERATA

Arthropods lacking mandibles and antennae; first pair of appendages pincer-like chelicerae; body of two main divisions.

Horseshoe crabs. Primitive marine chelicerates possessing heavily armoured exoskeleton. No British representatives.

Class **ARACHNIDA**

Terrestrial chelicerates possessing four pairs of walking legs; eyes simple.

Order **Pseudoscorpiones** False-scorpions. Small; secretive; with massive pincers but without 'tail'.

Neobisium	*Dactylochelifer*	*Lamprochernes*
Allochernes	*Cheiridium*	*Chelifer*

Order **Araneae** Spiders. With silk glands opening through spinnerets on abdomen.

Atypus	*Eresus*	*Amaurobius*
Hyptiotes	*Pholcus*	*Scytodes*
Segestria	*Oonops*	*Dysdera*
Harpactea	*Drassodes*	*Zelotes*
Clubiona	*Cheiracanthium*	*Anyphaena*
Marpissa	*Evarcha*	*Hyctia*
Synageles	*Thomisus*	*Misumena*
Xysticus	*Philodromus*	*Tibellus*
Micrommata	*Lycosa*	*Arctosa*
Argyroneta	*Tegenaria*	*Theridion*
Tetragnatha	*Meta*	*Araneus*
Zygiella	*Singa*	*Walckenaera*
Ero	*Drapetisca*	*Linyphia*

Order **Opiliones** Harvestmen. Body usually small and globular with long legs; paired eyes on tubercle.

Trogulus	*Megabunus*	*Phalangium*
		Leiobunum

Order **Acari** Ticks and mites. Generally small with globular body; includes free-living, parasitic, terrestrial and aquatic forms.

Steganacarus	*Rhodacarellus*	*Uropoda*
Halacarus	*Bdella*	*Halolaelaps*
Parasitus	*Demodex*	*Trombicula*
Halarachne	*Sarcoptes*	*Psoroptes*
Dermanyssus	*Ixodes*	*Argas*
Megnina	*Eutrombidium*	*Acarapis*
Dermacentor	*Spinturnix*	*Haemaphysalis*
Bryobia	*Panonychus*	*Steneotarsonemus*
Cecidophyopsis	*Hydrachna*	*Unionicola*
Arrhenurus	*Hygrobates*	*Mideopsis*
Limnesia	*Limnochares*	*Neumania*

Subphylum **PYCNOGONIDA**

Sea spiders. Marine arthropods with greatly reduced body; four, five or six pairs of legs, often extremely long; mouth on end of proboscis.

Nymphon	*Pycnogonum*

Subphylum **PENTASTOMIDA**

Worm-like animals of uncertain affinities but possessing certain arthropod features; parasitic in the respiratory tract of vertebrates.

Linguatula

Phylum **SIPUNCULOIDA**

Unsegmented worm-like animals with lophophore of tentacles each having a coelomic lumen and extended by water pressure.

Golfingia	*Phascolion*	*Aspidosiphon*

Phylum PHORONIDA

Slender worm-like animals similar to sipunculoids in organisation, but living permanently in secreted tubes.

Phoronis

Phylum BRYOZOA

The polyzoans or ectoprocts. Sessile, colonial animals living permanently in tubes or capsules; with lophophore that does not enclose the anus.

Class GYMNOLAEMATA

Lophophore a circle of tentacles; almost entirely marine.

Order Ctenostomata Capsules not calcareous.

Flustrella *Alcyonidium*

Order Cheilostomata Capsules calcareous; usually with an operculum.

Bugula *Membranipora* *Flustra*
Umbonula *Mucronella*

Class STENOLAEMATA

Lophophore circular; extrusion of polyp does not involve deformation of capsule. Marine.

Order Cyclostomata Capsules calcareous; opening round.

Crisia

Class PHYLACTOLAEMATA

Lophophore horseshoe-shaped; entirely freshwater.

Plumatella *Cristatella* *Lophopus*

Phylum BRACHIOPODA

The lamp shells. Superficially similar to bivalve molluscs, but stalked and possessing an internal lophophore.

Class INARTICULATA

No proper hinge; horny shell; gut ends in anus.

Crania

Class ARTICULATA

Hinged, calcareous shell; no anus.

`Terebratulina*

Phylum CHAETOGNATHA

The arrow worms. Small, pelagic marine predatory animals with powerful jaws bearing spines.

Sagitta *Spadella*

[Phylum POGONOPHORA]

Solitary tube-dwelling marine animals of deeper water; lacking gut; with lophophore of one or many tentacles.

Phylum HEMICHORDATA

Once thought to be primitive chordates, but now regarded as related to other phyla bearing a lophophore.

Class ENTEROPNEUSTA

The acorn worms. Solitary marine animals with body divided into three regions; many gill slits; lacking a lophophore.

Glossobalanus *Saccoglossus* *Protoglossus*

[Class PTEROBRANCHIA]

Small aggregated or colonial animals with double lophophore; usually from deeper water.

Phylum ECHINODERMATA

Wholly marine with mainly five-rayed symmetry, calcareous skeleton and protrusible hydraulic tube-feet.

Subphylum CRINOZOA

Sessile, with mouth directed upwards.

Class CRINOIDEA

Sea lilies and feather stars. Arms and pinnules form a feeding funnel.

Antedon

Subphylum ASTEROZOA

Sea stars. Mouth directed downwards.

Class ASTEROIDEA

Starfishes. Arms not clearly demarcated from central disk.

Order **Platyasterida** Ventral arm plates expanded laterally.

Luidia

Order **Phanerozonida** Conspicuous marginal plates.

Astropecten

Order **Spinulosida** Spines usually present on dorsal surface.

| *Henricia* | *Porania* | *Solaster* |
| | *Anseropoda* | *Asterina* |

Order **Forcipulata** With specially modified forceps-like spines.

Asterias *Marthasterias*

Class OPHIUROIDEA

Brittle-stars. Arms sharply demarcated from central disk.

Family **Ophiotrichidae**

Ophiothrix

Family **Ophiocomidae**

Ophiocomina *Ophiopsila*

Family **Ophiactidae**

Ophiactis

Family **Amphiuridae**

Amphiura *Acrocnida* *Amphipholis*

Family **Ophiolepidae**

Ophiura

Subphylum ECHINOZOA

Spherical or cylindrical; without arms.

Class ECHINOIDEA

Sea urchins. Mouth directed downwards, anus dorsal or ventral; protected by calcareous plated capsule or test.

Superorder **Echinacea** Regular echinoids.

Echinus *Psammechinus* *Paracentrotus*

Superorder **Gnathostomata** Irregular echinoids with mouth lantern.

Echinocyamus

Superorder **Atelostomata** Irregular echinoids without mouth lantern.

Spatangus *Brissopsis* *Echinocardium*

Class HOLOTHUROIDEA

Sea cucumbers. Elongated; mouth at one end, anus at the other; skeleton reduced to spicules.

Subclass **Dendrochirotacea** Possess retractible introvert.

| *Aslia* | *Pawsonia* | *Thyone* |
| | *Cucumaria* | *Leptopentacta* |

Subclass **Aspidochirotacea** No introvert.

Holothuria

Subclass **Apodacea** Tube-feet usually restricted to oral region only.

Leptosynapta *Labidoplax*

Phylum CHORDATA

Possessing a notochord at some stage in the life cycle.

Subphylum UROCHORDATA

Tunicates or ascidians. Notochord present only in larva.

Class **ASCIDIACEA**

Sea squirts. Adults usually sac-like, with inhalant and exhalant openings for feeding currents.

Order **Enterogona** Solitary tunicates.

Clavelina	*Aplydium*	*Sidnyum*
Perophora	*Ascidia*	*Phallusia*
		Ciona

Order **Pleurogona** Colonial or compound tunicates.

Botryllus	*Botrylloides*	*Molgula*
Dendrodoa (Styelopsis)		*Distomus*

Class **THALIACEA**

Pelagic, transparent, barrel-shaped animals lacking tail.

Salpa	*Doliolum*

Class **LARVACEA**

Adults retain larva-like character; live within a transparent mucous 'house'.

Oikopleura

Subphylum CEPHALOCHORDATA

Very simple fish-like chordates with segmental muscle blocks and gonads.

Amphioxus

GLOSSARY

Actinotrocha. The free-swimming larva of a phoronid.

Acontium. A stinging thread which arises in the gut cavity of sea-anemones and is discharged through the mouth or special pores.

Aestivation. Prolonged summer torpor.

Alimentary tract. The passage from mouth to anus in which food is broken down and digested.

Alternation of generations. The alternation of a generation reproducing sexually with one reproducing asexually, often having very different forms.

Amoeboid. Lacking a fixed form and creeping by means of protoplasmic projections, like *Amoeba*.

Antenna. One of the paired sensory appendages on the heads of certain arthropods, etc; usually composed of numerous segments.

Antennule. One of the small second pair of antennae occurring in Crustacea.

Anus. The orifice through which the alimentary tract empties.

Autotomy. Voluntary amputation of a part of the body, such as limb or tail, at a specialized point of weakness.

Avicularium. Modified zooid of some Bryozoa, shaped like a bird's head.

Benthic. Living on the sea bottom.

Bilateral symmetry. Mirror-image correspondence of opposite sides of the body.

Bursa. A blind-ending sac, as in ophiuroids.

Byssus. A thread or threads by which a bivalve attaches itself to a hard surface.

Calcareous. Containing lime (calcium carbonate).

Carapace. A hardened dorsal shield.

Carnivore. An organism which feeds on animal material.

Cephalothorax. The anterior division of an arachnid, consisting of the coalesced head and thorax.

Ceras (pl. **cerata**). A finger-like respiratory papilla of nudibranch molluscs, sometimes containing ingested nematocysts.

Cercaria. A larva of digenean flatworms.

Chaeta. A tough, embedded, ectodermal bristle.

Chelicerae. The first pair of arachnid appendages, modified as mouthparts, and bearing the poison fangs in spiders.

Chordate. Having a skeletal rod, the notochord.

Chromatophore. A pigment-bearing cell.

Cilia. Small whip-like processes from cells, that by beating in synchrony serve in propulsion.

Cirrus. A projection from a soft-bodied animal, such as an annelid.

Class. The major taxonomic subdivision of a phylum.

Clausilium. A sliding door to the mantle cavity in some gastropods.

Clitellum. The glandular region on the body of earthworms that secretes the cocoon.

Cloaca. The cavity into which the anus and other effluent pores open.

Coelom. The secondary body cavity in higher invertebrates.

Comb-plates. Rows of fused cilia that form the main organs of propulsion in ctenophores.

Commensalism. An association between two organisms in which one gains shelter and/or food; literally 'sharing a table'.

Convergence. The evolution of two groups along similar lines.

Coxa. The basal segment of an arthropod limb.

Cryptozoic. Living in dark places.

Cuticle. The outermost layers of the integument.

Cuvierian organ. A mass of sticky extrusible threads which emerge from the cloaca of some holothuroids.

Cysticercus. A larval stage of tapeworms.

Dextral. Rotating to the right.

Ectoparasite. Parasite that lives on the outside of its host.

Elytron. A plate-like extension of the dorsal body wall in polychaetes.

Endoparasite. Parasite that lives within the body of its host.

Endoskeleton. A skeleton formed within the body.

Ephippium. The thickened cuticle of the crustacean carapace that is shed at moulting and in waterfleas serves as an egg case.

Ephyra. The pelagic larva of a jellyfish.

Epiphragm. A mucoid seal across the shell-opening in some snails.

Errant. Moving freely (as opposed to sedentary); usually applied to annelids.

Exoskeleton. A skeleton formed outside the body, as in arthropods.

Faeces. The undigested excrement from an animal's gut.

Family. The major taxonomic subdivision of an order.

Fasciole. A region or band of ciliated spines, as in spatangoid echinoids.

Fission. Reproductive division by splitting.

Flagella. Long, whip-like processes used in propulsion.

Gemmule. The overwintering stage of a freshwater sponge.

Genus. The major taxonomic subdivision of a family; an assemblage of closely related species.

Glochidium. The larva of freshwater bivalves.

Gonad. An organ which produces male and/or female reproductive cells.

Gravid (of a female). Full of eggs.

Haemoglobin. An iron-based respiratory pigment.

Haptor. A complex attachment organ of some flukes.

Herbivore. A plant-eating organism.

Hermaphrodite. Having both sexes in one animal.

Holdfast. An organ of attachment.

Hydroid phase, hydranth. The polyp of a coelenterate.

Interstitial. Living in the spaces between sand grains or in small rock crevices.

Intromittent organ. A male copulatory structure that is inserted into the body of the female.

Introvert. An involuted portion of the body; especially the proboscis of worm-like animals.

Invertebrate. Lacking a backbone.

Larva. A free-living immature stage that is markedly distinct from the adult.

Littoral. Living in the area of seashore between the low and high tide marks.

Lophophore. Circlet of eversible ciliated tentacles surrounding the mouth of certain sedentary aquatic animals.

Luminescence. Chemical production of light.

Mantle. Lateral skirt-like fold of the molluscan body wall, usually having a spacious cavity beneath it.

Medusa. The free-living, sexual, jelly-fish-like stage of certain coelenterates.

Melanism. The abnormal condition in which the body is darkened by overproduction of the pigment melanin.

Metacercaria. An encysted cercaria larva of a fluke.

Metamorphosis. An abrupt change in body form from one larval stage to another or from larva to adult.

Metazoan. Composed of many cells.

Miracidium. A larva of some flukes.

Mucus. A viscous secretion.

Nematocyst. A stinging cell of coelenterates.

Notochord. A skeletal rod of turgid cells.

Nucleus. The controlling centre of a cell.

Nymph. An immature stage of certain insects in which there is some similarity to the adult form.

Onchosphere. The hooked embryo of some tapeworms.

Operculum. A calcified flap which closes off the opening in some gastropods.

Order. The major taxonomic subdivision of a class.

Osmosis. A process whereby water passes across a membrane from the weaker to the stronger solution.

Ovary. The female gamete-producing organ.

Paedomorphosis. The process in which an organism becomes sexually mature while still possessing larval features.

Palp. A fleshy process from the body, usually in annelids.

Parapodium. One of the paired fleshy projections from the body wall of polychaete worms, used in locomotion.

Parasite. An animal which lives at the expense of another living organism.

Parthenogenesis. Reproduction by means of ova that develop without fertilization.

Pedipalp. Each of the second pair of arachnid appendages; in spiders, leg-like and modified for sperm-transfer in males.

Pedunculate. With the body borne on a stalk.

Pelagic. Living in the upper waters of the open sea.

Pericardium. The part of the body cavity (coelom) which surrounds the heart.

Pharynx. 1. The first part of the gut from which the gill slits open. 2. The muscular and sometimes eversible first part of the gut.

Phoresy. Transportation by clinging to another animal.

Photosynthesis. The process whereby plants build up complex organic substances from simple ones, using the sun's energy.

Phylum. One of the main taxonomic subdivisions of the animal kingdom.

Pinnule. Lateral branch from the arms of crinoids.

Plankton. The small plants and animals living free in surface waters of lakes and oceans, taken collectively.

Planula. A larva of lowly invertebrates such as coelenterates and platyhelminths.

Pluteus. A larva of echinoids and ophiuroids, shaped like an inverted easel.

Pneumostome. The opening to the lung in pulmonate gastropods.

Polymorphism. The occurrence in any one animal species of more than one shape or colour.

Polyp. A simple animal with upwardly-directed mouth, fringe of tentacles, and simple gut.

Predator. An animal that preys on other organisms.

Primordium. The simple beginnings of an animal or organ.

Proboscis. A trunk-like process, sometimes eversible, usually arising from the head.

Proglottid. A compartment of a tapeworm's body.

Prostomium. The anterior part of an annelid, in front of the first segment.

Protoplasm. The material of which an animal's body is composed.

Radial symmetry. Mirror-image correspondence of the two halves of a body halved in any vertical plane.

Radula. The rasp-like feeding tongue of some molluscs.

Rectum. The terminal portion of the alimentary tract.

Redia. A larval stage in some flukes.

Respiratory tree. The internal organ for gaseous exchange in some holothuroids.

Scolex. The attachment end of a tapeworm.

Scute. A plate forming part of a skeletal cover.

Scyphistoma. The sessile stage in the life-history of a jellyfish.

Segment. A compartment of an animal's body, which is repeated almost identically throughout its length.

Segmented. Having a body divided into segments.

Septum. A partition separating two cavities.

Sessile. Settled, stationary, on a solid object.

Sinistral. Coiled to the left.

Spatulate. Having the shape of a flat spoon.

Species. The lowest taxonomic category into which living things are divided; members of the same species can interbreed.

Spermatophore. A capsule containing spermatozoa.

Spicules. Small pointed calcareous or siliceous bodies that make up the skeleton of certain animals.

Spinneret. A finger-like projection from the posterior end of the spider abdomen through which silk is discharged.

Sporocyst. The first sac-like larval stage of some flukes.

Spring tide. High tide occurring after full and new moon in each month.

Statoblast. The overwintering body of some bryozoans and sponges.

Stigma. In protozoans, an eye spot; in arthropods, one of the external openings to the tracheal system.

Stylet. A small needle-like spicule.

Sublittoral. The part of the shore just below the limit of the lowest tides.

Substratum. A solid bottom underlying a body of water.

Swimmerets. Swimming appendages, usually of arthropods.

Symbiosis. An association between two organisms which is of benefit to both.

Test. A skeletal covering which is itself covered by a living layer.

Testis. The male gamete-producing organ.

Tornaria. The larva of some hemichordates.

Torsion. A process of twisting during growth.

Trochophore. The larva of annelids and related groups.

Tube-foot. A hydraulic organ protruding from the bodies of all echinoderms.

Umbilicus. A navel-like orifice in the shells of some molluscs.

Vacuole. A fluid-filled cavity in a cell.

Veliger. A larva of some molluscs.

Vertebrate. Having a backbone.

Viviparous. Typically producing living young rather than eggs.

Whorl. A circlet of similar structures.

Zooid. An individual polyp-like organism.

SOURCES OF FURTHER INFORMATION

BOOKS : a selection appropriate for the general reader.

BARRETT, J. & YONGE, C. M., *Collins pocket guide to the sea shore*. 1958. Collins.
BRISTOWE, W. S., *The world of spiders*. 1958. Collins.
BUCHSBAUM, R., *Animals without backbones*. 1951. Pelican.
CLOUDSLEY-THOMPSON, J. L., *Spiders, Scorpions, Centipedes and Mites*. 2nd edn. 1968. Pergamon.
CLEGG, J., *The freshwater life of the British Isles*. 3rd edn. 1965. Warne.
CROFTON, H. D., *Nematodes*. 1966. Hutchinson.
DALES, R. P., *Annelids*. 2nd edn. 1967. Hutchinson.
EALES, N. B., *The littoral fauna of the British Isles*. 4th edn. 1967. C.U.P.
ELLIS, A. E., *British Snails*. 1926, reissue 1969. Clarendon Press.
GREEN, J., *A biology of Crustacea*. 1961. Witherby.
HARDY, A. C., *The open sea: the world of plankton*. 1956. Collins.
International Code of Zoological Nomenclature. 1961. International Trust for Zoological Nomenclature, London.
JANUS, H., *The young specialist looks at molluscs*. 1965. Burke.
KAESTNER, A., *Invertebrate Zoology*: Vol. 1, 1967; II, 1968; III & IV forthcoming. Interscience.
KERRICH, G. J., MEIKLE, R. D., & TEBBLE, N., *Bibliography of key works for the identification of the British fauna and flora*. 3rd edn. 1967. Systematics Association, London.
Larousse Encyclopedia of Animal Life. 1967. Hamlyn.
MACAN, T. T., *A guide to freshwater invertebrate animals*. 1959. Longman.
MACAN, T. T. & WORTHINGTON, E. B., *Life in lakes and rivers*. 2nd edn. 1968. Collins.
MARSHALL, S. M. & ORR, A. P., *The biology of a marine copepod*. 1955. Oliver & Boyd.
McMILLAN, N. F., *British shells*. 1968. Warne.
MELLANBY, H., *Animal life in freshwater*. 6th edn. 1963. Methuen.
MORTON, J. E., *Molluscs*. 4th edn. 1967. Hutchinson.
NICHOLS, D., *Echinoderms*. 4th edn. 1969. Hutchinson.
RUSSELL-HUNTER, W. D., *A biology of lower invertebrates*. 1968. Macmillan.
RYLAND, J. S., *Bryozoans*. 1970. Hutchinson.
SOUTHWARD, A. J., *Life on the sea-shore*. 1965. Heinemann.
WELLS, M., *Lower animals*. 1968. Weidenfeld & Nicholson.
YONGE, C. M., *The sea shore*. 1949. Collins.

SOCIETIES AND ASSOCIATIONS: most of these hold regular meetings and publish journals.
    The name of the honorary secretary or director is given.

British Arachnological Society. J. R. Parker, F.Z.S., Peare Tree House, The Green, Blennerhasset, Carlisle.
British Conchological Society. T. E. Crowley, Clonard Cottage, Rowlands Avenue, Hatch End, Middlesex.
British Malacological Society. Miss J. E. Rigby, Dept. of Biology, Queen Elizabeth College, London, W.8.
British Society for Parasitology. Dr. F. E. G. Cox, Dept. of Zoology, Kings College, Strand, London.
Fauna Preservation Society. c/o Zoological Society of London, Regent's Park, London N.W.1.
Field Studies Council. The Secretary, 9 Devereux Court, London W.C.2.
Freshwater Biological Association. H. C. Gilson, F.B.A. Laboratory, The Ferry House, Far Sawrey, Ambleside, Westmorland.
Institute of Biology. D. J. B. Copp, 41 Queen's Gate, London S.W.7.
Linnean Society of London. Burlington House, Piccadilly, London W1V 0LQ.
Marine Biological Association. Dr. J. E. Smith, F.R.S., The Laboratory, Citadel Hill, Plymouth, Devon.
The Ray Society. D. MacFarlane, c/o British Museum (Natural History), Cromwell Road, London S.W.7.
Royal Zoological Society of Scotland. Scottish National Zoological Park, Edinburgh 12.
Systematics Association. Dr. A. J. Boyce, Dept. of Biological Sciences, University of Surrey, 14 Falcon Road, London S.W.11.
Zoological Society of London. Regent's Park, London N.W.1.

INDEX

Aberrant annelid worms 120, **121**
Abida 54, 198; *secale* **55**
Abra 78, 199; *alba* **79**
Acanthinula 52, 198; *aculeata* **53**
Acanthobdellae 201
Acanthocardia 76, 199; *echinata, tuberculata*, **77**
Acanthochitona 32, 196; *crinitus* **33**
Acanthodoris 46, 197; *pillosa* **47**
Acarapis 146, 204; *woodi* **147**
Acari 138, 154, 160, 204
Acarine disease 146
Achaearanea 162; *saxatile* 162, **163**
Achatinacea 198
Acicula lineata (*Acme fusca*) 54, **55**, 197
Acid 4, 40
Acmaea 34, 196; *virginea* **35**
Acme 54, 197; *fusca* **55**
Acoela 22, 194
Acontium 14, 18
Acorn barnacle 128
Acorn worms 168, 205
Acrocnida 88, 184, **206**; *brachiata* **185**
Acroloxus 70, 198; *lacustris* **71**
Actaeon 44, 197; *tornatalis* **45**
Acteonia 48, 197; *senestra* **49**
Actinia 14, 194; *equina* **15**
Actiniaria 194
Actinopoda 192
Actinosphaerium viii, **1**, 192
Actinotrocha 170
Adalaria 46, 197; *proxima* **47**
Adamsia 18, 194; *palliata* **49**
Adapedonta 199
Adesmacea 199
Adhesive organ 22, 24
Aeolidia 48, 197; *papillosa* **49**
Aeolosoma 114, **115**, 201
Aeolosomatidae 114, 201
Agriolimax 62, 198; *reticulatus* **63**
Ague 148
Aiptasia 14, 194; *couchii* **15**
Akera 44, 197; *bullata* **45**
Alcyonaria 18, 194
Alcyonidium 172, 205; *hirsutum* **173**
Alcyonium 18, 194; *digitata* 19
Alderia 48, 197; *modesta* **49**
Alentia 94, 200; *gelatinosa* **95**
Alga viii
Alimentary tract vi
Allochernes 154, 204; *panzeri* **155**
Alloeocoela 195
Allolobophora 112, 201; *longa* **113**
Alternation of generations 2, 8, 188
Amber Snail 56, **57**, 198
American Hookworm 30
American Piddock 86, **87**
Ammonoidea 200
Amnicola 70, 197; *taylori* **71**
Amoeba viii, **1**, 192
Amoebida 192
Amoebulae 2
Amphictenidae 201
Amphilina 26, 195; *foliacea* **27**
Amphineura 196
Amphioxus 176, 207; *lanceolatus* **177**
Ampharpholis 180, 206; *squamata* **181**
Amphipod 124
Amphipoda 203
Amphiporus 28, 195; *lactifloreus* **29**
Amphitrite 102, 200; *johnstoni* **103**
Amphiura 182, 184, 206; *filiformis* **185**
Amphiuridae 206
Analasmocephalus 164
Analgesidae 146

Anaspidea 44, 197
Ancula 46, 197; *cristata* **47**
Ancylastrum 70, 198; *fluviatilis* **71**
Ancylidae 198
Ancylus 70, 198; *lacustris* **71**
Anemone *see* Sea Anemones
Anemonia 14, 194; *sulcata* **15**
Anisomyaria 198 – 199
Anisus 68, 198; *vortex* **69**
Annelida 94 – 121, 200 – 202, vi, 168
Anodonta 90, 138, 199; *anatina, cygnea* **91**
Anomalodesmacea 82, 199
Anomia 84, 199; *ephippium* **85**
Anomiacea 199
Anopla 195
Anostraca 202
Anseropoda 182, 206; *placenta* **183**
Antedon 120, 180, 206; *bifida* **181**
Antennae; of glycerids, 98; of *Leptodora* 132; of nereid 100; of *Palinurus* 122; of water-fleas 136
Antennules, of barnacles 136
Anthomedusae 193
Anthozoa 14, 18, 194
Anticoagulant, of leeches' saliva, 116, 118
Antipatharia 194
Anus 28, 195, 196, 197, 205
Anyphaena 158, 204; *accentuata* **159**
Aphrodita 94, 200; *aculeata* **95**
Aphroditidae 200
Aplacophora 196
Aplexa 70, 198; *hypnorum* **71**
Aplydium 190, 207; *proliferum* **191**
Aplysia 44, 197; *punctata* **45**
Aplysiomorpha 197
Apoda 186
Apodacea 206
Aporrhaidae 40
Aporrhais 40, 168, 197; *pespelicani* **41**
Arabellid 98
Arachnida 122, 130 – 133, 138 – 139, 142 – 167, 204
Arachnids: domestic 148; garden 152; grassland 164; sandhill 166; seashore 130
Araneae 204
Araneus 150, 158, 162, 204; *diadematus* **151**
Arca 84, 198; *tetragona* **85**
Arcacea 198
Arcella viii, **1**, 192
Archaeogastropoda 34, 36, 70 196
Archaeotaenioglossa 196
Archiannelida 94, 120, 202
Archidoris 46, 197; *pseudoargus* **47**
Arctosa 130, 162, 204; *perita* **163**
Arenicola 102, 200; *marina* **103**
Arenicolidae 200
Argas 142, 204; *vespertilionis* **143**
Argasids 142
Argulus 140, 203; *foliaceus* **141**
Argyroneta 132, 204; *aquatica* **133**
Arianta 54, 198; *arbustorum* **55**
Arion 62, 198; *hortensis* **63**
Ark Shell 84, **85**, 198
Armadillidium 150, 164, **165**; 166, 203
Armiger 198
Arrhenurus 138, 204; *caudatus* **139**
Arrow worms 176, **177**, 205
Arthropoda vii, 32, 202 – 204; seashore 122 – 131; freshwater 133 – 139; parasitic 140 – 147; terrestrial 148 – 167

Articulata 170, 205
Ascaris 30, 196; *suilla* **31**
Ascidia 188, 44, 207; *mentula* **189**
Ascidiacea 188, 207
Ascophyllum 8
Asellus 132, 203; *aquaticus* **133**
Ashfordia 52, 198; *granulata* **53**
Aslia 186, 206; *lefevrei* **187**
Aspidobothrea 24, 195
Aspidochirotacea 206
Aspidogaster 24, 195; *conchicola* **25**
Aspidosiphon 168, 204; *mulleri* **169**
Astacus 132, 203; *fluviatilis* **133**
Astartacea 199
Astarte 82, 199; *sulcata* **83**
Astartes 199
Asterias 40, 180, 206; *rubens* **181**
Asterina 180, 206; *gibbosa* **181**
Asteroidea 180, 206
Asterozoa 206
Astropecten 182, 206; *irregularis* **183**
Atelostomata 206
Atypus 162, 204; *affinis* **163**
Aurelia 12, 194; *aurita* **13**
Autolytus 96, **97**, 200
Autotomy 48
Avicularia, of bryozoans 172
Axial skeleton 18
Azeca 58, 198; *goodalli* **59**

Babesia 142
Backbone vi
Bacterium vi, 20
Bait, fisherman's 98, 102
Balanus 128, 140, 203; *balanoides* **129**
Balea 58, 198; *perversa* **59**
Baltic Tellin 78, **79**
Banded Carpet Shell 74, **75**
Banded Venus 74, **75**
Banded Wedge Shell 78, **79**
Barnacles 124, 128, **129**, 140, 203
Barnea 86, 199; *candida* **87**
Basket Shell 80, **81**, 199
Basommatophora 197 – 198
Bath sponge 4
Bathynella 203
Bathynella 134, 203; *natans* **135**
Bdella 130, **131**, 204
Beadlet Anemone 14, 15
Bean Horse Mussel 72, **73**
Bearded Horse Mussel 72, **73**
Beautiful Snail 60
Bêche de mer 186
Beetle mites 160
Big Bud 152
Bimastus 112, 201; *tenuis* **113**
Bipalium 20, 195; *kewense* **21**
Bithynia 66, 197; *tentaculata* **67**
Bittium 40, 197; *reticulatum* **41**
Bivalved shell 136, 170, 198
Bivalvia 72 – 91; 198 – 199; as prey, 42, 180
Bladder Snail 70, **71**, 198
Bladder-worm 26
Blaniulus 152, 203; *guttulatus* **153**
Blind Snail 54, **55**
Blood fluke 24
Blood system vi, 195; open 32; vascular, 28
Blue-rayed Limpet 34, **35**
Blunt Tellin 78, **79**
Bog or Marsh Snail 64, **65**
Bonellia 120, 124, 202; *viridis* **121**
Book Scorpion 148
Bootlace Worm 28, **29**

Bopyrus 124, 203; *squillarum* **125**
Boring polychaetes 108
Boring sponge 4
Botrylloides 190, 207; *leachi* **191**
Botryllus 190, 207; *schlosseri* **191**
Bougainvillea 8, 193; *ramosa* **9**
Bouyancy, of squid, 92
Brachionus **1**
Brachiopoda 170, 205
Bracken Bug 144, **145**
Brallion 80, **81**
Branchellion 118, 202; *boreale* **119**
Branchial basket 188
Branchiomma 28
Branchionus 196
Branchiopoda 202
Branchiostoma 176; *lanceolatus* **177**
Branchiura 140, 203
Bread-crumb Sponge 4, **5**, 22, **23**
Brissopsis 178, 206; *lyrifera* **179**
Bristle worms 94 – 111, 200
Brittle-stars 88, 120, 178, 180, **181**, 184, **185**, 206
Brood chamber, of *Ophiothrix* 184
Brood pouch: of *Leptodora*, 132; of *Oniscus* 152; of shrimp 134; of water-flea 136
Brooding, of *Henricia* 182
Bryobia 154, 204; *rubrioculus* **155**
Bryozoa 172 – 175, 205
Buccinacea 197
Buccinum 42, 110, 197; *undatum* **43**
Bud 12
Budding 6, 10, 188, 190
Bugula 172, 205; *turbinata* **173**
Bullomorpha 44, 197
Bunodactis 14, 194; *verrucosa* **15**
Bursae, of brittle-star, 184
Button Ramshorn 68, **69**
Buzzing Spider 158, **159**
Byssus threads 72
Bythinella 70, 197; *scholtzi* **71**
Bythotrephes 132, 202; *longimanus* **133**

Caddis fly larva 66
Caecilioides 54, 198; *acicula* **55**
Calanus 126, 202; *finmarchicus* **127**
Calcarea 193
Calcareous spicules 4; tubes, 110
Calliactis 18, 122, 194; *parasitica* **19**
Calliobdella 118, 202; *lophii* **119**
Calliostoma 36, 196; *zizyphinum* **37**
Calyptraea 34, 197; *chinensis* **35**
Calyptraeacea 197
Canal 6, 70
Canal system 12
Cancer 122, 203; *pagurus* **123**
Candidula 52, 198; *gigaxi* **53**
Cantharidus 36, 196; *exasperatus* **37**
Canthocamptus 134, 202; *staphylinus* **135**
Capitella 102, 200; *capitata* **103**
Capitellidae 102, 200
Caprella 126, 203; *linearis* **127**
Carcinus 124, 203; *maenas* **125**
Cardiacea 199
Cardiidae 76
Cardium 76, 199; *echinata, edule, exiguum, tuberculata* **77**
Carpet Shells 74, 199
Carychium 54, 197; *minimum* **55**
Caryophyllia 18, 194; *smithi* **19**
Castor Bean 142, **143**
Cecidophyopsis 152, 204; *ribis* **153**
Cell 192
Cellar Glass Snail 50, **51**
Cellulose viii
Centipede 130, 152, 154, 203
Cepaea 52, 60, 198; *hortensis, nemoralis* **53**
Cephalaspidea 44, 197
Cephalochordata 176, 207

Cephalopoda vii, 32, 92, 199 – 200
Cephalothorax 148, 158
Cephalothrix 28, 195; *rufifrons* **29**
Cerastoderma 76, 199; *edule* **77**
Cerata 48, 197
Ceratium 2, 192; *tripos* **3**
Cercaria 24, **25**, 64
Cereus 16, 194; *pedunculatus* **17**
Ceriantharia 194
Cerianthidea 16
Cerianthus 16, 194; *lloydi* **17**
Cerithiacea 197
Cerithiidae 40
Cestoda 20, 195
Cestodaria 20, 26, 195
Cestus 16, 194; *veneris* **17**
Chaetae 94, 112
Chaetogaster 114, 201; *limnaea* **115**
Chaetognatha 176, 205
Chaetonotus **1**, 196
Chaetopteridae 201
Chaetopterus 106, 201; *variopedatus* **107**
Chalk 2
Chamelea 74, 199; *striatula* **75**
Chartreuse Snail 50, **51**
Cheilostomata 205
Cheiracanthium 166, 204; *virescens* **167**
Cheiridium 148, 204
Cheirocephalus 134, 202; *grubei* **135**
Chelicera, of pseudoscorpion 154; of spider, 158
Chelicerata 203 – 204
Chelifer 148, 204; *cancroides* **149**
Chemical defense 44
Chilopoda 154, 203
Chinaman's Hat Shell 34, **35**
Chinese Liver Fluke 24
Chink Shells 38, **39**
Chitons 32, **33**, 196
Chlamys 84, 180, 199; *distorta, opercularis, varia* **85**
Chloraemidae 201
Chlorohydra 6, 193; *viridissima* **7**
Chondrophora 10, 193
Chordata vii, 207
Chrysalis Snail 52, **53**
Chrysaora 12, 194; *isosceles* **13**
Cilium viii, 12, 16, 20, 24, 36, 193, 194, 196
Ciliata 20, 193
Ciona 188, 207; *intestinalis* **189**
Cirratulidae 200
Cirratulus 102, 200; *cirratus* **103**
Cirripedia 124, 140, 203
Cirrus 94, 128, 180, 200
Cladocera 132, 202 – 203
Clam 80, **81**
Class 192
Classification 192
Clathrus 40, 197; *clathrus* **41**
Clausilia 58, 198; *bidentata* **59**
Clausilium, in snail, 58
Clausinella 74, 199; *fasciata* **75**
Clava 8, 193; *squamata* **9**
Clavelina 190, 207; *lepadiformis* **191**
Cliona 4, **5**, 193
Clione 44, 197; *limacina* **45**
Clitellata 118
Clitellum 112, 118
Clubiona 158, 166, 204; *corticalis* **159**
Clypeasteroida 178
Cnidospora 192
Coat-of-Mail Shells 32
Cochlicopa 52, 198; *lubrica* **53**
Cockles 76, 199; commensals of, 124
Cocoon 6, 20, 112, 118, 150
Coelenterata vi, 6 – 19, 193 – 194
Coelom 28, 94, 118
Coeloplana 16, **17**, 194
Coin Shells 199

Coleoidea 200
Collar, of enteropneust, 168
Colonies vi, 4, 8, 10, 172, 188, 190, 193, 194, 205, 207
Columella 56, 198; *edentula* **57**
Comb Jelly 16, **17**
Comb-row 16
Commensalism 88, 102, 124, 180, 184, 186
Common Cuttlefish 92, **93**
Common or Garden Snail 50, **51**
Common Lobster 122, **123**
Common Mussel 72, **73**
Common Octopus 92, **93**
Common Orb Mussel 90
Common Periwinkle 38, **39**
Common Piddock 86, **87**
Common Shrimp 122, **123**
Common Squid 92, **93**
Common Starfish 180, **181**
Common Valve Snail 66, **67**
Common Wendletrap 40, **41**
Compensation sac, of zooid, 172
Compound eye, in nauplius, 128
Compound Sea-squirts 190
Convoluta 22, 194; *roscoffensis* **23**
Copepoda 12, 26, 124, 134, 140, 202
Copse Snail 54, **55**
Coral 18, 194
Corallimorpha 18, 194
Corallina 88, 180
Corbula 80, 199; *gibba* **81**
Cordylophora 6, 193; *lacustris* **7**
Cortex 4
Corynactis 18, 194; *viridis* **19**
Coryne 8, 126, 193; *muscoides* **9**
Cotton Spinner 186, **187**
Courtship dance, spiders', 148, 150, 156, 166
Cowries 36, **37**, 197
Crabs 18, 122, **123**, 124, **125**, 203
Crabs, hermit 88, 100, 122, **123**, 203
Crab-spider 162, 164
Crangon 122, 203; *vulgaris* **123**
Crania 170, 205; *anomala* **171**
Craspedacusta 6, 7, 193
Crawfish 122, **123**
Crayfish 132, **133**
Crenella 72, 198; *decussata* **73**
Crepidula 34, 197; *fornicata* **35**
Crinoidea 120, 180, **181**, 206
Crinozoa 206
Crisia 172, **173**, 205
Cristatella 174, 205; *mucedo* **175**
Crosscut Carpet Shell 74, **75**
Cross-spider 150, **151**
Crumb-of-bread Sponge 46
Crustacea vii, 12, 20, 202 – 203; freshwater 132 – 137; parasitic 140 – 141; seashore 122 – 129
Cryptops 154, 203; *hortensis* **155**
Crystal Snail 56, **57**
Ctenophora 16, 194
Ctenostomata 205
Cucumaria 186, 206; *elongata* **187**
Curled Octopus 92, **93**
Cushion Star 182, **183**
Cuspidaria 82, 199; *cuspidata* **83**
Cusp Shells 199
Cuttlefish 92, **93**, 200
Cuvierian organs 186
Cyanea 12, 194; *capillata* **13**
Cyclophoracea 196
Cyclophyllidea 195
Cycloporus 22, 195; *papillosus* **23**
Cyclops 134, 202; *strenuus* **135**
Cyclostoma 54
Cyclostomata 205
Cylindroiulus 166, **167**, 203
Cypraeacea 197
Cypris 136, **137**, 202
Cypris larva 124, 128, **129**

Cysticercus 26
Cochlicella 60, 198; *acuta* **61**

Dactylochelifer 166, 204; *latreillei* **167**
Dahlia Anemone 14, **15**
Daisy Anemone 16, **17**
Dark-lipped Banded Snail 52, **53**
Dart, of snail, 50
Dead Men's Fingers 18, **19**
Decapoda 200, 203
Demodex 144, 204; *folliculorum* 144; *canis* **145**
Demospongiae 193
Dendrobaena 112, 201; *subrubicunda* **113**
Dendrochirotacea 206
Dendrocoelum 20, 195; *lacteum* **21**
Dendronotacea 197
Dendrodoa 190, 207; *grossularia* **191**
Dendronotacea 197
Dendronotus 197
Dentalium 32, 198; *entalis* **33**
Dermacentor 142, 204; *reticulatus* **143**
Dermanyssus 146, 204; *gallinae* **147**
Dero 114, **115**, 201
Derogenes 22, 195; *varicus* **23**
Devonia 88, 199; *perrieri* **89**
Devonshire Cup Coral 18, **19**
Diaptomus 134, 202; *gracilis* **135**
Difflugia viii, **1**, 192
Digenea 20, 22, 56, 195
Dinoflagellate 2, **3**
Dinophilus 120, **121**, 202
Diodora 34, 196; *apertura* **35**
Diotocardian 34
Diphyllobothrium 26, 195; *latum* **27**
Diplopoda 203
Dipylidium 26, 195; *canium* **27**
Discus Snails 198
Distomum (Fasciola) 24, 56, 64
Distomum 190, 207; *variolosus* **191**
Dog Cockle 82, **83**
Dog Whelk 42, **43**, 197
Doliolum 188, 207; *gegenbauri* **189**
Donax 78, 199; *vittatus* **79**
Doridacea 46, 197
Dorididae 46
Dosinia 74, 199; *exoleta* **75**
Doto 8, 197; *coronata* **9**
Drapetisca 156, 204; *socialis* **157**
Drassodes 150, 204; *lapidosus* **151**
Dreissena 90, 199; *polymorpha* **91**
Dreissenacea 199
Drills 42
Dublin Bay Prawn 122
Duck Mussel 90, **91**
Dugesia 20, 195; *subtentaculata* **21**
Dwarf Pond Snail 64, **65**
Dwarf Snail 58, **59**
Dynamena 8, 193; *pumila* **9**
Dysdera 150, 204; *crocata* **151**

Ear Pond Snail 64, **65**
Earthworm 20, 94, 112, **113**, 201
Echinacea 206
Echinocardium 178, 88, 182, 206; *cordatum* **179**
Echinococcus 26, 195; *granulosus* **27**
Echinocyamus 178, 206; *pusillus* **179**
Echinodermata vii, 42, 178 – 187, 205 – 206
Echinoidea 178, 206
Echinorhynchus 30, 195; *clavula* **31**
Echinozoa 206
Echinus 178, 206; *esculentus* **179**
Echiurida 120, 202
Ectoparasite 138, 140, 142, 146, 202

Ectoprocta 172 – 175, 205
Edible Cockle 76, **77**
Edible Crab 122, **123**
Edible Sea-urchin 178, **179**
Edible Snail 54, **55**
Eel-grass 36
Egg-sac, of *Calanus*, 126
Eisenia 112, 201; *rosea* **113**
Eiseniella 114, **115**, 201
Elasmobranchs 118
Eledone 92, 200; *cirrhosa* **93**
Elephantiasis 30
Ellobiidae 197
Elytra 94
Emarginula 34, 196; *reticulata* **35**
Emplectonema 28, 195; *neesii* **29**
Ena obscura 58, **59**, 198
Enchytraeidae 112, 201
Enchytraeus 112, **113**, 201
Encrustation 6, 190
Endodontacea 198
Endoparasite 128, 140, 142, 144, 195
Enopla 195
Ensis 80, 199; *arcuatus, ensis, siliqua* **81**
Enterogona 207
Enteropneusta 168, 205
Entoprocta 196
Eolidacea 46, 48, 197
Ephippium 136
Ephydatia 6, 193; *fluviatilis* **7**
Ephyrae 12
Epilepsy 26
Epipodial tentacles 36
Epithelium 12
Epitoniidae 40
Epizoanthus 18, 194; *incrustatus* **19**
Epizoic bivalves 88
Ergasilus 140, 202; *sieboldi* **141**
Erigone 130, 158; *arctica* **131**
Eriophyidae 152
Ero 162, 204; *furcata* **163**
Erpobdella 116, 202; *octoculata* **117**
Erpobdellidae 202
Errantia 94 – 101, 102, 200
Erycinacea 199
Escargots 54, **55**
Estuaries 100, 122
Euconulus 56, 198; *fulvus* **57**
Eulalia 96, 200; *viridis* **97**
Eunemertes 28, 195; *neesii* **29**
Eunereis 100, 200; *longissima* **101**
Eunice 98, 200; *harassii* **99**
Eunicella 18, 194; *verrucosa* **19**
Eunicidae 98, 106, 200
Eunoe 94, 200; *nodosa* **95**
Eupagurus, see *Pagurus*
Euparypha 60, 198; *pisana* **61**
Eupogebia 88
Euspongilla 6, 193; *lacustris* **7**
Eutrombidium 146, 204; *rostratus* **147**
Evarcha 162, 204; *arcuata* **163**
Excentric Snail 52, **53**
Excitatory organ, of snail, 52
Exoskeleton vii, 122
Eye: of arachnid 204; of branchiopod 202; of cephalopod 92, 199; of nereid 100; of pulmonate 197, 198; of scallop 84
Eye-spot 20, 104
Eye-Worm 30

Facelina 48, 197; *auriculata* **49**
Fairy-shrimp 134, **135**
False Scorpions 146, 154, 166, 204
Family 192
Fan Mussel 72, **73**, 198
Fasciola 24, 64, 195; *hepatica* **25**
Fascioles 178

Feather Mite 146, **147**
Feather-star 120, 178, 180, 206
Fertilization 6, 8
Filaria 30
File Shell 84, 199
Filograna 110, 201; *implexa* **111**
Fish Leech 116, **117**, 118, **119**, 202
Fish-louse 140, **141**
Fission 14, 20
Flabelligera 106, 201; *diplochaitus* **107**
Flabelligeridae 201
Flagellate 20, 22
Flagellum viii, 2, 4, 192
Flat Ramshorn 68, **69**
Flat Top 36, **37**
Flat Valve Snail 66, **67**
Flat Winkle 38, **39**
Flatworm vi, 16, 20, **21**, 22, 26, 56, 194 – 195
Float 10
Fluke 22 – 25, 56, 195
Flustra 172, 205; *foliacea* **173**
Flustrella 172, 205; *hispida* **173**
Follicle Mite 144, **145**
Follicular mange 144
Food webs vi
Foot, of mollusc, 32
Foraminifera 2, **3**
Foraminiferida 192
Forcipulata 206
Forest: mites, myriapods, woodlice 160 – 161
Formica rufa 154
Fragmentation, of oligochaetes, 114
Freshwater: arthropods 132 – 139; bivalves 90; bryozoans 174; hydroids 6; leeches 116; medusae 6; oligochaetes 114; shrimps 134; sponges 6; snails 64 – 71
Freshwater Nerite 70
Frill 84
Fruit-fly 20
Fruit Tree Red Spider Mite 152, **153**
Fucus 8, 110

Galathea 122, 203; *squamifera* **123**
Galba 64, 198; *truncatula* **65**
Gall 152
Gall Mite 152
Gametes 2, 12
Gammarus 124, 134, 203; *lacusta* **125**, *pulex* **135**
Gaper 80, 199
Gaping File-shell 84, **85**
Gari 78, 199; *fervensis* **79**
Garlic Glass Snail 50, **51**
Gastropoda 34 – 71, 196 – 198
Gastrotricha viii, **1**, 196
Gattyana 102, 200
Gem Anemone 14, **15**
Gemmules 6
Genus 192
Geophilomorpha 164
Geophilus 203
Germ cell 6
Gerris 138
Gibbous Starlet 180, **181**
Gibbula 36, 196; *cineraria, lineata, magus, umbilicalis* **37**
Gill Maggot 140, **141**
Gill slits 168, 176, 188
Gills 32, 36, 98, 196, 197, 198, 200
Glass Snail 198
Glass sponges 193
Globigerina 2, **3**, 192
Glochidia 90
Glomeris 164, **165**, 203
Glossiphonia 116, 201; *complanata* **117**
Glossiphoniidae 201

Glossobalanus 168, 205; *sarniensis* **169**
Glossy Snail 52, **53**
Glycera 98, 200; *convoluta* **99**
Glyceridae 98, 200
Glycymeris 82, 198; *glycymeris* **83**
Gnathobdellidae 116, 202
Gnathostomata 206
Golfingia 168, 204; *elongata* **169**
Gonads 6, 8, 12
Goniodiscus 58, 198; *rotundatus* **59**
Goniodoris 46, 197; *nodosa* **47**
Goose Barnacle 128, **129**
Goose Foot Star 182, **183**
Gordius 30, **31**, 196
Grantia 4, 193; *compressa* **5**
Great Grey Slug 62, **63**
Great Pond Snail 64, **65**
Great Ramshorn 68, **69**
Great Scallop 84, **85**
Grey Sea-slug 48, **49**
Grey Top 36, **37**
Grooved Razor 80, **81**
Grooved Top-shell 36, **37**
Grove Snail 52, **53**
Gulf Stream 10
Gut vi, 193
Gymnolaemata 174, 205
Gymnosomata 44, 197
Gyraulus 68, 198; *albus* **69**
Gyrocotyle 26, 195; *fimbriata* **27**
Gyrodactylus 22, **23**, 195

Haemaphysalis 142, 204; *punctata* **143**
Haemoglobin 68
Haemopis 116, 202; *sanguisuga* **117**
Hair Worms 30, 31
Hairy Snail 50, **51**, 198
Halacarus 130, 204; *bisculus* **131**
Halarachne 144, 204; *halichoeri* **145**
Halcampa 16, 194; *chrysanthemum* **17**
Halichondria 4, 2, 46, 193; *panicea* **5**
Haliclystus 12, 194; *auricula* **13**
Haliotis 34, 196; *tuberculata* **35**
Halolaelaps 130, 204; *marinus* **131**
Halosydna 94, 200; *gelatinosa* **95**
Haminoea 44, 197; *navicula* **45**
Haplophilus 164, 203; *subterraneus* **165**
Haptor 24
Hard Shell Clam 74, **75**
Harmothoe 184, 186, 200
Harpactea 156, 204; *hombergi* **157**
Harvest Mite 144, **145**
Harvestman 152, 156, 164, 204
Heart-urchin 88
Heath Snail 60, **61**
Heathland spiders 162
Helicacea 198
Helicella 60, 52, 198; *gigaxi* **53**;
  caperata, itala, virgata **61**
Heliozoida viii, 192
Helix 50, 54, 60, 198;
  aspersa **51**; *pomatia* **55**
Helobdella 116, 201; *stagnalis* **117**
Hemichordata 168, 205
Hemiclepsis 116, 201; *marginata* **117**
Henricia 182, 206; *oculata* **183**
Herald Snail 54, **55**
Hermit crabs 18, 40, 100, 122
Herring, food of, 126
Heterocoela 193
Heterodonta 199
Heteronereis 100
Hexacontium **3**, 192
Hexactinellida 193
Hiatella 86, 74, 199; *arctica* **87**
Hiatellacea 199
Hirudidae 202
Hirudinea 94, 116, 201 – 202
Hirudo 116, 202; *medicinalis* **117**
Histriobdella 98

Holdfast viii, 8, 18, 190, 193
Hollow-shelled snail 197
Holopedium 136, 202; *gibberum* **137**
Holothuria 186, 206; *forskali* **187**
Holothuroidea 88, 186, 206
Holotricha 193
Homarus 122, 203; *vulgaris* **123**
Homocoela 193
Hook 26
Horny Coral 18
Horse Leech 116, **117**, 202
Horse Mussel 72, **73**
Horseshoe Crab 204
Hunchbank Scallop 84, **85**
Hunting spiders 166
Hyalinoecia 106, 200; *tubicola* **107**
Hyctia 166, 204; *nivoyi* **167**
Hydra 6, 193; *oligactis* **7**
Hydrachna 138, 204; *globosa* **139**
Hydraulic pressure 172
Hydrobia 66, 197; *jenkinsi* **67**
Hydroida 6, 193
Hydroides 110, 201; *norvegica* **111**
Hydroschendyla 130, 203; *submarina* **131**
Hydrozoa 6, 8, 10, 193 – 194
Hygrobates 138, 204; *longipalpis* **139**
Hygromia 50, 198; *striolata* **51**
Hymeniacidon 4, **5**, 193
Hyperia 12
Hyptiotes 158, 204; *paradoxus* **159**

Ichthyobdellidae 202
Idotea 124, 203; *granulosa* **127**
Inarticulata 170, 205
Indicator species 176
'Ink', from squids, etc. 92
Insecta 122, 203
Intermediate host 26, 56, 64
Introvert 28
Ischnochiton 32, 196; *albus* **33**
Isle of Wight disease 146
Isopoda 124, 132, 140, 203
Itch Mite 144, **145**
Ixodes 142, 204; *ricinus* **143**
Ixodids 142, 204

Jack Sail-by-the-Wind 10, **11**
Janusion 118, 202; *scorpii* **119**
Jaw 100, 196, 203
Jellyfish 8, 12, **13**, 194
Jenkins's Spire Shell 66, **67**
Jet-propulsion 10, 32, 92, 188
Jewel Anemone 18, **19**
Jointed limbs, of arthropods 122
Jumping spiders 162, 166

Kamptozoa 196
Keeled Ramshorn 68, **69**
Keeled Slug 62, **63**
Kentish Snail 50, **51**
Keratosa 193
Kew Gardens 20
Keyhole Limpet 34, **35**
King Rag 100, **101**
Kinorhyncha 195

Labidoplax 186, 184, 206; *digitata* **187**
Lacuna 38, 197; *pallidula, vincta* **39**
Lagisca 94, 200; *extenuata* **95**
Lake Limpet 70, **71**
Lamellibranchia 72, 198
Laminaria 34, 172
Lamprochernes 146, 204; *nodosus* **147**
Lamp-shells 170 – 171, 205
Lancelets 176, **177**
Land planarians 20, **21**
Land Winkle 54, **55**

Langouste 122
Lanice 104, 102, 200; *conchilega* **105**
Lankester, E. Ray, 168
Lantern Shell 199
Large Chrysalis Snail 54, **55**
Large-mouthed Valve Snail 66, **67**
Larvacea 207
Larvae: actinotrocha 170; cypris 124,
  128, **129**; megalopa 126;
  miracidium 24; mysis 122;
  nauplius 124, 126, 128, **129**, 140;
  phyllosoma 122, 126; planula 8, 12;
  pycnogonid 126; redia 24;
  trochophore 100, 120; veliger 176;
  zoea 126
  of crinoid 180; of fluke 24, 56, 64;
  of *Hydra* 6; of *Natica* 40;
  of *Polystoma* 24; of *Rhabditis* 30;
  of sea-squirt 188; of top-shell 36;
  of water-mite 138
Lasaea 88, 199; *rubra* **89**
Laternulacea 82
Lauria 52, 198; *cylindracea* **53**
Leander 124; *serratus* **125**
Leeches 94, 201 – 202;
  freshwater 116 – 117; marine 118 – 119
Leiobunum 152, 204; *rotundum* **157**
Lepas 128, 203, 140; *anatifera* **129**
Lepeophtheirus 140, 202; *salmonis* **141**
Lepidochitona 32, 196; *cinerea* **33**
Lepidonotus 94, 200; *squamatus* **95**
Lepthyphantes 156, 158; *minutus* **157**
Leptodora 132, 202; *kindti* **133**
Leptolimnea 64, 198; *glabra* **65**
Leptomedusae 193
Lepton 88, 199; *squamosum* **89**
Leptopentacta 186, 206; *elongata* **187**
Leptoplana 22, 195; *tremellaris* **23**
Leptosynapta 88, 186, 206; *inhaerens* **187**
Lernaeocera 140, 202; *branchialis* **141**
Lesser Bulin 58, **59**
Lesser Octopus 92, **93**
Leucosolenia 4, **5**, 193
Lichina pygmaea 38, 88
Ligia 124
Lima 84, 199; *hians* **85**
Limapontia 48, 197; *capitata* **49**
Limax 62, 198; *maximus* **63**
Limnaea 64, 24, 198; *auricularia,*
  glabra, palustris, pereger, stagnalis,
  truncatula **65**
Limnaeidae 198
Limnesia 138, 204; *fulgida* **139**
Limnochares 138, 204; *aquatica* **139**
Limnomedusae 6, 193
Limnophilus flavicornis 66
Limpets 34, **35**, 36, 70
Lineus 28, 195; *longissimus* **29**
Linguatula 142, 204; *serrata* **143**
Linyphia 158, 204; *hortensis* **159**
Linyphiidae 130, 164
Lithobiomorpha 154
Lithobius 152, 154, 203; *forficatus* **153**,
  variegatus **155**
Litter Mites 160, **161**
Little Cockle 76, **77**
Little Cuttle 92, **93**
Littorina 38, 197; *littorea, neritoides,*
  obtusata, saxatilis (rudis) **39**
Littorinacea 197
Liver Fluke 24, **25**, 64
Lobster 98, 122, 203
Loliginacea 200
Loligo 92, 200; *forbesi* **93**
Lophophore 168, 170, 172, 204, 205
Lophopus 174, 205; *crystallina* **175**
Louping ill 142
Lugworm 102, **103**, 200
Luidia 182, 206; *sarsi* **183**
Lumbricidae 112, 201
Lumbriculidae 114, 201

Lumbriculus 114, 201; *rivulorum* **115**
Lumbricus 112, 201; *terrestris* **113**
Lumbrinereids 98
Luminescence 2
Lung: of frog 24; of pulmonate 50; of slug 62; of snail 70
Lutraria 80, 199; *lutraria* **81**
Lycosa 150, 130, 204; *amentata* **151**
Lyre Urchin 178, **179**

Macoma 78, 199; *balthica* **79**
Macrobiotus 136, **137**, 202
Macropipus 122, 203; *puber* **123**
Macropodia 124, 203; *rostrata* **125**
Mactra 76, 199; *corallina* **77**
Mactracea 199
Madreporaria 18, 194
Magelona 102, 200; *pappilicornis* **103**
Magelonidae 200
Malacostraca 203
Malaria 148
Malmgrenia 200
Mandibulata 203
Man-o'-War 10, **11**
Mantle, of mollusc, 32
Mantle cavity 32, 34, 36, 196, 197
Margaritifer 90, 199; *margaritifer* **91**
Marginal anchors 12
Marginal plates 182
Marine Leeches 118
Marpessa 58, 198; *laminata* **59**
Marphysa 98, 200; *sanguinea* **99**
Marpissa 156, 162, 204; *muscosa* **157**
Marshland snails 56
Marthasterias 180, 206; *glacialis* **181**
Mastigophora 192
Medicinal Leech 116, **117**, 202
Medusa 6, 8, 10, 12, 193 – 194
Megabunus 156, 204; *diadema* **157**
Megalomma 104, 201; *vesiculosum* **105**
Megalopa 126, **127**
Megaspheric 2
Megninia 146, 204; *cubitalis* **147**
Melanic forms 156, 162
Membranipora 172, 205; *membranipora* **173**
Mercenaria 74, 199; *mercenaria* **75**
Merostomata 204
Mesogastropoda 34, 36, 38, 40, 54, 70, 196 – 197
Mesostigmatida 144, 146
Mesostoma 20, 194; *tetragonum* **21**
Mesozoa 193
Meta 150, 204; *segmentata* **151**
Metacercaria 24
Metamorphosis 100, 168
Metastigmata 142
Metazoa viii, 4
Metridium 14, 194; *senile* **15**
Microhydra 6, **7**, 193
Micrommata 164, 204; *virescens* **165**
Microspheric 2
Mideopsis 138, 204; *orbicularis* **139**
Milax 62, 198; *sowerbyi* **63**
Millipede 154, 160, 203
Milne-Edwardsia 16, 194; *carnea* **17**
Miracidium 24, **25**
Misumena 164, 204; *vatia* **165**
Mites 204: forest 160; parasitic 144, 146; phoretic 146; woodland 154
Modiolus 72, 198; *barbatus, modiolus, phaseolinus* **73**
Molgula 188, 207; *citrina* **189**
Mollusca vii, 18, 22, 24, 122, 126, 196 – 200; freshwater 64 – 71, 90 – 91; marine 33 – 47, 72 – 89, 92 – 93; terrestrial 50 – 63
Monacha 50, 52, 198; *cartusiana, cantiana* **51**, *granulata* **53**
Monaxonida 193

Money-spider 130
Monodonta 36, 196; *lineata* **37**
Monogenea 20, 22, 195
Monopisthocotylea 195
Monoplacophora 196
Monotocardian 36
Montacuta 88, 199; *substriata* **89**
Moon Shell 40, **41**
Moss Bladder Snail 70, **71**
Moss Snail 52, **53**
Mother-of-pearl 34
Mucous: house 188, **189**; net 106; tube 28, 106
Mucronella 172, 205; *coccinea* **173**
Mud Snail 64, **65**
Muggiaea 10, 194; *atlantica* **11**
Muricacea 197
Musculus 72, 198; *discors* **73**
Mussels 198; sea 72; freshwater 90; commensals of 124
Mya 80, 199; *arenaria* **81**
Myacea 199
Mycetozoida 192
Myrianida 96, 200; *pinnigera* **97**
Myriapoda 203
Myriapods vii, 122, 203: domestic 148; forest 160; garden 152; grassland 164; sandhill 166; seashore 130; woodland 154
Myriothela 8, 193; *cocksi* **9**
Myrmarachne 166
Mysidacea 203
Mysis larva 122
Mysis 132, 203; *relicta* **133**
Mystacocarida 203
Mytilacea 198
Mytilus 72, 198; *edulis* **73**
Myxicola 106, 201; *infundibulum* **107**
Myzostoma 120, 180, 202; *cirriferum* **121**
Myzostomaria 202
Myzostomida 120

Naididae 201
Nais 114, **115**, 201
Naked Sea Butterfly 44, **45**
Nassarius 42, 197; *incrassatus, reticulatus* **43**
Natica 40, 197; *catena* **41**
Naticacea 197
Naticidae 40
Nauplius 124, 126, 128, **129**, 132, 136, 140
Nautiloidea 199
Neanthes 100, 200; *diversicolor, fucata, virens* **101**
Necklace Shell 40, **41**, 197
Necrophleophagus 130, 203; *longicornis* **131**
Needle-whelk 40, **41**
Nematocysts 48
Nematoda 12, 30, 196
Nematomorpha 30, 196
Nemertina vi, 28, 195
Nemertopsis 28, 195; *flavida* **29**
Neobisium 130, 204; *maritimum* **131**
Neogastropoda 38, 42, 197
Nephrops 122
Nephtyidae 98, 200
Nephtys 98, 200; *caeca, cirrosa, hombergi* **99**
Neptunea 42, 197; *antiqua* **43**
Nereidae 100, 200
Nereis 100, 200; *fucata* **101**, 122, *diversicolor, longissima, pelagica, virens* **101**
Nerilla 120, 202; *antennata* **121**
Nerine 102, 200; *cirratulus* **103**
Neritacea 196
Neritina 70, 196; *fluviatilis* **71**

Nerve cord, of protochordates, 176
Nervous system vi, 6
Netted Dog Whelk 42, **43**
Netted Slug 62, **63**
Neumania 138, 204; *spinipes* **139**
Noctiluca 2, **3**, 192
Nodosaria 2, **3**, 192
Norway Lobster 122
Notaspidea 197
Notochord vii, 168, 176, 188, 207
Notomastus 102, 200; *latericeus* **103**
Notostraca 202
Nucella 42, 197; *lapillus* **43**
Nucleus viii, 2
Nucula 82, 198; *turgida* **83**
Nuculacea 198
Nuda 194
Nudibranchia 8, 46, 48, 197
Nut Shell 82, **83**, 198
Nymph: of *Parasitus* 130; of pseudoscorpion 130; of tick 142; of water-mite 138
Nymphon 126, **127**, 204

Obelia 8, 193; *geniculata* **9**
Ocenebra 42, 197; *erinacea* **43**
Octolasium 112, 201; *cyaneum* **113**
Octopoda 200
Octopodacea 200
Octopus 92, 32, 199 – 200; *vulgaris* **93**
Oikopleura 188, **189**, 207
Oleacinacea 198
Oligochaeta 94, 112, 114, 201
Onchidella 48, 197; *celtica* **49**
Onchidiacea 48, 197
Onchidoris 46, 197; *oblonga* **47**
Onchosphere 26
Onchydium 48, 197
Oniscus 152, 160, 203; *asellus* **153**
Onychophora 202
Oonops 148, 204; *domesticus* **149**
Opelet Anemone 14, **15**
Operculum: false 54; of serpulid 110 of snail 54, 66; of whelk 42; of zooid 172
Ophiactidae 206
Ophiactis 180, 206; *balli* **181**
Ophiocomidae 206
Ophiocomina 184, 182, 206; *nigra* **185**
Ophiolepidae 206
Ophiopsila 180, 206; *aranea* **181**
Ophiothrix 184, 206; *fragilis* **185**
Ophiotrichidae 206
Ophiura 184, 182, 206; *texturata* **185**
Ophiuroidea 184, **185**, 88, 120, 206
Opiliones 204
Opisthobranchia 44, 48, 197
Opisthopora 201
Opossum Shrimp 124, 132, 203
Oral groove viii
Orb shells 90
Orchestia 124, 203; *gammarella* **125**
Order 192
Oribatidae 160
Ormer 34, **35**
Osmosis 100
Ostracoda 136, 202
Ostrea 84, 199; *virgata* **85**
Ostreacea 199
Otina 38; *otis* **39**
Otter Shell 80, **81**.
Oval Piddock 86, **87**
Oval Venus 74, **75**
Ovary 6
Oviger, of sea-spider, 126
Ovipositor, of *Phalangium*, 152
Owenia 104, 201; *fusiformis* **105**
Oweniidae 201
Oxychilus 50, 52, 198; *alliarius, cellarius* **51**, *helveticus* **53**

Oyster 4, 34, 42, 84, **85**, 199;
commensals of, 124
Oyster Drill 42, **43**

Paddle Worm 96, **97**
Paedomorphosis 188
Pagurus 122, 203; *bernhardus* 18, 100, **123**;
prideauxi 18
Painted Top-shell 36, **37**
Painter's Mussel 90, **91**
Palaemon 124, 203; *serratus* **125**
Palinurus 122, 203; *vulgaris* **123**
Palolos 98, 200
Palp proboscides 82
Palps, of bivalve, 82
Pandora 82, 199; *albida* **83**
Pandoracea 199
Pandora's Box Shell 82, **83**, 199
Panonychus 152, 204; *ulmi* **153**
Paracentrotus 178, 206; *lividus* **179**
Parapodia 44, 94, 200
Parasites, parasitism vi, 20, 88, 195, 196,
202, 204; crustaceans 124, 140;
flatworms 20 – 27, 195;
flukes 22, 24, 195; halacarines 130;
leeches 116, 94; mites 144, 146, 204;
round worms 30, 196; ticks 142, 204;
water-mites 138
of arthropods 30, 196;
of birds 30, 144, 146, 195;
of fish 22, 26, 30, 90, 116, 118, 140,
195, 203; of frogs 24; of insects 24,
26, 30, 138, 146; of mammals 26,
30, 144, 146, 195
Parasitus 130, 146, 204;
coleoptratorum **147**, *kempersi* **131**
Parthenogenesis 136, 140, 144
Parvicardium 76, 199; *exiguum* **77**
Patella 34, 196; *aspera, intermedia,*
vulgata **35**
Patellacea 196
Patina 34, 196; *pellucida* **35**
Pauropoda 154, 203
Pauropus 154, **155**, 203
Pea Crab 124, **125**
Pea Mussel 90, **91**, 199
Pea Urchin 178, **179**
Peachia 16, 194; *hastata* **17**
Peacock Worm 104, **105**, 201
Pearl 84, 90
Pearl Mussel 90, **91**
Pearly Nautilus 199
Pecten 84, 199; *maximus* **85**
Pectinacea 199
Pectinaria 104, 201; *belgica* **105**
Pectinariidae 201
Pedipalp, of spider, 158
Pelagia 12, 194; *noctiluca* **13**
Pelecypoda 198
Pelican's Foot Shell 40, **41**, 197
Pen, shell of squid, 92
Pentastomida 142, 204
Peppery Furrow Shell 78, **79**
Pergamasus 160, **161**
Perinereis 100, 200; *cultrifera* **101**
Peripatus 202
Peritricha 193
Periwinkle 36, 38, **39**, 197
Perophora 188, 207; *listeri* **189**
Petricola 86, 199; *pholadiformis* **87**
Phalangium 152, 204; *opilio* **153**
Phallusia 188, 207; *mammillata* **189**
Phanerozonida 206
Pharus 80, 199, 78; *legumen* **81**
Pharyngobdellae 202
Pharynx 20, 176, 188
Phascolion 168, 204; *strombi* **169**
Pheasant Shell 36, **37**
Philine 44, 197; *quadripartita* **45**
Philodromus 158, 204; *aureolus* **159**

Philoscia 164
Pholadacea 199
Pholas 86, 74, 168, 199; *dactylus* **87**
Pholcus 148, 204; *phalangioides* **149**
Phoresy 130, 146, 160
Phoronida 170, 205
Phoronis 170, 205; *hippocrepia* **171**
Phosphorescence 12, 94, 130, 180
Photosynthesis 2, 6, 22, 192
Phylactolaemata 174, 205
Phyllodoce 96, 200; *lamelligera* **97**
Phyllodocidae 96, 200
Phyllosoma 126, 122, **127**
Phylum vi, 192
Physa 70, 198; *fontinalis* **71**
Physalia 10, 194; *physalis* **11**
Physidae 198
Physophora 10, 194; *hydrostatica* **11**
Phytomastigophora 192
Piddock 86, 168, 180, 199
Pill Bug 164, **165**
Pill Millipede 164, **165**
Pink Ribbon Worm 28, **29**
Pinna 72, 198; *fragilis* **73**
Pinnotheres 124, 203; *pisum* **125**
Pinnules, of crinoids, 180
Pinworm 30
Pionocypris 136, 202; *vidua* **137**
Piscicola 116, 202; *geometra* **117**
Piscicolidae 202
Pisidium 90, 199; *amnicum* **91**
Plaited Door Snail 58, **59**
Planarian 20, 22
Plankton 2, 12, 36, 38, 96, 100, 128,
132, 136, 170, 176, 190
Planorbarius 68, 198; *corneus* **69**
Planorbidae 198
Planorbis 68, 198; *albus, carinatus,*
complanatus, corneus, planorbis
spirorbis, vortex **69**
Planula 8, 12
Platyasterida 206
Platyhelminthes 16, 20 – 27, 28, 194 – 195
Platynereis 100, 200; *dumerili* **101**
Plesiopora Plesiothecata 201
Plesiopora Prosothecata 201
Pleurobrachia 16, 194; *pileus* **17**
Pleurobranchomorpha 197
Pleurobranchus 44, 197;
membranaceus **45**
Pleurogona 207
Plumatella 174, 205; *fungosa* **175**
Plumose Anemone 14, **15**
Plumularia 8, 193; *setacea* **9**
Pluteus 176
Pneumonoeces 24, 195; *variegatus* **25**
Pneumostome 50, 62
Pod Razor 80, **81**
Pod Shell 80, **81**
Pogonophora 205
Point Shell 54, 55
Pointed Snail 60, **61**
Poison 48, 140; of centipede 154;
of spider 148, 162
Polybostrichus 96
Polycelis 20, 195; *nigra* **21**
Polycera 46, 197; *quadrilineata* **47**
Polyceridae 46
Polychaeta 94 – 110, 4, 200 – 201
Polycirrus 102, 200
Polycladida 22, 195
Polydesmus 152, 203; *complanatus* **153**
Polydora 108, **109**, 200
Polygordius 120, 202; *lacteus* **121**
Polymicrodon 160, 203;
polydesmoides **161**
Polymorphism 10, 52, 174
Polynoids 94, 184, 186
Polyopisthocotylea 195
Polyp 6 – 11, 12, 14, 193 – 194
Polyphemus 132, 202; *pediculus* **133**
Polyplacophora 196

Polystoma 24, 195; *integerrimum* **25**
Polystomella 2, **3**, 192
Polyxenus 154, 203; *lagurus* **155**
Polyzoa 46, 172 – 175, 205
Polyzonium 160, **161**, 203
Pomatias 54, 197; *elegans* **55**
Pomatoceros 110, 201; *triqueter* **111**
Pompilius 162
Pondskater 138
Pond Snails 64, 198
Pond Sponge 6
Pondworms 94
Pontobdella 118, 202; *muricata* **119**
Porania 182, 206; *pulvillus* **183**
Porcelain Crab 124, **125**
Porcellana 124, 203; *longicornis* **125**
Porcellio 166, 160, 203; *scaber* **167**
Porifera 4, **5**, 6, **7**, 193
Poromyacea 199
Portuguese Man-o'-War 10, **11**
Potamilla 108, 201; *reniformis* **109**
Potamopyrgus 66, 197; *jenkinsi* **67**
Pot Worm 112, **113**, 201
Praunus 124, **125**, 132, 203
Prawn 88, 124
Priapulida 28, 195
Priapulus 28, 195; *bicaudatus* **29**
Prickly Cockle 76, **77**
Prickly Snail 52, **53**
Proboscis 28, 30, 38, 42, 96, 98, 126,
140, 168, 197, 200
Proboscis Roundworm 30, **31**
Proboscis worms 28, 195
Procerodes 22, 195; *ulvae* **23**
Proglottid 26
Prosobranchia 34 – 43, 54, 66, 70,
196 – 197
Prosopora 201
Prostomium, of nereid, 100
Protein 4
Protobranchia 198
Protochordata vii, 168, 176
Protodrilus 120, 202; *flavocapitatus* **121**
Protoglossus 168, 205; *koehleri* **169**
Protonymphon 26
Protozoa vi, viii, **1**, **2**, **3**, 192 – 193
Psammechinus 178, 206; *miliaris* **179**
Pseudophyllidea 195
Pseudoscorpion 130, 146, 148, 154,
166, 204
Psoroptes 144, 204; *ovis* **145**
Ptenoglossa 197
Pteria 72, 198; *hirundo* **73**
Pteriacea 198
Pterobranchia 168, 205
Puddles 134
Pullet Carpet Shell 74, **75**
Pulmonata 38, 50, 54, 70, 197 – 198
Punctum 58, 198; *pygmaeum* **59**
Pupilla 60, 198; *muscorum* **61**
Purple Heart-urchin 178, **179**
Purple Sea-urchin 178, **179**
Purple Top 36, **37**
Purse sponges 4, **5**
Purse Web Spider 162, **163**
Pycnogonids 126, **127**, 204
Pycnogonum 126, 204; *littorale* **127**
Pyramidula 50, 198; *rupestris* **51**

Quahog 74, **75**
Queen Scallop 84, **85**
Quill Worm 106, **107**

Radial plates, of *Ophiothrix*, 184
Radiolaria 2, **3**, 192
Radix 64, 198;
auricularia, pereger **65**
Radula 40, 42, 44
Ragworm 100, 122, 200

Ramshorn organ 148
Ramshorn Snail 68, 198, **69**
Rayed Artemis 74, **75**
Rayed Snail 56, **57**
Rayed Trough Shell 76, **77**
Razor Shell 80, 199
Red Fowl Mite 146
Red mange 144
Red Rock Worm 98, **99**
Rediae 24, **25**
Redwater fever 142
Regeneration: in annelids 96, 114; in brittle-stars 184
Reproductive capsules 8
Respiratory tree 186
Retinella 56, 58, 198; *nitidula* **59**, *radiatula* **57**
Rhabditis 30, 196; *maupasi* **31**
Rhabdocoela 20, 194
Rhizopoda 2, 192
Rhizostoma 12, 194; *octopus* **13**
Rhizostomae 194
Rhodacarellus 160, **161**, 204
Rhynchobdellae 201 – 202
Rhynchocoela 28
Rhynchodemus 20, 195; *terrestris* **21**
Ribbed Grass Snail 60, **61**
Ribbon worm vi, 195
Rissoacea 197
River Limpet 70, 198
River Mussel 199
River Orb Mussel 90
River Snails 66, **67**, 196
River Sponge 6
Rock Borers 86, 178, 199
Rockling 22, **23**
Rock Lobster 122, **123**
Rock pool 8, 12, 14
Rock Snail 50, **51**
Rockworms 98, 200
Roman Snail 54, **55**
Rostanga 46, 197; *rufescens* **47**
Rosy Feather-star 180, **181**
Rotifera viii, **1**, 12, 196
Rough Winkle 38
Round Worms 30, 196
Rounded Snail 58, **59**
Round-mouthed Snail 54, **55**, 197

Sabella 104, 201; *pavonina* **105**
Sabellaria 108, 201; *alveolata* **109**
Sabellaridae 108, 201
Sabellidae 104 – 109, 201
Saccocirrus 120, **121**, 202
Saccoglossus 168, 205; *cambrensis* **169**
Sacconereis 96
Sacculina 124, 203, 128; *carcini* **125**
Sacoglossa 48, 197
Saddle: of hirudineans 118; of oligochaetes 112
Saddle Oyster 84, **85**, 199
Sagartia 14, 194; *elegans, troglodytes* **15**
Sagartiidae 14
Sagitta 176, 205; *elegans, setosa* **77**
Salmincola 140, 202; *salmonea* **141**
Salpa 188, 207; *fusiformis* **189**
Sand Dollar 178
Sand Egg 188, **189**
Sand Gaper 80, **81**
Sand gastropods 40
Sandhill: arachnids, myriapods, woodlice 166; molluscs 60
Sandhopper 124, 134, 203
Sand Mason 104, **105**
Sand-urchin 178, **179**
Sarcodina 192
Sarcoptes 144, 204; *scabei* **145**
Sarcoptic mange 144
Saxicavacea 199

Scab Mite 144
Scabies 144
Scale 200
Scaleworms 94, 102
Scallop 84, 199
Scalpellum 128, **129**, 203
Scaly leg 144
Scaphopoda 32, **33**, 198
Scarlet Starfish 182, **183**
Schistosoma 24, **25**, 195
Schizodonta 199
Scolex 26
Scolopendromorpha 154
Scrobicularia 78, 199; *plana* **79**
Scutigera 148, 203; *coleoptrata* **149**
Scutigerella 148, **149**, 203
Scutigeromorpha 148
Scyphistoma 12
Scyphozoa 12, 194
Scytodes 148, 204; *thoracica* **149**
Sea Anemones 14 – 19, 122, 193, 194
Sea-cucumbers 88, 178, 184, 186, 206
Sea Ear 34, **35**
Sea Fan 18, **19**
Sea Fern 8
Sea Gooseberry 16, **17**, 194
Sea-grass 12
Sea Hair 8, **9**
Sea Hare 44, **45**
Sea Lemon 46, **47**
Sea-lilies 120, 206
Sea Mats 172
Sea Mouse 94, **95**
Sea Nettle 12, **13**
Sea Potato 178, **179**
Sea-serpent 12
Sea-slater 124
Sea Slugs 46 – 47, 197
Sea Spiders 126, 204
Sea Squirts 207; simple 188; compound 190
Sea Urchins 178, 206
Sedentaria 102, 200 – 201
Sediments 2
Segestria 150, 204; *senoculata* **151**
Segment, segmentation: 94, 112, 122, 196, 200, 201, 202
Segmented worms vi, 94
Segmentina 68, 198; *complanatus, nitida* **69**
Semaeostomae 194
Sepia 92, 200; *officinalis* **93**
Sepiacea 200
Sepiola 92, 200; *atlantica* **93**
Septibranchia 82, 199
Septum 16, 18
Serpula 110, 201; *vermicularis* **111**
Serpulidae 110, 201
Serrated Wrack 110
Sertularia 8, 193, 48; *operculata* **9**
Sessile jellyfish 12, 194
Sex hormones 24
Sexual dimorphism 120, 128
Sheep scab 144
Sheep Tick 142, **143**
Shelled Slugs 198
Shield-shelled Slug 62, **63**
Shiny Ramshorn 68, **69**
Shiny Snail 56, **57**
Shipworms 86, 199
Shore Crab 124, **125**
Shrimp 122, 203
Sida 136, 202; *crystallina* **137**
Sidnyum 190, 207; *turbinatum* **191**
Silicon 4
Silk: spun by false scorpion 154; by spider 132, 148, 150, 152, 158; by millipede 160
Silky Snail 52, **53**
Silver Tommy 36, **37**
Simocephalus 136, 202; *vetulus* **137**
Singa 162, 204; *pygmaea* **163**

Sinistral coil 70
Sinistral shell 58
Siphon 40, 42, 74, 76, 188
Siphonophora 10, 194
Sipunculoida 168, 204
Sisyra 6, **7**
Skeleton 4
Skeleton shrimp 126, **127**
Slime Fungi 192
Slipper Limpet 34, **35**, 197
Slippery Snail 52, **53**
Slit Limpet 34, **35**
Slugs 196 – 198; land 62; sea 46 – 49
Smooth Snail 58, **59**
Snails 196 – 198: downland 54; garden 50; hedge 52; marshland 56; pond 64; ramshorn 68; river 66; sand 60; water 70; woodland 58 as prey of thrush 52
Snout Mite 130
Soft Coral 18
Soft Shell Clam 80, **81**
Solaster 182, 206; *papposus* **183**
Solen 80, 199; *marginatus* **81**
Solenacea 199
Spadella 176, **177**, 205
Spatangoida 88, 178
Spatangus 178, 88, 206; *purpureus* **179**
Species 192
Sperm 8
Sphaeriacea 199
Sphaerium 90, 199; *corneum, rivicola* **91**
Spicule 4, 6, 18, 186, 193, 206
Spider crab 124
Spiders: vii, 204; domestic 148; freshwater 132; garden 150; grassland 164; heathland 162; jumping 162; sand 166; seashore 130; tree 156; woodland 158
Spiders' webs: hammock 156, 158; orb 150, 156, 158, 162; purse 162; sheet 148; triangular 158
Spindle Shell 42, **43**
Spines: of brittle-star 184; of decapod larva 126; of *Palinurus* 122; of sea-urchin 178; of starfish 180
Spinneret 150, 158, 204
Spinther 4
Spinturnix 146, 204; *vespertilionis* **147**
Spinulosida 206
Spiny-headed worms, 195
Spiny Lobster 122, **123**
Spiny Starfish 180, **181**
Spionidae 102, 108, 200
Spire Shells 40, 197
Spire Snails 198
Spirorbis 110, 201; *borealis* **111**
Spirostomum viii, **1**, 193
Spirotricha 193
Spisula 76, 182, 199; *elliptica* **77**
Sponges vi, 22, 46, 96, 193; marine 4 – 5; freshwater 6 – 7
Spongilla Fly Larva 6
Sporocyst 24, **25**
Sporozoa 192
Spotted Snake Millipede 152, **153**
Spring-tail 20
Square Tailed Worm 114, **115**
Squat Lobster 122, **123**
Squid 32, 92, 199 – 200
Staircase Shell 40, **41**
Stalked barnacles 128
Starfish 40, 180 – 183, 206
Statoblast 192
Stauromedusae 12, 194
Steganacarus 160, **161**, 204
Steneotarsonemus 152, 204; *pallidus* **153**
Stenoglossa 197
Stenolaemata 205
Stenostomum 20, **21**, 194
Stentor viii, **1**, 193

Stickleback 140, **141**
Stinging cells 6, 8, 10, 12, 18, 193, 194
Sting Winkle 42, **43**
Strawberry Snail 50, **51**
Strigamia 130; *maritima* **131**
Striped Snail 60, **61**
Striped Venus 74, **75**
Strombacea 197
Styelopsis 190, 207; *grossularia* **191**
Stylaria 114, 201; *lacustris* **115**
Stylets 28, 146
Stylommatophora 198
Suberites 4, **5**, 193
Succinea 56, 198; *putris* **57**
Succineacea 198
Sucker: of fluke 24; of leech 116, 118;
    of tapeworm 26; of myzostomid 120
Sunset Shell 78, **79**
Sunstar 182, **183**
Swan Mussel 90, **91**
Swimmerets, of megalopa 126
Swimmer's itch 24
Swimming bell 10
Swollen River Mussel 90, **91**
Sycon 4, 193; *coronatum* **5**
Syllidae 96, 200
Syllis 96, 200; *hyalina* **97**
Symbiosis viii, 6, 18, 22, 88, 140
Symmetry, five-rayed 178, 205
Symphyla 148, 203
Synageles 166, 204; *venator* **167**
Syndosmya 78, 199; *alba* **79**

Tadpole: of *Polystoma* 24;
    of tunicate 188
Taenia 26, 195; *solium* **27**
Taeniarhynchus saginatus 26
Tapeworms 26 – 27, 195
Tardigrada 136, 202
Tawny Glass Snail 56, **57**
Taxodonta 198
Taylor's Spire Shell 70
Tealia 14, 194; *felina* **15**
Tegenaria 148, 204; *saeva* **149**
Tellina 78, 199;
    crassa, fabula, tenuis **79**
Tellinacea 199
Tentacles 6, 8, 10, 12, 14, 16, 102,
    194, 196, 197, 198, 200, 204
Tentaculata 194
Terebella 104, 102, 200; *lapidaria* **105**
Terebellidae 104, 200
Terebratulina 170, 205;
    caputserpentis **171**
Teredo 86, 199; *norvegica* **87**
Tergipes 48, 197; *despectus* **49**
Test 178
Testacella 62, 198; *haliotidea* **63**
Testacellidae 62
Testacida 192
Tetractinellida 193
Tetragnatha 158, 204; *extensa* **159**
Tetranychidae 152, 154
Tetrastemma 28, 195; *flavida* **29**
Thais 42, 197
Thalassema, 120, 200; *neptuni* **121**
Thaliacea 207
Theba 50, 60, 198; *cartusiana* **51**;
    pisana **51**
Thecosomata 197
Theodoxus 70, 196; *fluviatilis* **71**
Theridion 204
Theromyzon 116, 201; *tesselatum* **117**
Thersitina 140, 202; *gasterostei* **141**
Thick Top-shell 36, **37**
Thick Trough Shell 76, **77**
Thin Tellin 78, **79**

Thomisus 162, 158, 166, 204;
    onustus **163**
Threadworms 30, 196
Three-toothed Snail 58, **59**
Thyone 186, 206; *fusus* **187**
Tibellus 166, 158, 204; *maritimus* **167**
Ticks 142, 204
Timoclea 74, 199; *ovata* **75**
Tomopteridae 96, 200
Tomopteris 96, **97**, 200
Tongue-worms 142
Tonicella 32, 196; *marmorea* **33**
Toothed Snails 198
Toothless Chrysalis Snail 56, **57**
Topshell 34, 36
Torsion 34, 42, 44
Tortoiseshell Limpet 34, **35**
Trachylina 193
Tree Snail 58, **59**
Trematoda 22
Trepang 186
Trichia 50, 198; *hispida, striolata* **51**
Trichoniscus 160, 164, 203; *pusillus* **161**
Tricladida 20, 22, 195
Tricolia 36, 196; *pullus* **37**
Triops 134, 202; *cancriformis* **135**
Trivia 36, 197; *arctica, monacha* **37**
Trochacea 196
Trochophore 36, 100, 120, 176
Trogulus 164, 204; *tricarinatus* **165**
Trombicula 144, 204; *autumnalis* **145**
Trough Shells 76, 199
True Worms 200
Trunk, of enteropneust, 168
Tube-foot 184, 186, 205, 206
Tubifex 114, 201; *tubifex* **115**
Tubificidae 114, 201
Tubulanus 28, 195; *annulatus* **29**
Tubularia 8, 126, 193; *larynx* **9**
Tunicates 22, 96, 188 – 191, 207
Turbellaria 16, 20, 22, 24, 194 – 195
Turret-shell 40, **41**, 168, 197
Turritella 40, 197, 168, *communis* **41**
Turritellidae 40
Tusk-shells 32, **33**, 198
Two-toothed Door Snail 58, **59**
Typical Snails 198

Umbilicus, of snail 52
Umbonula 172, 205; *verrucosa* **173**
Unio 90, 138, 199; *margaritifer,
    pictorum, tumidus* **91**
Unionacea 199
Unionicola 138, 204; *intermidia* **139**
Urochordata 188 – 190, 207
Uropoda 160, **161**, 204
Uropodidae 160
Urosalpinx 42, 197; *cinerea* **43**

Vacuole viii
Vallonia 60, 198; *costata* **61**
Valvata 66, 196; *cristata, macrostoma,
    piscinalis* **67**
Valvatacea 196
Valve Snails 196
Variegated Scallop 84, **85**
Velella 10, 193; *velella* **11**
Veliger 36, 176
Velvet Mite 146, **147**
Velvet Swimming Crab 122, **123**
Veneracea 199
Venerupis 74, 199; *decussata, pullastra,
    saxatilis* **75**
Venom, of centipede 154

Venus 74, 182, 199; *fasciata,
    mercenaria, ovata, striatula* **75**
Venus Clams 74, 199
Venus's Girdle 16, **17**
Verruca 128, 203; *stroemia* **129**
Vertiginacea **198**
Vertigo 56, 198; *pygmaea* **57**
Violet snails 197
Virus vi
Vitrea 56, 198; *crystallina* **57**
Vitrina 198
Viviparity 14, 180, 190
Viviparus 66, 196; *viviparus* **67**
Volvox viii, 192, *globator* **1**
Vorticella viii, **1**, 193

Walckenaera 164, 204; *accuminata* **165**
Wandering Snail 64, **65**
Water-bear 136, 202
Water-flea 6, 136, 202
Water-louse 132, **133**
Water-mite 138
Water Spider 132, **133**
Wedge Shells 199
Wendletraps 197
Wheel animalcules 196
Whelks 42, 110, 197
Whipworm 30
Whirlpool Ramshorn 68, **69**
White-Lipped Banded Snail 52, **53**
White or Sandhill Snail 60, **61**
White Piddock 86, **87**
White Ramshorn 68, **69**
White Weed 8
White Worm 98, **99**
Whorl Snail 56, **57**
Wing Oyster 72, **73**, 198
Winter cysts, of *Lophopus*, 174
Wolf-spider 130, 150
Wood Ant 154
Woodland: false scorpions, mites,
    myriapods 154; snails 58; spiders 158
Woodlice 203: forest 160; garden 152;
    grassland 166; sandhill 166
Wood Piddock 86, **87**
Worm Leech 116, **117**
Worms: acorn 168; annelid 94 – 121;
    arrowworms 176; earth 112; errant
    polychaetes 96 – 101; flatworms 20, 22;
    higher 168; lowly 28 – 31;
    oligochaetes 114; proboscis 28;
    roundworms 30; scaleworms 94;
    sedentary polychaetes 102 – 111;
    tapeworms 26
Wrinkled Snail 60, **61**

Xenobalanus 140, **141**, 203
Xenocoeloma 140
Xylophaga 86, 199; *dorsalis* **87**
Xysticus 158, 166, 204; *lanio* **159**

Zebra Mussel 90, **91**, 199
Zelotes 164, **165**, 204
Zeugobranchia 196
Zirphaea 86, 199; *crispata* **87**
Zoantharia 194
Zoanthidea 18, 194
Zoea 126, **127**
Zonation 38
Zonitacea 198
Zonitoides 56, 198; *nitidus* **57**
Zooid 172, 174
Zoomastigophora 192
Zostera 12, 36
Zygiella 156, 204; *atrica* **157**
Zygote 2, 12

STATCRUNCH INTEGRATED THROUGHOUT

StatCrunch® is an online statistical software website that enables users to perform complex analyses, share data sets, and generate compelling reports of their data. Interactive graphics are embedded to help users understand statistical concepts; graphics are available for export to enrich reports with visual representations of data.

STATCRUNCH EXAMPLES

Examples with a **SC** icon indicate that you can log on to StatCrunch to learn how the example problem can be solved using the StatCrunch software. Note: accessing this content online requires a MyStatLab or StatCrunch account.

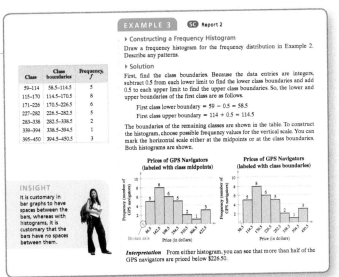

page 42

STATCRUNCH REPORTS

Data sets for 50 of the text's examples are provided in StatCrunch. Interactive, step-by-step instructions guide you through how to use StatCrunch to solve the example problem.

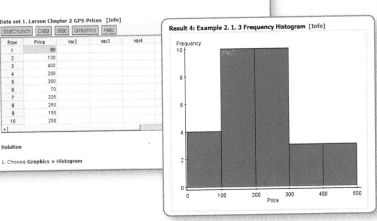

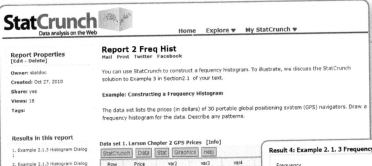

STATCRUNCH EXERCISES

Specific **SC** icon exercises direct you to solve the problem using StatCrunch, which allows you to practice using statistical software.

SC *In Exercises 27 and 28, use StatCrunch to construct 90%, 95%, and 99% confidence intervals for the population proportion. Interpret the results and compare the widths of the confidence intervals.*

27. Congress In a survey of 1025 U.S. adults, 802 disapprove of the job Congress is doing. *(Adapted from The Gallup Poll)*

28. UFOs In a survey of 2303 U.S. adults, 734 believe in UFOs. *(Adapted from Harris Interactive)*

page 335

CENTRAL TEXAS COLLEGE EDITION

ELEMENTARY STATISTICS
PICTURING THE WORLD
RON LARSON • BETSY FARBER

Taken from:

Elementary Statistics: Picturing the World, Fifth Edition
by Ron Larson and Betsy Farber

CENTRAL TEXAS COLLEGE

PEARSON

Cover Art: Courtesy of PhotoDisc/Getty Images.

Taken from:

Elementary Statistics: Picturing the World, Fifth Edition
by Ron Larson and Betsy Farber
Copyright © 2012, 2009, 2006, 2003 by Pearson Education, Inc.
Published by Prentice Hall
Upper Saddle River, New Jersey 07458

This special edition published in cooperation with Pearson Learning Solutions.

Pearson Learning Solutions, 501 Boylston Street, Suite 900, Boston, MA 02116
A Pearson Education Company
www.pearsoned.com

Printed in the United States of America

2 3 4 5 6 V0ZU 15 14 13 12 11

000200010270769385

SB

ISBN 10: 1-256-11595-9
ISBN 13: 978-1-256-11595-3

The Pathway to Success in Your College Courses

You've made the decision to take some college-level courses with Central Texas College. Your reason may be one of many: to earn a two-year degree and transfer to a four-year institution, to earn promotion points, or for self improvement. Regardless of why you are taking the courses, you want to do the best you can, and we at CTC want the same for you.

Before you attend that first lecture or complete your first reading assignment, go through this material to get on the right track and avoid speed bumps and dangerous detours.

Time Management

The time you spend preparing to study relates directly to a successful outcome. But time, or lack thereof, is a common complaint. In fact, can you identify with Louis? (See the quotation below.)

I am definitely going to take a course on time management . . . just as soon as I can work it into my schedule.   Louis E. Boone

How often do you comment that you don't have enough time to accomplish everything you wish to do? If you have a job and/or a family, school work may be a low priority. You can't add hours to the day or days to the week, but if you learn to plan your time wisely, you should be able to make better use of the time you have. As an added bonus, you should feel less stress.

The first step in time management is to see how you currently spend your time. You may spend 40 hours a week on your job, but what about commuting time? How much time do you give to your friends and family? Have you ever taken the time to add up all of the hours you spend on your regular activities? Is there any time left to sleep?

Go to the study skills self-help information site at http://www.ucc.vt.edu/stdysk/stdyhlp.html and take the survey "Where Does Time Go?" Then, view the suggestions for time scheduling:

* Time Scheduling Suggestions
* More Information on Time Scheduling

Additional information on time management is available at study skills and strategies (http://www.studygs.net/). You might want to examine some of the other links as well.

* Managing Time at http://www.studygs.net/timman.htm
* Scheduling and Setting Goals at http://www.studygs.net/timman.htm

How-to-Study.com offers hints on managing time on a term, weekly, and daily basis at http://www.how-to-study.com/Keeping%20Track%20of%20Assignments.htm.

Use the *Cliffs Notes* website (you're probably familiar with the books) at http://www.cliffsnotes.com/WileyCDA/Section/id-106283.html (see "How to Get a Grip on Stress and Time Management") for a cram session on time management.

Study Environment

Now that you have time under control, let's look at your study environment. Where you study and how you study is as important as how often you study.

You should identify a quiet place with a desk or table, a chair, and good lighting. Your bed might be inviting, but remember your goal is to stay awake and concentrate.

Although music or some type of background noise might be ok, avoid the television. It's too easy to get engrossed in a show rather than your course work. Watching *CSI* can be interesting, but it probably won't help you with your *Introduction to Criminal Justice* final exam.

Make sure you have everything you need: your textbooks, notes, paper and pencil, and a clock. Why a clock? To help you manage your time.

And, don't forget to take regular breaks.

Visit these websites for hints on preparing to study. Remember, these are just a few of the sites available to you on the Web. You can also search for sites related to "how to study" or "study guides."

- Read about just about everything related to how to study at http://www.how-to-study.com/.
- Examine hints for creating a good learning space at http://www.ablongman.com/textbooktips/index.html. Click the "Home Sweet Home" (Create a Good Learning Space) box.
- Find study skills self help information at http://www.ucc.vt.edu/stdysk/stdyhlp.html.
  - Take a study environment analysis survey at http://www.ucc.vt.edu/stdysk/studydis.html.
  - Learn how to control your environment at http://www.ucc.vt.edu/stdysk/control.html.
  - Evaluate your current study skills at http://www.ucc.vt.edu/stdysk/checklis.html.
  - Read some concentration guidelines at http://www.ucc.vt.edu/stdysk/concentr.html.
- Refer to the lists of study guides and strategies at http://www.studygs.net/.
- Improve your study skills and see if you are guilty of the ten traps of studying at http://caps.unc.edu/TenTraps.html.

Learning Styles

Do you have to *see* a name or address in writing in order to remember it?

Do you put instructions aside and refer to them only if you run into trouble putting a bookcase together or using a new computer program?

Do you enjoy audio books, or do you find your mind wandering as you listen?

Your answers to these questions relate to your preferred learning style and like clothing, one learning style doesn't fit all.

Take the learning style inventory at http://www.vark-learn.com/english/index.asp to identify your preferred learning style from among the following (note that some inventories show only three learning styles: visual, aural, and kinesthetic). Then, view the VARK helpsheets at http://www.vark-learn.com/english/page.asp?p=helpsheets to develop effective study skills for your preferred learning style(s).

Visual Learners Learn Through Seeing

These learners need to see the teacher's body language and facial expression to fully understand the content of a lesson. They tend to prefer sitting at the front of the classroom to avoid visual obstructions (e.g. people's heads). They may think in pictures and learn best from visual displays, including diagrams, illustrated text books, overhead transparencies, videos, flip charts, and hand-outs. During a lecture or classroom discussion, visual learners often prefer to take detailed notes to absorb the information.

If you are a visual learner, here are some suggestions just for you:

- Use visual materials such as pictures, charts, maps, graphs, etc.
- Have a clear view of your teachers when they are speaking so you can see their body language and facial expression
- Use color to highlight important points in text
- Take notes or ask your teacher to provide handouts
- Illustrate your ideas as a picture or brainstorming bubble before writing them down
- Write a story and illustrate it
- Use multi-media (e.g. computers, videos, and filmstrips)
- Study in a quiet place away from verbal disturbances
- Read illustrated books
- Visualize information as a picture to aid memorization

Aural Learners Learn Through Listening

They learn best through verbal lectures, discussions, talking things through, and listening to what others have to say. Auditory learners interpret the underlying meanings of speech through listening to tone of voice, pitch, speed, and other nuances. Written information may have little meaning until it is heard. These learners often benefit from reading text aloud and using a tape recorder.

If you are an aural learner, here are some suggestions just for you:

- Participate in class discussions/debates
- Make speeches and presentations
- Use a tape recorder during lectures instead of taking notes
- Read text out aloud
- Create musical jingles to aid memorization
- Create mnemonics to aid memorization
- Discuss your ideas verbally
- Dictate to someone while they write down your thoughts
- Use verbal analogies and story telling to demonstrate your point

Read/Write Learners Learn Through Reading and Writing

These learners learn best by reading and re-reading the textbook and their notes, writing and rewriting their notes, and in general, organizing items into lists.

Kinesthetic Learners Learn Through Moving, Doing, and Touching

Tactile/kinesthetic persons learn best through a hands-on approach, actively exploring the physical world around them. They may find it hard to sit still for long periods and may become distracted by their need for activity and exploration.

If you are a tactile/kinesthetic learner, here are some suggestions just for you:

- Take frequent study breaks
- Move around to learn new things (e.g. read while on an exercise bike, mold a piece of clay to learn a new concept)
- Work at a standing position
- Chew gum while studying
- Use bright colors to highlight reading material

- Dress up your work space with posters
- If you wish, listen to music while you study
- Skim through reading material to get a rough idea what it is about before settling down to read it in detail

Multimodal Learners

These learners don't have a single preferred learning style. They learn best through combinations. If you have multiple preferences you are in the majority—somewhere between fifty and seventy percent of any population seems to fit into that group.

VARK is one of many learning style inventories. Here are some other sites: http://www.Idpride.net/learningstyles.MI.htm and http://www.chaminade.org/inspire/learnstl.htm. If you want to try others, do a search on the Internet.

Activities

Now that you know your learning style(s), try these activities to see how your learning style can affect performance.

Activity A

1. Go to the Learning Resources site at http://literacynet.org/cnnsf/home.html.
2. Click on the Story Archives button.
3. Select a story that interests you.
4. Select Story and read the text.
5. After reading the story, click on Multiple Choice and answer the questions.

How did you do? Go to the next activity.

Activity B

1. Go to the Learning Resources site at http://literacynet.org/cnnsf/home.html.
2. Click on the Story Archives button.
3. Select a story that interests you.
4. Select the Movie option (you will need RealPlayer) to watch the video of the story. Do not read the text of the story.
5. After watching the video, click on Multiple Choice and answer the questions.

How did you do? Go to the next activity.

Activity C

1. Go to the Learning Resources site at http://literacynet.org/cnnsf/home.html.
2. Click on the Story Archives button.
3. Select a story that interests you.
4. Select the Movie option (you will need RealPlayer) to watch the video of the story.
5. After watching the video, go back and read the text of the story.
6. Click on Multiple Choice and answer the questions.

How did you do? Was your score higher on one of these activities than on the others? If so, it may be a result of your learning style. Learn to use this to your advantage.

Reading Skills and Strategies

You may be saying to yourself, "I learned to read in first grade! Why do I need to think about reading strategies in college?"

Yes, you can read the words in your textbooks, but how good is your comprehension of their meaning?

How often have you read a chapter, or even part of a chapter, in a textbook and then thought, "What did I just read?" Could you answer questions about the content without referring back to the pages?

And slowing your reading speed does not necessarily increase your comprehension. It only means that it takes you longer to read the assignment.

Experiment on Reading

Dr. Perry (psychologist), Director of the Harvard Reading-Study Center, gave 1500 first-year students a thirty-page chapter from a history book to read, with the explanation that in about twenty minutes they would be stopped and asked to identify the important details and to write an essay on what they had read.

The class scored well on a multiple-choice test on detail, but only fifteen students of 1500 (1%) were able to write a short statement on what the chapter was about in terms of its basic theme. Why? Only fifteen readers had thought of reading the last paragraph marked "Summary" or of skimming down the descriptive flags in the margin.

This demonstration of "obedient purposelessness" is evidence of "an enormous amount of wasted effort" in freshmen study. Dr. Perry suggested that students ask themselves what it is they want to get out of a reading assignment, then look around for those points. Students should "talk to themselves" while reading, asking "is this the point I'm looking for?"

Source: *A Harvard Report on Reading Improvement* at
http://www.dartmouth.edu/admin/acskills/lsg/harvard.html

Would you have been

- One of the 1% who skimmed the thirty pages to get the main idea and the important points, or
- One of the 99% who started reading page 1, continued to page 2, and so on until time was called?

Good reading skills are essential to your success in your college-level classes. Here are a couple of reasons why:

- In high school, you may have been able to get good grades without reading much of the text. Now that you're in college, professors will expect you to read the textbook and they may test you on information not discussed in class but covered in the reading. In fact, many professors test on assigned readings as a check to make sure students are using their texts.

- The average freshman is assigned over 250 pages of reading each week, so clearly you're going to need to keep up with your reading assignments. If you do not read during week one, that means that you will need to read 500 pages the next week—just to stay caught up! If you choose not to read during the second week either ... well, you can see how the work can just snowball.

Go to http://www.ablongman.com/textbooktips/index.html for quick and easy tips to improve your reading:

- Click on "No Dorothy, You're Not in High School Any More" (Be Responsible for Your Reading) for an explanation of why reading your textbook is so important as well as hints for how to read your textbook.

- Select "Keep Up With Your Reading" (Training Schedule) for tips on staying on track. Be sure to examine the sample schedule.

- Refer to "Do You Need Me to Draw You a Picture?" (Read Everything in the Text) for further information.

Another resource for improving your reading comprehension is at http://www.marin.cc.ca.us/%7Edon/Study/7read.html. These are the main ideas:

- Develop a broad background.
- Know the structure of paragraphs.
- Identify the type of reasoning.
- Anticipate and predict.
- Look for the method of organization.
- Create motivation and interest.
- Pay attention to supporting cues.
- Highlight, summarize, and review.
- Build a good vocabulary.
- Use a systematic reading technique like SQR3.
- Monitor effectiveness.

In order to know what you need to do, you must identify where you are. Here's how you can check your reading skills:

1. Go to the Learning Resources site at http://literacynet.org/cnnsf/home.html.
2. Click on the Current Story button.
3. Select Story.
4. Read the story.
5. After reading the story, click on Multiple Choice from the choices at the left side of the screen and answer the questions.

How did you do? If you missed any questions in the different areas listed below, you may want to spend some time honing your reading skills.

- Vocabulary: 10 questions
- Word Selection: 5 questions
- Multiple Choice: 5 questions
- Sequencing: 5 questions
- Conclusions: 4 questions

Improving Your Reading Skills and Applying Reading Strategies
A good reader:

- Seizes the main ideas
- Thinks about what the author is saying
- Is active, not passive
- Concentrates on what is being read
- Remembers as much as possible
- Applies what is being read to personal experience

Go to http://www.how-to-study.com/Improving%20Reading%20Skills.htm for more on reading skills.

SQ3R is one recommended method for improving your reading comprehension. The letters in the name stand for these five steps:

1. Survey: Before you read, scan the titles, headings, pictures, and summaries. Consider using the heading and subheadings as an outline for notes as you read.
2. Question: Ask yourself questions based on Step 1 and look for answers as you complete Step 3. For example, if a subheading is entitled "Basic Concepts of Reading," change it to read, "What are the Basic Concepts of Reading?"

3. Read: Read and take notes.

4. Recall: Without referring to the book or your notes, think about what you have read. See if your questions were answered. Could you explain the content to someone else? Try putting major concepts in your own words.

5. Review: Look at your questions, answers, notes, and book to see how well you did recall. Observe carefully the points stated incorrectly or omitted. Fix carefully in mind the logical sequence of the entire idea, concepts, or problem. Finish up with a mental picture of the whole.

Another method is **PQR3**, which stands for:

1. Preview: Preview what you are going to read.

2. Question: Question what you are going to learn after the preview.

3. Read: Read the assignment.

4. Recite: Stop every once in a while, look up from the book, and put in your own words what you have just read.

5. Review: After you have finished, review the main points.

(Sounds similar to **SQ3R**, doesn't it?) Go to http://www.how-to-study.com/pqr.htm to learn more about this method.

There is even a related study method known as **M.U.R.D.E.R.**, which stands for:

1. Mood: Set a *positive* mood for yourself to study in.

2. Understand: Mark any information *you don't understand* in a particular unit and keep a focus on one unit or a manageable group of exercise.

3. Recall: After studying the unit, stop and put what you have learned *into your own words*.

4. Digest: Go back to what you did not understand and reconsider the information. Contact external expert sources (e.g., other books or an instructor) if you still cannot understand it.

5. Expand: In this step, ask *three kinds of questions* concerning the studied material:

   • If I could speak to the author, what questions would I ask or what criticism would I offer?

   • How could I apply this material to what I am interested in?

   • How could I make this information interesting and understandable to other students?

6. Review: Go over the material you've covered. Review what strategies helped you understand and/or retain information in the past and apply these to your current studies.

Check this system out at http://www.studygs.net/murder.htm. Are you beginning to see similarities among the various systems?

If you search the Internet, you will find an abundance of sites dealing with reading strategies. Some are good; others are not so good. The sites listed below are a good place to start.

   • Read efficiently by reading intelligently at http://www.mindtools.com/rdstratg.html.

   • Make your textbook work for you at http://www.ablongman.com/textbooktips/index.html.

   • Make sense of reading strategies at http://www.suite101.com/article.cfm/1411/9169.

   • Vary your reading strategies by subject at http://www.utexas.edu/student/utlc/handouts/553.html.

As you visit the various websites (http://www.ucc.vt.edu/stdysk/stdyhlp.html, http://www.studygs.net/, and http://www.how-to-study.com/, for example), you might notice distinctions being made based on the type of material being read, i.e., a strategy for reading novels, how to read a difficult book, skimming and scanning scientific material, how to read essays you must analyze. This emphasizes the fact that how you read should depend on what you are reading and your purpose for reading it.

Note Taking

Why take notes?

- It triggers basic lecturing processes and helps you to remember information.
- It helps you to concentrate in class.
- It helps you prepare for tests.
- Your notes are often a source of valuable clues for what information the instructor thinks is most important (i.e., what will show up on the next test).
- Your notes often contain information that cannot be found elsewhere (i.e., in your textbook).

Evaluate your present note-taking system. Ask yourself:

- Did I use complete phrases or sentences that will mean something to me later?
- Did I use any form at all?
- Are my notes clear or confusing?
- Did I capture main points and all subpoints?
- Did I streamline using abbreviations and shortcuts?

If you answered no to any of these questions, you may need to develop some new note-taking skills!

Here are some guidelines for taking notes:

- Concentrate on the lecture or on the reading material.
- Take notes consistently.
- Take notes selectively. Do NOT try to write down every word. Remember that the average lecturer speaks approximately 125–140 words per minute, and the average note-taker writes at a rate of about 25 words per minute.
- Translate ideas into your own words.
- Organize notes into some sort of logical form.
- Be brief. Write down only the major points and important information.
- Write legibly. Notes are useless if you cannot read them later!
- Don't be concerned with spelling and grammar.

This information is from http://www.arc.sbc.edu/notes.html. Click on the link for more on how to streamline your notes. This document provides more information on note taking: http://www.dartmouth.edu/~acskills/docs/taking_notes.doc.

There are a number of different methods for taking notes and they may look very different.

These are some common notetaking methods:

- The Cornell Method (this MSWord document provides more on this method: http://www.dartmouth.edu/~acskills/docs/cornell_note_taking.doc)
- The Outline Method
- The Mapping Method
- The Charting Method
- The Sentence Method

For details on these methods, go to http://www.sas.calpoly.edu/asc/ssl/notetaking .systems.html. Also check out this resource about note taking: http://www.how-to-study.com/Taking%20Notes%20in%20Class.htm.

Go to http://www.ablongman.com/textbooktips/index.html and select the following for more on note taking:

- Zen and the Art of Note-Taking (Rely on Your Text and Class Notes)
- Don't Look Into the (High) Lighter (Take Notes in Your Text)
- Select the links related to note taking at http://www.ucc.vt.edu/stdysk/ stdyhlp.html.
- View http://www.how-to-study.com/Taking%20Notes%20in%20Class.htm at How-to-Study.com.
- Finally, check out these sites for hints on using your notes effectively to review: http://www.yorku.ca/cdc/lsp/notesonline/note5.htm and http://www.yorku.ca/ cdc/lsp/notesonline/note6.htm.

If some of these links don't work or if you are interested in further research, use a search tool and type in "note taking."

Practice, Practice, Practice

Practice makes perfect! If you practice enough, you won't have to think about applying reading strategies or taking notes. You will just do it.

Start small by using the short stories at http://literacynet.org/cnnsf/home.html to practice your reading skills and reading strategies.

1. Click on the Story Archives button.
2. Select a story that interests you.
3. Read the story.
4. After reading the story, complete the short quizzes listed at the left:
   - Vocabulary
   - Word Selection
   - Multiple Choice
   - Sequencing
   - Conclusions

Memory Techniques

We hope that this information on preparing to study has been helpful, but do you feel that your problem is remembering? Don't worry. There are ways to help you build your memory skills too. Some general hints on remembering are available at http://www.ucc.vt.edu/stdysk/remember.html, and a number of tools and techniques for improving your memory are presented at http://www.mindtools.com/ memory.html. These tools include acronyms and mnemonics.

An *acronym* is defined as "a word formed from the initial letters of a name," such as PCS for permanent change of station or SOC for Servicemembers Opportunity Colleges, "or by combining initial letters or parts of a series of words," as radar for radio detecting and ranging.

A *mnemonic* is defined as "a device, such as a formula or rhyme, used as an aid in remembering."

As a child, you might have determined the number of days in a given month by reciting the rhyme "Thirty days hath September, April, June, and November . . ." or by using your knuckles ("peaks" have 31 days and "valleys" have 30, except February, of course).

If you have studied music, you might have used these techniques for remembering the names of the notes: FACE represents the names of the notes in the spaces on the staff. The first letters of the words in sentence "Every good boy does fine" represent the names of the notes on the lines on the staff.

A mnemonic used to recall the steps for simplifying algebraic expressions is "Please excuse my dear Aunt Sally."

1. Perform operations within the innermost parentheses and work outward.
2. Evaluate all exponential expressions.
3. Perform multiplications and divisions as they occur, working from left to right.
4. Perform additions and subtractions as they occur, working from left to right.

We will illustrate the steps using this expression: $5 - 2(2^3 - 1)$

Step 1: Work inside the parentheses. This particular expression involves going through Steps 2–4 in order to simplify inside the parentheses: $5 - 2(8 - 1) = 5 - 2(7)$

Step 2: There are no exponents outside the parentheses—skip to Step 3.

Step 3: Multiply: $5 - 14$

Step 4: Subtract: -9

Following are a few more examples from other subject areas, but there is no law against creating your own memory aids, as needed.

Use the sentence "My Very Educated Mother Just Served Us Nine Pizzas" to recall the order of the planets from the sun (Mercury, Venus, Earth, Mars, Jupiter, Neptune, and Pluto).

or

Big Brown Rabbits Often Yield Great Big Vocal Groans When Gingerly Slapped for the color codes for resistors (black, brown, red, orange, yellow, green, blue, violet, gray, and white). The last two colors in the resistor sentence relate to gold and silver, which represent multipliers. Alternate sentences have been offered in the past for resistor color codes, but some are more "politically correct" than others.

More on acronyms is found at http://www.ucc.vt.edu/stdysk/acronyms.html and http://www.how-to-study.com/UsingAcronyms.htm.

Does the course involve vocabulary? Try the note or index card approach as described at http://www.how-to-study.com/Index%20Cards%20Vocabulary.htm.

Preparing For and Taking Tests

If you have practiced the strategies we have outlined in this material, you should be reviewing on a regular basis as you study rather than waiting to cram right before a test. Try to anticipate what is important and will be on the test, and use any review materials that are available, such as practice tests or review sheets. This doesn't mean that you don't need to study right before a test, but you shouldn't have to stay up all night to prepare for it, and you should feel more confident when you take the test.

These are a few resources to help you get ready. See the links for "Preparing for Tests" and "Taking Tests" at http://www.studygs.net/. Also, try "Strategies to Use With Questions You Cannot Answer Immediately" at http://www.ucc.vt.edu/stdysk/strategi.html and "A Strategy for Taking Tests" at http://www.how-to-study.com/A%20Strategy%20for%20Taking%20Tests.htm.

CliffsNotes comes through again at http://www.cliffsnotes.com/WileyCDA/Section/id-106284.html (Remember What Not to Forget at Test Time) and http://www.cliffsnotes.com/WileyCDA/Section/id-106287.html (Know Your Stuff for In-Class Exams).

Once you are sitting in the hot spot with your pencil in hand, use the **DETER** strategy for taking tests as described at http://www.how-to-study.com/A%20Strategy%20for%20Taking%20Tests.htm. DETER stands for

1. Directions: Read and understand the test directions.
2. Examine: Examine the entire test to see what is required.

3. <u>T</u>ime: Determine how much time to allow for each item.

4. <u>E</u>asiest: Answer the easiest items first.

5. <u>R</u>eview: Allow time to review the test to check your answers for accuracy and completeness.

Again, practice makes perfect. There are several web sites for taking practice tests. Here are a few:

- http://www.actstudent.org/testprep/index.html
- http://4tests.com/
- http://www.collegeboard.com/
- http://www.ets.org/

Computer Basics

For many classes, you need to know the basics about using a computer and possibly even surfing the Internet in order to complete certain assignments. If you are taking a distance learning class, you MUST have some basic knowledge of computers and the Internet. You must be able to prepare, save, and retrieve files; send and receive emails with attachments; deposit files in an electronic drop box; locate and navigate websites; download software and plug ins; and participate in discussion boards. A good resource for learning about these items is http://www.learnthenet.com/english/index.html. Once you have reached this site, note the "How To" list at the left side of the screen. If you are a novice, you might want to start with "How to Use this Site." Otherwise, start with "Master the Basics" and then work your way down the list. You will find information ranging from making the connection to the Internet to building your own website. Click on each underlined word or title to access the information. This information is also available as the "Animated Internet." Be sure to visit Netiquette, a primer on the do's and don'ts of communicating on the Internet. For example, did you know that you should avoid writing e-mail messages or posting in newsgroups using all caps. Why? Because IT LOOKS LIKE YOU'RE SHOUTING! The "Harness E-Mail" section has hints on E-Mail Etiquette. In the age of computer worms and viruses, we recommend you read the section on "Protect Yourself" to keep your computer from picking up a "bug"! And if you hear or see words that are unfamiliar to you, try the glossary.

Jan's Illustrated Computer Literacy 101 at http://www.jegsworks.com/Lessons/index.html includes lessons on the topics listed below, and the approach is very detailed yet easy to understand. Even if you have never touched a mouse before, you should be able to follow along.

- Computer Basics
  - Computer Types
  - Applications (includes word processing, e-mail, browser, Web pages, chats, and instant messaging)
  - Input (includes use of keyboard and mouse)
  - Processing (includes processor speed and physical components)
  - Output (includes printers and screen displays)
  - Storage (explains disks)
  - Computer to Computer (modems, software, and networks)
  - System Software (operating systems)
  - Programming
  - What You See—How the computer's parts all hook together
  - Hands on—Working with files and networks
  - On Your Own—Buying and managing your own computer

- Working with Windows
- Working with Words (Word Processing)
- Working with Numbers (Spreadsheets)
- Working with the Web (the Internet)
- Working with Presentations (MS PowerPoint)

Another "New User Tutorial" is available at http://northville.lib.mi.us/tech/tutor/welcome.htm.

The "Computers and Technology" section of the AARP website at http://www.aarp.org/computers/ covers topics from choosing a keyboard to anticipating and avoiding personal computer disasters.

Also consider these sites:

- Computer Basics for Newcomers to Personal Computers ("Newbies") at http://www.cyberwalker.net/basics/
- "How to" guides at http://www.pcnineoneone.com/howto.html
- Free career and computer training at http://www.gcflearnfree.org/en/course/course_detail.asp?Course_ID=17&Course_Title=Computer+Basics
- Computer Help A to Z at http://www.computerhelpatoz.com/how.html
- Internet Basics for ESL Students at http://iteslj.org/s/ib/

Do you want to learn about specific items; i.e., Windows XP or MSWord 2003? These are Microsoft products. You can go to http://www.microsoft.com/ and find training on just about every product produced by Microsoft—even older versions.

An important part of using the Internet is to be able to find and evaluate resources. This interactive tutorial provided by the University of Texas System Digital Library can help you build these skills: http://tilt.lib.utsystem.edu/.

Check out the research and study sites for college and graduate students, instructors, and advisors at http://www.education-world.com/higher_ed/study.shtml. This site includes information on copyright, citation and plagiarism, as well as online encyclopedias and other research tools.

Distant Learner Handbook

The content of this orientation is intended for all college students, regardless of whether they are taking courses in a classroom or as a distant learner. If you are taking courses online, on CD, or hybrid, we have developed a handbook just for you. This handbook addresses distant learner responsibilities and provides instructions and contact information on student and support services. You can find the handbook at our website at http://online.ctcd.edu.

ABOUT THE AUTHORS

Ron Larson

The Pennsylvania State University
The Behrend College

Ron Larson received his Ph.D. in mathematics from the University of Colorado in 1970. At that time he accepted a position with Penn State University, and he currently holds the rank of professor of mathematics at the university. Larson is the lead author of more than two dozen mathematics textbooks that range from sixth grade through calculus levels. Many of his texts, such as the eighth edition of his calculus text, are leaders in their markets. Larson is also one of the pioneers in the use of multimedia and the Internet to enhance the learning of mathematics. He has authored multimedia programs, extending from the elementary school through calculus levels. Larson is a member of several professional groups and is a frequent speaker at national and regional mathematics meetings.

Betsy Farber

Bucks County Community College

Betsy Farber received her Bachelor's degree in mathematics from Penn State University and her Master's degree in mathematics from the College of New Jersey. Since 1976, she has been teaching all levels of mathematics at Bucks County Community College in Newtown, Pennsylvania, where she currently holds the rank of professor. She is particularly interested in developing new ways to make statistics relevant and interesting to her students and has been teaching statistics in many different modes—with the TI-83 Plus, with MINITAB, and by distance learning as well as in the traditional classroom. A member of the American Mathematical Association of Two-Year Colleges (AMATYC), she is an author of *The Student Edition to MINITAB* and *A Guide to MINITAB*. She served as consulting editor for *Statistics, A First Course* and has written computer tutorials for the CD-ROM correlating to the texts in the Streeter Series in mathematics.

CONTENTS

Preface   X   How To Study Statistics   XV
Supplements   XII   Index of Applications   XVI
Acknowledgments   XIV

PART ONE DESCRIPTIVE STATISTICS

1 | INTRODUCTION TO STATISTICS

Where You've Been and *Where You're Going*   1
1.1   An Overview of Statistics   2
1.2   Data Classification   9
   ■ Case Study: *Rating Television Shows in the United States*   15
1.3   Data Collection and Experimental Design   16
   ■ Activity: *Random Numbers*   26
   ■ Uses and Abuses   27
   Chapter Summary   28
   Review Exercises   29
   Chapter Quiz   31
   ■ Real Statistics–Real Decisions—Putting It All Together   32
   ■ History of Statistics—Timeline   33
   ■ Technology: *Using Technology in Statistics*   34

2 | DESCRIPTIVE STATISTICS   36

Where You've Been and *Where You're Going*   37
2.1   Frequency Distributions and Their Graphs   38
2.2   More Graphs and Displays   53
2.3   Measures of Central Tendency   65
   ■ Activity: *Mean Versus Median*   79
2.4   Measures of Variation   80
   ■ Activity: *Standard Deviation*   98
   ■ Case Study: *Earnings of Athletes*   99
2.5   Measures of Position   100
   ■ Uses and Abuses   113
   Chapter Summary   114
   Review Exercises   115
   Chapter Quiz   119
   ■ Real Statistics–Real Decisions—Putting It All Together   120
   ■ Technology: *Monthly Milk Production*   121
   ■ Using Technology to Determine Descriptive Statistics   122
   Cumulative Review: *Chapters 1–2*   124

PART TWO PROBABILITY AND PROBABILITY DISTRIBUTIONS

3 PROBABILITY   126

Where You've Been and *Where You're Going*   127

3.1   Basic Concepts of Probability and Counting   128

■ Activity: *Simulating the Stock Market*   144

3.2   Conditional Probability and the Multiplication Rule   145

3.3   The Addition Rule   156

■ Activity: *Simulating the Probability of Rolling a 3 or 4*   166

■ Case Study: *United States Congress*   167

3.4   Additional Topics in Probability and Counting   168

■ Uses and Abuses   179

Chapter Summary   180

Review Exercises   181

Chapter Quiz   185

■ Real Statistics–Real Decisions—Putting It All Together   186

■ Technology: *Simulation: Composing Mozart Variations with Dice* 187

4 DISCRETE PROBABILITY DISTRIBUTIONS   188

Where You've Been and *Where You're Going*   189

4.1   Probability Distributions   190

4.2   Binomial Distributions   202

■ Activity: *Binomial Distribution*   216

■ Case Study: *Binomial Distribution of Airplane Accidents*   217

4.3   More Discrete Probability Distributions   218

■ Uses and Abuses   225

Chapter Summary   226

Review Exercises   227

Chapter Quiz   231

■ Real Statistics–Real Decisions—Putting It All Together   232

■ Technology: *Using Poisson Distributions as Queuing Models*   233

5   NORMAL PROBABILITY DISTRIBUTIONS   **234**

Where You've Been and *Where You're Going*   235

5.1   Introduction to Normal Distributions and   236
the Standard Normal Distribution

5.2   Normal Distributions: Finding Probabilities   249

5.3   Normal Distributions: Finding Values   257

■ Case Study: *Birth Weights in America*   265

5.4   Sampling Distributions and the Central Limit Theorem   266

■ Activity: *Sampling Distributions*   280

5.5   Normal Approximations to Binomial Distributions   281

■ Uses and Abuses   291

Chapter Summary   292

Review Exercises   293

Chapter Quiz   297

■ Real Statistics–Real Decisions—Putting It All Together   298

■ Technology: *Age Distribution in the United States*   299

Cumulative Review: *Chapters 3–5*   300

PART THREE   STATISTICAL INFERENCE

6   CONFIDENCE INTERVALS   **302**

Where You've Been and *Where You're Going*   303

6.1   Confidence Intervals for the Mean (Large Samples)   304

■ Case Study: *Marathon Training*   317

6.2   Confidence Intervals for the Mean (Small Samples)   318

■ Activity: *Confidence Intervals for a Mean*   326

6.3   Confidence Intervals for Population Proportions   327

■ Activity: *Confidence Intervals for a Proportion*   336

6.4   Confidence Intervals for Variance and Standard Deviation   337

■ Uses and Abuses   344

Chapter Summary   345

Review Exercises   346

Chapter Quiz   349

■ Real Statistics–Real Decisions—Putting It All Together   350

■ Technology: *Most Admired Polls*   351

■ Using Technology to Construct Confidence Intervals   352

HYPOTHESIS TESTING WITH ONE SAMPLE 354

Where You've Been and *Where You're Going* 355

7.1 **Introduction to Hypothesis Testing** 356

7.2 **Hypothesis Testing for the Mean (Large Samples)** 371

■ Case Study: *Human Body Temperature: What's Normal?* 386

7.3 **Hypothesis Testing for the Mean (Small Samples)** 387

■ Activity: *Hypothesis Tests for a Mean* 397

7.4 **Hypothesis Testing for Proportions** 398

■ Activity: *Hypothesis Tests for a Proportion* 403

7.5 **Hypothesis Testing for Variance and Standard Deviation** 404

■ **Uses and Abuses** 413

A Summary of Hypothesis Testing 414

Chapter Summary 416

Review Exercises 417

Chapter Quiz 421

■ **Real Statistics–Real Decisions—Putting It All Together** 422

■ Technology: *The Case of the Vanishing Women* 423

■ **Using Technology to Perform Hypothesis Tests** 424

PART FOUR MORE STATISTICAL INFERENCE

CORRELATION AND REGRESSION 482

Where You've Been and *Where You're Going* 483

9.1 **Correlation** 484

■ Activity: *Correlation by Eye* 500

9.2 **Linear Regression** 501

■ Activity: *Regression by Eye* 511

■ Case Study: *Correlation of Body Measurements* 512

9.3 **Measures of Regression and Prediction Intervals** 513

APPENDICES

APPENDIX A   **ALTERNATIVE PRESENTATION OF THE STANDARD NORMAL DISTRIBUTION**   **A1**

Standard Normal Distribution Table (0-to-z)   A1

Alternative Presentation of the Standard Normal Distribution   A2

APPENDIX B   **TABLES**   **A7**

TABLE 1   *Random Numbers*   A7

TABLE 2   *Binomial Distribution*   A8

TABLE 3   *Poisson Distribution*   A11

TABLE 4   *Standard Normal Distribution*   A16

TABLE 5   *t-Distribution*   A18

TABLE 6   *Chi-Square Distribution*   A19

TABLE 7   *F-Distribution*   A20

TABLE 8   *Critical Values for the Sign Test*   A25

TABLE 9   *Critical Values for the Wilcoxon Signed-Rank Test*   A25

TABLE 10   *Critical Values for the Spearman Rank Correlation*   A26

TABLE 11   *Critical Values for the Pearson Correlation Coefficient*   A26

TABLE 12   *Critical Values for the Number of Runs*   A27

APPENDIX C   **NORMAL PROBABILITY PLOTS AND THEIR GRAPHS**   A28

Answers to the Try It Yourself Exercises   A30

Answers to the Odd-Numbered Exercises   A50

Index   I1

Photo Credits

PREFACE

Welcome to *Elementary Statistics: Picturing the World,* Fifth Edition. You will find that this textbook is written with a balance of rigor and simplicity. It combines step-by-step instruction, real-life examples and exercises, carefully developed features, and technology that makes statistics accessible to all.

We are grateful for the overwhelming acceptance of the first four editions. It is gratifying to know that our vision of combining theory, pedagogy, and design to exemplify how statistics is used to picture and describe the world has helped students learn about statistics and make informed decisions.

WHAT'S NEW IN THIS EDITION

The goal of the Fifth Edition was a thorough update of the key features, examples, and exercises:

Examples This edition includes more than 210 examples, approximately 50% of which are new or revised.

Exercises Approximately 50% of the more than 2100 exercises are new or revised. We've also added 75 conceptual and critical thinking exercises throughout the text.

StatCrunch® Examples New to this edition are more than 50 StatCrunch Reports. These interactive reports, called out in the book with the **SC** icon, provide step-by-step instructions for how to use the online statistical software StatCrunch to solve the examples. *Note:* Accessing these reports requires a MyStatLab or StatCrunch account.

StatCrunch Exercises New to this edition are more than 80 exercises that instruct students to solve the exercise using StatCrunch. This allows students to practice the software skills learned in the StatCrunch Examples. *Note:* Solving the exercises using StatCrunch requires a MyStatLab or StatCrunch account.

Extensive Feature Updates Approximately 50% of the following key features have been replaced, making this edition fresh and relevant to today's students:

- Chapter Openers
- Case Studies
- Putting It All Together: Real Statistics—Real Decisions

Revised Content The following sections have been changed:

- **Section 2.2, More Graphs and Displays,** now defines misleading graphs.
- **Section 2.5, Measures of Position,** now defines the modified boxplot.

- **Section 9.1, Correlation,** now defines perfect positive linear correlation and perfect negative linear correlation.

FEATURES OF THE FIFTH EDITION

Guiding Student Learning

Where You've Been and **Where You're Going** Each chapter begins with a two-page visual description of a real-life problem. *Where You've Been* shows students how the chapter fits into the bigger picture of statistics by connecting it to topics learned in earlier chapters. *Where You're Going* gives students an overview of the chapter, exploring concepts in the context of real-world settings.

What You Should Learn Each section is organized by learning objectives, presented in everyday language in *What You Should Learn.* The same objectives are then used as subsection titles throughout the section.

Definitions and Formulas are clearly presented in easy-to-locate boxes. They are often followed by **Guidelines,** which explain *In Words* and *In Symbols* how to apply the formula or understand the definition.

Margin Features help reinforce understanding:

- **Study Tips** show how to read a table, use technology, or interpret a result or a graph. **Round-off Rules** guide the student during calculations.
- **Insights** help drive home an important interpretation or connect different concepts.
- **Picturing the World** Each section contains a real-life "mini case study" called *Picturing the World* illustrating important concepts in the section. Each feature concludes with a question and can be used for general class discussion or group work. The answers to these questions are included in the *Annotated Instructor's Edition.*

Examples and Exercises

Examples Every concept in the text is clearly illustrated with one or more step-by-step examples. Most examples have an interpretation step that shows the student how the solution may be interpreted within the real-life context of the example and promotes critical thinking and writing skills. Each example, which is numbered and titled for easy reference, is followed by a similar exercise called **Try It Yourself** so students can immediately practice the skill learned. The answers to these exercises are given in the back of the book, and the worked-out solutions are given in the *Student's Solutions Manual.* The Videos on DVD show clips of an instructor working out each *Try It Yourself* exercise.

StatCrunch Examples New to this edition are more than 50 StatCrunch Reports. These interactive reports, called out in the book with the **SC** icon, provide step-by-step instructions for how to use the online statistical software StatCrunch to solve the examples. Go to *www.stat-crunch.com*, choose **Explore ▼ Groups,** and search for "Larson Elementary Statistics 5/e" to access the StatCrunch Reports. *Note:* Accessing these reports requires a MyStatLab or StatCrunch account.

Technology Examples Many sections contain a worked example that shows how technology can be used to calculate formulas, perform tests, or display data. Screen displays from MINITAB®, Excel®, and the TI-83/84 Plus graphing calculator are given. Additional screen displays are presented at the ends of selected chapters, and detailed instructions are given in separate technology manuals available with the book.

Exercises The Fifth Edition includes more than 2100 exercises, giving students practice in performing calculations, making decisions, providing explanations, and applying results to a real-life setting. Approximately 50% of these exercises are new or revised. The exercises at the end of each section are divided into three parts:

- **Building Basic Skills and Vocabulary** are short answer, true or false, and vocabulary exercises carefully written to nurture student understanding.

- **Using and Interpreting Concepts** are skill or word problems that move from basic skill development to more challenging and interpretive problems.

- **Extending Concepts** go beyond the material presented in the section. They tend to be more challenging and are not required as prerequisites for subsequent sections.

For the sections that contain StatCrunch examples, there are corresponding **StatCrunch exercises** that direct students to use StatCrunch to solve the exercises. *Note:* Using StatCrunch requires a MyStatLab or StatCrunch account.

Technology Answers Answers in the back of the book are found using tables. Answers found using technology are also included when there are discrepancies due to rounding.

Review and Assessment

Chapter Summary Each chapter concludes with a Chapter Summary that answers the question *What did you learn?* The objectives listed are correlated to Examples in the section as well as to the Review Exercises.

Chapter Review Exercises A set of Review Exercises follows each Chapter Summary. The order of the exercises follows the chapter organization. Answers to all odd-numbered exercises are given in the back of the book.

Chapter Quizzes Each chapter ends with a Chapter Quiz. The answers to all quiz questions are provided in the back of the book. For additional help, see the step-by-step video solutions on the companion DVD-ROM.

Cumulative Review A Cumulative Review at the end of Chapters 2, 5, 8, and 11 concludes each part of the text. Exercises in the Cumulative Review are in random order and may incorporate multiple ideas. Answers to all odd-numbered exercises are given in the back of the book.

Statistics in the Real World

Uses and Abuses: Statistics in the Real World Each chapter features a discussion on how statistical techniques should be used, while cautioning students about common abuses. The discussion includes ethics, where appropriate. Exercises help students apply their knowledge.

Applet Activities Selected sections contain activities that encourage interactive investigation of concepts in the lesson with exercises that ask students to draw conclusions. The accompanying applets are contained on the DVD that accompanies new copies of the text.

Chapter Case Study Each chapter has a full-page Case Study featuring actual data from a real-world context and questions that illustrate the important concepts of the chapter.

Putting It All Together: Real Statistics–Real Decisions This feature encourages students to think critically and make informed decisions about real-world data. Exercises guide students from interpretation to drawing of conclusions.

Chapter Technology Project Each chapter has a Technology project using MINITAB, Excel, and the TI-83/84 Plus that gives students insight into how technology is used to handle large data sets or real-life questions.

CONTINUED STRONG PEDAGOGY FROM THE FOURTH EDITION

Versatile Course Coverage The table of contents was developed to give instructors many options. For instance, the *Extending Concepts* exercises, applet activities, Real Statistics–Real Decisions, and Uses and Abuses provide sufficient content for the text to be used in a two-semester course. More commonly, we expect the text to be used in a three-credit semester course or a four-credit semester course that includes a lab component. In such cases, instructors will have to pare down the text's 46 sections.

Graphical Approach As with most introductory statistics texts, we begin the descriptive statistics chapter (Chapter 2) with a survey of different ways to display data graphically. A difference between this text and many others is that **we continue to incorporate the graphical display of data throughout the text.** For example, see the use of stem-and-leaf plots to display data on pages 385 and 386. This emphasis on graphical displays is beneficial to all students, especially those utilizing visual learning strategies.

Balanced Approach The text strikes a **balance among computation, decision making, and conceptual understanding.** We have provided many Examples, Exercises, and Try It Yourself exercises that go beyond mere computation.

Variety of Real-Life Applications We have chosen real-life applications that are representative of the majors of students taking introductory statistics courses. We want statistics to come alive and appear relevant to students so they understand the importance of and rationale for studying statistics. We wanted the applications to be **authentic**—but they also need to be **accessible.** See the Index of Applications on page XVI.

Data Sets and Source Lines The data sets in the book were chosen for interest, variety, and their ability to illustrate concepts. Most of the **240-plus data sets** contain real data with source lines. The remaining data sets contain simulated data that are representative of real-life situations. All data sets containing 20 or more entries are available in a variety of formats; they are available electronically on the DVD and Internet. In the exercise sets, the data sets that are available electronically are indicated by the icon ⊙.

Flexible Technology Although most formulas in the book are illustrated with "hand" calculations, we assume that most students have access to some form of technology tool, such as MINITAB, Excel, the TI-83 Plus, or the TI-84 Plus. Because the use of technology varies widely, we have made the text flexible. **It can be used in courses with no more technology than a scientific calculator—or it can be used in courses that require sophisticated technology tools.** Whatever your use of technology, we are sure you agree with us that the goal of the course is not computation. Rather, it is to help students gain an understanding of the basic concepts and uses of statistics.

Prerequisites Algebraic manipulations are kept to a minimum—often we display informal versions of formulas using words in place of or in addition to variables.

Choice of Tables Our experience has shown that students find a **cumulative density function** (CDF) table easier to use than a "0-to-z" table. Using the CDF table to find the area under a normal curve is a topic of Section 5.1 on pages 239–243. Because we realize that some teachers prefer to use the "0-to-z" table, we have provided an alternative presentation of this topic using the "0-to-z" table in Appendix A.

Page Layout Statistics is more accessible when it is carefully formatted on each page with a consistent open layout. This text is the first college-level statistics book to be written so that its features are not split from one page to the next. Although this process requires extra planning, the result is a presentation that is clean and clear.

MEETING THE STANDARDS

MAA, AMATYC, NCTM Standards This text answers the call for a **student-friendly text that emphasizes the uses of statistics.** Our job as introductory instructors is not to produce statisticians but to produce informed consumers of statistical reports. For this reason, we have included exercises that require students to interpret results, provide written explanations, find patterns, and make decisions.

GAISE Recommendations Funded by the American Statistical Association, the Guidelines for Assessment and Instruction in Statistics Education (GAISE) Project developed six recommendations for teaching introductory statistics in a college course. These recommendations are:

- Emphasize statistical literacy and develop statistical thinking.
- Use real data.
- Stress conceptual understanding rather than mere knowledge of procedures.
- Foster active learning in the classroom.
- Use technology for developing conceptual understanding and analyzing data.
- Use assessments to improve and evaluate student learning.

The examples, exercises, and features in this text embrace all of these recommendations.

SUPPLEMENTS

STUDENT RESOURCES

Student Solutions Manual Includes complete worked-out solutions to all of the *Try It Yourself* exercises, the odd-numbered exercises, and all of the *Chapter Quiz* exercises.
(ISBN-13: 978-0-321-69373-0; ISBN-10: 0-321-69373-6)

Videos on DVD-ROM A comprehensive set of videos tied to the textbook, containing short video clips of an instructor working every *Try It Yourself* exercise. New to this edition are section lecture videos.
(ISBN-13: 978-0-321-69374-7; ISBN-10: 0-321-69374-4)

A **Companion DVD-ROM** is bound in new copies of *Elementary Statistics: Picturing the World*. The DVD holds a number of supporting materials, including:

- **Chapter Quiz Prep:** video solutions to Chapter Quiz questions in the text, with English and Spanish captions
- **Data Sets:** selected data sets from the text, available in Excel, MINITAB (v.14), TI-83 / TI-84 and txt (tab delimited)
- **Applets:** 15 applets by Webster West
- **DDXL:** an Excel add-in

Graphing Calculator Manual Tutorial instruction and worked out examples for the TI-83/84 Plus graphing calculator. (ISBN-13: 978-0-321-69379-2; ISBN-10: 0-321-69379-5)

Excel Manual Tutorial instruction and worked-out examples for Excel. (ISBN-13: 978-0-321-69380-8; ISBN-10: 0-321-69380-9)

Minitab Manual Tutorial instruction and worked-out examples for Minitab. (ISBN-13: 978-0-321-69377-8; ISBN-10: 0-321-69377-9)

Study Cards for the following statistical software products are available: Minitab, Excel, SPSS, JMP, R, StatCrunch, and the TI-83/84 Plus graphing calculator.

INSTRUCTOR RESOURCES

Annotated Instructor's Edition Includes suggested activities, additional ways to present material, common pitfalls, alternative formats or approaches, and other helpful teaching tips. All answers to the section and review exercises are provided with short answers appearing in the margin next to the exercise. (ISBN-13: 978-0-321-69365-5; ISBN-10: 0-321-69365-5)

Instructor Solutions Manual Includes complete solutions to all of the exercises, *Try It Yourself* exercises, Case Studies, Technology pages, Uses and Abuses exercises, and Real Statistics–Real Decisions exercises. (ISBN-13: 978-0-321-69366-2; ISBN-10: 0-321-69366-3)

TestGen® (*www.pearsoned.com/testgen*) Enables instructors to build, edit, and print, and administer tests using a computerized bank of questions developed to cover all the objectives of the text. TestGen is algorithmically based, allowing instructors to create multiple but equivalent versions of the same question or test with the click of a button. Instructors can also modify test bank questions or add new questions. The software and testbank are available for download from Pearson Education's online catalog (*www.pearsonhighered.com/irc*).

Online Test Bank A test bank derived from TestGen® available for download at *www.pearsonhighered.com/irc.*

PowerPoint Lecture Slides Fully editable and printable slides that follow the textbook. Use during lecture or post to a website in an online course. Most slides include notes offering suggestions for how the material may effectively be presented in class. These slides are available within MyStatLab or at *www.pearsonhighered.com/irc.*

Active Learning Questions Prepared in PowerPoint®, these questions are intended for use with classroom response systems. Several multiple-choice questions are available for each chapter of the book, allowing instructors to quickly assess mastery of material in class. The Active Learning Questions are available to download from within MyStatLab or at *www.pearsonhighered.com/irc.*

TECHNOLOGY SUPPLEMENTS

MyStatLab™ Online Course (access code required)

MyStatLab is a series of text-specific, easily customizable online courses for Pearson Education's textbooks in statistics. For students, MyStatLab™ provides students with a personalized interactive learning environment that adapts to each student's learning style and gives them immediate feedback and help. Because MyStatLab is delivered over the Internet, students can learn at their own pace and work whenever they want. MyStatLab provides instructors with a rich and flexible set of text-specific resources, including course management tools, to support online, hybrid, or traditional courses. MyStatLab is available to qualified adopters and includes access to **StatCrunch**. For more information, visit *www.mystatlab.com* or contact your Pearson representative.

MathXL® for Statistics Online Course (access code required)

MathXL® for Statistics is a powerful online homework, tutorial, and assessment system that accompanies Pearson textbooks in statistics. With MathXL for Statistics, instructors can:

- Create, edit, and assign online homework and tests using algorithmically generated exercises correlated at the objective level to the textbook.
- Create and assign their own online exercises and import TestGen tests for added flexibility.
- Maintain records of all student work, tracked in MathXL's online gradebook.

With MathXL for Statistics, students can:

- Take chapter tests in MathXL and receive personalized study plans and/or personalized homework assignments based on their test results.
- Use the study plan and/or the homework to link directly to tutorial exercises for the objectives they need to study.
- Students can also access supplemental animations and video clips directly from selected exercises.

MathXL for Statistics is available to qualified adopters. For more information, visit our website at *www.mathxl.com*, or contact your Pearson representative.

StatCrunch®

StatCrunch® is an online statistical software website that allows users to perform complex analyses, share data sets, and generate compelling reports of their data. Developed by programmers and statisticians, StatCrunch currently has more than twelve thousand data sets available for students to analyze, covering almost any topic of interest. Interactive graphics are embedded to help users understand statistical concepts and are available for export to enrich reports with visual representations of data. Additional features include:

- A full range of numerical and graphical methods that allow users to analyze and gain insights from any data set.
- Flexible upload options that allow users to work with their .txt or Excel® files, both online and offline.
- Reporting options that help users create a wide variety of visually appealing representations of their data.

StatCrunch is available to qualified adopters. For more information, visit our website at *www.statcrunch.com*, or contact your Pearson representative.

ACKNOWLEDGMENTS

We owe a debt of gratitude to the many reviewers who helped us shape and refine *Elementary Statistics: Picturing the World*, Fifth Edition.

REVIEWERS OF THE CURRENT EDITION

Dawn Dabney, Northeast State Community College
David Gilbert, Santa Barbara City College
Donna Gorton, Butler Community College
Dr. Larry Green, Lake Tahoe Community College
Lloyd Jaisingh, Morehead State
Austin Lovenstein, Pulaski Technical College
Lyn A. Noble, Florida Community College at Jacksonville—South Campus
Nishant Patel, Northwest Florida State
Jack Plaggemeyer, Little Big Horn College
Abdullah Shuaibi, Truman College
Cathleen Zucco-Teveloff, Rowan University

REVIEWERS OF THE PREVIOUS EDITIONS

Rosalie Abraham, Florida Community College at Jacksonville
Ahmed Adala, Metropolitan Community College
Olcay Akman, College of Charleston
Polly Amstutz, University of Nebraska, Kearney
John J. Avioli, Christopher Newport University
David P. Benzel, Montgomery College
John Bernard, University of Texas—Pan American
G. Andy Chang, Youngstown State University
Keith J. Craswell, Western Washington University
Carol Curtis, Fresno City College
Cara DeLong, Fayetteville Technical Community College
Ginger Dewey, York Technical College
David DiMarco, Neumann College
Gary Egan, Monroe Community College
Charles Ehler, Anne Arundel Community College
Harold W. Ellingsen, Jr., SUNY—Potsdam
Michael Eurgubian, Santa Rosa Jr. College
Jill Fanter, Walters State Community College
Douglas Frank, Indiana University of Pennsylvania
Frieda Ganter, California State University
Sonja Hensler, St. Petersburg Jr. College
Sandeep Holay, Southeast Community College, Lincoln Campus

Nancy Johnson, Manatee Community College
Martin Jones, College of Charleston
David Kay, Moorpark College
Mohammad Kazemi, University of North Carolina—Charlotte
Jane Keller, Metropolitan Community College
Susan Kellicut, Seminole Community College
Hyune-Ju Kim, Syracuse University
Rita Kolb, Cantonsville Community College
Rowan Lindley, Westchester Community College
Jeffrey Linek, St. Petersburg Jr. College
Benny Lo, DeVry University, Fremont
Diane Long, College of DuPage
Rhonda Magel, North Dakota State University
Mike McGann, Ventura Community College
Vicki McMillian, Ocean County College
Lynn Meslinsky, Erie Community College
Lyn A. Noble, Florida Community College at Jacksonville—South Campus
Julie Norton, California State University—Hayward
Lynn Onken, San Juan College
Lindsay Packer, College of Charleston
Eric Preibisius, Cuyamaca Community College
Melonie Rasmussen, Pierce College
Neal Rogness, Grand Valley State University
Elisabeth Schuster, Benedictine University
Jean Sells, Sacred Heart University
John Seppala, Valdosta State University
Carole Shapero, Oakton Community College
Aileen Solomon, Trident Technical College
Sandra L. Spain, Thomas Nelson Community College
Michelle Strager-McCarney, Penn State—Erie, The Behrend College
Deborah Swiderski, Macomb Community College
William J. Thistleton, SUNY—Institute of Technology, Utica
Agnes Tuska, California State University—Fresno
Clark Vangilder, DeVry University
Ting-Xiu Wang, Oakton Community College
Dex Whittinghall, Rowan University

We also give special thanks to the people at Pearson Education who worked with us in the development of *Elementary Statistics: Picturing the World*, Fifth Edition: Marianne Stepanian, Dana Jones, Chere Bemelmans, Alex Gay, Kathleen DeChavez, Audra Walsh, Tamela Ambush, Joyce Kneuer, Courtney Marsh, Andrea Sheehan and Rich Williams. We also thank the staff of Larson Texts, Inc., who assisted with the development and production of the book. On a personal level, we are grateful to our spouses, Deanna Gilbert Larson and Richard Farber, for their love, patience, and support. Also, a special thanks goes to R. Scott O'Neil.

We have worked hard to make *Elementary Statistics: Picturing the World*, Fifth Edition, a clean, clear, and enjoyable text from which to teach and learn statistics. Despite our best efforts to ensure accuracy and ease of use, many users will undoubtedly have suggestions for improvement. We welcome your suggestions.

Ron Larson

Ron Larson, odx@psu.edu

Betsy Farber

Betsy Farber, farberb@bucks.edu

HOW TO STUDY STATISTICS

FOR EXTRA HELP:
MyStatLab

STUDY STRATEGIES

Congratulations! You are about to begin your study of statistics. As you progress through the course, you should discover how to use statistics in your everyday life and in your career. The prerequisites for this course are two years of algebra, an open mind, and a willingness to study. When you are studying statistics, the material you learn each day builds on material you learned previously. There are no shortcuts—you must keep up with your studies every day. Before you begin, read through the following hints that will help you succeed.

Making a Plan   Make your own course plan right now! A good rule of thumb is to study at least two hours for every hour in class. After your first major exam, you will know if your efforts were sufficient. If you did not get the grade you wanted, then you should increase your study time, improve your study efficiency, or both.

Preparing for Class   Before every class, review your notes from the previous class and read the portion of the text that is to be covered. Pay special attention to the definitions and rules that are highlighted. Read the examples and work through the Try It Yourself exercises that accompany each example. These steps take self-discipline, but they will pay off because you will benefit much more from your instructor's presentation.

Attending Class   Attend every class. Arrive on time with your text, materials for taking notes, and your calculator. If you must miss a class, get the notes from another student, go to a tutor or your instructor for help, or view the appropriate Video on DVD. Try to learn the material that was covered in the class you missed before attending the next class.

Participating in Class   When reading the text before class, reviewing your notes from a previous class, or working on your homework, write down any questions you have about the material. Ask your instructor these questions during class. Doing so will help you (and others in your class) understand the material better.

Taking Notes
During class, be sure to take notes on definitions, examples, concepts, and rules. Focus on the instructor's cues to identify important material. Then, as soon after class

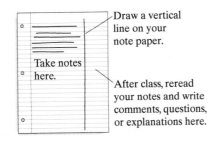

Draw a vertical line on your note paper.

Take notes here.

After class, reread your notes and write comments, questions, or explanations here.

as possible, review your notes and add any explanations that will help to make your notes more understandable to you.

Doing the Homework   Learning statistics is like learning to play the piano or to play basketball. You cannot develop skills just by watching someone do it; you must do it yourself. The best time to do your homework is right after class, when the concepts are still fresh in your mind. Doing homework at this time increases your chances of retaining the information in long-term memory.

Finding a Study Partner   When you get stuck on a problem, you may find that it helps to work with a partner. Even if you feel you are giving more help than you are getting, you will find that teaching others is an excellent way to learn.

Keeping Up with the Work   Don't let yourself fall behind in this course. If you are having trouble, seek help immediately—from your instructor, a statistics tutor, your study partner, or additional study aids such as the Chapter Quiz Prep videos on DVD-ROM and the Try It Yourself video clips on the videos on DVD-ROM. Remember: If you have trouble with one section of your statistics text, there's a good chance that you will have trouble with later sections unless you take steps to improve your understanding.

Getting Stuck   Every statistics student has had this experience: You work a problem and cannot solve it, or the answer you get does not agree with the one given in the text. When this happens, consider asking for help or taking a break to clear your thoughts. You might even want to sleep on it, or rework the problem, or reread the section in the text. Avoid getting frustrated or spending too much time on a single problem.

Preparing for Tests   Cramming for a statistics test seldom works. If you keep up with the work and follow the suggestions given here, you should be almost ready for the test. To prepare for the chapter test, review the Chapter Summary and work the Review Exercises and the Cumulative Review Exercises. Then set aside some time to take the sample Chapter Quiz. Analyze the results of your Chapter Quiz to locate and correct test-taking errors.

Taking a Test   Most instructors do not recommend studying right up to the minute the test begins. Doing so tends to make people anxious. The best cure for test-taking anxiety is to prepare well in advance. Once the test begins, read the directions carefully and work at a reasonable pace. (You might want to read the entire test first, then work the problems in the order in which you feel most comfortable.) Don't rush! People who hurry tend to make careless errors. If you finish early, take a few moments to clear your thoughts and then go over your work.

Learning from Mistakes   After your test is returned to you, go over any errors you might have made. Doing so will help you avoid repeating some systematic or conceptual errors. Don't dismiss any error as just a "dumb mistake." Take advantage of any mistakes by hunting for ways to improve your test-taking skills.

INDEX OF APPLICATIONS

Biology and Life Sciences

Air pollution, 31, 471
Air quality, 115
Alligator, 125
Atlantic croaker fish, 49
Beagle, 48, 253
Box turtle, 235
Brown trout, 220
Cats, 182, 259, 402, 447
Dogs, 142, 153, 182, 198, 200, 259, 402, 447
Elephants, 450, 527
Endangered species, 590
Environmentally friendly product, 287
Fish, 13, 526
Fisher's Iris data set, 58
Florida panther, 339
Fruit flies, 110
Green turtle migration, 294–295
House flies, 62, 276
Kitti's hog-nosed bat, 294, 296
Koalas, 30
Oats, 629
Ostrich, 481
Pets, 94, 215
Plants, 215
Rabbits, 220
Salmon, 135, 147
Sharks, 230
Snapdragon flowers, 142
Soil, 175, 570
Soybeans, 24
Swans, 369
Threatened species, 590
Trees, 13, 48, 170, 270, 368, 522, 527
Vertebrate groups, 590
Veterinarian, 447
Waste, 324, 389, 394
Water, 175, 385
    conductivity, 391
    consumption, 95
    hardness, 343
    pH level, 391
Wheat, 591, 629

Business

Advertisements, 228, 400, 585
Advertising and sales, 516
Bankruptcies, 223
Beverage company, 143
Board of directors, 169
Book sales, nonfiction, 14
Bookbinding defects, 154
Chief financial officers, survey of, 212
Clothing store purchases, 212
Consumer ratings, 459
Defective parts, 162, 176, 181, 185, 223
Executives, 111, 184
Fortune 500 companies, 29, 191
Free samples, 402
Inventory shrinkage, 57

Manufacturer
    claims, 246
    earnings, 109
Product assembly, 175
Quality control, 7, 31, 34, 35, 129, 198
Sales, 2, 50, 64, 119, 158, 192, 194, 195, 222, 375, 520, 521, 528, 578
Salesperson, 76, 107
Shipping errors, 368
Small business
    owners, 214
    websites, 208
Telemarketing, 190
Wal-Mart shareholder's equity, 528
Warehouses, 154, 177
Website costs, 343

Combinatorics

Answer guessing, 154, 212
Area code, 176
Letters, 172, 175
License plates, 131, 176, 181
Password, 174, 176
Security code, 169, 184, 185

Computers

Computer, 7, 8, 201, 209, 253–255, 332, 368
Computer software engineer earnings, 343
Disk drive, 586
Internet, 31, 70, 152, 183, 229, 287, 288, 369, 399, 429, 467, 497, 498, 506, 560
Microchips, 224
Monitor, 273, 445
Mozilla® Firefox®, 286
Operating system, 7
Printers, 96
Security, 297
Social networking sites, 61, 197, 203, 304, 306–308, 310, 605
Typing speed, 74
Videos, online, 349
Website, visitors per day, 600
Windows® Internet Explorer®, 286

Demographics

Age, 6, 25, 29, 31, 60, 76, 134, 157, 161, 496, 544, 546, 558
Birth weights in America, 265
Bride's age, 90, 608
Cars per household, 94
Children per household, 88
City rent, 296
Drive to work, 197
Ear wiggling, 153
Education, 593
Employee, 21, 134, 136, 141, 176, 177, 179, 181, 184, 198, 276, 524–526
Eye color, 13, 150, 157

Groom's age, 608
Height, 6, 13, 486, 496, 649
    of men, 77, 86, 110, 253, 263, 277, 507, 515
    of women, 48, 86, 253, 263, 277, 401
Home, 7
Household, 200, 300, 419, 473
Left-handed, 164
Marriage, 5, 29
Most admired polls, 351
Moving out, 468
New car, 130
New home prices, 119
Population
    Alaska, 87
    Brazil, 96
    cities, fastest growing, 1
    cities, largest numerical increase, 1
    Florida, 87
    U.S., 9, 95
    West Ridge County, 20–22
Retirement age, 51
Shoe size, 49, 507, 515
U.S. age distribution, 163, 181, 299
U.S. unemployment rate, 116
Weight of newborns, 13, 238, 480, A6
Zip codes, 29

Earth Science

Alternative energy, 333, 349
Carbon footprint, 300
Clear days, May, San Francisco, CA, 210
Climate conditions, 189
Cloudy days, June, Pittsburgh, PA, 210
Cyanide levels, 350
Earthquakes, 260, 499
Global warming, 3, 334
Green products, 75
Hurricane, 200, 223, 231
    relief efforts, 24
Ice thickness, 61
Lightning strikes, 231
Nitrogen dioxide, 384
Old Faithful, Yellowstone National Park, 44, 94, 279, 486, 489, 491, 503, 504, 514
Precipitation
    Baltimore, MD, 223
    Orlando, FL, 12
    San Francisco, CA, 343
    Tampa, FL, 222
Rain, 645
Saffir-Simpson Hurricane Scale, 231
Seawater, 312
Snowfall, 636
    January average, 14
    Mount Shasta, CA, 224
    New York county, 275
    Nome, AK, 197
Sunny and rainy days, 189, 193
    Seattle, WA, 140

Temperature, 63, 638
    Cleveland, OH, 47
    Denver, CO, 12
    Mohave, AZ, 29
    Pittsburgh, PA, 604
    San Diego, CA, 605
Tornadoes, 125, 197
UV Index, 62
Water contamination, 350
Water temperatures, 622
Wet or dry, Seattle, WA, 140
Wildland fires, 531

Economics and Finance

Account balance, 75
Accounting, 199
Allowance, 589
ATM machine, 50, 52
Audit, 133, 162, 335
Bill payment, 332, 550
Book spending, 49
Charitable donations, 332, 499
Children's savings accounts, 296
Commission, 113
Credit card, 29, 112, 213, 273, 347, 396, 432, 605
Credit score, 475, 641
Debit card, 31
Debt and income, 628
Depression, 14
Dividends and earnings, 497, 498
Dow Jones Industrial Average, 77
Economic power, 8
Emergency savings, 153
Executive compensation, 417
Financial advice, 213
Financial debt, 605
Financial shape, 177
Forecasting earnings, 5
Gross domestic product, 485, 488, 493, 502, 504, 514, 516, 518, 523
Home owner income, 7
Honeymoon financing, 213
Income, 125, 496, 592, 649
Investments, 63
IRAs, 521, 522
IRS tax filing wait times, 394
Manufacturing, 63
Missing tax deductions, 347, 348
Money managing, 7
Mortgages, 324
Mutual funds, 306, 648
Paycheck errors, 224
Primary investor in household, 7
Profit and loss analysis, 199
Raising a child, cost, 380, 418
Restaurant spending, 96, 437
Retirement income, 213
Salaries, 4, 6, 7, 29, 31, 48, 63, 64, 72, 75, 80–83, 91, 97, 117, 119, 124, 201, 276, 296, 349, 379, 383, 394, 395, 412, 431, 439, 440, 471, 508, 523, 524–526, 535, 572, 582, 583, 586, 595, 615, 616, 622, 623, 643
Baltimore, MD, 593
Boston, MA, 92

Chicago, IL, 83, 92
Dallas, TX, 92
Jacksonville, FL, 593
New York, NY, 92
San Francisco, CA, 593
Savings, 213, 548
more money, 348
Spending before traveling, 89
Stock, 113, 143, 181, 228, 229, 315,
519, 521, 570
McDonald's, 535
Stock market, 144
Tax preparation methods, 540,
541, 543
Taxes, 521, 522
U.S. exports, 77
U.S. income and economic
research, 647
Utility bills, 105, 254, 255
Vacation cost, 7, 395, 418, 421, 433

Education

Achievement, 558, 593
ACT, 8, 247, 253, 294, 437
Ages of students, 68, 291, 309,
313, 314, 480
Alumni, annual contributions by,
485, 489, 491, 503
Books, 197, 312
Business schools, 10
Career counselors, 124
Class size, 395, 618
Classes, 181
College costs, 548, 649
College graduates, 288
jobs, 17
College president, 31
College professors, 162
College students, 161
per faculty member, 115
Continuing education, 560
Day care, 439
Degrees, 56, 400
Degrees and gender, 185
Doctorate degree, 627, 642, 643
Dormitory room prices, 117
Education, study plans, 468
Educational attainment and
work location, 553
Elementary school students, 183,
431
Enrollment, 201, 227, 466,
626–627
Expenditure per student, 419
Extracurricular activities, 199,
358, 363, 364
Faculty hours, 394, 395
Final exam, 428
Final grade, 75, 76, 525, 526
Financial aid, 561
Genders of students, 634
GPA, 60, 74, 146, 324, 485, 494,
529, 579, 586
Health-related fields, study plans,
468
Highest level, 141
Homework, 325
Law school, 396, 642
Mathematics assessment test, 444
MCAT scores, 49, 73, 383
Medical school, 149
Midterm scores, 428, 525, 526

Musical training, 443
Nursing major, 152, 156, 164
Online classes, opinion, 30
Physics minors, 29
Plus/minus grading, 182
Preschool, 439
Public schools, 163, 535
Quiz, 139, 199, 203
Recess, 124
Reliability of testing, 155
SAT scores, 4, 52, 92, 104, 201,
247, 252, 254, 278, 324, 421,
434, 456, 481, 529, 591, 606,
607
Scholarship, 182
Science assessment tests, 411,
475, 572
Secondary school teachers, 431
Student advisory board, 172
Student-athletes, 197
Student ID numbers, 13, 137
Student loans, 496, A29
Student safety, 558
Student sleep habits, 326
Study habits, 30, 438, 497, 498,
506
Teaching experience, 301, 590
Teaching methods, 449, 472
Test grades/scores, 51, 60, 61, 63,
69, 72, 75, 76, 107, 109, 110,
116, 118, 125, 138, 238, 264,
428, 497, 498, 506, 550, 638,
649
Test scores and GNI, 629
Tuition, 73, 101, 102, 369
U.S. history assessment tests, 411,
572
Vocabulary, 496

Engineering

Aerospace engineers, 349
Bolts, 341, 420
Brick mortar, 50
Building heights
Atlanta, GA, 506
Houston, TX, 115
Cooling capacity, 505
Flow rate, 368
Gears, 256
Horsepower, 31
Liquid dispenser, 256, 631
Machine
calibrations, 278
part supplier, 140
Nails, 256
Nut, 438
Petroleum engineering, 4
Plastic injection mold, 592
Plastic sheet cutting, 314
Repairs, 176
Resistors, 293
Tensile strength, 448
Washers, 255, 438

Entertainment

Academy Award, winning, 133
Best-selling novel, 133
Blu-ray™ players, 342, 644
Broadway tickets, 14
Concert, 634
attendance, 197
tickets, 73, 289

Game show, 138
Games of chance, 143, 199, 200
Home theater system, 312, 358,
363, 572
Horse race, 176
Lottery, 139, 172, 175, 177, 179,
212, 224
Magazine, 8, 116, 184
Monopoly game, 146
Motion Picture Association,
ratings, 12, 161
Movie ticket prices, 117
Movies, 25, 31, 150, 183–184, 296,
558, 559
budget and gross, 497
on phone, 213
MP3 player, 13, 275, 368
Netbook, 185
New Year's Eve, 116
News, 289
Nielsen Company ratings, 15, 25
Oscar winners, ages, 106, 112
Political blog, 49
Powerball lottery, 186
Radio stations, 118
Raffle ticket, 135, 184, 196
Reading, 207, 333
Rock concert, fan age, 66
Satellite television, 117
Song lengths, 111
Summer vacation, 151
Television, 6, 10, 12, 108, 109, 118,
199, 228, 437, 439, 532, 533
3D TV, 282, 285
HDTV, 282, 284
late night, 582
LCD TV, 342
networks, Pittsburgh, PA, 10
The Price Is Right, 126, 127
top-ranked programs, 15
Video games, 31, 57, 174, 209, 368

Food and Nutrition

Apple, 61, 264
Beef, 627
Caffeine, 95, 384, 494, 624
Calories, 368, 507, 508, 572
Candy, 601
Carbohydrates, 316, 411, 573
Carrots, 264
Cereal, 247, 507, 537
Cheese, 314
Chicken, 627
Chicken wings, 228
Coffee, 77, 95, 159, 276, 320–321,
383, 546
Cookies, 213
Corn, toxin, 173
Dark chocolate, 413
Delivery, 547
Dried fruit, 417
Energy bar, 417
Fast food, 230, 384, 564
Fat, 505, 508
Fat substitute, 23
Food away from home, money
spent on, 419
Fruit consumption, 295
Hot chocolate, 508
Hot dogs, 206, 507
Ice cream, 264, 277, 333, 551, 552,
554–555, 572

Jelly beans, 182
Juice drinks, 312
Leftovers, 331
M&M's, 229, 544–545
Meat consumption, 295
Melons, 421
Menu, 139, 175
Milk
consumption, 246
containers, 277
processing, 407
production, 121, 532, 533
Multivitamin, 287
Oranges, 264
Peanuts, 255
Pepper pungencies, 50
Pizza, 176
Potatoes, 73, 527
Protein, 505
Restaurant, 466, 557, 569
Burger King, 573
Long John Silver's, 471
McDonald's, 573
serving, 420
Wendy's, 471
Rye, 527
Salmonella, 359
Saturated fat intake, 51
Sodium, 316, 417, 471, 507
Soft drinks, 255, 422
Sports drink, 369, 407
Storing fish, 4
Sugar, 507, 531, 532
Supermarket, 95, 250
Tea drinker, 143
Vegetables, 276, 421
Vending machine, 264
Water, 314, 383, 496

Government

Better Business Bureau, 57
Congress, 167, 335
gender profile, 14, 161
issue when voting, 7
Department of Energy, gas
prices, 3
Federal bailout, 417
Federal funding, alternative
energy, 349
Federal income tax, 474
Federal pension plan, 521, 522
Governor, Democrats, 8
Home Security Advisory System,
29
Legal system in U.S., 359
Registered voters, 6, 35
Securities and Exchange
Commission, 35
Senate, 637
Senators, years of service, 124
Tax cut, 556
U.S. Census
accuracy, 401
participation, 347
undercount, 4
U.S. government system, 152

Health and Medicine

Allergy medicines, 23, 340
Alzheimer's disease, 150
Appetite suppressant, 452
Arthritis, 24, 469

Assisted reproductive technology, 152, 232
Asthma, 401
Bacteria vaccine, 27
Bariatric surgery, 375
Blood, 6, 197
    donations, 156, 159, 198, 214
    pressure, 17, 31, 62, 150, 246, 357, 428, 440, 458, 460, 496, 505
    type, 139, 154, 214, 296
BMI, 75, 324
Body measurements, 512
Body temperature, 11, 13, 366, 386, 455, 496
BRCA gene, 151
Breast cancer, 27
Calcium supplements, 473, 615, 641
Cavities, 531, 532
Cholesterol, 6, 73, 113, 252, 254, 255, 261, 314, 459, 464, 573
Chronic medications, 481
Colds, 603
Cough syrup, 341, 343
Cyanosis, 212
Dentist, 213, 245, 332
Diabetes, 2, 150, 462
Diabetic, 16
Diet, 25, 31, 229, 440
Doctor, tell truth, 347
Drinking habits, 24
Drug testing, 140, 290, 460, 462, 466, 557, 559, 562
Emergency room visits, 333
Emphysema, 146
Exercise, 17, 24, 25, 119, 146, 485, 556, 628
Flu, 135
Fluorouracil, 440
Gastrointestinal stromal tumor, 466
Growth of a virus, 23
Headaches, 457, 643
Health care costs, 347, 348
Health care coverage, 420
Health care rating, 300
Health care reform, 124
Health care visits, 369, 541
Health club, 254, 419
Health improvement program, 298
Healthy foods, 25
Heart disease, 413
Heart medication, 369
Heart rate, 11, 75, 270, 322, 434, 616
Heart rhythm abnormality, 7
Heart transplant, 263, 279, 450
Herbal medicine, 460, 642
HIV, 420
Hospital, 52
Hospital beds, 76
Hospital costs, 412
Hospital length of stay, 77, 325, 411, 476, 583, 624
Hospital waiting times, 325
Influenza vaccine, 7, 19
Irinotecan, 440
Kidney cancer survival rate, 296
Kidney transplant, 23, 223, 265, 450
Knee surgery, 148, 204, 611
Lead levels, 31

Length of visit, physician's office, 589
Lower back pain, 606
Lung cancer, 368
Managed health care, 29
Maximal oxygen consumption, 447
Migraines, 466
MRI, 531, 532
Musculoskeletal injury, 559
Nausea, 557
No trouble sleeping, 214
Nutrients entering bloodstream, 569
Obesity, 8, 150
Organ donors, 287
Pain relievers, 577–578
Patient education material, 441
Personal hygiene, 23
Physical examination, 5
Physician assistant, 440
Physician's intake form, 14
Physicians, leaving medicine, 29
Placebo, 557, 559, 562
Plantar heel pain, 465
Plaque buildup in arteries, 458
Pregnancy study, Cebu, Philippines, 30
Prescription drugs, 24, 616
Prostate cancer, 31
Pulse rate, 51, 342, 471, 486
Recovery time, 366
Registered nurse, 119, 440, 508, 623
Rotator cuff surgery, 148
Sleep, 227, 263, 507, 523, 532, 533
    deprivation, 8, 24, 30
Sleep apnea, 150
Smoking, 19, 31, 143, 146, 230, 284, 368, 383, 401, 427, 463, 464, 534, 559
Stem cell research, 22
Stress, 150
Sudden infant death syndrome, 7
Surgery, 212
    corneal transplant, 231
    procedure, 203
    survival, 152
Triglyceride levels, 51, 251
Ulcers, 150
Vitamins, 124, 335, 341, 343
Weight, 72, 428
Weight loss, 18, 245, 384, 409, 452, 457, 459, 473, 496

Housing and Construction

City house value, 296
Construction, 322
Home insurance, 623
House size, 358, 363, 365, 549
Housing contract, 287
Monthly apartment rents, 120
Prices of condominiums, 67, 569
Prices of homes, 68, 272, 437, 506, 585, 600, 604
Realty, 139
Residence, rent or own, 24
Room and board, 272
Sales price, new apartments, 645
Security system, 131, 140, 174, 368

Square footage, 506, 605
Subdivision, 170
Tacoma Narrows Bridge, 163, 220
Unit size, 605

Law

Booster seat, 347, 348
California Peace Officer Standards and Training test, 260
Case of the vanishing women, 423
Child support, 270
Crime, 497, 498, 621–622
Fraud, 133, 594, 597
Jury selection, 149, 173, 175
Police officers, 368
Prison sentence, length of, 641
Repeat offenders, 603
Safe Drinking Water Act, 350
Seat belts, 463
Software piracy, 355
Speeding, 146, 395, 644
Theft, 645
    identity, 133, 301, 597

Miscellaneous

911 calls, 199
Aggressive behavior, children, 469
Air conditioners, 206
Appliances, 571
Archaeology, 91
Bacteria, 510
Badge numbers, police officers, 31
Ball, numbered, 150
Bank, 343, 557, 641
Barrel of oil, 62
Battery, 153, 255, 394
Birthday, 129, 154, 155, 161, 549
Births, 279
Bracelets, 175
Breaking up, 222
Calculators, defects, 184
Camcorder, 63
Camping chairs, 199
Car dealership, 321
Car wash, 174
Carbon dioxide emission, 227, 413, 485, 488, 493, 502, 504, 516, 518, 523, 536
Cards, 132, 138, 139, 143, 145–147, 150, 157, 158, 162, 173, 177, 181, 183, 184, 202, 204
Cellular phone, 7, 29, 59, 61, 67, 205, 228, 249, 269, 399, 401
Charitable donations, CEOs in Syracuse, NY, 30
Charity, 164
Chess, 368
Chlorine levels in a pool, 407
Chores, 229
Clocks, 368
Cloning, 400
Coffee shop, remodeling, 18
Coin toss, 35, 128, 129, 134, 137, 138, 139, 146, 147, 179, 181, 182, 229, 359, 477, 637
Consumption
    energy, 584, 624
    fuel, 572
Contact with parents, 607

Conversation, most annoying phrase, 347, 348
Cordless drills, 342
Crawling, 514
Customer service, 24
Die roll, 35, 72, 77, 128, 129, 132, 136, 138, 139, 140, 143, 146, 147, 150, 156–158, 162, 166, 181, 183, 185, 197
Digital camera, 342, 348
Digital photo frames, 66
Dog microchips, 638
DVDs, 170, 275
DVRs, 138, 313
Electricity consumption, 384
Electricity cost, 592
Employment and educational attainment, 562–563
Energy cost, 584, 586
Energy efficiency, 505
Eye survey, contacts, glasses, 73, 165
Farm values, 93, 94, 296, 504
Fire drill, 385
Floral arrangement, 301
Fluorescent lamps, 384
Full-body scanner, 413
Furnaces, 358, 363
Furniture store, 368
Garden hose, 368
Gas grill, 533–535
Gas station, 118, 227
Gasoline, volume of, 191
Gasoline consumption, 50
Gender of children, 181
Ghost sighting, 332
Global positioning system (GPS) navigators, 39, 41–46
Goals, 560
Grammatical errors, 456
Greeting cards, 227
Grip strength, 458
Grocery store waiting times, 13
Guitar, string tension, 197
Hats, 421
Hindenburg, 8
Hotel rooms, 74, 115, 313, 342, 412, 591
Impression of past decade, 133
Lawn mowers, 369
Life on other planets, 474
Light bulbs, 325, 368, 384
Liquid volume of cans, 115
Living on your own survey, 73
Marbles, 203
Memory, 8
Metacarpal bone length, 649
Metal detector, 227
Microwave, 421
Middle initial, 138
Mozart, 187
Museum attendance, 601
Music downloads, 355
Nail polish, 333
NASA budget, 62
Natural gas, 2, 648
Nitrogen oxides emission, 536
Nuclear power plants, 100, 102, 103
Number generator, 638
Oil, 29, 64, 520, 521
Opinion poll, 13
Pages, section, 228

Paint, 368, 644
    cans, 277, 314
    sprayer, 315
Parachute assembly, 359
Phone numbers, 10, 13
Photo printers, 276
Pilot's test, 222
Power failure, 73
Printing company departments,
    30
Product warning label, 23
Puzzle, 17
Questionnaire, 7
Queuing models, 233
Random number selection, 138,
    139
Recycling, 332
Refrigerator, 312, 313, 369
Salesperson, 13
Smartphones, 35
Social Security numbers, 29
Socks, 182
Space shuttle
    flights, 118
    fuel, 231
    menu, 175
    speed, 191
Speed of sound, 497
Spinner, 137, 140
Spray-on water repellent, 611
Spring break, 7, 30
Sprinkler system, 383
State troopers, 51
Statistics students, 20
Sudoku, 168
Survey of shoppers, 24
Survey of spectators, 6
Telephone, 76, 182, 408, 445
Terrorism, 475
Text messages, 25, 53–55, 205, 313
Toothbrush, 124
Toothpaste, 182, 225, 581
Transmission, 312, 313
Typographical error, 223, 231
UFO sighting, 133
UFOs, belief in, 335
Vacation, 230, 332
Vacuum cleaners, 629
Volunteering, 645
Waking times, 346
Washing cars, 295
Washing machine, 139
Wealthy people, 37, 40, 41, 43, 44,
    46, 54, 55, 70, 100, 102–104
Weigh station, 227
Well-being index, 583, 586
Wind energy, 6, 436
Winning a prize, 223
Yard sale, 77
Yoga, 64, 419

Mortality

Airplane accidents, 217
Alcohol-related accidents, 561
Emergency response time, 49,
    408
Heart disease, women, 125
Homicides, 547, 548
Motor vehicle
    casualties, 197, 562, 590
    crashes, 63

Shark attacks, 230
Tornado deaths, 230, 411

Motor Vehicles and
Transportation

Acceleration times, 348
Age of vehicles, 522, 601
Air travel, 30, 32, 65–67, 124
Airfare, 315
Airplanes, 17, 74, 109, 118
Airport scanners, 116
ATV, 368
Auto parts, 224, 301
Automobile insurance, 390, 614
Automobiles, 366
    battery, 301, 341, 358, 363, 581
Bicycle helmet, 285
Blood alcohol content, drivers, 30
Brakes, 278
Braking distance, 252, 260, 295,
    434, 436, 444, 498
Bumper, 447
Bus, 644
Car accident, 146, 193, 496
Car occupancy, 163, 200, 220
Carpooling, 200, 208
Carrying capacities, 13
Carry-on luggage, 74, 222
Compact cars, consumer testing,
    78
Crash test, 447, 539
Department of Motor Vehicles
    wait times, 392, 467
Diesel engines, 230
Drivers, 61, 330
Driver's license exam, 182
Driving time, 271–272
Engine, 256, 532, 533, 632
Flights, 155
    annoyances, 214
    bird-aircraft collisions, 328
Footwell intrusion, 447
Fuel additive, 617
Fuel economy, 76, 118, 582, 586
    ratings, 534
Fuel efficiency, 532, 533, 534
Gas prices, 276, 279, 312
Hybrid vehicle, 356, 556, 572
Intersection accidents, 219–220
Mileage, 117, 356, 369, 379, 385,
    396, 411, 480
Motor vehicles, 14, 61, 63, 522
Motorcycles, 343
    fuel economy, 118, 466
    helmet usage, 23, 466, 467
New highway, 171
Oil change, 346, 358, 363, 395
Oil tankers, 223
Parking ticket, 150
Pickup trucks, 151
Pit stop, 374
Power boats, 434
Price of a car, 9, 390, 533
Public transportation, 289
Speed of vehicles, 60, 73, 105,
    112, 249, 374
Sports cars, 73, 325, 347
Taxi cab, 369
Tires, 110, 198, 264, 278, 301, 649
Towing capacities, 118
Traffic congestion, 334

Traffic signal, 274
Traffic tickets, 227
Travel concerns, 552, 555
Used car cost, 394
Vehicle
    crashes, 560
    eco-friendly, 215
    manufacturers, 157
    owned, 179, 558, 590
    sales, 183, 585

Political Science

111th Congress, 13, 14
First Lady of the United States,
    124
Gubernatorial Election, Virginia,
    141
Officers, 176, 185, 300
Political analyst, 35
Political parties, 67
President's approval ratings, 17
Supreme Court justice, ages, 117
U.S. Presidents
    children, 51
    greatest, 327, 329
    political party, 6
Voters, 135, 141, 183, 216, 331,
    454–455, 521, 522

Psychology

Attention-deficit hyperactivity
    disorder, 428
Depression, 17, 78
Eating disorders, 74
Experimental group, 175
IQ, 107, 145, 297, 303, 434, 531,
    532
Mouthing behavior, 16
Obsessive-compulsive disorder,
    559, 562
Passive-aggressive traits, 192, 198,
    195
Psychological tests, 30, 198, 428
Reaction times, 50
Self-perception, age, 607
Training dogs, 30

Sports

400-meter dash, women's, 94
40-yard dash, 454
Baseball, 197, 212, 228, 369, 508
    batting averages, 92, 457
    games started, 125
    home run totals, 11
    Major League, 4, 29, 119, 124,
        251, 483, 486, 489, 493, 503,
        510
    World Series, 11, 199, 200, 333,
        638
Basketball, 10, 184, 300, 349, 459,
    460
    Howard, Dwight, 230
    heights, 6, 65, 96, A28–A29
    James, LeBron, 218–219
    NBA draft, 178
    vertical jumps, 116
    weights, 96
Bicycle race, 184
Boston marathon, 31
Curling, 402
Daytona 500, 169
Earnings, athlete, 99

Fishing, 207, 227, 409
Football, 322, 531
    college, 13, 142
    defensive player weights, 118
    Manning, Peyton, 223
    National Football League, 73,
        94, 160
    Super Bowl, 103, 636
    weight, 112, 118
    yards per carry, 325
Golf, 6, 78, 92, 185, 224, 368, 450,
    589, 607, 610, 638
Hockey, 125, 230, 481
Injury, 557
Jump height, 453, 478
Lacrosse, 174
Marathon training, 317
New York City marathon, 62
Olympics, 163
    100-meter times, 648
    Winter, 402
    women's hockey, 168
Running times, 52
Skiing, 174
Soccer, 13, 107, 315
Softball, 174
Sporting goods, 119
Sports, 288, 289, 549, 584
Strength shoes®, 453, 454
Swimming, 313
Tennis, 325
Volleyball, 75
Weightlifting, 325, 434

Work

Annual wage, 432, 448,
Career placement, 642
CEO
    ages, 61
    compensation, 30
Committees, 174, 225
Driving distance, 323
Earnings, 324, 421, 444, 481, 579,
    608, 613–614, 616
E-mail, 327, 329
Employee tenure, 261, 646
Employment, 14, 58
Genders of recent hires, 635–636
Happiness at work, 287
Hourly earnings, 62, 74, 109, 369,
    605, 606, 607, 645
Industry workers, 142
Injuries, 630
Job offers, 624
Job opening, 139
Job seekers, 2
Leaving job, 547
Lumber cutter, 277
Night shift, 347
Office rentals, 84
Overtime hours, 199, 295
Sick days, 74, 229
Time wasted, 578
Travel time, 47, 60, 74, 312, 323,
    346, 434
Union, 645, 648
Vacation days, 109
Work days, 379
Work environment, 177, 553
Work time and leisure time, 505,
    520, 521
Work weeks, 289

1
CHAPTER

INTRODUCTION TO STATISTICS

1.1 An Overview of Statistics

1.2 Data Classification

■ CASE STUDY

1.3 Data Collection and Experimental Design

■ ACTIVITY

■ USES AND ABUSES

■ REAL STATISTICS– REAL DECISIONS

■ HISTORY OF STATISTICS– TIMELINE

■ TECHNOLOGY

In 2008, the population of New Orleans, Louisiana grew faster than any other large city in the United States. Despite the increase, the population of 311,853 was still well below the pre-Hurricane Katrina population of 484,674.

You are already familiar with many of the practices of statistics, such as taking surveys, collecting data, and describing populations. What you may not know is that collecting accurate statistical data is often difficult and costly. Consider, for instance, the monumental task of counting and describing the entire population of the United States. If you were in charge of such a census, how would you do it? How would you ensure that your results are accurate? These and many more concerns are the responsibility of the United States Census Bureau, which conducts the census every decade.

In Chapter 1, you will be introduced to the basic concepts and goals of statistics. For instance, statistics were used to construct the following graphs, which show the fastest growing U.S. cities (population over 100,000) in 2008 by percent increase in population, U.S. cities with the largest numerical increases in population, and the regions where the cities are located.

For the 2010 Census, the Census Bureau sent short forms to every household. Short forms ask all members of every household such things as their gender, age, race, and ethnicity. Previously, a long form, which covered additional topics, was sent to about 17% of the population. But for the first time since 1940, the long form is being replaced by the American Community Survey, which will survey about 3 million households a year throughout the decade. These 3 million households will form a sample. In this course, you will learn how the data collected from a sample are used to infer characteristics about the entire population.

Fastest Growing U.S. Cities
(Population over 100,000)

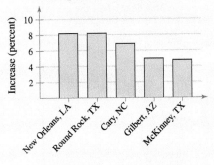

U.S. Cities with Largest
Numerical Increases

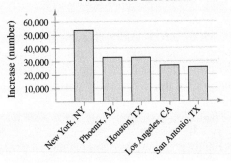

Location of the 25 Fastest
Growing U.S. Cities

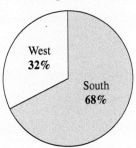

Location of the 25 U.S. Cities with
Largest Numerical Increases

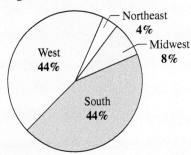

1.1 An Overview of Statistics

WHAT YOU SHOULD LEARN

▸ The definition of statistics

▸ How to distinguish between a population and a sample and between a parameter and a statistic

▸ How to distinguish between descriptive statistics and inferential statistics

A Definition of Statistics ▸ Data Sets ▸ Branches of Statistics

▸ A DEFINITION OF STATISTICS

As you begin this course, you may wonder: *What is statistics? Why should I study statistics? How can studying statistics help me in my profession?* Almost every day you are exposed to statistics. For instance, consider the following.

- "The number of Americans with diabetes will nearly double in the next 25 years." *(Source: Diabetes Care)*

- "The NRF expects holiday sales to decline 1% versus a 3.4% drop in holiday sales the previous year." *(Source: National Retail Federation)*

- "EIA projects total U.S. natural gas consumption will decline by 2.6 percent in 2009 and increase by 0.5 percent in 2010." *(Source: Energy Information Administration)*

The three statements you just read are based on the collection of *data*.

DEFINITION

Data consist of information coming from observations, counts, measurements, or responses.

Sometimes data are presented graphically. If you have ever read *USA TODAY*, you have certainly seen one of that newspaper's most popular features, *USA TODAY Snapshots*. Graphics such as this present information in a way that is easy to understand.

The use of statistics dates back to census taking in ancient Babylonia, Egypt, and later in the Roman Empire, when data were collected about matters concerning the state, such as births and deaths. In fact, the word *statistics* is derived from the Latin word *status*, meaning "state." So, what is statistics?

DEFINITION

Statistics is the science of collecting, organizing, analyzing, and interpreting data in order to make decisions.

▶ DATA SETS

There are two types of data sets you will use when studying statistics. These data sets are called *populations* and *samples*.

DEFINITION

A **population** is the collection of *all* outcomes, responses, measurements, or counts that are of interest.

A **sample** is a subset, or part, of a population.

A sample should be representative of a population so that sample data can be used to form conclusions about that population. Sample data must be collected using an appropriate method, such as *random sampling*. (You will learn more about random sampling in Section 1.3.) If they are not collected using an appropriate method, the data are of no value.

EXAMPLE 1

▶ **Identifying Data Sets**

In a recent survey, 1500 adults in the United States were asked if they thought there was solid evidence of global warming. Eight hundred fifty-five of the adults said yes. Identify the population and the sample. Describe the sample data set. *(Adapted from Pew Research Center)*

▶ **Solution**

The population consists of the responses of all adults in the United States, and the sample consists of the responses of the 1500 adults in the United States in the survey. The sample is a subset of the responses of all adults in the United States. The sample data set consists of 855 yes's and 645 no's.

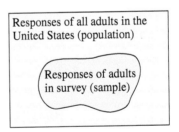

▶ **Try It Yourself 1**

The U.S. Department of Energy conducts weekly surveys of approximately 900 gasoline stations to determine the average price per gallon of regular gasoline. On January 11, 2010, the average price was $2.75 per gallon. Identify the population and the sample. Describe the sample data set. *(Source: Energy Information Administration)*

a. Identify the *population* and the *sample*.
b. What does the sample data set consist of?   *Answer: Page A30*

Whether a data set is a population or a sample usually depends on the context of the real-life situation. For instance, in Example 1, the population was the set of responses of all adults in the United States. Depending on the purpose of the survey, the population could have been the set of responses of all adults who live in California or who have cellular phones or who read a particular magazine.

Two important terms that are used throughout this course are *parameter* and *statistic*.

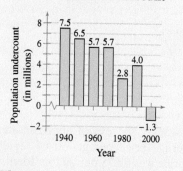

DEFINITION

A **parameter** is a numerical description of a *population* characteristic.

A **statistic** is a numerical description of a *sample* characteristic.

It is important to note that a sample statistic can differ from sample to sample whereas a population parameter is constant for a population.

EXAMPLE 2

▶ **Distinguishing Between a Parameter and a Statistic**

Decide whether the numerical value describes a population parameter or a sample statistic. Explain your reasoning.

1. A recent survey of 200 college career centers reported that the average starting salary for petroleum engineering majors is $83,121. *(Source: National Association of Colleges and Employers)*

2. The 2182 students who accepted admission offers to Northwestern University in 2009 have an average SAT score of 1442. *(Source: Northwestern University)*

3. In a random check of a sample of retail stores, the Food and Drug Administration found that 34% of the stores were not storing fish at the proper temperature.

▶ **Solution**

1. Because the average of $83,121 is based on a subset of the population, it is a sample statistic.

2. Because the SAT score of 1442 is based on all the students who accepted admission offers in 2009, it is a population parameter.

3. Because the percent of 34% is based on a subset of the population, it is a sample statistic.

▶ **Try It Yourself 2**

In 2009, Major League Baseball teams spent a total of $2,655,395,194 on players' salaries. Does this numerical value describe a population parameter or a sample statistic? *(Source: USA Today)*

a. Decide whether the numerical value is from a *population* or a *sample*.
b. Specify whether the numerical value is a *parameter* or a *statistic*.

Answer: Page A30

In this course, you will see how the use of statistics can help you make informed decisions that affect your life. Consider the census that the U.S. government takes every decade. When taking the census, the Census Bureau attempts to contact everyone living in the United States. Although it is impossible to count everyone, it is important that the census be as accurate as it can be, because public officials make many decisions based on the census information. Data collected in the 2010 census will determine how to assign congressional seats and how to distribute public funds.

▶ BRANCHES OF STATISTICS

The study of statistics has two major branches: *descriptive statistics* and *inferential statistics*.

DEFINITION

Descriptive statistics is the branch of statistics that involves the organization, summarization, and display of data.

Inferential statistics is the branch of statistics that involves using a sample to draw conclusions about a population. A basic tool in the study of inferential statistics is probability.

EXAMPLE 3

▶ Descriptive and Inferential Statistics

Decide which part of the study represents the descriptive branch of statistics. What conclusions might be drawn from the study using inferential statistics?

1. A large sample of men, aged 48, was studied for 18 years. For unmarried men, approximately 70% were alive at age 65. For married men, 90% were alive at age 65. *(Source: The Journal of Family Issues)*

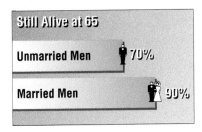

2. In a sample of Wall Street analysts, the percentage who incorrectly forecasted high-tech earnings in a recent year was 44%. *(Source: Bloomberg News)*

▶ Solution

1. Descriptive statistics involves statements such as "For unmarried men, approximately 70% were alive at age 65" and "For married men, 90% were alive at 65." A possible inference drawn from the study is that being married is associated with a longer life for men.

2. The part of this study that represents the descriptive branch of statistics involves the statement "the percentage [of Wall Street analysts] who incorrectly forecasted high-tech earnings in a recent year was 44%." A possible inference drawn from the study is that the stock market is difficult to forecast, even for professionals.

▶ Try It Yourself 3

A survey conducted among 1017 men and women by Opinion Research Corporation International found that 76% of women and 60% of men had a physical examination within the previous year. *(Source: Men's Health)*

a. Identify the *descriptive* aspect of the survey.
b. What *inferences* could be drawn from this survey?   *Answer: Page A30*

Throughout this course you will see applications of both branches. A major theme in this course will be how to use sample statistics to make inferences about unknown population parameters.

1.1 EXERCISES

■ BUILDING BASIC SKILLS AND VOCABULARY

1. How is a sample related to a population?

2. Why is a sample used more often than a population?

3. What is the difference between a parameter and a statistic?

4. What are the two main branches of statistics?

True or False? *In Exercises 5–10, determine whether the statement is true or false. If it is false, rewrite it as a true statement.*

5. A statistic is a measure that describes a population characteristic.

6. A sample is a subset of a population.

7. It is impossible for the Census Bureau to obtain all the census data about the population of the United States.

8. Inferential statistics involves using a population to draw a conclusion about a corresponding sample.

9. A population is the collection of some outcomes, responses, measurements, or counts that are of interest.

10. A sample statistic will not change from sample to sample.

Classifying a Data Set *In Exercises 11–20, determine whether the data set is a population or a sample. Explain your reasoning.*

11. The height of each player on a school's basketball team

12. The amount of energy collected from every wind turbine on a wind farm

13. A survey of 500 spectators from a stadium with 42,000 spectators

14. The annual salary of each pharmacist at a pharmacy

15. The cholesterol levels of 20 patients in a hospital with 100 patients

16. The number of televisions in each U.S. household

17. The final score of each golfer in a tournament

18. The age of every third person entering a clothing store

19. The political party of every U.S. president

20. The soil contamination levels at 10 locations near a landfill

Graphical Analysis *In Exercises 21–24, use the Venn diagram to identify the population and the sample.*

21.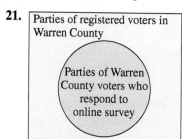

22.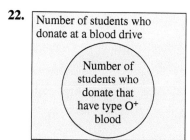

23.

Ages of adults in the United States who own cellular phones

Ages of adults in the U.S. who own Samsung cellular phones

24.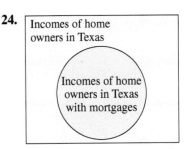

Incomes of home owners in Texas

Incomes of home owners in Texas with mortgages

■ USING AND INTERPRETING CONCEPTS

Identifying Populations and Samples   *In Exercises 25–34, identify the population and the sample.*

25. A survey of 1000 U.S. adults found that 59% think buying a home is the best investment a family can make. *(Source: Rasmussen Reports)*

26. A study of 33,043 infants in Italy was conducted to find a link between a heart rhythm abnormality and sudden infant death syndrome. *(Source: New England Journal of Medicine)*

27. A survey of 1442 U.S. adults found that 36% received an influenza vaccine for the current flu season. *(Source: Zogby International)*

28. A survey of 1600 people found that 76% plan on using the Microsoft Windows 7™ operating system at their businesses. *(Source: Information Technology Intelligence Corporation and Sunbelt Software)*

29. A survey of 800 registered voters found that 50% think economic stimulus is the most important issue to consider when voting for Congress. *(Source: Diageo/Hotline Poll)*

30. A survey of 496 students at a college found that 10% planned on traveling out of the country during spring break.

31. A survey of 546 U.S. women found that more than 56% are the primary investors in their households. *(Adapted from Roper Starch Worldwide for Intuit)*

32. A survey of 791 vacationers from the United States found that they planned on spending at least $2000 for their next vacation.

33. A magazine mails questionnaires to each company in Fortune magazine's top 100 best companies to work for and receives responses from 85 of them.

34. At the end of the day, a quality control inspector selects 20 light bulbs from the day's production and tests them.

Distinguishing Between a Parameter and a Statistic   *In Exercises 35–42, determine whether the numerical value is a parameter or a statistic. Explain your reasoning.*

35. The average annual salary for 35 of a company's 1200 accountants is $68,000.

36. In a survey of a sample of high school students, 43% said that their mothers had taught them the most about managing money. *(Source: Harris Poll for Girls Incorporated)*

37. Sixty-two of the 97 passengers aboard the Hindenburg airship survived its explosion.

38. In January 2010, 52% of the governors of the 50 states in the United States were Democrats.

39. In a survey of 300 computer users, 8% said their computers had malfunctions that needed to be repaired by service technicians.

40. In a recent year, the interest category for 12% of all new magazines was sports. *(Source: Oxbridge Communications)*

41. In a recent survey of 2000 people, 44% said China is the world's leading economic power. *(Source: Pew Research Center)*

42. In a recent year, the average math scores for all graduates on the ACT was 21.0. *(Source: ACT, Inc.)*

43. Which part of the survey described in Exercise 31 represents the descriptive branch of statistics? Make an inference based on the results of the survey.

44. Which part of the survey described in Exercise 32 represents the descriptive branch of statistics? Make an inference based on the results of the survey.

■ EXTENDING CONCEPTS

45. Identifying Data Sets in Articles Find a newspaper or magazine article that describes a survey.

    (a) Identify the sample used in the survey.

    (b) What is the sample's population?

    (c) Make an inference based on the results of the survey.

46. Sleep Deprivation In a recent study, volunteers who had 8 hours of sleep were three times more likely to answer questions correctly on a math test than were sleep-deprived participants. *(Source: CBS News)*

    (a) Identify the sample used in the study.

    (b) What is the sample's population?

    (c) Which part of the study represents the descriptive branch of statistics?

    (d) Make an inference based on the results of the study.

47. Living in Florida A study shows that senior citizens who live in Florida have better memories than senior citizens who do not live in Florida.

    (a) Make an inference based on the results of this study.

    (b) What is wrong with this type of reasoning?

48. Increase in Obesity Rates A study shows that the obesity rate among boys ages 2 to 19 has increased over the past several years. *(Source: Washington Post)*

    (a) Make an inference based on the results of this study.

    (b) What is wrong with this type of reasoning?

49. Writing Write an essay about the importance of statistics for one of the following.

    • A study on the effectiveness of a new drug

    • An analysis of a manufacturing process

    • Making conclusions about voter opinions using surveys

1.2 Data Classification

Types of Data ▸ Levels of Measurement

WHAT YOU SHOULD LEARN

▸ How to distinguish between qualitative data and quantitative data

▸ How to classify data with respect to the four levels of measurement: nominal, ordinal, interval, and ratio

▸ TYPES OF DATA

When doing a study, it is important to know the kind of data involved. The nature of the data you are working with will determine which statistical procedures can be used. In this section, you will learn how to classify data by type and by level of measurement. Data sets can consist of two types of data: *qualitative data* and *quantitative data*.

DEFINITION

Qualitative data consist of attributes, labels, or nonnumerical entries.

Quantitative data consist of numerical measurements or counts.

EXAMPLE 1

▸ Classifying Data by Type

The suggested retail prices of several Ford vehicles are shown in the table. Which data are qualitative data and which are quantitative data? Explain your reasoning. *(Source: Ford Motor Company)*

Model	Suggested retail price
Focus Sedan	$15,995
Fusion	$19,270
Mustang	$20,995
Edge	$26,920
Flex	$28,495
Escape Hybrid	$32,260
Expedition	$35,085
F-450	$44,145

▸ Solution

The information shown in the table can be separated into two data sets. One data set contains the names of vehicle models, and the other contains the suggested retail prices of vehicle models. The names are nonnumerical entries, so these are qualitative data. The suggested retail prices are numerical entries, so these are quantitative data.

▸ Try It Yourself 1

The populations of several U.S. cities are shown in the table. Which data are qualitative data and which are quantitative data? *(Source: U.S. Census Bureau)*

a. *Identify* the two data sets.
b. Decide whether each data set consists of *numerical* or *nonnumerical* entries.
c. Specify the *qualitative* data and the *quantitative* data.   *Answer: Page A30*

City	Population
Baltimore, MD	636,919
Jacksonville, FL	807,815
Memphis, TN	669,651
Pasadena, CA	143,080
San Antonio, TX	1,351,305
Seattle, WA	598,541

▶ LEVELS OF MEASUREMENT

Another characteristic of data is its level of measurement. The level of measurement determines which statistical calculations are meaningful. The four levels of measurement, in order from lowest to highest, are *nominal*, *ordinal*, *interval*, and *ratio*.

DEFINITION

Data at the **nominal level of measurement** are qualitative only. Data at this level are categorized using names, labels, or qualities. No mathematical computations can be made at this level.

Data at the **ordinal level of measurement** are qualitative or quantitative. Data at this level can be arranged in order, or ranked, but differences between data entries are not meaningful.

When numbers are at the nominal level of measurement, they simply represent a label. Examples of numbers used as labels include Social Security numbers and numbers on sports jerseys. For instance, it would not make sense to add the numbers on the players' jerseys for the Chicago Bears.

PICTURING THE WORLD

In 2009, Forbes Magazine chose the 75 best business schools in the United States. Forbes based their rankings on the return on investment achieved by the graduates from the class of 2004. Graduates of the top five M.B.A. programs typically earn more than $200,000 within five years.
(Source: Forbes)

Forbes Top Five U.S. Business Schools
1. Stanford
2. Dartmouth
3. Harvard
4. Chicago
5. Pennsylvania

In this list, what is the level of measurement?

EXAMPLE 2

▶ Classifying Data by Level

Two data sets are shown. Which data set consists of data at the nominal level? Which data set consists of data at the ordinal level? Explain your reasoning.
(Source: The Nielsen Company)

Top Five TV Programs (from 5/4/09 to 5/10/09)
1. American Idol–Wednesday
2. American Idol–Tuesday
3. Dancing with the Stars
4. NCIS
5. The Mentalist

Network Affiliates in Pittsburgh, PA	
WTAE	(ABC)
WPXI	(NBC)
KDKA	(CBS)
WPGH	(FOX)

▶ Solution

The first data set lists the ranks of five TV programs. The data set consists of the ranks 1, 2, 3, 4, and 5. Because the ranks can be listed in order, these data are at the ordinal level. Note that the difference between a rank of 1 and 5 has no mathematical meaning. The second data set consists of the call letters of each network affiliate in Pittsburgh. The call letters are simply the names of network affiliates, so these data are at the nominal level.

▶ Try It Yourself 2

Consider the following data sets. For each data set, decide whether the data are at the nominal level or at the ordinal level.

1. The final standings for the Pacific Division of the National Basketball Association

2. A collection of phone numbers

a. *Identify* what each data set represents.
b. Specify the *level of measurement* and justify your answer.

Answer: Page A30

The two highest levels of measurement consist of quantitative data only.

DEFINITION

Data at the **interval level of measurement** can be ordered, and meaningful differences between data entries can be calculated. At the interval level, a zero entry simply represents a position on a scale; the entry is not an inherent zero.

Data at the **ratio level of measurement** are similar to data at the interval level, with the added property that a zero entry is an inherent zero. A ratio of two data values can be formed so that one data value can be meaningfully expressed as a multiple of another.

An *inherent zero* is a zero that implies "none." For instance, the amount of money you have in a savings account could be zero dollars. In this case, the zero represents no money; it is an inherent zero. On the other hand, a temperature of 0°C does not represent a condition in which no heat is present. The 0°C temperature is simply a position on the Celsius scale; it is not an inherent zero.

To distinguish between data at the interval level and at the ratio level, determine whether the expression "twice as much" has any meaning in the context of the data. For instance, $2 is twice as much as $1, so these data are at the ratio level. On the other hand, 2°C is not twice as warm as 1°C, so these data are at the interval level.

**New York Yankees'
World Series Victories (Years)**

1923, 1927, 1928, 1932, 1936, 1937, 1938, 1939, 1941, 1943, 1947, 1949, 1950, 1951, 1952, 1953, 1956, 1958, 1961, 1962, 1977, 1978, 1996, 1998, 1999, 2000, 2009

**2009 American League
Home Run Totals (by Team)**

Team	Total
Baltimore	160
Boston	212
Chicago	184
Cleveland	161
Detroit	183
Kansas City	144
Los Angeles	173
Minnesota	172
New York	244
Oakland	135
Seattle	160
Tampa Bay	199
Texas	224
Toronto	209

EXAMPLE 3

▶ **Classifying Data by Level**

Two data sets are shown at the left. Which data set consists of data at the interval level? Which data set consists of data at the ratio level? Explain your reasoning. *(Source: Major League Baseball)*

▶ **Solution**

Both of these data sets contain quantitative data. Consider the dates of the Yankees' World Series victories. It makes sense to find differences between specific dates. For instance, the time between the Yankees' first and last World Series victories is

$$2009 - 1923 = 86 \text{ years.}$$

But it does not make sense to say that one year is a multiple of another. So, these data are at the interval level. However, using the home run totals, you can find differences *and* write ratios. From the data, you can see that Texas hit 63 more home runs than Cleveland hit and that New York hit about 1.5 times as many home runs as Seattle hit. So, these data are at the ratio level.

▶ **Try It Yourself 3**

Decide whether the data are at the interval level or at the ratio level.

1. The body temperatures (in degrees Fahrenheit) of an athlete during an exercise session

2. The heart rates (in beats per minute) of an athlete during an exercise session

   a. *Identify* what each data set represents.
   b. Specify the *level of measurement* and justify your answer.

Answer: Page A30

The following tables summarize which operations are meaningful at each of the four levels of measurement. When identifying a data set's level of measurement, use the highest level that applies.

Level of measurement	Put data in categories	Arrange data in order	Subtract data values	Determine if one data value is a multiple of another
Nominal	Yes	No	No	No
Ordinal	Yes	Yes	No	No
Interval	Yes	Yes	Yes	No
Ratio	Yes	Yes	Yes	Yes

Summary of Four Levels of Measurement

	Example of a Data Set	Meaningful Calculations
Nominal Level (Qualitative data)	*Types of Shows Televised by a Network* Comedy Documentaries Drama Cooking Reality Shows Soap Operas Sports Talk Shows	*Put in a category.* For instance, a show televised by the network could be put into one of the eight categories shown.

Ordinal Level (Qualitative or quantitative data)	*Motion Picture Association of America Ratings*
	
	Description
	     G General Audiences
	    PG Parental Guidance Suggested
	PG-13 Parents Strongly Cautioned
	     R Restricted
	NC-17 No One Under 17 Admitted
		Put in a category and *put in order.*
		
		For instance, a PG rating has a stronger restriction than a G rating.

Interval Level (Quantitative data)	*Average Monthly Temperatures (in degrees Fahrenheit) for Denver, CO* Jan 29.2 Jul 73.4 Feb 33.2 Aug 71.7 Mar 39.6 Sep 62.4 Apr 47.6 Oct 51.0 May 57.2 Nov 37.5 Jun 67.6 Dec 30.3 *(Source: National Climatic Data Center)*	Put in a category, put in order, and *find differences between values.* For instance, $57.2 - 47.6 = 9.6°F$. So, May is 9.6° warmer than April.
Ratio Level (Quantitative data)	*Average Monthly Precipitation (in inches) for Orlando, FL* Jan 2.4 Jul 7.2 Feb 2.4 Aug 6.3 Mar 3.5 Sep 5.8 Apr 2.4 Oct 2.7 May 3.7 Nov 2.3 Jun 7.4 Dec 2.3 *(Source: National Climatic Data Center)*	Put in a category, put in order, find differences between values, and *find ratios of values.* For instance, $\frac{7.4}{3.7} = 2$. So, there is twice as much rain in June as in May.

1.2 EXERCISES

■ BUILDING BASIC SKILLS AND VOCABULARY

1. Name each level of measurement for which data can be qualitative.

2. Name each level of measurement for which data can be quantitative.

True or False?   *In Exercises 3–6, determine whether the statement is true or false. If it is false, rewrite it as a true statement.*

3. Data at the ordinal level are quantitative only.

4. For data at the interval level, you cannot calculate meaningful differences between data entries.

5. More types of calculations can be performed with data at the nominal level than with data at the interval level.

6. Data at the ratio level cannot be put in order.

■ USING AND INTERPRETING CONCEPTS

Classifying Data by Type   *In Exercises 7–18, determine whether the data are qualitative or quantitative. Explain your reasoning.*

7. telephone numbers in a directory

8. heights of hot air balloons

9. body temperatures of patients

10. eye colors of models

11. lengths of songs on MP3 player

12. carrying capacities of pickups

13. player numbers for a soccer team

14. student ID numbers

15. weights of infants at a hospital

16. species of trees in a forest

17. responses on an opinion poll

18. wait times at a grocery store

Classifying Data by Level   *In Exercises 19–24, determine whether the data are qualitative or quantitative, and identify the data set's level of measurement. Explain your reasoning.*

19. Football   The top five teams in the final college football poll released in January 2010 are listed. *(Source: Associated Press)*

    1. Alabama   2. Texas   3. Florida   4. Boise State   5. Ohio State

20. Politics   The three political parties in the 111th Congress are listed below.

    Republican   Democrat   Independent

21. Top Salespeople   The regions representing the top salespeople in a corporation for the past six years are given.

    Southeast   Northwest   Northeast
    Southeast   Southwest   Southwest

22. Fish Lengths   The lengths (in inches) of a sample of striped bass caught in Maryland waters are listed. *(Adapted from National Marine Fisheries Service, Fisheries Statistics and Economics Division)*

    16   17.25   19   18.75   21   20.3   19.8   24   21.82

23. Best Seller List The top five hardcover nonfiction books on *The New York Times* Best Seller List on January 19, 2010 are shown. *(Source: The New York Times)*

1. Committed 2. Have a Little Faith 3. The Checklist Manifesto

4. Going Rogue 5. Stones Into Schools

24. Ticket Prices The average ticket prices for 10 Broadway shows in 2009 are listed. *(Adapted from The Broadway League)*

$149 $128 $124 $91 $96 $106 $112 $95 $86 $74

Graphical Analysis *In Exercises 25–28, identify the level of measurement of the data listed on the horizontal axis in the graph.*

25. Over the Next Few Years, How Likely Is It That the United States Will Enter a 1930s-Like Depression?

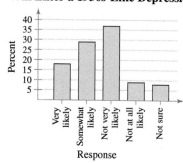

(Source: Rasmussen Reports)

26. Average January Snowfall for 15 Cities

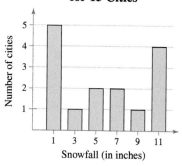

(Source: National Climatic Data Center)

27. Gender Profile of the 111th Congress

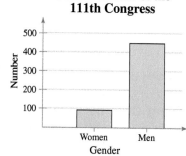

(Source: Congressional Research Service)

28. Motor Vehicle Accidents by Year

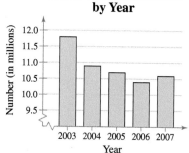

(Source: National Safety Council)

29. The following items appear on a physician's intake form. Identify the level of measurement of the data.

a. Temperature

b. Allergies

c. Weight

d. Pain level (scale of 0 to 10)

30. The following items appear on an employment application. Identify the level of measurement of the data.

a. Highest grade level completed

b. Gender

c. Year of college graduation

d. Number of years at last job

■ EXTENDING CONCEPTS

31. Writing What is an inherent zero? Describe three examples of data sets that have inherent zeros and three that do not.

32. Writing Describe two examples of data sets for each of the four levels of measurement. Justify your answer.

Rating Television Shows in the United States

The Nielsen Company has been rating television programs for more than 60 years. Nielsen uses several sampling procedures, but its main one is to track the viewing patterns of 20,000 households. These contain more than 45,000 people and are chosen to form a cross section of the overall population. The households represent various locations, ethnic groups, and income brackets. The data gathered from the Nielsen sample of 20,000 households are used to draw inferences about the population of all households in the United States.

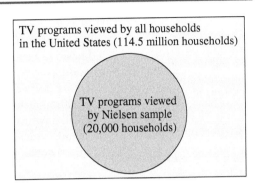

TV programs viewed by all households in the United States (114.5 million households)

TV programs viewed by Nielsen sample (20,000 households)

Top-Ranked Programs in Overall Viewing for the Week of 11/23/09–11/29/09

Rank	Rank Last Week	Program Name	Network	Day, Time	Rating	Share	Audience
1	2	Dancing with the Stars	ABC	Mon., 8:00 P.M.	12.9	19	20,411,000
2	1	NCIS	CBS	Tues., 8:00 P.M.	12.3	20	20,348,000
3	4	Dancing with the Stars Results	ABC	Tues., 9:00 P.M.	12.0	20	19,294,000
4	3	NBC Sunday Night Football	NBC	Sun., 8:15 P.M.	11.5	18	19,210,000
5	8	NCIS: Los Angeles	CBS	Tues., 9:00 P.M.	10.4	16	17,221,000
6	5	60 Minutes	CBS	Sun., 7:00 P.M.	9.0	14	14,377,000
7	15	The Big Bang Theory	CBS	Mon., 9:30 P.M.	8.4	13	14,129,000
8	16	Sunday Night NFL Pre-Kick	NBC	Sun., 8:00 P.M.	8.4	13	13,927,000
9	12	Two and a Half Men	CBS	Mon., 9:00 P.M.	8.3	12	13,877,000
10	11	Criminal Minds	CBS	Wed., 9:00 P.M.	8.2	14	13,605,000

EXERCISES

1. **Rating Points** Each rating point represents 1,145,000 households, or 1% of the households in the United States. Does a program with a rating of 8.4 have twice the number of households as a program with a rating of 4.2? Explain your reasoning.

2. **Sampling Percent** What percentage of the total number of U.S. households is used in the Nielsen sample?

3. **Nominal Level of Measurement** Which columns in the table contain data at the nominal level?

4. **Ordinal Level of Measurement** Which columns in the table contain data at the ordinal level? Describe two ways that the data can be ordered.

5. **Interval Level of Measurement** Which column in the table contains data at the interval level? How can these data be ordered?

6. **Ratio Level of Measurement** Which columns contain data at the ratio level?

7. **Rankings** The column listed as "Share" gives the percentage of televisions in use at a given time. The 11th ranked program for this week is CSI: Miami with a rating of 8.4 and share of 14. Using this information, how does Nielsen rank the programs? Why do you think they do it this way? Explain your reasoning.

8. **Inferences** What decisions (inferences) can be made on the basis of the Nielsen ratings?

1.3 Data Collection and Experimental Design

Design of a Statistical Study ▸ Data Collection ▸ Experimental Design ▸ Sampling Techniques

▸ DESIGN OF A STATISTICAL STUDY

The goal of every statistical study is to collect data and then use the data to make a decision. Any decision you make using the results of a statistical study is only as good as the process used to obtain the data. If the process is flawed, then the resulting decision is questionable.

Although you may never have to develop a statistical study, it is likely that you will have to interpret the results of one. And before you interpret the results of a study, you should determine whether the results are valid, as well as reliable. In other words, you should be familiar with how to design a statistical study.

GUIDELINES

Designing a Statistical Study

1. Identify the variable(s) of interest (the focus) and the population of the study.
2. Develop a detailed plan for collecting data. If you use a sample, make sure the sample is representative of the population.
3. Collect the data.
4. Describe the data, using descriptive statistics techniques.
5. Interpret the data and make decisions about the population using inferential statistics.
6. Identify any possible errors.

▸ DATA COLLECTION

There are several ways you can collect data. Often, the focus of the study dictates the best way to collect data. The following is a brief summary of four methods of data collection.

- *Do an observational study* In an **observational study,** a researcher observes and measures characteristics of interest of part of a population but does not change existing conditions. For instance, an observational study was performed in which researchers observed and recorded the mouthing behavior on nonfood objects of children up to three years old. *(Source: Pediatrics Magazine)*

- *Perform an experiment* In performing an **experiment,** a **treatment** is applied to part of a population and responses are observed. Another part of the population may be used as a **control group,** in which no treatment is applied. In many cases, subjects (sometimes called **experimental units**) in the control group are given a **placebo,** which is a harmless, unmedicated treatment, that is made to look like the real treatment. The responses of the treatment group and control group can then be compared and studied. In most cases, it is a good idea to use the same number of subjects for each treatment. For instance, an experiment was performed in which diabetics took cinnamon extract daily while a control group took none. After 40 days, the diabetics who took the cinnamon reduced their risk of heart disease while the control group experienced no change. *(Source: Diabetes Care)*

- *Use a simulation*   A **simulation** is the use of a mathematical or physical model to reproduce the conditions of a situation or process. Collecting data often involves the use of computers. Simulations allow you to study situations that are impractical or even dangerous to create in real life, and often they save time and money. For instance, automobile manufacturers use simulations with dummies to study the effects of crashes on humans. Throughout this course, you will have the opportunity to use applets that simulate statistical processes on a computer.

- *Use a survey*   A **survey** is an investigation of one or more characteristics of a population. Most often, surveys are carried out on *people* by asking them questions. The most common types of surveys are done by interview, mail, or telephone. In designing a survey, it is important to word the questions so that they do not lead to biased results, which are not representative of a population. For instance, a survey is conducted on a sample of female physicians to determine whether the primary reason for their career choice is financial stability. In designing the survey, it would be acceptable to make a list of reasons and ask each individual in the sample to select her first choice.

PICTURING THE WORLD

The Gallup Organization conducts many polls (or surveys) regarding the president, Congress, and political and nonpolitical issues. A commonly cited Gallup poll is the public approval rating of the president. For instance, the approval ratings for President Barack Obama throughout 2009 are shown in the following graph. (The rating is from the poll conducted at the end of each month.)

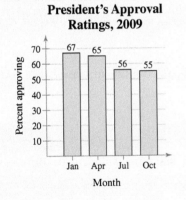

President's Approval Ratings, 2009

Discuss some ways that Gallup could select a biased sample to conduct a poll. How could Gallup select a sample that is unbiased?

EXAMPLE 1

▸ **Deciding on Methods of Data Collection**

Consider the following statistical studies. Which method of data collection would you use to collect data for each study? Explain your reasoning.

1. A study of the effect of changing flight patterns on the number of airplane accidents

2. A study of the effect of eating oatmeal on lowering blood pressure

3. A study of how fourth grade students solve a puzzle

4. A study of U.S. residents' approval rating of the U.S. president

▸ **Solution**

1. Because it is impractical to create this situation, use a simulation.

2. In this study, you want to measure the effect a treatment (eating oatmeal) has on patients. So, you would want to perform an experiment.

3. Because you want to observe and measure certain characteristics of part of a population, you could do an observational study.

4. You could use a survey that asks, "Do you approve of the way the president is handling his job?"

▸ Try It Yourself 1

Consider the following statistical studies. Which method of data collection would you use to collect data for each study?

1. A study of the effect of exercise on relieving depression

2. A study of the success of graduates of a large university in finding a job within one year of graduation

a. Identify the *focus* of the study.
b. Identify the *population* of the study.
c. Choose an appropriate *method of data collection*.   *Answer: Page A30*

▶ EXPERIMENTAL DESIGN

In order to produce meaningful unbiased results, experiments should be carefully designed and executed. It is important to know what steps should be taken to make the results of an experiment valid. Three key elements of a well-designed experiment are *control*, *randomization*, and *replication*.

Because experimental results can be ruined by a variety of factors, being able to control these influential factors is important. One such factor is a *confounding variable*.

> **DEFINITION**
>
> A **confounding variable** occurs when an experimenter cannot tell the difference between the effects of different factors on a variable.

For instance, to attract more customers, a coffee shop owner experiments by remodeling her shop using bright colors. At the same time, a shopping mall nearby has its grand opening. If business at the coffee shop increases, it cannot be determined whether it is because of the new colors or the new shopping mall. The effects of the colors and the shopping mall have been confounded.

Another factor that can affect experimental results is the *placebo effect*. The **placebo effect** occurs when a subject reacts favorably to a placebo when in fact the subject has been given no medicated treatment at all. To help control or minimize the placebo effect, a technique called *blinding* can be used.

> **DEFINITION**
>
> **Blinding** is a technique where the subjects do not know whether they are receiving a treatment or a placebo. In a **double-blind experiment,** neither the experimenter nor the subjects know if the subjects are receiving a treatment or a placebo. The experimenter is informed after all the data have been collected. This type of experimental design is preferred by researchers.

Another technique that can be used to obtain unbiased results is *randomization*.

> **DEFINITION**
>
> **Randomization** is a process of randomly assigning subjects to different treatment groups.

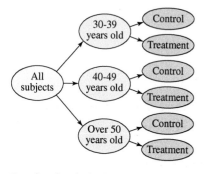

Randomized Block Design

In a **completely randomized design,** subjects are assigned to different treatment groups through random selection. In some experiments, it may be necessary for the experimenter to use **blocks,** which are groups of subjects with similar characteristics. A commonly used experimental design is a **randomized block design.** To use a randomized block design, you should divide subjects with similar characteristics into blocks, and then, within each block, randomly assign subjects to treatment groups. For instance, an experimenter who is testing the effects of a new weight loss drink may first divide the subjects into age categories such as 30–39 years old, 40–49 years old, and over 50 years old, and then, within each age group, randomly assign subjects to either the treatment group or the control group as shown.

Another type of experimental design is a **matched-pairs design,** where subjects are paired up according to a similarity. One subject in the pair is randomly selected to receive one treatment while the other subject receives a different treatment. For instance, two subjects may be paired up because of their age, geographical location, or a particular physical characteristic.

Sample size, which is the number of subjects, is another important part of experimental design. To improve the validity of experimental results, *replication* is required.

DEFINITION

Replication is the repetition of an experiment under the same or similar conditions.

For instance, suppose an experiment is designed to test a vaccine against a strain of influenza. In the experiment, 10,000 people are given the vaccine and another 10,000 people are given a placebo. Because of the sample size, the effectiveness of the vaccine would most likely be observed. But, if the subjects in the experiment are not selected so that the two groups are similar (according to age and gender), the results are of less value.

EXAMPLE 2

▶ **Analyzing an Experimental Design**

A company wants to test the effectiveness of a new gum developed to help people quit smoking. Identify a potential problem with the given experimental design and suggest a way to improve it.

1. The company identifies ten adults who are heavy smokers. Five of the subjects are given the new gum and the other five subjects are given a placebo. After two months, the subjects are evaluated and it is found that the five subjects using the new gum have quit smoking.

2. The company identifies one thousand adults who are heavy smokers. The subjects are divided into blocks according to gender. Females are given the new gum and males are given the placebo. After two months, a significant number of the female subjects have quit smoking.

▶ **Solution**

1. The sample size being used is not large enough to validate the results of the experiment. The experiment must be replicated to improve the validity.

2. The groups are not similar. The new gum may have a greater effect on women than on men, or vice versa. The subjects can be divided into blocks according to gender, but then, within each block, they must be randomly assigned to be in the treatment group or in the control group.

▶ **Try It Yourself 2**

Using the information in Example 2, suppose the company identifies 240 adults who are heavy smokers. The subjects are randomly assigned to be in a treatment group or in a control group. Each subject is also given a DVD featuring the dangers of smoking. After four months, most of the subjects in the treatment group have quit smoking.

a. Identify a *potential problem* with the experimental design.
b. How could the design be *improved*?   *Answer: Page A30*

▶ SAMPLING TECHNIQUES

A **census** is a count or measure of an *entire* population. Taking a census provides complete information, but it is often costly and difficult to perform. A **sampling** is a count or measure of *part* of a population, and is more commonly used in statistical studies. To collect unbiased data, a researcher must ensure that the sample is representative of the population. Appropriate sampling techniques must be used to ensure that inferences about the population are valid. Remember that when a study is done with faulty data, the results are questionable. Even with the best methods of sampling, a **sampling error** may occur. A sampling error is the difference between the results of a sample and those of the population. When you learn about inferential statistics, you will learn techniques of controlling sampling errors.

A **random sample** is one in which every member of the population has an equal chance of being selected. A **simple random sample** is a sample in which every possible sample of the same size has the same chance of being selected. One way to collect a simple random sample is to assign a different number to each member of the population and then use a random number table like the one in Appendix B. Responses, counts, or measures for members of the population whose numbers correspond to those generated using the table would be in the sample. Calculators and computer software programs are also used to generate random numbers (see page 34).

INSIGHT

A **biased sample** is one that is not representative of the population from which it is drawn. For instance, a sample consisting of only 18- to 22-year-old college students would not be representative of the entire 18- to 22-year-old population in the country.

To explore this topic further, see Activity 1.3 on page 26.

Table 1—Random Numbers

92630	78240	19267	95457	53497	23894	37708	79862
79445	78735	71549	44843	26104	67318	00701	34986
59654	71966	27386	50004	05358	94031	29281	18544
31524	49587	76612	39789	13537	48086	59483	60680
06348	76938	90379	51392	55887	71015	09209	79157

Portion of Table 1 found in Appendix B

Consider a study of the number of people who live in West Ridge County. To use a simple random sample to count the number of people who live in West Ridge County households, you could assign a different number to each household, use a technology tool or table of random numbers to generate a sample of numbers, and then count the number of people living in each selected household.

STUDY TIP

Here are instructions for using the random integer generator on a TI-83/84 Plus for Example 3.

MATH

Choose the PRB menu.

5: randInt(

1 , 7 3 1 , 8)

ENTER

randInt(1,731,8)
{537 33 249 728…

Continuing to press ENTER will generate more random samples of 8 integers.

EXAMPLE 3   Report 1

▶ Using a Simple Random Sample

There are 731 students currently enrolled in a statistics course at your school. You wish to form a sample of eight students to answer some survey questions. Select the students who will belong to the simple random sample.

▶ Solution

Assign numbers 1 to 731 to the students in the course. In the table of random numbers, choose a starting place at random and read the digits in groups of three (because 731 is a three-digit number). For instance, if you started in the third row of the table at the beginning of the second column, you would group the numbers as follows:

719|66 2|738|6 50|004| 053|58 9|403|1 29|281| 185|44

Ignoring numbers greater than 731, the first eight numbers are 719, 662, 650, 4, 53, 589, 403, and 129. The students assigned these numbers will make up the sample. To find the sample using a TI-83/84 Plus, follow the instructions in the margin.

▶ **Try It Yourself 3**

A company employs 79 people. Choose a simple random sample of five to survey.

a. In the table in Appendix B, randomly choose a *starting place*.
b. *Read the digits* in groups of two.
c. Write the five random numbers.

Answer: Page A30

When you choose members of a sample, you should decide whether it is acceptable to have the same population member selected more than once. If it is acceptable, then the sampling process is said to be *with replacement*. If it is not acceptable, then the sampling process is said to be *without replacement*.

There are several other commonly used sampling techniques. Each has advantages and disadvantages.

- **Stratified Sample**   When it is important for the sample to have members from each segment of the population, you should use a stratified sample. Depending on the focus of the study, members of the population are divided into two or more subsets, called *strata*, that share a similar characteristic such as age, gender, ethnicity, or even political preference. A sample is then randomly selected from each of the strata. Using a stratified sample ensures that each segment of the population is represented. For instance, to collect a stratified sample of the number of people who live in West Ridge County households, you could divide the households into socioeconomic levels, and then randomly select households from each level.

Group 1:
Low income

Group 2:
Middle income

Group 3:
High income

Stratified Sampling

INSIGHT

For stratified sampling, each of the strata contains members with a certain characteristic (for instance, a particular age group). In contrast, clusters consist of geographic groupings, and each cluster should contain members with all of the characteristics (for instance, all age groups). With stratified samples, some of the members of each group are used. In a cluster sampling, all of the members of one or more groups are used.

- **Cluster Sample**   When the population falls into naturally occurring subgroups, each having similar characteristics, a cluster sample may be the most appropriate. To select a cluster sample, divide the population into groups, called *clusters*, and select all of the members in one or more (but not all) of the clusters. Examples of clusters could be different sections of the same course or different branches of a bank. For instance, to collect a cluster sample of the number of people who live in West Ridge County households, divide the households into groups according to zip codes, then select all the households in one or more, but not all, zip codes and count the number of people living in each household. In using a cluster sample, care must be taken to ensure that all clusters have similar characteristics. For instance, if one of the zip code clusters has a greater proportion of high-income people, the data might not be representative of the population.

Zip Code Zones in West Ridge County

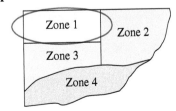

Cluster Sampling

- *Systematic Sample*   A systematic sample is a sample in which each member of the population is assigned a number. The members of the population are ordered in some way, a starting number is randomly selected, and then sample members are selected at regular intervals from the starting number. (For instance, every 3rd, 5th, or 100th member is selected.) For instance, to collect a systematic sample of the number of people who live in West Ridge County households, you could assign a different number to each household, randomly choose a starting number, select every 100th household, and count the number of people living in each. An advantage of systematic sampling is that it is easy to use. In the case of any regularly occurring pattern in the data, however, this type of sampling should be avoided.

Systematic Sampling

   A type of sample that often leads to biased studies (so it is not recommended) is a **convenience sample.** A convenience sample consists only of available members of the population.

EXAMPLE 4

▸ Identifying Sampling Techniques

You are doing a study to determine the opinions of students at your school regarding stem cell research. Identify the sampling technique you are using if you select the samples listed. Discuss potential sources of bias (if any). Explain.

1. You divide the student population with respect to majors and randomly select and question some students in each major.

2. You assign each student a number and generate random numbers. You then question each student whose number is randomly selected.

3. You select students who are in your biology class.

▸ Solution

1. Because students are divided into strata (majors) and a sample is selected from each major, this is a stratified sample.

2. Each sample of the same size has an equal chance of being selected and each student has an equal chance of being selected, so this is a simple random sample.

3. Because the sample is taken from students that are readily available, this is a convenience sample. The sample may be biased because biology students may be more familiar with stem cell research than other students and may have stronger opinions.

▸ Try It Yourself 4

You want to determine the opinions of students regarding stem cell research. Identify the sampling technique you are using if you select the samples listed.

1. You select a class at random and question each student in the class.

2. You assign each student a number and, after choosing a starting number, question every 25th student.

a. Determine *how* the sample is *selected* and identify the corresponding *sampling technique.*

b. Discuss potential sources of *bias* (if any). Explain.   *Answer: Page A30*

1.3 EXERCISES

1. In an experiment, a treatment is applied to part of a population and responses are observed. In an observational study, a researcher measures characteristics of interest of a part of a population but does not change existing conditions.

2. See Selected Answers, page A109.

3. See Odd Answers, page A51.

4. See Selected Answers, page A109.

5. True

6. See Selected Answers, page A109.

7. False. Using stratified sampling guarantees that members of each group within a population will be sampled.

8. See Selected Answers, page A109.

9. See Odd Answers, page A51.

10. True

11. Use a census because all the patients are accessible and the number of patients is not too large.

12. See Selected Answers, page A109.

13. Perform an experiment because you want to measure the effect of a treatment on the human digestive system.

14. See Selected Answers, page A109.

15. Use a simulation because the situation is impractical and dangerous to create in real life.

16. See Selected Answers, page A109.

17. See Odd Answers, page A51.

■ BUILDING BASIC SKILLS AND VOCABULARY

1. What is the difference between an observational study and an experiment?

2. What is the difference between a census and a sampling?

3. What is the difference between a random sample and a simple random sample?

4. What is replication in an experiment, and why is it important?

True or False?   *In Exercises 5–10, determine whether the statement is true or false. If it is false, rewrite it as a true statement.*

5. In a randomized block design, subjects with similar characteristics are divided into blocks, and then, within each block, randomly assigned to treatment groups.

6. A double-blind experiment is used to increase the placebo effect.

7. Using a systematic sample guarantees that members of each group within a population will be sampled.

8. A census is a count of part of a population.

9. The method for selecting a stratified sample is to order a population in some way and then select members of the population at regular intervals.

10. To select a cluster sample, divide a population into groups and then select all of the members in at least one (but not all) of the groups.

Deciding on the Method of Data Collection   *In Exercises 11–16, explain which method of data collection you would use to collect data for the study.*

11. A study of the health of 168 kidney transplant patients at a hospital

12. A study of motorcycle helmet usage in a city without a helmet law

13. A study of the effect on the human digestive system of potato chips made with a fat substitute

14. A study of the effect of a product's warning label to determine whether consumers will still buy the product

15. A study of how fast a virus would spread in a metropolitan area

16. A study of how often people wash their hands in public restrooms

■ USING AND INTERPRETING CONCEPTS

17. **Allergy Drug**   A pharmaceutical company wants to test the effectiveness of a new allergy drug. The company identifies 250 females 30–35 years old who suffer from severe allergies. The subjects are randomly assigned into two groups. One group is given the new allergy drug and the other is given a placebo that looks exactly like the new allergy drug. After six months, the subjects' symptoms are studied and compared.

    (a) Identify the experimental units and treatments used in this experiment.

    (b) Identify a potential problem with the experimental design being used and suggest a way to improve it.

    (c) How could this experiment be designed to be double-blind?

18. See Selected Answers, page A109.

19. Simple random sampling is used because each telephone number has an equal chance of being dialed, and all samples of 1400 phone numbers have an equal chance of being selected. The sample may be biased because only homes with telephones will be sampled.

20. Stratified sampling is used because the persons are divided into strata (rural and urban), and a sample is selected from each stratum.

21. Convenience sampling is used because the students are chosen due to their convenience of location. Bias may enter into the sample because the students sampled may not be representative of the population of students.

22. Cluster sampling is used because the disaster area is divided into grids, and 30 grids are then entirely selected. A possible source of bias is that certain grids may have been much more severely damaged than others.

23. Simple random sampling is used because each customer has an equal chance of being contacted, and all samples of 580 customers have an equal chance of being selected.

24. Systematic sampling is used because every tenth person entering the shopping mall is sampled. It is possible for bias to enter into the sample if, for some reason, there is a regular pattern to the people entering the shopping mall.

25. Stratified sampling is used because a sample is taken from each one-acre subplot.

26. See Selected Answers, page A109.

27. Answers will vary.

28. Answers will vary.

29. Answers will vary. *Sample answer:* Treatment group: Jake, Maria, Lucy, Adam, Bridget, Vanessa, Rick, Dan, and Mary. Control group: Mike, Ron, Carlos, Steve, Susan, Kate, Pete, Judy, and Connie. A random number table is used.

30. Answers will vary.

18. Sneakers Nike developed a new type of sneaker designed to help delay the onset of arthritis in the knee. Eighty people with early signs of arthritis volunteered for a study. One-half of the volunteers wore the experimental sneakers and the other half wore regular Nike sneakers that looked exactly like the experimental sneakers. The individuals wore the sneakers every day. At the conclusion of the study, their symptoms were evaluated and MRI tests were performed on their knees. *(Source: Washington Post)*

(a) Identify the experimental units and treatments used in this experiment.

(b) Identify a potential problem with the experimental design being used and suggest a way to improve it.

(c) The experiment is described as a placebo-controlled, double-blind study. Explain what this means.

(d) Of the 80 volunteers, suppose 40 are men and 40 are women. How could blocking be used in designing this experiment?

Identifying Sampling Techniques *In Exercises 19–26, identify the sampling technique used, and discuss potential sources of bias (if any). Explain.*

19. Using random digit dialing, researchers call 1400 people and ask what obstacles (such as childcare) keep them from exercising.

20. Chosen at random, 500 rural and 500 urban persons age 65 or older are asked about their health and their experience with prescription drugs.

21. Questioning students as they leave a university library, a researcher asks 358 students about their drinking habits.

22. After a hurricane, a disaster area is divided into 200 equal grids. Thirty of the grids are selected, and every occupied household in the grid is interviewed to help focus relief efforts on what residents require the most.

23. Chosen at random, 580 customers at a car dealership are contacted and asked their opinions of the service they received.

24. Every tenth person entering a mall is asked to name his or her favorite store.

25. Soybeans are planted on a 48-acre field. The field is divided into one-acre subplots. A sample is taken from each subplot to estimate the harvest.

26. From calls made with randomly generated telephone numbers, 1012 respondents are asked if they rent or own their residences.

27. Random Number Table Use the seventh row of Table 1 in Appendix B to generate 12 random numbers between 1 and 99.

28. Random Number Table Use the twelfth row of Table 1 in Appendix B to generate 10 random numbers between 1 and 920.

29. Sleep Deprivation A researcher wants to study the effects of sleep deprivation on motor skills. Eighteen people volunteer for the experiment: Jake, Maria, Mike, Lucy, Ron, Adam, Bridget, Carlos, Steve, Susan, Vanessa, Rick, Dan, Kate, Pete, Judy, Mary, and Connie. Use a random number generator to choose nine subjects for the treatment group. The other nine subjects will go into the control group. List the subjects in each group. Tell which method you would use to generate the random numbers.

30. Random Number Generation Volunteers for an experiment are numbered from 1 to 70. The volunteers are to be randomly assigned to two different treatment groups. Use a random number generator different from the one you used in Exercise 29 to choose 35 subjects for the treatment group. The other 35 subjects will go into the control group. List the subjects, according to number, in each group. Tell which method you used to generate the random numbers.

Choosing Between a Census and a Sampling   *In Exercises 31 and 32, determine whether you would take a census or use a sampling. If you would use a sampling, decide what sampling technique you would use. Explain your reasoning.*

31. The average age of the 115 residents of a retirement community

32. The most popular type of movie among 100,000 online movie rental subscribers

Recognizing a Biased Question   *In Exercises 33–36, determine whether the survey question is biased. If the question is biased, suggest a better wording.*

33. Why does eating whole-grain foods improve your health?

34. Why does text messaging while driving increase the risk of a crash?

35. How much do you exercise during an average week?

36. Why do you think the media have a negative effect on teen girls' dieting habits?

37. Writing   A sample of television program ratings by The Nielsen Company is described on page 15. Discuss the strata used in the sample. Why is it important to have a stratified sample for these ratings?

SC **38.** Use StatCrunch to generate the following random numbers.

    a. 8 numbers between 1 and 50

    b. 15 numbers between 1 and 150

    c. 16 numbers between 1 and 325

    d. 20 numbers between 1 and 1000

▓ EXTENDING CONCEPTS

39. Observational studies are sometimes referred to as *natural experiments*. Explain, in your own words, what this means.

40. Open and Closed Questions   Two types of survey questions are open questions and closed questions. An open question allows for any kind of response; a closed question allows for only a fixed response. An open question, and a closed question with its possible choices, are given below. List an advantage and a disadvantage of each question.

    Open Question   What can be done to get students to eat healthier foods?

    Closed Question   How would you get students to eat healthier foods?

      1. Mandatory nutrition course

      2. Offer only healthy foods in the cafeteria and remove unhealthy foods

      3. Offer more healthy foods in the cafeteria and raise the prices on unhealthy foods

41. Who Picked These People?   Some polling agencies ask people to call a telephone number and give their response to a question. (a) List an advantage and a disadvantage of a survey conducted in this manner. (b) What sampling technique is used in such a survey?

42. Give an example of an experiment where confounding may occur.

43. Why is it important to use blinding in an experiment?

44. How are the placebo effect and the Hawthorne effect similar? How are they different?

45. How is a randomized block design in experiments similar to a stratified sample?

ACTIVITY 1.3 Random Numbers

APPLET

The *random numbers* applet is designed to allow you to generate random numbers from a range of values. You can specify integer values for the minimum value, maximum value, and the number of samples in the appropriate fields. You should not use decimal points when filling in the fields. When SAMPLE is clicked, the applet generates random values, which are displayed as a list in the text field.

Minimum value: [   ]
Maximum value: [   ]
Number of samples: [   ]
[Sample]

■ Explore

Step 1 Specify a minimum value.
Step 2 Specify a maximum value.
Step 3 Specify the number of samples.
Step 4 Click SAMPLE to generate a list of random values.

■ Draw Conclusions

APPLET

1. Specify the minimum, maximum, and number of samples to be 1, 20, and 8, respectively, as shown. Run the applet. Continue generating lists until you obtain one that shows that the random sample is taken with replacement. Write down this list. How do you know that the list is a random sample taken with replacement?

Minimum value: [1]
Maximum value: [20]
Number of samples: [8]
[Sample]

2. Use the applet to repeat Example 3 on page 20. What values did you use for the minimum, maximum, and number of samples? Which method do you prefer? Explain.

USES AND ABUSES

Uses

Experiments with Favorable Results An experiment that began in March 2003 studied 321 women with advanced breast cancer. All of the women had been previously treated with other drugs, but the cancer had stopped responding to the medications. The women were then given the opportunity to take a new drug combined with a particular chemotherapy drug.

The subjects were divided into two groups, one that took the new drug combined with a chemotherapy drug, and one that took only the chemotherapy drug. After three years, results showed that the new drug in combination with the chemotherapy drug delayed the progression of cancer in the subjects. The results were so significant that the study was stopped, and the new drug was offered to all women in the study. The Food and Drug Administration has since approved use of the new drug in conjunction with a chemotherapy drug.

Abuses

Experiments with Unfavorable Results From 1988 to 1991, one hundred eighty thousand teenagers in Norway were used as subjects to test a new vaccine against the deadly bacteria *meningococcus b*. A brochure describing the possible effects of the vaccine stated, "it is unlikely to expect serious complications," while information provided to the Norwegian Parliament stated, "serious side effects can not be excluded." The vaccine trial had some disastrous results: More than 500 side effects were reported, with some considered serious, and several of the subjects developed serious neurological diseases. The results showed that the vaccine was providing immunity in only 57% of the cases. This result was not sufficient for the vaccine to be added to Norway's vaccination program. Compensations have since been paid to the vaccine victims.

Ethics

Experiments help us further understand the world that surrounds us. But, in some cases, they can do more harm than good. In the Norwegian experiments, several ethical questions arise. Was the Norwegian experiment unethical if the best interests of the subjects were neglected? When should the experiment have been stopped? Should it have been conducted at all? If serious side effects are not reported and are withheld from subjects, there is no ethical question here, it is just wrong.

On the other hand, the breast cancer researchers would not want to deny the new drug to a group of patients with a life-threatening disease. But again, questions arise. How long must a researcher continue an experiment that shows better-than-expected results? How soon can a researcher conclude a drug is safe for the subjects involved?

◼ EXERCISES

1. *Unfavorable Results* Find an example of a real-life experiment that had unfavorable results. What could have been done to avoid the outcome of the experiment?

2. *Stopping an Experiment* In your opinion, what are some problems that may arise if clinical trials of a new experimental drug or vaccine are stopped early and then the drug or vaccine is distributed to other subjects or patients?

1 CHAPTER SUMMARY

What did you learn?	EXAMPLE(S)	REVIEW EXERCISES
Section 1.1		
■ How to distinguish between a population and a sample	*1*	*1–4*
■ How to distinguish between a parameter and a statistic	*2*	*5–8*
■ How to distinguish between descriptive statistics and inferential statistics	*3*	*9, 10*
Section 1.2		
■ How to distinguish between qualitative data and quantitative data	*1*	*11–16*
■ How to classify data with respect to the four levels of measurement: nominal, ordinal, interval, and ratio	*2, 3*	*17–20*
Section 1.3		
■ How data are collected: by doing an observational study, performing an experiment, using a simulation, or using a survey	*1*	*21–24*
■ How to design an experiment	*2*	*25, 26*
■ How to create a sample using random sampling, simple random sampling, stratified sampling, cluster sampling, and systematic sampling	*3, 4*	*27–34*
■ How to identify a biased sample	*4*	*35–38*

1 REVIEW EXERCISES

■ SECTION 1.1

In Exercises 1–4, identify the population and the sample.

1. A survey of 1000 U.S. adults found that 83% think credit cards tempt people to buy things they cannot afford. *(Source: Rasmussen Reports)*

2. Thirty-eight nurses working in the San Francisco area were surveyed concerning their opinions of managed health care.

3. A survey of 39 credit cards found that the average annual percentage rate (APR) is 12.83%. *(Source: Consumer Action)*

4. A survey of 1205 physicians found that about 60% had considered leaving the practice of medicine because they were discouraged over the state of U.S. health care. *(Source: The Physician Executive Journal of Medical Management)*

In Exercises 5–8, determine whether the numerical value describes a parameter or a statistic.

5. The 2009 team payroll of the Philadelphia Phillies was $113,004,046. *(Source: USA Today)*

6. In a survey of 752 adults in the United States, 42% think there should be a law that prohibits people from talking on cell phones in public places. *(Source: University of Michigan)*

7. In a recent study of math majors at a university, 10 students were minoring in physics.

8. Fifty percent of a sample of 1508 U.S. adults say they oppose drilling for oil and gas in the Arctic National Wildlife Refuge. *(Source: Pew Research Center)*

9. Which part of the study described in Exercise 3 represents the descriptive branch of statistics? Make an inference based on the results of the study.

10. Which part of the survey described in Exercise 4 represents the descriptive branch of statistics? Make an inference based on the results of the survey.

■ SECTION 1.2

In Exercises 11–16, determine which data are qualitative data and which are quantitative data. Explain your reasoning.

11. The monthly salaries of the employees at an accounting firm

12. The Social Security numbers of the employees at an accounting firm

13. The ages of a sample of 350 employees of a software company

14. The zip codes of a sample of 350 customers at a sporting goods store

15. The 2010 revenues of the companies on the Fortune 500 list

16. The marital statuses of all professional golfers

In Exercises 17–20, identify the data set's level of measurement. Explain your reasoning.

17. The daily high temperatures (in degrees Fahrenheit) for Mohave, Arizona for a week in June are listed. *(Source: Arizona Meteorological Network)*

   93 91 86 94 103 104 103

18. The levels of the Homeland Security Advisory System are listed.

   Severe High Elevated Guarded Low

19. The four departments of a printing company are listed.

Administration   Sales   Production   Billing

20. The total compensations (in millions of dollars) of the top ten female CEOs in the United States are listed. *(Source: Forbes)*

9.4   5.3   11.8   11.1   9.4   4.1   6.6   5.7   4.6   4.5

■ SECTION 1.3

In Exercises 21–24, decide which method of data collection you would use to collect data for the study. Explain your reasoning.

21. A study of charitable donations of the CEOs in Syracuse, New York

22. A study of the effect of koalas on the ecosystem of Kangaroo Island, Australia

23. A study of how training dogs from animal shelters affects inmates at a prison

24. A study of college professors' opinions on teaching classes online

In Exercises 25 and 26, an experiment is being performed to test the effects of sleep deprivation on memory recall. Two hundred students volunteer for the experiment. The students will be placed in one of five different treatment groups, including the control group.

25. Explain how you could design an experiment so that it uses a randomized block design.

26. Explain how you could design an experiment so that it uses a completely randomized design.

27. Random Number Table   Use the fifth row of Table 1 in Appendix B to generate 8 random numbers between 1 and 650.

28. Census or Sampling?   You want to know the favorite spring break destination among 15,000 students at a university. Decide whether you would take a census or use a sampling. If you would use a sampling, decide what technique you would use. Explain your reasoning.

In Exercises 29–34, identify the sampling technique used in the study. Explain your reasoning.

29. Using random digit dialing, researchers ask 1003 U.S. adults their plans on working during retirement. *(Source: Princeton Survey Research Associates International)*

30. A student asks 18 friends to participate in a psychology experiment.

31. A pregnancy study in Cebu, Philippines randomly selects 33 communities from the Cebu metropolitan area, then interviews all available pregnant women in these communities. *(Adapted from Cebu Longitudinal Health and Nutrition Survey)*

32. Law enforcement officials stop and check the driver of every third vehicle for blood alcohol content.

33. Twenty-five students are randomly selected from each grade level at a high school and surveyed about their study habits.

34. A journalist interviews 154 people waiting at an airport baggage claim and asks them how safe they feel during air travel.

In Exercises 35–38, identify a bias or error that might occur in the indicated survey or study.

35. study in Exercise 29

36. experiment in Exercise 30

37. study in Exercise 31

38. sampling in Exercise 32

1 CHAPTER QUIZ

Take this quiz as you would take a quiz in class. After you are done, check your work against the answers given in the back of the book.

1. Identify the population and the sample in the following study.

   A study of the dietary habits of 20,000 men was conducted to find a link between high intakes of dairy products and prostate cancer. *(Source: Harvard School of Public Health)*

2. Determine whether the numerical value is a parameter or a statistic.

   (a) In a survey of 2253 Internet users, 19% use Twitter or another service to share social updates. *(Source: Pew Internet Project)*

   (b) At a college, 90% of the Board of Trustees members approved the contract of the new president.

   (c) A survey of 846 chief financial officers and senior comptrollers shows that 55% of U.S. companies are reducing bonuses. *(Source: Grant Thornton International)*

3. Determine whether the data are qualitative or quantitative.

   (a) A list of debit card pin numbers

   (b) The final scores on a video game

4. Identify each data set's level of measurement. Explain your reasoning.

   (a) A list of badge numbers of police officers at a precinct

   (b) The horsepowers of racing car engines

   (c) The top 10 grossing films released in 2010

   (d) The years of birth for the runners in the Boston marathon

5. Decide which method of data collection you would use to gather data for each study. Explain your reasoning.

   (a) A study on the effect of low dietary intake of vitamin C and iron on lead levels in adults

   (b) The ages of people living within 500 miles of your home

6. An experiment is being performed to test the effects of a new drug on high blood pressure. The experimenter identifies 320 people ages 35–50 years old with high blood pressure for participation in the experiment. The subjects are divided into equal groups according to age. Within each group, subjects are then randomly selected to be in either the treatment group or the control group. What type of experimental design is being used for this experiment?

7. Identify the sampling technique used in each study. Explain your reasoning.

   (a) A journalist goes to a campground to ask people how they feel about air pollution.

   (b) For quality assurance, every tenth machine part is selected from an assembly line and measured for accuracy.

   (c) A study on attitudes about smoking is conducted at a college. The students are divided by class (freshman, sophomore, junior, and senior). Then a random sample is selected from each class and interviewed.

8. Which sampling technique used in Exercise 7 could lead to a biased study?

PUTTING IT ALL TOGETHER

Real Statistics — Real Decisions

You are a researcher for a professional research firm. Your firm has won a contract to do a study for an air travel industry publication. The editors of the publication would like to know their readers' thoughts on air travel factors such as ticket purchase, services, safety, comfort, economic growth, and security. They would also like to know the thoughts of adults who use air travel for business as well as for recreation.

The editors have given you their readership database and 20 questions they would like to ask (two sample questions from a previous study are given at the right). You know that it is too expensive to contact all of the readers, so you need to determine a way to contact a representative sample of the entire readership population.

■ EXERCISES

1. How Would You Do It?

(a) What sampling technique would you use to select the sample for the study? Why?

(b) Will the technique you choose in part (a) give you a sample that is representative of the population?

(c) Describe the method for collecting data.

(d) Identify possible flaws or biases in your study.

2. Data Classification

(a) What type of data do you expect to collect: qualitative, quantitative, or both? Why?

(b) At what levels of measurement do you think the data in the study will be? Why?

(c) Will the data collected for the study represent a population or a sample?

(d) Will the numerical descriptions of the data be parameters or statistics?

3. How They Did It

When the *Resource Systems Group* did a similar study, they used an Internet survey. They sent out 1000 invitations to participate in the survey and received 621 completed surveys.

(a) Describe some possible errors in collecting data by Internet surveys.

(b) Compare your method for collecting data in Exercise 1 to this method.

How did you acquire your ticket?

Response	Percent
Travel agent	35.1%
Directly from airline	20.9%
Online, using the airline's website	21.0%
Online, from a travel site other than the airline	18.5%
Other	4.5%

(Source: Resource Systems Group)

How many associates, friends, or family members traveled together in your party?

Response	Percent
1 (traveled alone)	48.7%
2 (traveled with one other person)	29.7%
3 (traveled with 2 others)	7.1%
4 (traveled with 3 others)	7.7%
5 (traveled with 4 others)	3.0%
6 or more (traveled with 5 or more others)	3.8%

(Source: Resource Systems Group)

HISTORY OF STATISTICS - TIMELINE

CONTRIBUTOR	TIME	CONTRIBUTION

John Graunt (1620–1674)

Blaise Pascal (1623–1662)

Pierre de Fermat (1601–1665)

17th century

Studied records of deaths in London in the early 1600s. The first to make extensive statistical observations from massive amounts of data (Chapter 2), his work laid the foundation for modern statistics.

Pascal and Fermat corresponded about basic probability problems (Chapter 3)—especially those dealing with gaming and gambling.

Pierre Laplace (1749–1827)

Carl Friedrich Gauss (1777–1855)

18th century

Studied probability (Chapter 3) and is credited with putting probability on a sure mathematical footing.

Studied regression and the method of least squares (Chapter 9) through astronomy. In his honor, the normal distribution is sometimes called the Gaussian distribution.

Lambert Quetelet (1796–1874)

Francis Galton (1822–1911)

19th century

Used descriptive statistics (Chapter 2) to analyze crime and mortality data and studied census techniques. Described normal distributions (Chapter 5) in connection with human traits such as height.

Used regression and correlation (Chapter 9) to study genetic variation in humans. He is credited with the discovery of the Central Limit Theorem (Chapter 5).

Karl Pearson (1857–1936)

William Gosset (1876–1937)

Charles Spearman (1863–1945)

Ronald Fisher (1890–1962)

20th century

Studied natural selection using correlation (Chapter 9). Formed first academic department of statistics and helped develop chi-square analysis (Chapter 6).

Studied process of brewing and developed *t*-test to correct problems connected with small sample sizes (Chapter 6).

British psychologist who was one of the first to develop intelligence testing using factor analysis (Chapter 10).

Studied biology and natural selection and developed ANOVA (Chapter 10), stressed the importance of experimental design (Chapter 1), and was the first to identify the null and alternative hypotheses (Chapter 7).

Frank Wilcoxon (1892–1965)

John Tukey (1915–2000)

David Kendall (1918–2007)

20th century (later)

Biochemist who used statistics to study plant pathology. He introduced two-sample tests (Chapter 8), which led the way to the development of nonparametric statistics.

Worked at Princeton during World War II. Introduced exploratory data analysis techniques such as stem-and-leaf plots (Chapter 2). Also, worked at Bell Laboratories and is best known for his work in inferential statistics (Chapters 6–11).

Worked at Princeton and Cambridge. Was a leading authority on applied probability and data analysis (Chapters 2 and 3).

TECHNOLOGY

USING TECHNOLOGY IN STATISTICS

With large data sets, you will find that calculators or computer software programs can help perform calculations and create graphics. Of the many calculators and statistical software programs that are available, we have chosen to incorporate the TI-83/84 Plus graphing calculator, and MINITAB and Excel software into this text.

The following example shows how to use these three technologies to generate a list of random numbers. This list of random numbers can be used to select sample members or perform simulations.

EXAMPLE

▶ **Generating a List of Random Numbers**

A quality control department inspects a random sample of 15 of the 167 cars that are assembled at an auto plant. How should the cars be chosen?

▶ **Solution**

One way to choose the sample is to first number the cars from 1 to 167. Then you can use technology to form a list of random numbers from 1 to 167. Each of the technology tools shown requires different steps to generate the list. Each, however, does require that you identify the minimum value as 1 and the maximum value as 167. Check your user's manual for specific instructions.

MINITAB	
↓	C1
1	167
2	11
3	74
4	160
5	18
6	70
7	80
8	56
9	37
10	6
11	82
12	126
13	98
14	104
15	137

EXCEL	
	A
1	41
2	16
3	91
4	58
5	151
6	36
7	96
8	154
9	2
10	113
11	157
12	103
13	64
14	135
15	90

TI-83/84 PLUS

randInt(1, 167, 15)
{17 42 152 59 5 116 125
64 122 55 58 60 82 152
105}

Recall that when you generate a list of random numbers, you should decide whether it is acceptable to have numbers that repeat. If it is acceptable, then the sampling process is said to be with replacement. If it is not acceptable, then the sampling process is said to be without replacement.

With each of the three technology tools shown on page 34, you have the capability of sorting the list so that the numbers appear in order. Sorting helps you see whether any of the numbers in the list repeat. If it is not acceptable to have repeats, you should specify that the tool generate more random numbers than you need.

■ EXERCISES

1. The SEC (Securities and Exchange Commission) is investigating a financial services company. The company being investigated has 86 brokers. The SEC decides to review the records for a random sample of 10 brokers. Describe how this investigation could be done. Then use technology to generate a list of 10 random numbers from 1 to 86 and order the list.

2. A quality control department is testing 25 smartphones from a shipment of 300 smartphones. Describe how this test could be done. Then use technology to generate a list of 25 random numbers from 1 to 300 and order the list.

3. Consider the population of ten digits: 0, 1, 2, 3, 4, 5, 6, 7, 8, and 9. Select three random samples of five digits from this list. Find the average of each sample. Compare your results with the average of the entire population. Comment on your results. (*Hint:* To find the average, sum the data entries and divide the sum by the number of entries.)

4. Consider the population of 41 whole numbers from 0 to 40. What is the average of these numbers? Select three random samples of seven numbers from this list. Find the average of each sample. Compare your results with the average of the entire population. Comment on your results. (*Hint:* To find the average, sum the data entries and divide the sum by the number of entries.)

5. Use random numbers to simulate rolling a six-sided die 60 times. How many times did you obtain each number from 1 to 6? Are the results what you expected?

6. You rolled a six-sided die 60 times and got the following tally.

20 ones	20 twos	15 threes
3 fours	2 fives	0 sixes

Does this seem like a reasonable result? What inference might you draw from the result?

7. Use random numbers to simulate tossing a coin 100 times. Let 0 represent heads, and let 1 represent tails. How many times did you obtain each number? Are the results what you expected?

8. You tossed a coin 100 times and got 77 heads and 23 tails. Does this seem like a reasonable result? What inference might you draw from the result?

9. A political analyst would like to survey a sample of the registered voters in a county. The county has 47 election districts. How could the analyst use random numbers to obtain a cluster sample?

Extended solutions are given in the *Technology Supplement.*
Technical instruction is provided for MINITAB, Excel, and the TI-83/84 Plus.

2

CHAPTER

DESCRIPTIVE STATISTICS

2.1 Frequency Distributions and Their Graphs

2.2 More Graphs and Displays

2.3 Measures of Central Tendency

■ ACTIVITY

2.4 Measures of Variation

■ ACTIVITY

■ CASE STUDY

2.5 Measures of Position

■ USES AND ABUSES

■ REAL STATISTICS– REAL DECISIONS

■ TECHNOLOGY

Brothers Sam and Bud Walton opened the first Wal-Mart store in 1962. Today, the Walton family is one of the richest families in the world. Members of the Walton family held four spots in the top 50 richest people in the world in 2009.

In Chapter 1, you learned that there are many ways to collect data. Usually, researchers must work with sample data in order to analyze populations, but occasionally it is possible to collect all the data for a given population. For instance, the following represents the ages of the 50 richest people in the world in 2009.

89, 89, 87, 86, 86, 85, 83, 83, 82, 81, 80, 78, 78, 77, 76, 73, 73, 73, 72, 69, 69, 68, 67, 66, 66, 65, 65, 64, 63, 61, 61, 60, 59, 58, 57, 56, 54, 54, 53, 53, 51, 51, 49, 47, 46, 44, 43, 42, 36, 35

In Chapter 2, you will learn ways to organize and describe data sets. The goal is to make the data easier to understand by describing trends, averages, and variations. For instance, in the raw data showing the ages of the 50 richest people in the world in 2009, it is not easy to see any patterns or special characteristics. Here are some ways you can organize and describe the data.

Make a frequency distribution table.

Class	Frequency, f
35–41	2
42–48	5
49–55	7
56–62	7
63–69	10
70–76	5
77–83	8
84–90	6

Draw a histogram.

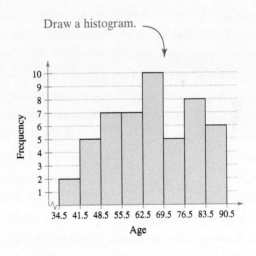

$$\text{Mean} = \frac{89 + 89 + 87 + 86 + 86 + \cdots + 43 + 42 + 36 + 35}{50}$$

$$= \frac{3263}{50}$$

$$= 65.26 \text{ years old}$$ ← Find an average.

$$\text{Range} = 89 - 35$$

$$= 54 \text{ years}$$ ← Find how the data vary.

2.1 Frequency Distributions and Their Graphs

Frequency Distributions ▸ Graphs of Frequency Distributions

▸ FREQUENCY DISTRIBUTIONS

WHAT YOU SHOULD LEARN

▸ How to construct a frequency distribution including limits, midpoints, relative frequencies, cumulative frequencies, and boundaries

▸ How to construct frequency histograms, frequency polygons, relative frequency histograms, and ogives

You will learn that there are many ways to organize and describe a data set. Important characteristics to look for when organizing and describing a data set are its **center,** its **variability** (or spread), and its **shape.** Measures of center and shapes of distributions are covered in Section 2.3.

When a data set has many entries, it can be difficult to see patterns. In this section, you will learn how to organize data sets by grouping the data into *intervals* called *classes* and forming a *frequency distribution.* You will also learn how to use frequency distributions to construct graphs.

DEFINITION

A **frequency distribution** is a table that shows **classes** or **intervals** of data entries with a count of the number of entries in each class. The **frequency** f of a class is the number of data entries in the class.

Example of a Frequency Distribution

Class	Frequency, f
1–5	5
6–10	8
11–15	6
16–20	8
21–25	5
26–30	4

In the frequency distribution shown at the left there are six classes. The frequencies for each of the six classes are 5, 8, 6, 8, 5, and 4. Each class has a **lower class limit,** which is the least number that can belong to the class, and an **upper class limit,** which is the greatest number that can belong to the class. In the frequency distribution shown, the lower class limits are 1, 6, 11, 16, 21, and 26, and the upper class limits are 5, 10, 15, 20, 25, and 30. The **class width** is the distance between lower (or upper) limits of consecutive classes. For instance, the class width in the frequency distribution shown is $6 - 1 = 5$.

The difference between the maximum and minimum data entries is called the **range.** In the frequency table shown, suppose the maximum data entry is 29, and the minimum data entry is 1. The range then is $29 - 1 = 28$. You will learn more about the range of a data set in Section 2.4.

STUDY TIP

In a frequency distribution, it is best if each class has the same width. Answers shown will use the minimum data value for the lower limit of the first class. Sometimes it may be more convenient to choose a lower limit that is slightly lower than the minimum value. The frequency distribution produced will vary slightly.

GUIDELINES

Constructing a Frequency Distribution from a Data Set

1. Decide on the number of classes to include in the frequency distribution. The number of classes should be between 5 and 20; otherwise, it may be difficult to detect any patterns.

2. Find the class width as follows. Determine the range of the data, divide the range by the number of classes, and *round up to the next convenient number.*

3. Find the class limits. You can use the minimum data entry as the lower limit of the first class. To find the remaining lower limits, add the class width to the lower limit of the preceding class. Then find the upper limit of the first class. Remember that classes cannot overlap. Find the remaining upper class limits.

4. Make a tally mark for each data entry in the row of the appropriate class.

5. Count the tally marks to find the total frequency f for each class.

EXAMPLE 1

▶ **Constructing a Frequency Distribution from a Data Set**

The following sample data set lists the prices (in dollars) of 30 portable global positioning system (GPS) navigators. Construct a frequency distribution that has seven classes.

90	130	400	200	350	70	325	250	150	250
275	270	150	130	59	200	160	450	300	130
220	100	200	400	200	250	95	180	170	150

▶ **Solution**

1. The number of classes (7) is stated in the problem.

2. The minimum data entry is 59 and the maximum data entry is 450, so the range is $450 - 59 = 391$. Divide the range by the number of classes and round up to find the class width.

$$\text{Class width} = \frac{391}{7} \qquad \frac{\text{Range}}{\text{Number of classes}}$$

$$\approx 55.86 \qquad \text{Round up to 56.}$$

3. The minimum data entry is a convenient lower limit for the first class. To find the lower limits of the remaining six classes, add the class width of 56 to the lower limit of each previous class. The upper limit of the first class is 114, which is one less than the lower limit of the second class. The upper limits of the other classes are $114 + 56 = 170$, $170 + 56 = 226$, and so on. The lower and upper limits for all seven classes are shown.

4. Make a tally mark for each data entry in the appropriate class. For instance, the data entry 130 is in the 115–170 class, so make a tally mark in that class. Continue until you have made a tally mark for each of the 30 data entries.

5. The number of tally marks for a class is the frequency of that class.

Lower limit	Upper limit
59	114
115	170
171	226
227	282
283	338
339	394
395	450

The frequency distribution is shown in the following table. The first class, 59–114, has five tally marks. So, the frequency of this class is 5. Notice that the sum of the frequencies is 30, which is the number of entries in the sample data set. The sum is denoted by Σf, where Σ is the uppercase Greek letter **sigma**.

Frequency Distribution for
Prices (in dollars) of GPS Navigators

Prices →   ← Number of
                    GPS navigators

Class	Tally	Frequency, f
59–114	ⅢⅢ	5
115–170	ⅢⅢ |||	8
171–226	ⅢⅢ		6
227–282	ⅢⅢ	5
283–338	||	2
339–394			1
395–450	|||	3
		$\Sigma f = 30$

Check that the sum of the frequencies equals the number in the sample.

▶ **Try It Yourself 1**

Construct a frequency distribution using the ages of the 50 richest people data set listed in the Chapter Opener on page 37. Use eight classes.

a. State the *number* of classes.
b. Find the minimum and maximum values and the *class width*.
c. Find the *class limits*.
d. *Tally* the data entries.
e. Write the *frequency f* of each class.

Answer: Page A30

After constructing a standard frequency distribution such as the one in Example 1, you can include several additional features that will help provide a better understanding of the data. These features (the *midpoint, relative frequency,* and *cumulative frequency* of each class) can be included as additional columns in your table.

DEFINITION

The **midpoint** of a class is the sum of the lower and upper limits of the class divided by two. The midpoint is sometimes called the *class mark*.

$$\text{Midpoint} = \frac{(\text{Lower class limit}) + (\text{Upper class limit})}{2}$$

The **relative frequency** of a class is the portion or percentage of the data that falls in that class. To find the relative frequency of a class, divide the frequency f by the sample size n.

$$\text{Relative frequency} = \frac{\text{Class frequency}}{\text{Sample size}} = \frac{f}{n}$$

The **cumulative frequency** of a class is the sum of the frequencies of that class and all previous classes. The cumulative frequency of the last class is equal to the sample size n.

After finding the first midpoint, you can find the remaining midpoints by adding the class width to the previous midpoint. For instance, if the first midpoint is 86.5 and the class width is 56, then the remaining midpoints are

$$86.5 + 56 = 142.5$$

$$142.5 + 56 = 198.5$$

$$198.5 + 56 = 254.5$$

$$254.5 + 56 = 310.5$$

and so on.

You can write the relative frequency as a fraction, decimal, or percent. The sum of the relative frequencies of all the classes should be equal to 1, or 100%. Due to rounding, the sum may be slightly less than or greater than 1. So, values such as 0.99 and 1.01 are sufficient.

EXAMPLE 2

▶ **Finding Midpoints, Relative Frequencies, and Cumulative Frequencies**

Using the frequency distribution constructed in Example 1, find the midpoint, relative frequency, and cumulative frequency of each class. Identify any patterns.

▶ **Solution**

The midpoints, relative frequencies, and cumulative frequencies of the first three classes are calculated as follows.

Class	f	Midpoint	Relative frequency	Cumulative frequency
59–114	5	$\dfrac{59 + 114}{2} = 86.5$	$\dfrac{5}{30} \approx 0.17$	5
115–170	8	$\dfrac{115 + 170}{2} = 142.5$	$\dfrac{8}{30} \approx 0.27$	$5 + 8 = 13$
171–226	6	$\dfrac{171 + 226}{2} = 198.5$	$\dfrac{6}{30} = 0.2$	$13 + 6 = 19$

The remaining midpoints, relative frequencies, and cumulative frequencies are shown in the following expanded frequency distribution.

Frequency Distribution for Prices (in dollars) of GPS Navigators

Prices

Number of GPS navigators

Portion of GPS navigators

Class	Frequency, f	Midpoint	Relative frequency	Cumulative frequency
59–114	5	86.5	0.17	5
115–170	8	142.5	0.27	13
171–226	6	198.5	0.2	19
227–282	5	254.5	0.17	24
283–338	2	310.5	0.07	26
339–394	1	366.5	0.03	27
395–450	3	422.5	0.1	30
	$\Sigma f = 30$		$\Sigma \dfrac{f}{n} \approx 1$	

Interpretation   There are several patterns in the data set. For instance, the most common price range for GPS navigators was $115 to $170.

▶ **Try It Yourself 2**

Using the frequency distribution constructed in Try It Yourself 1, find the midpoint, relative frequency, and cumulative frequency of each class. Identify any patterns.

a. Use the formulas to find each *midpoint*, *relative frequency*, and *cumulative frequency*.
b. *Organize* your results in a frequency distribution.
c. *Identify* patterns that emerge from the data.

Answer: Page A31

▶ GRAPHS OF FREQUENCY DISTRIBUTIONS

Sometimes it is easier to identify patterns of a data set by looking at a graph of the frequency distribution. One such graph is a *frequency histogram*.

DEFINITION

A **frequency histogram** is a bar graph that represents the frequency distribution of a data set. A histogram has the following properties.

1. The horizontal scale is quantitative and measures the data values.
2. The vertical scale measures the frequencies of the classes.
3. Consecutive bars must touch.

Because consecutive bars of a histogram must touch, bars must begin and end at class boundaries instead of class limits. **Class boundaries** are the numbers that separate classes *without* forming gaps between them. If data entries are integers, subtract 0.5 from each lower limit to find the lower class boundaries. To find the upper class boundaries, add 0.5 to each upper limit. The upper boundary of a class will equal the lower boundary of the next higher class.

EXAMPLE 3   Report 2

▶ Constructing a Frequency Histogram

Draw a frequency histogram for the frequency distribution in Example 2. Describe any patterns.

▶ Solution

First, find the class boundaries. Because the data entries are integers, subtract 0.5 from each lower limit to find the lower class boundaries and add 0.5 to each upper limit to find the upper class boundaries. So, the lower and upper boundaries of the first class are as follows.

First class lower boundary = 59 − 0.5 = 58.5

First class upper boundary = 114 + 0.5 = 114.5

The boundaries of the remaining classes are shown in the table. To construct the histogram, choose possible frequency values for the vertical scale. You can mark the horizontal scale either at the midpoints or at the class boundaries. Both histograms are shown.

Class	Class boundaries	Frequency, f
59–114	58.5–114.5	5
115–170	114.5–170.5	8
171–226	170.5–226.5	6
227–282	226.5–282.5	5
283–338	282.5–338.5	2
339–394	338.5–394.5	1
395–450	394.5–450.5	3

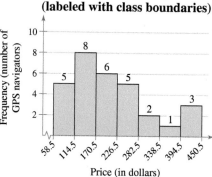

Interpretation   From either histogram, you can see that more than half of the GPS navigators are priced below $226.50.

▶ Try It Yourself 3

Use the frequency distribution from Try It Yourself 2 to construct a frequency histogram that represents the ages of the 50 richest people. Describe any patterns.

a. Find the *class boundaries*.
b. Choose appropriate *horizontal and vertical scales*.
c. Use the frequency distribution to *find the height of each bar*.
d. *Describe* any patterns in the data.   *Answer: Page A31*

Another way to graph a frequency distribution is to use a frequency polygon. A **frequency polygon** is a line graph that emphasizes the continuous change in frequencies.

EXAMPLE 4

▶ **Constructing a Frequency Polygon**

Draw a frequency polygon for the frequency distribution in Example 2. Describe any patterns.

▶ **Solution**

To construct the frequency polygon, use the same horizontal and vertical scales that were used in the histogram labeled with class midpoints in Example 3. Then plot points that represent the midpoint and frequency of each class and connect the points in order from left to right. Because the graph should begin and end on the horizontal axis, extend the left side to one class width before the first class midpoint and extend the right side to one class width after the last class midpoint.

Prices of GPS Navigators

Interpretation   You can see that the frequency of GPS navigators increases up to $142.50 and then decreases.

▶ Try It Yourself 4

Use the frequency distribution from Try It Yourself 2 to construct a frequency polygon that represents the ages of the 50 richest people. Describe any patterns.

a. Choose appropriate *horizontal and vertical scales*.
b. *Plot points* that represent the midpoint and frequency of each class.
c. *Connect the points* and extend the sides as necessary.
d. *Describe* any patterns in the data.   *Answer: Page A31*

STUDY TIP

A histogram and its corresponding frequency polygon are often drawn together. If you have not already constructed the histogram, begin constructing the frequency polygon by choosing appropriate horizontal and vertical scales. The horizontal scale should consist of the class midpoints, and the vertical scale should consist of appropriate frequency values.

A **relative frequency histogram** has the same shape and the same horizontal scale as the corresponding frequency histogram. The difference is that the vertical scale measures the *relative* frequencies, not frequencies.

EXAMPLE 5   SC Report 3

▶ **Constructing a Relative Frequency Histogram**

Draw a relative frequency histogram for the frequency distribution in Example 2.

▶ **Solution**

The relative frequency histogram is shown. Notice that the shape of the histogram is the same as the shape of the frequency histogram constructed in Example 3. The only difference is that the vertical scale measures the relative frequencies.

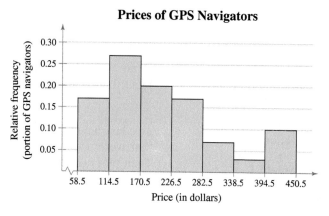

Prices of GPS Navigators

Interpretation   From this graph, you can quickly see that 0.27 or 27% of the GPS navigators are priced between $114.50 and $170.50, which is not as immediately obvious from the frequency histogram.

▶ **Try It Yourself 5**

Use the frequency distribution in Try It Yourself 2 to construct a relative frequency histogram that represents the ages of the 50 richest people.

a. *Use the same horizontal scale* that was used in the frequency histogram in the Chapter Opener.
b. *Revise the vertical scale* to reflect relative frequencies.
c. Use the relative frequencies to *find the height of each bar*.

Answer: Page A31

If you want to describe the number of data entries that are equal to or below a certain value, you can easily do so by constructing a *cumulative frequency graph*.

DEFINITION

A **cumulative frequency graph,** or **ogive** (pronounced ō′jīve), is a line graph that displays the cumulative frequency of each class at its upper class boundary. The upper boundaries are marked on the horizontal axis, and the cumulative frequencies are marked on the vertical axis.

GUIDELINES

Constructing an Ogive (Cumulative Frequency Graph)

1. Construct a frequency distribution that includes cumulative frequencies as one of the columns.
2. Specify the horizontal and vertical scales. The horizontal scale consists of upper class boundaries, and the vertical scale measures cumulative frequencies.
3. Plot points that represent the upper class boundaries and their corresponding cumulative frequencies.
4. Connect the points in order from left to right.
5. The graph should start at the lower boundary of the first class (cumulative frequency is zero) and should end at the upper boundary of the last class (cumulative frequency is equal to the sample size).

EXAMPLE 6

▶ **Constructing an Ogive**

Draw an ogive for the frequency distribution in Example 2. Estimate how many GPS navigators cost $300 or less. Also, use the graph to estimate when the greatest increase in price occurs.

▶ **Solution**

Using the cumulative frequencies, you can construct the ogive shown. The upper class boundaries, frequencies, and cumulative frequencies are shown in the table. Notice that the graph starts at 58.5, where the cumulative frequency is 0, and the graph ends at 450.5, where the cumulative frequency is 30.

Upper class boundary	f	Cumulative frequency
114.5	5	5
170.5	8	13
226.5	6	19
282.5	5	24
338.5	2	26
394.5	1	27
450.5	3	30

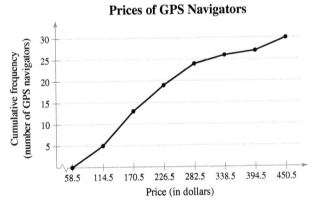

Prices of GPS Navigators

Interpretation From the ogive, you can see that about 25 GPS navigators cost $300 or less. It is evident that the greatest increase occurs between $114.50 and $170.50, because the line segment is steepest between these two class boundaries.

Another type of ogive uses percent as the vertical axis instead of frequency (see Example 5 in Section 2.5).

▶ Try It Yourself 6

Use the frequency distribution from Try It Yourself 2 to construct an ogive that represents the ages of the 50 richest people. Estimate the number of people who are 80 years old or younger.

a. Specify the *horizontal and vertical scales*.
b. *Plot* the points given by the upper class boundaries and the cumulative frequencies.
c. *Construct* the graph.
d. *Estimate* the number of people who are 80 years old or younger.
e. *Interpret* the results in the context of the data. *Answer: Page A31*

EXAMPLE 7

▶ **Using Technology to Construct Histograms**

Use a calculator or a computer to construct a histogram for the frequency distribution in Example 2.

▶ **Solution**

MINITAB, Excel, and the TI-83/84 Plus each have features for graphing histograms. Try using this technology to draw the histograms as shown.

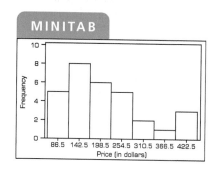

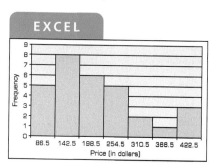

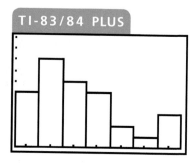

STUDY TIP

Detailed instructions for using MINITAB, Excel, and the TI-83/84 Plus are shown in the Technology Guide that accompanies this text. For instance, here are instructions for creating a histogram on a TI-83/84 Plus.

STAT   ENTER

Enter midpoints in L1.
Enter frequencies in L2.

2nd   STATPLOT

Turn on Plot 1.
Highlight Histogram.

   Xlist: L1
   Freq: L2

ZOOM   9

WINDOW

Xscl=56

GRAPH

▶ Try It Yourself 7

Use a calculator or a computer and the frequency distribution from Try It Yourself 2 to construct a frequency histogram that represents the ages of the 50 richest people.

a. *Enter* the data
b. *Construct* the histogram. *Answer: Page A31*

2.1 EXERCISES

■ BUILDING BASIC SKILLS AND VOCABULARY

1. What are some benefits of representing data sets using frequency distributions? What are some benefits of using graphs of frequency distributions?

2. Why should the number of classes in a frequency distribution be between 5 and 20?

3. What is the difference between class limits and class boundaries?

4. What is the difference between relative frequency and cumulative frequency?

5. After constructing an expanded frequency distribution, what should the sum of the relative frequencies be? Explain.

6. What is the difference between a frequency polygon and an ogive?

True or False?   *In Exercises 7–10, determine whether the statement is true or false. If it is false, rewrite it as a true statement.*

7. In a frequency distribution, the class width is the distance between the lower and upper limits of a class.

8. The midpoint of a class is the sum of its lower and upper limits divided by two.

9. An ogive is a graph that displays relative frequencies.

10. Class boundaries are used to ensure that consecutive bars of a histogram touch.

In Exercises 11–14, use the given minimum and maximum data entries and the number of classes to find the class width, the lower class limits, and the upper class limits.

11. min = 9, max = 64, 7 classes

12. min = 12, max = 88, 6 classes

13. min = 17, max = 135, 8 classes

14. min = 54, max = 247, 10 classes

Reading a Frequency Distribution   *In Exercises 15 and 16, use the given frequency distribution to find the (a) class width, (b) class midpoints, and (c) class boundaries.*

15. **Cleveland, OH High Temperatures (°F)**

Class	Frequency, f
20–30	19
31–41	43
42–52	68
53–63	69
64–74	74
75–85	68
86–96	24

16. **Travel Time to Work (in minutes)**

Class	Frequency, f
0–9	188
10–19	372
20–29	264
30–39	205
40–49	83
50–59	76
60–69	32

17. Use the frequency distribution in Exercise 15 to construct an expanded frequency distribution, as shown in Example 2.

18. Use the frequency distribution in Exercise 16 to construct an expanded frequency distribution, as shown in Example 2.

Graphical Analysis *In Exercises 19 and 20, use the frequency histogram to*

(a) *determine the number of classes.*

(b) *estimate the frequency of the class with the least frequency.*

(c) *estimate the frequency of the class with the greatest frequency.*

(d) *determine the class width.*

19.

Employee Salaries

20.

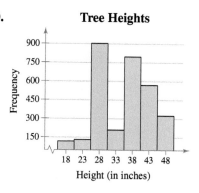

Tree Heights

Graphical Analysis *In Exercises 21 and 22, use the ogive to approximate*

(a) *the number in the sample.*

(b) *the location of the greatest increase in frequency.*

21.

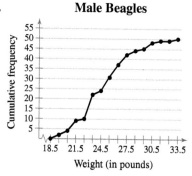

Male Beagles

22.

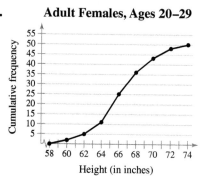

Adult Females, Ages 20–29

23. Use the ogive in Exercise 21 to approximate

(a) the cumulative frequency for a weight of 27.5 pounds.

(b) the weight for which the cumulative frequency is 45.

(c) the number of beagles that weigh between 22.5 pounds and 29.5 pounds.

(d) the number of beagles that weigh more than 30.5 pounds.

24. Use the ogive in Exercise 22 to approximate

(a) the cumulative frequency for a height of 72 inches.

(b) the height for which the cumulative frequency is 25.

(c) the number of adult females that are between 62 and 66 inches tall.

(d) the number of adult females that are taller than 70 inches.

Graphical Analysis   *In Exercises 25 and 26, use the relative frequency histogram to*

(a) *identify the class with the greatest, and the class with the least, relative frequency.*

(b) *approximate the greatest and least relative frequencies.*

(c) *approximate the relative frequency of the second class.*

25.

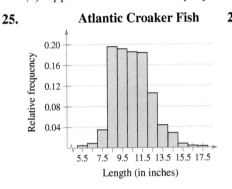

Atlantic Croaker Fish

26.

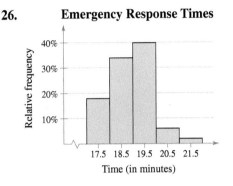

Emergency Response Times

Graphical Analysis   *In Exercises 27 and 28, use the frequency polygon to identify the class with the greatest, and the class with the least, frequency.*

27.

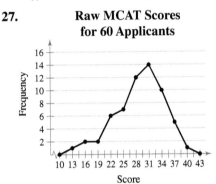

Raw MCAT Scores for 60 Applicants

28.

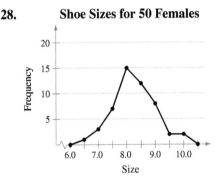

Shoe Sizes for 50 Females

■ USING AND INTERPRETING CONCEPTS

Constructing a Frequency Distribution   *In Exercises 29 and 30, construct a frequency distribution for the data set using the indicated number of classes. In the table, include the midpoints, relative frequencies, and cumulative frequencies. Which class has the greatest frequency and which has the least frequency?*

 29. Political Blog Reading Times
Number of classes: 5
Data set: Time (in minutes) spent reading a political blog in a day

7	39	13	9	25	8	22	0	2	18	2	30	7
35	12	15	8	6	5	29	0	11	39	16	15	

 **30. Book Spending**
Number of classes: 6
Data set: Amount (in dollars) spent on books for a semester

91	472	279	249	530	376	188	341	266	199
142	273	189	130	489	266	248	101	375	486
190	398	188	269	43	30	127	354	84	

 indicates that the data set for this exercise is available electronically.

Constructing a Frequency Distribution and a Frequency Histogram

In Exercises 31–34, construct a frequency distribution and a frequency histogram for the data set using the indicated number of classes. Describe any patterns.

31. Sales

Number of classes: 6

Data set: July sales (in dollars) for all sales representatives at a company

2114	2468	7119	1876	4105	3183	1932	1355
4278	1030	2000	1077	5835	1512	1697	2478
3981	1643	1858	1500	4608	1000		

32. Pepper Pungencies

Number of classes: 5

Data set: Pungencies (in 1000s of Scoville units) of 24 tabasco peppers

35	51	44	42	37	38	36	39
44	43	40	40	32	39	41	38
42	39	40	46	37	35	41	39

33. Reaction Times

Number of classes: 8

Data set: Reaction times (in milliseconds) of a sample of 30 adult females to an auditory stimulus

507	389	305	291	336	310	514	442
373	428	387	454	323	441	388	426
411	382	320	450	309	416	359	388
307	337	469	351	422	413		

34. Fracture Times

Number of classes: 5

Data set: Amounts of pressure (in pounds per square inch) at fracture time for 25 samples of brick mortar

2750	2862	2885	2490	2512	2456	2554
2872	2601	2877	2721	2692	2888	2755
2867	2718	2641	2834	2466	2596	2519
2532	2885	2853	2517			

Constructing a Frequency Distribution and a Relative Frequency Histogram

In Exercises 35–38, construct a frequency distribution and a relative frequency histogram for the data set using five classes. Which class has the greatest relative frequency and which has the least relative frequency?

35. Gasoline Consumption

Data set: Highway fuel consumptions (in miles per gallon) for a sample of cars

32	35	28	40	30	42	55	40	45	24
28	34	40	36	34	40	30	25	28	32
40	35	25	44	26	39	38	42	45	32

36. ATM Withdrawals

Data set: A sample of ATM withdrawals (in dollars)

35	10	30	25	75	10	30	20	20	10	40
50	40	30	60	70	25	40	10	60	20	80
40	25	20	10	20	25	30	50	80	20	

37. Triglyceride Levels

Data set: Triglyceride levels (in milligrams per deciliter of blood) of a sample of patients

209 140 155 170 265 138 180 295 250
320 270 225 215 390 420 462 150 200
400 295 240 200 190 145 160 175

38. Years of Service

Data set: Years of service of a sample of New York state troopers

12   7   9   8   9   8   12 10   9
10   6   8   13 12 10 11   7   14
12   9   8   10   9 11 13   8

Constructing a Cumulative Frequency Distribution and an Ogive

In Exercises 39 and 40, construct a cumulative frequency distribution and an ogive for the data set using six classes. Then describe the location of the greatest increase in frequency.

39. Retirement Ages

Data set: Retirement ages for a sample of doctors

70 54 55 71 57 58 63 65
60 66 57 62 63 60 63 60
66 60 67 69 69 52 61 73

40. Saturated Fat Intakes

Data set: Daily saturated fat intakes (in grams) of a sample of people

38 32 34 39 40 54 32 17 29 33
57 40 25 36 33 24 42 16 31 33

Constructing a Frequency Distribution and a Frequency Polygon

In Exercises 41 and 42, construct a frequency distribution and a frequency polygon for the data set. Describe any patterns.

41. Exam Scores

Number of classes: 5
Data set: Exam scores for all students in a statistics class

83 92 94 82 73 98 78 85 72 90
89 92 96 89 75 85 63 47 75 82

42. Children of the Presidents

Number of classes: 6
Data set: Number of children of the U.S. presidents

(Source: presidentschildren.com)

0 5 6 0 3 4 0 4 10 15 0 6 2 3 0
4 5 4 8 7 3 5 3   2   6 3 3 1 2
2 6 1 2 3 2 2 4   4   4 6 1 2 2

In Exercises 43 and 44, use the data set to construct (a) an expanded frequency distribution, (b) a frequency histogram, (c) a frequency polygon, (d) a relative frequency histogram, and (e) an ogive.

43. Pulse Rates

Number of classes: 6
Data set: Pulse rates of students in a class

68 105 95 80 90 100   75   70 84 98 102   70
65   88 90 75 78   94 110 120 95 80   76 108

44. Hospitals
Number of classes: 8
Data set: Number of hospitals in each state *(Source: American Hospital Directory)*

15	100	56	74	360	53	34	8	213	116
15	38	21	143	97	59	76	110	83	51
23	116	55	91	75	19	108	14	25	14
73	40	30	213	154	97	36	181	12	63
29	121	378	36	91	7	61	71	40	15

SC **45.** Use StatCrunch to construct a frequency histogram and a relative frequency histogram for the following data set that shows the finishing times (in minutes) for 25 runners in a marathon. Use seven classes.

159	164	165	170	215	200	167	225	192	185	235	240	225
191	194	175	167	234	158	172	180	240	176	159	231	

46. Writing What happens when the number of classes is increased for a frequency histogram? Use the data set listed and a technology tool to create frequency histograms with 5, 10, and 20 classes. Which graph displays the data best?

2	7	3	2	11	3	15	8	4	9	10	13	9
7	11	10	1	2	12	5	6	4	2	9	15

■ EXTENDING CONCEPTS

47. What Would You Do? You work at a bank and are asked to recommend the amount of cash to put in an ATM each day. You don't want to put in too much (security) or too little (customer irritation). Here are the daily withdrawals (in 100s of dollars) for 30 days.

72	84	61	76	104	76	86	92	80	88
98	76	97	82	84	67	70	81	82	89
74	73	86	81	85	78	82	80	91	83

(a) Construct a relative frequency histogram for the data using 8 classes.

(b) If you put $9000 in the ATM each day, what percent of the days in a month should you expect to run out of cash? Explain your reasoning.

(c) If you are willing to run out of cash for 10% of the days, how much cash should you put in the ATM each day? Explain your reasoning.

48. What Would You Do? You work in the admissions department for a college and are asked to recommend the minimum SAT scores that the college will accept for a position as a full-time student. Here are the SAT scores for a sample of 50 applicants.

1760	1502	1375	1310	1601	1942	1380	2211	1622	1771
1150	1351	1682	1618	2051	1742	1463	1395	1860	1918
1882	1996	1525	1510	2120	1700	1818	1869	1440	1235
976	1513	1790	2250	2102	1905	1979	1588	1420	1730
2175	1930	1965	1658	2005	2125	1260	1560	1635	1620

(a) Construct a relative frequency histogram for the data using 10 classes.

(b) If you set the minimum score at 1616, what percent of the applicants will meet this requirement? Explain your reasoning.

(c) If you want to accept the top 88% of the applicants, what should the minimum score be? Explain your reasoning.

2.2 More Graphs and Displays

Graphing Quantitative Data Sets ▸ Graphing Qualitative Data Sets ▸ Graphing Paired Data Sets

▸ GRAPHING QUANTITATIVE DATA SETS

In Section 2.1, you learned several traditional ways to display quantitative data graphically. In this section, you will learn a newer way to display quantitative data, called a **stem-and-leaf plot.** Stem-and-leaf plots are examples of **exploratory data analysis (EDA),** which was developed by John Tukey in 1977.

In a stem-and-leaf plot, each number is separated into a **stem** (for instance, the entry's leftmost digits) and a **leaf** (for instance, the rightmost digit). You should have as many leaves as there are entries in the original data set and the leaves should be single digits. A stem-and-leaf plot is similar to a histogram but has the advantage that the graph still contains the original data values. Another advantage of a stem-and-leaf plot is that it provides an easy way to sort data.

EXAMPLE 1   **SC** Report 4

▸ Constructing a Stem-and-Leaf Plot

The following are the numbers of text messages sent last week by the cellular phone users on one floor of a college dormitory. Display the data in a stem-and-leaf plot. What can you conclude?

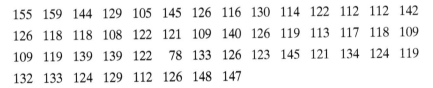

155 159 144 129 105 145 126 116 130 114 122 112 112 142
126 118 118 108 122 121 109 140 126 119 113 117 118 109
109 119 139 139 122 78 133 126 123 145 121 134 124 119
132 133 124 129 112 126 148 147

▸ **Solution**   Because the data entries go from a low of 78 to a high of 159, you should use stem values from 7 to 15. To construct the plot, list these stems to the left of a vertical line. For each data entry, list a leaf to the right of its stem. For instance, the entry 155 has a stem of 15 and a leaf of 5. The resulting stem-and-leaf plot will be unordered. To obtain an ordered stem-and-leaf plot, rewrite the plot with the leaves in increasing order from left to right. Be sure to include a key.

Number of Text Messages Sent

```
 7 | 8                 Key:  15|5 = 155
 8 |
 9 |
10 | 5 8 9 9 9
11 | 6 4 2 2 8 8 9 3 7 8 9 9 2
12 | 9 6 2 6 2 1 6 2 6 3 1 4 4 9 6
13 | 0 9 9 3 4 2 3
14 | 4 5 2 0 5 8 7
15 | 5 9
```

Unordered Stem-and-Leaf Plot

Number of Text Messages Sent

```
 7 | 8                 Key:  15|5 = 155
 8 |
 9 |
10 | 5 8 9 9 9
11 | 2 2 2 3 4 6 7 8 8 8 9 9 9
12 | 1 1 2 2 2 3 4 4 6 6 6 6 6 9 9
13 | 0 2 3 3 4 9 9
14 | 0 2 4 5 5 7 8
15 | 5 9
```

Ordered Stem-and-Leaf Plot

Interpretation   From the display, you can conclude that more than 50% of the cellular phone users sent between 110 and 130 text messages.

▶ Try It Yourself 1

Use a stem-and-leaf plot to organize the ages of the 50 richest people data set listed in the Chapter Opener on page 37. What can you conclude?

a. List all possible *stems*.
b. List the *leaf* of each data entry to the right of its stem and include a *key*.
c. *Rewrite* the stem-and-leaf plot so that the leaves are ordered.
d. Use the plot to make a *conclusion*.   *Answer: Page A31*

EXAMPLE 2

▶ Constructing Variations of Stem-and-Leaf Plots

Organize the data given in Example 1 using a stem-and-leaf plot that has two rows for each stem. What can you conclude?

▶ Solution

Use the stem-and-leaf plot from Example 1, except now list each stem twice. Use the leaves 0, 1, 2, 3, and 4 in the first stem row and the leaves 5, 6, 7, 8, and 9 in the second stem row. The revised stem-and-leaf plot is shown. Notice that by using two rows per stem, you obtain a more detailed picture of the data.

INSIGHT

You can use stem-and-leaf plots to identify unusual data values called *outliers*. In Examples 1 and 2, the data value 78 is an outlier. You will learn more about outliers in Section 2.3.

Number of Text Messages Sent

7		Key: 15|5 = 155
7	8	
8		
8		
9		
9		
10		
10	5 8 9 9 9	
11	4 2 2 3 2	
11	6 8 8 9 7 8 9 9	
12	2 2 1 2 3 1 4 4	
12	9 6 6 6 6 9 6	
13	0 3 4 2 3	
13	9 9	
14	4 2 0	
14	5 5 8 7	
15		
15	5 9	

Unordered Stem-and-Leaf Plot

Number of Text Messages Sent

7		Key: 15|5 = 155
7	8	
8		
8		
9		
9		
10		
10	5 8 9 9 9	
11	2 2 2 3 4	
11	6 7 8 8 8 9 9 9	
12	1 1 2 2 2 3 4 4	
12	6 6 6 6 9 9	
13	0 2 3 3 4	
13	9 9	
14	0 2 4	
14	5 5 7 8	
15		
15	5 9	

Ordered Stem-and-Leaf Plot

Interpretation From the display, you can conclude that most of the cellular phone users sent between 105 and 135 text messages.

▶ Try It Yourself 2

Using two rows for each stem, revise the stem-and-leaf plot you constructed in Try It Yourself 1. What can you conclude?

a. List each stem *twice*.
b. List all leaves *using the appropriate stem row*.
c. Use the plot to make a *conclusion*.   *Answer: Page A32*

You can also use a dot plot to graph quantitative data. In a **dot plot,** each data entry is plotted, using a point, above a horizontal axis. Like a stem-and-leaf plot, a dot plot allows you to see how data are distributed, determine specific data entries, and identify unusual data values.

> ### EXAMPLE 3   SC Report 5

> ▶ Constructing a Dot Plot

Use a dot plot to organize the text messaging data given in Example 1. What can you conclude from the graph?

155	159	144	129	105	145	126	116	130	114	122	112
112	142	126	118	118	108	122	121	109	140	126	119
113	117	118	109	109	119	139	139	122	78	133	126
123	145	121	134	124	119	132	133	124	129	112	126
148	147										

> ▶ Solution

So that each data entry is included in the dot plot, the horizontal axis should include numbers between 70 and 160. To represent a data entry, plot a point above the entry's position on the axis. If an entry is repeated, plot another point above the previous point.

Number of Text Messages Sent

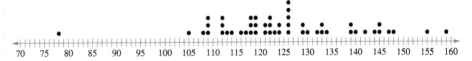

Interpretation   From the dot plot, you can see that most values cluster between 105 and 148 and the value that occurs the most is 126. You can also see that 78 is an unusual data value.

> ▶ Try It Yourself 3

Use a dot plot to organize the ages of the 50 richest people data set listed in the Chapter Opener on page 37. What can you conclude from the graph?

a. Choose an appropriate scale for the *horizontal axis*.
b. Represent each data entry by *plotting a point*.
c. *Describe* any patterns in the data.   *Answer: Page A32*

Technology can be used to construct stem-and-leaf plots and dot plots. For instance, a MINITAB dot plot for the text messaging data is shown below.

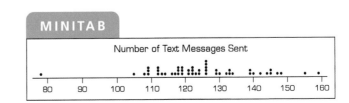

▶ GRAPHING QUALITATIVE DATA SETS

Pie charts provide a convenient way to present qualitative data graphically as percents of a whole. A **pie chart** is a circle that is divided into sectors that represent categories. The area of each sector is proportional to the frequency of each category. In most cases, you will be interpreting a pie chart or constructing one using technology. Example 4 shows how to construct a pie chart by hand.

EXAMPLE 4   Report 6

▶ Constructing a Pie Chart

The numbers of earned degrees conferred (in thousands) in 2007 are shown in the table. Use a pie chart to organize the data. What can you conclude? *(Source: U.S. National Center for Education Statistics)*

▶ Solution

Begin by finding the relative frequency, or percent, of each category. Then construct the pie chart using the central angle that corresponds to each category. To find the central angle, multiply 360° by the category's relative frequency. For instance, the central angle for associate's degrees is $360°(0.24) \approx 86°$. To construct a pie chart in Excel, follow the instructions in the margin.

Earned Degrees Conferred in 2007

Type of degree	Number (thousands)
Associate's	728
Bachelor's	1525
Master's	604
First professional	90
Doctoral	60

Type of degree	f	Relative frequency	Angle
Associate's	728	0.24	86°
Bachelor's	1525	0.51	184°
Master's	604	0.20	72°
First professional	90	0.03	11°
Doctoral	60	0.02	7°

Earned Degrees Conferred in 2007

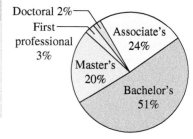

Interpretation   From the pie chart, you can see that over one half of the degrees conferred in 2007 were bachelor's degrees.

▶ Try It Yourself 4

The numbers of earned degrees conferred (in thousands) in 1990 are shown in the table. Use a pie chart to organize the data. Compare the 1990 data with the 2007 data. *(Source: U.S. National Center for Education Statistics)*

Earned Degrees Conferred in 1990

Type of degree	Number (thousands)
Associate's	455
Bachelor's	1052
Master's	325
First professional	71
Doctoral	38

a. Find the *relative frequency* and *central angle* of each category.
b. Use the *central angle* to find the portion that corresponds to each category.
c. *Compare* the 1990 data with the 2007 data.   *Answer: Page A32*

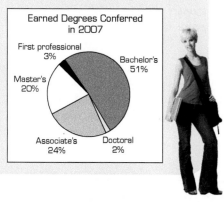

Another way to graph qualitative data is to use a Pareto chart. A **Pareto chart** is a vertical bar graph in which the height of each bar represents frequency or relative frequency. The bars are positioned in order of decreasing height, with the tallest bar positioned at the left. Such positioning helps highlight important data and is used frequently in business.

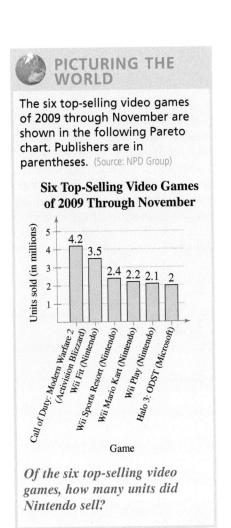
EXAMPLE 5   SC Report 7

▶ Constructing a Pareto Chart

In a recent year, the retail industry lost $36.5 billion in inventory shrinkage. Inventory shrinkage is the loss of inventory through breakage, pilferage, shoplifting, and so on. The main causes of inventory shrinkage are administrative error ($5.4 billion), employee theft ($15.9 billion), shoplifting ($12.7 billion), and vendor fraud ($1.4 billion). If you were a retailer, which causes of inventory shrinkage would you address first? *(Source: National Retail Federation and the University of Florida)*

▶ Solution

Using frequencies for the vertical axis, you can construct the Pareto chart as shown.

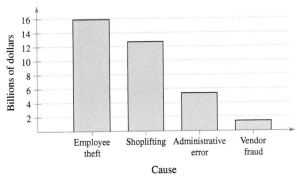

Main Causes of Inventory Shrinkage

Interpretation   From the graph, it is easy to see that the causes of inventory shrinkage that should be addressed first are employee theft and shoplifting.

▶ Try It Yourself 5

Every year, the Better Business Bureau (BBB) receives complaints from customers. In a recent year, the BBB received the following complaints.

> 7792 complaints about home furnishing stores
> 5733 complaints about computer sales and service stores
> 14,668 complaints about auto dealers
> 9728 complaints about auto repair shops
> 4649 complaints about dry cleaning companies

Use a Pareto chart to organize the data. What source is the greatest cause of complaints? *(Source: Council of Better Business Bureaus)*

a. Find the *frequency or relative frequency* for each data entry.
b. *Position the bars in decreasing order* according to frequency or relative frequency.
c. *Interpret* the results in the context of the data.   *Answer: Page A32*

▸ GRAPHING PAIRED DATA SETS

When each entry in one data set corresponds to one entry in a second data set, the sets are called **paired data sets.** For instance, suppose a data set contains the costs of an item and a second data set contains sales amounts for the item at each cost. Because each cost corresponds to a sales amount, the data sets are paired. One way to graph paired data sets is to use a **scatter plot,** where the ordered pairs are graphed as points in a coordinate plane. A scatter plot is used to show the relationship between two quantitative variables.

EXAMPLE 6

▸ Interpreting a Scatter Plot

The British statistician Ronald Fisher (see page 33) introduced a famous data set called Fisher's Iris data set. This data set describes various physical characteristics, such as petal length and petal width (in millimeters), for three species of iris. In the scatter plot shown, the petal lengths form the first data set and the petal widths form the second data set. As the petal length increases, what tends to happen to the petal width? *(Source: Fisher, R. A., 1936)*

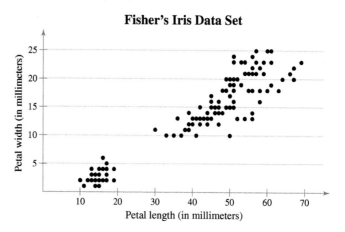

Fisher's Iris Data Set

Length of employment (in years)	Salary (in dollars)
5	32,000
4	32,500
8	40,000
4	27,350
2	25,000
10	43,000
7	41,650
6	39,225
9	45,100
3	28,000

▸ Solution

The horizontal axis represents the petal length, and the vertical axis represents the petal width. Each point in the scatter plot represents the petal length and petal width of one flower.

Interpretation From the scatter plot, you can see that as the petal length increases, the petal width also tends to increase.

▸ Try It Yourself 6

The lengths of employment and the salaries of 10 employees are listed in the table at the left. Graph the data using a scatter plot. What can you conclude?

a. Label the *horizontal and vertical axes.*
b. *Plot* the paired data.
c. *Describe* any trends.

Answer: Page A32

You will learn more about scatter plots and how to analyze them in Chapter 9.

A data set that is composed of quantitative entries taken at regular intervals over a period of time is called a **time series.** For instance, the amount of precipitation measured each day for one month is a time series. You can use a **time series chart** to graph a time series.

EXAMPLE 7   SC Report 8

▸ **Constructing a Time Series Chart**

See MINITAB and TI-83/84 Plus steps on pages 122 and 123.

The table lists the number of cellular telephone subscribers (in millions) and subscribers' average local monthly bills for service (in dollars) for the years 1998 through 2008. Construct a time series chart for the number of cellular subscribers. What can you conclude?
(Source: Cellular Telecommunications & Internet Association)

Year	Subscribers (in millions)	Average bill (in dollars)
1998	69.2	39.43
1999	86.0	41.24
2000	109.5	45.27
2001	128.4	47.37
2002	140.8	48.40
2003	158.7	49.91
2004	182.1	50.64
2005	207.9	49.98
2006	233.0	50.56
2007	255.4	49.79
2008	270.3	50.07

▸ **Solution**

Let the horizontal axis represent the years and let the vertical axis represent the number of subscribers (in millions). Then plot the paired data and connect them with line segments.

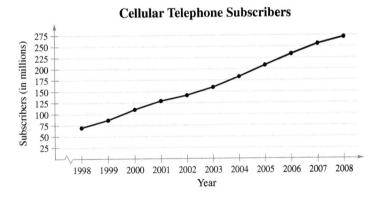

Cellular Telephone Subscribers

Interpretation The graph shows that the number of subscribers has been increasing since 1998.

▸ **Try It Yourself 7**

Use the table in Example 7 to construct a time series chart for subscribers' average local monthly cellular telephone bills for the years 1998 through 2008. What can you conclude?

a. Label the *horizontal and vertical axes*.
b. *Plot* the paired data and *connect* them with line segments.
c. *Describe* any patterns you see.

Answer: Page A32

2.2 EXERCISES

FOR EXTRA HELP:
MyStatLab

■ BUILDING BASIC SKILLS AND VOCABULARY

1. Name some ways to display quantitative data graphically. Name some ways to display qualitative data graphically.

2. What is an advantage of using a stem-and-leaf plot instead of a histogram? What is a disadvantage?

3. In terms of displaying data, how is a stem-and-leaf plot similar to a dot plot?

4. How is a Pareto chart different from a standard vertical bar graph?

Putting Graphs in Context *In Exercises 5–8, match the plot with the description of the sample.*

5.
```
0 | 8            Key: 0|8 = 0.8
1 | 5 6 8
2 | 1 3 4 5
3 | 0 9
4 | 0 0
```

6.
```
6 | 7 8            Key: 6|7 = 67
7 | 4 5 5 8 8 8
8 | 1 3 5 5 8 8 9
9 | 0 0 0 2 4
```

7.

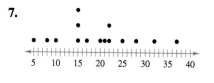

8.

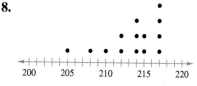

(a) Time (in minutes) it takes a sample of employees to drive to work

(b) Grade point averages of a sample of students with finance majors

(c) Top speeds (in miles per hour) of a sample of high-performance sports cars

(d) Ages (in years) of a sample of residents of a retirement home

Graphical Analysis *In Exercises 9–12, use the stem-and-leaf plot or dot plot to list the actual data entries. What is the maximum data entry? What is the minimum data entry?*

9. Key: 2|7 = 27
```
2 | 7
3 | 2
4 | 1 3 3 4 7 7 8
5 | 0 1 1 2 3 3 3 3 4 4 4 4 5 6 6 8 9
6 | 8 8 8
7 | 3 8 8
8 | 5
```

10. Key: 12|9 = 12.9
```
12 |
12 | 9
13 | 3
13 | 6 7 7
14 | 1 1 1 1 3 4 4
14 | 6 9 9
15 | 0 0 0 1 2 4
15 | 6 7 8 8 8 9
16 | 1
16 | 6 7
```

11.

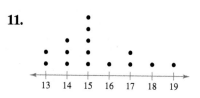

12.

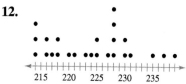

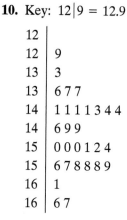

■ **USING AND INTERPRETING CONCEPTS**

Graphical Analysis   *In Exercises 13–16, give three conclusions that can be drawn from the graph.*

13.   **Average Time Spent on Top 5 Social Networking Sites**

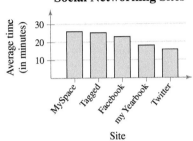

(Source: Experian Hitwise)

14.   **Motor Vehicle Thefts in U.S.**

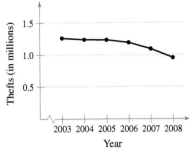

(Source: Federal Bureau of Investigation)

15.   **How Other Drivers Irk Us**

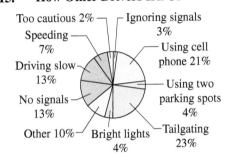

(Adapted from Reuters/Zogby)

16.   **Driving and Cell Phone Use**

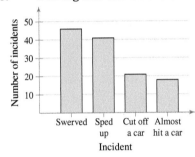

(Adapted from USA Today)

Graphing Data Sets   *In Exercises 17–30, organize the data using the indicated type of graph. What can you conclude about the data?*

17.   **Exam Scores**   Use a stem-and-leaf plot to display the data. The data represent the scores of a biology class on a midterm exam.

75   85   90   80   87   67   82   88   95   91   73   80
83   92   94   68   75   91   79   95   87   76   91   85

18.   **Highest Paid CEOs**   Use a stem-and-leaf plot that has two rows for each stem to display the data. The data represent the ages of the top 30 highest paid CEOs. *(Source: Forbes)*

64   74   55   55   62   63   50   67   51   59   50
52   50   59   62   64   57   61   49   63   62   60
55   56   48   58   64   60   60   57

19.   **Ice Thickness**   Use a stem-and-leaf plot to display the data. The data represent the thicknesses (in centimeters) of ice measured at 20 different locations on a frozen lake.

5.8   6.4   6.9   7.2   5.1   4.9   4.3   5.8   7.0   6.8
8.1   7.5   7.2   6.9   5.8   7.2   8.0   7.0   6.9   5.9

20.   **Apple Prices**   Use a stem-and-leaf plot to display the data. The data represent the prices (in cents per pound) paid to 28 farmers for apples.

19.2   19.6   16.4   17.1   19.0   17.4   17.3
20.1   19.0   17.5   17.6   18.6   18.4   17.7
19.5   18.4   18.9   17.5   19.3   20.8   19.3
18.6   18.6   18.3   17.1   18.1   16.8   17.9

21. Systolic Blood Pressures Use a dot plot to display the data. The data represent the systolic blood pressures (in millimeters of mercury) of 30 patients at a doctor's office.

120	135	140	145	130	150	120	170	145	125
130	110	160	180	200	150	200	135	140	120
120	130	140	170	120	165	150	130	135	140

22. Life Spans of Houseflies Use a dot plot to display the data. The data represent the life spans (in days) of 40 houseflies.

9	9	4	4	8	11	10	5	8	13	9
6	7	11	13	11	6	9	8	14	10	6
10	10	8	7	14	11	7	8	6	11	13
10	14	14	8	13	14	10				

23. New York City Marathon Use a pie chart to display the data. The data represent the number of men's New York City Marathon winners from each country through 2009. *(Source: New York Road Runners)*

United States	15	Mexico	4
Italy	4	Morocco	1
Ethiopia	1	Great Britain	1
South Africa	2	Brazil	2
Tanzania	1	New Zealand	1
Kenya	8		

24. NASA Budget Use a pie chart to display the data. The data represent the 2010 NASA budget request (in millions of dollars) divided among five categories. *(Source: NASA)*

Science, aeronautics, exploration	8947
Space operations	6176
Education	126
Cross-agency support	3401
Inspector general	36

25. Barrel of Oil Use a Pareto chart to display the data. The data represent how a 42-gallon barrel of crude oil is distributed. *(Adapted from American Petroleum Institute)*

Gasoline	43%
Kerosene-type jet fuel	9%
Distillate fuel oil (home heating, diesel fuel, etc.)	24%
Coke	5%
Residual fuel oil (industry, marine transportation, etc.)	4%
Liquefied refinery gases	3%
Other	12%

26. UV Index Use a Pareto chart to display the data. The data represent the ultraviolet indices for five cities at noon on a recent date. *(Source: National Oceanic and Atmospheric Administration)*

Atlanta, GA	Boise, ID	Concord, NH	Denver, CO	Miami, FL
9	7	8	7	10

27. Hourly Wages Use a scatter plot to display the data shown in the table. The data represent the number of hours worked and the hourly wages (in dollars) for a sample of 12 production workers. Describe any trends shown.

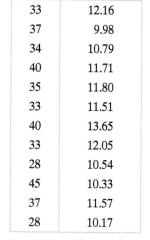

Hours	Hourly wage
33	12.16
37	9.98
34	10.79
40	11.71
35	11.80
33	11.51
40	13.65
33	12.05
28	10.54
45	10.33
37	11.57
28	10.17

TABLE FOR EXERCISE 27

Number of students per teacher	Average teacher's salary
17.1	28.7
17.5	47.5
18.9	31.8
17.1	28.1
20.0	40.3
18.6	33.8
14.4	49.8
16.5	37.5
13.3	42.5
18.4	31.9

TABLE FOR EXERCISE 28

 28. Salaries   Use a scatter plot to display the data shown in the table. The data represent the number of students per teacher and the average teacher salaries (in thousands of dollars) for a sample of 10 school districts. Describe any trends shown.

29. Daily High Temperatures   Use a time series chart to display the data. The data represent the daily high temperatures for a city for a period of 12 days.

May 1	May 2	May 3	May 4	May 5	May 6
77°	77°	79°	81°	82°	82°

May 7	May 8	May 9	May 10	May 11	May 12
85°	87°	90°	88°	89°	82°

 30. Manufacturing   Use a time series chart to display the data. The data represent the percentages of the U.S. gross domestic product (GDP) that come from the manufacturing sector. *(Source: U.S. Bureau of Economic Analysis)*

1997	1998	1999	2000	2001	2002
16.6%	15.4%	14.8%	14.5%	13.2%	12.9%

2003	2004	2005	2006	2007	2008
12.5%	12.2%	11.9%	12.0%	11.7%	11.5%

SC *In Exercises 31–34, use StatCrunch to organize the data using the indicated type of graph. What can you conclude about the data?*

31. Use a stem-and-leaf plot to display the data. The data represent the scores of an economics class on a final exam.

82 93 95 75 68 90 98 71 85 88 100 93
70 80 89 62 55 95 83 86 88 76   99 87

32. Use a dot plot to display the data. The data represent the screen sizes (in inches) of 20 DVD camcorders.

3.0 2.7 3.2 2.7 1.8 2.7 2.7 3.0 2.7 3.0
2.5 3.2 2.7 2.7 3.0 2.7 2.0 2.7 3.0 2.5

33. Use (a) a pie chart and (b) a Pareto chart to display the data. The data represent the results of an online survey that asked adults which type of investment they would focus on in 2010. *(Adapted from CNN)*

U.S. stocks	11,521	Emerging markets	5267
Bonds	3292	Commodities	1975
Bank accounts	10,533		

34. The data represent the number of motor vehicles (in millions) registered in the U.S. and the number of crashes (in millions). *(Source: U.S. National Highway Safety Traffic Administration)*

Year	2000	2001	2002	2003	2004	2005	2006	2007
Registrations	221	230	230	231	237	241	244	247
Crashes	6.4	6.3	6.3	6.3	6.2	6.2	6.0	6.0

(a) Use a scatter plot to display the number of registrations.

(b) Use a scatter plot to display the number of crashes.

(c) Construct a time series chart for the number of registrations.

(d) Construct a time series chart for the number of crashes.

■ **EXTENDING CONCEPTS**

A Misleading Graph? *A misleading graph is a statistical graph that is not drawn appropriately. This type of graph can misrepresent data and lead to false conclusions. In Exercises 35–38, (a) explain why the graph is misleading, and (b) redraw the graph so that it is not misleading.*

35.

Sales for Company A

Sales (in thousands of dollars)

120, 110, 100, 90

3rd 2nd 1st 4th

Quarter

36.

Results of a Survey

Percent that responded "yes"

72, 68, 64, 60, 56

Middle school, High school, College/ university

Type of student

37.

Sales for Company B

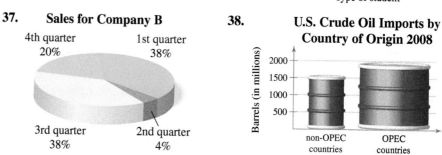

4th quarter 20%

1st quarter 38%

3rd quarter 38%

2nd quarter 4%

38.

U.S. Crude Oil Imports by Country of Origin 2008

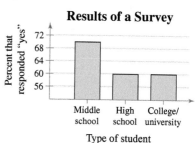

Barrels (in millions)

2000, 1500, 1000, 500

non-OPEC countries   OPEC countries

Law Firm A   **Law Firm B**

Law Firm A		Law Firm B
5 0	9	0 3
8 5 2 2 2	10	5 7
9 9 7 0 0	11	0 0 5
1 1	12	0 3 3 5
	13	2 2 5 9
	14	1 3 3 3 9
	15	5 5 5 6
	16	4 9 9
9 9 5 1 0	17	1 2 5
5 5 5 2 1	18	9
9 9 8 7 5	19	0
3	20	

Key: 5|19|0 = $195,000 for Law Firm A and $190,000 for Law Firm B

FIGURE FOR EXERCISE 39

39. Law Firm Salaries A **back-to-back stem-and-leaf plot** compares two data sets by using the same stems for each data set. Leaves for the first data set are on one side while leaves for the second data set are on the other side. The back-to-back stem-and-leaf plot shows the salaries (in thousands of dollars) of all lawyers at two small law firms.

(a) What are the lowest and highest salaries at Law Firm A? at Law Firm B?

(b) How many lawyers are in each firm?

(c) Compare the distribution of salaries at each law firm. What do you notice?

40. Yoga Classes The data sets show the ages of all participants in two yoga classes.

3:00 P.M. Class

40	60	73	77	51	68
68	35	68	53	64	75
76	69	59	55	38	57
68	84	75	62	73	75
85	77				

8:00 P.M. Class

19	18	20	29	39	43
71	56	44	44	18	19
19	18	18	20	25	29
25	22	31	24	24	23
19	19	18	28	20	31

(a) Make a back-to-back stem-and-leaf plot to display the data.

(b) What are the lowest and highest ages of participants in the 3:00 P.M. class? in the 8:00 P.M. class?

(c) How many participants are in each class?

(d) Compare the distribution of ages in each class. What conclusion(s) can you make based on your observations?

2.3   Measures of Central Tendency

WHAT YOU SHOULD LEARN

▸ How to find the mean, median, and mode of a population and of a sample

▸ How to find a weighted mean of a data set and the mean of a frequency distribution

▸ How to describe the shape of a distribution as symmetric, uniform, or skewed and how to compare the mean and median for each

Mean, Median, and Mode ▸ Weighted Mean and Mean of Grouped Data ▸ The Shapes of Distributions

▸ MEAN, MEDIAN, AND MODE

In Sections 2.1 and 2.2, you learned about the graphical representations of quantitative data. In Sections 2.3 and 2.4, you will learn how to supplement graphical representations with numerical statistics that describe the center and variability of a data set.

A **measure of central tendency** is a value that represents a typical, or central, entry of a data set. The three most commonly used measures of central tendency are the *mean*, the *median*, and the *mode*.

DEFINITION

The **mean** of a data set is the sum of the data entries divided by the number of entries. To find the mean of a data set, use one of the following formulas.

$$\text{Population Mean: } \mu = \frac{\sum x}{N} \qquad \text{Sample Mean: } \bar{x} = \frac{\sum x}{n}$$

The lowercase Greek letter μ (pronounced mu) represents the population mean and $\bar{x}$ (read as "x bar") represents the sample mean. Note that N represents the number of entries in a *population* and n represents the number of entries in a *sample*. Recall that the uppercase Greek letter sigma (Σ) indicates a summation of values.

EXAMPLE 1   **SC** Report 9

▸ **Finding a Sample Mean**

The prices (in dollars) for a sample of round-trip flights from Chicago, Illinois to Cancun, Mexico are listed. What is the mean price of the flights?

872   432   397   427   388   782   397

▸ **Solution**

The sum of the flight prices is

$$\sum x = 872 + 432 + 397 + 427 + 388 + 782 + 397 = 3695.$$

To find the mean price, divide the sum of the prices by the number of prices in the sample.

$$\bar{x} = \frac{\sum x}{n} = \frac{3695}{7} \approx 527.9$$

So, the mean price of the flights is about $527.90.

▸ **Try It Yourself 1**

The heights (in inches) of the players on the 2009–2010 Cleveland Cavaliers basketball team are listed. What is the mean height?

a. *Find the sum* of the data entries.
b. *Divide the sum* by the number of data entries.
c. *Interpret* the results in the context of the data.

Answer: Page A32

STUDY TIP

Notice that the mean in Example 1 has one more decimal place than the original set of data values. This *round-off rule* will be used throughout the text. Another important *round-off rule* is that rounding should not be done until the final answer of a calculation.

Heights of Players							
74	78	81	87	81	80	77	80
85	78	80	83	75	81	73	

DEFINITION

The **median** of a data set is the value that lies in the middle of the data when the data set is ordered. The median measures the center of an ordered data set by dividing it into two equal parts. If the data set has an odd number of entries, the median is the middle data entry. If the data set has an even number of entries, the median is the mean of the two middle data entries.

EXAMPLE 2   SC Report 10

▸ **Finding the Median**

Find the median of the flight prices given in Example 1.

▸ **Solution**

To find the median price, first order the data.

    388   397   397   427   432   782   872

Because there are seven entries (an odd number), the median is the middle, or fourth, data entry. So, the median flight price is $427.

▸ **Try It Yourself 2**

The ages of a sample of fans at a rock concert are listed. Find the median age.

    24   27   19   21   18   23   21   20   19   33   30   29   21
    18   24   26   38   19   35   34   33   30   21   27   30

a. *Order* the data entries.
b. *Find the middle* data entry.
c. *Interpret* the results in the context of the data.   *Answer: Page A32*

EXAMPLE 3

▸ **Finding the Median**

In Example 2, the flight priced at $432 is no longer available. What is the median price of the remaining flights?

▸ **Solution**

The remaining prices, in order, are 388, 397, 397, 427, 782, and 872.

Because there are six entries (an even number), the median is the mean of the two middle entries.

$$\text{Median} = \frac{397 + 427}{2} = 412$$

So, the median price of the remaining flights is $412.

▸ **Try It Yourself 3**

The prices (in dollars) of a sample of digital photo frames are listed. Find the median price of the digital photo frames.

    25   100   130   60   140   200   220   80   250   97

a. *Order* the data entries.
b. *Find the mean* of the two middle data entries.
c. *Interpret* the results in the context of the data.   *Answer: Page A32*

DEFINITION

The **mode** of a data set is the data entry that occurs with the greatest frequency. A data set can have one mode, more than one mode, or no mode. If no entry is repeated, the data set has no mode. If two entries occur with the same greatest frequency, each entry is a mode and the data set is called **bimodal.**

EXAMPLE 4   (SC) Report 11

▸ Finding the Mode

Find the mode of the flight prices given in Example 1.

▸ Solution

Ordering the data helps to find the mode.

388   397   397   427   432   782   872

From the ordered data, you can see that the entry 397 occurs twice, whereas the other data entries occur only once. So, the mode of the flight prices is $397.

▸ Try It Yourself 4

The prices (in dollars per square foot) for a sample of South Beach (Miami Beach, FL) condominiums are listed. Find the mode of the prices.

324	462	540	450	638	564	670	618	624	825
540	980	1650	1420	670	830	912	750	1260	450
975	670	1100	980	750	723	705	385	475	720

a. Write the data in *order*.
b. Identify the entry, or entries, that occur with the *greatest frequency*.
c. *Interpret* the results in the context of the data.   *Answer: Page A32*

EXAMPLE 5

▸ Finding the Mode

At a political debate, a sample of audience members were asked to name the political party to which they belonged. Their responses are shown in the table. What is the mode of the responses?

▸ Solution

The response occurring with the greatest frequency is Republican. So, the mode is Republican.

Interpretation   In this sample, there were more Republicans than people of any other single affiliation.

▸ Try It Yourself 5

In a survey, 1000 U.S. adults were asked if they thought public cellular phone conversations were rude. Of those surveyed, 510 responded "Yes," 370 responded "No," and 120 responded "Not sure." What is the mode of the responses? *(Adapted from Fox TV/Rasmussen Reports)*

a. Identify the entry that occurs with the *greatest frequency*.
b. *Interpret* the results in the context of the data.   *Answer: Page A32*

Political party	Frequency, f
Democrat	34
Republican	56
Other	21
Did not respond	9

Although the mean, the median, and the mode each describe a typical entry of a data set, there are advantages and disadvantages of using each. The mean is a reliable measure because it takes into account every entry of a data set. However, the mean can be greatly affected when the data set contains *outliers*.

DEFINITION

An **outlier** is a data entry that is far removed from the other entries in the data set.

A data set can have one or more outliers, causing **gaps** in a distribution. Conclusions that are drawn from a data set that contains outliers may be flawed.

Ages in a Class

20	20	20	20	20	20	21
21	21	21	22	22	22	23
23	23	23	24	24	65	

Outlier

EXAMPLE 6

▸ **Comparing the Mean, the Median, and the Mode**

Find the mean, the median, and the mode of the sample ages of students in a class shown at the left. Which measure of central tendency best describes a typical entry of this data set? Are there any outliers?

▸ **Solution**

Mean: $\bar{x} = \dfrac{\sum x}{n} = \dfrac{475}{20} \approx 23.8$ years

Median: Median $= \dfrac{21 + 22}{2} = 21.5$ years

Mode: The entry occurring with the greatest frequency is 20 years.

Interpretation The mean takes every entry into account but is influenced by the outlier of 65. The median also takes every entry into account, and it is not affected by the outlier. In this case the mode exists, but it doesn't appear to represent a typical entry. Sometimes a graphical comparison can help you decide which measure of central tendency best represents a data set. The histogram shows the distribution of the data and the locations of the mean, the median, and the mode. In this case, it appears that the median best describes the data set.

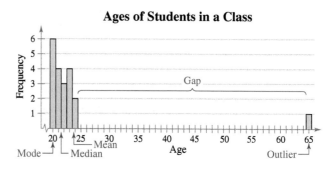

Ages of Students in a Class

▸ **Try It Yourself 6**

Remove the data entry 65 from the data set in Example 6. Then rework the example. How does the absence of this outlier change each of the measures?

a. Find the *mean*, the *median*, and the *mode*.
b. *Compare* these measures of central tendency with those found in Example 6.

Answer: Page A33

▶ WEIGHTED MEAN AND MEAN OF GROUPED DATA

Sometimes data sets contain entries that have a greater effect on the mean than do other entries. To find the mean of such a data set, you must find the *weighted mean*.

DEFINITION

A **weighted mean** is the mean of a data set whose entries have varying weights. A weighted mean is given by

$$\bar{x} = \frac{\sum (x \cdot w)}{\sum w}$$

where w is the weight of each entry x.

EXAMPLE 7

▶ **Finding a Weighted Mean**

You are taking a class in which your grade is determined from five sources: 50% from your test mean, 15% from your midterm, 20% from your final exam, 10% from your computer lab work, and 5% from your homework. Your scores are 86 (test mean), 96 (midterm), 82 (final exam), 98 (computer lab), and 100 (homework). What is the weighted mean of your scores? If the minimum average for an A is 90, did you get an A?

▶ **Solution**

Begin by organizing the scores and the weights in a table.

Source	Score, x	Weight, w	xw
Test mean	86	0.50	43.0
Midterm	96	0.15	14.4
Final exam	82	0.20	16.4
Computer lab	98	0.10	9.8
Homework	100	0.05	5.0
		$\sum w = 1$	$\sum (x \cdot w) = 88.6$

$$\bar{x} = \frac{\sum (x \cdot w)}{\sum w}$$

$$= \frac{88.6}{1}$$

$$= 88.6$$

Your weighted mean for the course is 88.6. So, you did not get an A.

▶ Try It Yourself 7

An error was made in grading your final exam. Instead of getting 82, you scored 98. What is your new weighted mean?

a. Multiply each score by its weight and *find the sum of these products.*
b. Find the *sum of the weights.*
c. Find the *weighted mean.*
d. *Interpret* the results in the context of the data.   *Answer: Page A33*

If data are presented in a frequency distribution, you can approximate the mean as follows.

DEFINITION

The **mean of a frequency distribution** for a sample is approximated by

$$\bar{x} = \frac{\Sigma\,(x \cdot f)}{n}$$ Note that $n = \Sigma f$.

where x and f are the midpoints and frequencies of a class, respectively.

GUIDELINES

Finding the Mean of a Frequency Distribution

IN WORDS	IN SYMBOLS
1. Find the midpoint of each class.	$x = \dfrac{(\text{Lower limit}) + (\text{Upper limit})}{2}$
2. Find the sum of the products of the midpoints and the frequencies.	$\Sigma\,(x \cdot f)$
3. Find the sum of the frequencies.	$n = \Sigma f$ inconsistence
4. Find the mean of the frequency distribution.	$\bar{x} = \dfrac{\Sigma\,(x \cdot f)}{n}$

EXAMPLE 8

▸ **Finding the Mean of a Frequency Distribution**

Use the frequency distribution at the left to approximate the mean number of minutes that a sample of Internet subscribers spent online during their most recent session.

Class midpoint, x	Frequency, f	xf
12.5	6	75.0
24.5	10	245.0
36.5	13	474.5
48.5	8	388.0
60.5	5	302.5
72.5	6	435.0
84.5	2	169.0
	$n = 50$	$\Sigma = 2089.0$

▸ **Solution**

$$\bar{x} = \frac{\Sigma\,(x \cdot f)}{n}$$

$$= \frac{2089.0}{50}$$

$$\approx 41.8$$

So, the mean time spent online was approximately 41.8 minutes.

▸ **Try It Yourself 8**

Use a frequency distribution to approximate the mean age of the 50 richest people. (See Try It Yourself 2 on page 41.)

a. Find the *midpoint* of each class.
b. Find the *sum of the products* of each midpoint and corresponding frequency.
c. Find the *sum of the frequencies*.
d. Find the *mean of the frequency distribution*.   *Answer: Page A33*

■ USING AND INTERPRETING CONCEPTS

Finding and Discussing the Mean, Median, and Mode   *In Exercises 17–34, find the mean, median, and mode of the data, if possible. If any of these measures cannot be found or a measure does not represent the center of the data, explain why.*

17. **Concert Tickets**   The number of concert tickets purchased online for the last 13 purchases

    4   2   5   8   6   6   4   3   2   4   7   8   5

18. **Tuition**   The 2009–2010 tuition and fees (in thousands of dollars) for the top 10 liberal arts colleges   *(Source: U.S. News and World Report)*

    39   39   38   51   38   40   37   40   35   39

19. **MCAT Scores**   The average medical college admission test (MCAT) scores for a sample of seven medical schools   *(Source: Association of American Medical Colleges)*

    11.0   11.7   10.3   11.7   11.7   10.7   9.7

20. **Cholesterol**   The cholesterol levels of a sample of 10 female employees

    154   240   171   188   235   203   184   173   181   275

21. **NFL**   The average points per game scored by each NFL team during the 2009 regular season   *(Source: National Football League)*

    20.4   19.7   17.5   26.7   22.7   21.8   16.6   29.4
    26.0   22.5   28.8   19.1   18.1   12.3   16.4   15.2
    16.1   23.4   20.6   18.4   23.0   25.1   26.8   31.9
    24.4   28.4   20.4   22.1   15.3   10.9   24.2   22.6

22. **Power Failures**   The durations (in minutes) of power failures at a residence in the last 10 years

    18   26   45   75   125   80   33   40   44   49
    89   80   96   125   12   61   31   63   103   28

23. **Eyeglasses and Contacts**   The responses of a sample of 1000 adults who were asked what type of corrective lenses they wore are shown in the table at the left.   *(Adapted from American Optometric Association)*

24. **Living on Your Own**   The responses of a sample of 1177 young adults who were asked what surprised them the most as they began to live on their own   *(Adapted from Charles Schwab)*

    Amount of first salary: 63   Trying to find a job: 125
    Number of decisions: 163   Money needed: 326
    Paying bills: 150   Trying to save: 275
    How hard it is breaking away from parents: 75

25. **Top Speeds**   The top speeds (in miles per hour) for a sample of seven sports cars

    187.3   181.8   180.0   169.3   162.2   158.1   155.7

26. **Potatoes**   The pie chart at the left shows the responses of a sample of 1000 adults who were asked their favorite way to eat potatoes.   *(Adapted from Idaho Potato Commission)*

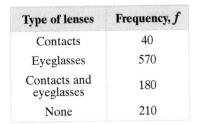

Type of lenses	Frequency, f
Contacts	40
Eyeglasses	570
Contacts and eyeglasses	180
None	210

TABLE FOR EXERCISE 23

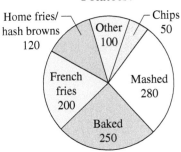

How Do People Eat Their Potatoes?

Home fries/hash browns 120
Other 100
Chips 50
French fries 200
Mashed 280
Baked 250

FIGURE FOR EXERCISE 26

27. Typing Speeds   The typing speeds (in words per minute) for several stenographers

125   140   170   155   132   175   225   210   125   230

28. Eating Disorders   The number of weeks it took to reach a target weight for a sample of five patients with eating disorders treated by psychodynamic psychotherapy *(Source: The Journal of Consulting and Clinical Psychology)*

15.0   31.5   10.0   25.5   1.0

29. Eating Disorders   The number of weeks it took to reach a target weight for a sample of 14 patients with eating disorders treated by psychodynamic psychotherapy and cognitive behavior techniques *(Source: The Journal of Consulting and Clinical Psychology)*

2.5   20.0   11.0   10.5   17.5   16.5   13.0
15.5   26.5   2.5   27.0   28.5   1.5   5.0

30. Aircraft   The number of aircraft that 15 airlines have in their fleets *(Source: Airline Transport Association)*

136   110   38   625   350   755   52   32
142   9   537   28   409   354   28

31.   **Weights (in pounds) of Carry-On Luggage on a Plane**

0	6 7
1	2 5 8 9
2	0 4 4 4 5 8 9
3	2 2 3 5 5 5 6 8 9
4	0 1 2 7 8
5	1

Key: $3|2 = 32$

32. Grade Point Averages of Students in a Class

0	8
1	5 6 8
2	1 3 4 5
3	0 9
4	0 0

Key: $0|8 = 0.8$

33.   **Time (in minutes) It Takes Employees to Drive to Work**

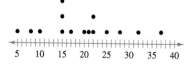

34.   **Prices (in dollars per night) of Hotel Rooms in a City**

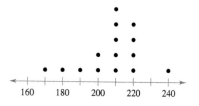

Graphical Analysis   *In Exercises 35 and 36, the letters A, B, and C are marked on the horizontal axis. Describe the shape of the data. Then determine which is the mean, which is the median, and which is the mode. Justify your answers.*

35.   **Sick Days Used by Employees**

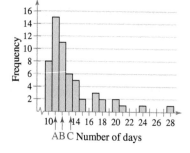

36.   **Hourly Wages of Employees**

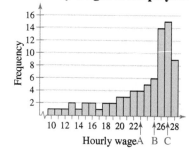

In Exercises 37–40, without performing any calculations, determine which measure of central tendency best represents the graphed data. Explain your reasoning.

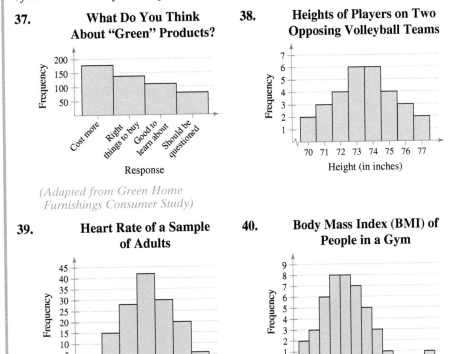

37. What Do You Think About "Green" Products?

(Adapted from Green Home Furnishings Consumer Study)

38. Heights of Players on Two Opposing Volleyball Teams

39. Heart Rate of a Sample of Adults

40. Body Mass Index (BMI) of People in a Gym

Finding the Weighted Mean   *In Exercises 41–46, find the weighted mean of the data.*

41. Final Grade   The scores and their percents of the final grade for a statistics student are given. What is the student's mean score?

	Score	Percent of final grade
Homework	85	5%
Quizzes	80	35%
Project	100	20%
Speech	90	15%
Final exam	93	25%

42. Salaries   The average starting salaries (by degree attained) for 25 employees at a company are given. What is the mean starting salary for these employees?

8 with MBAs: $92,500   17 with BAs in business: $68,000

43. Account Balance   For the month of April, a checking account has a balance of $523 for 24 days, $2415 for 2 days, and $250 for 4 days. What is the account's mean daily balance for April?

44. Account Balance   For the month of May, a checking account has a balance of $759 for 15 days, $1985 for 5 days, $1410 for 5 days, and $348 for 6 days. What is the account's mean daily balance for May?

45. Grades   A student receives the following grades, with an A worth 4 points, a B worth 3 points, a C worth 2 points, and a D worth 1 point. What is the student's mean grade point score?

B in 2 three-credit classes   D in 1 two-credit class
A in 1 four-credit class   C in 1 three-credit class

46. Scores The mean scores for students in a statistics course (by major) are given. What is the mean score for the class?

> 9 engineering majors: 85
> 5 math majors: 90
> 13 business majors: 81

47. Final Grade In Exercise 41, an error was made in grading your final exam. Instead of getting 93, you scored 85. What is your new weighted mean?

48. Grades In Exercise 45, one of the student's B grades gets changed to an A. What is the student's new mean grade point score?

Finding the Mean of Grouped Data *In Exercises 49–52, approximate the mean of the grouped data.*

49. Fuel Economy The highway mileage (in miles per gallon) for 30 small cars

Mileage (miles per gallon)	Frequency
29–33	11
34–38	12
39–43	2
44–48	5

50. Fuel Economy The city mileage (in miles per gallon) for 24 family sedans

Mileage (miles per gallon)	Frequency
22–27	16
28–33	2
34–39	2
40–45	3
46–51	1

51. Ages The ages of residents of a town

Age	Frequency
0–9	55
10–19	70
20–29	35
30–39	56
40–49	74
50–59	42
60–69	38
70–79	17
80–89	10

52. Phone Calls The lengths of calls (in minutes) made by a salesperson in one week

Length of call	Number of calls
1–5	12
6–10	26
11–15	20
16–20	7
21–25	11
26–30	7
31–35	4
36–40	4
41–45	1

Identifying the Shape of a Distribution *In Exercises 53–56, construct a frequency distribution and a frequency histogram of the data using the indicated number of classes. Describe the shape of the histogram as symmetric, uniform, negatively skewed, positively skewed, or none of these.*

 53. Hospital Beds
Number of classes: 5
Data set: The number of beds in a sample of 24 hospitals

> 149 167 162 127 130 180 160 167
> 221 145 137 194 207 150 254 262
> 244 297 137 204 166 174 180 151

54. Hospitalization

Number of classes: 6

Data set: The number of days 20 patients remained hospitalized

6 9 7 14 4 5 6 8 4 11
10 6 8 6 5 7 6 6 3 11

55. Heights of Males

Number of classes: 5

Data set: The heights (to the nearest inch) of 30 males

67 76 69 68 72 68 65 63 75 69
66 72 67 66 69 73 64 62 71 73
68 72 71 65 69 66 74 72 68 69

56. Six-Sided Die

Number of classes: 6

Data set: The results of rolling a six-sided die 30 times

1 4 6 1 5 3 2 5 4 6 1 2 4 3 5
6 3 2 1 1 5 6 2 4 4 3 1 6 2 4

57. Coffee Contents During a quality assurance check, the actual coffee contents (in ounces) of six jars of instant coffee were recorded as 6.03, 5.59, 6.40, 6.00, 5.99, and 6.02.

(a) Find the mean and the median of the coffee content.

(b) The third value was incorrectly measured and is actually 6.04. Find the mean and median of the coffee content again.

(c) Which measure of central tendency, the mean or the median, was affected more by the data entry error?

58. U.S. Exports The table at the left shows the U.S. exports (in billions of dollars) to 19 countries for a recent year. *(Source: U.S. Department of Commerce)*

(a) Find the mean and median.

(b) Find the mean and median without the U.S. exports to Canada. Which measure of central tendency, the mean or the median, was affected more by the elimination of the Canadian exports?

(c) The U.S. exports to India were $17.7 billion. Find the mean and median with the Indian exports added to the original data set. Which measure of central tendency was affected more by adding the Indian exports?

SC *In Exercises 59 and 60, use StatCrunch to find the sample size, mean, median, minimum data value, and maximum data value of the data.*

59. The data represent the amounts (in dollars) made by several families during a community yard sale.

95 120 125.50 105.25 82 102.75 130 151.50 145.25 79 97

60. The data represent the prices (in dollars) of the stocks in the Dow Jones Industrial Average during a recent session. *(Source: CNN Money)*

83.62 15.90 42.61 26.35 16.89 61.46 62.07 79.53 24.99 34.05
69.62 16.77 52.69 21.46 132.39 65.10 44.56 29.08 62.54 39.92
31.07 19.46 57.19 28.30 61.49 49.28 72.77 31.38 54.33 31.06

U.S. Exports (in billions of dollars)

Canada: 261.1	Japan: 65.1
Mexico: 151.2	South Korea: 34.7
Germany: 54.5	Singapore: 27.9
Taiwan: 24.9	France: 28.8
Netherlands: 39.7	Brazil: 32.3
China: 69.7	Belgium: 28.9
Australia: 22.2	Italy: 15.5
Malaysia: 12.9	Thailand: 9.1
Switzerland: 22.0	
Saudi Arabia: 12.5	
United Kingdom: 53.6	

TABLE FOR EXERCISE 58

■ EXTENDING CONCEPTS

61. Golf The distances (in yards) for nine holes of a golf course are listed.

  336 393 408 522 147 504 177 375 360

(a) Find the mean and median of the data.

(b) Convert the distances to feet. Then rework part (a).

(c) Compare the measures you found in part (b) with those found in part (a). What do you notice?

(d) Use your results from part (c) to explain how to find quickly the mean and median of the given data set if the distances are measured in inches.

62. Data Analysis A consumer testing service obtained the following mileages (in miles per gallon) in five test runs performed with three types of compact cars.

	Run 1	Run 2	Run 3	Run 4	Run 5
Car A:	28	32	28	30	34
Car B:	31	29	31	29	31
Car C:	29	32	28	32	30

(a) The manufacturer of Car A wants to advertise that its car performed best in this test. Which measure of central tendency—mean, median, or mode—should be used for its claim? Explain your reasoning.

(b) The manufacturer of Car B wants to advertise that its car performed best in this test. Which measure of central tendency—mean, median, or mode—should be used for its claim? Explain your reasoning.

(c) The manufacturer of Car C wants to advertise that its car performed best in this test. Which measure of central tendency—mean, median, or mode—should be used for its claim? Explain your reasoning.

63. Midrange Another measure of central tendency that is rarely used but is easy to calculate is the **midrange.** It can be found by the formula

$$\frac{(\text{Maximum data entry}) + (\text{Minimum data entry})}{2}.$$

Which of the manufacturers in Exercise 62 would prefer to use the midrange statistic in their ads? Explain your reasoning.

64. Data Analysis Students in an experimental psychology class did research on depression as a sign of stress. A test was administered to a sample of 30 students. The scores are given.

  44 51 11 90 76 36 64 37 43 72 53 62 36 74 51
  72 37 28 38 61 47 63 36 41 22 37 51 46 85 13

(a) Find the mean and median of the data.

(b) Draw a stem-and-leaf plot for the data using one row per stem. Locate the mean and median on the display.

(c) Describe the shape of the distribution.

65. Trimmed Mean To find the 10% **trimmed mean** of a data set, order the data, delete the lowest 10% of the entries and the highest 10% of the entries, and find the mean of the remaining entries.

(a) Find the 10% trimmed mean for the data in Exercise 64.

(b) Compare the four measures of central tendency, including the midrange.

(c) What is the benefit of using a trimmed mean versus using a mean found using all data entries? Explain your reasoning.

ACTIVITY 2.3   Mean Versus Median

APPLET

The *mean versus median* applet is designed to allow you to investigate interactively the mean and the median as measures of the center of a data set. Points can be added to the plot by clicking the mouse above the horizontal axis. The mean of the points is shown as a green arrow and the median is shown as a red arrow. If the two values are the same, then a single yellow arrow is displayed. Numeric values for the mean and median are shown above the plot. Points on the plot can be removed by clicking on the point and then dragging the point into the trash can. All of the points on the plot can be removed by simply clicking inside the trash can. The range of values for the horizontal axis can be specified by inputting lower and upper limits and then clicking UPDATE.

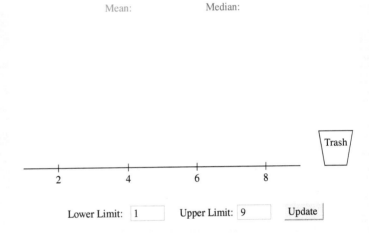

■ **Explore**

Step 1   Specify a lower limit.
Step 2   Specify an upper limit.
Step 3   Add 15 points to the plot.
Step 4   Remove all of the points from the plot.

■ **Draw Conclusions**

APPLET

1. Specify the lower limit to be 1 and the upper limit to be 50. Add at least 10 points that range from 20 to 40 so that the mean and the median are the same. What is the shape of the distribution? What happens at first to the mean and median when you add a few points that are less than 10? What happens over time as you continue to add points that are less than 10?

2. Specify the lower limit to be 0 and the upper limit to be 0.75. Place 10 points on the plot. Then change the upper limit to 25. Add 10 more points that are greater than 20 to the plot. Can the mean be any one of the points that were plotted? Can the median be any one of the points that were plotted? Explain.

2.4 Measures of Variation

WHAT YOU SHOULD LEARN

▸ How to find the range of a data set

▸ How to find the variance and standard deviation of a population and of a sample

▸ How to use the Empirical Rule and Chebychev's Theorem to interpret standard deviation

▸ How to approximate the sample standard deviation for grouped data

Range ▸ Deviation, Variance, and Standard Deviation ▸ Interpreting Standard Deviation ▸ Standard Deviation for Grouped Data

▸ RANGE

In this section, you will learn different ways to measure the variation of a data set. The simplest measure is the *range* of the set.

DEFINITION

The **range** of a data set is the difference between the maximum and minimum data entries in the set. To find the range, the data must be quantitative.

Range = (Maximum data entry) − (Minimum data entry)

EXAMPLE 1   **SC** Report 12

▸ **Finding the Range of a Data Set**

Two corporations each hired 10 graduates. The starting salaries for each graduate are shown. Find the range of the starting salaries for Corporation A.

Starting Salaries for Corporation A (1000s of dollars)

Salary	41	38	39	45	47	41	44	41	37	42

Starting Salaries for Corporation B (1000s of dollars)

Salary	40	23	41	50	49	32	41	29	52	58

▸ **Solution**

Ordering the data helps to find the least and greatest salaries.

37 38 39 41 41 41 42 44 45 47

Minimum ⟶   ⟵ Maximum

Range = (Maximum salary) − (Minimum salary)

= 47 − 37

= 10

So, the range of the starting salaries for Corporation A is 10, or $10,000.

▸ **Try It Yourself 1**

Find the range of the starting salaries for Corporation B.

a. Identify the *minimum* and *maximum* salaries.
b. Find the *range*.
c. *Compare* your answer with that for Example 1.

Answer: Page A33

INSIGHT

Both data sets in Example 1 have a mean of 41.5, or $41,500, a median of 41, or $41,000, and a mode of 41, or $41,000. And yet the two sets differ significantly.

The difference is that the entries in the second set have greater variation. Your goal in this section is to learn how to measure the variation of a data set.

▸ DEVIATION, VARIANCE, AND STANDARD DEVIATION

As a measure of variation, the range has the advantage of being easy to compute. Its disadvantage, however, is that it uses only two entries from the data set. Two measures of variation that use all the entries in a data set are the *variance* and the *standard deviation*. However, before you learn about these measures of variation, you need to know what is meant by the *deviation* of an entry in a data set.

DEFINITION

The **deviation** of an entry x in a population data set is the difference between the entry and the mean μ of the data set.

$$\text{Deviation of } x = x - \mu$$

Deviations of Starting Salaries for Corporation A

Salary (1000s of dollars) x	Deviation (1000s of dollars) $x - \mu$
41	−0.5
38	−3.5
39	−2.5
45	3.5
47	5.5
41	−0.5
44	2.5
41	−0.5
37	−4.5
42	0.5
$\Sigma x = 415$	$\Sigma(x - \mu) = 0$

EXAMPLE 2

▸ Finding the Deviations of a Data Set

Find the deviation of each starting salary for Corporation A given in Example 1.

▸ Solution

The mean starting salary is $\mu = 415/10 = 41.5$, or \$41,500. To find out how much each salary deviates from the mean, subtract 41.5 from the salary. For instance, the deviation of 41, or \$41,000 is

$$41 - 41.5 = -0.5, \text{ or } -\$500. \qquad \text{Deviation of } x = x - \mu$$

The table at the left lists the deviations of each of the 10 starting salaries.

▸ Try It Yourself 2

Find the deviation of each starting salary for Corporation B given in Example 1.

a. Find the *mean* of the data set.
b. *Subtract* the mean from each salary.   *Answer: Page A33*

In Example 2, notice that the sum of the deviations is zero. Because this is true for any data set, it doesn't make sense to find the average of the deviations. To overcome this problem, you can square each deviation. When you add the squares of the deviations, you compute a quantity called the **sum of squares,** denoted SS_x. In a population data set, the mean of the squares of the deviations is called the **population variance.**

DEFINITION

The **population variance** of a population data set of N entries is

$$\text{Population variance} = \sigma^2 = \frac{\Sigma(x - \mu)^2}{N}.$$

The symbol σ is the lowercase Greek letter sigma.

DEFINITION

The **population standard deviation** of a population data set of N entries is the square root of the population variance.

$$\text{Population standard deviation} = \sigma = \sqrt{\sigma^2} = \sqrt{\frac{\sum (x - \mu)^2}{N}}$$

GUIDELINES

Finding the Population Variance and Standard Deviation

IN WORDS	IN SYMBOLS
1. Find the mean of the population data set.	$\mu = \dfrac{\sum x}{N}$
2. Find the deviation of each entry.	$x - \mu$
3. Square each deviation.	$(x - \mu)^2$
4. Add to get the **sum of squares**.	$SS_x = \sum (x - \mu)^2$
5. Divide by N to get the **population variance**.	$\sigma^2 = \dfrac{\sum (x - \mu)^2}{N}$
6. Find the square root of the variance to get the **population standard deviation**.	$\sigma = \sqrt{\dfrac{\sum (x - \mu)^2}{N}}$

Sum of Squares of Starting Salaries for Corporation A

Salary x	Deviation $x - \mu$	Squares $(x - \mu)^2$
41	−0.5	0.25
38	−3.5	12.25
39	−2.5	6.25
45	3.5	12.25
47	5.5	30.25
41	−0.5	0.25
44	2.5	6.25
41	−0.5	0.25
37	−4.5	20.25
42	0.5	0.25
	$\Sigma = 0$	$SS_x = 88.5$

EXAMPLE 3

▶ **Finding the Population Standard Deviation**

Find the population standard deviation of the starting salaries for Corporation A given in Example 1.

▶ **Solution**

The table at the left summarizes the steps used to find SS_x.

$$SS_x = 88.5, \qquad N = 10, \qquad \sigma^2 = \frac{88.5}{10} \approx 8.9, \qquad \sigma = \sqrt{\frac{88.5}{10}} \approx 3.0$$

So, the population variance is about 8.9, and the population standard deviation is about 3.0, or $3000.

▶ **Try It Yourself 3**

Find the population variance and standard deviation of the starting salaries for Corporation B given in Example 1.

a. Find the *mean* and each *deviation*, as you did in Try It Yourself 2.
b. *Square* each deviation and *add* to get the sum of squares.
c. *Divide* by N to get the population variance.
d. Find the *square root* of the population variance to get the population standard deviation.
e. *Interpret* the results by giving the population standard deviation in dollars.

Answer: Page A33

STUDY TIP

Notice that the variance and standard deviation in Example 3 have one more decimal place than the original set of data values has. This is the same *round-off rule* that was used to calculate the mean.

DEFINITION

The **sample variance** and **sample standard deviation** of a sample data set of *n* entries are listed below.

$$\text{Sample variance} = s^2 = \frac{\sum (x - \overline{x})^2}{n - 1}$$

$$\text{Sample standard deviation} = s = \sqrt{s^2} = \sqrt{\frac{\sum (x - \overline{x})^2}{n - 1}}$$

Symbols in Variance and Standard Deviation Formulas

	Population	**Sample**
Variance	σ^2	s^2
Standard deviation	σ	s
Mean	μ	$\overline{x}$
Number of entries	N	n
Deviation	$x - \mu$	$x - \overline{x}$
Sum of squares	$\Sigma(x - \mu)^2$	$\Sigma(x - \overline{x})^2$

GUIDELINES

Finding the Sample Variance and Standard Deviation

IN WORDS	IN SYMBOLS
1. Find the mean of the sample data set.	$\overline{x} = \dfrac{\sum x}{n}$
2. Find the deviation of each entry.	$x - \overline{x}$
3. Square each deviation.	$(x - \overline{x})^2$
4. Add to get the **sum of squares**.	$SS_x = \Sigma (x - \overline{x})^2$
5. Divide by $n - 1$ to get the **sample variance**.	$s^2 = \dfrac{\Sigma (x - \overline{x})^2}{n - 1}$
6. Find the square root of the variance to get the **sample standard deviation**.	$s = \sqrt{\dfrac{\Sigma (x - \overline{x})^2}{n - 1}}$

See MINITAB and TI-83/84 Plus steps on pages 122 and 123.

EXAMPLE 4   Report 13

▸ **Finding the Sample Standard Deviation**

The starting salaries given in Example 1 are for the Chicago branches of Corporations A and B. Each corporation has several other branches, and you plan to use the starting salaries of the Chicago branches to estimate the starting salaries for the larger populations. Find the *sample* standard deviation of the starting salaries for the Chicago branch of Corporation A.

▸ **Solution**

$$SS_x = 88.5, \qquad n = 10, \qquad s^2 = \frac{88.5}{9} \approx 9.8, \qquad s = \sqrt{\frac{88.5}{9}} \approx 3.1$$

So, the sample variance is about 9.8, and the sample standard deviation is about 3.1, or $3100.

▸ **Try It Yourself 4**

Find the sample standard deviation of the starting salaries for the Chicago branch of Corporation B.

a. Find the *sum of squares*, as you did in Try It Yourself 3.
b. *Divide* by *n* − 1 to get the sample variance.
c. Find the *square root* of the sample variance to get the sample standard deviation.
d. *Interpret* the results by giving the sample standard deviation in dollars.

Answer: Page A33

Office Rental Rates		
35.00	33.50	37.00
23.75	26.50	31.25
36.50	40.00	32.00
39.25	37.50	34.75
37.75	37.25	36.75
27.00	35.75	26.00
37.00	29.00	40.50
24.50	33.00	38.00

EXAMPLE 5

▶ **Using Technology to Find the Standard Deviation**

Sample office rental rates (in dollars per square foot per year) for Miami's central business district are shown in the table. Use a calculator or a computer to find the mean rental rate and the sample standard deviation. *(Adapted from Cushman & Wakefield Inc.)*

▶ **Solution**

MINITAB, Excel, and the TI-83/84 Plus each have features that automatically calculate the means and the standard deviations of data sets. Try using this technology to find the mean and the standard deviation of the office rental rates. From the displays, you can see that $\bar{x} \approx 33.73$ and $s \approx 5.09$.

MINITAB

Descriptive Statistics: Rental Rates

Variable	N	Mean	SE Mean	StDev	Minimum
Rental Rates	24	33.73	1.04	5.09	23.75

Variable	Q1	Median	Q3	Maximum	
Rental Rates	29.56	35.38	37.44	40.50	

EXCEL

	A	B
1	Mean	33.72917
2	Standard Error	1.038864
3	Median	35.375
4	Mode	37
5	Standard Deviation	5.089373
6	Sample Variance	25.90172
7	Kurtosis	-0.74282
8	Skewness	-0.70345
9	Range	16.75
10	Minimum	23.75
11	Maximum	40.5
12	Sum	809.5
13	Count	24

TI-83/84 PLUS

```
1-Var Stats
x̄=33.72916667
Σx=809.5
Σx²=27899.5
Sx=5.089373342
σx=4.982216639
n=24
```

Sample Mean
Sample Standard Deviation

▶ **Try It Yourself 5**

Sample office rental rates (in dollars per square foot per year) for Seattle's central business district are listed. Use a calculator or a computer to find the mean rental rate and the sample standard deviation. *(Adapted from Cushman & Wakefield Inc.)*

40.00	43.00	46.00	40.50	35.75	39.75	32.75
36.75	35.75	38.75	38.75	36.75	38.75	39.00
29.00	35.00	42.75	32.75	40.75	35.25	

a. *Enter* the data.
b. *Calculate* the sample mean and the sample standard deviation.

Answer: Page A33

To explore this topic further, see Activity 2.4 on page 98.

▶ **INTERPRETING STANDARD DEVIATION**

When interpreting the standard deviation, remember that it is a measure of the typical amount an entry deviates from the mean. The more the entries are spread out, the greater the standard deviation.

INSIGHT

When all data values are equal, the standard deviation is 0. Otherwise, the standard deviation must be positive.

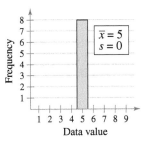 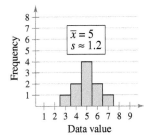 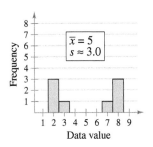

EXAMPLE 6

▶ **Estimating Standard Deviation**

Without calculating, estimate the population standard deviation of each data set.

1. 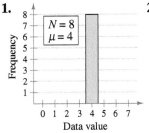 **2.** 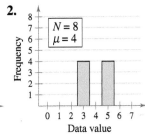 **3.**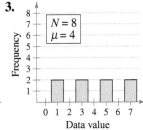

▶ **Solution**

1. Each of the eight entries is 4. So, each deviation is 0, which implies that

$$\sigma = 0.$$

2. Each of the eight entries has a deviation of ± 1. So, the population standard deviation should be 1. By calculating, you can see that

$$\sigma = 1.$$

3. Each of the eight entries has a deviation of ± 1 or ± 3. So, the population standard deviation should be about 2. By calculating, you can see that

$$\sigma \approx 2.24.$$

▶ **Try It Yourself 6**

Write a data set that has 10 entries, a mean of 10, and a population standard deviation that is approximately 3. (There are many correct answers.)

a. *Write* a data set that has five entries that are three units less than 10 and five entries that are three units more than 10.

b. *Calculate* the population standard deviation to check that σ is approximately 3.

Answer: Page A33

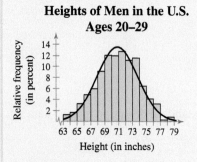

Heights of Men in the U.S. Ages 20–29

Roughly which two heights contain the middle 95% of the data?

Many real-life data sets have distributions that are approximately symmetric and bell-shaped. Later in the text, you will study this type of distribution in detail. For now, however, the following *Empirical Rule* can help you see how valuable the standard deviation can be as a measure of variation.

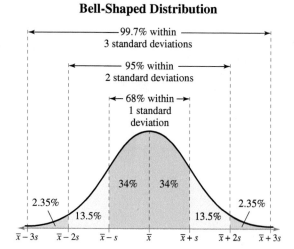

Bell-Shaped Distribution

EMPIRICAL RULE (OR 68–95–99.7 RULE)

For data with a (symmetric) bell-shaped distribution, the standard deviation has the following characteristics.

1. About 68% of the data lie within one standard deviation of the mean.
2. About 95% of the data lie within two standard deviations of the mean.
3. About 99.7% of the data lie within three standard deviations of the mean.

EXAMPLE 7

▶ **Using the Empirical Rule**

In a survey conducted by the National Center for Health Statistics, the sample mean height of women in the United States (ages 20–29) was 64.3 inches, with a sample standard deviation of 2.62 inches. Estimate the percent of women whose heights are between 59.06 inches and 64.3 inches. *(Adapted from National Center for Health Statistics)*

▶ **Solution**

The distribution of women's heights is shown. Because the distribution is bell-shaped, you can use the Empirical Rule. The mean height is 64.3, so when you subtract two standard deviations from the mean height, you get

$$\bar{x} - 2s = 64.3 - 2(2.62) = 59.06.$$

Because 59.06 is two standard deviations below the mean height, the percent of the heights between 59.06 and 64.3 inches is 13.5% + 34% = 47.5%.

Interpretation So, 47.5% of women are between 59.06 and 64.3 inches tall.

▶ **Try It Yourself 7**

Estimate the percent of women's heights that are between 64.3 and 66.92 inches tall.

a. How many *standard deviations* is 66.92 to the right of 64.3?
b. Use the *Empirical Rule* to estimate the percent of the data between $\bar{x}$ and $\bar{x} + s$.
c. *Interpret* the result in the context of the data. *Answer: Page A33*

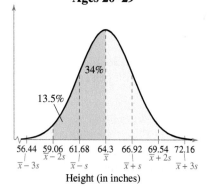

Heights of Women in the U.S. Ages 20–29

INSIGHT

Data values that lie more than two standard deviations from the mean are considered unusual. Data values that lie more than three standard deviations from the mean are very unusual.

The Empirical Rule applies only to (symmetric) bell-shaped distributions. What if the distribution is not bell-shaped, or what if the shape of the distribution is not known? The following theorem gives an inequality statement that applies to *all* distributions. It is named after the Russian statistician Pafnuti Chebychev (1821–1894).

CHEBYCHEV'S THEOREM

The portion of any data set lying within k standard deviations $(k > 1)$ of the mean is at least

$$1 - \frac{1}{k^2}.$$

- $k = 2$: In any data set, at least $1 - \frac{1}{2^2} = \frac{3}{4}$, or 75%, of the data lie within 2 standard deviations of the mean.

- $k = 3$: In any data set, at least $1 - \frac{1}{3^2} = \frac{8}{9}$, or 88.9%, of the data lie within 3 standard deviations of the mean.

EXAMPLE 8

▶ **Using Chebychev's Theorem**

The age distributions for Alaska and Florida are shown in the histograms. Decide which is which. Apply Chebychev's Theorem to the data for Florida using $k = 2$. What can you conclude?

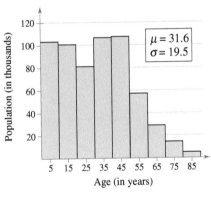
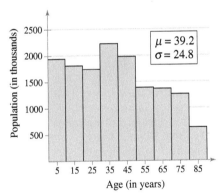

▶ **Solution**

The histogram on the right shows Florida's age distribution. You can tell because the population is greater and older. Moving two standard deviations to the left of the mean puts you below 0, because $\mu - 2\sigma = 39.2 - 2(24.8) = -10.4$. Moving two standard deviations to the right of the mean puts you at $\mu + 2\sigma = 39.2 + 2(24.8) = 88.8$. By Chebychev's Theorem, you can say that at least 75% of the population of Florida is between 0 and 88.8 years old.

▶ **Try It Yourself 8**

Apply Chebychev's Theorem to the data for Alaska using $k = 2$. What can you conclude?

a. *Subtract* two standard deviations from the mean.
b. *Add* two standard deviations to the mean.
c. *Apply* Chebychev's Theorem for $k = 2$ and *interpret* the results.

Answer: Page A33

INSIGHT

In Example 8, Chebychev's Theorem gives you an inequality statement that says that at least 75% of the population of Florida is under the age of 88.8. This is a true statement, but it is not nearly as strong a statement as could be made from reading the histogram.

In general, Chebychev's Theorem gives the minimum percent of data values that fall within the given number of standard deviations of the mean. Depending on the distribution, there is probably a higher percent of data falling in the given range.

▸ STANDARD DEVIATION FOR GROUPED DATA

In Section 2.1, you learned that large data sets are usually best represented by frequency distributions. The formula for the sample standard deviation for a frequency distribution is

$$\text{Sample standard deviation} = s = \sqrt{\frac{\Sigma(x - \bar{x})^2 f}{n - 1}}$$

where $n = \Sigma f$ is the number of entries in the data set.

EXAMPLE 9

▸ **Finding the Standard Deviation for Grouped Data**

You collect a random sample of the number of children per household in a region. The results are shown at the left. Find the sample mean and the sample standard deviation of the data set.

▸ **Solution**

These data could be treated as 50 individual entries, and you could use the formulas for mean and standard deviation. Because there are so many repeated numbers, however, it is easier to use a frequency distribution.

x	f	xf	$x - \bar{x}$	$(x - \bar{x})^2$	$(x - \bar{x})^2 f$
0	10	0	−1.8	3.24	32.40
1	19	19	−0.8	0.64	12.16
2	7	14	0.2	0.04	0.28
3	7	21	1.2	1.44	10.08
4	2	8	2.2	4.84	9.68
5	1	5	3.2	10.24	10.24
6	4	24	4.2	17.64	70.56
	$\Sigma = 50$	$\Sigma = 91$			$\Sigma = 145.40$

Number of Children in 50 Households

1	3	1	1	1
1	2	2	1	0
1	1	0	0	0
1	5	0	3	6
3	0	3	1	1
1	1	6	0	1
3	6	6	1	2
2	3	0	1	1
4	1	1	2	2
0	3	0	2	4

$$\bar{x} = \frac{\Sigma xf}{n} = \frac{91}{50} \approx 1.8 \qquad \text{Sample mean}$$

Use the sum of squares to find the sample standard deviation.

$$s = \sqrt{\frac{\Sigma(x - \bar{x})^2 f}{n - 1}} = \sqrt{\frac{145.4}{49}} \approx 1.7 \qquad \text{Sample standard deviation}$$

So, the sample mean is about 1.8 children, and the sample standard deviation is about 1.7 children.

▸ **Try It Yourself 9**

Change three of the 6's in the data set to 4's. How does this change affect the sample mean and sample standard deviation?

a. Write the first three columns of a *frequency distribution*.
b. Find the *sample mean*.
c. Complete the *last three columns* of the frequency distribution.
d. Find the *sample standard deviation*.

Answer: Page A33

When a frequency distribution has classes, you can estimate the sample mean and the sample standard deviation by using the midpoint of each class.

EXAMPLE 10

▶ **Using Midpoints of Classes**

The circle graph at the right shows the results of a survey in which 1000 adults were asked how much they spend in preparation for personal travel each year. Make a frequency distribution for the data. Then use the table to estimate the sample mean and the sample standard deviation of the data set. *(Adapted from Travel Industry Association of America)*

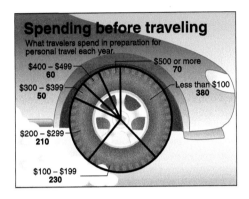

Spending before traveling
What travelers spend in preparation for personal travel each year.

$400 – $499 60
$500 or more 70
$300 – $399 50
Less than $100 380
$200 – $299 210
$100 – $199 230

▶ **Solution**

Begin by using a frequency distribution to organize the data.

Class	x	f	xf	$x - \bar{x}$	$(x - \bar{x})^2$	$(x - \bar{x})^2 f$
0–99	49.5	380	18,810	−142.5	20,306.25	7,716,375.0
100–199	149.5	230	34,385	−42.5	1806.25	415,437.5
200–299	249.5	210	52,395	57.5	3306.25	694,312.5
300–399	349.5	50	17,475	157.5	24,806.25	1,240,312.5
400–499	449.5	60	26,970	257.5	66,306.25	3,978,375.0
500+	599.5	70	41,965	407.5	166,056.25	11,623,937.5
		$\Sigma = 1000$	$\Sigma = 192,000$			$\Sigma = 25,668,750.0$

$$\bar{x} = \frac{\Sigma xf}{n} = \frac{192,000}{1000} = 192 \qquad \text{Sample mean}$$

Use the sum of squares to find the sample standard deviation.

$$s = \sqrt{\frac{\Sigma (x - \bar{x})^2 f}{n - 1}} = \sqrt{\frac{25,668,750}{999}} \approx 160.3 \qquad \text{Sample standard deviation}$$

So, the sample mean is $192 per year, and the sample standard deviation is about $160.30 per year.

▶ **Try It Yourself 10**

In the frequency distribution, 599.5 was chosen to represent the class of $500 or more. How would the sample mean and standard deviation change if you used 650 to represent this class?

a. Write the first four columns of a *frequency distribution*.
b. Find the *sample mean*.
c. Complete the *last three columns* of the frequency distribution.
d. Find the *sample standard deviation*.   *Answer: Page A34*

STUDY TIP

When a class is open, as in the last class, you must assign a single value to represent the midpoint. For this example, we selected 599.5.

2.4 EXERCISES

FOR EXTRA HELP:
MyStatLab

■ BUILDING BASIC SKILLS AND VOCABULARY

1. Explain how to find the range of a data set. What is an advantage of using the range as a measure of variation? What is a disadvantage?

2. Explain how to find the deviation of an entry in a data set. What is the sum of all the deviations in any data set?

3. Why is the standard deviation used more frequently than the variance? (*Hint:* Consider the units of the variance.)

4. Explain the relationship between variance and standard deviation. Can either of these measures be negative? Explain.

5. Construct a sample data set for which $n = 7$, $\bar{x} = 9$, and $s = 0$.

6. Construct a population data set for which $N = 6$, $\mu = 5$, and $\sigma = 2$.

7. Describe the difference between the calculation of population standard deviation and that of sample standard deviation.

8. Given a data set, how do you know whether to calculate σ or s?

9. Discuss the similarities and the differences between the Empirical Rule and Chebychev's Theorem.

10. What must you know about a data set before you can use the Empirical Rule?

In Exercises 11 and 12, find the range, mean, variance, and standard deviation of the population data set.

11. 9   5   9   10   11   12   7   7   8   12

12. 18   20   19   21   19   17   15
    17   25   22   19   20   16   18

In Exercises 13 and 14, find the range, mean, variance, and standard deviation of the sample data set.

13. 4   15   9   12   16   8   11   19   14

14. 28   25   21   15   7   14   9
    27   21   24   14   17   16

Graphical Reasoning *In Exercises 15–18, find the range of the data set represented by the display or graph.*

15.
2	3 9
3	0 0 2 3 6 7
4	0 1 2 3 3 8
5	0 1 1 9
6	1 2 9 9
7	5 9
8	4 8
9	0 2 5 6

Key: 2 | 3 = 23

16. **Bride's Age at First Marriage**

Age (in years)

17.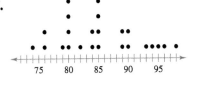

18.

0	5 5 9	Key: $0\vert5 = 0.5$
1	1 3 4 6 9	
2	2 5 7 9 9	
3	0 1 5 5 5 5	
4	7 7 9	
5		
6	3 4 7	

19. Archaeology   The depths (in inches) at which 10 artifacts are found are given below.

   20.7   24.8   30.5   26.2   36.0   34.3   30.3   29.5   27.0   38.5

(a) Find the range of the data set.

(b) Change 38.5 to 60.5 and find the range of the new data set.

20. In Exercise 19, compare your answer to part (a) with your answer to part (b). How do outliers affect the range of a data set?

▪ USING AND INTERPRETING CONCEPTS

21. Graphical Reasoning   Both data sets have a mean of 165. One has a standard deviation of 16, and the other has a standard deviation of 24. By looking at the graphs, which is which? Explain your reasoning.

(a)

12	8 9	Key: $12\vert8 = 128$
13	5 5 8	
14	1 2	
15	0 0 6 7	
16	4 5 9	
17	1 3 6 8	
18	0 8 9	
19	6	
20	3 5 7	

(b)

12		Key: $13\vert1 = 131$
13	1	
14	2 3 5	
15	0 4 5 6 8	
16	1 1 2 3 3 3	
17	1 5 8 8	
18	2 3 4 5	
19	0 2	
20		

22. Graphical Reasoning   Both data sets represented below have a mean of 50. One has a standard deviation of 2.4, and the other has a standard deviation of 5. By looking at the graphs, which is which? Explain your reasoning.

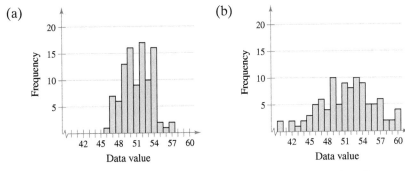

23. Salary Offers   You are applying for jobs at two companies. Company A offers starting salaries with $\mu = \$31{,}000$ and $\sigma = \$1000$. Company B offers starting salaries with $\mu = \$31{,}000$ and $\sigma = \$5000$. From which company are you more likely to get an offer of $\$33{,}000$ or more? Explain your reasoning.

24. Golf Strokes An Internet site compares the strokes per round for two professional golfers. Which golfer is more consistent: Player A with $\mu = 71.5$ strokes and $\sigma = 2.3$ strokes, or Player B with $\mu = 70.1$ strokes and $\sigma = 1.2$ strokes? Explain your reasoning.

Comparing Two Data Sets *In Exercises 25–28, you are asked to compare two data sets and interpret the results.*

25. Annual Salaries Sample annual salaries (in thousands of dollars) for accountants in Dallas and New York City are listed.

Dallas: 41.6   50.0   49.5   38.7   39.9   45.8   44.7   47.8   40.5
New York City: 45.6   41.5   57.6   55.1   59.3   59.0   50.6   47.2   42.3

(a) Find the mean, median, range, variance, and standard deviation of each data set.

(b) Interpret the results in the context of the real-life setting.

26. Annual Salaries Sample annual salaries (in thousands of dollars) for electrical engineers in Boston and Chicago are listed.

Boston: 70.4   84.2   58.5   64.5   71.6   79.9   88.3   80.1   69.9
Chicago: 69.4   71.5   65.4   59.9   70.9   68.5   62.9   70.1   60.9

(a) Find the mean, median, range, variance, and standard deviation of each data set.

(b) Interpret the results in the context of the real-life setting.

27. SAT Scores Sample SAT scores for eight males and eight females are listed.

Male SAT scores: 1520   1750   2120   1380   1982   1645   1033   1714
Female SAT scores: 1785   1507   1497   1952   2210   1871   1263   1588

(a) Find the mean, median, range, variance, and standard deviation of each data set.

(b) Interpret the results in the context of the real-life setting.

28. Batting Averages Sample batting averages for baseball players from two opposing teams are listed.

Team A: 0.295   0.310   0.325   0.272   0.256   0.297   0.320   0.384   0.235
Team B: 0.285   0.305   0.315   0.270   0.292   0.330   0.335   0.268   0.290

(a) Find the mean, median, range, variance, and standard deviation of each data set.

(b) Interpret the results in the context of the real-life setting.

Reasoning with Graphs *In Exercises 29–32, you are asked to compare three data sets. (a) Without calculating, determine which data set has the greatest sample standard deviation and which has the least sample standard deviation. Explain your reasoning. (b) How are the data sets the same? How do they differ?*

29. (i)   (ii)   (iii)

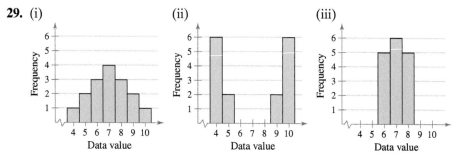

30. (i)

0	9
1	5 8
2	3 3 7 7
3	2 5
4	1

Key: 1|5 = 15

(ii)

0	9
1	5
2	3 3 3 7 7 7
3	5
4	1

Key: 1|5 = 15

(iii)

0	
1	5
2	3 3 3 3 7 7 7 7
3	5
4	

Key: 1|5 = 15

31. (i)   (ii)   (iii)

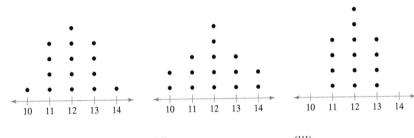

32. (i)   (ii)   (iii)

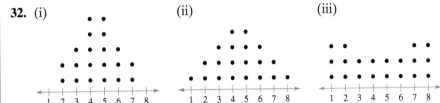

Using the Empirical Rule *In Exercises 33–38, you are asked to use the Empirical Rule.*

33. The mean value of land and buildings per acre from a sample of farms is $1500, with a standard deviation of $200. Estimate the percent of farms whose land and building values per acre are between $1300 and $1700. (Assume the data set has a bell-shaped distribution.)

34. The mean value of land and buildings per acre from a sample of farms is $2400, with a standard deviation of $450. Between what two values do about 95% of the data lie? (Assume the data set has a bell-shaped distribution.)

35. Using the sample statistics from Exercise 33, do the following. (Assume the number of farms in the sample is 75.)

(a) Estimate the number of farms whose land and building values per acre are between $1300 and $1700.

(b) If 25 additional farms were sampled, about how many of these farms would you expect to have land and building values between $1300 per acre and $1700 per acre?

36. Using the sample statistics from Exercise 34, do the following. (Assume the number of farms in the sample is 40.)

(a) Estimate the number of farms whose land and building values per acre are between $1500 and $3300.

(b) If 20 additional farms were sampled, about how many of these farms would you expect to have land and building values between $1500 per acre and $3300 per acre?

37. The land and building values per acre for eight more farms are listed. Using the sample statistics from Exercise 33, determine which of the data values are unusual. Are any of the data values very unusual? Explain.

$1150, $1775, $2180, $1000, $1475, $2000, $1850, $950

38. The land and building values per acre for eight more farms are listed. Using the sample statistics from Exercise 34, determine which of the data values are unusual. Are any of the data values very unusual? Explain.

$3325, $1045, $2450, $3200, $3800, $1490, $1675, $2950

39. Chebychev's Theorem   Old Faithful is a famous geyser at Yellowstone National Park. From a sample with $n = 32$, the mean duration of Old Faithful's eruptions is 3.32 minutes and the standard deviation is 1.09 minutes. Using Chebychev's Theorem, determine at least how many of the eruptions lasted between 1.14 minutes and 5.5 minutes. *(Source: Yellowstone National Park)*

40. Chebychev's Theorem   The mean time in a women's 400-meter dash is 57.07 seconds, with a standard deviation of 1.05 seconds. Apply Chebychev's Theorem to the data using $k = 2$. Interpret the results.

Calculating Using Grouped Data   *In Exercises 41–48, use the grouped data formulas to find the indicated mean and standard deviation.*

41. Pets per Household   The results of a random sample of the number of pets per household in a region are shown in the histogram. Estimate the sample mean and the sample standard deviation of the data set.

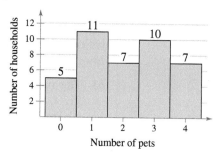

42. Cars per Household   The results of a random sample of the number of cars per household in a region are shown in the histogram. Estimate the sample mean and the sample standard deviation of the data set.

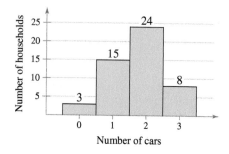

43. Football Wins   The number of regular season wins for each National Football League team in 2009 are listed. Make a frequency distribution (using five classes) for the data set. Then approximate the population mean and the population standard deviation of the data set. *(Source: National Football League)*

10	9	7	6	10	9	9	5	14	9	8	7
13	8	5	4	11	11	8	4	12	11	7	2
13	9	8	3	10	8	5	1				

44. Water Consumption   The number of gallons of water consumed per day by a small village are listed. Make a frequency distribution (using five classes) for the data set. Then approximate the population mean and the population standard deviation of the data set.

| 167 | 180 | 192 | 173 | 145 | 151 | 174 | 175 | 178 | 160 |
| 195 | 224 | 244 | 146 | 162 | 146 | 177 | 163 | 149 | 188 |

45. Amounts of Caffeine   The amounts of caffeine in a sample of five-ounce servings of brewed coffee are shown in the histogram. Make a frequency distribution for the data. Then use the table to estimate the sample mean and the sample standard deviation of the data set.

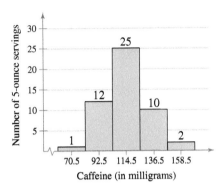

46. Supermarket Trips   Thirty people were randomly selected and asked how many trips to the supermarket they had made in the past week. The responses are shown in the histogram. Make a frequency distribution for the data. Then use the table to estimate the sample mean and the sample standard deviation of the data set.

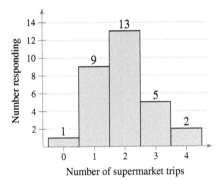

47. U.S. Population   The estimated distribution (in millions) of the U.S. population by age for the year 2015 is shown in the pie chart. Make a frequency distribution for the data. Then use the table to estimate the sample mean and the sample standard deviation of the data set. Use 70 as the midpoint for "65 years and over." *(Source: Population Division, U.S. Census Bureau)*

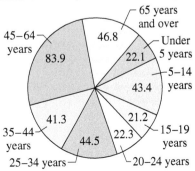

48. Brazil's Population

Brazil's estimated population for the year 2015 is shown in the histogram. Make a frequency distribution for the data. Then use the table to estimate the sample mean and the sample standard deviation of the data set. *(Adapted from U.S. Census Bureau, International Data Base)*

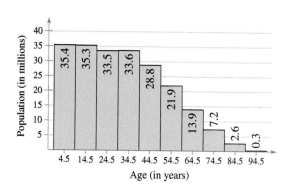

SC *In Exercises 49 and 50, use StatCrunch to find the sample size, mean, variance, standard deviation, median, range, minimum data value, and maximum data value of the data.*

49. The data represent the total amounts (in dollars) spent by several families at a restaurant.

49   56   75   64   55   49   62   89   30   34   60   52   60   72   75

50. The data represent the prices (in dollars) of several Hewlett-Packard office printers. *(Source: Hewlett-Packard)*

199.99   499.99   149.99   119.99   129.99   229.99
179.99   89.99   299.99   249.99   349.99   99.99

■ EXTENDING CONCEPTS

51. Coefficient of Variation The **coefficient of variation** CV describes the standard deviation as a percent of the mean. Because it has no units, you can use the coefficient of variation to compare data with different units.

$$CV = \frac{\text{Standard deviation}}{\text{Mean}} \times 100\%$$

The table at the left shows the heights (in inches) and weights (in pounds) of the members of a basketball team. Find the coefficient of variation for each data set. What can you conclude?

52. Shortcut Formula You used $SS_x = \sum (x - \bar{x})^2$ when calculating variance and standard deviation. An alternative formula that is sometimes more convenient for hand calculations is

$$SS_x = \sum x^2 - \frac{(\sum x)^2}{n}.$$

You can find the sample variance by dividing the sum of squares by $n - 1$ and the sample standard deviation by finding the square root of the sample variance.

(a) Use the shortcut formula to calculate the sample standard deviations for the data sets given in Exercise 27.

(b) Compare your results with those obtained in Exercise 27.

Heights	Weights
72	180
74	168
68	225
76	201
74	189
69	192
72	197
79	162
70	174
69	171
77	185
73	210

TABLE FOR EXERCISE 51

53. **Scaling Data** Sample annual salaries (in thousands of dollars) for employees at a company are listed.

42 36 48 51 39 39 42 36 48 33 39 42 45

   (a) Find the sample mean and sample standard deviation.
   (b) Each employee in the sample is given a 5% raise. Find the sample mean and sample standard deviation for the revised data set.
   (c) To calculate the monthly salary, divide each original salary by 12. Find the sample mean and sample standard deviation for the revised data set.
   (d) What can you conclude from the results of (a), (b), and (c)?

54. **Shifting Data** Sample annual salaries (in thousands of dollars) for employees at a company are listed.

40 35 49 53 38 39 40 37 49 34 38 43 47

   (a) Find the sample mean and sample standard deviation.
   (b) Each employee in the sample is given a $1000 raise. Find the sample mean and sample standard deviation for the revised data set.
   (c) Each employee in the sample takes a pay cut of $2000 from their original salary. Find the sample mean and sample standard deviation for the revised data set.
   (d) What can you conclude from the results of (a), (b), and (c)?

55. **Mean Absolute Deviation** Another useful measure of variation for a data set is the **mean absolute deviation** (*MAD*). It is calculated by the formula

$$\frac{\sum |x - \bar{x}|}{n}.$$

   (a) Find the mean absolute deviations of the data sets in Exercise 27. Compare your results with the sample standard deviation.
   (b) Find the mean absolute deviations of the data sets in Exercise 28. Compare your results with the sample standard deviation.

56. **Chebychev's Theorem** At least 99% of the data in any data set lie within how many standard deviations of the mean? Explain how you obtained your answer.

57. **Pearson's Index of Skewness** The English statistician Karl Pearson (1857–1936) introduced a formula for the skewness of a distribution.

$$P = \frac{3(\bar{x} - \text{median})}{s} \qquad \text{Pearson's index of skewness}$$

Most distributions have an index of skewness between −3 and 3. When $P > 0$, the data are skewed right. When $P < 0$, the data are skewed left. When $P = 0$, the data are symmetric. Calculate the coefficient of skewness for each distribution. Describe the shape of each.

   (a) $\bar{x} = 17$, $s = 2.3$, median $= 19$
   (b) $\bar{x} = 32$, $s = 5.1$, median $= 25$
   (c) $\bar{x} = 9.2$, $s = 1.8$, median $= 9.2$
   (d) $\bar{x} = 42$, $s = 6.0$, median $= 40$

ACTIVITY 2.4 Standard Deviation

 APPLET

The *standard deviation* applet is designed to allow you to investigate interactively the standard deviation as a measure of spread for a data set. Points can be added to the plot by clicking the mouse above the horizontal axis. The mean of the points is shown as a green arrow. A numeric value for the standard deviation is shown above the plot. Points on the plot can be removed by clicking on the point and then dragging the point into the trash can. All of the points on the plot can be removed by simply clicking inside the trash can. The range of values for the horizontal axis can be specified by inputting lower and upper limits and then clicking UPDATE.

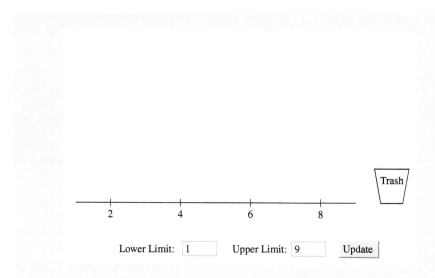

■ Explore

Step 1 Specify a lower limit.
Step 2 Specify an upper limit.
Step 3 Add 15 points to the plot.
Step 4 Remove all of the points from the plot.

■ Draw Conclusions

 APPLET

1. Specify the lower limit to be 10 and the upper limit to be 20. Plot 10 points that have a mean of about 15 and a standard deviation of about 3. Write the estimates of the values of the points. Plot a point with a value of 15. What happens to the mean and standard deviation? Plot a point with a value of 20. What happens to the mean and standard deviation?

2. Specify the lower limit to be 30 and the upper limit to be 40. How can you plot eight points so that the points have the largest possible standard deviation? Use the applet to plot the set of points and then use the formula for standard deviation to confirm the value given in the applet. How can you plot eight points so that the points have the lowest possible standard deviation? Explain.

Earnings of Athletes

The earnings of professional athletes in different sports can vary. An athlete can be paid a base salary, earn signing bonuses upon signing a new contract, or even earn money by finishing in a certain position in a race or tournament. The data shown below are the earnings (for performance only, no endorsements) from Major League Baseball (MLB), Major League Soccer (MLS), the National Basketball Association (NBA), the National Football League (NFL), the National Hockey League (NHL), the National Association for Stock Car Auto Racing (NASCAR), and the Professional Golf Association Tour (PGA) for a recent year.

Organization	Number of players
MLB	858
MLS	410
NBA	463
NFL	1861
NHL	722
NASCAR	76
PGA	262

Number of Players Separated into Earnings Ranges

Organization	$0–$500,000	$500,001–$2,000,000	$2,000,001–$6,000,000	$6,000,001–$10,000,000	$10,000,001 +
MLB	353	182	164	85	74
MLS	403	5	1	1	0
NBA	35	157	137	77	57
NFL	554	746	438	85	38
NHL	42	406	237	37	0
NASCAR	23	16	31	6	0
PGA	110	115	36	1	0

■ EXERCISES

1. **Revenue** Which organization had the greatest total player earnings? Explain your reasoning.

2. **Mean Earnings** Estimate the mean earnings of a player in each organization. Use $19,000,000 as the midpoint for $10,000,001+.

3. **Revenue** Which organization had the greatest earnings per player? Explain your reasoning.

4. **Standard Deviation** Estimate the standard deviation for the earnings of a player in each organization. Use $19,000,000 as the midpoint for $10,000,001+.

5. **Standard Deviation** Which organization had the greatest standard deviation? Explain your reasoning.

6. **Bell-Shaped Distribution** Of the seven organizations, which is most bell-shaped? Explain your reasoning.

2.5 Measures of Position

Quartiles ▸ Percentiles and Other Fractiles ▸ The Standard Score

WHAT YOU SHOULD LEARN

▸ How to find the first, second, and third quartiles of a data set

▸ How to find the interquartile range of a data set

▸ How to represent a data set graphically using a box-and-whisker plot

▸ How to interpret other fractiles such as percentiles

▸ How to find and interpret the standard score (*z*-score)

▸ QUARTILES

In this section, you will learn how to use fractiles to specify the position of a data entry within a data set. **Fractiles** are numbers that partition, or divide, an ordered data set into equal parts. For instance, the median is a fractile because it divides an ordered data set into two equal parts.

DEFINITION

The three **quartiles**, Q_1, Q_2, and Q_3, approximately divide an ordered data set into four equal parts. About one quarter of the data fall on or below the **first quartile** Q_1. About one half of the data fall on or below the **second quartile** Q_2 (the second quartile is the same as the median of the data set). About three quarters of the data fall on or below the **third quartile** Q_3.

EXAMPLE 1   SC Report 14

▸ **Finding the Quartiles of a Data Set**

The number of nuclear power plants in the top 15 nuclear power-producing countries in the world are listed. Find the first, second, and third quartiles of the data set. What can you conclude? *(Source: International Atomic Energy Agency)*

| 7 | 18 | 11 | 6 | 59 | 17 | 18 | 54 | 104 | 20 | 31 | 8 | 10 | 15 | 19 |

▸ **Solution**

First, order the data set and find the median Q_2. Once you find Q_2, divide the data set into two halves. The first and third quartiles are the medians of the lower and upper halves of the data set.

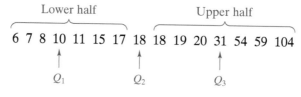

Lower half   Upper half

6 7 8 10 11 15 17 18 18 19 20 31 54 59 104

Q_1 Q_2 Q_3

Interpretation   About one fourth of the countries have 10 or fewer nuclear power plants; about one half have 18 or fewer; and about three fourths have 31 or fewer.

▸ **Try It Yourself 1**

Find the first, second, and third quartiles for the ages of the 50 richest people using the data set listed in the Chapter Opener on page 37. What can you conclude?

a. *Order* the data set.
b. Find the *median* Q_2.
c. Find the *first and third quartiles*, Q_1 and Q_3.
d. *Interpret* the results in the context of the data.   *Answer: Page A34*

EXAMPLE 2

▸ Using Technology to Find Quartiles

The tuition costs (in thousands of dollars) for 25 liberal arts colleges are listed. Use a calculator or a computer to find the first, second, and third quartiles. What can you conclude?

23 25 30 23 20 22 21 15 25 24 30 25 30
20 23 29 20 19 22 23 29 23 28 22 28

▸ Solution

MINITAB, Excel, and the TI-83/84 Plus each have features that automatically calculate quartiles. Try using this technology to find the first, second, and third quartiles of the tuition data. From the displays, you can see that $Q_1 = 21.5$, $Q_2 = 23$, and $Q_3 = 28$.

STUDY TIP

There are several ways to find the quartiles of a data set. Regardless of how you find the quartiles, the results are rarely off by more than one data entry. For instance, in Example 2, the first quartile, as determined by Excel, is 22 instead of 21.5.

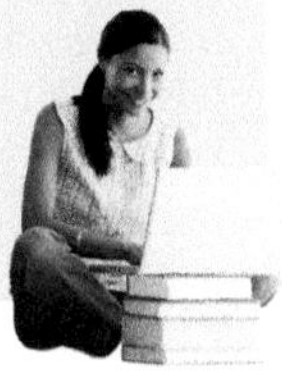

MINITAB

Descriptive Statistics: Tuition

Variable	N	Mean	SE Mean	StDev	Minimum
Tuition	25	23.960	0.788	3.942	15.000

Variable	Q1	Median	Q3	Maximum
Tuition	21.500	23.000	28.000	30.000

EXCEL

	A	B	C	D
1	23			
2	25		Quartile(A1:A25,1)	
3	30		22	
4	23			
5	20		Quartile(A1:A25,2)	
6	22		23	
7	21			
8	15		Quartile(A1:A25,3)	
9	25		28	
10	24			
11	30			
12	25			
13	30			
14	20			
15	23			
16	29			
17	20			
18	19			
19	22			
20	23			
21	29			
22	23			
23	28			
24	22			
25	28			

TI-83/84 PLUS

1-Var Stats
↑n=25
minX=15
Q_1=21.5
Med=23
Q_3=28
maxX=30

Interpretation About one quarter of these colleges charge tuition of \$21,500 or less; one half charge \$23,000 or less; and about three quarters charge \$28,000 or less.

▶ Try It Yourself 2

The tuition costs (in thousands of dollars) for 25 universities are listed. Use a calculator or a computer to find the first, second, and third quartiles. What can you conclude?

20 26 28 25 31 14 23 15 12 26 29 24 31
19 31 17 15 17 20 31 32 16 21 22 28

a. *Enter* the data.
b. Calculate the *first, second, and third quartiles.*
c. *Interpret* the results in the context of the data.   *Answer: Page A34*

After finding the quartiles of a data set, you can find the *interquartile range.*

DEFINITION

The **interquartile range (IQR)** of a data set is a measure of variation that gives the range of the middle 50% of the data. It is the difference between the third and first quartiles.

$$\text{Interquartile range (IQR)} = Q_3 - Q_1$$

EXAMPLE 3

▶ **Finding the Interquartile Range**

Find the interquartile range of the data set given in Example 1. What can you conclude from the result?

▶ **Solution**

From Example 1, you know that $Q_1 = 10$ and $Q_3 = 31$. So, the interquartile range is

$$\text{IQR} = Q_3 - Q_1 = 31 - 10 = 21.$$

Interpretation The number of power plants in the middle portion of the data set vary by at most 21.

▶ Try It Yourself 3

Find the interquartile range for the ages of the 50 richest people listed in the Chapter Opener on page 37.

a. Find the *first and third quartiles*, Q_1 and Q_3.
b. *Subtract* Q_1 from Q_3.
c. *Interpret* the result in the context of the data.   *Answer: Page A34*

The IQR can also be used to identify outliers. First, multiply the IQR by 1.5. Then subtract that value from Q_1, and add that value to Q_3. Any data value that is smaller than $Q_1 - 1.5(\text{IQR})$ or larger than $Q_3 + 1.5(\text{IQR})$ is an outlier. For instance, the IQR in Example 1 is $31 - 10 = 21$ and $1.5(21) = 31.5$. So, adding 31.5 to Q_3 gives $Q_3 + 31.5 = 31 + 31.5 = 62.5$. Because $104 > 62.5$, 104 is an outlier.

Another important application of quartiles is to represent data sets using box-and-whisker plots. A **box-and-whisker plot** (or **boxplot**) is an exploratory data analysis tool that highlights the important features of a data set. To graph a box-and-whisker plot, you must know the following values.

PICTURING THE WORLD

Of the first 44 Super Bowls played, Super Bowl XIV had the highest attendance at about 104,000. Super Bowl I had the lowest attendance at about 62,000. The box-and-whisker plot summarizes the attendances (in thousands of people) at the 44 Super Bowls. (Source: National Football League)

Super Bowl Attendance

About how many Super Bowl attendances are represented by the right whisker? About how many are represented by the left whisker?

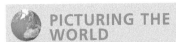

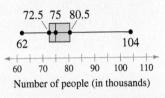

1. The minimum entry
2. The first quartile Q_1
3. The median Q_2
4. The third quartile Q_3
5. The maximum entry

These five numbers are called the **five-number summary** of the data set.

GUIDELINES

Drawing a Box-and-Whisker Plot

1. Find the five-number summary of the data set.
2. Construct a horizontal scale that spans the range of the data.
3. Plot the five numbers above the horizontal scale.
4. Draw a box above the horizontal scale from Q_1 to Q_3 and draw a vertical line in the box at Q_2.
5. Draw whiskers from the box to the minimum and maximum entries.

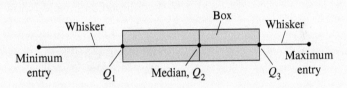

EXAMPLE 4

SC Report 15

▸ Drawing a Box-and-Whisker Plot

Draw a box-and-whisker plot that represents the data set given in Example 1. What can you conclude from the display?

See MINITAB and TI-83/84 Plus steps on pages 122 and 123.

▸ Solution

The five-number summary of the data set is displayed below. Using these five numbers, you can construct the box-and-whisker plot shown.

Min = 6, $Q_1 = 10$, $Q_2 = 18$, $Q_3 = 31$, Max = 104,

Number of Power Plants

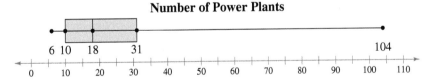

Interpretation You can make several conclusions from the display. One is that about half the data values are between 10 and 31. By looking at the length of the right whisker, you can also conclude that the data value of 104 is a possible outlier.

INSIGHT

You can use a box-and-whisker plot to determine the shape of a distribution. Notice that the box-and-whisker plot in Example 4 represents a distribution that is skewed right.

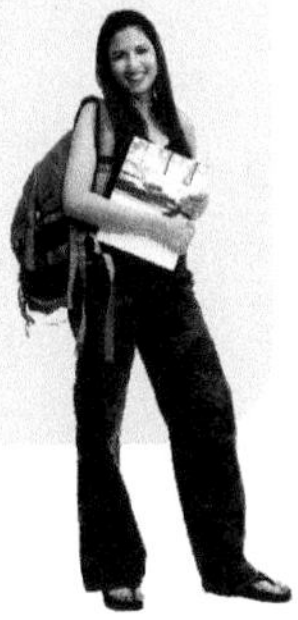

▸ Try It Yourself 4

Draw a box-and-whisker plot that represents the ages of the 50 richest people listed in the Chapter Opener on page 37. What can you conclude?

a. Find the *five-number summary* of the data set.
b. Construct a *horizontal scale* and *plot* the five numbers above it.
c. Draw the *box*, the *vertical line*, and the *whiskers*.
d. Make some *conclusions*.

Answer: Page A34

▶ PERCENTILES AND OTHER FRACTILES

In addition to using quartiles to specify a measure of position, you can also use percentiles and deciles. These common fractiles are summarized as follows.

Fractiles	Summary	Symbols
Quartiles	Divide a data set into 4 equal parts.	Q_1, Q_2, Q_3
Deciles	Divide a data set into 10 equal parts.	$D_1, D_2, D_3, \ldots, D_9$
Percentiles	Divide a data set into 100 equal parts.	$P_1, P_2, P_3, \ldots, P_{99}$

Percentiles are often used in education and health-related fields to indicate how one individual compares with others in a group. They can also be used to identify unusually high or unusually low values. For instance, test scores and children's growth measurements are often expressed in percentiles. Scores or measurements in the 95th percentile and above are unusually high, while those in the 5th percentile and below are unusually low.

EXAMPLE 5

▶ Interpreting Percentiles

The ogive at the right represents the cumulative frequency distribution for SAT test scores of college-bound students in a recent year. What test score represents the 62nd percentile? How should you interpret this? *(Source: The College Board)*

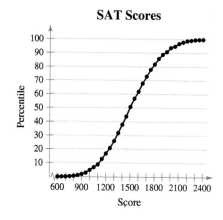

SAT Scores

▶ Solution

From the ogive, you can see that the 62nd percentile corresponds to a test score of 1600.

Interpretation This means that approximately 62% of the students had an SAT score of 1600 or less.

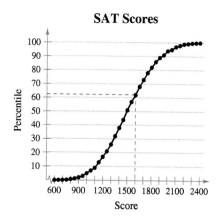

SAT Scores

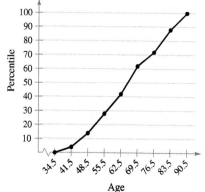

Ages of the 50 Richest People

▶ Try It Yourself 5

The ages of the 50 richest people are represented in the cumulative frequency graph at the left. At what percentile is someone who is 66 years old? How should you interpret this?

a. *Use the graph* to find the percentile that corresponds to the given age.
b. *Interpret* the results in the context of the data. *Answer: Page A34*

▸ **THE STANDARD SCORE**

When you know the mean and standard deviation of a data set, you can measure a data value's position in the data set with a *standard score*, or *z-score*.

> **DEFINITION**
>
> The **standard score**, or **z-score**, represents the number of standard deviations a given value x falls from the mean μ. To find the z-score for a given value, use the following formula.
>
> $$z = \frac{\text{Value} - \text{Mean}}{\text{Standard deviation}} = \frac{x - \mu}{\sigma}$$

A z-score can be negative, positive, or zero. If z is negative, the corresponding x-value is less than the mean. If z is positive, the corresponding x-value is greater than the mean. And if $z = 0$, the corresponding x-value is equal to the mean. A z-score can be used to identify an unusual value of a data set that is approximately bell-shaped.

EXAMPLE 6

▸ **Finding z-Scores**

The mean speed of vehicles along a stretch of highway is 56 miles per hour with a standard deviation of 4 miles per hour. You measure the speeds of three cars traveling along this stretch of highway as 62 miles per hour, 47 miles per hour, and 56 miles per hour. Find the z-score that corresponds to each speed. What can you conclude?

▸ **Solution**

The z-score that corresponds to each speed is calculated below.

$x = 62$ mph	$x = 47$ mph	$x = 56$ mph
$z = \dfrac{62 - 56}{4} = 1.5$	$z = \dfrac{47 - 56}{4} = -2.25$	$z = \dfrac{56 - 56}{4} = 0$

Interpretation   From the z-scores, you can conclude that a speed of 62 miles per hour is 1.5 standard deviations above the mean; a speed of 47 miles per hour is 2.25 standard deviations below the mean; and a speed of 56 miles per hour is equal to the mean. If the distribution of the speeds is approximately bell-shaped, the car traveling 47 miles per hour is said to be traveling unusually slowly, because its speed corresponds to a z-score of -2.25.

▸ **Try It Yourself 6**

The monthly utility bills in a city have a mean of $70 and a standard deviation of $8. Find the z-scores that correspond to utility bills of $60, $71, and $92. What can you conclude?

a. *Identify* μ and σ. *Transform* each value to a z-score.
b. *Interpret* the results.

Answer: Page A34

When a distribution is approximately bell-shaped, you know from the Empirical Rule that about 95% of the data lie within 2 standard deviations of the mean. So, when this distribution's values are transformed to z-scores, about 95% of the z-scores should fall between -2 and 2. A z-score outside of this range will occur about 5% of the time and would be considered unusual. So, according to the Empirical Rule, a z-score less than -3 or greater than 3 would be very unusual, with such a score occurring about 0.3% of the time.

In Example 6, you used *z*-scores to compare data values within the same data set. You can also use *z*-scores to compare data values from different data sets.

EXAMPLE 7

▶ **Comparing *z*-Scores from Different Data Sets**

In 2009, Heath Ledger won the Oscar for Best Supporting Actor at age 29 for his role in the movie *The Dark Knight*. Penelope Cruz won the Oscar for Best Supporting Actress at age 34 for her role in *Vicky Cristina Barcelona*. The mean age of all Best Supporting Actor winners is 49.5, with a standard deviation of 13.8. The mean age of all Best Supporting Actress winners is 39.9, with a standard deviation of 14.0. Find the *z*-scores that correspond to the ages of Ledger and Cruz. Then compare your results.

▶ **Solution**

The *z*-scores that correspond to the ages of the two performers are calculated below.

$$\textbf{\textit{Heath Ledger}} \quad z = \frac{x - \mu}{\sigma}$$

$$= \frac{29 - 49.5}{13.8}$$

$$\approx -1.49$$

$$\textbf{\textit{Penelope Cruz}} \quad z = \frac{x - \mu}{\sigma}$$

$$= \frac{34 - 39.9}{14.0}$$

$$\approx -0.42$$

The age of Heath Ledger was 1.49 standard deviations below the mean, and the age of Penelope Cruz was 0.42 standard deviation below the mean.

Interpretation   Compared with other Best Supporting Actor winners, Heath Ledger was relatively younger, whereas the age of Penelope Cruz was only slightly lower than the average age of other Best Supporting Actress winners. Both *z*-scores fall between −2 and 2, so neither score would be considered unusual.

▶ **Try It Yourself 7**

In 2009, Sean Penn won the Oscar for Best Actor at age 48 for his role in the movie *Milk*. Kate Winslet won the Oscar for Best Actress at age 33 for her role in *The Reader*. The mean age of all Best Actor winners is 43.7, with a standard deviation of 8.7. The mean age of all Best Actress winners is 35.9, with a standard deviation of 11.4. Find the *z*-scores that correspond to the ages of Penn and Winslet. Then compare your results.

a. *Identify* μ and σ for each data set.
b. *Transform* each value to a *z*-score.
c. *Compare* your results.

Answer: Page A34

2.5 EXERCISES

■ BUILDING BASIC SKILLS AND VOCABULARY

1. The goals scored per game by a soccer team represent the first quartile for all teams in a league. What can you conclude about the team's goals scored per game?

2. A salesperson at a company sold \$6,903,435 of hardware equipment last year, a figure that represented the eighth decile of sales performance at the company. What can you conclude about the salesperson's performance?

3. A student's score on an actuarial exam is in the 78th percentile. What can you conclude about the student's exam score?

4. A counselor tells a child's parents that their child's IQ is in the 93rd percentile for the child's age group. What can you conclude about the child's IQ?

5. Explain how the interquartile range of a data set can be used to identify outliers.

6. Describe the relationship between quartiles and percentiles.

True or False?   *In Exercises 7–14, determine whether the statement is true or false. If it is false, rewrite it as a true statement.*

7. The mean and median of a data set are both fractiles.

8. About one quarter of a data set falls below Q_1.

9. The second quartile is the median of an ordered data set.

10. The five numbers you need to graph a box-and-whisker plot are the minimum, the maximum, Q_1, Q_3, and the mean.

11. The 50th percentile is equivalent to Q_1.

12. It is impossible to have a z-score of 0.

13. A z-score of -2.5 is considered very unusual.

14. A z-score of 1.99 is considered usual.

■ USING AND INTERPRETING CONCEPTS

Graphical Analysis   *In Exercises 15–20, use the box-and-whisker plot to identify (a) the five-number summary, and (b) the interquartile range.*

15.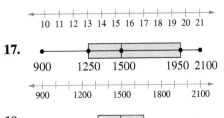

16.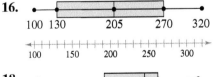

17.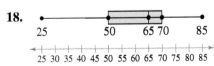

18.

19.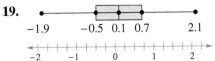

20.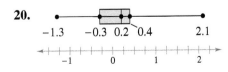

In Exercises 21–24, (a) find the five-number summary, and (b) draw a box-and-whisker plot of the data.

21. 39 36 30 27 26 24 28 35 39 60 50 41 35 32 51

22. 171 176 182 150 178 180 173 170 174 178 181 180

 23. 4 7 7 5 2 9 7 6 8 5 8 4 1 5 2 8 7 6 6 9

24. 2 7 1 3 1 2 8 9 9 2 5 4 7 3 7 5 4 7
    2 3 5 9 5 6 3 9 3 4 9 8 8 2 3 9 5

Interpreting Graphs *In Exercises 25–28, use the box-and-whisker plot to determine if the shape of the distribution represented is symmetric, skewed left, skewed right, or none of these. Justify your answer.*

25.

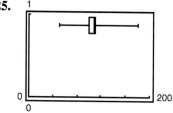

26.

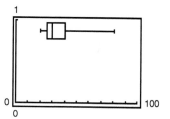

27.

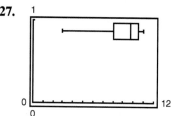

28.

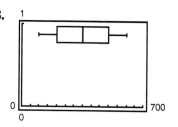

29. Graphical Analysis The letters A, B, and C are marked on the histogram. Match them with Q_1, Q_2 (the median), and Q_3. Justify your answer.

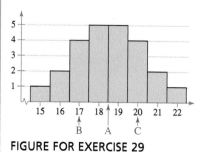

FIGURE FOR EXERCISE 29

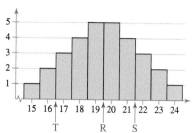

FIGURE FOR EXERCISE 30

30. Graphical Analysis The letters R, S, and T are marked on the histogram. Match them with P_{10}, P_{50}, and P_{80}. Justify your answer.

Using Technology to Find Quartiles and Draw Graphs *In Exercises 31–34, use a calculator or a computer to (a) find the data set's first, second, and third quartiles, and (b) draw a box-and-whisker plot that represents the data set.*

31. TV Viewing The number of hours of television watched per day by a sample of 28 people

    2 4 1 5 7 2 5 4 4 2 3 6 4 3
    5 2 0 3 5 9 4 5 2 1 3 6 7 2

32. Vacation Days The number of vacation days used by a sample of 20 employees in a recent year

3 9   2 1 7 5 3 2 2 6
4 0 10 0 3 5 7 8 6 5

33. Airplane Distances The distances (in miles) from an airport of a sample of 22 inbound and outbound airplanes

2.8 2.0 3.0 3.0 3.2 5.9 3.5 3.6
1.8 5.5 3.7 5.2 3.8 3.9 6.0 2.5
4.0 4.1 4.6 5.0 5.5 6.0

34. Hourly Earnings The hourly earnings (in dollars) of a sample of 25 railroad equipment manufacturers

15.60 18.75 14.60 15.80 14.35 13.90 17.50 17.55 13.80
14.20 19.05 15.35 15.20 19.45 15.95 16.50 16.30 15.25
15.05 19.10 15.20 16.22 17.75 18.40 15.25

35. TV Viewing Refer to the data set given in Exercise 31 and the box-and-whisker plot you drew that represents the data set.

(a) About 75% of the people watched no more than how many hours of television per day?

(b) What percent of the people watched more than 4 hours of television per day?

(c) If you randomly selected one person from the sample, what is the likelihood that the person watched less than 2 hours of television per day? Write your answer as a percent.

36. Manufacturer Earnings Refer to the data set given in Exercise 34 and the box-and-whisker plot you drew that represents the data set.

(a) About 75% of the manufacturers made less than what amount per hour?

(b) What percent of the manufacturers made more than $15.80 per hour?

(c) If you randomly selected one manufacturer from the sample, what is the likelihood that the manufacturer made less than $15.80 per hour? Write your answer as a percent.

Graphical Analysis *In Exercises 37 and 38, the midpoints A, B, and C are marked on the histogram. Match them with the indicated z-scores. Which z-scores, if any, would be considered unusual?*

37. $z = 0$ **38.** $z = 0.77$

$z = 2.14$ $z = 1.54$

$z = -1.43$ $z = -1.54$

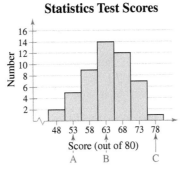

Statistics Test Scores

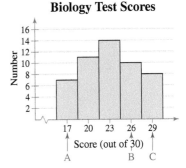

Biology Test Scores

Comparing Test Scores *For the statistics test scores in Exercise 37, the mean is 63 and the standard deviation is 7.0, and for the biology test scores in Exercise 38, the mean is 23 and the standard deviation is 3.9. In Exercises 39–42, you are given the test scores of a student who took both tests.*

(a) *Transform each test score to a z-score.*

(b) *Determine on which test the student had a better score.*

39. A student gets a 75 on the statistics test and a 25 on the biology test.

40. A student gets a 60 on the statistics test and a 22 on the biology test.

41. A student gets a 78 on the statistics test and a 29 on the biology test.

42. A student gets a 63 on the statistics test and a 23 on the biology test.

43. Life Spans of Tires A certain brand of automobile tire has a mean life span of 35,000 miles, with a standard deviation of 2250 miles. (Assume the life spans of the tires have a bell-shaped distribution.)

(a) The life spans of three randomly selected tires are 34,000 miles, 37,000 miles, and 30,000 miles. Find the z-score that corresponds to each life span. According to the z-scores, would the life spans of any of these tires be considered unusual?

(b) The life spans of three randomly selected tires are 30,500 miles, 37,250 miles, and 35,000 miles. Using the Empirical Rule, find the percentile that corresponds to each life span.

44. Life Spans of Fruit Flies The life spans of a species of fruit fly have a bell-shaped distribution, with a mean of 33 days and a standard deviation of 4 days.

(a) The life spans of three randomly selected fruit flies are 34 days, 30 days, and 42 days. Find the z-score that corresponds to each life span and determine if any of these life spans are unusual.

(b) The life spans of three randomly selected fruit flies are 29 days, 41 days, and 25 days. Using the Empirical Rule, find the percentile that corresponds to each life span.

Interpreting Percentiles *In Exercises 45–50, use the cumulative frequency distribution to answer the questions. The cumulative frequency distribution represents the heights of males in the United States in the 20–29 age group. The heights have a bell-shaped distribution (see Picturing the World, page 86) with a mean of 69.9 inches and a standard deviation of 3.0 inches. (Adapted from National Center for Health Statistics)*

45. What height represents the 60th percentile? How should you interpret this?

46. What percentile is a height of 77 inches? How should you interpret this?

47. Three adult males in the 20–29 age group are randomly selected. Their heights are 74 inches, 62 inches, and 80 inches. Use z-scores to determine which heights, if any, are unusual.

48. Three adult males in the 20–29 age group are randomly selected. Their heights are 70 inches, 66 inches, and 68 inches. Use z-scores to determine which heights, if any, are unusual.

49. Find the z-score for a male in the 20–29 age group whose height is 71.1 inches. What percentile is this?

50. Find the z-score for a male in the 20–29 age group whose height is 66.3 inches. What percentile is this?

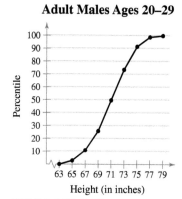

Adult Males Ages 20–29

Percentile / Height (in inches)

FIGURE FOR EXERCISES 45–50

Over the hill or on top?

Number of 100 top executives
in the following age groups:

TOP EXECUTIVES

		36	31			

13

16

2   1   1

24.5 34.5 44.5 54.5 64.5 74.5 84.5
Age

FIGURE FOR EXERCISE 51

■ EXTENDING CONCEPTS

51. Ages of Executives The ages of a sample of 100 executives are listed.

```
31  62  51  44  61  47  49  45  40  52  60  51  67
47  63  54  59  43  63  52  50  54  61  41  48  49
51  54  39  54  47  52  36  53  74  33  53  68  44
40  60  42  50  48  42  42  36  57  42  48  56  51
54  42  27  43  43  41  54  49  49  47  51  28  54
36  36  41  60  55  42  59  35  65  48  56  82  39
54  49  61  56  57  32  38  48  64  51  45  46  62
63  59  63  32  47  40  37  49  57
```

(a) Find the five-number summary.

(b) Draw a box-and-whisker plot that represents the data set.

(c) Interpret the results in the context of the data.

(d) On the basis of this sample, at what age would you expect to be an executive? Explain your reasoning.

(e) Which age groups, if any, can be considered unusual? Explain your reasoning.

Midquartile *Another measure of position is called the* **midquartile.** *You can find the midquartile of a data set by using the following formula.*

$$Midquartile = \frac{Q_1 + Q_3}{2}$$

In Exercises 52–55, find the midquartile of the given data set.

52. 5 7 1 2 3 10 8 7 5 3

53. 23 36 47 33 34 40 39 24 32 22 38 41

54. 12.3 9.7 8.0 15.4 16.1 11.8 12.7 13.4
12.2 8.1 7.9 10.3 11.2

55. 21.4 20.8 19.7 15.2 31.9 18.7 15.6 16.7
19.8 13.4 22.9 28.7 19.8 17.2 30.1

56. Song Lengths **Side-by-side box-and-whisker plots** can be used to compare two or more different data sets. Each box-and-whisker plot is drawn on the same number line to compare the data sets more easily. The lengths (in seconds) of songs played at two different concerts are shown.

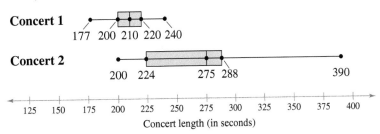

(a) Describe the shape of each distribution. Which concert has less variation in song lengths?

(b) Which distribution is more likely to have outliers? Explain your reasoning.

(c) Which concert do you think has a standard deviation of 16.3? Explain your reasoning.

(d) Can you determine which concert lasted longer? Explain.

57. Credit Card Purchases   The monthly credit card purchases (rounded to the nearest dollar) over the last two years for you and a friend are listed.

You: 60   95   102   110   130   130   162   200   215   120   124   28
          58   40   102   105   141   160   130   210   145   90   46   76

Friend: 100   125   132   90   85   75   140   160   180   190   160   105
             145   150   151   82   78   115   170   158   140   130   165   125

Use a calculator or a computer to draw a side-by-side box-and-whisker plot that represents the data sets. Then describe the shapes of the distributions.

Finding Percentiles   *You can find the percentile that corresponds to a specific data value x by using the following formula, then rounding the result to the nearest whole number.*

$$Percentile \ of \ x = \frac{number \ of \ data \ values \ less \ than \ x}{total \ number \ of \ data \ values} \cdot 100$$

In Exercises 58 and 59, use the information from Example 7 and the fact that there have been 73 Oscars for Best Supporting Actor and 73 Oscars for Best Supporting Actress awarded.

58. Only three winners were younger than Heath Ledger when they won the Oscar for Best Supporting Actor. Find the percentile that corresponds to Heath Ledger's age.

59. Forty-three winners were older than Penelope Cruz when they won the Oscar for Best Supporting Actress. Find the percentile that corresponds to Penelope Cruz's age.

Modified Boxplot   *A **modified boxplot** is a boxplot that uses symbols to identify outliers. The horizontal line of a modified boxplot extends as far as the minimum data value that is not an outlier and the maximum data value that is not an outlier. In Exercises 60 and 61, (a) identify any outliers (using the 1.5 × IQR rule), and (b) draw a modified boxplot that represents the data set. Use asterisks (*) to identify outliers.*

60. 16   9   11   12   8   10   12   13   11   10   24   9   2   15   7

61. 75   78   80   75   62   72   74   75   80   95   76   72

SC *In Exercises 62 and 63, use StatCrunch to (a) find the five-number summary, (b) construct a regular boxplot, and (c) construct a modified boxplot for the data.*

62. The data represent the speeds (in miles per hour) of several vehicles.

   68   88   70   72   70   69   72   62   65   70   75   52   65

63. The data represent the weights (in pounds) of several professional football players.

   225   250   305   285   275   265   290   310   290   250   210   225
   308   325   260   165   195   245   235   298   395   255   268   190

USES AND ABUSES

Uses

Descriptive statistics help you see trends or patterns in a set of raw data. A good description of a data set consists of (1) a measure of the center of the data, (2) a measure of the variability (or spread) of the data, and (3) the shape (or distribution) of the data. When you read reports, news items, or advertisements prepared by other people, you are seldom given the raw data used for a study. Instead you see graphs, measures of central tendency, and measures of variability. To be a discerning reader, you need to understand the terms and techniques of descriptive statistics.

Procter & Gamble's Stock Price

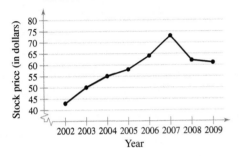

Abuses

Knowing how statistics are calculated can help you analyze questionable statistics. For instance, suppose you are interviewing for a sales position and the company reports that the average yearly commission earned by the five people in its sales force is $60,000. This is a misleading statement if it is based on four commissions of $25,000 and one of $200,000. The median would more accurately describe the yearly commission, but the company used the mean because it is a greater amount.

Statistical graphs can also be misleading. Compare the two time series charts at the left, which show the year-end stock prices for the Procter & Gamble Corporation. The data are the same for each chart. The first graph, however, has a cropped vertical axis, which makes it appear that the stock price increased greatly from 2002 to 2007, then decreased greatly from 2007 to 2009. In the second graph, the scale on the vertical axis begins at zero. This graph correctly shows that the stock price changed modestly during this time period. *(Source: Procter & Gamble Corporation)*

Procter & Gamble's Stock Price

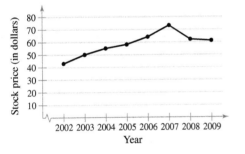

Ethics

Mark Twain helped popularize the saying, "There are three kinds of lies: lies, damned lies, and statistics." In short, even the most accurate statistics can be used to support studies or statements that are incorrect. Unscrupulous people can use misleading statistics to "prove" their point. Being informed about how statistics are calculated and questioning the data are ways to avoid being misled.

■ EXERCISES

1. Use the Internet or some other resource to find an example of a graph that might lead to incorrect conclusions.

2. You are publishing an article that discusses how eating oatmeal can help lower cholesterol. Because eating oatmeal might help people with high cholesterol, you include a graph that exaggerates the effects of eating oatmeal on lowering cholesterol. Do you think it is ethical to publish this graph? Explain.

2 CHAPTER SUMMARY

What did you **learn?**	EXAMPLE(S)	REVIEW EXERCISES
Section 2.1		
■ How to construct a frequency distribution including limits, midpoints, relative frequencies, cumulative frequencies, and boundaries	*1, 2*	*1*
■ How to construct frequency histograms, frequency polygons, relative frequency histograms, and ogives	*3–7*	*2–6*
Section 2.2		
■ How to graph quantitative data sets using stem-and-leaf plots and dot plots	*1–3*	*7, 8*
■ How to graph and interpret paired data sets using scatter plots and time series charts	*6, 7*	*9, 10*
■ How to graph qualitative data sets using pie charts and Pareto charts	*4, 5*	*11, 12*
Section 2.3		
■ How to find the mean, median, and mode of a population and a sample	*1–6*	*13, 14*
■ How to find a weighted mean of a data set and the mean of a frequency distribution	*7, 8*	*15–18*
■ How to describe the shape of a distribution as symmetric, uniform, or skewed and how to compare the mean and median for each		*19–24*
Section 2.4		
■ How to find the range of a data set	*1*	*25, 26*
■ How to find the variance and standard deviation of a population and a sample	*2–5*	*27–30*
■ How to use the Empirical Rule and Chebychev's Theorem to interpret standard deviation	*6–8*	*31–34*
■ How to approximate the sample standard deviation for grouped data	*9, 10*	*35, 36*
Section 2.5		
■ How to find the quartiles and interquartile range of a data set	*1–3*	*37, 38, 41*
■ How to draw a box-and-whisker plot	*4*	*39, 40, 42*
■ How to interpret other fractiles such as percentiles	*5*	*43, 44*
■ How to find and interpret the standard score (z-score)	*6, 7*	*45–48*

2 REVIEW EXERCISES

■ SECTION 2.1

In Exercises 1 and 2, use the following data set. The data set represents the number of students per faculty member for 20 public colleges. (Source: Kiplinger)

13 15 15   8 16 20 28 19 18 15
21 23 30 17 10 16 15 16 20 15

1. Make a frequency distribution of the data set using five classes. Include the class limits, midpoints, boundaries, frequencies, relative frequencies, and cumulative frequencies.

2. Make a relative frequency histogram using the frequency distribution in Exercise 1. Then determine which class has the greatest relative frequency and which has the least relative frequency.

In Exercises 3 and 4, use the following data set. The data represent the actual liquid volumes (in ounces) in 24 twelve-ounce cans.

11.95 11.91 11.86 11.94 12.00 11.93 12.00 11.94
12.10 11.95 11.99 11.94 11.89 12.01 11.99 11.94
11.92 11.98 11.88 11.94 11.98 11.92 11.95 11.93

3. Make a frequency histogram of the data set using seven classes.

4. Make a relative frequency histogram of the data set using seven classes.

In Exercises 5 and 6, use the following data set. The data represent the number of rooms reserved during one night's business at a sample of hotels.

153 104 118 166   89 104 100   79
  93   96 116   94 140   84   81   96
108 111   87 126 101 111 122 108
126   93 108   87 103   95 129   93

5. Make a frequency distribution of the data set with six classes and draw a frequency polygon.

6. Make an ogive of the data set using six classes.

■ SECTION 2.2

In Exercises 7 and 8, use the following data set. The data represent the air quality indices for 30 U.S. cities. (Source: AIRNow)

25 35 20 75 10 10 61 89 44 22
34 33 38 30 47 53 44 57 71 20
42 52 48 41 35 59 53 61 65 25

7. Make a stem-and-leaf plot of the data set. Use one line per stem.

8. Make a dot plot of the data set.

9. The following are the heights (in feet) and the number of stories of nine notable buildings in Houston. Use the data to construct a scatter plot. What type of pattern is shown in the scatter plot? (Source: Emporis Corporation)

Height (in feet)	992	780	762	756	741	732	714	662	579
Number of stories	71	56	53	55	47	53	50	49	40

10. The U.S. unemployment rate over a 12-year period is given. Use the data to construct a time series chart. *(Source: U.S. Bureau of Labor Statistics)*

Year	1998	1999	2000	2001	2002	2003
Unemployment rate	4.5	4.2	4.0	4.7	5.8	6.0

Year	2004	2005	2006	2007	2008	2009
Unemployment rate	5.5	5.1	4.6	4.6	5.8	9.3

In Exercises 11 and 12, use the following data set. The data set represents the results of a survey that asked U.S. adults where they would be at midnight when the new year arrived. *(Adapted from Rasmussen Reports)*

Response	At home	At friend's home	At restaurant or bar	Somewhere else	Not sure
Number	620	110	50	100	130

11. Make a Pareto chart of the data set.

12. Make a pie chart of the data set.

■ SECTION 2.3

In Exercises 13 and 14, find the mean, median, and mode of the data, if possible. If any of these measures cannot be found or a measure does not represent the center of the data, explain why.

13. Vertical Jumps   The vertical jumps (in inches) of a sample of 10 college basketball players at the 2009 NBA Draft Combine *(Source: Sports Phenoms, Inc.)*

26.0   29.5   27.0   30.5   29.5   25.0   31.5   33.0   32.0   27.5

14. Airport Scanners   The responses of 542 adults who were asked whether they approved the use of full-body scanners at airport security checkpoints *(Adapted from USA Today/Gallup Poll)*

Approved: 423   Did not approve: 108   No opinion: 11

15. Estimate the mean of the frequency distribution you made in Exercise 1.

16. The following frequency distribution shows the number of magazine subscriptions per household for a sample of 60 households. Find the mean number of subscriptions per household.

Number of magazines	0	1	2	3	4	5	6
Frequency	13	9	19	8	5	2	4

17. Six test scores are given. The first 5 test scores are 15% of the final grade, and the last test score is 25% of the final grade. Find the weighted mean of the test scores.

78   72   86   91   87   80

18. Four test scores are given. The first 3 test scores are 20% of the final grade, and the last test score is 40% of the final grade. Find the weighted mean of the test scores.

96   85   91   86

19. Describe the shape of the distribution in the histogram you made in Exercise 3. Is the distribution symmetric, uniform, or skewed?

20. Describe the shape of the distribution in the histogram you made in Exercise 4. Is the distribution symmetric, uniform, or skewed?

In Exercises 21 and 22, determine whether the approximate shape of the distribution in the histogram is symmetric, uniform, skewed left, skewed right, or none of these. Justify your answer.

21.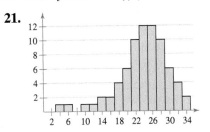

22.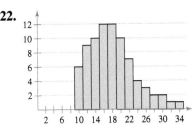

23. For the histogram in Exercise 21, which is greater, the mean or the median? Explain your reasoning.

24. For the histogram in Exercise 22, which is greater, the mean or the median? Explain your reasoning.

■ SECTION 2.4

25. The data set represents the mean prices of movie tickets (in U.S. dollars) for a sample of 12 U.S. cities. Find the range of the data set.

    7.82 7.38 6.42 6.76 6.34 7.44 6.15 5.46 7.92 6.58 8.26 7.17

26. The data set represents the mean prices of movie tickets (in U.S. dollars) for a sample of 12 Japanese cities. Find the range of the data set.

    19.73 16.48 19.10 18.56 17.68 17.19
    16.63 15.99 16.66 19.59 15.89 16.49

27. The mileages (in thousands of miles) for a rental car company's fleet are listed. Find the population mean and the population standard deviation of the data.

    4 2 9 12 15 3 6 8 1 4 14 12 3 3

28. The ages of the Supreme Court justices as of January 27, 2010 are listed. Find the population mean and the population standard deviation of the data. *(Source: Supreme Court of the United States)*

    55 89 73 73 61 76 71 59 55

29. Dormitory room prices (in dollars) for one school year for a sample of four-year universities are listed. Find the sample mean and the sample standard deviation of the data.

    2445 2940 2399 1960 2421 2940 2657 2153
    2430 2278 1947 2383 2710 2761 2377

30. Sample salaries (in dollars) of high school teachers are listed. Find the sample mean and the sample standard deviation of the data.

    49,632 54,619 58,298 48,250 51,842 50,875 53,219 49,924

31. The mean rate for satellite television for a sample of households was $49.00 per month, with a standard deviation of $2.50 per month. Between what two values do 99.7% of the data lie? (Assume the data set has a bell-shaped distribution.)

32. The mean rate for satellite television for a sample of households was $49.50 per month, with a standard deviation of $2.75 per month. Estimate the percent of satellite television rates between $46.75 and $52.25. (Assume the data set has a bell-shaped distribution.)

33. The mean sale per customer for 40 customers at a gas station is $36.00, with a standard deviation of $8.00. Using Chebychev's Theorem, determine at least how many of the customers spent between $20.00 and $52.00.

34. The mean length of the first 20 space shuttle flights was about 7 days, and the standard deviation was about 2 days. Using Chebychev's Theorem, determine at least how many of the flights lasted between 3 days and 11 days. *(Source: NASA)*

35. From a random sample of households, the number of televisions are listed. Find the sample mean and the sample standard deviation of the data.

Number of televisions	0	1	2	3	4	5
Number of households	1	8	13	10	5	3

36. From a random sample of airplanes, the number of defects found in their fuselages are listed. Find the sample mean and the sample standard deviation of the data.

Number of defects	0	1	2	3	4	5	6
Number of airplanes	4	5	2	9	1	3	1

■ SECTION 2.5

In Exercises 37–40, use the following data set. The data represent the fuel economies (in highway miles per gallon) of several Harley-Davidson motorcycles. (Source: Total Motorcycle)

53 57 60 57 54 53 54 53 54 42 48
53 47 47 50 48 42 42 54 54 60

37. Find the five-number summary of the data set.

38. Find the interquartile range.

39. Make a box-and-whisker plot of the data.

40. About how many motorcycles fall on or below the third quartile?

41. Find the interquartile range of the data from Exercise 13.

42. The weights (in pounds) of the defensive players on a high school football team are given. Draw a box-and-whisker plot of the data and describe the shape of the distribution.

173 145 205 192 197 227 156 240 172 185
208 185 190 167 212 228 190 184 195

43. A student's test grade of 75 represents the 65th percentile of the grades. What percent of students scored higher than 75?

44. As of January 2010, there were 755 "oldies" radio stations in the United States. If one station finds that 104 stations have a larger daily audience than it has, what percentile does this station come closest to in the daily audience rankings? *(Source: Radio-locator.com)*

In Exercises 45–48, use the following information. The towing capacities (in pounds) of 25 four-wheel drive pickup trucks have a bell-shaped distribution, with a mean of 11,830 pounds and a standard deviation of 2370 pounds. Use z-scores to determine if the towing capacities of the following randomly selected four-wheel drive pickup trucks are unusual.

45. 16,500 pounds

46. 5500 pounds

47. 18,000 pounds

48. 11,300 pounds

2 CHAPTER QUIZ

Take this quiz as you would take a quiz in class. After you are done, check your work against the answers given in the back of the book.

1. The data set represents the number of minutes a sample of 25 people exercise each week.

108	139	120	123	120	132	123	131	131
157	150	124	111	101	135	119	116	117
127	128	139	119	118	114	127		

(a) Make a frequency distribution of the data set using five classes. Include class limits, midpoints, boundaries, frequencies, relative frequencies, and cumulative frequencies.

(b) Display the data using a frequency histogram and a frequency polygon on the same axes.

(c) Display the data using a relative frequency histogram.

(d) Describe the distribution's shape as symmetric, uniform, or skewed.

(e) Display the data using a stem-and-leaf plot. Use one line per stem.

(f) Display the data using a box-and-whisker plot.

(g) Display the data using an ogive.

2. Use frequency distribution formulas to approximate the sample mean and the sample standard deviation of the data set in Exercise 1.

3. U.S. sporting goods sales (in billions of dollars) can be classified in four areas: clothing (10.6), footwear (17.2), equipment (24.9), and recreational transport (27.0). Display the data using (a) a pie chart and (b) a Pareto chart. *(Source: National Sporting Goods Association)*

4. Weekly salaries (in dollars) for a sample of registered nurses are listed.

774 446 1019 795 908 667 444 960

(a) Find the mean, median, and mode of the salaries. Which best describes a typical salary?

(b) Find the range, variance, and standard deviation of the data set. Interpret the results in the context of the real-life setting.

5. The mean price of new homes from a sample of houses is $155,000 with a standard deviation of $15,000. The data set has a bell-shaped distribution. Between what two prices do 95% of the houses fall?

6. Refer to the sample statistics from Exercise 5 and use z-scores to determine which, if any, of the following house prices is unusual.

(a) $200,000   (b) $55,000   (c) $175,000   (d) $122,000

7. The number of regular season wins for each Major League Baseball team in 2009 are listed. *(Source: Major League Baseball)*

103	95	84	75	64	87	86	79	65	65	97	87	85	75	93
87	86	70	59	91	83	80	78	74	62	95	92	88	75	70

(a) Find the five-number summary of the data set.

(b) Find the interquartile range.

(c) Display the data using a box-and-whisker plot.

PUTTING IT ALL TOGETHER

Real Statistics — Real Decisions

You are a member of your local apartment association. The association represents rental housing owners and managers who operate residential rental property throughout the greater metropolitan area. Recently, the association has received several complaints from tenants in a particular area of the city who feel that their monthly rental fees are much higher compared to other parts of the city.

You want to investigate the rental fees. You gather the data shown in the table at the right. Area A represents the area of the city where tenants are unhappy about their monthly rents. The data represent the monthly rents paid by a random sample of tenants in Area A and three other areas of similar size. Assume all the apartments represented are approximately the same size with the same amenities.

NATIONAL APARTMENT ASSOCIATION

AMERICA'S LEADING ADVOCATE FOR QUALITY RENTAL HOUSING

■ EXERCISES

1. How Would You Do It?

(a) How would you investigate the complaints from renters who are unhappy about their monthly rents?

(b) Which statistical measure do you think would best represent the data sets for the four areas of the city?

(c) Calculate the measure from part (b) for each of the four areas.

2. Displaying the Data

(a) What type of graph would you choose to display the data? Explain your reasoning.

(b) Construct the graph from part (a).

(c) Based on your data displays, does it appear that the monthly rents in Area A are higher than the rents in the other areas of the city? Explain.

3. Measuring the Data

(a) What other statistical measures in this chapter could you use to analyze the monthly rent data?

(b) Calculate the measures from part (a).

(c) Compare the measures from part (b) with the graph you constructed in Exercise 2. Do the measurements support your conclusion in Exercise 2? Explain.

4. Discussing the Data

(a) Do you think the complaints in Area A are legitimate? How do you think they should be addressed?

(b) What reasons might you give as to why the rents vary among different areas of the city?

The Monthly Rents (in dollars) Paid by 12 Randomly Selected Apartment Tenants in 4 Areas of Your City

Area A	Area B	Area C	Area D
1275	1124	1085	928
1110	954	827	1096
975	815	793	862
862	1078	1170	735
1040	843	919	798
997	745	943	812
1119	796	756	1232
908	816	765	1036
890	938	809	998
1055	1082	1020	914
860	750	710	1005
975	703	775	930

Highest Monthly Rents

AVERAGE PER CITY

New York, NY	$2922
San Francisco, CA	$1904
Boston, MA	$1658
San Jose, CA	$1612
Los Angeles, CA	$1452

(Source: Forbes)

TECHNOLOGY

MINITAB   EXCEL   TI-83/84 PLUS

Dairy Farmers of America is an association that provides help to dairy farmers. Part of this help is gathering and distributing statistics on milk production.

MONTHLY MILK PRODUCTION

The following data set was supplied by a dairy farmer. It lists the monthly milk productions (in pounds) for 50 Holstein dairy cows. *(Source: Matlink Dairy, Clymer, NY)*

2825	2072	2733	2069	2484
4285	2862	3353	1449	2029
1258	2982	2045	1677	1619
2597	3512	2444	1773	2284
1884	2359	2046	2364	2669
3109	2804	1658	2207	2159
2207	2882	1647	2051	2202
3223	2383	1732	2230	1147
2711	1874	1979	1319	2923
2281	1230	1665	1294	2936

www.dfamilk.com

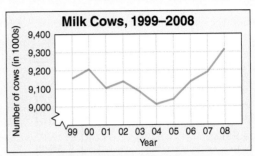

(Source: National Agricultural Statistics Service)

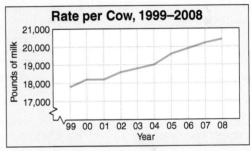

(Source: National Agricultural Statistics Service)

From 1999 to 2008, the number of dairy cows in the United States increased by only 1.7% while the yearly milk production per cow increased by almost 15%.

EXERCISES

In Exercises 1–4, use a computer or calculator. If possible, print your results.

1. Find the sample mean of the data.

2. Find the sample standard deviation of the data.

3. Make a frequency distribution for the data. Use a class width of 500.

4. Draw a histogram for the data. Does the distribution appear to be bell-shaped?

5. What percent of the distribution lies within one standard deviation of the mean? Within two standard deviations of the mean? How do these results agree with the Empirical Rule?

In Exercises 6–8, use the frequency distribution found in Exercise 3.

6. Use the frequency distribution to estimate the sample mean of the data. Compare your results with Exercise 1.

7. Use the frequency distribution to find the sample standard deviation for the data. Compare your results with Exercise 2.

8. **Writing** Use the results of Exercises 6 and 7 to write a general statement about the mean and standard deviation for grouped data. Do the formulas for grouped data give results that are as accurate as the individual entry formulas?

Extended solutions are given in the *Technology Supplement.*
Technical instruction is provided for MINITAB, Excel, and the TI-83/84 Plus.

2 USING TECHNOLOGY TO DETERMINE DESCRIPTIVE STATISTICS

Here are some MINITAB and TI-83/84 Plus printouts for three examples in this chapter.

(See Example 7, page 59.)

Bar Chart...
Pie Chart...
Time Series Plot...
Area Graph...
Contour Plot...
3D Scatterplot...
3D Surface Plot...

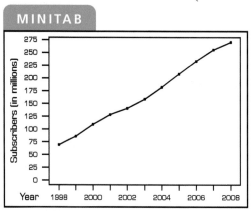

(See Example 4, page 83.)

Display Descriptive Statistics...
Store Descriptive Statistics...
Graphical Summary...

1-Sample Z...
1-Sample t...
2-Sample t...
Paired t...

MINITAB

Descriptive Statistics: Salaries

Variable	N	Mean	SE Mean	StDev	Minimum
Salaries	10	41.500	0.992	3.136	37.000

Variable	Q1	Median	Q3	Maximum
Salaries	38.750	41.000	44.250	47.000

(See Example 4, page 103.)

Empirical CDF...
Probability Distribution Plot ...

Boxplot...
Interval Plot...
Individual Value Plot...
Line Plot...

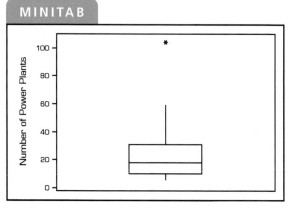

(See Example 7, page 59.)

TI-83/84 PLUS

STAT PLOTS
1: Plot1...Off
      L1   L2   □
2: Plot2...Off
      L1   L2   □
3: Plot3...Off
      L1   L2   □

↓

TI-83/84 PLUS

Plot1 Plot2 Plot3
On Off
Type:
Xlist: L1
Ylist: L2
Mark: □ + .

↓

TI-83/84 PLUS

ZOOM MEMORY
4↑ ZDecimal
5: ZSquare
6: ZStandard
7: ZTrig
8: ZInteger
9: ZoomStat
0: ZoomFit

↓

TI-83/84 PLUS

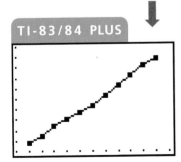

(See Example 4, page 83.)

TI-83/84 PLUS

EDIT CALC TESTS
1: 1-Var Stats
2: 2-Var Stats
3: Med-Med
4: LinReg(ax+b)
5: QuadReg
6: CubicReg
7↓ QuartReg

↓

TI-83/84 PLUS

1-Var Stats L1

↓

TI-83/84 PLUS

1-Var Stats
$\bar{x}$= 41.5
Σx= 415
Σx^2= 17311
Sx= 3.13581462
σx= 2.974894956
↓n= 10

(See Example 4, page 103.)

TI-83/84 PLUS

STAT PLOTS
1: Plot1...Off
      L1   L2   □
2: Plot2...Off
      L1   L2   □
3: Plot3...Off
      L1   L2   □
4↓ PlotsOff

↓

TI-83/84 PLUS

Plot1 Plot2 Plot3
On Off
Type:
Xlist: L1
Freq: 1

↓

TI-83/84 PLUS

ZOOM MEMORY
4↑ ZDecimal
5: ZSquare
6: ZStandard
7: ZTrig
8: ZInteger
9: ZoomStat
0: ZoomFit

↓

TI-83/84 PLUS

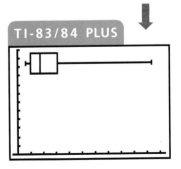

CUMULATIVE REVIEW
Chapters 1 and 2

In Exercises 1 and 2, identify the sampling technique used and discuss potential sources of bias (if any). Explain.

1. For quality assurance, every fortieth toothbrush is taken from each of four assembly lines and tested to make sure the bristles stay in the toothbrush.

2. Using random digit dialing, researchers asked 1200 U.S. adults their thoughts on health care reform.

3. In 2008, a worldwide study of all airlines found that baggage delays were caused by transfer baggage mishandling (49%), failure to load at originating airport (16%), arrival station mishandling (8%), space-weight restriction (6%), loading/offloading error (5%), tagging error (3%), and ticketing error/bag switch/security/other (13%). Use a Pareto chart to organize the data. *(Source: Société International de Télécommunications Aéronautiques)*

In Exercises 4 and 5, determine whether the numerical value is a parameter or a statistic. Explain your reasoning.

4. In 2009, the average salary of a Major League Baseball player was $2,996,106. *(Source: Major League Baseball)*

5. In a recent survey of 1000 voters, 19% said that First Lady of the United States Michelle Obama will be very involved in policy decisions. *(Source: Rasmussen Reports)*

6. The mean annual salary for a sample of electrical engineers is $83,500, with a standard deviation of $1500. The data set has a bell-shaped distribution.

   (a) Use the Empirical Rule to estimate the number of electrical engineers whose annual salaries are between $80,500 and $86,500.

   (b) If 40 additional electrical engineers were sampled, about how many of these electrical engineers would you expect to have annual salaries between $80,500 and $86,500?

   (c) The salaries of three randomly selected electrical engineers are $90,500, $79,750, and $82,600. Find the z-score that corresponds to each salary. According to the z-scores, would the salaries of any of these engineers be considered unusual?

In Exercises 7 and 8, identify the population and the sample.

7. A survey of career counselors at 195 colleges and universities found that 90% of the students working with their offices were interested in federal jobs or internships. *(Source: Partnership for Public Service Survey)*

8. A study of 232,606 people was conducted to find a link between taking antioxidant vitamins and living a longer life. *(Source: Journal of the American Medical Association)*

In Exercises 9 and 10, decide which method of data collection you would use to collect data for the study. Explain.

9. A study of the years of service of the 100 members of the Senate

10. A study of the effects of removing recess from schools

In Exercises 11 and 12, determine whether the data are qualitative or quantitative and identify the data set's level of measurement.

11. The number of games started by pitchers with at least one start for the New York Yankees in 2009 are listed. *(Source: Major League Baseball)*

9 34 1 33 32 31 7 9 6

12. The five top-earning states in 2008 by median income are listed. *(Source: U.S. Census Bureau)*

1. Maryland   2. New Jersey   3. Connecticut   4. Alaska   5. Hawaii

13. The number of tornadoes by state in a recent year is listed. (a) Find the data set's five-number summary, (b) draw a box-and-whisker plot that represents the data set, and (c) describe the shape of the distribution. *(Source: National Climatic Data Center)*

81	1	8	69	30	34	0	0	56	54
2	6	21	14	46	136	17	23	2	0
1	5	71	105	39	10	40	1	0	7
4	0	23	53	4	27	1	11	0	14
19	23	105	4	0	24	4	0	63	6

14. Five test scores are given. The first four test scores are 15% of the final grade, and the last test score is 40% of the final grade. Find the weighted mean of the test scores.

85 92 84 89 91

15. Tail lengths (in feet) for a sample of American alligators are listed.

6.5 3.4 4.2 7.1 5.4 6.8 7.5 3.9 4.6

(a) Find the mean, median, and mode of the tail lengths. Which best describes a typical American alligator tail length? Explain your reasoning.

(b) Find the range, variance, and standard deviation of the data set. Interpret the results in the context of the real-life setting.

16. A study shows that the number of deaths due to heart disease for women has decreased every year for the past five years.

(a) Make an inference based on the results of the study.

(b) What is wrong with this type of reasoning?

In Exercises 17–19, use the following data set. The data represent the points scored by each player on the Montreal Canadiens in a recent NHL season. (Source: National Hockey League)

5	64	50	1	41	0	39	23	32	28
26	23	33	23	22	1	17	18	12	11
11	9	65	3	2	41	21	1	0	39

17. Make a frequency distribution using eight classes. Include the class limits, midpoints, boundaries, frequencies, relative frequencies, and cumulative frequencies.

18. Describe the shape of the distribution.

19. Make a relative frequency histogram using the frequency distribution in Exercise 17. Then determine which class has the greatest relative frequency and which has the least relative frequency.

3 PROBABILITY

3.1 Basic Concepts of Probability and Counting

■ ACTIVITY

3.2 Conditional Probability and the Multiplication Rule

3.3 The Addition Rule

■ ACTIVITY

■ CASE STUDY

3.4 Additional Topics in Probability and Counting

■ USES AND ABUSES

■ REAL STATISTICS– REAL DECISIONS

■ TECHNOLOGY

The television game show *The Price Is Right* presents a wide range of pricing games in which contestants compete for prizes using strategy, probability, and their knowledge of prices. One popular game is *Spelling Bee*.

In Chapters 1 and 2, you learned how to collect and describe data. Once the data are collected and described, you can use the results to write summaries, form conclusions, and make decisions. For instance, in *Spelling Bee*, contestants have a chance to win a car by choosing lettered cards that spell CAR or by choosing a single card that displays the entire word CAR. By collecting and analyzing data, you can determine the chances of winning the car.

To play *Spelling Bee*, contestants choose from 30 cards. Eleven cards display the letter C, eleven cards display A, six cards display R, and two cards display CAR. Depending on how well contestants play the game, they can choose two, three, four, or five cards.

Before the chosen cards are displayed, contestants are offered $1000 for each card. If contestants choose the money, the game is over. If contestants choose to try to win the car, the host displays one card. After a card is displayed, contestants are offered $1000 for each remaining card. If they do not accept the money, the host continues displaying cards. Play continues until contestants take the money, spell the word CAR, display the word CAR, or display all cards and do not spell CAR.

In Chapter 3, you will learn how to determine the probability of an event. For instance, the following table shows the four ways that contestants on *Spelling Bee* can win a car and the corresponding probabilities.

You can see from the table that choosing more cards gives you a better chance of winning. These probabilities can be found using *combinations*, which will be discussed in Section 3.4.

Event	Probability
Winning by selecting two cards	$\frac{57}{435} \approx 0.131$
Winning by selecting three cards	$\frac{151}{406} \approx 0.372$
Winning by selecting four cards	$\frac{1067}{1827} \approx 0.584$
Winning by selecting five cards	$\frac{52{,}363}{71{,}253} \approx 0.735$

3.1 Basic Concepts of Probability and Counting

WHAT YOU SHOULD LEARN

▸ How to identify the sample space of a probability experiment and how to identify simple events

▸ How to use the Fundamental Counting Principle to find the number of ways two or more events can occur

▸ How to distinguish among classical probability, empirical probability, and subjective probability

▸ How to find the probability of the complement of an event

▸ How to use a tree diagram and the Fundamental Counting Principle to find more probabilities

Probability Experiments ▸ The Fundamental Counting Principle ▸ Types of Probability ▸ Complementary Events ▸ Probability Applications

▸ PROBABILITY EXPERIMENTS

When weather forecasters say that there is a 90% chance of rain or a physician says there is a 35% chance for a successful surgery, they are stating the likelihood, or *probability*, that a specific event will occur. Decisions such as "should you go golfing" or "should you proceed with surgery" are often based on these probabilities. In the previous chapter, you learned about the role of the descriptive branch of statistics. Because probability is the foundation of inferential statistics, it is necessary to learn about probability before proceeding to the second branch—inferential statistics.

DEFINITION

A **probability experiment** is an action, or trial, through which specific results (counts, measurements, or responses) are obtained. The result of a single trial in a probability experiment is an **outcome.** The set of all possible outcomes of a probability experiment is the **sample space.** An **event** is a subset of the sample space. It may consist of one or more outcomes.

STUDY TIP

Here is a simple example of the use of the terms *probability experiment, sample space, event,* and *outcome.*

Probability Experiment:
  Roll a six-sided die.

Sample Space:
  {1, 2, 3, 4, 5, 6}

Event:
  Roll an even number, {2, 4, 6}.

Outcome:
  Roll a 2, {2}.

EXAMPLE 1

▸ **Identifying the Sample Space of a Probability Experiment**

A probability experiment consists of tossing a coin and then rolling a six-sided die. Determine the number of outcomes and identify the sample space.

▸ **Solution**

There are two possible outcomes when tossing a coin: a head (H) or a tail (T). For each of these, there are six possible outcomes when rolling a die: 1, 2, 3, 4, 5, or 6. **A tree diagram** gives a visual display of the outcomes of a probability experiment by using branches that originate from a starting point. It can be used to find the number of possible outcomes in a sample space as well as individual outcomes.

Tree Diagram for Coin and Die Experiment

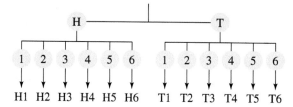

From the tree diagram, you can see that the sample space has 12 outcomes.

{H1, H2, H3, H4, H5, H6, T1, T2, T3, T4, T5, T6}

▶ **Try It Yourself 1**

For each probability experiment, determine the number of outcomes and identify the sample space.

1. A probability experiment consists of recording a response to the survey statement at the left *and* the gender of the respondent.
2. A probability experiment consists of recording a response to the survey statement at the left *and* the geographic location (Northeast, South, Midwest, West) of the respondent.

a. Start a *tree diagram* by forming a branch for each possible response to the survey.
b. At the end of each survey response branch, draw a *new branch* for each possible outcome.
c. Find the *number of outcomes* in the sample space.
d. List the *sample space*.

Answer: Page A34

In the rest of this chapter, you will learn how to calculate the probability or likelihood of an event. Events are often represented by uppercase letters, such as *A*, *B*, and *C*. An event that consists of a single outcome is called a **simple event.** In Example 1, the event "tossing heads and rolling a 3" is a simple event and can be represented as $A = \{H3\}$. In contrast, the event "tossing heads and rolling an even number" is not simple because it consists of three possible outcomes $B = \{H2, H4, H6\}$.

EXAMPLE 2

▶ **Identifying Simple Events**

Determine the number of outcomes in each event. Then decide whether each event is simple or not. Explain your reasoning.

1. For quality control, you randomly select a machine part from a batch that has been manufactured that day. Event *A* is selecting a specific defective machine part.

2. You roll a six-sided die. Event *B* is rolling at least a 4.

▶ **Solution**

1. Event *A* has only one outcome: choosing the specific defective machine part. So, the event is a simple event.

2. Event *B* has three outcomes: rolling a 4, a 5, or a 6. Because the event has more than one outcome, it is not simple.

▶ **Try It Yourself 2**

You ask for a student's age at his or her last birthday. Determine the number of outcomes in each event. Then decide whether each event is simple or not. Explain your reasoning.

1. Event *C*: The student's age is between 18 and 23, inclusive.
2. Event *D*: The student's age is 20.

a. Determine the number of *outcomes* in the event.
b. State whether the event is *simple* or not. Explain your reasoning.

Answer: Page A34

▶ THE FUNDAMENTAL COUNTING PRINCIPLE

In some cases, an event can occur in so many different ways that it is not practical to write out all the outcomes. When this occurs, you can rely on the Fundamental Counting Principle. The Fundamental Counting Principle can be used to find the number of ways two or more events can occur in sequence.

THE FUNDAMENTAL COUNTING PRINCIPLE

If one event can occur in m ways and a second event can occur in n ways, the number of ways the two events can occur in sequence is $m \cdot n$. This rule can be extended to any number of events occurring in sequence.

In words, the number of ways that events can occur in sequence is found by multiplying the number of ways one event can occur by the number of ways the other event(s) can occur.

EXAMPLE 3

▶ Using the Fundamental Counting Principle

You are purchasing a new car. The possible manufacturers, car sizes, and colors are listed.

Manufacturer: Ford, GM, Honda
Car size: compact, midsize
Color: white (W), red (R), black (B), green (G)

How many different ways can you select one manufacturer, one car size, and one color? Use a tree diagram to check your result.

▶ Solution

There are three choices of manufacturers, two choices of car sizes, and four choices of colors. Using the Fundamental Counting Principle, you can conclude that the number of ways to select one manufacturer, one car size, and one color is

$3 \cdot 2 \cdot 4 = 24$ ways.

Using a tree diagram, you can see why there are 24 options.

Tree Diagram for Car Selections

▶ Try It Yourself 3

Your choices now include a Toyota and a tan car. How many different ways can you select one manufacturer, one car size, and one color? Use a tree diagram to check your result.

a. Find the *number of ways* each event can occur.
b. Use the *Fundamental Counting Principle*.
c. Use a *tree diagram* to check your result.

Answer: Page A35

EXAMPLE 4

▶ **Using the Fundamental Counting Principle**

The access code for a car's security system consists of four digits. Each digit can be any number from 0 through 9.

Access Code

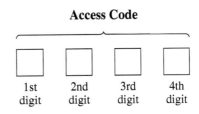

| 1st digit | 2nd digit | 3rd digit | 4th digit |

How many access codes are possible if

1. each digit can be used only once and not repeated?

2. each digit can be repeated?

3. each digit can be repeated but the first digit cannot be 0 or 1?

▶ **Solution**

1. Because each digit can be used only once, there are 10 choices for the first digit, 9 choices left for the second digit, 8 choices left for the third digit, and 7 choices left for the fourth digit. Using the Fundamental Counting Principle, you can conclude that there are

$$10 \cdot 9 \cdot 8 \cdot 7 = 5040$$

possible access codes.

2. Because each digit can be repeated, there are 10 choices for each of the four digits. So, there are

$$10 \cdot 10 \cdot 10 \cdot 10 = 10^4$$
$$= 10{,}000$$

possible access codes.

3. Because the first digit cannot be 0 or 1, there are 8 choices for the first digit. Then there are 10 choices for each of the other three digits. So, there are

$$8 \cdot 10 \cdot 10 \cdot 10 = 8000$$

possible access codes.

▶ Try It Yourself 4

How many license plates can you make if a license plate consists of

1. six (out of 26) alphabetical letters each of which can be repeated?
2. six (out of 26) alphabetical letters each of which cannot be repeated?
3. six (out of 26) alphabetical letters each of which can be repeated but the first letter cannot be A, B, C, or D?

a. *Identify* each event and the *number of ways* each event can occur.
b. Use the *Fundamental Counting Principle*.   *Answer: Page A35*

▶ TYPES OF PROBABILITY

The method you will use to calculate a probability depends on the type of probability. There are three types of probability: *classical probability*, *empirical probability*, and *subjective probability*. The probability that event E will occur is written as $P(E)$ and is read "the probability of event E."

DEFINITION

Classical (or **theoretical**) **probability** is used when each outcome in a sample space is equally likely to occur. The classical probability for an event E is given by

$$P(E) = \frac{\text{Number of outcomes in event } E}{\text{Total number of outcomes in sample space}}.$$

EXAMPLE 5

▶ **Finding Classical Probabilities**

You roll a six-sided die. Find the probability of each event.

1. Event A: rolling a 3

2. Event B: rolling a 7

3. Event C: rolling a number less than 5

▶ **Solution**

When a six-sided die is rolled, the sample space consists of six outcomes: $\{1, 2, 3, 4, 5, 6\}$.

1. There is one outcome in event $A = \{3\}$. So,

$$P(\text{rolling a 3}) = \frac{1}{6} \approx 0.167.$$

2. Because 7 is not in the sample space, there are no outcomes in event B. So,

$$P(\text{rolling a 7}) = \frac{0}{6} = 0.$$

3. There are four outcomes in event $C = \{1, 2, 3, 4\}$. So,

$$P(\text{rolling a number less than 5}) = \frac{4}{6} = \frac{2}{3} \approx 0.667.$$

▶ **Try It Yourself 5**

You select a card from a standard deck. Find the probability of each event.

1. Event D: Selecting a nine of clubs
2. Event E: Selecting a heart
3. Event F: Selecting a diamond, heart, club, or spade

a. Identify the *total number of outcomes* in the sample space.
b. Find the *number of outcomes* in the event.
c. Use the *classical probability formula*. *Answer: Page A35*

Standard Deck of Playing Cards

Hearts	Diamonds	Spades	Clubs
A ♥	A ♦	A ♠	A ♣
K ♥	K ♦	K ♠	K ♣
Q ♥	Q ♦	Q ♠	Q ♣
J ♥	J ♦	J ♠	J ♣
10 ♥	10 ♦	10 ♠	10 ♣
9 ♥	9 ♦	9 ♠	9 ♣
8 ♥	8 ♦	8 ♠	8 ♣
7 ♥	7 ♦	7 ♠	7 ♣
6 ♥	6 ♦	6 ♠	6 ♣
5 ♥	5 ♦	5 ♠	5 ♣
4 ♥	4 ♦	4 ♠	4 ♣
3 ♥	3 ♦	3 ♠	3 ♣
2 ♥	2 ♦	2 ♠	2 ♣

When an experiment is repeated many times, regular patterns are formed. These patterns make it possible to find empirical probability. Empirical probability can be used even if each outcome of an event is not equally likely to occur.

PICTURING THE WORLD

It seems as if no matter how strange an event is, somebody wants to know the probability that it will occur. The following table lists the probabilities that some intriguing events will happen. (Adapted from Life: The Odds)

Event	Probability
Being audited by the IRS	0.6%
Writing a *New York Times* best seller	0.0045
Winning an Academy Award	0.000087
Having your identity stolen	0.5%
Spotting a UFO	0.0000003

Which of these events is most likely to occur? Least likely?

To explore this topic further, see Activity 3.1 on page 144.

DEFINITION

Empirical (or **statistical**) **probability** is based on observations obtained from probability experiments. The empirical probability of an event E is the relative frequency of event E.

$$P(E) = \frac{\text{Frequency of event } E}{\text{Total frequency}} = \frac{f}{n}$$

EXAMPLE 6

Finding Empirical Probabilities

A company is conducting a telephone survey of randomly selected individuals to get their overall impressions of the past decade (2000s). So far, 1504 people have been surveyed. The frequency distribution shows the results. What is the probability that the next person surveyed has a positive overall impression of the 2000s? *(Adapted from Princeton Survey Research Associates International)*

Response	Number of times, f
Positive	406
Negative	752
Neither	316
Don't know	30
	$\Sigma f = 1504$

Solution

The event is a response of "positive." The frequency of this event is 406. Because the total of the frequencies is 1504, the empirical probability of the next person having a positive overall impression of the 2000s is

$$P(\text{positive}) = \frac{406}{1504} \approx 0.270.$$

Try It Yourself 6

An insurance company determines that in every 100 claims, 4 are fraudulent. What is the probability that the next claim the company processes will be fraudulent?

a. *Identify* the event. Find the *frequency* of the event.
b. Find the *total frequency* for the experiment.
c. Find the *empirical probability* of the event.

Answer: Page A35

As you increase the number of times a probability experiment is repeated, the empirical probability (relative frequency) of an event approaches the theoretical probability of the event. This is known as the **law of large numbers.**

LAW OF LARGE NUMBERS

As an experiment is repeated over and over, the empirical probability of an event approaches the theoretical (actual) probability of the event.

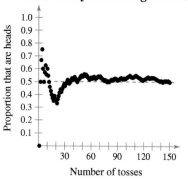

Probability of Tossing a Head

Proportion that are heads
Number of tosses

As an example of this law, suppose you want to determine the probability of tossing a head with a fair coin. If you toss the coin 10 times and get only 3 heads, you obtain an empirical probability of $\frac{3}{10}$. Because you tossed the coin only a few times, your empirical probability is not representative of the theoretical probability, which is $\frac{1}{2}$. If, however, you toss the coin several thousand times, then the law of large numbers tells you that the empirical probability will be very close to the theoretical or actual probability.

The scatter plot at the left shows the results of simulating a coin toss 150 times. Notice that, as the number of tosses increases, the probability of tossing a head gets closer and closer to the theoretical probability of 0.5.

EXAMPLE 7

▶ **Using Frequency Distributions to Find Probabilities**

You survey a sample of 1000 employees at a company and record the age of each. The results are shown in the frequency distribution at the left. If you randomly select another employee, what is the probability that the employee will be between 25 and 34 years old?

▶ **Solution**

The event is selecting an employee who is between 25 and 34 years old. The frequency of this event is 366. Because the total of the frequencies is 1000, the empirical probability of selecting an employee between the ages of 25 and 34 years old is

$$P(\text{age 25 to 34}) = \frac{366}{1000}$$

$$= 0.366.$$

Employee ages	Frequency, f
15 to 24	54
25 to 34	366
35 to 44	233
45 to 54	180
55 to 64	125
65 and over	42
	$\Sigma f = 1000$

▶ **Try It Yourself 7**

Find the probability that an employee chosen at random will be between 15 and 24 years old.

a. Find the *frequency* of the event.
b. Find the *total of the frequencies.*
c. Find the *empirical probability* of the event.   *Answer: Page A35*

The third type of probability is **subjective probability.** Subjective probabilities result from intuition, educated guesses, and estimates. For instance, given a patient's health and extent of injuries, a doctor may feel that the patient has a 90% chance of a full recovery. Or a business analyst may predict that the chance of the employees of a certain company going on strike is 0.25.

EXAMPLE 8

▶ **Classifying Types of Probability**

Classify each statement as an example of classical probability, empirical probability, or subjective probability. Explain your reasoning.

1. The probability that you will get the flu this year is 0.1.

2. The probability that a voter chosen at random will be younger than 35 years old is 0.3.

3. The probability of winning a 1000-ticket raffle with one ticket is $\frac{1}{1000}$.

▶ **Solution**

1. This probability is most likely based on an educated guess. It is an example of subjective probability.

2. This statement is most likely based on a survey of a sample of voters, so it is an example of empirical probability.

3. Because you know the number of outcomes and each is equally likely, this is an example of classical probability.

▶ **Try It Yourself 8**

Based on previous counts, the probability of a salmon successfully passing through a dam on the Columbia River is 0.85. Is this statement an example of classical probability, empirical probability, or subjective probability? *(Source: Army Corps of Engineers)*

a. Identify the *event*.
b. Decide whether the probability is *determined* by knowing all possible outcomes, whether the probability is *estimated* from the results of an experiment, or whether the probability is an *educated guess*.
c. Make a *conclusion*.   *Answer: Page A35*

A probability cannot be negative or greater than 1. So, the probability of an event E is between 0 and 1, inclusive, as stated in the following rule.

RANGE OF PROBABILITIES RULE

The probability of an event E is between 0 and 1, inclusive. That is,

$$0 \le P(E) \le 1.$$

If the probability of an event is 1, the event is certain to occur. If the probability of an event is 0, the event is impossible. A probability of 0.5 indicates that an event has an even chance of occurring.

The following graph shows the possible range of probabilities and their meanings.

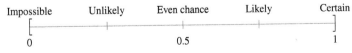

An event that occurs with a probability of 0.05 or less is typically considered unusual. Unusual events are highly unlikely to occur. Later in this course you will identify unusual events when studying inferential statistics.

▶ COMPLEMENTARY EVENTS

The sum of the probabilities of all outcomes in a sample space is 1 or 100%. An important result of this fact is that if you know the probability of an event E, you can find the probability of the *complement of event E*.

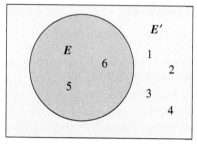

The area of the rectangle represents the total probability of the sample space (1 = 100%). The area of the circle represents the probability of event E, and the area outside the circle represents the probability of the complement of event E.

DEFINITION

The **complement of event E** is the set of all outcomes in a sample space that are not included in event E. The complement of event E is denoted by E' and is read as "E prime."

For instance, if you roll a die and let E be the event "the number is at least 5," then the complement of E is the event "the number is less than 5." In symbols, $E = \{5, 6\}$ and $E' = \{1, 2, 3, 4\}$.

Using the definition of the complement of an event and the fact that the sum of the probabilities of all outcomes is 1, you can determine the following formulas.

$$P(E) + P(E') = 1 \qquad P(E) = 1 - P(E') \qquad P(E') = 1 - P(E)$$

The Venn diagram at the left illustrates the relationship between the sample space, an event E, and its complement E'.

EXAMPLE 9

▶ **Finding the Probability of the Complement of an Event**

Use the frequency distribution in Example 7 to find the probability of randomly choosing an employee who is not between 25 and 34 years old.

▶ **Solution**

From Example 7, you know that

$$P(\text{age 25 to 34}) = \frac{366}{1000}$$
$$= 0.366.$$

So, the probability that an employee is not between 25 and 34 years old is

$$P(\text{age is not 25 to 34}) = 1 - \frac{366}{1000}$$
$$= \frac{634}{1000}$$
$$= 0.634.$$

▶ **Try It Yourself 9**

Use the frequency distribution in Example 7 to find the probability of randomly choosing an employee who is not between 45 and 54 years old.

a. Find the *probability* of randomly choosing an employee who is between 45 and 54 years old.
b. *Subtract* the resulting probability from 1.
c. *State the probability* as a fraction and as a decimal.   *Answer: Page A35*

▸ PROBABILITY APPLICATIONS

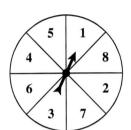

Tree Diagram for Coin and Spinner Experiment

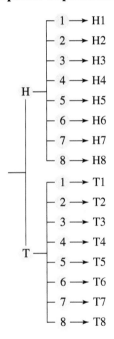

> ### EXAMPLE 10

▸ **Using a Tree Diagram**

A probability experiment consists of tossing a coin and spinning the spinner shown at the left. The spinner is equally likely to land on each number. Use a tree diagram to find the probability of each event.

1. Event A: tossing a tail and spinning an odd number

2. Event B: tossing a head or spinning a number greater than 3

▸ **Solution**   From the tree diagram at the left, you can see that there are 16 outcomes.

1. There are four outcomes in event $A = \{T1, T3, T5, T7\}$. So,

$$P(\text{tossing a tail and spinning an odd number}) = \frac{4}{16} = \frac{1}{4} = 0.25.$$

2. There are 13 outcomes in event $B = \{H1, H2, H3, H4, H5, H6, H7, H8, T4, T5, T6, T7, T8\}$. So,

$$P(\text{tossing a head or spinning a number greater than 3}) = \frac{13}{16} \approx 0.813.$$

▸ Try It Yourself 10

Find the probability of tossing a tail and spinning a number less than 6.

a. Find the *number of outcomes* in the event.
b. Find the *probability of the event.*   *Answer: Page A35*

> ### EXAMPLE 11

▸ **Using the Fundamental Counting Principle**

Your college identification number consists of eight digits. Each digit can be 0 through 9 and each digit can be repeated. What is the probability of getting your college identification number when randomly generating eight digits?

▸ **Solution**   Because each digit can be repeated, there are 10 choices for each of the 8 digits. So, using the Fundamental Counting Principle, there are $10 \cdot 10 \cdot 10 \cdot 10 \cdot 10 \cdot 10 \cdot 10 \cdot 10 = 10^8 = 100,000,000$ possible identification numbers. But only one of those numbers corresponds to your college identification number. So, the probability of randomly generating 8 digits and getting your college identification number is 1/100,000,000.

▸ Try It Yourself 11

Your college identification number consists of nine digits. The first two digits of each number will be the last two digits of the year you are scheduled to graduate. The other digits can be any number from 0 through 9, and each digit can be repeated. What is the probability of getting your college identification number when randomly generating the other seven digits?

a. Find the *total number* of possible identification numbers. Assume that you are scheduled to graduate in 2015.
b. Find the *probability* of randomly generating your identification number.
                                                                Answer: Page A35

3.1 EXERCISES

■ BUILDING BASIC SKILLS AND VOCABULARY

1. What is the difference between an outcome and an event?

2. Determine which of the following numbers could not represent the probability of an event. Explain your reasoning.

   (a) 33.3%   (b) −1.5   (c) 0.0002   (d) 0   (e) $\frac{320}{1058}$ (f) $\frac{64}{25}$

3. Explain why the following statement is incorrect: *The probability of rain tomorrow is 150%.*

4. When you use the Fundamental Counting Principle, what are you counting?

5. Use your own words to describe the law of large numbers. Give an example.

6. List the three formulas that can be used to describe complementary events.

True or False? *In Exercises 7–10, determine whether the statement is true or false. If it is false, rewrite it as a true statement.*

7. If you roll a six-sided die six times, you will roll an even number at least once.

8. You toss a fair coin nine times and it lands tails up each time. The probability it will land heads up on the tenth flip is greater than 0.5.

9. A probability of $\frac{1}{10}$ indicates an unusual event.

10. If an event is almost certain to happen, its complement will be an unusual event.

Matching Probabilities *In Exercises 11–14, match the event with its probability.*

(a) 0.95   *(b) 0.05*   *(c) 0.25*   *(d) 0*

11. You toss a coin and randomly select a number from 0 to 9. What is the probability of getting tails and selecting a 3?

12. A random number generator is used to select a number from 1 to 100. What is the probability of selecting the number 153?

13. A game show contestant must randomly select a door. One door doubles her money while the other three doors leave her with no winnings. What is the probability she selects the door that doubles her money?

14. Five of the 100 digital video recorders (DVRs) in an inventory are known to be defective. What is the probability you randomly select an item that is not defective?

■ USING AND INTERPRETING CONCEPTS

Identifying a Sample Space *In Exercises 15–20, identify the sample space of the probability experiment and determine the number of outcomes in the sample space. Draw a tree diagram if it is appropriate.*

15. Guessing the initial of a student's middle name

16. Guessing a student's letter grade (A, B, C, D, F) in a class

17. Drawing one card from a standard deck of cards

18. Tossing three coins

19. Determining a person's blood type (A, B, AB, O) and Rh-factor (positive, negative)

20. Rolling a pair of six-sided dice

Recognizing Simple Events   *In Exercises 21–24, determine the number of outcomes in each event. Then decide whether the event is a simple event or not. Explain your reasoning.*

21. A computer is used to randomly select a number between 1 and 4000. Event *A* is selecting 253.

22. A computer is used to randomly select a number between 1 and 4000. Event *B* is selecting a number less than 500.

23. You randomly select one card from a standard deck. Event *A* is selecting an ace.

24. You randomly select one card from a standard deck. Event *B* is selecting a ten of diamonds.

25. Job Openings   A software company is hiring for two positions: a software development engineer and a sales operations manager. How many ways can these positions be filled if there are 12 people applying for the engineering position and 17 people applying for the managerial position?

26. Menu   A restaurant offers a $12 dinner special that has 5 choices for an appetizer, 10 choices for entrées, and 4 choices for dessert. How many different meals are available if you select an appetizer, an entrée, and a dessert?

27. Realty   A realtor uses a lock box to store the keys for a house that is for sale. The access code for the lock box consists of four digits. The first digit cannot be zero and the last digit must be even. How many different codes are available?

28. True or False Quiz   Assuming that no questions are left unanswered, in how many ways can a six-question true-false quiz be answered?

Classical Probabilities   *In Exercises 29–34, a probability experiment consists of rolling a 12-sided die. Find the probability of each event.*

29. Event *A*: rolling a 2

30. Event *B*: rolling a 10

31. Event *C*: rolling a number greater than 4

32. Event *D*: rolling an even number

33. Event *E*: rolling a prime number

34. Event *F*: rolling a number divisible by 5

Classifying Types of Probability   *In Exercises 35 and 36, classify the statement as an example of classical probability, empirical probability, or subjective probability. Explain your reasoning.*

35. According to company records, the probability that a washing machine will need repairs during a six-year period is 0.10.

36. The probability of choosing 6 numbers from 1 to 40 that match the 6 numbers drawn by a state lottery is $1/3,838,380 \approx 0.00000026$.

Finding Probabilities   *In Exercises 37–40, consider a company that selects employees for random drug tests. The company uses a computer to randomly select employee numbers that range from 1 to 6296.*

37. Find the probability of selecting a number less than 1000.

38. Find the probability of selecting a number greater than 1000.

39. Find the probability of selecting a number divisible by 1000.

40. Find the probability of selecting a number that is not divisible by 1000.

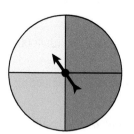

FIGURE FOR EXERCISES 41–44

Probability Experiment   *In Exercises 41–44, a probability experiment consists of rolling a six-sided die and spinning the spinner shown at the left. The spinner is equally likely to land on each color. Use a tree diagram to find the probability of each event. Then tell whether the event can be considered unusual.*

41. Event *A*: rolling a 5 and the spinner landing on blue

42. Event *B*: rolling an odd number and the spinner landing on green

43. Event *C*: rolling a number less than 6 and the spinner landing on yellow

44. Event *D*: not rolling a number less than 6 and the spinner landing on yellow

45. Security System   The access code for a garage door consists of three digits. Each digit can be any number from 0 through 9, and each digit can be repeated.

(a) Find the number of possible access codes.

(b) What is the probability of randomly selecting the correct access code on the first try?

(c) What is the probability of not selecting the correct access code on the first try?

46. Security System   An access code consists of a letter followed by four digits. Any letter can be used, the first digit cannot be 0, and the last digit must be even.

(a) Find the number of possible access codes.

(b) What is the probability of randomly selecting the correct access code on the first try?

(c) What is the probability of not selecting the correct access code on the first try?

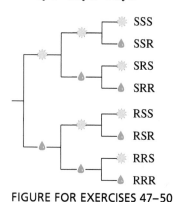

Day 1   Day 2   Day 3

SSS
SSR
SRS
SRR
RSS
RSR
RRS
RRR

FIGURE FOR EXERCISES 47–50

Wet or Dry?   *You are planning a three-day trip to Seattle, Washington in October. In Exercises 47–50, use the tree diagram shown at the left to answer each question.*

47. List the sample space.

48. List the outcome(s) of the event "It rains all three days."

49. List the outcome(s) of the event "It rains on exactly one day."

50. List the outcome(s) of the event "It rains on at least one day."

51. Sunny and Rainy Days   You are planning a four-day trip to Seattle, Washington in October.

(a) Make a sunny day/rainy day tree diagram for your trip.

(b) List the sample space.

(c) List the outcome(s) of the event "It rains on exactly one day."

52. Machine Part Suppliers   Your company buys machine parts from three different suppliers. Make a tree diagram that shows the three suppliers and whether the parts they supply are defective.

Graphical Analysis   *In Exercises 53 and 54, use the diagram to answer the question.*

53. What is the probability that a registered voter in Virginia voted in the 2009 gubernatorial election? *(Source: Commonwealth of Virginia State Board of Elections)*

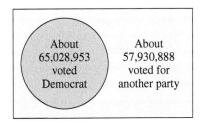

FIGURE FOR EXERCISE 53   FIGURE FOR EXERCISE 54

54. What is the probability that a voter chosen at random did not vote for a Democratic representative in the 2008 election? *(Source: Federal Election Commission)*

Using a Frequency Distribution to Find Probabilities   *In Exercises 55–58, use the frequency distribution at the left, which shows the number of American voters (in millions) according to age, to find the probability that a voter chosen at random is in the given age range.* *(Source: U.S. Census Bureau)*

55. between 18 and 20 years old   **56.** between 35 and 44 years old

57. not between 21 and 24 years old   **58.** not between 45 and 64 years old

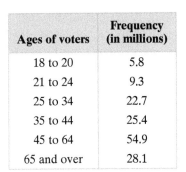

Ages of voters	Frequency (in millions)
18 to 20	5.8
21 to 24	9.3
25 to 34	22.7
35 to 44	25.4
45 to 64	54.9
65 and over	28.1

TABLE FOR EXERCISES 55–58

Using a Bar Graph to Find Probabilities   *In Exercises 59–62, use the following bar graph, which shows the highest level of education received by employees of a company.*

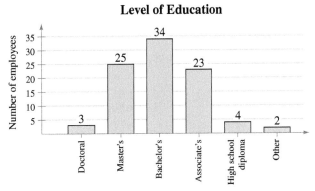

Find the probability that the highest level of education for an employee chosen at random is

59. a doctorate.   **60.** an associate's degree.

61. a master's degree.   **62.** a high school diploma.

63. Can any of the events in Exercises 55–58 be considered unusual? Explain.

64. Can any of the events in Exercises 59–62 be considered unusual? Explain.

Parents
Ssmm and SsMm

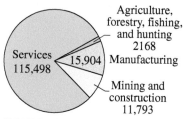

	SM	**Sm**
Sm	SSMm	SSmm
Sm	SSMm	SSmm
sm	SsMm	Ssmm
sm	SsMm	Ssmm

	sM	**sm**
Sm	SsMm	Ssmm
Sm	SsMm	Ssmm
sm	ssMm	ssmm
sm	ssMm	ssmm

TABLE FOR EXERCISE 66

Workers (in thousands) by Industry for the U.S.

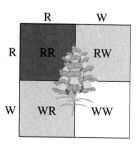

Agriculture, forestry, fishing, and hunting 2168

Services 115,498 15,904 Manufacturing

Mining and construction 11,793

FIGURE FOR EXERCISES 67–70

65. Genetics A *Punnett square* is a diagram that shows all possible gene combinations in a cross of parents whose genes are known. When two pink snapdragon flowers (RW) are crossed, there are four equally likely possible outcomes for the genetic makeup of the offspring: red (RR), pink (RW), pink (WR), and white (WW), as shown in the Punnett square. If two pink snapdragons are crossed, what is the probability that the offspring will be (a) pink, (b) red, and (c) white?

	R	W
R	RR	RW
W	WR	WW

66. Genetics There are six basic types of coloring in registered collies: sable (SSmm), tricolor (ssmm), trifactored sable (Ssmm), blue merle (ssMm), sable merle (SSMm), and trifactored sable merle (SsMm). The Punnett square at the left shows the possible coloring of the offspring of a trifactored sable merle collie and a trifactored sable collie. What is the probability that the offspring will have the same coloring as one of its parents?

Using a Pie Chart to Find Probabilities *In Exercises 67–70, use the pie chart at the left, which shows the number of workers (in thousands) by industry for the United States.* (*Source: U.S. Bureau of Labor Statistics*)

67. Find the probability that a worker chosen at random was employed in the services industry.

68. Find the probability that a worker chosen at random was employed in the manufacturing industry.

69. Find the probability that a worker chosen at random was not employed in the services industry.

70. Find the probability that a worker chosen at random was not employed in the agriculture, forestry, fishing, and hunting industry.

71. College Football A stem-and-leaf plot for the number of touchdowns scored by all NCAA Division I Football Bowl Subdivision teams is shown. If a team is selected at random, find the probability the team scored (a) at least 51 touchdowns, (b) between 20 and 30 touchdowns, inclusive, and (c) more than 69 touchdowns. Are any of these events unusual? Explain. (*Source: NCAA*)

```
1 | 8 8 9                    Key: 1|8 = 18
2 | 1 1 3 4 4 5 5 6 6 7 7 8 8 9 9
3 | 0 1 1 2 3 3 3 3 3 4 4 4 4 4 5 5 5 5 5 7 7 7 7 8 8 9 9 9
4 | 0 0 0 0 1 2 2 2 2 3 3 4 4 4 4 4 4 4 5 5 5 5 5 5 6 6 6 6 6 7 7 7 8 8 8 8 8 9 9 9
5 | 0 0 0 0 1 1 2 2 2 2 4 5 5 5 6 6 7 9
6 | 0 0 1 2 2 3 4 5 6 8 9
7 | 6 7
```

72. Individual Stock Price An individual stock is selected at random from the portfolio represented by the box-and-whisker plot shown. Find the probability that the stock price is (a) less than $21, (b) between $21 and $50, and (c) $30 or more.

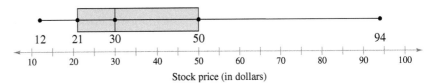

Writing *In Exercises 73 and 74, write a statement that represents the complement of the given probability.*

73. The probability of randomly choosing a tea drinker who has a college degree (Assume that you are choosing from the population of all tea drinkers.)

74. The probability of randomly choosing a smoker whose mother also smoked (Assume that you are choosing from the population of all smokers.)

■ EXTENDING CONCEPTS

75. Rolling a Pair of Dice You roll a pair of six-sided dice and record the sum.

   (a) List all of the possible sums and determine the probability of rolling each sum.

   (b) Use a technology tool to simulate rolling a pair of dice and recording the sum 100 times. Make a tally of the 100 sums and use these results to list the probability of rolling each sum.

   (c) Compare the probabilities in part (a) with the probabilities in part (b). Explain any similarities or differences.

Odds *In Exercises 76–81, use the following information. The chances of winning are often written in terms of odds rather than probabilities. The **odds of winning** is the ratio of the number of successful outcomes to the number of unsuccessful outcomes. The **odds of losing** is the ratio of the number of unsuccessful outcomes to the number of successful outcomes. For example, if the number of successful outcomes is 2 and the number of unsuccessful outcomes is 3, the odds of winning are 2 : 3 (read "2 to 3") or $\frac{2}{3}$.*

76. A beverage company puts game pieces under the caps of its drinks and claims that one in six game pieces wins a prize. The official rules of the contest state that the odds of winning a prize are 1 : 6. Is the claim "one in six game pieces wins a prize" correct? Why or why not?

77. The probability of winning an instant prize game is $\frac{1}{10}$. The odds of winning a different instant prize game are 1 : 10. If you want the best chance of winning, which game should you play? Explain your reasoning.

78. The odds of an event occurring are 4 : 5. Find (a) the probability that the event will occur and (b) the probability that the event will not occur.

79. A card is picked at random from a standard deck of 52 playing cards. Find the odds that it is a spade.

80. A card is picked at random from a standard deck of 52 playing cards. Find the odds that it is not a spade.

81. The odds of winning an event A are $p : q$. Show that the probability of event A is given by $P(A) = \dfrac{p}{p + q}$.

ACTIVITY 3.1 Simulating the Stock Market

APPLET

The *simulating the stock market* applet allows you to investigate the probability that the stock market will go up on any given day. The plot at the top left corner shows the probability associated with each outcome. In this case, the market has a 50% chance of going up on any given day. When SIMULATE is clicked, outcomes for *n* days are simulated. The results of the simulations are shown in the frequency plot. If the animate option is checked, the display will show each outcome dropping into the frequency plot as the simulation runs. The individual outcomes are shown in the text field at the far right of the applet. The center plot shows in red the cumulative proportion of times that the market went up. The green line in the plot reflects the true probability of the market going up. As the experiment is conducted over and over, the cumulative proportion should converge to the true value.

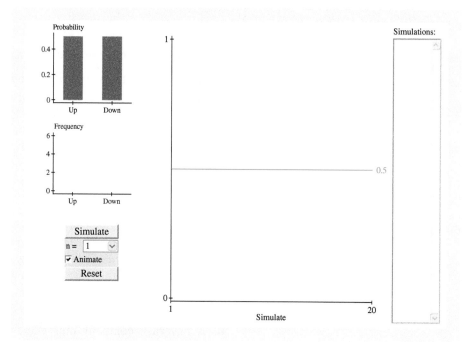

■ Explore

Step 1 Specify a value for *n*.
Step 2 Click SIMULATE four times.
Step 3 Click RESET.
Step 4 Specify another value for *n*.
Step 5 Click SIMULATE.

■ Draw Conclusions

APPLET

1. Run the simulation using $n = 1$ without clicking RESET. How many days did it take until there were three straight days on which the stock market went up? three straight days on which the stock market went down?

2. Run the applet to simulate the stock market activity over the last 35 business days. Find the empirical probability that the market goes up on day 36.

3.2   Conditional Probability and the Multiplication Rule

WHAT YOU SHOULD LEARN

▸ How to find the probability of an event given that another event has occurred

▸ How to distinguish between independent and dependent events

▸ How to use the Multiplication Rule to find the probability of two events occurring in sequence

▸ How to use the Multiplication Rule to find conditional probabilities

Conditional Probability ▸ Independent and Dependent Events ▸ The Multiplication Rule

▸ CONDITIONAL PROBABILITY

In this section, you will learn how to find the probability that two events occur in sequence. Before you can find this probability, however, you must know how to find *conditional probabilities*.

DEFINITION

A **conditional probability** is the probability of an event occurring, given that another event has already occurred. The conditional probability of event B occurring, given that event A has occurred, is denoted by $P(B|A)$ and is read as "probability of B, given A."

EXAMPLE 1

▸ **Finding Conditional Probabilities**

1. Two cards are selected in sequence from a standard deck. Find the probability that the second card is a queen, given that the first card is a king. (Assume that the king is not replaced.)

2. The table at the left shows the results of a study in which researchers examined a child's IQ and the presence of a specific gene in the child. Find the probability that a child has a high IQ, given that the child has the gene.

	Gene present	Gene not present	Total
High IQ	33	19	52
Normal IQ	39	11	50
Total	72	30	102

▸ Solution

1. Because the first card is a king and is not replaced, the remaining deck has 51 cards, 4 of which are queens. So,

$$P(B|A) = \frac{4}{51} \approx 0.078.$$

So, the probability that the second card is a queen, given that the first card is a king, is about 0.078.

2. There are 72 children who have the gene. So, the sample space consists of these 72 children, as shown at the left. Of these, 33 have a high IQ. So,

$$P(B|A) = \frac{33}{72} \approx 0.458.$$

So, the probability that a child has a high IQ, given that the child has the gene, is about 0.458.

Sample Space

	Gene present
High IQ	33
Normal IQ	39
Total	72

▸ Try It Yourself 1

1. Find the probability that a child does not have the gene.
2. Find the probability that a child does not have the gene, given that the child has a normal IQ.

a. Find the *number of outcomes* in the event and in the sample space.
b. *Divide* the number of outcomes in the event by the number of outcomes in the sample space.

Answer: Page A35

▶ INDEPENDENT AND DEPENDENT EVENTS

In some experiments, one event does not affect the probability of another. For instance, if you roll a die and toss a coin, the outcome of the roll of the die does not affect the probability of the coin landing on heads. These two events are *independent*. The question of the independence of two or more events is important to researchers in fields such as marketing, medicine, and psychology. You can use conditional probabilities to determine whether events are *independent*.

DEFINITION

Two events are **independent** if the occurrence of one of the events does not affect the probability of the occurrence of the other event. Two events A and B are independent if

$$P(B|A) = P(B) \quad \text{or if} \quad P(A|B) = P(A).$$

Events that are not independent are **dependent.**

To determine if A and B are independent, first calculate $P(B)$, the probability of event B. Then calculate $P(B|A)$, the probability of B, given A. If the values are equal, the events are independent. If $P(B) \neq P(B|A)$, then A and B are dependent events.

EXAMPLE 2

▶ **Classifying Events as Independent or Dependent**

Decide whether the events are independent or dependent.

1. Selecting a king from a standard deck (A), not replacing it, and then selecting a queen from the deck (B)

2. Tossing a coin and getting a head (A), and then rolling a six-sided die and obtaining a 6 (B)

3. Driving over 85 miles per hour (A), and then getting in a car accident (B)

▶ **Solution**

1. $P(B|A) = \frac{4}{51}$ and $P(B) = \frac{4}{52}$. The occurrence of A changes the probability of the occurrence of B, so the events are dependent.

2. $P(B|A) = \frac{1}{6}$ and $P(B) = \frac{1}{6}$. The occurrence of A does not change the probability of the occurrence of B, so the events are independent.

3. If you drive over 85 miles per hour, the chances of getting in a car accident are greatly increased, so these events are dependent.

▶ **Try It Yourself 2**

Decide whether the events are independent or dependent.

1. Smoking a pack of cigarettes per day (A) and developing emphysema, a chronic lung disease (B)

2. Exercising frequently (A) and having a 4.0 grade point average (B)

a. *Decide* whether the occurrence of the first event affects the probability of the second event.

b. *State* if the events are *independent* or *dependent*.   *Answer: Page A35*

▶ THE MULTIPLICATION RULE

To find the probability of two events occurring in sequence, you can use the Multiplication Rule.

THE MULTIPLICATION RULE FOR THE PROBABILITY OF *A* AND *B*

The probability that two events *A* and *B* will occur in sequence is

$$P(A \text{ and } B) = P(A) \cdot P(B|A).$$

If events *A* and *B* are independent, then the rule can be simplified to $P(A \text{ and } B) = P(A) \cdot P(B)$. This simplified rule can be extended to any number of independent events.

EXAMPLE 3

▶ **Using the Multiplication Rule to Find Probabilities**

1. Two cards are selected, without replacing the first card, from a standard deck. Find the probability of selecting a king and then selecting a queen.

2. A coin is tossed and a die is rolled. Find the probability of tossing a head and then rolling a 6.

▶ **Solution**

1. Because the first card is not replaced, the events are dependent.

$$P(K \text{ and } Q) = P(K) \cdot P(Q|K)$$

$$= \frac{4}{52} \cdot \frac{4}{51}$$

$$= \frac{16}{2652}$$

$$\approx 0.006$$

So, the probability of selecting a king and then a queen is about 0.006.

2. The events are independent.

$$P(H \text{ and } 6) = P(H) \cdot P(6)$$

$$= \frac{1}{2} \cdot \frac{1}{6}$$

$$= \frac{1}{12}$$

$$\approx 0.083$$

So, the probability of tossing a head and then rolling a 6 is about 0.083.

▶ **Try It Yourself 3**

1. The probability that a salmon swims successfully through a dam is 0.85. Find the probability that two salmon swim successfully through the dam.

2. Two cards are selected from a standard deck without replacement. Find the probability that they are both hearts.

   a. Decide if the events are *independent* or *dependent*.

   b. Use the *Multiplication Rule* to find the probability.   *Answer: Page A35*

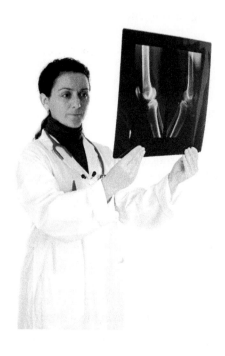

EXAMPLE 4

▶ **Using the Multiplication Rule to Find Probabilities**

The probability that a particular knee surgery is successful is 0.85.

1. Find the probability that three knee surgeries are successful.

2. Find the probability that none of the three knee surgeries are successful.

3. Find the probability that at least one of the three knee surgeries is successful.

▶ **Solution**

1. The probability that each knee surgery is successful is 0.85. The chance of success for one surgery is independent of the chances for the other surgeries.

$$P(\text{three surgeries are successful}) = (0.85)(0.85)(0.85)$$

$$\approx 0.614$$

So, the probability that all three surgeries are successful is about 0.614.

2. Because the probability of success for one surgery is 0.85, the probability of failure for one surgery is $1 - 0.85 = 0.15$.

$$P(\text{none of the three are successful}) = (0.15)(0.15)(0.15)$$

$$\approx 0.003$$

So, the probability that none of the surgeries are successful is about 0.003. Because 0.003 is less than 0.05, this can be considered an unusual event.

3. The phrase "at least one" means one or more. The complement to the event "at least one is successful" is the event "none are successful." Use the complement to find the probability.

$$P(\text{at least one is successful}) = 1 - P(\text{none are successful})$$

$$\approx 1 - 0.003$$

$$= 0.997.$$

So, the probability that at least one of the three surgeries is successful is about 0.997.

▶ **Try It Yourself 4**

The probability that a particular rotator cuff surgery is successful is 0.9. *(Source: The Orthopedic Center of St. Louis)*

1. Find the probability that three rotator cuff surgeries are successful.
2. Find the probability that none of the three rotator cuff surgeries are successful.
3. Find the probability that at least one of the three rotator cuff surgeries is successful.

a. Decide whether to find the probability of the event or its complement.
b. Use the *Multiplication Rule* to find the probability. If necessary, use the *complement*.
c. Determine if the event is *unusual*. Explain.   *Answer: Page A35*

In Example 4, you were asked to find a probability using the phrase "at least one." Notice that it was easier to find the probability of its complement, "none," and then subtract the probability of its complement from 1.

EXAMPLE 5

▶ **Using the Multiplication Rule to Find Probabilities**

More than 15,000 U.S. medical school seniors applied to residency programs in 2009. Of those, 93% were matched with residency positions. Eighty-two percent of the seniors matched with residency positions were matched with one of their top three choices. Medical students electronically rank the residency programs in their order of preference, and program directors across the United States do the same. The term "match" refers to the process whereby a student's preference list and a program director's preference list overlap, resulting in the placement of the student in a residency position. *(Source: National Resident Matching Program)*

1. Find the probability that a randomly selected senior was matched with a residency position *and* it was one of the senior's top three choices.

2. Find the probability that a randomly selected senior who was matched with a residency position did *not* get matched with one of the senior's top three choices.

3. Would it be unusual for a randomly selected senior to be matched with a residency position *and* that it was one of the senior's top three choices?

▶ **Solution**

Let A = {matched with residency position} and B = {matched with one of top three choices}. So, $P(A) = 0.93$ and $P(B|A) = 0.82$.

1. The events are dependent.

$$P(A \text{ and } B) = P(A) \cdot P(B|A) = (0.93) \cdot (0.82) \approx 0.763$$

So, the probability that a randomly selected senior was matched with one of the senior's top three choices is about 0.763.

2. To find this probability, use the complement.

$$P(B'|A) = 1 - P(B|A) = 1 - 0.82 = 0.18$$

So, the probability that a randomly selected senior was matched with a residency position that was not one of the senior's top three choices is 0.18.

3. It is not unusual because the probability of a senior being matched with a residency position that was one of the senior's top three choices is about 0.763, which is greater than 0.05.

▶ **Try It Yourself 5**

In a jury selection pool, 65% of the people are female. Of these 65%, one out of four works in a health field.

1. Find the probability that a randomly selected person from the jury pool is female and works in a health field.

2. Find the probability that a randomly selected person from the jury pool is female and does not work in a health field.

a. Determine *events A* and *B*.

b. Use the *Multiplication Rule* to write a formula to find the probability. If necessary, use the *complement*.

c. *Calculate* the probability. *Answer: Page A35*

Medical School

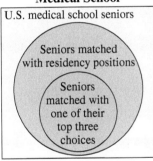

U.S. medical school seniors

Seniors matched with residency positions

Seniors matched with one of their top three choices

Jury Selection

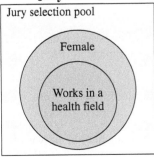

Jury selection pool

Female

Works in a health field

3.2) EXERCISES

■ BUILDING BASIC SKILLS AND VOCABULARY

1. What is the difference between independent and dependent events?

2. List examples of

  (a) two events that are independent.

  (b) two events that are dependent.

3. What does the notation $P(B|A)$ mean?

4. Explain how the complement can be used to find the probability of getting at least one item of a particular type.

True or False?   *In Exercises 5 and 6, determine whether the statement is true or false. If it is false, rewrite it as a true statement.*

5. If two events are independent, $P(A|B) = P(B)$.

6. If events A and B are dependent, then $P(A \text{ and } B) = P(A) \cdot P(B)$.

Classifying Events   *In Exercises 7–12, decide whether the events are independent or dependent. Explain your reasoning.*

7. Selecting a king from a standard deck, replacing it, and then selecting a queen from the deck

8. Returning a rented movie after the due date and receiving a late fee

9. A father having hazel eyes and a daughter having hazel eyes

10. Not putting money in a parking meter and getting a parking ticket

11. Rolling a six-sided die and then rolling the die a second time so that the sum of the two rolls is five

12. A ball numbered from 1 through 52 is selected from a bin, replaced, and then a second numbered ball is selected from the bin.

Classifying Events Based on Studies   *In Exercises 13–16, identify the two events described in the study. Do the results indicate that the events are independent or dependent? Explain your reasoning.*

13. A study found that people who suffer from moderate to severe sleep apnea are at increased risk of having high blood pressure. *(Source: Journal of the American Medical Association)*

14. Stress causes the body to produce higher amounts of acid, which can irritate already existing ulcers. But, stress does not cause stomach ulcers. *(Source: Baylor College of Medicine)*

15. Studies found that exposure to everyday sources of aluminum does not cause Alzheimer's disease. *(Source: Alzheimer's Association)*

16. According to researchers, diabetes is rare in societies in which obesity is rare. In societies in which obesity has been common for at least 20 years, diabetes is also common. *(Source: American Diabetes Association)*

■ USING AND INTERPRETING CONCEPTS

17. BRCA Gene   In the general population, one woman in eight will develop breast cancer. Research has shown that approximately 1 woman in 600 carries a mutation of the BRCA gene. About 6 out of 10 women with this mutation develop breast cancer. *(Adapted from Susan G. Komen Breast Cancer Foundation)*

(a) Find the probability that a randomly selected woman will develop breast cancer, given that she has a mutation of the BRCA gene.

(b) Find the probability that a randomly selected woman will carry the mutation of the BRCA gene and will develop breast cancer.

(c) Are the events "carrying this mutation" and "developing breast cancer" independent or dependent? Explain.

Breast Cancer and the BRCA Gene

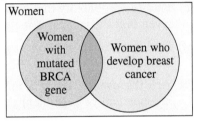

FIGURE FOR EXERCISE 17

What Do You Drive?

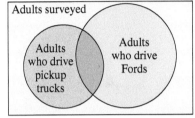

FIGURE FOR EXERCISE 18

18. Pickup Trucks   In a survey, 510 adults were asked if they drive a pickup truck and if they drive a Ford. The results showed that one in six adults surveyed drives a pickup truck, and three in ten adults surveyed drive a Ford. Of the adults surveyed that drive Fords, two in nine drive a pickup truck.

(a) Find the probability that a randomly selected adult drives a pickup truck, given that the adult drives a Ford.

(b) Find the probability that a randomly selected adult drives a Ford and drives a pickup truck.

(c) Are the events "driving a Ford" and "driving a pickup truck" independent or dependent? Explain.

19. Summer Vacation   The table shows the results of a survey in which 146 families were asked if they own a computer and if they will be taking a summer vacation during the current year.

		Summer Vacation This Year		
		Yes	**No**	**Total**
Own a Computer	**Yes**	87	28	115
	No	14	17	31
	Total	101	45	146

(a) Find the probability that a randomly selected family is not taking a summer vacation this year.

(b) Find the probability that a randomly selected family owns a computer.

(c) Find the probability that a randomly selected family is taking a summer vacation this year, given that they own a computer.

(d) Find the probability that a randomly selected family is taking a summer vacation this year and owns a computer.

(e) Are the events "owning a computer" and "taking a summer vacation this year" independent or dependent events? Explain.

20. Nursing Majors The table shows the number of male and female students enrolled in nursing at the University of Oklahoma Health Sciences Center for a recent semester. *(Source: University of Oklahoma Health Sciences Center Office of Institutional Research)*

	Nursing majors	Non-nursing majors	Total
Males	151	1104	1255
Females	1016	1693	2709
Total	1167	2797	3964

(a) Find the probability that a randomly selected student is a nursing major.

(b) Find the probability that a randomly selected student is male.

(c) Find the probability that a randomly selected student is a nursing major, given that the student is male.

(d) Find the probability that a randomly selected student is a nursing major and male.

(e) Are the events "being a male student" and "being a nursing major" independent or dependent events? Explain.

21. Assisted Reproductive Technology A study found that 37% of the assisted reproductive technology (ART) cycles resulted in pregnancies. Twenty-five percent of the ART pregnancies resulted in multiple births. *(Source: National Center for Chronic Disease Prevention and Health Promotion)*

(a) Find the probability that a randomly selected ART cycle resulted in a pregnancy *and* produced a multiple birth.

(b) Find the probability that a randomly selected ART cycle that resulted in a pregnancy did *not* produce a multiple birth.

(c) Would it be unusual for a randomly selected ART cycle to result in a pregnancy and produce a multiple birth? Explain.

22. Government According to a survey, 86% of adults in the United States think the U.S. government system is broken. Of these 86%, about 8 out of 10 think the government can be fixed. *(Adapted from CNN/Opinion Research Corporation)*

(a) Find the probability that a randomly selected adult thinks the U.S. government system is broken *and* thinks the government can be fixed.

(b) Given that a randomly selected adult thinks the U.S. government system is broken, find the probability that he or she thinks the government *cannot* be fixed.

(c) Would it be unusual for a randomly selected adult to think the U.S. government system is broken and think the government can be fixed? Explain.

23. Computers and Internet Access A study found that 81% of households in the United States have computers. Of those 81%, 92% have Internet access. Find the probability that a U.S. household selected at random has a computer and has Internet access. *(Source: The Nielsen Company)*

24. Surviving Surgery A doctor gives a patient a 60% chance of surviving bypass surgery after a heart attack. If the patient survives the surgery, he has a 50% chance that the heart damage will heal. Find the probability that the patient survives surgery and the heart damage heals.

Pregnancies

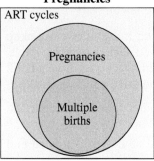

FIGURE FOR EXERCISE 21

Government

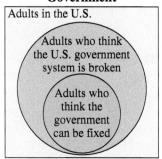

FIGURE FOR EXERCISE 22

25. People Who Can Wiggle Their Ears In a sample of 1000 people, 130 can wiggle their ears. Two unrelated people are selected at random without replacement.

(a) Find the probability that both people can wiggle their ears.

(b) Find the probability that neither person can wiggle his or her ears.

(c) Find the probability that at least one of the two people can wiggle his or her ears.

(d) Which of the events can be considered unusual? Explain.

26. Batteries Sixteen batteries are tested to see if they last as long as the manufacturer claims. Four batteries fail the test. Two batteries are selected at random without replacement.

(a) Find the probability that both batteries fail the test.

(b) Find the probability that both batteries pass the test.

(c) Find the probability that at least one battery fails the test.

(d) Which of the events can be considered unusual? Explain.

27. Emergency Savings The table shows the results of a survey in which 142 male and 145 female workers ages 25 to 64 were asked if they had at least one month's income set aside for emergencies.

	Male	Female	Total
Less than one month's income	66	83	149
One month's income or more	76	62	138
Total	142	145	287

(a) Find the probability that a randomly selected worker has one month's income or more set aside for emergencies.

(b) Given that a randomly selected worker is a male, find the probability that the worker has less than one month's income.

(c) Given that a randomly selected worker has one month's income or more, find the probability that the worker is a female.

(d) Are the events "having less than one month's income saved" and "being male" independent or dependent? Explain.

28. Health Care for Dogs The table shows the results of a survey in which 90 dog owners were asked how much they had spent in the last year for their dog's health care, and whether their dogs were purebred or mixed breeds.

		Type of Dog		
		Purebred	Mixed breed	Total
Health Care	**Less than $100**	19	21	40
	$100 or more	35	15	50
	Total	54	36	90

(a) Find the probability that $100 or more was spent on a randomly selected dog's health care in the last year.

(b) Given that a randomly selected dog owner spent less than $100, find the probability that the dog was a mixed breed.

(c) Find the probability that a randomly selected dog owner spent $100 or more on health care and the dog was a mixed breed.

(d) Are the events "spending $100 or more on health care" and "having a mixed breed dog" independent or dependent? Explain.

29. Blood Types The probability that a person in the United States has type B^+ blood is 9%. Five unrelated people in the United States are selected at random. *(Source: American Association of Blood Banks)*

(a) Find the probability that all five have type B^+ blood.

(b) Find the probability that none of the five have type B^+ blood.

(c) Find the probability that at least one of the five has type B^+ blood.

30. Blood Types The probability that a person in the United States has type A^+ blood is 31%. Three unrelated people in the United States are selected at random. *(Source: American Association of Blood Banks)*

(a) Find the probability that all three have type A^+ blood.

(b) Find the probability that none of the three have type A^+ blood.

(c) Find the probability that at least one of the three has type A^+ blood.

31. Guessing A multiple-choice quiz has five questions, each with four answer choices. Only one of the choices is correct. You have no idea what the answer is to any question and have to guess each answer.

(a) Find the probability of answering the first question correctly.

(b) Find the probability of answering the first two questions correctly.

(c) Find the probability of answering all five questions correctly.

(d) Find the probability of answering none of the questions correctly.

(e) Find the probability of answering at least one of the questions correctly.

32. Bookbinding Defects A printing company's bookbinding machine has a probability of 0.005 of producing a defective book. This machine is used to bind three books.

(a) Find the probability that none of the books are defective.

(b) Find the probability that at least one of the books is defective.

(c) Find the probability that all of the books are defective.

33. Warehouses A distribution center receives shipments of a product from three different factories in the following quantities: 50, 35, and 25. Three times a product is selected at random, each time without replacement. Find the probability that (a) all three products came from the third factory and (b) none of the three products came from the third factory.

34. Birthdays Three people are selected at random. Find the probability that (a) all three share the same birthday and (b) none of the three share the same birthday. Assume 365 days in a year.

■ **EXTENDING CONCEPTS**

*According to **Bayes' Theorem**, the probability of event A, given that event B has occurred, is*

$$P(A|B) = \frac{P(A) \cdot P(B|A)}{P(A) \cdot P(B|A) + P(A') \cdot P(B|A')}.$$

In Exercises 35–38, use Bayes' Theorem to find $P(A|B)$.

35. $P(A) = \frac{2}{3}$, $P(A') = \frac{1}{3}$, $P(B|A) = \frac{1}{5}$, and $P(B|A') = \frac{1}{2}$

36. $P(A) = \frac{3}{8}$, $P(A') = \frac{5}{8}$, $P(B|A) = \frac{2}{3}$, and $P(B|A') = \frac{3}{5}$

37. $P(A) = 0.25$, $P(A') = 0.75$, $P(B|A) = 0.3$, and $P(B|A') = 0.5$

38. $P(A) = 0.62$, $P(A') = 0.38$, $P(B|A) = 0.41$, and $P(B|A') = 0.17$

39. Reliability of Testing   A certain virus infects one in every 200 people. A test used to detect the virus in a person is positive 80% of the time if the person has the virus and 5% of the time if the person does not have the virus. (This 5% result is called a *false positive*.) Let A be the event "the person is infected" and B be the event "the person tests positive."

(a) Using Bayes' Theorem, if a person tests positive, determine the probability that the person is infected.

(b) Using Bayes' Theorem, if a person tests negative, determine the probability that the person is *not* infected.

40. Birthday Problem   You are in a class that has 24 students. You want to find the probability that at least two of the students share the same birthday.

(a) First, find the probability that each student has a different birthday.

$$P(\text{different birthdays}) = \overbrace{\frac{365}{365} \cdot \frac{364}{365} \cdot \frac{363}{365} \cdot \frac{362}{365} \cdots \frac{343}{365} \cdot \frac{342}{365}}^{24 \text{ factors}}$$

(b) The probability that at least two students have the same birthday is the complement of the probability in part (a). What is this probability?

(c) We used a technology tool to generate 24 random numbers between 1 and 365. Each number represents a birthday. Did we get at least two people with the same birthday?

228	348	181	317	81	183
52	346	177	118	315	273
252	168	281	266	285	13
118	360	8	193	57	107

(d) Use a technology tool to simulate the "Birthday Problem." Repeat the simulation 10 times. How many times did you get at least two people with the same birthday?

The Multiplication Rule and Conditional Probability   *By rewriting the formula for the Multiplication Rule, you can write a formula for finding conditional probabilities. The conditional probability of event B occurring, given that event A has occurred, is*

$$P(B|A) = \frac{P(A \text{ and } B)}{P(A)}.$$

In Exercises 41 and 42, use the following information.

- *The probability that an airplane flight departs on time is 0.89.*

- *The probability that a flight arrives on time is 0.87.*

- *The probability that a flight departs and arrives on time is 0.83.*

41. Find the probability that a flight departed on time given that it arrives on time.

42. Find the probability that a flight arrives on time given that it departed on time.

3.3 The Addition Rule

- How to determine if two events are mutually exclusive

- How to use the Addition Rule to find the probability of two events

Mutually Exclusive Events ▸ The Addition Rule ▸ A Summary of Probability

▸ MUTUALLY EXCLUSIVE EVENTS

In Section 3.2, you learned how to find the probability of two events, A and B, occurring in sequence. Such probabilities are denoted by $P(A \text{ and } B)$. In this section, you will learn how to find the probability that at least one of two events will occur. Probabilities such as these are denoted by $P(A \text{ or } B)$ and depend on whether the events are *mutually exclusive*.

> ### DEFINITION
>
> Two events A and B are **mutually exclusive** if A and B cannot occur at the same time.

STUDY TIP

In probability and statistics, the word *or* is usually used as an "inclusive or" rather than an "exclusive or." For instance, there are three ways for "event A or B" to occur.

(1) A occurs and B does not occur.

(2) B occurs and A does not occur.

(3) A and B both occur.

The Venn diagrams show the relationship between events that are mutually exclusive and events that are not mutually exclusive.

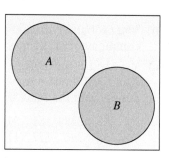

A and B are mutually exclusive.

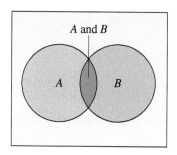

A and B are not mutually exclusive.

EXAMPLE 1

▸ **Mutually Exclusive Events**

Decide if the events are mutually exclusive. Explain your reasoning.

1. Event A: Roll a 3 on a die.
   Event B: Roll a 4 on a die.

2. Event A: Randomly select a male student.
   Event B: Randomly select a nursing major.

3. Event A: Randomly select a blood donor with type O blood.
   Event B: Randomly select a female blood donor.

▸ **Solution**

1. The first event has one outcome, a 3. The second event also has one outcome, a 4. These outcomes cannot occur at the same time, so the events are mutually exclusive.

2. Because the student can be a male nursing major, the events are not mutually exclusive.

3. Because the donor can be a female with type O blood, the events are not mutually exclusive.

▶ Try It Yourself 1

Decide if the events are mutually exclusive. Explain your reasoning.

1. Event *A*: Randomly select a jack from a standard deck of cards.
   Event *B*: Randomly select a face card from a standard deck of cards.
2. Event *A*: Randomly select a 20-year-old student.
   Event *B*: Randomly select a student with blue eyes.
3. Event *A*: Randomly select a vehicle that is a Ford.
   Event *B*: Randomly select a vehicle that is a Toyota.

a. Decide if one of the following statements is true.
   - Events *A* and *B* cannot occur at the same time.
   - Events *A* and *B* have no outcomes in common.
   - $P(A \text{ and } B) = 0$

b. Make a *conclusion*.   *Answer: Page A35*

▶ THE ADDITION RULE

STUDY TIP

By subtracting $P(A \text{ and } B)$ you avoid double counting the probability of outcomes that occur in both *A* and *B*.

To explore this topic further, see Activity 3.3 on page 166.

THE ADDITION RULE FOR THE PROBABILITY OF *A* OR *B*

The probability that events *A* or *B* will occur, $P(A \text{ or } B)$, is given by

$$P(A \text{ or } B) = P(A) + P(B) - P(A \text{ and } B).$$

If events *A* and *B* are mutually exclusive, then the rule can be simplified to $P(A \text{ or } B) = P(A) + P(B)$. This simplified rule can be extended to any number of mutually exclusive events.

In words, to find the probability that one event or the other will occur, add the individual probabilities of each event and subtract the probability that they both occur.

EXAMPLE 2

▶ Using the Addition Rule to Find Probabilities

1. You select a card from a standard deck. Find the probability that the card is a 4 or an ace.

2. You roll a die. Find the probability of rolling a number less than 3 or rolling an odd number.

▶ Solution

1. If the card is a 4, it cannot be an ace. So, the events are mutually exclusive, as shown in the Venn diagram. The probability of selecting a 4 or an ace is

$$P(4 \text{ or ace}) = P(4) + P(\text{ace}) = \frac{4}{52} + \frac{4}{52} = \frac{8}{52} = \frac{2}{13} \approx 0.154.$$

2. The events are not mutually exclusive because 1 is an outcome of both events, as shown in the Venn diagram. So, the probability of rolling a number less than 3 or an odd number is

$$P(\text{less than 3 or odd}) = P(\text{less than 3}) + P(\text{odd})$$
$$- P(\text{less than 3 and odd})$$
$$= \frac{2}{6} + \frac{3}{6} - \frac{1}{6} = \frac{4}{6} = \frac{2}{3} \approx 0.667.$$

Deck of 52 Cards

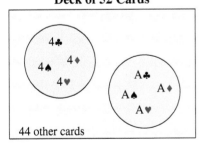

44 other cards

Roll a Die

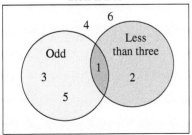

▶ Try It Yourself 2

1. A die is rolled. Find the probability of rolling a 6 or an odd number.
2. A card is selected from a standard deck. Find the probability that the card is a face card or a heart.

a. Decide whether the events are *mutually exclusive*.
b. Find $P(A)$, $P(B)$, and, if necessary, $P(A \text{ and } B)$.
c. Use the *Addition Rule* to find the probability.   *Answer: Page A35*

EXAMPLE 3

▶ **Finding Probabilities of Mutually Exclusive Events**

The frequency distribution shows volumes of sales (in dollars) and the number of months in which a sales representative reached each sales level during the past three years. If this sales pattern continues, what is the probability that the sales representative will sell between $75,000 and $124,999 next month?

Sales volume ($)	Months
0–24,999	3
25,000–49,999	5
50,000–74,999	6
75,000–99,999	7
100,000–124,999	9
125,000–149,999	2
150,000–174,999	3
175,000–199,999	1

▶ **Solution**

To solve this problem, define events A and B as follows.

$A = \{$monthly sales between $75,000 and $99,999$\}$

$B = \{$monthly sales between $100,000 and $124,999$\}$

Because events A and B are mutually exclusive, the probability that the sales representative will sell between $75,000 and $124,999 next month is

$$P(A \text{ or } B) = P(A) + P(B)$$

$$= \frac{7}{36} + \frac{9}{36}$$

$$= \frac{16}{36}$$

$$= \frac{4}{9} \approx 0.444.$$

▶ **Try It Yourself 3**

Find the probability that the sales representative will sell between $0 and $49,999.

a. Identify *events A and B*.
b. Decide if the events are *mutually exclusive*.
c. Find the *probability* of each event.
d. Use the *Addition Rule* to find the probability.   *Answer: Page A35*

EXAMPLE 4

▸ **Using the Addition Rule to Find Probabilities**

A blood bank catalogs the types of blood, including positive or negative Rh-factor, given by donors during the last five days. The number of donors who gave each blood type is shown in the table. A donor is selected at random.

1. Find the probability that the donor has type O or type A blood.

2. Find the probability that the donor has type B blood or is Rh-negative.

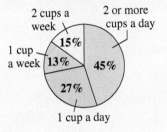
		Blood Type				
		O	**A**	**B**	**AB**	**Total**
Rh-factor	**Positive**	156	139	37	12	344
	Negative	28	25	8	4	65
	Total	184	164	45	16	409

▸ **Solution**

1. Because a donor cannot have type O blood and type A blood, these events are mutually exclusive. So, using the Addition Rule, the probability that a randomly chosen donor has type O or type A blood is

$$P(\text{type O or type A}) = P(\text{type O}) + P(\text{type A})$$

$$= \frac{184}{409} + \frac{164}{409}$$

$$= \frac{348}{409}$$

$$\approx 0.851.$$

2. Because a donor can have type B blood and be Rh-negative, these events are not mutually exclusive. So, using the Addition Rule, the probability that a randomly chosen donor has type B blood or is Rh-negative is

$$P(\text{type B or Rh-neg}) = P(\text{type B}) + P(\text{Rh-neg}) - P(\text{type B and Rh-neg})$$

$$= \frac{45}{409} + \frac{65}{409} - \frac{8}{409}$$

$$= \frac{102}{409}$$

$$\approx 0.249.$$

▸ **Try It Yourself 4**

1. Find the probability that the donor has type B or type AB blood.
2. Find the probability that the donor has type O blood or is Rh-positive.

a. Identify *events A* and *B*.
b. Decide if the events are *mutually exclusive*.
c. Find the *probability* of each event.
d. Use the *Addition Rule* to find the probability.   *Answer: Page A35*

▸ A SUMMARY OF PROBABILITY

Type of Probability and Probability Rules	In Words	In Symbols
Classical Probability	The number of outcomes in the sample space is known and each outcome is equally likely to occur.	$P(E) = \dfrac{\text{Number of outcomes in event } E}{\text{Number of outcomes in sample space}}$
Empirical Probability	The frequency of outcomes in the sample space is estimated from experimentation.	$P(E) = \dfrac{\text{Frequency of event } E}{\text{Total frequency}} = \dfrac{f}{n}$
Range of Probabilities Rule	The probability of an event is between 0 and 1, inclusive.	$0 \le P(E) \le 1$
Complementary Events	The complement of event E is the set of all outcomes in a sample space that are not included in E, denoted by E'.	$P(E') = 1 - P(E)$
Multiplication Rule	The Multiplication Rule is used to find the probability of two events occurring in a sequence.	$P(A \text{ and } B) = P(A) \cdot P(B \mid A)$ $P(A \text{ and } B) = P(A) \cdot P(B)$ *Independent events*
Addition Rule	The Addition Rule is used to find the probability of at least one of two events occurring.	$P(A \text{ or } B) = P(A) + P(B) - P(A \text{ and } B)$ $P(A \text{ or } B) = P(A) + P(B)$ *Mutually exclusive events*

EXAMPLE 5

▸ **Combining Rules to Find Probabilities**

Use the graph at the right to find the probability that a randomly selected draft pick is not a running back or a wide receiver.

(Source: National Football League)

▸ **Solution**

Define events A and B.

　　A: Draft pick is a running back.
　　B: Draft pick is a wide receiver.

These events are mutually exclusive, so the probability that the draft pick is a running back or wide receiver is

$$P(A \text{ or } B) = P(A) + P(B) = \tfrac{22}{256} + \tfrac{34}{256} = \tfrac{56}{256} = \tfrac{7}{32} \approx 0.219.$$

By taking the complement of $P(A \text{ or } B)$, you can determine that the probability of randomly selecting a draft pick who is not a running back or wide receiver is

$$1 - P(A \text{ or } B) = 1 - \tfrac{7}{32} = \tfrac{25}{32} \approx 0.781.$$

▸ **Try it Yourself 5**

Find the probability that a randomly selected draft pick is not a linebacker or a quarterback.

a. Find the *probability* that the draft pick is a linebacker or a quarterback.
b. Find the *complement* of the event.

Answer: Page A35

3.3 EXERCISES

■ BUILDING BASIC SKILLS AND VOCABULARY

1. If two events are mutually exclusive, why is $P(A \text{ and } B) = 0$?

2. List examples of

(a) two events that are mutually exclusive.

(b) two events that are not mutually exclusive.

True or False?   *In Exercises 3–6, determine whether the statement is true or false. If it is false, explain why.*

3. If two events are mutually exclusive, they have no outcomes in common.

4. If two events are independent, then they are also mutually exclusive.

5. The probability that event A or event B will occur is

$$P(A \text{ or } B) = P(A) + P(B) - P(A \text{ or } B).$$

6. If events A and B are mutually exclusive, then

$$P(A \text{ or } B) = P(A) + P(B).$$

Graphical Analysis   *In Exercises 7 and 8, decide if the events shown in the Venn diagram are mutually exclusive. Explain your reasoning.*

7.

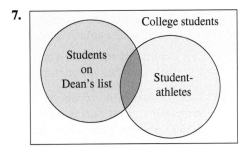

8.

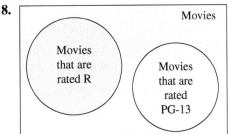

Recognizing Mutually Exclusive Events   *In Exercises 9–12, decide if the events are mutually exclusive. Explain your reasoning.*

9. Event A: Randomly select a female public school teacher.
   Event B: Randomly select a public school teacher who is 25 years old.

10. Event A: Randomly select a member of the U.S. Congress.
    Event B: Randomly select a male U.S. Senator.

11. Event A: Randomly select a student with a birthday in April.
    Event B: Randomly select a student with a birthday in May.

12. Event A: Randomly select a person between 18 and 24 years old.
    Event B: Randomly select a person who drives a convertible.

■ USING AND INTERPRETING CONCEPTS

13. Audit During a 52-week period, a company paid overtime wages for 18 weeks and hired temporary help for 9 weeks. During 5 weeks, the company paid overtime *and* hired temporary help.

(a) Are the events "selecting a week in which overtime wages were paid" and "selecting a week in which temporary help wages were paid" mutually exclusive? Explain.

(b) If an auditor randomly examined the payroll records for only one week, what is the probability that the payroll for that week contained overtime wages or temporary help wages?

14. Conference A math conference has an attendance of 4950 people. Of these, 2110 are college professors and 2575 are female. Of the college professors, 960 are female.

(a) Are the events "selecting a female" and "selecting a college professor" mutually exclusive? Explain.

(b) The conference selects people at random to win prizes. Find the probability that a selected person is a female or a college professor.

15. Carton Defects A company that makes cartons finds that the probability of producing a carton with a puncture is 0.05, the probability that a carton has a smashed corner is 0.08, and the probability that a carton has a puncture and has a smashed corner is 0.004.

(a) Are the events "selecting a carton with a puncture" and "selecting a carton with a smashed corner" mutually exclusive? Explain.

(b) If a quality inspector randomly selects a carton, find the probability that the carton has a puncture or has a smashed corner.

16. Can Defects A company that makes soda pop cans finds that the probability of producing a can without a puncture is 0.96, the probability that a can does not have a smashed edge is 0.93, and the probability that a can does not have a puncture and does not have a smashed edge is 0.893.

(a) Are the events "selecting a can without a puncture" and "selecting a can without a smashed edge" mutually exclusive? Explain.

(b) If a quality inspector randomly selects a can, find the probability that the can does not have a puncture or does not have a smashed edge.

17. Selecting a Card A card is selected at random from a standard deck. Find each probability.

(a) Randomly selecting a club or a 3

(b) Randomly selecting a red suit or a king

(c) Randomly selecting a 9 or a face card

18. Rolling a Die You roll a die. Find each probability.

(a) Rolling a 5 or a number greater than 3

(b) Rolling a number less than 4 or an even number

(c) Rolling a 2 or an odd number

19. U.S. Age Distribution The estimated percent distribution of the U.S. population for 2020 is shown in the pie chart. Find each probability. *(Source: U.S. Census Bureau)*

(a) Randomly selecting someone who is under 5 years old

(b) Randomly selecting someone who is not 65 years or over

(c) Randomly selecting someone who is between 20 and 34 years old

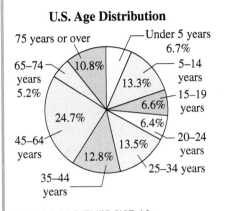

U.S. Age Distribution

75 years or over — Under 5 years 6.7%
65–74 years 5.2% — 10.8%
— 5–14 years 13.3%
— 15–19 years 6.6%
24.7% — 6.4%
45–64 years — 20–24 years
13.5%
12.8% — 25–34 years
35–44 years

FIGURE FOR EXERCISE 19

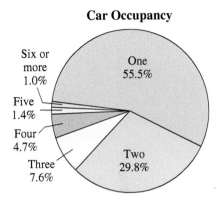

Car Occupancy

Six or more 1.0%
One 55.5%
Five 1.4%
Four 4.7%
Three 7.6%
Two 29.8%

FIGURE FOR EXERCISE 20

20. Tacoma Narrows Bridge The percent distribution of the number of occupants in vehicles crossing the Tacoma Narrows Bridge in Washington is shown in the pie chart. Find each probability. *(Source: Washington State Department of Transportation)*

(a) Randomly selecting a car with two occupants

(b) Randomly selecting a car with two or more occupants

(c) Randomly selecting a car with between two and five occupants, inclusive

21. Education The number of responses to a survey are shown in the Pareto chart. The survey asked 1026 U.S. adults how they would grade the quality of public schools in the United States. Each person gave one response. Find each probability. *(Adapted from CBS News Poll)*

(a) Randomly selecting a person from the sample who did not give the public schools an A

(b) Randomly selecting a person from the sample who gave the public schools a D or an F

22. Olympics The number of responses to a survey are shown in the Pareto chart. The survey asked 1000 U.S. adults if they would watch a large portion of the 2010 Winter Olympics. Each person gave one response. Find each probability. *(Adapted from Rasmussen Reports)*

(a) Randomly selecting a person from the sample who is not at all likely to watch a large portion of the Winter Olympics

(b) Randomly selecting a person from the sample who is not sure whether they will watch a large portion of the Winter Olympics

(c) Randomly selecting a person from the sample who is neither somewhat likely nor very likely to watch a large portion of the Winter Olympics

How Would You Grade the Quality of Public Schools in the U.S.?

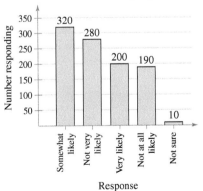

Number responding

| C | D | B | F | A |
335, 272, 241, 126, 52

Response

FIGURE FOR EXERCISE 21

Will You Watch a Large Portion of the Winter Olympics?

Number responding

320, 280, 200, 190, 10

Somewhat likely | Not very likely | Very likely | Not at all likely | Not sure

Response

FIGURE FOR EXERCISE 22

23. Nursing Majors The table shows the number of male and female students enrolled in nursing at the University of Oklahoma Health Sciences Center for a recent semester. A student is selected at random. Find the probability of each event. *(Adapted from University of Oklahoma Health Sciences Center Office of Institutional Research)*

	Nursing majors	Non-nursing majors	Total
Males	151	1104	1255
Females	1016	1693	2709
Total	1167	2797	3964

(a) The student is male or a nursing major.

(b) The student is female or not a nursing major.

(c) The student is not female or is a nursing major.

(d) Are the events "being male" and "being a nursing major" mutually exclusive? Explain.

24. Left-Handed People In a sample of 1000 people (525 men and 475 women), 113 are left-handed (63 men and 50 women). The results of the sample are shown in the table. A person is selected at random from the sample. Find the probability of each event.

		Gender		
		Male	Female	Total
Dominant Hand	Left	63	50	113
	Right	462	425	887
	Total	525	475	1000

(a) The person is left-handed or female.

(b) The person is right-handed or male.

(c) The person is not right-handed or is a male.

(d) The person is right-handed and is a female.

(e) Are the events "being right-handed" and "being female" mutually exclusive? Explain.

25. Charity The table shows the results of a survey that asked 2850 people whether they were involved in any type of charity work. A person is selected at random from the sample. Find the probability of each event.

	Frequently	Occasionally	Not at all	Total
Male	221	456	795	1472
Female	207	430	741	1378
Total	428	886	1536	2850

(a) The person is frequently or occasionally involved in charity work.

(b) The person is female or not involved in charity work at all.

(c) The person is male or frequently involved in charity work.

(d) The person is female or not frequently involved in charity work.

(e) Are the events "being female" and "being frequently involved in charity work" mutually exclusive? Explain.

26. **Eye Survey**   The table shows the results of a survey that asked 3203 people whether they wore contacts or glasses. A person is selected at random from the sample. Find the probability of each event.

	Only contacts	Only glasses	Both	Neither	Total
Male	64	841	177	456	1538
Female	189	427	368	681	1665
Total	253	1268	545	1137	3203

   (a) The person wears only contacts or only glasses.

   (b) The person is male or wears both contacts and glasses.

   (c) The person is female or wears neither contacts nor glasses.

   (d) The person is male or does not wear glasses.

   (e) Are the events "wearing only contacts" and "wearing both contacts and glasses" mutually exclusive? Explain.

■ EXTENDING CONCEPTS

27. **Writing**   Is there a relationship between independence and mutual exclusivity? To decide, find examples of the following, if possible.

   (a) Describe two events that are dependent and mutually exclusive.

   (b) Describe two events that are independent and mutually exclusive.

   (c) Describe two events that are dependent and not mutually exclusive.

   (d) Describe two events that are independent and not mutually exclusive.

Use your results to write a conclusion about the relationship between independence and mutual exclusivity.

Addition Rule for Three Events   *The Addition Rule for the probability that event A or B or C will occur, $P(A \text{ or } B \text{ or } C)$, is given by*

$$P(A \text{ or } B \text{ or } C) = P(A) + P(B) + P(C) - P(A \text{ and } B) - P(A \text{ and } C) \\ - P(B \text{ and } C) + P(A \text{ and } B \text{ and } C).$$

In the Venn diagram shown, $P(A \text{ or } B \text{ or } C)$ is represented by the blue areas.

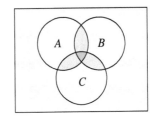

In Exercises 28 and 29, find $P(A \text{ or } B \text{ or } C)$ for the given probabilities.

28. $P(A) = 0.40$, $P(B) = 0.10$, $P(C) = 0.50$,
$P(A \text{ and } B) = 0.05$, $P(A \text{ and } C) = 0.25$, $P(B \text{ and } C) = 0.10$,
$P(A \text{ and } B \text{ and } C) = 0.03$

29. $P(A) = 0.38$, $P(B) = 0.26$, $P(C) = 0.14$,
$P(A \text{ and } B) = 0.12$, $P(A \text{ and } C) = 0.03$, $P(B \text{ and } C) = 0.09$,
$P(A \text{ and } B \text{ and } C) = 0.01$

30. Explain, in your own words, why in the Addition Rule for $P(A \text{ or } B \text{ or } C)$, $P(A \text{ and } B \text{ and } C)$ is added at the end of the formula.

ACTIVITY 3.3 Simulating the Probability of Rolling a 3 or 4

APPLET

The *simulating the probability of rolling a 3 or 4* applet allows you to investigate the probability of rolling a 3 or 4 on a fair die. The plot at the top left corner shows the probability associated with each outcome of a die roll. When ROLL is clicked, *n* simulations of the experiment of rolling a die are performed. The results of the simulations are shown in the frequency plot. If the animate option is checked, the display will show each outcome dropping into the frequency plot as the simulation runs. The individual outcomes are shown in the text field at the far right of the applet. The center plot shows in blue the cumulative proportion of times that an event of rolling a 3 or 4 occurs. The green line in the plot reflects the true probability of rolling a 3 or 4. As the experiment is conducted over and over, the cumulative proportion should converge to the true value.

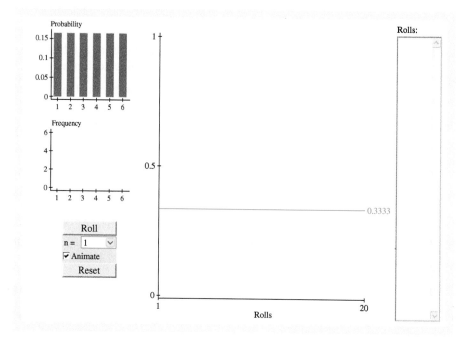

■ Explore

Step 1 Specify a value for *n*.
Step 2 Click ROLL four times.
Step 3 Click RESET.
Step 4 Specify another value for *n*.
Step 5 Click ROLL.

■ Draw Conclusions

APPLET

1. What is the theoretical probability of rolling a 3 or 4?

2. Run the simulation using each value of *n* one time. Clear the results after each trial. Compare the cumulative proportion of rolling a 3 or 4 for each trial with the theoretical probability of rolling a 3 or 4.

3. Suppose you want to modify the applet so you can find the probability of rolling a number less than 4. Describe the placement of the green line.

United States Congress

Congress is made up of the House of Representatives and the Senate. Members of the House of Representatives serve two-year terms and represent a district in a state. The number of representatives each state has is determined by population. States with larger populations have more representatives than states with smaller populations. The total number of representatives is set by law at 435 members. Members of the Senate serve six-year terms and represent a state. Each state has 2 senators, for a total of 100. The tables show the makeup of the 111th Congress by gender and political party. There are two vacant seats in the House of Representatives.

House of Representatives

		Political Party			
		Republican	Democrat	Independent	Total
Gender	Male	161	196	0	357
	Female	17	59	0	76
	Total	178	255	0	433

Senate

		Political Party			
		Republican	Democrat	Independent	Total
Gender	Male	37	44	2	83
	Female	4	13	0	17
	Total	41	57	2	100

■ EXERCISES

1. Find the probability that a randomly selected representative is female. Find the probability that a randomly selected senator is female.

2. Compare the probabilities from Exercise 1.

3. A representative is selected at random. Find the probability of each event.

   (a) The representative is male.
   (b) The representative is a Republican.
   (c) The representative is male given that the representative is a Republican.
   (d) The representative is female and a Democrat.
   (e) Are the events "being female" and "being a Democrat" independent or dependent events? Explain.

4. A senator is selected at random. Find the probability of each event.

   (a) The senator is male.
   (b) The senator is not a Democrat.
   (c) The senator is female or a Republican.
   (d) The senator is male or a Democrat.
   (e) Are the events "being female" and "being an Independent" mutually exclusive? Explain.

5. Using the same row and column headings as the tables above, create a combined table for Congress.

6. A member of Congress is selected at random. Use the table from Exercise 5 to find the probability of each event.

   (a) The member is Independent.
   (b) The member is female and a Republican.
   (c) The member is male or a Democrat.

3.4   Additional Topics in Probability and Counting

Permutations ▸ Combinations ▸ Applications of Counting Principles

▸ PERMUTATIONS

In Section 3.1, you learned that the Fundamental Counting Principle is used to find the number of ways two or more events can occur in sequence. In this section, you will study several other techniques for counting the number of ways an event can occur. An important application of the Fundamental Counting Principle is determining the number of ways that n objects can be arranged in order or in a *permutation*.

DEFINITION

A **permutation** is an ordered arrangement of objects. The number of different permutations of n distinct objects is $n!$.

The expression $n!$ is read as **n factorial** and is defined as follows.

$$n! = n \cdot (n - 1) \cdot (n - 2) \cdot (n - 3) \cdots 3 \cdot 2 \cdot 1$$

As a special case, $0! = 1$. Here are several other values of $n!$.

$$1! = 1, 2! = 2 \cdot 1 = 2, 3! = 3 \cdot 2 \cdot 1 = 6, 4! = 4 \cdot 3 \cdot 2 \cdot 1 = 24$$

EXAMPLE 1

▸ **Finding the Number of Permutations of n Objects**

The objective of a 9×9 Sudoku number puzzle is to fill the grid so that each row, each column, and each 3×3 grid contain the digits 1 to 9. How many different ways can the first row of a blank 9×9 Sudoku grid be filled?

▸ **Solution**

The number of permutations is $9! = 9 \cdot 8 \cdot 7 \cdot 6 \cdot 5 \cdot 4 \cdot 3 \cdot 2 \cdot 1 = 362{,}880$. So, there are 362,880 different ways the first row can be filled.

▸ **Try It Yourself 1**

The women's hockey teams for the 2010 Olympics are Canada, Sweden, Switzerland, Slovakia, United States, Finland, Russia, and China. How many different final standings are possible?

a. Determine the total number of women's hockey teams n that are in the 2010 Olympics.

b. Evaluate $n!$.   *Answer: Page A35*

Sudoku Number Puzzle

6	7	1				2	4	9
8			7		2			1
2				6				3
	5		6		3		2	
		8				7		
	1		8		4		6	
9				1				6
1			5		9			7
5	8	7				9	1	2

Suppose you want to choose some of the objects in a group and put them in order. Such an ordering is called a **permutation of n objects taken r at a time.**

PERMUTATIONS OF n OBJECTS TAKEN r AT A TIME

The number of permutations of n distinct objects taken r at a time is

$$_nP_r = \frac{n!}{(n - r)!}, \text{ where } r \leq n.$$

Detailed instructions for using MINITAB, Excel, and the TI-83/84 Plus are shown in the Technology Guide that accompanies this text. For instance, here are instructions for finding the number of permutations of *n* objects taken *r* at a time on a TI-83/84 Plus.

Enter the total number of objects *n*.

MATH

Choose the PRB menu.

2: nPr

Enter the number of objects *r* taken.

ENTER

EXAMPLE 2

▶ Finding $_nP_r$

Find the number of ways of forming four-digit codes in which no digit is repeated.

▶ Solution

To form a four-digit code with no repeating digits, you need to select 4 digits from a group of 10, so $n = 10$ and $r = 4$.

$$_nP_r = {}_{10}P_4 = \frac{10!}{(10-4)!}$$

$$= \frac{10!}{6!} = \frac{10 \cdot 9 \cdot 8 \cdot 7 \cdot \cancel{6} \cdot \cancel{5} \cdot \cancel{4} \cdot \cancel{3} \cdot \cancel{2} \cdot \cancel{1}}{\cancel{6} \cdot \cancel{5} \cdot \cancel{4} \cdot \cancel{3} \cdot \cancel{2} \cdot \cancel{1}} = 5040$$

So, there are 5040 possible four-digit codes that do not have repeating digits.

▶ Try It Yourself 2

A psychologist shows a list of eight activities to her subject. How many ways can the subject pick a first, second, and third activity?

a. Find the quotient of $n!$ and $(n-r)!$. (List the factors and divide out.)
b. *Write* the result as a sentence.   *Answer: Page A35*

Notice that the Fundamental Counting Principle can be used in Example 3 to obtain the same result. There are 43 choices for first place, 42 choices for second place, and 41 choices for third place. So, there are

$$43 \cdot 42 \cdot 41 = 74,046$$

ways the cars can finish first, second, and third.

EXAMPLE 3

▶ Finding $_nP_r$

Forty-three race cars started the 2010 Daytona 500. How many ways can the cars finish first, second, and third?

▶ Solution

You need to select three race cars from a group of 43, so $n = 43$ and $r = 3$. Because the order is important, the number of ways the cars can finish first, second, and third is

$$_nP_r = {}_{43}P_3 = \frac{43!}{(43-3)!} = \frac{43!}{40!} = 43 \cdot 42 \cdot 41 = 74{,}046.$$

▶ Try It Yourself 3

The board of directors of a company has 12 members. One member is the president, another is the vice president, another is the secretary, and another is the treasurer. How many ways can these positions be assigned?

a. *Identify* the total number of objects *n* and the number of objects *r* being chosen in order.
b. *Evaluate* $_nP_r$.   *Answer: Page A35*

The letters *AAAABBC* can be rearranged in 7! orders, but many of these are not distinguishable. The number of distinguishable orders is

$$\frac{7!}{4! \cdot 2! \cdot 1!} = \frac{7 \cdot 6 \cdot 5}{2}$$

$$= 105.$$

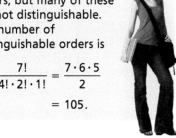

You may want to order a group of *n* objects in which some of the objects are the same. For instance, consider a group of letters consisting of four As, two Bs, and one C. How many ways can you order such a group? Using the previous formula, you might conclude that there are $_7P_7 = 7!$ possible orders. However, because some of the objects are the same, not all of these permutations are *distinguishable*. How many distinguishable permutations are possible? The answer can be found using the formula for the number of distinguishable permutations.

DISTINGUISHABLE PERMUTATIONS

The number of **distinguishable permutations** of n objects, where n_1 are of one type, n_2 are of another type, and so on, is

$$\frac{n!}{n_1! \cdot n_2! \cdot n_3! \cdots n_k!}, \text{ where } n_1 + n_2 + n_3 + \cdots + n_k = n.$$

EXAMPLE 4

▶ **Finding the Number of Distinguishable Permutations**

A building contractor is planning to develop a subdivision. The subdivision is to consist of 6 one-story houses, 4 two-story houses, and 2 split-level houses. In how many distinguishable ways can the houses be arranged?

▶ **Solution**

There are to be 12 houses in the subdivision, 6 of which are of one type (one-story), 4 of another type (two-story), and 2 of a third type (split-level). So, there are

$$\frac{12!}{6! \cdot 4! \cdot 2!} = \frac{12 \cdot 11 \cdot 10 \cdot 9 \cdot 8 \cdot 7 \cdot 6!}{6! \cdot 4! \cdot 2!}$$

$$= 13,860 \text{ distinguishable ways.}$$

Interpretation There are 13,860 distinguishable ways to arrange the houses in the subdivision.

▶ **Try It Yourself 4**

The contractor wants to plant six oak trees, nine maple trees, and five poplar trees along the subdivision street. The trees are to be spaced evenly. In how many distinguishable ways can they be planted?

a. *Identify* the total number of objects n and the number of each type of object in the groups n_1, n_2, and n_3.

b. *Evaluate* $\dfrac{n!}{n_1! \cdot n_2! \cdots n_k!}$.

Answer: Page A36

▶ COMBINATIONS

You want to buy three DVDs from a selection of five DVDs labeled $A, B, C, D,$ and E. There are 10 ways to make your selections.

ABC, ABD, ABE, ACD, ACE, ADE, BCD, BCE, BDE, CDE

In each selection, order does not matter (ABC is the same set as BAC). The number of ways to choose r objects from n objects without regard to order is called the number of **combinations of n objects taken r at a time.**

INSIGHT

You can think of a combination of n objects chosen r at a time as a permutation of n objects in which the r selected objects are alike and the remaining $n - r$ (not selected) objects are alike.

COMBINATIONS OF n OBJECTS TAKEN r AT A TIME

A combination is a selection of r objects from a group of n objects without regard to order and is denoted by $_nC_r$. The number of combinations of r objects selected from a group of n objects is

$$_nC_r = \frac{n!}{(n - r)! r!}.$$

EXAMPLE 5

▶ Finding the Number of Combinations

A state's department of transportation plans to develop a new section of interstate highway and receives 16 bids for the project. The state plans to hire four of the bidding companies. How many different combinations of four companies can be selected from the 16 bidding companies?

▶ Solution

The state is selecting four companies from a group of 16, so $n = 16$ and $r = 4$. Because order is not important, there are

$$_nC_r = {}_{16}C_4 = \frac{16!}{(16 - 4)!4!}$$

$$= \frac{16!}{12!4!}$$

$$= \frac{16 \cdot 15 \cdot 14 \cdot 13 \cdot 12!}{12! \cdot 4!}$$

$$= 1820 \text{ different combinations.}$$

Interpretation   There are 1820 different combinations of four companies that can be selected from the 16 bidding companies.

▶ Try It Yourself 5

The manager of an accounting department wants to form a three-person advisory committee from the 20 employees in the department. In how many ways can the manager form this committee?

a. *Identify* the number of objects in the group n and the number of objects r to be selected.
b. *Evaluate* $_nC_r$.
c. *Write* the result as a sentence.   *Answer: Page A36*

Answer: Page A36

The table summarizes the counting principles.

Principle	Description	Formula
Fundamental Counting Principle	If one event can occur in m ways and a second event can occur in n ways, the number of ways the two events can occur in sequence is $m \cdot n$.	$m \cdot n$
Permutations	The number of different ordered arrangements of n distinct objects	$n!$
	The number of permutations of n distinct objects taken r at a time, where $r \leq n$	$_nP_r = \dfrac{n!}{(n - r)!}$
	The number of distinguishable permutations of n objects where n_1 are of one type, n_2 are of another type, and so on	$\dfrac{n!}{n_1! \cdot n_2! \cdots n_k!}$
Combinations	The number of combinations of r objects selected from a group of n objects without regard to order	$_nC_r = \dfrac{n!}{(n - r)!r!}$

▶ APPLICATIONS OF COUNTING PRINCIPLES

EXAMPLE 6

▶ **Finding Probabilities**

A student advisory board consists of 17 members. Three members serve as the board's chair, secretary, and webmaster. Each member is equally likely to serve in any of the positions. What is the probability of selecting at random the three members who currently hold the three positions?

▶ **Solution**   There is one favorable outcome and there are

$$_{17}P_3 = \frac{17!}{(17-3)!} = \frac{17!}{14!} = \frac{17 \cdot 16 \cdot 15 \cdot 14!}{14!} = 17 \cdot 16 \cdot 15 = 4080$$

ways the three positions can be filled. So, the probability of correctly selecting the three members who hold each position is

$$P(\text{selecting the three members}) = \frac{1}{4080} \approx 0.0002.$$

▶ **Try It Yourself 6**

A student advisory board consists of 20 members. Two members serve as the board's chair and secretary. Each member is equally likely to serve in either of the positions. What is the probability of selecting at random the two members who currently hold the two positions?

a. Find the *number of ways* the two positions can be filled.
b. Find the *probability* of correctly selecting the two members.

Answer: Page A36

EXAMPLE 7

▶ **Finding Probabilities**

You have 11 letters consisting of one M, four I's, four S's, and two P's. If the letters are randomly arranged in order, what is the probability that the arrangement spells the word *Mississippi*?

▶ **Solution**   There is one favorable outcome and there are

$$\frac{11!}{1! \cdot 4! \cdot 4! \cdot 2!} = 34{,}650 \qquad \text{11 letters with 1, 4, 4, and 2 like letters}$$

distinguishable permutations of the given letters. So, the probability that the arrangement spells the word *Mississippi* is

$$P(\text{Mississippi}) = \frac{1}{34{,}650} \approx 0.00003.$$

▶ **Try It Yourself 7**

You have 6 letters consisting of one L, two E's, two T's, and one R. If the letters are randomly arranged in order, what is the probability that the arrangement spells the word *letter*?

a. Find the *number of favorable outcomes* and the *number of distinguishable permutations*.
b. Find the *probability* that the arrangement spells the word *letter*.

Answer: Page A36

EXAMPLE 8

▶ **Finding Probabilities**

Find the probability of picking five diamonds from a standard deck of playing cards.

▶ **Solution**

The possible number of ways of choosing 5 diamonds out of 13 is $_{13}C_5$. The number of possible five-card hands is $_{52}C_5$. So, the probability of being dealt 5 diamonds is

$$P(5 \text{ diamonds}) = \frac{_{13}C_5}{_{52}C_5} = \frac{1287}{2,598,960} \approx 0.0005.$$

▶ Try It Yourself 8

Find the probability of being dealt five diamonds from a standard deck of playing cards that also includes two jokers. In this case, the joker is considered to be a wild card that can be used to represent any card in the deck.

a. Find the *number of ways* of choosing 5 diamonds.
b. Find the *number of possible five-card hands*.
c. Find the *probability* of being dealt five diamonds.   *Answer: Page A36*

EXAMPLE 9

▶ **Finding Probabilities**

A food manufacturer is analyzing a sample of 400 corn kernels for the presence of a toxin. In this sample, three kernels have dangerously high levels of the toxin. If four kernels are randomly selected from the sample, what is the probability that exactly one kernel contains a dangerously high level of the toxin?

▶ **Solution**

The possible number of ways of choosing one toxic kernel out of three toxic kernels is $_3C_1$. The possible number of ways of choosing 3 nontoxic kernels from 397 nontoxic kernels is $_{397}C_3$. So, using the Fundamental Counting Principle, the number of ways of choosing one toxic kernel and three nontoxic kernels is

$$_3C_1 \cdot _{397}C_3 = 3 \cdot 10,349,790$$

$$= 31,049,370.$$

The number of possible ways of choosing 4 kernels from 400 kernels is $_{400}C_4 = 1,050,739,900$. So, the probability of selecting exactly 1 toxic kernel is

$$P(1 \text{ toxic kernel}) = \frac{_3C_1 \cdot _{397}C_3}{_{400}C_4} = \frac{31,049,370}{1,050,739,900} \approx 0.030.$$

▶ Try It Yourself 9

A jury consists of five men and seven women. Three jury members are selected at random for an interview. Find the probability that all three are men.

a. *Find* the product of the number of ways to choose three men from five and the number of ways to choose zero women from seven.
b. Find the *number of ways* to choose 3 jury members from 12.
c. Find the *probability* that all three are men.   *Answer: Page A36*

3.4 EXERCISES

FOR EXTRA HELP:
MyStatLab

■ BUILDING BASIC SKILLS AND VOCABULARY

1. When you calculate the number of permutations of n distinct objects taken r at a time, what are you counting? Give an example.

2. When you calculate the number of combinations of r objects taken from a group of n objects, what are you counting? Give an example.

True or False?   *In Exercises 3–6, determine whether the statement is true or false. If it is false, rewrite it as a true statement.*

3. A combination is an ordered arrangement of objects.

4. The number of different ordered arrangements of n distinct objects is $n!$.

5. If you divide the number of permutations of 11 objects taken 3 at a time by 3!, you will get the number of combinations of 11 objects taken 3 at a time.

6. $_7C_5 = {_7}C_2$

In Exercises 7–14, perform the indicated calculation.

7. $_9P_5$

8. $_{16}P_2$

9. $_8C_3$

10. $_7P_4$

11. $_{21}C_8$

12. $\dfrac{_8C_4}{_{12}C_6}$

13. $\dfrac{_6P_2}{_{11}P_3}$

14. $\dfrac{_{10}C_7}{_{14}C_7}$

In Exercises 15–18, decide if the situation involves permutations, combinations, or neither. Explain your reasoning.

15. The number of ways eight cars can line up in a row for a car wash

16. The number of ways a four-member committee can be chosen from 10 people

17. The number of ways 2 captains can be chosen from 28 players on a lacrosse team

18. The number of four-letter passwords that can be created when no letter can be repeated

■ USING AND INTERPRETING CONCEPTS

19. **Video Games**   You have seven different video games. How many different ways can you arrange the games side by side on a shelf?

20. **Skiing**   Eight people compete in a downhill ski race. Assuming that there are no ties, in how many different orders can the skiers finish?

21. **Security Code**   In how many ways can the letters A, B, C, D, E, and F be arranged for a six-letter security code?

22. **Starting Lineup**   The starting lineup for a softball team consists of 10 players. How many different batting orders are possible using the starting lineup?

23. Lottery Number Selection   A lottery has 52 numbers. In how many different ways can 6 of the numbers be selected? (Assume that order of selection is not important.)

24. Assembly Process   There are four processes involved in assembling a certain product. These processes can be performed in any order. Management wants to find which order is the least time-consuming. How many different orders will have to be tested?

25. Bracelets   You are putting 4 spacers, 10 gold charms, and 8 silver charms on a bracelet. In how many distinguishable ways can the spacers and charms be put on the bracelet?

26. Experimental Group   In order to conduct an experiment, 4 subjects are randomly selected from a group of 20 subjects. How many different groups of four subjects are possible?

27. Letters   In how many distinguishable ways can the letters in the word *statistics* be written?

28. Jury Selection   From a group of 40 people, a jury of 12 people is selected. In how many different ways can a jury of 12 people be selected?

29. Space Shuttle Menu   Space shuttle astronauts each consume an average of 3000 calories per day. One meal normally consists of a main dish, a vegetable dish, and two different desserts. The astronauts can choose from 10 main dishes, 8 vegetable dishes, and 13 desserts. How many different meals are possible? *(Source: NASA)*

30. Menu   A restaurant offers a dinner special that has 12 choices for entrées, 10 choices for side dishes, and 6 choices for dessert. For the special, you can choose one entrée, two side dishes, and one dessert. How many different meals are possible?

31. Water Samples   An environmental agency is analyzing water samples from 80 lakes for pollution. Five of the lakes have dangerously high levels of dioxin. If six lakes are randomly selected from the sample, how many ways could one polluted lake and five non-polluted lakes be chosen? Use a technology tool.

32. Soil Samples   An environmental agency is analyzing soil samples from 50 farms for lead contamination. Eight of the farms have dangerously high levels of lead. If 10 farms are randomly selected from the sample, how many ways could 2 contaminated farms and 8 noncontaminated farms be chosen? Use a technology tool.

Word Jumble   *In Exercises 33–38, do the following.*

  (a) *Find the number of distinguishable ways the letters can be arranged.*

  (b) *There is one arrangement that spells an important term used throughout the course. Find the term.*

  (c) *If the letters are randomly arranged in order, what is the probability that the arrangement spells the word from part (b)? Can this event be considered unusual? Explain.*

33. palmes

34. nevte

35. etre

36. rnctee

37. unoppolati

38. sidtbitoiurn

39. Horse Race A horse race has 12 entries. Assuming that there are no ties, what is the probability that the three horses owned by one person finish first, second, and third?

40. Pizza Toppings A pizza shop offers nine toppings. No topping is used more than once. What is the probability that the toppings on a three-topping pizza are pepperoni, onions, and mushrooms?

41. Jukebox You look over the songs on a jukebox and determine that you like 15 of the 56 songs.

(a) What is the probability that you like the next three songs that are played? (Assume a song cannot be repeated.)

(b) What is the probability that you do not like the next three songs that are played? (Assume a song cannot be repeated.)

42. Officers The offices of president, vice president, secretary, and treasurer for an environmental club will be filled from a pool of 14 candidates. Six of the candidates are members of the debate team.

(a) What is the probability that all of the offices are filled by members of the debate team?

(b) What is the probability that none of the offices are filled by members of the debate team?

43. Employee Selection Four sales representatives for a company are to be chosen to participate in a training program. The company has eight sales representatives, two in each of four regions. In how many ways can the four sales representatives be chosen if (a) there are no restrictions and (b) the selection must include a sales representative from each region? (c) What is the probability that the four sales representatives chosen to participate in the training program will be from only two of the four regions if they are chosen at random?

44. License Plates In a certain state, each automobile license plate number consists of two letters followed by a four-digit number. How many distinct license plate numbers can be formed if (a) there are no restrictions and (b) the letters O and I are not used? (c) What is the probability of selecting at random a license plate that ends in an even number?

45. Password A password consists of two letters followed by a five-digit number. How many passwords are possible if (a) there are no restrictions and (b) none of the letters or digits can be repeated? (c) What is the probability of guessing the password in one trial if there are no restrictions?

46. Area Code An area code consists of three digits. How many area codes are possible if (a) there are no restrictions and (b) the first digit cannot be a 1 or a 0? (c) What is the probability of selecting an area code at random that ends in an odd number if the first digit cannot be a 1 or a 0?

47. Repairs In how many orders can three broken computers and two broken printers be repaired if (a) there are no restrictions, (b) the printers must be repaired first, and (c) the computers must be repaired first? (d) If the order of repairs has no restrictions and the order of repairs is done at random, what is the probability that a printer will be repaired first?

48. Defective Units A shipment of 10 microwave ovens contains two defective units. In how many ways can a restaurant buy three of these units and receive (a) no defective units, (b) one defective unit, and (c) at least two nondefective units? (d) What is the probability of the restaurant buying at least two nondefective units?

Rate Your Financial Shape

FIGURE FOR EXERCISES 49–52

Financial Shape *In Exercises 49–52, use the pie chart, which shows how U.S. adults rate their financial shape.* (*Source: Pew Research Center*)

49. Suppose 4 people are chosen at random from a group of 1200. What is the probability that all four would rate their financial shape as excellent? (Make the assumption that the 1200 people are represented by the pie chart.)

50. Suppose 10 people are chosen at random from a group of 1200. What is the probability that all 10 would rate their financial shape as poor? (Make the assumption that the 1200 people are represented by the pie chart.)

51. Suppose 80 people are chosen at random from a group of 500. What is the probability that none of the 80 people would rate their financial shape as fair? (Make the assumption that the 500 people are represented by the pie chart.)

52. Suppose 55 people are chosen at random from a group of 500. What is the probability that none of the 55 people would rate their financial shape as good? (Make the assumption that the 500 people are represented by the pie chart.)

53. Probability In a state lottery, you must correctly select 5 numbers (in any order) out of 40 to win the top prize.

(a) How many ways can 5 numbers be chosen from 40 numbers?

(b) You purchase one lottery ticket. What is the probability that you will win the top prize?

54. Probability A company that has 200 employees chooses a committee of 15 to represent employee retirement issues. When the committee is formed, none of the 56 minority employees are selected.

(a) Use a technology tool to find the number of ways 15 employees can be chosen from 200.

(b) Use a technology tool to find the number of ways 15 employees can be chosen from 144 nonminorities.

(c) If the committee is chosen randomly (without bias), what is the probability that it contains no minorities?

(d) Does your answer to part (c) indicate that the committee selection is biased? Explain your reasoning.

55. Cards You are dealt a hand of five cards from a standard deck of playing cards. Find the probability of being dealt a hand consisting of

(a) four-of-a-kind.

(b) a full house, which consists of three of one kind and two of another kind.

(c) three-of-a-kind. (The other two cards are different from each other.)

(d) two clubs and one of each of the other three suits.

56. Warehouse A warehouse employs 24 workers on first shift and 17 workers on second shift. Eight workers are chosen at random to be interviewed about the work environment. Find the probability of choosing

(a) all first-shift workers.

(b) all second-shift workers.

(c) six first-shift workers.

(d) four second-shift workers.

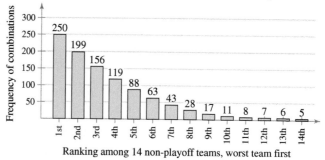

■ EXTENDING CONCEPTS

NBA Draft Lottery *In Exercises 57–62, use the following information. The National Basketball Association (NBA) uses a lottery to determine which team gets the first pick in its annual draft. The teams eligible for the lottery are the 14 non-playoff teams. Fourteen Ping-Pong balls numbered 1 through 14 are placed in a drum. Each of the 14 teams is assigned a certain number of possible four-number combinations that correspond to the numbers on the Ping-Pong balls, such as 3, 8, 10, and 12, as shown. Four balls are then drawn out to determine the first pick in the draft. The order in which the balls are drawn is not important. All of the four-number combinations are assigned to the 14 teams by computer except for one four-number combination. When this four-number combination is drawn, the balls are put back in the drum and another drawing takes place. For instance, if Team A has been assigned the four-number combination 3, 8, 10, 12 and the balls shown at the left are drawn, then Team A wins the first pick.*

After the first pick of the draft is determined, the process continues to choose the teams that will select second and third picks. A team may not win the lottery more than once. If the four-number combination belonging to a team that has already won is drawn, the balls are put back in the drum and another drawing takes place. The remaining order of the draft is determined by the number of losses of each team.

57. In how many ways can 4 of the numbers 1 to 14 be selected if order is not important? How many sets of 4 numbers are assigned to the 14 teams?

58. In how many ways can four of the numbers be selected if order is important?

In the Pareto chart, the number of combinations assigned to each of the 14 teams is shown. The team with the most losses (the worst team) gets the most chances to win the lottery. So, the worst team receives the greatest frequency of four-number combinations, 250. The team with the best record of the 14 non-playoff teams has the fewest chances, with 5 four-number combinations.

Frequency of Four-Number Combinations Assigned in the NBA Draft Lottery

Frequency of combinations: 1st: 250, 2nd: 199, 3rd: 156, 4th: 119, 5th: 88, 6th: 63, 7th: 43, 8th: 28, 9th: 17, 10th: 11, 11th: 8, 12th: 7, 13th: 6, 14th: 5

Ranking among 14 non-playoff teams, worst team first

59. For each team, find the probability that the team will win the first pick. Which of these events would be considered unusual? Explain.

60. What is the probability that the team with the worst record will win the second pick, given that the team with the best record, ranked 14th, wins the first pick?

61. What is the probability that the team with the worst record will win the third pick, given that the team with the best record, ranked 14th, wins the first pick and the team ranked 2nd wins the second pick?

62. What is the probability that neither the first- nor the second-worst team will get the first pick?

USES AND ABUSES

Uses

Probability affects decisions when the weather is forecast, when marketing strategies are determined, when medications are selected, and even when players are selected for professional sports teams. Although intuition is often used for determining probabilities, you will be better able to assess the likelihood that an event will occur by applying the rules of classical probability and empirical probability.

For instance, suppose you work for a real estate company and are asked to estimate the likelihood that a particular house will sell for a particular price within the next 90 days. You could use your intuition, but you could better assess the probability by looking at sales records for similar houses.

Abuses

One common abuse of probability is thinking that probabilities have "memories." For instance, if a coin is tossed eight times, the probability that it will land heads up all eight times is only about 0.004. However, if the coin has already been tossed seven times and has landed heads up each time, the probability that it will land heads up on the eighth time is 0.5. Each toss is independent of all other tosses. The coin does not "remember" that it has already landed heads up seven times.

Ethics

A human resources director for a company with 100 employees wants to show that her company is an equal opportunity employer of women and minorities. There are 40 women employees and 20 minority employees in the company. Nine of the women employees are minorities. Despite this fact, the director reports that 60% of the company is either a woman or a minority. If one employee is selected at random, the probability that the employee is a woman is 0.4 and the probability that the employee is a minority is 0.2. This does not mean, however, that the probability that a randomly selected employee is a woman or a minority is $0.4 + 0.2 = 0.6$, because nine employees belong to both groups. In this case, it would be ethically incorrect to omit this information from her report because these individuals would have been counted twice.

▮ EXERCISES

1. *Assuming That Probability Has a "Memory"*   A "Daily Number" lottery has a three-digit number from 000 to 999. You buy one ticket each day. Your number is 389.

   a. What is the probability of winning next Tuesday and Wednesday?

   b. You won on Tuesday. What's the probability of winning on Wednesday?

   c. You didn't win on Tuesday. What's the probability of winning on Wednesday?

2. *Adding Probabilities Incorrectly*   A town has a population of 500 people. Suppose that the probability that a randomly chosen person owns a pickup truck is 0.25 and the probability that a randomly chosen person owns an SUV is 0.30. What can you say about the probability that a randomly chosen person owns a pickup or an SUV? Could this probability be 0.55? Could it be 0.60? Explain your reasoning.

3 CHAPTER SUMMARY

What did you **learn?**	EXAMPLE(S)	REVIEW EXERCISES
Section 3.1		
■ How to identify the sample space of a probability experiment and how to identify simple events	*1, 2*	*1–4*
■ How to use the Fundamental Counting Principle to find the number of ways two or more events can occur	*3, 4*	*5, 6*
■ How to distinguish among classical probability, empirical probability, and subjective probability	*5–8*	*7–12*
■ How to find the probability of the complement of an event and how to find other probabilities using the Fundamental Counting Principle	*9–11*	*13–16*
Section 3.2		
■ How to find conditional probabilities	*1*	*17, 18*
■ How to distinguish between independent and dependent events	*2*	*19–21*
■ How to use the Multiplication Rule to find the probability of two events occurring in sequence $P(A \text{ and } B) = P(A) \cdot P(B\|A)$ if events are dependent $P(A \text{ and } B) = P(A) \cdot P(B)$ if events are independent	*3–5*	*22–24*
Section 3.3		
■ How to determine if two events are mutually exclusive	*1*	*25–27*
■ How to use the Addition Rule to find the probability of two events $P(A \text{ or } B) = P(A) + P(B) - P(A \text{ and } B)$ $P(A \text{ or } B) = P(A) + P(B)$ if events are mutually exclusive	*2–5*	*28–40*
Section 3.4		
■ How to find the number of ways a group of objects can be arranged in order and the number of ways to choose several objects from a group without regard to order $_nP_r = \dfrac{n!}{(n-r)!}$ permutations of n objects taken r at a time $\dfrac{n!}{n_1! \cdot n_2! \cdot n_3! \cdots n_k!}$ distinguishable permutations $_nC_r = \dfrac{n!}{(n-r)!r!}$ combinations of n objects taken r at a time	*1–5*	*41–50*
■ How to use counting principles to find probabilities	*6–9*	*51–55*

3 REVIEW EXERCISES

■ SECTION 3.1

In Exercises 1–4, identify the sample space of the probability experiment and determine the number of outcomes in the event. Draw a tree diagram if it is appropriate.

1. *Experiment:* Tossing four coins
   Event: Getting three heads

2. *Experiment:* Rolling 2 six-sided dice
   Event: Getting a sum of 4 or 5

3. *Experiment:* Choosing a month of the year
   Event: Choosing a month that begins with the letter J

4. *Experiment:* Guessing the gender(s) of the three children in a family
   Event: The family has two boys

In Exercises 5 and 6, use the Fundamental Counting Principle.

5. A student must choose from 7 classes to take at 8:00 A.M., 4 classes to take at 9:00 A.M., and 3 classes to take at 10:00 A.M. How many ways can the student arrange the schedule?

6. The state of Virginia's license plates have three letters followed by four digits. Assuming that any letter or digit can be used, how many different license plates are possible?

In Exercises 7–12, classify the statement as an example of classical probability, empirical probability, or subjective probability. Explain your reasoning.

7. On the basis of prior counts, a quality control officer says there is a 0.05 probability that a randomly chosen part is defective.

8. The probability of randomly selecting five cards of the same suit from a standard deck is about 0.0005.

9. The chance that Corporation A's stock price will fall today is 75%.

10. The probability that a person can roll his or her tongue is 70%.

11. The probability of rolling 2 six-sided dice and getting a sum greater than 9 is $\frac{1}{6}$.

12. The chance that a randomly selected person in the United States is between 15 and 29 years old is about 21%. *(Source: U.S. Census Bureau)*

In Exercises 13 and 14, the table shows the approximate distribution of the sizes of firms for a recent year. Use the table to determine the probability of the event.
(Adapted from U.S. Small Business Administration)

Number of employees	0 to 4	5 to 9	10 to 19	20 to 99	100 or more
Percent of firms	60.9%	17.6%	10.7%	9.0%	1.8%

13. What is the probability that a randomly selected firm will have at least 10 employees?

14. What is the probability that a randomly selected firm will have fewer than 20 employees?

Telephone Numbers   *The telephone numbers for a region of a state have an area code of 570. The next seven digits represent the local telephone numbers for that region. A local telephone number cannot begin with a 0 or 1. Your cousin lives within the given area code.*

15. What is the probability of randomly generating your cousin's telephone number?

16. What is the probability of not randomly generating your cousin's telephone number?

■ SECTION 3.2

For Exercises 17 and 18, the two statements below summarize the results of a study on the use of plus/minus grading at North Carolina State University. It shows the percents of graduate and undergraduate students who received grades with pluses and minuses (for example, C+, A−, etc.). (Source: North Carolina State University)

- *Of all students who received one or more plus grades, 92% were undergraduates and 8% were graduates.*

- *Of all students who received one or more minus grades, 93% were undergraduates and 7% were graduates.*

17. Find the probability that a student is an undergraduate student, given that the student received a plus grade.

18. Find the probability that a student is a graduate student, given that the student received a minus grade.

In Exercises 19–21, decide whether the events are independent or dependent. Explain your reasoning.

19. Tossing a coin four times, getting four heads, and tossing it a fifth time and getting a head

20. Taking a driver's education course and passing the driver's license exam

21. Getting high grades and being awarded an academic scholarship

22. You are given that $P(A) = 0.35$ and $P(B) = 0.25$. Do you have enough information to find $P(A \text{ and } B)$? Explain.

In Exercises 23 and 24, find the probability of the sequence of events.

23. You are shopping, and your roommate has asked you to pick up toothpaste and dental rinse. However, your roommate did not tell you which brands to get. The store has eight brands of toothpaste and five brands of dental rinse. What is the probability that you will purchase the correct brands of both products? Is this an unusual event? Explain.

24. Your sock drawer has 18 folded pairs of socks, with 8 pairs of white, 6 pairs of black, and 4 pairs of blue. What is the probability, without looking in the drawer, that you will first select and remove a black pair, then select either a blue or a white pair? Is this an unusual event? Explain.

■ SECTION 3.3

In Exercises 25–27, decide if the events are mutually exclusive. Explain your reasoning.

25. Event *A*: Randomly select a red jelly bean from a jar.
Event *B*: Randomly select a yellow jelly bean from the same jar.

26. Event *A*: Randomly select a person who loves cats.
Event *B*: Randomly select a person who owns a dog.

27. Event *A*: Randomly select a U.S. adult registered to vote in Illinois.
Event *B*: Randomly select a U.S. adult registered to vote in Florida.

28. You are given that $P(A) = 0.15$ and $P(B) = 0.40$. Do you have enough information to find $P(A \text{ or } B)$? Explain.

29. A random sample of 250 working adults found that 37% access the Internet at work, 44% access the Internet at home, and 21% access the Internet at both work and home. What is the probability that a person in this sample selected at random accesses the Internet at home or at work?

30. A sample of automobile dealerships found that 19% of automobiles sold are silver, 22% of automobiles sold are sport utility vehicles (SUVs), and 16% of automobiles sold are silver SUVs. What is the probability that a randomly chosen sold automobile from this sample is silver or an SUV?

In Exercises 31–34, determine the probability.

31. A card is randomly selected from a standard deck. Find the probability that the card is between 4 and 8, inclusive, or is a club.

32. A card is randomly selected from a standard deck. Find the probability that the card is red or a queen.

33. A 12-sided die, numbered 1 to 12, is rolled. Find the probability that the roll results in an odd number or a number less than 4.

34. An 8-sided die, numbered 1 to 8, is rolled. Find the probability that the roll results in an even number or a number greater than 6.

In Exercises 35 and 36, use the pie chart, which shows the percent distribution of the number of students in traditional U.S. elementary schools. (Source: U.S. National Center for Education Statistics)

35. Find the probability of randomly selecting a school with 600 or more students.

36. Find the probability of randomly selecting a school with between 300 and 999 students, inclusive.

In Exercises 37–40, use the Pareto chart, which shows the results of a survey in which 874 adults were asked which genre of movie they preferred. (Adapted from Rasmussen Reports)

Students in Elementary Schools

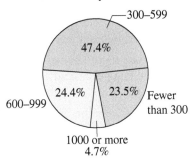

FIGURE FOR EXERCISES 35 AND 36

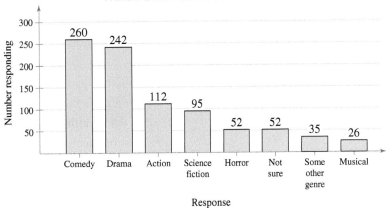

Which Genre of Movie Do You Prefer?

37. Find the probability of randomly selecting an adult from the sample who prefers an action movie or a horror movie.

38. Find the probability of randomly selecting an adult from the sample who prefers a drama or a musical.

39. Find the probability of randomly selecting an adult from the sample who does not prefer a comedy.

40. Find the probability of randomly selecting an adult from the sample who does not prefer a science fiction movie or an action movie.

■ SECTION 3.4

In Exercises 41–44, perform the indicated calculation.

41. $_{11}P_2$ **42.** $_8P_6$ **43.** $_7C_4$ **44.** $\dfrac{_5C_3}{_{10}C_3}$

45. Use a technology tool to find $_{50}P_5$.

46. Use a technology tool to find $_{38}C_{25}$.

In Exercises 47–50, use combinations and permutations.

47. Fifteen cyclists enter a race. In how many ways can they finish first, second, and third?

48. Five players on a basketball team must each choose a player on the opposing team to defend. In how many ways can they choose their defensive assignments?

49. A literary magazine editor must choose 4 short stories for this month's issue from 17 submissions. In how many ways can the editor choose this month's stories?

50. An employer must hire 2 people from a list of 13 applicants. In how many ways can the employer choose to hire the 2 people?

In Exercises 51–55, use counting principles to find the probability. Then tell whether the event can be considered unusual.

51. A full house consists of a three of one kind and two of another kind. Find the probability of a full house consisting of three kings and two queens.

52. A security code consists of three letters followed by one digit. The first letter cannot be an A, B, or C. What is the probability of guessing the security code in one trial?

53. A batch of 200 calculators contains 3 defective units. What is the probability that a sample of three calculators will have

(a) no defective calculators?

(b) all defective calculators?

(c) at least one defective calculator?

(d) at least one nondefective calculator?

54. A batch of 350 raffle tickets contains four winning tickets. You buy four tickets. What is the probability that you have

(a) no winning tickets?

(b) all of the winning tickets?

(c) at least one winning ticket?

(d) at least one nonwinning ticket?

55. A corporation has six male senior executives and four female senior executives. Four senior executives are chosen at random to attend a technology seminar. What is the probability of choosing

(a) four men?

(b) four women?

(c) two men and two women?

(d) one man and three women?

3 CHAPTER QUIZ

Take this quiz as you would take a quiz in class. After you are done, check your work against the answers given in the back of the book.

1. The table shows the number (in thousands) of earned degrees, by level and gender, conferred in the United States in a recent year. *(Source: U.S. National Center for Education Statistics)*

		Gender		
		Male	Female	Total
Level of Degree	Associate's	275	453	728
	Bachelor's	650	875	1525
	Master's	238	366	604
	Doctoral	30	30	60
	Total	1193	1724	2917

A person who earned a degree in the year is randomly selected. Find the probability of selecting someone who

   (a) earned a bachelor's degree.

   (b) earned a bachelor's degree given that the person is a female.

   (c) earned a bachelor's degree given that the person is not a female.

   (d) earned an associate's degree or a bachelor's degree.

   (e) earned a doctorate given that the person is a male.

   (f) earned a master's degree or is a female.

   (g) earned an associate's degree and is a male.

   (h) is a female given that the person earned a bachelor's degree.

2. Which event(s) in Exercise 1 can be considered unusual? Explain your reasoning.

3. Decide if the events are mutually exclusive. Then decide if the events are independent or dependent. Explain your reasoning.

   Event *A*: A golfer scoring the best round in a four-round tournament

   Event *B*: Losing the golf tournament

4. A shipment of 250 netbooks contains 3 defective units. Determine how many ways a vending company can buy three of these units and receive

   (a) no defective units.

   (b) all defective units.

   (c) at least one good unit.

5. In Exercise 4, find the probability of the vending company receiving

   (a) no defective units.

   (b) all defective units.

   (c) at least one good unit.

6. The access code for a warehouse's security system consists of six digits. The first digit cannot be 0 and the last digit must be even. How many different codes are available?

7. From a pool of 30 candidates, the offices of president, vice president, secretary, and treasurer will be filled. In how many different ways can the offices be filled?

PUTTING IT ALL TOGETHER

Real Statistics — Real Decisions

You work for the company that runs the Powerball® lottery. Powerball is a lottery game in which five white balls are chosen from a drum containing 59 balls and one red ball is chosen from a drum containing 39 balls. To win the jackpot, a player must match all five white balls and the red ball. Other winners and their prizes are also shown in the table.

Working in the public relations department, you handle many inquiries from the media and from lottery players. You receive the following e-mail.

You list the probability of matching only the red ball as 1/62. I know from my statistics class that the probability of winning is the ratio of the number of successful outcomes to the total number of outcomes. Could you please explain why the probability of matching only the red ball is 1/62?

Your job is to answer this question, using the probability techniques you have learned in this chapter to justify your answer. In answering the question, assume only one ticket is purchased.

■ EXERCISES

1. How Would You Do It?

(a) How would you investigate the question about the probability of matching only the red ball?

(b) What statistical methods taught in this chapter would you use?

2. Answering the Question

Write an explanation that answers the question about the probability of matching only the red ball. Include in your explanation any probability formulas that justify your explanation.

3. Another Question

You receive another question asking how the overall probability of winning a prize in the Powerball lottery is determined. The overall probability of winning a prize in the Powerball lottery is 1/35. Write an explanation that answers the question and include any probability formulas that justify your explanation.

www.musl.com

Reprinted with permission from the MultiState Lottery Association.

Powerball Winners and Prizes

Match	Prize	Approximate probability
5 white, 1 red	Jackpot	1/195,249,054
5 white	$200,000	1/5,138,133
4 white, 1 red	$10,000	1/723,145
4 white	$100	1/19,030
3 white, 1 red	$100	1/13,644
3 white	$7	1/359
2 white, 1 red	$7	1/787
1 white, 1 red	$4	1/123
1 red	$3	1/62

(Source: Multi-State Lottery Association)

Where Is Powerball Played?
Powerball is played in 42 states, Washington, D.C., and the U.S. Virgin Islands

U.S Virgin Islands

(Source: Multi-State Lottery Association)

TECHNOLOGY

SIMULATION: COMPOSING MOZART VARIATIONS WITH DICE

Wolfgang Mozart (1756–1791) composed a wide variety of musical pieces. In his Musical Dice Game, he wrote a Wiener minuet with an almost endless number of variations. Each minuet has 16 bars. In the eighth and sixteenth bars, the player has a choice of two musical phrases. In each of the other 14 bars, the player has a choice of 11 phrases.

To create a minuet, Mozart suggested that the player toss 2 six-sided dice 16 times. For the eighth and sixteenth bars, choose Option 1 if the dice total is odd and Option 2 if it is even. For each of the other 14 bars, subtract 1 from the dice total. The following minuet is the result of the following sequence of numbers.

5	7	1	6	4	10	5	1
6	6	2	4	6	8	8	2

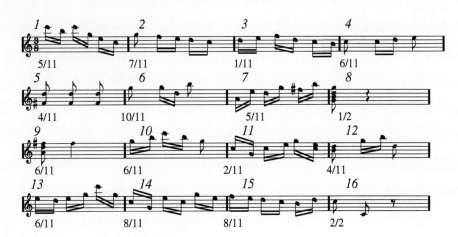

■ EXERCISES

1. How many phrases did Mozart write to create the Musical Dice Game minuet? Explain.

2. How many possible variations are there in Mozart's Musical Dice Game minuet? Explain.

3. Use technology to randomly select a number from 1 to 11.

  (a) What is the theoretical probability of each number from 1 to 11 occurring?

  (b) Use this procedure to select 100 integers from 1 to 11. Tally your results and compare them with the probabilities in part (a).

4. What is the probability of randomly selecting option 6, 7, or 8 for the first bar? For all 14 bars? Find each probability using (a) theoretical probability and (b) the results of Exercise 3(b).

5. Use technology to randomly select two numbers from 1, 2, 3, 4, 5, and 6. Find the sum and subtract 1 to obtain a total.

  (a) What is the theoretical probability of each total from 1 to 11?

  (b) Use this procedure to select 100 totals from 1 to 11. Tally your results and compare them with the probabilities in part (a).

6. Repeat Exercise 4 using the results of Exercise 5(b).

Extended solutions are given in the *Technology Supplement*.
Technical instruction is provided for MINITAB, Excel, and the TI-83/84 Plus.

4 DISCRETE PROBABILITY DISTRIBUTIONS

4.1 Probability Distributions

4.2 Binomial Distributions

■ ACTIVITY

■ CASE STUDY

4.3 More Discrete Probability Distributions

■ USES AND ABUSES

■ REAL STATISTICS–
REAL DECISIONS

■ TECHNOLOGY

The National Climatic Data Center (NCDC) is the world's largest active archive of weather data. NCDC archives weather data from the Coast Guard, Federal Aviation Administration, Military Services, the National Weather Service, and voluntary observers.

In Chapters 1 through 3, you learned how to collect and describe data and how to find the probability of an event. These skills are used in many different types of careers. For instance, data about climatic conditions are used to analyze and forecast the weather throughout the world. On a typical day, aircraft, National Weather Service cooperative observers, radar, remote sensing systems, satellites, ships, weather balloons, wind profilers, and a variety of other data-collection devices work together to provide meteorologists with data that are used to forecast the weather. Even with this much data, meteorologists cannot forecast the weather with certainty. Instead, they assign probabilities to certain weather conditions. For instance, a meteorologist might determine that there is a 40% chance of rain (based on the relative frequency of rain under similar weather conditions).

In Chapter 4, you will learn how to create and use probability distributions. Knowing the shape, center, and variability of a probability distribution will enable you to make decisions in inferential statistics. You are a meteorologist working on a three-day forecast. Assuming that having rain on one day is independent of having rain on another day, you have determined that there is a 40% probability of rain

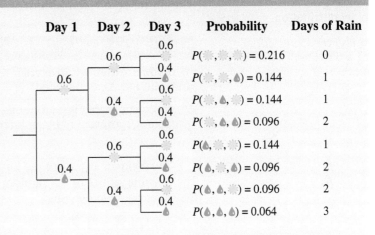

(and a 60% probability of no rain) on each of the three days. What is the probability that it will rain on 0, 1, 2, or 3 of the days? To answer this, you can create a probability distribution for the possible outcomes.

Using the *Addition Rule* with the probabilities in the tree diagram, you can determine the probabilities of having rain on various numbers of days. You can then use this information to graph a probability distribution.

Probability Distribution

Days of rain	Tally	Probability
0	1	0.216
1	3	0.432
2	3	0.288
3	1	0.064

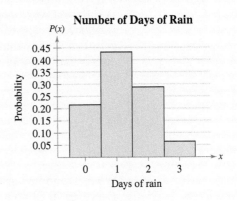

Number of Days of Rain

4.1 Probability Distributions

WHAT YOU SHOULD LEARN

▸ How to distinguish between discrete random variables and continuous random variables

▸ How to construct a discrete probability distribution and its graph

▸ How to determine if a distribution is a probability distribution

▸ How to find the mean, variance, and standard deviation of a discrete probability distribution

▸ How to find the expected value of a discrete probability distribution

Random Variables ▸ Discrete Probability Distributions ▸ Mean, Variance, and Standard Deviation ▸ Expected Value

▸ RANDOM VARIABLES

The outcome of a probability experiment is often a count or a measure. When this occurs, the outcome is called a *random variable*.

DEFINITION

A **random variable** x represents a numerical value associated with each outcome of a probability experiment.

The word *random* indicates that x is determined by chance. There are two types of random variables: *discrete* and *continuous*.

DEFINITION

A random variable is **discrete** if it has a finite or countable number of possible outcomes that can be listed.

A random variable is **continuous** if it has an uncountable number of possible outcomes, represented by an interval on the number line.

You conduct a study of the number of calls a telemarketer makes in one day. The possible values of the random variable x are 0, 1, 2, 3, 4, and so on. Because the set of possible outcomes

$$\{0, 1, 2, 3, \dots\}$$

can be listed, x is a discrete random variable. You can represent its values as points on a number line.

Number of Calls (Discrete)

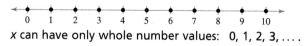

x can have only whole number values: 0, 1, 2, 3, ...

A different way to conduct the study would be to measure the time (in hours) a telemarketer spends making calls in one day. Because the time spent making calls can be any number from 0 to 24 (including fractions and decimals), x is a continuous random variable. You can represent its values with an interval on a number line.

Hours Spent on Calls (Continuous)

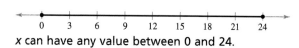

x can have any value between 0 and 24.

When a random variable is discrete, you can list the possible values it can assume. However, it is impossible to list all values for a continuous random variable.

EXAMPLE 1

▸ **Discrete Variables and Continuous Variables**

Decide whether the random variable x is discrete or continuous. Explain your reasoning.

1. Let x represent the number of Fortune 500 companies that lost money in the previous year.

2. Let x represent the volume of gasoline in a 21-gallon tank.

▸ **Solution**

1. The number of companies that lost money in the previous year can be counted.

$$\{0, 1, 2, 3, \ldots, 500\}$$

So, x is a *discrete* random variable.

2. The amount of gasoline in the tank can be any volume between 0 gallons and 21 gallons. So, x is a *continuous* random variable.

▸ **Try It Yourself 1**

Decide whether the random variable x is discrete or continuous. Explain your reasoning.

1. Let x represent the speed of a Space Shuttle.
2. Let x represent the number of calves born on a farm in one year.

a. Decide if x represents *counted* data or *measured* data.
b. Make a *conclusion* and *explain* your reasoning. *Answer: Page A36*

It is important that you can distinguish between discrete and continuous random variables because different statistical techniques are used to analyze each. The remainder of this chapter focuses on discrete random variables and their probability distributions. You will study continuous distributions later.

▸ DISCRETE PROBABILITY DISTRIBUTIONS

Each value of a discrete random variable can be assigned a probability. By listing each value of the random variable with its corresponding probability, you are forming a *discrete probability distribution*.

DEFINITION

A **discrete probability distribution** lists each possible value the random variable can assume, together with its probability. A discrete probability distribution must satisfy the following conditions.

IN WORDS	IN SYMBOLS
1. The probability of each value of the discrete random variable is between 0 and 1, inclusive.	$0 \leq P(x) \leq 1$
2. The sum of all the probabilities is 1.	$\sum P(x) = 1$

Because probabilities represent relative frequencies, a discrete probability distribution can be graphed with a relative frequency histogram.

Constructing a Discrete Probability Distribution

Let x be a discrete random variable with possible outcomes $x_1, x_2, \ldots, x_n$.

1. Make a frequency distribution for the possible outcomes.
2. Find the sum of the frequencies.
3. Find the probability of each possible outcome by dividing its frequency by the sum of the frequencies.
4. Check that each probability is between 0 and 1, inclusive, and that the sum of all the probabilities is 1.

Frequency Distribution

Score, x	Frequency, f
1	24
2	33
3	42
4	30
5	21

Passive-Aggressive Traits

Frequency Distribution

Sales per day, x	Number of days, f
0	16
1	19
2	15
3	21
4	9
5	10
6	8
7	2

EXAMPLE 2   SC Report 16

▶ **Constructing and Graphing a Discrete Probability Distribution**

An industrial psychologist administered a personality inventory test for passive-aggressive traits to 150 employees. Each individual was given a score from 1 to 5, where 1 was extremely passive and 5 extremely aggressive. A score of 3 indicated neither trait. The results are shown at the left. Construct a probability distribution for the random variable x. Then graph the distribution using a histogram.

▶ **Solution**

Divide the frequency of each score by the total number of individuals in the study to find the probability for each value of the random variable.

$$P(1) = \frac{24}{150} = 0.16 \qquad P(2) = \frac{33}{150} = 0.22 \qquad P(3) = \frac{42}{150} = 0.28$$

$$P(4) = \frac{30}{150} = 0.20 \qquad P(5) = \frac{21}{150} = 0.14$$

The discrete probability distribution is shown in the following table.

x	1	2	3	4	5
$P(x)$	0.16	0.22	0.28	0.20	0.14

Note that $0 \leq P(x) \leq 1$ and $\sum P(x) = 1$.

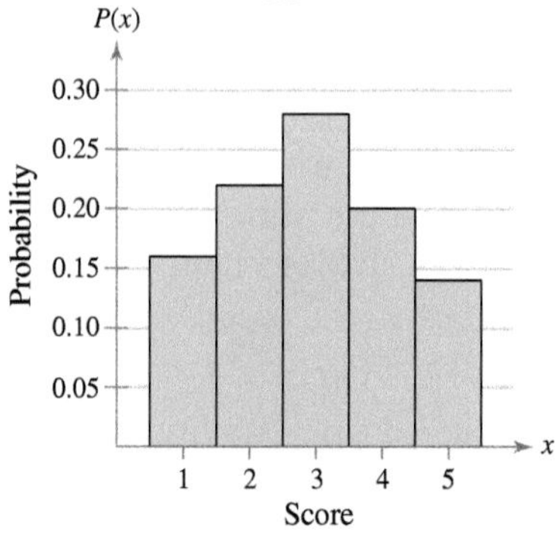

The histogram is shown at the left. Because the width of each bar is one, the area of each bar is equal to the probability of a particular outcome. Also, the probability of an event corresponds to the sum of the areas of the outcomes included in the event. For instance, the probability of the event "having a score of 2 or 3" is equal to the sum of the areas of the second and third bars,

$$(1)(0.22) + (1)(0.28) = 0.22 + 0.28 = 0.50.$$

Interpretation You can see that the distribution is approximately symmetric.

▶ **Try It Yourself 2**

A company tracks the number of sales new employees make each day during a 100-day probationary period. The results for one new employee are shown at the left. Construct and graph a probability distribution.

a. Find the *probability* of each outcome.
b. Organize the probabilities in a *probability distribution*.
c. Graph the probability distribution using a *histogram*. *Answer: Page A36*

Probability Distribution

Days of rain, x	Probability, $P(x)$
0	0.216
1	0.432
2	0.288
3	0.064

PICTURING THE WORLD

In a recent year in the United States, nearly 11 million traffic accidents were reported to the police. A histogram of traffic accidents for various age groups from 16 to 84 is shown. (Adapted from National Safety Council)

U.S. Traffic Accidents by Age

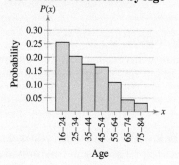

Estimate the probability that a randomly selected person involved in a traffic accident is in the 16 to 34 age group.

EXAMPLE 3

▶ **Verifying Probability Distributions**

Verify that the distribution at the left (see page 189) is a probability distribution.

▶ **Solution**

If the distribution is a probability distribution, then (1) each probability is between 0 and 1, inclusive, and (2) the sum of the probabilities equals 1.

1. Each probability is between 0 and 1.

2. $\sum P(x) = 0.216 + 0.432 + 0.288 + 0.064$

$\qquad = 1.$

Interpretation Because both conditions are met, the distribution is a probability distribution.

▶ **Try It Yourself 3**

Verify that the distribution you constructed in Try It Yourself 2 is a probability distribution.

a. Verify that the *probability* of each outcome is between 0 and 1, inclusive.
b. Verify that the *sum* of all the probabilities is 1.
c. Make a *conclusion*.   *Answer: Page A36*

EXAMPLE 4

▶ **Identifying Probability Distributions**

Decide whether the distribution is a probability distribution. Explain your reasoning.

1.

x	5	6	7	8
$P(x)$	0.28	0.21	0.43	0.15

2.

x	1	2	3	4
$P(x)$	$\frac{1}{2}$	$\frac{1}{4}$	$\frac{5}{4}$	-1

▶ **Solution**

1. Each probability is between 0 and 1, but the sum of all the probabilities is 1.07, which is greater than 1. So, it is *not* a probability distribution.

2. The sum of all the probabilities is equal to 1, but $P(3)$ and $P(4)$ are not between 0 and 1. So, it is *not* a probability distribution. Probabilities can never be negative or greater than 1.

▶ **Try It Yourself 4**

Decide whether the distribution is a probability distribution. Explain your reasoning.

1.

x	5	6	7	8
$P(x)$	$\frac{1}{16}$	$\frac{5}{8}$	$\frac{1}{4}$	$\frac{1}{16}$

2.

x	1	2	3	4
$P(x)$	0.09	0.36	0.49	0.06

a. Verify that the *probability* of each outcome is between 0 and 1.
b. Verify that the *sum* of all the probabilities is 1.
c. Make a *conclusion*.   *Answer: Page A36*

MEAN, VARIANCE, AND STANDARD DEVIATION

You can measure the center of a probability distribution with its mean and measure the variability with its variance and standard deviation. The mean of a discrete random variable is defined as follows.

MEAN OF A DISCRETE RANDOM VARIABLE

The **mean** of a discrete random variable is given by

$$\mu = \Sigma x P(x).$$

Each value of x is multiplied by its corresponding probability and the products are added.

The mean of a random variable represents the "theoretical average" of a probability experiment and sometimes is not a possible outcome. If the experiment were performed many thousands of times, the mean of all the outcomes would be close to the mean of the random variable.

x	$P(x)$
1	0.16
2	0.22
3	0.28
4	0.20
5	0.14

EXAMPLE 5

Finding the Mean of a Probability Distribution

The probability distribution for the personality inventory test for passive-aggressive traits discussed in Example 2 is given at the left. Find the mean score. What can you conclude?

Solution

Use a table to organize your work, as shown below. From the table, you can see that the mean score is approximately 2.9. A score of 3 represents an individual who exhibits neither passive nor aggressive traits. The mean is slightly under 3.

x	$P(x)$	$xP(x)$	
1	0.16	$1(0.16) = 0.16$	
2	0.22	$2(0.22) = 0.44$	
3	0.28	$3(0.28) = 0.84$	
4	0.20	$4(0.20) = 0.80$	
5	0.14	$5(0.14) = 0.70$	
	$\Sigma P(x) = 1$	$\Sigma xP(x) = 2.94$	← Mean

Interpretation You can conclude that the mean personality trait is neither extremely passive nor extremely aggressive, but is slightly closer to passive.

Try It Yourself 5

Find the mean of the probability distribution you constructed in Try It Yourself 2. What can you conclude?

a. Find the *product* of each random outcome and its corresponding probability.
b. Find the *sum* of the products.
c. Make a *conclusion*.

Answer: Page A36

STUDY TIP

Notice that the mean in Example 5 is rounded to one decimal place. This rounding was done because the mean of a probability distribution should be rounded to one more decimal place than was used for the random variable x. This *round-off rule* is also used for the variance and standard deviation of a probability distribution.

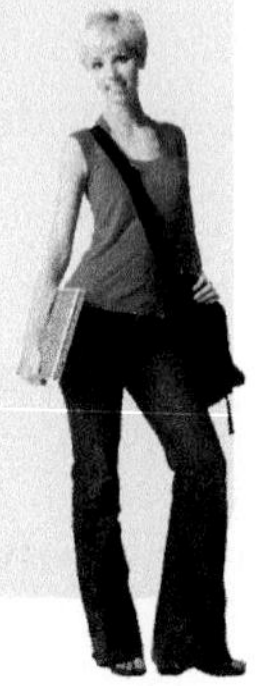

Although the mean of the random variable of a probability distribution describes a typical outcome, it gives no information about how the outcomes vary. To study the variation of the outcomes, you can use the variance and standard deviation of the random variable of a probability distribution.

STUDY TIP

A shortcut formula for the variance of a probability distribution is

$$\sigma^2 = [\Sigma x^2 P(x)] - \mu^2.$$

VARIANCE AND STANDARD DEVIATION OF A DISCRETE RANDOM VARIABLE

The **variance** of a discrete random variable is

$$\sigma^2 = \Sigma(x - \mu)^2 P(x).$$

The **standard deviation** is

$$\sigma = \sqrt{\sigma^2} = \sqrt{\Sigma(x - \mu)^2 P(x)}.$$

EXAMPLE 6

▶ **Finding the Variance and Standard Deviation**

The probability distribution for the personality inventory test for passive-aggressive traits discussed in Example 2 is given at the left. Find the variance and standard deviation of the probability distribution.

x	P(x)
1	0.16
2	0.22
3	0.28
4	0.20
5	0.14

▶ **Solution**

From Example 5, you know that before rounding, the mean of the distribution is $\mu = 2.94$. Use a table to organize your work, as shown below.

x	P(x)	x − μ	(x − μ)²	P(x)(x − μ)²
1	0.16	−1.94	3.764	0.602
2	0.22	−0.94	0.884	0.194
3	0.28	0.06	0.004	0.001
4	0.20	1.06	1.124	0.225
5	0.14	2.06	4.244	0.594
	$\Sigma P(x) = 1$			$\Sigma P(x)(x - \mu)^2 = 1.616$

Variance

So, the variance is

$$\sigma^2 = 1.616 \approx 1.6$$

and the standard deviation is

$$\sigma = \sqrt{\sigma^2} = \sqrt{1.616} \approx 1.3.$$

Interpretation Most of the data values differ from the mean by no more than 1.3.

STUDY TIP

Detailed instructions for using MINITAB, Excel, and the TI-83/84 Plus are shown in the Technology Guide that accompanies this text. Here are instructions for finding the mean and standard deviation of a discrete random variable of a probability distribution on a TI-83/84 Plus.

STAT

Choose the EDIT menu.

1: Edit

Enter the possible values of the discrete random variable x in L1. Enter the probabilities P(x) in L2.

STAT

Choose the CALC menu.

1: 1–Var Stats

ENTER

2nd L1 , 2nd L2

ENTER

▶ **Try It Yourself 6**

Find the variance and standard deviation of the probability distribution constructed in Try It Yourself 2.

a. For each value of x, find the *square of the deviation* from the mean and multiply that value by the corresponding probability of x.
b. Find the *sum of the products* found in part (a) for the variance.
c. Take the *square root of the variance* to find the standard deviation.
d. *Interpret* the results.

Answer: Page A36

▶ EXPECTED VALUE

The mean of a random variable represents what you would expect to happen over thousands of trials. It is also called the *expected value*.

DEFINITION

The **expected value** of a discrete random variable is equal to the mean of the random variable.

$$\text{Expected Value} = E(x) = \mu = \Sigma\, xP(x)$$

Although probabilities can never be negative, the expected value of a random variable can be negative.

INSIGHT

In most applications, an expected value of 0 has a practical interpretation. For instance, in games of chance, an expected value of 0 implies that a game is fair (an unlikely occurrence!). In a profit and loss analysis, an expected value of 0 represents the break-even point.

EXAMPLE 7

▶ Finding an Expected Value

At a raffle, 1500 tickets are sold at $2 each for four prizes of $500, $250, $150, and $75. You buy one ticket. What is the expected value of your gain?

▶ Solution

To find the gain for each prize, subtract the price of the ticket from the prize. For instance, your gain for the $500 prize is

$500 − $2 = $498

and your gain for the $250 prize is

$250 − $2 = $248.

Write a probability distribution for the possible gains (or outcomes).

Gain, x	$498	$248	$148	$73	−$2
Probability, $P(x)$	$\frac{1}{1500}$	$\frac{1}{1500}$	$\frac{1}{1500}$	$\frac{1}{1500}$	$\frac{1496}{1500}$

Then, using the probability distribution, you can find the expected value.

$$E(x) = \Sigma\, xP(x)$$

$$= \$498 \cdot \frac{1}{1500} + \$248 \cdot \frac{1}{1500} + \$148 \cdot \frac{1}{1500} + \$73 \cdot \frac{1}{1500} + (-\$2) \cdot \frac{1496}{1500}$$

$$= -\$1.35$$

Interpretation Because the expected value is negative, you can expect to lose an average of $1.35 for each ticket you buy.

▶ Try It Yourself 7

At a raffle, 2000 tickets are sold at $5 each for five prizes of $2000, $1000, $500, $250, and $100. You buy one ticket. What is the expected value of your gain?

a. Find the *gain* for each prize.
b. Write a *probability distribution* for the possible gains.
c. Find the *expected value*.
d. *Interpret* the results. *Answer: Page A36*

4.1 EXERCISES

■ BUILDING BASIC SKILLS AND VOCABULARY

1. What is a random variable? Give an example of a discrete random variable and a continuous random variable. Justify your answer.

2. What is a discrete probability distribution? What are the two conditions that determine a probability distribution?

3. Is the expected value of the probability distribution of a random variable always one of the possible values of *x*? Explain.

4. What is the significance of the mean of a probability distribution?

True or False?   *In Exercises 5–8, determine whether the statement is true or false. If it is false, rewrite it as a true statement.*

5. In most applications, continuous random variables represent counted data, while discrete random variables represent measured data.

6. For a random variable *x*, the word *random* indicates that the value of *x* is determined by chance.

7. The mean of a random variable represents the "theoretical average" of a probability experiment and sometimes is not a possible outcome.

8. The expected value of a discrete random variable is equal to the standard deviation of the random variable.

Graphical Analysis   *In Exercises 9–12, decide whether the graph represents a discrete random variable or a continuous random variable. Explain your reasoning.*

9. The attendance at concerts for a rock group

40,000   45,000   50,000

10. The length of time student-athletes practice each week

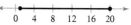
0   4   8   12   16   20

11. The distance a baseball travels after being hit

0   200   400   600

12. The annual traffic fatalities in the United States *(Source: U.S. National Highway Traffic Safety Administration)*

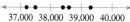
37,000 38,000 39,000 40,000

Distinguishing Between Discrete and Continuous Random Variables
In Exercises 13–20, decide whether the random variable x is discrete or continuous. Explain your reasoning.

13. Let *x* represent the number of books in a university library.

14. Let *x* represent the length of time it takes to get to work.

15. Let *x* represent the volume of blood drawn for a blood test.

16. Let *x* represent the number of tornadoes in the month of June in Oklahoma.

17. Let *x* represent the number of messages posted each month on a social networking website.

18. Let *x* represent the tension at which a randomly selected guitar's strings have been strung.

19. Let *x* represent the amount of snow (in inches) that fell in Nome, Alaska last winter.

20. Let *x* represent the total number of die rolls required for an individual to roll a five.

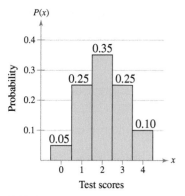

FIGURE FOR EXERCISE 21

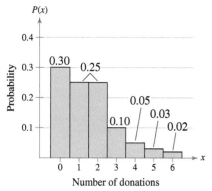

FIGURE FOR EXERCISE 22

USING AND INTERPRETING CONCEPTS

21. Employee Testing A company gave psychological tests to prospective employees. The random variable x represents the possible test scores. Use the histogram to find the probability that a person selected at random from the survey's sample had a test score of (a) more than two and (b) less than four.

22. Blood Donations A survey asked a sample of people how many times they donate blood each year. The random variable x represents the number of donations in one year. Use the histogram to find the probability that a person selected at random from the survey's sample donated blood (a) more than once in a year and (b) less than three times in a year.

Determining a Missing Probability *In Exercises 23 and 24, determine the probability distribution's missing probability value.*

23.

x	0	1	2	3	4
$P(x)$	0.07	0.20	0.38	?	0.13

24.

x	0	1	2	3	4	5	6
$P(x)$	0.05	?	0.23	0.21	0.17	0.11	0.08

Identifying Probability Distributions *In Exercises 25 and 26, decide whether the distribution is a probability distribution. If it is not a probability distribution, identify the property (or properties) that are not satisfied.*

25. Tires A mechanic checked the tire pressures on each car that he worked on for one week. The random variable x represents the number of tires that were underinflated.

x	0	1	2	3	4
$P(x)$	0.30	0.25	0.25	0.15	0.05

26. Quality Control A quality inspector checked for imperfections in rolls of fabric for one week. The random variable x represents the number of imperfections found.

x	0	1	2	3	4	5
$P(x)$	$\frac{3}{4}$	$\frac{1}{10}$	$\frac{1}{20}$	$\frac{1}{25}$	$\frac{1}{50}$	$\frac{1}{100}$

Constructing Probability Distributions *In Exercises 27–32, (a) use the frequency distribution to construct a probability distribution, (b) graph the probability distribution using a histogram and describe its shape, (c) find the mean, variance, and standard deviation of the probability distribution, and (d) interpret the results in the context of the real-life situation.*

27. Dogs The number of dogs per household in a small town

Dogs	0	1	2	3	4	5
Households	1491	425	168	48	29	14

28. Baseball   The number of games played in the World Series from 1903 to 2009 *(Source: Major League Baseball)*

Games played	4	5	6	7	8
Frequency	20	23	23	36	3

29. Televisions   The number of televisions per household in a small town

Televisions	0	1	2	3
Households	26	442	728	1404

30. Camping Chairs   The number of defects per batch of camping chairs inspected

Defects	0	1	2	3	4	5
Batches	95	113	87	64	13	8

31. Overtime Hours   The number of overtime hours worked in one week per employee

Overtime hours	0	1	2	3	4	5	6
Employees	6	12	29	57	42	30	16

32. Extracurricular Activities   The number of school-related extracurricular activities per student

Activities	0	1	2	3	4	5	6	7
Students	19	39	52	57	68	41	27	17

33. Writing   The expected value of an accountant's profit and loss analysis is 0. Explain what this means.

34. Writing   In a game of chance, what is the relationship between a "fair bet" and its expected value? Explain.

Finding Expected Value   *In Exercises 35–40, use the probability distribution or histogram to find the (a) mean, (b) variance, (c) standard deviation, and (d) expected value of the probability distribution, and (e) interpret the results.*

35. Quiz   Students in a class take a quiz with eight questions. The random variable x represents the number of questions answered correctly.

x	0	1	2	3	4	5	6	7	8
$P(x)$	0.02	0.02	0.06	0.06	0.08	0.22	0.30	0.16	0.08

36. 911 Calls   A 911 service center recorded the number of calls received per hour. The random variable x represents the number of calls per hour for one week.

x	0	1	2	3	4	5	6	7
$P(x)$	0.01	0.10	0.26	0.33	0.18	0.06	0.03	0.03

37. Hurricanes The histogram shows the distribution of hurricanes that have hit the U.S. mainland by category, with 1 the weakest level and 5 the strongest. (*Source: Weather Research Center*)

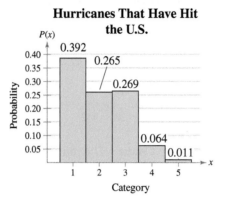

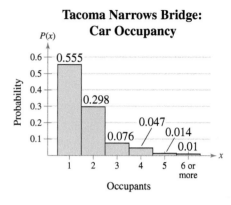

FIGURE FOR EXERCISE 37   FIGURE FOR EXERCISE 38

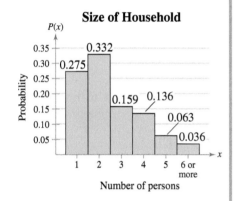

FIGURE FOR EXERCISE 39

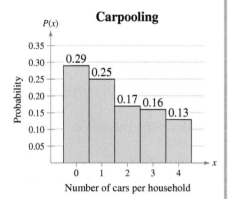

FIGURE FOR EXERCISE 40

38. Car Occupancy The histogram shows the distribution of occupants in cars crossing the Tacoma Narrows Bridge in Washington each week. (*Adapted from Washington State Department of Transportation*)

39. Household Size The histogram shows the distribution of household sizes in the United States for a recent year. (*Adapted from U.S. Census Bureau*)

40. Carpooling The histogram shows the distribution of carpooling by the number of cars per household. (*Adapted from Federal Highway Administration*)

41. Finding Probabilities Use the probability distribution you made for Exercise 27 to find the probability of randomly selecting a household that has (a) fewer than two dogs, (b) at least one dog, and (c) between one and three dogs, inclusive.

42. Finding Probabilities Use the probability distribution you made for Exercise 28 to find the probability of randomly selecting a World Series that consisted of (a) four games, (b) at least five games, and (c) between four and six games, inclusive.

43. Unusual Values A person lives in a household with three dogs and claims that having three dogs is not unusual. Use the information in Exercise 27 to determine if this person is correct. Explain your reasoning.

44. Unusual Values A person randomly chooses a World Series in which eight games were played and claims that this is an unusual event. Use the information in Exercise 28 to determine if this person is correct. Explain you reasoning.

Games of Chance *In Exercises 45 and 46, find the expected net gain to the player for one play of the game. If x is the net gain to a player in a game of chance, then E(x) is usually negative. This value gives the average amount per game the player can expect to lose.*

45. In American roulette, the wheel has the 38 numbers

$$00, 0, 1, 2, \ldots, 34, 35, \text{ and } 36$$

marked on equally spaced slots. If a player bets $1 on a number and wins, then the player keeps the dollar and receives an additional 35 dollars. Otherwise, the dollar is lost.

46. A charity organization is selling $5 raffle tickets as part of a fund-raising program. The first prize is a trip to Mexico valued at $3450, and the second prize is a weekend spa package valued at $750. The remaining 20 prizes are $25 gas cards. The number of tickets sold is 6000.

SC *In Exercises 47 and 48, use StatCrunch to (a) construct and graph a probability distribution and (b) describe its shape.*

47. Computers   The number of computers per household in a small town

Computers	0	1	2	3
Households	300	280	95	20

48. Students   The enrollments (in thousands) for grades 1 through 8 in the United States for a recent year *(Source: U.S. National Center for Education Statistics)*

Grade	1	2	3	4	5	6	7	8
Enrollment	3750	3640	3627	3585	3601	3660	3715	3765

■ EXTENDING CONCEPTS

Linear Transformation of a Random Variable   *In Exercises 49 and 50, use the following information. For a random variable x, a new random variable y can be created by applying a **linear transformation** $y = a + bx$, where a and b are constants. If the random variable x has mean μ_x and standard deviation σ_x, then the mean, variance, and standard deviation of y are given by the following formulas.*

$$\mu_y = a + b\mu_x \qquad \sigma_y^{\,2} = b^2\sigma_x^{\,2} \qquad \sigma_y = |b|\sigma_x$$

49. The mean annual salary of employees at a company is \$36,000. At the end of the year, each employee receives a \$1000 bonus and a 5% raise (based on salary). What is the new mean annual salary (including the bonus and raise) of the employees?

50. The mean annual salary of employees at a company is \$36,000 with a variance of 15,202,201. At the end of the year, each employee receives a \$2000 bonus and a 4% raise (based on salary). What is the standard deviation of the new salaries?

Independent and Dependent Random Variables   *Two random variables x and y are **independent** if the value of x does not affect the value of y. If the variables are not independent, they are **dependent**. A new random variable can be formed by finding the sum or difference of random variables. If a random variable x has mean μ_x and a random variable y has mean μ_y, then the means of the sum and difference of the variables are given by the following equations.*

$$\mu_{x+y} = \mu_x + \mu_y \qquad\qquad \mu_{x-y} = \mu_x - \mu_y$$

If random variables are independent, then the variance and standard deviation of the sum or difference of the random variables can be found. So, if a random variable x has variance σ^2_x and a random variable y has variance σ^2_y, then the variances of the sum and difference of the variables are given by the following equations. Note that the variance of the difference is the sum of the variances.

$$\sigma^2_{x+y} = \sigma^2_x + \sigma^2_y \qquad\qquad \sigma^2_{x-y} = \sigma^2_x + \sigma^2_y$$

In Exercises 51 and 52, the distribution of SAT scores for college-bound male seniors has a mean of 1524 and a standard deviation of 317. The distribution of SAT scores for college-bound female seniors has a mean of 1496 and a standard deviation of 307. One male and one female are randomly selected. Assume their scores are independent. (Source: The College Board)

51. What is the average sum of their scores? What is the average difference of their scores?

52. What is the standard deviation of the difference in their scores?

4.2 Binomial Distributions

WHAT YOU SHOULD LEARN

▸ How to determine if a probability experiment is a binomial experiment

▸ How to find binomial probabilities using the binomial probability formula

▸ How to find binomial probabilities using technology, formulas, and a binomial probability table

▸ How to graph a binomial distribution

▸ How to find the mean, variance, and standard deviation of a binomial probability distribution

Binomial Experiments ▸ Binomial Probability Formula ▸ Finding Binomial Probabilities ▸ Graphing Binomial Distributions ▸ Mean, Variance, and Standard Deviation

▸ BINOMIAL EXPERIMENTS

There are many probability experiments for which the results of each trial can be reduced to two outcomes: success and failure. For instance, when a basketball player attempts a free throw, he or she either makes the basket or does not. Probability experiments such as these are called *binomial experiments*.

DEFINITION

A **binomial experiment** is a probability experiment that satisfies the following conditions.

1. The experiment is repeated for a fixed number of trials, where each trial is independent of the other trials.
2. There are only two possible outcomes of interest for each trial. The outcomes can be classified as a success (S) or as a failure (F).
3. The probability of a success $P(S)$ is the same for each trial.
4. The random variable x counts the number of successful trials.

NOTATION FOR BINOMIAL EXPERIMENTS

SYMBOL	DESCRIPTION
n	The number of times a trial is repeated
$p = P(S)$	The probability of success in a single trial
$q = P(F)$	The probability of failure in a single trial ($q = 1 - p$)
x	The random variable represents a count of the number of successes in n trials: $x = 0, 1, 2, 3, \ldots, n$.

Here is a simple example of a binomial experiment. From a standard deck of cards, you pick a card, note whether it is a club or not, and replace the card. You repeat the experiment five times, so $n = 5$. The outcomes of each trial can be classified in two categories: S = selecting a club and F = selecting another suit. The probabilities of success and failure are

$$p = P(S) = \frac{1}{4} \quad \text{and} \quad q = P(F) = \frac{3}{4}.$$

The random variable x represents the number of clubs selected in the five trials. So, the possible values of the random variable are

0, 1, 2, 3, 4, and 5.

For instance, if $x = 2$, then exactly two of the five cards are clubs and the other three are not clubs. An example of an experiment with $x = 2$ is shown at the left. Note that x is a discrete random variable because its possible values can be listed.

Trial	Outcome	S or F?
1		F
2		S
3		F
4		F
5		S

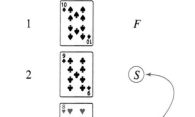

There are two successful outcomes. So, $x = 2$.

PICTURING THE WORLD

In a recent survey of U.S. adults who used the social networking website Twitter were asked if they had ever posted comments about their personal lives. The respondents' answers were either yes or no. (Adapted from Zogby International)

Survey question: Have you ever posted comments about your personal life on Twitter?

Why is this a binomial experiment? Identify the probability of success p. Identify the probability of failure q.

EXAMPLE 1

▶ **Identifying and Understanding Binomial Experiments**

Decide whether the experiment is a binomial experiment. If it is, specify the values of $n, p,$ and $q,$ and list the possible values of the random variable x. If it is not, explain why.

1. A certain surgical procedure has an 85% chance of success. A doctor performs the procedure on eight patients. The random variable represents the number of successful surgeries.

2. A jar contains five red marbles, nine blue marbles, and six green marbles. You randomly select three marbles from the jar, *without replacement*. The random variable represents the number of red marbles.

▶ **Solution**

1. The experiment is a binomial experiment because it satisfies the four conditions of a binomial experiment. In the experiment, each surgery represents one trial. There are eight surgeries, and each surgery is independent of the others. There are only two possible outcomes for each surgery—either the surgery is a success or it is a failure. Also, the probability of success for each surgery is 0.85. Finally, the random variable x represents the number of successful surgeries.

$$n = 8$$
$$p = 0.85$$
$$q = 1 - 0.85$$
$$\quad = 0.15$$
$$x = 0, 1, 2, 3, 4, 5, 6, 7, 8$$

2. The experiment is not a binomial experiment because it does not satisfy all four conditions of a binomial experiment. In the experiment, each marble selection represents one trial, and selecting a red marble is a success. When the first marble is selected, the probability of success is 5/20. However, because the marble is not replaced, the probability of success for subsequent trials is no longer 5/20. So, the trials are not independent, and the probability of a success is not the same for each trial.

▶ **Try It Yourself 1**

Decide whether the following is a binomial experiment. If it is, specify the values of $n, p,$ and $q,$ and list the possible values of the random variable x. If it is not, explain why.

> You take a multiple-choice quiz that consists of 10 questions. Each question has four possible answers, only one of which is correct. To complete the quiz, you randomly guess the answer to each question. The random variable represents the number of correct answers.

a. Identify a *trial* of the experiment and what is a success.
b. Decide if the experiment *satisfies the four conditions* of a binomial experiment.
c. Make a *conclusion* and *identify* $n, p, q,$ and the possible values of x, if possible.

Answer: Page A36

▶ **BINOMIAL PROBABILITY FORMULA**

There are several ways to find the probability of x successes in n trials of a binomial experiment. One way is to use a tree diagram and the Multiplication Rule. Another way is to use the binomial probability formula.

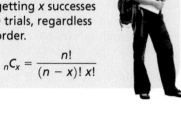
> ### BINOMIAL PROBABILITY FORMULA
>
> In a binomial experiment, the probability of exactly x successes in n trials is
>
> $$P(x) = {}_nC_x\, p^x q^{n-x} = \frac{n!}{(n-x)!\,x!}\, p^x q^{n-x}.$$

EXAMPLE 2   SC Report 17

▶ **Finding Binomial Probabilities**

Microfracture knee surgery has a 75% chance of success on patients with degenerative knees. The surgery is performed on three patients. Find the probability of the surgery being successful on exactly two patients. *(Source: Illinois Sportsmedicine and Orthopedic Center)*

▶ **Solution**   **Method 1:** Draw a tree diagram and use the Multiplication Rule.

1st Surgery	2nd Surgery	3rd Surgery	Outcome	Number of Successes	Probability
		S	SSS	3	$\frac{3}{4} \cdot \frac{3}{4} \cdot \frac{3}{4} = \frac{27}{64}$
	S	F	SSF	2	$\frac{3}{4} \cdot \frac{3}{4} \cdot \frac{1}{4} = \frac{9}{64}$
S		S	SFS	2	$\frac{3}{4} \cdot \frac{1}{4} \cdot \frac{3}{4} = \frac{9}{64}$
	F	F	SFF	1	$\frac{3}{4} \cdot \frac{1}{4} \cdot \frac{1}{4} = \frac{3}{64}$
		S	FSS	2	$\frac{1}{4} \cdot \frac{3}{4} \cdot \frac{3}{4} = \frac{9}{64}$
	S	F	FSF	1	$\frac{1}{4} \cdot \frac{3}{4} \cdot \frac{1}{4} = \frac{3}{64}$
F		S	FFS	1	$\frac{1}{4} \cdot \frac{1}{4} \cdot \frac{3}{4} = \frac{3}{64}$
	F	F	FFF	0	$\frac{1}{4} \cdot \frac{1}{4} \cdot \frac{1}{4} = \frac{1}{64}$

There are three outcomes that have exactly two successes, and each has a probability of $\frac{9}{64}$. So, the probability of a successful surgery on exactly two patients is $3\left(\frac{9}{64}\right) \approx 0.422$.

Method 2: Use the binomial probability formula.

In this binomial experiment, the values of n, p, q, and x are $n = 3$, $p = \frac{3}{4}$, $q = \frac{1}{4}$, and $x = 2$. The probability of exactly two successful surgeries is

$$P(2 \text{ successful surgeries}) = \frac{3!}{(3-2)!\,2!}\left(\frac{3}{4}\right)^2\left(\frac{1}{4}\right)^1$$

$$= 3\left(\frac{9}{16}\right)\left(\frac{1}{4}\right) = 3\left(\frac{9}{64}\right) = \frac{27}{64} \approx 0.422.$$

▶ **Try It Yourself 2**

A card is selected from a standard deck and replaced. This experiment is repeated a total of five times. Find the probability of selecting exactly three clubs.

a. *Identify* a trial, a success, and a failure.
b. *Identify* n, p, q, and x.
c. Use the *binomial probability formula*.   *Answer: Page A36*

By listing the possible values of *x* with the corresponding probabilities, you can construct a **binomial probability distribution.**

EXAMPLE 3

▸ **Constructing a Binomial Distribution**

Why Do We Like Texting?

Convenient for basic information — 79%
Works where talking won't do — 75%
Quicker than calling — 56%
Easier when facing arguments — 37%
Dislike phone conversations — 27%
Great for flirting — 25%

(Source: GfK Roper for Best Buy Mobile)

In a survey, U.S. adults were asked to give reasons why they liked texting on their cellular phones. The results are shown in the graph. Seven adults who participated in the survey are randomly selected and asked whether they like texting because it is quicker than calling. Create a binomial probability distribution for the number of adults who respond yes.

▸ **Solution**

From the graph, you can see that 56% of adults like texting because it is quicker than calling. So, $p = 0.56$ and $q = 0.44$. Because $n = 7$, the possible values of *x* are 0, 1, 2, 3, 4, 5, 6, and 7.

$$P(0) = {}_7C_0(0.56)^0(0.44)^7 = 1(0.56)^0(0.44)^7 \approx 0.0032$$

$$P(1) = {}_7C_1(0.56)^1(0.44)^6 = 7(0.56)^1(0.44)^6 \approx 0.0284$$

$$P(2) = {}_7C_2(0.56)^2(0.44)^5 = 21(0.56)^2(0.44)^5 \approx 0.1086$$

$$P(3) = {}_7C_3(0.56)^3(0.44)^4 = 35(0.56)^3(0.44)^4 \approx 0.2304$$

$$P(4) = {}_7C_4(0.56)^4(0.44)^3 = 35(0.56)^4(0.44)^3 \approx 0.2932$$

$$P(5) = {}_7C_5(0.56)^5(0.44)^2 = 21(0.56)^5(0.44)^2 \approx 0.2239$$

$$P(6) = {}_7C_6(0.56)^6(0.44)^1 = 7(0.56)^6(0.44)^1 \approx 0.0950$$

$$P(7) = {}_7C_7(0.56)^7(0.44)^0 = 1(0.56)^7(0.44)^0 \approx 0.0173$$

x	*P(x)*
0	0.0032
1	0.0284
2	0.1086
3	0.2304
4	0.2932
5	0.2239
6	0.0950
7	0.0173
	$\Sigma P(x) = 1$

Notice in the table at the left that all the probabilities are between 0 and 1 and that the sum of the probabilities is 1.

▸ **Try It Yourself 3**

Seven adults who participated in the survey are randomly selected and asked whether they like texting because it works where talking won't do. Create a binomial distribution for the number of adults who respond yes.

a. *Identify* a trial, a success, and a failure.
b. *Identify* n, p, q, and possible values for *x*.
c. Use the *binomial probability formula* for each value of *x*.
d. Use a *table* to show that the properties of a probability distribution are satisfied.

Answer: Page A37

STUDY TIP

When probabilities are rounded to a fixed number of decimal places, the sum of the probabilities may differ slightly from 1.

▶ FINDING BINOMIAL PROBABILITIES

In Examples 2 and 3, you used the binomial probability formula to find the probabilities. A more efficient way to find binomial probabilities is to use a calculator or a computer. For instance, you can find binomial probabilities using MINITAB, Excel, and the TI-83/84 Plus.

EXAMPLE 4

> #### ▶ Finding a Binomial Probability Using Technology

The results of a recent survey indicate that 67% of U.S. adults consider air conditioning a necessity. If you randomly select 100 adults, what is the probability that exactly 75 adults consider air conditioning a necessity? Use a technology tool to find the probability. *(Source: Opinion Research Corporation)*

> #### ▶ Solution

MINITAB, Excel, and the TI-83/84 Plus each have features that allow you to find binomial probabilities automatically. Try using these technologies. You should obtain results similar to the following.

MINITAB
Probability Distribution Function
Binomial with n = 100 and p = 0.67
x   P[X=x]
75   0.0201004

TI-83/84 PLUS
binompdf(100,.67,75)
.0201004116

EXCEL

	A	B	C	D
1	BINOMDIST(75,100,0.67,FALSE)			
2				0.020100412

Interpretation From these displays, you can see that the probability that exactly 75 adults consider air conditioning a necessity is about 0.02. Because 0.02 is less than 0.05, this can be considered an unusual event.

> #### ▶ Try It Yourself 4

The results of a recent survey indicate that 71% of people in the United States use more than one topping on their hot dogs. If you randomly select 250 people, what is the probability that exactly 178 of them will use more than one topping? Use a technology tool to find the probability. *(Source: ICR Survey Research Group for Hebrew International)*

a. *Identify n, p,* and *x.*
b. Calculate the *binomial probability.*
c. *Interpret* the results.
d. Determine if the event is *unusual.* Explain.   *Answer: Page A37*

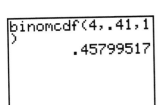

Using a TI-83/84 Plus, you can find the probability in part (1) automatically.

The cumulative distribution function (CDF) computes the probability of "x or fewer" successes. The CDF adds the areas for the given x-value and all those to its left.

EXAMPLE 5   **SC** Report 18

▶ **Finding Binomial Probabilities Using Formulas**

A survey indicates that 41% of women in the United States consider reading their favorite leisure-time activity. You randomly select four U.S. women and ask them if reading is their favorite leisure-time activity. Find the probability that (1) exactly two of them respond yes, (2) at least two of them respond yes, and (3) fewer than two of them respond yes. *(Source: Louis Harris & Associates)*

▶ **Solution**

1. Using $n = 4$, $p = 0.41$, $q = 0.59$, and $x = 2$, the probability that exactly two women will respond yes is

$$P(2) = {}_4C_2(0.41)^2(0.59)^2 = 6(0.41)^2(0.59)^2 \approx 0.351.$$

2. To find the probability that at least two women will respond yes, find the sum of $P(2)$, $P(3)$, and $P(4)$.

$$P(2) = {}_4C_2(0.41)^2(0.59)^2 = 6(0.41)^2(0.59)^2 \approx 0.351094$$
$$P(3) = {}_4C_3(0.41)^3(0.59)^1 = 4(0.41)^3(0.59)^1 \approx 0.162654$$
$$P(4) = {}_4C_4(0.41)^4(0.59)^0 = 1(0.41)^4(0.59)^0 \approx 0.028258$$

So, the probability that at least two will respond yes is

$$P(x \geq 2) = P(2) + P(3) + P(4)$$
$$\approx 0.351094 + 0.162654 + 0.028258$$
$$\approx 0.542.$$

3. To find the probability that fewer than two women will respond yes, find the sum of $P(0)$ and $P(1)$.

$$P(0) = {}_4C_0(0.41)^0(0.59)^4 = 1(0.41)^0(0.59)^4 \approx 0.121174$$
$$P(1) = {}_4C_1(0.41)^1(0.59)^3 = 4(0.41)^1(0.59)^3 \approx 0.336822$$

So, the probability that fewer than two will respond yes is

$$P(x < 2) = P(0) + P(1)$$
$$\approx 0.121174 + 0.336822$$
$$\approx 0.458.$$

▶ **Try It Yourself 5**

A survey indicates that 21% of men in the United States consider fishing their favorite leisure-time activity. You randomly select five U.S. men and ask them if fishing is their favorite leisure-time activity. Find the probability that (1) exactly two of them respond yes, (2) at least two of them respond yes, and (3) fewer than two of them respond yes. *(Source: Louis Harris & Associates)*

a. Determine the appropriate *value of x* for each situation.
b. Find the *binomial probability* for each value of x. Then find the *sum*, if necessary.
c. *Write* the result as a sentence.

Answer: Page A37

Finding binomial probabilities with the binomial probability formula can be a tedious process. To make this process easier, you can use a binomial probability table. Table 2 in Appendix B lists the binomial probabilities for selected values of n and p.

EXAMPLE 6

▶ Finding a Binomial Probability Using a Table

About ten percent of workers (16 years and over) in the United States commute to their jobs by carpooling. You randomly select eight workers. What is the probability that exactly four of them carpool to work? Use a table to find the probability. *(Source: American Community Survey)*

▶ Solution

A portion of Table 2 in Appendix B is shown here. Using the distribution for $n = 8$ and $p = 0.1$, you can find the probability that $x = 4$, as shown by the highlighted areas in the table.

														p
n	*x*	.01	.05	.10	.15	.20	.25	.30	.35	.40	.45	.50	.55	.60
2	0	.980	.902	.810	.723	.640	.563	.490	.423	.360	.303	.250	.203	.160
	1	.020	.095	.180	.255	.320	.375	.420	.455	.480	.495	.500	.495	.480
	2	.000	.002	.010	.023	.040	.063	.090	.123	.160	.203	.250	.303	.360
3	0	.970	.857	.729	.614	.512	.422	.343	.275	.216	.166	.125	.091	.064
	1	.029	.135	.243	.325	.384	.422	.441	.444	.432	.408	.375	.334	.288
	2	.000	.007	.027	.057	.096	.141	.189	.239	.288	.334	.375	.408	.432
	3	.000	.000	.001	.003	.008	.016	.027	.043	.064	.091	.125	.166	.216
8	0	.923	.663	.430	.272	.168	.100	.058	.032	.017	.008	.004	.002	.001
	1	.075	.279	.383	.385	.336	.267	.198	.137	.090	.055	.031	.016	.008
	2	.003	.051	.149	.238	.294	.311	.296	.259	.209	.157	.109	.070	.041
	3	.000	.005	.033	.084	.147	.208	.254	.279	.279	.257	.219	.172	.124
	4	.000	.000	(.005)	.018	.046	.087	.136	.188	.232	.263	.273	.263	.232
	5	.000	.000	.000	.003	.009	.023	.047	.081	.124	.172	.219	.257	.279
	6	.000	.000	.000	.000	.001	.004	.010	.022	.041	.070	.109	.157	.209
	7	.000	.000	.000	.000	.000	.000	.001	.003	.008	.016	.031	.055	.090
	8	.000	.000	.000	.000	.000	.000	.000	.000	.001	.002	.004	.008	.017

Interpretation So, the probability that exactly four of the eight workers carpool to work is 0.005. Because 0.005 is less than 0.05, this can be considered an unusual event.

To explore this topic further, see Activity 4.2 on page 216.

▶ Try It Yourself 6

About fifty-five percent of all small businesses in the United States have a website. If you randomly select 10 small businesses, what is the probability that exactly four of them have a website? Use a table to find the probability. *(Adapted from Webvisible/Nielsen Online)*

a. *Identify* a trial, a success, and a failure.
b. Identify n, p, and x.
c. Use Table 2 in Appendix B to find the *binomial probability*.
d. *Interpret* the results.
e. Determine if the event is *unusual*. Explain.

Answer: Page A37

▶ GRAPHING BINOMIAL DISTRIBUTIONS

In Section 4.1, you learned how to graph discrete probability distributions. Because a binomial distribution is a discrete probability distribution, you can use the same process.

EXAMPLE 7

▶ Graphing a Binomial Distribution

Sixty percent of households in the United States own a video game console. You randomly select six households and ask them if they own a video game console. Construct a probability distribution for the random variable x. Then graph the distribution. *(Source: Deloitte LLP)*

▶ Solution

To construct the binomial distribution, find the probability for each value of x. Using $n = 6$, $p = 0.6$, and $q = 0.4$, you can obtain the following.

x	0	1	2	3	4	5	6
$P(x)$	0.004	0.037	0.138	0.276	0.311	0.187	0.047

You can graph the probability distribution using a histogram as shown below.

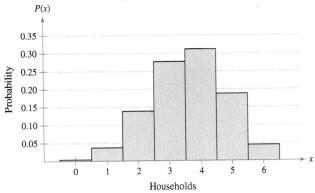

Interpretation   From the histogram, you can see that it would be unusual if none, only one, or all six of the households owned a video game console because of the low probabilities.

▶ Try It Yourself 7

Eighty-one percent of households in the United States own a computer. You randomly select four households and ask if they own a computer. Construct a probability distribution for the random variable x. Then graph the distribution. *(Source: Nielsen)*

a. Find the *binomial probability* for each value of the random variable x.
b. *Organize* the values of x and their corresponding probabilities in a table.
c. Use a *histogram* to graph the binomial distribution. Then describe its shape.
d. Are any of the events *unusual*? Explain.   *Answer: Page A37*

Notice in Example 7 that the histogram is skewed left. The graph of a binomial distribution with $p > 0.5$ is skewed left, whereas the graph of a binomial distribution with $p < 0.5$ is skewed right. The graph of a binomial distribution with $p = 0.5$ is symmetric.

▶ MEAN, VARIANCE, AND STANDARD DEVIATION

Although you can use the formulas you learned in Section 4.1 for mean, variance, and standard deviation of a discrete probability distribution, the properties of a binomial distribution enable you to use much simpler formulas.

POPULATION PARAMETERS OF A BINOMIAL DISTRIBUTION

$$\text{Mean: } \mu = np$$
$$\text{Variance: } \sigma^2 = npq$$
$$\text{Standard deviation: } \sigma = \sqrt{npq}$$

EXAMPLE 8

▶ Finding and Interpreting Mean, Variance, and Standard Deviation

In Pittsburgh, Pennsylvania, about 56% of the days in a year are cloudy. Find the mean, variance, and standard deviation for the number of cloudy days during the month of June. Interpret the results and determine any unusual values. *(Source: National Climatic Data Center)*

▶ Solution

There are 30 days in June. Using $n = 30$, $p = 0.56$, and $q = 0.44$, you can find the mean, variance, and standard deviation as shown below.

$$\mu = np = 30 \cdot 0.56$$
$$= 16.8$$
$$\sigma^2 = npq = 30 \cdot 0.56 \cdot 0.44$$
$$\approx 7.4$$
$$\sigma = \sqrt{npq} = \sqrt{30 \cdot 0.56 \cdot 0.44}$$
$$\approx 2.7$$

Interpretation On average, there are 16.8 cloudy days during the month of June. The standard deviation is about 2.7 days. Values that are more than two standard deviations from the mean are considered unusual. Because $16.8 - 2(2.7) = 11.4$, a June with 11 cloudy days or less would be unusual. Similarly, because $16.8 + 2(2.7) = 22.2$, a June with 23 cloudy days or more would also be unusual.

▶ Try It Yourself 8

In San Francisco, California, 44% of the days in a year are clear. Find the mean, variance, and standard deviation for the number of clear days during the month of May. Interpret the results and determine any unusual events. *(Source: National Climatic Data Center)*

a. *Identify* a success and the values of n, p, and q.
b. Find the *product* of n and p to calculate the mean.
c. Find the *product* of n, p, and q for the variance.
d. Find the *square root* of the variance to find the standard deviation.
e. *Interpret* the results.
f. Determine any *unusual* events.

Answer: Page A37

4.2 EXERCISES

■ BUILDING BASIC SKILLS AND VOCABULARY

1. In a binomial experiment, what does it mean to say that each trial is independent of the other trials?

2. In a binomial experiment with n trials, what does the random variable measure?

Graphical Analysis   *In Exercises 3 and 4, match each given probability with the correct graph. The histograms represent binomial distributions. Each distribution has the same number of trials n but different probabilities of success p.*

3. $p = 0.20$, $p = 0.50$, $p = 0.80$

(a) 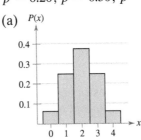   (b) 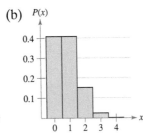   (c)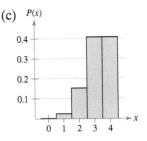

4. $p = 0.25$, $p = 0.50$, $p = 0.75$

(a) 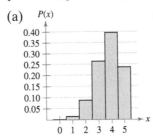   (b) 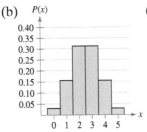   (c)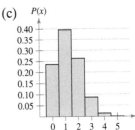

Graphical Analysis   *In Exercises 5 and 6, match each given value of n with the correct graph. Each histogram shown represents part of a binomial distribution. Each distribution has the same probability of success p but different numbers of trials n. What happens as the value of n increases and p remains the same?*

5. $n = 4$, $n = 8$, $n = 12$

(a) 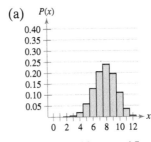   (b) 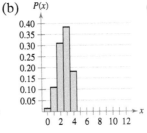   (c)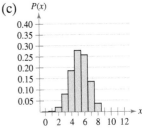

6. $n = 5$, $n = 10$, $n = 15$

(a) 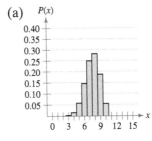   (b) 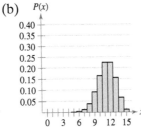   (c)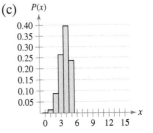

7. Identify the unusual values of x in each histogram in Exercise 5.

8. Identify the unusual values of x in each histogram in Exercise 6.

Identifying and Understanding Binomial Experiments
In Exercises 9–12, decide whether the experiment is a binomial experiment. If it is, identify a success, specify the values of n, p, and q, and list the possible values of the random variable x. If it is not a binomial experiment, explain why.

9. Cyanosis Cyanosis is the condition of having bluish skin due to insufficient oxygen in the blood. About 80% of babies born with cyanosis recover fully. A hospital is caring for five babies born with cyanosis. The random variable represents the number of babies that recover fully. *(Source: The World Book Encyclopedia)*

10. Clothing Store Purchases From past records, a clothing store finds that 26% of the people who enter the store will make a purchase. During a one-hour period, 18 people enter the store. The random variable represents the number of people who do not make a purchase.

11. Survey A survey asks 1400 chief financial officers, "Has the economy forced you to postpone or reduce the amount of vacation you plan to take this year?" Thirty-one percent of those surveyed say they are postponing or reducing the amount of vacation. Twenty officers participating in the survey are randomly selected. The random variable represents the number of officers who are postponing or reducing the amount of vacation. *(Source: Robert Half Management Resources)*

12. Lottery A state lottery randomly chooses 6 balls numbered from 1 through 40. You choose six numbers and purchase a lottery ticket. The random variable represents the number of matches on your ticket to the numbers drawn in the lottery.

Mean, Variance, and Standard Deviation
In Exercises 13–16, find the mean, variance, and standard deviation of the binomial distribution with the given values of n and p.

13. $n = 50$, $p = 0.4$

14. $n = 84$, $p = 0.65$

15. $n = 124$, $p = 0.26$

16. $n = 316$, $p = 0.82$

■ USING AND INTERPRETING CONCEPTS

Finding Binomial Probabilities
In Exercises 17–26, find the indicated probabilities. If convenient, use technology to find the probabilities.

17. Answer Guessing You are taking a multiple-choice quiz that consists of five questions. Each question has four possible answers, only one of which is correct. To complete the quiz, you randomly guess the answer to each question. Find the probability of guessing (a) exactly three answers correctly, (b) at least three answers correctly, and (c) less than three answers correctly.

18. Surgery Success A surgical technique is performed on seven patients. You are told there is a 70% chance of success. Find the probability that the surgery is successful for (a) exactly five patients, (b) at least five patients, and (c) less than five patients.

19. Baseball Fans Fifty-nine percent of men consider themselves fans of professional baseball. You randomly select 10 men and ask each if he considers himself a fan of professional baseball. Find the probability that the number who consider themselves baseball fans is (a) exactly eight, (b) at least eight, and (c) less than eight. *(Source: Gallup Poll)*

20. **Favorite Cookie**   Ten percent of adults say oatmeal raisin is their favorite cookie. You randomly select 12 adults and ask them to name their favorite cookie. Find the probability that the number who say oatmeal raisin is their favorite cookie is (a) exactly four, (b) at least four, and (c) less than four. *(Source: WEAREVER)*

21. **Savings**   Fifty-five percent of U.S. households say they would feel secure if they had $50,000 in savings. You randomly select 8 households and ask them if they would feel secure if they had $50,000 in savings. Find the probability that the number that say they would feel secure is (a) exactly five, (b) more than five, and (c) at most five. *(Source: HSBC Consumer Survey)*

22. **Honeymoon Financing**   Seventy percent of married couples paid for their honeymoon themselves. You randomly select 20 married couples and ask them if they paid for their honeymoon themselves. Find the probability that the number of couples who say they paid for their honeymoon themselves is (a) exactly one, (b) more than one, and (c) at most one. *(Source: Bride's Magazine)*

23. **Financial Advice**   Forty-three percent of adults say they get their financial advice from family members. You randomly select 14 adults and ask them if they get their financial advice from family members. Find the probability that the number who say they get their financial advice from family members is (a) exactly five, (b) at least six, and (c) at most three. *(Source: Sun Life Unretirement Index)*

24. **Retirement**   Fourteen percent of workers believe they will need less than $250,000 when they retire. You randomly select 10 workers and ask them how much money they think they will need for retirement. Find the probability that the number of workers who say they will need less than $250,000 when they retire is (a) exactly two, (b) more than six, and (c) at most five. *(Source: Retirement Corporation of America)*

25. **Credit Cards**   Twenty-eight percent of college students say they use credit cards because of the rewards program. You randomly select 10 college students and ask them to name the reason they use credit cards. Find the probability that the number of college students who say they use credit cards because of the rewards program is (a) exactly two, (b) more than two, and (c) between two and five, inclusive. *(Source: Experience.com)*

26. **Movies on Phone**   Twenty-five percent of adults say they would watch streaming movies on their phone at work. You randomly select 12 adults and ask them if they would watch streaming movies on their phone at work. Find the probability that the number who say they would watch streaming movies on their phone at work is (a) exactly four, (b) more than four, and (c) between four and eight, inclusive. *(Source: mSpot)*

Constructing Binomial Distributions   *In Exercises 27–30, (a) construct a binomial distribution, (b) graph the binomial distribution using a histogram and describe its shape, (c) find the mean, variance, and standard deviation of the binomial distribution, and (d) interpret the results in the context of the real-life situation. What values of the random variable x would you consider unusual? Explain your reasoning.*

27. **Visiting the Dentist**   Sixty-three percent of adults say they are visiting the dentist less because of the economy. You randomly select six adults and ask them if they are visiting the dentist less because of the economy. *(Source: American Optometric Association)*

28. **No Trouble Sleeping** One in four adults claims to have no trouble sleeping at night. You randomly select five adults and ask them if they have any trouble sleeping at night. *(Source: Marist Institute for Public Opinion)*

29. **Blood Donors** Five percent of people in the United States eligible to donate blood actually do. You randomly select four eligible blood donors and ask them if they donate blood. *(Source: MetLife Consumer Education Center)*

30. **Blood Types** Thirty-nine percent of people in the United States have type O^+ blood. You randomly select five Americans and ask them if their blood type is O^+. *(Source: American Association of Blood Banks)*

31. **Annoying Flights** The graph shows the results of a survey of travelers who were asked to name what they found most annoying on a flight. You randomly select six people who participated in the survey and ask them to name what they find most annoying on a flight. Let x represent the number who name crying kids as the most annoying. *(Source: USA Today)*

    (a) Construct a binomial distribution.

    (b) Find the probability that exactly two people name "crying kids."

    (c) Find the probability that at least five people name "crying kids."

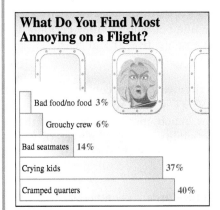

FIGURE FOR EXERCISE 31

FIGURE FOR EXERCISE 32

32. **Small-Business Owners** The graph shows the results of a survey of small-business owners who were asked which business skills they would like to develop further. You randomly select five owners who participated in the survey and ask them which business skills they want to develop further. Let x represent the number who said financial management was the skill they wanted to develop further. *(Source: American Express)*

    (a) Construct a binomial distribution.

    (b) Find the probability that exactly two owners say "financial management."

    (c) Find the probability that fewer than four owners say "financial management."

33. Find the mean and standard deviation of the binomial distribution in Exercise 31 and interpret the results in the context of the real-life situation. What values of x would you consider unusual? Explain your reasoning.

34. Find the mean and standard deviation of the binomial distribution in Exercise 32 and interpret the results in the context of the real-life situation. What values of x would you consider unusual? Explain your reasoning.

SC *In Exercises 35 and 36, use the StatCrunch binomial calculator to find the indicated probabilities. Then determine if the event is unusual. Explain your reasoning.*

35. Pet Owners Sixty-six percent of pet owners say they consider their pet to be their best friend. You randomly select 10 pet owners and ask them if they consider their pet to be their best friend. Find the probability that the number who say their pet is their best friend is (a) exactly nine, (b) at least seven, and (c) at most three. *(Adapted from Kelton Research)*

36. Eco-Friendly Vehicles Fifty-three percent of 18- to 30-year-olds say they would pay more for an eco-friendly vehicle. You randomly select eight 18- to 30-year-olds and ask each if they would pay more for an eco-friendly vehicle. Find the probability that the number who say they would pay more for an eco-friendly vehicle is (a) exactly four, (b) at least five, and (c) less than two. *(Source: Deloitte LLP and Michigan State University)*

◼ EXTENDING CONCEPTS

Multinomial Experiments *In Exercises 37 and 38, use the following information.*

A **multinomial experiment** is a probability experiment that satisfies the following conditions.

1. The experiment is repeated a fixed number of times n where each trial is independent of the other trials.

2. Each trial has k possible mutually exclusive outcomes: $E_1, E_2, E_3, \ldots, E_k$.

3. Each outcome has a fixed probability. So, $P(E_1) = p_1$, $P(E_2) = p_2$, $P(E_3) = p_3, \ldots, P(E_k) = p_k$. The sum of the probabilities for all outcomes is

$$p_1 + p_2 + p_3 + \cdots + p_k = 1.$$

4. x_1 is the number of times E_1 will occur, x_2 is the number of times E_2 will occur, x_3 is the number of times E_3 will occur, and so on.

5. The discrete random variable x counts the number of times $x_1, x_2, x_3, \ldots, x_k$ occurs in n independent trials where

$$x_1 + x_2 + x_3 + \cdots + x_k = n.$$

The probability that x will occur is

$$P(x) = \frac{n!}{x_1! x_2! x_3! \cdots x_k!} \, p_1^{x_1} p_2^{x_2} p_3^{x_3} \cdots p_k^{x_k}.$$

37. Genetics According to a theory in genetics, if tall and colorful plants are crossed with short and colorless plants, four types of plants will result: tall and colorful, tall and colorless, short and colorful, and short and colorless, with corresponding probabilities of $\frac{9}{16}$, $\frac{3}{16}$, $\frac{3}{16}$, and $\frac{1}{16}$. If 10 plants are selected, find the probability that 5 will be tall and colorful, 2 will be tall and colorless, 2 will be short and colorful, and 1 will be short and colorless.

38. Genetics Another proposed theory in genetics gives the corresponding probabilities for the four types of plants described in Exercise 37 as $\frac{5}{16}$, $\frac{4}{16}$, $\frac{1}{16}$, and $\frac{6}{16}$. If 10 plants are selected, find the probability that 5 will be tall and colorful, 2 will be tall and colorless, 2 will be short and colorful, and 1 will be short and colorless.

ACTIVITY 4.2   Binomial Distribution

APPLET

The *binomial distribution* applet allows you to simulate values from a binomial distribution. You can specify the parameters for the binomial distribution (n and p) and the number of values to be simulated (N). When you click SIMULATE, N values from the specified binomial distribution will be plotted at the right. The frequency of each outcome is shown in the plot.

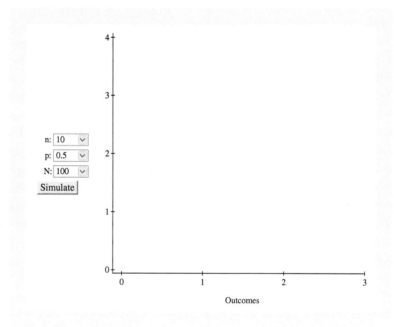

■ Explore

Step 1   Specify a value of n.
Step 2   Specify a value of p.
Step 3   Specify a value of N.
Step 4   Click SIMULATE.

■ Draw Conclusions

APPLET

1. During a presidential election year, 70% of a county's eligible voters actually vote. Simulate selecting $n = 10$ eligible voters $N = 10$ times (for 10 communities in the county). Use the results to estimate the probability that the number who voted in this election is (a) exactly 5, (b) at least 8, and (c) at most 7.

2. During a non-presidential election year, 20% of the eligible voters in the same county as in Exercise 1 actually vote. Simulate selecting $n = 10$ eligible voters $N = 10$ times (for 10 communities in the county). Use the results to estimate the probability that the number who voted in this election is (a) exactly 4, (b) at least 5, and (c) less than 4.

3. Suppose in Exercise 1 you select $n = 10$ eligible voters $N = 100$ times. Estimate the probability that the number who voted in this election is exactly 5. Compare this result with the result in Exercise 1 part (a). Which of these is closer to the probability found using the binomial probability formula?

Binomial Distribution of Airplane Accidents

The Air Transport Association of America (ATA) is a support organization for the principal U.S. airlines. Some of the ATA's activities include promoting the air transport industry and conducting industry-wide studies.

The ATA also keeps statistics about commercial airline flights, including those that involve accidents. From 1979 through 2008 for aircraft with 10 or more seats, there were 76 fatal commercial airplane accidents involving U.S. airlines. The distribution of these accidents is shown in the histogram at the right.

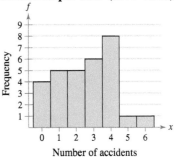

Fatal Commercial Airplane Accidents per Year (1979–2008)

Number of accidents

Year	1979	1980	1981	1982	1983	1984	1985	1986	1987	1988	1989	1990	1991	1992	1993
Accidents	4	0	4	4	4	1	4	2	4	3	5	4	3	3	1

Year	1994	1995	1996	1997	1998	1999	2000	2001	2002	2003	2004	2005	2006	2007	2008
Accidents	4	1	3	3	1	2	2	6	0	2	1	3	2	0	0

▆ EXERCISES

1. In 2006, there were about 11 million commercial flights in the United States. If one is selected at random, what is the probability that it involved a fatal accident?

2. Suppose that the probability of a fatal accident in a given year is 0.0000004. A binomial probability distribution for $n = 11,000,000$ and $p = 0.0000004$ with $x = 0$ to 12 is shown.

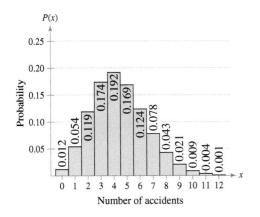

Number of accidents

   What is the probability that there will be (a) 4 fatal accidents in a year? (b) 10 fatal accidents? (c) between 1 and 5, inclusive?

3. Construct a binomial distribution for $n = 11,000,000$ and $p = 0.0000008$ with $x = 0$ to 12. Compare your results with the distribution in Exercise 2.

4. Is a binomial distribution a good model for determining the probabilities of various numbers of fatal accidents during a year? Explain your reasoning and include a discussion of the four criteria for a binomial experiment.

5. According to analysis by *USA TODAY*, air flight is so safe that a person "would have to fly every day for more than 64,000 years before dying in an accident." How can such a statement be justified?

4.3 More Discrete Probability Distributions

WHAT YOU SHOULD LEARN

▸ How to find probabilities using the geometric distribution

▸ How to find probabilities using the Poisson distribution

The Geometric Distribution ▸ The Poisson Distribution ▸ Summary of Discrete Probability Distributions

▸ THE GEOMETRIC DISTRIBUTION

In this section, you will study two more discrete probability distributions—the geometric distribution and the Poisson distribution.

Many actions in life are repeated until a success occurs. For instance, a CPA candidate might take the CPA exam several times before receiving a passing score, or you might have to send an e-mail several times before it is successfully sent. Situations such as these can be represented by a *geometric distribution*.

DEFINITION

A **geometric distribution** is a discrete probability distribution of a random variable x that satisfies the following conditions.

1. A trial is repeated until a success occurs.
2. The repeated trials are independent of each other.
3. The probability of success p is constant for each trial.
4. The random variable x represents the number of the trial in which the first success occurs.

The probability that the first success will occur on trial number x is

$$P(x) = pq^{x-1}, \text{ where } q = 1 - p.$$

In other words, when the first success occurs on the third trial, the outcome is *FFS*, and the probability is $P(3) = q \cdot q \cdot p$, or $P(3) = p \cdot q^2$.

STUDY TIP

Here are instructions for finding a geometric probability on a TI-83/84 Plus.

2nd DISTR

D: geometpdf(

Enter the values of p and x separated by commas.

ENTER

```
geometpdf(.74,3)
           .050024
geometpdf(.74,4)
          .01300624
```

Using a TI-83/84 Plus, you can find the probabilities used in Example 1 automatically.

EXAMPLE 1

▸ **Finding Probabilities Using the Geometric Distribution**

Basketball player LeBron James makes a free throw shot about 74% of the time. Find the probability that the first free throw shot LeBron makes occurs on the third or fourth attempt. *(Source: ESPN)*

▸ **Solution** To find the probability that LeBron makes his first free throw shot on the third or fourth attempt, first find the probability that the first shot he makes will occur on the third attempt and the probability that the first shot he makes will occur on the fourth attempt. Then, find the sum of the resulting probabilities. Using $p = 0.74$, $q = 0.26$, and $x = 3$, you have

$$P(3) = 0.74 \cdot (0.26)^2 = 0.050024.$$

Using $p = 0.74$, $q = 0.26$, and $x = 4$, you have

$$P(4) = 0.74 \cdot (0.26)^3 \approx 0.013006.$$

So, the probability that LeBron makes his first free throw shot on the third or fourth attempt is

$$P(\text{shot made on third or fourth attempt}) = P(3) + P(4)$$

$$\approx 0.050024 + 0.013006 \approx 0.063.$$

▶ **Try It Yourself 1**

Find the probability that LeBron makes his first free throw shot before his third attempt.

a. Use the *geometric distribution* to find $P(1)$ and $P(2)$.
b. Find the *sum* of $P(1)$ and $P(2)$.
c. *Write* the result as a sentence.   *Answer: Page A37*

Even though theoretically a success may never occur, the geometric distribution is a discrete probability distribution because the values of x can be listed—1, 2, 3, ... Notice that as x becomes larger, $P(x)$ gets closer to zero. For instance,

$$P(15) = 0.74(0.26)^{14} \approx 0.0000000048.$$

▶ THE POISSON DISTRIBUTION

In a binomial experiment, you are interested in finding the probability of a specific number of successes in a given number of trials. Suppose instead that you want to know the probability that a specific number of occurrences takes place within a given unit of time or space. For instance, to determine the probability that an employee will take 15 sick days within a year, you can use the *Poisson distribution*.

DEFINITION

The **Poisson distribution** is a discrete probability distribution of a random variable x that satisfies the following conditions.

1. The experiment consists of counting the number of times x an event occurs in a given interval. The interval can be an interval of time, area, or volume.
2. The probability of the event occurring is the same for each interval.
3. The number of occurrences in one interval is independent of the number of occurrences in other intervals.

The probability of exactly x occurrences in an interval is

$$P(x) = \frac{\mu^x e^{-\mu}}{x!}$$

where e is an irrational number approximately equal to 2.71828 and μ is the mean number of occurrences per interval unit.

STUDY TIP

Here are instructions for finding a Poisson probability on a TI-83/84 Plus.

| 2nd | DISTR

B: poissonpdf(

Enter the values of μ and x separated by commas.

| ENTER |

```
poissonpdf(3,4)
       .1680313557
```

Using a TI-83/84 Plus, you can find the probability in Example 2 automatically.

EXAMPLE 2   SC Report 19

▶ **Using the Poisson Distribution**

The mean number of accidents per month at a certain intersection is three. What is the probability that in any given month four accidents will occur at this intersection?

▶ **Solution**

Using $x = 4$ and $\mu = 3$, the probability that 4 accidents will occur in any given month at the intersection is

$$P(4) \approx \frac{3^4(2.71828)^{-3}}{4!} \approx 0.168.$$

▶ **Try It Yourself 2**

What is the probability that more than four accidents will occur in any given month at the intersection?

a. Use the *Poisson distribution* to find $P(0)$, $P(1)$, $P(2)$, $P(3)$, and $P(4)$.
b. Find the *sum* of $P(0)$, $P(1)$, $P(2)$, $P(3)$, and $P(4)$.
c. *Subtract* the sum from 1.
d. *Write* the result as a sentence.   *Answer: Page A37*

In Example 2, you used a formula to determine a Poisson probability. You can also use a table to find Poisson probabilities. Table 3 in Appendix B lists the Poisson probabilities for selected values of x and μ. You can use technology tools, such as MINITAB, Excel, and the TI-83/84 Plus, to find Poisson probabilities as well.

EXAMPLE 3

▶ **Finding Poisson Probabilities Using a Table**

A population count shows that the average number of rabbits per acre living in a field is 3.6. Use a table to find the probability that seven rabbits are found on any given acre of the field.

▶ **Solution**

A portion of Table 3 in Appendix B is shown here. Using the distribution for $\mu = 3.6$ and $x = 7$, you can find the Poisson probability as shown by the highlighted areas in the table.

x	3.1	3.2	3.3	3.4	3.5	3.6	3.7
0	.0450	.0408	.0369	.0334	.0302	.0273	.0247
1	.1397	.1304	.1217	.1135	.1057	.0984	.0915
2	.2165	.2087	.2008	.1929	.1850	.1771	.1692
3	.2237	.2226	.2209	.2186	.2158	.2125	.2087
4	.1734	.1781	.1823	.1858	.1888	.1912	.1931
5	.1075	.1140	.1203	.1264	.1322	.1377	.1429
6	.0555	.0608	.0662	.0716	.0771	.0826	.0881
7	.0246	.0278	.0312	.0348	.0385	.0425	.0466
8	.0095	.0111	.0129	.0148	.0169	.0191	.0215
9	.0033	.0040	.0047	.0056	.0066	.0076	.0089
10	.0010	.0013	.0016	.0019	.0023	.0028	.0033

(column header group: μ)

Interpretation So, the probability that seven rabbits are found on any given acre is 0.0425. Because 0.0425 is less than 0.05, this can be considered an unusual event.

▶ **Try It Yourself 3**

Two thousand brown trout are introduced into a small lake. The lake has a volume of 20,000 cubic meters. Use a table to find the probability that three brown trout are found in any given cubic meter of the lake.

a. Find the *average* number of brown trout per cubic meter.
b. *Identify* μ and x.
c. *Use* Table 3 in Appendix B to find the Poisson probability.
d. *Interpret* the results.
e. Determine if the event is *unusual*. Explain.   *Answer: Page A37*

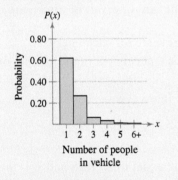

▶ SUMMARY OF DISCRETE PROBABILITY
DISTRIBUTIONS

The following table summarizes the discrete probability distributions discussed in this chapter.

Distribution	Summary	Formulas
Binomial Distribution	A binomial experiment satisfies the following conditions. 1. The experiment is repeated for a fixed number n of independent trials. 2. There are only two possible outcomes for each trial. Each outcome can be classified as a success or as a failure. 3. The probability of a success must remain constant for each trial. 4. The random variable x counts the number of successful trials. The parameters of a binomial distribution are n and p.	n = the number of times a trial repeats x = the number of successes in n trials p = probability of success in a single trial q = probability of failure in a single trial $q = 1 - p$ The probability of exactly x successes in n trials is $$P(x) = {}_nC_x\, p^x q^{n-x}$$ $$= \frac{n!}{(n-x)!x!}\, p^x q^{n-x}.$$
Geometric Distribution	A geometric distribution is a discrete probability distribution of a random variable x that satisfies the following conditions. 1. A trial is repeated until a success occurs. 2. The repeated trials are independent of each other. 3. The probability of success p is constant for each trial. 4. The random variable x represents the number of the trial in which the first success occurs. The parameter of a geometric distribution is p.	x = the number of the trial in which the first success occurs p = probability of success in a single trial q = probability of failure in a single trial $q = 1 - p$ The probability that the first success occurs on trial number x is $$P(x) = pq^{x-1}.$$
Poisson Distribution	The Poisson distribution is a discrete probability distribution of a random variable x that satisfies the following conditions. 1. The experiment consists of counting the number of times x an event occurs over a specified interval of time, area, or volume. 2. The probability of the event occurring is the same for each interval. 3. The number of occurrences in one interval is independent of the number of occurrences in other intervals. The parameter of a Poisson distribution is μ.	x = the number of occurrences in the given interval μ = the mean number of occurrences in a given time or space unit The probability of exactly x occurrences in an interval is $$P(x) = \frac{\mu^x e^{-\mu}}{x!}.$$

4.3 EXERCISES

■ BUILDING BASIC SKILLS AND VOCABULARY

In Exercises 1–4, assume the geometric distribution applies. Use the given probability of success p to find the indicated probability.

1. Find $P(3)$ when $p = 0.65$. **2.** Find $P(1)$ when $p = 0.45$.

3. Find $P(5)$ when $p = 0.09$. **4.** Find $P(8)$ when $p = 0.28$.

In Exercises 5–8, assume the Poisson distribution applies. Use the given mean μ to find the indicated probability.

5. Find $P(4)$ when $\mu = 5$. **6.** Find $P(3)$ when $\mu = 6$.

7. Find $P(2)$ when $\mu = 1.5$. **8.** Find $P(5)$ when $\mu = 9.8$.

9. In your own words, describe the difference between the value of x in a binomial distribution and in a geometric distribution.

10. In your own words, describe the difference between the value of x in a binomial distribution and in a Poisson distribution.

Deciding on a Distribution *In Exercises 11–14, decide which probability distribution—binomial, geometric, or Poisson—applies to the question. You do not need to answer the question. Instead, justify your choice.*

11. Pilot's Test *Given:* The probability that a student passes the written test for a private pilot's license is 0.75. *Question:* What is the probability that a student will fail on the first attempt and pass on the second attempt?

12. Precipitation *Given:* In Tampa, Florida, the mean number of days in July with 0.01 inch or more precipitation is 16. *Question:* What is the probability that Tampa has 20 days with 0.01 inch or more precipitation next July? *(Source: National Climatic Data Center)*

13. Carry-On Luggage *Given:* Fifty-four percent of U.S. adults think Congress should place size limits on carry-on bags. In a survey of 110 randomly chosen adults, people are asked, "Do you think Congress should place size limits on carry-on bags?" *Question:* What is the probability that exactly 60 of the people answer yes? *(Source: TripAdvisor)*

14. Breaking Up *Given:* Twenty-nine percent of Americans ages 16 to 21 years old say that they would break up with their boyfriend/girlfriend for $10,000. You select at random twenty 16- to 21-year-olds. *Question:* What is the probability that the first person who says he or she would break up with their boyfriend/girlfriend for $10,000 is the fifth person selected? *(Source: Bank of America Student Banking & Seventeen)*

■ USING AND INTERPRETING CONCEPTS

Using a Distribution to Find Probabilities *In Exercises 15–22, find the indicated probabilities using the geometric distribution or the Poisson distribution. Then determine if the events are unusual. If convenient, use a Poisson probability table or technology to find the probabilities.*

15. Telephone Sales Assume the probability that you will make a sale on any given telephone call is 0.19. Find the probability that you (a) make your first sale on the fifth call, (b) make your first sale on the first, second, or third call, and (c) do not make a sale on the first three calls.

16. **Bankruptcies** The mean number of bankruptcies filed per minute in the United States in a recent year was about two. Find the probability that (a) exactly five businesses will file bankruptcy in any given minute, (b) at least five businesses will file bankruptcy in any given minute, and (c) more than five businesses will file bankruptcy in any given minute. *(Source: Administrative Office of the U.S. Courts)*

17. **Typographical Errors** A newspaper finds that the mean number of typographical errors per page is four. Find the probability that (a) exactly three typographical errors are found on a page, (b) at most three typographical errors are found on a page, and (c) more than three typographical errors are found on a page.

18. **Pass Completions** Football player Peyton Manning completes a pass 64.8% of the time. Find the probability that (a) the first pass Peyton completes is the second pass, (b) the first pass Peyton completes is the first or second pass, and (c) Peyton does not complete his first two passes. *(Source: National Football League)*

19. **Major Hurricanes** A major hurricane is a hurricane with wind speeds of 111 miles per hour or greater. During the 20th century, the mean number of major hurricanes to strike the U.S. mainland per year was about 0.6. Find the probability that in a given year (a) exactly one major hurricane strikes the U.S. mainland, (b) at most one major hurricane strikes the U.S. mainland, and (c) more than one major hurricane strikes the U.S. mainland. *(Source: National Hurricane Center)*

20. **Glass Manufacturer** A glass manufacturer finds that 1 in every 500 glass items produced is warped. Find the probability that (a) the first warped glass item is the tenth item produced, (b) the first warped glass item is the first, second, or third item produced, and (c) none of the first 10 glass items produced are defective.

21. **Winning a Prize** A cereal maker places a game piece in each of its cereal boxes. The probability of winning a prize in the game is 1 in 4. Find the probability that you (a) win your first prize with your fourth purchase, (b) win your first prize with your first, second, or third purchase, and (c) do not win a prize with your first four purchases.

22. **Precipitation** The mean number of days with 0.01 inch or more precipitation per month in Baltimore, Maryland, is about 9.5. Find the probability that in a given month, (a) there are exactly 10 days with 0.01 inch or more precipitation, (b) there are at most 10 days with 0.01 inch or more precipitation, and (c) there are more than 10 days with 0.01 inch or more precipitation. *(Source: National Climatic Data Center)*

SC *In Exercises 23 and 24, use the StatCrunch Poisson calculator to find the indicated probabilities. Then determine if the events are unusual. Explain your reasoning.*

23. **Oil Tankers** The mean number of oil tankers at a port city is 8 per day. The port has facilities to handle up to 12 oil tankers in a day. Find the probability that on a given day, (a) eight oil tankers will arrive, (b) at most three oil tankers will arrive, and (c) too many oil tankers will arrive.

24. **Kidney Transplants** The mean number of kidney transplants performed per day in the United States in a recent year was about 45. Find the probability that on a given day, (a) exactly 50 kidney transplants will be performed, (b) at least 65 kidney transplants will be performed, and (c) no more than 40 kidney transplants will be performed. *(Source: U.S. Department of Health and Human Services)*

▦ EXTENDING CONCEPTS

25. Comparing Binomial and Poisson Distributions An automobile manufacturer finds that 1 in every 2500 automobiles produced has a particular manufacturing defect. (a) Use a binomial distribution to find the probability of finding 4 cars with the defect in a random sample of 6000 cars. (b) The Poisson distribution can be used to approximate the binomial distribution for large values of n and small values of p. Repeat (a) using a Poisson distribution and compare the results.

26. Hypergeometric Distribution Binomial experiments require that any sampling be done with replacement because each trial must be independent of the others. The **hypergeometric distribution** also has two outcomes: success and failure. However, the sampling is done without replacement. Given a population of N items having k successes and $N - k$ failures, the probability of selecting a sample of size n that has x successes and $n - x$ failures is given by

$$P(x) = \frac{(_kC_x)(_{N-k}C_{n-x})}{_NC_n}.$$

In a shipment of 15 microchips, 2 are defective and 13 are not defective. A sample of three microchips is chosen at random. Find the probability that (a) all three microchips are not defective, (b) one microchip is defective and two are not defective, and (c) two microchips are defective and one is not defective.

Geometric Distribution: Mean and Variance
In Exercises 27 and 28, use the fact that the mean of a geometric distribution is $\mu = 1/p$ and the variance is $\sigma^2 = q/p^2$.

27. Daily Lottery A daily number lottery chooses three balls numbered 0 to 9. The probability of winning the lottery is 1/1000. Let x be the number of times you play the lottery before winning the first time. (a) Find the mean, variance, and standard deviation. Interpret the results. (b) How many times would you expect to have to play the lottery before winning? Assume that it costs $1 to play and winners are paid $500. Would you expect to make or lose money playing this lottery? Explain.

28. Paycheck Errors A company assumes that 0.5% of the paychecks for a year were calculated incorrectly. The company has 200 employees and examines the payroll records from one month. (a) Find the mean, variance, and standard deviation. Interpret the results. (b) How many employee payroll records would you expect to examine before finding one with an error?

Poisson Distribution: Variance
In Exercises 29 and 30, use the fact that the variance of a Poisson distribution is $\sigma^2 = \mu$.

29. Golf In a recent year, the mean number of strokes per hole for golfer Phil Mickelson was about 3.9. (a) Find the variance and standard deviation. Interpret the results. (b) How likely is Phil to play an 18-hole round and have more than 72 strokes? *(Source: PGATour.com)*

30. Snowfall The mean snowfall in January in Mount Shasta, California is 29.9 inches. (a) Find the variance and standard deviation. Interpret the results. (b) Find the probability that the snowfall in January in Mount Shasta, California will exceed 3 feet. *(Source: National Climatic Data Center)*

USES AND ABUSES

Uses

There are countless occurrences of binomial probability distributions in business, science, engineering, and many other fields.

For instance, suppose you work for a marketing agency and are in charge of creating a television ad for Brand A toothpaste. The toothpaste manufacturer claims that 40% of toothpaste buyers prefer its brand. To check whether the manufacturer's claim is reasonable, your agency conducts a survey. Of 100 toothpaste buyers selected at random, you find that only 35 (or 35%) prefer Brand A. Could the manufacturer's claim still be true? What if your random sample of 100 found only 25 people (or 25%) who express a preference for Brand A? Would you still be justified in running the advertisement?

Knowing the characteristics of binomial probability distributions will help you answer this type of question. By the time you have completed this course, you will be able make educated decisions about the reasonableness of the manufacturer's claim.

Ethics

Suppose the toothpaste manufacturer also claims that four out of five dentists recommend Brand A toothpaste. Your agency wants to mention this fact in the television ad, but when determining how the sample of dentists was formed, you find that the dentists were paid to recommend the toothpaste. Including this statement when running the advertisement would be unethical.

Abuses

Interpreting the "Most Likely" Outcome A common misuse of binomial probability distributions is to think that the "most likely" outcome is the outcome that will occur most of the time. For instance, suppose you randomly choose a committee of four from a large population that is 50% women and 50% men. The most likely composition of the committee will be two men and two women. Although this is the most likely outcome, the probability that it will occur is only 0.375. There is a 0.5 chance that the committee will contain one man and three women or three men and one woman. So, if either of these outcomes occurs, you should not assume that the selection was unusual or biased.

▦ EXERCISES

In Exercises 1–4, suppose that the manufacturer's claim is true—40% of toothpaste buyers prefer Brand A toothpaste. Use the graph and technology to answer the questions. Explain your reasoning.

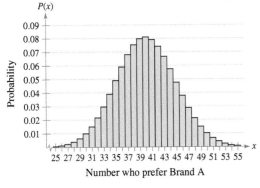

Number who prefer Brand A

1. ***Interpreting the "Most Likely" Outcome*** In a random sample of 100, what is the most likely outcome? How likely is it?

2. ***Interpreting the "Most Likely" Outcome*** In a random sample of 100, what is the probability that between 35 and 45 people, inclusive, prefer Brand A?

3. Suppose in a random sample of 100, you found 36 who prefer Brand A. Would the manufacturer's claim be believable?

4. Suppose in a random sample of 100, you found 25 who prefer Brand A. Would the manufacturer's claim be believable?

4 CHAPTER SUMMARY

What did you **learn?**	EXAMPLE(S)	REVIEW EXERCISES
Section 4.1		
▪ How to distinguish between discrete random variables and continuous random variables	*1*	*1–6*
▪ How to determine if a distribution is a probability distribution	*3–4*	*7–10*
▪ How to construct a discrete probability distribution and its graph and find the mean, variance, and standard deviation of a discrete probability distribution	*2, 5, 6*	*11–14*
$\mu = \Sigma x P(x)$ Mean of a discrete random variable $\sigma^2 = \Sigma (x - \mu)^2 P(x)$ Variance of a discrete random variable $\sigma = \sqrt{\sigma^2} = \sqrt{\Sigma (x - \mu)^2 P(x)}$ Standard deviation of a discrete random variable		
▪ How to find the expected value of a discrete probability distribution	*7*	*15–16*
Section 4.2		
▪ How to determine if a probability experiment is a binomial experiment	*1*	*17–20*
▪ How to find binomial probabilities using the binomial probability formula, a binomial probability table, and technology	*2, 4–6*	*21–24*
$P(x) = {}_nC_x p^x q^{n-x} = \dfrac{n!}{(n-x)!x!} p^x q^{n-x}$ Binomial probability formula		
▪ How to construct a binomial distribution and its graph and find the mean, variance, and standard deviation of a binomial probability distribution	*3, 7, 8*	*25–28*
$\mu = np$ Mean of a binomial distribution $\sigma^2 = npq$ Variance of a binomial distribution $\sigma = \sqrt{npq}$ Standard deviation of a binomial distribution		
Section 4.3		
▪ How to find probabilities using the geometric distribution $P(x) = pq^{x-1}$ Probability that the first success will occur on trial number x	*1*	*29, 30*
▪ How to find probabilities using the Poisson distribution $P(x) = \dfrac{\mu^x e^{-\mu}}{x!}$ Probability of exactly x occurrences in an interval	*2, 3*	*31–33*

4 REVIEW EXERCISES

■ SECTION 4.1

In Exercises 1 and 2, decide whether the graph represents a discrete random variable or a continuous random variable. Explain your reasoning.

1. The number of hours spent sleeping each day

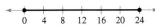

2. The number of fish caught during a fishing tournament

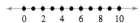

In Exercises 3–6, decide whether the random variable x is discrete or continuous.

3. Let x represent the number of pumps in use at a gas station.

4. Let x represent the weight of a truck at a weigh station.

5. Let x represent the amount of carbon dioxide emitted from a car's tailpipe each day.

6. Let x represent the number of people that activate a metal detector at an airport each hour.

In Exercises 7–10, decide whether the distribution is a probability distribution. If it is not, identify the property that is not satisfied.

7. The daily limit for catching bass at a lake is four. The random variable x represents the number of fish caught in a day.

x	0	1	2	3	4
$P(x)$	0.36	0.23	0.08	0.14	0.29

8. The random variable x represents the number of tickets a police officer writes out each shift.

x	0	1	2	3	4	5
$P(x)$	0.09	0.23	0.29	0.16	0.21	0.02

9. A greeting card shop keeps records of customers' buying habits. The random variable x represents the number of cards sold to an individual customer in a shopping visit.

x	1	2	3	4	5	6	7
$P(x)$	0.68	0.14	0.08	0.05	0.02	0.02	0.01

10. The random variable x represents the number of classes in which a student is enrolled in a given semester at a university.

x	1	2	3	4	5	6	7	8
$P(x)$	$\frac{1}{80}$	$\frac{2}{75}$	$\frac{1}{10}$	$\frac{12}{25}$	$\frac{27}{20}$	$\frac{1}{5}$	$\frac{2}{25}$	$\frac{1}{120}$

In Exercises 11–14,

(a) use the frequency distribution table to construct a probability distribution,

(b) graph the probability distribution using a histogram and describe its shape,

(c) find the mean, variance, and standard deviation of the probability distribution, and

(d) interpret the results in the context of the real-life situation.

11. The number of pages in a section from 10 statistics texts

Pages	Sections
2	3
3	12
4	72
5	115
6	169
7	120
8	83
9	48
10	22
11	6

12. The number of hits per game played by a baseball player during a recent season

Hits	Games
0	29
1	62
2	33
3	12
4	3
5	1

13. The distribution of the number of cellular phones per household in a small town is given.

Cellphones	Families
0	5
1	35
2	68
3	73
4	42
5	19
6	8

14. A television station sells advertising in 15-, 30-, 60-, 90-, and 120-second blocks. The distribution of sales for one 24-hour day is given.

Length (in seconds)	Number
15	76
30	445
60	30
90	3
120	12

In Exercises 15 and 16, find the expected value of the random variable.

15. A person has shares of eight different stocks. The random variable x represents the number of stocks showing a loss on a selected day.

x	0	1	2	3	4	5	6	7	8
$P(x)$	0.02	0.11	0.18	0.32	0.15	0.09	0.05	0.05	0.03

16. A local pub has a chicken wing special on Tuesdays. The pub owners purchase wings in cases of 300. The random variable x represents the number of cases used during the special.

x	1	2	3	4
$P(x)$	$\frac{1}{9}$	$\frac{1}{3}$	$\frac{1}{2}$	$\frac{1}{18}$

■ SECTION 4.2

In Exercises 17 and 18, use the following information.

A probability experiment has n independent trials. Each trial has three possible outcomes: A, B, and C. For each trial, P(A) = 0.30, P(B) = 0.50, and P(C) = 0.20. There are 20 trials.

17. Can a binomial experiment be used to find the probability of 6 outcomes of *A*, 10 outcomes of *B*, and 4 outcomes of *C*? Explain your reasoning.

18. Can a binomial experiment be used to find the probability of 4 outcomes of *C* and 16 outcomes that are *not C*? Explain your reasoning. What is the probability of success for each trial?

In Exercises 19 and 20, decide whether the experiment is a binomial experiment. If it is not, identify the property that is not satisfied. If it is, list the values of n, p, and q and the values that x can assume.

19. Bags of plain M&M's contain 24% blue candies. One candy is selected from each of 12 bags. The random variable represents the number of blue candies selected. *(Source: Mars, Incorporated)*

20. A fair coin is tossed repeatedly until 15 heads are obtained. The random variable *x* counts the number of tosses.

In Exercises 21–24, find the indicated probabilities.

21. One in four adults is currently on a diet. You randomly select eight adults and ask them if they are currently on a diet. Find the probability that the number who say they are currently on a diet is (a) exactly three, (b) at least three, and (c) more than three. *(Source: Wirthlin Worldwide)*

22. One in four people in the United States owns individual stocks. You randomly select 12 people and ask them if they own individual stocks. Find the probability that the number who say they own individual stocks is (a) exactly two, (b) at least two, and (c) more than two. *(Source: Pew Research Center)*

23. Forty-three percent of businesses in the United States require a doctor's note when an employee takes sick time. You randomly select nine businesses and ask each if it requires a doctor's note when an employee takes sick time. Find the probability that the number who say they require a doctor's note is (a) exactly five, (b) at least five, and (c) more than five. *(Source: Harvard School of Public Health)*

24. In a typical day, 31% of people in the United States with Internet access go online to get news. You randomly select five people in the United States with Internet access and ask them if they go online to get news. Find the probability that the number who say they go online to get news is (a) exactly two, (b) at least two, and (c) more than two. *(Source: Pew Research Center)*

In Exercises 25–28,

(a) construct a binomial distribution,

(b) graph the binomial distribution using a histogram and describe its shape,

(c) find the mean, variance, and standard deviation of the binomial distribution and interpret the results in the context of the real-life situation, and

(d) determine the values of the random variable x that you would consider unusual.

25. Thirty-four percent of women in the United States say their spouses never help with household chores. You randomly select five U.S. women and ask if their spouses help with household chores. *(Source: Boston Consulting Group)*

26. Sixty-eight percent of families say that their children have an influence on their vacation destinations. You randomly select six families and ask if their children have an influence on their vacation destinations. *(Source: YPB&R)*

27. In a recent year, forty percent of trucks sold by a company had diesel engines. You randomly select four trucks sold by the company and check if they have diesel engines.

28. Sixty-three percent of U.S. mothers with school-age children choose fast food as a dining option for their families one to three times a week. You randomly select five U.S. mothers with school-age children and ask if they choose fast food as a dining option for their families one to three times a week. *(Adapted from Market Day)*

■ SECTION 4.3

In Exercises 29–32, find the indicated probabilities using the geometric distribution or the Poisson distribution. Then determine if the events are unusual. If convenient, use a Poisson probability table or technology to find the probabilities.

29. Twenty-two percent of former smokers say they tried to quit four or more times before they were habit-free. You randomly select 10 former smokers. Find the probability that the first person who tried to quit four or more times is (a) the third person selected, (b) the fourth or fifth person selected, and (c) not one of the first seven people selected. *(Source: Porter Novelli Health Styles)*

30. In a recent season, hockey player Sidney Crosby scored 33 goals in 77 games he played. Assume that his goal production stayed at that level the following season. What is the probability that he would get his first goal

   (a) in the first game of the season?

   (b) in the second game of the season?

   (c) in the first or second game of the season?

   (d) within the first three games of the season? *(Source: ESPN)*

31. During a 69-year period, tornadoes killed 6755 people in the United States. Assume this rate holds true today and is constant throughout the year. Find the probability that tomorrow

   (a) no one in the U. S. is killed by a tornado,

   (b) one person in the U.S. is killed by a tornado,

   (c) at most two people in the U.S. are killed by a tornado, and

   (d) more than one person in the U.S. is killed by a tornado. *(Source: National Weather Service)*

32. It is estimated that sharks kill 10 people each year worldwide. Find the probability that at least 3 people are killed by sharks this year

   (a) assuming that this rate is true,

   (b) if the rate is actually 5 people a year, and

   (c) if the rate is actually 15 people a year. *(Source: International Shark Attack File)*

33. In Exercise 32, describe what happens to the probability of at least three people being killed by sharks this year as the rate increases and decreases.

4 CHAPTER QUIZ

Take this quiz as you would take a quiz in class. After you are done, check your work against the answers given in the back of the book.

1. Decide if the random variable x is discrete or continuous. Explain your reasoning.

   (a) Let x represent the number of lightning strikes that occur in Wyoming during the month of June.

   (b) Let x represent the amount of fuel (in gallons) used by the Space Shuttle during takeoff.

2. The table lists the number of U.S. mainland hurricane strikes (from 1851 to 2008) for various intensities according to the Saffir-Simpson Hurricane Scale. *(Source: National Oceanic and Atmospheric Administration)*

   (a) Construct a probability distribution of the data.

   (b) Graph the discrete probability distribution using a probability histogram. Then describe its shape.

   (c) Find the mean, variance, and standard deviation of the probability distribution and interpret the results.

   (d) Find the probability that a hurricane selected at random for further study has an intensity of at least four.

3. The success rate of corneal transplant surgery is 85%. The surgery is performed on six patients. *(Adapted from St. Luke's Cataract & Laser Institute)*

   (a) Construct a binomial distribution.

   (b) Graph the binomial distribution using a probability histogram. Then describe its shape.

   (c) Find the mean, variance, and standard deviation of the probability distribution and interpret the results.

   (d) Find the probability that the surgery is successful for exactly three patients. Is this an unusual event? Explain.

   (e) Find the probability that the surgery is successful for fewer than four patients. Is this an unusual event? Explain.

4. A newspaper finds that the mean number of typographical errors per page is five. Find the probability that

   (a) exactly five typographical errors will be found on a page,

   (b) fewer than five typographical errors will be found on a page, and

   (c) no typographical errors will be found on a page.

In Exercises 5 and 6, use the following information. Basketball player Dwight Howard makes a free throw shot about 60.2% of the time. *(Source: ESPN)*

5. Find the probability that the first free throw shot Dwight makes is the fourth shot. Is this an unusual event? Explain.

6. Find the probability that the first free throw shot Dwight makes is the second or third shot. Is this an unusual event? Explain.

Intensity	Number of hurricanes
1	114
2	74
3	76
4	18
5	3

TABLE FOR EXERCISE 2

PUTTING IT ALL TOGETHER

Real Statistics — Real Decisions

The Centers for Disease Control and Prevention (CDC) is required by law to publish a report on assisted reproductive technologies (ART). ART includes all fertility treatments in which both the egg and the sperm are used. These procedures generally involve removing eggs from a woman's ovaries, combining them with sperm in the laboratory, and returning them to the woman's body or giving them to another woman.

You are helping to prepare the CDC report and select at random 10 ART cycles for a special review. None of the cycles resulted in a clinical pregnancy. Your manager feels it is impossible to select at random 10 ART cycles that did not result in a clinical pregnancy. Use the information provided at the right and your knowledge of statistics to determine if your manager is correct.

**Results of ART Cycles
Using Fresh Nondonor
Eggs or Embryos**

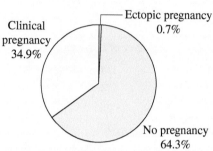

(Source: Centers for Disease Control and Prevention)

■ EXERCISES

1. How Would You Do It?

(a) How would you determine if your manager's view is correct, that it is impossible to select at random 10 ART cycles that did not result in a clinical pregnancy?

(b) What probability distribution do you think best describes the situation? Do you think the distribution of the number of clinical pregnancies is discrete or continuous? Why?

2. Answering the Question

Write an explanation that answers the question, "Is it possible to select at random 10 ART cycles that did not result in a clinical pregnancy?" Include in your explanation the appropriate probability distribution and your calculation of the probability of no clinical pregnancies in 10 ART cycles.

3. Suspicious Samples?

Which of the following samples would you consider suspicious if someone told you that the sample was selected at random? Would you believe that the samples were selected at random? Why or why not?

(a) Selecting at random 10 ART cycles among women of age 40, eight of which resulted in clinical pregnancies.

(b) Selecting at random 10 ART cycles among women of age 41, none of which resulted in clinical pregnancies.

**Pregnancy and Live Birth Rates for
ART Cycles Among Women
of Age 40 and Older**

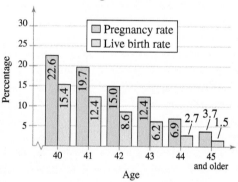

(Source: Centers for Disease Control and Prevention)

TECHNOLOGY   MINITAB   EXCEL   TI-83/84 PLUS

USING POISSON DISTRIBUTIONS AS QUEUING MODELS

Queuing means waiting in line to be served. There are many examples of queuing in everyday life: waiting at a traffic light, waiting in line at a grocery checkout counter, waiting for an elevator, holding for a telephone call, and so on.

Poisson distributions are used to model and predict the number of people (calls, computer programs, vehicles) arriving at the line. In the following exercises, you are asked to use Poisson distributions to analyze the queues at a grocery store checkout counter.

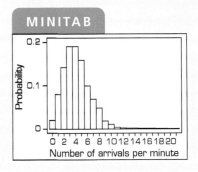

■ EXERCISES

In Exercises 1–7, consider a grocery store that can process a total of four customers at its checkout counters each minute.

1. Suppose that the mean number of customers who arrive at the checkout counters each minute is 4. Create a Poisson distribution with $\mu = 4$ for $x = 0$ to 20. Compare your results with the histogram shown at the upper right.

2. MINITAB was used to generate 20 random numbers with a Poisson distribution for $\mu = 4$. Let the random number represent the number of arrivals at the checkout counter each minute for 20 minutes.

       3 3 3 3 5 5 6 7 3 6
       3 5 6 3 4 6 2 2 4 1

   During each of the first four minutes, only three customers arrived. These customers could all be processed, so there were no customers waiting after four minutes.

   (a) How many customers were waiting after 5 minutes? 6 minutes? 7 minutes? 8 minutes?

   (b) Create a table that shows the number of customers waiting at the end of 1 through 20 minutes.

3. Generate a list of 20 random numbers with a Poisson distribution for $\mu = 4$. Create a table that shows the number of customers waiting at the end of 1 through 20 minutes.

4. Suppose that the mean increases to 5 arrivals per minute. You can still process only four per minute. How many would you expect to be waiting in line after 20 minutes?

5. Simulate the setting in Exercise 4. Do this by generating a list of 20 random numbers with a Poisson distribution for $\mu = 5$. Then create a table that shows the number of customers waiting at the end of 20 minutes.

6. Suppose that the mean number of arrivals per minute is 5. What is the probability that 10 customers will arrive during the first minute?

7. Suppose that the mean number of arrivals per minute is 4.

   (a) What is the probability that three, four, or five customers will arrive during the third minute?

   (b) What is the probability that more than four customers will arrive during the first minute?

   (c) What is the probability that more than four customers will arrive during each of the first four minutes?

Extended solutions are given in the *Technology Supplement.*
Technical instruction is provided for MINITAB, Excel, and the TI-83/84 Plus.

5 NORMAL PROBABILITY DISTRIBUTIONS

5.1 Introduction to Normal Distributions and the Standard Normal Distribution

5.2 Normal Distributions: Finding Probabilities

5.3 Normal Distributions: Finding Values

■ CASE STUDY

5.4 Sampling Distributions and the Central Limit Theorem

■ ACTIVITY

5.5 Normal Approximations to Binomial Distributions

■ USES AND ABUSES

■ REAL STATISTICS–REAL DECISIONS

■ TECHNOLOGY

The bottom shell of an Eastern Box Turtle has hinges so the turtle can retract its head, tail, and legs into the shell. The shell can also regenerate if it has been damaged.

In Chapters 1 through 4, you learned how to collect and describe data, find the probability of an event, and analyze discrete probability distributions. You also learned that if a sample is used to make inferences about a population, then it is critical that the sample not be biased. Suppose, for instance, that you wanted to determine the rate of clinical mastitis (infections caused by bacteria that can alter milk production) in dairy herds. How would you organize the study? When the Animal Health Service performed this study, it used random sampling and then classified the results according to breed, housing, hygiene, health, milking management, and milking machine. One conclusion from the study was that herds with Red and White cows as the predominant breed had a higher rate of clinical mastitis than herds with Holstein-Friesian cows as the main breed.

In Chapter 5, you will learn how to recognize normal (bell-shaped) distributions and how to use their properties in real-life applications. Suppose that you worked for the North Carolina Zoo and were collecting data about various physical traits of Eastern Box Turtles at the zoo. Which of the following would you expect to have bell-shaped, symmetric distributions: carapace (top shell) length, plastral (bottom shell) length, carapace width, plastral width, weight, total length? For instance, the four graphs below show the carapace length and plastral length of male and female Eastern Box Turtles in the North Carolina Zoo. Notice that the male Eastern Box Turtle carapace length distribution is bell-shaped, but the other three distributions are skewed left.

Female Eastern Box Turtle Carapace Length

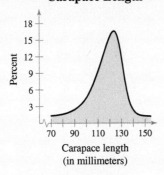

Male Eastern Box Turtle Carapace Length

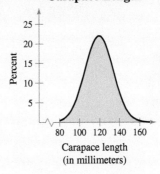

Female Eastern Box Turtle Plastral Length

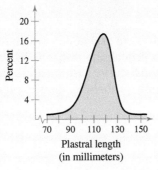

Male Eastern Box Turtle Plastral Length

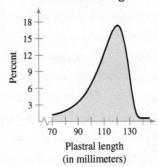

5.1 Introduction to Normal Distributions and the Standard Normal Distribution

Properties of a Normal Distribution ▸ The Standard Normal Distribution

▶ PROPERTIES OF A NORMAL DISTRIBUTION

In Section 4.1, you distinguished between discrete and continuous random variables, and learned that a continuous random variable has an infinite number of possible values that can be represented by an interval on the number line. Its probability distribution is called a **continuous probability distribution.** In this chapter, you will study the most important continuous probability distribution in statistics—the *normal distribution*. Normal distributions can be used to model many sets of measurements in nature, industry, and business. For instance, the systolic blood pressures of humans, the lifetimes of plasma televisions, and even housing costs are all normally distributed random variables.

DEFINITION

A **normal distribution** is a continuous probability distribution for a random variable x. The graph of a normal distribution is called the **normal curve.** A normal distribution has the following properties.

1. The mean, median, and mode are equal.
2. The normal curve is bell-shaped and is symmetric about the mean.
3. The total area under the normal curve is equal to 1.
4. The normal curve approaches, but never touches, the x-axis as it extends farther and farther away from the mean.
5. Between $\mu - \sigma$ and $\mu + \sigma$ (in the center of the curve), the graph curves downward. The graph curves upward to the left of $\mu - \sigma$ and to the right of $\mu + \sigma$. The points at which the curve changes from curving upward to curving downward are called *inflection points*.

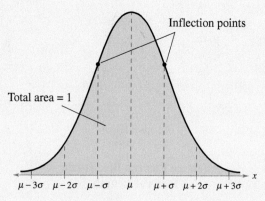

INSIGHT

To learn how to determine if a random sample is taken from a normal distribution, see Appendix C.

You have learned that a discrete probability distribution can be graphed with a histogram. For a continuous probability distribution, you can use a **probability density function (pdf).** A normal curve with mean μ and standard deviation σ can be graphed using the normal probability density function.

INSIGHT

A probability density function has two requirements.

1. The total area under the curve is equal to 1.
2. The function can never be negative.

$$y = \frac{1}{\sigma\sqrt{2\pi}}\, e^{-(x-\mu)^2/2\sigma^2}.$$

A normal curve depends completely on the two parameters μ and σ because $e \approx 2.718$ and $\pi \approx 3.14$ are constants.

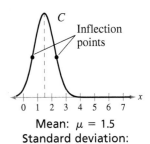

A normal distribution can have any mean and any positive standard deviation. These two parameters, μ and σ, completely determine the shape of the normal curve. The mean gives the location of the line of symmetry, and the standard deviation describes how much the data are spread out.

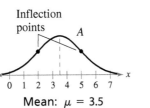

Inflection points

A

0 1 2 3 4 5 6 7

Mean: $\mu = 3.5$
Standard deviation:
$\sigma = 1.5$

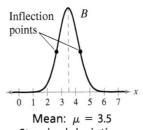

Inflection points

B

0 1 2 3 4 5 6 7

Mean: $\mu = 3.5$
Standard deviation:
$\sigma = 0.7$

C

Inflection points

0 1 2 3 4 5 6 7

Mean: $\mu = 1.5$
Standard deviation:
$\sigma = 0.7$

Notice that curve A and curve B above have the same mean, and curve B and curve C have the same standard deviation. The total area under each curve is 1.

EXAMPLE 1

▶ **Understanding Mean and Standard Deviation**

1. Which normal curve has a greater mean?

2. Which normal curve has a greater standard deviation?

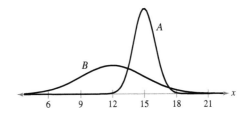

▶ **Solution**

1. The line of symmetry of curve A occurs at $x = 15$. The line of symmetry of curve B occurs at $x = 12$. So, curve A has a greater mean.

2. Curve B is more spread out than curve A. So, curve B has a greater standard deviation.

▶ **Try It Yourself 1**

Consider the normal curves shown at the right. Which normal curve has the greatest mean? Which normal curve has the greatest standard deviation?

a. Find the location of the *line of symmetry* of each curve. Make a conclusion about which mean is greatest.

b. Determine which normal curve is *more spread out*. Make a conclusion about which standard deviation is greatest.   *Answer: Page A37*

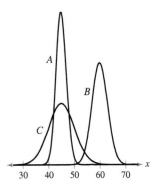

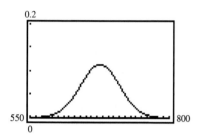

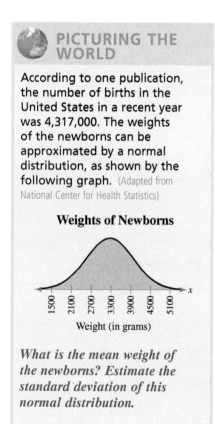

Once you determine the mean and standard deviation, you can use a TI-83/84 Plus to graph the normal curve in Example 2.

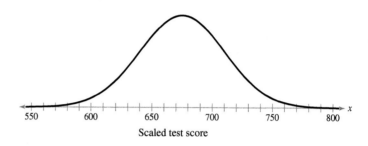

EXAMPLE 2

▸ Interpreting Graphs of Normal Distributions

The scaled test scores for the New York State Grade 8 Mathematics Test are normally distributed. The normal curve shown below represents this distribution. What is the mean test score? Estimate the standard deviation of this normal distribution. *(Adapted from New York State Education Department)*

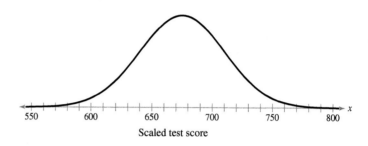

Scaled test score

▸ Solution

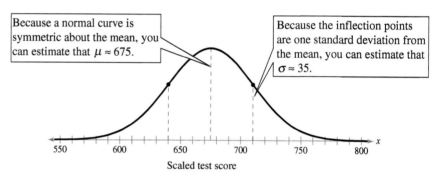

Because a normal curve is symmetric about the mean, you can estimate that $\mu \approx 675$.

Because the inflection points are one standard deviation from the mean, you can estimate that $\sigma \approx 35$.

Scaled test score

Interpretation The scaled test scores for the New York State Grade 8 Mathematics Test are normally distributed with a mean of about 675 and a standard deviation of about 35.

▸ Try It Yourself 2

The scaled test scores for the New York State Grade 8 English Language Arts Test are normally distributed. The normal curve shown below represents this distribution. What is the mean test score? Estimate the standard deviation of this normal distribution. *(Adapted from New York State Education Department)*

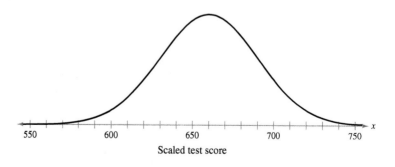

Scaled test score

a. Find the *line of symmetry* and identify the *mean*.
b. Estimate the *inflection points* and identify the *standard deviation*.

Answer: Page A37

▸ THE STANDARD NORMAL DISTRIBUTION

There are infinitely many normal distributions, each with its own mean and standard deviation. The normal distribution with a mean of 0 and a standard deviation of 1 is called the *standard normal distribution*. The horizontal scale of the graph of the standard normal distribution corresponds to z-scores. In Section 2.5, you learned that a z-score is a measure of position that indicates the number of standard deviations a value lies from the mean. Recall that you can transform an x-value to a z-score using the formula

$$z = \frac{\text{Value} - \text{Mean}}{\text{Standard deviation}}$$

$$= \frac{x - \mu}{\sigma}. \qquad \text{Round to the nearest hundredth.}$$

DEFINITION

The **standard normal distribution** is a normal distribution with a mean of 0 and a standard deviation of 1.

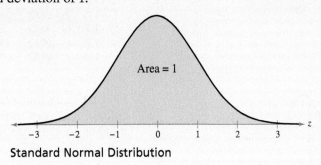

Area = 1

Standard Normal Distribution

If each data value of a normally distributed random variable x is transformed into a z-score, the result will be the standard normal distribution. When this transformation takes place, the area that falls in the interval under the nonstandard normal curve is the *same* as that under the standard normal curve within the corresponding z-boundaries.

In Section 2.4, you learned to use the Empirical Rule to approximate areas under a normal curve when the values of the random variable x corresponded to −3, −2, −1, 0, 1, 2, or 3 standard deviations from the mean. Now, you will learn to calculate areas corresponding to other x-values. After you use the formula given above to transform an x-value to a z-score, you can use the Standard Normal Table in Appendix B. The table lists the cumulative area under the standard normal curve to the left of z for z-scores from −3.49 to 3.49. As you examine the table, notice the following.

PROPERTIES OF THE STANDARD NORMAL DISTRIBUTION

1. The cumulative area is close to 0 for z-scores close to $z = -3.49$.

2. The cumulative area increases as the z-scores increase.

3. The cumulative area for $z = 0$ is 0.5000.

4. The cumulative area is close to 1 for z-scores close to $z = 3.49$.

EXAMPLE 3

▶ **Using the Standard Normal Table**

1. Find the cumulative area that corresponds to a z-score of 1.15.

2. Find the cumulative area that corresponds to a z-score of -0.24.

▶ **Solution**

1. Find the area that corresponds to $z = 1.15$ by finding 1.1 in the left column and then moving across the row to the column under 0.05. The number in that row and column is 0.8749. So, the area to the left of $z = 1.15$ is 0.8749.

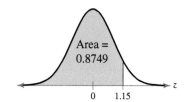
Area = 0.8749

z	.00	.01	.02	.03	.04	.05	.06
0.0	.5000	.5040	.5080	.5120	.5160	.5199	.5239
0.1	.5398	.5438	.5478	.5517	.5557	.5596	.5636
0.2	.5793	.5832	.5871	.5910	.5948	.5987	.6026
0.9	.8159	.8186	.8212	.8238	.8264	.8289	.8315
1.0	.8413	.8438	.8461	.8485	.8508	.8531	.8554
1.1	.8643	.8665	.8686	.8708	.8729	(.8749)	.8770
1.2	.8849	.8869	.8888	.8907	.8925	.8944	.8962
1.3	.9032	.9049	.9066	.9082	.9099	.9115	.9131
1.4	.9192	.9207	.9222	.9236	.9251	.9265	.9279

2. Find the area that corresponds to $z = -0.24$ by finding -0.2 in the left column and then moving across the row to the column under 0.04. The number in that row and column is 0.4052. So, the area to the left of $z = -0.24$ is 0.4052.

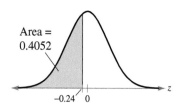
Area = 0.4052

z	.09	.08	.07	.06	.05	.04	.03
−3.4	.0002	.0003	.0003	.0003	.0003	.0003	.0003
−3.3	.0003	.0004	.0004	.0004	.0004	.0004	.0004
−3.2	.0005	.0005	.0005	.0006	.0006	.0006	.0006
−0.5	.2776	.2810	.2843	.2877	.2912	.2946	.2981
−0.4	.3121	.3156	.3192	.3228	.3264	.3300	.3336
−0.3	.3483	.3520	.3557	.3594	.3632	.3669	.3707
−0.2	.3859	.3897	.3936	.3974	.4013	(.4052)	.4090
−0.1	.4247	.4286	.4325	.4364	.4404	.4443	.4483
−0.0	.4641	.4681	.4721	.4761	.4801	.4840	.4880

You can also use a computer or calculator to find the cumulative area that corresponds to a z-score, as shown in the margin.

▶ **Try It Yourself 3**

1. Find the cumulative area that corresponds to a z-score of -2.19.

2. Find the cumulative area that corresponds to a z-score of 2.17.

Locate the given z-score and *find the area* that corresponds to it in the Standard Normal Table. *Answer: Page A37*

When the z-score is not in the table, use the entry closest to it. If the given z-score is exactly midway between two z-scores, then use the area midway between the corresponding areas.

STUDY TIP

Here are instructions for finding the area that corresponds to $z = -0.24$ on a TI-83/84 Plus.

To specify the lower bound in this case, use $-10,000$.

2nd DISTR

2: normalcdf(

−10000, −.24)

ENTER

```
normalcdf(-10000
,-.24)
          .405165175
```

You can use the following guidelines to find various types of areas under the standard normal curve.

GUIDELINES

Finding Areas Under the Standard Normal Curve

1. Sketch the standard normal curve and shade the appropriate area under the curve.
2. Find the area by following the directions for each case shown.

   a. To find the area to the *left* of z, find the area that corresponds to z in the Standard Normal Table.

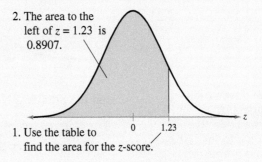

   2. The area to the left of $z = 1.23$ is 0.8907.

   1. Use the table to find the area for the z-score.

   b. To find the area to the *right* of z, use the Standard Normal Table to find the area that corresponds to z. Then subtract the area from 1.

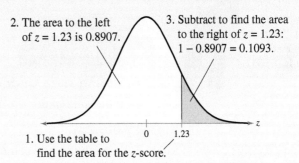

   2. The area to the left of $z = 1.23$ is 0.8907.

   3. Subtract to find the area to the right of $z = 1.23$:
   $1 - 0.8907 = 0.1093$.

   1. Use the table to find the area for the z-score.

   c. To find the area *between* two z-scores, find the area corresponding to each z-score in the Standard Normal Table. Then subtract the smaller area from the larger area.

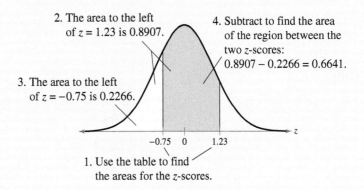

   2. The area to the left of $z = 1.23$ is 0.8907.

   4. Subtract to find the area of the region between the two z-scores:
   $0.8907 - 0.2266 = 0.6641$.

   3. The area to the left of $z = -0.75$ is 0.2266.

   1. Use the table to find the areas for the z-scores.

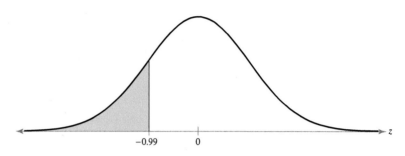

Using a TI-83/84 Plus, you can find the area automatically.

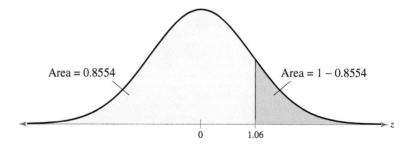

Use 10,000 for the upper bound.

EXAMPLE 4

▶ **Finding Area Under the Standard Normal Curve**

Find the area under the standard normal curve to the left of $z = -0.99$.

▶ **Solution**

The area under the standard normal curve to the left of $z = -0.99$ is shown.

From the Standard Normal Table, this area is equal to 0.1611.

▶ **Try It Yourself 4**

Find the area under the standard normal curve to the left of $z = 2.13$.

a. *Draw* the standard normal curve and shade the area under the curve and to the left of $z = 2.13$.
b. Use the Standard Normal Table to *find the area* that corresponds to $z = 2.13$.   *Answer: Page A38*

EXAMPLE 5

▶ **Finding Area Under the Standard Normal Curve**

Find the area under the standard normal curve to the right of $z = 1.06$.

▶ **Solution**

The area under the standard normal curve to the right of $z = 1.06$ is shown.

From the Standard Normal Table, the area to the left of $z = 1.06$ is 0.8554. Because the total area under the curve is 1, the area to the right of $z = 1.06$ is

$$\text{Area} = 1 - 0.8554$$
$$= 0.1446.$$

▶ Try It Yourself 5

Find the area under the standard normal curve to the right of $z = -2.16$.

a. *Draw* the standard normal curve and shade the area below the curve and to the right of $z = -2.16$.
b. Use the Standard Normal Table to *find the area* to the left of $z = -2.16$.
c. *Subtract* the area from 1.   *Answer: Page A38*

EXAMPLE 6

▶ **Finding Area Under the Standard Normal Curve**

Find the area under the standard normal curve between $z = -1.5$ and $z = 1.25$.

▶ **Solution**

The area under the standard normal curve between $z = -1.5$ and $z = 1.25$ is shown.

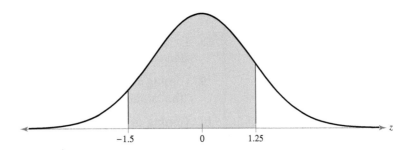

From the Standard Normal Table, the area to the left of $z = 1.25$ is 0.8944 and the area to the left of $z = -1.5$ is 0.0668. So, the area between $z = -1.5$ and $z = 1.25$ is

$$\text{Area} = 0.8944 - 0.0668$$
$$= 0.8276.$$

Interpretation   So, 82.76% of the area under the curve falls between $z = -1.5$ and $z = 1.25$.

▶ Try It Yourself 6

Find the area under the standard normal curve between $z = -2.165$ and $z = -1.35$.

a. Use the Standard Normal Table to *find the area* to the left of $z = -1.35$.
b. Use the Standard Normal Table to *find the area* to the left of $z = -2.165$.
c. *Subtract* the smaller area from the larger area.
d. *Interpret* the results.   *Answer: Page A38*

Recall that in Section 2.4 you learned, using the Empirical Rule, that values lying more than two standard deviations from the mean are considered unusual. Values lying more than three standard deviations from the mean are considered *very* unusual. So, if a z-score is greater than 2 or less than -2, it is unusual. If a z-score is greater than 3 or less than -3, it is *very* unusual.

When using technology, your answers may differ slightly from those found using the Standard Normal Table.

5.1 EXERCISES

■ BUILDING BASIC SKILLS AND VOCABULARY

1. Find three real-life examples of a continuous variable. Which do you think may be normally distributed? Why?

2. In a normal distribution, which is greater, the mean or the median? Explain.

3. What is the total area under the normal curve?

4. What do the inflection points on a normal distribution represent? Where do they occur?

5. Draw two normal curves that have the same mean but different standard deviations. Describe the similarities and differences.

6. Draw two normal curves that have different means but the same standard deviation. Describe the similarities and differences.

7. What is the mean of the standard normal distribution? What is the standard deviation of the standard normal distribution?

8. Describe how you can transform a nonstandard normal distribution to a standard normal distribution.

9. **Getting at the Concept** Why is it correct to say "a" normal distribution and "the" standard normal distribution?

10. **Getting at the Concept** If a z-score is 0, which of the following must be true? Explain your reasoning.

    (a) The mean is 0.
    (b) The corresponding x-value is 0.
    (c) The corresponding x-value is equal to the mean.

Graphical Analysis *In Exercises 11–16, determine whether the graph could represent a variable with a normal distribution. Explain your reasoning.*

11.

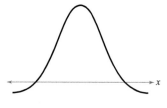

12.

13.

14.

15.

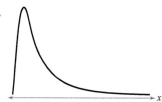

16.

Graphical Analysis   *In Exercises 17 and 18, determine whether the histogram represents data with a normal distribution. Explain your reasoning.*

17. **Waiting Time in a Dentist's Office**

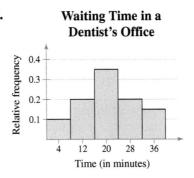

Time (in minutes)

18. **Weight Loss**

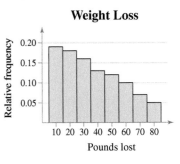

Pounds lost

■ USING AND INTERPRETING CONCEPTS

Graphical Analysis   *In Exercises 19–24, find the area of the indicated region under the standard normal curve. If convenient, use technology to find the area.*

19.

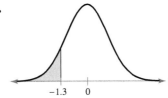

20.

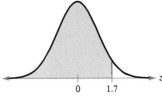

21.

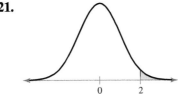

22.

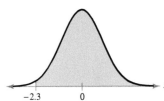

23.

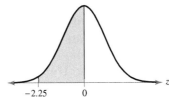

24.

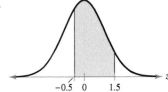

Finding Area   *In Exercises 25–38, find the indicated area under the standard normal curve. If convenient, use technology to find the area.*

25. To the left of $z = 0.08$

26. To the right of $z = -3.16$

27. To the left of $z = -2.575$

28. To the left of $z = 1.365$

29. To the right of $z = -0.65$

30. To the right of $z = 3.25$

31. To the right of $z = -0.355$

32. To the right of $z = 1.615$

33. Between $z = 0$ and $z = 2.86$

34. Between $z = -1.53$ and $z = 0$

35. Between $z = -1.96$ and $z = 1.96$

36. Between $z = -2.33$ and $z = 2.33$

37. To the left of $z = -1.28$ and to the right of $z = 1.28$

38. To the left of $z = -1.96$ and to the right of $z = 1.96$

39. Manufacturer Claims You work for a consumer watchdog publication and are testing the advertising claims of a tire manufacturer. The manufacturer claims that the life spans of the tires are normally distributed, with a mean of 40,000 miles and a standard deviation of 4000 miles. You test 16 tires and get the following life spans.

48,778 41,046 29,083 36,394 32,302 42,787 41,972 37,229
25,314 31,920 38,030 38,445 30,750 38,886 36,770 46,049

(a) Draw a frequency histogram to display these data. Use five classes. Is it reasonable to assume that the life spans are normally distributed? Why?

(b) Find the mean and standard deviation of your sample.

(c) Compare the mean and standard deviation of your sample with those in the manufacturer's claim. Discuss the differences.

40. Milk Consumption You are performing a study about weekly per capita milk consumption. A previous study found weekly per capita milk consumption to be normally distributed, with a mean of 48.7 fluid ounces and a standard deviation of 8.6 fluid ounces. You randomly sample 30 people and find their weekly milk consumptions to be as follows.

40 45 54 41 43 31 47 30 33 37 48 57 52 45 38
65 25 39 53 51 58 52 40 46 44 48 61 47 49 57

(a) Draw a frequency histogram to display these data. Use seven classes. Is it reasonable to assume that the consumptions are normally distributed? Why?

(b) Find the mean and standard deviation of your sample.

(c) Compare the mean and standard deviation of your sample with those of the previous study. Discuss the differences.

Computing and Interpreting z-Scores of Normal Distributions *In Exercises 41–44, you are given a normal distribution, the distribution's mean and standard deviation, four values from that distribution, and a graph of the standard normal distribution. (a) Without converting to z-scores, match the values with the letters A, B, C, and D on the given graph of the standard normal distribution. (b) Find the z-score that corresponds to each value and check your answers to part (a). (c) Determine whether any of the values are unusual.*

41. Blood Pressure The systolic blood pressures of a sample of adults are normally distributed, with a mean pressure of 115 millimeters of mercury and a standard deviation of 3.6 millimeters of mercury. The systolic blood pressures of four adults selected at random are 121 millimeters of mercury, 113 millimeters of mercury, 105 millimeters of mercury, and 127 millimeters of mercury.

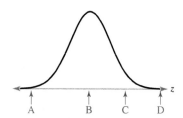

42. Cereal Boxes   The weights of the contents of cereal boxes are normally distributed, with a mean weight of 12 ounces and a standard deviation of 0.05 ounce. The weights of the contents of four cereal boxes selected at random are 12.01 ounces, 11.92 ounces, 12.12 ounces, and 11.99 ounces.

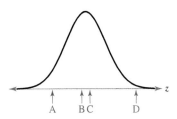

43. SAT Scores   The SAT is an exam used by colleges and universities to evaluate undergraduate applicants. The test scores are normally distributed. In a recent year, the mean test score was 1509 and the standard deviation was 312. The test scores of four students selected at random are 1924, 1241, 2202, and 1392. *(Source: The College Board)*

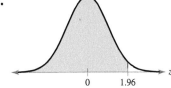

FIGURE FOR EXERCISE 43

FIGURE FOR EXERCISE 44

44. ACT Scores   The ACT is an exam used by colleges and universities to evaluate undergraduate applicants. The test scores are normally distributed. In a recent year, the mean test score was 21.1 and the standard deviation was 5.0. The test scores of four students selected at random are 15, 22, 9, and 35. *(Source: ACT, Inc.)*

Graphical Analysis   *In Exercises 45–50, find the probability of z occurring in the indicated region. If convenient, use technology to find the probability.*

45.

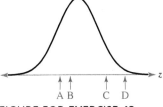

46.

47.

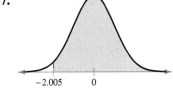

48.

49.

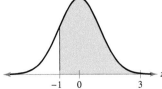

50.

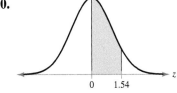

Finding Probabilities *In Exercises 51–60, find the indicated probability using the standard normal distribution. If convenient, use technology to find the probability.*

51. $P(z < 1.45)$ **52.** $P(z < -0.18)$ **53.** $P(z > 2.175)$

54. $P(z > -1.85)$ **55.** $P(-0.89 < z < 0)$ **56.** $P(0 < z < 0.525)$

57. $P(-1.65 < z < 1.65)$ **58.** $P(-1.54 < z < 1.54)$

59. $P(z < -2.58 \text{ or } z > 2.58)$ **60.** $P(z < -1.54 \text{ or } z > 1.54)$

■ EXTENDING CONCEPTS

61. Writing Draw a normal curve with a mean of 60 and a standard deviation of 12. Describe how you constructed the curve and discuss its features.

62. Writing Draw a normal curve with a mean of 450 and a standard deviation of 50. Describe how you constructed the curve and discuss its features.

63. Uniform Distribution Another continuous distribution is the **uniform distribution.** An example is $f(x) = 1$ for $0 \le x \le 1$. The mean of the distribution for this example is 0.5 and the standard deviation is approximately 0.29. The graph of the distribution for this example is a square with the height and width both equal to 1 unit. In general, the density function for a uniform distribution on the interval from $x = a$ to $x = b$ is given by

$$f(x) = \frac{1}{b - a}.$$

The mean is

$$\frac{a + b}{2}$$

and the standard deviation is

$$\sqrt{\frac{(b - a)^2}{12}}.$$

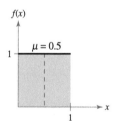

(a) Verify that the area under the curve is 1.

(b) Find the probability that x falls between 0.25 and 0.5.

(c) Find the probability that x falls between 0.3 and 0.7.

64. Uniform Distribution Consider the uniform density function $f(x) = 0.1$ for $10 \le x \le 20$. The mean of this distribution is 15 and the standard deviation is about 2.89.

(a) Draw a graph of the distribution and show that the area under the curve is 1.

(b) Find the probability that x falls between 12 and 15.

(c) Find the probability that x falls between 13 and 18.

5.2 Normal Distributions: Finding Probabilities

Probability and Normal Distributions

▸ PROBABILITY AND NORMAL DISTRIBUTIONS

If a random variable x is normally distributed, you can find the probability that x will fall in a given interval by calculating the area under the normal curve for the given interval. To find the area under any normal curve, you can first convert the upper and lower bounds of the interval to z-scores. Then use the standard normal distribution to find the area. For instance, consider a normal curve with $\mu = 500$ and $\sigma = 100$, as shown at the upper left. The value of x one standard deviation above the mean is $\mu + \sigma = 500 + 100 = 600$. Now consider the standard normal curve shown at the lower left. The value of z one standard deviation above the mean is $\mu + \sigma = 0 + 1 = 1$. Because a z-score of 1 corresponds to an x-value of 600, and areas are not changed with a transformation to a standard normal curve, the shaded areas in the graphs are equal.

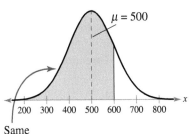

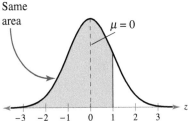

Same area

EXAMPLE 1   **SC** Report 20

▸ **Finding Probabilities for Normal Distributions**

A survey indicates that people use their cellular phones an average of 1.5 years before buying a new one. The standard deviation is 0.25 year. A cellular phone user is selected at random. Find the probability that the user will use their current phone for less than 1 year before buying a new one. Assume that the variable x is normally distributed. *(Adapted from Fonebak)*

▸ **Solution**

The graph shows a normal curve with $\mu = 1.5$ and $\sigma = 0.25$ and a shaded area for x less than 1. The z-score that corresponds to 1 year is

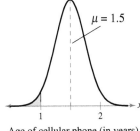

$$z = \frac{x - \mu}{\sigma} = \frac{1 - 1.5}{0.25} = -2.$$

The Standard Normal Table shows that $P(z < -2) = 0.0228$. The probability that the user will use their cellular phone for less than 1 year before buying a new one is 0.0228.

Age of cellular phone (in years)

Interpretation So, 2.28% of cellular phone users will use their cellular phone for less than 1 year before buying a new one. Because 2.28% is less than 5%, this is an unusual event.

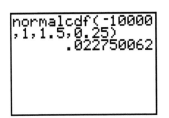

In Example 1, you can use a TI-83/84 Plus to find the probability automatically.

▸ **Try It Yourself 1**

The average speed of vehicles traveling on a stretch of highway is 67 miles per hour with a standard deviation of 3.5 miles per hour. A vehicle is selected at random. What is the probability that it is violating the 70 mile per hour speed limit? Assume the speeds are normally distributed.

a. *Sketch* a graph.
b. *Find the z-score* that corresponds to 70 miles per hour.
c. *Find the area* to the right of that z-score.
d. *Interpret* the results.

Answer: Page A38

EXAMPLE 2

▶ **Finding Probabilities for Normal Distributions**

A survey indicates that for each trip to the supermarket, a shopper spends an average of 45 minutes with a standard deviation of 12 minutes in the store. The lengths of time spent in the store are normally distributed and are represented by the variable x. A shopper enters the store. (a) Find the probability that the shopper will be in the store for each interval of time listed below. (b) Interpret your answer if 200 shoppers enter the store. How many shoppers would you expect to be in the store for each interval of time listed below?

1. Between 24 and 54 minutes   **2.** More than 39 minutes

▶ **Solution**

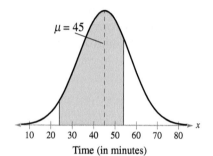

$\mu = 45$

10 20 30 40 50 60 70 80
Time (in minutes)

1. (a) The graph at the left shows a normal curve with $\mu = 45$ minutes and $\sigma = 12$ minutes. The area for x between 24 and 54 minutes is shaded. The z-scores that correspond to 24 minutes and to 54 minutes are

$$z_1 = \frac{24 - 45}{12} = -1.75 \quad \text{and} \quad z_2 = \frac{54 - 45}{12} = 0.75.$$

So, the probability that a shopper will be in the store between 24 and 54 minutes is

$$\begin{aligned} P(24 < x < 54) &= P(-1.75 < z < 0.75) \\ &= P(z < 0.75) - P(z < -1.75) \\ &= 0.7734 - 0.0401 = 0.7333. \end{aligned}$$

(b) *Interpretation*   If 200 shoppers enter the store, then you would expect $200(0.7333) = 146.66$, or about 147, shoppers to be in the store between 24 and 54 minutes.

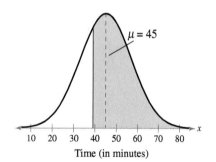

$\mu = 45$

10 20 30 40 50 60 70 80
Time (in minutes)

2. (a) The graph at the left shows a normal curve with $\mu = 45$ minutes and $\sigma = 12$ minutes. The area for x greater than 39 minutes is shaded. The z-score that corresponds to 39 minutes is

$$z = \frac{39 - 45}{12} = -0.5.$$

So, the probability that a shopper will be in the store more than 39 minutes is

$$P(x > 39) = P(z > -0.5) = 1 - P(z < -0.5) = 1 - 0.3085 = 0.6915.$$

(b) *Interpretation*   If 200 shoppers enter the store, then you would expect $200(0.6915) = 138.3$, or about 138, shoppers to be in the store more than 39 minutes.

▶ **Try It Yourself 2**

What is the probability that the shopper in Example 2 will be in the supermarket between 33 and 60 minutes?

a. *Sketch* a graph.
b. *Find the z-scores* that correspond to 33 minutes and 60 minutes.
c. *Find the cumulative area* for each z-score and *subtract* the smaller area from the larger area.
d. *Interpret* your answer if 150 shoppers enter the store. How many shoppers would you expect to be in the store between 33 and 60 minutes?

Answer: Page A38

Another way to find normal probabilities is to use a calculator or a computer. You can find normal probabilities using MINITAB, Excel, and the TI-83/84 Plus.

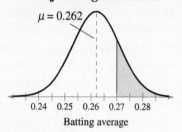
EXAMPLE 3

▶ Using Technology to Find Normal Probabilities

Triglycerides are a type of fat in the bloodstream. The mean triglyceride level in the United States is 134 milligrams per deciliter. Assume the triglyceride levels of the population of the United States are normally distributed, with a standard deviation of 35 milligrams per deciliter. You randomly select a person from the United States. What is the probability that the person's triglyceride level is less than 80? Use a technology tool to find the probability. *(Adapted from University of Maryland Medical Center)*

▶ Solution

MINITAB, Excel, and the TI-83/84 Plus each have features that allow you to find normal probabilities without first converting to standard z-scores. For each, you must specify the mean and standard deviation of the population, as well as the x-value(s) that determine the interval.

MINITAB

Cumulative Distribution Function

Normal with mean = 134 and standard deviation = 35

x	P[X <= x]
80	0.0614327

EXCEL

	A	B	C
1	NORMDIST(80,134,35,TRUE)		
2			0.06143272

TI-83/84 PLUS

normalcdf(-10000,80,134,35)
        .0614327356

From the displays, you can see that the probability that the person's triglyceride level is less than 80 is about 0.0614, or 6.14%.

▶ Try It Yourself 3

A person from the United States is selected at random. What is the probability that the person's triglyceride level is between 100 and 150? Use a technology tool.

a. *Read the user's guide* for the technology tool you are using.
b. *Enter the appropriate data* to obtain the probability.
c. *Write* the result as a sentence.   *Answer: Page A38*

Example 3 shows only one of several ways to find normal probabilities using MINITAB, Excel, and the TI-83/84 Plus.

5.2 EXERCISES

■ BUILDING BASIC SKILLS AND VOCABULARY

Computing Probabilities *In Exercises 1–6, assume the random variable x is normally distributed with mean* $\mu = 174$ *and standard deviation* $\sigma = 20$. *Find the indicated probability.*

1. $P(x < 170)$

2. $P(x < 200)$

3. $P(x > 182)$

4. $P(x > 155)$

5. $P(160 < x < 170)$

6. $P(172 < x < 192)$

Graphical Analysis *In Exercises 7–12, assume a member is selected at random from the population represented by the graph. Find the probability that the member selected at random is from the shaded area of the graph. Assume the variable x is normally distributed.*

7.   **SAT Writing Scores**

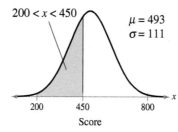

$200 < x < 450$

$\mu = 493$
$\sigma = 111$

Score

(Source: The College Board)

8.   **SAT Math Scores**

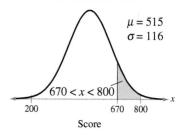

$670 < x < 800$

$\mu = 515$
$\sigma = 116$

Score

(Source: The College Board)

9.   **U.S. Men Ages 35–44:**
Total Cholesterol

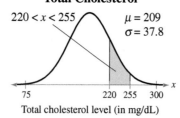

$220 < x < 255$

$\mu = 209$
$\sigma = 37.8$

Total cholesterol level (in mg/dL)

(Adapted from National Center for Health Statistics)

10.   **U.S. Women Ages 35–44:**
Total Cholesterol

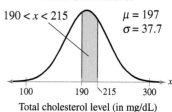

$190 < x < 215$

$\mu = 197$
$\sigma = 37.7$

Total cholesterol level (in mg/dL)

(Adapted from National Center for Health Statistics)

11.   **Ford Fusion:**
Braking Distance

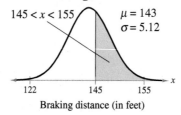

$145 < x < 155$

$\mu = 143$
$\sigma = 5.12$

Braking distance (in feet)

(Adapted from Consumer Reports)

12.   **Hyundai Elantra:**
Braking Distance

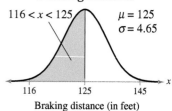

$116 < x < 125$

$\mu = 125$
$\sigma = 4.65$

Braking distance (in feet)

(Adapted from Consumer Reports)

■ USING AND INTERPRETING CONCEPTS

Finding Probabilities   *In Exercises 13–20, find the indicated probabilities. If convenient, use technology to find the probabilities.*

13. **Heights of Men**   A survey was conducted to measure the heights of U.S. men. In the survey, respondents were grouped by age. In the 20–29 age group, the heights were normally distributed, with a mean of 69.9 inches and a standard deviation of 3.0 inches. A study participant is randomly selected. *(Adapted from U.S. National Center for Health Statistics)*

    (a) Find the probability that his height is less than 66 inches.

    (b) Find the probability that his height is between 66 and 72 inches.

    (c) Find the probability that his height is more than 72 inches.

    (d) Can any of these events be considered unusual? Explain your reasoning.

14. **Heights of Women**   A survey was conducted to measure the heights of U.S. women. In the survey, respondents were grouped by age. In the 20–29 age group, the heights were normally distributed, with a mean of 64.3 inches and a standard deviation of 2.6 inches. A study participant is randomly selected. *(Adapted from U.S. National Center for Health Statistics)*

    (a) Find the probability that her height is less than 56.5 inches.

    (b) Find the probability that her height is between 61 and 67 inches.

    (c) Find the probability that her height is more than 70.5 inches.

    (d) Can any of these events be considered unusual? Explain your reasoning.

15. **ACT English Scores**   In a recent year, the ACT scores for the English portion of the test were normally distributed, with a mean of 20.6 and a standard deviation of 6.3. A high school student who took the English portion of the ACT is randomly selected. *(Source: ACT, Inc.)*

    (a) Find the probability that the student's ACT score is less than 15.

    (b) Find the probability that the student's ACT score is between 18 and 25.

    (c) Find the probability that the student's ACT score is more than 34.

    (d) Can any of these events be considered unusual? Explain your reasoning.

16. **Beagles**   The weights of adult male beagles are normally distributed, with a mean of 25 pounds and a standard deviation of 3 pounds. A beagle is randomly selected.

    (a) Find the probability that the beagle's weight is less than 23 pounds.

    (b) Find the probability that the weight is between 24.5 and 25 pounds.

    (c) Find the probability that the beagle's weight is more than 30 pounds.

    (d) Can any of these events be considered unusual? Explain your reasoning.

17. **Computer Usage**   A survey was conducted to measure the number of hours per week adults in the United States spend on their computers. In the survey, the numbers of hours were normally distributed, with a mean of 7 hours and a standard deviation of 1 hour. A survey participant is randomly selected.

    (a) Find the probability that the number of hours spent on the computer by the participant is less than 5 hours per week.

    (b) Find the probability that the number of hours spent on the computer by the participant is between 5.5 and 9.5 hours per week.

    (c) Find the probability that the number of hours spent on the computer by the participant is more than 10 hours per week.

18. Utility Bills The monthly utility bills in a city are normally distributed, with a mean of $100 and a standard deviation of $12. A utility bill is randomly selected.

(a) Find the probability that the utility bill is less than $70.

(b) Find the probability that the utility bill is between $90 and $120.

(c) Find the probability that the utility bill is more than $140.

19. Computer Lab Schedule The times per week a student uses a lab computer are normally distributed, with a mean of 6.2 hours and a standard deviation of 0.9 hour. A student is randomly selected.

(a) Find the probability that the student uses a lab computer less than 4 hours per week.

(b) Find the probability that the student uses a lab computer between 5 and 7 hours per week.

(c) Find the probability that the student uses a lab computer more than 8 hours per week.

20. Health Club Schedule The times per workout an athlete uses a stairclimber are normally distributed, with a mean of 20 minutes and a standard deviation of 5 minutes. An athlete is randomly selected.

(a) Find the probability that the athlete uses a stairclimber for less than 17 minutes.

(b) Find the probability that the athlete uses a stairclimber between 20 and 28 minutes.

(c) Find the probability that the athlete uses a stairclimber for more than 30 minutes.

Using Normal Distributions *In Exercises 21–28, answer the questions about the specified normal distribution.*

21. SAT Writing Scores Use the normal distribution of SAT writing scores in Exercise 7 for which the mean is 493 and the standard deviation is 111.

(a) What percent of the SAT writing scores are less than 600?

(b) If 1000 SAT writing scores are randomly selected, about how many would you expect to be greater than 550?

22. SAT Math Scores Use the normal distribution of SAT math scores in Exercise 8 for which the mean is 515 and the standard deviation is 116.

(a) What percent of the SAT math scores are less than 500?

(b) If 1500 SAT math scores are randomly selected, about how many would you expect to be greater than 600?

23. Cholesterol Use the normal distribution of men's total cholesterol levels in Exercise 9 for which the mean is 209 milligrams per deciliter and the standard deviation is 37.8 milligrams per deciliter.

(a) What percent of the men have a total cholesterol level less than 225 milligrams per deciliter of blood?

(b) If 250 U.S. men in the 35–44 age group are randomly selected, about how many would you expect to have a total cholesterol level greater than 260 milligrams per deciliter of blood?

24. **Cholesterol** Use the normal distribution of women's total cholesterol levels in Exercise 10 for which the mean is 197 milligrams per deciliter and the standard deviation is 37.7 milligrams per deciliter.

    (a) What percent of the women have a total cholesterol level less than 217 milligrams per deciliter of blood?

    (b) If 200 U.S. women in the 35–44 age group are randomly selected, about how many would you expect to have a total cholesterol level greater than 185 milligrams per deciliter of blood?

25. **Computer Usage** Use the normal distribution of computer usage in Exercise 17 for which the mean is 7 hours and the standard deviation is 1 hour.

    (a) What percent of the adults spend more than 4 hours per week on their computer?

    (b) If 35 adults in the United States are randomly selected, about how many would you expect to say they spend less than 5 hours per week on their computer?

26. **Utility Bills** Use the normal distribution of utility bills in Exercise 18 for which the mean is $100 and the standard deviation is $12.

    (a) What percent of the utility bills are more than $125?

    (b) If 300 utility bills are randomly selected, about how many would you expect to be less than $90?

27. **Battery Life Spans** The life spans of batteries are normally distributed, with a mean of 2000 hours and a standard deviation of 30 hours. What percent of batteries have a life span that is more than 2065 hours? Would it be unusual for a battery to have a life span that is more than 2065 hours? Explain your reasoning.

28. **Peanuts** Assume the mean annual consumptions of peanuts are normally distributed, with a mean of 5.9 pounds per person and a standard deviation of 1.8 pounds per person. What percent of people annually consume less than 3.1 pounds of peanuts per person? Would it be unusual for a person to consume less than 3.1 pounds of peanuts in a year? Explain your reasoning.

SC *In Exercises 29 and 30, use the StatCrunch normal calculator to find the indicated probabilities.*

29. **Soft Drink Machine** The amounts a soft drink machine is designed to dispense for each drink are normally distributed, with a mean of 12 fluid ounces and a standard deviation of 0.2 fluid ounce. A drink is randomly selected.

    (a) Find the probability that the drink is less than 11.9 fluid ounces.

    (b) Find the probability that the drink is between 11.8 and 11.9 fluid ounces.

    (c) Find the probability that the drink is more than 12.3 fluid ounces. Can this be considered an unusual event? Explain your reasoning.

30. **Machine Parts** The thicknesses of washers produced by a machine are normally distributed, with a mean of 0.425 inch and a standard deviation of 0.005 inch. A washer is randomly selected.

    (a) Find the probability that the washer is less than 0.42 inch thick.

    (b) Find the probability that the washer is between 0.40 and 0.42 inch thick.

    (c) Find the probability that the washer is more than 0.44 inch thick. Can this be considered an unusual event? Explain your reasoning.

■ **EXTENDING CONCEPTS**

Control Charts *Statistical process control (SPC) is the use of statistics to monitor and improve the quality of a process, such as manufacturing an engine part. In SPC, information about a process is gathered and used to determine if a process is meeting all of the specified requirements. One tool used in SPC is a* **control chart.** *When individual measurements of a variable x are normally distributed, a control chart can be used to detect processes that are possibly out of statistical control. Three warning signals that a control chart uses to detect a process that may be out of control are as follows.*

(1) A point lies beyond three standard deviations of the mean.

(2) There are nine consecutive points that fall on one side of the mean.

(3) At least two of three consecutive points lie more than two standard deviations from the mean.

In Exercises 31–34, a control chart is shown. Each chart has horizontal lines drawn at the mean μ, at $\mu \pm 2\sigma$, and at $\mu \pm 3\sigma$. Determine if the process shown is in control or out of control. Explain.

31. A gear has been designed to have a diameter of 3 inches. The standard deviation of the process is 0.2 inch.

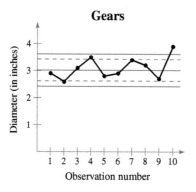

Gears

32. A nail has been designed to have a length of 4 inches. The standard deviation of the process is 0.12 inch.

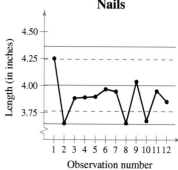

Nails

33. A liquid-dispensing machine has been designed to fill bottles with 1 liter of liquid. The standard deviation of the process is 0.1 liter.

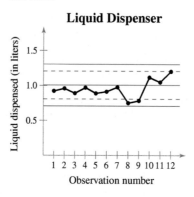

Liquid Dispenser

34. An engine part has been designed to have a diameter of 55 millimeters. The standard deviation of the process is 0.001 millimeter.

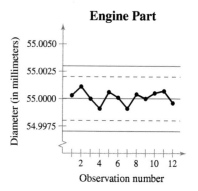

Engine Part

| 5.3 | **Normal Distributions: Finding Values** |

Finding z-Scores ▸ Transforming a z-Score to an x-Value ▸ Finding a Specific Data Value for a Given Probability

▸ FINDING z-SCORES

In Section 5.2, you were given a normally distributed random variable x and you found the probability that x would fall in a given interval by calculating the area under the normal curve for the given interval.

But what if you are given a probability and want to find a value? For instance, a university might want to know the lowest test score a student can have on an entrance exam and still be in the top 10%, or a medical researcher might want to know the cutoff values for selecting the middle 90% of patients by age. In this section, you will learn how to find a value given an area under a normal curve (or a probability), as shown in the following example.

EXAMPLE 1

▸ **Finding a z-Score Given an Area**

1. Find the z-score that corresponds to a cumulative area of 0.3632.

2. Find the z-score that has 10.75% of the distribution's area to its right.

▸ **Solution**

1. Find the z-score that corresponds to an area of 0.3632 by locating 0.3632 in the Standard Normal Table. The values at the beginning of the corresponding row and at the top of the corresponding column give the z-score. For this area, the row value is -0.3 and the column value is 0.05. So, the z-score is -0.35.

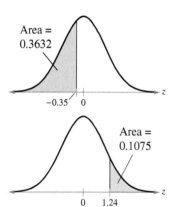

Area = 0.3632

z	.09	.08	.07	.06	.05	.04	.03
−3.4	.0002	.0003	.0003	.0003	.0003	.0003	.0003
−0.5	.2776	.2810	.2843	.2877	.2912	.2946	.2981
−0.4	.3121	.3156	.3192	.3228	.3264	.3300	.3336
−0.3	.3483	.3520	.3557	.3594	.3632	.3669	.3707
−0.2	.3859	.3897	.3936	.3974	.4013	.4052	.4090

2. Because the area to the right is 0.1075, the cumulative area is $1 - 0.1075 = 0.8925$. Find the z-score that corresponds to an area of 0.8925 by locating 0.8925 in the Standard Normal Table. For this area, the row value is 1.2 and the column value is 0.04. So, the z-score is 1.24.

z	.00	.01	.02	.03	.04	.05	.06
0.0	.5000	.5040	.5080	.5120	.5160	.5199	.5239
1.0	.8413	.8438	.8461	.8485	.8508	.8531	.8554
1.1	.8643	.8665	.8686	.8708	.8729	.8749	.8770
1.2	.8849	.8869	.8888	.8907	.8925	.8944	.8962
1.3	.9032	.9049	.9066	.9082	.9099	.9115	.9131

You can also use a computer or calculator to find the z-scores that correspond to the given cumulative areas, as shown in the margin.

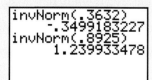

▶ **Try It Yourself 1**

1. Find the z-score that has 96.16% of the distribution's area to the right.
2. Find the z-score for which 95% of the distribution's area lies between $-z$ and z.

   a. *Determine* the cumulative area.
   b. *Locate* the area in the Standard Normal Table.
   c. *Find the z-score* that corresponds to the area.   *Answer: Page A38*

In Section 2.5, you learned that percentiles divide a data set into 100 equal parts. To find a z-score that corresponds to a percentile, you can use the Standard Normal Table. Recall that if a value x represents the 83rd percentile P_{83}, then 83% of the data values are below x and 17% of the data values are above x.

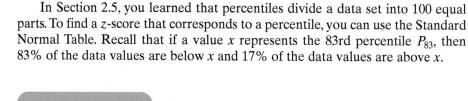

EXAMPLE 2

▶ **Finding a z-Score Given a Percentile**

Find the z-score that corresponds to each percentile.

1. P_5

2. P_{50}

3. P_{90}

▶ **Solution**

1. To find the z-score that corresponds to P_5, find the z-score that corresponds to an area of 0.05 (see upper figure) by locating 0.05 in the Standard Normal Table. The areas closest to 0.05 in the table are 0.0495 ($z = -1.65$) and 0.0505 ($z = -1.64$). Because 0.05 is halfway between the two areas in the table, use the z-score that is halfway between -1.64 and -1.65. So, the z-score that corresponds to an area of 0.05 is -1.645.

2. To find the z-score that corresponds to P_{50}, find the z-score that corresponds to an area of 0.5 (see middle figure) by locating 0.5 in the Standard Normal Table. The area closest to 0.5 in the table is 0.5000, so the z-score that corresponds to an area of 0.5 is 0.

3. To find the z-score that corresponds to P_{90}, find the z-score that corresponds to an area of 0.9 (see lower figure) by locating 0.9 in the Standard Normal Table. The area closest to 0.9 in the table is 0.8997, so the z-score that corresponds to an area of 0.9 is about 1.28.

▶ **Try It Yourself 2**

Find the z-score that corresponds to each percentile.

1. P_{10} 2. P_{20} 3. P_{99}

a. *Write* the percentile as an area. If necessary, draw a graph of the area to visualize the problem.
b. *Locate* the area in the Standard Normal Table. If the area is not in the table, use the closest area. (See Study Tip above.)
c. *Identify* the z-score that corresponds to the area.   *Answer: Page A38*

▶ TRANSFORMING A z-SCORE TO AN x-VALUE

Recall that to transform an x-value to a z-score, you can use the formula

$$z = \frac{x - \mu}{\sigma}.$$

This formula gives z in terms of x. If you solve this formula for x, you get a new formula that gives x in terms of z.

$z = \dfrac{x - \mu}{\sigma}$	Formula for z in terms of x
$z\sigma = x - \mu$	Multiply each side by σ.
$\mu + z\sigma = x$	Add μ to each side.
$x = \mu + z\sigma$	Interchange sides.

TRANSFORMING A z-SCORE TO AN x-VALUE

To transform a standard z-score to a data value x in a given population, use the formula

$$x = \mu + z\sigma.$$

EXAMPLE 3

▶ Finding an x-Value Corresponding to a z-Score

A veterinarian records the weights of cats treated at a clinic. The weights are normally distributed, with a mean of 9 pounds and a standard deviation of 2 pounds. Find the weights x corresponding to z-scores of 1.96, −0.44, and 0. Interpret your results.

▶ Solution

The x-value that corresponds to each standard z-score is calculated using the formula $x = \mu + z\sigma$.

$z = 1.96$:	$x = 9 + 1.96(2) = 12.92$ pounds
$z = -0.44$:	$x = 9 + (-0.44)(2) = 8.12$ pounds
$z = 0$:	$x = 9 + 1.96(0) = 9$ pounds

Interpretation You can see that 12.92 pounds is above the mean, 8.12 pounds is below the mean, and 9 pounds is equal to the mean.

▶ Try It Yourself 3

A veterinarian records the weights of dogs treated at a clinic. The weights are normally distributed, with a mean of 52 pounds and a standard deviation of 15 pounds. Find the weights x corresponding to z-scores of −2.33, 3.10, and 0.58. Interpret your results.

a. *Identify* μ and σ of the normal distribution.
b. *Transform* each z-score to an x-value.
c. *Interpret* the results. *Answer: Page A38*

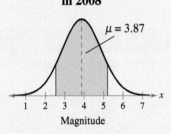
▸ FINDING A SPECIFIC DATA VALUE FOR A GIVEN PROBABILITY

You can also use the normal distribution to find a specific data value (*x*-value) for a given probability, as shown in Examples 4 and 5.

EXAMPLE 4   Report 21

▸ Finding a Specific Data Value

Scores for the California Peace Officer Standards and Training test are normally distributed, with a mean of 50 and a standard deviation of 10. An agency will only hire applicants with scores in the top 10%. What is the lowest score you can earn and still be eligible to be hired by the agency? *(Source: State of California)*

▸ Solution

Exam scores in the top 10% correspond to the shaded region shown.

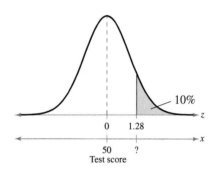

A test score in the top 10% is any score above the 90th percentile. To find the score that represents the 90th percentile, you must first find the *z*-score that corresponds to a cumulative area of 0.9. From the Standard Normal Table, you can find that the area closest to 0.9 is 0.8997. So, the *z*-score that corresponds to an area of 0.9 is $z = 1.28$. Using the equation $x = \mu + z\sigma$, you have

$$x = \mu + z\sigma$$
$$= 50 + 1.28(10)$$
$$\approx 62.8.$$

Interpretation   The lowest score you can earn and still be eligible to be hired by the agency is about 63.

▸ Try It Yourself 4

The braking distances of a sample of Nissan Altimas are normally distributed, with a mean of 129 feet and a standard deviation of 5.18 feet. What is the longest braking distance one of these Nissan Altimas could have and still be in the bottom 1%? *(Adapted from Consumer Reports)*

a. *Sketch* a graph.
b. *Find the z-score* that corresponds to the given area.
c. *Find x* using the equation $x = \mu + z\sigma$.
d. *Interpret* the result.

Answer: Page A38

EXAMPLE 5   **SC** Report 22

▶ **Finding a Specific Data Value**

In a randomly selected sample of women ages 20–34, the mean total cholesterol level is 188 milligrams per deciliter with a standard deviation of 41.3 milligrams per deciliter. Assume the total cholesterol levels are normally distributed. Find the highest total cholesterol level a woman in this 20–34 age group can have and still be in the bottom 1%. *(Adapted from National Center for Health Statistics)*

▶ **Solution**

Total cholesterol levels in the lowest 1% correspond to the shaded region shown.

Total Cholesterol Levels in Women Ages 20–34

1%

−2.33   0   *z*

?   188   *x*

Total cholesterol level (in mg/dL)

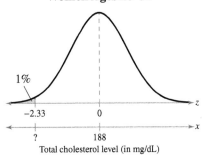

Using a TI-83/84 Plus, you can find the highest total cholesterol level automatically.

A total cholesterol level in the lowest 1% is any level below the 1st percentile. To find the level that represents the 1st percentile, you must first find the *z*-score that corresponds to a cumulative area of 0.01. From the Standard Normal Table, you can find that the area closest to 0.01 is 0.0099. So, the *z*-score that corresponds to an area of 0.01 is $z = -2.33$. Using the equation $x = \mu + z\sigma$, you have

$$x = \mu + z\sigma$$

$$= 188 + (-2.33)(41.3)$$

$$\approx 91.77.$$

Interpretation   The value that separates the lowest 1% of total cholesterol levels for women in the 20–34 age group from the highest 99% is about 92 milligrams per deciliter.

▶ **Try It Yourself 5**

The lengths of time employees have worked at a corporation are normally distributed, with a mean of 11.2 years and a standard deviation of 2.1 years. In a company cutback, the lowest 10% in seniority are laid off. What is the maximum length of time an employee could have worked and still be laid off?

a. *Sketch* a graph.
b. *Find the z-score* that corresponds to the given area.
c. *Find x* using the equation $x = \mu + z\sigma$.
d. *Interpret* the result.

Answer: Page A38

5.3 EXERCISES

■ BUILDING BASIC SKILLS AND VOCABULARY

In Exercises 1–16, use the Standard Normal Table to find the z-score that corresponds to the given cumulative area or percentile. If the area is not in the table, use the entry closest to the area. If the area is halfway between two entries, use the z-score halfway between the corresponding z-scores. If convenient, use technology to find the z-score.

1. 0.2090   **2.** 0.4364   **3.** 0.9916   **4.** 0.7995

5. 0.05   **6.** 0.85   **7.** 0.94   **8.** 0.0046

9. P_{15} **10.** P_{30} **11.** P_{88} **12.** P_{67}

13. P_{25} **14.** P_{40} **15.** P_{75} **16.** P_{80}

Graphical Analysis *In Exercises 17–22, find the indicated z-score(s) shown in the graph. If convenient, use technology to find the z-score(s).*

17.

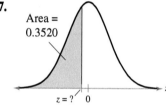

Area = 0.3520

$z = ?$ 0

18.

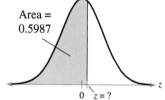

Area = 0.5987

0   $z = ?$

19.

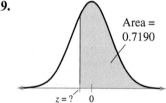

Area = 0.7190

$z = ?$ 0

20.

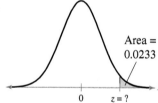

Area = 0.0233

0   $z = ?$

21.

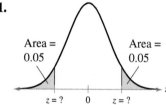

Area = 0.05   Area = 0.05

$z = ?$ 0 $z = ?$

22.

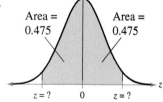

Area = 0.475   Area = 0.475

$z = ?$ 0 $z = ?$

In Exercises 23–30, find the indicated z-score.

23. Find the z-score that has 11.9% of the distribution's area to its left.

24. Find the z-score that has 78.5% of the distribution's area to its left.

25. Find the z-score that has 11.9% of the distribution's area to its right.

26. Find the z-score that has 78.5% of the distribution's area to its right.

27. Find the z-score for which 80% of the distribution's area lies between $-z$ and z.

28. Find the z-score for which 99% of the distribution's area lies between $-z$ and z.

29. Find the z-score for which 5% of the distribution's area lies between $-z$ and z.

30. Find the z-score for which 12% of the distribution's area lies between $-z$ and z.

■ USING AND INTERPRETING CONCEPTS

Using Normal Distributions   *In Exercises 31–36, answer the questions about the specified normal distribution.*

31. Heights of Women   In a survey of women in the United States (ages 20–29), the mean height was 64.3 inches with a standard deviation of 2.6 inches. *(Adapted from National Center for Health Statistics)*

(a) What height represents the 95th percentile?

(b) What height represents the first quartile?

32. Heights of Men   In a survey of men in the United States (ages 20–29), the mean height was 69.9 inches with a standard deviation of 3.0 inches. *(Adapted from National Center for Health Statistics)*

(a) What height represents the 90th percentile?

(b) What height represents the first quartile?

33. Heart Transplant Waiting Times   The time spent (in days) waiting for a heart transplant for people ages 35–49 in a recent year can be approximated by a normal distribution, as shown in the graph. *(Adapted from Organ Procurement and Transplantation Network)*

(a) What waiting time represents the 5th percentile?

(b) What waiting time represents the third quartile?

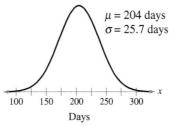

Time Spent Waiting for a Heart

$\mu = 204$ days
$\sigma = 25.7$ days

FIGURE FOR EXERCISE 33

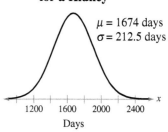

Time Spent Waiting for a Kidney

$\mu = 1674$ days
$\sigma = 212.5$ days

FIGURE FOR EXERCISE 34

34. Kidney Transplant Waiting Times   The time spent (in days) waiting for a kidney transplant for people ages 35–49 in a recent year can be approximated by a normal distribution, as shown in the graph. *(Adapted from Organ Procurement and Transplantation Network)*

(a) What waiting time represents the 80th percentile?

(b) What waiting time represents the first quartile?

35. Sleeping Times of Medical Residents   The average time spent sleeping (in hours) for a group of medical residents at a hospital can be approximated by a normal distribution, as shown in the graph. *(Source: National Institute of Occupational Safety and Health, Japan)*

(a) What is the shortest time spent sleeping that would still place a resident in the top 5% of sleeping times?

(b) Between what two values does the middle 50% of the sleep times lie?

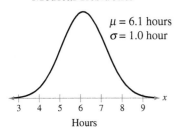

Sleeping Times of Medical Residents

$\mu = 6.1$ hours
$\sigma = 1.0$ hour

Hours

FIGURE FOR EXERCISE 35

Annual U.S. per Capita Ice Cream Consumption

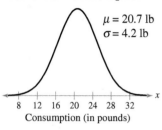

$\mu = 20.7$ lb
$\sigma = 4.2$ lb

Consumption (in pounds)

FIGURE FOR EXERCISE 36

36. Ice Cream The annual per capita consumption of ice cream (in pounds) in the United States can be approximated by a normal distribution, as shown in the graph. *(Adapted from U.S. Department of Agriculture)*

(a) What is the largest annual per capita consumption of ice cream that can be in the bottom 10% of consumptions?

(b) Between what two values does the middle 80% of the consumptions lie?

37. Bags of Baby Carrots The weights of bags of baby carrots are normally distributed, with a mean of 32 ounces and a standard deviation of 0.36 ounce. Bags in the upper 4.5% are too heavy and must be repackaged. What is the most a bag of baby carrots can weigh and not need to be repackaged?

38. Vending Machine A vending machine dispenses coffee into an eight-ounce cup. The amounts of coffee dispensed into the cup are normally distributed, with a standard deviation of 0.03 ounce. You can allow the cup to overfill 1% of the time. What amount should you set as the mean amount of coffee to be dispensed?

SC *In Exercises 39 and 40, use the StatCrunch normal calculator to find the indicated values.*

39. Apples The annual per capita consumption of fresh apples (in pounds) in the United States can be approximated by a normal distribution, with a mean of 16.2 pounds and a standard deviation of 4 pounds. *(Adapted from U.S. Department of Agriculture)*

(a) What is the smallest annual per capita consumption of apples that can be in the top 25% of consumptions?

(b) What is the largest annual per capita consumption of apples that can be in the bottom 15% of consumptions?

40. Oranges The annual per capita consumption of fresh oranges (in pounds) in the United States can be approximated by a normal distribution, with a mean of 9.9 pounds and a standard deviation of 2.5 pounds. *(Adapted from U.S. Department of Agriculture)*

(a) What is the smallest annual per capita consumption of oranges that can be in the top 10% of consumptions?

(b) What is the largest annual per capita consumption of oranges that can be in the bottom 5% of consumptions?

■ EXTENDING CONCEPTS

41. Writing a Guarantee You sell a brand of automobile tire that has a life expectancy that is normally distributed, with a mean life of 30,000 miles and a standard deviation of 2500 miles. You want to give a guarantee for free replacement of tires that don't wear well. How should you word your guarantee if you are willing to replace approximately 10% of the tires?

42. Statistics Grades In a large section of a statistics class, the points for the final exam are normally distributed, with a mean of 72 and a standard deviation of 9. Grades are to be assigned according to the following rule: the top 10% receive A's, the next 20% receive B's, the middle 40% receive C's, the next 20% receive D's, and the bottom 10% receive F's. Find the lowest score on the final exam that would qualify a student for an A, a B, a C, and a D.

Final Exam Grades

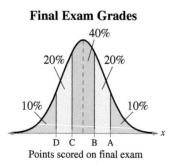

40%

20%   20%

10%   10%

D C   B A
Points scored on final exam

FIGURE FOR EXERCISE 42

Birth Weights in America

The National Center for Health Statistics (NCHS) keeps records of many health-related aspects of people, including the birth weights of all babies born in the United States.

The birth weight of a baby is related to its gestation period (the time between conception and birth). For a given gestation period, the birth weights can be approximated by a normal distribution. The means and standard deviations of the birth weights for various gestation periods are shown in the table below.

One of the many goals of the NCHS is to reduce the percentage of babies born with low birth weights. As you can see from the graph below, the problem of low birth weights increased from 1992 to 2006.

Gestation period	Mean birth weight	Standard deviation
Under 28 weeks	1.90 lb	1.22 lb
28 to 31 weeks	4.12 lb	1.87 lb
32 to 33 weeks	5.14 lb	1.57 lb
34 to 36 weeks	6.19 lb	1.29 lb
37 to 39 weeks	7.29 lb	1.08 lb
40 weeks	7.66 lb	1.04 lb
41 weeks	7.75 lb	1.07 lb
42 weeks and over	7.57 lb	1.11 lb

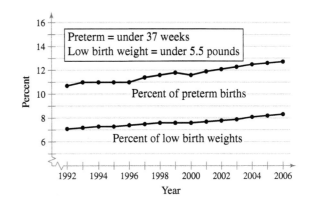

Preterm = under 37 weeks
Low birth weight = under 5.5 pounds

Percent of preterm births

Percent of low birth weights

▪ EXERCISES

1. The distributions of birth weights for three gestation periods are shown. Match the curves with the gestation periods. Explain your reasoning.

(a)

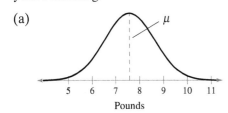

Pounds

(b)

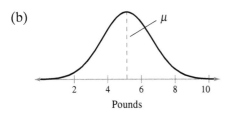

Pounds

(c)

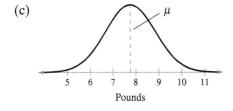

Pounds

2. What percent of the babies born within each gestation period have a low birth weight (under 5.5 pounds)? Explain your reasoning.

(a) Under 28 weeks (b) 32 to 33 weeks
(c) 40 weeks (d) 42 weeks and over

3. Describe the weights of the top 10% of the babies born within each gestation period. Explain your reasoning.

(a) Under 28 weeks (b) 34 to 36 weeks
(c) 41 weeks (d) 42 weeks and over

4. For each gestation period, what is the probability that a baby will weigh between 6 and 9 pounds at birth?

(a) Under 28 weeks (b) 28 to 31 weeks
(c) 34 to 36 weeks (d) 37 to 39 weeks

5. A birth weight of less than 3.25 pounds is classified by the NCHS as a "very low birth weight." What is the probability that a baby has a very low birth weight for each gestation period?

(a) Under 28 weeks (b) 28 to 31 weeks
(c) 32 to 33 weeks (d) 37 to 39 weeks

5.4   Sampling Distributions and the Central Limit Theorem

WHAT YOU SHOULD LEARN

▸ How to find sampling distributions and verify their properties

▸ How to interpret the Central Limit Theorem

▸ How to apply the Central Limit Theorem to find the probability of a sample mean

Sampling Distributions ▸ The Central Limit Theorem ▸ Probability and the Central Limit Theorem

▸ SAMPLING DISTRIBUTIONS

In previous sections, you studied the relationship between the mean of a population and values of a random variable. In this section, you will study the relationship between a population mean and the means of samples taken from the population.

DEFINITION

A **sampling distribution** is the probability distribution of a sample statistic that is formed when samples of size n are repeatedly taken from a population. If the sample statistic is the sample mean, then the distribution is the **sampling distribution of sample means.** Every sample statistic has a sampling distribution.

For instance, consider the following Venn diagram. The rectangle represents a large population, and each circle represents a sample of size n. Because the sample entries can differ, the sample means can also differ. The mean of Sample 1 is $\overline{x}_1$; the mean of Sample 2 is $\overline{x}_2$; and so on. The sampling distribution of the sample means for samples of size n for this population consists of $\overline{x}_1$, $\overline{x}_2$, $\overline{x}_3$, and so on. If the samples are drawn with replacement, an infinite number of samples can be drawn from the population.

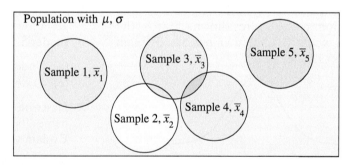

INSIGHT

Sample means can vary from one another and can also vary from the population mean. This type of variation is to be expected and is called *sampling error*.

PROPERTIES OF SAMPLING DISTRIBUTIONS OF SAMPLE MEANS

1. The mean of the sample means $\mu_{\overline{x}}$ is equal to the population mean μ.

$$\mu_{\overline{x}} = \mu$$

2. The standard deviation of the sample means $\sigma_{\overline{x}}$ is equal to the population standard deviation σ divided by the square root of the sample size n.

$$\sigma_{\overline{x}} = \frac{\sigma}{\sqrt{n}}$$

The standard deviation of the sampling distribution of the sample means is called the **standard error of the mean.**

Probability Histogram of Population of x

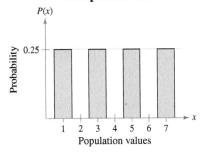

Probability Distribution of Sample Means

$\bar{x}$	f	Probability
1	1	$1/16 = 0.0625$
2	2	$2/16 = 0.1250$
3	3	$3/16 = 0.1875$
4	4	$4/16 = 0.2500$
5	3	$3/16 = 0.1875$
6	2	$2/16 = 0.1250$
7	1	$1/16 = 0.0625$

Probability Histogram of Sampling Distribution of $\bar{x}$

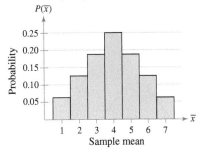
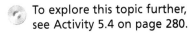

To explore this topic further, see Activity 5.4 on page 280.

STUDY TIP

Review Section 4.1 to find the mean and standard deviation of a probability distribution.

EXAMPLE 1

▶ **A Sampling Distribution of Sample Means**

You write the population values $\{1, 3, 5, 7\}$ on slips of paper and put them in a box. Then you randomly choose two slips of paper, with replacement. List all possible samples of size $n = 2$ and calculate the mean of each. These means form the sampling distribution of the sample means. Find the mean, variance, and standard deviation of the sample means. Compare your results with the mean $\mu = 4$, variance $\sigma^2 = 5$, and standard deviation $\sigma = \sqrt{5} \approx 2.236$ of the population.

▶ **Solution**

List all 16 samples of size 2 from the population and the mean of each sample.

Sample	Sample mean, $\bar{x}$	Sample	Sample mean, $\bar{x}$
1, 1	1	5, 1	3
1, 3	2	5, 3	4
1, 5	3	5, 5	5
1, 7	4	5, 7	6
3, 1	2	7, 1	4
3, 3	3	7, 3	5
3, 5	4	7, 5	6
3, 7	5	7, 7	7

After constructing a probability distribution of the sample means, you can graph the sampling distribution using a probability histogram as shown at the left. Notice that the shape of the histogram is bell-shaped and symmetric, similar to a normal curve. The mean, variance, and standard deviation of the 16 sample means are

$$\mu_{\bar{x}} = 4$$

$$(\sigma_{\bar{x}})^2 = \frac{5}{2} = 2.5 \quad \text{and} \quad \sigma_{\bar{x}} = \sqrt{\frac{5}{2}} = \sqrt{2.5} \approx 1.581.$$

These results satisfy the properties of sampling distributions because

$$\mu_{\bar{x}} = \mu = 4 \quad \text{and} \quad \sigma_{\bar{x}} = \frac{\sigma}{\sqrt{n}} = \frac{\sqrt{5}}{\sqrt{2}} \approx 1.581.$$

▶ **Try It Yourself 1**

List all possible samples of $n = 3$, with replacement, from the population $\{1, 3, 5, 7\}$. Calculate the mean, variance, and standard deviation of the sample means. Compare these values with the corresponding population parameters.

a. Form all possible *samples of size 3* and find the *mean* of each.
b. Make a *probability distribution* of the sample means and find the *mean*, *variance*, and *standard deviation*.
c. *Compare* the mean, variance, and standard deviation of the sample means with those of the population.

Answer: Page A38

▶ THE CENTRAL LIMIT THEOREM

The Central Limit Theorem forms the foundation for the inferential branch of statistics. This theorem describes the relationship between the sampling distribution of sample means and the population that the samples are taken from. The Central Limit Theorem is an important tool that provides the information you'll need to use sample statistics to make inferences about a population mean.

THE CENTRAL LIMIT THEOREM

1. If samples of size n, where $n \geq 30$, are drawn from any population with a mean μ and a standard deviation σ, then the sampling distribution of sample means approximates a normal distribution. The greater the sample size, the better the approximation.

2. If the population itself is normally distributed, then the sampling distribution of sample means is normally distributed for *any* sample size n.

In either case, the sampling distribution of sample means has a mean equal to the population mean.

$$\mu_{\bar{x}} = \mu \qquad \text{Mean}$$

The sampling distribution of sample means has a variance equal to $1/n$ times the variance of the population and a standard deviation equal to the population standard deviation divided by the square root of n.

$$\sigma_{\bar{x}}^2 = \frac{\sigma^2}{n} \qquad \text{Variance}$$

$$\sigma_{\bar{x}} = \frac{\sigma}{\sqrt{n}} \qquad \text{Standard deviation}$$

Recall that the standard deviation of the sampling distribution of the sample means, $\sigma_{\bar{x}}$, is also called the standard error of the mean.

1. Any Population Distribution

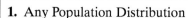

Distribution of Sample Means, $n \geq 30$

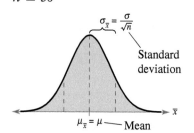

2. Normal Population Distribution

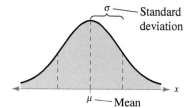

Distribution of Sample Means (any n)

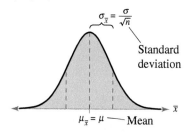

EXAMPLE 2

▶ **Interpreting the Central Limit Theorem**

Cellular phone bills for residents of a city have a mean of $63 and a standard deviation of $11, as shown in the following graph. Random samples of 100 cellular phone bills are drawn from this population and the mean of each sample is determined. Find the mean and standard error of the mean of the sampling distribution. Then sketch a graph of the sampling distribution of sample means. *(Adapted from JD Power and Associates)*

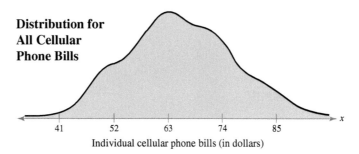

Distribution for All Cellular Phone Bills

Individual cellular phone bills (in dollars)

▶ **Solution**

The mean of the sampling distribution is equal to the population mean, and the standard error of the mean is equal to the population standard deviation divided by $\sqrt{n}$. So,

$$\mu_{\bar{x}} = \mu = 63 \quad \text{and} \quad \sigma_{\bar{x}} = \frac{\sigma}{\sqrt{n}} = \frac{11}{\sqrt{100}} = 1.1.$$

Interpretation   From the Central Limit Theorem, because the sample size is greater than 30, the sampling distribution can be approximated by a normal distribution with $\mu = \$63$ and $\sigma = \$1.10$, as shown in the graph below.

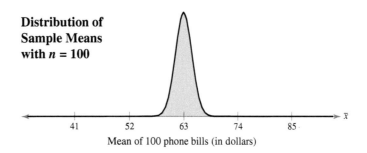

Distribution of Sample Means with *n* = 100

Mean of 100 phone bills (in dollars)

▶ **Try It Yourself 2**

Suppose random samples of size 64 are drawn from the population in Example 2. Find the mean and standard error of the mean of the sampling distribution. Sketch a graph of the sampling distribution and compare it with the sampling distribution in Example 2.

a. *Find* $\mu_{\bar{x}}$ *and* $\sigma_{\bar{x}}$.
b. *Identify* the sample size. If $n \geq 30$, *sketch* a normal curve with mean $\mu_{\bar{x}}$ and standard deviation $\sigma_{\bar{x}}$.
c. *Compare* the results with those in Example 2.   *Answer: Page A39*

PICTURING THE WORLD

In a recent year, there were about 4.8 million parents in the United States who received child support payments. The following histogram shows the distribution of children per custodial parent. The mean number of children was 1.7 and the standard deviation was 0.8. (Adapted from U.S. Census Bureau)

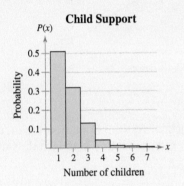

Child Support

You randomly select 35 parents who receive child support and ask how many children in their custody are receiving child support payments. What is the probability that the mean of the sample is between 1.5 and 1.9 children?

EXAMPLE 3

▶ **Interpreting the Central Limit Theorem**

Suppose the training heart rates of all 20-year-old athletes are normally distributed, with a mean of 135 beats per minute and standard deviation of 18 beats per minute, as shown in the following graph. Random samples of size 4 are drawn from this population, and the mean of each sample is determined. Find the mean and standard error of the mean of the sampling distribution. Then sketch a graph of the sampling distribution of sample means.

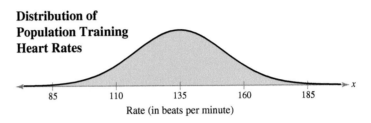

Distribution of Population Training Heart Rates

Rate (in beats per minute)

▶ **Solution**

The mean of the sampling distribution is equal to the population mean, and the standard error of the mean is equal to the population standard deviation divided by $\sqrt{n}$. So,

$$\mu_{\bar{x}} = \mu = 135 \text{ beats per minute} \quad \text{and} \quad \sigma_{\bar{x}} = \frac{\sigma}{\sqrt{n}} = \frac{18}{\sqrt{4}} = 9 \text{ beats per minute.}$$

Interpretation   From the Central Limit Theorem, because the population is normally distributed, the sampling distribution of the sample means is also normally distributed, as shown in the graph below.

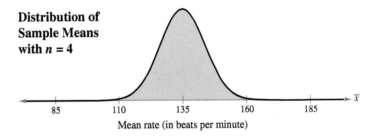

Distribution of Sample Means with $n = 4$

Mean rate (in beats per minute)

▶ **Try It Yourself 3**

The diameters of fully grown white oak trees are normally distributed, with a mean of 3.5 feet and a standard deviation of 0.2 foot, as shown in the graph below. Random samples of size 16 are drawn from this population, and the mean of each sample is determined. Find the mean and standard error of the mean of the sampling distribution. Then sketch a graph of the sampling distribution.

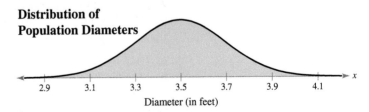

Distribution of Population Diameters

Diameter (in feet)

a. *Find* $\mu_{\bar{x}}$ *and* $\sigma_{\bar{x}}$.
b. *Sketch* a normal curve with mean $\mu_{\bar{x}}$ and standard deviation $\sigma_{\bar{x}}$.

Answer: Page A39

▶ PROBABILITY AND THE CENTRAL LIMIT THEOREM

In Section 5.2, you learned how to find the probability that a random variable x will fall in a given interval of population values. In a similar manner, you can find the probability that a sample mean $\bar{x}$ will fall in a given interval of the $\bar{x}$ sampling distribution. To transform $\bar{x}$ to a z-score, you can use the formula

$$z = \frac{\text{Value} - \text{Mean}}{\text{Standard error}} = \frac{\bar{x} - \mu_{\bar{x}}}{\sigma_{\bar{x}}} = \frac{\bar{x} - \mu}{\sigma/\sqrt{n}}.$$

EXAMPLE 4

▶ Finding Probabilities for Sampling Distributions

The graph at the right shows the lengths of time people spend driving each day. You randomly select 50 drivers ages 15 to 19. What is the probability that the mean time they spend driving each day is between 24.7 and 25.5 minutes? Assume that $\sigma = 1.5$ minutes.

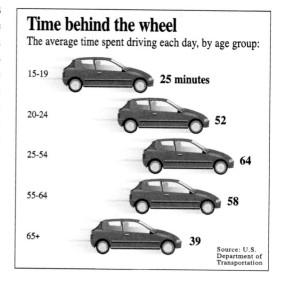

Time behind the wheel

The average time spent driving each day, by age group:

15-19	25 minutes
20-24	52
25-54	64
55-64	58
65+	39

Source: U.S. Department of Transportation

▶ Solution

The sample size is greater than 30, so you can use the Central Limit Theorem to conclude that the distribution of sample means is approximately normal, with a mean and a standard deviation of

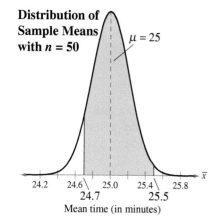

Distribution of Sample Means with $n = 50$

$\mu = 25$

24.2   24.6   25.0   25.4   25.8 $\bar{x}$
  24.7 25.5
Mean time (in minutes)

$$\mu_{\bar{x}} = \mu = 25 \text{ minutes} \quad \text{and} \quad \sigma_{\bar{x}} = \frac{\sigma}{\sqrt{n}} = \frac{1.5}{\sqrt{50}} \approx 0.21213 \text{ minute}.$$

The graph of this distribution is shown at the left with a shaded area between 24.7 and 25.5 minutes. The z-scores that correspond to sample means of 24.7 and 25.5 minutes are

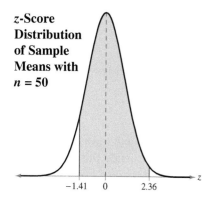

z-Score Distribution of Sample Means with $n = 50$

−1.41   0   2.36 z

$$z_1 = \frac{24.7 - 25}{1.5/\sqrt{50}} \approx \frac{-0.3}{0.21213} \approx -1.41 \quad \text{and}$$

$$z_2 = \frac{25.5 - 25}{1.5/\sqrt{50}} \approx \frac{0.5}{0.21213} \approx 2.36.$$

So, the probability that the mean time the 50 people spend driving each day is between 24.7 and 25.5 minutes is

$$P(24.7 < \bar{x} < 25.5) = P(-1.41 < z < 2.36)$$
$$= P(z < 2.36) - P(z < -1.41)$$
$$= 0.9909 - 0.0793 = 0.9116.$$

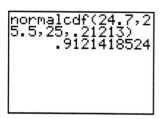

```
normalcdf(24.7,2
5.5,25,.21213)
       .9121418524
```

In Example 4, you can use a TI-83/84 Plus to find the probability automatically once the standard error of the mean is calculated.

Interpretation   Of the samples of 50 drivers ages 15 to 19, 91.16% will have a mean driving time that is between 24.7 and 25.5 minutes, as shown in the graph at the left. This implies that, assuming the value of $\mu = 25$ is correct, only 8.84% of such sample means will lie outside the given interval.

STUDY TIP

Before you find probabilities for intervals of the sample mean $\bar{x}$, use the Central Limit Theorem to determine the mean and the standard deviation of the sampling distribution of the sample means. That is, calculate $\mu_{\bar{x}}$ and $\sigma_{\bar{x}}$.

▶ **Try It Yourself 4**

You randomly select 100 drivers ages 15 to 19 from Example 4. What is the probability that the mean time they spend driving each day is between 24.7 and 25.5 minutes? Use $\mu = 25$ and $\sigma = 1.5$ minutes.

a. Use the Central Limit Theorem to *find* $\mu_{\bar{x}}$ and $\sigma_{\bar{x}}$ and *sketch* the sampling distribution of the sample means.
b. *Find the z-scores* that correspond to $\bar{x} = 24.7$ minutes and $\bar{x} = 25.5$ minutes.
c. *Find the cumulative area* that corresponds to each z-score and calculate the probability.
d. *Interpret* the results. *Answer: Page A39*

EXAMPLE 5

▶ **Finding Probabilities for Sampling Distributions**

The mean room and board expense per year at four-year colleges is $7540. You randomly select 9 four-year colleges. What is the probability that the mean room and board is less than $7800? Assume that the room and board expenses are normally distributed with a standard deviation of $1245. *(Adapted from National Center for Education Statistics)*

▶ **Solution**

Because the population is normally distributed, you can use the Central Limit Theorem to conclude that the distribution of sample means is normally distributed, with a mean of $7540 and a standard deviation of $415.

$$\mu_{\bar{x}} = \mu = 7540 \quad \text{and} \quad \sigma_{\bar{x}} = \frac{\sigma}{\sqrt{n}} = \frac{1245}{\sqrt{9}} = 415$$

The graph of this distribution is shown at the left. The area to the left of $7800 is shaded. The z-score that corresponds to $7800 is

$$z = \frac{7800 - 7540}{1245/\sqrt{9}} = \frac{260}{415} \approx 0.63.$$

So, the probability that the mean room and board expense is less than $7800 is

$$P(\bar{x} < 7800) = P(z < 0.63)$$
$$= 0.7357.$$

Interpretation So, 73.57% of such samples with $n = 9$ will have a mean less than $7800 and 26.43% of these sample means will lie outside this interval.

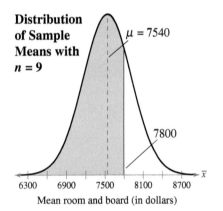

Distribution of Sample Means with $n = 9$

$\mu = 7540$

7800

6300   6900   7500   8100   8700 $\bar{x}$

Mean room and board (in dollars)

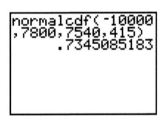

```
normalcdf(-10000
,7800,7540,415)
      .7345085183
```

In Example 5, you can use a TI-83/84 Plus to find the probability automatically.

▶ **Try It Yourself 5**

The average sales price of a single-family house in the United States is $290,600. You randomly select 12 single-family houses. What is the probability that the mean sales price is more than $265,000? Assume that the sales prices are normally distributed with a standard deviation of $36,000. *(Adapted from The U.S. Commerce Department)*

a. Use the Central Limit Theorem to *find* $\mu_{\bar{x}}$ and $\sigma_{\bar{x}}$ and *sketch* the sampling distribution of the sample means.
b. *Find the z-score* that corresponds to $\bar{x} = \$265,000$.
c. *Find the cumulative area* that corresponds to the z-score and calculate the probability.
d. *Interpret* the results. *Answer: Page A39*

The Central Limit Theorem can also be used to investigate unusual events. An unusual event is one that occurs with a probability of less than 5%.

EXAMPLE 6

▶ **Finding Probabilities for *x* and $\bar{x}$**

An education finance corporation claims that the average credit card debts carried by undergraduates are normally distributed, with a mean of $3173 and a standard deviation of $1120. *(Adapted from Sallie Mae)*

1. What is the probability that a randomly selected undergraduate, who is a credit card holder, has a credit card balance less than $2700?

2. You randomly select 25 undergraduates who are credit card holders. What is the probability that their mean credit card balance is less than $2700?

3. Compare the probabilities from (1) and (2) and interpret your answer in terms of the corporation's claim.

▶ **Solution**

1. In this case, you are asked to find the probability associated with a certain value of the random variable *x*. The *z*-score that corresponds to *x* = $2700 is

$$z = \frac{x - \mu}{\sigma} = \frac{2700 - 3173}{1120} \approx -0.42.$$

So, the probability that the card holder has a balance less than $2700 is

$$P(x < 2700) = P(z < -0.42) = 0.3372.$$

2. Here, you are asked to find the probability associated with a sample mean $\bar{x}$. The *z*-score that corresponds to $\bar{x}$ = $2700 is

$$z = \frac{\bar{x} - \mu_{\bar{x}}}{\sigma_{\bar{x}}} = \frac{\bar{x} - \mu}{\sigma/\sqrt{n}} = \frac{2700 - 3173}{1120/\sqrt{25}} = \frac{-473}{224} \approx -2.11.$$

So, the probability that the mean credit card balance of the 25 card holders is less than $2700 is

$$P(\bar{x} < 2700) = P(z < -2.11) = 0.0174.$$

3. *Interpretation* Although there is about a 34% chance that an undergraduate will have a balance less than $2700, there is only about a 2% chance that the mean of a sample of 25 will have a balance less than $2700. Because there is only a 2% chance that the mean of a sample of 25 will have a balance less than $2700, this is an unusual event. So, it is possible that the corporation's claim that the mean is $3173 is incorrect.

▶ **Try It Yourself 6**

A consumer price analyst claims that prices for liquid crystal display (LCD) computer monitors are normally distributed, with a mean of $190 and a standard deviation of $48. (1) What is the probability that a randomly selected LCD computer monitor costs less than $200? (2) You randomly select 10 LCD computer monitors. What is the probability that their mean cost is less than $200? (3) Compare these two probabilities.

a. *Find the z-scores* that correspond to *x* and $\bar{x}$.
b. Use the Standard Normal Table to *find the probability* associated with each *z*-score.
c. *Compare* the probabilities and *interpret* your answer.   *Answer: Page A39*

5.4 EXERCISES

■ BUILDING BASIC SKILLS AND VOCABULARY

In Exercises 1–4, a population has a mean $\mu = 150$ and a standard deviation $\sigma = 25$. Find the mean and standard deviation of a sampling distribution of sample means with the given sample size n.

1. $n = 50$

2. $n = 100$

3. $n = 250$

4. $n = 1000$

True or False? *In Exercises 5–8, determine whether the statement is true or false. If it is false, rewrite it as a true statement.*

5. As the size of a sample increases, the mean of the distribution of sample means increases.

6. As the size of a sample increases, the standard deviation of the distribution of sample means increases.

7. A sampling distribution is normal only if the population is normal.

8. If the size of a sample is at least 30, you can use z-scores to determine the probability that a sample mean falls in a given interval of the sampling distribution.

Graphical Analysis *In Exercises 9 and 10, the graph of a population distribution is shown with its mean and standard deviation. Assume that a sample size of 100 is drawn from each population. Decide which of the graphs labeled (a)–(c) would most closely resemble the sampling distribution of the sample means for each graph. Explain your reasoning.*

9. The waiting time (in seconds) at a traffic signal during a red light

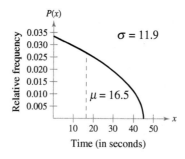

(a)

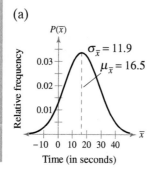

(b)

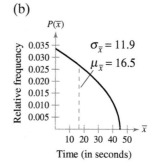

(c)

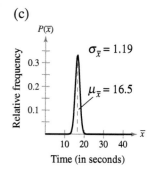

10. The annual snowfall (in feet) for a central New York state county

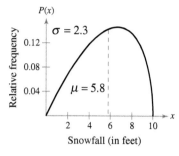

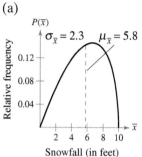

(a)

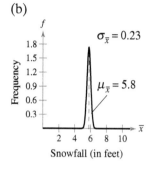

(b)

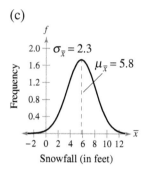

(c)

Verifying Properties of Sampling Distributions   *In Exercises 11 and 12, find the mean and standard deviation of the population. List all samples (with replacement) of the given size from that population. Find the mean and standard deviation of the sampling distribution and compare them with the mean and standard deviation of the population.*

11. The number of DVDs rented by each of four families in the past month is 8, 4, 16, and 2. Use a sample size of 3.

12. Four friends paid the following amounts for their MP3 players: $200, $130, $270, and $230. Use a sample size of 2.

Finding Probabilities   *In Exercises 13–16, the population mean and standard deviation are given. Find the required probability and determine whether the given sample mean would be considered unusual. If convenient, use technology to find the probability.*

13. For a sample of $n = 64$, find the probability of a sample mean being less than 24.3 if $\mu = 24$ and $\sigma = 1.25$.

14. For a sample of $n = 100$, find the probability of a sample mean being greater than 24.3 if $\mu = 24$ and $\sigma = 1.25$.

15. For a sample of $n = 45$, find the probability of a sample mean being greater than 551 if $\mu = 550$ and $\sigma = 3.7$.

16. For a sample of $n = 36$, find the probability of a sample mean being less than 12,750 or greater than 12,753 if $\mu = 12,750$ and $\sigma = 1.7$.

◼ USING AND INTERPRETING CONCEPTS

Using the Central Limit Theorem   *In Exercises 17–22, use the Central Limit Theorem to find the mean and standard error of the mean of the indicated sampling distribution. Then sketch a graph of the sampling distribution.*

17. **Employed Persons** The amounts of time employees at a large corporation work each day are normally distributed, with a mean of 7.6 hours and a standard deviation of 0.35 hour. Random samples of size 12 are drawn from the population and the mean of each sample is determined.

18. **Fly Eggs** The numbers of eggs female house flies lay during their lifetimes are normally distributed, with a mean of 800 eggs and a standard deviation of 100 eggs. Random samples of size 15 are drawn from this population and the mean of each sample is determined.

19. **Photo Printers** The mean price of photo printers on a website is $235 with a standard deviation of $62. Random samples of size 20 are drawn from this population and the mean of each sample is determined.

20. **Employees' Ages** The mean age of employees at a large corporation is 47.2 years with a standard deviation of 3.6 years. Random samples of size 36 are drawn from this population and the mean of each sample is determined.

21. **Fresh Vegetables** The per capita consumption of fresh vegetables by people in the United States in a recent year was normally distributed, with a mean of 188.4 pounds and a standard deviation of 54.5 pounds. Random samples of 25 are drawn from this population and the mean of each sample is determined. *(Adapted from U.S. Department of Agriculture)*

22. **Coffee** The per capita consumption of coffee by people in the United States in a recent year was normally distributed, with a mean of 24.2 gallons and a standard deviation of 8.1 gallons. Random samples of 30 are drawn from this population and the mean of each sample is determined. *(Adapted from U.S. Department of Agriculture)*

23. Repeat Exercise 17 for samples of size 24 and 36. What happens to the mean and the standard deviation of the distribution of sample means as the size of the sample increases?

24. Repeat Exercise 18 for samples of size 30 and 45. What happens to the mean and the standard deviation of the distribution of sample means as the size of the sample increases?

Finding Probabilities *In Exercises 25–30, find the probabilities and interpret the results. If convenient, use technology to find the probabilities.*

25. **Salaries** The population mean annual salary for environmental compliance specialists is about $63,500. A random sample of 35 specialists is drawn from this population. What is the probability that the mean salary of the sample is less than $60,000? Assume $\sigma = \$6100$. *(Adapted from Salary.com)*

26. **Salaries** The population mean annual salary for flight attendants is $56,275. A random sample of 48 flight attendants is selected from this population. What is the probability that the mean annual salary of the sample is less than $56,100? Assume $\sigma = \$1800$. *(Adapted from Salary.com)*

27. **Gas Prices: New England** During a certain week the mean price of gasoline in the New England region was $2.714 per gallon. A random sample of 32 gas stations is drawn from this population. What is the probability that the mean price for the sample was between $2.695 and $2.725 that week? Assume $\sigma = \$0.045$. *(Adapted from U.S. Energy Information Administration)*

28. **Gas Prices: California** During a certain week the mean price of gasoline in California was $2.999 per gallon. A random sample of 38 gas stations is drawn from this population. What is the probability that the mean price for the sample was between $3.010 and $3.025 that week? Assume $\sigma = \$0.049$. *(Adapted from U.S. Energy Information Administration)*

29. **Heights of Women** The mean height of women in the United States (ages 20–29) is 64.3 inches. A random sample of 60 women in this age group is selected. What is the probability that the mean height for the sample is greater than 66 inches? Assume $\sigma = 2.6$ inches. *(Source: National Center for Health Statistics)*

30. **Heights of Men** The mean height of men in the United States (ages 20–29) is 69.9 inches. A random sample of 60 men in this age group is selected. What is the probability that the mean height for the sample is greater than 70 inches? Assume $\sigma = 3.0$ inches. *(Source: National Center for Health Statistics)*

31. **Which Is More Likely?** Assume that the heights given in Exercise 29 are normally distributed. Are you more likely to randomly select 1 woman with a height less than 70 inches or are you more likely to select a sample of 20 women with a mean height less than 70 inches? Explain.

32. **Which Is More Likely?** Assume that the heights given in Exercise 30 are normally distributed. Are you more likely to randomly select 1 man with a height less than 65 inches or are you more likely to select a sample of 15 men with a mean height less than 65 inches? Explain.

33. **Make a Decision** A machine used to fill gallon-sized paint cans is regulated so that the amount of paint dispensed has a mean of 128 ounces and a standard deviation of 0.20 ounce. You randomly select 40 cans and carefully measure the contents. The sample mean of the cans is 127.9 ounces. Does the machine need to be reset? Explain your reasoning.

34. **Make a Decision** A machine used to fill half-gallon-sized milk containers is regulated so that the amount of milk dispensed has a mean of 64 ounces and a standard deviation of 0.11 ounce. You randomly select 40 containers and carefully measure the contents. The sample mean of the containers is 64.05 ounces. Does the machine need to be reset? Explain your reasoning.

35. **Lumber Cutter** Your lumber company has bought a machine that automatically cuts lumber. The seller of the machine claims that the machine cuts lumber to a mean length of 8 feet (96 inches) with a standard deviation of 0.5 inch. Assume the lengths are normally distributed. You randomly select 40 boards and find that the mean length is 96.25 inches.

    (a) Assuming the seller's claim is correct, what is the probability that the mean of the sample is 96.25 inches or more?

    (b) Using your answer from part (a), what do you think of the seller's claim?

    (c) Would it be unusual to have an individual board with a length of 96.25 inches? Why or why not?

36. **Ice Cream Carton Weights** A manufacturer claims that the mean weight of its ice cream cartons is 10 ounces with a standard deviation of 0.5 ounce. Assume the weights are normally distributed. You test 25 cartons and find their mean weight is 10.21 ounces.

    (a) Assuming the manufacturer's claim is correct, what is the probability that the mean of the sample is 10.21 ounces or more?

    (b) Using your answer from part (a), what do you think of the manufacturer's claim?

    (c) Would it be unusual to have an individual carton with a weight of 10.21 ounces? Why or why not?

37. Life of Tires   A manufacturer claims that the life span of its tires is 50,000 miles. You work for a consumer protection agency and you are testing this manufacturer's tires. Assume the life spans of the tires are normally distributed. You select 100 tires at random and test them. The mean life span is 49,721 miles. Assume $\sigma = 800$ miles.

(a) Assuming the manufacturer's claim is correct, what is the probability that the mean of the sample is 49,721 miles or less?

(b) Using your answer from part (a), what do you think of the manufacturer's claim?

(c) Would it be unusual to have an individual tire with a life span of 49,721 miles? Why or why not?

38. Brake Pads   A brake pad manufacturer claims its brake pads will last for 38,000 miles. You work for a consumer protection agency and you are testing this manufacturer's brake pads. Assume the life spans of the brake pads are normally distributed. You randomly select 50 brake pads. In your tests, the mean life of the brake pads is 37,650 miles. Assume $\sigma = 1000$ miles.

(a) Assuming the manufacturer's claim is correct, what is the probability that the mean of the sample is 37,650 miles or less?

(b) Using your answer from part (a), what do you think of the manufacturer's claim?

(c) Would it be unusual to have an individual brake pad last for 37,650 miles? Why or why not?

■ EXTENDING CONCEPTS

39. SAT Scores   The mean critical reading SAT score is 501, with a standard deviation of 112. A particular high school claims that its students have unusually high critical reading SAT scores. A random sample of 50 students from this school was selected, and the mean critical reading SAT score was 515. Is the high school justified in its claim? Explain.   *(Source: The College Board)*

40. Machine Calibrations   A machine in a manufacturing plant is calibrated to produce a bolt that has a mean diameter of 4 inches and a standard deviation of 0.5 inch. An engineer takes a random sample of 100 bolts from this machine and finds the mean diameter is 4.2 inches. What are some possible consequences of these findings?

Finite Correction Factor   *The formula for the standard error of the mean*

$$\sigma_{\overline{x}} = \frac{\sigma}{\sqrt{n}}$$

*given in the Central Limit Theorem is based on an assumption that the population has infinitely many members. This is the case whenever sampling is done with replacement (each member is put back after it is selected), because the sampling process could be continued indefinitely. The formula is also valid if the sample size is small in comparison with the population. However, when sampling is done without replacement and the sample size n is more than 5% of the finite population of size N ($n/N > 0.05$), there is a finite number of possible samples. A **finite correction factor,***

$$\sqrt{\frac{N - n}{N - 1}}$$

should be used to adjust the standard error. The sampling distribution of the sample means will be normal with a mean equal to the population mean, and the standard error of the mean will be

$$\sigma_{\bar{x}} = \frac{\sigma}{\sqrt{n}} \sqrt{\frac{N-n}{N-1}}.$$

In Exercises 41 and 42, determine if the finite correction factor should be used. If so, use it in your calculations when you find the probability.

41. Gas Prices   In a sample of 900 gas stations, the mean price of regular gasoline at the pump was $2.702 per gallon and the standard deviation was $0.009 per gallon. A random sample of size 55 is drawn from this population. What is the probability that the mean price per gallon is less than $2.698? *(Adapted from U.S. Department of Energy)*

42. Old Faithful   In a sample of 500 eruptions of the Old Faithful geyser at Yellowstone National Park, the mean duration of the eruptions was 3.32 minutes and the standard deviation was 1.09 minutes. A random sample of size 30 is drawn from this population. What is the probability that the mean duration of eruptions is between 2.5 minutes and 4 minutes? *(Adapted from Yellowstone National Park)*

Sampling Distribution of Sample Proportions   *The sample mean is not the only statistic with a sampling distribution. Every sample statistic, such as the sample median, the sample standard deviation, and the sample proportion, has a sampling distribution. For a random sample of size n, the **sample proportion** is the number of individuals in the sample with a specified characteristic divided by the sample size. The **sampling distribution of sample proportions** is the distribution formed when sample proportions of size n are repeatedly taken from a population where the probability of an individual with a specified characteristic is p.*

In Exercises 43–46, suppose three births are randomly selected. There are two equally possible outcomes for each birth, a boy (b) or a girl (g). The number of boys can equal 0, 1, 2, or 3. These correspond to sample proportions of 0, 1/3, 2/3, and 1.

43. List the eight possible samples that can result from randomly selecting three births. For instance, let bbb represent a sample of three boys. Make a table that shows each sample, the number of boys in each sample, and the proportion of boys in each sample.

44. Use the table from Exercise 43 to construct the sampling distribution of the sample proportion of boys from three births. Graph the sampling distribution using a probability histogram. What do you notice about the spread of the histogram as compared to the binomial probability distribution for the number of boys in each sample?

45. Let $x = 1$ represent a boy and $x = 0$ represent a girl. Using these values, find the sample mean for each sample. What do you notice?

46. Construct a sampling distribution of the sample proportion of boys from four births.

47. Heart Transplants   About 77% of all female heart transplant patients will survive for at least 3 years. One hundred five female heart transplant patients are randomly selected. What is the probability that the sample proportion surviving for at least 3 years will be less than 70%? Interpret your results. Assume the sampling distribution of sample proportions is a normal distribution. The mean of the sample proportion is equal to the population proportion p, and the standard deviation is equal to $\sqrt{\dfrac{pq}{n}}$. *(Source: American Heart Association)*

ACTIVITY 5.4   Sampling Distributions

APPLET

The *sampling distributions* applet allows you to investigate sampling distributions by repeatedly taking samples from a population. The top plot displays the distribution of a population. Several options are available for the population distribution (Uniform, Bell-shaped, Skewed, Binary, and Custom). When SAMPLE is clicked, N random samples of size n will be repeatedly selected from the population. The sample statistics specified in the bottom two plots will be updated for each sample. If N is set to 1 and n is less than or equal to 50, the display will show, in an animated fashion, the points selected from the population dropping into the second plot and the corresponding summary statistic values dropping into the third and fourth plots. Click RESET to stop an animation and clear existing results. Summary statistics for each plot are shown in the panel at the left of the plot.

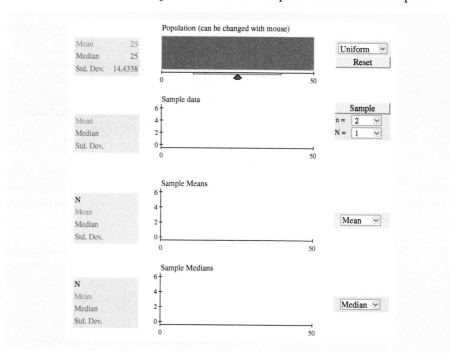

■ Explore

Step 1   Specify a distribution.
Step 2   Specify values of n and N.
Step 3   Specify what to display in the bottom two graphs.
Step 4   Click SAMPLE to generate the sampling distributions.

■ Draw Conclusions

APPLET

1. Run the simulation using $n = 30$ and $N = 10$ for a uniform, a bell-shaped, and a skewed distribution. What is the mean of the sampling distribution of the sample means for each distribution? For each distribution, is this what you would expect?

2. Run the simulation using $n = 50$ and $N = 10$ for a bell-shaped distribution. What is the standard deviation of the sampling distribution of the sample means? According to the formula, what should the standard deviation of the sampling distribution of the sample means be? Is this what you would expect?

5.5   Normal Approximations to Binomial Distributions

WHAT YOU SHOULD LEARN

▸ How to decide when a normal distribution can approximate a binomial distribution

▸ How to find the continuity correction

▸ How to use a normal distribution to approximate binomial probabilities

Approximating a Binomial Distribution ▸ Continuity Correction
▸ Approximating Binomial Probabilities

▸ APPROXIMATING A BINOMIAL DISTRIBUTION

In Section 4.2, you learned how to find binomial probabilities. For instance, if a surgical procedure has an 85% chance of success and a doctor performs the procedure on 10 patients, it is easy to find the probability of exactly two successful surgeries.

But what if the doctor performs the surgical procedure on 150 patients and you want to find the probability of *fewer than 100* successful surgeries? To do this using the techniques described in Section 4.2, you would have to use the binomial formula 100 times and find the sum of the resulting probabilities. This approach is not practical, of course. A better approach is to use a normal distribution to approximate the binomial distribution.

NORMAL APPROXIMATION TO A BINOMIAL DISTRIBUTION

If $np \geq 5$ and $nq \geq 5$, then the binomial random variable x is approximately normally distributed, with mean

$$\mu = np$$

and standard deviation

$$\sigma = \sqrt{npq}$$

where n is the number of independent trials, p is the probability of success in a single trial, and q is the probability of failure in a single trial.

STUDY TIP

Properties of a binomial experiment

• n independent trials

• Two possible outcomes: success or failure

• Probability of success is p; probability of failure is $q = 1 - p$

• p is constant for each trial

To see why this result is valid, look at the following binomial distributions for $p = 0.25$, $q = 1 - 0.25 = 0.75$, and $n = 4$, $n = 10$, $n = 25$, and $n = 50$. Notice that as n increases, the histogram approaches a normal curve.

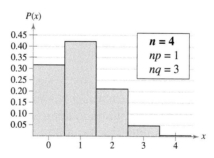

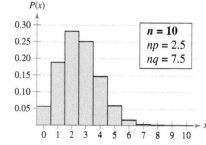

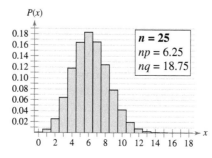

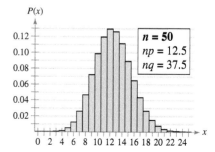

EXAMPLE 1

▶ **Approximating a Binomial Distribution**

Two binomial experiments are listed. Decide whether you can use the normal distribution to approximate x, the number of people who reply yes. If you can, find the mean and standard deviation. If you cannot, explain why. *(Source: Opinion Research Corporation)*

1. Sixty-two percent of adults in the United States have an HDTV in their home. You randomly select 45 adults in the United States and ask them if they have an HDTV in their home.

2. Twelve percent of adults in the United States who do not have an HDTV in their home are planning to purchase one in the next two years. You randomly select 30 adults in the United States who do not have an HDTV and ask them if they are planning to purchase one in the next two years.

▶ **Solution**

1. In this binomial experiment, $n = 45$, $p = 0.62$, and $q = 0.38$. So,

$$np = 45(0.62) = 27.9$$

and

$$nq = 45(0.38) = 17.1.$$

Because np and nq are greater than 5, you can use a normal distribution with

$$\mu = np = 27.9$$

and

$$\sigma = \sqrt{npq} = \sqrt{45 \cdot 0.62 \cdot 0.38} \approx 3.26$$

to approximate the distribution of x.

2. In this binomial experiment, $n = 30$, $p = 0.12$, and $q = 0.88$. So,

$$np = 30(0.12) = 3.6$$

and

$$nq = 30(0.88) = 26.4.$$

Because $np < 5$, you cannot use a normal distribution to approximate the distribution of x.

▶ **Try It Yourself 1**

Consider the following binomial experiment. Decide whether you can use the normal distribution to approximate x, the number of people who reply yes. If you can, find the mean and standard deviation. If you cannot, explain why. *(Source: Opinion Research Corporation)*

Five percent of adults in the United States are planning to purchase a 3D TV in the next two years. You randomly select 125 adults in the United States and ask them if they are planning to purchase a 3D TV in the next two years.

a. *Identify* n, p, and q.
b. *Find* the products np and nq.
c. *Decide* whether you can use a normal distribution to approximate x.
d. *Find* the mean μ and standard deviation σ, if appropriate.

Answer: Page A39

▶ CONTINUITY CORRECTION

A binomial distribution is discrete and can be represented by a probability histogram. To calculate *exact* binomial probabilities, you can use the binomial formula for each value of x and add the results. Geometrically, this corresponds to adding the areas of bars in the probability histogram. Remember that each bar has a width of one unit and x is the midpoint of the interval.

When you use a *continuous* normal distribution to approximate a binomial probability, you need to move 0.5 unit to the left and right of the midpoint to include all possible x-values in the interval. When you do this, you are making a **continuity correction.**

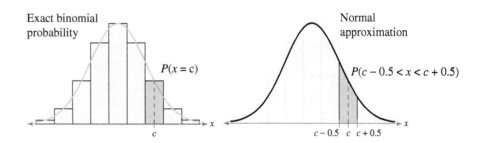

EXAMPLE 2

▶ Using a Continuity Correction

Use a continuity correction to convert each of the following binomial intervals to a normal distribution interval.

1. The probability of getting between 270 and 310 successes, inclusive

2. The probability of getting at least 158 successes

3. The probability of getting fewer than 63 successes

▶ Solution

1. The discrete midpoint values are 270, 271, ..., 310. The corresponding interval for the continuous normal distribution is

$$269.5 < x < 310.5.$$

2. The discrete midpoint values are 158, 159, 160, ... The corresponding interval for the continuous normal distribution is

$$x > 157.5.$$

3. The discrete midpoint values are ..., 60, 61, 62. The corresponding interval for the continuous normal distribution is

$$x < 62.5.$$

▶ Try It Yourself 2

Use a continuity correction to convert each of the following binomial intervals to a normal distribution interval.

1. The probability of getting between 57 and 83 successes, inclusive
2. The probability of getting at most 54 successes

a. List the *midpoint values* for the binomial probability.
b. Use a *continuity correction* to write the normal distribution interval.

Answer: Page A39

PICTURING THE WORLD

In a survey of U.S. adults, people were asked if there should be a nationwide ban on smoking in all public places. The results of the survey are shown in the following pie chart. (Adapted from Rasmussen Reports)

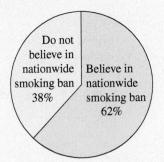

Do not believe in nationwide smoking ban 38%

Believe in nationwide smoking ban 62%

Assume that this survey is a true indication of the proportion of the population who say there should be a nationwide ban on smoking in all public places. If you sampled 50 adults at random, what is the probability that between 25 and 30, inclusive, would say there should be a nationwide ban on smoking in all public places?

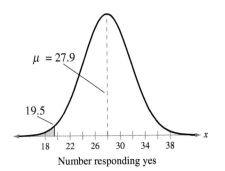

$\mu = 27.9$

19.5

18   22   26   30   34   38

Number responding yes

▶ **APPROXIMATING BINOMIAL PROBABILITIES**

GUIDELINES

Using a Normal Distribution to Approximate Binomial Probabilities

IN WORDS	IN SYMBOLS
1. Verify that a binomial distribution applies.	Specify $n, p,$ and q.
2. Determine if you can use a normal distribution to approximate x, the binomial variable.	Is $np \geq 5$? Is $nq \geq 5$?
3. Find the mean μ and standard deviation σ for the distribution.	$\mu = np$ $\sigma = \sqrt{npq}$
4. Apply the appropriate continuity correction. Shade the corresponding area under the normal curve.	Add or subtract 0.5 from endpoints.
5. Find the corresponding z-score(s).	$z = \dfrac{x - \mu}{\sigma}$
6. Find the probability.	Use the Standard Normal Table.

EXAMPLE 3

▶ **Approximating a Binomial Probability**

Sixty-two percent of adults in the United States have an HDTV in their home. You randomly select 45 adults in the United States and ask them if they have an HDTV in their home. What is the probability that fewer than 20 of them respond yes? *(Source: Opinion Research Corporation)*

▶ **Solution**

From Example 1, you know that you can use a normal distribution with $\mu = 27.9$ and $\sigma \approx 3.26$ to approximate the binomial distribution. Remember to apply the continuity correction for the value of x. In the binomial distribution, the possible midpoint values for "fewer than 20" are

    ... 17, 18, 19.

To use a normal distribution, add 0.5 to the right-hand boundary 19 to get $x = 19.5$. The graph at the left shows a normal curve with $\mu = 27.9$ and $\sigma \approx 3.26$ and a shaded area to the left of 19.5. The z-score that corresponds to $x = 19.5$ is

$$z = \frac{19.5 - 27.9}{3.26}$$

$$\approx -2.58.$$

Using the Standard Normal Table,

    $P(z < -2.58) = 0.0049.$

Interpretation The probability that fewer than 20 people respond yes is approximately 0.0049, or about 0.49%.

▸ **Try It Yourself 3**

Five percent of adults in the United States are planning to purchase a 3D TV in the next two years. You randomly select 125 adults in the United States and ask them if they are planning to purchase a 3D TV in the next two years. What is the probability that more than 9 respond yes? (See Try It Yourself 1.)
(Source: Opinion Research Corporation)

a. *Determine* whether you can use a normal distribution to approximate the binomial variable (see part (c) of Try It Yourself 1).
b. Find the *mean* μ and the *standard deviation* σ for the distribution (see part (d) of Try It Yourself 1).
c. Apply a *continuity correction* to rewrite $P(x > 9)$ and sketch a graph.
d. *Find* the corresponding z-score.
e. Use the Standard Normal Table to *find the area* to the left of z and *calculate the probability*. *Answer: Page A39*

EXAMPLE 4

▸ **Approximating a Binomial Probability**

Fifty-eight percent of adults say that they never wear a helmet when riding a bicycle. You randomly select 200 adults in the United States and ask them if they wear a helmet when riding a bicycle. What is the probability that at least 120 adults will say they never wear a helmet when riding a bicycle? *(Source: Consumer Reports National Research Center)*

▸ **Solution**   Because $np = 200 \cdot 0.58 = 116$ and $nq = 200 \cdot 0.42 = 84$, the binomial variable x is approximately normally distributed, with

$$\mu = np = 116 \quad \text{and} \quad \sigma = \sqrt{npq} = \sqrt{200 \cdot 0.58 \cdot 0.42} \approx 6.98.$$

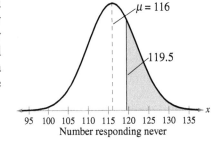

Using the continuity correction, you can rewrite the discrete probability $P(x \geq 120)$ as the continuous probability $P(x \geq 119.5)$. The graph shows a normal curve with $\mu = 116$, $\sigma = 6.98$, and a shaded area to the right of 119.5. The z-score that corresponds to 119.5 is

$$z = \frac{119.5 - 116}{6.98} \approx 0.50.$$

So, the probability that at least 120 will say yes is approximately

$$P(x \geq 119.5) = P(z \geq 0.50)$$
$$= 1 - P(z \leq 0.50) = 1 - 0.6915 = 0.3085.$$

▸ **Try It Yourself 4**

In Example 4, what is the probability that at most 100 adults will say they never wear a helmet when riding a bicycle?

a. *Determine* whether you can use a normal distribution to approximate the binomial variable (see Example 4).
b. Find the *mean* μ and the *standard deviation* σ for the distribution (see Example 4).
c. Apply a *continuity correction* to rewrite $P(x \leq 100)$ and sketch a graph.
d. *Find* the corresponding z-score.
e. Use the Standard Normal Table to *find the area* to the left of z and *calculate the probability*. *Answer: Page A39*

STUDY TIP

In a discrete distribution, there is a difference between $P(x \geq c)$ and $P(x > c)$. This is true because the probability that x is exactly c is not 0. In a continuous distribution, however, there is no difference between $P(x \geq c)$ and $P(x > c)$ because the probability that x is exactly c is 0.

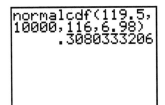

```
normalcdf(119.5,
10000,116,6.98)
      .3080333206
```

In Example 4, you can use a TI-83/84 Plus to find the probability once the mean, standard deviation, and continuity correction are calculated. Use 10,000 for the upper bound.

EXAMPLE 5

> ▸ **Approximating a Binomial Probability**

A survey reports that 62% of Internet users use Windows® Internet Explorer® as their browser. You randomly select 150 Internet users and ask them whether they use Internet Explorer® as their browser. What is the probability that exactly 96 will say yes? *(Source: Net Applications)*

▸ **Solution**

Because $np = 150 \cdot 0.62 = 93$ and $nq = 150 \cdot 0.38 = 57$, the binomial variable x is approximately normally distributed, with

$$\mu = np = 93 \quad \text{and} \quad \sigma = \sqrt{npq} = \sqrt{150 \cdot 0.62 \cdot 0.38} \approx 5.94.$$

Using the continuity correction, you can rewrite the discrete probability $P(x = 96)$ as the continuous probability $P(95.5 < x < 96.5)$. The graph shows a normal curve with $\mu = 93$, $\sigma = 5.94$, and a shaded area between 95.5 and 96.5.

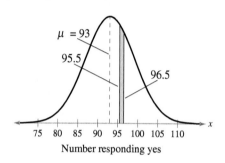

Number responding yes

The z-scores that correspond to 95.5 and 96.5 are

$$z_1 = \frac{95.5 - 93}{5.94} \approx 0.42 \quad \text{and} \quad z_2 = \frac{96.5 - 93}{5.94} \approx 0.59.$$

So, the probability that exactly 96 Internet users will say they use Internet Explorer® is

$$\begin{aligned}
P(95.5 < x < 96.5) &= P(0.42 < z < 0.59) \\
&= P(z < 0.59) - P(z < 0.42) \\
&= 0.7224 - 0.6628 \\
&= 0.0596.
\end{aligned}$$

Interpretation The probability that exactly 96 of the Internet users will say they use Internet Explorer® is approximately 0.0596, or about 6%.

▸ **Try It Yourself 5**

A survey reports that 24% of Internet users use Mozilla® Firefox® as their browser. You randomly select 150 Internet users and ask them whether they use Firefox® as their browser. What is the probability that exactly 27 will say yes? *(Source: Net Applications)*

a. *Determine* whether you can use a normal distribution to approximate the binomial variable.
b. Find the *mean* μ and the *standard deviation* σ for the distribution.
c. Apply a *continuity correction* to rewrite $P(x = 27)$ and sketch a graph.
d. *Find* the corresponding z-scores.
e. Use the Standard Normal Table to *find the area* to the left of each z-score and *calculate the probability*.

Answer: Page A39

```
binompdf(150,.62
,96)
       .0595828329
```

The approximation in Example 5 is almost exactly equal to the exact probability found using the binompdf(command on a TI-83/84 Plus.

5.5 EXERCISES

■ BUILDING BASIC SKILLS AND VOCABULARY

1. What are the properties of a binomial experiment?

2. What are the conditions for using a normal distribution to approximate a binomial distribution?

In Exercises 3–6, the sample size n, probability of success p, and probability of failure q are given for a binomial experiment. Decide whether you can use a normal distribution to approximate the random variable x.

3. $n = 24$, $p = 0.85$, $q = 0.15$

4. $n = 15$, $p = 0.70$, $q = 0.30$

5. $n = 18$, $p = 0.90$, $q = 0.10$

6. $n = 20$, $p = 0.65$, $q = 0.35$

Approximating a Binomial Distribution *In Exercises 7–12, a binomial experiment is given. Decide whether you can use a normal distribution to approximate the binomial distribution. If you can, find the mean and standard deviation. If you cannot, explain why.*

7. House Contract A survey of U.S. adults found that 85% read every word or at least enough to understand a contract for buying or selling a home before signing. You randomly select 10 adults and ask them if they read every word or at least enough to understand a contract for buying or selling a home before signing. *(Source: FindLaw.com)*

8. Organ Donors A survey of U.S. adults found that 63% would want their organs transplanted into a patient who needs them if they were killed in an accident. You randomly select 20 adults and ask them if they would want their organs transplanted into a patient who needs them if they were killed in an accident. *(Source: USA Today)*

9. Multivitamins A survey of U.S. adults found that 55% have used a multivitamin in the past 12 months. You randomly select 50 adults and ask them if they have used a multivitamin in the past 12 months. *(Source: Harris Interactive)*

10. Happiness at Work A survey of U.S. adults found that 19% are happy with their current employer. You randomly select 30 adults and ask them if they are happy with their current employer. *(Source: Opinion Research Corporation)*

11. Going Green A survey of U.S. adults found that 76% would pay more for an environmentally friendly product. You randomly select 20 adults and ask them if they would pay more for an environmentally friendly product. *(Source: Opinion Research Corporation)*

12. Online Habits A survey of U.S. adults found that 61% look online for health information. You randomly select 15 adults and ask them if they look online for health information. *(Source: Pew Research Center)*

In Exercises 13–16, use a continuity correction and match the binomial probability statement with the corresponding normal distribution statement.

Binomial Probability	Normal Probability
13. $P(x > 109)$	(a) $P(x > 109.5)$
14. $P(x \geq 109)$	(b) $P(x < 108.5)$
15. $P(x \leq 109)$	(c) $P(x \leq 109.5)$
16. $P(x < 109)$	(d) $P(x \geq 108.5)$

In Exercises 17–22, a binomial probability is given. Write the probability in words. Then, use a continuity correction to convert the binomial probability to a normal distribution probability.

17. $P(x < 25)$ **18.** $P(x \geq 110)$ **19.** $P(x = 33)$

20. $P(x > 65)$ **21.** $P(x \leq 150)$ **22.** $P(55 < x < 60)$

■ USING AND INTERPRETING CONCEPTS

Approximating Binomial Probabilities   *In Exercises 23–30, decide whether you can use a normal distribution to approximate the binomial distribution. If you can, use the normal distribution to approximate the indicated probabilities and sketch their graphs. If you cannot, explain why and use a binomial distribution to find the indicated probabilities.*

23. Internet Use   A survey of U.S. adults ages 18–29 found that 93% use the Internet. You randomly select 100 adults ages 18–29 and ask them if they use the Internet. *(Source: Pew Research Center)*

    (a) Find the probability that exactly 90 people say they use the Internet.

    (b) Find the probability that at least 90 people say they use the Internet.

    (c) Find the probability that fewer than 90 people say they use the Internet.

    (d) Are any of the probabilities in parts (a)–(c) unusual? Explain.

24. Internet Use   A survey of U.S. adults ages 50–64 found that 70% use the Internet. You randomly select 80 adults ages 50–64 and ask them if they use the Internet. *(Source: Pew Research Center)*

    (a) Find the probability that at least 70 people say they use the Internet.

    (b) Find the probability that exactly 50 people say they use the Internet.

    (c) Find the probability that more than 60 people say they use the Internet.

    (d) Are any of the probabilities in parts (a)–(c) unusual? Explain.

25. Favorite Sport   A survey of U.S. adults found that 35% say their favorite sport is professional football. You randomly select 150 adults and ask them if their favorite sport is professional football. *(Source: Harris Interactive)*

    (a) Find the probability that at most 75 people say their favorite sport is professional football.

    (b) Find the probability that more than 40 people say their favorite sport is professional football.

    (c) Find the probability that between 50 and 60 people, inclusive, say their favorite sport is professional football.

    (d) Are any of the probabilities in parts (a)–(c) unusual? Explain.

26. College Graduates   About 34% of workers in the United States are college graduates. You randomly select 50 workers and ask them if they are a college graduate. *(Source: U.S. Bureau of Labor Statistics)*

    (a) Find the probability that exactly 12 workers are college graduates.

    (b) Find the probability that more than 23 workers are college graduates.

    (c) Find the probability that at most 18 workers are college graduates.

    (d) A committee is looking for 30 working college graduates to volunteer at a career fair. The committee randomly selects 125 workers. What is the probability that there will not be enough college graduates?

27. Public Transportation   Five percent of workers in the United States use public transportation to get to work. You randomly select 250 workers and ask them if they use public transportation to get to work. *(Source: U.S. Census Bureau)*

(a) Find the probability that exactly 16 workers will say yes.

(b) Find the probability that at least 9 workers will say yes.

(c) Find the probability that fewer than 16 workers will say yes.

(d) A transit authority offers discount rates to companies that have at least 30 employees who use public transportation to get to work. There are 500 employees in a company. What is the probability that the company will not get the discount?

28. Concert Tickets   A survey of U.S. adults who attend at least one music concert a year found that 67% say concert tickets are too expensive. You randomly select 12 adults who attend at least one music concert a year and ask them if concert tickets are too expensive. *(Source: Rasmussen Reports)*

(a) Find the probability that fewer than 4 people say that concert tickets are too expensive.

(b) Find the probability that between 7 and 9 people, inclusive, say that concert tickets are too expensive.

(c) Find the probability that at most 10 people say that concert tickets are too expensive.

(d) Are any of the probabilities in parts (a)–(c) unusual? Explain.

29. News   A survey of U.S. adults ages 18–24 found that 34% get no news on an average day. You randomly select 200 adults ages 18–24 and ask them if they get news on an average day. *(Source: Pew Research Center)*

(a) Find the probability that at least 85 people say they get no news on an average day.

(b) Find the probability that fewer than 66 people say they get no news on an average day.

(c) Find the probability that exactly 68 people say they get no news on an average day.

(d) A college English teacher wants students to discuss current events. The teacher randomly selects six students from the class. What is the probability that none of the students can talk about current events because they get no news on an average day.

30. Long Work Weeks   A survey of U.S. workers found that 2.9% work more than 70 hours per week. You randomly select 10 workers in the United States and ask them if they work more than 70 hours per week.

(a) Find the probability that at most 3 people say they work more than 70 hours per week.

(b) Find the probability that at least 1 person says he or she works more than 70 hours per week.

(c) Find the probability that more than 2 people say they work more than 70 hours per week.

(d) A large company is concerned about overworked employees who work more than 70 hours per week. The company randomly selects 50 employees. What is the probability there will be no employee working more than 70 hours?

Graphical Analysis   *In Exercises 31 and 32, write the binomial probability and the normal probability for the shaded region of the graph. Find the value of each probability and compare the results.*

31.

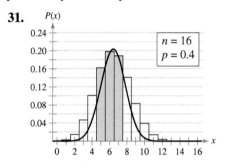

32.

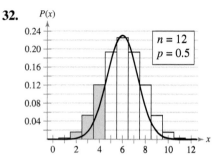

■ EXTENDING CONCEPTS

Getting Physical   *In Exercises 33 and 34, use the following information. The graph shows the results of a survey of adults in the United States ages 33 to 51 who were asked if they participated in a sport. Seventy percent of adults said they regularly participated in at least one sport, and they gave their favorite sport.*

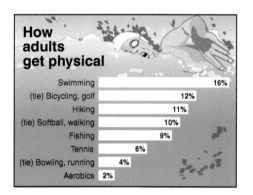

33. You randomly select 250 people in the United States ages 33 to 51 and ask them if they regularly participate in at least one sport. You find that 60% say no. How likely is this result? Do you think this sample is a good one? Explain your reasoning.

34. You randomly select 300 people in the United States ages 33 to 51 and ask them if they regularly participate in at least one sport. Of the 200 who say yes, 9% say they participate in hiking. How likely is this result? Do you think this sample is a good one? Explain your reasoning.

Testing a Drug   *In Exercises 35 and 36, use the following information. A drug manufacturer claims that a drug cures a rare skin disease 75% of the time. The claim is checked by testing the drug on 100 patients. If at least 70 patients are cured, this claim will be accepted.*

35. Find the probability that the claim will be rejected assuming that the manufacturer's claim is true.

36. Find the probability that the claim will be accepted assuming that the actual probability that the drug cures the skin disease is 65%.

USES AND ABUSES

Uses

Normal Distributions   Normal distributions can be used to describe many real-life situations and are widely used in the fields of science, business, and psychology. They are the most important probability distributions in statistics and can be used to approximate other distributions, such as discrete binomial distributions.

The most incredible application of the normal distribution lies in the Central Limit Theorem. This theorem states that no matter what type of distribution a population may have, as long as the sample size is at least 30, the distribution of sample means will be approximately normal. If the population is itself normal, then the distribution of sample means will be normal no matter how small the sample is.

The normal distribution is essential to sampling theory. Sampling theory forms the basis of statistical inference, which you will begin to study in the next chapter.

Abuses

Unusual Events   Suppose a population is normally distributed, with a mean of 100 and standard deviation of 15. It would not be unusual for an individual value taken from this population to be 115 or more. In fact, this will happen almost 16% of the time. It *would* be, however, highly unusual to take random samples of 100 values from that population and obtain a sample with a mean of 115 or more. Because the population is normally distributed, the mean of the sample distribution will be 100, and the standard deviation will be 1.5. A sample mean of 115 lies 10 standard deviations above the mean. This would be an extremely unusual event. When an event this unusual occurs, it is a good idea to question the original claimed value of the mean.

Although normal distributions are common in many populations, people try to make *non-normal* statistics fit a normal distribution. The statistics used for normal distributions are often inappropriate when the distribution is obviously non-normal.

■ EXERCISES

1. ***Is It Unusual?***   A population is normally distributed, with a mean of 100 and a standard deviation of 15. Determine if either of the following events is unusual. Explain your reasoning.

   a. The mean of a sample of 3 is 115 or more.

   b. The mean of a sample of 20 is 105 or more.

2. ***Find the Error***   The mean age of students at a high school is 16.5, with a standard deviation of 0.7. You use the Standard Normal Table to help you determine that the probability of selecting one student at random and finding his or her age to be more than 17.5 years is about 8%. What is the error in this problem?

3. Give an example of a distribution that might be non-normal.

5 CHAPTER SUMMARY

What did you **learn?**	EXAMPLE(S)	REVIEW EXERCISES
Section 5.1		
■ How to interpret graphs of normal probability distributions	*1, 2*	*1–6*
■ How to find areas under the standard normal curve	*3–6*	*7–28*
Section 5.2		
■ How to find probabilities for normally distributed variables	*1–3*	*29–38*
Section 5.3		
■ How to find a *z*-score given the area under the normal curve	*1, 2*	*39–46*
■ How to transform a *z*-score to an *x*-value $x = \mu + z\sigma$	*3*	*47, 48*
■ How to find a specific data value of a normal distribution given the probability	*4, 5*	*49–52*
Section 5.4		
■ How to find sampling distributions and verify their properties	*1*	*53, 54*
■ How to interpret the Central Limit Theorem $\mu_{\bar{x}} = \mu$ Mean $\sigma_{\bar{x}} = \dfrac{\sigma}{\sqrt{n}}$ Standard deviation	*2, 3*	*55, 56*
■ How to apply the Central Limit Theorem to find the probability of a sample mean	*4–6*	*57–62*
Section 5.5		
■ How to decide when a normal distribution can approximate a binomial distribution $\mu = np$ Mean $\sigma = \sqrt{npq}$ Standard deviation	*1*	*63, 64*
■ How to find the continuity correction	*2*	*65–68*
■ How to use a normal distribution to approximate binomial probabilities	*3–5*	*69, 70*

5 REVIEW EXERCISES

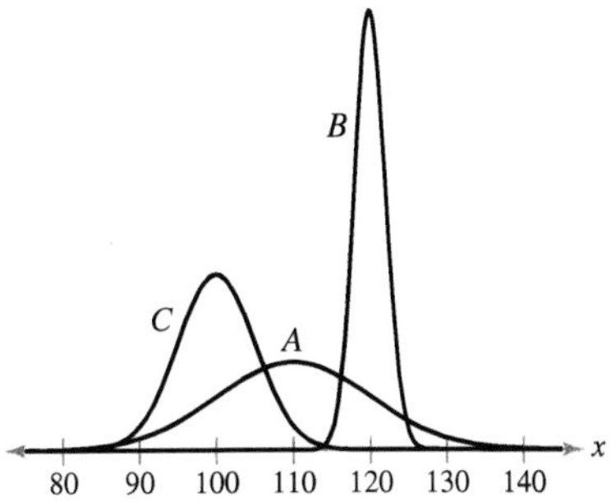

FIGURE FOR EXERCISES 3 AND 4

SECTION 5.1

In Exercises 1 and 2, use the graph to estimate μ and σ.

1.

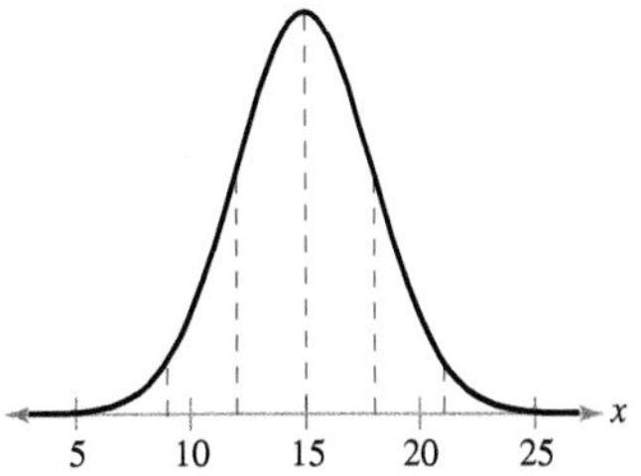

2.

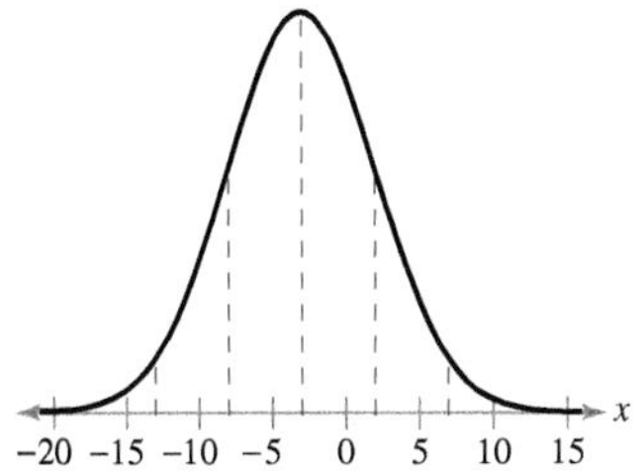

In Exercises 3 and 4, use the normal curves shown.

3. Which normal curve has the greatest mean? Explain your reasoning.

4. Which normal curve has the greatest standard deviation? Explain your reasoning.

In Exercises 5 and 6, use the following information and standard scores to investigate observations about a normal population. A batch of 2500 resistors is normally distributed, with a mean resistance of 1.5 ohms and a standard deviation of 0.08 ohm. Four resistors are randomly selected and tested. Their resistances are measured at 1.32, 1.54, 1.66, and 1.78 ohms.

5. How many standard deviations from the mean are these observations?

6. Are there any unusual observations?

In Exercises 7 and 8, find the area of the indicated region under the standard normal curve. If convenient, use technology to find the area.

7.

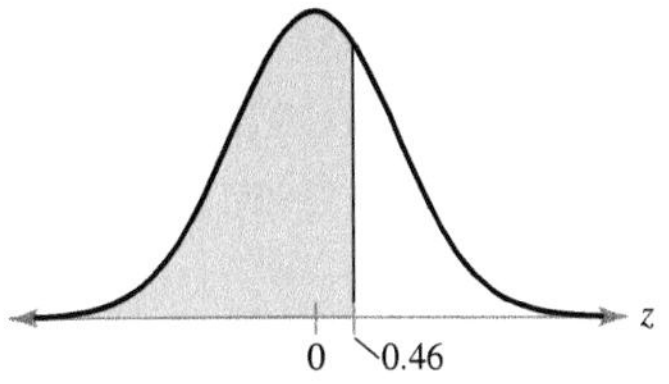

8.

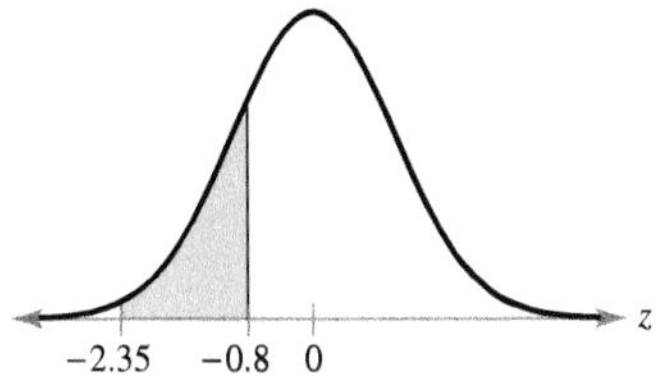

In Exercises 9–20, find the indicated area under the standard normal curve. If convenient, use technology to find the area.

9. To the left of $z = 0.33$

10. To the left of $z = -1.95$

11. To the right of $z = -0.57$

12. To the right of $z = 3.22$

13. To the left of $z = -2.825$

14. To the right of $z = 0.015$

15. Between $z = -1.64$ and the mean

16. Between $z = -1.55$ and $z = 1.04$

17. Between $z = 0.05$ and $z = 1.71$

18. Between $z = -2.68$ and $z = 2.68$

19. To the left of $z = -1.5$ and to the right of $z = 1.5$

20. To the left of $z = 0.64$ and to the right of $z = 3.415$

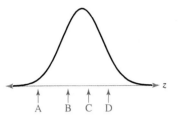

FIGURE FOR EXERCISES 21 AND 22

In Exercises 21 and 22, use the following information. In a recent year, the ACT scores for the reading portion of the test were normally distributed, with a mean of 21.4 and a standard deviation of 6.2. The test scores of four students selected at random are 17, 29, 8, and 23. *(Source: ACT, Inc.)*

21. Without converting to z-scores, match the values with the letters A, B, C, and D on the given graph.

22. Find the z-score that corresponds to each value and check your answers in Exercise 21. Are any of the values unusual? Explain.

In Exercises 23–28, find the indicated probabilities. If convenient, use technology to find the probability.

23. $P(z < 1.28)$

24. $P(z > -0.74)$

25. $P(-2.15 < z < 1.55)$

26. $P(0.42 < z < 3.15)$

27. $P(z < -2.50 \text{ or } z > 2.50)$

28. $P(z < 0 \text{ or } z > 1.68)$

▪ SECTION 5.2

In Exercises 29–34, assume the random variable x is normally distributed, with mean $\mu = 74$ and standard deviation $\sigma = 8$. Find the indicated probability.

29. $P(x < 84)$

30. $P(x < 55)$

31. $P(x > 80)$

32. $P(x > 71.6)$

33. $P(60 < x < 70)$

34. $P(72 < x < 82)$

In Exercises 35 and 36, find the indicated probabilities.

35. A study found that the mean migration distance of the green turtle was 2200 kilometers and the standard deviation was 625 kilometers. Assuming that the distances are normally distributed, find the probability that a randomly selected green turtle migrates a distance of

(a) less than 1900 kilometers.

(b) between 2000 kilometers and 2500 kilometers.

(c) greater than 2450 kilometers.

(Adapted from Dorling Kindersley Visual Encyclopedia)

36. The world's smallest mammal is the Kitti's hog-nosed bat, with a mean weight of 1.5 grams and a standard deviation of 0.25 gram. Assuming that the weights are normally distributed, find the probability of randomly selecting a bat that weighs

(a) between 1.0 gram and 2.0 grams.

(b) between 1.6 grams and 2.2 grams.

(c) more than 2.2 grams.

(Adapted from Dorling Kindersley Visual Encyclopedia)

37. Can any of the events in Exercise 35 be considered unusual? Explain your reasoning.

38. Can any of the events in Exercise 36 be considered unusual? Explain your reasoning.

▪ SECTION 5.3

In Exercises 39–44, use the Standard Normal Table to find the z-score that corresponds to the given cumulative area or percentile. If the area is not in the table, use the entry closest to the area. If convenient, use technology to find the z-score.

39. 0.4721

40. 0.1

41. 0.8708

42. P_2

43. P_{85}

44. P_{46}

45. Find the z-score that has 30.5% of the distribution's area to its right.

46. Find the z-score for which 94% of the distribution's area lies between $-z$ and z.

In Exercises 47–52, use the following information. On a dry surface, the braking distance (in meters) of a Cadillac Escalade can be approximated by a normal distribution, as shown in the graph at the left. (Adapted from Consumer Reports)

47. Find the braking distance of a Cadillac Escalade that corresponds to $z = -2.5$.

48. Find the braking distance of a Cadillac Escalade that corresponds to $z = 1.2$.

49. What braking distance of a Cadillac Escalade represents the 95th percentile?

50. What braking distance of a Cadillac Escalade represents the third quartile?

51. What is the shortest braking distance of a Cadillac Escalade that can be in the top 10% of braking distances?

52. What is the longest braking distance of a Cadillac Escalade that can be in the bottom 5% of braking distances?

▨ SECTION 5.4

In Exercises 53 and 54, use the given population to find the mean and standard deviation of the population and the mean and standard deviation of the sampling distribution. Compare the values.

53. A corporation has four executives. The number of minutes of overtime per week reported by each is 90, 120, 160, and 210. Draw three executives' names from this population, with replacement.

54. There are four residents sharing a house. The number of times each washes their car each month is 1, 2, 0, and 3. Draw two names from this population, with replacement.

In Exercises 55 and 56, use the Central Limit Theorem to find the mean and standard error of the mean of the indicated sampling distribution. Then sketch a graph of the sampling distribution.

55. The per capita consumption of citrus fruits by people in the United States in a recent year was normally distributed, with a mean of 76.0 pounds and a standard deviation of 20.5 pounds. Random samples of 35 people are drawn from this population and the mean of each sample is determined. *(Adapted from U.S. Department of Agriculture)*

56. The per capita consumption of red meat by people in the United States in a recent year was normally distributed, with a mean of 108.3 pounds and a standard deviation of 35.1 pounds. Random samples of 40 people are drawn from this population and the mean of each sample is determined. *(Adapted from U.S. Department of Agriculture)*

In Exercises 57–62, find the probabilities for the sampling distributions. Interpret the results.

57. Refer to Exercise 35. A sample of 12 green turtles is randomly selected. Find the probability that the sample mean of the distance migrated is (a) less than 1900 kilometers, (b) between 2000 kilometers and 2500 kilometers, and (c) greater than 2450 kilometers. Compare your answers with those in Exercise 35.

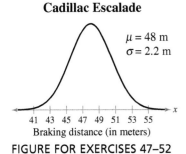

Braking Distance of a Cadillac Escalade

$\mu = 48$ m
$\sigma = 2.2$ m

Braking distance (in meters)

FIGURE FOR EXERCISES 47–52

58. Refer to Exercise 36. A sample of seven Kitti's hog-nosed bats is randomly selected. Find the probability that the sample mean is (a) between 1.0 gram and 2.0 grams, (b) between 1.6 grams and 2.2 grams, and (c) more than 2.2 grams. Compare your answers with those in Exercise 36.

59. The mean annual salary for chauffeurs is $29,200. A sample of 45 chauffeurs is randomly selected. What is the probability that the mean annual salary is (a) less than $29,000 and (b) more than $31,000? Assume $\sigma = \$1500$. *(Source: Salary.com)*

60. The mean value of land and buildings per acre for farms is $1300. A sample of 36 farms is randomly selected. What is the probability that the mean value of land and buildings per acre is (a) less than $1400 and (b) more than $1150? Assume $\sigma = \$250$.

61. The mean price of houses in a city is $1.5 million with a standard deviation of $500,000. The house prices are normally distributed. You randomly select 15 houses in this city. What is the probability that the mean price will be less than $1.125 million?

62. Mean rent in a city is $500 per month with a standard deviation of $30. The rents are normally distributed. You randomly select 15 apartments in this city. What is the probability that the mean rent will be more than $525?

▧ SECTION 5.5

In Exercises 63 and 64, a binomial experiment is given. Decide whether you can use a normal distribution to approximate the binomial distribution. If you can, find the mean and standard deviation. If you cannot, explain why.

63. In a recent year, the American Cancer Society said that the five-year survival rate for new cases of stage 1 kidney cancer is 96%. You randomly select 12 men who were new stage 1 kidney cancer cases this year and calculate the five-year survival rate of each. *(Source: American Cancer Society, Inc.)*

64. A survey indicates that 75% of U.S. adults who go to the theater at least once a month think movie tickets are too expensive. You randomly select 30 adults and ask them if they think movie tickets are too expensive. *(Source: Rasmussen Reports)*

In Exercises 65–68, write the binomial probability as a normal probability using the continuity correction.

65. $P(x \geq 25)$	**66.** $P(x \leq 36)$
67. $P(x = 45)$	**68.** $P(x = 50)$

In Exercises 69 and 70, decide whether you can use a normal distribution to approximate the binomial distribution. If you can, use the normal distribution to approximate the indicated probabilities and sketch their graphs. If you cannot, explain why and use a binomial distribution to find the indicated probabilities.

69. Seventy percent of children ages 12 to 17 keep at least part of their savings in a savings account. You randomly select 45 children and ask them if they keep at least part of their savings in a savings account. Find the probability that at most 20 children will say yes. *(Source: International Communications Research for Merrill Lynch)*

70. Thirty-one percent of people in the United States have type A⁺ blood. You randomly select 15 people in the United States and ask them if their blood type is A⁺. Find the probability that more than 8 adults say they have A⁺ blood. *(Source: American Association of Blood Banks)*

5 ⟩ CHAPTER QUIZ

Take this quiz as you would take a quiz in class. After you are done, check your work against the answers given in the back of the book.

1. Find each standard normal probability.
   (a) $P(z > -2.54)$
   (b) $P(z < 3.09)$
   (c) $P(-0.88 < z < 0.88)$
   (d) $P(z < -1.445 \text{ or } z > -0.715)$

2. Find each normal probability for the given parameters.
   (a) $\mu = 5.5, \sigma = 0.08, P(5.36 < x < 5.64)$
   (b) $\mu = -8.2, \sigma = 7.84, P(-5.00 < x < 0)$
   (c) $\mu = 18.5, \sigma = 9.25, P(x < 0 \text{ or } x > 37)$

In Exercises 3–10, use the following information. Students taking a standardized IQ test had a mean score of 100 with a standard deviation of 15. Assume that the scores are normally distributed. (Adapted from Audiblox)

3. Find the probability that a student had a score higher than 125. Is this an unusual event? Explain.

4. Find the probability that a student had a score between 95 and 105. Is this an unusual event? Explain.

5. What percent of the students had an IQ score that is greater than 112?

6. If 2000 students are randomly selected, how many would be expected to have an IQ score that is less than 90?

7. What is the lowest score that would still place a student in the top 5% of the scores?

8. What is the highest score that would still place a student in the bottom 10% of the scores?

9. A random sample of 60 students is drawn from this population. What is the probability that the mean IQ score is greater than 105? Interpret your result.

10. Are you more likely to randomly select one student with an IQ score greater than 105 or are you more likely to randomly select a sample of 15 students with a mean IQ score greater than 105? Explain.

In Exercises 11 and 12, use the following information. In a survey of adults under age 65, 81% say they are concerned about the amount and security of personal online data that can be accessed by cybercriminals and hackers. You randomly select 35 adults and ask them if they are concerned about the amount and security of personal online data that can be accessed by cybercriminals and hackers. (Source: Financial Times/Harris Poll)

11. Decide whether you can use a normal distribution to approximate the binomial distribution. If you can, find the mean and standard deviation. If you cannot, explain why.

12. Find the probability that at most 20 adults say they are concerned about the amount and security of personal online data that can be accessed by cybercriminals and hackers. Interpret the result.

PUTTING IT ALL TOGETHER

Real Statistics — Real Decisions

You are the human resources director for a corporation and want to implement a health improvement program for employees to decrease employee medical absences. You perform a six-month study with a random sample of employees. Your goal is to decrease absences by 50%. (Assume all data are normally distributed.)

■ EXERCISES

1. *Preliminary Thoughts*

You got the idea for this health improvement program from a national survey in which 75% of people who responded said they would participate in such a program if offered by their employer. You randomly select 60 employees and ask them whether they would participate in such a program.

(a) Find the probability that exactly 35 will say yes.

(b) Find the probability that at least 40 will say yes.

(c) Find the probability that fewer than 20 will say yes.

(d) Based on the results in parts (a)–(c), explain why you chose to perform the study.

2. *Before the Program*

Before the study, the mean number of absences during a six-month period of the participants was 6, with a standard deviation of 1.5. An employee is randomly selected.

(a) Find the probability that the employee's number of absences is less than 5.

(b) Find the probability that the employee's number of absences is between 5 and 7.

(c) Find the probability that the employee's number of absences is more than 7.

3. *After the Program*

The graph at the right represents the results of the study.

(a) What is the mean number of absences for employees? Explain how you know.

(b) Based on the results, was the goal of decreasing absences by 50% reached?

(c) Describe how you would present your results to the board of directors of the corporation.

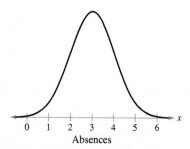

Absences

FIGURE FOR EXERCISE 3

TECHNOLOGY

 U.S. Census Bureau

www.census.gov

AGE DISTRIBUTION IN THE UNITED STATES

One of the jobs of the U.S. Census Bureau is to keep track of the age distribution in the country. The age distribution in 2009 is shown below.

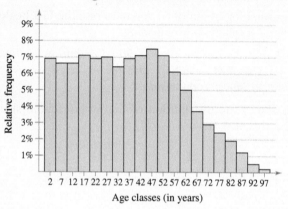

Age Distribution in the U.S.

Class	Class midpoint	Relative frequency
0–4	2	6.9%
5–9	7	6.6%
10–14	12	6.6%
15–19	17	7.1%
20–24	22	6.9%
25–29	27	7.0%
30–34	32	6.4%
35–39	37	6.9%
40–44	42	7.1%
45–49	47	7.5%
50–54	52	7.1%
55–59	57	6.1%
60–64	62	5.0%
65–69	67	3.7%
70–74	72	2.9%
75–79	77	2.4%
80–84	82	1.9%
85–89	87	1.2%
90–94	92	0.5%
95–99	97	0.2%

■ EXERCISES

We used a technology tool to select random samples with $n = 40$ from the age distribution of the United States. The means of the 36 samples were as follows.

 28.14, 31.56, 36.86, 32.37, 36.12, 39.53,
36.19, 39.02, 35.62, 36.30, 34.38, 32.98,
36.41, 30.24, 34.19, 44.72, 38.84, 42.87,
38.90, 34.71, 34.13, 38.25, 38.04, 34.07,
39.74, 40.91, 42.63, 35.29, 35.91, 34.36,
36.51, 36.47, 32.88, 37.33, 31.27, 35.80

1. Enter the age distribution of the United States into a technology tool. Use the tool to find the mean age in the United States.

2. Enter the set of sample means into a technology tool. Find the mean of the set of sample means. How does it compare with the mean age in the United States? Does this agree with the result predicted by the Central Limit Theorem?

3. Are the ages of people in the United States normally distributed? Explain your reasoning.

4. Sketch a relative frequency histogram for the 36 sample means. Use nine classes. Is the histogram approximately bell-shaped and symmetric? Does this agree with the result predicted by the Central Limit Theorem?

5. Use a technology tool to find the standard deviation of the ages of people in the United States.

6. Use a technology tool to find the standard deviation of the set of 36 sample means. How does it compare with the standard deviation of the ages? Does this agree with the result predicted by the Central Limit Theorem?

Extended solutions are given in the *Technology Supplement*.
Technical instruction is provided for MINITAB, Excel, and the TI-83/84 Plus.

CUMULATIVE REVIEW
Chapters 3 – 5

1. A survey of voters in the United States found that 15% rate the U.S. health care system as excellent. You randomly select 50 voters and ask them how they rate the U.S. health care system. *(Source: Rasmussen Reports)*

   (a) Verify that the normal distribution can be used to approximate the binomial distribution.

   (b) Find the probability that at most 14 voters rate the U.S. health care system as excellent.

   (c) Is it unusual for 14 out of 50 voters to rate the U.S. health care system as excellent? Explain your reasoning.

In Exercises 2 and 3, use the probability distribution to find the (a) mean, (b) variance, (c) standard deviation, and (d) expected value of the probability distribution, and (e) interpret the results.

2. The table shows the distribution of family household sizes in the United States for a recent year. *(Source: U.S. Census Bureau)*

x	2	3	4	5	6	7
$P(x)$	0.427	0.227	0.200	0.093	0.034	0.018

3. The table shows the distribution of fouls per game for a player in a recent NBA season. *(Source: NBA.com)*

x	0	1	2	3	4	5	6
$P(x)$	0.012	0.049	0.159	0.256	0.244	0.195	0.085

4. Use the probability distribution in Exercise 3 to find the probability of randomly selecting a game in which the player had (a) fewer than four fouls, (b) at least three fouls, and (c) between two and four fouls, inclusive.

5. From a pool of 16 candidates, 9 men and 7 women, the offices of president, vice president, secretary, and treasurer will be filled. (a) In how many different ways can the offices be filled? (b) What is the probability that all four of the offices are filled by women?

In Exercises 6–11, use the Standard Normal Table to find the indicated area under the standard normal curve.

6. To the left of $z = 0.72$

7. To the left of $z = -3.08$

8. To the right of $z = -0.84$

9. Between $z = 0$ and $z = 2.95$

10. Between $z = -1.22$ and $z = -0.26$

11. To the left of $z = 0.12$ or to the right of $z = 1.72$

12. Forty-five percent of adults say they are interested in regularly measuring their carbon footprint. You randomly select 11 adults and ask them if they are interested in regularly measuring their carbon footprint. Find the probability that the number of adults who say they are interested is (a) exactly eight, (b) at least five, and (c) less than two. Are any of these events unusual? Explain your reasoning. *(Source: Sacred Heart University Polling)*

13. An auto parts seller finds that 1 in every 200 parts sold is defective. Use the geometric distribution to find the probability that (a) the first defective part is the tenth part sold, (b) the first defective part is the first, second, or third part sold, and (c) none of the first 10 parts sold are defective.

14. The table shows the results of a survey in which 2,944,100 public and 401,900 private school teachers were asked about their full-time teaching experience. *(Adapted from U.S. National Center for Education Statistics)*

	Public	Private	Total
Less than 3 years	177,300	27,600	204,900
3 to 9 years	995,800	154,500	1,150,300
10 to 20 years	906,300	111,600	1,017,900
20 years or more	864,700	108,200	972,900
Total	2,944,100	401,900	3,346,000

(a) Find the probability that a randomly selected private school teacher has 10 to 20 years of full-time teaching experience.

(b) Given that a randomly selected teacher has 3 to 9 years of full-time experience, find the probability that the teacher is at a public school.

(c) Are the events "being a public school teacher" and "having 20 years or more of full-time teaching experience" independent? Explain.

(d) Find the probability that a randomly selected teacher is either at a public school or has less than 3 years of full-time teaching experience.

(e) Find the probability that a randomly selected teacher has 3 to 9 years of full-time teaching experience or is at a private school.

15. The initial pressures for bicycle tires when first filled are normally distributed, with a mean of 70 pounds per square inch (psi) and a standard deviation of 1.2 psi.

(a) Random samples of size 40 are drawn from this population and the mean of each sample is determined. Use the Central Limit Theorem to find the mean and standard error of the mean of the sampling distribution. Then sketch a graph of the sampling distribution of sample means.

(b) A random sample of 15 tires is drawn from this population. What is the probability that the mean tire pressure of the sample $\bar{x}$ is less than 69 psi?

16. The life spans of car batteries are normally distributed, with a mean of 44 months and a standard deviation of 5 months.

(a) A car battery is selected at random. Find the probability that the life span of the battery is less than 36 months.

(b) A car battery is selected at random. Find the probability that the life span of the battery is between 42 and 60 months.

(c) What is the shortest life expectancy a car battery can have and still be in the top 5% of life expectancies?

17. A florist has 12 different flowers from which floral arrangements can be made. (a) If a centerpiece is to be made using four different flowers, how many different centerpieces can be made? (b) What is the probability that the four flowers in the centerpiece are roses, gerbers, hydrangeas, and callas?

18. About fifty percent of adults say they feel vulnerable to identity theft. You randomly select 16 adults and ask them if they feel vulnerable to identity theft. Find the probability that the number who say they feel vulnerable is (a) exactly 12, (b) no more than 6, and (c) more than 7. Are any of these events unusual? Explain your reasoning. *(Adapted from KRC Research for Fellowes)*

6 CONFIDENCE INTERVALS

CHAPTER

6.1 Confidence Intervals for the Mean (Large Samples)

■ CASE STUDY

6.2 Confidence Intervals for the Mean (Small Samples)

■ ACTIVITY

6.3 Confidence Intervals for Population Proportions

■ ACTIVITY

6.4 Confidence Intervals for Variance and Standard Deviation

■ USES AND ABUSES

■ REAL STATISTICS– REAL DECISIONS

■ TECHNOLOGY

David Wechsler was one of the most influential psychologists of the 20th century. He is known for developing intelligence tests, such as the Wechsler Adult Intelligence Scale and the Wechsler Intelligence Scale for Children.

In Chapters 1 through 5, you studied descriptive statistics (how to collect and describe data) and probability (how to find probabilities and analyze discrete and continuous probability distributions). For instance, psychologists use descriptive statistics to analyze the data collected during experiments and trials.

One of the most commonly administered psychological tests is the Wechsler Adult Intelligence Scale. It is an intelligence quotient (IQ) test that is standardized to have a normal distribution with a mean of 100 and a standard deviation of 15.

In this chapter, you will begin your study of inferential statistics—the second major branch of statistics. For instance, a chess club wants to estimate the mean IQ of its members. The mean of a random sample of members is 115. Because this estimate consists of a single number represented by a point on a number line, it is called a point estimate. The problem with using a point estimate is that it is rarely equal to the exact parameter (mean, standard deviation, or proportion) of the population.

In this chapter, you will learn how to make a more meaningful estimate by specifying an interval of values on a number line, together with a statement of how confident you are that your interval contains the population parameter. Suppose the club wants to be 90% confident of its estimate for the mean IQ of its members. Here is an overview of how to construct an interval estimate.

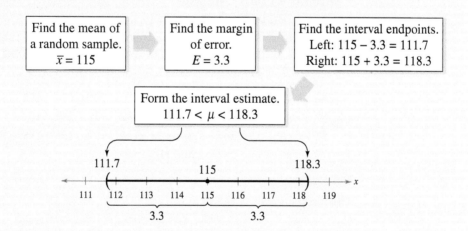

So, the club can be 90% confident that the mean IQ of its members is between 111.7 and 118.3.

6.1 Confidence Intervals for the Mean (Large Samples)

WHAT YOU SHOULD LEARN

▸ How to find a point estimate and a margin of error

▸ How to construct and interpret confidence intervals for the population mean

▸ How to determine the minimum sample size required when estimating μ

Estimating Population Parameters ▸ Confidence Intervals for the Population Mean ▸ Sample Size

▸ ESTIMATING POPULATION PARAMETERS

In this chapter, you will learn an important technique of statistical inference—to use sample statistics to estimate the value of an unknown population parameter. In this section, you will learn how to use sample statistics to make an estimate of the population parameter μ when the sample size is at least 30 or when the population is normally distributed and the standard deviation σ is known. To make such an inference, begin by finding a *point estimate*.

DEFINITION

A **point estimate** is a single value estimate for a population parameter. The most unbiased point estimate of the population mean μ is the sample mean $\bar{x}$.

The validity of an estimation method is increased if a sample statistic is unbiased and has low variability. A statistic is unbiased if it does not overestimate or underestimate the population parameter. In Chapter 5, you learned that the mean of all possible sample means of the same size equals the population mean. As a result, $\bar{x}$ is an unbiased estimator of μ. When the standard error $\sigma/\sqrt{n}$ of a sample mean is decreased by increasing n, it becomes less variable.

EXAMPLE 1

▸ **Finding a Point Estimate**

A social networking website allows its users to add friends, send messages, and update their personal profiles. The following represents a random sample of the number of friends for 40 users of the website. Find a point estimate of the population mean μ. *(Adapted from Facebook)*

140	105	130	97	80	165	232	110	214	201	122
98	65	88	154	133	121	82	130	211	153	114
58	77	51	247	236	109	126	132	125	149	122
74	59	218	192	90	117	105				

▸ **Solution**

The sample mean of the data is

$$\bar{x} = \frac{\Sigma x}{n} = \frac{5232}{40} = 130.8.$$

So, the point estimate for the mean number of friends for all users of the website is 130.8 friends.

▸ **Try It Yourself 1**

Another random sample of the number of friends for 30 users of the website is shown at the left. Use this sample to find another point estimate for μ.

a. *Find* the sample mean.
b. *Estimate* the mean number of friends of the population. *Answer: Page A39*

Sample Data

\multicolumn{6}{c}{**Number of Friends**}
162	114	131	87	108	63
249	135	172	196	127	100
146	214	80	55	71	130
95	156	201	227	137	125
145	179	74	215	137	124

In Example 1, the probability that the population mean is exactly 130.8 is virtually zero. So, instead of estimating μ to be exactly 130.8 using a point estimate, you can estimate that μ lies in an interval. This is called making an *interval estimate*.

DEFINITION

An **interval estimate** is an interval, or range of values, used to estimate a population parameter.

Although you can assume that the point estimate in Example 1 is not equal to the actual population mean, it is probably close to it. To form an interval estimate, use the point estimate as the center of the interval, then add and subtract a margin of error. For instance, if the margin of error is 15.7, then an interval estimate would be given by 130.8 ± 15.7 or $115.1 < \mu < 146.5$. The point estimate and interval estimate are as follows.

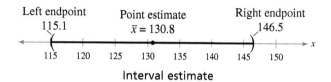

Left endpoint
115.1

Point estimate
$\bar{x} = 130.8$

Right endpoint
146.5

Interval estimate

Before finding a margin of error for an interval estimate, you should first determine how confident you need to be that your interval estimate contains the population mean μ.

DEFINITION

The **level of confidence** c is the probability that the interval estimate contains the population parameter.

STUDY TIP

In this course, you will usually use 90%, 95%, and 99% levels of confidence. The following z-scores correspond to these levels of confidence.

Level of Confidence	z_c
90%	1.645
95%	1.96
99%	2.575

You know from the Central Limit Theorem that when $n \geq 30$, the sampling distribution of sample means is a normal distribution. The level of confidence c is the area under the standard normal curve between the *critical values*, $-z_c$ and z_c. **Critical values** are values that separate sample statistics that are probable from sample statistics that are improbable, or unusual. You can see from the graph that c is the percent of the area under the normal curve between $-z_c$ and z_c. The area remaining is $1 - c$, so the area in each tail is $\frac{1}{2}(1 - c)$. For instance, if $c = 90\%$, then 5% of the area lies to the left of $-z_c = -1.645$ and 5% lies to the right of $z_c = 1.645$.

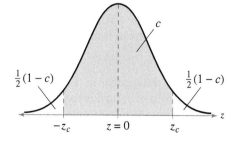

$\frac{1}{2}(1-c)$

c

$\frac{1}{2}(1-c)$

$-z_c$ $z = 0$ z_c z

If $c = 90\%$:	
$c = 0.90$	Area in blue region
$1 - c = 0.10$	Area in yellow regions
$\frac{1}{2}(1 - c) = 0.05$	Area in each tail
$-z_c = -1.645$	Critical value separating left tail
$z_c = 1.645$	Critical value separating right tail

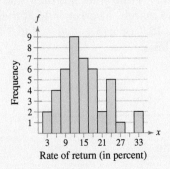

f

Frequency

Rate of return (in percent)

For a 95% confidence interval, what would be the margin of error for the population mean rate of return?

STUDY TIP

Remember that you can calculate the sample standard deviation s using the formula

$$s = \sqrt{\frac{\Sigma(x - \bar{x})^2}{n - 1}}$$

or the shortcut formula

$$s = \sqrt{\frac{\Sigma x^2 - (\Sigma x)^2/n}{n - 1}}.$$

However, the most convenient way to find the sample standard deviation is to use the *1-Var Stats* feature of a graphing calculator.

The difference between the point estimate and the actual parameter value is called the **sampling error.** When μ is estimated, the sampling error is the difference $\bar{x} - \mu$. In most cases, of course, μ is unknown, and $\bar{x}$ varies from sample to sample. However, you can calculate a maximum value for the error if you know the level of confidence and the sampling distribution.

DEFINITION

Given a level of confidence c, the **margin of error E** (sometimes also called the maximum error of estimate or error tolerance) is the greatest possible distance between the point estimate and the value of the parameter it is estimating.

$$E = z_c \sigma_{\bar{x}} = z_c \frac{\sigma}{\sqrt{n}}$$

In order to use this technique, it is assumed that the population standard deviation is known. This is rarely the case, but when $n \geq 30$, the sample standard deviation s can be used in place of σ.

EXAMPLE 2

▶ **Finding the Margin of Error**

Use the data given in Example 1 and a 95% confidence level to find the margin of error for the mean number of friends for all users of the website. Assume that the sample standard deviation is about 53.0.

▶ **Solution**

The z-score that corresponds to a 95% confidence level is 1.96. This implies that 95% of the area under the standard normal curve falls within 1.96 standard deviations of the mean. (You can approximate the distribution of the sample means with a normal curve by the Central Limit Theorem because $n = 40 \geq 30$.) You don't know the population standard deviation σ. But because $n \geq 30$, you can use s in place of σ.

Using the values $z_c = 1.96$, $\sigma \approx s \approx 53.0$, and $n = 40$,

$$E = z_c \frac{\sigma}{\sqrt{n}}$$

$$\approx 1.96 \cdot \frac{53.0}{\sqrt{40}}$$

$$\approx 16.4.$$

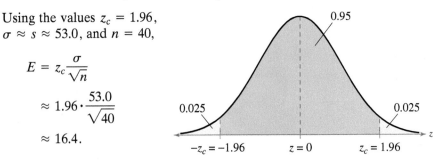

Interpretation You are 95% confident that the margin of error for the population mean is about 16.4 friends.

▶ **Try It Yourself 2**

Use the data given in Try It Yourself 1 and a 95% confidence level to find the margin of error for the mean number of friends for all users of the website.

a. *Identify z_c, n, and s.*
b. *Find E using z_c, $\sigma \approx s$, and n.*
c. *Interpret the results.*

Answer: Page A39

▸ CONFIDENCE INTERVALS FOR THE POPULATION MEAN

Using a point estimate and a margin of error, you can construct an interval estimate of a population parameter such as μ. This interval estimate is called a *confidence interval*.

DEFINITION

A *c*-**confidence interval for the population mean** μ is

$$\overline{x} - E < \mu < \overline{x} + E.$$

The probability that the confidence interval contains μ is c.

GUIDELINES

Finding a Confidence Interval for a Population Mean ($n \geq 30$ or σ known with a normally distributed population)

IN WORDS	IN SYMBOLS
1. Find the sample statistics n and $\overline{x}$.	$\overline{x} = \dfrac{\sum x}{n}$
2. Specify σ, if known. Otherwise, if $n \geq 30$, find the sample standard deviation s and use it as an estimate for σ.	$s = \sqrt{\dfrac{\sum (x - \overline{x})^2}{n - 1}}$
3. Find the critical value z_c that corresponds to the given level of confidence.	Use the Standard Normal Table or technology.
4. Find the margin of error E.	$E = z_c \dfrac{\sigma}{\sqrt{n}}$
5. Find the left and right endpoints and form the confidence interval.	Left endpoint: $\overline{x} - E$ Right endpoint: $\overline{x} + E$ Interval: $\overline{x} - E < \mu < \overline{x} + E$

EXAMPLE 3   **SC** Report 23

See MINITAB steps on page 352.

▸ **Constructing a Confidence Interval**

Use the data given in Example 1 to construct a 95% confidence interval for the mean number of friends for all users of the website.

▸ **Solution**   In Examples 1 and 2, you found that $\overline{x} = 130.8$ and $E \approx 16.4$. The confidence interval is as follows.

Left Endpoint   Right Endpoint

$$\overline{x} - E \approx 130.8 - 16.4 = 114.4 \qquad \overline{x} + E \approx 130.8 + 16.4 = 147.2$$

$$114.4 < \mu < 147.2$$

```
        114.4          130.8        147.2
    ←+—+—(—+—+—+—•—+—+—+—+—)—+—+→ x
     110 115 120 125 130 135 140 145 150
```

Interpretation   With 95% confidence, you can say that the population mean number of friends is between 114.4 and 147.2.

Answer: Page A39

INSIGHT

The width of a confidence interval is 2E. Examine the formula for E to see why a larger sample size tends to give you a narrower confidence interval for the same level of confidence.

STUDY TIP

Using a TI-83/84 Plus, you can either enter the original data into a list to construct the confidence interval or enter the descriptive statistics.

[STAT]

Choose the TESTS menu.

   7: ZInterval…

Select the *Data* input option if you use the original data. Select the *Stats* input option if you use the descriptive statistics. In each case, enter the appropriate values, then select *Calculate*. Your results may differ slightly depending on the method you use. For Example 4, the original data values were entered.

```
ZInterval
 (109.21,152.39)
 x̄=130.8
 Sx=52.63234844
 n=40
```

▶ **Try It Yourself 3**

Use the data given in Try It Yourself 1 to construct a 95% confidence interval for the mean number of friends for all users of the website. Compare your result with the interval found in Example 3.

a. *Find* $\bar{x}$ and *E*.
b. Find the *left* and *right endpoints* of the confidence interval.
c. *Interpret* the results and compare them with Example 3.   *Answer: Page A39*

EXAMPLE 4

▶ **Constructing a Confidence Interval Using Technology**

Use a technology tool to construct a 99% confidence interval for the mean number of friends for all users of the website using the sample in Example 1.

▶ **Solution**

To use a technology tool to solve the problem, enter the data and recall that the sample standard deviation is $s \approx 53.0$. Then, use the confidence interval command to calculate the confidence interval (*1-Sample Z* for MINITAB). The display should look like the one shown below. To construct a confidence interval using a TI-83/84 Plus, follow the instructions in the margin.

MINITAB

One-Sample Z: Friends

The assumed standard deviation = 53

Variable	N	Mean	StDev	SE Mean	99% CI
Friends	40	130.80	52.63	8.38	(109.21, 152.39)

So, a 99% confidence interval for μ is $(109.2, 152.4)$.

Interpretation   With 99% confidence, you can say that the population mean number of friends is between 109.2 and 152.4.

▶ **Try It Yourself 4**

Use the sample data in Example 1 and a technology tool to construct 75%, 85%, and 99% confidence intervals for the mean number of friends for all users of the website. How does the width of the confidence interval change as the level of confidence increases?

a. *Enter* the data.
b. *Use* the appropriate command to construct each confidence interval.
c. *Compare* the widths of the confidence intervals for $c = 0.75, 0.85,$ and 0.99.
    Answer: Page A39

In Example 4 and Try It Yourself 4, the same sample data were used to construct confidence intervals with different levels of confidence. Notice that as the level of confidence increases, the width of the confidence interval also increases. In other words, when the same sample data are used, *the greater the level of confidence, the wider the interval.*

If the population is normally distributed and the population standard deviation σ is known, you may use the normal sampling distribution for any sample size, as shown in Example 5.

See TI-83/84 Plus steps on page 353.

EXAMPLE 5

▶ **Constructing a Confidence Interval, σ Known**

A college admissions director wishes to estimate the mean age of all students currently enrolled. In a random sample of 20 students, the mean age is found to be 22.9 years. From past studies, the standard deviation is known to be 1.5 years, and the population is normally distributed. Construct a 90% confidence interval of the population mean age.

▶ **Solution**

Using $n = 20$, $\bar{x} = 22.9$, $\sigma = 1.5$, and $z_c = 1.645$, the margin of error at the 90% confidence level is

$$E = z_c \frac{\sigma}{\sqrt{n}}$$

$$= 1.645 \cdot \frac{1.5}{\sqrt{20}} \approx 0.6.$$

The 90% confidence interval can be written as $\bar{x} \pm E \approx 22.9 \pm 0.6$ or as follows.

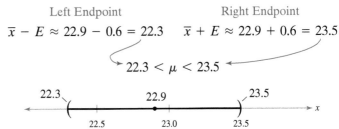

Left Endpoint   Right Endpoint

$\bar{x} - E \approx 22.9 - 0.6 = 22.3$ $\bar{x} + E \approx 22.9 + 0.6 = 23.5$

$22.3 < \mu < 23.5$

Interpretation   With 90% confidence, you can say that the mean age of all the students is between 22.3 and 23.5 years.

▶ **Try It Yourself 5**

Construct a 90% confidence interval of the population mean age for the college students in Example 5 with the sample size increased to 30 students. Compare your answer with Example 5.

a. *Identify* n, $\bar{x}$, σ, and z_c, and *find* E.
b. *Find* the *left* and *right endpoints* of the confidence interval.
c. *Interpret* the results and compare them with Example 5.   *Answer: Page A39*

After constructing a confidence interval, it is important that you interpret the results correctly. Consider the 90% confidence interval constructed in Example 5. Because μ is a fixed value predetermined by the population, it is either in the interval or not. It is *not* correct to say "There is a 90% probability that the actual mean will be in the interval $(22.3, 23.5)$." This statement is wrong because it suggests that the value of μ can vary, which is not true. The correct way to interpret your confidence interval is "If a large number of samples is collected and a confidence interval is created for each sample, approximately 90% of these intervals will contain μ."

STUDY TIP

Here are instructions for constructing a confidence interval in Excel. First, click *Insert* at the top of the screen and select *Function*. Select the category *Statistical* and select the *Confidence* function. In the dialog box, enter the values of alpha, the standard deviation, and the sample size. Then click OK. The value returned is the margin of error, which is used to construct the confidence interval.

	A	B
1	=CONFIDENCE(0.1,1.5,20)	
2		0.55170068

Alpha is the *level of significance*, which will be explained in Chapter 7. When using Excel in Chapter 6, you can think of alpha as the complement of the level of confidence. So, for a 90% confidence interval, alpha is equal to $1 - 0.90 = 0.10$.

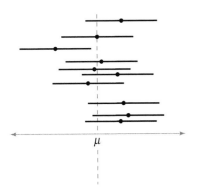

The horizontal segments represent 90% confidence intervals for different samples of the same size. In the long run, 9 of every 10 such intervals will contain μ.

▶ SAMPLE SIZE

For the same sample statistics, as the level of confidence increases, the confidence interval widens. As the confidence interval widens, the precision of the estimate decreases. One way to improve the precision of an estimate without decreasing the level of confidence is to increase the sample size. But how large a sample size is needed to guarantee a certain level of confidence for a given margin of error?

FIND A MINIMUM SAMPLE SIZE TO ESTIMATE μ

Given a c-confidence level and a margin of error E, the minimum sample size n needed to estimate the population mean μ is

$$n = \left(\frac{z_c\sigma}{E}\right)^2.$$

If σ is unknown, you can estimate it using s, provided you have a preliminary sample with at least 30 members.

EXAMPLE 6

▶ **Determining a Minimum Sample Size**

You want to estimate the mean number of friends for all users of the website. How many users must be included in the sample if you want to be 95% confident that the sample mean is within seven friends of the population mean?

▶ **Solution**

Using $c = 0.95$, $z_c = 1.96$, $\sigma \approx s \approx 53.0$ (from Example 2), and $E = 7$, you can solve for the minimum sample size n.

$$n = \left(\frac{z_c\sigma}{E}\right)^2$$
$$\approx \left(\frac{1.96 \cdot 53.0}{7}\right)^2$$
$$\approx 220.23$$

When necessary, round up to obtain a whole number. So, you should include at least 221 users in your sample.

Interpretation   You already have 40, so you need 181 more. Note that 221 is the *minimum* number of users to include in the sample. You could include more, if desired.

▶ **Try It Yourself 6**

How many users must be included in the sample if you want to be 95% confident that the sample mean is within 10 users of the population mean? Compare your answer with Example 6.

a. *Identify* z_c, E, and s.
b. *Use* z_c, E, and $\sigma \approx s$ to find the *minimum sample size n*.
c. *Interpret* the results and compare them with Example 6.

Answer: Page A40

6.1 EXERCISES

FOR EXTRA HELP:
MyStatLab

■ BUILDING BASIC SKILLS AND VOCABULARY

1. When estimating a population mean, are you more likely to be correct if you use a point estimate or an interval estimate? Explain your reasoning.

2. A news reporter reports the results of a survey and states that 45% of those surveyed responded "yes" with a margin of error of "plus or minus 5%." Explain what this means.

3. Given the same sample statistics, which level of confidence would produce the widest confidence interval? Explain your reasoning.

(a) 90%   (b) 95%   (c) 98%   (d) 99%

4. You construct a 95% confidence interval for a population mean using a random sample. The confidence interval is $24.9 < \mu < 31.5$. Is the probability that μ is in this interval 0.95? Explain.

In Exercises 5–8, find the critical value z_c necessary to construct a confidence interval at the given level of confidence.

5. $c = 0.80$ **6.** $c = 0.85$ **7.** $c = 0.75$ **8.** $c = 0.97$

Graphical Analysis   *In Exercises 9–12, use the values on the number line to find the sampling error.*

9. $\bar{x} = 3.8$ $\mu = 4.27$
        3.4 3.6 3.8 4.0 4.2 4.4 4.6

10. $\mu = 8.76$ $\bar{x} = 9.5$
        8.6 8.8 9.0 9.2 9.4 9.6 9.8

11. $\mu = 24.67$ $\bar{x} = 26.43$
        24   25   26   27

12. $\bar{x} = 46.56$ $\mu = 48.12$
        46   47   48   49

In Exercises 13–16, find the margin of error for the given values of c, s, and n.

13. $c = 0.95, s = 5.2, n = 30$ **14.** $c = 0.90, s = 2.9, n = 50$

15. $c = 0.80, s = 1.3, n = 75$ **16.** $c = 0.975, s = 4.6, n = 100$

Matching   *In Exercises 17–20, match the level of confidence c with its representation on the number line, given $\bar{x} = 57.2$, $s = 7.1$, and $n = 50$.*

17. $c = 0.88$ **18.** $c = 0.90$ **19.** $c = 0.95$ **20.** $c = 0.98$

(a)   54.9   57.2   59.5
        54 55 56 57 58 59 60

(b)   55.2   57.2   59.2
        54 55 56 57 58 59 60

(c)   55.6   57.2   58.8
        54 55 56 57 58 59 60

(d)   55.5   57.2   58.9
        54 55 56 57 58 59 60

In Exercises 21–24, construct the indicated confidence interval for the population mean μ. If convenient, use technology to construct the confidence interval.

21. $c = 0.90, \bar{x} = 12.3, s = 1.5, n = 50$

22. $c = 0.95, \bar{x} = 31.39, s = 0.8, n = 82$

23. $c = 0.99, \bar{x} = 10.5, s = 2.14, n = 45$

24. $c = 0.80, \bar{x} = 20.6, s = 4.7, n = 100$

In Exercises 25–28, use the given confidence interval to find the margin of error and the sample mean.

25. $(12.0, 14.8)$

26. $(21.61, 30.15)$

27. $(1.71, 2.05)$

28. $(3.144, 3.176)$

In Exercises 29–32, determine the minimum sample size n needed to estimate μ for the given values of c, s, and E.

29. $c = 0.90, s = 6.8, E = 1$

30. $c = 0.95, s = 2.5, E = 1$

31. $c = 0.80, s = 4.1, E = 2$

32. $c = 0.98, s = 10.1, E = 2$

■ USING AND INTERPRETING CONCEPTS

Finding the Margin of Error   *In Exercises 33 and 34, use the given confidence interval to find the estimated margin of error. Then find the sample mean.*

33. Commute Times A government agency reports a confidence interval of $(26.2, 30.1)$ when estimating the mean commute time (in minutes) for the population of workers in a city.

34. Book Prices A store manager reports a confidence interval of $(44.07, 80.97)$ when estimating the mean price (in dollars) for the population of textbooks.

Constructing Confidence Intervals   *In Exercises 35–38, you are given the sample mean and the sample standard deviation. Use this information to construct the 90% and 95% confidence intervals for the population mean. Interpret the results and compare the widths of the confidence intervals. If convenient, use technology to construct the confidence intervals.*

35. Home Theater Systems A random sample of 34 home theater systems has a mean price of $452.80 and a standard deviation of $85.50.

36. Gasoline Prices From a random sample of 48 days in a recent year, U.S. gasoline prices had a mean of $2.34 and a standard deviation of $0.32. *(Source: U.S. Energy Information Administration)*

37. Juice Drinks A random sample of 31 eight-ounce servings of different juice drinks has a mean of 99.3 calories and a standard deviation of 41.5 calories. *(Adapted from The Beverage Institute for Health and Wellness)*

38. Sodium Chloride Concentration In 36 randomly selected seawater samples, the mean sodium chloride concentration was 23 cubic centimeters per cubic meter and the standard deviation was 6.7 cubic centimeters per cubic meter. *(Adapted from Dorling Kindersley Visual Encyclopedia)*

39. Replacement Costs: Transmissions You work for a consumer advocate agency and want to estimate the population mean cost of replacing a car's transmission. As part of your study, you randomly select 50 replacement costs and find the mean to be $2650.00. The sample standard deviation is $425.00. Construct a 95% confidence interval for the population mean replacement cost. Interpret the results. *(Adapted from CostHelper)*

40. Repair Costs: Refrigerators In a random sample of 60 refrigerators, the mean repair cost was $150.00 and the standard deviation was $15.50. Construct a 99% confidence interval for the population mean repair cost. Interpret the results. *(Adapted from Consumer Reports)*

41. Repeat Exercise 39, changing the sample size to $n = 80$. Which confidence interval is wider? Explain.

42. Repeat Exercise 40, changing the sample size to $n = 40$. Which confidence interval is wider? Explain.

43. Swimming Times A random sample of forty-eight 200-meter swims has a mean time of 3.12 minutes and a standard deviation of 0.09 minute. Construct a 95% confidence interval for the population mean time. Interpret the results.

44. Hotels A random sample of 55 standard hotel rooms in the Philadelphia, PA area has a mean nightly cost of $154.17 and a standard deviation of $38.60. Construct a 99% confidence interval for the population mean cost. Interpret the results.

45. Repeat Exercise 43, using a standard deviation of $s = 0.06$ minute. Which confidence interval is wider? Explain.

46. Repeat Exercise 44, using a standard deviation of $s = \$42.50$. Which confidence interval is wider? Explain.

47. If all other quantities remain the same, how does the indicated change affect the width of a confidence interval?

(a) Increase in the level of confidence

(b) Increase in the sample size

(c) Increase in the standard deviation

48. Describe how you would construct a 90% confidence interval to estimate the population mean age for students at your school.

Constructing Confidence Intervals *In Exercises 49 and 50, use the given information to construct the 90% and 99% confidence intervals for the population mean. Interpret the results and compare the widths of the confidence intervals. If convenient, use technology to construct the confidence intervals.*

49. DVRs A research council wants to estimate the mean length of time (in minutes) the average U.S. adult spends watching TVs using digital video recorders (DVRs) each day. To determine this estimate, the research council takes a random sample of 20 U.S. adults and obtains the following results.

    15, 18, 17, 20, 24, 12, 9, 15, 14, 25, 8, 6, 10, 14, 16, 20, 27, 10, 9, 13

From past studies, the research council assumes that σ is 1.3 minutes and that the population of times is normally distributed. *(Adapted from the Council for Research Excellence)*

50. Text Messaging A telecommunications company wants to estimate the mean length of time (in minutes) that 18- to 24-year-olds spend text messaging each day. In a random sample of twenty-seven 18- to 24-year-olds, the mean length of time spent text messaging was 29 minutes. From past studies, the company assumes that σ is 4.5 minutes and that the population of times is normally distributed. *(Adapted from the Council for Research Excellence)*

51. Minimum Sample Size Determine the minimum required sample size if you want to be 95% confident that the sample mean is within one unit of the population mean given $\sigma = 4.8$. Assume the population is normally distributed.

52. Minimum Sample Size Determine the minimum required sample size if you want to be 99% confident that the sample mean is within two units of the population mean given $\sigma = 1.4$. Assume the population is normally distributed.

53. Cholesterol Contents of Cheese A cheese processing company wants to estimate the mean cholesterol content of all one-ounce servings of cheese. The estimate must be within 0.5 milligram of the population mean.

(a) Determine the minimum required sample size to construct a 95% confidence interval for the population mean. Assume the population standard deviation is 2.8 milligrams.

(b) Repeat part (a) using a 99% confidence interval.

(c) Which level of confidence requires a larger sample size? Explain.

54. Ages of College Students An admissions director wants to estimate the mean age of all students enrolled at a college. The estimate must be within 1 year of the population mean. Assume the population of ages is normally distributed.

(a) Determine the minimum required sample size to construct a 90% confidence interval for the population mean. Assume the population standard deviation is 1.2 years.

(b) Repeat part (a) using a 99% confidence interval.

(c) Which level of confidence requires a larger sample size? Explain.

55. Paint Can Volumes A paint manufacturer uses a machine to fill gallon cans with paint (see figure).

(a) The manufacturer wants to estimate the mean volume of paint the machine is putting in the cans within 0.25 ounce. Determine the minimum sample size required to construct a 90% confidence interval for the population mean. Assume the population standard deviation is 0.85 ounce.

(b) Repeat part (a) using an error tolerance of 0.15 ounce. Which error tolerance requires a larger sample size? Explain.

56. Water Dispensing Machine A beverage company uses a machine to fill one-liter bottles with water (see figure). Assume that the population of volumes is normally distributed.

(a) The company wants to estimate the mean volume of water the machine is putting in the bottles within 1 milliliter. Determine the minimum sample size required to construct a 95% confidence interval for the population mean. Assume the population standard deviation is 3 milliliters.

(b) Repeat part (a) using an error tolerance of 2 milliliters. Which error tolerance requires a larger sample size? Explain.

57. Plastic Sheet Cutting A machine cuts plastic into sheets that are 50 feet (600 inches) long. Assume that the population of lengths is normally distributed.

(a) The company wants to estimate the mean length of the sheets within 0.125 inch. Determine the minimum sample size required to construct a 95% confidence interval for the population mean. Assume the population standard deviation is 0.25 inch.

(b) Repeat part (a) using an error tolerance of 0.0625 inch. Which error tolerance requires a larger sample size? Explain.

Error tolerance = 0.25 oz

FIGURE FOR EXERCISE 55

Error tolerance = 1 mL

FIGURE FOR EXERCISE 56

58. Paint Sprayer A company uses an automated sprayer to apply paint to metal furniture. The company sets the sprayer to apply the paint one mil (1/1000 of an inch) thick.

(a) The company wants to estimate the mean thickness of paint the sprayer is applying within 0.0425 mil. Determine the minimum sample size required to construct a 90% confidence interval for the population mean. Assume the population standard deviation is 0.15 mil.

(b) Repeat part (a) using an error tolerance of 0.02125 mil. Which error tolerance requires a larger sample size? Explain.

59. Soccer Balls A soccer ball manufacturer wants to estimate the mean circumference of soccer balls within 0.1 inch.

(a) Determine the minimum sample size required to construct a 99% confidence interval for the population mean. Assume the population standard deviation is 0.25 inch.

(b) Repeat part (a) using a standard deviation of 0.3 inch. Which standard deviation requires a larger sample size? Explain.

60. Mini-Soccer Balls A soccer ball manufacturer wants to estimate the mean circumference of mini-soccer balls within 0.15 inch. Assume that the population of circumferences is normally distributed.

(a) Determine the minimum sample size required to construct a 99% confidence interval for the population mean. Assume the population standard deviation is 0.20 inch.

(b) Repeat part (a) using a standard deviation of 0.10 inch. Which standard deviation requires a larger sample size? Explain.

61. If all other quantities remain the same, how does the indicated change affect the minimum sample size requirement?

(a) Increase in the level of confidence

(b) Increase in the error tolerance

(c) Increase in the standard deviation

62. When estimating the population mean, why not construct a 99% confidence interval every time?

Using Technology *In Exercises 63 and 64, you are given a data sample. Use a technology tool to construct a 95% confidence interval for the population mean. Interpret your answer.*

63. Airfare The stem-and-leaf plot shows the results of a random sample of airfare prices (in dollars) for a one-way ticket from Boston, MA to Chicago, IL *(Adapted from Expedia, Inc.)*

64. Stock Prices A random sample of the closing stock prices for the Oracle Corporation for a recent year *(Source: Yahoo! Inc.)*

18.41	16.91	16.83	17.72	15.54	15.56	18.01	19.11	19.79
18.32	18.65	20.71	20.66	21.04	21.74	22.13	21.96	22.16
22.86	20.86	20.74	22.05	21.42	22.34	22.83	24.34	17.97
14.47	19.06	18.42	20.85	21.43	21.97	21.81

18	3 3
19	7
20	9 9
21	2 2 2 3 3 3 3 3 3 6 6
22	2 2 2 3 6 6 8 8 8 8 9
23	8 8

Key: 18|3 = 183

FIGURE FOR EXERCISE 63

SC *In Exercises 65 and 66, use StatCrunch to construct the 80%, 90%, and 95% confidence intervals for the population mean. Interpret the results.*

65. Sodium   A random sample of 30 sandwiches from a fast food restaurant has a mean of 1042.7 milligrams of sodium and a standard deviation of 344.9 milligrams of sodium. *(Source: McDonald's Corporation)*

66. Carbohydrates   The following represents a random sample of the amounts of carbohydrates (in grams) for 30 sandwiches from a fast food restaurant. *(Source: McDonald's Corporation)*

31	33	34	33	37	40	40	45	37	38	63	61	59	38	40
44	51	59	52	60	54	62	39	33	26	34	27	35	28	26

■ EXTENDING CONCEPTS

Finite Population Correction Factor   *In Exercises 67 and 68, use the following information.*

In this section, you studied the construction of a confidence interval to estimate a population mean when the population is large or infinite. When a population is finite, the formula that determines the standard error of the mean $\sigma_{\bar{x}}$ needs to be adjusted. If N is the size of the population and n is the size of the sample (where $n \geq 0.05N$), the standard error of the mean is

$$\sigma_{\bar{x}} = \frac{\sigma}{\sqrt{n}} \sqrt{\frac{N - n}{N - 1}}.$$

The expression $\sqrt{(N - n)/(N - 1)}$ is called the *finite population correction factor.* The margin of error is

$$E = z_c \frac{\sigma}{\sqrt{n}} \sqrt{\frac{N - n}{N - 1}}.$$

67. Determine the finite population correction factor for each of the following.

(a) $N = 1000$ and $n = 500$ (b) $N = 1000$ and $n = 100$

(c) $N = 1000$ and $n = 75$ (d) $N = 1000$ and $n = 50$

(e) What happens to the finite population correction factor as the sample size n decreases but the population size N remains the same?

68. Determine the finite population correction factor for each of the following.

(a) $N = 100$ and $n = 50$ (b) $N = 400$ and $n = 50$

(c) $N = 700$ and $n = 50$ (d) $N = 1000$ and $n = 50$

(e) What happens to the finite population correction factor as the population size N increases but the sample size n remains the same?

69. Sample Size   The equation for determining the sample size

$$n = \left(\frac{z_c \sigma}{E} \right)^2$$

can be obtained by solving the equation for the margin of error

$$E = \frac{z_c \sigma}{\sqrt{n}}$$

for n. Show that this is true and justify each step.

Marathon Training

A marathon is a foot race with a distance of 26.22 miles. It was one of the original events of the modern Olympics, where it was a men's-only event. The women's marathon did not become an Olympic event until 1984. The Olympic record for the men's marathon was set during the 2008 Olympics by Samuel Kamau Wanjiru of Kenya, with a time of 2 hours, 6 minutes, 32 seconds. The Olympic record for the women's marathon was set during the 2000 Olympics by Naoko Takahashi of Japan, with a time of 2 hours, 23 minutes, 14 seconds.

 Training for a marathon typically lasts at least 6 months. The training is gradual, with increases in distance about every 2 weeks. About 1 to 3 weeks before the race, the distance run is decreased slightly. The stem-and-leaf plots below show the marathon training times (in minutes) for a sample of 30 male runners and 30 female runners.

**Training Times (in minutes)
of Male Runners**

15	5 8 9 9 9	Key: 15|5 = 155
16	0 0 0 0 1 2 3 4 4 5 8 9	
17	0 1 1 3 5 6 6 7 7 9	
18	0 1 5	

**Training Times (in minutes)
of Female Runners**

17	8 9 9	Key: 17|8 = 178
18	0 0 0 0 1 2 3 4 6 6 7 9	
19	0 0 0 1 3 4 5 5 6 6	
20	0 0 1 2 3	

▇ EXERCISES

1. Use the sample to find a point estimate for the mean training time of the
   (a) male runners.
   (b) female runners.

2. Find the standard deviation of the training times for the
   (a) male runners.
   (b) female runners.

3. Use the sample to construct a 95% confidence interval for the population mean training time of the
   (a) male runners.
   (b) female runners.

4. Interpret the results of Exercise 3.

5. Use the sample to construct a 95% confidence interval for the population mean training time of all runners. How do your results differ from those in Exercise 3? Explain.

6. A trainer wants to estimate the population mean running times for both male and female runners within 2 minutes. Determine the minimum sample size required to construct a 99% confidence interval for the population mean training time of
   (a) male runners. Assume the population standard deviation is 8.9 minutes.
   (b) female runners. Assume the population standard deviation is 8.4 minutes.

6.2   Confidence Intervals for the Mean (Small Samples)

WHAT YOU SHOULD LEARN

▸ How to interpret the t-distribution and use a t-distribution table

▸ How to construct confidence intervals when $n < 30$, the population is normally distributed, and σ is unknown

The t-Distribution ▸ Confidence Intervals and t-Distributions

▸ THE t-DISTRIBUTION

In many real-life situations, the population standard deviation is unknown. Moreover, because of various constraints such as time and cost, it is often not practical to collect samples of size 30 or more. So, how can you construct a confidence interval for a population mean given such circumstances? If the random variable is normally distributed (or approximately normally distributed), you can use a *t-distribution*.

DEFINITION

If the distribution of a random variable x is approximately normal, then

$$t = \frac{\bar{x} - \mu}{\dfrac{s}{\sqrt{n}}}$$

follows a **t-distribution.**

Critical values of t are denoted by t_c. Several properties of the t-distribution are as follows.

1. The t-distribution is bell-shaped and symmetric about the mean.
2. The t-distribution is a family of curves, each determined by a parameter called the *degrees of freedom*. The **degrees of freedom** are the number of free choices left after a sample statistic such as $\bar{x}$ is calculated. When you use a t-distribution to estimate a population mean, the degrees of freedom are equal to one less than the sample size.

$$\text{d.f.} = n - 1 \qquad \text{Degrees of freedom}$$

3. The total area under a t-curve is 1 or 100%.
4. The mean, median, and mode of the t-distribution are equal to 0.
5. As the degrees of freedom increase, the t-distribution approaches the normal distribution. After 30 d.f., the t-distribution is very close to the standard normal z-distribution.

INSIGHT

The following example illustrates the concept of degrees of freedom. Suppose the number of chairs in a classroom equals the number of students: 25 chairs and 25 students. Each of the first 24 students to enter the classroom has a choice as to which chair he or she will sit in. There is no freedom of choice, however, for the 25th student who enters the room.

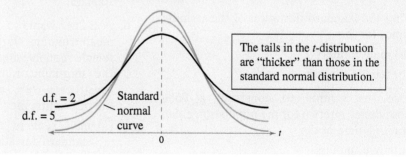

The tails in the t-distribution are "thicker" than those in the standard normal distribution.

d.f. = 2
d.f. = 5
Standard normal curve

Table 5 in Appendix B lists critical values of t for selected confidence intervals and degrees of freedom.

EXAMPLE 1

▶ **Finding Critical Values of *t***

Find the critical value t_c for a 95% confidence level when the sample size is 15.

▶ **Solution**

Because $n = 15$, the degrees of freedom are

$$\text{d.f.} = n - 1$$
$$= 15 - 1$$
$$= 14.$$

A portion of Table 5 is shown. Using d.f. $= 14$ and $c = 0.95$, you can find the critical value t_c, as shown by the highlighted areas in the table.

STUDY TIP

Unlike the *z*-table, critical values for a specific confidence interval can be found in the column headed by *c* in the appropriate d.f. row. (The symbol α will be explained in Chapter 7.)

	Level of confidence, *c*	0.50	0.80	0.90	0.95	0.98
	One tail, α	0.25	0.10	0.05	0.025	0.01
d.f.	Two tails, α	0.50	0.20	0.10	0.05	0.02
1		1.000	3.078	6.314	12.706	31.821
2		.816	1.886	2.920	4.303	6.965
3		.765	1.638	2.353	3.182	4.541
12		.695	1.356	1.782	2.179	2.681
13		.694	1.350	1.771	2.160	2.650
14		.692	1.345	1.761	2.145	2.624
15		.691	1.341	1.753	2.131	2.602
16		.690	1.337	1.746	2.120	2.583
28		.683	1.313	1.701	2.048	2.467
29		.683	1.311	1.699	2.045	2.462
∞		.674	1.282	1.645	1.960	2.326

From the table, you can see that $t_c = 2.145$. The graph shows the *t*-distribution for 14 degrees of freedom, $c = 0.95$, and $t_c = 2.145$.

INSIGHT

For 30 or more degrees of freedom, the critical values for the *t*-distribution are close to the corresponding critical values for the normal distribution. Moreover, the values in the last row of the table marked ∞ d.f. correspond *exactly* to the normal distribution values.

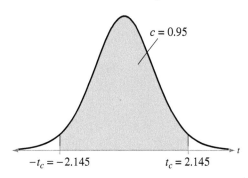

Interpretation So, 95% of the area under the *t*-distribution curve with 14 degrees of freedom lies between $t = \pm 2.145$.

▶ **Try It Yourself 1**

Find the critical value t_c for a 90% confidence level when the sample size is 22.

a. Identify the *degrees of freedom.*
b. Identify the *level of confidence c.*
c. *Use* Table 5 in Appendix B to find t_c. *Answer: Page A40*

▶ CONFIDENCE INTERVALS AND *t*-DISTRIBUTIONS

Constructing a confidence interval using the *t*-distribution is similar to constructing a confidence interval using the normal distribution—both use a point estimate $\bar{x}$ and a margin of error E.

GUIDELINES

Constructing a Confidence Interval for the Mean: *t*-Distribution

IN WORDS	IN SYMBOLS
1. Find the sample statistics n, $\bar{x}$, and s.	$\bar{x} = \dfrac{\sum x}{n},\ \ s = \sqrt{\dfrac{\sum (x - \bar{x})^2}{n - 1}}$
2. Identify the degrees of freedom, the level of confidence c, and the critical value t_c.	d.f. $= n - 1$
3. Find the margin of error E.	$E = t_c \dfrac{s}{\sqrt{n}}$
4. Find the left and right endpoints and form the confidence interval.	Left endpoint: $\bar{x} - E$ Right endpoint: $\bar{x} + E$ Interval: $\bar{x} - E < \mu < \bar{x} + E$

EXAMPLE 2 ⬤ SC Report 24

See MINITAB steps on page 352.

▶ Constructing a Confidence Interval

You randomly select 16 coffee shops and measure the temperature of the coffee sold at each. The sample mean temperature is 162.0°F with a sample standard deviation of 10.0°F. Construct a 95% confidence interval for the population mean temperature. Assume the temperatures are approximately normally distributed.

▶ Solution

Because the sample size is less than 30, σ is unknown, and the temperatures are approximately normally distributed, you can use the *t*-distribution. Using $n = 16$, $\bar{x} = 162.0$, $s = 10.0$, $c = 0.95$, and d.f. $= 15$, you can use Table 5 to find that $t_c = 2.131$. The margin of error at the 95% confidence level is

$$E = t_c \frac{s}{\sqrt{n}} = 2.131 \cdot \frac{10.0}{\sqrt{16}} \approx 5.3.$$

The confidence interval is as follows.

Left Endpoint   Right Endpoint

$\bar{x} - E \approx 162 - 5.3 = 156.7$ $\bar{x} + E \approx 162 + 5.3 = 167.3$

$156.7 < \mu < 167.3$

Interpretation   With 95% confidence, you can say that the population mean temperature of coffee sold is between 156.7°F and 167.3°F.

STUDY TIP

For a TI-83/84 Plus, constructing a confidence interval using the *t*-distribution is similar to constructing a confidence interval using the normal distribution.

[STAT]

Choose the TESTS menu.

   8: TInterval...

Select the *Data* input option if you use the original data. Select the *Stats* input option if you use the descriptive statistics. In each case, enter the appropriate values, then select *Calculate*. Your results may vary slightly depending on the method you use. For Example 2, the descriptive statistics were entered.

```
TInterval
 (156.67,167.33)
x̄=162
Sx=10
n=16
```

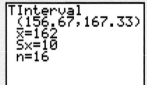

▶ Try It Yourself 2

Construct 90% and 99% confidence intervals for the population mean temperature.

a. *Find t_c and E for each level of confidence.*
b. *Use $\bar{x}$ and E to find the left and right endpoints of the confidence interval.*
c. *Interpret the results.*   *Answer: Page A40*

To explore this topic further, see Activity 6.2 on page 326.

EXAMPLE 3   Report 25

See TI-83/84 Plus steps on page 353.

▶ **Constructing a Confidence Interval**

You randomly select 20 cars of the same model that were sold at a car dealership and determine the number of days each car sat on the dealership's lot before it was sold. The sample mean is 9.75 days, with a sample standard deviation of 2.39 days. Construct a 99% confidence interval for the population mean number of days the car model sits on the dealership's lot. Assume the days on the lot are normally distributed.

▶ **Solution**

Because the sample size is less than 30, σ is unknown, and the days on the lot are normally distributed, you can use the t-distribution. Using $n = 20$, $\bar{x} = 9.75$, $s = 2.39$, $c = 0.99$, and d.f. = 19, you can use Table 5 to find that $t_c = 2.861$. The margin of error at the 99% confidence level is

$$E = t_c \frac{s}{\sqrt{n}}$$

$$= 2.861 \cdot \frac{2.39}{\sqrt{20}}$$

$$\approx 1.53.$$

The confidence interval is as follows.

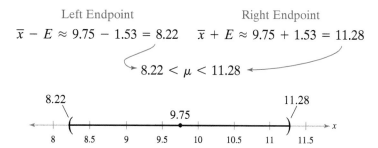

Left Endpoint
$\bar{x} - E \approx 9.75 - 1.53 = 8.22$

Right Endpoint
$\bar{x} + E \approx 9.75 + 1.53 = 11.28$

$8.22 < \mu < 11.28$

Interpretation   With 99% confidence, you can say that the population mean number of days the car model sits on the dealership's lot is between 8.22 and 11.28.

▶ Try It Yourself 3

Construct 90% and 95% confidence intervals for the population mean number of days the car model sits on the dealership's lot. Compare the widths of the confidence intervals.

a. *Find t_c and E for each level of confidence.*
b. *Use $\bar{x}$ and E to find the left and right endpoints of the confidence interval.*
c. *Interpret the results and compare the widths of the confidence intervals.*
                                                      Answer: Page A40

HISTORICAL REFERENCE

William S. Gosset (1876–1937)

Developed the *t*-distribution while employed by the Guinness Brewing Company in Dublin, Ireland. Gosset published his findings using the pseudonym Student. The *t*-distribution is sometimes referred to as Student's *t*-distribution. (See page 33 for others who were important in the history of statistics.)

The following flowchart describes when to use the normal distribution and when to use a *t*-distribution to construct a confidence interval for the population mean.

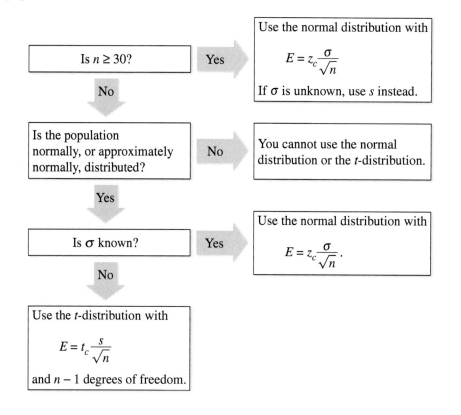

▶ **Choosing the Normal Distribution or the *t*-Distribution**

You randomly select 25 newly constructed houses. The sample mean construction cost is $181,000 and the population standard deviation is $28,000. Assuming construction costs are normally distributed, should you use the normal distribution, the *t*-distribution, or neither to construct a 95% confidence interval for the population mean construction cost? Explain your reasoning.

▶ **Solution**

Because the population is normally distributed and the population standard deviation is known, you should use the normal distribution.

▶ **Try It Yourself 4**

You randomly select 18 adult male athletes and measure the resting heart rate of each. The sample mean heart rate is 64 beats per minute, with a sample standard deviation of 2.5 beats per minute. Assuming the heart rates are normally distributed, should you use the normal distribution, the *t*-distribution, or neither to construct a 90% confidence interval for the population mean heart rate? Explain your reasoning.

Use the flowchart above to determine which distribution you should use to construct the 90% confidence interval for the population mean heart rate.

Answer: Page A40

6.2 EXERCISES

■ BUILDING BASIC SKILLS AND VOCABULARY

In Exercises 1–4, find the critical value t_c for the given confidence level c and sample size n.

1. $c = 0.90$, $n = 10$

2. $c = 0.95$, $n = 12$

3. $c = 0.99$, $n = 16$

4. $c = 0.98$, $n = 20$

In Exercises 5–8, find the margin of error for the given values of c, s, and n.

5. $c = 0.95$, $s = 5$, $n = 16$

6. $c = 0.99$, $s = 3$, $n = 6$

7. $c = 0.90$, $s = 2.4$, $n = 12$

8. $c = 0.98$, $s = 4.7$, $n = 9$

In Exercises 9–12, (a) construct the indicated confidence interval for the population mean μ using a t-distribution. (b) If you had incorrectly used a normal distribution, which interval would be wider?

9. $c = 0.90$, $\bar{x} = 12.5$, $s = 2.0$, $n = 6$

10. $c = 0.95$, $\bar{x} = 13.4$, $s = 0.85$, $n = 8$

11. $c = 0.98$, $\bar{x} = 4.3$, $s = 0.34$, $n = 14$

12. $c = 0.99$, $\bar{x} = 24.7$, $s = 4.6$, $n = 10$

In Exercises 13–16, use the given confidence interval to find the margin of error and the sample mean.

13. $(14.7, 22.1)$

14. $(6.17, 8.53)$

15. $(64.6, 83.6)$

16. $(16.2, 29.8)$

■ USING AND INTERPRETING CONCEPTS

Constructing Confidence Intervals *In Exercises 17 and 18, you are given the sample mean and the sample standard deviation. Assume the random variable is normally distributed and use a t-distribution to find the margin of error and construct a 95% confidence interval for the population mean. Interpret the results. If convenient, use technology to construct the confidence interval.*

17. Commute Time to Work In a random sample of eight people, the mean commute time to work was 35.5 minutes and the sample standard deviation was 7.2 minutes.

18. Driving Distance to Work In a random sample of five people, the mean driving distance to work was 22.2 miles and the sample standard deviation was 5.8 miles.

19. You research commute times to work and find that the population standard deviation was 9.3 minutes. Repeat Exercise 17, using a normal distribution with the appropriate calculations for a standard deviation that is known. Compare the results.

20. You research driving distances to work and find that the population standard deviation was 5.2 miles. Repeat Exercise 18, using a normal distribution with the appropriate calculations for a standard deviation that is known. Compare the results.

Constructing Confidence Intervals *In Exercises 21 and 22, you are given the sample mean and the sample standard deviation. Assume the random variable is normally distributed and use a normal distribution or a t-distribution to construct a 90% confidence interval for the population mean. If convenient, use technology to construct the confidence interval.*

21. **Waste Generated** (a) In a random sample of 10 adults from the United States, the mean waste generated per person per day was 4.50 pounds and the standard deviation was 1.21 pounds. (b) Repeat part (a), assuming the same statistics came from a sample size of 500. Compare the results. *(Adapted from U.S. Environmental Protection Agency)*

22. **Waste Recycled** (a) In a random sample of 12 adults from the United States, the mean waste recycled per person per day was 1.50 pounds and the standard deviation was 0.28 pound. (b) Repeat part (a), assuming the same statistics came from a sample size of 600. Compare the results. *(Adapted from U.S. Environmental Protection Agency)*

Constructing Confidence Intervals *In Exercises 23–26, a data set is given. For each data set, (a) find the sample mean, (b) find the sample standard deviation, and (c) construct a 99% confidence interval for the population mean. Assume the population of each data set is normally distributed. If convenient, use a technology tool.*

23. **Earnings** The annual earnings of 16 randomly selected computer software engineers *(Adapted from U.S. Bureau of Labor Statistics)*

    92,184 86,919 90,176 91,740 95,535 90,108 94,815 88,114
    85,406 90,197 89,944 93,950 84,116 96,054 85,119 88,549

24. **Earnings** The annual earnings of 14 randomly selected physical therapists *(Adapted from U.S. Bureau of Labor Statistics)*

    63,118 65,740 72,899 68,500 66,726 65,554 69,247
    64,963 68,627 70,448 71,842 66,873 74,103 71,138

25. **SAT Scores** The SAT scores of 12 randomly selected high school seniors

    1704 1940 1518 2005 1432 1872
    1998 1658 1825 1670 2210 1380

26. **GPA** The grade point averages (GPA) of 15 randomly selected college students

    2.3 3.3 2.6 1.8 0.2 3.1 4.0 0.7
    2.3 2.0 3.1 3.4 1.3 2.6 2.6

Choosing a Distribution *In Exercises 27–32, use a normal distribution or a t-distribution to construct a 95% confidence interval for the population mean. Justify your decision. If neither distribution can be used, explain why. Interpret the results. If convenient, use technology to construct the confidence interval.*

27. **Body Mass Index** In a random sample of 50 people, the mean body mass index (BMI) was 27.7 and the standard deviation was 6.12. Assume the body mass indexes are normally distributed. *(Adapted from Centers for Disease Control)*

28. **Mortgages** In a random sample of 15 mortgage institutions, the mean interest rate was 4.99% and the standard deviation was 0.36%. Assume the interest rates are normally distributed. *(Adapted from Federal Reserve)*

29. Sports Cars: Miles per Gallon   You take a random survey of 25 sports cars and record the miles per gallon for each. The data are listed below. Assume the miles per gallon are normally distributed.

15   27   24   24   20   21   24   14   21   25   21   13   21
25   22   21   25   24   22   24   24   22   21   24   24

30. Yards Per Carry   In a recent season, the standard deviation of the yards per carry for all running backs was 1.34. The yards per carry of 20 randomly selected running backs are listed below. Assume the yards per carry are normally distributed. *(Source: National Football League)*

5.6   4.4   3.8   4.5   3.3   5.0   3.6   3.7   4.8   3.5
5.6   3.0   6.8   4.7   2.2   3.3   5.7   3.0   5.0   4.5

31. Hospital Waiting Times   In a random sample of 19 patients at a hospital's minor emergency department, the mean waiting time before seeing a medical professional was 23 minutes and the standard deviation was 11 minutes. Assume the waiting times are not normally distributed.

32. Hospital Length of Stay   In a random sample of 13 people, the mean length of stay at a hospital was 6.3 days and the standard deviation was 1.7 days. Assume the lengths of stay are normally distributed. *(Adapted from American Hospital Association)*

SC *In Exercises 33 and 34, use StatCrunch to construct the 90%, 95%, and 99% confidence intervals for the population mean. Interpret the results and compare the widths of the confidence intervals. Assume the random variable is normally distributed.*

33. Homework   The weekly time spent (in hours) on homework for 18 randomly selected high school students

12.0   11.3   13.5   11.7   12.0   13.0   15.5   10.8   12.5
12.3   14.0   9.5   8.8   10.0   12.8   15.0   11.8   13.0

34. Weight Lifting   In a random sample of 11 college football players, the mean weekly time spent weight lifting was 7.2 hours and the standard deviation was 1.9 hours.

■ EXTENDING CONCEPTS

35. Tennis Ball Manufacturing   A company manufactures tennis balls. When its tennis balls are dropped onto a concrete surface from a height of 100 inches, the company wants the mean height the balls bounce upward to be 55.5 inches. This average is maintained by periodically testing random samples of 25 tennis balls. If the *t*-value falls between $-t_{0.99}$ and $t_{0.99}$, the company will be satisfied that it is manufacturing acceptable tennis balls. A sample of 25 balls is randomly selected and tested. The mean bounce height of the sample is 56.0 inches and the standard deviation is 0.25 inch. Assume the bounce heights are approximately normally distributed. Is the company making acceptable tennis balls? Explain your reasoning.

36. Light Bulb Manufacturing   A company manufactures light bulbs. The company wants the bulbs to have a mean life span of 1000 hours. This average is maintained by periodically testing random samples of 16 light bulbs. If the *t*-value falls between $-t_{0.99}$ and $t_{0.99}$, the company will be satisfied that it is manufacturing acceptable light bulbs. A sample of 16 light bulbs is randomly selected and tested. The mean life span of the sample is 1015 hours and the standard deviation is 25 hours. Assume the life spans are approximately normally distributed. Is the company making acceptable light bulbs? Explain your reasoning.

ACTIVITY 6.2

Confidence Intervals for a Mean (the impact of not knowing the standard deviation)

APPLET

The *confidence intervals for a mean (the impact of not knowing the standard deviation)* applet allows you to visually investigate confidence intervals for a population mean. You can specify the sample size *n*, the shape of the distribution (Normal or Right-skewed), the population mean (Mean), and the true population standard deviation (Std. Dev.). When you click SIMULATE, 100 separate samples of size *n* will be selected from a population with these population parameters. For each of the 100 samples, a 95% Z confidence interval (known standard deviation) and a 95% T confidence interval (unknown standard deviation) are displayed in the plot at the right. The 95% Z confidence interval is displayed in green and the 95% T confidence interval is displayed in blue. If an interval does not contain the population mean, it is displayed in red. Additional simulations can be carried out by clicking SIMULATE multiple times. The cumulative number of times that each type of interval contains the population mean is also shown. Press CLEAR to clear existing results and start a new simulation.

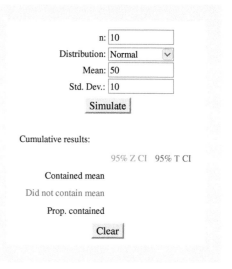

■ Explore

Step 1 Specify a value for *n*.
Step 2 Specify a distribution.
Step 3 Specify a value for the mean.
Step 4 Specify a value for the standard deviation.
Step 5 Click SIMULATE to generate the confidence intervals.

■ Draw Conclusions

APPLET

1. Set *n* = 30, Mean = 25, Std. Dev. = 5, and the distribution to Normal. Run the simulation so that at least 1000 confidence intervals are generated. Compare the proportion of the 95% Z confidence intervals and 95% T confidence intervals that contain the population mean. Is this what you would expect? Explain.

2. In a random sample of 24 high school students, the mean number of hours of sleep per night during the school week was 7.26 hours and the standard deviation was 1.19 hours. Assume the sleep times are normally distributed. Run the simulation for *n* = 10 so that at least 500 confidence intervals are generated. What proportion of the 95% Z confidence intervals and 95% T confidence intervals contain the population mean? Should you use a Z confidence interval or a T confidence interval for the mean number of hours of sleep? Explain.

6.3 Confidence Intervals for Population Proportions

WHAT YOU SHOULD LEARN

▸ How to find a point estimate for a population proportion

▸ How to construct a confidence interval for a population proportion

▸ How to determine the minimum sample size required when estimating a population proportion

Point Estimate for a Population Proportion ▸ Confidence Intervals for a Population Proportion ▸ Finding a Minimum Sample Size

▸ POINT ESTIMATE FOR A POPULATION PROPORTION

Recall from Section 4.2 that the probability of success in a single trial of a binomial experiment is p. This probability is a **population proportion.** In this section, you will learn how to estimate a population proportion p using a confidence interval. As with confidence intervals for μ, you will start with a point estimate.

DEFINITION

The **point estimate for p,** the population proportion of successes, is given by the proportion of successes in a sample and is denoted by

$$\hat{p} = \frac{x}{n} \qquad \text{Sample proportion}$$

where x is the number of successes in the sample and n is the sample size. The point estimate for the population proportion of failures is $\hat{q} = 1 - \hat{p}$. The symbols $\hat{p}$ and $\hat{q}$ are read as "p hat" and "q hat."

EXAMPLE 1

▸ Finding a Point Estimate for p

In a survey of 1000 U.S. adults, 662 said that it is acceptable to check personal e-mail while at work. Find a point estimate for the population proportion of U.S. adults who say it is acceptable to check personal e-mail while at work. *(Adapted from Liberty Mutual)*

▸ Solution

Using $n = 1000$ and $x = 662$,

$$\hat{p} = \frac{x}{n}$$
$$= \frac{662}{1000}$$
$$= 0.662$$
$$= 66.2\%$$

So, the point estimate for the population proportion of U.S. adults who say it is acceptable to check personal e-mail while at work is 66.2%.

▸ Try It Yourself 1

In a survey of 1006 U.S. adults, 181 said that Abraham Lincoln was the greatest president. Find a point estimate for the population proportion of U.S. adults who say Abraham Lincoln was the greatest president. *(Adapted from The Gallup Poll)*

a. *Identify* x and n.
b. Use x and n to find $\hat{p}$.

Answer: Page A40

INSIGHT

In the first two sections, estimates were made for quantitative data. In this section, sample proportions are used to make estimates for qualitative data.

▸ CONFIDENCE INTERVALS FOR A POPULATION PROPORTION

Constructing a confidence interval for a population proportion p is similar to constructing a confidence interval for a population mean. You start with a point estimate and calculate a margin of error.

DEFINITION

A *c*-**confidence interval for a population proportion** p **is**

$$\hat{p} - E < p < \hat{p} + E$$

where

$$E = z_c \sqrt{\frac{\hat{p}\hat{q}}{n}}.$$

The probability that the confidence interval contains p is c.

In Section 5.5, you learned that a binomial distribution can be approximated by a normal distribution if $np \geq 5$ and $nq \geq 5$. When $n\hat{p} \geq 5$ and $n\hat{q} \geq 5$, the sampling distribution of $\hat{p}$ is approximately normal with a mean of

$$\mu_{\hat{p}} = p$$

and a standard error of

$$\sigma_{\hat{p}} = \sqrt{\frac{pq}{n}}.$$

GUIDELINES

Constructing a Confidence Interval for a Population Proportion

IN WORDS	IN SYMBOLS
1. Identify the sample statistics n and x.	
2. Find the point estimate $\hat{p}$.	$\hat{p} = \dfrac{x}{n}$
3. Verify that the sampling distribution of $\hat{p}$ can be approximated by a normal distribution.	$n\hat{p} \geq 5$, $n\hat{q} \geq 5$
4. Find the critical value z_c that corresponds to the given level of confidence c.	Use the Standard Normal Table or technology.
5. Find the margin of error E.	$E = z_c \sqrt{\dfrac{\hat{p}\hat{q}}{n}}$
6. Find the left and right endpoints and form the confidence interval.	Left endpoint: $\hat{p} - E$ Right endpoint: $\hat{p} + E$ Interval: $\hat{p} - E < p < \hat{p} + E$

MINITAB and TI-83/84 Plus steps are shown on pages 352 and 353.

EXAMPLE 2   SC Report 26

▶ **Constructing a Confidence Interval for *p***

Use the data given in Example 1 to construct a 95% confidence interval for the population proportion of U.S. adults who say that it is acceptable to check personal e-mail while at work.

▶ **Solution**

From Example 1, $\hat{p} = 0.662$. So,

$$\hat{q} = 1 - 0.662 = 0.338.$$

Using $n = 1000$, you can verify that the sampling distribution of $\hat{p}$ can be approximated by a normal distribution.

$$n\hat{p} = 1000 \cdot 0.662 = 662 > 5$$

$$n\hat{q} = 1000 \cdot 0.338 = 338 > 5$$

Using $z_c = 1.96$, the margin of error is

$$E = z_c \sqrt{\frac{\hat{p}\hat{q}}{n}} = 1.96 \sqrt{\frac{(0.662)(0.338)}{1000}} \approx 0.029.$$

The 95% confidence interval is as follows.

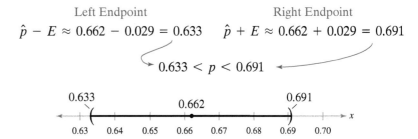

Left Endpoint	Right Endpoint
$\hat{p} - E \approx 0.662 - 0.029 = 0.633$	$\hat{p} + E \approx 0.662 + 0.029 = 0.691$

$$0.633 < p < 0.691$$

STUDY TIP

Notice in Example 2 that the confidence interval for the population proportion *p* is rounded to three decimal places. This *round-off rule* will be used throughout the text.

Interpretation   With 95% confidence, you can say that the population proportion of U.S. adults who say that it is acceptable to check personal e-mail while at work is between 63.3% and 69.1%.

▶ **Try It Yourself 2**

Use the data given in Try It Yourself 1 to construct a 90% confidence interval for the population proportion of U.S. adults who say that Abraham Lincoln was the greatest president.

a. *Find* $\hat{p}$ and $\hat{q}$.
b. *Verify* that the sampling distribution of $\hat{p}$ can be approximated by a normal distribution.
c. *Find* z_c and *E*.
d. *Use* $\hat{p}$ and *E* to find the *left* and *right endpoints* of the confidence interval.
e. *Interpret* the results.   *Answer: Page A40*

The confidence level of 95% used in Example 2 is typical of opinion polls. The result, however, is usually not stated as a confidence interval. Instead, the result of Example 2 would be stated as "66.2% with a margin of error of ±2.9%."

EXAMPLE 3 **SC** Report 27

▶ **Constructing a Confidence Interval for p**

The graph shown at the right is from a survey of 498 U.S. adults. Construct a 99% confidence interval for the population proportion of U.S. adults who think that teenagers are the more dangerous drivers.
(Source: The Gallup Poll)

Who are the more dangerous drivers?

71% Teenagers

People over 65 **25%**

4% No opinion

> **INSIGHT**
>
> In Example 3, note that $n\hat{p} \geq 5$ and $n\hat{q} \geq 5$. So, the sampling distribution of $\hat{p}$ is approximately normal.

To explore this topic further, see Activity 6.3 on page 336.

▶ **Solution**

From the graph, $\hat{p} = 0.71$. So,

$$\hat{q} = 1 - 0.71$$
$$= 0.29.$$

Using these values and the values $n = 498$ and $z_c = 2.575$, the margin of error is

$$E = z_c \sqrt{\frac{\hat{p}\hat{q}}{n}}$$

$$\approx 2.575\sqrt{\frac{(0.71)(0.29)}{498}}$$ Use Table 4 in Appendix B to estimate that z_c is halfway between 2.57 and 2.58.

$$\approx 0.052.$$

The 99% confidence interval is as follows.

Left Endpoint   Right Endpoint

$\hat{p} - E \approx 0.71 - 0.052 = 0.658$ $\hat{p} + E \approx 0.71 + 0.052 = 0.762$

$0.658 < p < 0.762$

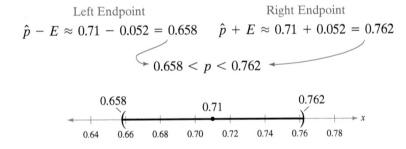

Interpretation With 99% confidence, you can say that the population proportion of U.S. adults who think that teenagers are the more dangerous drivers is between 65.8% and 76.2%.

▶ **Try It Yourself 3**

Use the data given in Example 3 to construct a 99% confidence interval for the population proportion of adults who think that people over 65 are the more dangerous drivers.

a. *Find* $\hat{p}$ and $\hat{q}$.
b. *Verify* that the sampling distribution of $\hat{p}$ can be approximated by a normal distribution.
c. *Find* z_c and E.
d. *Use* $\hat{p}$ and E to find the *left* and *right endpoints* of the confidence interval.
e. *Interpret* the results.

Answer: Page A40

▶ FINDING A MINIMUM SAMPLE SIZE

One way to increase the precision of a confidence interval without decreasing the level of confidence is to increase the sample size.

INSIGHT

The reason for using 0.5 as the values of $\hat{p}$ and $\hat{q}$ when no preliminary estimate is available is that these values yield a maximum value of the product $\hat{p}\hat{q} = \hat{p}(1 - \hat{p})$. In other words, if you don't estimate the values of $\hat{p}$ and $\hat{q}$, you must pay the penalty of using a larger sample.

FINDING A MINIMUM SAMPLE SIZE TO ESTIMATE p

Given a c-confidence level and a margin of error E, the minimum sample size n needed to estimate p is

$$n = \hat{p}\hat{q}\left(\frac{z_c}{E}\right)^2.$$

This formula assumes that you have preliminary estimates of $\hat{p}$ and $\hat{q}$. If not, use $\hat{p} = 0.5$ and $\hat{q} = 0.5$.

EXAMPLE 4

▶ **Determining a Minimum Sample Size**

You are running a political campaign and wish to estimate, with 95% confidence, the population proportion of registered voters who will vote for your candidate. Your estimate must be accurate within 3% of the population proportion. Find the minimum sample size needed if (1) no preliminary estimate is available and (2) a preliminary estimate gives $\hat{p} = 0.31$. Compare your results.

▶ **Solution**

1. Because you do not have a preliminary estimate of $\hat{p}$, use $\hat{p} = 0.5$ and $\hat{q} = 0.5$. Using $z_c = 1.96$ and $E = 0.03$, you can solve for n.

$$n = \hat{p}\hat{q}\left(\frac{z_c}{E}\right)^2 = (0.5)(0.5)\left(\frac{1.96}{0.03}\right)^2 \approx 1067.11$$

Because n is a decimal, round up to the nearest whole number, 1068.

2. You have a preliminary estimate of $\hat{p} = 0.31$. So, $\hat{q} = 0.69$. Using $z_c = 1.96$ and $E = 0.03$, you can solve for n.

$$n = \hat{p}\hat{q}\left(\frac{z_c}{E}\right)^2 = (0.31)(0.69)\left(\frac{1.96}{0.03}\right)^2 \approx 913.02$$

Because n is a decimal, round up to the nearest whole number, 914.

Interpretation With no preliminary estimate, the minimum sample size should be at least 1068 registered voters. With a preliminary estimate of $\hat{p} = 0.31$, the sample size should be at least 914 registered voters. So, you will need a larger sample size if no preliminary estimate is available.

▶ **Try It Yourself 4**

You wish to estimate, with 90% confidence, the population proportion of females who refuse to eat leftovers. Your estimate must be accurate within 2% of the population proportion. Find the minimum sample size needed if (1) no preliminary estimate is available and (2) a previous survey found that 11% of females refuse to eat leftovers. *(Source: Consumer Reports National Research Center)*

a. *Identify* $\hat{p}$, $\hat{q}$, z_c, and E. If $\hat{p}$ is unknown, use 0.5.
b. *Use* $\hat{p}$, $\hat{q}$, z_c, and E to find the *minimum sample size n*.
c. *Determine* how many females should be included in the sample.

Answer: Page A40

6.3 EXERCISES

■ BUILDING BASIC SKILLS AND VOCABULARY

True or False?   *In Exercises 1 and 2, determine whether the statement is true or false. If it is false, rewrite it as a true statement.*

1. To estimate the value of p, the population proportion of successes, use the point estimate x.

2. The point estimate for the proportion of failures is $1 - \hat{p}$.

Finding $\hat{p}$ and $\hat{q}$ *In Exercises 3–6, let p be the population proportion for the given condition. Find point estimates of p and q.*

3. **Recycling**   In a survey of 1002 U.S. adults, 752 say they recycle. *(Adapted from ABC News Poll)*

4. **Charity**   In a survey of 2939 U.S. adults, 2439 say they have contributed to a charity in the past 12 months. *(Adapted from Harris Interactive)*

5. **Computers**   In a survey of 11,605 parents, 4912 think that the government should subsidize the costs of computers for lower-income families. *(Adapted from DisneyFamily.com)*

6. **Vacation**   In a survey of 1003 U.S. adults, 110 say they would go on vacation to Europe if cost did not matter. *(Adapted from The Gallup Poll)*

In Exercises 7–10, use the given confidence interval to find the margin of error and the sample proportion.

7. $(0.905, 0.933)$

8. $(0.245, 0.475)$

9. $(0.512, 0.596)$

10. $(0.087, 0.263)$

■ USING AND INTERPRETING CONCEPTS

Constructing Confidence Intervals   *In Exercises 11 and 12, construct 90% and 95% confidence intervals for the population proportion. Interpret the results and compare the widths of the confidence intervals. If convenient, use technology to construct the confidence intervals.*

11. **Dental Visits**   In a survey of 674 U.S. males ages 18–64, 396 say they have gone to the dentist in the past year. *(Adapted from National Center for Health Statistics)*

12. **Dental Visits**   In a survey of 420 U.S. females ages 18–64, 279 say they have gone to the dentist in the past year. *(Adapted from National Center for Health Statistics)*

Constructing Confidence Intervals   *In Exercises 13 and 14, construct a 99% confidence interval for the population proportion. Interpret the results.*

13. **Going Green**   In a survey of 3110 U.S. adults, 1435 say they have started paying bills online in the last year. *(Adapted from Harris Interactive)*

14. **Seen a Ghost**   In a survey of 4013 U.S. adults, 722 say they have seen a ghost. *(Adapted from Pew Research Center)*

15. Nail Polish   In a survey of 7000 women, 4431 say they change their nail polish once a week. Construct a 95% confidence interval for the population proportion of women who change their nail polish once a week. *(Adapted from Essie Cosmetics)*

16. World Series   In a survey of 891 U.S. adults who follow baseball in a recent year, 184 said that the Boston Red Sox would win the World Series. Construct a 90% confidence interval for the population proportion of U.S. adults who follow baseball who in a recent year said that the Boston Red Sox would win the World Series. *(Adapted from Harris Interactive)*

17. Alternative Energy   You wish to estimate, with 95% confidence, the population proportion of U.S. adults who want more funding for alternative energy. Your estimate must be accurate within 4% of the population proportion.

(a) No preliminary estimate is available. Find the minimum sample size needed.

(b) Find the minimum sample size needed, using a prior study that found that 78% of U.S. adults want more funding for alternative energy. *(Source: Pew Research Center)*

(c) Compare the results from parts (a) and (b).

18. Reading Fiction   You wish to estimate, with 99% confidence, the population proportion of U.S. adults who read fiction books. Your estimate must be accurate within 2% of the population proportion.

(a) No preliminary estimate is available. Find the minimum sample size needed.

(b) Find the minimum sample size needed, using a prior study that found that 47% of U.S. adults read fiction books. *(Source: National Endowment for the Arts)*

(c) Compare the results from parts (a) and (b).

19. Emergency Room Visits   You wish to estimate, with 90% confidence, the population proportion of U.S. adults who made one or more emergency room visits in the past year. Your estimate must be accurate within 3% of the population proportion.

(a) No preliminary estimate is available. Find the minimum sample size needed.

(b) Find the minimum sample size needed, using a prior study that found that 20.1% of U.S. adults made one or more emergency room visits in the past year. *(Source: National Center for Health Statistics)*

(c) Compare the results from parts (a) and (b).

20. Ice Cream   You wish to estimate, with 95% confidence, the population proportion of U.S. adults who say chocolate is their favorite ice cream flavor. Your estimate must be accurate within 5% of the population proportion.

(a) No preliminary estimate is available. Find the minimum sample size needed.

(b) Find the minimum sample size needed, using a prior study that found that 27% of U.S. adults say that chocolate is their favorite ice cream flavor. *(Source: Harris Interactive)*

(c) Compare the results from parts (a) and (b).

Constructing Confidence Intervals

In Exercises 21 and 22, use the following information. The graph shows the results of a survey in which 1017 adults from the United States, 1060 adults from Italy, and 1126 adults from Great Britain were asked if they believe climate change poses a large threat to the world. (Source: Harris Interactive)

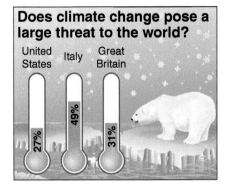

Does climate change pose a large threat to the world?

United States 27%
Italy 49%
Great Britain 31%

21. Global Warming Construct a 99% confidence interval for

   (a) the population proportion of adults from the United States who say that climate change poses a large threat to the world.

   (b) the population proportion of adults from Italy who say that climate change poses a large threat to the world.

   (c) the population proportion of adults from Great Britain who say that climate change poses a large threat to the world.

22. Global Warming Determine whether it is possible that the following proportions are equal and explain your reasoning.

   (a) The proportion of adults from Exercise 21(a) and the proportion of adults from Exercise 21(b).

   (b) The proportion of adults from Exercise 21(b) and the proportion of adults from Exercise 21(c).

   (c) The proportion of adults from Exercise 21(a) and the proportion of adults from Exercise 21(c).

Constructing Confidence Intervals

In Exercises 23 and 24, use the following information. The table shows the results of a survey in which separate samples of 400 adults each from the East, South, Midwest, and West were asked if traffic congestion is a serious problem in their community. (Adapted from Harris Interactive)

Bad Traffic Congestion?
Adults who say that traffic congestion is a serious problem

East	36%
South	32%
Midwest	26%
West	56%

23. South and West Construct a 95% confidence interval for the population proportion of adults

   (a) from the South who say traffic congestion is a serious problem.

   (b) from the West who say traffic congestion is a serious problem.

24. East and Midwest Construct a 95% confidence interval for the population proportion of adults

   (a) from the East who say traffic congestion is a serious problem.

   (b) from the Midwest who say traffic congestion is a serious problem.

25. Writing Is it possible that the proportions in Exercise 23 are equal? What if you used a 99% confidence interval? Explain your reasoning.

26. Writing Is it possible that the proportions in Exercise 24 are equal? What if you used a 99% confidence interval? Explain your reasoning.

SC *In Exercises 27 and 28, use StatCrunch to construct 90%, 95%, and 99% confidence intervals for the population proportion. Interpret the results and compare the widths of the confidence intervals.*

27. **Congress** In a survey of 1025 U.S. adults, 802 disapprove of the job Congress is doing. *(Adapted from The Gallup Poll)*

28. **UFOs** In a survey of 2303 U.S. adults, 734 believe in UFOs. *(Adapted from Harris Interactive)*

■ EXTENDING CONCEPTS

Newspaper Surveys *In Exercises 29 and 30, translate the newspaper excerpt into a confidence interval for p. Approximate the level of confidence.*

29. In a survey of 8451 U.S. adults, 31.4% said they were taking vitamin E as a supplement. The survey's margin of error is plus or minus 1%. *(Source: Decision Analyst, Inc.)*

30. In a survey of 1000 U.S. adults, 19% are concerned that their taxes will be audited by the Internal Revenue Service. The survey's margin of error is plus or minus 3%. *(Source: Rasmussen Reports)*

31. **Why Check It?** Why is it necessary to check that $n\hat{p} \geq 5$ and $n\hat{q} \geq 5$?

32. **Sample Size** The equation for determining the sample size

$$n = \hat{p}\hat{q}\left(\frac{z_c}{E}\right)^2$$

can be obtained by solving the equation for the margin of error

$$E = z_c \sqrt{\frac{\hat{p}\hat{q}}{n}}$$

for n. Show that this is true and justify each step.

33. **Maximum Value of $\hat{p}\hat{q}$** Complete the tables for different values of $\hat{p}$ and $\hat{q} = 1 - \hat{p}$. From the tables, which value of $\hat{p}$ appears to give the maximum value of the product $\hat{p}\hat{q}$?

$\hat{p}$	$\hat{q} = 1 - \hat{p}$	$\hat{p}\hat{q}$	$\hat{p}$	$\hat{q} = 1 - \hat{p}$	$\hat{p}\hat{q}$
0.0	1.0	0.00	0.45		
0.1	0.9	0.09	0.46		
0.2	0.8		0.47		
0.3			0.48		
0.4			0.49		
0.5			0.50		
0.6			0.51		
0.7			0.52		
0.8			0.53		
0.9			0.54		
1.0			0.55		

ACTIVITY 6.3 Confidence Intervals for a Proportion

APPLET

The *confidence intervals for a proportion* applet allows you to visually investigate confidence intervals for a population proportion. You can specify the sample size n and the population proportion p. When you click SIMULATE, 100 separate samples of size n will be selected from a population with a proportion of successes equal to p. For each of the 100 samples, a 95% confidence interval (in green) and a 99% confidence interval (in blue) are displayed in the plot at the right. Each of these intervals is computed using the standard normal approximation. If an interval does not contain the population proportion, it is displayed in red. Note that the 99% confidence interval is always wider than the 95% confidence interval. Additional simulations can be carried out by clicking SIMULATE multiple times. The cumulative number of times that each type of interval contains the population proportion is also shown. Press CLEAR to clear existing results and start a new simulation.

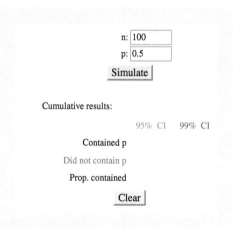

■ Explore

Step 1 Specify a value for n.
Step 2 Specify a value for p.
Step 3 Click SIMULATE to generate the confidence intervals.

■ Draw Conclusions

APPLET

1. Run the simulation for $p = 0.6$ and $n = 10, 20, 40$, and 100. Clear the results after each trial. What proportion of the confidence intervals for each confidence level contains the population proportion? What happens to the proportion of confidence intervals that contains the population proportion for each confidence level as the sample size increases?

2. Run the simulation for $p = 0.4$ and $n = 100$ so that at least 1000 confidence intervals are generated. Compare the proportion of confidence intervals that contains the population proportion for each confidence level. Is this what you would expect? Explain.

6.4 Confidence Intervals for Variance and Standard Deviation

WHAT YOU SHOULD LEARN

▸ How to interpret the chi-square distribution and use a chi-square distribution table

▸ How to use the chi-square distribution to construct a confidence interval for the variance and standard deviation

The Chi-Square Distribution ▸ Confidence Intervals for σ^2 and σ

▸ THE CHI-SQUARE DISTRIBUTION

In manufacturing, it is necessary to control the amount that a process varies. For instance, an automobile part manufacturer must produce thousands of parts to be used in the manufacturing process. It is important that the parts vary little or not at all. How can you measure, and consequently control, the amount of variation in the parts? You can start with a point estimate.

> **DEFINITION**
>
> The **point estimate for σ^2** is s^2 and the **point estimate for σ** is s. The most unbiased estimate for σ^2 is s^2.

You can use a *chi-square distribution* to construct a confidence interval for the variance and standard deviation.

STUDY TIP

The Greek letter χ is pronounced "$k\bar{\iota}$," which rhymes with the more familiar Greek letter π.

> **DEFINITION**
>
> If a random variable x has a normal distribution, then the distribution of
>
> $$\chi^2 = \frac{(n-1)s^2}{\sigma^2}$$
>
> forms a **chi-square distribution** for samples of any size $n > 1$. Four properties of the chi-square distribution are as follows.
>
> **1.** All chi-square values χ^2 are greater than or equal to 0.
>
> **2.** The chi-square distribution is a family of curves, each determined by the degrees of freedom. To form a confidence interval for σ^2, use the χ^2-distribution with degrees of freedom equal to one less than the sample size.
>
> $$\text{d.f.} = n - 1 \qquad \text{Degrees of freedom}$$
>
> **3.** The area under each curve of the chi-square distribution equals 1.
>
> **4.** Chi-square distributions are positively skewed.

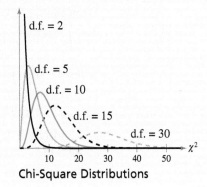

Chi-Square Distributions

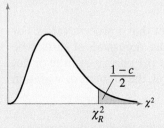
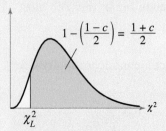
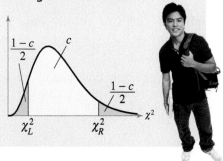
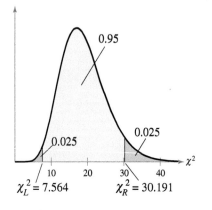

There are two critical values for each level of confidence. The value χ_R^2 represents the right-tail critical value and χ_L^2 represents the left-tail critical value. Table 6 in Appendix B lists critical values of χ^2 for various degrees of freedom and areas. Each area in the table represents the region under the chi-square curve to the *right* of the critical value.

EXAMPLE 1

▶ **Finding Critical Values for χ^2**

Find the critical values χ_R^2 and χ_L^2 for a 95% confidence interval when the sample size is 18.

▶ **Solution**

Because the sample size is 18, there are

$$\text{d.f.} = n - 1 = 18 - 1 = 17 \text{ degrees of freedom.}$$

The areas to the right of χ_R^2 and χ_L^2 are

$$\text{Area to right of } \chi_R^2 = \frac{1 - c}{2} = \frac{1 - 0.95}{2} = 0.025$$

and

$$\text{Area to right of } \chi_L^2 = \frac{1 + c}{2} = \frac{1 + 0.95}{2} = 0.975.$$

Part of Table 6 is shown. Using d.f. = 17 and the areas 0.975 and 0.025, you can find the critical values, as shown by the highlighted areas in the table.

Degrees of freedom	α							
	0.995	0.99	0.975	0.95	0.90	0.10	0.05	0.025
1	—	—	0.001	0.004	0.016	2.706	3.841	5.024
2	0.010	0.020	0.051	0.103	0.211	4.605	5.991	7.378
3	0.072	0.115	0.216	0.352	0.584	6.251	7.815	9.348
15	4.601	5.229	6.262	7.261	8.547	22.307	24.996	27.488
16	5.142	5.812	6.908	7.962	9.312	23.542	26.296	28.845
17	5.697	6.408	7.564	8.672	10.085	24.769	27.587	30.191
18	6.265	7.015	8.231	9.390	10.865	25.989	28.869	31.526
19	6.844	7.633	8.907	10.117	11.651	27.204	30.144	32.852
20	7.434	8.260	9.591	10.851	12.443	28.412	31.410	34.170

χ_L^2 χ_R^2

From the table, you can see that $\chi_R^2 = 30.191$ and $\chi_L^2 = 7.564$.

Interpretation So, 95% of the area under the curve lies between 7.564 and 30.191.

▶ **Try It Yourself 1**

Find the critical values χ_R^2 and χ_L^2 for a 90% confidence interval when the sample size is 30.

a. Identify the *degrees of freedom* and the *level of confidence*.
b. *Find the areas* to the right of χ_R^2 and χ_L^2.
c. *Use* Table 6 in Appendix B to find χ_R^2 and χ_L^2.
d. *Interpret* the results.   *Answer: Page A40*

▶ CONFIDENCE INTERVALS FOR σ^2 AND σ

You can use the critical values χ_R^2 and χ_L^2 to construct confidence intervals for a population variance and standard deviation. The best point estimate for the variance is s^2 and the best point estimate for the standard deviation is s.

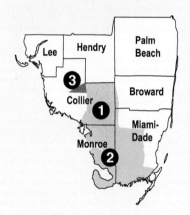
DEFINITION

The c-confidence intervals for the population variance and standard deviation are as follows.

Confidence Interval for σ^2:

$$\frac{(n-1)s^2}{\chi_R^2} < \sigma^2 < \frac{(n-1)s^2}{\chi_L^2}$$

Confidence Interval for σ:

$$\sqrt{\frac{(n-1)s^2}{\chi_R^2}} < \sigma < \sqrt{\frac{(n-1)s^2}{\chi_L^2}}$$

The probability that the confidence intervals contain σ^2 or σ is c.

GUIDELINES

Constructing a Confidence Interval for a Variance and Standard Deviation

IN WORDS	IN SYMBOLS
1. Verify that the population has a normal distribution.	
2. Identify the sample statistic n and the degrees of freedom.	d.f. $= n - 1$
3. Find the point estimate s^2.	$s^2 = \dfrac{\Sigma(x - \bar{x})^2}{n-1}$
4. Find the critical values χ_R^2 and χ_L^2 that correspond to the given level of confidence c.	Use Table 6 in Appendix B.

Left Endpoint   Right Endpoint

5. Find the left and right endpoints and form the confidence interval for the population variance.

$$\frac{(n-1)s^2}{\chi_R^2} < \sigma^2 < \frac{(n-1)s^2}{\chi_L^2}$$

6. Find the confidence interval for the population standard deviation by taking the square root of each endpoint.

$$\sqrt{\frac{(n-1)s^2}{\chi_R^2}} < \sigma < \sqrt{\frac{(n-1)s^2}{\chi_L^2}}$$

EXAMPLE 2   SC   Report 28

▶ **Constructing a Confidence Interval**

You randomly select and weigh 30 samples of an allergy medicine. The sample standard deviation is 1.20 milligrams. Assuming the weights are normally distributed, construct 99% confidence intervals for the population variance and standard deviation.

▶ **Solution**

The areas to the right of χ_R^2 and χ_L^2 are

$$\text{Area to right of } \chi_R^2 = \frac{1-c}{2} = \frac{1-0.99}{2} = 0.005$$

and

$$\text{Area to right of } \chi_L^2 = \frac{1+c}{2} = \frac{1+0.99}{2} = 0.995.$$

Using the values $n = 30$, d.f. = 29, and $c = 0.99$, the critical values χ_R^2 and χ_L^2 are

$$\chi_R^2 = 52.336 \quad \text{and} \quad \chi_L^2 = 13.121.$$

Using these critical values and $s = 1.20$, the confidence interval for σ^2 is as follows.

Left Endpoint

$$\frac{(n-1)s^2}{\chi_R^2} = \frac{(30-1)(1.20)^2}{52.336} \approx 0.80$$

Right Endpoint

$$\frac{(n-1)s^2}{\chi_L^2} = \frac{(30-1)(1.20)^2}{13.121} \approx 3.18$$

$$0.80 < \sigma^2 < 3.18$$

The confidence interval for σ is

$$\sqrt{\frac{(30-1)(1.20)^2}{52.336}} < \sigma < \sqrt{\frac{(30-1)(1.20)^2}{13.121}}$$

$$0.89 < \sigma < 1.78.$$

Interpretation With 99% confidence, you can say that the population variance is between 0.80 and 3.18, and the population standard deviation is between 0.89 and 1.78 milligrams.

▶ **Try It Yourself 2**

Find the 90% and 95% confidence intervals for the population variance and standard deviation of the medicine weights.

a. Find the *critical values* χ_R^2 and χ_L^2 for each confidence interval.
b. Use n, s, χ_R^2, and χ_L^2 to find the *left* and *right endpoints* for each confidence interval for the population variance.
c. Find the *square roots* of the endpoints of each confidence interval.
d. *Specify* the 90% and 95% confidence intervals for the population variance and standard deviation.

Answer: Page A40

STUDY TIP

When a confidence interval for a population variance or standard deviation is computed, the general *round-off rule* is to round off to the same number of decimal places given for the sample variance or standard deviation.

6.4 EXERCISES

BUILDING BASIC SKILLS AND VOCABULARY

1. Does a population have to be normally distributed in order to use the chi-square distribution?

2. What happens to the shape of the chi-square distribution as the degrees of freedom increase?

In Exercises 3–8, find the critical values χ_R^2 and χ_L^2 for the given confidence level c and sample size n.

3. $c = 0.90$, $n = 8$

4. $c = 0.99$, $n = 15$

5. $c = 0.95$, $n = 20$

6. $c = 0.98$, $n = 26$

7. $c = 0.99$, $n = 30$

8. $c = 0.80$, $n = 51$

USING AND INTERPRETING CONCEPTS

Constructing Confidence Intervals   *In Exercises 9–24, assume each sample is taken from a normally distributed population and construct the indicated confidence intervals for (a) the population variance σ^2 and (b) the population standard deviation σ. Interpret the results.*

9. Vitamins   To analyze the variation in weights of vitamin supplement tablets, you randomly select and weigh 14 tablets. The results (in milligrams) are shown. Use a 90% level of confidence.

500.000	499.995	500.010	499.997	500.015
499.988	500.000	499.996	500.020	500.002
499.998	499.996	500.003	500.000	

10. Cough Syrup   You randomly select and measure the volumes of the contents of 15 bottles of cough syrup. The results (in fluid ounces) are shown. Use a 90% level of confidence.

4.211	4.246	4.269	4.241	4.260
4.293	4.189	4.248	4.220	4.239
4.253	4.209	4.300	4.256	4.290

11. Car Batteries   The reserve capacities (in hours) of 18 randomly selected automotive batteries are shown. Use a 99% level of confidence. *(Adapted from Consumer Reports)*

1.70	1.60	1.94	1.58	1.74	1.60
1.86	1.72	1.38	1.46	1.64	1.49
1.55	1.70	1.75	0.88	1.77	2.07

12. Bolts   You randomly select and measure the lengths of 17 bolts. The results (in inches) are shown. Use a 95% level of confidence.

1.286	1.138	1.240	1.132	1.381	1.137
1.300	1.167	1.240	1.401	1.241	1.171
1.217	1.360	1.302	1.331	1.383	

13. **LCD TVs** A magazine includes a report on the energy costs per year for 32-inch liquid crystal display (LCD) televisions. The article states that 14 randomly selected 32-inch LCD televisions have a sample standard deviation of $3.90. Use a 99% level of confidence. *(Adapted from Consumer Reports)*

14. **Digital Cameras** A magazine includes a report on the prices of subcompact digital cameras. The article states that 11 randomly selected subcompact digital cameras have a sample standard deviation of $109. Use an 80% level of confidence. *(Adapted from Consumer Reports)*

15. **Spring Break** As part of your spring break planning, you randomly select 10 hotels in Cancun, Mexico, and record the room rate for each hotel. The results are shown in the stem-and-leaf plot. Use a 98% level of confidence. *(Source: Expedia, Inc.)*

```
 6 | 9        Key: 7|4 = 74
 7 | 4
 8 |
 9 | 0 9 9
10 |
11 | 2
12 |
13 | 6 9
14 | 9
15 | 0
```

16. **Cordless Drills** The weights (in pounds) of a random sample of 14 cordless drills are shown in the stem-and-leaf plot. Use a 99% level of confidence. *(Adapted from Consumer Reports)*

```
3 | 4 6 9      Key: 3|4 = 3.4
4 | 6 8 9
5 | 1 3 4 5 7 9
6 | 0 1
```

17. **Pulse Rates** The pulse rates of a random sample of 16 adults are shown in the dot plot. Use a 95% level of confidence.

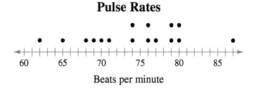

Pulse Rates

Beats per minute

18. **Blu-Ray™ Players** The prices of a random sample of 27 Blu-ray™ players are shown in the dot plot. Use a 98% level of confidence. *(Adapted from Consumer Reports)*

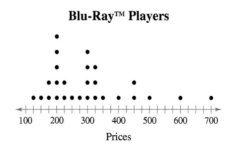

Blu-Ray™ Players

Prices

19. Water Quality As part of a water quality survey, you test the water hardness in several randomly selected streams. The results are shown in the figure. Use a 95% level of confidence.

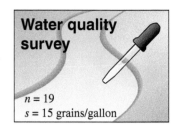

Water quality survey

$n = 19$
$s = 15$ grains/gallon

20. Website Costs As part of a survey, you ask a random sample of business owners how much they would be willing to pay for a website for their company. The results are shown in the figure. Use a 90% level of confidence.

How much will you pay for your site?

$n = 30$
$s = \$3600$

21. Annual Earnings The annual earnings of 14 randomly selected computer software engineers have a sample standard deviation of $3725. Use an 80% level of confidence. *(Adapted from U.S. Bureau of Labor Statistics)*

22. Annual Precipitation The average annual precipitations (in inches) of a random sample of 30 years in San Francisco, California have a sample standard deviation of 8.18 inches. Use a 98% level of confidence. *(Source: Golden Gate Weather Services)*

23. Waiting Times The waiting times (in minutes) of a random sample of 22 people at a bank have a sample standard deviation of 3.6 minutes. Use a 98% level of confidence.

24. Motorcycles The prices of a random sample of 20 new motorcycles have a sample standard deviation of $3900. Use a 90% level of confidence.

SC *In Exercises 25–28, use StatCrunch to help you construct the indicated confidence intervals for the population variance σ^2 and the population standard deviation σ. Assume each sample is taken from a normally distributed population.*

25. $c = 0.95, s^2 = 11.56, n = 30$ **26.** $c = 0.99, s^2 = 0.64, n = 7$

27. $c = 0.90, s = 35, n = 18$ **28.** $c = 0.97, s = 278.1, n = 45$

■ EXTENDING CONCEPTS

29. Vitamin Tablet Weights You are analyzing the sample of vitamin supplement tablets in Exercise 9. The population standard deviation of the tablets' weights should be less than 0.015 milligram. Does the confidence interval you constructed for σ suggest that the variation in the tablets' weights is at an acceptable level? Explain your reasoning.

30. Cough Syrup Bottle Contents You are analyzing the sample of cough syrup bottles in Exercise 10. The population standard deviation of the volumes of the bottles' contents should be less than 0.025 fluid ounce. Does the confidence interval you constructed for σ suggest that the variation in the volumes of the bottles' contents is at an acceptable level? Explain your reasoning.

31. In your own words, explain how finding a confidence interval for a population variance is different from finding a confidence interval for a population mean or proportion.

USES AND ABUSES

Uses

By now, you know that complete information about population parameters is often not available. The techniques of this chapter can be used to make interval estimates of these parameters so that you can make informed decisions.

From what you learned in this chapter, you know that point estimates (sample statistics) of population parameters are usually close but rarely equal to the actual values of the parameters they are estimating. Remembering this can help you make good decisions in your career and in everyday life. For instance, suppose the results of a survey tell you that 52% of the population plans to vote in favor of the rezoning of a portion of a town from residential to commercial use. You know that this is only a point estimate of the actual proportion that will vote in favor of rezoning. If the interval estimate is $0.49 < p < 0.55$, then you know this means it is possible that the item will not receive a majority vote.

Abuses

Unrepresentative Samples There are many ways that surveys can result in incorrect predictions. When you read the results of a survey, remember to question the sample size, the sampling technique, and the questions asked. For instance, suppose you want to know the proportion of people who will vote in favor of rezoning. From the diagram below, you can see that even if your sample is large enough, it may not consist of actual voters.

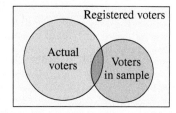

Using a small sample might be the only way to make an estimate, but be aware that a change in one data value may completely change the results. Generally, the larger the sample size, the more accurate the results will be.

Biased Survey Questions In surveys, it is also important to analyze the wording of the questions. For instance, the question about rezoning might be presented as: "Knowing that rezoning will result in more businesses contributing to school taxes, would you support the rezoning?"

■ EXERCISES

1. ***Unrepresentative Samples*** Find an example of a survey that is reported in a newspaper, magazine, or on a website. Describe different ways that the sample could have been unrepresentative of the population.

2. ***Biased Survey Questions*** Find an example of a survey that is reported in a newspaper, magazine, or on a website. Describe different ways that the survey questions could have been biased.

6 CHAPTER SUMMARY

What did you **learn?**	EXAMPLE(S)	REVIEW EXERCISES
Section 6.1		
■ How to find a point estimate and a margin of error $E = z_c \dfrac{\sigma}{\sqrt{n}}$ Margin of error	1, 2	1, 2
■ How to construct and interpret confidence intervals for the population mean $\bar{x} - E < \mu < \bar{x} + E$	3–5	3–6
■ How to determine the minimum sample size required when estimating μ	6	7–10
Section 6.2		
■ How to interpret the t-distribution and use a t-distribution table $t = \dfrac{(\bar{x} - \mu)}{(s/\sqrt{n})}$	1	11–16
■ How to construct confidence intervals when $n < 30$, the population is normally distributed, and σ is unknown $\bar{x} - E < \mu < \bar{x} + E, \quad E = t_c \dfrac{s}{\sqrt{n}}$	2–4	17–26
Section 6.3		
■ How to find a point estimate for a population proportion $\hat{p} = \dfrac{x}{n}$	1	27–34
■ How to construct a confidence interval for a population proportion $\hat{p} - E < p < \hat{p} + E, \quad E = z_c \sqrt{\dfrac{\hat{p}\hat{q}}{n}}$	2, 3	35–42
■ How to determine the minimum sample size required when estimating a population proportion	4	43, 44
Section 6.4		
■ How to interpret the chi-square distribution and use a chi-square distribution table $\chi^2 = \dfrac{(n-1)s^2}{\sigma^2}$	1	45–48
■ How to use the chi-square distribution to construct a confidence interval for the variance and standard deviation $\dfrac{(n-1)s^2}{\chi_R^2} < \sigma^2 < \dfrac{(n-1)s^2}{\chi_L^2}, \quad \sqrt{\dfrac{(n-1)s^2}{\chi_R^2}} < \sigma < \sqrt{\dfrac{(n-1)s^2}{\chi_L^2}}$	2	49–52

6 REVIEW EXERCISES

■ SECTION 6.1

In Exercises 1 and 2, find (a) the point estimate of the population mean μ and (b) the margin of error for a 90% confidence interval.

1. Waking times of 40 people who start work at 8:00 A.M. (in minutes past 5:00 A.M.)

135	145	95	140	135	95	110	50
90	165	110	125	80	125	130	110
25	75	65	100	60	125	115	135
95	90	140	40	75	50	130	85
100	160	135	45	135	115	75	130

2. Lengths of commutes to work of 32 people (in miles)

12	9	7	2	8	7	3	27
21	10	13	3	7	2	30	7
6	13	6	14	4	1	10	3
13	6	2	9	2	12	16	18

In Exercises 3 and 4, construct the indicated confidence interval for the population mean μ. If convenient, use technology to construct the confidence interval.

3. $c = 0.99, \bar{x} = 15.8, s = 0.85, n = 80$

4. $c = 0.95, \bar{x} = 7.675, s = 0.105, n = 55$

In Exercises 5 and 6, use the given confidence interval to find the margin of error and the sample mean.

5. (20.75, 24.10)

6. (7.428, 7.562)

In Exercises 7–10, determine the minimum sample size n needed to estimate μ.

7. Use the results of Exercise 1. Determine the minimum survey size that is necessary to be 95% confident that the sample mean waking time is within 10 minutes of the actual mean waking time.

8. Use the results of Exercise 1. Now suppose you want 99% confidence with a margin of error of 2 minutes. How many people would you need to survey?

9. Use the results of Exercise 2. Determine the minimum survey size that is necessary to be 95% confident that the sample mean length of commutes to work is within 2 miles of the actual mean length of commutes to work.

10. Use the results of Exercise 2. Now suppose you want 98% confidence with a margin of error of 0.5 mile. How many people would you need to survey?

■ SECTION 6.2

In Exercises 11–14, find the critical value t_c for the given confidence level c and sample size n.

11. $c = 0.80, n = 10$
12. $c = 0.95, n = 24$
13. $c = 0.98, n = 15$
14. $c = 0.99, n = 30$

15. Consider a 90% confidence interval for μ. Assume σ is not known. For which sample size, $n = 20$ or $n = 30$, is the critical value t_c larger?

16. Consider a 90% confidence interval for μ. Assume σ is not known. For which sample size, $n = 20$ or $n = 30$, is the confidence interval wider?

In Exercises 17–20, find the margin of error for μ.

17. $c = 0.90, s = 25.6, n = 16, \bar{x} = 72.1$

18. $c = 0.95, s = 1.1, n = 25, \bar{x} = 3.5$

19. $c = 0.98, s = 0.9, n = 12, \bar{x} = 6.8$

20. $c = 0.99, s = 16.5, n = 20, \bar{x} = 25.2$

In Exercises 21–24, construct the confidence interval for μ using the statistics from the given exercise. If convenient, use technology to construct the confidence interval.

21. Exercise 17   **22.** Exercise 18

23. Exercise 19   **24.** Exercise 20

25. In a random sample of 28 sports cars, the average annual fuel cost was $2218 and the standard deviation was $523. Construct a 90% confidence interval for μ. Assume the annual fuel costs are normally distributed. *(Adapted from U.S. Department of Energy)*

26. Repeat Exercise 25 using a 99% confidence interval.

■ SECTION 6.3

In Exercises 27–34, let p be the proportion of the population who respond yes. Use the given information to find $\hat{p}$ and $\hat{q}$.

27. A survey asks 1500 U.S. adults if they will participate in the 2010 Census. The results are shown in the pie chart. *(Adapted from Pew Research Center)*

28. In a survey of 500 U.S. adults, 425 say they would trust doctors to tell the truth. *(Adapted from Harris Interactive)*

29. In a survey of 1023 U.S. adults, 552 say they have worked the night shift at some point in their lives. *(Adapted from CNN/Opinion Research)*

30. In a survey of 800 U.S. adults, 90 are making the minimum payment(s) on their credit card(s). *(Adapted from Cambridge Consumer Credit Index)*

31. In a survey of 1008 U.S. adults, 141 say the cost of health care is the most important financial problem facing their family today. *(Adapted from Gallup, Inc.)*

32. In a survey of 938 U.S. adults, 235 say the phrase "you know" is the most annoying conversational phrase. *(Adapted from Marist Poll)*

33. In a survey of 706 parents with kids 4 to 8 years old, 346 say that they know their state booster seat law. *(Adapted from Knowledge Networks, Inc.)*

34. In a survey of 2365 U.S. adults, 1230 say they worry most about missing deductions when filing their taxes. *(Adapted from USA TODAY)*

In Exercises 35–42, construct the indicated confidence interval for the population proportion p. If convenient, use technology to construct the confidence interval. Interpret the results.

35. Use the sample in Exercise 27 with $c = 0.95$.

36. Use the sample in Exercise 28 with $c = 0.99$.

37. Use the sample in Exercise 29 with $c = 0.90$.

38. Use the sample in Exercise 30 with $c = 0.98$.

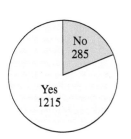

FIGURE FOR EXERCISE 27

39. Use the sample in Exercise 31 with $c = 0.99$.

40. Use the sample in Exercise 32 with $c = 0.90$.

41. Use the sample in Exercise 33 with $c = 0.80$.

42. Use the sample in Exercise 34 with $c = 0.98$.

43. You wish to estimate, with 95% confidence, the population proportion of U.S. adults who think they should be saving more money. Your estimate must be accurate within 5% of the population proportion.

   (a) No preliminary estimate is available. Find the minimum sample size needed.

   (b) Find the minimum sample size needed, using a prior study that found that 63% of U.S. adults think that they should be saving more money. *(Source: Pew Research Center)*

   (c) Compare the results from parts (a) and (b).

44. Repeat Exercise 43 part (b), using a 99% confidence level and a margin of error of 2.5%. How does this sample size compare with your answer from Exercise 43 part (b)?

■ SECTION 6.4

In Exercises 45–48, find the critical values χ_R^2 and χ_L^2 for the given confidence level c and sample size n.

45. $c = 0.95, n = 13$ **46.** $c = 0.98, n = 25$

47. $c = 0.90, n = 8$ **48.** $c = 0.99, n = 10$

In Exercises 49–52, construct the indicated confidence intervals for the population variance σ^2 and the population standard deviation σ. Assume each sample is taken from a normally distributed population.

49. A random sample of the weights (in ounces) of 17 superzoom digital cameras is shown in the stem-and-leaf plot. Use a 95% level of confidence. *(Adapted from Consumer Reports)*

```
0 | 7 8 8 9          Key: 1|3 = 13
1 | 0 1 3 4 5 5 5 7 7 9
2 | 1 4
3 | 5
```

50. Repeat Exercise 49 using a 99% level of confidence. Interpret the results and compare with Exercise 49.

51. A random sample of the acceleration times (in seconds) from 0 to 60 miles per hour for 26 sedans is shown in the dot plot. Use a 98% level of confidence. *(Adapted from Consumer Reports)*

Acceleration Times From 0–60 Miles Per Hour for Sedans

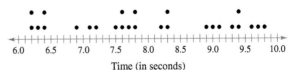

Time (in seconds)

52. Repeat Exercise 51 using a 90% level of confidence. Interpret the results and compare with Exercise 51.

6 CHAPTER QUIZ

Take this quiz as you would take a quiz in class. After you are done, check your work against the answers given in the back of the book.

 1. The following data set represents the amounts of time (in minutes) spent watching online videos each day for a random sample of 30 college students. *(Adapted from the Council for Research Excellence)*

5.0	6.25	8.0	5.5	4.75	4.5	7.2	6.6	5.8	5.5
4.2	5.4	6.75	9.8	8.2	6.4	7.8	6.5	5.5	6.0
3.8	6.75	9.25	10.0	9.6	7.2	6.4	6.8	9.8	10.2

(a) Find the point estimate of the population mean.

(b) Find the margin of error for a 95% level of confidence. Interpret the result.

(c) Construct a 95% confidence interval for the population mean. Interpret the results.

2. You want to estimate the mean time college students spend watching online videos each day. The estimate must be within 1 minute of the population mean. Determine the required sample size to construct a 99% confidence interval for the population mean. Assume the population standard deviation is 2.4 minutes.

3. The following data set represents the average number of minutes played for a random sample of professional basketball players in a recent season. *(Source: ESPN)*

35.9 33.8 34.7 31.5 33.2 29.1 30.7 31.2 36.1 34.9

(a) Find the sample mean and the sample standard deviation.

(b) Construct a 90% confidence interval for the population mean and interpret the results. Assume the population of the data set is normally distributed.

(c) Repeat part (b), assuming $\sigma = 5.25$ minutes per game. Interpret and compare the results.

4. In a random sample of seven aerospace engineers, the mean monthly income was $6824 and the standard deviation was $340. Assume the monthly incomes are normally distributed and construct a 95% confidence interval for the population mean monthly income for aerospace engineers. *(Adapted from U.S. Bureau of Labor Statistics)*

5. In a survey of 1383 U.S. adults, 1079 favor increasing federal funding for research on wind, solar, and hydrogen technology. *(Adapted from Pew Research Center)*

(a) Find a point estimate for the population proportion p of those in favor of increasing federal funding for research on wind, solar, and hydrogen technology.

(b) Construct a 90% confidence interval for the population proportion.

(c) Find the minimum sample size needed to estimate the population proportion at the 99% confidence level in order to ensure that the estimate is accurate within 4% of the population proportion.

6. Refer to the data set in Exercise 1. Assume the population of times spent watching online videos each day is normally distributed.

(a) Construct a 95% confidence interval for the population variance.

(b) Construct a 95% confidence interval for the population standard deviation.

PUTTING IT ALL TOGETHER

Real Statistics — Real Decisions

In 1974, the Safe Drinking Water Act was passed "to protect public health by regulating the nation's public drinking water supply." In accordance with the act, the Environmental Protection Agency (EPA) has regulations that limit the levels of contaminants in drinking water supplied by water utilities. These utilities are required to supply water quality reports to their customers annually. These reports discuss the source of the water, its treatment, and the results of water quality monitoring that is performed daily. The results of this monitoring indicate whether or not drinking water is healthy enough for consumption.

A water department tests for contaminants at water treatment plants and at customers' taps. These regulated parameters include microorganisms, organic chemicals, and inorganic chemicals. For instance, cyanide is an inorganic chemical that is regulated. Its presence in drinking water is the result of discharges from steel, plastics, and fertilizer factories. The maximum contaminant level for cyanide is set at 0.2 part per million.

You work for a city's water department and are interpreting the results shown in the graph at the right. The graph shows the point estimates for the population mean concentration and the 95% confidence intervals for μ for cyanide over a three-year period. The data are based on random water samples taken by the city's three water treatment plants.

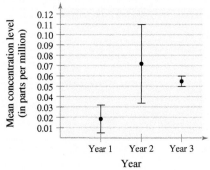

Cyanide

■ EXERCISES

1. *Interpreting the Results*

Use the graph to decide if there has been a change in the mean concentration level of cyanide for the given years. Explain your reasoning.

(a) From Year 1 to Year 2   (b) From Year 2 to Year 3

(c) From Year 1 to Year 3

2. *What Can You Conclude?*

Using the results of Exercise 1, what can you conclude about the concentrations of cyanide in the drinking water?

3. *What Do You Think?*

The confidence interval for Year 2 is much larger than the other years. What do you think may have caused this larger confidence level?

4. *How Do You Think They Did It?*

How do you think the water department constructed the 95% confidence intervals for the population mean concentration of cyanide in the water? Do the following to answer the question. (You do not need to make any calculations.)

(a) What sampling distribution do you think they used? Why?

(b) Do you think they used the population standard deviation in calculating the margin of error? Why or why not? If not, what could they have used?

TECHNOLOGY   MINITAB   EXCEL   TI-83/84 PLUS

THE GALLUP ORGANIZATION
WWW.GALLUP.COM

MOST ADMIRED POLLS

Since 1946, the Gallup Organization has conducted a "most admired" poll. The methodology for the 2009 poll is described at the right.

> **Survey Question**
> *What man* that you have heard or read about, living today in any part of the world, do you admire most? And who is your second choice?*

Reprinted with permission from GALLUP.

*Survey respondents are asked an identical question about most admired woman.

"Results are based on telephone interviews with 1,025 national adults, aged 18 and older, conducted Dec. 11–13, 2009. For results based on the total sample of national adults, one can say with 95% confidence that the maximum margin of sampling error is ±4 percentage points. Interviews are conducted with respondents on land-line telephones (for respondents with a land-line telephone) and cellular phones (for respondents who are cell-phone only). In addition to sampling error, question wording and practical difficulties in conducting surveys can introduce error or bias into the findings of public opinion polls."

■ EXERCISES

1. In 2009, the most named man was Barack Obama at 30%. Use a technology tool to find a 95% confidence interval for the population proportion that would have chosen Barack Obama.

2. In 2009, the most named woman was Hillary Clinton at 16%. Use a technology tool to find a 95% confidence interval for the population proportion that would have chosen Hillary Clinton.

3. Do the confidence intervals you obtained in Exercises 1 and 2 agree with the statement issued by the Gallup Organization that the margin of error is ±4%? Explain.

4. The second most named woman was Sarah Palin, who was named by 15% of the people in the sample. Use a technology tool to find a 95% confidence interval for the population proportion that would have chosen Sarah Palin.

5. Use a technology tool to simulate a most admired poll. Assume that the actual population proportion who most admire Sarah Palin is 18%. Run the simulation several times using $n = 1025$.

   (a) What was the least value you obtained for $\hat{p}$?

   (b) What was the greatest value you obtained for $\hat{p}$?

   > **MINITAB**
   >
   > Number of rows of data to generate: 200
   >
   > Store in column(s): C1
   >
   > Number of trials: 1025
   >
   > Event probability: 0.18

6. Is it probable that the population proportion who most admire Sarah Palin is 18% or greater? Explain your reasoning.

Extended solutions are given in the *Technology Supplement*.
Technical instruction is provided for MINITAB, Excel, and the TI-83/84 Plus.

6 USING TECHNOLOGY TO CONSTRUCT CONFIDENCE INTERVALS

Here are some MINITAB and TI-83/84 Plus printouts for some examples in this chapter. Answers may be slightly different because of rounding.

(See Example 3, page 307.)

140	105	130	97	80	165	232	110	214	201	122	98	65	88
154	133	121	82	130	211	153	114	58	77	51	247	236	109
126	132	125	149	122	74	59	218	192	90	117	105		

Display Descriptive Statistics...
Store Descriptive Statistics...
Graphical Summary...

1-Sample Z...
1-Sample t...
2-Sample t...
Paired t...

1 Proportion...
2 Proportions...

MINITAB

One-Sample Z: Friends

The assumed standard deviation = 53

Variable	N	Mean	StDev	SE Mean	95% CI
Friends	40	130.80	52.63	8.38	(114.38, 147.22)

(See Example 2, page 320.)

Display Descriptive Statistics...
Store Descriptive Statistics...
Graphical Summary...

1-Sample Z...
1-Sample t...
2-Sample t...
Paired t...

1 Proportion...
2 Proportions...

MINITAB

One-Sample T

N	Mean	StDev	SE Mean	95% CI
16	162.00	10.00	2.50	(156.67, 167.33)

(See Example 2, page 329.)

Display Descriptive Statistics...
Store Descriptive Statistics...
Graphical Summary...

1-Sample Z...
1-Sample t...
2-Sample t...
Paired t...

1 Proportion...
2 Proportions...

MINITAB

Test and CI for One Proportion

Sample	X	N	Sample p	95% CI
1	662	1000	0.662000	(0.631738, 0.691305)

(See Example 5, page 309.)

TI-83/84 PLUS

EDIT CALC **TESTS**
1: Z-Test...
2: T-Test...
3: 2-SampZTest...
4: 2-SampTTest...
5: 1-PropZTest...
6: 2-PropZTest...
7↓ ZInterval...

⬇

TI-83/84 PLUS

ZInterval
 Inpt: Data **Stats**
 s: 1.5
 x̄: 22.9
 n: 20
 C-Level: .9
 Calculate

⬇

TI-83/84 PLUS

ZInterval
 (22.348, 23.452)
 x̄= 22.9
 n= 20

(See Example 3, page 321.)

TI-83/84 PLUS

EDIT CALC **TESTS**
2↑ T-Test...
3: 2-SampZTest...
4: 2-SampTTest...
5: 1-PropZTest...
6: 2-PropZTest...
7: ZInterval...
8↓ TInterval...

⬇

TI-83/84 PLUS

TInterval
 Inpt: Data **Stats**
 x̄: 9.75
 Sx: 2.39
 n: 20
 C-Level: .99
 Calculate

⬇

TI-83/84 PLUS

TInterval
 (8.2211, 11.279)
 x̄= 9.75
 Sx= 2.39
 n= 20

(See Example 2, page 329.)

TI-83/84 PLUS

EDIT CALC **TESTS**
5↑ 1-PropZTest...
6: 2-PropZTest...
7: ZInterval...
8: TInterval...
9: 2-SampZInt...
0: 2-SampTInt...
A↓ 1-PropZInt...

⬇

TI-83/84 PLUS

1-PropZInt
 x: 662
 n: 1000
 C-Level: .95
 Calculate

⬇

TI-83/84 PLUS

1-PropZInt
 (.63268, .69132)
 p̂= 0.662
 n= 1000

7 HYPOTHESIS TESTING WITH ONE SAMPLE

7.1 Introduction to Hypothesis Testing

7.2 Hypothesis Testing for the Mean (Large Samples)
- CASE STUDY

7.3 Hypothesis Testing for the Mean (Small Samples)
- ACTIVITY

7.4 Hypothesis Testing for Proportions
- ACTIVITY

7.5 Hypothesis Testing for Variance and Standard Deviation
- USES AND ABUSES
- REAL STATISTICS–REAL DECISIONS
- TECHNOLOGY

Computer software is protected by federal copyright laws. Each year, software companies lose billions of dollars because of pirated software. Federal criminal penalties for software piracy can include fines of up to $250,000 and jail terms of up to five years.

In Chapter 6, you began your study of inferential statistics. There, you learned how to form a confidence interval estimate about a population parameter, such as the proportion of people in the United States who agree with a certain statement. For instance, in a nationwide poll conducted by *Harris Interactive* on behalf of the Business Software Alliance (BSA), U.S. students ages 8 to 18 years were asked several questions about their attitudes toward copyright law and Internet behavior. Here are some of the results.

Survey Question	Number Surveyed	Number Who Said Yes
Have you ever downloaded music from the Internet without paying for it?	1196	361
Have you ever downloaded movies from the Internet without paying for them?	1196	95
Have you ever downloaded software from the Internet without paying for it?	1196	133

In this chapter, you will continue your study of inferential statistics. But now, instead of making an estimate about a population parameter, you will learn how to test a claim about a parameter.

For instance, suppose that you work for *Harris Interactive* and are asked to test a claim that the proportion of U.S. students ages 8 to 18 who download music without paying for it is $p = 0.25$. To test the claim, you take a random sample of $n = 1196$ students and find that 361 of them download music without paying for it. Your sample statistic is $\hat{p} \approx 0.302$.

Is your sample statistic different enough from the claim ($p = 0.25$) to decide that the claim is false? The answer lies in the sampling distribution of sample proportions taken from a population in which $p = 0.25$. The graph below shows that your sample statistic is more than 4 standard errors from the claimed value. If the claim is true, the probability of the sample statistic being 4 standard errors or more from the claimed value is extremely small. Something is wrong! If your sample was truly random, then you can conclude that the actual proportion of the student population is not 0.25. In other words, you tested the original claim (hypothesis), and you decided to reject it.

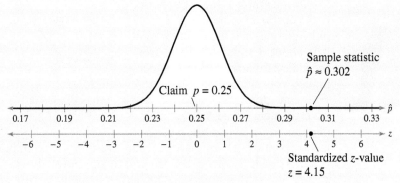

Sampling Distribution

7.1 Introduction to Hypothesis Testing

WHAT YOU SHOULD LEARN

▸ A practical introduction to hypothesis tests

▸ How to state a null hypothesis and an alternative hypothesis

▸ How to identify type I and type II errors and interpret the level of significance

▸ How to know whether to use a one-tailed or two-tailed statistical test and find a *P*-value

▸ How to make and interpret a decision based on the results of a statistical test

▸ How to write a claim for a hypothesis test

Hypothesis Tests ▸ Stating a Hypothesis ▸ Types of Errors and Level of Significance ▸ Statistical Tests and *P*-Values ▸ Making a Decision and Interpreting the Decision ▸ Strategies for Hypothesis Testing

▸ HYPOTHESIS TESTS

Throughout the remainder of this course, you will study an important technique in inferential statistics called hypothesis testing. A **hypothesis test** is a process that uses sample statistics to test a claim about the value of a population parameter. Researchers in fields such as medicine, psychology, and business rely on hypothesis testing to make informed decisions about new medicines, treatments, and marketing strategies.

For instance, suppose an automobile manufacturer advertises that its new hybrid car has a mean gas mileage of 50 miles per gallon. If you suspect that the mean mileage is not 50 miles per gallon, how could you show that the advertisement is false?

Obviously, you cannot test *all* the vehicles, but you can still make a reasonable decision about the mean gas mileage by taking a random sample from the population of vehicles and measuring the mileage of each. If the sample mean differs enough from the advertisement's mean, you can decide that the advertisement is wrong.

For instance, to test that the mean gas mileage of all hybrid vehicles of this type is $\mu = 50$ miles per gallon, you could take a random sample of $n = 30$ vehicles and measure the mileage of each. Suppose you obtain a sample mean of $\bar{x} = 47$ miles per gallon with a sample standard deviation of $s = 5.5$ miles per gallon. Does this indicate that the manufacturer's advertisement is false?

To decide, you do something unusual—*you assume the advertisement is correct!* That is, you assume that $\mu = 50$. Then, you examine the sampling distribution of sample means (with $n = 30$) taken from a population in which $\mu = 50$ and $\sigma = 5.5$. From the Central Limit Theorem, you know this sampling distribution is normal with a mean of 50 and standard error of

$$\frac{5.5}{\sqrt{30}} \approx 1.$$

In the graph at the right, notice that your sample mean of $\bar{x} = 47$ miles per gallon is highly unlikely—it is about 3 standard errors from the claimed mean! Using the techniques you studied in Chapter 5, you can determine that if the advertisement is true, the probability of obtaining a sample mean of 47 or

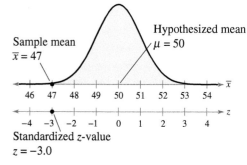

Sampling Distribution of $\bar{x}$

Sample mean $\bar{x} = 47$

Hypothesized mean $\mu = 50$

Standardized *z*-value $z = -3.0$

less is about 0.0013. This is an unusual event! Your assumption that the company's advertisement is correct has led you to an improbable result. So, either you had a very unusual sample, or the advertisement is probably false. The logical conclusion is that the advertisement is probably false.

▶ STATING A HYPOTHESIS

A statement about a population parameter is called a **statistical hypothesis.** To test a population parameter, you should carefully state a pair of hypotheses—one that represents the claim and the other, its complement. When one of these hypotheses is false, the other must be true. Either hypothesis—the *null hypothesis* or the *alternative hypothesis*—may represent the original claim.

DEFINITION

1. A **null hypothesis H_0** is a statistical hypothesis that contains a statement of equality, such as $\leq$, $=$, or $\geq$.
2. The **alternative hypothesis H_a** is the complement of the null hypothesis. It is a statement that must be true if H_0 is false and it contains a statement of strict inequality, such as $>$, $\neq$, or $<$.

H_0 is read as "H sub-zero" or "H naught" and H_a is read as "H sub-a."

To write the null and alternative hypotheses, translate the claim made about the population parameter from a verbal statement to a mathematical statement. Then, write its complement. For instance, if the claim value is k and the population parameter is μ, then some possible pairs of null and alternative hypotheses are

$$\begin{cases} H_0: \mu \leq k \\ H_a: \mu > k \end{cases} \qquad \begin{cases} H_0: \mu \geq k \\ H_a: \mu < k \end{cases} \qquad \begin{cases} H_0: \mu = k \\ H_a: \mu \neq k \end{cases}.$$

Regardless of which of the three pairs of hypotheses you use, you always assume $\mu = k$ and examine the sampling distribution on the basis of this assumption. Within this sampling distribution, you will determine whether or not a sample statistic is unusual.

The following table shows the relationship between possible verbal statements about the parameter μ and the corresponding null and alternative hypotheses. Similar statements can be made to test other population parameters, such as p, σ, or σ^2.

Verbal Statement H_0 *The mean is . . .*	Mathematical Statements	Verbal Statement H_a *The mean is . . .*
. . . greater than or equal to k. . . . at least k. . . . not less than k.	$\begin{cases} H_0: \mu \geq k \\ H_a: \mu < k \end{cases}$	. . . less than k. . . . below k. . . . fewer than k.
. . . less than or equal to k. . . . at most k. . . . not more than k.	$\begin{cases} H_0: \mu \leq k \\ H_a: \mu > k \end{cases}$	. . . greater than k. . . . above k. . . . more than k.
. . . equal to k. . . . k. . . . exactly k.	$\begin{cases} H_0: \mu = k \\ H_a: \mu \neq k \end{cases}$	. . . not equal to k. . . . different from k. . . . not k.

EXAMPLE 1

▶ **Stating the Null and Alternative Hypotheses**

Write the claim as a mathematical sentence. State the null and alternative hypotheses, and identify which represents the claim.

1. A school publicizes that the proportion of its students who are involved in at least one extracurricular activity is 61%.

2. A car dealership announces that the mean time for an oil change is less than 15 minutes.

3. A company advertises that the mean life of its furnaces is more than 18 years.

▶ **Solution**

1. The claim "the proportion ... is 61%" can be written as $p = 0.61$. Its complement is $p \neq 0.61$. Because $p = 0.61$ contains the statement of equality, it becomes the null hypothesis. In this case, the null hypothesis represents the claim.

$$H_0: p = 0.61 \quad \text{(Claim)}$$

$$H_a: p \neq 0.61$$

2. The claim "the mean ... is less than 15 minutes" can be written as $\mu < 15$. Its complement is $\mu \geq 15$. Because $\mu \geq 15$ contains the statement of equality, it becomes the null hypothesis. In this case, the alternative hypothesis represents the claim.

$$H_0: \mu \geq 15 \text{ minutes}$$

$$H_a: \mu < 15 \text{ minutes} \quad \text{(Claim)}$$

3. The claim "the mean ... is more than 18 years" can be written as $\mu > 18$. Its complement is $\mu \leq 18$. Because $\mu \leq 18$ contains the statement of equality, it becomes the null hypothesis. In this case, the alternative hypothesis represents the claim.

$$H_0: \mu \leq 18 \text{ years}$$

$$H_a: \mu > 18 \text{ years} \quad \text{(Claim)}$$

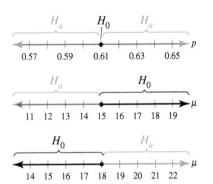

In each of these graphs, notice that each point on the number line is in H_0 or H_a, but no point is in both.

▶ **Try It Yourself 1**

Write the claim as a mathematical sentence. State the null and alternative hypotheses, and identify which represents the claim.

1. A consumer analyst reports that the mean life of a certain type of automobile battery is not 74 months.
2. An electronics manufacturer publishes that the variance of the life of its home theater systems is less than or equal to 2.7.
3. A realtor publicizes that the proportion of homeowners who feel their house is too small for their family is more than 24%.

a. Identify the *verbal claim* and write it as a *mathematical statement*.
b. Write the *complement* of the claim.
c. Identify the *null* and *alternative hypotheses* and determine which one represents the claim.   *Answer: Page A40*

▶ TYPES OF ERRORS AND LEVEL OF SIGNIFICANCE

No matter which hypothesis represents the claim, you always begin a hypothesis test by assuming that the equality condition in the null hypothesis is true. So, when you perform a hypothesis test, you make one of two decisions:

1. reject the null hypothesis or

2. fail to reject the null hypothesis.

Because your decision is based on a sample rather than the entire population, there is always the possibility you will make the wrong decision.

For instance, suppose you claim that a certain coin is not fair. To test your claim, you flip the coin 100 times and get 49 heads and 51 tails. You would probably agree that you do not have enough evidence to support your claim. Even so, it is possible that the coin is actually not fair and you had an unusual sample.

But what if you flip the coin 100 times and get 21 heads and 79 tails? It would be a rare occurrence to get only 21 heads out of 100 tosses with a fair coin. So, you probably have enough evidence to support your claim that the coin is not fair. However, you can't be 100% sure. It is possible that the coin is fair and you had an unusual sample.

If p represents the proportion of heads, the claim that "the coin is not fair" can be written as the mathematical statement $p \neq 0.5$. Its complement, "the coin is fair," is written as $p = 0.5$. So, your null hypothesis and alternative hypothesis are

$$H_0: p = 0.5$$

and

$$H_a: p \neq 0.5. \quad \text{(Claim)}$$

Remember, the only way to be absolutely certain of whether H_0 is true or false is to test the entire population. Because your decision—to reject H_0 or to fail to reject H_0—is based on a sample, you must accept the fact that your decision might be incorrect. You might reject a null hypothesis when it is actually true. Or, you might fail to reject a null hypothesis when it is actually false.

DEFINITION

A **type I error** occurs if the null hypothesis is rejected when it is true.

A **type II error** occurs if the null hypothesis is not rejected when it is false.

The following table shows the four possible outcomes of a hypothesis test.

Decision	Truth of H_0	
	H_0 is true.	H_0 is false.
Do not reject H_0.	Correct decision	Type II error
Reject H_0.	Type I error	Correct decision

	Truth About Defendant	
Verdict	Innocent	Guilty
Not guilty	Justice	Type II error
Guilty	Type I error	Justice

Hypothesis testing is sometimes compared to the legal system used in the United States. Under this system, the following steps are used.

1. A carefully worded accusation is written.

2. The defendant is assumed innocent (H_0) until proven guilty. The burden of proof lies with the prosecution. If the evidence is not strong enough, there is no conviction. A "not guilty" verdict does not prove that a defendant is innocent.

3. The evidence needs to be conclusive beyond a reasonable doubt. The system assumes that more harm is done by convicting the innocent (type I error) than by not convicting the guilty (type II error).

EXAMPLE 2

▶ **Identifying Type I and Type II Errors**

The USDA limit for salmonella contamination for chicken is 20%. A meat inspector reports that the chicken produced by a company exceeds the USDA limit. You perform a hypothesis test to determine whether the meat inspector's claim is true. When will a type I or type II error occur? Which is more serious? *(Source: U.S. Department of Agriculture)*

▶ **Solution**

Let p represent the proportion of the chicken that is contaminated. The meat inspector's claim is "more than 20% is contaminated." You can write the null and alternative hypotheses as follows.

H_0: $p \le 0.2$ — The proportion is less than or equal to 20%.

H_a: $p > 0.2$ (Claim) — The proportion is greater than 20%.

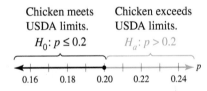

A type I error will occur if the actual proportion of contaminated chicken is less than or equal to 0.2, but you reject H_0. A type II error will occur if the actual proportion of contaminated chicken is greater than 0.2, but you do not reject H_0. With a type I error, you might create a health scare and hurt the sales of chicken producers who were actually meeting the USDA limits. With a type II error, you could be allowing chicken that exceeded the USDA contamination limit to be sold to consumers. A type II error is more serious because it could result in sickness or even death.

▶ **Try It Yourself 2**

A company specializing in parachute assembly states that its main parachute failure rate is not more than 1%. You perform a hypothesis test to determine whether the company's claim is false. When will a type I or type II error occur? Which is more serious?

a. State the *null* and *alternative hypotheses*.
b. Write the possible *type I* and *type II* errors.
c. *Determine* which error is more serious. *Answer: Page A40*

You will reject the null hypothesis when the sample statistic from the sampling distribution is unusual. You have already identified unusual events to be those that occur with a probability of 0.05 or less. When statistical tests are used, an unusual event is sometimes required to have a probability of 0.10 or less, 0.05 or less, or 0.01 or less. Because there is variation from sample to sample, there is always a possibility that you will reject a null hypothesis when it is actually true. In other words, although the null hypothesis is true, your sample statistic is determined to be an unusual event in the sampling distribution. You can decrease the probability of this happening by lowering the *level of significance*.

> **INSIGHT**
>
> When you decrease α (the maximum allowable probability of making a type I error), you are likely to be increasing β. The value $1 - \beta$ is called the **power of the test**. It represents the probability of rejecting the null hypothesis when it is false. The value of the power is difficult (and sometimes impossible) to find in most cases.

DEFINITION

In a hypothesis test, the **level of significance** is your maximum allowable probability of making a type I error. It is denoted by α, the lowercase Greek letter alpha.

The probability of a type II error is denoted by β, the lowercase Greek letter beta.

By setting the level of significance at a small value, you are saying that you want the probability of rejecting a true null hypothesis to be small. Three commonly used levels of significance are $\alpha = 0.10$, $\alpha = 0.05$, and $\alpha = 0.01$.

▶ STATISTICAL TESTS AND *P*-VALUES

After stating the null and alternative hypotheses and specifying the level of significance, the next step in a hypothesis test is to obtain a random sample from the population and calculate sample statistics such as the mean and the standard deviation. The statistic that is compared with the parameter in the null hypothesis is called the **test statistic.** The type of test used and the sampling distribution are based on the test statistic.

In this chapter, you will learn about several one-sample statistical tests. The following table shows the relationships between population parameters and their corresponding test statistics and standardized test statistics.

Population parameter	Test statistic	Standardized test statistic
μ	$\overline{x}$	z (Section 7.2, $n \geq 30$), t (Section 7.3, $n < 30$)
p	$\hat{p}$	z (Section 7.4)
σ^2	s^2	χ^2 (Section 7.5)

One way to decide whether to reject the null hypothesis is to determine whether the probability of obtaining the standardized test statistic (or one that is more extreme) is less than the level of significance.

DEFINITION

If the null hypothesis is true, a **_P_-value** (or **probability value**) of a hypothesis test is the probability of obtaining a sample statistic with a value as extreme or more extreme than the one determined from the sample data.

The *P*-value of a hypothesis test depends on the nature of the test. There are three types of hypothesis tests—left-tailed, right-tailed, and two-tailed. The type of test depends on the location of the region of the sampling distribution that favors a rejection of H_0. This region is indicated by the alternative hypothesis.

DEFINITION

1. If the alternative hypothesis H_a contains the less-than inequality symbol ($<$), the hypothesis test is a **left-tailed test.**

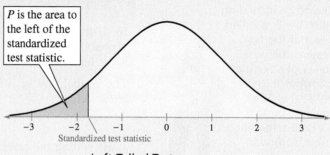

$H_0: \mu \geq k$
$H_a: \mu < k$

P is the area to the left of the standardized test statistic.

Standardized test statistic

Left-Tailed Test

2. If the alternative hypothesis H_a contains the greater-than inequality symbol ($>$), the hypothesis test is a **right-tailed test.**

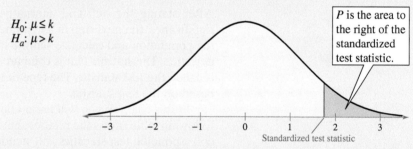

$H_0: \mu \leq k$
$H_a: \mu > k$

P is the area to the right of the standardized test statistic.

Standardized test statistic

Right-Tailed Test

3. If the alternative hypothesis H_a contains the not-equal-to symbol ($\neq$), the hypothesis test is a **two-tailed test.** In a two-tailed test, each tail has an area of $\frac{1}{2}P$.

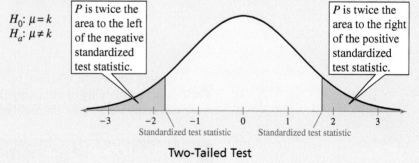
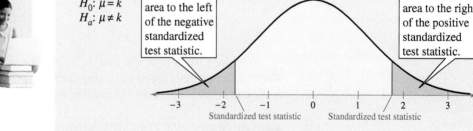

$H_0: \mu = k$
$H_a: \mu \neq k$

P is twice the area to the left of the negative standardized test statistic.

P is twice the area to the right of the positive standardized test statistic.

Standardized test statistic   Standardized test statistic

Two-Tailed Test

The smaller the *P*-value of the test, the more evidence there is to reject the null hypothesis. A very small *P*-value indicates an unusual event. Remember, however, that even a very low *P*-value does not constitute proof that the null hypothesis is false, only that it is probably false.

EXAMPLE 3

▶ **Identifying the Nature of a Hypothesis Test**

For each claim, state H_0 and H_a in words and in symbols. Then determine whether the hypothesis test is a left-tailed test, right-tailed test, or two-tailed test. Sketch a normal sampling distribution and shade the area for the P-value.

1. A school publicizes that the proportion of its students who are involved in at least one extracurricular activity is 61%.

2. A car dealership announces that the mean time for an oil change is less than 15 minutes.

3. A company advertises that the mean life of its furnaces is more than 18 years.

▶ **Solution**

In Symbols	*In Words*

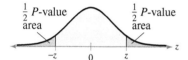

1. H_0: $p = 0.61$ The proportion of students who are involved in at least one extracurricular activity is 61%.

H_a: $p \neq 0.61$ The proportion of students who are involved in at least one extracurricular activity is not 61%.

Because H_a contains the $\neq$ symbol, the test is a two-tailed hypothesis test. The graph of the normal sampling distribution at the left shows the shaded area for the P-value.

2. H_0: $\mu \geq 15$ min   The mean time for an oil change is greater than or equal to 15 minutes.

H_a: $\mu < 15$ min   The mean time for an oil change is less than 15 minutes.

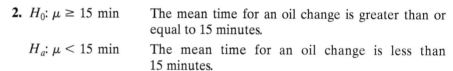

Because H_a contains the $<$ symbol, the test is a left-tailed hypothesis test. The graph of the normal sampling distribution at the left shows the shaded area for the P-value.

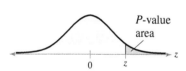

3. H_0: $\mu \leq 18$ yr   The mean life of the furnaces is less than or equal to 18 years.

H_a: $\mu > 18$ yr   The mean life of the furnaces is more than 18 years.

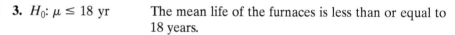

Because H_a contains the $>$ symbol, the test is a right-tailed hypothesis test. The graph of the normal sampling distribution at the left shows the shaded area for the P-value.

▶ **Try It Yourself 3**

For each claim, state H_0 and H_a in words and in symbols. Then determine whether the hypothesis test is a left-tailed test, right-tailed test, or two-tailed test. Sketch a normal sampling distribution and shade the area for the P-value.

1. A consumer analyst reports that the mean life of a certain type of automobile battery is not 74 months.
2. An electronics manufacturer publishes that the variance of the life of its home theater systems is less than or equal to 2.7.
3. A realtor publicizes that the proportion of homeowners who feel their house is too small for their family is more than 24%.

a. *Write* H_0 and H_a in words and in symbols.
b. *Determine* whether the test is *left-tailed*, *right-tailed*, or *two-tailed*.
c. *Sketch* the sampling distribution and *shade* the area for the P-value.

Answer: Page A40

MAKING A DECISION AND INTERPRETING THE DECISION

To conclude a hypothesis test, you make a decision and interpret that decision. There are only two possible outcomes to a hypothesis test: (1) reject the null hypothesis and (2) fail to reject the null hypothesis.

INSIGHT

In this chapter, you will learn that there are two basic types of **decision rules** for deciding whether to reject H_0 or fail to reject H_0. The decision rule described on this page is based on *P*-values. The second basic type of decision rule is based on rejection regions. When the standardized test statistic falls in the rejection region, the observed probability (*P*-value) of a type I error is less than α. You will learn more about rejection regions in the next section.

DECISION RULE BASED ON *P*-VALUE

To use a *P*-value to make a conclusion in a hypothesis test, compare the *P*-value with α.

1. If $P \leq \alpha$, then reject H_0.
2. If $P > \alpha$, then fail to reject H_0.

Failing to reject the null hypothesis does not mean that you have accepted the null hypothesis as true. It simply means that there is not enough evidence to reject the null hypothesis. If you want to support a claim, state it so that it becomes the alternative hypothesis. If you want to reject a claim, state it so that it becomes the null hypothesis. The following table will help you interpret your decision.

	Claim	
Decision	Claim is H_0.	Claim is H_a.
Reject H_0.	There is enough evidence to reject the claim.	There is enough evidence to support the claim.
Fail to reject H_0.	There is not enough evidence to reject the claim.	There is not enough evidence to support the claim.

EXAMPLE 4

Interpreting a Decision

You perform a hypothesis test for each of the following claims. How should you interpret your decision if you reject H_0? If you fail to reject H_0?

1. H_0 (Claim): A school publicizes that the proportion of its students who are involved in at least one extracurricular activity is 61%.
2. H_a (Claim): A car dealership announces that the mean time for an oil change is less than 15 minutes.

Solution

1. The claim is represented by H_0. If you reject H_0, then you should conclude "there is enough evidence to reject the school's claim that the proportion of students who are involved in at least one extracurricular activity is 61%." If you fail to reject H_0, then you should conclude "there is not enough evidence to reject the school's claim that the proportion of students who are involved in at least one extracurricular activity is 61%."
2. The claim is represented by H_a, so the null hypothesis is "the mean time for an oil change is greater than or equal to 15 minutes." If you reject H_0, then you should conclude "there is enough evidence to support the dealership's claim that the mean time for an oil change is less than 15 minutes." If you fail to reject H_0, then you should conclude "there is not enough evidence to support the dealership's claim that the mean time for an oil change is less than 15 minutes."

▶ Try It Yourself 4

You perform a hypothesis test for the following claim. How should you interpret your decision if you reject H_0? If you fail to reject H_0?

H_a (Claim): A realtor publicizes that the proportion of homeowners who feel their house is too small for their family is more than 24%.

a. Interpret your decision if you *reject* the null hypothesis.
b. Interpret your decision if you *fail to reject* the null hypothesis.

Answer: Page A41

The general steps for a hypothesis test using *P*-values are summarized below.

STEPS FOR HYPOTHESIS TESTING

1. State the claim mathematically and verbally. Identify the null and alternative hypotheses.

$$H_0: \quad ? \qquad H_a: \quad ?$$

2. Specify the level of significance.

$$\alpha = \quad ?$$

3. Determine the standardized sampling distribution and sketch its graph.

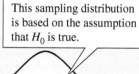

This sampling distribution is based on the assumption that H_0 is true.

4. Calculate the test statistic and its corresponding standardized test statistic. Add it to your sketch.

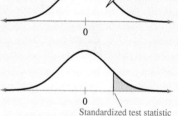

Standardized test statistic

5. Find the *P*-value.
6. Use the following decision rule.

| Is the *P*-value less than or equal to the level of significance? | → No → | Fail to reject H_0. |

↓ Yes

Reject H_0.

7. Write a statement to interpret the decision in the context of the original claim.

In the steps above, the graphs show a right-tailed test. However, the same basic steps also apply to left-tailed and two-tailed tests.

▶ STRATEGIES FOR HYPOTHESIS TESTING

In a courtroom, the strategy used by an attorney depends on whether the attorney is representing the defense or the prosecution. In a similar way, the strategy that you will use in hypothesis testing should depend on whether you are trying to support or reject a claim. Remember that you cannot use a hypothesis test to support your claim if your claim is the null hypothesis. So, as a researcher, if you want a conclusion that supports your claim, word your claim so it is the alternative hypothesis. If you want to reject a claim, word it so it is the null hypothesis.

EXAMPLE 5

▶ **Writing the Hypotheses**

A medical research team is investigating the benefits of a new surgical treatment. One of the claims is that the mean recovery time for patients after the new treatment is less than 96 hours. How would you write the null and alternative hypotheses if (1) you are on the research team and want to support the claim? (2) you are on an opposing team and want to reject the claim?

▶ **Solution**

1. To answer the question, first think about the context of the claim. Because you want to support this claim, make the alternative hypothesis state that the mean recovery time for patients is less than 96 hours. So, $H_a: \mu < 96$ hours. Its complement, $\mu \geq 96$ hours, would be the null hypothesis.

$$H_0: \mu \geq 96$$
$$H_a: \mu < 96 \text{ (Claim)}$$

2. First think about the context of the claim. As an opposing researcher, you do not want the recovery time to be less than 96 hours. Because you want to reject this claim, make it the null hypothesis. So, $H_0: \mu \leq 96$ hours. Its complement, $\mu > 96$ hours, would be the alternative hypothesis.

$$H_0: \mu \leq 96 \text{ (Claim)}$$
$$H_a: \mu > 96$$

▶ **Try It Yourself 5**

1. You represent a chemical company that is being sued for paint damage to automobiles. You want to support the claim that the mean repair cost per automobile is less than $650. How would you write the null and alternative hypotheses?

2. You are on a research team that is investigating the mean temperature of adult humans. The commonly accepted claim is that the mean temperature is about 98.6°F. You want to show that this claim is false. How would you write the null and alternative hypotheses?

a. *Determine* whether you want to support or reject the claim.
b. Write the *null* and *alternative hypotheses*. *Answer: Page A41*

7.1 EXERCISES

■ BUILDING BASIC SKILLS AND VOCABULARY

1. What are the two types of hypotheses used in a hypothesis test? How are they related?

2. Describe the two types of error possible in a hypothesis test decision.

3. What are the two decisions that you can make from performing a hypothesis test?

4. Does failing to reject the null hypothesis mean that the null hypothesis is true? Explain.

True or False? *In Exercises 5–10, determine whether the statement is true or false. If it is false, rewrite it as a true statement.*

5. In a hypothesis test, you assume the alternative hypothesis is true.

6. A statistical hypothesis is a statement about a sample.

7. If you decide to reject the null hypothesis, you can support the alternative hypothesis.

8. The level of significance is the maximum probability you allow for rejecting a null hypothesis when it is actually true.

9. A large *P*-value in a test will favor rejection of the null hypothesis.

10. If you want to support a claim, write it as your null hypothesis.

Stating Hypotheses *In Exercises 11–16, use the given statement to represent a claim. Write its complement and state which is H_0 and which is H_a.*

11. $\mu \leq 645$

12. $\mu < 128$

13. $\sigma \neq 5$

14. $\sigma^2 \geq 1.2$

15. $p < 0.45$

16. $p = 0.21$

Graphical Analysis *In Exercises 17–20, match the alternative hypothesis with its graph. Then state the null hypothesis and sketch its graph.*

17. $H_a: \mu > 3$ (a)

18. $H_a: \mu < 3$ (b)

19. $H_a: \mu \neq 3$ (c)

20. $H_a: \mu > 2$ (d)

Identifying Tests *In Exercises 21–24, determine whether the hypothesis test with the given null and alternative hypotheses is left-tailed, right-tailed, or two-tailed.*

21. $H_0: \mu \leq 8.0$
    $H_a: \mu > 8.0$

22. $H_0: \sigma \geq 5.2$
    $H_a: \sigma < 5.2$

23. $H_0: \sigma^2 = 142$
    $H_a: \sigma^2 \neq 142$

24. $H_0: p = 0.25$
    $H_a: p \neq 0.25$

■ USING AND INTERPRETING CONCEPTS

Stating the Hypotheses *In Exercises 25–30, write the claim as a mathematical sentence. State the null and alternative hypotheses, and identify which represents the claim.*

25. **Light Bulbs** A light bulb manufacturer claims that the mean life of a certain type of light bulb is more than 750 hours.

26. **Shipping Errors** As stated by a company's shipping department, the number of shipping errors per million shipments has a standard deviation that is less than 3.

27. **Base Price of an ATV** The standard deviation of the base price of a certain type of all-terrain vehicle is no more than $320.

28. **Oak Trees** A state park claims that the mean height of the oak trees in the park is at least 85 feet.

29. **Drying Time** A company claims that its brands of paint have a mean drying time of less than 45 minutes.

30. **MP3 Players** According to a recent survey, 74% of college students own an MP3 player. *(Source: Harris Interactive)*

Identifying Errors *In Exercises 31–36, write sentences describing type I and type II errors for a hypothesis test of the indicated claim.*

31. **Repeat Buyers** A furniture store claims that at least 60% of its new customers will return to buy their next piece of furniture.

32. **Flow Rate** A garden hose manufacturer advertises that the mean flow rate of a certain type of hose is 16 gallons per minute.

33. **Chess** A local chess club claims that the length of time to play a game has a standard deviation of more than 12 minutes.

34. **Video Game Systems** A researcher claims that the proportion of adults in the United States who own a video game system is not 26%.

35. **Police** A police station publicizes that at most 20% of applicants become police officers.

36. **Computers** A computer repairer advertises that the mean cost of removing a virus infection is less than $100.

Identifying Tests *In Exercises 37–42, state H_0 and H_a in words and in symbols. Then determine whether the hypothesis test is left-tailed, right-tailed, or two-tailed. Explain your reasoning.*

37. **Security Alarms** At least 14% of all homeowners have a home security alarm.

38. **Clocks** A manufacturer of grandfather clocks claims that the mean time its clocks lose is no more than 0.02 second per day.

39. **Golf** The standard deviation of the 18-hole scores for a golfer is less than 2.1 strokes.

40. **Lung Cancer** A government report claims that the proportion of lung cancer cases that are due to smoking is 87%. *(Source: LungCancer.org)*

41. **Baseball**   A baseball team claims that the mean length of its games is less than 2.5 hours.

42. **Tuition**   A state claims that the mean tuition of its universities is no more than $25,000 per year.

Interpreting a Decision   *In Exercises 43–48, consider each claim. If a hypothesis test is performed, how should you interpret a decision that*

(a) *rejects the null hypothesis?*

(b) *fails to reject the null hypothesis?*

43. **Swans**   A scientist claims that the mean incubation period for swan eggs is less than 40 days.

44. **Lawn Mowers**   The standard deviation of the life of a certain type of lawn mower is at most 2.8 years.

45. **Hourly Wages**   The U.S. Department of Labor claims that the proportion of full-time workers earning over $450 per week is greater than 75%. *(Adapted from U.S. Bureau of Labor Statistics)*

46. **Gas Mileage**   An automotive manufacturer claims the standard deviation for the gas mileage of its models is 3.9 miles per gallon.

47. **Health Care Visits**   A researcher claims that the proportion of people who have had no health care visits in the past year is less than 17%. *(Adapted from National Center for Health Statistics)*

48. **Calories**   A sports drink maker claims the mean calorie content of its beverages is 72 calories per serving.

49. **Writing Hypotheses: Medicine**   Your medical research team is investigating the mean cost of a 30-day supply of a certain heart medication. A pharmaceutical company thinks that the mean cost is less than $60. You want to support this claim. How would you write the null and alternative hypotheses?

50. **Writing Hypotheses: Taxicab Company**   A taxicab company claims that the mean travel time between two destinations is about 21 minutes. You work for the bus company and want to reject this claim. How would you write the null and alternative hypotheses?

51. **Writing Hypotheses: Refrigerator Manufacturer**   A refrigerator manufacturer claims that the mean life of its competitor's refrigerators is less than 15 years. You are asked to perform a hypothesis test to test this claim. How would you write the null and alternative hypotheses if

   (a) you represent the manufacturer and want to support the claim?

   (b) you represent the competitor and want to reject the claim?

52. **Writing Hypotheses: Internet Provider**   An Internet provider is trying to gain advertising deals and claims that the mean time a customer spends online per day is greater than 28 minutes. You are asked to test this claim. How would you write the null and alternative hypotheses if

   (a) you represent the Internet provider and want to support the claim?

   (b) you represent a competing advertiser and want to reject the claim?

■ **EXTENDING CONCEPTS**

53. Getting at the Concept Why can decreasing the probability of a type I error cause an increase in the probability of a type II error?

54. Getting at the Concept Explain why a level of significance of $\alpha = 0$ is not used.

55. Writing A null hypothesis is rejected with a level of significance of 0.05. Is it also rejected at a level of significance of 0.10? Explain.

56. Writing A null hypothesis is rejected with a level of significance of 0.10. Is it also rejected at a level of significance of 0.05? Explain.

Graphical Analysis *In Exercises 57–60, you are given a null hypothesis and three confidence intervals that represent three samplings. Decide whether each confidence interval indicates that you should reject H_0. Explain your reasoning.*

57.

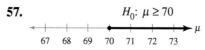

$H_0: \mu \geq 70$

(a) $67 < \mu < 71$

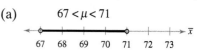

(b) $67 < \mu < 69$

(c) $69.5 < \mu < 72.5$

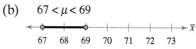

58.

$H_0: \mu \leq 54$

(a) $53.5 < \mu < 56.5$

(b) $51.5 < \mu < 54.5$

(c) $54.5 < \mu < 55.5$

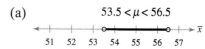

59.

$H_0: p \leq 0.20$

(a) $0.21 < p < 0.23$

(b) $0.19 < p < 0.23$

(c) $0.175 < p < 0.205$

60.

$H_0: p \geq 0.73$

(a) $0.73 < p < 0.75$

(b) $0.715 < p < 0.725$

(c) $0.695 < p < 0.745$

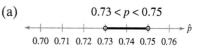

7.2 Hypothesis Testing for the Mean (Large Samples)

WHAT YOU SHOULD LEARN

▸ How to find *P*-values and use them to test a mean μ

▸ How to use *P*-values for a *z*-test

▸ How to find critical values and rejection regions in a normal distribution

▸ How to use rejection regions for a *z*-test

Using *P*-Values to Make Decisions ▸ Using *P*-Values for a *z*-Test ▸ Rejection Regions and Critical Values ▸ Using Rejection Regions for a *z*-Test

▸ USING *P*-VALUES TO MAKE DECISIONS

In Chapter 5, you learned that when the sample size is at least 30, the sampling distribution for $\overline{x}$ (the sample mean) is normal. In Section 7.1, you learned that a way to reach a conclusion in a hypothesis test is to use a *P*-value for the sample statistic, such as $\overline{x}$. Recall that when you assume the null hypothesis is true, a *P*-value (or probability value) of a hypothesis test is the probability of obtaining a sample statistic with a value as extreme or more extreme than the one determined from the sample data. The decision rule for a hypothesis test based on a *P*-value is as follows.

DECISION RULE BASED ON *P*-VALUE

To use a *P*-value to make a conclusion in a hypothesis test, compare the *P*-value with α.

1. If $P \leq \alpha$, then reject H_0.
2. If $P > \alpha$, then fail to reject H_0.

EXAMPLE 1

▸ **Interpreting a *P*-Value**

The *P*-value for a hypothesis test is $P = 0.0237$. What is your decision if the level of significance is (1) $\alpha = 0.05$ and (2) $\alpha = 0.01$?

▸ **Solution**

1. Because $0.0237 < 0.05$, you should reject the null hypothesis.

2. Because $0.0237 > 0.01$, you should fail to reject the null hypothesis.

▸ **Try It Yourself 1**

The *P*-value for a hypothesis test is $P = 0.0347$. What is your decision if the level of significance is (1) $\alpha = 0.01$ and (2) $\alpha = 0.05$?

a. *Compare* the *P*-value with the level of significance.
b. *Make* a decision.   *Answer: Page A41*

INSIGHT

The lower the *P*-value, the more evidence there is in favor of rejecting H_0. The *P*-value gives you the lowest level of significance for which the sample statistic allows you to reject the null hypothesis. In Example 1, you would reject H_0 at any level of significance greater than or equal to 0.0237.

FINDING THE *P*-VALUE FOR A HYPOTHESIS TEST

After determining the hypothesis test's standardized test statistic and the test statistic's corresponding area, do one of the following to find the *P*-value.

a. For a left-tailed test, $P = $ (Area in left tail).
b. For a right-tailed test, $P = $ (Area in right tail).
c. For a two-tailed test, $P = 2$(Area in tail of test statistic).

EXAMPLE 2

▶ **Finding a *P*-Value for a Left-Tailed Test**

Find the *P*-value for a left-tailed hypothesis test with a test statistic of $z = -2.23$. Decide whether to reject H_0 if the level of significance is $\alpha = 0.01$.

▶ **Solution**

The graph shows a standard normal curve with a shaded area to the left of $z = -2.23$. For a left-tailed test,

$$P = (\text{Area in left tail}).$$

From Table 4 in Appendix B, the area corresponding to $z = -2.23$ is 0.0129, which is the area in the left tail. So, the *P*-value for a left-tailed hypothesis test with a test statistic of $z = -2.23$ is $P = 0.0129$.

Interpretation Because the *P*-value of 0.0129 is greater than 0.01, you should fail to reject H_0.

▶ **Try It Yourself 2**

Find the *P*-value for a left-tailed hypothesis test with a test statistic of $z = -1.71$. Decide whether to reject H_0 if the level of significance is $\alpha = 0.05$.

a. *Use* Table 4 in Appendix B to find the area that corresponds to $z = -1.71$.
b. *Calculate* the *P*-value for a left-tailed test, the area in the left tail.
c. *Compare* the *P*-value with α and *decide* whether to reject H_0.

Answer: Page A41

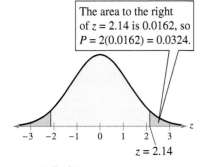

The area to the left of $z = -2.23$ is $P = 0.0129$.

$z = -2.23$

Left-Tailed Test

EXAMPLE 3

▶ **Finding a *P*-Value for a Two-Tailed Test**

Find the *P*-value for a two-tailed hypothesis test with a test statistic of $z = 2.14$. Decide whether to reject H_0 if the level of significance is $\alpha = 0.05$.

▶ **Solution**

The graph shows a standard normal curve with shaded areas to the left of $z = -2.14$ and to the right of $z = 2.14$. For a two-tailed test,

$$P = 2(\text{Area in tail of test statistic}).$$

From Table 4, the area corresponding to $z = 2.14$ is 0.9838. The area in the right tail is $1 - 0.9838 = 0.0162$. So, the *P*-value for a two-tailed hypothesis test with a test statistic of $z = 2.14$ is

$$P = 2(0.0162) = 0.0324.$$

Interpretation Because the *P*-value of 0.0324 is less than 0.05, you should reject H_0.

▶ **Try It Yourself 3**

Find the *P*-value for a two-tailed hypothesis test with a test statistic of $z = 1.64$. Decide whether to reject H_0 if the level of significance is $\alpha = 0.10$.

a. *Use* Table 4 to find the area that corresponds to $z = 1.64$.
b. *Calculate* the *P*-value for a two-tailed test, twice the area in the tail of the test statistic.
c. *Compare* the *P*-value with α and *decide* whether to reject H_0.

Answer: Page A41

The area to the right of $z = 2.14$ is 0.0162, so $P = 2(0.0162) = 0.0324$.

$z = 2.14$

Two-Tailed Test

▶ USING *P*-VALUES FOR A *z*-TEST

The *z*-test for the mean is used in populations for which the sampling distribution of sample means is normal. To use the *z*-test, you need to find the standardized value for your test statistic $\bar{x}$.

$$z = \frac{(\text{Sample mean}) - (\text{Hypothesized mean})}{\text{Standard error}}$$

z-TEST FOR A MEAN μ

The **z-test for a mean** is a statistical test for a population mean. The *z*-test can be used when the population is normal and σ is known, or for any population when the sample size *n* is at least 30. The **test statistic** is the sample mean $\bar{x}$ and the **standardized test statistic** is

$$z = \frac{\bar{x} - \mu}{\sigma/\sqrt{n}}.$$

Recall that $\frac{\sigma}{\sqrt{n}}$ = standard error = $\sigma_{\bar{x}}$.

When $n \geq 30$, you can use the sample standard deviation *s* in place of σ.

GUIDELINES

Using *P*-Values for a *z*-Test for Mean μ

IN WORDS	IN SYMBOLS
1. State the claim mathematically and verbally. Identify the null and alternative hypotheses.	State H_0 and H_a.
2. Specify the level of significance.	Identify α.
3. Determine the standardized test statistic.	$z = \frac{\bar{x} - \mu}{\sigma/\sqrt{n}}$ or, if $n \geq 30$, use $\sigma \approx s$.
4. Find the area that corresponds to *z*.	Use Table 4 in Appendix B.

5. Find the *P*-value.
   a. For a left-tailed test, *P* = (Area in left tail).
   b. For a right-tailed test, *P* = (Area in right tail).
   c. For a two-tailed test, *P* = 2(Area in tail of test statistic).

6. Make a decision to reject or fail to reject the null hypothesis.	Reject H_0 if *P*-value is less than or equal to α. Otherwise, fail to reject H_0.

7. Interpret the decision in the context of the original claim.

EXAMPLE 4

▶ **Hypothesis Testing Using P-Values**

In auto racing, a pit stop is where a racing vehicle stops for new tires, fuel, repairs, and other mechanical adjustments. The efficiency of a pit crew that makes these adjustments can affect the outcome of a race. A pit crew claims that its mean pit stop time (for 4 new tires and fuel) is less than 13 seconds. A random selection of 32 pit stop times has a sample mean of 12.9 seconds and a standard deviation of 0.19 second. Is there enough evidence to support the claim at $\alpha = 0.01$? Use a P-value.

▶ **Solution**

The claim is "the mean pit stop time is less than 13 seconds." So, the null and alternative hypotheses are

$$H_0: \mu \geq 13 \text{ seconds} \quad \text{and} \quad H_a: \mu < 13 \text{ seconds.} \text{ (Claim)}$$

The level of significance is $\alpha = 0.01$. The standardized test statistic is

$$z = \frac{\bar{x} - \mu}{\sigma/\sqrt{n}} \qquad \text{Because } n \geq 30, \text{ use the } z\text{-test.}$$

$$\approx \frac{12.9 - 13}{0.19/\sqrt{32}} \qquad \text{Because } n \geq 30, \text{ use } \sigma \approx s = 0.19. \text{ Assume } \mu = 13.$$

$$\approx -2.98.$$

In Table 4 in Appendix B, the area corresponding to $z = -2.98$ is 0.0014. Because this test is a left-tailed test, the P-value is equal to the area to the left of $z = -2.98$. So, $P = 0.0014$. Because the P-value is less than $\alpha = 0.01$, you should decide to reject the null hypothesis.

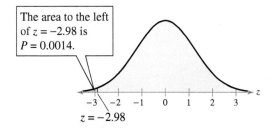

The area to the left of $z = -2.98$ is $P = 0.0014$.

$z = -2.98$

Left-Tailed Test

Interpretation There is enough evidence at the 1% level of significance to support the claim that the mean pit stop time is less than 13 seconds.

▶ **Try It Yourself 4**

Homeowners claim that the mean speed of automobiles traveling on their street is greater than the speed limit of 35 miles per hour. A random sample of 100 automobiles has a mean speed of 36 miles per hour and a standard deviation of 4 miles per hour. Is there enough evidence to support the claim at $\alpha = 0.05$? Use a P-value.

a. Identify the *claim*. Then state the *null* and *alternative hypotheses*.
b. Identify the *level of significance*.
c. Find the *standardized test statistic z*.
d. Find the *P-value*.
e. *Decide* whether to reject the null hypothesis.
f. *Interpret* the decision in the context of the original claim.

Answer: Page A41

EXAMPLE 5   SC Report 29   See MINITAB steps on page 424.

▶ Hypothesis Testing Using *P*-Values

The National Institute of Diabetes and Digestive and Kidney Diseases reports that the average cost of bariatric (weight loss) surgery is about $22,500. You think this information is incorrect. You randomly select 30 bariatric surgery patients and find that the average cost for their surgeries is $21,545 with a standard deviation of $3015. Is there enough evidence to support your claim at $\alpha = 0.05$? Use a *P*-value. *(Adapted from National Institute of Diabetes and Digestive and Kidney Diseases)*

▶ Solution

The claim is "the mean is different from $22,500." So, the null and alternative hypotheses are

$$H_0\text{: } \mu = \$22,500$$

and

$$H_a\text{: } \mu \neq \$22,500. \text{ (Claim)}$$

The level of significance is $\alpha = 0.05$. The standardized test statistic is

$$z = \frac{\overline{x} - \mu}{\sigma/\sqrt{n}}$$ Because $n \geq 30$, use the z-test.

$$\approx \frac{21,545 - 22,500}{3015/\sqrt{30}}$$ Because $n \geq 30$, use $\sigma \approx s = 3015$. Assume $\mu = 22,500$.

$$\approx -1.73.$$

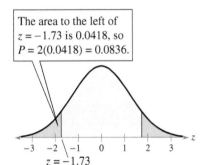

The area to the left of $z = -1.73$ is 0.0418, so $P = 2(0.0418) = 0.0836$.

$z = -1.73$

Two-Tailed Test

In Table 4, the area corresponding to $z = -1.73$ is 0.0418. Because the test is a two-tailed test, the *P*-value is equal to twice the area to the left of $z = -1.73$. So,

$$P = 2(0.0418)$$

$$= 0.0836.$$

Because the *P*-value is greater than α, you should fail to reject the null hypothesis.

Interpretation   There is not enough evidence at the 5% level of significance to support the claim that the mean cost of bariatric surgery is different from $22,500.

▶ Try It Yourself 5

One of your distributors reports an average of 150 sales per day. You suspect that this average is not accurate, so you randomly select 35 days and determine the number of sales each day. The sample mean is 143 daily sales with a standard deviation of 15 sales. At $\alpha = 0.01$, is there enough evidence to doubt the distributor's reported average? Use a *P*-value.

a. Identify the *claim*. Then state the *null* and *alternative hypotheses*.
b. Identify the *level of significance*.
c. Find the *standardized test statistic z*.
d. Find the *P-value*.
e. *Decide* whether to reject the null hypothesis.
f. *Interpret* the decision in the context of the original claim.

Answer: Page A41

EXAMPLE 6

▶ **Using a Technology Tool to Find a *P*-Value**

What decision should you make for the following TI-83/84 Plus displays, using a level of significance of $\alpha = 0.05$?

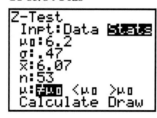

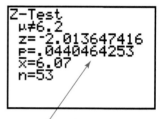

▶ **Solution**

The *P*-value for this test is given as 0.0440464253. Because the *P*-value is less than 0.05, you should reject the null hypothesis.

▶ **Try It Yourself 6**

For the TI-83/84 Plus hypothesis test shown in Example 6, make a decision at the $\alpha = 0.01$ level of significance.

a. *Compare* the *P*-value with the level of significance.
b. *Make* your decision.   *Answer: Page A41*

▶ REJECTION REGIONS AND CRITICAL VALUES

Another method to decide whether to reject the null hypothesis is to determine whether the standardized test statistic falls within a range of values called the rejection region of the sampling distribution.

DEFINITION

A **rejection region** (or **critical region**) of the sampling distribution is the range of values for which the null hypothesis is not probable. If a test statistic falls in this region, the null hypothesis is rejected. A **critical value** z_0 separates the rejection region from the nonrejection region.

GUIDELINES

Finding Critical Values in a Normal Distribution

1. Specify the level of significance α.
2. Decide whether the test is left-tailed, right-tailed, or two-tailed.
3. Find the critical value(s) z_0. If the hypothesis test is
   a. *left-tailed*, find the *z*-score that corresponds to an area of α.
   b. *right-tailed*, find the *z*-score that corresponds to an area of $1 - \alpha$.
   c. *two-tailed*, find the *z*-scores that correspond to $\frac{1}{2}\alpha$ and $1 - \frac{1}{2}\alpha$.
4. Sketch the standard normal distribution. Draw a vertical line at each critical value and shade the rejection region(s).

If you cannot find the exact area in Table 4, use the area that is closest. When the area is exactly midway between two areas in the table, use the z-score midway between the corresponding z-scores.

EXAMPLE 7

▶ **Finding a Critical Value for a Left-Tailed Test**

Find the critical value and rejection region for a left-tailed test with $\alpha = 0.01$.

▶ **Solution**

The graph shows a standard normal curve with a shaded area of 0.01 in the left tail. In Table 4, the z-score that is closest to an area of 0.01 is -2.33. So, the critical value is $z_0 = -2.33$. The rejection region is to the left of this critical value.

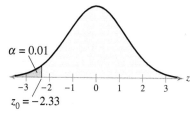

$\alpha = 0.01$

$z_0 = -2.33$

1% Level of Significance

▶ **Try It Yourself 7**

Find the critical value and rejection region for a left-tailed test with $\alpha = 0.10$.

a. *Draw* a graph of the standard normal curve with an area of α in the left tail.
b. *Use* Table 4 to find the area that is closest to α.
c. *Find* the z-score that corresponds to this area.
d. *Identify* the rejection region.   *Answer: Page A41*

EXAMPLE 8

▶ **Finding a Critical Value for a Two-Tailed Test**

Find the critical values and rejection regions for a two-tailed test with $\alpha = 0.05$.

▶ **Solution**

The graph shows a standard normal curve with shaded areas of $\frac{1}{2}\alpha = 0.025$ in each tail. The area to the left of $-z_0$ is $\frac{1}{2}\alpha = 0.025$, and the area to the left of z_0 is $1 - \frac{1}{2}\alpha = 0.975$. In Table 4, the z-scores that correspond to the areas 0.025 and 0.975 are -1.96 and 1.96, respectively. So, the critical values are $-z_0 = -1.96$ and $z_0 = 1.96$. The rejection regions are to the left of -1.96 and to the right of 1.96.

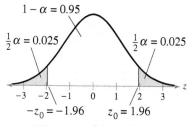

$1 - \alpha = 0.95$

$\frac{1}{2}\alpha = 0.025$ $\frac{1}{2}\alpha = 0.025$

$-z_0 = -1.96$ $z_0 = 1.96$

5% Level of Significance

▶ **Try It Yourself 8**

Find the critical values and rejection regions for a two-tailed test with $\alpha = 0.08$.

a. *Draw* a graph of the standard normal curve with an area of $\frac{1}{2}\alpha$ in each tail.
b. *Use* Table 4 to find the areas that are closest to $\frac{1}{2}\alpha$ and $1 - \frac{1}{2}\alpha$.
c. *Find* the z-scores that correspond to these areas.
d. *Identify* the rejection regions.   *Answer: Page A41*

STUDY TIP

Notice in Example 8 that the critical values are opposites. This is always true for two-tailed z-tests.

The table lists the critical values for commonly used levels of significance.

Alpha	Tail	z
0.10	Left	-1.28
	Right	1.28
	Two	± 1.645
0.05	Left	-1.645
	Right	1.645
	Two	± 1.96
0.01	Left	-2.33
	Right	2.33
	Two	± 2.575

▶ USING REJECTION REGIONS FOR A z-TEST

To conclude a hypothesis test using rejection region(s), you make a decision and interpret the decision as follows.

DECISION RULE BASED ON REJECTION REGION

To use a rejection region to conduct a hypothesis test, calculate the standardized test statistic z. If the standardized test statistic

1. is in the rejection region, then reject H_0.
2. is *not* in the rejection region, then fail to reject H_0.

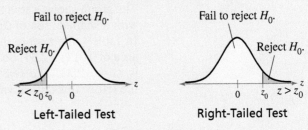

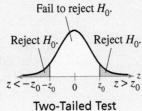

Failing to reject the null hypothesis does not mean that you have accepted the null hypothesis as true. It simply means that there is not enough evidence to reject the null hypothesis.

GUIDELINES

Using Rejection Regions for a z-Test for a Mean μ

IN WORDS	IN SYMBOLS
1. State the claim mathematically and verbally. Identify the null and alternative hypotheses.	State H_0 and H_a.
2. Specify the level of significance.	Identify α.
3. Determine the critical value(s).	Use Table 4 in Appendix B.
4. Determine the rejection region(s).	
5. Find the standardized test statistic and sketch the sampling distribution.	$z = \dfrac{\bar{x} - \mu}{\sigma/\sqrt{n}}$, or, if $n \geq 30$, use $\sigma \approx s$.
6. Make a decision to reject or fail to reject the null hypothesis.	If z is in the rejection region, reject H_0. Otherwise, fail to reject H_0.
7. Interpret the decision in the context of the original claim.	

EXAMPLE 9

See TI-83/84 Plus
steps on page 425.

▶ **Testing μ with a Large Sample**

Employees at a construction and mining company claim that the mean salary of the company's mechanical engineers is less than that of one of its competitors, which is $68,000. A random sample of 30 of the company's mechanical engineers has a mean salary of $66,900 with a standard deviation of $5500. At $\alpha = 0.05$, test the employees' claim.

▶ **Solution**

The claim is "the mean salary is less than $68,000." So, the null and alternative hypotheses can be written as

$$H_0:\ \mu \geq \$68,000 \qquad \text{and} \qquad H_a:\ \mu < \$68,000. \ \text{(Claim)}$$

Because the test is a left-tailed test and the level of significance is $\alpha = 0.05$, the critical value is $z_0 = -1.645$ and the rejection region is $z < -1.645$. The standardized test statistic is

$$z = \frac{\bar{x} - \mu}{\sigma/\sqrt{n}} \qquad \text{Because } n \geq 30, \text{ use the } z\text{-test.}$$

$$\approx \frac{66,900 - 68,000}{5500/\sqrt{30}} \qquad \begin{array}{l}\text{Because } n \geq 30, \text{ use } \sigma \approx s = 5500.\\ \text{Assume } \mu = 68,000.\end{array}$$

$$\approx -1.10.$$

The graph shows the location of the rejection region and the standardized test statistic z. Because z is not in the rejection region, you fail to reject the null hypothesis.

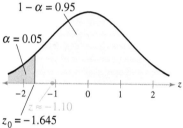

5% Level of Significance

Interpretation There is not enough evidence at the 5% level of significance to support the employees' claim that the mean salary is less than $68,000.

Be sure you understand the decision made in this example. Even though your sample has a mean of $66,900, you cannot (at a 5% level of significance) support the claim that the mean of all the mechanical engineers' salaries is less than $68,000. The difference between your test statistic and the hypothesized mean is probably due to sampling error.

▶ **Try It Yourself 9**

The CEO of the company claims that the mean work day of the company's mechanical engineers is less than 8.5 hours. A random sample of 35 of the company's mechanical engineers has a mean work day of 8.2 hours with a standard deviation of 0.5 hour. At $\alpha = 0.01$, test the CEO's claim.

a. Identify the *claim* and state H_0 and H_a.
b. Identify the *level of significance* α.
c. Find the *critical value* z_0 and identify the *rejection region*.
d. Find the *standardized test statistic z*. *Sketch* a graph.
e. *Decide* whether to reject the null hypothesis.
f. *Interpret* the decision in the context of the original claim.

Answer: Page A41

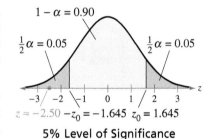

Using a TI-83/84 Plus, you can find the standardized test statistic automatically.

EXAMPLE 10

▶ **Testing μ with a Large Sample**

The U.S. Department of Agriculture claims that the mean cost of raising a child from birth to age 2 by husband-wife families in the United States is $13,120. A random sample of 500 children (age 2) has a mean cost of $12,925 with a standard deviation of $1745. At $\alpha = 0.10$, is there enough evidence to reject the claim? *(Adapted from U.S. Department of Agriculture Center for Nutrition Policy and Promotion)*

▶ **Solution**

The claim is "the mean cost is $13,120." So, the null and alternative hypotheses are

$$H_0: \mu = \$13,120 \text{ (Claim)}$$

and

$$H_a: \mu \neq \$13,120.$$

Because the test is a two-tailed test and the level of significance is $\alpha = 0.10$, the critical values are $-z_0 = -1.645$ and $z_0 = 1.645$. The rejection regions are $z < -1.645$ and $z > 1.645$. The standardized test statistic is

$$z = \frac{\bar{x} - \mu}{\sigma/\sqrt{n}}$$ 　Because $n \geq 30$, use the z-test.

$$\approx \frac{12,925 - 13,120}{1745/\sqrt{500}}$$ 　Because $n \geq 30$, use $\sigma \approx s = 1745$.
　Assume $\mu = 13,120$.

$$\approx -2.50.$$

The graph shows the location of the rejection regions and the standardized test statistic z. Because z is in the rejection region, you should reject the null hypothesis.

Interpretation There is enough evidence at the 10% level of significance to reject the claim that the mean cost of raising a child from birth to age 2 by husband-wife families in the United States is $13,120.

5% Level of Significance

▶ **Try It Yourself 10**

Using the information and results of Example 10, determine whether there is enough evidence to reject the claim that the mean cost of raising a child from birth to age 2 by husband-wife families in the United States is $13,120. Use $\alpha = 0.01$.

a. Identify the *level of significance* α.
b. Find the *critical values* $-z_0$ and z_0 and identify the *rejection regions*.
c. *Sketch* a graph. *Decide* whether to reject the null hypothesis.
d. *Interpret* the decision in the context of the original claim.

Answer: Page A41

7.2 EXERCISES

■ BUILDING BASIC SKILLS AND VOCABULARY

1. Explain the difference between the z-test for μ using rejection region(s) and the z-test for μ using a *P*-value.

2. In hypothesis testing, does choosing between the critical value method or the *P*-value method affect your conclusion? Explain.

In Exercises 3–8, find the P-value for the indicated hypothesis test with the given standardized test statistic z. Decide whether to reject H_0 for the given level of significance α.

3. Left-tailed test, $z = -1.32$, $\alpha = 0.10$

4. Left-tailed test, $z = -1.55$, $\alpha = 0.05$

5. Right-tailed test, $z = 2.46$, $\alpha = 0.01$

6. Right-tailed test, $z = 1.23$, $\alpha = 0.10$

7. Two-tailed test, $z = -1.68$, $\alpha = 0.05$

8. Two-tailed test, $z = 2.30$, $\alpha = 0.01$

Graphical Analysis   *In Exercises 9–12, match each P-value with the graph that displays its area. The graphs are labeled (a)–(d).*

9. $P = 0.0089$

10. $P = 0.3050$

11. $P = 0.0688$

12. $P = 0.0287$

(a)

$z = 1.90$

(b)

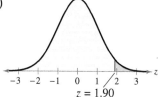
$z = -2.37$

(c)

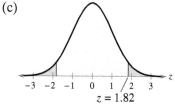
$z = 1.82$

(d)

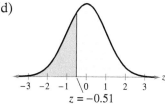
$z = -0.51$

13. Given H_0: $\mu = 100$, H_a: $\mu \neq 100$, and $P = 0.0461$.

    (a) Do you reject or fail to reject H_0 at the 0.01 level of significance?

    (b) Do you reject or fail to reject H_0 at the 0.05 level of significance?

14. Given H_0: $\mu \geq 8.5$, H_a: $\mu < 8.5$, and $P = 0.0691$.

    (a) Do you reject or fail to reject H_0 at the 0.01 level of significance?

    (b) Do you reject or fail to reject H_0 at the 0.05 level of significance?

In Exercises 15 and 16, use the TI-83/84 Plus displays to make a decision to reject or fail to reject the null hypothesis at the given level of significance.

15. $\alpha = 0.05$

```
Z-Test
 Inpt:Data Stats
 μ0:60
 σ:4.25
 x̄:58.75
 n:40
 μ:≠μ0 <μ0 >μ0
 Calculate Draw
```

```
Z-Test
 μ≠60
 z=-1.86016333
 p=.0628622957
 x̄=58.75
 n=40
```

16. $\alpha = 0.01$

```
Z-Test
 Inpt:Data Stats
 μ0:742
 σ:68.1
 x̄:763
 n:65
 μ:≠μ0 <μ0 >μ0
 Calculate Draw
```

```
Z-Test
 μ>742
 z=2.486158777
 p=.0064565285
 x̄=763
 n=65
```

Finding Critical Values *In Exercises 17–22, find the critical value(s) for the indicated type of test and level of significance α. Include a graph with your answer.*

17. Right-tailed test, $\alpha = 0.05$

18. Right-tailed test, $\alpha = 0.08$

19. Left-tailed test, $\alpha = 0.03$

20. Left-tailed test, $\alpha = 0.09$

21. Two-tailed test, $\alpha = 0.02$

22. Two-tailed test, $\alpha = 0.10$

Graphical Analysis *In Exercises 23 and 24, state whether each standardized test statistic z allows you to reject the null hypothesis. Explain your reasoning.*

23. (a) $z = -1.301$
   (b) $z = 1.203$
   (c) $z = 1.280$
   (d) $z = 1.286$

24. (a) $z = 1.98$
   (b) $z = -1.89$
   (c) $z = 1.65$
   (d) $z = -1.99$

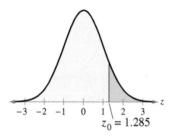

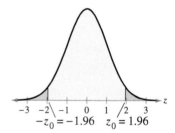

In Exercises 25–28, test the claim about the population mean μ at the given level of significance α using the given sample statistics.

25. Claim: $\mu = 40$; $\alpha = 0.05$.
   Sample statistics: $\bar{x} = 39.2$, $s = 3.23$, $n = 75$

26. Claim: $\mu > 1745$; $\alpha = 0.10$.
   Sample statistics: $\bar{x} = 1752$, $s = 38$, $n = 44$

27. Claim: $\mu \neq 8550$; $\alpha = 0.02$.
   Sample statistics: $\bar{x} = 8420$, $s = 314$, $n = 38$

28. Claim: $\mu \leq 22{,}500$; $\alpha = 0.01$.
   Sample statistics: $\bar{x} = 23{,}250$, $s = 1200$, $n = 45$

■ USING AND INTERPRETING CONCEPTS

Testing Claims Using P-Values   *In Exercises 29–34,*

(a) *write the claim mathematically and identify H_0 and H_a.*

(b) *find the standardized test statistic z and its corresponding area. If convenient, use technology.*

(c) *find the P-value. If convenient, use technology.*

(d) *decide whether to reject or fail to reject the null hypothesis.*

(e) *interpret the decision in the context of the original claim.*

29. MCAT Scores   A random sample of 50 medical school applicants at a university has a mean raw score of 31 with a standard deviation of 2.5 on the multiple choice portions of the Medical College Admission Test (MCAT). A student says that the mean raw score for the school's applicants is more than 30. At $\alpha = 0.01$, is there enough evidence to support the student's claim? *(Adapted from Association of American Medical Colleges)*

30. Sprinkler Systems   A manufacturer of sprinkler systems designed for fire protection claims that the average activating temperature is at least 135°F. To test this claim, you randomly select a sample of 32 systems and find the mean activation temperature to be 133°F with a standard deviation of 3.3°F. At $\alpha = 0.10$, do you have enough evidence to reject the manufacturer's claim?

31. Bottled Water Consumption   The U.S. Department of Agriculture claims that the mean consumption of bottled water by a person in the United States is 28.5 gallons per year. A random sample of 100 people in the United States has a mean bottled water consumption of 27.8 gallons per year with a standard deviation of 4.1 gallons. At $\alpha = 0.08$, can you reject the claim? *(Adapted from U.S. Department of Agriculture)*

32. Coffee Consumption   The U.S. Department of Agriculture claims that the mean consumption of coffee by a person in the United States is 24.2 gallons per year. A random sample of 120 people in the United States shows that the mean coffee consumption is 23.5 gallons per year with a standard deviation of 3.2 gallons. At $\alpha = 0.05$, can you reject the claim? *(Adapted from U.S. Department of Agriculture)*

33. Quitting Smoking   The lengths of time (in years) it took a random sample of 32 former smokers to quit smoking permanently are listed. At $\alpha = 0.05$, is there enough evidence to reject the claim that the mean time it takes smokers to quit smoking permanently is 15 years? *(Adapted from The Gallup Organization)*

15.7	13.2	22.6	13.0	10.7	18.1	14.7	7.0	17.3	7.5	21.8
12.3	19.8	13.8	16.0	15.5	13.1	20.7	15.5	9.8	11.9	16.9
7.0	19.3	13.2	14.6	20.9	15.4	13.3	11.6	10.9	21.6	

34. Salaries   An analyst claims that the mean annual salary for advertising account executives in Denver, Colorado is more than the national mean, $66,200. The annual salaries (in dollars) for a random sample of 35 advertising account executives in Denver are listed. At $\alpha = 0.09$, is there enough evidence to support the analyst's claim? *(Adapted from Salary.com)*

69,450	65,910	68,780	66,724	64,125	67,561	62,419
70,375	65,835	62,653	65,090	67,997	65,176	64,936
66,716	69,832	63,111	64,550	63,512	65,800	66,150
68,587	68,276	65,902	63,415	64,519	70,275	70,102
67,230	65,488	66,225	69,879	69,200	65,179	69,755

Testing Claims Using Critical Values *In Exercises 35–42, (a) write the claim mathematically and identify H_0 and H_a, (b) find the critical values and identify the rejection regions, (c) find the standardized test statistic, (d) decide whether to reject or fail to reject the null hypothesis, and (e) interpret the decision in the context of the original claim.*

35. Caffeine Content in Colas A company that makes cola drinks states that the mean caffeine content per 12-ounce bottle of cola is 40 milligrams. You want to test this claim. During your tests, you find that a random sample of thirty 12-ounce bottles of cola has a mean caffeine content of 39.2 milligrams with a standard deviation of 7.5 milligrams. At $\alpha = 0.01$, can you reject the company's claim? *(Adapted from American Beverage Association)*

36. Electricity Consumption The U.S. Energy Information Association claims that the mean monthly residential electricity consumption in your town is 874 kilowatt-hours (kWh). You want to test this claim. You find that a random sample of 64 residential customers has a mean monthly electricity consumption of 905 kWh and a standard deviation of 125 kWh. At $\alpha = 0.05$, do you have enough evidence to reject the association's claim? *(Adapted from U.S. Energy Information Association)*

37. Light Bulbs A light bulb manufacturer guarantees that the mean life of a certain type of light bulb is at least 750 hours. A random sample of 36 light bulbs has a mean life of 745 hours with a standard deviation of 60 hours. At $\alpha = 0.02$, do you have enough evidence to reject the manufacturer's claim?

38. Fast Food A fast food restaurant estimates that the mean sodium content in one of its breakfast sandwiches is no more than 920 milligrams. A random sample of 44 breakfast sandwiches has a mean sodium content of 925 with a standard deviation of 18 milligrams. At $\alpha = 0.10$, do you have enough evidence to reject the restaurant's claim?

39. Nitrogen Dioxide Levels A scientist estimates that the mean nitrogen dioxide level in Calgary is greater than 32 parts per billion. You want to test this estimate. To do so, you determine the nitrogen dioxide levels for 34 randomly selected days. The results (in parts per billion) are listed below. At $\alpha = 0.06$, can you support the scientist's estimate? *(Adapted from Clean Air Strategic Alliance)*

24	36	44	35	44	34	29	40	39	43	41	32
33	29	29	43	25	39	25	42	29	22	22	25
14	15	14	29	25	27	22	24	18	17		

40. Fluorescent Lamps A fluorescent lamp manufacturer guarantees that the mean life of a certain type of lamp is at least 10,000 hours. You want to test this guarantee. To do so, you record the lives of a random sample of 32 fluorescent lamps. The results (in hours) are shown below. At $\alpha = 0.09$, do you have enough evidence to reject the manufacturer's claim?

8,800	9,155	13,001	10,250	10,002	11,413	8,234	10,402
10,016	8,015	6,110	11,005	11,555	9,254	6,991	12,006
10,420	8,302	8,151	10,980	10,186	10,003	8,814	11,445
6,277	8,632	7,265	10,584	9,397	11,987	7,556	10,380

41. Weight Loss A weight loss program claims that program participants have a mean weight loss of at least 10 pounds after 1 month. You work for a medical association and are asked to test this claim. A random sample of 30 program participants and their weight losses (in pounds) after 1 month is listed in the stem-and-leaf plot at the left. At $\alpha = 0.03$, do you have enough evidence to reject the program's claim?

Weight Loss (in pounds) after One Month

5	7 7	Key: 5|7 = 5.7
6	6 7	
7	0 1 9	
8	2 2 7 9	
9	0 3 5 6 8	
10	2 5 6 6	
11	1 2 5 7 8	
12	0 7 8	
13	8	
14		
15	0	

FIGURE FOR EXERCISE 41

Evacuation Time (in seconds)

```
 0 | 7 9                    Key: 0|7 = 7
 1 | 1 9 9
 2 | 2 6 7 9 9
 3 | 1 1 6 7 7 9 9
 4 | 1 1 3 3 3 4 6 6 7
 5 | 2 3 4 5 7 8 8 8 9 9
 6 | 1 3 3 4 6 6 7
 7 | 4 6 9
 8 | 4 6
 9 | 4
10 | 2
```

FIGURE FOR EXERCISE 42

 42. Fire Drill An engineering company claims that the mean time it takes an employee to evacuate a building during a fire drill is less than 60 seconds. You want to test this claim. A random sample of 50 employees and their evacuation times (in seconds) is listed in the stem-and-leaf plot at the left. At $\alpha = 0.01$, can you support the company's claim?

SC *In Exercises 43–46, use StatCrunch to help you test the claim about the population mean μ at the given level of significance α using the given sample statistics. For each claim, assume the population is normally distributed.*

43. Claim: $\mu = 58$; $\alpha = 0.10$. Sample statistics: $\bar{x} = 57.6$, $s = 2.35$, $n = 80$

44. Claim: $\mu > 495$; $\alpha = 0.05$. Sample statistics: $\bar{x} = 498.4$, $s = 17.8$, $n = 65$

45. Claim: $\mu \leq 1210$; $\alpha = 0.08$. Sample statistics: $\bar{x} = 1234.21$, $s = 205.87$, $n = 250$

46. Claim: $\mu \neq 28{,}750$; $\alpha = 0.01$. Sample statistics: $\bar{x} = 29{,}130$, $s = 3200$, $n = 600$

▆ EXTENDING CONCEPTS

47. Water Usage You believe the mean annual water usage of U.S. households is less than 127,400 gallons. You find that a random sample of 30 households has a mean water usage of 125,270 gallons with a standard deviation of 6275 gallons. You conduct a statistical experiment where $H_0: \mu \geq 127{,}400$ and $H_a: \mu < 127{,}400$. At $\alpha = 0.01$, explain why you cannot reject H_0. *(Adapted from American Water Works Association)*

48. Vehicle Miles of Travel You believe the annual mean vehicle miles of travel (VMT) per U.S. household is greater than 22,000 miles. You do some research and find that a random sample of 36 U.S. households has a mean annual VMT of 22,200 miles with a standard deviation of 775 miles. You conduct a statistical experiment where $H_0: \mu \leq 22{,}000$ and $H_a: \mu > 22{,}000$. At $\alpha = 0.05$, explain why you cannot reject H_0. *(Adapted from U.S. Federal Highway Administration)*

49. Using Different Values of α and n In Exercise 47, you believe that H_0 is not valid. Which of the following allows you to reject H_0? Explain your reasoning.

(a) Use the same values but increase α from 0.01 to 0.02.

(b) Use the same values but increase α from 0.01 to 0.05.

(c) Use the same values but increase n from 30 to 40.

(d) Use the same values but increase n from 30 to 50.

50. Using Different Values of α and n In Exercise 48, you believe that H_0 is not valid. Which of the following allows you to reject H_0? Explain your reasoning.

(a) Use the same values but increase α from 0.05 to 0.06.

(b) Use the same values but increase α from 0.05 to 0.07.

(c) Use the same values but increase n from 36 to 40.

(d) Use the same values but increase n from 36 to 80.

CASE STUDY

Human Body Temperature: What's Normal?

In an article in the *Journal of Statistics Education* (vol. 4, no. 2), Allen Shoemaker describes a study that was reported in the Journal of the American Medical Association (JAMA).* It is generally accepted that the mean body temperature of an adult human is 98.6°F. In his article, Shoemaker uses the data from the JAMA article to test this hypothesis. Here is a summary of his test.

Claim: The body temperature of adults is 98.6°F.

$H_0: \mu = 98.6°F$ (Claim) $H_a: \mu \neq 98.6°F$

Sample Size: $n = 130$

Population: Adult human temperatures (Fahrenheit)

Distribution: Approximately normal

Test Statistics: $\bar{x} = 98.25$, $s = 0.73$

* Data for the JAMA article were collected from healthy men and women, ages 18 to 40, at the University of Maryland Center for Vaccine Development, Baltimore.

Men's Temperatures (in degrees Fahrenheit)

```
96 | 3
96 | 7 9
97 | 0 1 1 1 2 3 4 4 4 4
97 | 5 5 6 6 6 7 8 8 8 8 9 9
98 | 0 0 0 0 0 0 1 1 2 2 2 2 3 3 4 4 4 4
98 | 5 5 6 6 6 6 6 6 7 7 8 8 8 9
99 | 0 0 0 1 2 3 4
99 | 5
100 |
100 |
```
Key: 96|3 = 96.3

Women's Temperatures (in degrees Fahrenheit)

```
96 | 4
96 | 7 8
97 | 2 2 4
97 | 6 7 7 8 8 8 9 9 9
98 | 0 0 0 0 0 1 2 2 2 2 2 2 3 3 3 4 4 4 4 4
98 | 5 6 6 6 6 7 7 7 7 7 7 8 8 8 8 8 8 8 9
99 | 0 0 1 1 2 2 3 4
99 | 9
100 | 0
100 | 8
```
Key: 96|4 = 96.4

■ EXERCISES

1. Complete the hypothesis test for all adults (men and women) by performing the following steps. Use a level of significance of $\alpha = 0.05$.

   (a) Sketch the sampling distribution.

   (b) Determine the critical values and add them to your sketch.

   (c) Determine the rejection regions and shade them in your sketch.

   (d) Find the standardized test statistic. Add it to your sketch.

   (e) Make a decision to reject or fail to reject the null hypothesis.

   (f) Interpret the decision in the context of the original claim.

2. If you lower the level of significance to $\alpha = 0.01$, does your decision change? Explain your reasoning.

3. Test the hypothesis that the mean temperature of men is 98.6°F. What can you conclude at a level of significance of $\alpha = 0.01$?

4. Test the hypothesis that the mean temperature of women is 98.6°F. What can you conclude at a level of significance of $\alpha = 0.01$?

5. Use the sample of 130 temperatures to form a 99% confidence interval for the mean body temperature of adult humans.

6. The conventional "normal" body temperature was established by Carl Wunderlich over 100 years ago. What were possible sources of error in Wunderlich's sampling procedure?

7.3 Hypothesis Testing for the Mean (Small Samples)

WHAT YOU SHOULD LEARN

▸ How to find critical values in a t-distribution

▸ How to use the t-test to test a mean μ

▸ How to use technology to find P-values and use them with a t-test to test a mean μ

Critical Values in a t-Distribution ▸ The t-Test for a Mean μ ($n < 30$, σ unknown) ▸ Using P-Values with t-Tests

▸ CRITICAL VALUES IN A t-DISTRIBUTION

In Section 7.2, you learned how to perform a hypothesis test for a population mean when the sample size was at least 30. In real life, it is often not practical to collect samples of size 30 or more. However, if the population has a normal, or nearly normal, distribution, you can still test the population mean μ. To do so, you can use the t-sampling distribution with $n - 1$ degrees of freedom.

GUIDELINES

Finding Critical Values in a t-Distribution

1. Identify the level of significance α.
2. Identify the degrees of freedom d.f. $= n - 1$.
3. Find the critical value(s) using Table 5 in Appendix B in the row with $n - 1$ degrees of freedom. If the hypothesis test is
   a. *left-tailed*, use the "One Tail, α" column with a negative sign.
   b. *right-tailed*, use the "One Tail, α" column with a positive sign.
   c. *two-tailed*, use the "Two Tails, α" column with a negative and a positive sign.

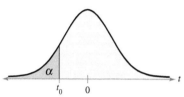

Left-Tailed Test

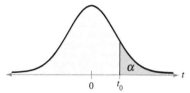

Right-Tailed Test

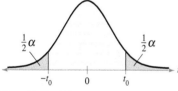

Two-Tailed Test

EXAMPLE 1

▸ **Finding Critical Values for t**

Find the critical value t_0 for a left-tailed test with $\alpha = 0.05$ and $n = 21$.

▸ **Solution**

The degrees of freedom are

$$\text{d.f.} = n - 1$$
$$= 21 - 1$$
$$= 20.$$

To find the critical value, use Table 5 in Appendix B with d.f. $= 20$ and $\alpha = 0.05$ in the "One Tail, α" column. Because the test is a left-tailed test, the critical value is negative. So,

$$t_0 = -1.725.$$

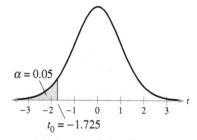

5% Level of Significance

▸ **Try It Yourself 1**

Find the critical value t_0 for a left-tailed test with $\alpha = 0.01$ and $n = 14$.

a. Identify the *degrees of freedom*.
b. *Use* the "One Tail, α" column in Table 5 in Appendix B to find t_0.

Answer: Page A41

EXAMPLE 2

▶ **Finding Critical Values for *t***

Find the critical value t_0 for a right-tailed test with $\alpha = 0.01$ and $n = 17$.

▶ **Solution**

The degrees of freedom are

> d.f. = $n - 1$
> = $17 - 1$
> = 16.

To find the critical value, use Table 5 with d.f. = 16 and $\alpha = 0.01$ in the "One Tail, α" column. Because the test is right-tailed, the critical value is positive. So,

> $t_0 = 2.583$.

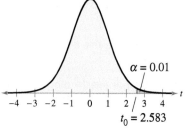

1% Level of Significance

▶ **Try It Yourself 2**

Find the critical value t_0 for a right-tailed test with $\alpha = 0.10$ and $n = 9$.

a. *Identify* the *degrees of freedom.*
b. *Use* the "One Tail, α" column in Table 5 in Appendix B to find t_0.

Answer: Page A41

EXAMPLE 3

▶ **Finding Critical Values for *t***

Find the critical values $-t_0$ and t_0 for a two-tailed test with $\alpha = 0.10$ and $n = 26$.

▶ **Solution**

The degrees of freedom are

> d.f. = $n - 1$
> = $26 - 1$
> = 25.

To find the critical values, use Table 5 with d.f. = 25 and $\alpha = 0.10$ in the "Two Tails, α" column. Because the test is two-tailed, one critical value is negative and one is positive. So,

> $-t_0 = -1.708$ and $t_0 = 1.708$.

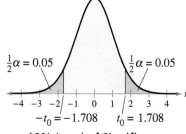

10% Level of Significance

▶ **Try It Yourself 3**

Find the critical values $-t_0$ and t_0 for a two-tailed test with $\alpha = 0.05$ and $n = 16$.

a. *Identify* the *degrees of freedom.*
b. *Use* the "Two Tails, α" column in Table 5 in Appendix B to find t_0.

Answer: Page A41

▶ **THE t-TEST FOR A MEAN μ ($n < 30$, σ UNKNOWN)**

To test a claim about a mean μ using a small sample ($n < 30$) from a normal, or nearly normal, distribution when σ is unknown, you can use a t-sampling distribution.

$$t = \frac{(\text{Sample mean}) - (\text{Hypothesized mean})}{\text{Standard error}}$$

t-TEST FOR A MEAN μ

The **t-test for a mean** is a statistical test for a population mean. The t-test can be used when the population is normal or nearly normal, σ is unknown, and $n < 30$. The **test statistic** is the sample mean $\bar{x}$ and the **standardized test statistic** is

$$t = \frac{\bar{x} - \mu}{s/\sqrt{n}}.$$

The degrees of freedom are

d.f. $= n - 1$.

PICTURING THE WORLD

On the basis of a t-test, a decision was made whether to send truckloads of waste contaminated with cadmium to a sanitary landfill or a hazardous waste landfill. The trucks were sampled to determine if the mean level of cadmium exceeded the allowable amount of 1 milligram per liter for a sanitary landfill. Assume the null hypothesis was $\mu \le 1$.
(Adapted from Pacific Northwest National Laboratory)

	H_0 True	H_0 False
Fail to reject H_0.		
Reject H_0.		

Describe the possible type I and type II errors of this situation.

GUIDELINES

Using the t-Test for a Mean μ (Small Sample)

IN WORDS	IN SYMBOLS
1. State the claim mathematically and verbally. Identify the null and alternative hypotheses.	State H_0 and H_a.
2. Specify the level of significance.	Identify α.
3. Identify the degrees of freedom.	d.f. $= n - 1$
4. Determine the critical value(s).	Use Table 5 in Appendix B.
5. Determine the rejection region(s).	
6. Find the standardized test statistic and sketch the sampling distribution.	$t = \dfrac{\bar{x} - \mu}{s/\sqrt{n}}$
7. Make a decision to reject or fail to reject the null hypothesis.	If t is in the rejection region, reject H_0. Otherwise, fail to reject H_0.
8. Interpret the decision in the context of the original claim.	

Remember that when you make a decision, the possibility of a type I or a type II error exists.

If you prefer using P-values, turn to page 392 to learn how to use P-values for a t-test for a mean μ (small sample).

EXAMPLE 4

See MINITAB steps on page 424.

▶ **Testing μ with a Small Sample**

A used car dealer says that the mean price of a 2008 Honda CR-V is at least $20,500. You suspect this claim is incorrect and find that a random sample of 14 similar vehicles has a mean price of $19,850 and a standard deviation of $1084. Is there enough evidence to reject the dealer's claim at $\alpha = 0.05$? Assume the population is normally distributed. *(Adapted from Kelley Blue Book)*

▶ **Solution**

The claim is "the mean price is at least $20,500." So, the null and alternative hypotheses are

$$H_0: \mu \geq \$20,500 \quad \text{(Claim)}$$

and

$$H_a: \mu < \$20,500.$$

The test is a left-tailed test, the level of significance is $\alpha = 0.05$, and the degrees of freedom are d.f. $= 14 - 1 = 13$. So, the critical value is $t_0 = -1.771$. The rejection region is $t < -1.771$. The standardized test statistic is

$$t = \frac{\overline{x} - \mu}{s/\sqrt{n}} \qquad \text{Because } n < 30, \text{ use the } t\text{-test.}$$

$$= \frac{19,850 - 20,500}{1084/\sqrt{14}} \qquad \text{Assume } \mu = 20,500.$$

$$\approx -2.244.$$

To explore this topic further, see Activity 7.3 on page 397.

The graph shows the location of the rejection region and the standardized test statistic t. Because t is in the rejection region, you should reject the null hypothesis.

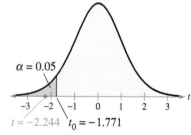

$\alpha = 0.05$

$t \approx -2.244$ $t_0 = -1.771$

5% Level of Significance

Interpretation There is enough evidence at the 5% level of significance to reject the claim that the mean price of a 2008 Honda CR-V is at least $20,500.

▶ **Try It Yourself 4**

An insurance agent says that the mean cost of insuring a 2008 Honda CR-V is less than $1200. A random sample of 7 similar insurance quotes has a mean cost of $1125 and a standard deviation of $55. Is there enough evidence to support the agent's claim at $\alpha = 0.10$? Assume the population is normally distributed.

a. Identify the *claim* and state H_0 and H_a.
b. Identify the *level of significance* α and the *degrees of freedom*.
c. Find the *critical value* t_0 and identify the *rejection region*.
d. Find the *standardized test statistic t. Sketch* a graph.
e. *Decide* whether to reject the null hypothesis.
f. *Interpret* the decision in the context of the original claim.

Answer: Page A41

EXAMPLE 5

See TI-83/84 Plus steps on page 425.

▸ **Testing μ with a Small Sample**

An industrial company claims that the mean pH level of the water in a nearby river is 6.8. You randomly select 19 water samples and measure the pH of each. The sample mean and standard deviation are 6.7 and 0.24, respectively. Is there enough evidence to reject the company's claim at $\alpha = 0.05$? Assume the population is normally distributed.

▸ **Solution**

The claim is "the mean pH level is 6.8." So, the null and alternative hypotheses are

$$H_0: \mu = 6.8 \quad \text{(Claim)}$$

and

$$H_a: \mu \neq 6.8.$$

The test is a two-tailed test, the level of significance is $\alpha = 0.05$, and the degrees of freedom are d.f. $= 19 - 1 = 18$. So, the critical values are $-t_0 = -2.101$ and $t_0 = 2.101$. The rejection regions are $t < -2.101$ and $t > 2.101$. The standardized test statistic is

$$t = \frac{\bar{x} - \mu}{s/\sqrt{n}} \qquad \text{Because } n < 30, \text{ use the } t\text{-test.}$$

$$= \frac{6.7 - 6.8}{0.24/\sqrt{19}} \qquad \text{Assume } \mu = 6.8.$$

$$\approx -1.816.$$

The graph shows the location of the rejection region and the standardized test statistic t. Because t is not in the rejection region, you fail to reject the null hypothesis.

Interpretation There is not enough evidence at the 5% level of significance to reject the claim that the mean pH is 6.8.

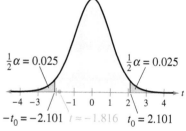

$\frac{1}{2}\alpha = 0.025 \qquad \frac{1}{2}\alpha = 0.025$

$-t_0 = -2.101 \quad t \approx -1.816 \quad t_0 = 2.101$

5% Level of Significance

▸ **Try It Yourself 5**

The company also claims that the mean conductivity of the river is 1890 milligrams per liter. The conductivity of a water sample is a measure of the total dissolved solids in the sample. You randomly select 19 water samples and measure the conductivity of each. The sample mean and standard deviation are 2500 milligrams per liter and 700 milligrams per liter, respectively. Is there enough evidence to reject the company's claim at $\alpha = 0.01$? Assume the population is normally distributed.

a. Identify the *claim* and state H_0 and H_a.
b. Identify the *level of significance α* and the *degrees of freedom*.
c. Find the *critical values* $-t_0$ and t_0 and identify the *rejection region*.
d. Find the *standardized test statistic t. Sketch* a graph.
e. *Decide* whether to reject the null hypothesis.
f. *Interpret* the decision in the context of the original claim.

Answer: Page A42

▶ USING *P*-VALUES WITH *t*-TESTS

Suppose you wanted to find a *P*-value given $t = 1.98$, 15 degrees of freedom, and a right-tailed test. Using Table 5 in Appendix B, you can determine that *P* falls between $\alpha = 0.025$ and $\alpha = 0.05$, but you cannot determine an exact value for *P*. In such cases, you can use technology to perform a hypothesis test and find exact *P*-values.

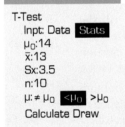
| EXAMPLE 6 | Report 30 |

▶ **Using *P*-Values with a *t*-Test**

A Department of Motor Vehicles office claims that the mean wait time is less than 14 minutes. A random sample of 10 people has a mean wait time of 13 minutes with a standard deviation of 3.5 minutes. At $\alpha = 0.10$, test the office's claim. Assume the population is normally distributed.

▶ **Solution**

The claim is "the mean wait time is less than 14 minutes." So, the null and alternative hypotheses are

$$H_0: \mu \geq 14 \text{ minutes}$$

and

$$H_a: \mu < 14 \text{ minutes. (Claim)}$$

The TI-83/84 Plus display at the far left shows how to set up the hypothesis test. The two displays on the right show the possible results, depending on whether you select "Calculate" or "Draw."

TI-83/84 PLUS

```
T-Test
 Inpt: Data Stats
 μ₀:14
 x̄:13
 Sx:3.5
 n:10
 μ: ≠μ₀  <μ₀  >μ₀
 Calculate Draw
```

TI-83/84 PLUS

```
T-Test
 μ<14
 t=-.9035079029
 p=.1948994027
 x̄=13
 Sx=3.5
 n=10
```

TI-83/84 PLUS

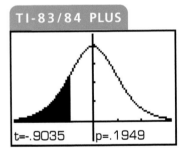

```
t=-.9035    p=.1949
```

From the displays, you can see that $P \approx 0.1949$. Because the *P*-value is greater than $\alpha = 0.10$, you fail to reject the null hypothesis.

Interpretation There is not enough evidence at the 10% level of significance to support the office's claim that the mean wait time is less than 14 minutes.

▶ **Try It Yourself 6**

Another Department of Motor Vehicles office claims that the mean wait time is at most 18 minutes. A random sample of 12 people has a mean wait time of 15 minutes with a standard deviation of 2.2 minutes. At $\alpha = 0.05$, test the office's claim. Assume the population is normally distributed.

a. Identify the *claim* and state H_0 and H_a.
b. *Use* a TI-83/84 Plus to find the *P*-value.
c. *Compare* the *P*-value with the level of significance α and *make* a decision.
d. *Interpret* the decision in the context of the original claim.

Answer: Page A42

7.3 EXERCISES

BUILDING BASIC SKILLS AND VOCABULARY

1. Explain how to find critical values for a *t*-sampling distribution.

2. Explain how to use a *t*-test to test a hypothesized mean μ given a small sample ($n < 30$). What assumption about the population is necessary?

In Exercises 3–8, find the critical value(s) for the indicated t-test, level of significance α, and sample size n.

3. Right-tailed test, $\alpha = 0.05$, $n = 23$ **4.** Right-tailed test, $\alpha = 0.01$, $n = 11$

5. Left-tailed test, $\alpha = 0.10$, $n = 20$ **6.** Left-tailed test, $\alpha = 0.01$, $n = 28$

7. Two-tailed test, $\alpha = 0.05$, $n = 27$ **8.** Two-tailed test, $\alpha = 0.10$, $n = 22$

Graphical Analysis   *In Exercises 9–12, state whether the standardized test statistic t indicates that you should reject the null hypothesis. Explain.*

9. (a) $t = 2.091$
   (b) $t = 0$
   (c) $t = -1.08$
   (d) $t = -2.096$

10. (a) $t = 1.308$
   (b) $t = -1.389$
   (c) $t = 1.650$
   (d) $t = -0.998$

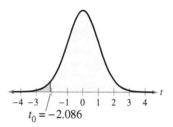

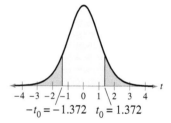

11. (a) $t = -2.502$
   (b) $t = 2.203$
   (c) $t = 2.680$
   (d) $t = -2.703$

12. (a) $t = 1.705$
   (b) $t = -1.755$
   (c) $t = -1.585$
   (d) $t = 1.745$

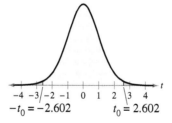

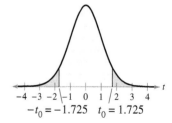

In Exercises 13–16, use a t-test to test the claim about the population mean μ at the given level of significance α using the given sample statistics. For each claim, assume the population is normally distributed.

13. Claim: $\mu = 15$; $\alpha = 0.01$. Sample statistics: $\bar{x} = 13.9$, $s = 3.23$, $n = 6$

14. Claim: $\mu > 25$; $\alpha = 0.05$. Sample statistics: $\bar{x} = 26.2$, $s = 2.32$, $n = 17$

15. Claim: $\mu \geq 8000$; $\alpha = 0.01$. Sample statistics: $\bar{x} = 7700$, $s = 450$, $n = 25$

16. Claim: $\mu \neq 52{,}200$; $\alpha = 0.10$. Sample statistics: $\bar{x} = 53{,}220$, $s = 2700$, $n = 18$

■ USING AND INTERPRETING CONCEPTS

Testing Claims *In Exercises 17–24, (a) write the claim mathematically and identify H_0 and H_a, (b) find the critical value(s) and identify the rejection region(s), (c) find the standardized test statistic t, (d) decide whether to reject or fail to reject the null hypothesis, and (e) interpret the decision in the context of the original claim. If convenient, use technology. For each claim, assume the population is normally distributed.*

17. **Used Car Cost** A used car dealer says that the mean price of a 2008 Subaru Forester is $18,000. You suspect this claim is incorrect and find that a random sample of 15 similar vehicles has a mean price of $18,550 and a standard deviation of $1767. Is there enough evidence to reject the claim at $\alpha = 0.05$? *(Adapted from Kelley Blue Book)*

18. **IRS Wait Times** The Internal Revenue Service claims that the mean wait time for callers during a recent tax filing season was at most 7 minutes. A random sample of 11 callers has a mean wait time of 8.7 minutes and a standard deviation of 2.7 minutes. Is there enough evidence to reject the claim at $\alpha = 0.10$? *(Adapted from Internal Revenue Service)*

19. **Work Hours** A medical board claims that the mean number of hours worked per week by surgical faculty who teach at an academic institution is more than 60 hours. The hours worked include teaching hours as well as regular working hours. A random sample of 7 surgical faculty has a mean hours worked per week of 70 hours and a standard deviation of 12.5 hours. At $\alpha = 0.05$, do you have enough evidence to support the board's claim? *(Adapted from Journal of the American College of Surgeons)*

20. **Battery Life** A company claims that the mean battery life of their MP3 player is at least 30 hours. You suspect this claim is incorrect and find that a random sample of 18 MP3 players has a mean battery life of 28.5 hours and a standard deviation of 1.7 hours. Is there enough evidence to reject the claim at $\alpha = 0.01$?

21. **Waste Recycled** An environmentalist estimates that the mean amount of waste recycled by adults in the United States is more than 1 pound per person per day. You want to test this claim. You find that the mean waste recycled per person per day for a random sample of 13 adults in the United States is 1.50 pounds and the standard deviation is 0.28 pound. At $\alpha = 0.10$, can you support the claim? *(Adapted from U.S. Environmental Protection Agency)*

22. **Waste Generated** As part of your work for an environmental awareness group, you want to test a claim that the mean amount of waste generated by adults in the United States is more than 4 pounds per day. In a random sample of 22 adults in the United States, you find that the mean waste generated per person per day is 4.50 pounds with a standard deviation of 1.21 pounds. At $\alpha = 0.01$, can you support the claim? *(Adapted from U.S. Environmental Protection Agency)*

23. **Annual Pay** An employment information service claims the mean annual salary for full-time male workers over age 25 and without a high school diploma is $26,000. The annual salaries for a random sample of 10 full-time male workers without a high school diploma are listed. At $\alpha = 0.05$, test the claim that the mean salary is $26,000. *(Adapted from U.S. Bureau of Labor Statistics)*

| 26,185 | 23,814 | 22,374 | 25,189 | 26,318 |
| 20,767 | 30,782 | 29,541 | 24,597 | 28,955 |

24. Annual Pay   An employment information service claims the mean annual salary for full-time female workers over age 25 and without a high school diploma is more than $18,500. The annual salaries for a random sample of 12 full-time female workers without a high school diploma are listed. At $\alpha = 0.10$, is there enough evidence to support the claim that the mean salary is more than $18,500? *(Adapted from U.S. Bureau of Labor Statistics)*

> 18,665   16,312   18,794   19,403   20,864   19,177
> 17,328   21,445   20,354   19,143   18,316   19,237

Testing Claims Using P-Values   *In Exercises 25–30, (a) write the claim mathematically and identify H_0 and H_a, (b) use technology to find the P-value, (c) decide whether to reject or fail to reject the null hypothesis, and (d) interpret the decision in the context of the original claim. Assume the population is normally distributed.*

25. Speed Limit   A county is considering raising the speed limit on a road because they claim that the mean speed of vehicles is greater than 45 miles per hour. A random sample of 25 vehicles has a mean speed of 48 miles per hour and a standard deviation of 5.4 miles per hour. At $\alpha = 0.10$, do you have enough evidence to support the county's claim?

26. Oil Changes   A repair shop believes that people travel more than 3500 miles between oil changes. A random sample of 8 cars getting an oil change has a mean distance of 3375 miles since having an oil change with a standard deviation of 225 miles. At $\alpha = 0.05$, do you have enough evidence to support the shop's claim?

27. Meal Cost   A travel association claims that the mean daily meal cost for two adults traveling together on vacation in San Francisco is $105. A random sample of 20 such groups of adults has a mean daily meal cost of $110 and a standard deviation of $8.50. Is there enough evidence to reject the claim at $\alpha = 0.01$? *(Adapted from American Automobile Association)*

28. Lodging Cost   A travel association claims that the mean daily lodging cost for two adults traveling together on vacation in San Francisco is at least $240. A random sample of 24 such groups of adults has a mean daily lodging cost of $233 and a standard deviation of $12.50. Is there enough evidence to reject the claim at $\alpha = 0.10$? *(Adapted from American Automobile Association)*

29. Class Size   You receive a brochure from a large university. The brochure indicates that the mean class size for full-time faculty is fewer than 32 students. You want to test this claim. You randomly select 18 classes taught by full-time faculty and determine the class size of each. The results are listed below. At $\alpha = 0.05$, can you support the university's claim?

> 35   28   29   33   32   40   26   25   29
> 28   30   36   33   29   27   30   28   25

30. Faculty Classroom Hours   The dean of a university estimates that the mean number of classroom hours per week for full-time faculty is 11.0. As a member of the student council, you want to test this claim. A random sample of the number of classroom hours for eight full-time faculty for one week is listed below. At $\alpha = 0.01$, can you reject the dean's claim?

> 11.8   8.6   12.6   7.9   6.4   10.4   13.6   9.1

SC *In Exercises 31–34, use StatCrunch and a t-test to help you test the claim about the population mean μ at the given level of significance α using the given sample statistics. For each claim, assume the population is normally distributed.*

31. Claim: $\mu \leq 75$; $\alpha = 0.05$. Sample statistics: $\bar{x} = 73.6$, $s = 3.2$, $n = 26$

32. Claim: $\mu \neq 27$; $\alpha = 0.01$. Sample statistics: $\bar{x} = 31.5$, $s = 4.7$, $n = 12$

33. Claim: $\mu < 188$; $\alpha = 0.05$. Sample statistics: $\bar{x} = 186$, $s = 12$, $n = 9$

34. Claim: $\mu \geq 2118$; $\alpha = 0.10$. Sample statistics: $\bar{x} = 1787$, $s = 384$, $n = 17$

■ EXTENDING CONCEPTS

35. Credit Card Balances To test the claim that the mean credit card debt for individuals is greater than \$5000, you do some research and find that a random sample of 6 cardholders has a mean credit card balance of \$5434 with a standard deviation of \$625. You conduct a statistical experiment where H_0: $\mu \leq \$5000$ and H_a: $\mu > \$5000$. At $\alpha = 0.05$, explain why you cannot reject H_0. Assume the population is normally distributed. *(Adapted from TransUnion)*

36. Using Different Values of α and n In Exercise 35, you believe that H_0 is not valid. Which of the following allows you to reject H_0? Explain your reasoning.

(a) Use the same values but decrease α from 0.05 to 0.01.

(b) Use the same values but increase α from 0.05 to 0.10.

(c) Use the same values but increase n from 6 to 8.

(d) Use the same values but increase n from 6 to 24.

Deciding on a Distribution *In Exercises 37 and 38, decide whether you should use a normal sampling distribution or a t-sampling distribution to perform the hypothesis test. Justify your decision. Then use the distribution to test the claim. Write a short paragraph about the results of the test and what you can conclude about the claim.*

37. Gas Mileage A car company says that the mean gas mileage for its luxury sedan is at least 23 miles per gallon (mpg). You believe the claim is incorrect and find that a random sample of 5 cars has a mean gas mileage of 22 mpg and a standard deviation of 4 mpg. At $\alpha = 0.05$, test the company's claim. Assume the population is normally distributed.

38. Private Law School An education publication claims that the average in-state tuition for one year of law school at a private institution is more than \$35,000. A random sample of 50 private law schools has a mean in-state tuition of \$34,967 and a standard deviation of \$5933 for one year. At $\alpha = 0.01$, test the publication's claim. Assume the population is normally distributed. *(Adapted from U.S. News and World Report)*

39. Writing You are testing a claim and incorrectly use the normal sampling distribution instead of the *t*-sampling distribution. Does this make it more or less likely to reject the null hypothesis? Is this result the same no matter whether the test is left-tailed, right-tailed, or two-tailed? Explain your reasoning.

ACTIVITY 7.3   Hypothesis Tests for a Mean

APPLET

The *hypothesis tests for a mean* applet allows you to visually investigate hypothesis tests for a mean. You can specify the sample size n, the shape of the distribution (Normal or Right skewed), the true population mean (Mean), the true population standard deviation (Std. Dev.), the null value for the mean (Null mean), and the alternative for the test (Alternative). When you click SIMULATE, 100 separate samples of size n will be selected from a population with these population parameters. For each of the 100 samples, a hypothesis test based on the T statistic is performed, and the results from each test are displayed in the plots at the right. The test statistic for each test is shown in the top plot and the P-value is shown in the bottom plot. The green and blue lines represent the cutoffs for rejecting the null hypothesis with the 0.05 and 0.01 level tests, respectively. Additional simulations can be carried out by clicking SIMULATE multiple times. The cumulative number of times that each test rejects the null hypothesis is also shown. Press CLEAR to clear existing results and start a new simulation.

■ Explore

Step 1   Specify a value for n.
Step 2   Specify a distribution.
Step 3   Specify a value for the mean.
Step 4   Specify a value for the standard deviation.
Step 5   Specify a value for the null mean.
Step 6   Specify an alternative hypothesis.
Step 7   Click SIMULATE to generate the hypothesis tests.

n: `100`
Distribution: `Normal ▾`
Mean: `50`
Std. Dev.: `10`
Null mean: `50`
Alternative: `< ▾`

Simulate

Cumulative results:

	0.05 level	0.01 level
Reject null		
Fail to reject null		
Prop. rejected		

Clear

■ Draw Conclusions

APPLET

1. Set $n = 15$, Mean $= 40$, Std. Dev. $= 5$, Null mean $= 40$, alternative hypothesis to "not equal," and the distribution to "Normal." Run the simulation so that at least 1000 hypothesis tests are run. Compare the proportion of null hypothesis rejections for the 0.05 level and the 0.01 level. Is this what you would expect? Explain.

2. Suppose a null hypothesis is rejected at the 0.01 level. Will it be rejected at the 0.05 level? Explain. Suppose a null hypothesis is rejected at the 0.05 level. Will it be rejected at the 0.01 level? Explain.

3. Set $n = 25$, Mean $= 25$, Std. Dev. $= 3$, Null mean $= 27$, alternative hypothesis to "<," and the distribution to "Normal." What is the null hypothesis? Run the simulation so that at least 1000 hypothesis tests are run. Compare the proportion of null hypothesis rejections for the 0.05 level and the 0.01 level. Is this what you would expect? Explain.

7.4 Hypothesis Testing for Proportions

WHAT YOU SHOULD LEARN

▸ How to use the z-test to test a population proportion p

Hypothesis Test for Proportions

▸ HYPOTHESIS TEST FOR PROPORTIONS

In Sections 7.2 and 7.3, you learned how to perform a hypothesis test for a population mean. In this section, you will learn how to test a population proportion p.

Hypothesis tests for proportions can be used when politicians want to know the proportion of their constituents who favor a certain bill or when quality assurance engineers test the proportion of parts that are defective.

If $np \geq 5$ and $nq \geq 5$ for a binomial distribution, then the sampling distribution for $\hat{p}$ is approximately normal with a mean of

$$\mu_{\hat{p}} = p$$

and a standard error of

$$\sigma_{\hat{p}} = \sqrt{pq/n}.$$

z-TEST FOR A PROPORTION p

The **z-test for a proportion** is a statistical test for a population proportion p. The z-test can be used when a binomial distribution is given such that $np \geq 5$ and $nq \geq 5$. The **test statistic** is the sample proportion $\hat{p}$ and the **standardized test statistic** is

$$z = \frac{\hat{p} - \mu_{\hat{p}}}{\sigma_{\hat{p}}} = \frac{\hat{p} - p}{\sqrt{pq/n}}.$$

GUIDELINES

Using a z-Test for a Proportion p

Verify that $np \geq 5$ and $nq \geq 5$.

IN WORDS	IN SYMBOLS
1. State the claim mathematically and verbally. Identify the null and alternative hypotheses.	State H_0 and H_a.
2. Specify the level of significance.	Identify α.
3. Determine the critical value(s).	Use Table 4 in Appendix B.
4. Determine the rejection region(s).	
5. Find the standardized test statistic and sketch the sampling distribution.	$z = \dfrac{\hat{p} - p}{\sqrt{pq/n}}$
6. Make a decision to reject or fail to reject the null hypothesis.	If z is in the rejection region, reject H_0. Otherwise, fail to reject H_0.
7. Interpret the decision in the context of the original claim.	

INSIGHT

A hypothesis test for a proportion p can also be performed using P-values. Use the guidelines on page 373 for using P-values for a z-test for a mean μ, but in Step 3 find the standardized test statistic by using the formula

$$z = \frac{\hat{p} - p}{\sqrt{pq/n}}.$$

The other steps in the test are the same.

See TI-83/84 Plus steps on page 425.

EXAMPLE 1

To explore this topic further, see Activity 7.4 on page 403.

▶ **Hypothesis Test for a Proportion**

A research center claims that less than 50% of U.S. adults have accessed the Internet over a wireless network with a laptop computer. In a random sample of 100 adults, 39% say they have accessed the Internet over a wireless network with a laptop computer. At $\alpha = 0.01$, is there enough evidence to support the researcher's claim? *(Adapted from Pew Research Center)*

▶ **Solution**   The products $np = 100(0.50) = 50$ and $nq = 100(0.50) = 50$ are both greater than 5. So, you can use a z-test. The claim is "less than 50% have accessed the Internet over a wireless network with a laptop computer." So, the null and alternative hypotheses are

$$H_0: p \geq 0.5 \qquad \text{and} \qquad H_a: p < 0.5. \quad \text{(Claim)}$$

Because the test is a left-tailed test and the level of significance is $\alpha = 0.01$, the critical value is $z_0 = -2.33$ and the rejection region is $z < -2.33$. The standardized test statistic is

$$z = \frac{\hat{p} - p}{\sqrt{pq/n}} \qquad \text{Because } np \geq 5 \text{ and } nq \geq 5, \text{ you can use the z-test.}$$

$$= \frac{0.39 - 0.5}{\sqrt{(0.5)(0.5)/100}} \qquad \text{Assume } p = 0.5.$$

$$= -2.2.$$

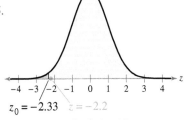

$z_0 = -2.33$ $z = -2.2$

1% Level of Significance

The graph shows the location of the rejection region and the standardized test statistic *z*. Because *z* is not in the rejection region, you should fail to reject the null hypothesis.

Interpretation   There is not enough evidence at the 1% level of significance to support the claim that less than 50% of U.S. adults have accessed the Internet over a wireless network with a laptop computer.

▶ **Try It Yourself 1**

A research center claims that more than 25% of U.S. adults have used a cellular phone to access the Internet. In a random sample of 125 adults, 32% say they have used a cellular phone to access the Internet. At $\alpha = 0.05$, is there enough evidence to support the researcher's claim? *(Adapted from Pew Research Center)*

a. *Verify* that $np \geq 5$ and $nq \geq 5$.
b. Identify the *claim* and state H_0 and H_a.
c. Identify the *level of significance* α.
d. Find the *critical value* z_0 and identify the *rejection region*.
e. Find the *standardized test statistic z*. *Sketch* a graph.
f. *Decide* whether to reject the null hypothesis.
g. *Interpret* the decision in the context of the original claim.

Answer: Page A42

To use a *P*-value to perform the hypothesis test in Example 1, use Table 4 to find the area corresponding to $z = -2.2$. The area is 0.0139. Because this is a left-tailed test, the *P*-value is equal to the area to the left of $z = -2.2$. So, $P = 0.0139$. Because the *P*-value is greater than $\alpha = 0.01$, you should fail to reject the null hypothesis. Note that this is the same result obtained in Example 1.

See MINITAB
steps on page 424.

EXAMPLE 2

▶ **Hypothesis Test for a Proportion**

A research center claims that 25% of college graduates think a college degree is not worth the cost. You decide to test this claim and ask a random sample of 200 college graduates whether they think a college degree is not worth the cost. Of those surveyed, 21% reply yes. At $\alpha = 0.10$, is there enough evidence to reject the claim? *(Adapted from Zogby International)*

▶ **Solution**

The products $np = 200(0.25) = 50$ and $nq = 200(0.75) = 150$ are both greater than 5. So, you can use a z-test. The claim is "25% of college graduates think a college degree is not worth the cost." So, the null and alternative hypotheses are

$$H_0: p = 0.25 \text{ (Claim)} \quad \text{and} \quad H_a: p \neq 0.25.$$

Because the test is a two-tailed test and the level of significance is $\alpha = 0.10$, the critical values are $-z_0 = -1.645$ and $z_0 = 1.645$. The rejection regions are $z < -1.645$ and $z > 1.645$. The standardized test statistic is

$$z = \frac{\hat{p} - p}{\sqrt{pq/n}} \qquad \text{Because } np \geq 5 \text{ and } nq \geq 5, \text{ you can use the } z\text{-test.}$$

$$= \frac{0.21 - 0.25}{\sqrt{(0.25)(0.75)/200}} \qquad \text{Assume } p = 0.25.$$

$$= -1.31.$$

The graph shows the location of the rejection regions and the standardized test statistic z. Because z is not in the rejection region, you should fail to reject the null hypothesis.

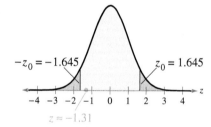

$-z_0 = -1.645$ $z_0 = 1.645$

$z \approx -1.31$

10% Level of Significance

Interpretation There is not enough evidence at the 10% level of significance to reject the claim that 25% of college graduates think a college degree is not worth the cost.

▶ **Try It Yourself 2**

A research center claims that 30% of U.S. adults have not purchased a certain brand because they found the advertisements distasteful. You decide to test this claim and ask a random sample of 250 U.S. adults whether they have not purchased a certain brand because they found the advertisements distasteful. Of those surveyed, 36% reply yes. At $\alpha = 0.10$, is there enough evidence to reject the claim? *(Adapted from Harris Interactive)*

a. *Verify* that $np \geq 5$ and $nq \geq 5$.
b. Identify the *claim* and state H_0 and H_a.
c. Identify the *level of significance* α.
d. Find the *critical values* $-z_0$ and z_0 and identify the *rejection regions*.
e. Find the *standardized test statistic z. Sketch* a graph.
f. *Decide* whether to reject the null hypothesis.
g. *Interpret* the decision in the context of the original claim.

Answer: Page A42

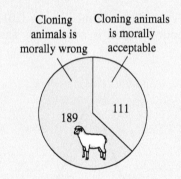

7.4 EXERCISES

■ BUILDING BASIC SKILLS AND VOCABULARY

1. Explain how to decide when a normal distribution can be used to approximate a binomial distribution.

2. Explain how to test a population proportion p.

In Exercises 3–8, decide whether the normal sampling distribution can be used. If it can be used, test the claim about the population proportion p at the given level of significance α using the given sample statistics.

3. Claim: $p < 0.12$; $\alpha = 0.01$. Sample statistics: $\hat{p} = 0.10$, $n = 40$

4. Claim: $p \geq 0.48$; $\alpha = 0.08$. Sample statistics: $\hat{p} = 0.40$, $n = 90$

5. Claim: $p \neq 0.15$; $\alpha = 0.05$. Sample statistics: $\hat{p} = 0.12$, $n = 500$

6. Claim: $p > 0.70$; $\alpha = 0.04$. Sample statistics: $\hat{p} = 0.64$, $n = 225$

7. Claim: $p \leq 0.45$; $\alpha = 0.05$. Sample statistics: $\hat{p} = 0.52$, $n = 100$

8. Claim: $p = 0.95$; $\alpha = 0.10$. Sample statistics: $\hat{p} = 0.875$, $n = 50$

■ USING AND INTERPRETING CONCEPTS

Testing Claims   *In Exercises 9–16, (a) write the claim mathematically and identify H_0 and H_a, (b) find the critical value(s) and identify the rejection region(s), (c) find the standardized test statistic z, (d) decide whether to reject or fail to reject the null hypothesis, and (e) interpret the decision in the context of the original claim. If convenient, use technology to find the standardized test statistic.*

9. **Smokers**   A medical researcher says that less than 25% of U.S. adults are smokers. In a random sample of 200 U.S. adults, 18.5% say that they are smokers. At $\alpha = 0.05$, is there enough evidence to reject the researcher's claim? *(Adapted from National Center for Health Statistics)*

10. **Census**   A research center claims that at least 40% of U.S. adults think the Census count is accurate. In a random sample of 600 U.S. adults, 35% say that the Census count is accurate. At $\alpha = 0.02$, is there enough evidence to reject the center's claim? *(Adapted from Rasmussen Reports)*

11. **Cellular Phones and Driving**   A research center claims that at most 50% of people believe that drivers should be allowed to use cellular phones with hands-free devices while driving. In a random sample of 150 U.S. adults, 58% say that drivers should be allowed to use cellular phones with hands-free devices while driving. At $\alpha = 0.01$, is there enough evidence to reject the center's claim? *(Adapted from Rasmussen Reports)*

12. **Asthma**   A medical researcher claims that 5% of children under 18 years of age have asthma. In a random sample of 250 children under 18 years of age, 9.6% say they have asthma. At $\alpha = 0.08$, is there enough evidence to reject the researcher's claim? *(Adapted from National Center for Health Statistics)*

13. **Female Height**   A research center claims that more than 75% of females ages 20–29 are taller than 62 inches. In a random sample of 150 females ages 20–29, 82% are taller than 62 inches. At $\alpha = 0.10$, is there enough evidence to support the center's claim? *(Adapted from National Center for Health Statistics)*

14. **Curling** A research center claims that 16% of U.S. adults say that curling is the Winter Olympic sport they would like to try the most. In a random sample of 300 U.S. adults, 20% say that curling is the Winter Olympic sport they would like to try the most. At $\alpha = 0.05$, is there enough evidence to reject the researcher's claim? *(Adapted from Zogby International)*

15. **Dog Ownership** A humane society claims that less than 35% of U.S. households own a dog. In a random sample of 400 U.S. households, 156 say they own a dog. At $\alpha = 0.10$, is there enough evidence to support the society's claim? *(Adapted from The Humane Society of the United States)*

16. **Cat Ownership** A humane society claims that 30% of U.S. households own a cat. In a random sample of 200 U.S. households, 72 say they own a cat. At $\alpha = 0.05$, is there enough evidence to reject the society's claim? *(Adapted from The Humane Society of the United States)*

Free Samples *In Exercises 17 and 18, use the graph, which shows what adults think about the effectiveness of free samples.*

17. **Do Free Samples Work?** You interview a random sample of 50 adults. The results of the survey show that 48% of the adults said they were more likely to buy a product when there are free samples. At $\alpha = 0.05$, can you reject the claim that at least 52% of adults are more likely to buy a product when there are free samples?

18. **Should Free Samples Be Used?** Use your conclusion from Exercise 17 to write a paragraph on the use of free samples. Do you think a company should use free samples to get people to buy a product? Explain.

■ EXTENDING CONCEPTS

Alternative Formula *In Exercises 19 and 20, use the following information. When you know the number of successes x, the sample size n, and the population proportion p, it can be easier to use the formula*

$$z = \frac{x - np}{\sqrt{npq}}$$

to find the standardized test statistic when using a z-test for a population proportion p.

19. Rework Exercise 15 using the alternative formula and compare the results.

20. The alternative formula is derived from the formula

$$z = \frac{\hat{p} - p}{\sqrt{pq/n}} = \frac{(x/n) - p}{\sqrt{pq/n}}.$$

Use this formula to derive the alternative formula. Justify each step.

ACTIVITY 7.4   Hypothesis Tests for a Proportion

APPLET

The *hypothesis tests for a proportion* applet allows you to visually investigate hypothesis tests for a population proportion. You can specify the sample size n, the population proportion (True p), the null value for the proportion (Null p), and the alternative for the test (Alternative). When you click SIMULATE, 100 separate samples of size n will be selected from a population with a proportion of successes equal to True p. For each of the 100 samples, a hypothesis test based on the Z statistic is performed, and the results from each test are displayed in plots at the right. The standardized test statistic for each test is shown in the top plot and the P-value is shown in the bottom plot. The green and blue lines represent the cutoffs for rejecting the null hypothesis with the 0.05 and 0.01 level tests, respectively. Additional simulations can be carried out by clicking SIMULATE multiple times. The cumulative number of times that each test rejects the null hypothesis is also shown. Press CLEAR to clear existing results and start a new simulation.

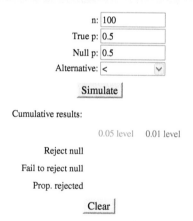

■ Explore

Step 1   Specify a value for n.
Step 2   Specify a value for True p.
Step 3   Specify a value for Null p.
Step 4   Specify an alternative hypothesis.
Step 5   Click SIMULATE to generate the hypothesis tests.

■ Draw Conclusions

APPLET

1. Set $n = 25$, True $p = 0.35$, Null $p = 0.35$, and the alternative hypothesis to "not equal." Run the simulation so that at least 1000 hypothesis tests are run. Compare the proportion of null hypothesis rejections for the 0.05 level and the 0.01 level. Is this what you would expect? Explain.

2. Set $n = 50$, True $p = 0.6$, Null $p = 0.4$, and the alternative hypothesis to "<." What is the null hypothesis? Run the simulation so that at least 1000 hypothesis tests are run. Compare the proportion of null hypothesis rejections for the 0.05 level and the 0.01 level. Perform a hypothesis test for each level. Use the results of the hypothesis tests to explain the results of the simulation.

7.5 Hypothesis Testing for Variance and Standard Deviation

WHAT YOU SHOULD LEARN

▸ How to find critical values for a χ^2-test

▸ How to use the χ^2-test to test a variance or a standard deviation

Critical Values for a χ^2-Test ▸ The Chi-Square Test

▸ CRITICAL VALUES FOR A χ^2-TEST

In real life, it is often important to produce consistent predictable results. For instance, consider a company that manufactures golf balls. The manufacturer must produce millions of golf balls, each having the same size and the same weight. There is a very low tolerance for variation. If the population is normal, you can test the variance and standard deviation of the process using the chi-square distribution with $n - 1$ degrees of freedom.

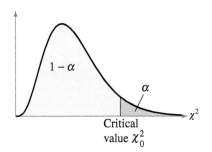

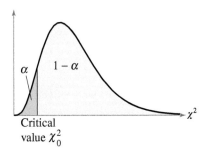

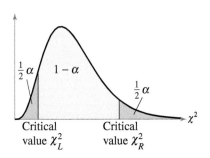

GUIDELINES

Finding Critical Values for the χ^2-Test

1. Specify the level of significance α.
2. Determine the degrees of freedom d.f. $= n - 1$.
3. The critical values for the χ^2-distribution are found in Table 6 in Appendix B. To find the critical value(s) for a
   a. *right-tailed test*, use the value that corresponds to d.f. and α.
   b. *left-tailed test*, use the value that corresponds to d.f. and $1 - \alpha$.
   c. *two-tailed test*, use the values that correspond to d.f. and $\frac{1}{2}\alpha$, and d.f. and $1 - \frac{1}{2}\alpha$.

EXAMPLE 1

▸ **Finding Critical Values for χ^2**

Find the critical χ^2-value for a right-tailed test when $n = 26$ and $\alpha = 0.10$.

▸ **Solution**

The degrees of freedom are

$$\text{d.f.} = n - 1 = 26 - 1 = 25.$$

The graph at the right shows a χ^2-distribution with 25 degrees of freedom and a shaded area of $\alpha = 0.10$ in the right tail. In Table 6 in Appendix B with d.f. $= 25$ and $\alpha = 0.10$, the critical value is

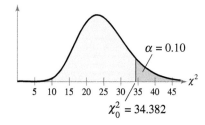

$$\chi_0^2 = 34.382.$$

▸ **Try It Yourself 1**

Find the critical χ^2-value for a right-tailed test when $n = 18$ and $\alpha = 0.01$.

a. *Identify* the degrees of freedom and the level of significance.
b. *Use* Table 6 in Appendix B to find the critical χ^2-value. *Answer: Page A42*

EXAMPLE 2

▶ **Finding Critical Values for χ^2**

Find the critical χ^2-value for a left-tailed test when $n = 11$ and $\alpha = 0.01$.

▶ **Solution**

The degrees of freedom are

$$\text{d.f.} = n - 1 = 11 - 1 = 10.$$

The graph shows a χ^2-distribution with 10 degrees of freedom and a shaded area of $\alpha = 0.01$ in the left tail. The area to the right of the critical value is

$$1 - \alpha = 1 - 0.01 = 0.99.$$

In Table 6 with d.f. $= 10$ and the area $1 - \alpha = 0.99$, the critical value is $\chi_0^2 = 2.558$.

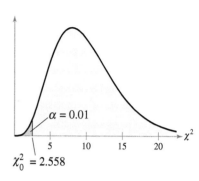

$\chi_0^2 = 2.558$

▶ **Try It Yourself 2**

Find the critical χ^2-value for a left-tailed test when $n = 30$ and $\alpha = 0.05$.

a. *Identify* the degrees of freedom and the level of significance.
b. *Use* Table 6 in Appendix B to find the critical χ^2-value. *Answer: Page A42*

EXAMPLE 3

▶ **Finding Critical Values for χ^2**

Find the critical χ^2-values for a two-tailed test when $n = 9$ and $\alpha = 0.05$.

▶ **Solution**

The degrees of freedom are

$$\text{d.f.} = n - 1 = 9 - 1 = 8.$$

The graph shows a χ^2-distribution with 8 degrees of freedom and a shaded area of $\frac{1}{2}\alpha = 0.025$ in each tail. The areas to the right of the critical values are

$$\tfrac{1}{2}\alpha = 0.025$$

and

$$1 - \tfrac{1}{2}\alpha = 0.975.$$

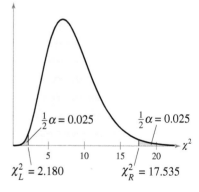

$\chi_L^2 = 2.180$ $\chi_R^2 = 17.535$

In Table 6 with d.f. $= 8$ and the areas 0.025 and 0.975, the critical values are $\chi_L^2 = 2.180$ and $\chi_R^2 = 17.535$.

▶ **Try It Yourself 3**

Find the critical χ^2-values for a two-tailed test when $n = 51$ and $\alpha = 0.01$.

a. *Identify* the degrees of freedom and the level of significance.
b. *Find* the first critical value χ_R^2 using Table 6 in Appendix B and the area $\frac{1}{2}\alpha$.
c. *Find* the second critical value χ_L^2 using Table 6 in Appendix B and the area $1 - \frac{1}{2}\alpha$. *Answer: Page A42*

▶ **THE CHI-SQUARE TEST**

To test a variance σ^2 or a standard deviation σ of a population that is normally distributed, you can use the χ^2-test. The χ^2-test for a variance or standard deviation is not as robust as the tests for the population mean μ or the population proportion p. So, it is essential in performing a χ^2-test for a variance or standard deviation that the population be normally distributed. The results can be misleading if the population is not normal.

χ^2-TEST FOR A VARIANCE σ^2 OR STANDARD DEVIATION σ

The **χ^2-test for a variance or standard deviation** is a statistical test for a population variance or standard deviation. The χ^2-test can be used when the population is normal. The **test statistic** is s^2 and the **standardized test statistic**

$$\chi^2 = \frac{(n-1)s^2}{\sigma^2}$$

follows a chi-square distribution with degrees of freedom

d.f. $= n - 1$.

GUIDELINES

Using the χ^2-Test for a Variance or Standard Deviation

IN WORDS	IN SYMBOLS
1. State the claim mathematically and verbally. Identify the null and alternative hypotheses.	State H_0 and H_a.
2. Specify the level of significance.	Identify α.
3. Determine the degrees of freedom.	d.f. $= n - 1$
4. Determine the critical value(s).	Use Table 6 in Appendix B.
5. Determine the rejection region(s).	
6. Find the standardized test statistic and sketch the sampling distribution.	$\chi^2 = \dfrac{(n-1)s^2}{\sigma^2}$
7. Make a decision to reject or fail to reject the null hypothesis.	If χ^2 is in the rejection region, reject H_0. Otherwise, fail to reject H_0.
8. Interpret the decision in the context of the original claim.	

EXAMPLE 4   SC Report 31

▸ **Using a Hypothesis Test for the Population Variance**

A dairy processing company claims that the variance of the amount of fat in the whole milk processed by the company is no more than 0.25. You suspect this is wrong and find that a random sample of 41 milk containers has a variance of 0.27. At $\alpha = 0.05$, is there enough evidence to reject the company's claim? Assume the population is normally distributed.

▸ **Solution**

The claim is "the variance is no more than 0.25." So, the null and alternative hypotheses are

$$H_0: \sigma^2 \leq 0.25 \text{ (Claim)} \quad \text{and} \quad H_a: \sigma^2 > 0.25.$$

The test is a right-tailed test, the level of significance is $\alpha = 0.05$, and the degrees of freedom are d.f. $= 41 - 1 = 40$. So, the critical value is

$$\chi_0^2 = 55.758.$$

The rejection region is $\chi^2 > 55.758$. The standardized test statistic is

$$\chi^2 = \frac{(n-1)s^2}{\sigma^2} \qquad \text{Use the chi-square test.}$$

$$= \frac{(41-1)(0.27)}{0.25} \qquad \text{Assume } \sigma^2 = 0.25.$$

$$= 43.2.$$

The graph shows the location of the rejection region and the standardized test statistic χ^2. Because χ^2 is not in the rejection region, you should fail to reject the null hypothesis.

Interpretation There is not enough evidence at the 5% level of significance to reject the company's claim that the variance of the amount of fat in the whole milk is no more than 0.25.

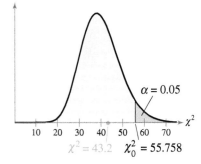

▸ **Try It Yourself 4**

A bottling company claims that the variance of the amount of sports drink in a 12-ounce bottle is no more than 0.40. A random sample of 31 bottles has a variance of 0.75. At $\alpha = 0.01$, is there enough evidence to reject the company's claim? Assume the population is normally distributed.

a. Identify the *claim* and state H_0 and H_a.
b. Identify the *level of significance* α and the *degrees of freedom*.
c. Find the *critical value* and identify the *rejection region*.
d. Find the *standardized test statistic* χ^2.
e. *Decide* whether to reject the null hypothesis. Use a graph if necessary.
f. *Interpret* the decision in the context of the original claim.

Answer: Page A42

EXAMPLE 5   SC   Report 32

▶ **Using a Hypothesis Test for the Standard Deviation**

A company claims that the standard deviation of the lengths of time it takes an incoming telephone call to be transferred to the correct office is less than 1.4 minutes. A random sample of 25 incoming telephone calls has a standard deviation of 1.1 minutes. At $\alpha = 0.10$, is there enough evidence to support the company's claim? Assume the population is normally distributed.

▶ **Solution**

The claim is "the standard deviation is less than 1.4 minutes." So, the null and alternative hypotheses are

$$H_0: \sigma \geq 1.4 \text{ minutes} \quad \text{and} \quad H_a: \sigma < 1.4 \text{ minutes}. \text{ (Claim)}$$

The test is a left-tailed test, the level of significance is $\alpha = 0.10$, and the degrees of freedom are

$$\text{d.f.} = 25 - 1$$
$$= 24.$$

So, the critical value is

$$\chi_0^2 = 15.659.$$

The rejection region is $\chi^2 < 15.659$. The standardized test statistic is

$$\chi^2 = \frac{(n-1)s^2}{\sigma^2} \qquad \text{Use the chi-square test.}$$

$$= \frac{(25-1)(1.1)^2}{1.4^2} \qquad \text{Assume } \sigma = 1.4.$$

$$\approx 14.816.$$

The graph shows the location of the rejection region and the standardized test statistic χ^2. Because χ^2 is in the rejection region, you should reject the null hypothesis.

Interpretation There is enough evidence at the 10% level of significance to support the claim that the standard deviation of the lengths of time it takes an incoming telephone call to be transferred to the correct office is less than 1.4 minutes.

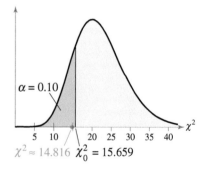

▶ **Try It Yourself 5**

A police chief claims that the standard deviation of the lengths of response times is less than 3.7 minutes. A random sample of 9 response times has a standard deviation of 3.0 minutes. At $\alpha = 0.05$, is there enough evidence to support the police chief's claim? Assume the population is normally distributed.

a. Identify the *claim* and state H_0 and H_a.
b. Identify the *level of significance* α and the *degrees of freedom*.
c. Find the *critical value* and identify the *rejection region*.
d. Find the *standardized test statistic* χ^2.
e. *Decide* whether to reject the null hypothesis. Use a graph if necessary.
f. *Interpret* the decision in the context of the original claim.

Answer: Page A42

EXAMPLE 6

▶ **Using a Hypothesis Test for the Population Variance**

A sporting goods manufacturer claims that the variance of the strengths of a certain fishing line is 15.9. A random sample of 15 fishing line spools has a variance of 21.8. At $\alpha = 0.05$, is there enough evidence to reject the manufacturer's claim? Assume the population is normally distributed.

▶ **Solution**

The claim is "the variance is 15.9." So, the null and alternative hypotheses are

$$H_0: \sigma^2 = 15.9 \text{ (Claim)}$$

and

$$H_a: \sigma^2 \neq 15.9.$$

The test is a two-tailed test, the level of significance is $\alpha = 0.05$, and the degrees of freedom are

$$\text{d.f.} = 15 - 1$$
$$= 14.$$

So, the critical values are $\chi_L^2 = 5.629$ and $\chi_R^2 = 26.119$.

The rejection regions are $\chi^2 < 5.629$ and $\chi^2 > 26.119$. The standardized test statistic is

$$\chi^2 = \frac{(n-1)s^2}{\sigma^2} \qquad \text{Use the chi-square test.}$$

$$= \frac{(15-1)(21.8)}{15.9} \qquad \text{Assume } \sigma^2 = 15.9.$$

$$\approx 19.195.$$

The graph shows the location of the rejection regions and the standardized test statistic χ^2. Because χ^2 is not in the rejection regions, you should fail to reject the null hypothesis.

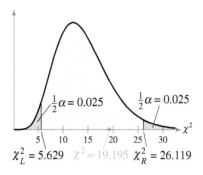

$\chi_L^2 = 5.629$ $\chi^2 \approx 19.195$ $\chi_R^2 = 26.119$

Interpretation There is not enough evidence at the 5% level of significance to reject the claim that the variance of the strengths of the fishing line is 15.9.

▶ **Try It Yourself 6**

A company that offers dieting products and weight loss services claims that the variance of the weight losses of their users is 25.5. A random sample of 13 users has a variance of 10.8. At $\alpha = 0.10$, is there enough evidence to reject the company's claim? Assume the population is normally distributed.

a. Identify the *claim* and state H_0 and H_a.
b. Identify the *level of significance* α and the *degrees of freedom*.
c. Find the *critical values* and identify the *rejection regions*.
d. Find the *standardized test statistic* χ^2.
e. *Decide* whether to reject the null hypothesis. Use a graph if necessary.
f. *Interpret* the decision in the context of the original claim.

Answer: Page A42

7.5 EXERCISES

FOR EXTRA HELP:
MyStatLab

■ BUILDING BASIC SKILLS AND VOCABULARY

1. Explain how to find critical values in a χ^2- sampling distribution.

2. Can a critical value for the χ^2-test be negative? Explain.

3. When testing a claim about a population mean or a population standard deviation, a requirement is that the sample is from a population that is normally distributed. How is this requirement different between the two tests?

4. Explain how to test a population variance or a population standard deviation.

In Exercises 5–10, find the critical value(s) for the indicated test for a population variance, sample size n, and level of significance α.

5. Right-tailed test,
   $n = 27, \alpha = 0.05$

6. Right-tailed test,
   $n = 10, \alpha = 0.10$

7. Left-tailed test,
   $n = 7, \alpha = 0.01$

8. Left-tailed test,
   $n = 24, \alpha = 0.05$

9. Two-tailed test,
   $n = 81, \alpha = 0.10$

10. Two-tailed test,
    $n = 61, \alpha = 0.01$

Graphical Analysis   *In Exercises 11–14, state whether the standardized test statistic χ^2 allows you to reject the null hypothesis.*

11. (a) $\chi^2 = 2.091$
    (b) $\chi^2 = 0$
    (c) $\chi^2 = 1.086$
    (d) $\chi^2 = 6.3471$

12. (a) $\chi^2 = 0.771$
    (b) $\chi^2 = 9.486$
    (c) $\chi^2 = 0.701$
    (d) $\chi^2 = 9.508$

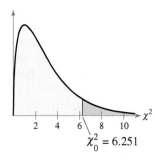

$\chi_0^2 = 6.251$

$\chi_L^2 = 0.711$ $\chi_R^2 = 9.488$

13. (a) $\chi^2 = 22.302$
    (b) $\chi^2 = 23.309$
    (c) $\chi^2 = 8.457$
    (d) $\chi^2 = 8.577$

14. (a) $\chi^2 = 10.065$
    (b) $\chi^2 = 10.075$
    (c) $\chi^2 = 10.585$
    (d) $\chi^2 = 10.745$

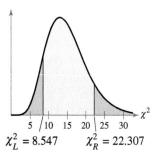

$\chi_L^2 = 8.547$ $\chi_R^2 = 22.307$

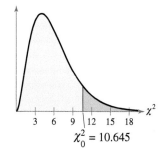

$\chi_0^2 = 10.645$

In Exercises 15–18, use a χ^2-test to test the claim about the population variance σ^2 or standard deviation σ at the given level of significance α using the given sample statistics. For each claim, assume the population is normally distributed.

15. Claim: $\sigma^2 = 0.52$; $\alpha = 0.05$. Sample statistics: $s^2 = 0.508$, $n = 18$

16. Claim: $\sigma^2 \geq 8.5$; $\alpha = 0.05$. Sample statistics: $s^2 = 7.45$, $n = 23$

17. Claim: $\sigma = 24.9$; $\alpha = 0.10$. Sample statistics: $s = 29.1$, $n = 51$

18. Claim: $\sigma < 40$; $\alpha = 0.01$. Sample statistics: $s = 40.8$, $n = 12$

■ USING AND INTERPRETING CONCEPTS

Testing Claims   *In Exercises 19–28, (a) write the claim mathematically and identify H_0 and H_a, (b) find the critical value(s) and identify the rejection region(s), (c) find the standardized test statistic χ^2, (d) decide whether to reject or fail to reject the null hypothesis, and (e) interpret the decision in the context of the original claim. For each claim, assume the population is normally distributed.*

19. Carbohydrates   A snack food manufacturer estimates that the variance of the number of grams of carbohydrates in servings of its tortilla chips is 1.25. A dietician is asked to test this claim and finds that a random sample of 22 servings has a variance of 1.35. At $\alpha = 0.05$, is there enough evidence to reject the manufacturer's claim?

20. Hybrid Vehicle Gas Mileage   An auto manufacturer believes that the variance of the gas mileages of its hybrid vehicles is 1.0. You work for an energy conservation agency and want to test this claim. You find that a random sample of the gas mileages of 25 of the manufacturer's hybrid vehicles has a variance of 1.65. At $\alpha = 0.05$, do you have enough evidence to reject the manufacturer's claim? *(Adapted from Green Hybrid)*

21. Science Assessment Tests   On a science assessment test, the scores of a random sample of 22 eighth grade students have a standard deviation of 33.4 points. This result prompts a test administrator to claim that the standard deviation for eighth graders on the examination is less than 36 points. At $\alpha = 0.10$, is there enough evidence to support the administrator's claim? *(Adapted from National Center for Educational Statistics)*

22. U.S. History Assessment Tests   A state school administrator says that the standard deviation of test scores for eighth grade students who took a U.S. history assessment test is less than 30 points. You work for the administrator and are asked to test this claim. You randomly select 18 tests and find that the tests have a standard deviation of 33.6 points. At $\alpha = 0.01$, is there enough evidence to support the administrator's claim? *(Adapted from National Center for Educational Statistics)*

23. Tornadoes   A weather service claims that the standard deviation of the number of fatalities per year from tornadoes is no more than 25. A random sample of the number of deaths for 28 years has a standard deviation of 31 fatalities. At $\alpha = 0.10$, is there enough evidence to reject the weather service's claim? *(Source: NOAA Weather Partners)*

24. Lengths of Stay   A doctor says the standard deviation of the lengths of stay for patients involved in a crash in which the vehicle struck a tree is 6.14 days. A random sample of 20 lengths of stay for patients involved in this type of crash has a standard deviation of 6.5 days. At $\alpha = 0.05$, can you reject the doctor's claim? *(Adapted from National Highway Traffic Safety Administration)*

25. Total Charges An insurance agent says the standard deviation of the total hospital charges for patients involved in a crash in which the vehicle struck a construction barricade is less than $3500. A random sample of 28 total hospital charges for patients involved in this type of crash has a standard deviation of $4100. At $\alpha = 0.10$, can you support the agent's claim? *(Adapted from National Highway Traffic Safety Administration)*

26. Hotel Room Rates A travel agency estimates that the standard deviation of the room rates of hotels in a certain city is no more than $30. You work for a consumer advocacy group and are asked to test this claim. You find that a random sample of 21 hotels has a standard deviation of $35.25. At $\alpha = 0.01$, do you have enough evidence to reject the agency's claim?

27. Salaries The annual salaries (in dollars) of 18 randomly chosen environmental engineers are listed. At $\alpha = 0.05$, can you conclude that the standard deviation of the annual salaries is greater than $6100? *(Adapted from Salary.com)*

63,125	59,749	52,369	55,979	61,550	54,644	50,420
47,291	51,357	56,901	53,499	49,998	69,712	64,575
45,850	46,297	63,770	71,589			

28. Salaries A staffing organization states that the standard deviation of the annual salaries of commodity buyers is at least $10,600. The annual salaries (in dollars) of 20 randomly chosen commodity buyers are listed. At $\alpha = 0.10$, can you reject the organization's claim? *(Adapted from Salary.com)*

79,319	68,825	65,129	75,899	85,070	76,270	68,750
70,982	69,237	63,470	79,025	55,880	80,985	75,264
66,918	65,459	70,598	86,579	71,225	57,311	

SC *In Exercises 29–32, use StatCrunch to help you test the claim about the population variance σ^2 or standard deviation σ at the given level of significance α using the given sample statistics. For each claim, assume the population is normally distributed.*

29. Claim: $\sigma^2 \geq 9$; $\alpha = 0.01$. Sample statistics: $s^2 = 2.03$, $n = 10$

30. Claim: $\sigma^2 = 14.85$; $\alpha = 0.05$. Sample statistics: $s^2 = 28.75$, $n = 17$

31. Claim: $\sigma > 4.5$; $\alpha = 0.05$. Sample statistics: $s = 5.8$, $n = 15$

32. Claim: $\sigma \neq 418$; $\alpha = 0.10$. Sample statistics: $s = 305$, $n = 24$

■ EXTENDING CONCEPTS

P-Values *You can calculate the P-value for a χ^2-test using technology. After calculating the χ^2-test value, you can use the cumulative density function (CDF) to calculate the area under the curve. From Example 4 on page 407, $\chi^2 = 43.2$. Using a TI-83/84 Plus (choose 7 from the DISTR menu), enter 0 for the lower bound, 43.2 for the upper bound, and 40 for the degrees of freedom, as shown at the left.*

The P-value is approximately $1 - 0.6638 = 0.3362$. Because $P > \alpha = 0.05$, the conclusion is to fail to reject H_0.

In Exercises 33–36, use the P-value method to perform the hypothesis test for the indicated exercise.

33. Exercise 25   **34.** Exercise 26

35. Exercise 27   **36.** Exercise 28

TI-83/84 PLUS

χ^2 cdf (0, 43.2, 40)
                .6637768667

USES AND ABUSES

Uses

Hypothesis Testing Hypothesis testing is important in many different fields because it gives a scientific procedure for assessing the validity of a claim about a population. Some of the concepts in hypothesis testing are intuitive, but some are not. For instance, the *American Journal of Clinical Nutrition* suggests that eating dark chocolate can help prevent heart disease. A random sample of healthy volunteers were assigned to eat 3.5 ounces of dark chocolate each day for 15 days. After 15 days, the mean systolic blood pressure of the volunteers was 6.4 millimeters of mercury lower. A hypothesis test could show if this drop in systolic blood pressure is significant or simply due to sampling error.

Careful inferences must be made concerning the results. In another part of the study, it was found that white chocolate did not result in similar benefits. So, the inference of health benefits cannot be extended to all types of chocolate. You also would not infer that you should eat large quantities of chocolate because the benefits must be weighed against known risks, such as weight gain, acne, and acid reflux.

Abuses

Not Using a Random Sample The entire theory of hypothesis testing is based on the fact that the sample is randomly selected. If the sample is not random, then you cannot use it to infer anything about a population parameter.

Attempting to Prove the Null Hypothesis If the P-value for a hypothesis test is greater than the level of significance, you have not proven the null hypothesis is true—only that there is not enough evidence to reject it. For instance, with a P-value higher than the level of significance, a researcher could not prove that there is no benefit to eating dark chocolate—only that there is not enough evidence to support the claim that there is a benefit.

Making Type I or Type II Errors Remember that a type I error is rejecting a null hypothesis that is true and a type II error is failing to reject a null hypothesis that is false. You can decrease the probability of a type I error by lowering the level of significance. Generally, if you decrease the probability of making a type I error, you increase the probability of making a type II error. You can decrease the chance of making both types of errors by increasing the sample size.

EXERCISES

In Exercises 1–4, assume that you work in a transportation department. You are asked to write a report about the claim that 73% of U.S. adults who fly at least once a year favor full-body scanners at airports. (Adapted from Rasmussen Reports)

Do You Favor the Use of Full-Body Scanners at Airports in the U.S.?

No 16%
Don't know **11%**
Yes **73%**

1. ***Not Using a Random Sample*** How could you choose a random sample to test this hypothesis?

2. ***Attempting to Prove the Null Hypothesis*** What is the null hypothesis in this situation? Describe how your report could be incorrect by trying to prove the null hypothesis.

3. ***Making a Type I Error*** Describe how your report could make a type I error.

4. ***Making a Type II Error*** Describe how your report could make a type II error.

7 ▸ A SUMMARY OF HYPOTHESIS TESTING

With hypothesis testing, perhaps more than any other area of statistics, it can be difficult to see the forest for all the trees. To help you see the forest—the overall picture—a summary of what you studied in this chapter is provided.

Writing the Hypotheses

- You are given a claim about a population parameter μ, p, σ^2, or σ.
- Rewrite the claim and its complement using $\underbrace{\leq, \geq, =}_{H_0}$ and $\underbrace{>, <, \neq}_{H_a}$.
- Identify the claim. Is it H_0 or H_a?

Specifying a Level of Significance

- Specify α, the maximum acceptable probability of rejecting a valid H_0 (a type I error).

Specifying the Sample Size

- Specify your sample size n.

INSIGHT

Large sample sizes will usually increase the cost and effort of testing a hypothesis, but they also tend to make your decision more reliable.

Choosing the Test ▪ Any population ▪ Normally distributed population

- **Mean:** H_0 describes a hypothesized population mean μ.
  - Use a **z-test** for *any* population if $n \geq 30$.
  - Use a **z-test** if the population is normal and σ is known for any n.
  - Use a **t-test** if the population is normal and $n < 30$, but σ is unknown.
- **Proportion:** H_0 describes a hypothesized population proportion p.
  - Use a **z-test** for any population if $np \geq 5$ and $nq \geq 5$.
- **Variance or Standard Deviation:** H_0 describes a hypothesized population variance σ^2 or standard deviation σ.
  - Use a χ^2-**test** if the population is normal.

Sketching the Sampling Distribution

- Use H_a to decide if the test is left-tailed, right-tailed, or two-tailed.

Finding the Standardized Test Statistic

- Take a random sample of size n from the population.
- Compute the test statistic $\bar{x}$, $\hat{p}$, or s^2.
- Find the standardized test statistic z, t, or χ^2.

Making a Decision

Option 1. Decision based on rejection region

- Use α to find the critical value(s) z_0, t_0, or χ_0^2 and rejection region(s).
- **Decision Rule:**

  Reject H_0 if the standardized test statistic is in the rejection region.
  Fail to reject H_0 if the standardized test statistic is not in the rejection region.

Option 2. Decision based on P-value

- Use the standardized test statistic or a technology tool to find the P-value.
- **Decision Rule:**

  Reject H_0 if $P \leq \alpha$.
  Fail to reject H_0 if $P > \alpha$.

z-Test for a Hypothesized Mean μ *(Section 7.2)*

Test statistic: $\overline{x}$

Critical value: z_0 (Use Table 4.)

If $n \geq 30$, s can be used in place of σ. Sampling distribution of sample means is a normal distribution.

Standardized test statistic: z

$$z = \frac{\overline{x} - \mu}{\sigma/\sqrt{n}}$$

Sample mean — , Hypothesized mean, Population standard deviation, Sample size

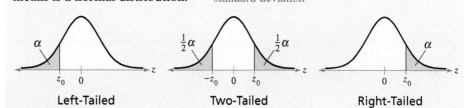

Left-Tailed   Two-Tailed   Right-Tailed

z-Test for a Hypothesized Proportion *p*   *(Section 7.4)*

Test statistic: $\hat{p}$

Critical value: z_0 (Use Table 4.)

Sampling distribution of sample proportions is a normal distribution.

Standardized test statistic: z

$$z = \frac{\hat{p} - p}{\sqrt{pq/n}}$$

$q = 1 - p$

Sample proportion, Hypothesized proportion, Sample size

t-Test for a Hypothesized Mean μ *(Section 7.3)*

Test statistic: $\overline{x}$

Critical value: t_0 (Use Table 5.)

Sampling distribution of sample means is approximated by a *t*-distribution with d.f. $= n - 1$.

Standardized test statistic: t

$$t = \frac{\overline{x} - \mu}{s/\sqrt{n}}$$

Sample mean — , Hypothesized mean, Sample standard deviation, Sample size

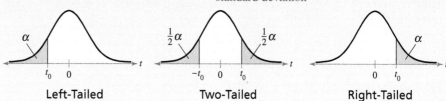

Left-Tailed   Two-Tailed   Right-Tailed

χ^2-Test for a Hypothesized Variance σ^2 or Standard Deviation σ *(Section 7.5)*

Test statistic: s^2

Critical value: χ_0^2 (Use Table 6.)

Sampling distribution is approximated by a chi-square distribution with d.f. $= n - 1$.

Standardized test statistic: χ^2

$$\chi^2 = \frac{(n - 1)s^2}{\sigma^2}$$

Sample size — , Sample variance, Hypothesized variance

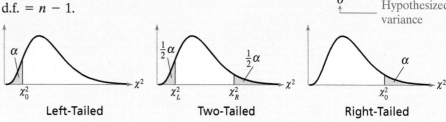

Left-Tailed   Two-Tailed   Right-Tailed

7 CHAPTER SUMMARY

What did you **learn?**	EXAMPLE(S)	REVIEW EXERCISES
Section 7.1		
■ How to state a null hypothesis and an alternative hypothesis	*1*	*1–6*
■ How to identify type I and type II errors	*2*	*7–10*
■ How to know whether to use a one-tailed or a two-tailed statistical test	*3*	*7–10*
■ How to interpret a decision based on the results of a statistical test	*4*	*7–10*
Section 7.2		
■ How to find *P*-values and use them to test a mean μ	*1–3*	*11, 12*
■ How to use *P*-values for a *z*-test	*4–6*	*13, 14, 23–28*
■ How to find critical values and rejection regions in a normal distribution	*7, 8*	*15–18*
■ How to use rejection regions for a *z*-test	*9, 10*	*19–28*
Section 7.3		
■ How to find critical values in a *t*-distribution	*1–3*	*29–32*
■ How to use the *t*-test to test a mean μ	*4, 5*	*33–40*
■ How to use technology to find *P*-values and use them with a *t*-test to test a mean μ	*6*	*41, 42*
Section 7.4		
■ How to use the *z*-test to test a population proportion *p*	*1, 2*	*43–52*
Section 7.5		
■ How to find critical values for a χ^2-test	*1–3*	*53–56*
■ How to use the χ^2-test to test a variance or a standard deviation	*4–6*	*57–63*

■ SECTION 7.1

In Exercises 1–6, use the given statement to represent a claim. Write its complement and state which is H_0 and which is H_a.

1. $\mu \le 375$ **2.** $\mu = 82$

3. $p < 0.205$ **4.** $\mu \ne 150{,}020$

5. $\sigma > 1.9$ **6.** $p \ge 0.64$

In Exercises 7–10, do the following.

(a) State the null and alternative hypotheses, and identify which represents the claim.

(b) Determine when a type I or type II error occurs for a hypothesis test of the claim.

(c) Determine whether the hypothesis test is left-tailed, right-tailed, or two-tailed. Explain your reasoning.

(d) Explain how you should interpret a decision that rejects the null hypothesis.

(e) Explain how you should interpret a decision that fails to reject the null hypothesis.

7. A news outlet reports that the proportion of Americans who support plans to order deep cuts in executive compensation at companies that have received federal bailout funds is 71%. *(Source: ABC News)*

8. An agricultural cooperative guarantees that the mean shelf life of a certain type of dried fruit is at least 400 days.

9. A soup maker says that the standard deviation of the sodium content in one serving of a certain soup is no more than 50 milligrams. *(Adapted from Consumer Reports)*

10. An energy bar maker claims that the mean number of grams of carbohydrates in one bar is less than 25.

■ SECTION 7.2

In Exercises 11 and 12, find the P-value for the indicated hypothesis test with the given standardized test statistic z. Decide whether to reject H_0 for the given level of significance α.

11. Left-tailed test, $z = -0.94$, $\alpha = 0.05$

12. Two-tailed test, $z = 2.57$, $\alpha = 0.10$

In Exercises 13 and 14, use a P-value to test the claim about the population mean μ using the given sample statistics. State your decision for $\alpha = 0.10$, $\alpha = 0.05$, and $\alpha = 0.01$ levels of significance. If convenient, use technology.

13. Claim: $\mu \le 0.05$; Sample statistics: $\bar{x} = 0.057$, $s = 0.018$, $n = 32$

14. Claim: $\mu \ne 230$; Sample statistics: $\bar{x} = 216.5$, $s = 17.3$, $n = 48$

In Exercises 15–18, find the critical value(s) for the indicated z-test and level of significance α. Include a graph with your answer.

15. Left-tailed test, $\alpha = 0.02$ **16.** Two-tailed test, $\alpha = 0.005$

17. Right-tailed test, $\alpha = 0.025$ **18.** Two-tailed test, $\alpha = 0.08$

In Exercises 19–22, state whether each standardized test statistic z allows you to reject the null hypothesis. Explain your reasoning.

19. $z = 1.631$

20. $z = 1.723$

21. $z = -1.464$

22. $z = -1.655$

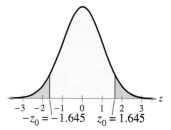

$-z_0 = -1.645 \quad z_0 = 1.645$

In Exercises 23–26, use a z-test to test the claim about the population mean μ at the given level of significance α using the given sample statistics. If convenient, use technology.

23. Claim: $\mu \le 45$; $\alpha = 0.05$. Sample statistics: $\bar{x} = 47.2$, $s = 6.7$, $n = 42$

24. Claim: $\mu \ne 8.45$; $\alpha = 0.03$. Sample statistics: $\bar{x} = 7.88$, $s = 1.75$, $n = 60$

25. Claim: $\mu < 5.500$; $\alpha = 0.01$. Sample statistics: $\bar{x} = 5.497$, $s = 0.011$, $n = 36$

26. Claim: $\mu = 7450$; $\alpha = 0.10$. Sample statistics: $\bar{x} = 7495$, $s = 243$, $n = 57$

In Exercises 27 and 28, test the claim about the population mean μ using rejection region(s) or a P-value. Interpret your decision in the context of the original claim. If convenient, use technology.

27. The U.S. Department of Agriculture claims that the mean cost of raising a child from birth to age 2 by husband-wife families in rural areas is $10,380. A random sample of 800 children (age 2) has a mean cost of $10,240 with a standard deviation of $1561. At $\alpha = 0.01$, is there enough evidence to reject the claim? *(Adapted from U.S. Department of Agriculture Center for Nutrition Policy and Promotion)*

28. A tourist agency in Hawaii claims the mean daily cost of meals and lodging for a family of 4 traveling in Hawaii is at most $650. You work for a consumer protection advocate and want to test this claim. In a random sample of 45 families of 4 traveling in Hawaii, the mean daily cost of meals and lodging is $657 with a standard deviation of $40. At $\alpha = 0.05$, do you have enough evidence to reject the tourist agency's claim? *(Adapted from American Automobile Association)*

■ SECTION 7.3

In Exercises 29–32, find the critical value(s) for the indicated t-test, level of significance α, and sample size n.

29. Two-tailed test, $\alpha = 0.05$, $n = 20$ **30.** Right-tailed test, $\alpha = 0.01$, $n = 8$

31. Left-tailed test, $\alpha = 0.005$, $n = 15$ **32.** Two-tailed test, $\alpha = 0.02$, $n = 12$

In Exercises 33–38, use a t-test to test the claim about the population mean μ at the given level of significance α using the given sample statistics. For each claim, assume the population is normally distributed. If convenient, use technology.

33. Claim: $\mu \ne 95$; $\alpha = 0.05$. Sample statistics: $\bar{x} = 94.1$, $s = 1.53$, $n = 12$

34. Claim: $\mu > 12,700$; $\alpha = 0.005$. Sample statistics: $\bar{x} = 12,855$, $s = 248$, $n = 21$

35. Claim: $\mu \ge 0$; $\alpha = 0.10$. Sample statistics: $\bar{x} = -0.45$, $s = 1.38$, $n = 16$

36. Claim: $\mu = 4.20$; $\alpha = 0.02$. Sample statistics: $\bar{x} = 4.61$, $s = 0.33$, $n = 9$

37. Claim: $\mu \le 48$; $\alpha = 0.01$. Sample statistics: $\bar{x} = 52$, $s = 2.5$, $n = 7$

38. Claim: $\mu < 850$; $\alpha = 0.025$. Sample statistics: $\bar{x} = 875$, $s = 25$, $n = 14$

In Exercises 39 and 40, use a t-test to test the claim. Interpret your decision in the context of the original claim. For each claim, assume the population is normally distributed. If convenient, use technology.

39. A fitness magazine advertises that the mean monthly cost of joining a health club is $25. You work for a consumer advocacy group and are asked to test this claim. You find that a random sample of 18 clubs has a mean monthly cost of $26.25 and a standard deviation of $3.23. At $\alpha = 0.10$, do you have enough evidence to reject the advertisement's claim?

40. A fitness magazine claims that the mean cost of a yoga session is no more than $14. You work for a consumer advocacy group and are asked to test this claim. You find that a random sample of 29 yoga sessions has a mean cost of $15.59 and a standard deviation of $2.60. At $\alpha = 0.025$, do you have enough evidence to reject the magazine's claim?

In Exercises 41 and 42, use a t-statistic and its P-value to test the claim about the population mean μ using the given data. Interpret your decision in the context of the original claim. For each claim, assume the population is normally distributed. If convenient, use technology.

41. An education publication claims that the mean expenditure per student in public elementary and secondary schools is at least $10,200. You want to test this claim. You randomly select 16 school districts and find the average expenditure per student. The results are listed below. At $\alpha = 0.01$, can you reject the publication's claim? *(Adapted from National Center for Education Statistics)*

9,242	10,857	10,377	8,935	9,545	9,974
9,847	10,641	9,364	10,157	9,784	9,962
10,065	9,851	9,763	9,969		

42. A restaurant association says the typical household in the United States spends a mean amount of $2698 per year on food away from home. You are a consumer reporter for a national publication and want to test this claim. A random sample of 28 U.S. households has a mean amount spent on food away from home of $2764 and a standard deviation of $322. At $\alpha = 0.05$, do you have enough evidence to reject the association's claim? *(Adapted from U.S. Bureau of Labor Statistics)*

SECTION 7.4

In Exercises 43–50, decide whether the normal sampling distribution can be used to approximate the binomial distribution. If it can, use the z-test to test the claim about the population proportion p at the given level of significance α using the given sample statistics. If convenient, use technology.

43. Claim: $p = 0.15$; $\alpha = 0.05$. Sample statistics: $\hat{p} = 0.09$, $n = 40$

44. Claim: $p < 0.70$; $\alpha = 0.01$. Sample statistics: $\hat{p} = 0.50$, $n = 68$

45. Claim: $p < 0.09$; $\alpha = 0.08$. Sample statistics: $\hat{p} = 0.07$, $n = 75$

46. Claim: $p = 0.65$; $\alpha = 0.03$. Sample statistics: $\hat{p} = 0.76$, $n = 116$

47. Claim: $p \geq 0.04$; $\alpha = 0.10$. Sample statistics: $\hat{p} = 0.03$, $n = 30$

48. Claim: $p \neq 0.34$; $\alpha = 0.01$. Sample statistics: $\hat{p} = 0.29$, $n = 60$

49. Claim: $p \neq 0.24$; $\alpha = 0.02$. Sample statistics: $\hat{p} = 0.32$, $n = 50$

50. Claim: $p \leq 0.80$; $\alpha = 0.10$. Sample statistics: $\hat{p} = 0.85$, $n = 43$

In Exercises 51 and 52, test the claim about the population proportion p. Interpret your decision in the context of the original claim. If convenient, use technology.

51. A polling agency reports that over 16% of U.S. adults are without health care coverage. In a random survey of 1420 U.S. adults, 256 said they did not have health care coverage. At $\alpha = 0.02$, is there enough evidence to support the agency's claim? *(Source: The Gallup Poll)*

52. The Western blot assay is a blood test for the presence of HIV. It has been found that this test sometimes gives false positive results for HIV. A medical researcher claims that the rate of false positives is 2%. A recent study of 300 randomly selected U.S. blood donors who do not have HIV found that 3 received a false positive test result. At $\alpha = 0.05$, is there enough evidence to reject the researcher's claim? *(Adapted from Centers for Disease Control and Prevention)*

■ SECTION 7.5

In Exercises 53–56, find the critical value(s) for the indicated χ^2-test for a population variance, sample size n, and level of significance α.

53. Right-tailed test, $n = 20$, $\alpha = 0.05$

54. Two-tailed test, $n = 14$, $\alpha = 0.01$

55. Right-tailed test, $n = 51$, $\alpha = 0.10$

56. Left-tailed test, $n = 6$, $\alpha = 0.05$

In Exercises 57–60, use a χ^2-test to test the claim about the population variance σ^2 or standard deviation σ at the given level of significance α and using the given sample statistics. For each claim, assume the population is normally distributed.

57. Claim: $\sigma^2 > 2$; $\alpha = 0.10$. Sample statistics: $s^2 = 2.95$, $n = 18$

58. Claim: $\sigma^2 \leq 60$; $\alpha = 0.025$. Sample statistics: $s^2 = 72.7$, $n = 15$

59. Claim: $\sigma = 1.25$; $\alpha = 0.05$. Sample statistics: $s = 1.03$, $n = 6$

60. Claim: $\sigma \neq 0.035$; $\alpha = 0.01$. Sample statistics: $s = 0.026$, $n = 16$

In Exercises 61 and 62, test the claim about the population variance or standard deviation. Interpret your decision in the context of the original claim. For each claim, assume the population is normally distributed.

61. A bolt manufacturer makes a type of bolt to be used in airtight containers. The manufacturer needs to be sure that all of its bolts are very similar in width, so it sets an upper tolerance limit for the variance of bolt width at 0.01. A random sample of the widths of 28 bolts has a variance of 0.064. At $\alpha = 0.005$, is there enough evidence to reject the manufacturer's claim?

62. A restaurant claims that the standard deviation of the lengths of serving times is 3 minutes. A random sample of 27 serving times has a standard deviation of 3.9 minutes. At $\alpha = 0.01$, is there enough evidence to reject the restaurant's claim?

63. In Exercise 62, is there enough evidence to reject the restaurant's claim at the $\alpha = 0.05$ level? Explain.

7 CHAPTER QUIZ

Take this quiz as you would take a quiz in class. After you are done, check your work against the answers given in the back of the book. If convenient, use technology.

For this quiz, do the following.

(a) *Write the claim mathematically. Identify H_0 and H_a.*

(b) *Determine whether the hypothesis test is one-tailed or two-tailed and whether to use a z-test, a t-test, or a χ^2-test. Explain your reasoning.*

(c) *If necessary, find the critical value(s) and identify the rejection region(s).*

(d) *Find the appropriate test statistic. If necessary, find the P-value.*

(e) *Decide whether to reject or fail to reject the null hypothesis.*

(f) *Interpret the decision in the context of the original claim.*

1. A research service estimates that the mean annual consumption of vegetables and melons by people in the United States is at least 170 pounds per person. A random sample of 360 people in the United States has a mean consumption of vegetables and melons of 168.5 pounds per year and a standard deviation of 11 pounds. At $\alpha = 0.03$, is there enough evidence to reject the service's claim that the mean consumption of vegetables and melons by people in the United States is at least 170 pounds per person? *(Adapted from U.S. Department of Agriculture)*

2. A hat company states that the mean hat size for a male is at least 7.25. A random sample of 12 hat sizes has a mean of 7.15 and a standard deviation of 0.27. At $\alpha = 0.05$, can you reject the company's claim that the mean hat size for a male is at least 7.25? Assume the population is normally distributed.

3. A maker of microwave ovens advertises that no more than 10% of its microwaves need repair during the first 5 years of use. In a random sample of 57 microwaves that are 5 years old, 13% needed repairs. At $\alpha = 0.04$, can you reject the maker's claim that no more than 10% of its microwaves need repair during the first five years of use? *(Adapted from Consumer Reports)*

4. A state school administrator says that the standard deviation of SAT critical reading test scores is 112. A random sample of 19 SAT critical reading test scores has a standard deviation of 143. At $\alpha = 0.10$, test the administrator's claim. What can you conclude? Assume the population is normally distributed. *(Adapted from The College Board)*

5. A government agency reports that the mean amount of earnings for full-time workers ages 25 to 34 with a master's degree is \$62,569. In a random sample of 15 full-time workers ages 25 to 34 with a master's degree, the mean amount of earnings is \$59,231 and the standard deviation is \$5945. Is there enough evidence to reject the agency's claim? Use a *P*-value and $\alpha = 0.05$. Assume the population is normally distributed. *(Adapted from U.S. Census Bureau)*

6. A tourist agency in Kansas claims the mean daily cost of meals and lodging for a family of 4 traveling in the state is \$201. You work for a consumer protection advocate and want to test this claim. In a random sample of 35 families of 4 traveling in Kansas, the mean daily cost of meals and lodging is \$216 and the standard deviation is \$30. Do you have enough evidence to reject the agency's claim? Use a *P*-value and $\alpha = 0.05$. *(Adapted from American Automobile Association)*

PUTTING IT ALL TOGETHER

Real Statistics — Real Decisions

In the 1970s and 1980s, PepsiCo, maker of Pepsi®, began airing television commercials in which it claimed more cola drinkers preferred Pepsi® over Coca-Cola® in a blind taste test. The Coca-Cola Company, maker of Coca-Cola®, was the market leader in soda sales. After the television ads began airing, Pepsi® sales increased and began rivaling Coca-Cola® sales.

   Assume the claim is that more than 50% of cola drinkers preferred Pepsi® over Coca-Cola®. You work for an independent market research firm and are asked to test this claim.

■ EXERCISES

1. *How Would You Do It?*

  (a) When PepsiCo performed this challenge, PepsiCo representatives went to shopping malls to obtain their sample. Do you think this type of sampling is representative of the population? Explain.

  (b) What sampling technique would you use to select the sample for your study?

  (c) Identify possible flaws or biases in your study.

2. *Testing a Proportion*

  In your study, 280 out of 560 cola drinkers prefer Pepsi® over Coca-Cola®. Using these results, test the claim that more than 50% of cola drinkers prefer Pepsi® over Coca-Cola®. Use $\alpha = 0.05$. Interpret your decision in the context of the original claim. Does the decision support PepsiCo's claim?

3. *Labeling Influence*

  The Baylor College of Medicine decided to replicate this taste test by monitoring brain activity while conducting the test on participants. They also wanted to see if brand labeling would affect the results. When participants were shown which cola they were sampling, Coca-Cola® was preferred by 75% of the participants. What conclusions can you draw from this study?

4. *Your Conclusions*

  (a) Why do you think PepsiCo used a blind taste test?

  (b) Do you think brand image or taste has more influence on consumer preferences for cola?

  (c) What other factors may influence consumer preferences besides taste and branding?

TECHNOLOGY   MINITAB   EXCEL   TI-83/84 PLUS

THE CASE OF THE VANISHING WOMEN

53% ➡ **29%** ➡ **9%** ➡ **0%**

From 1966 to 1968, Dr. Benjamin Spock and others were tried for conspiracy to violate the Selective Service Act by encouraging resistance to the Vietnam War. By a series of three selections, no women ended up being on the jury. In 1969, Hans Zeisel wrote an article in *The University of Chicago Law Review* using statistics and hypothesis testing to argue that the jury selection was biased against Dr. Spock. Dr. Spock was a well-known pediatrician and author of books about raising children. Millions of mothers had read his books and followed his advice. Zeisel argued that, by keeping women off the jury, the court prejudiced the verdict.

The jury selection process for Dr. Spock's trial is shown at the right.

Stage 1. The clerk of the Federal District Court selected 350 people "at random" from the Boston City Directory. The directory contained several hundred names, 53% of whom were women. However, only 102 of the 350 people selected were women.

Stage 2. The trial judge, Judge Ford, selected 100 people "at random" from the 350 people. This group was called a venire and it contained only nine women.

Stage 3. The court clerk assigned numbers to the members of the venire and, one by one, they were interrogated by the attorneys for the prosecution and defense until 12 members of the jury were chosen. At this stage, only one potential female juror was questioned, and she was eliminated by the prosecutor under his quota of peremptory challenges (for which he did not have to give a reason).

▩ EXERCISES

1. The MINITAB display below shows a hypothesis test for a claim that the proportion of women in the city directory is $p = 0.53$. In the test, $n = 350$ and $\hat{p} \approx 0.2914$. Should you reject the claim? What is the level of significance? Explain.

2. In Exercise 1, you rejected the claim that $p = 0.53$. But this claim was true. What type of error is this?

3. If you reject a true claim with a level of significance that is virtually zero, what can you infer about the randomness of your sampling process?

4. Describe a hypothesis test for Judge Ford's "random" selection of the venire. Use a claim of

$$p = \frac{102}{350} \approx 0.2914.$$

(a) Write the null and alternative hypotheses.

(b) Use a technology tool to perform the test.

(c) Make a decision.

(d) Interpret the decision in the context of the original claim. Could Judge Ford's selection of 100 venire members have been random?

MINITAB

Test and CI for One Proportion

Test of p = 0.53 vs p not = 0.53

Sample	X	N	Sample p	99 % CI	Z-Value	P-Value
1	102	350	0.291429	(0.228862, 0.353995)	−8.94	0.000

Using the normal approximation.

Extended solutions are given in the *Technology Supplement*.
Technical instruction is provided for MINITAB, Excel, and the TI-83/84 Plus.

7 USING TECHNOLOGY TO PERFORM HYPOTHESIS TESTS

Here are some MINITAB and TI-83/84 Plus printouts for some of the examples in this chapter.

(See Example 5, page 375.)

Display Descriptive Statistics...
Store Descriptive Statistics...
Graphical Summary...

1-Sample Z...
1-Sample t...
2-Sample t...
Paired t...

1 Proportion...
2 Proportions...

MINITAB

One-Sample Z

Test of mu = 22500 vs not = 22500
The assumed standard deviation = 3015

N	Mean	SE Mean	95% CI	Z	P
30	21545	550	(20466, 22624)	−1.73	0.083

(See Example 4, page 390.)

Display Descriptive Statistics...
Store Descriptive Statistics...
Graphical Summary...

1-Sample Z...
1-Sample t...
2-Sample t...
Paired t...

1 Proportion...
2 Proportions...

MINITAB

One-Sample T

Test of mu = 20500 vs < 20500

N	Mean	StDev	SE Mean	95% Upper Bound	T	P
14	19850	1084	290	20363	−2.24	0.021

(See Example 2, page 400.)

Display Descriptive Statistics...
Store Descriptive Statistics...
Graphical Summary...

1-Sample Z...
1-Sample t...
2-Sample t...
Paired t...

1 Proportion...
2 Proportions...

MINITAB

Test and CI for One Proportion

Test of p = 0.25 vs p not = 0.25

Sample	X	N	Sample p	90% CI	Z-Value	P-Value
1	42	200	0.210000	(0.162627, 0.257373)	−1.31	0.191

Using the normal approximation.

(See Example 9, page 379.)   (See Example 5, page 391.)   (See Example 1, page 399.)

TI-83/84 PLUS

EDIT CALC **TESTS**

1: Z-Test...
2: T-Test...
3: 2-SampZTest...
4: 2-SampTTest...
5: 1-PropZTest...
6: 2-PropZTest...
7↓ZInterval...

TI-83/84 PLUS

EDIT CALC **TESTS**

1: Z-Test...
2: T-Test...
3: 2-SampZTest...
4: 2-SampTTest...
5: 1-PropZTest...
6: 2-PropZTest...
7↓ZInterval...

TI-83/84 PLUS

EDIT CALC **TESTS**

1: Z-Test...
2: T-Test...
3: 2-SampZTest...
4: 2-SampTTest...
5: 1-PropZTest...
6: 2-PropZTest...
7↓ZInterval...

⬇   ⬇   ⬇

TI-83/84 PLUS

Z-Test
  Inpt: Data **Stats**
  μ_0: 68000
  σ: 5500
  $\bar{x}$: 66900
  n: 30
  μ: $\neq \mu_0$ **$<\mu_0$** $>\mu_0$
Calculate Draw

TI-83/84 PLUS

T-Test
  Inpt: Data **Stats**
  μ_0: 6.8
  $\bar{x}$: 6.7
  Sx: .24
  n: 19
  μ: **$\neq \mu_0$** $<\mu_0$ $>\mu_0$
Calculate Draw

TI-83/84 PLUS

1-PropZTest
  p_0: .5
  x: 39
  n: 100
  prop$\neq p_0$ **$<p_0$** $>p_0$
Calculate Draw

⬇   ⬇   ⬇

TI-83/84 PLUS

Z-Test
  $\mu < 68000$
  z= −1.095445115
  p= .1366608782
  $\bar{x}$= 66900
  n= 30

TI-83/84 PLUS

T-Test
  $\mu \neq 6.8$
  t= −1.816207893
  p= .0860316039
  $\bar{x}$= 6.7
  Sx= .24
  n= 19

TI-83/84 PLUS

1-PropZTest
  prop< .5
  z= −2.2
  p= .0139033989
  $\hat{p}$= .39
  n= 100

⬇   ⬇   ⬇

TI-83/84 PLUS

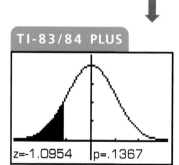

z=−1.0954 | p=.1367

TI-83/84 PLUS

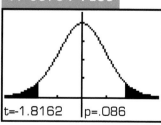

t=−1.8162 | p=.086

TI-83/84 PLUS

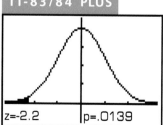

z=−2.2 | p=.0139

9

CHAPTER

CORRELATION AND REGRESSION

9.1 Correlation

■ ACTIVITY

9.2 Linear Regression

■ ACTIVITY

■ CASE STUDY

9.3 Measures of Regression and Prediction Intervals

9.4 Multiple Regression

■ USES AND ABUSES

■ REAL STATISTICS–
　REAL DECISIONS

■ TECHNOLOGY

In 2009, the New York Yankees had the highest team salary in Major League Baseball at $201.4 million and the Florida Marlins had the lowest team salary at $36.8 million. In the same year, the Los Angeles Dodgers had the highest average attendance at 46,440 and the Oakland Athletics had the lowest average attendance at 17,392.

In Chapters 1–8, you studied descriptive statistics, probability, and inferential statistics. One of the techniques you learned in descriptive statistics was graphing paired data with a scatter plot (Section 2.2). For instance, the salaries and average attendances at home games for the teams in Major League Baseball in 2009 are shown in graphical form at the right and in tabular form below.

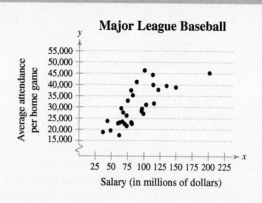

Major League Baseball

Salary (in millions of dollars)	73.5	96.7	67.1	121.7	134.8	96.1	73.6	81.6	75.2	115.1
Average Attendance Per Home Game	26,281	29,304	23,545	37,811	39,610	28,199	21,579	22,492	32,902	31,693

Salary (in millions of dollars)	36.8	103.0	70.5	113.7	100.4	80.2	65.3	149.4	201.4	62.3
Average Attendance Per Home Game	18,770	31,124	22,473	40,004	46,440	37,499	29,466	38,941	45,364	17,392

Salary (in millions of dollars)	113.0	48.7	43.7	82.6	98.9	88.5	63.3	68.2	80.5	60.3
Average Attendance Per Home Game	44,453	19,479	23,735	35,322	27,116	41,274	23,147	27,641	23,162	22,715

In this chapter, you will study how to describe and test the significance of relationships between two variables when data are presented as ordered pairs. For instance, in the scatter plot above, it appears that higher team salaries tend to correspond to higher average attendances and lower team salaries tend to correspond to lower average attendances. This relationship is described by saying that the team salaries are positively correlated to the average attendances. Graphically, the relationship can be described by drawing a line, called a regression line, that fits the points as closely as possible, as shown below. The second scatter plot below shows the salaries and wins for the teams in Major League Baseball in 2009. From the scatter plot, it appears that there is a weak positive correlation between the team salaries and wins.

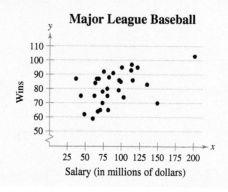

Major League Baseball

Salary (in millions of dollars)

Major League Baseball

Salary (in millions of dollars)

483

9.1 Correlation

WHAT YOU SHOULD LEARN

▸ An introduction to linear correlation, independent and dependent variables, and the types of correlation

▸ How to find a correlation coefficient

▸ How to test a population correlation coefficient ρ using a table

▸ How to perform a hypothesis test for a population correlation coefficient ρ

▸ How to distinguish between correlation and causation

An Overview of Correlation ▸ Correlation Coefficient ▸ Using a Table to Test a Population Correlation Coefficient ρ ▸ Hypothesis Testing for a Population Correlation Coefficient ρ ▸ Correlation and Causation

▸ AN OVERVIEW OF CORRELATION

Suppose a safety inspector wants to determine whether a relationship exists between the number of hours of training for an employee and the number of accidents involving that employee. Or suppose a psychologist wants to know whether a relationship exists between the number of hours a person sleeps each night and that person's reaction time. How would he or she determine if any relationship exists?

In this section, you will study how to describe what type of relationship, or correlation, exists between two quantitative variables and how to determine whether the correlation is significant.

DEFINITION

A **correlation** is a relationship between two variables. The data can be represented by the ordered pairs (x, y), where x is the **independent** (or **explanatory**) **variable** and y is the **dependent** (or **response**) **variable.**

In Section 2.2, you learned that the graph of ordered pairs (x, y) is called a *scatter plot*. In a scatter plot, the ordered pairs (x, y) are graphed as points in a coordinate plane. The independent (explanatory) variable x is measured by the horizontal axis, and the dependent (response) variable y is measured by the vertical axis. A scatter plot can be used to determine whether a linear (straight line) correlation exists between two variables. The following scatter plots show several types of correlation.

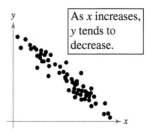

Negative Linear Correlation

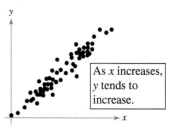

Positive Linear Correlation

No Correlation

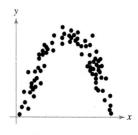

Nonlinear Correlation

GDP (trillions of \$), x	CO_2 emissions (millions of metric tons), y
1.6	428.2
3.6	828.8
4.9	1214.2
1.1	444.6
0.9	264.0
2.9	415.3
2.7	571.8
2.3	454.9
1.6	358.7
1.5	573.5

Number of years out of school, x	Annual contribution (1000s of \$), y
1	12.5
10	8.7
5	14.6
15	5.2
3	9.9
24	3.1
30	2.7

EXAMPLE 1

▶ **Constructing a Scatter Plot**

An economist wants to determine whether there is a linear relationship between a country's gross domestic product (GDP) and carbon dioxide (CO_2) emissions. The data are shown in the table at the left. Display the data in a scatter plot and determine whether there appears to be a positive or negative linear correlation or no linear correlation. *(Source: World Bank and U.S. Energy Information Administration)*

▶ **Solution**

The scatter plot is shown at the right. From the scatter plot, it appears that there is a positive linear correlation between the variables.

Interpretation Reading from left to right, as the gross domestic products increase, the carbon dioxide emissions tend to increase.

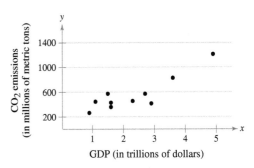

▶ **Try It Yourself 1**

A director of alumni affairs at a small college wants to determine whether there is a linear relationship between the number of years alumni classes have been out of school and their annual contributions (in thousands of dollars). The data are shown in the table at the left. Display the data in a scatter plot and determine the type of correlation.

a. *Draw* and *label* the x- and y-axes.
b. *Plot* each ordered pair.
c. Does there appear to be a linear correlation? If so, *interpret* the correlation in the context of the data.   *Answer: Page A44*

EXAMPLE 2

▶ **Constructing a Scatter Plot**

A student conducts a study to determine whether there is a linear relationship between the number of hours a student exercises each week and the student's grade point average (GPA). The data are shown in the following table. Display the data in a scatter plot and describe the type of correlation.

Hours of exercise, x	12	3	0	6	10	2	20	14	15	5
GPA, y	3.6	4.0	3.9	2.5	2.4	2.2	3.7	3.0	1.8	3.1

▶ **Solution**

The scatter plot is shown at the right. From the scatter plot, it appears that there is no linear correlation between the variables.

Interpretation The number of hours a student exercises each week does not appear to be related to the student's grade point average.

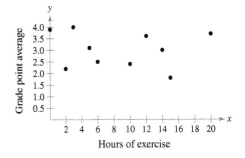

Duration, x	Time, y	Duration, x	Time, y
1.80	56	3.78	79
1.82	58	3.83	85
1.90	62	3.88	80
1.93	56	4.10	89
1.98	57	4.27	90
2.05	57	4.30	89
2.13	60	4.43	89
2.30	57	4.47	86
2.37	61	4.53	89
2.82	73	4.55	86
3.13	76	4.60	92
3.27	77	4.63	91
3.65	77		

▶ **Try It Yourself 2**

A researcher conducts a study to determine whether there is a linear relationship between a person's height (in inches) and pulse rate (in beats per minute). The data are shown in the following table. Display the data in a scatter plot and describe the type of correlation.

Height, x	68	72	65	70	62	75	78	64	68
Pulse rate, y	90	85	88	100	105	98	70	65	72

a. *Draw* and *label* the x- and y-axes.
b. *Plot* each ordered pair.
c. Does there appear to be a linear correlation? If so, *interpret* the correlation in the context of the data.

Answer: Page A44

EXAMPLE 3   Report 37

▶ **Constructing a Scatter Plot Using Technology**

Old Faithful, located in Yellowstone National Park, is the world's most famous geyser. The durations (in minutes) of several of Old Faithful's eruptions and the times (in minutes) until the next eruption are shown in the table at the left. Using a TI-83/84 Plus, display the data in a scatter plot. Describe the type of correlation.

▶ **Solution**

Begin by entering the x-values into List 1 and the y-values into List 2. Use *Stat Plot* to construct the scatter plot. The plot should look similar to the one shown below. From the scatter plot, it appears that the variables have a positive linear correlation.

 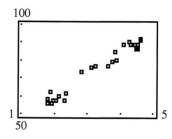

Interpretation You can conclude that the longer the duration of the eruption, the longer the time before the next eruption begins.

▶ **Try It Yourself 3**

Consider the data from the Chapter Opener on page 483 on the salaries and average attendances at home games for the teams in Major League Baseball. Use a technology tool to display the data in a scatter plot. Describe the type of correlation.

a. *Enter* the data into List 1 and List 2.
b. *Construct* the scatter plot.
c. Does there appear to be a linear correlation? If so, *interpret* the correlation in the context of the data.

Answer: Page A44

▶ CORRELATION COEFFICIENT

Interpreting correlation using a scatter plot can be subjective. A more precise way to measure the type and strength of a linear correlation between two variables is to calculate the *correlation coefficient*. Although a formula for the sample correlation coefficient is given, it is more convenient to use a technology tool to calculate this value.

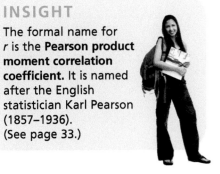

> ### DEFINITION
>
> The **correlation coefficient** is a measure of the strength and the direction of a linear relationship between two variables. The symbol *r* represents the sample correlation coefficient. A formula for *r* is
>
> $$r = \frac{n\sum xy - (\sum x)(\sum y)}{\sqrt{n\sum x^2 - (\sum x)^2}\sqrt{n\sum y^2 - (\sum y)^2}}$$
>
> where *n* is the number of pairs of data.
>
> The population correlation coefficient is represented by ρ (the lowercase Greek letter rho, pronounced "row").

The range of the correlation coefficient is −1 to 1, inclusive. If *x* and *y* have a strong positive linear correlation, *r* is close to 1. If *x* and *y* have a strong negative linear correlation, *r* is close to −1. If *x* and *y* have perfect positive linear correlation or perfect negative linear correlation, *r* is equal to 1 or −1, respectively. If there is no linear correlation or a weak linear correlation, *r* is close to 0. It is important to remember that if *r* is close to 0, it does not mean that there is no relation between *x* and *y*, just that there is no linear relation. Several examples are shown below.

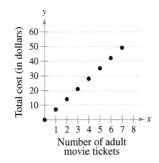

Perfect positive correlation
r = 1

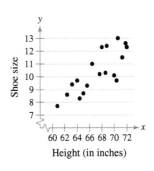

Strong positive correlation
r = 0.81

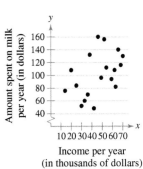

Weak positive correlation
r = 0.45

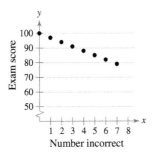

Perfect negative correlation
r = −1

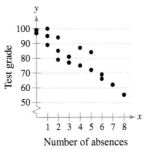

Strong negative correlation
r = −0.92

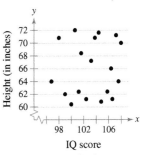

No correlation
r = 0.04

GUIDELINES

Calculating a Correlation Coefficient

IN WORDS	IN SYMBOLS
1. Find the sum of the x-values.	$\sum x$
2. Find the sum of the y-values.	$\sum y$
3. Multiply each x-value by its corresponding y-value and find the sum.	$\sum xy$
4. Square each x-value and find the sum.	$\sum x^2$
5. Square each y-value and find the sum.	$\sum y^2$
6. Use these five sums to calculate the correlation coefficient.	$r = \dfrac{n\sum xy - (\sum x)(\sum y)}{\sqrt{n\sum x^2 - (\sum x)^2}\sqrt{n\sum y^2 - (\sum y)^2}}$

EXAMPLE 4

▶ **Finding the Correlation Coefficient**

Calculate the correlation coefficient for the gross domestic products and carbon dioxide emissions data given in Example 1. What can you conclude?

▶ **Solution** Use a table to help calculate the correlation coefficient.

GDP (trillions of \$), x	CO$_2$ emissions (millions of metric tons), y	xy	x^2	y^2
1.6	428.2	685.12	2.56	183,355.24
3.6	828.8	2983.68	12.96	686,909.44
4.9	1214.2	5949.58	24.01	1,474,281.64
1.1	444.6	489.06	1.21	197,669.16
0.9	264.0	237.6	0.81	69,696
2.9	415.3	1204.37	8.41	172,474.09
2.7	571.8	1543.86	7.29	326,955.24
2.3	454.9	1046.27	5.29	206,934.01
1.6	358.7	573.92	2.56	128,665.69
1.5	573.5	860.25	2.25	328,902.25
$\sum x = 23.1$	$\sum y = 5554$	$\sum xy = 15{,}573.71$	$\sum x^2 = 67.35$	$\sum y^2 = 3{,}775{,}842.76$

With these sums and $n = 10$, the correlation coefficient is

$$r = \frac{n\sum xy - (\sum x)(\sum y)}{\sqrt{n\sum x^2 - (\sum x)^2}\sqrt{n\sum y^2 - (\sum y)^2}}$$

$$= \frac{10(15{,}573.71) - (23.1)(5554)}{\sqrt{10(67.35) - 23.1^2}\sqrt{10(3{,}775{,}842.76) - 5554^2}}$$

$$= \frac{27{,}439.7}{\sqrt{139.89}\sqrt{6{,}911{,}511.6}} \approx 0.882.$$

The result $r \approx 0.882$ suggests a strong positive linear correlation.

Interpretation As the gross domestic product increases, the carbon dioxide emissions also increase.

Number of years out of school, x	Annual contribution (1000s of $), y
1	12.5
10	8.7
5	14.6
15	5.2
3	9.9
24	3.1
30	2.7

▶ **Try It Yourself 4**

Calculate the correlation coefficient for the number of years out of school and annual contribution data given in Try It Yourself 1. What can you conclude?

a. *Identify* n and *use a table* to help calculate $\sum x$, $\sum y$, $\sum xy$, $\sum x^2$, and $\sum y^2$.
b. *Use the resulting sums and* n *to calculate* r.
c. What can you conclude?   *Answer: Page A44*

EXAMPLE 5   Report 38

▶ **Using Technology to Find a Correlation Coefficient**

Use a technology tool to calculate the correlation coefficient for the Old Faithful data given in Example 3. What can you conclude?

▶ **Solution**

MINITAB, Excel, and the TI-83/84 Plus each have features that allow you to calculate a correlation coefficient for paired data sets. Try using this technology to find r. You should obtain results similar to the following.

MINITAB

Correlations: C1, C2

Pearson correlation of C1 and C2 = 0.979

EXCEL

	A	B	C
26	CORREL(A1:A25,B1:B25)		
27			0.978659

Before using the TI-83/84 Plus to calculate r, you must enter the Diagnostic On command. To do so, enter the following keystrokes:

[2nd] [0] cursor to *DiagnosticOn* [ENTER] [ENTER].

The following screens show how to find r using a TI-83/84 Plus with the data stored in List 1 and List 2. To begin, use the STAT keystroke.

To explore this topic further, see Activity 9.1 on page 500.

TI-83/84 PLUS

EDIT **CALC** TESTS
1: 1-Var Stats
2: 2-Var Stats
3: Med-Med
4: LinReg(ax+b)
5: QuadReg
6: CubicReg
7↓QuartReg

TI-83/84 PLUS

LinReg(ax+b) L1,
L2

TI-83/84 PLUS

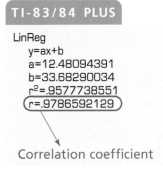

LinReg
 y=ax+b
 a=12.48094391
 b=33.68290034
 r²=.9577738551
 r=.9786592129

Correlation coefficient

The result $r \approx 0.979$ suggests a strong positive linear correlation.

▶ **Try It Yourself 5**

Calculate the correlation coefficient for the data from the Chapter Opener on page 483 on the salaries and average attendances at home games for the teams in Major League Baseball. What can you conclude?

a. *Enter* the data.
b. Use the appropriate feature to *calculate* r.
c. What can you conclude?   *Answer: Page A44*

▶ USING A TABLE TO TEST A POPULATION CORRELATION COEFFICIENT ρ

Once you have calculated r, the sample correlation coefficient, you will want to determine whether there is enough evidence to decide that the population correlation coefficient ρ is significant. In other words, based on a few pairs of data, can you make an inference about the population of all such data pairs? Remember that you are using sample data to make a decision about population data, so it is always possible that your inference may be wrong. In correlation studies, the small percentage of times when you decide that the correlation is significant when it is really not is called the *level of significance*. It is typically set at $\alpha = 0.01$ or 0.05. When $\alpha = 0.05$, you will probably decide that the population correlation coefficient is significant when it is really not 5% of the time. (Of course, 95% of the time, you will correctly determine that a correlation coefficient is significant.) When $\alpha = 0.01$, you will make this type of error only 1% of the time. When using a lower level of significance, however, you may fail to identify some significant correlations.

In order for a correlation coefficient to be significant, its absolute value must be close to 1. To determine whether the population correlation coefficient ρ is significant, use the critical values given in Table 11 in Appendix B. A portion of the table is shown below. If $|r|$ is greater than the critical value, there is enough evidence to decide that the correlation is significant. Otherwise, there is *not* enough evidence to say that the correlation is significant. For instance, to determine whether ρ is significant for five pairs of data ($n = 5$) at a level of significance of $\alpha = 0.01$, you need to compare $|r|$ with a critical value of 0.959, as shown in the table.

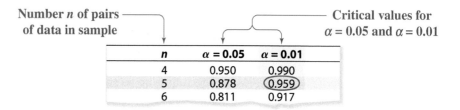

n	$\alpha = 0.05$	$\alpha = 0.01$
4	0.950	0.990
5	0.878	0.959
6	0.811	0.917

If $|r| > 0.959$, the correlation is significant. Otherwise, there is *not* enough evidence to conclude that the correlation is significant. The guidelines for this process are as follows.

GUIDELINES

Using Table 11 for the Correlation Coefficient ρ

IN WORDS	IN SYMBOLS
1. Determine the number of pairs of data in the sample.	Determine n.
2. Specify the level of significance.	Identify α.
3. Find the critical value.	Use Table 11 in Appendix B.
4. Decide if the correlation is significant.	If $|r| >$ critical value, the correlation is significant. Otherwise, there is *not* enough evidence to conclude that the correlation is significant.
5. Interpret the decision in the context of the original claim.	

EXAMPLE 6

▶ **Using Table 11 for a Correlation Coefficient**

In Example 5, you used 25 pairs of data to find $r \approx 0.979$. Is the correlation coefficient significant? Use $\alpha = 0.05$.

▶ **Solution**

The number of pairs of data is 25, so $n = 25$. The level of significance is $\alpha = 0.05$. Using Table 11, find the critical value in the $\alpha = 0.05$ column that corresponds to the row with $n = 25$. The number in that column and row is 0.396.

INSIGHT

Notice that the fewer the data points in your study, the stronger the evidence has to be to conclude that the correlation coefficient is significant.

n	α = 0.05	α = 0.01
4	0.950	0.990
5	0.878	0.959
6	0.811	0.917
7	0.754	0.875
8	0.707	0.834
9	0.666	0.798
10	0.632	0.765
11	0.602	0.735
12	0.576	0.708
13	0.553	0.684
14	0.532	0.661
19	0.456	0.575
20	0.444	0.561
21	0.433	0.549
22	0.423	0.537
23	0.413	0.526
24	0.404	0.515
25	(0.396)	0.505
26	0.388	0.496
27	0.381	0.487
28	0.374	0.479
29	0.367	0.471

Because $|r| \approx 0.979 > 0.396$, you can decide that the population correlation is significant.

Interpretation   There is enough evidence at the 5% level of significance to conclude that there is a significant linear correlation between the duration of Old Faithful's eruptions and the time between eruptions.

▶ **Try It Yourself 6**

In Try It Yourself 4, you calculated the correlation coefficient of the number of years out of school and annual contribution data to be $r \approx -0.908$. Is the correlation coefficient significant? Use $\alpha = 0.01$.

a. Determine the *number of pairs of data* in the sample.
b. Identify the *level of significance*.
c. Find the *critical value*. Use Table 11 in Appendix B.
d. *Compare* $|r|$ with the critical value and *decide* if the correlation is significant.
e. *Interpret* the decision in the context of the original claim.

Answer: Page A44

▶ HYPOTHESIS TESTING FOR A POPULATION CORRELATION COEFFICIENT ρ

You can also use a hypothesis test to determine whether the sample correlation coefficient r provides enough evidence to conclude that the population correlation coefficient ρ is significant. A hypothesis test for ρ can be one-tailed or two-tailed. The null and alternative hypotheses for these tests are as follows.

$$\begin{cases} H_0: \rho \geq 0 \ \text{(no significant negative correlation)} \\ H_a: \rho < 0 \ \text{(significant negative correlation)} \end{cases} \quad \textbf{Left-tailed test}$$

$$\begin{cases} H_0: \rho \leq 0 \ \text{(no significant positive correlation)} \\ H_a: \rho > 0 \ \text{(significant positive correlation)} \end{cases} \quad \textbf{Right-tailed test}$$

$$\begin{cases} H_0: \rho = 0 \ \text{(no significant correlation)} \\ H_a: \rho \neq 0 \ \text{(significant correlation)} \end{cases} \quad \textbf{Two-tailed test}$$

In this text, you will consider only two-tailed hypothesis tests for ρ.

THE t-TEST FOR THE CORRELATION COEFFICIENT

A **t-test** can be used to test whether the correlation between two variables is significant. The **test statistic** is r and the **standardized test statistic**

$$t = \frac{r}{\sigma_r} = \frac{r}{\sqrt{\dfrac{1 - r^2}{n - 2}}}$$

follows a t-distribution with $n - 2$ degrees of freedom.

GUIDELINES

Using the t-Test for the Correlation Coefficient ρ

IN WORDS	IN SYMBOLS
1. Identify the null and alternative hypotheses.	State H_0 and H_a.
2. Specify the level of significance.	Identify α.
3. Identify the degrees of freedom.	d.f. $= n - 2$
4. Determine the critical value(s) and the rejection region(s).	Use Table 5 in Appendix B.
5. Find the standardized test statistic.	$t = \dfrac{r}{\sqrt{\dfrac{1 - r^2}{n - 2}}}$
6. Make a decision to reject or fail to reject the null hypothesis.	If t is in the rejection region, reject H_0. Otherwise, fail to reject H_0.
7. Interpret the decision in the context of the original claim.	

EXAMPLE 7 SC Report 39

▶ The *t*-Test for a Correlation Coefficient

In Example 4, you used 10 pairs of data to find $r \approx 0.882$. Test the significance of this correlation coefficient. Use $\alpha = 0.05$.

▶ Solution

The null and alternative hypotheses are

$$H_0: \rho = 0 \text{ (no correlation)} \quad \text{and} \quad H_a: \rho \neq 0 \text{ (significant correlation)}.$$

Because there are 10 pairs of data in the sample, there are $10 - 2 = 8$ degrees of freedom. Because the test is a two-tailed test, $\alpha = 0.05$, and d.f. = 8, the critical values are $-t_0 = -2.306$ and $t_0 = 2.306$. The rejection regions are $t < -2.306$ and $t > 2.306$. Using the *t*-test, the standardized test statistic is

$$t = \frac{r}{\sqrt{\dfrac{1 - r^2}{n - 2}}}$$

$$\approx \frac{0.882}{\sqrt{\dfrac{1 - (0.882)^2}{10 - 2}}} \approx 5.294.$$

The following graph shows the location of the rejection regions and the standardized test statistic.

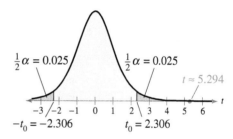

$\frac{1}{2}\alpha = 0.025$ $\frac{1}{2}\alpha = 0.025$

$t \approx 5.294$

$-t_0 = -2.306$ $t_0 = 2.306$

Because t is in the rejection region, you should decide to reject the null hypothesis.

Interpretation There is enough evidence at the 5% level of significance to conclude that there is a significant linear correlation between gross domestic products and carbon dioxide emissions.

▶ Try It Yourself 7

In Try It Yourself 5, you calculated the correlation coefficient of the salaries and average attendances at home games for the teams in Major League Baseball to be $r \approx 0.74972$. Test the significance of this correlation coefficient. Use $\alpha = 0.01$.

a. State the *null* and *alternative hypotheses*.
b. Identify the *level of significance*.
c. Identify the *degrees of freedom*.
d. Determine the *critical values* and the *rejection regions*.
e. Find the *standardized test statistic*.
f. *Make a decision* to reject or fail to reject the null hypothesis.
g. *Interpret* the decision in the context of the original claim.

Answer: Page A45

INSIGHT

In Example 7, you can use Table 11 in Appendix B to test the population correlation coefficient ρ. Given $n = 10$ and $\alpha = 0.05$, the critical value from Table 11 is 0.632. Because

$$|r| \approx 0.882 > 0.632,$$

the correlation is significant. Note that this is the same result you obtained using a *t*-test for the population correlation coefficient ρ.

STUDY TIP

Be sure you see in Example 7 that rejecting the null hypothesis means that there is enough evidence that the correlation is significant.

PICTURING THE WORLD

The following scatter plot shows the results of a survey conducted as a group project by students in a high school statistics class in the San Francisco area. In the survey, 125 high school students were asked their grade point average (GPA) and the number of caffeine drinks they consumed each day.

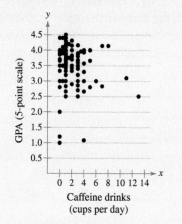

What type of correlation, if any, does the scatter plot show between caffeine consumption and GPA?

▶ **CORRELATION AND CAUSATION**

The fact that two variables are strongly correlated does not in itself imply a cause-and-effect relationship between the variables. More in-depth study is usually needed to determine whether there is a causal relationship between the variables.

   If there is a significant correlation between two variables, a researcher should consider the following possibilities.

1. **Is there a direct cause-and-effect relationship between the variables?**

   That is, does x cause y? For instance, consider the relationship between gross domestic products and carbon dioxide emissions that has been discussed throughout this section. It is reasonable to conclude that an increase in a country's gross domestic product will result in higher carbon dioxide emissions.

2. **Is there a reverse cause-and-effect relationship between the variables?**

   That is, does y cause x? For instance, consider the Old Faithful data that have been discussed throughout this section. These variables have a positive linear correlation, and it is possible to conclude that the duration of an eruption affects the time before the next eruption. However, it is also possible that the time between eruptions affects the duration of the next eruption.

3. **Is it possible that the relationship between the variables can be caused by a third variable or perhaps a combination of several other variables?**

   For instance, consider the salaries and average attendances per home game for the teams in Major League Baseball listed in the Chapter Opener. Although these variables have a positive linear correlation, it is doubtful that just because a team's salary decreases, the average attendance per home game will also decrease. The relationship is probably due to several other variables, such as the economy, the players on the team, and whether or not the team is winning games.

4. **Is it possible that the relationship between two variables may be a coincidence?**

   For instance, although it may be possible to find a significant correlation between the number of animal species living in certain regions and the number of people who own more than two cars in those regions, it is highly unlikely that the variables are directly related. The relationship is probably due to coincidence.

   Determining which of the cases above is valid for a data set can be difficult. For instance, consider the following example. Suppose a person breaks out in a rash each time he eats shrimp at a certain restaurant. The natural conclusion is that the person is allergic to shrimp. However, upon further study by an allergist, it is found that the person is not allergic to shrimp, but to a type of seasoning the chef is putting into the shrimp.

9.1 EXERCISES

■ BUILDING BASIC SKILLS AND VOCABULARY

1. Two variables have a positive linear correlation. Does the dependent variable increase or decrease as the independent variable increases?

2. Two variables have a negative linear correlation. Does the dependent variable increase or decrease as the independent variable increases?

3. Describe the range of values for the correlation coefficient.

4. What does the sample correlation coefficient r measure? Which value indicates a stronger correlation: $r = 0.918$ or $r = -0.932$? Explain your reasoning.

5. Give examples of two variables that have perfect positive linear correlation and two variables that have perfect negative linear correlation.

6. Explain how to decide whether a sample correlation coefficient indicates that the population correlation coefficient is significant.

7. Discuss the difference between r and ρ.

8. In your own words, what does it mean to say "correlation does not imply causation"?

Graphical Analysis   *In Exercises 9–14, the scatter plots of paired data sets are shown. Determine whether there is a perfect positive linear correlation, a strong positive linear correlation, a perfect negative linear correlation, a strong negative linear correlation, or no linear correlation between the variables.*

9.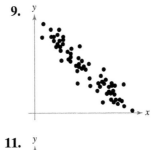

10.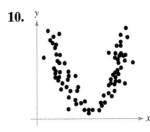

11.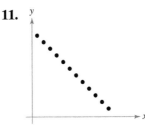

12.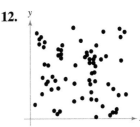

13.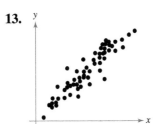

14.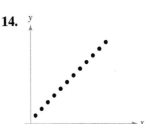

Graphical Analysis   *In Exercises 15–18, the scatter plots show the results of a survey of 20 randomly selected males ages 24–35. Using age as the explanatory variable, match each graph with the appropriate description. Explain your reasoning.*

(a) *Age and body temperature*   (b) *Age and balance on student loans*
(c) *Age and income*   (d) *Age and height*

15.

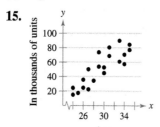

16.

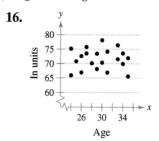

17.

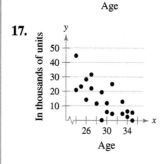

18.

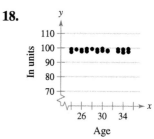

In Exercises 19 and 20, identify the explanatory variable and the response variable.

19. A nutritionist wants to determine if the amounts of water consumed each day by persons of the same weight and on the same diet can be used to predict individual weight loss.

20. An insurance company hires an actuary to determine whether the number of hours of safety driving classes can be used to predict the number of driving accidents for each driver.

■ USING AND INTERPRETING CONCEPTS

Constructing a Scatter Plot and Determining Correlation   *In Exercises 21–28, (a) display the data in a scatter plot, (b) calculate the sample correlation coefficient r, and (c) make a conclusion about the type of correlation.*

21. Age and Blood Pressure   The ages (in years) of 10 men and their systolic blood pressures

Age, x	16	25	39	45	49	64	70	29	57	22
Systolic blood pressure, y	109	122	143	132	199	185	199	130	175	118

22. Age and Vocabulary   The ages (in years) of 11 children and the number of words in their vocabulary

Age, x	1	2	3	4	5	6
Vocabulary size, y	3	440	1200	1500	2100	2600

Age, x	3	5	2	4	6
Vocabulary size, y	1100	2000	500	1525	2500

23. Hours Studying and Test Scores The number of hours 13 students spent studying for a test and their scores on that test

Hours spent studying, x	0	1	2	4	4	5	5	5	6	6	7	7	8
Test score, y	40	41	51	48	64	69	73	75	68	93	84	90	95

24. Hours Online and Test Scores The number of hours 12 students spent online during the weekend and the scores of each student who took a test the following Monday

Hours spent online, x	0	1	2	3	3	5	5	5	6	7	7	10
Test score, y	96	85	82	74	95	68	76	84	58	65	75	50

25. Movie Budgets and Grosses The budget (in millions of dollars) and worldwide gross (in millions of dollars) for eight of the most expensive movies ever made *(Adapted from The Numbers)*

Budget, x	300	258	250	210	232	230	225	207
Gross, y	961	891	937	836	391	576	419	551

26. Speed of Sound The altitude (in thousands of feet) and speed of sound (in feet per second)

Altitude, x	0	5	10	15	20	25
Speed of sound, y	1116.3	1096.9	1077.3	1057.2	1036.8	1015.8

Altitude, x	30	35	40	45	50
Speed of sound, y	994.5	969.0	967.7	967.7	967.7

27. Earnings and Dividends The earnings per share and dividends per share for 12 medical supplies companies in a recent year *(Source: The Value Line Investment Survey)*

Earnings per share, x	6.00	1.44	4.44	3.38	3.63	4.46
Dividends per share, y	2.45	0.15	0.62	0.91	0.68	1.14

Earnings per share, x	3.80	1.43	1.88	4.57	4.28	2.92
Dividends per share, y	0.52	0.06	0.19	1.80	0.48	0.63

28. Crimes and Arrests The number of crimes reported (in millions) and the number of arrests reported (in millions) by the U.S. Department of Justice for 14 years *(Adapted from the National Crime Victimization Survey and Uniform Crime Reports)*

Crimes, x	1.60	1.55	1.44	1.40	1.32	1.23	1.22
Arrests, y	0.78	0.80	0.73	0.72	0.68	0.64	0.63

Crimes, x	1.23	1.22	1.18	1.16	1.19	1.21	1.20
Arrests, y	0.63	0.62	0.60	0.59	0.60	0.61	0.58

29. A student spends 1 hour studying and gets a test score of 99. Add this data entry to the current data set in Exercise 23. Describe how adding this data entry changes the correlation coefficient r. Why do you think it changed?

30. A student spends 12 hours online during the weekend and gets a test score of 98. Add this data entry to the current data set in Exercise 24. Describe how adding this data entry changes the correlation coefficient r. Why do you think it changed?

Testing Claims   *In Exercises 31–36, use Table 11 in Appendix B as shown in Example 6, or perform a hypothesis test using Table 5 in Appendix B as shown in Example 7, to make a conclusion about the indicated correlation coefficient. If convenient, use technology to solve the problem.*

31. Braking Distances: Dry Surface   The weights (in pounds) of eight vehicles and the variability of their braking distances (in feet) when stopping on a dry surface are shown in the table. Can you conclude that there is a significant linear correlation between vehicle weight and variability in braking distance on a dry surface? Use $\alpha = 0.01$. *(Adapted from National Highway Traffic Safety Administration)*

Weight, x	5940	5340	6500	5100	5850	4800	5600	5890
Variability in braking distance, y	1.78	1.93	1.91	1.59	1.66	1.50	1.61	1.70

32. Braking Distances: Wet Surface   The weights (in pounds) of eight vehicles and the variability of their braking distances (in feet) when stopping on a wet surface are shown in the table. At $\alpha = 0.05$, can you conclude that there is a significant linear correlation between vehicle weight and variability in braking distance on a wet surface? *(Adapted from National Highway Traffic Safety Administration)*

Weight, x	5890	5340	6500	4800	5940	5600	5100	5850
Variability in braking distance, y	2.92	2.40	4.09	1.72	2.88	2.53	2.32	2.78

33. Hours Studying and Test Scores   The table in Exercise 23 shows the number of hours 13 students spent studying for a test and their scores on that test. At $\alpha = 0.01$, is there enough evidence to conclude that there is a significant linear correlation between the data? (Use the value of r found in Exercise 23.)

34. Hours Online and Test Scores   The table in Exercise 24 shows the number of hours spent online and the test scores for 12 randomly selected students. At $\alpha = 0.05$, is there enough evidence to conclude that there is a significant linear correlation between the data? (Use the value of r found in Exercise 24.)

35. Earnings and Dividends   The table in Exercise 27 shows the earnings per share and dividends per share for 12 medical supplies companies in a recent year. At $\alpha = 0.01$, can you conclude that there is a significant linear correlation between earnings per share and dividends per share? (Use the value of r found in Exercise 27.)

36. Crimes and Arrests   The table in Exercise 28 shows the number of crimes reported (in millions) and the number of arrests reported (in millions) by the U.S. Department of Justice for 14 years. At $\alpha = 0.05$, can you conclude that there is a significant linear correlation between the number of crimes and the number of arrests? (Use the value of r found in Exercise 28.)

SC *In Exercises 37 and 38, use StatCrunch to (a) display the data in a scatter plot, (b) calculate the correlation coefficient r, and (c) test the significance of the correlation coefficient.*

37. Earthquakes   A researcher wants to determine if there is a linear relationship between the magnitudes of earthquakes and their depths below the surface at the epicenter. The magnitudes and depths (in kilometers) of eight recent earthquakes are shown in the table. Use $\alpha = 0.01$. *(Source: U.S. Geological Survey)*

Magnitude, x	7.7	6.7	6.9	6.8	4.0	3.8	7.1	5.9
Depth, y	35	18	17	26	5	10	25	10

38. Income Level and Charitable Donations   A sociologist wants to determine if there is a linear relationship between family income level and percent of income donated to charities. The income levels (in thousands of dollars) and percents of income donated to charities for seven families are shown in the table. Use $\alpha = 0.05$.

Income level, x	50	65	48	42	59	72	60
Donating percent, y	4	8	5	5	10	7	6

39. An earthquake is recorded with a magnitude of 6.3 and a depth of 620 kilometers. Add this data entry to the current data set in Exercise 37. Describe how adding this data entry changes the correlation coefficient *r* and your decision to reject or fail to reject the null hypothesis.

40. A family has an income level of $75,000 and donates 1% of their income to charities. Add this data entry to the current data set in Exercise 38. Describe how adding this data entry changes the correlation coefficient *r* and your decision to reject or fail to reject the null hypothesis.

■ EXTENDING CONCEPTS

Interchanging x and y   *In Exercises 41 and 42, calculate the correlation coefficient r, letting Row 1 represent the x-values and Row 2 the y-values. Then calculate the correlation coefficient r, letting Row 2 represent the x-values and Row 1 the y-values. What effect does switching the explanatory and response variables have on the correlation coefficient?*

41.

Row 1	16	25	39	45	49	64	70
Row 2	109	122	143	132	199	185	199

42.

Row 1	0	1	2	3	3	5	5	5	6	7
Row 2	96	85	82	74	95	68	76	84	58	65

43. Writing   Use your school's library, the Internet, or some other reference source to find a real-life data set with the indicated cause-and-effect relationship. Write a paragraph describing each variable and explain why you think the variables have the indicated cause-and-effect relationship.

   (a) *Direct Cause-and-Effect:* Changes in one variable cause changes in the other variable.

   (b) *Other Factors:* The relationship between the variables is caused by a third variable.

   (c) *Coincidence:* The relationship between the variables is a coincidence.

ACTIVITY 9.1 Correlation by Eye

APPLET

The *correlation by eye* applet allows you to guess the sample correlation coefficient *r* for a data set. When the applet loads, a data set consisting of 20 points is displayed. Points can be added to the plot by clicking the mouse. Points on the plot can be removed by clicking on the point and then dragging the point into the trash can. All of the points on the plot can be removed by simply clicking inside the trash can. You can enter your guess for *r* in the "Guess" field, and then click SHOW R! to see if you are within 0.1 of the true value. When you click NEW DATA, a new data set is generated.

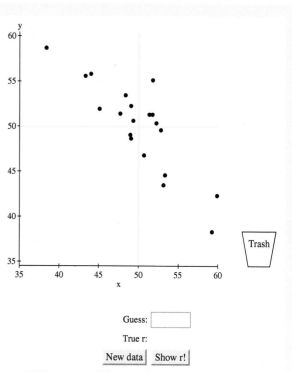

■ Explore

Step 1 Add five points to the plot.
Step 2 Enter a guess for *r*.
Step 3 Click SHOW R!.
Step 4 Click NEW DATA.
Step 5 Remove five points from the plot.
Step 6 Enter a guess for *r*.
Step 7 Click SHOW R!.

■ Draw Conclusions

APPLET

1. Generate a new data set. Using your knowledge of correlation, try to guess the value of *r* for the data set. Repeat this 10 times. How many times were you correct? Describe how you chose each *r* value.

2. Describe how to create a data set with a value of *r* that is approximately 1.

3. Describe how to create a data set with a value of *r* that is approximately 0.

4. Try to create a data set with a value of *r* that is approximately −0.9. Then try to create a data set with a value of *r* that is approximately 0.9. What did you do differently to create the two data sets?

9.2 Linear Regression

WHAT YOU SHOULD LEARN

▸ How to find the equation of a regression line

▸ How to predict y-values using a regression equation

Regression Lines ▸ Applications of Regression Lines

▸ REGRESSION LINES

After verifying that the linear correlation between two variables is significant, the next step is to determine the equation of the line that best models the data. This line is called a *regression line*, and its equation can be used to predict the value of y for a given value of x. Although many lines can be drawn through a set of points, a regression line is determined by specific criteria.

Consider the scatter plot and the line shown below. For each data point, d_i represents the difference between the observed y-value and the predicted y-value for a given x-value on the line. These differences are called **residuals** and can be positive, negative, or zero. When the point is above the line, d_i is positive. When the point is below the line, d_i is negative. If the observed y-value equals the predicted y-value, $d_i = 0$. Of all possible lines that can be drawn through a set of points, the regression line is the line for which the sum of the squares of all the residuals

$$\sum d_i^2$$

is a minimum.

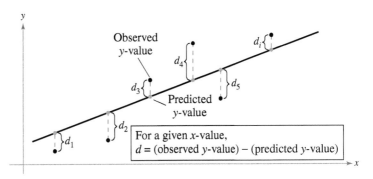

For a given x-value,
$d = $ (observed y-value) − (predicted y-value)

DEFINITION

A **regression line,** also called a **line of best fit,** is the line for which the sum of the squares of the residuals is a minimum.

STUDY TIP

When determining the equation of a regression line, it is helpful to construct a scatter plot of the data to check for outliers, which can greatly influence a regression line. You should also check for gaps and clusters in the data.

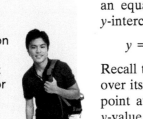

In algebra, you learned that you can write an equation of a line by finding its slope m and y-intercept b. The equation has the form

$$y = mx + b.$$

Recall that the slope of a line is the ratio of its rise over its run and the y-intercept is the y-value of the point at which the line crosses the y-axis. It is the y-value when $x = 0$.

In algebra, you used two points to determine the equation of a line. In statistics, you will use every point in the data set to determine the equation of the regression line.

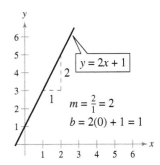

$y = 2x + 1$

$m = \frac{2}{1} = 2$

$b = 2(0) + 1 = 1$

The equation of a regression line allows you to use the independent (explanatory) variable x to make predictions for the dependent (response) variable y.

THE EQUATION OF A REGRESSION LINE

The equation of a regression line for an independent variable x and a dependent variable y is

$$\hat{y} = mx + b$$

where $\hat{y}$ is the predicted y-value for a given x-value. The slope m and y-intercept b are given by

$$m = \frac{n\Sum xy - (\Sum x)(\Sum y)}{n\Sum x^2 - (\Sum x)^2} \quad \text{and} \quad b = \bar{y} - m\bar{x} = \frac{\Sum y}{n} - m\frac{\Sum x}{n}$$

where $\bar{y}$ is the mean of the y-values in the data set and $\bar{x}$ is the mean of the x-values. The regression line always passes through the point $(\bar{x}, \bar{y})$.

EXAMPLE 1

▶ **Finding the Equation of a Regression Line**

Find the equation of the regression line for the gross domestic products and carbon dioxide emissions data used in Section 9.1.

▶ **Solution**

In Example 4 of Section 9.1, you found that $n = 10$, $\Sum x = 23.1$, $\Sum y = 5554$, $\Sum xy = 15{,}573.71$, and $\Sum x^2 = 67.35$. You can use these values to calculate the slope and y-intercept of the regression line as shown.

GDP (trillions of \$), x	CO_2 emissions (millions of metric tons), y
1.6	428.2
3.6	828.8
4.9	1214.2
1.1	444.6
0.9	264.0
2.9	415.3
2.7	571.8
2.3	454.9
1.6	358.7
1.5	573.5

$$m = \frac{n\Sum xy - (\Sum x)(\Sum y)}{n\Sum x^2 - (\Sum x)^2}$$

$$= \frac{10(15{,}573.71) - (23.1)(5554)}{10(67.35) - 23.1^2}$$

$$= \frac{27{,}439.7}{139.89} \approx 196.151977$$

$$b = \bar{y} - m\bar{x} \approx \frac{5554}{10} - (196.151977)\frac{23.1}{10}$$

$$= 555.4 - (196.151977)(2.31)$$

$$\approx 102.2889$$

So, the equation of the regression line is

$$\hat{y} = 196.152x + 102.289.$$

To sketch the regression line, use any two x-values within the range of data and calculate their corresponding y-values from the regression line. Then draw a line through the two points. The regression line and scatter plot of the data are shown at the right. If you plot the point $(\bar{x}, \bar{y}) = (2.31, 555.4)$, you will notice that the line passes through this point.

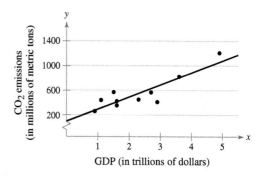

▶ Try It Yourself 1

Find the equation of the regression line for the number of years out of school and annual contribution data used in Section 9.1.

a. *Identify n, $\sum x$, $\sum y$, $\sum xy$, and $\sum x^2$ from Try It Yourself 4 in Section 9.1.*
b. Calculate the *slope m* and the *y-intercept b*.
c. Write the *equation* of the regression line.   *Answer: Page A45*

Duration, x	Time, y	Duration, x	Time, y
1.80	56	3.78	79
1.82	58	3.83	85
1.90	62	3.88	80
1.93	56	4.10	89
1.98	57	4.27	90
2.05	57	4.30	89
2.13	60	4.43	89
2.30	57	4.47	86
2.37	61	4.53	89
2.82	73	4.55	86
3.13	76	4.60	92
3.27	77	4.63	91
3.65	77		

EXAMPLE 2   SC Report 40

▶ **Using Technology to Find a Regression Equation**

Use a technology tool to find the equation of the regression line for the Old Faithful data used in Section 9.1.

▶ **Solution**

MINITAB, Excel, and the TI-83/84 Plus each have features that automatically calculate a regression equation. Try using this technology to find the regression equation. You should obtain results similar to the following.

MINITAB

Regression Analysis: C2 versus C1

The regression equation is
C2 = 33.7 + 12.5 C1

Predictor	Coef	SE Coef	T	P
Constant	33.683	1.894	17.79	0.000
C1	12.4809	0.5464	22.84	0.000

S = 2.88153   R-Sq = 95.8%   R-Sq(adj) = 95.6%

To explore this topic further, see Activity 9.2 on page 511.

EXCEL

	A	B	C	D
1	Slope:			
2	INDEX(LINEST(known_y's,known_x's),1)			
3				12.48094
4				
5	Y-intercept:			
6	INDEX(LINEST(known_y's,known_x's),2)			
7				33.6829

TI-83/84 PLUS

LinReg
y=ax+b
a=12.48094391
b=33.68290034
r^2=.9577738551
r=.9786592129

From the displays, you can see that the regression equation is

$$\hat{y} = 12.481x + 33.683.$$

The TI-83/84 Plus display at the left shows the regression line and a scatter plot of the data in the same viewing window. To do this, use *Stat Plot* to construct the scatter plot and enter the regression equation as y_1.

▶ Try It Yourself 2

Use a technology tool to find the equation of the regression line for the salaries and average attendances at home games for the teams in Major League Baseball given in the Chapter Opener on page 483.

a. *Enter* the data.
b. Perform the necessary steps to calculate the *slope* and *y-intercept*.
c. Specify the *regression equation*.   *Answer: Page A45*

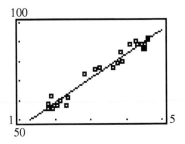

▶ APPLICATIONS OF REGRESSION LINES

After finding the equation of a regression line, you can use the equation to predict y-values over the range of the data *if the correlation between x and y is significant.* For instance, an environmentalist could forecast carbon dioxide emissions on the basis of gross domestic products. To predict y-values, substitute the given x-value into the regression equation, then calculate $\hat{y}$, the predicted y-value.

EXAMPLE 3

▶ Predicting y-Values Using Regression Equations

The regression equation for the gross domestic products (in trillions of dollars) and carbon dioxide emissions (in millions of metric tons) data is

$$\hat{y} = 196.152x + 102.289.$$

Use this equation to predict the *expected* carbon dioxide emissions for the following gross domestic products. (Recall from Section 9.1, Example 7, that x and y have a significant linear correlation.)

1. 1.2 trillion dollars   **2.** 2.0 trillion dollars   **3.** 2.5 trillion dollars

▶ Solution

To predict the expected carbon dioxide emissions, substitute each gross domestic product for x in the regression equation. Then calculate $\hat{y}$.

1. $\hat{y} = 196.152x + 102.289$
$= 196.152(1.2) + 102.289$
≈ 337.671

Interpretation   When the gross domestic product is \$1.2 trillion, the CO_2 emissions are about 337.671 million metric tons.

2. $\hat{y} = 196.152x + 102.289$
$= 196.152(2.0) + 102.289$
$= 494.593$

Interpretation   When the gross domestic product is \$2.0 trillion, the CO_2 emissions are 494.593 million metric tons.

3. $\hat{y} = 196.152x + 102.289$
$= 196.152(2.5) + 102.289$
$= 592.669$

Interpretation   When the gross domestic product is \$2.5 trillion, the CO_2 emissions are 592.669 million metric tons.

Prediction values are meaningful only for x-values in (or close to) the range of the data. The x-values in the original data set range from 0.9 to 4.9. So, it would not be appropriate to use the regression line $\hat{y} = 196.152x + 102.289$ to predict carbon dioxide emissions for gross domestic products such as \$0.2 or \$14.5 trillion dollars.

▶ Try It Yourself 3

The regression equation for the Old Faithful data is $\hat{y} = 12.481x + 33.683$. Use this to predict the time until the next eruption for each of the following eruption durations. (Recall from Section 9.1, Example 6, that x and y have a significant linear correlation.)

1. 2 minutes
2. 3.32 minutes

a. *Substitute* each value of x into the regression equation.
b. *Calculate* $\hat{y}$.
c. *Specify* the predicted time until the next eruption for each eruption duration.

Answer: Page A45

▶ PICTURING THE WORLD

The following scatter plot shows the relationship between the number of farms (in thousands) in a state and the total value of the farms (in billions of dollars).
(Source: U.S. Department of Agriculture and National Agriculture Statistics Service)

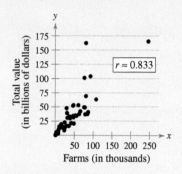

Describe the correlation between these two variables in words. Use the scatter plot to predict the total value of farms in a state that has 150,000 farms. The regression line for this scatter plot is $\hat{y} = 0.714x + 3.367$. Use this equation to make a prediction. (Assume x and y have a significant linear correlation.) How does your algebraic prediction compare with your graphical one?

9.2 EXERCISES

■ BUILDING BASIC SKILLS AND VOCABULARY

1. What is a residual? Explain when a residual is positive, negative, and zero.

2. Two variables have a positive linear correlation. Is the slope of the regression line for the variables positive or negative?

3. Explain how to predict y-values using the equation of a regression line.

4. Given a set of data and a corresponding regression line, describe all values of x that provide meaningful predictions for y.

5. In order to predict y-values using the equation of a regression line, what must be true about the correlation coefficient of the variables?

6. Why is it not appropriate to use a regression line to predict y-values for x-values that are not in (or close to) the range of x-values found in the data?

In Exercises 7–12, match the description in the left column with its symbol(s) in the right column.

7. The y-value of a data point corresponding to x_i

8. The y-value for a point on the regression line corresponding to x_i

9. Slope

10. y-intercept

11. The mean of the y-values

12. The point a regression line always passes through

a. $\hat{y}_i$

b. y_i

c. b

d. $(\bar{x}, \bar{y})$

e. m

f. $\bar{y}$

Graphical Analysis   *In Exercises 13–16, match the regression equation with the appropriate graph. (Note that the x- and y-axes are broken.)*

13. $\hat{y} = -1.04x + 50.3$

14. $\hat{y} = 1.662x + 83.34$

15. $\hat{y} = 0.00114x + 2.53$

16. $\hat{y} = -0.667x + 52.6$

a.

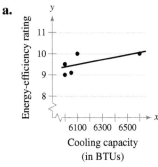

Cooling capacity
(in BTUs)

b.

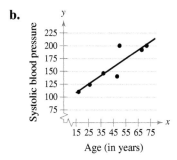

Age (in years)

c.

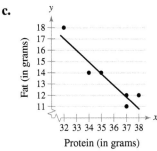

Protein (in grams)

d.

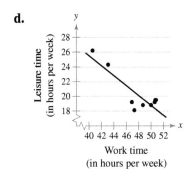

Work time
(in hours per week)

■ USING AND INTERPRETING CONCEPTS

Finding the Equation of a Regression Line   *In Exercises 17–24, find the equation of the regression line for the given data. Then construct a scatter plot of the data and draw the regression line. (Each pair of variables has a significant correlation.) Then use the regression equation to predict the value of y for each of the given x-values, if meaningful. If the x-value is not meaningful to predict the value of y, explain why not. If convenient, use technology to solve the problem.*

17. Atlanta Building Heights   The heights (in feet) and the number of stories of nine notable buildings in Atlanta *(Source: Emporis Corporation)*

Height, x	869	820	771	696	692	676	656	492	486
Stories, y	60	50	50	52	40	47	41	39	26

(a) $x = 800$ feet   (b) $x = 750$ feet
(c) $x = 400$ feet   (d) $x = 625$ feet

18. Square Footages and Home Sale Prices   The square footages and sale prices (in thousands of dollars) of seven homes *(Source: Howard Hanna)*

Square footage, x	1924	1592	2413	2332	1552	1312	1278
Sale price, y	174.9	136.9	275.0	219.9	120.0	99.9	145.0

(a) $x = 1450$ square feet   (b) $x = 2720$ square feet
(c) $x = 2175$ square feet   (d) $x = 1890$ square feet

19. Hours Studying and Test Scores   The number of hours 13 students spent studying for a test and their scores on that test

Hours spent studying, x	0	1	2	4	4	5	5
Test score, y	40	41	51	48	64	69	73

Hours spent studying, x	5	6	6	7	7	8
Test score, y	75	68	93	84	90	95

(a) $x = 3$ hours   (b) $x = 6.5$ hours
(c) $x = 13$ hours   (d) $x = 4.5$ hours

20. Hours Online   The number of hours 12 students spent online during the weekend and the scores of each student who took a test the following Monday

Hours spent online, x	0	1	2	3	3	5
Test score, y	96	85	82	74	95	68

Hours spent online, x	5	5	6	7	7	10
Test score, y	76	84	58	65	75	50

(a) $x = 4$ hours   (b) $x = 8$ hours
(c) $x = 9$ hours   (d) $x = 15$ hours

21. Hot Dogs: Caloric and Sodium Content The caloric contents and the sodium contents (in milligrams) of 10 beef hot dogs *(Source: Consumer Reports)*

Calories, x	150	170	120	120	90
Sodium, y	420	470	350	360	270

Calories, x	180	170	140	90	110
Sodium, y	550	530	460	380	330

(a) $x = 170$ calories   (b) $x = 100$ calories

(c) $x = 140$ calories   (d) $x = 210$ calories

22. High-Fiber Cereals: Caloric and Sugar Content The caloric contents and the sugar contents (in grams) of 11 high-fiber breakfast cereals *(Source: Consumer Reports)*

Calories, x	140	200	160	170	170	190
Sugar, y	6	9	6	9	10	17

Calories, x	190	210	190	170	160
Sugar, y	13	18	19	10	10

(a) $x = 150$ calories   (b) $x = 90$ calories

(c) $x = 175$ calories   (d) $x = 208$ calories

23. Shoe Size and Height The shoe sizes and heights (in inches) of 14 men

Shoe size, x	8.5	9.0	9.0	9.5	10.0	10.0	10.5
Height, y	66.0	68.5	67.5	70.0	70.0	72.0	71.5

Shoe size, x	10.5	11.0	11.0	11.0	12.0	12.0	12.5
Height, y	69.5	71.5	72.0	73.0	73.5	74.0	74.0

(a) $x = $ size 11.5   (b) $x = $ size 8.0

(c) $x = $ size 15.5   (d) $x = $ size 10.0

24. Age and Hours Slept The ages (in years) of 10 infants and the number of hours each slept in a day

Age, x	0.1	0.2	0.4	0.7	0.6	0.9
Hours slept, y	14.9	14.5	13.9	14.1	13.9	13.7

Age, x	0.1	0.2	0.4	0.9
Hours slept, y	14.3	13.9	14.0	14.1

(a) $x = 0.3$ year   (b) $x = 3.9$ years

(c) $x = 0.6$ year   (d) $x = 0.8$ year

Years of experience, x	Annual salary (in thousands), y
0.5	40.2
2	42.9
4	45.1
5	46.7
7	50.2
9	53.6
10	54.0
12.5	58.4
13	61.8
16	63.9
18	67.5
20	64.3
22	60.1
25	59.9

TABLE FOR EXERCISES 25–29

25. Strong positive linear correlation; As the years of experience of the registered nurses increase, their salaries tend to increase.

26. $\hat{y} = 0.998x + 43.214$

See Selected Answers, page A129.

27. No, it is not meaningful to predict a salary for a registered nurse with 28 years of experience because x = 28 is outside the range of the original data.

28. $|r| \approx 0.880 > 0.661$ (the critical value), so the population has a significant correlation.

29. Answers will vary. *Sample answer:* Although it is likely that there is a cause-and-effect relationship between a registered nurse's years of experience and salary, you cannot use significant correlation to claim cause and effect. The relationship between the variables may also be influenced by other factors, such as work performance, level of education, or the number of years with an employer.

30. (a) $\hat{y} = 0.024x + 0.181$

(b) 0.773

(c) See Selected Answers, page A129.

31. (a) $\hat{y} = -0.159x + 5.827$

(b) −0.852

(c) See Odd Answers, page A95.

Registered Nurse Salaries *In Exercises 25–29, use the following information. You work for a salary analyst and gather the data shown in the table. The table shows the years of experience of 14 registered nurses and their annual salaries.* (Adapted from Payscale, Inc.)

25. Correlation Using the scatter plot of the registered nurse salary data shown, what type of correlation, if any, do you think the data have? Explain.

26. Regression Line Find an equation of the regression line for the data. Sketch a scatter plot of the data and draw the regression line.

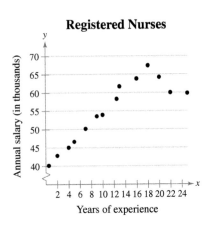

27. Using the Regression Line The analyst used the regression line you found in Exercise 26 to predict the annual salary for a registered nurse with 28 years of experience. Is this a valid prediction? Explain your reasoning.

28. Significant Correlation? The analyst claims that the population has a significant correlation for $\alpha = 0.01$. Verify this claim.

29. Cause and Effect Write a paragraph describing the cause-and-effect relationship between the years of experience and the annual salaries of registered nurses.

SC *In Exercises 30 and 31, use StatCrunch to (a) find the equation of the regression line for the data, (b) find the correlation coefficient r, and (c) construct the scatter plot of the data that also shows the regression line. (Each pair of variables has a significant correlation.)*

30. Hot Chocolates: Caloric and Fat Contents The caloric contents and the fat contents (in grams) of 6- to 8-ounce servings for 10 hot chocolate products (Source: Consumer Reports)

Calories, x	262	140	150	159	120	140	185	150	80	80
Fat, y	6.3	2.0	3.5	3.5	2.5	3.5	6.8	3	3	2.5

31. Wins and Earned Run Averages The number of wins and the earned run averages (mean number of earned runs allowed per nine innings pitched) for eight professional baseball pitchers in the 2009 regular season (Source: ESPN)

Wins, x	19	17	16	15	15	14	12	9
Earned run average, y	2.63	2.79	3.75	3.23	3.47	3.96	4.05	4.12

■ **EXTENDING CONCEPTS**

Interchanging x and y *In Exercises 32 and 33, do the following.*

(a) Find the equation of the regression line for the given data, letting Row 1 represent the x-values and Row 2 the y-values. Sketch a scatter plot of the data and draw the regression line.

(b) *Find the equation of the regression line for the given data, letting Row 2 represent the x-values and Row 1 the y-values. Sketch a scatter plot of the data and draw the regression line.*

(c) *What effect does switching the explanatory and response variables have on the regression line?*

32.

Row 1	16	25	39	45	49	64	70
Row 2	109	122	143	132	199	185	199

33.

Row 1	0	1	2	3	3	5	5	5	6	7
Row 2	96	85	82	74	95	68	76	84	58	65

Residual Plots   *A **residual plot** allows you to assess correlation data and check for possible problems with a regression model. To construct a residual plot, make a scatter plot of $(x, y - \hat{y})$, where $y - \hat{y}$ is the residual of each y-value. If the resulting plot shows any type of pattern, the regression line is not a good representation of the relationship between the two variables. If it does not show a pattern—that is, if the residuals fluctuate about 0—then the regression line is a good representation. Be aware that if a point on the residual plot appears to be outside the pattern of the other points, then it may be an outlier.*

In Exercises 34 and 35, (a) find the equation of the regression line, (b) construct a scatter plot of the data and draw the regression line, (c) construct a residual plot, and (d) determine if there are any patterns in the residual plot and explain what they suggest about the relationship between the variables.

34.

x	8	4	15	7	6	3	12	10	5
y	18	11	29	18	14	8	25	20	12

35.

x	38	34	40	46	43	48	60	55	52
y	24	22	27	32	30	31	27	26	28

Influential Points   *An **influential point** is a point in a data set that can greatly affect the graph of a regression line. An outlier may or may not be an influential point. To determine if a point is influential, find two regression lines: one including all the points in the data set, and the other excluding the possible influential point. If the slope or y-intercept of the regression line shows significant changes, the point can be considered influential. An influential point can be removed from a data set only if there is proper justification.*

In Exercises 36 and 37, (a) construct a scatter plot of the data, (b) identify any possible outliers, and (c) determine if the point is influential. Explain your reasoning.

36.

x	1	3	6	8	12	14
y	4	7	10	9	15	3

37.

x	5	6	9	10	14	17	19	44
y	32	33	28	26	25	23	23	8

Number of hours, x	Number of bacteria, y
1	165
2	280
3	468
4	780
5	1310
6	1920
7	4900

TABLE FOR EXERCISES 39–42

x	y
1	695
2	410
3	256
4	110
5	80
6	75
7	68
8	74

TABLE FOR EXERCISES 43–46

38. Chapter Opener Consider the data from the Chapter Opener on page 483 on the salaries and average attendances at home games for the teams in Major League Baseball. Is the data point (201.4, 45,364) an outlier? If so, is it influential? Explain.

Transformations to Achieve Linearity *When a linear model is not appropriate for representing data, other models can be used. In some cases, the values of x or y must be transformed to find an appropriate model. In a **logarithmic transformation,** the logarithms of the variables are used instead of the original variables when creating a scatter plot and calculating the regression line.*

In Exercises 39–42, use the data shown in the table, which shows the number of bacteria present after a certain number of hours.

39. Find the equation of the regression line for the data. Then construct a scatter plot of (x, y) and sketch the regression line with it.

40. Replace each y-value in the table with its logarithm, $\log y$. Find the equation of the regression line for the transformed data. Then construct a scatter plot of $(x, \log y)$ and sketch the regression line with it. What do you notice?

41. An **exponential equation** is a nonlinear regression equation of the form $y = ab^x$. Use a technology tool to find and graph the exponential equation for the original data. Include a scatter plot in your graph. Note that you can also find this model by solving the equation $\log y = mx + b$ from Exercise 40 for y.

42. Compare your results in Exercise 41 with the equation of the regression line and its graph in Exercise 39. Which equation is a better model for the data? Explain.

In Exercises 43–46, use the data shown in the table.

43. Find the equation of the regression line for the data. Then construct a scatter plot of (x, y) and sketch the regression line with it.

44. Replace each x-value and y-value in the table with its logarithm. Find the equation of the regression line for the transformed data. Then construct a scatter plot of $(\log x, \log y)$ and sketch the regression line with it. What do you notice?

45. A **power equation** is a nonlinear regression equation of the form $y = ax^b$. Use a technology tool to find and graph the power equation for the original data. Include a scatter plot in your graph. Note that you can also find this model by solving the equation $\log y = m(\log x) + b$ from Exercise 44 for y.

46. Compare your results in Exercise 45 with the equation of the regression line and its graph in Exercise 43. Which equation is a better model for the data? Explain.

Logarithmic Equation *In Exercises 47–50, use the following information and a technology tool. The **logarithmic equation** is a nonlinear regression equation of the form $y = a + b \ln x$.*

47. Find and graph the logarithmic equation for the data given in Exercise 23.

48. Find and graph the logarithmic equation for the data given in Exercise 24.

49. Compare your results in Exercise 47 with the equation of the regression line and its graph. Which equation is a better model for the data? Explain.

50. Compare your results in Exercise 48 with the equation of the regression line and its graph. Which equation is a better model for the data? Explain.

ACTIVITY 9.2 Regression by Eye

APPLET

The *regression by eye* applet allows you to interactively estimate the regression line for a data set. When the applet loads, a data set consisting of 20 points is displayed. Points on the plot can be added to the plot by clicking the mouse. Points on the plot can be removed by clicking on the point and then dragging the point into the trash can. All of the points on the plot can be removed by simply clicking inside the trash can. You can move the green line on the plot by clicking and dragging the endpoints. You should try to move the line in order to minimize the sum of the squares of the residuals, also known as the sum of square error (SSE). Note that the regression line minimizes SSE. The SSE for the green line and for the regression line are given below the plot. The equations of each line are given above the plot. Click SHOW REGRESSION LINE! to see the regression line in the plot. Click NEW DATA to generate a new data set.

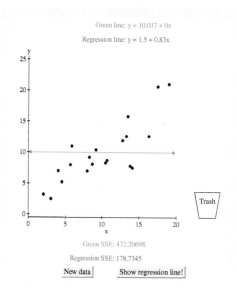

Green line: $y = 10.017 + 0x$

Regression line: $y = 1.5 + 0.83x$

Green SSE: 472.20698

Regression SSE: 178.7345

New data | Show regression line!

■ Explore

Step 1 Move the endpoints of the green line to try to approximate the regression line.

Step 2 Click SHOW REGRESSION LINE!.

■ Draw Conclusions

APPLET

1. Click NEW DATA to generate a new data set. Try to move the green line to where the regression line should be. Then click SHOW REGRESSION LINE!. Repeat this five times. Describe how you moved each green line.

2. On a blank plot, place 10 points so that they have a strong positive correlation. Record the equation of the regression line. Then, add a point in the upper left corner of the plot and record the equation of the regression line. How does the regression line change?

3. Remove the point from the upper-left corner of the plot. Add 10 more points so that there is still a strong positive correlation. Record the equation of the regression line. Add a point in the upper-left corner of the plot and record the equation of the regression line. How does the regression line change?

4. Use the results of Exercises 2 and 3 to describe what happens to the slope of the regression line when an outlier is added as the sample size increases.

Correlation of Body Measurements

In a study published in *Medicine and Science in Sports and Exercise* (volume 17, no. 2, page 189) the measurements of 252 men (ages 22–81) are given. Of the 14 measurements taken of each man, some have significant correlations and others don't. For instance, the scatter plot at the right shows that the hip and abdomen circumferences of the men have a strong linear correlation ($r = 0.85$). The partial table shown here lists only the first nine rows of the data.

Hip and Abdomen Circumferences

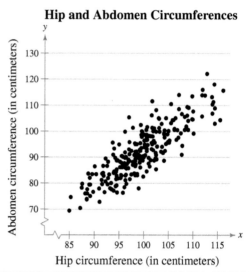

Hip circumference (in centimeters)

Age (yr)	Weight (lb)	Height (in.)	Neck (cm)	Chest (cm)	Abdom. (cm)	Hip (cm)	Thigh (cm)	Knee (cm)	Ankle (cm)	Bicep (cm)	Forearm (cm)	Wrist (cm)	Body fat %
22	173.25	72.25	38.5	93.6	83.0	98.7	58.7	37.3	23.4	30.5	28.9	18.2	6.1
22	154.00	66.25	34.0	95.8	87.9	99.2	59.6	38.9	24.0	28.8	25.2	16.6	25.3
23	154.25	67.75	36.2	93.1	85.2	94.5	59.0	37.3	21.9	32.0	27.4	17.1	12.3
23	198.25	73.50	42.1	99.6	88.6	104.1	63.1	41.7	25.0	35.6	30.0	19.2	11.7
23	159.75	72.25	35.5	92.1	77.1	93.9	56.1	36.1	22.7	30.5	27.2	18.2	9.4
23	188.15	77.50	38.0	96.6	85.3	102.5	59.1	37.6	23.2	31.8	29.7	18.3	10.3
24	184.25	71.25	34.4	97.3	100.0	101.9	63.2	42.2	24.0	32.2	27.7	17.7	28.7
24	210.25	74.75	39.0	104.5	94.4	107.8	66.0	42.0	25.6	35.7	30.6	18.8	20.9
24	156.00	70.75	35.7	92.7	81.9	95.3	56.4	36.5	22.0	33.5	28.3	17.3	14.2

Source: "Generalized Body Composition Prediction Equation for Men Using Simple Measurement Techniques" by K.W. Penrose et al. (1985). MEDICINE AND SCIENCE IN SPORTS AND EXERCISE, vol. 17, no.2, p. 189.

■ EXERCISES

1. Using your intuition, classify the following (x, y) pairs as having a weak correlation ($0 < r < 0.5$), a moderate correlation ($0.5 < r < 0.8$), or a strong correlation ($0.8 < r < 1.0$).

   (a) (weight, neck)   (b) (weight, height)
   (c) (age, body fat)   (d) (chest, hip)
   (e) (age, wrist)   (f) (ankle, wrist)
   (g) (forearm, height) (h) (bicep, forearm)
   (i) (weight, body fat) (j) (knee, thigh)
   (k) (hip, abdomen)   (l) (abdomen, hip)

2. Now, use a technology tool to find the correlation coefficient for each pair in Exercise 1. Compare your results with those obtained by intuition.

3. Use a technology tool to find the regression line for each pair in Exercise 1 that has a strong correlation.

4. Use the results of Exercise 3 to predict the following.

   (a) The neck circumference of a man whose weight is 180 pounds

   (b) The abdomen circumference of a man whose hip circumference is 100 centimeters

5. Are there pairs of measurements that have stronger correlation coefficients than 0.85? Use a technology tool and intuition to reach a conclusion.

9.3   Measures of Regression and Prediction Intervals

WHAT YOU SHOULD LEARN

▸ How to interpret the three types of variation about a regression line

▸ How to find and interpret the coefficient of determination

▸ How to find and interpret the standard error of estimate for a regression line

▸ How to construct and interpret a prediction interval for *y*

Variation about a Regression Line ▸ The Coefficient of Determination ▸ The Standard Error of Estimate ▸ Prediction Intervals

▸ VARIATION ABOUT A REGRESSION LINE

In this section, you will study two measures used in correlation and regression studies—the coefficient of determination and the standard error of estimate. You will also learn how to construct a prediction interval for *y* using a regression line and a given value of *x*. Before studying these concepts, you need to understand the three types of variation about a regression line.

To find the total variation, the explained variation, and the unexplained variation about a regression line, you must first calculate the **total deviation,** the **explained deviation,** and the **unexplained deviation** for each ordered pair (x_i, y_i) in a data set. These deviations are shown in the graph.

Total deviation $= y_i - \bar{y}$

Explained deviation $= \hat{y}_i - \bar{y}$

Unexplained deviation $= y_i - \hat{y}_i$

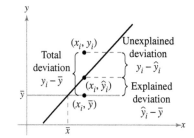

After calculating the deviations for each data point (x_i, y_i), you can find the *total variation*, the *explained variation*, and the *unexplained variation*.

STUDY TIP

Consider the gross domestic products and carbon dioxide emissions data used throughout this chapter with a regression line of

$$\hat{y} = 196.152x + 102.289.$$

Using the data point (2.7, 571.8), you can find the total, explained, and unexplained deviations as follows.

Total deviation:

$y_i - \bar{y} = 571.8 - 555.4$

$\quad\quad = 16.4$

Explained deviation:

$\hat{y}_i - \bar{y} = 631.8994 - 555.4$

$\quad\quad = 76.4994$

Unexplained deviation:

$y_i - \hat{y}_i = 571.8 - 631.8994$

$\quad\quad = -60.0994$

DEFINITION

The **total variation** about a regression line is the sum of the squares of the differences between the *y*-value of each ordered pair and the mean of *y*.

Total variation $= \Sigma(y_i - \bar{y})^2$

The **explained variation** is the sum of the squares of the differences between each predicted *y*-value and the mean of *y*.

Explained variation $= \Sigma(\hat{y}_i - \bar{y})^2$

The **unexplained variation** is the sum of the squares of the differences between the *y*-value of each ordered pair and each corresponding predicted *y*-value.

Unexplained variation $= \Sigma(y_i - \hat{y}_i)^2$

The sum of the explained and unexplained variations is equal to the total variation.

Total variation = Explained variation + Unexplained variation

As its name implies, the *explained variation* can be explained by the relationship between *x* and *y*. The *unexplained variation* cannot be explained by the relationship between *x* and *y* and is due to chance or other variables.

▸ THE COEFFICIENT OF DETERMINATION

You already know how to calculate the correlation coefficient r. The square of this coefficient is called the *coefficient of determination*. It can be shown that the coefficient of determination is equal to the ratio of the explained variation to the total variation.

DEFINITION

The **coefficient of determination** r^2 is the ratio of the explained variation to the total variation. That is,

$$r^2 = \frac{\text{Explained variation}}{\text{Total variation}}.$$

It is important that you interpret the coefficient of determination correctly. For instance, if the correlation coefficient is $r = 0.90$, then the coefficient of determination is

$$r^2 = 0.90^2$$
$$= 0.81.$$

This means that 81% of the variation in y can be explained by the relationship between x and y. The remaining 19% of the variation is unexplained and is due to other factors or to sampling error.

EXAMPLE 1   Report 41

▸ Finding the Coefficient of Determination

The correlation coefficient for the gross domestic products and carbon dioxide emissions data as calculated in Example 4 in Section 9.1 is $r \approx 0.882$. Find the coefficient of determination. What does this tell you about the explained variation of the data about the regression line? About the unexplained variation?

▸ Solution

The coefficient of determination is

$$r^2 \approx (0.882)^2$$
$$\approx 0.778.$$

Interpretation   About 77.8% of the variation in the carbon dioxide emissions can be explained by the variation in the gross domestic products. About 22.2% of the variation is unexplained and is due to chance or other variables.

▸ Try It Yourself 1

The correlation coefficient for the Old Faithful data as calculated in Example 5 in Section 9.1 is $r \approx 0.979$. Find the coefficient of determination. What does this tell you about the explained variation of the data about the regression line? About the unexplained variation?

a. Identify the *correlation coefficient r*.
b. Calculate the *coefficient of determination* r^2.
c. What percent of the variation in the times is *explained*? What percent is *unexplained*?

Answer: Page A45

PICTURING THE WORLD

Janette Benson (Psychology Department, University of Denver) performed a study relating the age at which infants crawl (in weeks after birth) with the average monthly temperature six months after birth. Her results are based on a sample of 414 infants. Janette Benson believes that the reason for the correlation of temperature and crawling age is that parents tend to bundle infants in more restrictive clothing and blankets during cold months. This bundling doesn't allow the infant as much opportunity to move and experiment with crawling.

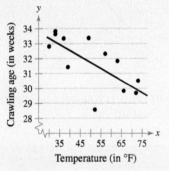

The correlation coefficient is $r = -0.70$. What percent of the variation in the data can be explained? What percent is due to chance, sampling error, or other factors?

▶ THE STANDARD ERROR OF ESTIMATE

When a $\hat{y}$-value is predicted from an x-value, the prediction is a point estimate. You can construct an interval estimate for $\hat{y}$, but first you need to calculate the *standard error of estimate*.

DEFINITION

The **standard error of estimate** s_e is the standard deviation of the observed y_i-values about the predicted $\hat{y}$-value for a given x_i-value. It is given by

$$s_e = \sqrt{\frac{\sum(y_i - \hat{y}_i)^2}{n - 2}}$$

where n is the number of ordered pairs in the data set.

From this formula, you can see that the standard error of estimate is the square root of the unexplained variation divided by $n - 2$. So, the closer the observed y-values are to the predicted y-values, the smaller the standard error of estimate will be.

GUIDELINES

Finding the Standard Error of Estimate s_e

IN WORDS	IN SYMBOLS
1. Make a table that includes the column headings shown at the right.	$x_i, y_i, \hat{y}_i, (y_i - \hat{y}_i),$ $(y_i - \hat{y}_i)^2$
2. Use the regression equation to calculate the predicted y-values.	$\hat{y}_i = mx_i + b$
3. Calculate the sum of the squares of the differences between each observed y-value and the corresponding predicted y-value.	$\sum(y_i - \hat{y}_i)^2$
4. Find the standard error of estimate.	$s_e = \sqrt{\dfrac{\sum(y_i - \hat{y}_i)^2}{n - 2}}$

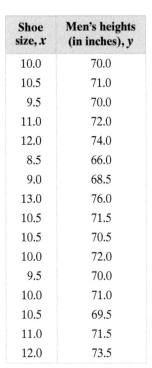

Shoe size, x	Men's heights (in inches), y
10.0	70.0
10.5	71.0
9.5	70.0
11.0	72.0
12.0	74.0
8.5	66.0
9.0	68.5
13.0	76.0
10.5	71.5
10.5	70.5
10.0	72.0
9.5	70.0
10.0	71.0
10.5	69.5
11.0	71.5
12.0	73.5

You can also find the standard error of estimate using the following formula.

$$s_e = \sqrt{\frac{\sum y^2 - b\sum y - m\sum xy}{n - 2}}$$

This formula is easy to use if you have already calculated the slope m, the y-intercept b, and several of the sums. For instance, the regression line for the data set given at the left is $\hat{y} = 1.84247x + 51.77413$, and the values of the sums are $\sum y^2 = 80{,}877.5$, $\sum y = 1137$, and $\sum xy = 11{,}940.25$. When the alternative formula is used, the standard error of estimate is

$$s_e = \sqrt{\frac{\sum y^2 - b\sum y - m\sum xy}{n - 2}}$$

$$= \sqrt{\frac{80{,}877.5 - 51.77413(1137) - 1.84247(11{,}940.25)}{16 - 2}}$$

$$\approx 0.877.$$

EXAMPLE 2   SC Report 42

▶ Finding the Standard Error of Estimate

The regression equation for the gross domestic products and carbon dioxide emissions data as calculated in Example 1 in Section 9.2 is

$$\hat{y} = 196.152x + 102.289.$$

Find the standard error of estimate.

▶ Solution

Use a table to calculate the sum of the squared differences of each observed y-value and the corresponding predicted y-value.

x_i	y_i	$\hat{y}_i$	$y_i - \hat{y}_i$	$(y_i - \hat{y}_i)^2$
1.6	428.2	416.1322	12.0678	145.63179684
3.6	828.8	808.4362	20.3638	414.68435044
4.9	1214.2	1063.4338	150.7662	22,730.44706244
1.1	444.6	318.0562	126.5438	16,013.33331844
0.9	264.0	278.8258	−14.8258	219.80434564
2.9	415.3	671.1298	−255.8298	65,448.88656804
2.7	571.8	631.8994	−60.0994	3611.93788036
2.3	454.9	553.4386	−98.5386	9709.85568996
1.6	358.7	416.1322	−57.4322	3298.45759684
1.5	573.5	396.517	176.983	31,322.982289
				$\Sigma = 152,916.020898$

Unexplained variation

When $n = 10$ and $\Sigma (y_i - \hat{y}_i)^2 = 152,916.020898$ are used, the standard error of estimate is

$$s_e = \sqrt{\frac{\Sigma (y_i - \hat{y}_i)^2}{n - 2}}$$

$$= \sqrt{\frac{152,916.020898}{10 - 2}}$$

$$\approx 138.255.$$

Interpretation The standard error of estimate of the carbon dioxide emissions for a specific gross domestic product is about 138.255 million metric tons.

▶ Try It Yourself 2

A researcher collects the data shown at the left and concludes that there is a significant relationship between the amount of radio advertising time (in minutes per week) and the weekly sales of a product (in hundreds of dollars). Find the standard error of estimate. Use the regression equation

$$\hat{y} = 1.405x + 7.311.$$

a. *Use a table* to calculate the sum of the squared differences of each observed y-value and the corresponding predicted y-value.
b. Identify the *number n of ordered pairs* in the data set.
c. *Calculate* s_e.
d. *Interpret* the results.

Answer: Page A45

Radio ad time	Weekly sales
15	26
20	32
20	38
30	56
40	54
45	78
50	80
60	88

▶ **PREDICTION INTERVALS**

Two variables have a **bivariate normal distribution** if for any fixed values of x the corresponding values of y are normally distributed, and for any fixed values of y the corresponding values of x are normally distributed.

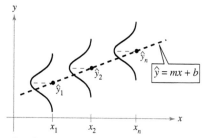

Bivariate Normal Distribution

Because regression equations are determined using sample data and because x and y are assumed to have a bivariate normal distribution, you can construct a *prediction interval* for the true value of y. To construct the prediction interval, use a t-distribution with $n - 2$ degrees of freedom.

DEFINITION

Given a linear regression equation $\hat{y} = mx + b$ and x_0, a specific value of x, a *c*-**prediction interval** for y is

$$\hat{y} - E < y < \hat{y} + E$$

where

$$E = t_c s_e \sqrt{1 + \frac{1}{n} + \frac{n(x_0 - \overline{x})^2}{n\Sigma x^2 - (\Sigma x)^2}}.$$

The point estimate is $\hat{y}$ and the margin of error is E. The probability that the prediction interval contains y is c.

GUIDELINES

Construct a Prediction Interval for y for a Specific Value of x

IN WORDS	IN SYMBOLS
1. Identify the number of ordered pairs in the data set n and the degrees of freedom.	d.f. $= n - 2$
2. Use the regression equation and the given x-value to find the point estimate $\hat{y}$.	$\hat{y}_i = mx_i + b$
3. Find the critical value t_c that corresponds to the given level of confidence c.	Use Table 5 in Appendix B.
4. Find the standard error of estimate s_e.	$s_e = \sqrt{\dfrac{\Sigma (y_i - \hat{y}_i)^2}{n - 2}}$
5. Find the margin of error E.	$E = t_c s_e \sqrt{1 + \dfrac{1}{n} + \dfrac{n(x_0 - \overline{x})^2}{n\Sigma x^2 - (\Sigma x)^2}}$
6. Find the left and right endpoints and form the prediction interval.	Left endpoint: $\hat{y} - E$ Right endpoint: $\hat{y} + E$ Interval: $\hat{y} - E < y < \hat{y} + E$

STUDY TIP

The formulas for s_e and E use the quantities $\Sigma(y_i - \hat{y}_i)^2$, $(\Sigma x)^2$, and Σx^2. Use a table to calculate these quantities.

EXAMPLE 3   **SC** Report 43

▶ **Constructing a Prediction Interval**

Using the results of Example 2, construct a 95% prediction interval for the carbon dioxide emissions when the gross domestic product is $3.5 trillion. What can you conclude?

▶ **Solution**

Because $n = 10$, there are

$$10 - 2 = 8$$

degrees of freedom. Using the regression equation

$$\hat{y} = 196.152x + 102.289$$

and

$$x = 3.5,$$

the point estimate is

$$\hat{y} = 196.152x + 102.289$$
$$= 196.152(3.5) + 102.289$$
$$= 788.821.$$

From Table 5, the critical value is $t_c = 2.306$, and from Example 2, $s_e \approx 138.255$. Using these values, the margin of error is

$$E = t_c s_e \sqrt{1 + \frac{1}{n} + \frac{n(x_0 - \bar{x})^2}{n(\Sigma x^2) - (\Sigma x)^2}}$$

$$\approx (2.306)(138.255)\sqrt{1 + \frac{1}{10} + \frac{10(3.5 - 2.31)^2}{10(67.35) - (23.1)^2}}$$

$$\approx 349.424.$$

Using $\hat{y} = 788.821$ and $E = 349.424$, the prediction interval is

Left Endpoint	Right Endpoint
$788.821 - 349.424 = 439.397$	$788.821 + 349.424 = 1138.245$

$$439.397 < y < 1138.245.$$

Interpretation   You can be 95% confident that when the gross domestic product is $3.5 trillion, the carbon dioxide emissions will be between 439.397 and 1138.245 million metric tons.

▶ **Try It Yourself 3**

Construct a 95% prediction interval for the carbon dioxide emissions when the gross domestic product is $4 trillion. What can you conclude?

a. *Specify* n, d.f., t_c, s_e.
b. *Calculate* $\hat{y}$ when $x = 4$.
c. Calculate the *margin of error E*.
d. Construct the *prediction interval*.
e. *Interpret* the results.

Answer: Page A45

INSIGHT

The greater the difference between x and $\bar{x}$, the wider the prediction interval is. For instance, in Example 3 the 95% prediction intervals for $0.9 < x < 4.9$ are shown below. Notice how the bands curve away from the regression line as x gets close to 0.9 and 4.9.

9.3 EXERCISES

■ BUILDING BASIC SKILLS AND VOCABULARY

Graphical Analysis *In Exercises 1–3, use the graph to answer the question.*

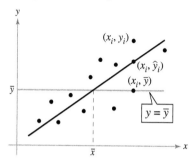

1. Describe the total variation about a regression line in words and in symbols.

2. Describe the explained variation about a regression line in words and in symbols.

3. Describe the unexplained variation about a regression line in words and in symbols.

4. The coefficient of determination r^2 is the ratio of which two types of variations? What does r^2 measure? What does $1 - r^2$ measure?

5. What is the coefficient of determination for two variables that have perfect positive linear correlation or perfect negative linear correlation? Interpret your answer.

6. Two variables have a bivariate normal distribution. Explain what this means.

In Exercises 7–10, use the value of the linear correlation coefficient to calculate the coefficient of determination. What does this tell you about the explained variation of the data about the regression line? About the unexplained variation?

7. $r = 0.465$

8. $r = -0.328$

9. $r = -0.957$

10. $r = 0.881$

■ USING AND INTERPRETING CONCEPTS

Finding Types of Variation and the Coefficient of Determination
In Exercises 11–18, use the data to find (a) the coefficient of determination and interpret the result, and (b) the standard error of estimate s_e and interpret the result.

11. Stock Offerings The number of initial public offerings of stock issued in a recent 12-year period and the total proceeds of these offerings (in millions of U.S. dollars) are shown in the table. The equation of the regression line is $\hat{y} = 104.982x + 14,128.671$. *(Source: University of Florida)*

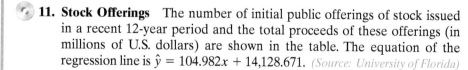

Number of issues, x	318	486	382	79	70	67
Proceeds, y	34,614	64,927	65,088	34,241	22,136	10,068

Number of issues, x	184	168	162	162	21	43
Proceeds, y	32,269	28,593	30,648	35,762	22,762	13,307

12. Crude Oil The table shows the amounts of crude oil (in thousands of barrels per day) produced by the United States and the amounts of crude oil (in thousands of barrels per day) imported by the United States for seven years. The equation of the regression line is $\hat{y} = -2.735x + 27{,}657.823$. (*Source: Energy Information Administration*)

Crude oil produced by U.S., x	5801	5746	5681	5419	5178	5102	5064
Crude oil imported by U.S., y	11,871	11,530	12,264	13,145	13,714	13,707	13,468

13. Retail Space and Sales The table shows the total square footage (in billions) of retailing space at shopping centers and their sales (in billions) of U.S. dollars) for 11 years. The equation of the regression line is $\hat{y} = 549.448x - 1881.694$. (*Adapted from International Council of Shopping Centers*)

Total square footage, x	5.0	5.1	5.2	5.3	5.5	5.6
Sales, y	893.8	933.9	980.0	1032.4	1105.3	1181.1

Total square footage, x	5.7	5.8	5.9	6.0	6.1
Sales, y	1221.7	1277.2	1339.2	1432.6	1530.4

14. Work and Leisure Time The median number of work hours per week and the median number of leisure hours per week for people in the United States for 10 recent years are shown in the table. The equation of the regression line is $\hat{y} = -0.646x + 50.734$. (*Source: Louis Harris & Associates*)

Median number of work hours per week, x	40.6	43.1	46.9	47.3	46.8
Median number of leisure hours per week, y	26.2	24.3	19.2	18.1	16.6

Median number of work hours per week, x	48.7	50.0	50.7	50.6	50.8
Median number of leisure hours per week, y	18.8	18.8	19.5	19.2	19.5

15. State and Federal Government Wages The table shows the average weekly wages for state government employees and federal government employees for six years. The equation of the regression line is $\hat{y} = 1.900x - 411.976$. (*Source: U.S. Bureau of Labor Statistics*)

Average weekly wages (state), x	754	770	791	812	844	883
Average weekly wages (federal), y	1001	1043	1111	1151	1198	1248

16. **Voter Turnout** The U.S. voting age population (in millions) and the turnout of the voting age population (in millions) for federal elections for eight nonpresidential election years are shown in the table. The equation of the regression line is $\hat{y} = 0.333x + 7.580$. *(Adapted from Federal Election Commission)*

Voting age population, x	158.4	169.9	178.6	185.8
Turnout in federal elections, y	58.9	67.6	65.0	67.9

Voting age population, x	193.7	200.9	215.5	220.6
Turnout in federal elections, y	75.1	73.1	79.8	80.6

17. **Taxes** The table shows the gross collections (in billions of dollars) of individual income taxes and corporate income taxes by the U.S. Internal Revenue Service for seven years. The equation of the regression line is $\hat{y} = 0.415x - 186.626$. *(Source: Internal Revenue Service)*

Individual income taxes, x	1038	987	990	1108	1236	1366	1426
Corporate income taxes, y	211	194	231	307	381	396	354

18. **Fund Assets** The table shows the total assets (in billions of U.S. dollars) of individual retirement accounts (IRAs) and federal pension plans for nine years. The equation of the regression line is $\hat{y} = 0.174x + 432.225$. *(Source: Investment Company Institute)*

IRAs, x	2629	2619	2533	2993	3299
Federal pension plans, y	797	860	894	958	1023

IRAs, x	3652	4220	4736	3572
Federal pension plans, y	1072	1141	1197	1221

Constructing and Interpreting Prediction Intervals *In Exercises 19–26, construct the indicated prediction interval and interpret the results.*

19. **Proceeds** Construct a 95% prediction interval for the proceeds from initial public offerings in Exercise 11 when the number of issues is 450.

20. **Crude Oil** Construct a 95% prediction interval for the amount of crude oil imported by the United States in Exercise 12 when the amount of crude oil (in thousands of barrels per day) produced by the United States is 5500.

21. **Retail Sales** Using the results of Exercise 13, construct a 90% prediction interval for shopping center sales when the total square footage of shopping centers is 5.75 billion.

22. **Leisure Hours** Using the results of Exercise 14, construct a 90% prediction interval for the median number of leisure hours per week when the median number of work hours per week is 45.1.

23. **Federal Government Wages** When the average weekly wages of state government employees is $800, find a 99% prediction interval for the average weekly wages of federal government employees. Use the results of Exercise 15.

24. **Predicting Voter Turnout**   When the voting age population is 210 million, construct a 99% prediction interval for the voter turnout in federal elections. Use the results of Exercise 16.

25. **Taxes**   The U.S. Internal Revenue Service collects $1250 billion in individual income taxes for a given year. Construct a prediction interval for the corporate income taxes collected by the U.S. Internal Revenue Service. Use the results of Exercise 17 and $c = 0.95$.

26. **Total Assets**   The total assets in IRAs is $3800 billion. Construct a prediction interval for the total assets in federal pension plans. Use the results of Exercise 18 and $c = 0.90$.

Old Vehicles   *In Exercises 27–33, use the information shown at the left.*

27. **Scatter Plot**   Construct a scatter plot of the data. Show $\bar{y}$ and $\bar{x}$ on the graph.

28. **Regression Line**   Find and graph the regression line.

29. **Deviation**   Calculate the explained deviation, the unexplained deviation, and the total deviation for each data point.

30. **Variation**   Find (a) the explained variation, (b) the unexplained variation, and (c) the total variation.

31. **Coefficient of Determination**   Find the coefficient of determination. What can you conclude?

32. **Error of Estimate**   Find the standard error of estimate s_e and interpret the results.

33. **Prediction Interval**   Construct and interpret a 95% prediction interval for the median age of trucks in use when the median age of cars in use is 7.0 years.

34. **Correlation Coefficient and Slope**   Recall the formula for the correlation coefficient r and the formula for the slope m of a regression line. Given a set of data, why must the slope m of the data's regression line always have the same sign as the data's correlation coefficient r?

SC   *In Exercises 35 and 36, use StatCrunch and the given data to (a) find the coefficient of determination, (b) find the standard error of estimate s_e, and (c) construct a 95% prediction interval for y using the given value of x.*

35. **Trees**   The table shows the heights (in feet) and trunk diameters (in inches) of eight trees. The equation of the regression line is $\hat{y} = 0.479x - 24.086$.

$x = 80$ feet

Height, x	70	72	75	76	85	78	77	82
Trunk diameter, y	8.3	10.5	11.0	11.4	14.9	14.0	16.3	15.8

36. **Motor Vehicles**   The table shows the number of motor vehicle registrations (in millions) and the number of motor vehicle accidents (in millions) in the United States for six years. The equation of the regression line is $\hat{y} = -0.314x + 87.116$.

$x = 235$ million registrations

Registrations, x	229.6	231.4	237.2	241.2	244.2	247.3
Accidents, y	18.3	11.8	10.9	10.7	10.4	10.6

(Adapted from U.S. Federal Highway Administration and National Safety Council)

Keeping cars longer
The median age of vehicles on U.S. roads for eight different years:

Median age in years	
Cars, x	Trucks, y
9.4	7.6
9.2	6.9
8.9	6.6
8.4	6.8
8.3	6.9
6.5	6.5
6.0	6.3
4.9	5.9

(Source: Polk Co.)

FIGURE FOR EXERCISES 27–33

■ **EXTENDING CONCEPTS**

Hypothesis Testing for Slope   *In Exercises 37 and 38, use the following information.*

When testing the slope M of the regression line for the population, you usually test that the slope is 0, or $H_0: M = 0$. A slope of 0 indicates that there is no linear relationship between x and y. To perform the t-test for the slope M, use the standardized test statistic

$$t = \frac{m}{s_e} \sqrt{\Sigma x^2 - \frac{(\Sigma x)^2}{n}}$$

with $n - 2$ degrees of freedom. Then, using the critical values found in Table 5 in Appendix B, make a decision whether to reject or fail to reject the null hypothesis. You can also use the LinRegTTest feature on a TI-83/84 Plus to calculate the standardized test statistic as well as the corresponding P-value. If $P \leq \alpha$, then reject the null hypothesis. If $P > \alpha$, then do not reject H_0.

37. The following table shows the weights (in pounds) and the number of hours slept in a day by a random sample of infants. Test the claim that $M \neq 0$. Use $\alpha = 0.01$. Then interpret the results in the context of the problem. If convenient, use technology to solve the problem.

Weight, x	8.1	10.2	9.9	7.2	6.9	11.2	11	15
Hours slept, y	14.8	14.6	14.1	14.2	13.8	13.2	13.9	12.5

38. The following table shows the ages (in years) and salaries (in thousands of dollars) of a random sample of engineers at a company. Test the claim that $M \neq 0$. Use $\alpha = 0.05$. Then interpret the results in the context of the problem. If convenient, use technology to solve the problem.

Age, x	25	34	29	30	42	38	49	52	35	40
Salary, y	57.5	61.2	59.9	58.7	87.5	67.4	89.2	85.3	69.5	75.1

Confidence Intervals for *y*-Intercept and Slope   *You can construct confidence intervals for the y-intercept B and slope M of the regression line $y = Mx + B$ for the population by using the following inequalities.*

***y*-intercept B:** $b - E < B < b + E$

$$\text{where } E = t_c s_e \sqrt{\frac{1}{n} + \frac{\overline{x}^2}{\Sigma x^2 - \frac{(\Sigma x)^2}{n}}} \quad \text{and}$$

slope M: $m - E < M < m + E$

$$\text{where } E = \frac{t_c s_e}{\sqrt{\Sigma x^2 - \frac{(\Sigma x)^2}{n}}}$$

The values of m and b are obtained from the sample data, and the critical value t_c is found using Table 5 in Appendix B with $n - 2$ degrees of freedom.

In Exercises 39 and 40, construct the indicated confidence interval for B and M using the gross domestic products and carbon dioxide emissions data found in Example 2.

39. 95% confidence interval   40. 99% confidence interval

Statistics in the Real World

USES AND ABUSES

Uses

Correlation and Regression   Correlation and regression analysis can be used to determine whether there is a significant relationship between two variables. If there is, you can use one of the variables to predict the value of the other variable. For instance, educators have used correlation and regression analysis to determine that there is a significant correlation between a student's SAT score and the grade point average from a student's freshman year at college. Consequently, many colleges and universities use SAT scores of high school applicants as a predictor of the applicant's initial success at college.

Abuses

Confusing Correlation and Causation   The most common abuse of correlation in studies is to confuse the concepts of correlation with those of causation (see page 494). Good SAT scores do not cause good college grades. Rather, there are other variables, such as good study habits and motivation, that contribute to both. When a strong correlation is found between two variables, look for other variables that are correlated with both.

Considering Only Linear Correlation   The correlation studied in this chapter is linear correlation. When the correlation coefficient is close to 1 or close to -1, the data points can be modeled by a straight line. It is possible that a correlation coefficient is close to 0 but there is still a strong correlation of a different type. Consider the data listed in the table at the left. The value of the correlation coefficient is 0; however, the data are perfectly correlated with the equation $x^2 + y^2 = 1$, as shown in the graph.

x	1	0	-1	0
y	0	1	0	-1

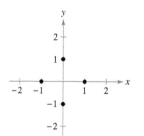

Ethics

When data are collected, all of the data should be used when calculating statistics. In this chapter, you learned that before finding the equation of a regression line, it is helpful to construct a scatter plot of the data to check for outliers, gaps, and clusters in the data. Researchers cannot use only those data points that fit their hypotheses or those that show a significant correlation. Although eliminating outliers may help a data set coincide with predicted patterns or fit a regression line, it is unethical to amend data in such a way. An outlier or any other point that influences a regression model can be removed only if it is properly justified.

In most cases, the best and sometimes safest approach for presenting statistical measurements is with and without an outlier being included. By doing this, the decision as to whether or not to recognize the outlier is left to the reader.

■ EXERCISES

1. ***Confusing Correlation and Causation***   Find an example of an article that confuses correlation and causation. Discuss other variables that could contribute to the relationship between the variables.

2. ***Considering Only Linear Correlation***   Find an example of two real-life variables that have a nonlinear correlation.

9 CHAPTER SUMMARY

What did you **learn?**	EXAMPLE(S)	REVIEW EXERCISES
Section 9.1		
■ How to construct a scatter plot	*1–3*	*1–4*
■ How to find a correlation coefficient	*4, 5*	*1–4*
$$r = \frac{n\Sigma xy - (\Sigma x)(\Sigma y)}{\sqrt{n\Sigma x^2 - (\Sigma x)^2}\sqrt{n\Sigma y^2 - (\Sigma y)^2}}$$		
■ How to perform a hypothesis test for a population correlation coefficient ρ	*7*	*5–10*
$$t = \frac{r}{\sqrt{\dfrac{1 - r^2}{n - 2}}}$$		
Section 9.2		
■ How to find the equation of a regression line, $\hat{y} = mx + b$	*1, 2*	*11–14*
$$m = \frac{n\Sigma xy - (\Sigma x)(\Sigma y)}{n\Sigma x^2 - (\Sigma x)^2}$$ $$b = \bar{y} - m\bar{x}$$ $$= \frac{\Sigma y}{n} - m\frac{\Sigma x}{n}$$		
■ How to predict y-values using a regression equation	*3*	*15–18*
Section 9.3		
■ How to find and interpret the coefficient of determination r^2	*1*	*19–24*
■ How to find and interpret the standard error of estimate for a regression line	*2*	*23, 24*
$$s_e = \sqrt{\frac{\Sigma(y_i - \hat{y}_i)^2}{n - 2}} = \sqrt{\frac{\Sigma y^2 - b\Sigma y - m\Sigma xy}{n - 2}}$$		
■ How to construct and interpret a prediction interval for y, $\hat{y} - E < y < \hat{y} + E$	*3*	*25–30*
$$E = t_c s_e \sqrt{1 + \frac{1}{n} + \frac{n(x_0 - \bar{x})^2}{n\Sigma x^2 - (\Sigma x)^2}}$$		
Section 9.4		
■ How to use technology to find a multiple regression equation, the standard error of estimate, and the coefficient of determination	*1*	*31, 32*
■ How to use a multiple regression equation to predict y-values	*2*	*33, 34*
$$\hat{y} = b + m_1 x_1 + m_2 x_2 + m_3 x_3 + \cdots + m_k x_k$$		

9 REVIEW EXERCISES

■ SECTION 9.1

In Exercises 1–4, display the data in a scatter plot. Then calculate the sample correlation coefficient r. Determine whether there is a positive linear correlation, a negative linear correlation, or no linear correlation between the variables. What can you conclude?

1. The number of pass attempts and passing yards for seven professional quarterbacks for a recent regular season *(Source: National Football League)*

Pass attempts, x	583	571	550	541	506	514	486
Passing yards, y	4770	4500	4483	4434	4328	4388	4254

2. The number of wildland fires (in thousands) and the number of wildland acres burned (in millions) in the United States for eight years *(Source: National Interagency Coordinate Center)*

Fires, x	84.1	73.5	63.6	65.5	66.8	96.4	85.7	79.0
Acres, y	3.6	7.2	4.0	8.1	8.7	9.9	9.3	5.3

3. The IQ and brain size, as measured by the total pixel count (in thousands) from an MRI scan, for nine female college students *(Adapted from Intelligence)*

IQ, x	138	140	96	83	101	135	85	77	88
Pixel count, y	991	856	879	865	808	791	799	794	894

4. The annual per capita sugar consumption (in kilograms) and the average number of cavities of 11- and 12-year-old children in seven countries

Sugar consumption, x	2.1	5.0	6.3	6.5	7.7	8.7	11.6
Cavities, y	0.59	1.51	1.55	1.70	2.18	2.10	2.73

In Exercises 5 and 6, use the given sample statistics to test the claim about the population correlation coefficient ρ at the indicated level of significance α.

5. Claim: $\rho \neq 0$; $\alpha = 0.01$. Sample statistics: $r = 0.24$, $n = 26$

6. Claim: $\rho \neq 0$; $\alpha = 0.05$. Sample statistics: $r = -0.55$, $n = 22$

In Exercises 7–10, test the claim about the population correlation coefficient ρ at the indicated level of significance α. Then interpret the decision in the context of the original claim.

7. Refer to the data in Exercise 1. At $\alpha = 0.05$, test the claim that there is a significant linear correlation between a quarterback's pass attempts and passing yards.

8. Refer to the data in Exercise 2. At $\alpha = 0.05$, is there enough evidence to conclude that there is a significant linear correlation between the number of wildland fires and the number of acres burned?

9. Refer to the data in Exercise 3. At $\alpha = 0.01$, test the claim that there is a significant linear correlation between a female college student's IQ and brain size.

10. Refer to the data in Exercise 4. At $\alpha = 0.01$, is there enough evidence to conclude that there is a significant linear correlation between sugar consumption and tooth decay?

■ SECTION 9.2

In Exercises 11–14, find the equation of the regression line for the given data. Then construct a scatter plot of the data and draw the regression line. Can you make a guess about the sign and magnitude of r? Calculate r and check your guess. If convenient, use technology to solve the problem.

11. The amounts of milk (in billions of pounds) produced in the United States and the average prices per gallon of milk for nine years *(Adapted from U.S. Department of Agriculture and U.S. Bureau of Labor Statistics)*

Milk produced, x	167.6	165.3	170.1	170.4	170.9
Price per gallon, y	2.79	2.90	2.68	2.95	3.23

Milk produced, x	177.0	181.8	185.7	190.0
Price per gallon, y	3.24	3.00	3.87	3.68

12. The average times (in hours) per day spent watching television for men and women for the last 10 years *(Adapted from The Nielsen Company)*

Men, x	4.03	4.18	4.32	4.37	4.48	4.43	4.52	4.58	4.65	4.82
Women, y	4.67	4.77	4.85	4.97	5.08	5.12	5.28	5.28	5.32	5.42

13. The ages (in years) and the number of hours of sleep in one night for seven adults

Age, x	35	20	59	42	68	38	75
Hours of sleep, y	7	9	5	6	5	8	4

14. The engine displacements (in cubic inches) and the fuel efficiencies (in miles per gallon) of seven automobiles

Displacement, x	170	134	220	305	109	256	322
Fuel efficiency, y	29.5	34.5	23.0	17.0	33.5	23.0	15.5

In Exercises 15–18, use the regression equations found in Exercises 11–14 to predict the value of y for each value of x, if meaningful. If not, explain why not. (Each pair of variables has a significant correlation.)

15. Refer to Exercise 11. What price per gallon would you predict for a milk production of (a) 160 billion pounds? (b) 175 billion pounds? (c) 180 billion pounds? (d) 200 billion pounds?

16. Refer to Exercise 12. What average time per day spent watching television for women would you predict when the average time per day for men is (a) 4.2 hours? (b) 4.5 hours? (c) 4.75 hours? (d) 5 hours?

17. Refer to Exercise 13. How many hours of sleep would you predict for an adult of age (a) 18 years? (b) 25 years? (c) 85 years? (d) 50 years?

18. Refer to Exercise 14. What fuel efficiency rating would you predict for a car with an engine displacement of (a) 86 cubic inches? (b) 198 cubic inches? (c) 289 cubic inches? (d) 407 cubic inches?

■ SECTION 9.3

In Exercises 19–22, use the value of the linear correlation coefficient to calculate the coefficient of determination. What does this tell you about the explained variation of the data about the regression line? About the unexplained variation?

19. $r = -0.450$ **20.** $r = -0.937$

21. $r = 0.642$ **22.** $r = 0.795$

In Exercises 23 and 24, use the data to find the (a) coefficient of determination r^2 and interpret the result, and (b) standard error of estimate s_e and interpret the result.

23. The table shows the prices (in thousands of dollars) and fuel efficiencies (in miles per gallon) for nine compact sports sedans. The regression equation is $\hat{y} = -0.414x + 37.147$. *(Adapted from Consumer Reports)*

Price, x	37.2	40.8	29.7	33.7	37.5	32.7	39.2	37.3	31.6
Fuel efficiency, x	21	19	25	24	22	24	23	21	23

 24. The table shows the cooking areas (in square inches) of 18 gas grills and their prices (in dollars). The regression equation is $\hat{y} = 1.454x - 532.053$. *(Source: Lowe's)*

Area, x	780	530	942	660	600	732	660	640	869
Price, y	359	98	547	299	449	799	699	199	1049

Area, x	860	700	942	890	733	732	464	869	600
Price, y	499	248	597	999	428	849	99	999	399

In Exercises 25–30, construct the indicated prediction interval and interpret the results.

25. Construct a 90% prediction interval for the price per gallon of milk in Exercise 11 when 185 billion pounds of milk is produced.

26. Construct a 90% prediction interval for the average time women spend per day watching television in Exercise 12 when the average time men spend per day watching television is 4.25 hours.

27. Construct a 95% prediction interval for the number of hours of sleep for an adult in Exercise 13 who is 45 years old.

28. Construct a 95% prediction interval for the fuel efficiency of an automobile in Exercise 14 that has an engine displacement of 265 cubic inches.

29. Construct a 99% prediction interval for the fuel efficiency of a compact sports sedan in Exercise 23 that costs $39,900.

30. Construct a 99% prediction interval for the price of a gas grill in Exercise 24 with a usable cooking area of 900 square inches.

9 CHAPTER QUIZ

Take this quiz as you would take a quiz in class. After you are done, check your work against the answers given in the back of the book.

For Exercises 1–8, use the data in the table, which shows the average annual salaries (both in thousands of dollars) for public school principals and public school classroom teachers in the United States for 11 years. *(Adapted from Educational Research Service)*

Principals, x	Classroom teachers, y
62.5	37.3
71.9	41.4
74.4	42.2
77.8	43.7
78.4	43.8
80.8	45.0
80.5	45.6
81.5	45.9
84.8	48.2
87.7	49.3
91.6	51.3

1. Construct a scatter plot for the data. Do the data appear to have a positive linear correlation, a negative linear correlation, or no linear correlation? Explain.

2. Calculate the correlation coefficient r. What can you conclude?

3. Test the level of significance of the correlation coefficient r. Use $\alpha = 0.05$.

4. Find the equation of the regression line for the data. Draw the regression line on the scatter plot.

5. Use the regression equation to predict the average annual salary of public school classroom teachers when the average annual salary of public school principals is \$90,500.

6. Find the coefficient of determination r^2 and interpret the result.

7. Find the standard error of estimate s_e and interpret the result.

8. Construct a 95% prediction interval for the average annual salary of public school classroom teachers when the average annual salary of public school principals is \$85,750. Interpret the results.

9. **Stock Price** The equation used to predict the stock price (in dollars) at the end of the year for McDonald's Corporation is

$$\hat{y} = -47 + 5.91x_1 - 1.99x_2$$

where x_1 is the total revenue (in billions of dollars) and x_2 is the shareholders' equity (in billions of dollars). Use the multiple regression equation to predict the y-values for the given values of the independent variables. *(Adapted from McDonald's Corporation)*

(a) $x_1 = 22.7$, $x_2 = 14.0$

(b) $x_1 = 17.9$, $x_2 = 14.2$

(c) $x_1 = 20.9$, $x_2 = 15.5$

(d) $x_1 = 19.1$, $x_2 = 15.1$

PUTTING IT ALL TOGETHER

Real Statistics — Real Decisions

Acid rain affects the environment by increasing the acidity of lakes and streams to dangerous levels, damaging trees and soil, accelerating the decay of building materials and paint, and destroying national monuments. The goal of the Environmental Protection Agency's (EPA) Acid Rain Program is to achieve environmental health benefits by reducing the emissions of the primary causes of acid rain: sulfur dioxide and nitrogen oxides.

You work for the EPA and you want to determine if there is a significant correlation between sulfur dioxide emissions and nitrogen oxides emissions.

■ EXERCISES

1. *Analyzing the Data*

(a) The data in the table show the sulfur dioxide emissions (in millions of tons) and the nitrogen oxides emissions (in millions of tons) for 14 years. Construct a scatter plot of the data and make a conclusion about the type of correlation between sulfur dioxide emissions and nitrogen oxides emissions.

(b) Calculate the correlation coefficient r and verify your conclusion in part (a).

(c) Test the significance of the correlation coefficient found in part (b). Use $\alpha = 0.05$.

(d) Find the equation of the regression line for sulfur dioxide emissions and nitrogen oxides emissions. Add the graph of the regression line to your scatter plot in part (a). Does the regression line appear to be a good fit?

(e) Can you use the equation of the regression line to predict the nitrogen oxides emission given the sulfur dioxide emission? Why or why not?

(f) Find the coefficient of determination r^2 and the standard error of estimate s_e. Interpret your results.

2. *Making Predictions*

The EPA set a goal of reducing sulfur dioxide emissions levels by 10 million tons from 1980 levels of 17.3 million tons. Construct a 95% prediction interval for the nitrogen oxides emissions for this sulfur dioxide emissions goal level. Interpret the results.

Sulfur dioxide emissions, x	Nitrogen oxides emissions, y
11.8	5.8
12.5	6.0
12.9	6.0
13.1	6.0
12.5	5.5
11.2	5.1
10.6	4.7
10.2	4.5
10.6	4.2
10.3	3.8
10.2	3.6
9.4	3.4
8.9	3.3
7.6	3.0

Source: Environmental Protection Agency

TECHNOLOGY

MINITAB   EXCEL   TI-83/84 PLUS

 U.S. Food and Drug Administration

NUTRIENTS IN BREAKFAST CEREALS

The U.S. Food and Drug Administration (FDA) requires nutrition labeling for most foods. Under FDA regulations, manufacturers are required to list the amounts of certain nutrients in their foods, such as calories, sugar, fat, and carbohydrates. This nutritional information is displayed in the "Nutrition Facts" panel on the food's package.

The table shows the following nutritional content for one cup of each of 21 different breakfast cereals.

C = calories

S = sugar in grams

F = fat in grams

R = carbohydrates in grams

Cereal	C	S	F	R
Apple Jacks®	100	12	0.5	25
Berry Burst Cheerios®	130	11	1.5	29
Cheerios®	100	1	2	20
Cocoa Puffs®	130	15	2	31
Cookie Crisp®	130	13	1.5	29
Corn Chex®	120	3	0.5	26
Corn Flakes®	100	2	0	24
Corn Pops®	120	10	0	29
Count Chocula®	150	16	1.5	31
Crispix®	110	4	0	25
Froot Loops®	110	12	1	25
Frosted Flakes®	150	15	0	36
Golden Grahams®	160	15	1.5	35
Honey Nut Cheerios®	150	12	2	29
Lucky Charms®	150	15	1.5	29
Multi Grain Cheerios®	110	6	1	23
Raisin Bran®	190	19	1.5	45
Rice Krispies®	100	3	0	23
Special K®	120	4	0.5	23
Trix®	120	11	1.5	28
Wheaties®	130	5	0.5	29

■ EXERCISES

1. Use a technology tool to draw a scatter plot of the following (x, y) pairs in the data set.

(a) (calories, sugar)

(b) (calories, fat)

(c) (calories, carbohydrates)

(d) (sugar, fat)

(e) (sugar, carbohydrates)

(f) (fat, carbohydrates)

2. From the scatter plots in Exercise 1, which pairs of variables appear to have a strong linear correlation?

3. Use a technology tool to find the correlation coefficient for each pair of variables in Exercise 1. Which has the strongest linear correlation?

4. Use a technology tool to find an equation of a regression line for the following variables.

(a) (calories, sugar)

(b) (calories, carbohydrates)

5. Use the results of Exercise 4 to predict the following.

(a) The sugar content of one cup of cereal that has a caloric content of 120 calories

(b) The carbohydrate content of one cup of cereal that has a caloric content of 120 calories

6. Use a technology tool to find the multiple regression equations of the following forms.

(a) $C = b + m_1S + m_2F + m_3R$

(b) $C = b + m_1S + m_2R$

7. Use the equations from Exercise 6 to predict the caloric content of 1 cup of cereal that has 7 grams of sugar, 0.5 gram of fat, and 31 grams of carbohydrates.

Extended solutions are given in the *Technology Supplement*.
Technical instruction is provided for MINITAB, Excel, and the TI-83/84 Plus.

In this appendix, we use a 0-to-z table as an alternative development of the standard normal distribution. It is intended that this appendix be used after completion of the "Properties of a Normal Distribution" subsection of Section 5.1 in the text. If used, this appendix should replace the material in the "Standard Normal Distribution" subsection of Section 5.1 except for the exercises.

Standard Normal Distribution (0-to-z)

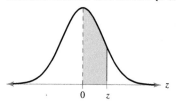

z	.00	.01	.02	.03	.04	.05	.06	.07	.08	.09
0.0	.0000	.0040	.0080	.0120	.0160	.0199	.0239	.0279	.0319	.0359
0.1	.0398	.0438	.0478	.0517	.0557	.0596	.0636	.0675	.0714	.0753
0.2	.0793	.0832	.0871	.0910	.0948	.0987	.1026	.1064	.1103	.1141
0.3	.1179	.1217	.1255	.1293	.1331	.1368	.1406	.1443	.1480	.1517
0.4	.1554	.1591	.1628	.1664	.1700	.1736	.1772	.1808	.1844	.1879
0.5	.1915	.1950	.1985	.2019	.2054	.2088	.2123	.2157	.2190	.2224
0.6	.2257	.2291	.2324	.2357	.2389	.2422	.2454	.2486	.2517	.2549
0.7	.2580	.2611	.2642	.2673	.2704	.2734	.2764	.2794	.2823	.2852
0.8	.2881	.2910	.2939	.2967	.2995	.3023	.3051	.3078	.3106	.3133
0.9	.3159	.3186	.3212	.3238	.3264	.3289	.3315	.3340	.3365	.3389
1.0	.3413	.3438	.3461	.3485	.3508	.3531	.3554	.3577	.3599	.3621
1.1	.3643	.3665	.3686	.3708	.3729	.3749	.3770	.3790	.3810	.3830
1.2	.3849	.3869	.3888	.3907	.3925	.3944	.3962	.3980	.3997	.4015
1.3	.4032	.4049	.4066	.4082	.4099	.4115	.4131	.4147	.4162	.4177
1.4	.4192	.4207	.4222	.4236	.4251	.4265	.4279	.4292	.4306	.4319
1.5	.4332	.4345	.4357	.4370	.4382	.4394	.4406	.4418	.4429	.4441
1.6	.4452	.4463	.4474	.4484	.4495	.4505	.4515	.4525	.4535	.4545
1.7	.4554	.4564	.4573	.4582	.4591	.4599	.4608	.4616	.4625	.4633
1.8	.4641	.4649	.4656	.4664	.4671	.4678	.4686	.4693	.4699	.4706
1.9	.4713	.4719	.4726	.4732	.4738	.4744	.4750	.4756	.4761	.4767
2.0	.4772	.4778	.4783	.4788	.4793	.4798	.4803	.4808	.4812	.4817
2.1	.4821	.4826	.4830	.4834	.4838	.4842	.4846	.4850	.4854	.4857
2.2	.4861	.4864	.4868	.4871	.4875	.4878	.4881	.4884	.4887	.4890
2.3	.4893	.4896	.4898	.4901	.4904	.4906	.4909	.4911	.4913	.4916
2.4	.4918	.4920	.4922	.4925	.4927	.4929	.4931	.4932	.4934	.4936
2.5	.4938	.4940	.4941	.4943	.4945	.4946	.4948	.4949	.4951	.4952
2.6	.4953	.4955	.4956	.4957	.4959	.4960	.4961	.4962	.4963	.4964
2.7	.4965	.4966	.4967	.4968	.4969	.4970	.4971	.4972	.4973	.4974
2.8	.4974	.4975	.4976	.4977	.4977	.4978	.4979	.4979	.4980	.4981
2.9	.4981	.4982	.4982	.4983	.4984	.4984	.4985	.4985	.4986	.4986
3.0	.4987	.4987	.4987	.4988	.4988	.4989	.4989	.4989	.4990	.4990
3.1	.4990	.4991	.4991	.4991	.4992	.4992	.4992	.4992	.4993	.4993
3.2	.4993	.4993	.4994	.4994	.4994	.4994	.4994	.4995	.4995	.4995
3.3	.4995	.4995	.4995	.4996	.4996	.4996	.4996	.4996	.4996	.4997
3.4	.4997	.4997	.4997	.4997	.4997	.4997	.4997	.4997	.4997	.4998

Reprinted with permission of Gale Mosteller, executor of estate of Frederick Mosteller, 3830 13th Street North, Arlington, VA 22201   mosteller.g@ei.com.

Alternative Presentation of the Standard Normal Distribution

The Standard Normal Distribution

▸ THE STANDARD NORMAL DISTRIBUTION

There are infinitely many normal distributions, each with its own mean and standard deviation. The normal distribution with a mean of 0 and a standard deviation of 1 is called the *standard normal distribution*. The horizontal scale of the graph of the standard normal distribution corresponds to z-scores. In Section 2.5, you learned that a z-score is a measure of position that indicates the number of standard deviations a value lies from the mean. Recall that you can transform an x-value to a z-score using the formula

$$z = \frac{\text{Value} - \text{Mean}}{\text{Standard deviation}} = \frac{x - \mu}{\sigma}.$$

DEFINITION

The **standard normal distribution** is a normal distribution with a mean of 0 and a standard deviation of 1.

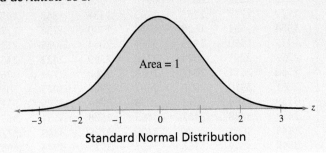

Area = 1

Standard Normal Distribution

If each data value of a normally distributed random variable x is transformed into a z-score, the result will be the standard normal distribution. When this transformation takes place, the area that falls in the interval under the nonstandard normal curve is the *same* as that under the standard normal curve within the corresponding z-boundaries.

In Section 2.4, you learned to use the Empirical Rule to approximate areas under a normal curve when values of the random variable x corresponded to −3, −2, −1, 0, 1, 2, or 3 standard deviations from the mean. Now, you will learn to calculate areas corresponding to other x-values. After you transform an x-value to a z-score, you can use the Standard Normal Table (0-to-z) on page A1. The table lists the area under the standard normal curve between 0 and the given z-score. As you examine the table, notice the following.

PROPERTIES OF THE STANDARD NORMAL DISTRIBUTION

1. The distribution is symmetric about the mean ($z = 0$).
2. The area under the standard normal curve to the left of $z = 0$ is 0.5 and the area to the right of $z = 0$ is 0.5.
3. The area under the standard normal curve increases as the distance between 0 and z increases.

At first glance, the table on page A1 appears to give areas for positive z-scores only. However, because of the symmetry of the standard normal curve, the table also gives areas for negative z-scores (see Example 1).

EXAMPLE 1

▸ **Using the Standard Normal Table (0-to-z)**

1. Find the area under the standard normal curve between $z = 0$ and $z = 1.15$.
2. Find the z-scores that correspond to an area of 0.0948.

▸ **Solution**

1. Find the area that corresponds to $z = 1.15$ by finding 1.1 in the left column and then moving across the row to the column under 0.05. The number in that row and column is 0.3749. So, the area between $z = 0$ and $z = 1.15$ is 0.3749.

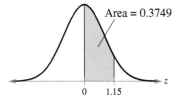

Area = 0.3749

z	.00	.01	.02	.03	.04	.05	.06
0.0	.0000	.0040	.0080	.0120	.0160	.0199	.0239
0.1	.0398	.0438	.0478	.0517	.0557	.0596	.0636
0.2	.0793	.0832	.0871	.0910	.0948	.0987	.1026
0.3	.1179	.1217	.1255	.1293	.1331	.1368	.1406
0.9	.3159	.3186	.3212	.3238	.3264	.3289	.3315
1.0	.3413	.3438	.3461	.3485	.3508	.3531	.3554
1.1	.3643	.3665	.3686	.3708	.3729	.3749	.3770
1.2	.3849	.3869	.3888	.3907	.3925	.3944	.3962
1.3	.4032	.4049	.4066	.4082	.4099	.4115	.4131
1.4	.4192	.4207	.4222	.4236	.4251	.4265	.4279

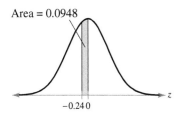

Area = 0.0948

2. Find the z-scores that correspond to an area of 0.0948 by locating 0.0948 in the table. The values at the beginning of the corresponding row and at the top of the corresponding column give the z-score. For an area of 0.0948, the row value is 0.2 and the column value is 0.04. So, the z-scores are $z = -0.24$ and $z = 0.24$.

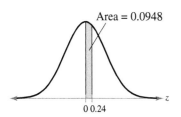

Area = 0.0948

z	.00	.01	.02	.03	.04	.05	.06
0.0	.0000	.0040	.0080	.0120	.0160	.0199	.0239
0.1	.0398	.0438	.0478	.0517	.0557	.0596	.0636
0.2	.0793	.0832	.0871	.0910	.0948	.0987	.1026
0.3	.1179	.1217	.1255	.1293	.1331	.1368	.1406
0.4	.1554	.1591	.1628	.1664	.1700	.1736	.1772
0.5	.1915	.1950	.1985	.2019	.2054	.2088	.2123

▸ **Try It Yourself 1**

1. Find the area under the standard normal curve between $z = 0$ and $z = 2.19$.

   Locate the given z-score and *find the corresponding area* in the Standard Normal Table (0-to-z) on page A1.

2. Find the z-scores that correspond to an area of 0.4850.

   Locate the given area in the Standard Normal Table (0-to-z) on page A1 and *find the corresponding z-score*. *Answer: Page A49*

Use the following guidelines to find various types of areas under the standard normal curve.

GUIDELINES

Finding Areas Under the Standard Normal Curve

1. Sketch the standard normal curve and shade the appropriate area under the curve.
2. Use the Standard Normal Table (0-to-z) on page A1 to find the area that corresponds to the given z-score(s).
3. Find the desired area by following the directions for each case shown.

   a. Area to the left of z

   i. When $z < 0$, *subtract* the area from 0.5.

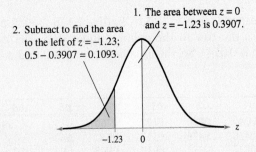

2. Subtract to find the area to the left of $z = -1.23$; $0.5 - 0.3907 = 0.1093$.

1. The area between $z = 0$ and $z = -1.23$ is 0.3907.

   ii. When $z > 0$, *add* 0.5 to the area.

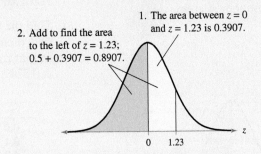

2. Add to find the area to the left of $z = 1.23$; $0.5 + 0.3907 = 0.8907$.

1. The area between $z = 0$ and $z = 1.23$ is 0.3907.

   b. Area to the right of z

   i. When $z < 0$, *add* 0.5 to the area.

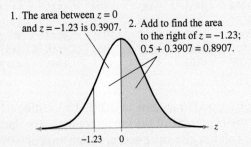

1. The area between $z = 0$ and $z = -1.23$ is 0.3907.

2. Add to find the area to the right of $z = -1.23$; $0.5 + 0.3907 = 0.8907$.

   ii. When $z > 0$, *subtract* the area from 0.5.

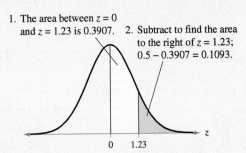

1. The area between $z = 0$ and $z = 1.23$ is 0.3907.

2. Subtract to find the area to the right of $z = 1.23$; $0.5 - 0.3907 = 0.1093$.

   c. Area between two z-scores

   i. When the two z-scores have the same sign (both positive or both negative), *subtract* the smaller area from the larger area.

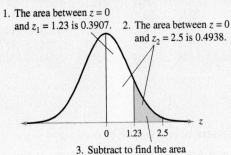

1. The area between $z = 0$ and $z_1 = 1.23$ is 0.3907.

2. The area between $z = 0$ and $z_2 = 2.5$ is 0.4938.

3. Subtract to find the area between $z_1 = 1.23$ and $z_2 = 2.5$; $0.4938 - 0.3907 = 0.1031$.

   ii. When the two z-scores have opposite signs (one negative and one positive), *add* the areas.

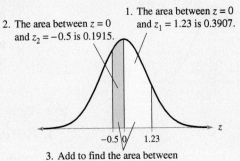

2. The area between $z = 0$ and $z_2 = -0.5$ is 0.1915.

1. The area between $z = 0$ and $z_1 = 1.23$ is 0.3907.

3. Add to find the area between $z_1 = 1.23$ and $z_2 = -0.5$; $0.3907 + 0.1915 = 0.5822$.

EXAMPLE 2

▶ **Finding Area Under the Standard Normal Curve**

Find the area under the standard normal curve to the left of $z = -0.99$.

▶ **Solution**

The area under the standard normal curve to the left of $z = -0.99$ is shown.

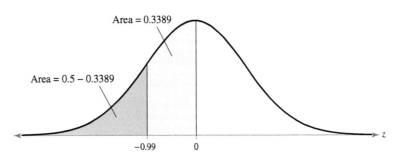

INSIGHT

Because the normal distribution is a continuous probability distribution, the area under the standard normal curve to the left of a z-score gives the probability that z is less than that z-score. For instance, in Example 2, the area to the left of $z = -0.99$ is 0.1611. So, $P(z < -0.99) = 0.1611$, which is read as "the probability that z is less than -0.99 is 0.1611."

From the Standard Normal Table (0-to-z), the area corresponding to $z = -0.99$ is 0.3389. Because the area to the left of $z = 0$ is 0.5, the area to the left of $z = -0.99$ is $0.5 - 0.3389 = 0.1611$.

▶ **Try It Yourself 2**

Find the area under the standard normal curve to the left of $z = 2.13$.

a. *Draw* the standard normal curve and shade the area under the curve and to the left of $z = 2.13$.
b. Use the Standard Normal Table (0-to-z) on page A1 to *find the area* that corresponds to $z = 2.13$.
c. *Add* 0.5 to the resulting area.   *Answer: Page A49*

EXAMPLE 3

▶ **Finding Area Under the Standard Normal Curve**

Find the area under the standard normal curve to the right of $z = 1.06$.

▶ **Solution**

The area under the standard normal curve to the right of $z = 1.06$ is shown.

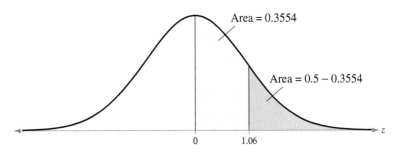

From the Standard Normal Table (0-to-z), the area corresponding to $z = 1.06$ is 0.3554. Because the area to the right of $z = 0$ is 0.5, the area to the right of $z = 1.06$ is $0.5 - 0.3554 = 0.1446$.

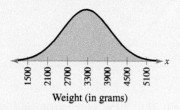
▶ **Try It Yourself 3**

Find the area under the standard normal curve to the right of $z = -2.16$.

a. *Draw* the standard normal curve and shade the area below the curve and to the right of $z = -2.16$.
b. Use the Standard Normal Table (0-to-z) on page A1 to *find the area* that corresponds to $z = -2.16$.
c. *Add* 0.5 to the resulting area.   *Answer: Page A49*

EXAMPLE 4

▶ **Finding Area Under the Standard Normal Curve**

Find the area under the standard normal curve between $z = -1.5$ and $z = 1.25$.

▶ **Solution**

The area under the standard normal curve between $z = -1.5$ and $z = 1.25$ is shown.

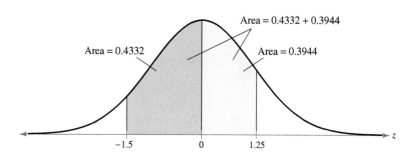

From the Standard Normal Table, the area corresponding to $z = -1.5$ is 0.4332 and the area corresponding to $z = 1.25$ is 0.3944. To find the area between these two z-scores, add the resulting areas.

   Area $= 0.4332 + 0.3944 = 0.8276$

Interpretation So, 82.76% of the area under the curve falls between $z = -1.5$ and $z = 1.25$.

▶ **Try It Yourself 4**

Find the area under the standard normal curve between $z = -2.165$ and $z = -1.35$.

a. *Draw* the standard normal curve and shade the area below the curve that is between $z = -2.165$ and $z = -1.35$.
b. Use the Standard Normal Table (0-to-z) on page A1 to *find the areas* that correspond to $z = -2.165$ and to $z = -1.35$.
c. *Subtract* the smaller area from the larger area.   *Answer: Page A49*

   Recall that in Section 2.5 you learned, using the Empirical Rule, that values lying more than two standard deviations from the mean are considered unusual. Values lying more than three standard deviations from the mean are considered *very* unusual. So, if a z-score is greater than 2 or less than -2, it is unusual. If a z-score is greater than 3 or less than -3, it is *very* unusual.

Table 1— Random Numbers

92630	78240	19267	95457	53497	23894	37708	79862	76471	66418
79445	78735	71549	44843	26104	67318	00701	34986	66751	99723
59654	71966	27386	50004	05358	94031	29281	18544	52429	06080
31524	49587	76612	39789	13537	48086	59483	60680	84675	53014
06348	76938	90379	51392	55887	71015	09209	79157	24440	30244
28703	51709	94456	48396	73780	06436	86641	69239	57662	80181
68108	89266	94730	95761	75023	48464	65544	96583	18911	16391
99938	90704	93621	66330	33393	95261	95349	51769	91616	33238
91543	73196	34449	63513	83834	99411	58826	40456	69268	48562
42103	02781	73920	56297	72678	12249	25270	36678	21313	75767
17138	27584	25296	28387	51350	61664	37893	05363	44143	42677
28297	14280	54524	21618	95320	38174	60579	08089	94999	78460
09331	56712	51333	06289	75345	08811	82711	57392	25252	30333
31295	04204	93712	51287	05754	79396	87399	51773	33075	97061
36146	15560	27592	42089	99281	59640	15221	96079	09961	05371
29553	18432	13630	05529	02791	81017	49027	79031	50912	09399
23501	22642	63081	08191	89420	67800	55137	54707	32945	64522
57888	85846	67967	07835	11314	01545	48535	17142	08552	67457
55336	71264	88472	04334	63919	36394	11196	92470	70543	29776
10087	10072	55980	64688	68239	20461	89381	93809	00796	95945
34101	81277	66090	88872	37818	72142	67140	50785	21380	16703
53362	44940	60430	22834	14130	96593	23298	56203	92671	15925
82975	66158	84731	19436	55790	69229	28661	13675	99318	76873
54827	84673	22898	08094	14326	87038	42892	21127	30712	48489
25464	59098	27436	89421	80754	89924	19097	67737	80368	08795
67609	60214	41475	84950	40133	02546	09570	45682	50165	15609
44921	70924	61295	51137	47596	86735	35561	76649	18217	63446
33170	30972	98130	95828	49786	13301	36081	80761	33985	68621
84687	85445	06208	17654	51333	02878	35010	67578	61574	20749
71886	56450	36567	09395	96951	35507	17555	35212	69106	01679
00475	02224	74722	14721	40215	21351	08596	45625	83981	63748
25993	38881	68361	59560	41274	69742	40703	37993	03435	18873
92882	53178	99195	93803	56985	53089	15305	50522	55900	43026
25138	26810	07093	15677	60688	04410	24505	37890	67186	62829
84631	71882	12991	83028	82484	90339	91950	74579	03539	90122
34003	92326	12793	61453	48121	74271	28363	66561	75220	35908
53775	45749	05734	86169	42762	70175	97310	73894	88606	19994
59316	97885	72807	54966	60859	11932	35265	71601	55577	67715
20479	66557	50705	26999	09854	52591	14063	30214	19890	19292
86180	84931	25455	26044	02227	52015	21820	50599	51671	65411
21451	68001	72710	40261	61281	13172	63819	48970	51732	54113
98062	68375	80089	24135	72355	95428	11808	29740	81644	86610
01788	64429	14430	94575	75153	94576	61393	96192	03227	32258
62465	04841	43272	68702	01274	05437	22953	18946	99053	41690
94324	31089	84159	92933	99989	89500	91586	02802	69471	68274
05797	43984	21575	09908	70221	19791	51578	36432	33494	79888
10395	14289	52185	09721	25789	38562	54794	04897	59012	89251
35177	56986	25549	59730	64718	52630	31100	62384	49483	11409
25633	89619	75882	98256	02126	72099	57183	55887	09320	73463
16464	48280	94254	45777	45150	68865	11382	11782	22695	41988

Table 2— Binomial Distribution

This table shows the probability of x successes in n independent trials, each with probability of success p.

												p									
n	x	.01	.05	.10	.15	.20	.25	.30	.35	.40	.45	.50	.55	.60	.65	.70	.75	.80	.85	.90	.95
2	0	.980	.902	.810	.723	.640	.563	.490	.423	.360	.303	.250	.203	.160	.123	.090	.063	.040	.023	.010	.002
	1	.020	.095	.180	.255	.320	.375	.420	.455	.480	.495	.500	.495	.480	.455	.420	.375	.320	.255	.180	.095
	2	.000	.002	.010	.023	.040	.063	.090	.123	.160	.203	.250	.303	.360	.423	.490	.563	.640	.723	.810	.902
3	0	.970	.857	.729	.614	.512	.422	.343	.275	.216	.166	.125	.091	.064	.043	.027	.016	.008	.003	.001	.000
	1	.029	.135	.243	.325	.384	.422	.441	.444	.432	.408	.375	.334	.288	.239	.189	.141	.096	.057	.027	.007
	2	.000	.007	.027	.057	.096	.141	.189	.239	.288	.334	.375	.408	.432	.444	.441	.422	.384	.325	.243	.135
	3	.000	.000	.001	.003	.008	.016	.027	.043	.064	.091	.125	.166	.216	.275	.343	.422	.512	.614	.729	.857
4	0	.961	.815	.656	.522	.410	.316	.240	.179	.130	.092	.062	.041	.026	.015	.008	.004	.002	.001	.000	.000
	1	.039	.171	.292	.368	.410	.422	.412	.384	.346	.300	.250	.200	.154	.112	.076	.047	.026	.011	.004	.000
	2	.001	.014	.049	.098	.154	.211	.265	.311	.346	.368	.375	.368	.346	.311	.265	.211	.154	.098	.049	.014
	3	.000	.000	.004	.011	.026	.047	.076	.112	.154	.200	.250	.300	.346	.384	.412	.422	.410	.368	.292	.171
	4	.000	.000	.000	.001	.002	.004	.008	.015	.026	.041	.062	.092	.130	.179	.240	.316	.410	.522	.656	.815
5	0	.951	.774	.590	.444	.328	.237	.168	.116	.078	.050	.031	.019	.010	.005	.002	.001	.000	.000	.000	.000
	1	.048	.204	.328	.392	.410	.396	.360	.312	.259	.206	.156	.113	.077	.049	.028	.015	.006	.002	.000	.000
	2	.001	.021	.073	.138	.205	.264	.309	.336	.346	.337	.312	.276	.230	.181	.132	.088	.051	.024	.008	.001
	3	.000	.001	.008	.024	.051	.088	.132	.181	.230	.276	.312	.337	.346	.336	.309	.264	.205	.138	.073	.021
	4	.000	.000	.000	.002	.006	.015	.028	.049	.077	.113	.156	.206	.259	.312	.360	.396	.410	.392	.328	.204
	5	.000	.000	.000	.000	.000	.001	.002	.005	.010	.019	.031	.050	.078	.116	.168	.237	.328	.444	.590	.774
6	0	.941	.735	.531	.377	.262	.178	.118	.075	.047	.028	.016	.008	.004	.002	.001	.000	.000	.000	.000	.000
	1	.057	.232	.354	.399	.393	.356	.303	.244	.187	.136	.094	.061	.037	.020	.010	.004	.002	.000	.000	.000
	2	.001	.031	.098	.176	.246	.297	.324	.328	.311	.278	.234	.186	.138	.095	.060	.033	.015	.006	.001	.000
	3	.000	.002	.015	.042	.082	.132	.185	.236	.276	.303	.312	.303	.276	.236	.185	.132	.082	.042	.015	.002
	4	.000	.000	.001	.006	.015	.033	.060	.095	.138	.186	.234	.278	.311	.328	.324	.297	.246	.176	.098	.031
	5	.000	.000	.000	.000	.002	.004	.010	.020	.037	.061	.094	.136	.187	.244	.303	.356	.393	.399	.354	.232
	6	.000	.000	.000	.000	.000	.000	.001	.002	.004	.008	.016	.028	.047	.075	.118	.178	.262	.377	.531	.735
7	0	.932	.698	.478	.321	.210	.133	.082	.049	.028	.015	.008	.004	.002	.001	.000	.000	.000	.000	.000	.000
	1	.066	.257	.372	.396	.367	.311	.247	.185	.131	.087	.055	.032	.017	.008	.004	.001	.000	.000	.000	.000
	2	.002	.041	.124	.210	.275	.311	.318	.299	.261	.214	.164	.117	.077	.047	.025	.012	.004	.001	.000	.000
	3	.000	.004	.023	.062	.115	.173	.227	.268	.290	.292	.273	.239	.194	.144	.097	.058	.029	.011	.003	.000
	4	.000	.000	.003	.011	.029	.058	.097	.144	.194	.239	.273	.292	.290	.268	.227	.173	.115	.062	.023	.004
	5	.000	.000	.000	.001	.004	.012	.025	.047	.077	.117	.164	.214	.261	.299	.318	.311	.275	.210	.124	.041
	6	.000	.000	.000	.000	.000	.001	.004	.008	.017	.032	.055	.087	.131	.185	.247	.311	.367	.396	.372	.257
	7	.000	.000	.000	.000	.000	.000	.000	.001	.002	.004	.008	.015	.028	.049	.082	.133	.210	.321	.478	.698
8	0	.923	.663	.430	.272	.168	.100	.058	.032	.017	.008	.004	.002	.001	.000	.000	.000	.000	.000	.000	.000
	1	.075	.279	.383	.385	.336	.267	.198	.137	.090	.055	.031	.016	.008	.003	.001	.000	.000	.000	.000	.000
	2	.003	.051	.149	.238	.294	.311	.296	.259	.209	.157	.109	.070	.041	.022	.010	.004	.001	.000	.000	.000
	3	.000	.005	.033	.084	.147	.208	.254	.279	.279	.257	.219	.172	.124	.081	.047	.023	.009	.003	.000	.000
	4	.000	.000	.005	.018	.046	.087	.136	.188	.232	.263	.273	.263	.232	.188	.136	.087	.046	.018	.005	.000
	5	.000	.000	.000	.003	.009	.023	.047	.081	.124	.172	.219	.257	.279	.279	.254	.208	.147	.084	.033	.005
	6	.000	.000	.000	.000	.001	.004	.010	.022	.041	.070	.109	.157	.209	.259	.296	.311	.294	.238	.149	.051
	7	.000	.000	.000	.000	.000	.000	.001	.003	.008	.016	.031	.055	.090	.137	.198	.267	.336	.385	.383	.279
	8	.000	.000	.000	.000	.000	.000	.000	.000	.001	.002	.004	.008	.017	.032	.058	.100	.168	.272	.430	.663
9	0	.914	.630	.387	.232	.134	.075	.040	.021	.010	.005	.002	.001	.000	.000	.000	.000	.000	.000	.000	.000
	1	.083	.299	.387	.368	.302	.225	.156	.100	.060	.034	.018	.008	.004	.001	.000	.000	.000	.000	.000	.000
	2	.003	.063	.172	.260	.302	.300	.267	.216	.161	.111	.070	.041	.021	.010	.004	.001	.000	.000	.000	.000
	3	.000	.008	.045	.107	.176	.234	.267	.272	.251	.212	.164	.116	.074	.042	.021	.009	.003	.001	.000	.000
	4	.000	.001	.007	.028	.066	.117	.172	.219	.251	.260	.246	.213	.167	.118	.074	.039	.017	.005	.001	.000
	5	.000	.000	.001	.005	.017	.039	.074	.118	.167	.213	.246	.260	.251	.219	.172	.117	.066	.028	.007	.001
	6	.000	.000	.000	.001	.003	.009	.021	.042	.074	.116	.164	.212	.251	.272	.267	.234	.176	.107	.045	.008
	7	.000	.000	.000	.000	.000	.001	.004	.010	.021	.041	.070	.111	.161	.216	.267	.300	.302	.260	.172	.063
	8	.000	.000	.000	.000	.000	.000	.000	.001	.004	.008	.018	.034	.060	.100	.156	.225	.302	.368	.387	.299
	9	.000	.000	.000	.000	.000	.000	.000	.000	.001	.002	.005	.010	.021	.040	.075	.134	.232	.387	.630	.630

Table 2—Binomial Distribution *(continued)*

												p									
n	x	.01	.05	.10	.15	.20	.25	.30	.35	.40	.45	.50	.55	.60	.65	.70	.75	.80	.85	.90	.95
10	0	.904	.599	.349	.197	.107	.056	.028	.014	.006	.003	.001	.000	.000	.000	.000	.000	.000	.000	.000	.000
	1	.091	.315	.387	.347	.268	.188	.121	.072	.040	.021	.010	.004	.002	.000	.000	.000	.000	.000	.000	.000
	2	.004	.075	.194	.276	.302	.282	.233	.176	.121	.076	.044	.023	.011	.004	.001	.000	.000	.000	.000	.000
	3	.000	.010	.057	.130	.201	.250	.267	.252	.215	.166	.117	.075	.042	.021	.009	.003	.001	.000	.000	.000
	4	.000	.001	.011	.040	.088	.146	.200	.238	.251	.238	.205	.160	.111	.069	.037	.016	.006	.001	.000	.000
	5	.000	.000	.001	.008	.026	.058	.103	.154	.201	.234	.246	.234	.201	.154	.103	.058	.026	.008	.001	.000
	6	.000	.000	.000	.001	.006	.016	.037	.069	.111	.160	.205	.238	.251	.238	.200	.146	.088	.040	.011	.001
	7	.000	.000	.000	.000	.001	.003	.009	.021	.042	.075	.117	.166	.215	.252	.267	.250	.201	.130	.057	.010
	8	.000	.000	.000	.000	.000	.000	.001	.004	.011	.023	.044	.076	.121	.176	.233	.282	.302	.276	.194	.075
	9	.000	.000	.000	.000	.000	.000	.000	.000	.002	.004	.010	.021	.040	.072	.121	.188	.268	.347	.387	.315
	10	.000	.000	.000	.000	.000	.000	.000	.000	.000	.000	.001	.003	.006	.014	.028	.056	.107	.197	.349	.599
11	0	.895	.569	.314	.167	.086	.042	.020	.009	.004	.001	.000	.000	.000	.000	.000	.000	.000	.000	.000	.000
	1	.099	.329	.384	.325	.236	.155	.093	.052	.027	.013	.005	.002	.001	.000	.000	.000	.000	.000	.000	.000
	2	.005	.087	.213	.287	.295	.258	.200	.140	.089	.051	.027	.013	.005	.002	.001	.000	.000	.000	.000	.000
	3	.000	.014	.071	.152	.221	.258	.257	.225	.177	.126	.081	.046	.023	.010	.004	.001	.000	.000	.000	.000
	4	.000	.001	.016	.054	.111	.172	.220	.243	.236	.206	.161	.113	.070	.038	.017	.006	.002	.000	.000	.000
	5	.000	.000	.002	.013	.039	.080	.132	.183	.221	.236	.226	.193	.147	.099	.057	.027	.010	.002	.000	.000
	6	.000	.000	.000	.002	.010	.027	.057	.099	.147	.193	.226	.236	.221	.183	.132	.080	.039	.013	.002	.000
	7	.000	.000	.000	.000	.002	.006	.017	.038	.070	.113	.161	.206	.236	.243	.220	.172	.111	.054	.016	.001
	8	.000	.000	.000	.000	.000	.001	.004	.010	.023	.046	.081	.126	.177	.225	.257	.258	.221	.152	.071	.014
	9	.000	.000	.000	.000	.000	.000	.001	.002	.005	.013	.027	.051	.089	.140	.200	.258	.295	.287	.213	.087
	10	.000	.000	.000	.000	.000	.000	.000	.000	.001	.002	.005	.013	.027	.052	.093	.155	.236	.325	.384	.329
	11	.000	.000	.000	.000	.000	.000	.000	.000	.000	.000	.000	.001	.004	.009	.020	.042	.086	.167	.314	.569
12	0	.886	.540	.282	.142	.069	.032	.014	.006	.002	.001	.000	.000	.000	.000	.000	.000	.000	.000	.000	.000
	1	.107	.341	.377	.301	.206	.127	.071	.037	.017	.008	.003	.001	.000	.000	.000	.000	.000	.000	.000	.000
	2	.006	.099	.230	.292	.283	.232	.168	.109	.064	.034	.016	.007	.002	.001	.000	.000	.000	.000	.000	.000
	3	.000	.017	.085	.172	.236	.258	.240	.195	.142	.092	.054	.028	.012	.005	.001	.000	.000	.000	.000	.000
	4	.000	.002	.021	.068	.133	.194	.231	.237	.213	.170	.121	.076	.042	.020	.008	.002	.001	.000	.000	.000
	5	.000	.000	.004	.019	.053	.103	.158	.204	.227	.223	.193	.149	.101	.059	.029	.011	.003	.001	.000	.000
	6	.000	.000	.000	.004	.016	.040	.079	.128	.177	.212	.226	.212	.177	.128	.079	.040	.016	.004	.000	.000
	7	.000	.000	.000	.001	.003	.011	.029	.059	.101	.149	.193	.223	.227	.204	.158	.103	.053	.019	.004	.000
	8	.000	.000	.000	.000	.001	.002	.008	.020	.042	.076	.121	.170	.213	.237	.231	.194	.133	.068	.021	.002
	9	.000	.000	.000	.000	.000	.000	.001	.005	.012	.028	.054	.092	.142	.195	.240	.258	.236	.172	.085	.017
	10	.000	.000	.000	.000	.000	.000	.000	.001	.002	.007	.016	.034	.064	.109	.168	.232	.283	.292	.230	.099
	11	.000	.000	.000	.000	.000	.000	.000	.000	.000	.001	.003	.008	.017	.037	.071	.127	.206	.301	.377	.341
	12	.000	.000	.000	.000	.000	.000	.000	.000	.000	.000	.000	.001	.002	.006	.014	.032	.069	.142	.282	.540
15	0	.860	.463	.206	.087	.035	.013	.005	.002	.000	.000	.000	.000	.000	.000	.000	.000	.000	.000	.000	.000
	1	.130	.366	.343	.231	.132	.067	.031	.013	.005	.002	.000	.000	.000	.000	.000	.000	.000	.000	.000	.000
	2	.009	.135	.267	.286	.231	.156	.092	.048	.022	.009	.003	.001	.000	.000	.000	.000	.000	.000	.000	.000
	3	.000	.031	.129	.218	.250	.225	.170	.111	.063	.032	.014	.005	.002	.000	.000	.000	.000	.000	.000	.000
	4	.000	.005	.043	.116	.188	.225	.219	.179	.127	.078	.042	.019	.007	.002	.001	.000	.000	.000	.000	.000
	5	.000	.001	.010	.045	.103	.165	.206	.212	.186	.140	.092	.051	.024	.010	.003	.001	.000	.000	.000	.000
	6	.000	.000	.002	.013	.043	.092	.147	.191	.207	.191	.153	.105	.061	.030	.012	.003	.001	.000	.000	.000
	7	.000	.000	.000	.003	.014	.039	.081	.132	.177	.201	.196	.165	.118	.071	.035	.013	.003	.001	.000	.000
	8	.000	.000	.000	.001	.003	.013	.035	.071	.118	.165	.196	.201	.177	.132	.081	.039	.014	.003	.000	.000
	9	.000	.000	.000	.000	.001	.003	.012	.030	.061	.105	.153	.191	.207	.191	.147	.092	.043	.013	.002	.000
	10	.000	.000	.000	.000	.000	.001	.003	.010	.024	.051	.092	.140	.186	.212	.206	.165	.103	.045	.010	.001
	11	.000	.000	.000	.000	.000	.000	.001	.002	.007	.019	.042	.078	.127	.179	.219	.225	.188	.116	.043	.005
	12	.000	.000	.000	.000	.000	.000	.000	.000	.002	.005	.014	.032	.063	.111	.170	.225	.250	.218	.129	.031
	13	.000	.000	.000	.000	.000	.000	.000	.000	.000	.001	.003	.009	.022	.048	.092	.156	.231	.286	.267	.135
	14	.000	.000	.000	.000	.000	.000	.000	.000	.000	.000	.000	.002	.005	.013	.031	.067	.132	.231	.343	.366
	15	.000	.000	.000	.000	.000	.000	.000	.000	.000	.000	.000	.000	.000	.002	.005	.013	.035	.087	.206	.463

Table 2 — Binomial Distribution *(continued)*

n	x	.01	.05	.10	.15	.20	.25	.30	.35	.40	.45	.50	.55	.60	.65	.70	.75	.80	.85	.90	.95
16	0	.851	.440	.185	.074	.028	.010	.003	.001	.000	.000	.000	.000	.000	.000	.000	.000	.000	.000	.000	.000
	1	.138	.371	.329	.210	.113	.053	.023	.009	.003	.001	.000	.000	.000	.000	.000	.000	.000	.000	.000	.000
	2	.010	.146	.275	.277	.211	.134	.073	.035	.015	.006	.002	.001	.000	.000	.000	.000	.000	.000	.000	.000
	3	.000	.036	.142	.229	.246	.208	.146	.089	.047	.022	.009	.003	.001	.000	.000	.000	.000	.000	.000	.000
	4	.000	.006	.051	.131	.200	.225	.204	.155	.101	.057	.028	.011	.004	.001	.000	.000	.000	.000	.000	.000
	5	.000	.001	.014	.056	.120	.180	.210	.201	.162	.112	.067	.034	.014	.005	.001	.000	.000	.000	.000	.000
	6	.000	.000	.003	.018	.055	.110	.165	.198	.198	.168	.122	.075	.039	.017	.006	.001	.000	.000	.000	.000
	7	.000	.000	.000	.005	.020	.052	.101	.152	.189	.197	.175	.132	.084	.044	.019	.006	.001	.000	.000	.000
	8	.000	.000	.000	.001	.006	.020	.049	.092	.142	.181	.196	.181	.142	.092	.049	.020	.006	.001	.000	.000
	9	.000	.000	.000	.000	.001	.006	.019	.044	.084	.132	.175	.197	.189	.152	.101	.052	.020	.005	.000	.000
	10	.000	.000	.000	.000	.000	.001	.006	.017	.039	.075	.122	.168	.198	.198	.165	.110	.055	.018	.003	.000
	11	.000	.000	.000	.000	.000	.000	.001	.005	.014	.034	.067	.112	.162	.201	.210	.180	.120	.056	.014	.001
	12	.000	.000	.000	.000	.000	.000	.000	.001	.004	.011	.028	.057	.101	.155	.204	.225	.200	.131	.051	.006
	13	.000	.000	.000	.000	.000	.000	.000	.000	.001	.003	.009	.022	.047	.089	.146	.208	.246	.229	.142	.036
	14	.000	.000	.000	.000	.000	.000	.000	.000	.000	.001	.002	.006	.015	.035	.073	.134	.211	.277	.275	.146
	15	.000	.000	.000	.000	.000	.000	.000	.000	.000	.000	.000	.001	.003	.009	.023	.053	.113	.210	.329	.371
	16	.000	.000	.000	.000	.000	.000	.000	.000	.000	.000	.000	.000	.001	.003	.010	.028	.074	.185	.440
20	0	.818	.358	.122	.039	.012	.003	.001	.000	.000	.000	.000	.000	.000	.000	.000	.000	.000	.000	.000	.000
	1	.165	.377	.270	.137	.058	.021	.007	.002	.000	.000	.000	.000	.000	.000	.000	.000	.000	.000	.000	.000
	2	.016	.189	.285	.229	.137	.067	.028	.010	.003	.001	.000	.000	.000	.000	.000	.000	.000	.000	.000	.000
	3	.001	.060	.190	.243	.205	.134	.072	.032	.012	.004	.001	.000	.000	.000	.000	.000	.000	.000	.000	.000
	4	.000	.013	.090	.182	.218	.190	.130	.074	.035	.014	.005	.001	.000	.000	.000	.000	.000	.000	.000	.000
	5	.000	.002	.032	.103	.175	.202	.179	.127	.075	.036	.015	.005	.001	.000	.000	.000	.000	.000	.000	.000
	6	.000	.000	.009	.045	.109	.169	.192	.171	.124	.075	.036	.015	.005	.001	.000	.000	.000	.000	.000	.000
	7	.000	.000	.002	.016	.055	.112	.164	.184	.166	.122	.074	.037	.015	.005	.001	.000	.000	.000	.000	.000
	8	.000	.000	.000	.005	.022	.061	.114	.161	.180	.162	.120	.073	.035	.014	.004	.001	.000	.000	.000	.000
	9	.000	.000	.000	.001	.007	.027	.065	.116	.160	.177	.160	.119	.071	.034	.012	.003	.000	.000	.000	.000
	10	.000	.000	.000	.000	.002	.010	.031	.069	.117	.159	.176	.159	.117	.069	.031	.010	.002	.000	.000	.000
	11	.000	.000	.000	.000	.000	.003	.012	.034	.071	.119	.160	.177	.160	.116	.065	.027	.007	.001	.000	.000
	12	.000	.000	.000	.000	.000	.001	.004	.014	.035	.073	.120	.162	.180	.161	.114	.061	.022	.005	.000	.000
	13	.000	.000	.000	.000	.000	.000	.001	.005	.015	.037	.074	.122	.166	.184	.164	.112	.055	.016	.002	.000
	14	.000	.000	.000	.000	.000	.000	.000	.001	.005	.015	.037	.075	.124	.171	.192	.169	.109	.045	.009	.000
	15	.000	.000	.000	.000	.000	.000	.000	.001	.005	.015	.036	.075	.127	.179	.202	.175	.103	.032	.002
	16	.000	.000	.000	.000	.000	.000	.000	.000	.001	.005	.014	.035	.074	.130	.190	.218	.182	.090	.013
	17	.000	.000	.000	.000	.000	.000	.000	.000	.000	.000	.001	.004	.012	.032	.072	.134	.205	.243	.190	.060
	18	.000	.000	.000	.000	.000	.000	.000	.000	.000	.000	.000	.001	.003	.010	.028	.067	.137	.229	.285	.189
	19	.000	.000	.000	.000	.000	.000	.000	.000	.000	.000	.000	.000	.000	.002	.007	.021	.058	.137	.270	.377
	20	.000	.000	.000	.000	.000	.000	.000	.000	.000	.000	.000	.000	.000	.000	.001	.003	.012	.039	.122	.358

Table 3 — Poisson Distribution

x	μ 0.1	0.2	0.3	0.4	0.5	0.6	0.7	0.8	0.9	1.0
0	.9048	.8187	.7408	.6703	.6065	.5488	.4966	.4493	.4066	.3679
1	.0905	.1637	.2222	.2681	.3033	.3293	.3476	.3595	.3659	.3679
2	.0045	.0164	.0333	.0536	.0758	.0988	.1217	.1438	.1647	.1839
3	.0002	.0011	.0033	.0072	.0126	.0198	.0284	.0383	.0494	.0613
4	.0000	.0001	.0003	.0007	.0016	.0030	.0050	.0077	.0111	.0153
5	.0000	.0000	.0000	.0001	.0002	.0004	.0007	.0012	.0020	.0031
6	.0000	.0000	.0000	.0000	.0000	.0000	.0001	.0002	.0003	.0005
7	.0000	.0000	.0000	.0000	.0000	.0000	.0000	.0000	.0000	.0001

x	μ 1.1	1.2	1.3	1.4	1.5	1.6	1.7	1.8	1.9	2.0
0	.3329	.3012	.2725	.2466	.2231	.2019	.1827	.1653	.1496	.1353
1	.3662	.3614	.3543	.3452	.3347	.3230	.3106	.2975	.2842	.2707
2	.2014	.2169	.2303	.2417	.2510	.2584	.2640	.2678	.2700	.2707
3	.0738	.0867	.0998	.1128	.1255	.1378	.1496	.1607	.1710	.1804
4	.0203	.0260	.0324	.0395	.0471	.0551	.0636	.0723	.0812	.0902
5	.0045	.0062	.0084	.0111	.0141	.0176	.0216	.0260	.0309	.0361
6	.0008	.0012	.0018	.0026	.0035	.0047	.0061	.0078	.0098	.0120
7	.0001	.0002	.0003	.0005	.0008	.0011	.0015	.0020	.0027	.0034
8	.0000	.0000	.0001	.0001	.0001	.0002	.0003	.0005	.0006	.0009
9	.0000	.0000	.0000	.0000	.0000	.0000	.0001	.0001	.0001	.0002

x	μ 2.1	2.2	2.3	2.4	2.5	2.6	2.7	2.8	2.9	3.0
0	.1225	.1108	.1003	.0907	.0821	.0743	.0672	.0608	.0550	.0498
1	.2572	.2438	.2306	.2177	.2052	.1931	.1815	.1703	.1596	.1494
2	.2700	.2681	.2652	.2613	.2565	.2510	.2450	.2384	.2314	.2240
3	.1890	.1966	.2033	.2090	.2138	.2176	.2205	.2225	.2237	.2240
4	.0992	.1082	.1169	.1254	.1336	.1414	.1488	.1557	.1622	.1680
5	.0417	.0476	.0538	.0602	.0668	.0735	.0804	.0872	.0940	.1008
6	.0146	.0174	.0206	.0241	.0278	.0319	.0362	.0407	.0455	.0504
7	.0044	.0055	.0068	.0083	.0099	.0118	.0139	.0163	.0188	.0216
8	.0011	.0015	.0019	.0025	.0031	.0038	.0047	.0057	.0068	.0081
9	.0003	.0004	.0005	.0007	.0009	.0011	.0014	.0018	.0022	.0027
10	.0001	.0001	.0001	.0002	.0002	.0003	.0004	.0005	.0006	.0008
11	.0000	.0000	.0000	.0000	.0000	.0001	.0001	.0001	.0002	.0002
12	.0000	.0000	.0000	.0000	.0000	.0000	.0000	.0000	.0000	.0001

x	μ 3.1	3.2	3.3	3.4	3.5	3.6	3.7	3.8	3.9	4.0
0	.0450	.0408	.0369	.0334	.0302	.0273	.0247	.0224	.0202	.0183
1	.1397	.1304	.1217	.1135	.1057	.0984	.0915	.0850	.0789	.0733
2	.2165	.2087	.2008	.1929	.1850	.1771	.1692	.1615	.1539	.1465
3	.2237	.2226	.2209	.2186	.2158	.2125	.2087	.2046	.2001	.1954
4	.1734	.1781	.1823	.1858	.1888	.1912	.1931	.1944	.1951	.1954
5	.1075	.1140	.1203	.1264	.1322	.1377	.1429	.1477	.1522	.1563
6	.0555	.0608	.0662	.0716	.0771	.0826	.0881	.0936	.0989	.1042
7	.0246	.0278	.0312	.0348	.0385	.0425	.0466	.0508	.0551	.0595
8	.0095	.0111	.0129	.0148	.0169	.0191	.0215	.0241	.0269	.0298
9	.0033	.0040	.0047	.0056	.0066	.0076	.0089	.0102	.0116	.0132
10	.0010	.0013	.0016	.0019	.0023	.0028	.0033	.0039	.0045	.0053
11	.0003	.0004	.0005	.0006	.0007	.0009	.0011	.0013	.0016	.0019
12	.0001	.0001	.0001	.0002	.0002	.0003	.0003	.0004	.0005	.0006
13	.0000	.0000	.0000	.0000	.0001	.0001	.0001	.0001	.0002	.0002
14	.0000	.0000	.0000	.0000	.0000	.0000	.0000	.0000	.0000	.0001

Reprinted with permission from W. H. Beyer, *Handbook of Tables for Probability and Statistics*, 2e, CRC Press, Boca Raton, Florida, 1986.

Table 3— Poisson Distribution *(continued)*

					μ					
x	4.1	4.2	4.3	4.4	4.5	4.6	4.7	4.8	4.9	5.0
0	.0166	.0150	.0136	.0123	.0111	.0101	.0091	.0082	.0074	.0067
1	.0679	.0630	.0583	.0540	.0500	.0462	.0427	.0395	.0365	.0337
2	.1393	.1323	.1254	.1188	.1125	.1063	.1005	.0948	.0894	.0842
3	.1904	.1852	.1798	.1743	.1687	.1631	.1574	.1517	.1460	.1404
4	.1951	.1944	.1933	.1917	.1898	.1875	.1849	.1820	.1789	.1755
5	.1600	.1633	.1662	.1687	.1708	.1725	.1738	.1747	.1753	.1755
6	.1093	.1143	.1191	.1237	.1281	.1323	.1362	.1398	.1432	.1462
7	.0640	.0686	.0732	.0778	.0824	.0869	.0914	.0959	.1002	.1044
8	.0328	.0360	.0393	.0428	.0463	.0500	.0537	.0575	.0614	.0653
9	.0150	.0168	.0188	.0209	.0232	.0255	.0280	.0307	.0334	.0363
10	.0061	.0071	.0081	.0092	.0104	.0118	.0132	.0147	.0164	.0181
11	.0023	.0027	.0032	.0037	.0043	.0049	.0056	.0064	.0073	.0082
12	.0008	.0009	.0011	.0014	.0016	.0019	.0022	.0026	.0030	.0034
13	.0002	.0003	.0004	.0005	.0006	.0007	.0008	.0009	.0011	.0013
14	.0001	.0001	.0001	.0001	.0002	.0002	.0003	.0003	.0004	.0005
15	.0000	.0000	.0000	.0000	.0001	.0001	.0001	.0001	.0001	.0002

					μ					
x	5.1	5.2	5.3	5.4	5.5	5.6	5.7	5.8	5.9	6.0
0	.0061	.0055	.0050	.0045	.0041	.0037	.0033	.0030	.0027	.0025
1	.0311	.0287	.0265	.0244	.0225	.0207	.0191	.0176	.0162	.0149
2	.0793	.0746	.0701	.0659	.0618	.0580	.0544	.0509	.0477	.0446
3	.1348	.1293	.1239	.1185	.1133	.1082	.1033	.0985	.0938	.0892
4	.1719	.1681	.1641	.1600	.1558	.1515	.1472	.1428	.1383	.1339
5	.1753	.1748	.1740	.1728	.1714	.1697	.1678	.1656	.1632	.1606
6	.1490	.1515	.1537	.1555	.1571	.1584	.1594	.1601	.1605	.1606
7	.1086	.1125	.1163	.1200	.1234	.1267	.1298	.1326	.1353	.1377
8	.0692	.0731	.0771	.0810	.0849	.0887	.0925	.0962	.0998	.1033
9	.0392	.0423	.0454	.0486	.0519	.0552	.0586	.0620	.0654	.0688
10	.0200	.0220	.0241	.0262	.0285	.0309	.0334	.0359	.0386	.0413
11	.0093	.0104	.0116	.0129	.0143	.0157	.0173	.0190	.0207	.0225
12	.0039	.0045	.0051	.0058	.0065	.0073	.0082	.0092	.0102	.0113
13	.0015	.0018	.0021	.0024	.0028	.0032	.0036	.0041	.0046	.0052
14	.0006	.0007	.0008	.0009	.0011	.0013	.0015	.0017	.0019	.0022
15	.0002	.0002	.0003	.0003	.0004	.0005	.0006	.0007	.0008	.0009
16	.0001	.0001	.0001	.0001	.0001	.0002	.0002	.0002	.0003	.0003
17	.0000	.0000	.0000	.0000	.0000	.0000	.0001	.0001	.0001	.0001

Table 3 — Poisson Distribution *(continued)*

x	6.1	6.2	6.3	6.4	6.5	6.6	6.7	6.8	6.9	7.0
0	.0022	.0020	.0018	.0017	.0015	.0014	.0012	.0011	.0010	.0009
1	.0137	.0126	.0116	.0106	.0098	.0090	.0082	.0076	.0070	.0064
2	.0417	.0390	.0364	.0340	.0318	.0296	.0276	.0258	.0240	.0223
3	.0848	.0806	.0765	.0726	.0688	.0652	.0617	.0584	.0552	.0521
4	.1294	.1249	.1205	.1162	.1118	.1076	.1034	.0992	.0952	.0912
5	.1579	.1549	.1519	.1487	.1454	.1420	.1385	.1349	.1314	.1277
6	.1605	.1601	.1595	.1586	.1575	.1562	.1546	.1529	.1511	.1490
7	.1399	.1418	.1435	.1450	.1462	.1472	.1480	.1486	.1489	.1490
8	.1066	.1099	.1130	.1160	.1188	.1215	.1240	.1263	.1284	.1304
9	.0723	.0757	.0791	.0825	.0858	.0891	.0923	.0954	.0985	.1014
10	.0441	.0469	.0498	.0528	.0558	.0588	.0618	.0649	.0679	.0710
11	.0245	.0265	.0285	.0307	.0330	.0353	.0377	.0401	.0426	.0452
12	.0124	.0137	.0150	.0164	.0179	.0194	.0210	.0227	.0245	.0264
13	.0058	.0065	.0073	.0081	.0089	.0098	.0108	.0119	.0130	.0142
14	.0025	.0029	.0033	.0037	.0041	.0046	.0052	.0058	.0064	.0071
15	.0010	.0012	.0014	.0016	.0018	.0020	.0023	.0026	.0029	.0033
16	.0004	.0005	.0005	.0006	.0007	.0008	.0010	.0011	.0013	.0014
17	.0001	.0002	.0002	.0002	.0003	.0003	.0004	.0004	.0005	.0006
18	.0000	.0001	.0001	.0001	.0001	.0001	.0001	.0002	.0002	.0002
19	.0000	.0000	.0000	.0000	.0000	.0000	.0000	.0001	.0001	.0001

μ

x	7.1	7.2	7.3	7.4	7.5	7.6	7.7	7.8	7.9	8.0
0	.0008	.0007	.0007	.0006	.0006	.0005	.0005	.0004	.0004	.0003
1	.0059	.0054	.0049	.0045	.0041	.0038	.0035	.0032	.0029	.0027
2	.0208	.0194	.0180	.0167	.0156	.0145	.0134	.0125	.0116	.0107
3	.0492	.0464	.0438	.0413	.0389	.0366	.0345	.0324	.0305	.0286
4	.0874	.0836	.0799	.0764	.0729	.0696	.0663	.0632	.0602	.0573
5	.1241	.1204	.1167	.1130	.1094	.1057	.1021	.0986	.0951	.0916
6	.1468	.1445	.1420	.1394	.1367	.1339	.1311	.1282	.1252	.1221
7	.1489	.1486	.1481	.1474	.1465	.1454	.1442	.1428	.1413	.1396
8	.1321	.1337	.1351	.1363	.1373	.1382	.1388	.1392	.1395	.1396
9	.1042	.1070	.1096	.1121	.1144	.1167	.1187	.1207	.1224	.1241
10	.0740	.0770	.0800	.0829	.0858	.0887	.0914	.0941	.0967	.0993
11	.0478	.0504	.0531	.0558	.0585	.0613	.0640	.0667	.0695	.0722
12	.0283	.0303	.0323	.0344	.0366	.0388	.0411	.0434	.0457	.0481
13	.0154	.0168	.0181	.0196	.0211	.0227	.0243	.0260	.0278	.0296
14	.0078	.0086	.0095	.0104	.0113	.0123	.0134	.0145	.0157	.0169
15	.0037	.0041	.0046	.0051	.0057	.0062	.0069	.0075	.0083	.0090
16	.0016	.0019	.0021	.0024	.0026	.0030	.0033	.0037	.0041	.0045
17	.0007	.0008	.0009	.0010	.0012	.0013	.0015	.0017	.0019	.0021
18	.0003	.0003	.0004	.0004	.0005	.0006	.0006	.0007	.0008	.0009
19	.0001	.0001	.0001	.0002	.0002	.0002	.0003	.0003	.0003	.0004
20	.0000	.0000	.0001	.0001	.0001	.0001	.0001	.0001	.0001	.0002
21	.0000	.0000	.0000	.0000	.0000	.0000	.0000	.0000	.0001	.0001

Table 3— Poisson Distribution *(continued)*

x	8.1	8.2	8.3	8.4	8.5	8.6	8.7	8.8	8.9	9.0
0	.0003	.0003	.0002	.0002	.0002	.0002	.0002	.0002	.0001	.0001
1	.0025	.0023	.0021	.0019	.0017	.0016	.0014	.0013	.0012	.0011
2	.0100	.0092	.0086	.0079	.0074	.0068	.0063	.0058	.0054	.0050
3	.0269	.0252	.0237	.0222	.0208	.0195	.0183	.0171	.0160	.0150
4	.0544	.0517	.0491	.0466	.0443	.0420	.0398	.0377	.0357	.0337
5	.0882	.0849	.0816	.0784	.0752	.0722	.0692	.0663	.0635	.0607
6	.1191	.1160	.1128	.1097	.1066	.1034	.1003	.0972	.0941	.0911
7	.1378	.1358	.1338	.1317	.1294	.1271	.1247	.1222	.1197	.1171
8	.1395	.1392	.1388	.1382	.1375	.1366	.1356	.1344	.1332	.1318
9	.1256	.1269	.1280	.1290	.1299	.1306	.1311	.1315	.1317	.1318
10	.1017	.1040	.1063	.1084	.1104	.1123	.1140	.1157	.1172	.1186
11	.0749	.0776	.0802	.0828	.0853	.0878	.0902	.0925	.0948	.0970
12	.0505	.0530	.0555	.0579	.0604	.0629	.0654	.0679	.0703	.0728
13	.0315	.0334	.0354	.0374	.0395	.0416	.0438	.0459	.0481	.0504
14	.0182	.0196	.0210	.0225	.0240	.0256	.0272	.0289	.0306	.0324
15	.0098	.0107	.0116	.0126	.0136	.0147	.0158	.0169	.0182	.0194
16	.0050	.0055	.0060	.0066	.0072	.0079	.0086	.0093	.0101	.0109
17	.0024	.0026	.0029	.0033	.0036	.0040	.0044	.0048	.0053	.0058
18	.0011	.0012	.0014	.0015	.0017	.0019	.0021	.0024	.0026	.0029
19	.0005	.0005	.0006	.0007	.0008	.0009	.0010	.0011	.0012	.0014
20	.0002	.0002	.0002	.0003	.0003	.0004	.0004	.0005	.0005	.0006
21	.0001	.0001	.0001	.0001	.0001	.0002	.0002	.0002	.0002	.0003
22	.0000	.0000	.0000	.0000	.0001	.0001	.0001	.0001	.0001	.0001

x	9.1	9.2	9.3	9.4	9.5	9.6	9.7	9.8	9.9	10.0
0	.0001	.0001	.0001	.0001	.0001	.0001	.0001	.0001	.0001	.0000
1	.0010	.0009	.0009	.0008	.0007	.0007	.0006	.0005	.0005	.0005
2	.0046	.0043	.0040	.0037	.0034	.0031	.0029	.0027	.0025	.0023
3	.0140	.0131	.0123	.0115	.0107	.0100	.0093	.0087	.0081	.0076
4	.0319	.0302	.0285	.0269	.0254	.0240	.0226	.0213	.0201	.0189
5	.0581	.0555	.0530	.0506	.0483	.0460	.0439	.0418	.0398	.0378
6	.0881	.0851	.0822	.0793	.0764	.0736	.0709	.0682	.0656	.0631
7	.1145	.1118	.1091	.1064	.1037	.1010	.0982	.0955	.0928	.0901
8	.1302	.1286	.1269	.1251	.1232	.1212	.1191	.1170	.1148	.1126
9	.1317	.1315	.1311	.1306	.1300	.1293	.1284	.1274	.1263	.1251
10	.1198	.1210	.1219	.1228	.1235	.1241	.1245	.1249	.1250	.1251
11	.0991	.1012	.1031	.1049	.1067	.1083	.1098	.1112	.1125	.1137
12	.0752	.0776	.0799	.0822	.0844	.0866	.0888	.0908	.0928	.0948
13	.0526	.0549	.0572	.0594	.0617	.0640	.0662	.0685	.0707	.0729
14	.0342	.0361	.0380	.0399	.0419	.0439	.0459	.0479	.0500	.0521
15	.0208	.0221	.0235	.0250	.0265	.0281	.0297	.0313	.0330	.0347
16	.0118	.0127	.0137	.0147	.0157	.0168	.0180	.0192	.0204	.0217
17	.0063	.0069	.0075	.0081	.0088	.0095	.0103	.0111	.0119	.0128
18	.0032	.0035	.0039	.0042	.0046	.0051	.0055	.0060	.0065	.0071
19	.0015	.0017	.0019	.0021	.0023	.0026	.0028	.0031	.0034	.0037
20	.0007	.0008	.0009	.0010	.0011	.0012	.0014	.0015	.0017	.0019
21	.0003	.0003	.0004	.0004	.0005	.0006	.0006	.0007	.0008	.0009
22	.0001	.0001	.0002	.0002	.0002	.0002	.0003	.0003	.0004	.0004
23	.0000	.0001	.0001	.0001	.0001	.0001	.0001	.0001	.0002	.0002
24	.0000	.0000	.0000	.0000	.0000	.0000	.0000	.0001	.0001	.0001

Table 3— Poisson Distribution *(continued)*

x	11	12	13	14	15	16	17	18	19	20
0	.0000	.0000	.0000	.0000	.0000	.0000	.0000	.0000	.0000	.0000
1	.0002	.0001	.0000	.0000	.0000	.0000	.0000	.0000	.0000	.0000
2	.0010	.0004	.0002	.0001	.0000	.0000	.0000	.0000	.0000	.0000
3	.0037	.0018	.0008	.0004	.0002	.0001	.0000	.0000	.0000	.0000
4	.0102	.0053	.0027	.0013	.0006	.0003	.0001	.0001	.0000	.0000
5	.0224	.0127	.0070	.0037	.0019	.0010	.0005	.0002	.0001	.0001
6	.0411	.0255	.0152	.0087	.0048	.0026	.0014	.0007	.0004	.0002
7	.0646	.0437	.0281	.0174	.0104	.0060	.0034	.0018	.0010	.0005
8	.0888	.0655	.0457	.0304	.0194	.0120	.0072	.0042	.0024	.0013
9	.1085	.0874	.0661	.0473	.0324	.0213	.0135	.0083	.0050	.0029
10	.1194	.1048	.0859	.0663	.0486	.0341	.0230	.0150	.0095	.0058
11	.1194	.1144	.1015	.0844	.0663	.0496	.0355	.0245	.0164	.0106
12	.1094	.1144	.1099	.0984	.0829	.0661	.0504	.0368	.0259	.0176
13	.0926	.1056	.1099	.1060	.0956	.0814	.0658	.0509	.0378	.0271
14	.0728	.0905	.1021	.1060	.1024	.0930	.0800	.0655	.0514	.0387
15	.0534	.0724	.0885	.0989	.1024	.0992	.0906	.0786	.0650	.0516
16	.0367	.0543	.0719	.0866	.0960	.0992	.0963	.0884	.0772	.0646
17	.0237	.0383	.0550	.0713	.0847	.0934	.0963	.0936	.0863	.0760
18	.0145	.0256	.0397	.0554	.0706	.0830	.0909	.0936	.0911	.0844
19	.0084	.0161	.0272	.0409	.0557	.0699	.0814	.0887	.0911	.0888
20	.0046	.0097	.0177	.0286	.0418	.0559	.0692	.0798	.0866	.0888

x	11	12	13	14	15	16	17	18	19	20
21	.0024	.0055	.0109	.0191	.0299	.0426	.0560	.0684	.0783	.0846
22	.0012	.0030	.0065	.0121	.0204	.0310	.0433	.0560	.0676	.0769
23	.0006	.0016	.0037	.0074	.0133	.0216	.0320	.0438	.0559	.0669
24	.0003	.0008	.0020	.0043	.0083	.0144	.0226	.0328	.0442	.0557
25	.0001	.0004	.0010	.0024	.0050	.0092	.0154	.0237	.0336	.0446
26	.0000	.0002	.0005	.0013	.0029	.0057	.0101	.0164	.0246	.0343
27	.0000	.0001	.0002	.0007	.0016	.0034	.0063	.0109	.0173	.0254
28	.0000	.0000	.0001	.0003	.0009	.0019	.0038	.0070	.0117	.0181
29	.0000	.0000	.0001	.0002	.0004	.0011	.0023	.0044	.0077	.0125
30	.0000	.0000	.0000	.0001	.0002	.0006	.0013	.0026	.0049	.0083
31	.0000	.0000	.0000	.0000	.0001	.0003	.0007	.0015	.0030	.0054
32	.0000	.0000	.0000	.0000	.0001	.0001	.0004	.0009	.0018	.0034
33	.0000	.0000	.0000	.0000	.0000	.0001	.0002	.0005	.0010	.0020
34	.0000	.0000	.0000	.0000	.0000	.0000	.0001	.0002	.0006	.0012
35	.0000	.0000	.0000	.0000	.0000	.0000	.0000	.0001	.0003	.0007
36	.0000	.0000	.0000	.0000	.0000	.0000	.0000	.0001	.0002	.0004
37	.0000	.0000	.0000	.0000	.0000	.0000	.0000	.0000	.0001	.0002
38	.0000	.0000	.0000	.0000	.0000	.0000	.0000	.0000	.0000	.0001
39	.0000	.0000	.0000	.0000	.0000	.0000	.0000	.0000	.0000	.0001

Table 4 — Standard Normal Distribution

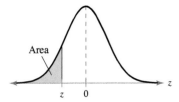

Area

z	.09	.08	.07	.06	.05	.04	.03	.02	.01	.00
− 3.4	.0002	.0003	.0003	.0003	.0003	.0003	.0003	.0003	.0003	.0003
− 3.3	.0003	.0004	.0004	.0004	.0004	.0004	.0004	.0005	.0005	.0005
− 3.2	.0005	.0005	.0005	.0006	.0006	.0006	.0006	.0006	.0007	.0007
− 3.1	.0007	.0007	.0008	.0008	.0008	.0008	.0009	.0009	.0009	.0010
− 3.0	.0010	.0010	.0011	.0011	.0011	.0012	.0012	.0013	.0013	.0013
− 2.9	.0014	.0014	.0015	.0015	.0016	.0016	.0017	.0018	.0018	.0019
− 2.8	.0019	.0020	.0021	.0021	.0022	.0023	.0023	.0024	.0025	.0026
− 2.7	.0026	.0027	.0028	.0029	.0030	.0031	.0032	.0033	.0034	.0035
− 2.6	.0036	.0037	.0038	.0039	.0040	.0041	.0043	.0044	.0045	.0047
− 2.5	.0048	.0049	.0051	.0052	.0054	.0055	.0057	.0059	.0060	.0062
− 2.4	.0064	.0066	.0068	.0069	.0071	.0073	.0075	.0078	.0080	.0082
− 2.3	.0084	.0087	.0089	.0091	.0094	.0096	.0099	.0102	.0104	.0107
− 2.2	.0110	.0113	.0116	.0119	.0122	.0125	.0129	.0132	.0136	.0139
− 2.1	.0143	.0146	.0150	.0154	.0158	.0162	.0166	.0170	.0174	.0179
− 2.0	.0183	.0188	.0192	.0197	.0202	.0207	.0212	.0217	.0222	.0228
− 1.9	.0233	.0239	.0244	.0250	.0256	.0262	.0268	.0274	.0281	.0287
− 1.8	.0294	.0301	.0307	.0314	.0322	.0329	.0336	.0344	.0351	.0359
− 1.7	.0367	.0375	.0384	.0392	.0401	.0409	.0418	.0427	.0436	.0446
− 1.6	.0455	.0465	.0475	.0485	.0495	.0505	.0516	.0526	.0537	.0548
− 1.5	.0559	.0571	.0582	.0594	.0606	.0618	.0630	.0643	.0655	.0668
− 1.4	.0681	.0694	.0708	.0721	.0735	.0749	.0764	.0778	.0793	.0808
− 1.3	.0823	.0838	.0853	.0869	.0885	.0901	.0918	.0934	.0951	.0968
− 1.2	.0985	.1003	.1020	.1038	.1056	.1075	.1093	.1112	.1131	.1151
− 1.1	.1170	.1190	.1210	.1230	.1251	.1271	.1292	.1314	.1335	.1357
− 1.0	.1379	.1401	.1423	.1446	.1469	.1492	.1515	.1539	.1562	.1587
− 0.9	.1611	.1635	.1660	.1685	.1711	.1736	.1762	.1788	.1814	.1841
− 0.8	.1867	.1894	.1922	.1949	.1977	.2005	.2033	.2061	.2090	.2119
− 0.7	.2148	.2177	.2206	.2236	.2266	.2296	.2327	.2358	.2389	.2420
− 0.6	.2451	.2483	.2514	.2546	.2578	.2611	.2643	.2676	.2709	.2743
− 0.5	.2776	.2810	.2843	.2877	.2912	.2946	.2981	.3015	.3050	.3085
− 0.4	.3121	.3156	.3192	.3228	.3264	.3300	.3336	.3372	.3409	.3446
− 0.3	.3483	.3520	.3557	.3594	.3632	.3669	.3707	.3745	.3783	.3821
− 0.2	.3859	.3897	.3936	.3974	.4013	.4052	.4090	.4129	.4168	.4207
− 0.1	.4247	.4286	.4325	.4364	.4404	.4443	.4483	.4522	.4562	.4602
− 0.0	.4641	.4681	.4721	.4761	.4801	.4840	.4880	.4920	.4960	.5000

Critical Values

Level of Confidence c	z_c
0.80	1.28
0.90	1.645
0.95	1.96
0.99	2.575

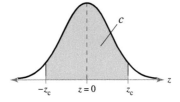

Table A-3, pp. 681–682 from *Probability and Statistics for Engineers and Scientists*, 6e by Walpole, Meyers, and Myers. Copyright 1997. Reprinted by permission of Pearson Prentice Hall, Upper Saddle River, N.J.

Table 4 — Standard Normal Distribution *(continued)*

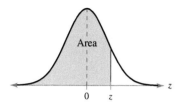

z	.00	.01	.02	.03	.04	.05	.06	.07	.08	.09
0.0	.5000	.5040	.5080	.5120	.5160	.5199	.5239	.5279	.5319	.5359
0.1	.5398	.5438	.5478	.5517	.5557	.5596	.5636	.5675	.5714	.5753
0.2	.5793	.5832	.5871	.5910	.5948	.5987	.6026	.6064	.6103	.6141
0.3	.6179	.6217	.6255	.6293	.6331	.6368	.6406	.6443	.6480	.6517
0.4	.6554	.6591	.6628	.6664	.6700	.6736	.6772	.6808	.6844	.6879
0.5	.6915	.6950	.6985	.7019	.7054	.7088	.7123	.7157	.7190	.7224
0.6	.7257	.7291	.7324	.7357	.7389	.7422	.7454	.7486	.7517	.7549
0.7	.7580	.7611	.7642	.7673	.7704	.7734	.7764	.7794	.7823	.7852
0.8	.7881	.7910	.7939	.7967	.7995	.8023	.8051	.8078	.8106	.8133
0.9	.8159	.8186	.8212	.8238	.8264	.8289	.8315	.8340	.8365	.8389
1.0	.8413	.8438	.8461	.8485	.8508	.8531	.8554	.8577	.8599	.8621
1.1	.8643	.8665	.8686	.8708	.8729	.8749	.8770	.8790	.8810	.8830
1.2	.8849	.8869	.8888	.8907	.8925	.8944	.8962	.8980	.8997	.9015
1.3	.9032	.9049	.9066	.9082	.9099	.9115	.9131	.9147	.9162	.9177
1.4	.9192	.9207	.9222	.9236	.9251	.9265	.9279	.9292	.9306	.9319
1.5	.9332	.9345	.9357	.9370	.9382	.9394	.9406	.9418	.9429	.9441
1.6	.9452	.9463	.9474	.9484	.9495	.9505	.9515	.9525	.9535	.9545
1.7	.9554	.9564	.9573	.9582	.9591	.9599	.9608	.9616	.9625	.9633
1.8	.9641	.9649	.9656	.9664	.9671	.9678	.9686	.9693	.9699	.9706
1.9	.9713	.9719	.9726	.9732	.9738	.9744	.9750	.9756	.9761	.9767
2.0	.9772	.9778	.9783	.9788	.9793	.9798	.9803	.9808	.9812	.9817
2.1	.9821	.9826	.9830	.9834	.9838	.9842	.9846	.9850	.9854	.9857
2.2	.9861	.9864	.9868	.9871	.9875	.9878	.9881	.9884	.9887	.9890
2.3	.9893	.9896	.9898	.9901	.9904	.9906	.9909	.9911	.9913	.9916
2.4	.9918	.9920	.9922	.9925	.9927	.9929	.9931	.9932	.9934	.9936
2.5	.9938	.9940	.9941	.9943	.9945	.9946	.9948	.9949	.9951	.9952
2.6	.9953	.9955	.9956	.9957	.9959	.9960	.9961	.9962	.9963	.9964
2.7	.9965	.9966	.9967	.9968	.9969	.9970	.9971	.9972	.9973	.9974
2.8	.9974	.9975	.9976	.9977	.9977	.9978	.9979	.9979	.9980	.9981
2.9	.9981	.9982	.9982	.9983	.9984	.9984	.9985	.9985	.9986	.9986
3.0	.9987	.9987	.9987	.9988	.9988	.9989	.9989	.9989	.9990	.9990
3.1	.9990	.9991	.9991	.9991	.9992	.9992	.9992	.9992	.9993	.9993
3.2	.9993	.9993	.9994	.9994	.9994	.9994	.9994	.9995	.9995	.9995
3.3	.9995	.9995	.9995	.9996	.9996	.9996	.9996	.9996	.9996	.9997
3.4	.9997	.9997	.9997	.9997	.9997	.9997	.9997	.9997	.9997	.9998

Table 5 — *t*-Distribution

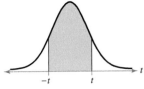

c-confidence interval

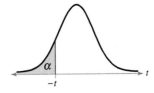

Left-tailed test

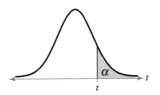

Right-tailed test

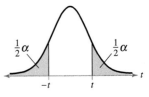

Two-tailed test

	Level of confidence, c	0.50	0.80	0.90	0.95	0.98	0.99
	One tail, α	0.25	0.10	0.05	0.025	0.01	0.005
d.f.	Two tails, α	0.50	0.20	0.10	0.05	0.02	0.01
1		1.000	3.078	6.314	12.706	31.821	63.657
2		.816	1.886	2.920	4.303	6.965	9.925
3		.765	1.638	2.353	3.182	4.541	5.841
4		.741	1.533	2.132	2.776	3.747	4.604
5		.727	1.476	2.015	2.571	3.365	4.032
6		.718	1.440	1.943	2.447	3.143	3.707
7		.711	1.415	1.895	2.365	2.998	3.499
8		.706	1.397	1.860	2.306	2.896	3.355
9		.703	1.383	1.833	2.262	2.821	3.250
10		.700	1.372	1.812	2.228	2.764	3.169
11		.697	1.363	1.796	2.201	2.718	3.106
12		.695	1.356	1.782	2.179	2.681	3.055
13		.694	1.350	1.771	2.160	2.650	3.012
14		.692	1.345	1.761	2.145	2.624	2.977
15		.691	1.341	1.753	2.131	2.602	2.947
16		.690	1.337	1.746	2.120	2.583	2.921
17		.689	1.333	1.740	2.110	2.567	2.898
18		.688	1.330	1.734	2.101	2.552	2.878
19		.688	1.328	1.729	2.093	2.539	2.861
20		.687	1.325	1.725	2.086	2.528	2.845
21		.686	1.323	1.721	2.080	2.518	2.831
22		.686	1.321	1.717	2.074	2.508	2.819
23		.685	1.319	1.714	2.069	2.500	2.807
24		.685	1.318	1.711	2.064	2.492	2.797
25		.684	1.316	1.708	2.060	2.485	2.787
26		.684	1.315	1.706	2.056	2.479	2.779
27		.684	1.314	1.703	2.052	2.473	2.771
28		.683	1.313	1.701	2.048	2.467	2.763
29		.683	1.311	1.699	2.045	2.462	2.756
∞		.674	1.282	1.645	1.960	2.326	2.576

Adapted from W. H. Beyer, *Handbook of Tables of Probability and Statistics*, 2e, CRC Press, Boca Raton, Florida, 1986. Reprinted with permission.

Table 6 — Chi-Square Distribution

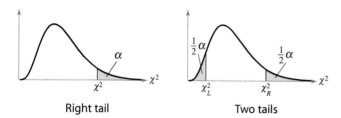

Right tail

Two tails

Degrees of freedom	α									
	0.995	0.99	0.975	0.95	0.90	0.10	0.05	0.025	0.01	0.005
1	—	—	0.001	0.004	0.016	2.706	3.841	5.024	6.635	7.879
2	0.010	0.020	0.051	0.103	0.211	4.605	5.991	7.378	9.210	10.597
3	0.072	0.115	0.216	0.352	0.584	6.251	7.815	9.348	11.345	12.838
4	0.207	0.297	0.484	0.711	1.064	7.779	9.488	11.143	13.277	14.860
5	0.412	0.554	0.831	1.145	1.610	9.236	11.071	12.833	15.086	16.750
6	0.676	0.872	1.237	1.635	2.204	10.645	12.592	14.449	16.812	18.548
7	0.989	1.239	1.690	2.167	2.833	12.017	14.067	16.013	18.475	20.278
8	1.344	1.646	2.180	2.733	3.490	13.362	15.507	17.535	20.090	21.955
9	1.735	2.088	2.700	3.325	4.168	14.684	16.919	19.023	21.666	23.589
10	2.156	2.558	3.247	3.940	4.865	15.987	18.307	20.483	23.209	25.188
11	2.603	3.053	3.816	4.575	5.578	17.275	19.675	21.920	24.725	26.757
12	3.074	3.571	4.404	5.226	6.304	18.549	21.026	23.337	26.217	28.299
13	3.565	4.107	5.009	5.892	7.042	19.812	22.362	24.736	27.688	29.819
14	4.075	4.660	5.629	6.571	7.790	21.064	23.685	26.119	29.141	31.319
15	4.601	5.229	6.262	7.261	8.547	22.307	24.996	27.488	30.578	32.801
16	5.142	5.812	6.908	7.962	9.312	23.542	26.296	28.845	32.000	34.267
17	5.697	6.408	7.564	8.672	10.085	24.769	27.587	30.191	33.409	35.718
18	6.265	7.015	8.231	9.390	10.865	25.989	28.869	31.526	34.805	37.156
19	6.844	7.633	8.907	10.117	11.651	27.204	30.144	32.852	36.191	38.582
20	7.434	8.260	9.591	10.851	12.443	28.412	31.410	34.170	37.566	39.997
21	8.034	8.897	10.283	11.591	13.240	29.615	32.671	35.479	38.932	41.401
22	8.643	9.542	10.982	12.338	14.042	30.813	33.924	36.781	40.289	42.796
23	9.260	10.196	11.689	13.091	14.848	32.007	35.172	38.076	41.638	44.181
24	9.886	10.856	12.401	13.848	15.659	33.196	36.415	39.364	42.980	45.559
25	10.520	11.524	13.120	14.611	16.473	34.382	37.652	40.646	44.314	46.928
26	11.160	12.198	13.844	15.379	17.292	35.563	38.885	41.923	45.642	48.290
27	11.808	12.879	14.573	16.151	18.114	36.741	40.113	43.194	46.963	49.645
28	12.461	13.565	15.308	16.928	18.939	37.916	41.337	44.461	48.278	50.993
29	13.121	14.257	16.047	17.708	19.768	39.087	42.557	45.722	49.588	52.336
30	13.787	14.954	16.791	18.493	20.599	40.256	43.773	46.979	50.892	53.672
40	20.707	22.164	24.433	26.509	29.051	51.805	55.758	59.342	63.691	66.766
50	27.991	29.707	32.357	34.764	37.689	63.167	67.505	71.420	76.154	79.490
60	35.534	37.485	40.482	43.188	46.459	74.397	79.082	83.298	88.379	91.952
70	43.275	45.442	48.758	51.739	55.329	85.527	90.531	95.023	100.425	104.215
80	51.172	53.540	57.153	60.391	64.278	96.578	101.879	106.629	112.329	116.321
90	59.196	61.754	65.647	69.126	73.291	107.565	113.145	118.136	124.116	128.299
100	67.328	70.065	74.222	77.929	82.358	118.498	124.342	129.561	135.807	140.169

Table 7—*F*-Distribution

$\alpha = 0.005$

d.f.$_D$: Degrees of freedom, denominator	d.f.$_N$: Degrees of freedom, numerator																		
	1	**2**	**3**	**4**	**5**	**6**	**7**	**8**	**9**	**10**	**12**	**15**	**20**	**24**	**30**	**40**	**60**	**120**	**∞**
1	16211	20000	21615	22500	23056	23437	23715	23925	24091	24224	24426	24630	24836	24940	25044	25148	25253	25359	25465
2	198.5	199.0	199.2	199.2	199.3	199.3	199.4	199.4	199.4	199.4	199.4	199.4	199.4	199.5	199.5	199.5	199.5	199.5	199.5
3	55.55	49.80	47.47	46.19	45.39	44.84	44.43	44.13	43.88	43.69	43.39	43.08	42.78	42.62	42.47	42.31	42.15	41.99	41.83
4	31.33	26.28	24.26	23.15	22.46	21.97	21.62	21.35	21.14	20.97	20.70	20.44	20.17	20.03	19.89	19.75	19.61	19.47	19.32
5	22.78	18.31	16.53	15.56	14.94	14.51	14.20	13.96	13.77	13.62	13.38	13.15	12.90	12.78	12.66	12.53	12.40	12.27	12.14
6	18.63	14.54	12.92	12.03	11.46	11.07	10.79	10.57	10.39	10.25	10.03	9.81	9.59	9.47	9.36	9.24	9.12	9.00	8.88
7	16.24	12.40	10.88	10.05	9.52	9.16	8.89	8.68	8.51	8.38	8.18	7.97	7.75	7.65	7.53	7.42	7.31	7.19	7.08
8	14.69	11.04	9.60	8.81	8.30	7.95	7.69	7.50	7.34	7.21	7.01	6.81	6.61	6.50	6.40	6.29	6.18	6.06	5.95
9	13.61	10.11	8.72	7.96	7.47	7.13	6.88	6.69	6.54	6.42	6.23	6.03	5.83	5.73	5.62	5.52	5.41	5.30	5.19
10	12.83	9.43	8.08	7.34	6.87	6.54	6.30	6.12	5.97	5.85	5.66	5.47	5.27	5.17	5.07	4.97	4.86	4.75	4.64
11	12.73	8.91	7.60	6.88	6.42	6.10	5.86	5.68	5.54	5.42	5.24	5.05	4.86	4.76	4.65	4.55	4.44	4.34	4.23
12	11.75	8.51	7.23	6.52	6.07	5.76	5.52	5.35	5.20	5.09	4.91	4.72	4.53	4.43	4.33	4.23	4.12	4.01	3.90
13	11.37	8.19	6.93	6.23	5.79	5.48	5.25	5.08	4.94	4.82	4.64	4.46	4.27	4.17	4.07	3.97	3.87	3.76	3.65
14	11.06	7.92	6.68	6.00	5.56	5.26	5.03	4.86	4.72	4.60	4.43	4.25	4.06	3.96	3.86	3.76	3.66	3.55	3.44
15	10.80	7.70	6.48	5.80	5.37	5.07	4.85	4.67	4.54	4.42	4.25	4.07	3.88	3.79	3.69	3.58	3.48	3.37	3.26
16	10.58	7.51	6.30	5.64	5.21	4.91	4.69	4.52	4.38	4.27	4.10	3.92	3.73	3.64	3.54	3.44	3.33	3.22	3.11
17	10.38	7.35	6.16	5.50	5.07	4.78	4.56	4.39	4.25	4.14	3.97	3.79	3.61	3.51	3.41	3.31	3.21	3.10	2.98
18	10.22	7.21	6.03	5.37	4.96	4.66	4.44	4.28	4.14	4.03	3.86	3.68	3.50	3.40	3.30	3.20	3.10	2.99	2.87
19	10.07	7.09	5.92	5.27	4.85	4.56	4.34	4.18	4.04	3.93	3.76	3.59	3.40	3.31	3.21	3.11	3.00	2.89	2.78
20	9.94	6.99	5.82	5.17	4.76	4.47	4.26	4.09	3.96	3.85	3.68	3.50	3.32	3.22	3.12	3.02	2.92	2.81	2.69
21	9.83	6.89	5.73	5.09	4.68	4.39	4.18	4.01	3.88	3.77	3.60	3.43	3.24	3.15	3.05	2.95	2.84	2.73	2.61
22	9.73	6.81	5.65	5.02	4.61	4.32	4.11	3.94	3.81	3.70	3.54	3.36	3.18	3.08	2.98	2.88	2.77	2.66	2.55
23	9.63	6.73	5.58	4.95	4.54	4.26	4.05	3.88	3.75	3.64	3.47	3.30	3.12	3.02	2.92	2.82	2.71	2.60	2.48
24	9.55	6.66	5.52	4.89	4.49	4.20	3.99	3.83	3.69	3.59	3.42	3.25	3.06	2.97	2.87	2.77	2.66	2.55	2.43
25	9.48	6.60	5.46	4.84	4.43	4.15	3.94	3.78	3.64	3.54	3.37	3.20	3.01	2.92	2.82	2.72	2.61	2.50	2.38
26	9.41	6.54	5.41	4.79	4.38	4.10	3.89	3.73	3.60	3.49	3.33	3.15	2.97	2.87	2.77	2.67	2.56	2.45	2.33
27	9.34	6.49	5.36	4.74	4.34	4.06	3.85	3.69	3.56	3.45	3.28	3.11	2.93	2.83	2.73	2.63	2.52	2.41	2.29
28	9.28	6.44	5.32	4.70	4.30	4.02	3.81	3.65	3.52	3.41	3.25	3.07	2.89	2.79	2.69	2.59	2.48	2.37	2.25
29	9.23	6.40	5.28	4.66	4.26	3.98	3.77	3.61	3.48	3.38	3.21	3.04	2.86	2.76	2.66	2.56	2.45	2.33	2.24
30	9.18	6.35	5.24	4.62	4.23	3.95	3.74	3.58	3.45	3.34	3.18	3.01	2.82	2.73	2.63	2.52	2.42	2.30	2.18
40	8.83	6.07	4.98	4.37	3.99	3.71	3.51	3.35	3.22	3.12	2.95	2.78	2.60	2.50	2.40	2.30	2.18	2.06	1.93
60	8.49	5.79	4.73	4.14	3.76	3.49	3.29	3.13	3.01	2.90	2.74	2.57	2.39	2.29	2.19	2.08	1.96	1.83	1.69
120	8.18	5.54	4.50	3.92	3.55	3.28	3.09	2.93	2.81	2.71	2.54	2.37	2.19	2.09	1.98	1.87	1.75	1.61	1.43
∞	7.88	5.30	4.28	3.72	3.35	3.09	2.90	2.74	2.62	2.52	2.36	2.19	2.00	1.90	1.79	1.67	1.53	1.36	1.00

Table 7— F-Distribution (continued)

$\alpha = 0.01$

d.f._D: Degrees of freedom, denominator	\multicolumn{19}{c}{d.f._N: Degrees of freedom, numerator}																		
	1	2	3	4	5	6	7	8	9	10	12	15	20	24	30	40	60	120	∞
1	4052	4999.5	5403	5625	5764	5859	5928	5982	6022	6056	6106	6157	6209	6235	6261	6287	6313	6339	6366
2	98.50	99.00	99.17	99.25	99.30	99.33	99.36	99.37	99.39	99.40	99.42	99.43	99.45	99.46	99.47	99.47	99.48	99.49	99.50
3	34.12	30.82	29.46	28.71	28.24	27.91	27.67	27.49	27.35	27.23	27.05	26.87	26.69	26.60	26.50	26.41	26.32	26.22	26.13
4	21.20	18.00	16.69	15.98	15.52	15.21	14.98	14.80	14.66	14.55	14.37	14.20	14.02	13.93	13.84	13.75	13.65	13.56	13.46
5	16.26	13.27	12.06	11.39	10.97	10.67	10.46	10.29	10.16	10.05	9.89	9.72	9.55	9.47	9.38	9.29	9.20	9.11	9.02
6	13.75	10.92	9.78	9.15	8.75	8.47	8.26	8.10	7.98	7.87	7.72	7.56	7.40	7.31	7.23	7.14	7.06	6.97	6.88
7	12.25	9.55	8.45	7.85	7.46	7.19	6.99	6.84	6.72	6.62	6.47	6.31	6.16	6.07	5.99	5.91	5.82	5.74	5.65
8	11.26	8.65	7.59	7.01	6.63	6.37	6.18	6.03	5.91	5.81	5.67	5.52	5.36	5.28	5.20	5.12	5.03	4.95	4.86
9	10.56	8.02	6.99	6.42	6.06	5.80	5.61	5.47	5.35	5.26	5.11	4.96	4.81	4.73	4.65	4.57	4.48	4.40	4.31
10	10.04	7.56	6.55	5.99	5.64	5.39	5.20	5.06	4.94	4.85	4.71	4.56	4.41	4.33	4.25	4.17	4.08	4.00	3.91
11	9.65	7.21	6.22	5.67	5.32	5.07	4.89	4.74	4.63	4.54	4.40	4.25	4.10	4.02	3.94	3.86	3.78	3.69	3.60
12	9.33	6.93	5.95	5.41	5.06	4.82	4.64	4.50	4.39	4.30	4.16	4.01	3.86	3.78	3.70	3.62	3.54	3.45	3.36
13	9.07	6.70	5.74	5.21	4.86	4.62	4.44	4.30	4.19	4.10	3.96	3.82	3.66	3.59	3.51	3.43	3.34	3.25	3.17
14	8.86	6.51	5.56	5.04	4.69	4.46	4.28	4.14	4.03	3.94	3.80	3.66	3.51	3.43	3.35	3.27	3.18	3.09	3.00
15	8.68	6.36	5.42	4.89	4.56	4.32	4.14	4.00	3.89	3.80	3.67	3.52	3.37	3.29	3.21	3.13	3.05	2.96	2.87
16	8.53	6.23	5.29	4.77	4.44	4.20	4.03	3.89	3.78	3.69	3.55	3.41	3.26	3.18	3.10	3.02	2.93	2.84	2.75
17	8.40	6.11	5.18	4.67	4.34	4.10	3.93	3.79	3.68	3.59	3.46	3.31	3.16	3.08	3.00	2.92	2.83	2.75	2.65
18	8.29	6.01	5.09	4.58	4.25	4.01	3.84	3.71	3.60	3.51	3.37	3.23	3.08	3.00	2.92	2.84	2.75	2.66	2.57
19	8.18	5.93	5.01	4.50	4.17	3.94	3.77	3.63	3.52	3.43	3.30	3.15	3.00	2.92	2.84	2.76	2.67	2.58	2.49
20	8.10	5.85	4.94	4.43	4.10	3.87	3.70	3.56	3.46	3.37	3.23	3.09	2.94	2.86	2.78	2.69	2.61	2.52	2.42
21	8.02	5.78	4.87	4.37	4.04	3.81	3.64	3.51	3.40	3.31	3.17	3.03	2.88	2.80	2.72	2.64	2.55	2.46	2.36
22	7.95	5.72	4.82	4.31	3.99	3.76	3.59	3.45	3.35	3.26	3.12	2.98	2.83	2.75	2.67	2.58	2.50	2.40	2.31
23	7.88	5.66	4.76	4.26	3.94	3.71	3.54	3.41	3.30	3.21	3.07	2.93	2.78	2.70	2.62	2.54	2.45	2.35	2.26
24	7.82	5.61	4.72	4.22	3.90	3.67	3.50	3.36	3.26	3.17	3.03	2.89	2.74	2.66	2.58	2.49	2.40	2.31	2.21
25	7.77	5.57	4.68	4.18	3.85	3.63	3.46	3.32	3.22	3.13	2.99	2.85	2.70	2.62	2.54	2.45	2.36	2.27	2.17
26	7.72	5.53	4.64	4.14	3.82	3.59	3.42	3.29	3.18	3.09	2.96	2.81	2.66	2.58	2.50	2.42	2.33	2.23	2.13
27	7.68	5.49	4.60	4.11	3.78	3.56	3.39	3.26	3.15	3.06	2.93	2.78	2.63	2.55	2.47	2.38	2.29	2.20	2.10
28	7.64	5.45	4.57	4.07	3.75	3.53	3.36	3.23	3.12	3.03	2.90	2.75	2.60	2.52	2.44	2.35	2.26	2.17	2.06
29	7.60	5.42	4.54	4.04	3.73	3.50	3.33	3.20	3.09	3.00	2.87	2.73	2.57	2.49	2.41	2.33	2.23	2.14	2.03
30	7.56	5.39	4.51	4.02	3.70	3.47	3.30	3.17	3.07	2.98	2.84	2.70	2.55	2.47	2.39	2.30	2.21	2.11	2.01
40	7.31	5.18	4.31	3.83	3.51	3.29	3.12	2.99	2.89	2.80	2.66	2.52	2.37	2.29	2.20	2.11	2.02	1.92	1.80
60	7.08	4.98	4.13	3.65	3.34	3.12	2.95	2.82	2.72	2.63	2.50	2.35	2.20	2.12	2.03	1.94	1.84	1.73	1.60
120	6.85	4.79	3.95	3.48	3.17	2.96	2.79	2.66	2.56	2.47	2.34	2.19	2.03	1.95	1.86	1.76	1.66	1.53	1.38
∞	6.63	4.61	3.78	3.32	3.02	2.80	2.64	2.51	2.41	2.32	2.18	2.04	1.88	1.79	1.70	1.59	1.47	1.32	1.00

Table 7 — F-Distribution (continued)

$\alpha = 0.025$

d.f._D: Degrees of freedom, denominator	d.f._N: Degrees of freedom, numerator																		
	1	2	3	4	5	6	7	8	9	10	12	15	20	24	30	40	60	120	∞
1	647.8	799.5	864.2	899.6	921.8	937.1	948.2	956.7	963.3	968.6	976.7	984.9	993.1	997.2	1001	1006	1010	1014	1018
2	38.51	39.00	39.17	39.25	39.30	39.33	39.36	39.37	39.39	39.40	39.41	39.43	39.45	39.46	39.46	39.47	39.48	39.49	39.50
3	17.44	16.04	15.44	15.10	14.88	14.73	14.62	14.54	14.47	14.42	14.34	14.25	14.17	14.12	14.08	14.04	13.99	13.95	13.90
4	12.22	10.65	9.98	9.60	9.36	9.20	9.07	8.98	8.90	8.84	8.75	8.66	8.56	8.51	8.46	8.41	8.36	8.31	8.26
5	10.01	8.43	7.76	7.39	7.15	6.98	6.85	6.76	6.68	6.62	6.52	6.43	6.33	6.28	6.23	6.18	6.12	6.07	6.02
6	8.81	7.26	6.60	6.23	5.99	5.82	5.70	5.60	5.52	5.46	5.37	5.27	5.17	5.12	5.07	5.01	4.96	4.90	4.85
7	8.07	6.54	5.89	5.52	5.29	5.12	4.99	4.90	4.82	4.76	4.67	4.57	4.47	4.42	4.36	4.31	4.25	4.20	4.14
8	7.57	6.06	5.42	5.05	4.82	4.65	4.53	4.43	4.36	4.30	4.20	4.10	4.00	3.95	3.89	3.84	3.78	3.73	3.67
9	7.21	5.71	5.08	4.72	4.48	4.32	4.20	4.10	4.03	3.96	3.87	3.77	3.67	3.61	3.56	3.51	3.45	3.39	3.33
10	6.94	5.46	4.83	4.47	4.24	4.07	3.95	3.85	3.78	3.72	3.62	3.52	3.42	3.37	3.31	3.26	3.20	3.14	3.08
11	6.72	5.26	4.63	4.28	4.04	3.88	3.76	3.66	3.59	3.53	3.43	3.33	3.23	3.17	3.12	3.06	3.00	2.94	2.88
12	6.55	5.10	4.47	4.12	3.89	3.73	3.61	3.51	3.44	3.37	3.28	3.18	3.07	3.02	2.96	2.91	2.85	2.79	2.72
13	6.41	4.97	4.35	4.00	3.77	3.60	3.48	3.39	3.31	3.25	3.15	3.05	2.95	2.89	2.84	2.78	2.72	2.66	2.60
14	6.30	4.86	4.24	3.89	3.66	3.50	3.38	3.29	3.21	3.15	3.05	2.95	2.84	2.79	2.73	2.67	2.61	2.55	2.49
15	6.20	4.77	4.15	3.80	3.58	3.41	3.29	3.20	3.12	3.06	2.96	2.86	2.76	2.70	2.64	2.59	2.52	2.46	2.40
16	6.12	4.69	4.08	3.73	3.50	3.34	3.22	3.12	3.05	2.99	2.89	2.79	2.68	2.63	2.57	2.51	2.45	2.38	2.32
17	6.04	4.62	4.01	3.66	3.44	3.28	3.16	3.06	2.98	2.92	2.82	2.72	2.62	2.56	2.50	2.44	2.38	2.32	2.25
18	5.98	4.56	3.95	3.61	3.38	3.22	3.10	3.01	2.93	2.87	2.77	2.67	2.56	2.50	2.44	2.38	2.32	2.26	2.19
19	5.92	4.51	3.90	3.56	3.33	3.17	3.05	2.96	2.88	2.82	2.72	2.62	2.51	2.45	2.39	2.33	2.27	2.20	2.13
20	5.87	4.46	3.86	3.51	3.29	3.13	3.01	2.91	2.84	2.77	2.68	2.57	2.46	2.41	2.35	2.29	2.22	2.16	2.09
21	5.83	4.42	3.82	3.48	3.25	3.09	2.97	2.87	2.80	2.73	2.64	2.53	2.42	2.37	2.31	2.25	2.18	2.11	2.04
22	5.79	4.38	3.78	3.44	3.22	3.05	2.93	2.84	2.76	2.70	2.60	2.50	2.39	2.33	2.27	2.21	2.14	2.08	2.00
23	5.75	4.35	3.75	3.41	3.18	3.02	2.90	2.81	2.73	2.67	2.57	2.47	2.36	2.30	2.24	2.18	2.11	2.04	1.97
24	5.72	4.32	3.72	3.38	3.15	2.99	2.87	2.78	2.70	2.64	2.54	2.44	2.33	2.27	2.21	2.15	2.08	2.01	1.94
25	5.69	4.29	3.69	3.35	3.13	2.97	2.85	2.75	2.68	2.61	2.51	2.41	2.30	2.24	2.18	2.12	2.05	1.98	1.91
26	5.66	4.27	3.67	3.33	3.10	2.94	2.82	2.73	2.65	2.59	2.49	2.39	2.28	2.22	2.16	2.09	2.03	1.95	1.88
27	5.63	4.24	3.65	3.31	3.08	2.92	2.80	2.71	2.63	2.57	2.47	2.36	2.25	2.19	2.13	2.07	2.00	1.93	1.85
28	5.61	4.22	3.63	3.29	3.06	2.90	2.78	2.69	2.61	2.55	2.45	2.34	2.23	2.17	2.11	2.05	1.98	1.91	1.83
29	5.59	4.20	3.61	3.27	3.04	2.88	2.76	2.67	2.59	2.53	2.43	2.32	2.21	2.15	2.09	2.03	1.96	1.89	1.81
30	5.57	4.18	3.59	3.25	3.03	2.87	2.75	2.65	2.57	2.51	2.41	2.31	2.20	2.14	2.07	2.01	1.94	1.87	1.79
40	5.42	4.05	3.46	3.13	2.90	2.74	2.62	2.53	2.45	2.39	2.29	2.18	2.07	2.01	1.94	1.88	1.80	1.72	1.64
60	5.29	3.93	3.34	3.01	2.79	2.63	2.51	2.41	2.33	2.27	2.17	2.06	1.94	1.88	1.82	1.74	1.67	1.58	1.48
120	5.15	3.80	3.23	2.89	2.67	2.52	2.39	2.30	2.22	2.16	2.05	1.94	1.82	1.76	1.69	1.61	1.53	1.43	1.31
∞	5.02	3.69	3.12	2.79	2.57	2.41	2.29	2.19	2.11	2.05	1.94	1.83	1.71	1.64	1.57	1.48	1.39	1.27	1.00

Table 7— F-Distribution (continued)

$\alpha = 0.05$

d.f.$_N$: Degrees of freedom, numerator

d.f.$_D$: Degrees of freedom, denominator	1	2	3	4	5	6	7	8	9	10	12	15	20	24	30	40	60	120	∞
1	161.4	199.5	215.7	224.6	230.2	234.0	236.8	238.9	240.5	241.9	243.9	245.9	248.0	249.1	250.1	251.1	252.2	253.3	254.3
2	18.51	19.00	19.16	19.25	19.30	19.33	19.35	19.37	19.38	19.40	19.41	19.43	19.45	19.45	19.46	19.47	19.48	19.49	19.50
3	10.13	9.55	9.28	9.12	9.01	8.94	8.89	8.85	8.81	8.79	8.74	8.70	8.66	8.64	8.62	8.59	8.57	8.55	8.53
4	7.71	6.94	6.59	6.39	6.26	6.16	6.09	6.04	6.00	5.96	5.91	5.86	5.80	5.77	5.75	5.72	5.69	5.66	5.63
5	6.61	5.79	5.41	5.19	5.05	4.95	4.88	4.82	4.77	4.74	4.68	4.62	4.56	4.53	4.50	4.46	4.43	4.40	4.36
6	5.99	5.14	4.76	4.53	4.39	4.28	4.21	4.15	4.10	4.06	4.00	3.94	3.87	3.84	3.81	3.77	3.74	3.70	3.67
7	5.59	4.74	4.35	4.12	3.97	3.87	3.79	3.73	3.68	3.64	3.57	3.51	3.44	3.41	3.38	3.34	3.30	3.27	3.23
8	5.32	4.46	4.07	3.84	3.69	3.58	3.50	3.44	3.39	3.35	3.28	3.22	3.15	3.12	3.08	3.04	3.01	2.97	2.93
9	5.12	4.26	3.86	3.63	3.48	3.37	3.29	3.23	3.18	3.14	3.07	3.01	2.94	2.90	2.86	2.83	2.79	2.75	2.71
10	4.96	4.10	3.71	3.48	3.33	3.22	3.14	3.07	3.02	2.98	2.91	2.85	2.77	2.74	2.70	2.66	2.62	2.58	2.54
11	4.84	3.98	3.59	3.36	3.20	3.09	3.01	2.95	2.90	2.85	2.79	2.72	2.65	2.61	2.57	2.53	2.49	2.45	2.40
12	4.75	3.89	3.49	3.26	3.11	3.00	2.91	2.85	2.80	2.75	2.69	2.62	2.54	2.51	2.47	2.43	2.38	2.34	2.30
13	4.67	3.81	3.41	3.18	3.03	2.92	2.83	2.77	2.71	2.67	2.60	2.53	2.46	2.42	2.38	2.34	2.30	2.25	2.21
14	4.60	3.74	3.34	3.11	2.96	2.85	2.76	2.70	2.65	2.60	2.53	2.46	2.39	2.35	2.31	2.27	2.22	2.18	2.13
15	4.54	3.68	3.29	3.06	2.90	2.79	2.71	2.64	2.59	2.54	2.48	2.40	2.33	2.29	2.25	2.20	2.16	2.11	2.07
16	4.49	3.63	3.24	3.01	2.85	2.74	2.66	2.59	2.54	2.49	2.42	2.35	2.28	2.24	2.19	2.15	2.11	2.06	2.01
17	4.45	3.59	3.20	2.96	2.81	2.70	2.61	2.55	2.49	2.45	2.38	2.31	2.23	2.19	2.15	2.10	2.06	2.01	1.96
18	4.41	3.55	3.16	2.93	2.77	2.66	2.58	2.51	2.46	2.41	2.34	2.27	2.19	2.15	2.11	2.06	2.02	1.97	1.92
19	4.38	3.52	3.13	2.90	2.74	2.63	2.54	2.48	2.42	2.38	2.31	2.23	2.16	2.11	2.07	2.03	1.98	1.93	1.88
20	4.35	3.49	3.10	2.87	2.71	2.60	2.51	2.45	2.39	2.35	2.28	2.20	2.12	2.08	2.04	1.99	1.95	1.90	1.84
21	4.32	3.47	3.07	2.84	2.68	2.57	2.49	2.42	2.37	2.32	2.25	2.18	2.10	2.05	2.01	1.96	1.92	1.87	1.81
22	4.30	3.44	3.05	2.82	2.66	2.55	2.46	2.40	2.34	2.30	2.23	2.15	2.07	2.03	1.98	1.94	1.89	1.84	1.78
23	4.28	3.42	3.03	2.80	2.64	2.53	2.44	2.37	2.32	2.27	2.20	2.13	2.05	2.01	1.96	1.91	1.86	1.81	1.76
24	4.26	3.40	3.01	2.78	2.62	2.51	2.42	2.36	2.30	2.25	2.18	2.11	2.03	1.98	1.94	1.89	1.84	1.79	1.73
25	4.24	3.39	2.99	2.76	2.60	2.49	2.40	2.34	2.28	2.24	2.16	2.09	2.01	1.96	1.92	1.87	1.82	1.77	1.71
26	4.23	3.37	2.98	2.74	2.59	2.47	2.39	2.32	2.27	2.22	2.15	2.07	1.99	1.95	1.90	1.85	1.80	1.75	1.69
27	4.21	3.35	2.96	2.73	2.57	2.46	2.37	2.31	2.25	2.20	2.13	2.06	1.97	1.93	1.88	1.84	1.79	1.73	1.67
28	4.20	3.34	2.95	2.71	2.56	2.45	2.36	2.29	2.24	2.19	2.12	2.04	1.96	1.91	1.87	1.82	1.77	1.71	1.65
29	4.18	3.33	2.93	2.70	2.55	2.43	2.35	2.28	2.22	2.18	2.10	2.03	1.94	1.90	1.85	1.81	1.75	1.70	1.64
30	4.17	3.32	2.92	2.69	2.53	2.42	2.33	2.27	2.21	2.16	2.09	2.01	1.93	1.89	1.84	1.79	1.74	1.68	1.62
40	4.08	3.23	2.84	2.61	2.45	2.34	2.25	2.18	2.12	2.08	2.00	1.92	1.84	1.79	1.74	1.69	1.64	1.58	1.51
60	4.00	3.15	2.76	2.53	2.37	2.25	2.17	2.10	2.04	1.99	1.92	1.84	1.75	1.70	1.65	1.59	1.53	1.47	1.39
120	3.92	3.07	2.68	2.45	2.29	2.17	2.09	2.02	1.96	1.91	1.83	1.75	1.66	1.61	1.55	1.50	1.43	1.35	1.25
∞	3.84	3.00	2.60	2.37	2.21	2.10	2.01	1.94	1.88	1.83	1.75	1.67	1.57	1.52	1.46	1.39	1.32	1.22	1.00

Table 7— F-Distribution (continued)

$\alpha = 0.10$

d.f.$_D$: Degrees of freedom, denominator	d.f.$_N$: Degrees of freedom, numerator																		
	1	2	3	4	5	6	7	8	9	10	12	15	20	24	30	40	60	120	∞
1	39.86	49.50	53.59	55.83	57.24	58.20	58.91	59.44	59.86	60.19	60.71	61.22	61.74	62.00	62.26	62.53	62.79	63.06	63.33
2	8.53	9.00	9.16	9.24	9.29	9.33	9.35	9.37	9.38	9.39	9.41	9.42	9.44	9.45	9.46	9.47	9.47	9.48	9.49
3	5.54	5.46	5.39	5.34	5.31	5.28	5.27	5.25	5.24	5.23	5.22	5.20	5.18	5.18	5.17	5.16	5.15	5.14	5.13
4	4.54	4.32	4.19	4.11	4.05	4.01	3.98	3.95	3.94	3.92	3.90	3.87	3.84	3.83	3.82	3.80	3.79	3.78	3.76
5	4.06	3.78	3.62	3.52	3.45	3.40	3.37	3.34	3.32	3.30	3.27	3.24	3.21	3.19	3.17	3.16	3.14	3.12	3.10
6	3.78	3.46	3.29	3.18	3.11	3.05	3.01	2.98	2.96	2.94	2.90	2.87	2.84	2.82	2.80	2.78	2.76	2.74	2.72
7	3.59	3.26	3.07	2.96	2.88	2.83	2.78	2.75	2.72	2.70	2.67	2.63	2.59	2.58	2.56	2.54	2.51	2.49	2.47
8	3.46	3.11	2.92	2.81	2.73	2.67	2.62	2.59	2.56	2.54	2.50	2.46	2.42	2.40	2.38	2.36	2.34	2.32	2.29
9	3.36	3.01	2.81	2.69	2.61	2.55	2.51	2.47	2.44	2.42	2.38	2.34	2.30	2.28	2.25	2.23	2.21	2.18	2.16
10	3.29	2.92	2.73	2.61	2.52	2.46	2.41	2.38	2.35	2.32	2.28	2.24	2.20	2.18	2.16	2.13	2.11	2.08	2.06
11	3.23	2.86	2.66	2.54	2.45	2.39	2.34	2.30	2.27	2.25	2.21	2.17	2.12	2.10	2.08	2.05	2.03	2.00	1.97
12	3.18	2.81	2.61	2.48	2.39	2.33	2.28	2.24	2.21	2.19	2.15	2.10	2.06	2.04	2.01	1.99	1.96	1.93	1.90
13	3.14	2.76	2.56	2.43	2.35	2.28	2.23	2.20	2.16	2.14	2.10	2.05	2.01	1.98	1.96	1.93	1.90	1.88	1.85
14	3.10	2.73	2.52	2.39	2.31	2.24	2.19	2.15	2.12	2.10	2.05	2.01	1.96	1.94	1.91	1.89	1.86	1.83	1.80
15	3.07	2.70	2.49	2.36	2.27	2.21	2.16	2.12	2.09	2.06	2.02	1.97	1.92	1.90	1.87	1.85	1.82	1.79	1.76
16	3.05	2.67	2.46	2.33	2.24	2.18	2.13	2.09	2.06	2.03	1.99	1.94	1.89	1.87	1.84	1.81	1.78	1.75	1.72
17	3.03	2.64	2.44	2.31	2.22	2.15	2.10	2.06	2.03	2.00	1.96	1.91	1.86	1.84	1.81	1.78	1.75	1.72	1.69
18	3.01	2.62	2.42	2.29	2.20	2.13	2.08	2.04	2.00	1.98	1.93	1.89	1.84	1.81	1.78	1.75	1.72	1.69	1.66
19	2.99	2.61	2.40	2.27	2.18	2.11	2.06	2.02	1.98	1.96	1.91	1.86	1.81	1.79	1.76	1.73	1.70	1.67	1.63
20	2.97	2.59	2.38	2.25	2.16	2.09	2.04	2.00	1.96	1.94	1.89	1.84	1.79	1.77	1.74	1.71	1.68	1.64	1.61
21	2.96	2.57	2.36	2.23	2.14	2.08	2.02	1.98	1.95	1.92	1.87	1.83	1.78	1.75	1.72	1.69	1.66	1.62	1.59
22	2.95	2.56	2.35	2.22	2.13	2.06	2.01	1.97	1.93	1.90	1.86	1.81	1.76	1.73	1.70	1.67	1.64	1.60	1.57
23	2.94	2.55	2.34	2.21	2.11	2.05	1.99	1.95	1.92	1.89	1.84	1.80	1.74	1.72	1.69	1.66	1.62	1.59	1.55
24	2.93	2.54	2.33	2.19	2.10	2.04	1.98	1.94	1.91	1.88	1.83	1.78	1.73	1.70	1.67	1.64	1.61	1.57	1.53
25	2.92	2.53	2.32	2.18	2.09	2.02	1.97	1.93	1.89	1.87	1.82	1.77	1.72	1.69	1.66	1.63	1.59	1.56	1.52
26	2.91	2.52	2.31	2.17	2.08	2.01	1.96	1.92	1.88	1.86	1.81	1.76	1.71	1.68	1.65	1.61	1.58	1.54	1.50
27	2.90	2.51	2.30	2.17	2.07	2.00	1.95	1.91	1.87	1.85	1.80	1.75	1.70	1.67	1.64	1.60	1.57	1.53	1.49
28	2.89	2.50	2.29	2.16	2.06	2.00	1.94	1.90	1.87	1.84	1.79	1.74	1.69	1.66	1.63	1.59	1.56	1.52	1.48
29	2.89	2.50	2.28	2.15	2.06	1.99	1.93	1.89	1.86	1.83	1.78	1.73	1.68	1.65	1.62	1.58	1.55	1.51	1.47
30	2.88	2.49	2.28	2.14	2.05	1.98	1.93	1.88	1.85	1.82	1.77	1.72	1.67	1.64	1.61	1.57	1.54	1.50	1.46
40	2.84	2.44	2.23	2.09	2.00	1.93	1.87	1.83	1.79	1.76	1.71	1.66	1.61	1.57	1.54	1.51	1.47	1.42	1.38
60	2.79	2.39	2.18	2.04	1.95	1.87	1.82	1.77	1.74	1.71	1.66	1.60	1.54	1.51	1.48	1.44	1.40	1.35	1.29
120	2.75	2.35	2.13	1.99	1.90	1.82	1.77	1.72	1.68	1.65	1.60	1.55	1.48	1.45	1.41	1.37	1.32	1.26	1.19
∞	2.71	2.30	2.08	1.94	1.85	1.77	1.72	1.67	1.63	1.60	1.55	1.49	1.42	1.38	1.34	1.30	1.24	1.17	1.00

From M. Merrington and C.M. THompson, "Table of Percentage Points of the Inverted Beta (F) Distribution", *Biometrika* 33 (1943), pp. 74-87, by permission of Oxford University Press.

Table 8 — Critical Values for the Sign Test

Reject the null hypothesis if the test statistic is less than or equal to the value in the table.

	One-tailed, $\alpha = 0.005$	$\alpha = 0.01$	$\alpha = 0.025$	$\alpha = 0.05$
n	Two-tailed, $\alpha = 0.01$	$\alpha = 0.02$	$\alpha = 0.05$	$\alpha = 0.10$
8	0	0	0	1
9	0	0	1	1
10	0	0	1	1
11	0	1	1	2
12	1	1	2	2
13	1	1	2	3
14	1	2	3	3
15	2	2	3	3
16	2	2	3	4
17	2	3	4	4
18	3	3	4	5
19	3	4	4	5
20	3	4	5	5
21	4	4	5	6
22	4	5	5	6
23	4	5	6	7
24	5	5	6	7
25	5	6	6	7

Note: Table 8 is for one-tailed or two-tailed tests. The sample size n represents the total number of + and − signs. The test value is the smaller number of + or − signs.

From *Journal of American Statistical Association* Vol. 41 (1946), pp. 557–66. W. J. Dixon and A. M. Mood. Reprinted with permission.

Table 9 — Critical Values for the Wilcoxon Signed-Rank Test

Reject the null hypothesis if the value of the test statistic w_s is less than or equal to the value given in the table.

	One-tailed, $\alpha = 0.05$	$\alpha = 0.025$	$\alpha = 0.01$	$\alpha = 0.005$
n	Two-tailed, $\alpha = 0.10$	$\alpha = 0.05$	$\alpha = 0.02$	$\alpha = 0.01$
5	1	—	—	—
6	2	1	—	—
7	4	2	0	—
8	6	4	2	0
9	8	6	3	2
10	11	8	5	3
11	14	11	7	5
12	17	14	10	7
13	21	17	13	10
14	26	21	16	13
15	30	25	20	16
16	36	30	24	19
17	41	35	28	23
18	47	40	33	28
19	54	46	38	32
20	60	52	43	37
21	68	59	49	43
22	75	66	56	49
23	83	73	62	55
24	92	81	69	61
25	101	90	77	68
26	110	98	85	76
27	120	107	93	84
28	130	117	102	92
29	141	127	111	100
30	152	137	120	109

From *Some Rapid Approximate Statistical Procedures.* Copyright 1949, 1964 Lederle Laboratories, American Cyanamid Co., Wayne, N.J. Reprinted with permission.

Table 10— Critical Values for the Spearman Rank Correlation

Reject H_0: $\rho_s = 0$ if the absolute value of r_s is greater than the value given in the table.

n	$\alpha = 0.10$	$\alpha = 0.05$	$\alpha = 0.01$
5	0.900	—	—
6	0.829	0.886	—
7	0.714	0.786	0.929
8	0.643	0.738	0.881
9	0.600	0.700	0.833
10	0.564	0.648	0.794
11	0.536	0.618	0.818
12	0.497	0.591	0.780
13	0.475	0.566	0.745
14	0.457	0.545	0.716
15	0.441	0.525	0.689
16	0.425	0.507	0.666
17	0.412	0.490	0.645
18	0.399	0.476	0.625
19	0.388	0.462	0.608
20	0.377	0.450	0.591
21	0.368	0.438	0.576
22	0.359	0.428	0.562
23	0.351	0.418	0.549
24	0.343	0.409	0.537
25	0.336	0.400	0.526
26	0.329	0.392	0.515
27	0.323	0.385	0.505
28	0.317	0.377	0.496
29	0.311	0.370	0.487
30	0.305	0.364	0.478

Reprinted with permission from the Institute of Mathematical Statistics.

Table 11— Critical Values for the Pearson Correlation Coefficient

Reject H_0: $\rho = 0$ if the absolute value of r is greater than the value given in the table.

n	$\alpha = 0.05$	$\alpha = 0.01$
4	0.950	0.990
5	0.878	0.959
6	0.811	0.917
7	0.754	0.875
8	0.707	0.834
9	0.666	0.798
10	0.632	0.765
11	0.602	0.735
12	0.576	0.708
13	0.553	0.684
14	0.532	0.661
15	0.514	0.641
16	0.497	0.623
17	0.482	0.606
18	0.468	0.590
19	0.456	0.575
20	0.444	0.561
21	0.433	0.549
22	0.423	0.537
23	0.413	0.526
24	0.404	0.515
25	0.396	0.505
26	0.388	0.496
27	0.381	0.487
28	0.374	0.479
29	0.367	0.471
30	0.361	0.463
35	0.334	0.430
40	0.312	0.403
45	0.294	0.380
50	0.279	0.361
55	0.266	0.345
60	0.254	0.330
65	0.244	0.317
70	0.235	0.306
75	0.227	0.296
80	0.220	0.286
85	0.213	0.278
90	0.207	0.270
95	0.202	0.263
100	0.197	0.256

The critical values in Table 11 were generated using Excel.

Table 12 — Critical Values for the Number of Runs

Reject the null hypothesis if the test statistic G is less than or equal to the smaller entry or greater than or equal to the larger entry.

Value of n_1		2	3	4	5	6	7	8	9	10	11	12	13	14	15	16	17	18	19	20
	2	1	1	1	1	1	1	1	1	1	1	2	2	2	2	2	2	2	2	2
		6	6	6	6	6	6	6	6	6	6	6	6	6	6	6	6	6	6	6
	3	1	1	1	1	2	2	2	2	2	2	2	2	2	3	3	3	3	3	3
		6	8	8	8	8	8	8	8	8	8	8	8	8	8	8	8	8	8	8
	4	1	1	1	2	2	2	3	3	3	3	3	3	3	3	4	4	4	4	4
		6	8	9	9	9	10	10	10	10	10	10	10	10	10	10	10	10	10	10
	5	1	1	2	2	3	3	3	3	3	4	4	4	4	4	4	4	5	5	5
		6	8	9	10	10	11	11	12	12	12	12	12	12	12	12	12	12	12	12
	6	1	2	2	3	3	3	3	4	4	4	4	5	5	5	5	5	5	6	6
		6	8	9	10	11	12	12	13	13	13	13	14	14	14	14	14	14	14	14
	7	1	2	2	3	3	3	4	4	5	5	5	5	5	6	6	6	6	6	6
		6	8	10	11	12	13	13	14	14	14	14	15	15	15	16	16	16	16	16
	8	1	2	3	3	3	4	4	5	5	5	6	6	6	6	6	7	7	7	7
		6	8	10	11	12	13	14	14	15	15	16	16	16	16	17	17	17	17	17
	9	1	2	3	3	4	4	5	5	5	6	6	6	7	7	7	7	8	8	8
		6	8	10	12	13	14	14	15	16	16	16	17	17	18	18	18	18	18	18
	10	1	2	3	3	4	5	5	5	6	6	7	7	7	7	8	8	8	8	9
		6	8	10	12	13	14	15	16	16	17	17	18	18	18	19	19	19	20	20
	11	1	2	3	4	4	5	5	6	6	7	7	7	8	8	8	9	9	9	9
		6	8	10	12	13	14	15	16	17	17	18	19	19	19	20	20	20	21	21
	12	2	2	3	4	4	5	6	6	7	7	7	8	8	8	9	9	9	10	10
		6	8	10	12	13	14	16	16	17	18	19	19	20	20	21	21	21	22	22
	13	2	2	3	4	5	5	6	6	7	7	8	8	9	9	9	10	10	10	10
		6	8	10	12	14	15	16	17	18	19	19	20	20	21	21	22	22	23	23
	14	2	2	3	4	5	5	6	7	7	8	8	9	9	9	10	10	10	11	11
		6	8	10	12	14	15	16	17	18	19	20	20	21	22	22	23	23	23	24
	15	2	3	3	4	5	6	6	7	7	8	8	9	9	10	10	11	11	11	12
		6	8	10	12	14	15	16	18	18	19	20	21	22	22	23	23	24	24	25
	16	2	3	4	4	5	6	6	7	8	8	9	9	10	10	11	11	11	12	12
		6	8	10	12	14	16	17	18	19	20	21	21	22	23	23	24	25	25	25
	17	2	3	4	4	5	6	7	7	8	9	9	10	10	11	11	11	12	12	13
		6	8	10	12	14	16	17	18	19	20	21	22	23	23	24	25	25	26	26
	18	2	3	4	5	5	6	7	8	8	9	9	10	10	11	11	12	12	13	13
		6	8	10	12	14	16	17	18	19	20	21	22	23	24	25	25	26	26	27
	19	2	3	4	5	6	6	7	8	8	9	10	10	11	11	12	12	13	13	13
		6	8	10	12	14	16	17	18	20	21	22	23	23	24	25	26	26	27	27
	20	2	3	4	5	6	6	7	8	9	9	10	10	11	12	12	13	13	13	14
		6	8	10	12	14	16	17	18	20	21	22	23	24	25	25	26	27	27	28

Note: Table 12 is for a two-tailed test with $\alpha = 0.05$.
Reprinted with permission from the Institute of Mathematical Statistics.

APPENDIX C

C Normal Probability Plots and Their Graphs

Normal Probability Plots

▸ NORMAL PROBABILITY PLOTS

For the majority of problems throughout this book, it has been assumed that a random sample of data is selected from a population that has a normal distribution. Suppose you select a random sample from a population with an unknown distribution. How can you determine if the sample was selected from a population that has a normal distribution?

You have already learned that a histogram or stem-and-leaf plot can reveal the shape of a distribution and any outliers, clusters, or gaps in a distribution. These data displays are useful for assessing large sets of data, but assessing small data sets in this manner can be difficult and unreliable. A reliable method for assessing normality in small data sets is to use a graph called a *normal probability plot*.

> **DEFINITION**
>
> A **normal probability plot** is a graph that plots each observed value from the data set along with its corresponding z-score. The observed values are usually plotted along the horizontal axis while the corresponding z-scores are plotted along the vertical axis.

INSIGHT

A normal probability plot is also called a **normal quantile plot**.

If the plotted points in a normal probability plot are approximately linear, then you can conclude that the data come from a normal distribution. If the plotted points are not approximately linear or follow some type of pattern that is not linear, you can conclude that the data come from a distribution that is not normal. When examining a normal probability plot, look for deviations or clusters of points that stray from the line, which indicate a distribution that is not normal. Individual points that stray from the line in a normal probability plot may be outliers.

Constructing a normal probability plot by hand can be rather tedious. Technology tools such as MINITAB or a TI-83/84 Plus can be used to construct normal probability plots, as shown in Example 1.

> **EXAMPLE 1**

▸ Constructing a Normal Probability Plot

The heights (in inches) of 12 current National Basketball Association players are listed. Use a technology tool to construct a normal probability plot to determine if the data come from a population that has a normal distribution. Identify any possible outliers.

74, 69, 78, 75, 73, 71, 80, 82, 81, 76, 86, 77

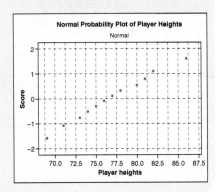

▸ **Solution**

Using a TI-83/84 Plus, begin by entering the data into List 1. Then use *Stat Plot* to construct the normal probability plot. The plot should look similar to the one shown below. From the scatter plot, it appears that there are no outliers and the points are approximately linear. To construct a normal probability plot using MINITAB, follow the instructions in the margin.

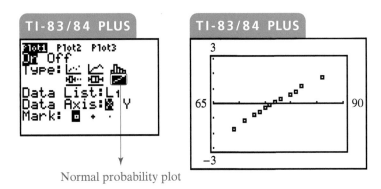

Normal probability plot

Interpretation   Because the points are approximately linear, you can conclude that the sample data come from a population that has a normal distribution.

▸ **Try It Yourself 1**

The balances (in dollars) on student loans for 18 randomly selected college seniors are listed.

29,150	16,980	12,470	19,235	15,875	8,960
16,105	14,575	39,860	20,170	9,710	19,650
21,590	8,200	18,100	25,530	9,285	10,075

a. *Use* a technology tool to construct a normal probability plot. Are the points approximately linear?
b. *Identify* any possible outliers.
c. *Interpret* your answer.   *Answer: Page A49*

To see that the points are approximately linear, you can graph the regression line for the original data values and their corresponding z-scores. The regression line for the heights and z-scores from Example 1 is shown in the graph. From the graph, you can see that the points lie along the regression line. You can also approximate the mean of the data set by determining where the line crosses the x-axis.

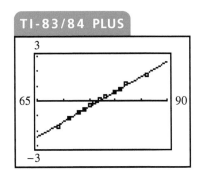

Try It Yourself Answers

CHAPTER 1

Section 1.1

1a. The population consists of the prices per gallon of regular gasoline at all gasoline stations in the United States. The sample consists of the prices per gallon of regular gasoline at the 900 surveyed stations.

b. The data set consists of the 900 prices.

2a. Population   **b.** Parameter

3a. Descriptive statistics involve the statement "76% of women and 60% of men had a physical examination within the previous year."

b. An inference drawn from the study is that a higher percentage of women had a physical examination within the previous year.

Section 1.2

1a. City names and city populations

b. City names: Nonnumerical
City populations: Numerical

c. City names: Qualitative
City populations: Quantitative

2a. (1) The final standings represent a ranking of basketball teams.

(2) The collection of phone numbers represents labels.

b. (1) Ordinal, because the data can be put in order.

(2) Nominal, because no mathematical computations can be made.

3a. (1) The data set is the collection of body temperatures.

(2) The data set is the collection of heart rates.

b. (1) Interval, because the data can be ordered and meaningful differences can be calculated, but it does not make sense to write a ratio using the temperatures.

(2) Ratio, because the data can be ordered, meaningful differences can be calculated, the data can be written as a ratio, and the data set contains an inherent zero.

Section 1.3

1a. (1) Focus: Effect of exercise on relieving depression

(2) Focus: Success rates of graduates of a large university in finding a job within one year of graduation

b. (1) Population: Collection of all people with depression

(2) Population: The employment status of all graduates of a large university one year after graduation

c. (1) Experiment

(2) Survey

2a. There is no way to tell why the people quit smoking. They could have quit smoking as a result of either chewing the gum or watching the DVD.

b. Two experiments could be done; one using the gum and the other using the DVD.

3a. Answers will vary. *Sample answer:* Start with the first digits 92630782 . . . .

b. 92|63|07|82|40|19|26

c. 63, 7, 40, 19, 26

4a. Sample selection:

(1) The sample was selected by using the students in a randomly chosen class.

(2) The sample was selected by numbering each student in the school, randomly choosing a starting number, and selecting students at regular intervals from the starting number.

Sampling technique:

(1) Cluster sampling

(2) Systematic sampling

b. (1) The sample may be biased because some classes may be more familiar with stem cell research than other classes and have stronger opinions.

(2) The sample may be biased if there is any regularly occurring pattern in the data.

CHAPTER 2

Section 2.1

1a. 8 classes

b. Min = 35; Max = 89; Class width = 7

c.

Lower limit	Upper limit
35	41
42	48
49	55
56	62
63	69
70	76
77	83
84	90

d. See part (e).

e.

Class	Frequency, f
35–41	2
42–48	5
49–55	7
56–62	7
63–69	10
70–76	5
77–83	8
84–90	6

2 a. See part (b).

b.

Class	Frequency, f	Midpoint	Relative frequency	Cumulative frequency
35–41	2	38	0.04	2
42–48	5	45	0.10	7
49–55	7	52	0.14	14
56–62	7	59	0.14	21
63–69	10	66	0.20	31
70–76	5	73	0.10	36
77–83	8	80	0.16	44
84–90	6	87	0.12	50
	$\Sigma f = 50$		$\Sigma \dfrac{f}{n} = 1$	

c. The most common age bracket for the 50 richest people is 63–69. 72% of the 50 richest people are older than 55. 4% of the 50 richest people are younger than 42. (Answers will vary.)

3 a.

Class boundaries
34.5–41.5
41.5–48.5
48.5–55.5
55.5–62.5
62.5–69.5
69.5–76.5
76.5–83.5
83.5–90.5

b. Use class midpoints for the horizontal scale and frequency for the vertical scale. (Class boundaries can also be used for the horizontal scale.)

c.

Ages of the 50 Richest People

d. Same as 2(c).

4 a. Same as 3(b).

b. See part (c).

c.

Ages of the 50 Richest People

d. The frequency of ages increases up to 66 and then decreases.

5 abc.

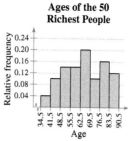

Ages of the 50 Richest People

6 a. Use upper class boundaries for the horizontal scale and cumulative frequency for the vertical scale.

b. See part (c).

c.

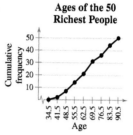

Ages of the 50 Richest People

d. Approximately 40 of the 50 richest people are 80 years old or younger.

e. Answers will vary.

7 a. Enter data.

b.

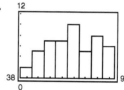

Section 2.2

1 a.

```
3 |
4 |
5 |
6 |
7 |
8 |
```

b.

```
3 | 6 5                    Key: 3|6 = 36
4 | 9 7 6 4 3 2
5 | 9 8 7 6 4 4 3 3 1 1
6 | 9 9 8 7 6 6 5 5 4 3 1 1 0
7 | 8 8 7 6 3 3 3 2
8 | 9 9 7 6 6 5 3 3 2 1 0
```

c.

```
3 | 5 6                    Key: 3|5 = 35
4 | 2 3 4 6 7 9
5 | 1 1 3 3 4 4 6 7 8 9
6 | 0 1 1 3 4 5 5 6 6 7 8 9 9
7 | 2 3 3 3 6 7 8 8
8 | 0 1 2 3 3 5 6 6 7 9 9
```

d. More than 50% of the 50 richest people are older than 60. (Answers will vary.)

2 ab.

```
3 |          Key: 3|5 = 35
3 | 5 6
4 | 2 3 4
4 | 6 7 9
5 | 1 1 3 3 4 4
5 | 6 7 8 9
6 | 0 1 1 3 4
6 | 5 5 6 6 7 8 9 9
7 | 2 3 3 3
7 | 6 7 8 8
8 | 0 1 2 3 3
8 | 5 6 6 7 9 9
```

c. Most of the 50 richest people are older than 60. (Answers will vary.)

3 a. Use age for the horizontal axis.

b.

Ages of the 50 Richest People

c. A large percentage of the ages are over 60. (Answers will vary.)

4 a.

Type of degree	f	Relative frequency	Angle
Associate's	455	0.23	82.8°
Bachelor's	1052	0.54	194.4°
Master's	325	0.17	61.2°
First professional	71	0.04	14.4°
Doctoral	38	0.02	7.2°
	$\Sigma f = 1941$	$\Sigma \dfrac{f}{n} = 1$	$\Sigma = 360°$

b.

Earned Degrees Conferred in 1990

Doctoral 2%
First professional 4%
Associate's 23%
Master's 17%
Bachelor's 54%

c. From 1990 to 2007, as percentages of the total degrees conferred, associate's degrees increased by 1%, bachelor's degrees decreased by 3%, master's degrees increased by 3%, first professional degrees decreased by 1%, and doctoral degrees remained unchanged.

5 a.

Cause	Frequency, f
Auto Dealers	14,668
Auto Repair	9728
Home Furnishing	7792
Computer Sales	5733
Dry Cleaning	4649

b.

Causes of BBB Complaints

c. It appears that the auto industry (dealers and repair shops) account for the largest portion of complaints filed at the BBB. (Answers will vary.)

6 ab.

Salaries

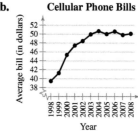

c. It appears that the longer an employee is with the company, the larger the employee's salary will be.

7 ab.

Cellular Phone Bills

c. The average bill increased from 1998 to 2004, then it hovered around $50.00 from 2004 to 2008.

Section 2.3

1 a. 1193   **b.** 79.5

c. The mean height of the players is about 79.5 inches.

2 a. 18, 18, 19, 19, 19, 20, 21, 21, 21, 21, 23, 24, 24, 26, 27, 27, 29, 30, 30, 30, 33, 33, 34, 35, 38

b. 24

c. The median age of the sample of fans at the concert is 24.

3 a. 25, 60, 80, 97, 100, 130, 140, 200, 220, 250   **b.** 115

c. The median price of the sample of digital photo frames is $115.

4 a. 324, 385, 450, 450, 462, 475, 540, 540, 564, 618, 624, 638, 670, 670, 670, 705, 720, 723, 750, 750, 825, 830, 912, 975, 980, 980, 1100, 1260, 1420, 1650

b. 670

c. The mode of the prices for the sample of South Beach, FL condominiums is $670.

5 a. Yes

b. In this sample, there were more people who thought public cell phone conversations were rude than people who did not or had no opinion.

6 a. 21.6; 21; 20

b. The mean in Example 6 ($\bar{x} \approx 23.8$) was heavily influenced by the entry 65. Neither the median nor the mode was affected as much by the entry 65.

7 ab.

Source	Score, x	Weight, w	xw
Test mean	86	0.50	43.0
Midterm	96	0.15	14.4
Final exam	98	0.20	19.6
Computer lab	98	0.10	9.8
Homework	100	0.05	5.0
		$\sum w = 1.00$	$\sum(x \cdot w) = 91.8$

c. 91.8

d. The weighted mean for the course is 91.8. So you did get an A.

8 abc.

Class	Midpoint, x	Frequency, f	xf
35–41	38	2	76
42–48	45	5	225
49–55	52	7	364
56–62	59	7	413
63–69	66	10	660
70–76	73	5	365
77–83	80	8	640
84–90	87	6	522
		$N = 50$	$\sum(x \cdot f) = 3265$

d. 65.3

Section 2.4

1 a. Min = 23, or \$23,000; Max = 58, or \$58,000

b. 35, or \$35,000

c. The range of the starting salaries for Corporation B, which is 35, or \$35,000, is much larger than the range of Corporation A.

2 a. 41.5, or \$41,500

b.

Salary, x (1000s of dollars)	Deviation, $x - \mu$ (1000s of dollars)
23	−18.5
29	−12.5
32	−9.5
40	−1.5
41	−0.5
41	−0.5
49	7.5
50	8.5
52	10.5
58	16.5
$\sum x = 415$	$\sum(x - \mu) = 0$

3 ab. $\mu = 41.5$, or \$41,500

Salary, x	$x - \mu$	$(x - \mu)^2$
23	−18.5	342.25
29	−12.5	156.25
32	−9.5	90.25
40	−1.5	2.25
41	−0.5	0.25
41	−0.5	0.25
49	7.5	56.25
50	8.5	72.25
52	10.5	110.25
58	16.5	272.25
$\sum x = 415$	$\sum(x - \mu) = 0$	$\sum(x - \mu)^2 = 1102.5$

c. 110.3 **d.** 10.5, or \$10,500

e. The population standard deviation is 10.5, or \$10,500.

4 a. See 3ab. **b.** 122.5 **c.** 11.1, or \$11,100

d. The population standard deviation is 11.1, or \$11,100.

5 a. Enter data. **b.** 37.89; 3.98

6 a. 7, 7, 7, 7, 7, 13, 13, 13, 13, 13 **b.** 3

7 a. 1 standard deviation **b.** 34%

c. Approximately 34% of women ages 20–29 are between 64.3 and 66.92 inches tall.

8 a. 0 **b.** 70.6

c. At least 75% of the data lie within 2 standard deviations of the mean. At least 75% of the population of Alaska is between 0 and 70.6 years old.

9 a.

x	f	xf
0	10	0
1	19	19
2	7	14
3	7	21
4	5	20
5	1	5
6	1	6
	$n = 50$	$\sum xf = 85$

b. 1.7

c.

$x - \bar{x}$	$(x - \bar{x})^2$	$(x - \bar{x})^2 f$
−1.7	2.89	28.90
−0.7	0.49	9.31
0.3	0.09	0.63
1.3	1.69	11.83
2.3	5.29	26.45
3.3	10.89	10.89
4.3	18.49	18.49
		$\sum(x - \bar{x})^2 f = 106.5$

d. 1.5

10 a.

Class	x	f	xf
0–99	49.5	380	18,810
100–199	149.5	230	34,385
200–299	249.5	210	52,395
300–399	349.5	50	17,475
400–499	449.5	60	26,970
500+	650.0	70	45,500
		$n = 1000$	$\Sigma xf = 195{,}535$

b. 195.5

c.

$x - \bar{x}$	$(x - \bar{x})^2$	$(x - \bar{x})^2 f$
−146.0	21,316	8,100,080
−46.0	2116	486,680
54.0	2916	612,360
154.0	23,716	1,185,800
254.0	64,516	3,870,960
454.5	206,570.25	14,459,917.5
		$\Sigma(x - \bar{x})^2 f = 28{,}715{,}797.5$

d. 169.5

Section 2.5

1 a. 35, 36, 42, 43, 44, 46, 47, 49, 51, 51, 53, 53, 54, 54, 56, 57, 58, 59, 60, 61, 61, 63, 64, 65, 65, 66, 66, 67, 68, 69, 69, 72, 73, 73, 73, 76, 77, 78, 78, 80, 81, 82, 83, 83, 85, 86, 86, 87, 89, 89

b. 65.5   **c.** 54, 78

d. About one fourth of the 50 richest people are 54 years old or younger; one half are 65.5 years old or younger; and about three fourths of the 50 richest people are 78 years old or younger.

2 a. Enter data.   **b.** 17, 23, 28.5

c. One quarter of the tuition costs is $17,000 or less, one half is $23,000 or less, and three quarters is $28,500 or less.

3 a. 54, 78   **b.** 24

c. The ages of the 50 richest people in the middle portion of the data set vary by at most 24 years.

4 a. Min = 35, Q_1 = 54, Q_2 = 65.5, Q_3 = 78, Max = 89

bc. Ages of the 50 Richest People

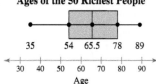

d. It appears that half of the ages are between 54 and 78.

5 a. 50th percentile

b. 50% of the 50 richest people are younger than 66.

6 a. $\mu = 70$, $\sigma = 8$

$$z_1 = \frac{60 - 70}{8} = -1.25$$

$$z_2 = \frac{71 - 70}{8} = 0.125$$

$$z_3 = \frac{92 - 70}{8} = 2.75$$

b. From the z-scores, $60 is 1.25 standard deviations below the mean, $71 is 0.125 standard deviation above the mean, and $92 is 2.75 standard deviations above the mean.

7 a. Best Actor: $\mu = 43.7$, $\sigma = 8.7$

Best Actress: $\mu = 35.9$, $\sigma = 11.4$

b. Sean Penn: $z = 0.49$

Kate Winslet: $z = -0.25$

c. The age of Sean Penn is 0.49 standard deviation above the mean and the age of Kate Winslet is 0.25 standard deviation below the mean. Both z-scores fall between -2 and 2, so neither would be considered unusual. Comparing the two measures indicates that Sean Penn is further above the average age of actors than Kate Winslet is below the average age of actresses. (Answers will vary.)

CHAPTER 3

Section 3.1

1ab. (1)

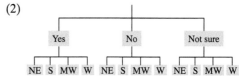

(2)

Yes	No	Not sure

| NE S MW W | NE S MW W | NE S MW W |

c. (1) 6   (2) 12

d. (1) Let Y = Yes, N = No, NS = Not sure, M = Male, F = Female.

Sample space =

    {YM, YF, NM, NF, NSM, NSF}

(2) Let Y = Yes, N = No, NS = Not sure, NE = Northeast, S = South, MW = Midwest, W = West.

Sample space =

    {$YNE, YS, YMW, YW, NNE, NS, NMW, NW,$ $NSNE, NSS, NSMW, NSW$}

2 a. (1) 6   (2) 1

b. (1) Not a simple event because it is an event that consists of more than a single outcome.

(2) Simple event because it is an event that consists of a single outcome.

3 a. Manufacturer: 4, Size: 2, Color: 5   **b.** 40

c.

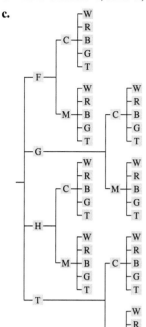

4 a. (1) Each letter is an event (26 choices for each).

(2) Each letter is an event (26, 25, 24, 23, 22, and 21 choices).

(3) Each letter is an event (22, 26, 26, 26, 26, and 26 choices).

b. (1) 308,915,776   (2) 165,765,600   (3) 261,390,272

5 a. (1) 52   (2) 52   (3) 52

b. (1) 1   (2) 13   (3) 52

c. (1) 0.019   (2) 0.25   (3) 1

6 a. The event is "the next claim processed is fraudulent." The frequency is 4.

b. 100   **c.** 0.04

7 a. 54   **b.** 1000   **c.** 0.054

8 a. The event is "salmon successfully passing through a dam on the Columbia River."

b. Estimated   **c.** Empirical probability

9 a. 0.18   **b.** 0.82   **c.** $\frac{41}{50}$ or 0.82

10 a. 5   **b.** 0.313

11 a. 10,000,000   **b.** $\dfrac{1}{10,000,000}$

Section 3.2

1 a. (1) 30 and 102   (2) 11 and 50

b. (1) 0.294   (2) 0.22

2 a. (1) Yes   (2) No

b. (1) Dependent   (2) Independent

3 a. (1) Independent   (2) Dependent

b. (1) 0.723   (2) 0.059

4 a. (1) Event   (2) Event   (3) Complement

b. (1) 0.729   (2) 0.001   (3) 0.999

c. (1) The event cannot be considered unusual because its probability is not less than or equal to 0.05.

(2) The event can be considered unusual because its probability is less than or equal to 0.05.

(3) The event cannot be considered unusual because its probability is not less than or equal to 0.05.

5 a. (1) and (2) $A = \{$is female$\}$,
$B = \{$works in health field$\}$

b. (1) $P(A \text{ and } B) = P(A) \cdot P(B|A) = (0.65) \cdot (0.25)$

(2) $P(A \text{ and } B') = P(A) \cdot (1 - P(B|A))$
$= (0.65) \cdot (0.75)$

c. (1) 0.163   (2) 0.488

Section 3.3

1 a. (1) None are true.   (2) None are true.

(3) All are true.

b. (1) Not mutually exclusive   (2) Not mutually exclusive

(3) Mutually exclusive

2 a. (1) Mutually exclusive   (2) Not mutually exclusive

b. (1) $\frac{1}{6}, \frac{1}{2}$ (2) $\frac{12}{52}, \frac{13}{52}, \frac{3}{52}$ **c.** (1) 0.667   (2) 0.423

3 a. $A = \{$sales between \$0 and \$24,999$\}$

$B = \{$sales between \$25,000 and \$49,999$\}$

b. A and B cannot occur at the same time.

A and B are mutually exclusive.

c. $\frac{3}{36}, \frac{5}{36}$ **d.** 0.222

4 a. (1) $A = \{$type B$\}$

$B = \{$type AB$\}$

(2) $A = \{$type O$\}$

$B = \{$Rh-positive$\}$

b. (1) A and B cannot occur at the same time.

A and B are mutually exclusive.

(2) A and B can occur at the same time.

A and B are not mutually exclusive.

c. (1) $\frac{45}{409}, \frac{16}{409}$ (2) $\frac{184}{409}, \frac{344}{409}$

d. (1) 0.149   (2) 0.910

5 a. 0.141   **b.** 0.859

Section 3.4

1 a. 8   **b.** 40,320

2 a. 336

b. There are 336 possible ways that the subject can pick a first, second, and third activity.

3 a. $n = 12, r = 4$ **b.** 11,880

4a. $n = 20$, $n_1 = 6$, $n_2 = 9$, $n_3 = 5$ **b.** 77,597,520

5a. $n = 20$, $r = 3$ **b.** 1140

c. There are 1140 different possible three-person committees that can be selected from 20 employees.

6a. 380   **b.** 0.003

7a. 1 outcome and 180 distinguishable permutations

b. 0.006

8a. 3003   **b.** 3,162,510   **c.** 0.0009

9a. 10   **b.** 220   **c.** 0.045

CHAPTER 4

Section 4.1

1a. (1) Measured   (2) Counted

b. (1) The random variable is continuous because x can be any speed up to the maximum speed of a Space Shuttle.

(2) The random variable is discrete because the number of calves born on a farm in one year is countable.

2ab.

x	f	$P(x)$
0	16	0.16
1	19	0.19
2	15	0.15
3	21	0.21
4	9	0.09
5	10	0.10
6	8	0.08
7	2	0.02
	$n = 100$	$\Sigma P(x) = 1$

c.

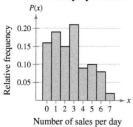

New Employee Sales

Number of sales per day

3a. Each $P(x)$ is between 0 and 1.   **b.** $\Sigma P(x) = 1$

c. Because both conditions are met, the distribution is a probability distribution.

4a. (1) Yes, each outcome is between 0 and 1.

(2) Yes, each outcome is between 0 and 1.

b. (1) Yes   (2) Yes

c. (1) A probability distribution

(2) A probability distribution

5ab.

x	$P(x)$	$xP(x)$
0	0.16	0.00
1	0.19	0.19
2	0.15	0.30
3	0.21	0.63
4	0.09	0.36
5	0.10	0.50
6	0.08	0.48
7	0.02	0.14
	$\Sigma P(x) = 1$	$\Sigma xP(x) = 2.60$

c. $\mu = 2.6$

On average, a new employee makes 2.6 sales per day.

6ab.

x	$P(x)$	$x - \mu$	$(x - \mu)^2$	$P(x)(x - \mu)^2$
0	0.16	−2.6	6.76	1.0816
1	0.19	−1.6	2.56	0.4864
2	0.15	−0.6	0.36	0.0540
3	0.21	0.4	0.16	0.0336
4	0.09	1.4	1.96	0.1764
5	0.10	2.4	5.76	0.5760
6	0.08	3.4	11.56	0.9248
7	0.02	4.4	19.36	0.3872
	$\Sigma P(x) = 1$			$\Sigma P(x)(x - \mu)^2 = 3.72$

c. 1.9

d. Most of the data values differ from the mean by no more than 1.9 sales per day.

7ab.

Gain, x	$1995	$995	$495	$245	$95	−$5
Probability, $P(x)$	$\frac{1}{2000}$	$\frac{1}{2000}$	$\frac{1}{2000}$	$\frac{1}{2000}$	$\frac{1}{2000}$	$\frac{1995}{2000}$

c. −$3.08

d. Because the expected value is negative, you can expect to lose an average of $3.08 for each ticket you buy.

Section 4.2

1a. Trial: answering a question

Success: question answered correctly

b. Yes

c. It is a binomial experiment; $n = 10$, $p = 0.25$, $q = 0.75$, $x = 0, 1, 2, 3, 4, 5, 6, 7, 8, 9, 10$

2a. Trial: drawing a card with replacement

Success: card drawn is a club

Failure: card drawn is not a club

b. $n = 5$, $p = 0.25$, $q = 0.75$, $x = 3$

c. $P(3) = \dfrac{5!}{2!\,3!}(0.25)^3(0.75)^2 \approx 0.088$

3a. Trial: selecting an adult and asking a question

Success: selecting an adult who likes texting because it works where talking won't do

Failure: selecting an adult who does not like texting because it works where talking won't do

b. $n = 7$, $p = 0.75$, $q = 0.25$, $x = 0, 1, 2, 3, 4, 5, 6, 7$

c. $P(0) = {_7}C_0\,(0.75)^0\,(0.25)^7 \approx 0.00006$

$P(1) = {_7}C_1\,(0.75)^1\,(0.25)^6 \approx 0.00128$

$P(2) = {_7}C_2\,(0.75)^2\,(0.25)^5 \approx 0.01154$

$P(3) = {_7}C_3\,(0.75)^3\,(0.25)^4 \approx 0.05768$

$P(4) = {_7}C_4\,(0.75)^4\,(0.25)^3 \approx 0.17303$

$P(5) = {_7}C_5\,(0.75)^5\,(0.25)^2 \approx 0.31146$

$P(6) = {_7}C_6\,(0.75)^6\,(0.25)^1 \approx 0.31146$

$P(7) = {_7}C_7\,(0.75)^7\,(0.25)^0 \approx 0.13348$

d.

x	$P(x)$
0	0.00006
1	0.00128
2	0.01154
3	0.05768
4	0.17303
5	0.31146
6	0.31146
7	0.13348
	$\sum P(x) \approx 1$

4a. $n = 250$, $p = 0.71$, $x = 178$　　**b.** 0.056

c. The probability that exactly 178 people from a random sample of 250 people in the United States will use more than one topping on their hotdogs is about 0.056.

d. Because 0.056 is not less than or equal to 0.05, this event is not unusual.

5a. (1) $x = 2$　(2) $x = 2, 3, 4$, or 5　(3) $x = 0$ or 1

b. (1) 0.217　(2) 0.217, 0.058, 0.008, 0.0004; 0.283

(3) 0.308, 0.409; 0.717

c. (1) The probability that exactly two of the five men consider fishing their favorite leisure-time activity is about 0.217.

(2) The probability that at least two of the five men consider fishing their favorite leisure-time activity is about 0.283.

(3) The probability that fewer than two of the five men consider fishing their favorite leisure-time activity is about 0.717.

6a. Trial: selecting a business and asking if it has a website

Success: selecting a business with a website

Failure: selecting a business without a website

b. $n = 10$, $p = 0.55$, $x = 4$　　**c.** 0.160

d. The probability that exactly 4 of the 10 small businesses have websites is 0.160.

e. Because 0.160 is greater than 0.05, this event is not unusual.

7a. 0.001, 0.022, 0.142, 0.404, 0.430

b.

x	$P(x)$
0	0.001
1	0.022
2	0.142
3	0.404
4	0.430

c.

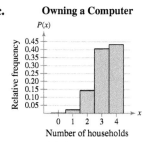

Owning a Computer

Skewed left

d. Yes, it would be unusual if exactly zero or exactly one of the four households owned a computer, because each of these events has a probability that is less than 0.05.

8a. Success: selecting a clear day

$n = 31$, $p = 0.44$, $q = 0.56$

b. 13.6　**c.** 7.6　**d.** 2.8

e. On average, there are about 14 clear days during the month of May.

f. A May with fewer than 8 clear days or more than 19 clear days would be unusual.

Section 4.3

1a. 0.74; 0.192　**b.** 0.932

c. The probability that LeBron makes his first free throw shot before his third attempt is 0.932.

2a. $P(0) \approx 0.050$

$P(1) \approx 0.149$

$P(2) \approx 0.224$

$P(3) \approx 0.224$

$P(4) \approx 0.168$

b. 0.815　**c.** 0.185

d. The probability that more than four accidents will occur in any given month at the intersection is 0.185.

3a. 0.10　**b.** 0.10, 3　**c.** 0.0002

d. The probability of finding three brown trout in any given cubic meter of the lake is 0.0002.

e. Because 0.0002 is less than 0.05, this can be considered an unusual event.

CHAPTER 5

Section 5.1

1a. A: $x = 45$, B: $x = 60$, C: $x = 45$; B has the greatest mean.

b. Curve C is more spread out, so curve C has the greatest standard deviation.

2a. $x = 660$　**b.** 630, 690; 30

3. (1) 0.0143　(2) 0.9850

4 a.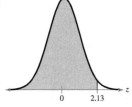
b. 0.9834

5 a.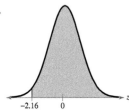
b. 0.0154   **c.** 0.9846

6 a. 0.0885   **b.** 0.0152   **c.** 0.0733

d. 7.33% of the area under the curve falls between $z = -2.165$ and $z = -1.35$.

Section 5.2

1 a.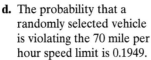
- $\mu = 67$
- $x = 70$
- Speed (in miles per hour)

b. 0.86
c. 0.1949
d. The probability that a randomly selected vehicle is violating the 70 mile per hour speed limit is 0.1949.

2 a.
- $\mu = 45$
- $x = 33$
- $x = 60$
- Time (in minutes)

b. -1, 1.25
c. 0.1587; 0.8944; 0.7357
d. If 150 shoppers enter the store, then you would expect $150(0.7357) = 110.355$, or about 110, shoppers to be in the store between 33 and 60 minutes.

3 a. Read user's guide for the technology tool.

b. 0.5105

c. The probability that a randomly selected U.S. person's triglyceride level is between 100 and 150 is 0.5105.

Section 5.3

1 a. (1) 0.0384   (2) 0.0250 and 0.9750
bc. (1) -1.77 (2) ± 1.96
2 a. (1) Area $= 0.10$ (2) Area $= 0.20$
   (3) Area $= 0.99$
bc. (1) -1.28 (2) -0.84 (3) 2.33
3 a. $\mu = 52$, $\sigma = 15$ **b.** 17.05; 98.5; 60.7
c. 17.05 pounds is below the mean, 60.7 pounds and 98.5 pounds are above the mean.

4ab.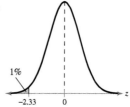
- 1%
- -2.33 0

c. 116.93
d. So, the longest braking distance a Nissan Altima could have and still be in the bottom 1% is about 117 feet.

5ab.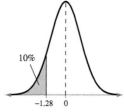
- 10%
- -1.28 0

c. 8.512
d. So, the maximum length of time an employee could have worked and still be laid off is about 8.5 years.

Section 5.4

1a.

Sample	Mean	Sample	Mean	Sample	Mean
1, 1, 1	1	3, 3, 5	3.67	5, 7, 1	4.33
1, 1 ,3	1.67	3, 3, 7	4.33	5, 7, 3	5
1, 1, 5	2.33	3, 5, 1	3	5, 7, 5	5.67
1, 1, 7	3	3, 5, 3	3.67	5, 7, 7	6.33
1, 3, 1	1.67	3, 5, 5	4.33	7, 1, 1	3
1, 3, 3	2.33	3, 5, 7	5	7, 1, 3	3.67
1, 3, 5	3	3, 7, 1	3.67	7, 1, 5	4.33
1, 3, 7	3.67	3, 7, 3	4.33	7 ,1, 7	5
1, 5, 1	2.33	3, 7, 5	5	7, 3, 1	3.67
1, 5, 3	3	3, 7, 7	5.67	7, 3, 3	4.33
1, 5, 5	3.67	5, 1, 1	2.33	7, 3, 5	5
1, 5, 7	4.33	5, 1, 3	3	7, 3, 7	5.67
1, 7, 1	3	5, 1, 5	3.67	7, 5, 1	4.33
1, 7, 3	3.67	5, 1, 7	4.33	7, 5, 3	5
1, 7, 5	4.33	5, 3, 1	3	7, 5, 5	5.67
1, 7, 7	5	5, 3, 3	3.67	7, 5, 7	6.33
3, 1, 1	1.67	5, 3, 5	4.33	7, 7, 1	5
3, 1, 3	2.33	5, 3, 7	5	7, 7, 3	5.67
3, 1, 5	3	5, 5, 1	3.67	7, 7, 5	6.33
3, 1, 7	3.67	5, 5, 3	4.33	7, 7, 7	7
3, 3, 1	2.33	5, 5, 5	5		
3, 3, 3	3	5, 5, 7	5.67		

b.

$\bar{x}$	f	Probability
1	1	0.0156
1.67	3	0.0469
2.33	6	0.0938
3	10	0.1563
3.67	12	0.1875
4.33	12	0.1875
5	10	0.1563
5.67	6	0.0938
6.33	3	0.0469
7	1	0.0156

$\mu_{\bar{x}} = 4$
$(\sigma_{\bar{x}})^2 \approx 1.667$
$\sigma_{\bar{x}} \approx 1.291$

c. $\mu_{\bar{x}} = \mu = 4$

$$(\sigma_{\bar{x}})^2 = \frac{\sigma^2}{n} = \frac{5}{3} \approx 1.667; \; \sigma_{\bar{x}} = \frac{\sigma}{\sqrt{n}} = \frac{\sqrt{5}}{\sqrt{3}} \approx 1.291$$

2a. 63, 1.4

b. $n = 64$

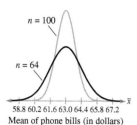

Mean of phone bills (in dollars)

c. With a smaller sample size, the mean stays the same but the standard deviation increases.

3a. 3.5, 0.05

b.

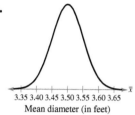

Mean diameter (in feet)

4a. 25; 0.15   **b.** $-2, 3.33$ **c.** 0.0228; 0.996; 0.9768

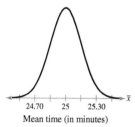

Mean time (in minutes)

d. Of the samples of 100 drivers ages 15 to 19, 97.68% will have a mean driving time between 24.7 and 25.5 minutes.

5a. 290,600; 10,392.30

Mean sales price (in dollars)

b. -2.46 **c.** 0.0069; 0.9931

d. 99.31% of samples of 12 single-family houses will have a mean sales price greater than $265,000.

6a. 0.21; 0.66   **b.** 0.5832; 0.7454

c. There is about a 58% chance that an LCD computer monitor will cost less than $200. There is about a 75% chance that the mean of a sample of 10 LCD computer monitors is less than $200.

Section 5.5

1a. $n = 125$, $p = 0.05$, $q = 0.95$ **b.** 6.25, 118.75

c. Normal distribution can be used.   **d.** 6.25, 2.44

2a. (1) 57, 58, ..., 83   (2) ..., 52, 53, 54

b. (1) $56.5 < x < 83.5$ (2) $x < 54.5$

3a. Normal distribution can be used.   **b.** 6.25, 2.44

c.

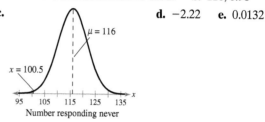

Number responding yes

d. 1.33   **e.** 0.9082; 0.0918

4a. Normal distribution can be used.   **b.** 116, 6.98

c.

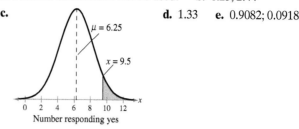

Number responding never

d. -2.22 **e.** 0.0132

5a. Normal distribution can be used.   **b.** 36, 5.23

c.

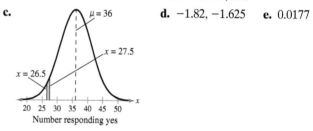

Number responding yes

d. $-1.82, -1.625$ **e.** 0.0177

CHAPTER 6

Section 6.1

1a. $\bar{x} = 138.5$

b. A point estimate for the population mean number of friends is 138.5.

2a. $z_c = 1.96$, $n = 30$, $s \approx 51.0$ **b.** $E \approx 18.3$

c. You are 95% confident that the maximum error of the estimate is about 18.3 friends.

3a. $\bar{x} = 138.5$, $E \approx 18.3$ **b.** 120.2, 156.8

c. With 95% confidence, you can say that the population mean number of friends is between 120.2 and 156.8. This confidence interval is wider than the one found in Example 3.

4a. Enter the data.

b. (121.2, 140.4); (118.7, 142.9); (109.2, 152.4)

c. As the confidence level increases, so does the width of the interval.

5a. $n = 30$, $\bar{x} = 22.9$, $\sigma = 1.5$, $z_c = 1.645$, $E \approx 0.5$

b. (22.4, 23.4) [*Tech:* (22.5, 23.4)]

c. With 90% confidence, you can say that the mean age of the students is between 22.4 (*Tech:* 22.5) and 23.4 years. Because of the larger sample size, the confidence interval is slightly narrower.

6 a. $z_c = 1.96$, $E = 10$, $s \approx 53.0$ **b.** $n = 108$

 c. You should have at least 108 users in your sample. Because of the larger margin of error, the sample size needed is much smaller.

Section 6.2

1 a. d.f. $= 21$ **b.** $c = 0.90$ **c.** $t_c = 1.721$

2 a. $t_c = 1.753$, $E \approx 4.4$; $t_c = 2.947$, $E \approx 7.4$

 b. $(157.6, 166.4)$; $(154.6, 169.4)$

 c. With 90% confidence, you can say that the population mean temperature of coffee sold is between 157.6°F and 166.4°F.

 With 99% confidence, you can say that the population mean temperature of coffee sold is between 154.6°F and 169.4°F.

3 a. $t_c = 1.729$, $E \approx 0.92$; $t_c = 2.093$, $E \approx 1.12$

 b. $(8.83, 10.67)$; $(8.63, 10.87)$

 c. With 90% confidence, you can say that the population mean number of days the car model sits on the lot is between 8.83 and 10.67; with 95% confidence, you can say that the population mean number of days the car model sits on the lot is between 8.63 and 10.87. The 90% confidence interval is slightly narrower.

4. Use a t-distribution because the sample size is small ($n < 30$), the population is normally distributed, and the population standard deviation is unknown.

Section 6.3

1 a. $x = 181$, $n = 1006$ **b.** $\hat{p} \approx 0.180$

2 a. $\hat{p} \approx 0.180$, $\hat{q} \approx 0.820$

 b. $n\hat{p} \approx 181 > 5$ and $n\hat{q} \approx 825 > 5$

 c. $z_c = 1.645$, $E = 0.020$ **d.** $(0.160, 0.200)$

 e. With 90% confidence, you can say that the proportion of adults who think Abraham Lincoln was the greatest president is between 16.0% and 20.0%.

3 a. $\hat{p} = 0.25$, $\hat{q} = 0.75$

 b. $n\hat{p} = 124.5 > 5$ and $n\hat{q} = 373.5 > 5$

 c. $z_c = 2.575$, $E \approx 0.050$ **d.** $(0.20, 0.30)$

 e. With 99% confidence, you can say that the proportion of U.S. adults who think that people over 65 are the more dangerous drivers is between 20% and 30%.

4 a. (1) $\hat{p} = 0.5$, $\hat{q} = 0.5$, $z_c = 1.645$, $E = 0.02$

 (2) $\hat{p} = 0.11$, $\hat{q} = 0.89$, $z_c = 1.645$, $E = 0.02$

 b. (1) 1691.27   (2) 662.30

 c. (1) 1692 females   (2) 663 females

Section 6.4

1 a. d.f. $= 29$, $c = 0.90$ **b.** $0.05, 0.95$ **c.** $42.557, 17.708$

 d. 90% of the area under the curve lies between 17.708 and 42.557.

2 a. $42.557, 17.708$; $45.722, 16.047$

 b. $(0.98, 2.36)$; $(0.91, 2.60)$ **c.** $(0.99, 1.54)$; $(0.96, 1.61)$

 d. With 90% confidence, you can say that the population variance is between 0.98 and 2.36 and the population standard deviation is between 0.99 and 1.54. With 95% confidence, you can say that the population variance is between 0.91 and 2.60 and the population standard deviation is between 0.96 and 1.61.

CHAPTER 7

Section 7.1

1 a. (1) The mean is not 74 months.

 $\mu \neq 74$

 (2) The variance is less than or equal to 2.7.

 $\sigma^2 \leq 2.7$

 (3) The proportion is more than 24%.

 $p > 0.24$

 b. (1) $\mu = 74$ (2) $\sigma^2 > 2.7$ (3) $p \leq 0.24$

 c. (1) H_0: $\mu = 74$; H_a: $\mu \neq 74$ (claim)

 (2) H_0: $\sigma^2 \leq 2.7$ (claim); H_a: $\sigma^2 > 2.7$

 (3) H_0: $p \leq 0.24$; H_a: $p > 0.24$ (claim)

2 a. H_0: $p \leq 0.01$; H_a: $p > 0.01$

 b. A type I error will occur if the actual proportion is less than or equal to 0.01, but you reject H_0.

 A type II error will occur if the actual proportion is greater than 0.01, but you fail to reject H_0.

 c. A type II error is more serious because you would be misleading the consumer, possibly causing serious injury or death.

3 a. (1) H_0: The mean life of a certain type of automobile battery is 74 months.

 H_a: The mean life of a certain type of automobile battery is not 74 months.

 H_0: $\mu = 74$; H_a: $\mu \neq 74$

 (2) H_0: The variance of the life of the home theater systems is less than or equal to 2.7.

 H_a: The variance of the life of the home theater systems is greater than 2.7.

 H_0: $\sigma^2 \leq 2.7$; H_a: $\sigma^2 > 2.7$

 (3) H_0: The proportion of homeowners who feel their house is too small for their family is less than or equal to 24%.

 H_a: The proportion of homeowners who feel their house is too small for their family is greater than 24%.

 H_0: $p \leq 0.24$; H_a: $p > 0.24$

 b. (1) Two-tailed   (2) Right-tailed   (3) Right-tailed

c. (1)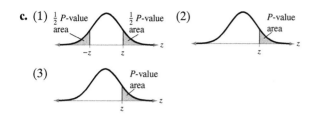

(2)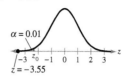

(3)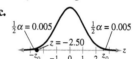

4 a. There is enough evidence to support the realtor's claim that the proportion of homeowners who feel their house is too small for their family is more than 24%.

b. There is not enough evidence to support the realtor's claim that the proportion of homeowners who feel their house is too small for their family is more than 24%.

5 a. (1) Support claim. (2) Reject claim.

b. (1) $H_0: \mu \geq 650$; $H_a: \mu < 650$ (claim)

(2) $H_0: \mu = 98.6$ (claim); $H_a: \mu \neq 98.6$

Section 7.2

1 a. (1) $0.0347 > 0.01$ (2) $0.0347 < 0.05$

b. (1) Fail to reject H_0. (2) Reject H_0.

2 a b. 0.0436

  c. Reject H_0 because $0.0436 < 0.05$.

3 a. 0.9495 **b.** 0.1010

  c. Fail to reject H_0 because $0.1010 > 0.01$.

4 a. The claim is "the mean speed is greater than 35 miles per hour."

$H_0: \mu \leq 35$; $H_a: \mu > 35$ (claim)

b. $\alpha = 0.05$ **c.** 2.5 **d.** 0.0062 **e.** Reject H_0.

f. There is enough evidence at the 5% level of significance to support the claim that the average speed is greater than 35 miles per hour.

5 a. The claim is "one of your distributors reports an average of 150 sales per day."

$H_0: \mu = 150$ (claim); $H_a: \mu \neq 150$

b. $\alpha = 0.01$ **c.** −2.76 **d.** 0.0058

e. Reject H_0 because $0.0058 < 0.01$.

f. There is enough evidence at the 1% level of significance to reject the claim that the distributorship averages 150 sales per day.

6 a. $0.0440 > 0.01$ **b.** Fail to reject H_0.

7 a.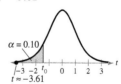

b. 0.1003

c. $z_0 = -1.28$

d. Rejection region: $z < -1.28$

8 a.

b. 0.0401, 0.9599

c. $-z_0 = -1.75$, $z_0 = 1.75$

d. Rejection regions: $z < -1.75$, $z > 1.75$

9 a. The claim is "the mean work day of the company's mechanical engineers is less than 8.5 hours."

$H_0: \mu \geq 8.5$; $H_a: \mu < 8.5$ (claim)

b. $\alpha = 0.01$

c. $z_0 = -2.33$; Rejection region: $z < -2.33$

d. −3.55

e. Because $-3.55 < -2.33$, reject H_0.

f. There is enough evidence at the 1% level of significance to support the claim that the mean work day is less than 8.5 hours.

10 a. $\alpha = 0.01$

b. $-z_0 = -2.575$, $z_0 = 2.575$;

Rejection regions: $z < -2.575$, $z > 2.575$

c.

Fail to reject H_0.

d. There is not enough evidence at the 1% level of significance to reject the claim that the mean cost of raising a child from birth to age 2 by husband-wife families in the United States is $13,120.

Section 7.3

1 a. 13 **b.** −2.650

2 a. 8 **b.** 1.397

3 a. 15 **b.** −2.131, 2.131

4 a. The claim is "the mean cost of insuring a 2008 Honda CR-V is less than $1200."

$H_0: \mu \geq \$1200$; $H_a: \mu < \$1200$ (claim)

b. $\alpha = 0.10$, d.f. = 6

c. $t_0 = -1.440$; Rejection region: $t < -1.440$

d. −3.61

e. Reject H_0.

f. There is enough evidence at the 10% level of significance to support the insurance agent's claim that the mean cost of insuring a 2008 Honda CR-V is less than $1200.

5a. The claim is "the mean conductivity of the river is 1890 milligrams per liter."

$H_0: \mu = 1890$ (claim); $H_a: \mu \neq 1890$

b. $\alpha = 0.01$, d.f. $= 18$

c. $-t_0 = -2.878$, $t_0 = 2.878$

Rejection regions: $t < -2.878$, $t > 2.878$

d. 3.798

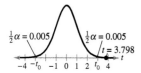

e. Reject H_0.

f. There is enough evidence at the 1% level of significance to reject the company's claim that the mean conductivity of the river is 1890 milligrams per liter.

6a. The claim is "the mean wait time is at most 18 minutes."

$H_0: \mu \leq 18$ minutes (claim); $H_a: \mu > 18$ minutes

b. 0.9997

c. $0.9997 > 0.05$; Fail to reject H_0.

d. There is not enough evidence at the 5% level of significance to reject the office's claim that the mean wait time is at most 18 minutes..

Section 7.4

1a. $np = 31.25 > 5$, $nq = 93.75 > 5$

b. The claim is "more than 25% of U.S. adults have used a cellular phone to access the Internet."

$H_0: p \leq 0.25$; $H_a: p > 0.25$ (claim)

c. $\alpha = 0.05$

d. $z_0 = 1.645$; Rejection region: $z > 1.645$

e. 1.81

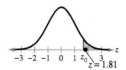

f. Reject H_0.

g. There is enough evidence at the 5% level of significance to support the research center's claim that more than 25% of U.S. adults have used a cellular phone to access the Internet.

2a. $np = 75 > 5$, $nq = 175 > 5$

b. The claim is "30% of U.S. adults have not purchased a certain brand because they found the advertisements distasteful."

$H_0: p = 0.30$ (claim); $H_a: p \neq 0.30$

c. $\alpha = 0.10$

d. $-z_0 = -1.645$, $z_0 = 1.645$;

Rejection regions: $z < -1.645$, $z > 1.645$

e. 2.07

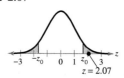

f. Reject H_0.

g. There is enough evidence at the 10% level of significance to reject the claim that 30% of U.S. adults have not purchased a certain brand because they found the advertisements distasteful.

Section 7.5

1a. d.f. $= 17$, $\alpha = 0.01$ **b.** 33.409

2a. d.f. $= 29$, $\alpha = 0.05$ **b.** 17.708

3a. d.f. $= 50$, $\alpha = 0.01$ **b.** 79.490 **c.** 27.991

4a. The claim is "the variance of the amount of sports drink in a 12-ounce bottle is no more than 0.40."

$H_0: \sigma^2 \leq 0.40$ (claim); $H_a: \sigma^2 > 0.40$

b. $\alpha = 0.01$, d.f. $= 30$

c. $\chi_0^2 = 50.892$; Rejection region: $\chi^2 > 50.892$

d. 56.250 **e.** Reject H_0.

f. There is enough evidence at the 1% level of significance to reject the bottling company's claim that the variance of the amount of sports drink in a 12-ounce bottle is no more than 0.40.

5a. The claim is "the standard deviation of the lengths of response times is less than 3.7 minutes."

$H_0: \sigma \geq 3.7$; $H_a: \sigma < 3.7$ (claim)

b. $\alpha = 0.05$, d.f. $= 8$

c. $\chi_0^2 = 2.733$; Rejection region: $\chi^2 < 2.733$

d. 5.259 **e.** Fail to reject H_0.

f. There is not enough evidence at the 5% level of significance to support the police chief's claim that the standard deviation of the lengths of response times is less than 3.7 minutes.

6a. The claim is "the variance of the weight losses is 25.5."

$H_0: \sigma^2 = 25.5$ (claim); $H_a: \sigma^2 \neq 25.5$

b. $\alpha = 0.10$, d.f. $= 12$

c. $\chi_L^2 = 5.226$, $\chi_R^2 = 21.026$;

Rejection regions: $\chi^2 < 5.226$, $\chi^2 > 21.026$

d. 5.082 **e.** Reject H_0.

f. There is enough evidence at the 10% level of significance to reject the company's claim that the variance of the weight losses of the users is 25.5.

CHAPTER 8

Section 8.1

1a. (1) Independent (2) Dependent

b. (1) Because each sample represents blood pressures of different individuals, and it is not possible to form a pairing between the members of the samples.

(2) Because the samples represent exam scores of the same students, the samples can be paired with respect to each student.

2a. The claim is "there is a difference in the mean annual wages for forensic science technicians working for local and state governments."

$H_0: \mu_1 = \mu_2$; $H_a: \mu_1 \neq \mu_2$ (claim)

b. $\alpha = 0.10$

c. $-z_0 = -1.645$, $z_0 = 1.645$;

Rejection regions: $z < -1.645$, $z > 1.645$

d. 1.667

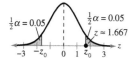

e. Reject H_0.

f. There is enough evidence at the 10% level of significance to support the claim that there is a difference in the mean annual wages for forensic science technicians working for local and state governments.

3a. $z \approx 1.36$; $P \approx 0.0865$

b. Fail to reject H_0.

c. There is not enough evidence at the 5% level of significance to support the travel agency's claim that the average daily cost of meals and lodging for vacationing in Alaska is greater than the same average cost for vacationing in Colorado.

Section 8.2

1a. The claim is "there is a difference in the mean annual earnings based on level of education."

$H_0: \mu_1 = \mu_2$; $H_a: \mu_1 \neq \mu_2$ (claim)

b. $\alpha = 0.01$; d.f. = 11

c. $-t_0 = -3.106$, $t_0 = 3.106$; Rejection regions: $t < -3.106$, $t > 3.106$

d. -4.63

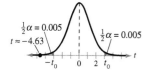

e. Reject H_0.

f. There is enough evidence at the 1% level of significance to support the claim that there is a difference in the mean annual earnings based on level of education.

2a. The claim is "the watt usage of a manufacturer's 17-inch flat panel monitors is less than that of its leading competitor."

$H_0: \mu_1 \geq \mu_2$; $H_a: \mu_1 < \mu_2$ (claim)

b. $\alpha = 0.10$; d.f. = 25

c. $t_0 = -1.316$; Rejection region: $t < -1.316$

d. -3.997

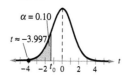

e. Reject H_0.

f. There is enough evidence at the 10% level of significance to support the manufacturer's claim that the watt usage of its monitors is less than that of its leading competitor.

Section 8.3

1a. The claim is "athletes can decrease their times in the 40-yard dash."

$H_0: \mu_d \leq 0$; $H_a: \mu_d > 0$ (claim)

b. $\alpha = 0.05$; d.f. = 11

c. $t_0 = 1.796$; Rejection region: $t > 1.796$

d. $\overline{d} \approx 0.0233$; $s_d \approx 0.0607$

e. 1.333

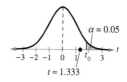

f. Fail to reject H_0.

g. There is not enough evidence at the 5% level of significance to support the claim that athletes can decrease their times in the 40-yard dash.

2a. The claim is "the drug changes the body's temperature."

$H_0: \mu_d = 0$; $H_a: \mu_d \neq 0$ (claim)

b. $\alpha = 0.05$; d.f. = 6

c. $-t_0 = -2.447$, $t_0 = 2.447$; Rejection regions: $t < -2.447$, $t > 2.447$

d. $\overline{d} \approx 0.5571$; $s_d \approx 0.9235$

e. 1.596

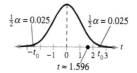

f. Fail to reject H_0.

g. There is not enough evidence at the 5% level of significance to support the claim that the drug changes the body's temperature.

Section 8.4

1a. The claim is "there is a difference between the proportion of male high school students who smoke cigarettes and the proportion of female high school students who smoke cigarettes."

$H_0: p_1 = p_2$; $H_a: p_1 \neq p_2$ (claim)

b. $\alpha = 0.05$

c. $-z_0 = -1.96$, $z_0 = 1.96$;

Rejection regions: $z < -1.96$, $z > 1.96$

d. $\bar{p} \approx 0.1975$; $\bar{q} \approx 0.8025$

e. $n_1\bar{p} \approx 1382.5 > 5$, $n_1\bar{q} \approx 5617.5 > 5$, $n_2\bar{p} \approx 1479.1 > 5$, and $n_2\bar{q} \approx 6009.9 > 5$.

f. 4.23

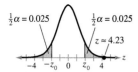

g. Reject H_0.

h. There is enough evidence at the 5% level of significance to support the claim that there is a difference between the proportion of male high school students who smoke cigarettes and the proportion of female high school students who smoke cigarettes.

2a. The claim is "the proportion of male high school students who smoke cigars is greater than the proportion of female high school students who smoke cigars."

$H_0: p_1 \leq p_2$; $H_a: p_1 > p_2$ (claim)

b. $\alpha = 0.05$ **c.** $z_0 = 1.645$; Rejection region: $z > 1.645$

d. $\bar{p} \approx 0.1174$; $\bar{q} \approx 0.8826$

e. $n_1\bar{p} \approx 821.8 > 5$, $n_1\bar{q} \approx 6178.2 > 5$, $n_2\bar{p} \approx 879.2 > 5$, and $n_2\bar{q} = 6609.8 > 5$

f. 17.565

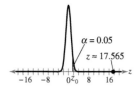

g. Reject H_0.

h. There is enough evidence at the 5% level of significance to support the claim that the proportion of male high school students who smoke cigars is greater than the proportion of female high school students who smoke cigars.

CHAPTER 9

Section 9.1

1ab.

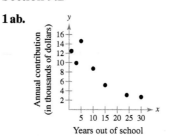

c. Yes, it appears that there is a negative linear correlation. As the number of years out of school increases, the annual contribution decreases.

2ab.

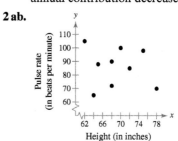

c. No, it appears that there is no linear correlation between height and pulse rate.

3ab.

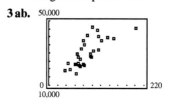

c. Yes, it appears that there is a positive linear correlation. As the team salary increases, the average attendance per home game increases.

4a. 7; $\Sigma x = 88$, $\Sigma y = 56.7$, $\Sigma xy = 435.6$, $\Sigma x^2 = 1836$, $\Sigma y^2 = 587.05$

b. -0.908

c. Because r is close to -1, this suggests a strong negative linear correlation between years out of school and annual contribution.

5ab. 0.750

c. Because r is close to 1, this suggests a strong positive linear correlation between the salaries and the average attendances at home games.

6a. 7   **b.** 0.01   **c.** 0.875

d. $|r| \approx 0.908 > 0.875$; The correlation is significant.

e. There is enough evidence at the 1% level of significance to conclude that there is a significant linear correlation between the number of years out of school and the annual contribution.

7 a. $H_0: \rho = 0;\ H_a: \rho \neq 0$ **b.** 0.01 **c.** 28

d. $-t_0 = -2.763,\ t_0 = 2.763$;

Rejection regions: $t < -2.763,\ t > 2.763$

e. 5.995 **f.** Reject H_0.

g. There is enough evidence at the 1% level of significance to conclude that there is a significant linear correlation between the salaries and average attendances at home games for the teams in Major League Baseball.

Section 9.2

1 a. $n = 7,\ \Sigma x = 88,\ \Sigma y = 56.7,\ \Sigma xy = 435.6,\ \Sigma x^2 = 1836$

b. $m \approx -0.379875;\ b \approx 12.8756$

c. $\hat{y} = -0.380x + 12.876$

2 a. Enter the data. **b.** $m \approx 189.038015;\ b \approx 13,497.9583$

c. $\hat{y} = 189.038x + 13,497.958$

3 a. (1) $\hat{y} = 12.481(2) + 33.683$

(2) $\hat{y} = 12.481(3.32) + 33.683$

b. (1) 58.645 (2) 75.120

c. (1) 58.645 minutes (2) 75.120 minutes

Section 9.3

1 a. 0.979 **b.** 0.958

c. About 95.8% of the variation in the times is explained. About 4.2% of the variation is unexplained.

2 a.

x_i	y_i	$\hat{y}_i$	$y_i - \hat{y}_i$	$(y_i - \hat{y}_i)^2$
15	26	28.386	−2.386	5.692996
20	32	35.411	−3.411	11.634921
20	38	35.411	2.589	6.702921
30	56	49.461	6.539	42.758521
40	54	63.511	−9.511	90.459121
45	78	70.536	7.464	55.711296
50	80	77.561	2.439	5.948721
60	88	91.611	−3.611	13.039321
				$\Sigma = 231.947818$

b. 8 **c.** 6.218

d. The standard error of estimate of the weekly sales for a specific radio ad time is about $621.80.

3 a. $n = 10,\ \text{d.f.} = 8,\ t_c = 2.306,\ s_e \approx 138.255$

b. 886.897 **c.** 364.088

d. $522.809 < y < 1250.985$

e. You can be 95% confident that when the gross domestic product is $4 trillion, the carbon dioxide emissions will be between 522.809 and 1250.985 million metric tons.

Section 9.4

1 a. Enter the data.

b. $\hat{y} = 46.385 + 0.540x_1 - 4.897x_2$

2 ab. (1) $\hat{y} = 46.385 + 0.540(89) - 4.897(1)$

(2) $\hat{y} = 46.385 + 0.540(78) - 4.897(3)$

(3) $\hat{y} = 46.385 + 0.540(83) - 4.897(2)$

c. (1) $\hat{y} = 89.548$ (2) $\hat{y} = 73.814$ (3) $\hat{y} = 81.411$

d. (1) 90 (2) 74 (3) 81

CHAPTER 10

Section 10.1

1.

Tax preparation method	% of people	Expected frequency
Accountant	25%	125
By hand	20%	100
Computer software	35%	175
Friend/family	5%	25
Tax preparation service	15%	75

2 a. The expected frequencies are 64, 80, 32, 56, 60, 48, 40, and 20, all of which are at least 5.

b. Claimed distribution:

Ages	Distribution
0–9	16%
10–19	20%
20–29	8%
30–39	14%
40–49	15%
50–59	12%
60–69	10%
70+	5%

H_0: The distribution of ages is as shown in table above.

H_a: The distribution of ages differs from the claimed distribution. (claim)

c. 0.05 **d.** 7

e. $\chi_0^2 = 14.067$; Rejection region: $\chi^2 > 14.067$

f. 6.694

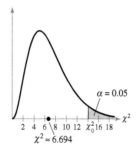

$\alpha = 0.05$

$\chi^2 \approx 6.694$

g. Fail to reject H_0.

h. There is not enough evidence at the 5% level of significance to support the sociologist's claim that the age distribution differs from the age distribution 10 years ago.

3 a. The expected frequency for each category is 30, which is at least 5.

b. Claimed distribution:

Color	Distribution
Brown	$16.\overline{6}\%$
Yellow	$16.\overline{6}\%$
Red	$16.\overline{6}\%$
Blue	$16.\overline{6}\%$
Orange	$16.\overline{6}\%$
Green	$16.\overline{6}\%$

H_0: The distribution of colors is uniform, as shown in the table above. (claim)

H_a: The distribution of colors is not uniform.

c. 0.05　　**d.** 5

e. $\chi_0^2 = 11.071$; Rejection region: $\chi^2 > 11.071$

f. 12.933

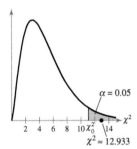
$\alpha = 0.05$
$\chi^2 \approx 12.933$

g. Reject H_0.

h. There is enough evidence at the 5% level of significance to reject the claim that the distribution of different-colored candies in bags of peanut M&M's is uniform.

Section 10.2

1 a. Marginal frequencies: Row 1: 180; Row 2: 120; Column 1: 74; Column 2: 162; Column 3: 28; Column 4: 36

b. 300

c. $E_{1,1} = 44.4$, $E_{1,2} = 97.2$, $E_{1,3} = 16.8$, $E_{1,4} = 21.6$, $E_{2,1} = 29.6$, $E_{2,2} = 64.8$, $E_{2,3} = 11.2$, $E_{2,4} = 14.4$

2 a. H_0: Travel concern is independent of travel purpose.

H_a: Travel concern is dependent on travel purpose. (claim)

b. 0.01　　**c.** 3

d. $\chi_0^2 = 11.345$; Rejection region: $\chi^2 > 11.345$

e. 8.158

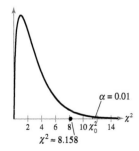
$\alpha = 0.01$
$\chi^2 \approx 8.158$

f. Fail to reject H_0.

g. There is not enough evidence at the 1% level of significance for the consultant to conclude that travel concern is dependent on travel purpose.

3 a. H_0: Whether or not a tax cut would influence an adult to purchase a hybrid vehicle is independent of age.

H_a: Whether or not a tax cut would influence an adult to purchase a hybrid vehicle is dependent on age. (claim)

b. Enter the data.

c. $\chi_0^2 = 9.210$; Rejection region: $\chi^2 > 9.210$

d. 15.306　　**e.** Reject H_0.

f. There is enough evidence at the 1% level of significance to conclude that whether or not a tax cut would influence an adult to purchase a hybrid vehicle is dependent on age.

Section 10.3

1 a. 0.05　　**b.** 2.45

2 a. 0.01　　**b.** 18.31

3 a. $H_0: \sigma_1^2 \leq \sigma_2^2$; $H_a: \sigma_1^2 > \sigma_2^2$ (claim)

b. 0.01　　**c.** d.f.$_N$ = 24, d.f.$_D$ = 19

d. $F_0 = 2.92$; Rejection region: $F > 2.92$

e. 3.21

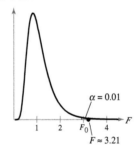
$\alpha = 0.01$
$F \approx 3.21$

f. Reject H_0.

g. There is enough evidence at the 1% level of significance to support the researcher's claim that a specially treated intravenous solution decreases the variance of the time required for nutrients to enter the bloodstream.

4 a. $H_0: \sigma_1 = \sigma_2$ (claim); $H_a: \sigma_1 \neq \sigma_2$

b. 0.01　　**c.** d.f.$_N$ = 15, d.f.$_D$ = 21

d. $F_0 = 3.43$; Rejection region: $F > 3.43$

e. 1.48　　**f.** Fail to reject H_0.

g. There is not enough evidence at the 1% level of significance to reject the biologist's claim that the pH levels of the soil in the two geographic locations have equal standard deviations.

Section 10.4

1 a. $H_0: \mu_1 = \mu_2 = \mu_3 = \mu_4$

H_a: At least one mean is different from the others. (claim)

b. 0.05　　**c.** d.f.$_N$ = 3, d.f.$_D$ = 14

d. $F_0 = 3.34$; Rejection region: $F > 3.34$

e. 4.22

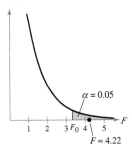

$\alpha = 0.05$

$F \approx 4.22$

f. Reject H_0.

g. There is enough evidence at the 5% level of significance for the analyst to conclude that there is a difference in the mean monthly sales among the sales regions.

2 a. H_0: $\mu_1 = \mu_2 = \mu_3 = \mu_4$

H_a: At least one mean is different from the others. (claim)

b. Enter the data.

c. $F \approx 1.34$; P-value ≈ 0.280

d. Fail to reject H_0.

e. There is not enough evidence at the 5% level of significance to conclude that there is a difference in the means of the GPAs.

CHAPTER 11

Section 11.1

1 a. H_0: median ≤ 120; H_a: median > 120 (claim)

b. 0.025   **c.** 23   **d.** 6   **e.** 6   **f.** Reject H_0.

g. There is enough evidence at the 2.5% level of significance to support the agency's claim that the median number of days a home is on the market in its city is greater than 120.

2 a. H_0: median $= 9.4$ (claim); H_a: median $\neq 9.4$

b. 0.10   **c.** 92   **d.** -1.645 **e.** -0.94

f. Fail to reject H_0.

g. There is not enough evidence at the 10% level of significance to reject the organization's claim that the median age of automobiles in operation in the United States is 9.4 years.

3 a. H_0: The number of colds will not decrease.

H_a: The number of colds will decrease. (claim)

b. 0.05   **c.** 11   **d.** 2   **e.** 2   **f.** Reject H_0.

g. There is enough evidence at the 5% level of significance to support the researcher's claim that a new vaccine will decrease the number of colds in adults.

Section 11.2

1 a. H_0: There is no difference in the amounts of water repelled.

H_a: There is a difference in the amounts of water repelled. (claim)

b. 0.01   **c.** 11   **d.** 5

e.

No repellent	Repellent applied	Differ-ence	Absolute value	Rank	Signed rank
8	15	−7	7	11	−11
7	12	−5	5	9	−9
7	11	−4	4	7.5	−7.5
4	6	−2	2	3.5	−3.5
6	6	0	0		
10	8	2	2	3.5	3.5
9	8	1	1	1.5	1.5
5	6	−1	1	1.5	−1.5
9	12	−3	3	5.5	−5.5
11	8	3	3	5.5	5.5
8	14	−6	6	10	−10
4	8	−4	4	7.5	−7.5

$w_s = 10.5$

f. Fail to reject H_0.

g. There is not enough evidence at the 1% level of significance for the quality control inspector to conclude that the spray-on water repellent is effective.

2 a. H_0: There is no difference in the claims paid by the companies.

H_a: There is a difference in the claims paid by the companies. (claim)

b. 0.05

c. $-z_0 = -1.96$, $z_0 = 1.96$;

Rejection regions: $z < -1.96$, $z > 1.96$

d. $n_1 = 12$ and $n_2 = 12$

e.

Ordered data	Sample	Rank
1.7	B	1
1.8	B	2
2.2	B	3
2.5	A	4
3.0	A	5.5
3.0	B	5.5
3.4	B	7
3.9	A	8
4.1	B	9
4.4	B	10
4.5	A	11
4.7	B	12

Ordered data	Sample	Rank
5.3	B	13
5.6	B	14
5.8	A	15
6.0	A	16
6.2	A	17
6.3	A	18
6.5	A	19
7.3	B	20
7.4	A	21
9.9	A	22
10.6	A	23
10.8	B	24

$R = 120.5$ (or $R = 179.5$)

f. -1.703 (or 1.703)

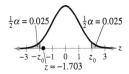

$\frac{1}{2}\alpha = 0.025$ $\frac{1}{2}\alpha = 0.025$

$z \approx -1.703$

g. Fail to reject H_0.

h. There is not enough evidence at the 5% level of significance to conclude that there is a difference in the claims paid by the companies.

A47

Section 11.3

1 a. H_0: There is no difference in the salaries in the three states.

H_a: There is a difference in the salaries in the three states. (claim)

b. 0.05   **c.** 2

d. $\chi_0^2 = 5.991$; Rejection region: $\chi^2 > 5.991$

e.

Ordered data	State	Rank
88.28	CA	1
88.80	PA	2
92.50	NY	3
93.10	NY	4
94.40	NY	5
95.15	PA	6
96.25	CA	7
97.25	PA	8
97.44	PA	9
97.50	CA	10.5
97.50	NY	10.5
97.89	NY	12
98.85	CA	13
99.20	PA	14

Ordered data	State	Rank
99.70	CA	15
99.75	NY	16
99.95	CA	17
99.99	PA	18
100.55	PA	19
100.75	CA	20
101.20	CA	21
101.55	NY	22
101.97	NY	23
102.35	NY	24
103.20	CA	25
103.70	PA	26
110.45	PA	27
113.90	CA	28

$R_1 = 157.5$

$R_2 = 129$

$R_3 = 119.5$

f. 0.433

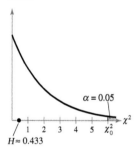

$\alpha = 0.05$

$H \approx 0.433$

g. Fail to reject H_0.

h. There is not enough evidence at the 5% level of significance to conclude that the distributions of the veterinarians' salaries in these three states are different.

Section 11.4

1 a. H_0: $\rho_s = 0$; H_a: $\rho_s \neq 0$ (claim)

b. 0.01   **c.** 0.929

d.

Male	Rank	Female	Rank	d	d^2
25	3.5	20	1.5	2	4
24	1.5	20	1.5	0	0
24	1.5	22	3.0	−1.5	2.25
25	3.5	23	4.0	−0.5	0.25
27	5.0	26	5.0	0	0
29	6.0	27	6.0	0	0
30	7.0	30	7.0	0	0
					$\sum d^2 = 6.5$

$\sum d^2 = 6.5$

e. 0.884   **f.** Fail to reject H_0.

g. There is not enough evidence at the 1% level of significance to conclude that a significant correlation exists between the number of males and females who received doctoral degrees.

Section 11.5

1 a. $P\ P\ P\quad F\quad P\quad F\quad P\ P\ P\ P\quad F\ F\quad P\quad F\quad P\ P$
$F\ F\ F\quad P\ P\ P\quad F\quad P\ P\ P$

b. 13

c. 3, 1, 1, 1, 4, 2, 1, 1, 2, 3, 3, 1, 3

2 a. H_0: The sequence of genders is random.

H_a: The sequence of genders is not random. (claim)

b. 0.05

c. n_1 = number of F's = 9

n_2 = number of M's = 6

G = number of runs = 8

d. lower critical value = 4

upper critical value = 13

e. 8   **f.** Fail to reject H_0.

g. There is not enough evidence at the 5% level of significance to support the claim that the sequence of genders is not random.

3 a. H_0: The sequence of weather conditions is random.

H_a: The sequence of weather conditions is not random. (claim)

b. 0.05

c. n_1 = number of N's = 21

n_2 = number of S's = 10

G = number of runs = 17

d. ±1.96   **e.** 1.03   **f.** Fail to reject H_0.

g. There is not enough evidence at the 5% level of significance to support the claim that the sequence of weather conditions is not random.

APPENDIX A

1. (1) 0.4857

(2) $z = \pm 2.17$

2 a.

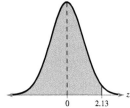

b. 0.4834   **c.** 0.9834

3 a.

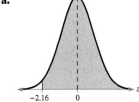

b. 0.4846   **c.** 0.9846

4 a.

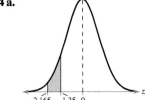

b. 0.4848; 0.4115   **c.** 0.0733

APPENDIX C

1 a.

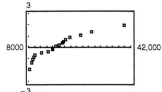

The points do not appear to be approximately linear.

b. 39,860 is a possible outlier because it is far removed from the other entries in the data set.

c. Because the points do not appear to be approximately linear and there is an outlier, you can conclude that the sample data do not come from a population that has a normal distribution.

Odd Answers

CHAPTER 1

Section 1.1 (page 6)

1. A sample is a subset of a population.

3. A parameter is a numerical description of a population characteristic. A statistic is a numerical description of a sample characteristic.

5. False. A statistic is a numerical measure that describes a sample characteristic.

7. True

9. False. A population is the collection of *all* outcomes, responses, measurements, or counts that are of interest.

11. Population, because it is a collection of the heights of all the players on the school's basketball team.

13. Sample, because the collection of the 500 spectators is a subset of the population.

15. Sample, because the collection of the 20 patients is a subset of the population.

17. Population, because it is a collection of all the golfers' scores in the tournament.

19. Population, because it is a collection of all the U.S. presidents' political parties.

21. Population: Parties of registered voters in Warren County
   Sample: Parties of Warren County voters who respond to online survey

23. Population: Ages of adults in the United States who own cellular phones
   Sample: Ages of adults in the United States who own Samsung cellular phones

25. Population: Collection of the responses of all adults in the United States
   Sample: Collection of the responses of the 1000 adults surveyed

27. Population: Collection of the immunization status of all adults in the United States
   Sample: Collection of the immunization status of the 1442 adults surveyed

29. Population: Collection of the opinions of all registered voters
   Sample: Collection of the opinions of the 800 registered voters surveyed

31. Population: Collection of the investor role of all women in the United States
   Sample: Collection of the investor role of the 546 U.S. women surveyed

33. Population: Collection of the responses of all Fortune magazine's top 100 best companies to work for
   Sample: Collection of the responses of the 85 companies who responded to the questionnaire

35. Statistic. The value $68,000 is a numerical description of a sample of annual salaries.

37. Parameter. The 62 surviving passengers out of 97 total passengers is a numerical description of all of the passengers of the Hindenburg that survived.

39. Statistic. 8% is a numerical description of a sample of computer users.

41. Statistic. 44% is a numerical description of a sample of all people.

43. The statement "more than 56% are the primary investors in their households" is an example of descriptive statistics.
   An inference drawn from the sample is that an association exists between U.S. women and being the primary investors in their households.

45. Answers will vary.

47. (a) An inference drawn from the sample is that senior citizens who live in Florida have better memories than senior citizens who do not live in Florida.
   (b) This inference may incorrectly imply that if you live in Florida you will have a better memory.

49. Answers will vary.

Section 1.2 (page 13)

1. Nominal and ordinal

3. False. Data at the ordinal level can be qualitative or quantitative.

5. False. More types of calculations can be performed with data at the interval level than with data at the nominal level.

7. Qualitative, because telephone numbers are labels.

9. Quantitative, because body temperatures are numerical measurements.

11. Quantitative, because song lengths are numerical measurements.

13. Qualitative, because player numbers are labels.

15. Quantitative, because infant weights are numerical measurements.

17. Qualitative, because the poll responses are attributes.

19. Qualitative. Ordinal. Data can be arranged in order, but the differences between data entries make no sense.

21. Qualitative. Nominal. No mathematical computations can be made and data are categorized by region.

23. Qualitative. Ordinal. Data can be arranged in order, but the differences between data entries are not meaningful.

25. Ordinal   27. Nominal

29. (a) Interval   (b) Nominal   (c) Ratio   (d) Ordinal

31. An inherent zero is a zero that implies "none." Answers will vary.

Section 1.3 *(page 23)*

1. In an experiment, a treatment is applied to part of a population and responses are observed. In an observational study, a researcher measures characteristics of interest of a part of a population but does not change existing conditions.

3. In a random sample, every member of the population has an equal chance of being selected. In a simple random sample, every possible sample of the same size has an equal chance of being selected.

5. True

7. False. Using stratified sampling guarantees that members of each group within a population will be sampled.

9. False. A systematic sample is selected by ordering a population in some way and then selecting members of the population at regular intervals.

11. Use a census because all the patients are accessible and the number of patients is not too large.

13. Perform an experiment because you want to measure the effect of a treatment on the human digestive system.

15. Use a simulation because the situation is impractical and dangerous to create in real life.

17. (a) The experimental units are the 30- to 35-year-old females being given the treatment. The treatment is the new allergy drug.

    (b) A problem with the design is that there may be some bias on the part of the researcher if the researcher knows which patients were given the real drug. A way to eliminate this problem would be to make the study into a double-blind experiment.

    (c) The study would be a double-blind study if the researcher did not know which patients received the real drug or the placebo.

19. Simple random sampling is used because each telephone number has an equal chance of being dialed, and all samples of 1400 phone numbers have an equal chance of being selected. The sample may be biased because only homes with telephones will be sampled.

21. Convenience sampling is used because the students are chosen due to their convenience of location. Bias may enter into the sample because the students sampled may not be representative of the population of students.

23. Simple random sampling is used because each customer has an equal chance of being contacted, and all samples of 580 customers have an equal chance of being selected.

25. Stratified sampling is used because a sample is taken from each one-acre subplot.

27. Answers will vary.

29. Answers will vary. *Sample answer:* Treatment group: Jake, Maria, Lucy, Adam, Bridget, Vanessa, Rick, Dan, and Mary. Control group: Mike, Ron, Carlos, Steve, Susan, Kate, Pete, Judy, and Connie. A random number table is used.

31. Census, because it is relatively easy to obtain the ages of the 115 residents.

33. The question is biased because it already suggests that eating whole-grain foods improves your health. The question might be rewritten as "How does eating whole-grain foods affect your health?"

35. The survey question is unbiased.

37. The households sampled represent various locations, ethnic groups, and income brackets. Each of these variables is considered a stratum. Stratified sampling ensures that each segment of the population is represented.

39. Observational studies may be referred to as natural experiments because they involve observing naturally occurring events that are not influenced by the study.

41. (a) Advantage: Usually results in a savings in the survey cost.

    Disadvantage: There tends to be a lower response rate and this can introduce a bias into the sample. Only a certain segment of the population might respond.

    (b) Sampling technique: Convenience sampling

43. If blinding is not used, then the placebo effect is more likely to occur.

45. Both a randomized block design and a stratified sample split their members into groups based on similar characteristics.

Section 1.3 Activity *(page 26)*

1. Answers will vary. The list contains one number at least twice.

2. The minimum is 1, the maximum is 731, and the number of samples is 8. Answers will vary.

Uses and Abuses for Chapter 1 *(page 27)*

1. Answers will vary.   2. Answers will vary.

Review Answers for Chapter 1 *(page 29)*

1. Population: Collection of the opinions of all U.S. adults about credit cards

   Sample: Collection of the opinions of the 1000 U.S. adults surveyed about credit cards

3. Population: Collection of the average annual percentage rates of all credit cards

   Sample: Collection of the average annual percentage rates of the 39 credit cards sampled

5. Parameter   7. Parameter

9. The statement "the average annual percentage rate (APR) [charged by credit cards] is 12.83%" is an example of descriptive statistics.

   An inference drawn from the sample is that all credit cards have an annual percentage rate of 12.83%.

11. Quantitative, because monthly salaries are numerical measurements.

13. Quantitative, because ages are numerical measurements.

15. Quantitative, because revenues are numerical measurements.

17. Interval. The data can be ordered and meaningful differences can be calculated, but it makes no sense to say that 100 degrees is twice as hot as 50 degrees.

19. Nominal. The data are qualitative and cannot be arranged in a meaningful order.

21. Take a census because CEOs keep accurate records of charitable donations.

23. Perform an experiment because you want to measure the effect of training dogs from animal shelters on inmates.

25. The subjects could be split into male and female and then be randomly assigned to each of the five treatment groups.

27. Answers will vary.

29. Simple random sampling is used because random telephone numbers were generated and called.

31. Cluster sampling is used because each community is considered a cluster and every pregnant woman in a selected community is surveyed.

33. Stratified sampling is used because 25 students are randomly selected from each grade level.

35. Telephone sampling samples only individuals who have telephones, who are available, and who are willing to respond.

37. The selected communities may not be representative of the entire area.

Chapter Quiz for Chapter 1   *(page 31)*

1. Population: Collection of the prostate conditions of all men

Sample: Collection of the prostate conditions of 20,000 men in study

2. (a) Statistic   (b) Parameter   (c) Statistic

3. (a) Qualitative   (b) Quantitative

4. (a) Ordinal, because badge numbers can be ordered and often indicate seniority of service, but no meaningful mathematical computation can be performed.

(b) Ratio, because one data value can be expressed as a multiple of another.

(c) Ordinal, because data can be arranged in order, but the differences between data entries make no sense.

(d) Interval, because meaningful differences between entries can be calculated but a zero entry is not an inherent zero.

5. (a) Perform an experiment because you want to measure the effect of a treatment on lead levels in adults.

(b) Use a survey because it would be impossible to question everyone in the population.

6. Randomized block design

7. (a) Convenience sampling, because all of the people sampled are in one convenient location.

(b) Systematic sampling, because every tenth machine part is sampled.

(c) Stratified sampling, because the population is first stratified and then a sample is collected from each stratum.

8. Convenience sampling

Real Statistics—Real Decisions for Chapter 1   *(page 32)*

1. (a) Answers will vary.   (b) Yes   (c) Use surveys.

(d) You may take too large a percentage of your sample from a subgroup of the population that is relatively small.

2. (a) Both, because questions will ask for demographics (qualitative) as well as cost (quantitative).

(b) Gender, business/recreational: nominal
Cost of ticket: ratio
Comfort, safety: ordinal

(c) Sample   (d) Statistics

3. (a) Answers will vary. *Sample answer:* Sample includes only members of the population with access to the Internet.

(b) Answers will vary.

CHAPTER 2

Section 2.1   *(page 47)*

1. Organizing the data into a frequency distribution may make patterns within the data more evident. Sometimes it is easier to identify patterns of a data set by looking at a graph of the frequency distribution.

3. Class limits determine which numbers can belong to each class.

Class boundaries are the numbers that separate classes without forming gaps between them.

5. The sum of the relative frequencies must be 1 or 100% because it is the sum of all portions or percentages of the data.

7. False. Class width is the difference between lower or upper limits of consecutive classes.

9. False. An ogive is a graph that displays cumulative frequencies.

11. Class width = 8; Lower class limits: 9, 17, 25, 33, 41, 49, 57; Upper class limits: 16, 24, 32, 40, 48, 56, 64

13. Class width = 15; Lower class limits: 17, 32, 47, 62, 77, 92, 107, 122; Upper class limits: 31, 46, 61, 76, 91, 106, 121, 136

15. (a) Class width = 11

(b) and (c)

Class	Midpoint	Class boundaries
20–30	25	19.5–30.5
31–41	36	30.5–41.5
42–52	47	41.5–52.5
53–63	58	52.5–63.5
64–74	69	63.5–74.5
75–85	80	74.5–85.5
86–96	91	85.5–96.5

17.

Class	Frequency, f	Midpoint	Relative frequency	Cumulative frequency
20–30	19	25	0.05	19
31–41	43	36	0.12	62
42–52	68	47	0.19	130
53–63	69	58	0.19	199
64–74	74	69	0.20	273
75–85	68	80	0.19	341
86–96	24	91	0.07	365
	$\Sigma f = 365$		$\Sigma \dfrac{f}{n} \approx 1$	

19. (a) Number of classes = 7   (b) Least frequency $\approx$ 10
   (c) Greatest frequency $\approx$ 300   (d) Class width = 10

21. (a) 50   (b) 22.5–23.5 pounds

23. (a) 42   (b) 29.5 pounds   (c) 35   (d) 2

25. (a) Class with greatest relative frequency: 8–9 inches
       Class with least relative frequency: 17–18 inches
   (b) Greatest relative frequency $\approx$ 0.195
       Least relative frequency $\approx$ 0.005
   (c) Approximately 0.01

27. Class with greatest frequency: 29.5–32.5
   Classes with least frequency: 11.5–14.5 and 38.5–41.5

29.

Class	Frequency, f	Midpoint	Relative frequency	Cumulative frequency
0–7	8	3.5	0.32	8
8–15	8	11.5	0.32	16
16–23	3	19.5	0.12	19
24–31	3	27.5	0.12	22
32–39	3	35.5	0.12	25
	$\Sigma f = 25$		$\Sigma \dfrac{f}{n} = 1$	

Classes with greatest frequency: 0–7, 8–15
Classes with least frequency: 16–23, 24–31, 32–39

31.

Class	Frequency, f	Mid-point	Relative frequency	Cumulative frequency
1000–2019	12	1509.5	0.5455	12
2020–3039	3	2529.5	0.1364	15
3040–4059	2	3549.5	0.0909	17
4060–5079	3	4569.5	0.1364	20
5080–6099	1	5589.5	0.0455	21
6100–7119	1	6609.5	0.0455	22
	$\Sigma f = 22$		$\Sigma \dfrac{f}{N} \approx 1$	

July Sales for Representatives

The graph shows that most of the sales representatives at the company sold between $1000 and $2019. (Answers will vary.)

33.

Class	Frequency, f	Mid-point	Relative frequency	Cumulative frequency
291–318	5	304.5	0.1667	5
319–346	4	332.5	0.1333	9
347–374	3	360.5	0.1000	12
375–402	5	388.5	0.1667	17
403–430	6	416.5	0.2000	23
431–458	4	444.5	0.1333	27
459–486	1	472.5	0.0333	28
487–514	2	500.5	0.0667	30
	$\Sigma f = 30$		$\Sigma \dfrac{f}{n} = 1$	

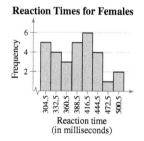

Reaction Times for Females

The graph shows that the most frequent reaction times were between 403 and 430 milliseconds. (Answers will vary.)

35.

Class	Frequency, f	Mid-point	Relative frequency	Cumulative frequency
24–30	9	27	0.30	9
31–37	8	34	0.27	17
38–44	10	41	0.33	27
45–51	2	48	0.07	29
52–58	1	55	0.03	30
	$\Sigma f = 30$		$\Sigma \dfrac{f}{n} = 1$	

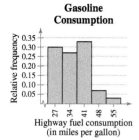

Gasoline Consumption

Class with greatest relative frequency: 38–44
Class with least relative frequency: 52–58

37.

Class	Frequency, f	Mid-point	Relative frequency	Cumulative frequency
138–202	12	170	0.46	12
203–267	6	235	0.23	18
268–332	4	300	0.15	22
333–397	1	365	0.04	23
398–462	3	430	0.12	26
	$\Sigma f = 26$		$\Sigma \dfrac{f}{n} = 1$	

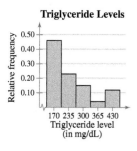

Triglyceride Levels

Class with greatest relative frequency: 138–202
Class with least relative frequency: 333–397

39.

Class	Frequency, f	Relative frequency	Cumulative frequency
52–55	3	0.125	3
56–59	3	0.125	6
60–63	9	0.375	15
64–67	4	0.167	19
68–71	4	0.167	23
72–75	1	0.042	24
	$\Sigma f = 24$	$\Sigma \dfrac{f}{n} \approx 1$	

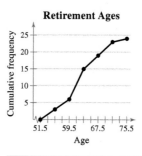

Retirement Ages

Location of the greatest increase in frequency: 60–63

41.

Class	Frequency, f	Mid-point	Relative frequency	Cumulative frequency
47–57	1	52	0.05	1
58–68	1	63	0.05	2
69–79	5	74	0.25	7
80–90	8	85	0.40	15
91–101	5	96	0.25	20
	$\Sigma f = 20$		$\Sigma \dfrac{f}{N} = 1$	

Exam Scores

The graph shows that the most frequent exam scores were between 80 and 90. (Answers will vary.)

43. (a)

Class	Frequency, f	Midpoint	Relative frequency	Cumulative frequency
65–74	4	69.5	0.17	4
75–84	7	79.5	0.29	11
85–94	4	89.5	0.17	15
95–104	5	99.5	0.21	20
105–114	3	109.5	0.13	23
115–124	1	119.5	0.04	24
	$\Sigma f = 24$		$\Sigma \dfrac{f}{n} \approx 1$	

(b)

Pulse Rates

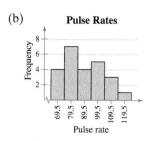

(c)

Pulse Rates

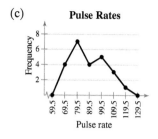

(d)

Pulse Rates

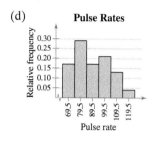

(e)

Pulse Rates

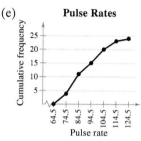

45.

Finishing Times of Marathon Runners

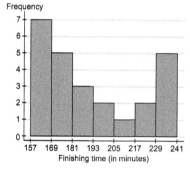

Finishing Times of Marathon Runners

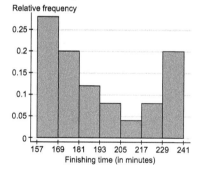

47. (a)

Daily Withdrawals

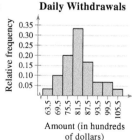

(b) 16.7%, because the sum of the relative frequencies for the last three classes is 0.167.

(c) $9600, because the sum of the relative frequencies for the last two classes is 0.10.

Section 2.2 *(page 60)*

1. Quantitative: stem-and-leaf plot, dot plot, histogram, scatter plot, time series chart

Qualitative: pie chart, Pareto chart

3. Both the stem-and-leaf plot and the dot plot allow you to see how data are distributed, to determine specific data entries, and to identify unusual data values.

5. b   **6.** d   **7.** a   **8.** c

9. 27, 32, 41, 43, 43, 44, 47, 47, 48, 50, 51, 51, 52, 53, 53, 53, 54, 54, 54, 54, 55, 56, 56, 58, 59, 68, 68, 68, 73, 78, 78, 85

Max: 85; Min: 27

11. 13, 13, 14, 14, 14, 15, 15, 15, 15, 15, 16, 17, 17, 18, 19

Max: 19; Min: 13

13. Answers will vary. *Sample answer:* Users spend the most amount of time on MySpace and the least amount of time on Twitter.

15. Answers will vary. *Sample answer:* Tailgaters irk drivers the most, and too-cautious drivers irk drivers the least.

17. Key: $6|7 = 67$

```
6 | 7 8
7 | 3 5 5 6 9
8 | 0 0 2 3 5 5 7 7 8
9 | 0 1 1 1 2 4 5 5
```

It appears that most grades for the biology midterm were in the 80s or 90s. (Answers will vary.)

19. Key: $4|3 = 4.3$

```
4 | 3 9
5 | 1 8 8 8 9
6 | 4 8 9 9 9
7 | 0 0 2 2 2 5
8 | 0 1
```

It appears that most ice had a thickness of 5.8 centimeters to 7.2 centimeters. (Answers will vary.)

21.   **Systolic Blood Pressures**

```
. I.IIIII ..I .    I
100 110 120 130 140 150 160 170 180 190 200
Systolic blood pressure (in mmHg)
```

It appears that systolic blood pressure tends to be between 120 and 150 millimeters of mercury. (Answers will vary.)

23.   **Marathon Winners' Countries of Origin**

United States 37.5%, Kenya 20%, Mexico 10%, Italy 10%, Brazil 5%, South Africa 5%, Morocco 2.5%, Tanzania 2.5%, Ethiopia 2.5%, New Zealand 2.5%, Great Britain 2.5%

Most of the New York City Marathon winners are from the United States and Kenya. (Answers will vary.)

25.

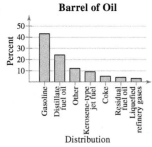

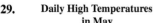

Barrel of Oil

It appears that the largest portion of a 42-gallon barrel of crude oil is used for making gasoline. (Answers will vary.)

27.

Hourly Wages

It appears that there is no relation between wages and hours worked. (Answers will vary.)

29.

Daily High Temperatures in May

It appears that it was hottest from May 7th to May 11th. (Answers will vary.)

31. Variable: Scores

Decimal point is 1 digit(s) to the right of the colon.

    5 : 5
    6 : 2
    6 : 8
    7 : 0 1
    7 : 5 6
    8 : 0 2 3
    8 : 5 6 7 8 8 9
    9 : 0 3 3
    9 : 5 5 8 9
   10 : 0

It appears that most scores on the final exam in economics were in the 80s and 90s. (Answers will vary.)

33. (a)

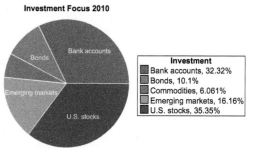

Investment Focus 2010

Investment
■ Bank accounts, 32.32%
■ Bonds, 10.1%
■ Commodities, 6.061%
■ Emerging markets, 16.16%
■ U.S. stocks, 35.35%

It appears a large portion of adults said that the type of investment that they would focus on in 2010 was U.S. stocks or bank accounts. (Answers will vary.)

(b)

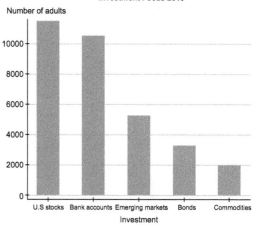

Investment Focus 2010

It appears that most adults said that the type of investment that they would focus on in 2010 was U.S. stocks or bank accounts. (Answers will vary.)

35. (a) The graph is misleading because the large gap from 0 to 90 makes it appear that the sales for the 3rd quarter are disproportionately larger than the other quarters. (Answers will vary.)

(b)

Sales for Company A

37. (a) The graph is misleading because the angle makes it appear as though the 3rd quarter had a larger percent of sales than the others, when the 1st and 3rd quarters have the same percent.

(b) Sales for Company B

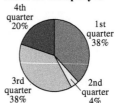

39. (a) At Law Firm A, the lowest salary was $90,000 and the highest salary was $203,000; at Law Firm B, the lowest salary was $90,000 and the highest salary was $190,000.

(b) There are 30 lawyers at Law Firm A and 32 lawyers at Law Firm B.

(c) At Law Firm A, the salaries tend to be clustered at the far ends of the distribution range and at Law Firm B, the salaries tend to fall in the middle of the distribution range.

Section 2.3 *(page 72)*

1. True　**3.** True　**5.** 1, 2, 2, 2, 3 (Answers will vary.)

7. 2, 5, 7, 9, 35 (Answers will vary.)

9. The shape of the distribution is skewed right because the bars have a "tail" to the right.

11. The shape of the distribution is uniform because the bars are approximately the same height.

13. (11), because the distribution of values ranges from 1 to 12 and has (approximately) equal frequencies.

15. (12), because the distribution has a maximum value of 90 and is skewed left due to a few students scoring much lower than the majority of the students.

17. $\bar{x} \approx 4.9$; median = 5; mode = 4

19. $\bar{x} \approx 11.0$; median = 11.0; mode = 11.7; The mode does not represent the center of the data because 11.7 is the largest number in the data set.

21. $\bar{x} \approx 21.46$; median = 21.95; mode = 20.4

23. $\bar{x}$ = not possible; median = not possible; mode = "Eyeglasses"; The mean and median cannot be found because the data are at the nominal level of measurement.

25. $\bar{x} \approx 170.63$; median = 169.3; mode = none; There is no mode because no data point is repeated.

27. $\bar{x} \approx 168.7$; median = 162.5; mode = 125; The mode does not represent the center of the data because 125 is the smallest number in the data set.

29. $\bar{x} \approx 14.11$; median = 14.25; mode = 2.5; The mode does not represent the center of the data because 2.5 is much smaller than most of the data in the set.

31. $\bar{x} \approx 29.82$; median = 32; mode = 24, 35

33. $\bar{x} \approx 19.5$; median = 20; mode = 15

35. The data are skewed right.

A = mode, because it is the data entry that occurred most often.

B = median, because the median is to the left of the mean in a skewed right distribution.

C = mean, because the mean is to the right of the median in a skewed right distribution.

37. Mode, because the data are at the nominal level of measurement.

39. Mean, because there are no outliers.

41. 89　**43.** $612.73　**45.** 2.8　**47.** 87

49. 36.2 miles per gallon　**51.** 35.8 years old

53.

Class	Frequency, f	Midpoint
127–161	9	144
162–196	8	179
197–231	3	214
232–266	3	249
267–301	1	284
	$\Sigma f = 24$	

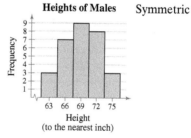

Hospital Beds　Positively skewed

55.

Class	Frequency, f	Midpoint
62–64	3	63
65–67	7	66
68–70	9	69
71–73	8	72
74–76	3	75
	$\Sigma f = 30$	

Heights of Males　Symmetric

57. (a) $\bar{x}$ = 6.005　(b) $\bar{x}$ = 5.945
median = 6.01　median = 6.01
(c) Mean

59. Summary statistics:

Column	n	Mean	Median	Min	Max
Amount (in dollars)	11	112.11364	105.25	79	151.5

61. (a) $\bar{x}$ = 358, median = 375
(b) $\bar{x}$ = 1074, median = 1125
(c) The mean and median in part (b) are three times the mean and median in part (a).
(d) If you multiply the mean and median from part (b) by 12, you will get the mean and median of the data set in inches.

63. Car A, because the midrange is the largest.

65. (a) 49.2
(b) $\bar{x} \approx 49.2$; median = 46.5; mode = 36, 37, 51; midrange = 50.5

(c) Using the trimmed mean eliminates potential outliers that could affect the mean of the entries.

Section 2.3 Activity *(page 79)*

1. The distribution is symmetric. The mean and median both decrease slightly. Over time, the median will decrease dramatically and the mean will also decrease, but to a lesser degree.

2. Neither the mean nor the median can be any of the points that were plotted. Because there are 10 points in each output region, the mean will fall somewhere between the two regions. By the same logic, the median will be the average of the greatest point between 0 and 0.75 and the least point between 20 and 25.

Section 2.4 *(page 90)*

1. The range is the difference between the maximum and minimum values of a data set. The advantage of the range is that it is easy to calculate. The disadvantage is that it uses only two entries from the data set.

3. The units of variance are squared. Its units are meaningless. (Example: dollars^2)

5. $\{9, 9, 9, 9, 9, 9, 9\}$

7. When calculating the population standard deviation, you divide the sum of the squared deviations by N, then take the square root of that value. When calculating the sample standard deviation, you divide the sum of the squared deviations by $n - 1$, then take the square root of that value.

9. Similarity: Both estimate proportions of the data contained within k standard deviations of the mean.

   Difference: The Empirical Rule assumes the distribution is bell-shaped; Chebychev's Theorem makes no such assumption.

11. Range $= 7$, $\mu = 9$, $\sigma^2 = 4.8$, $\sigma \approx 2.2$

13. Range $= 15$, $\bar{x} = 12$, $s^2 \approx 21$, $s \approx 4.6$

15. 73   **17.** 24

19. (a) Range $= 17.8$ (b) Range $= 39.8$

21. The data set in (a) has a standard deviation of 24 and the data set in (b) has a standard deviation of 16, because the data in (a) have more variability.

23. Company B; An offer of $33,000 is two standard deviations from the mean of Company A's starting salaries, which makes it unlikely. The same offer is within one standard deviation of the mean of Company B's starting salaries, which makes the offer likely.

25. (a) Dallas: $\bar{x} \approx 44.28$; median $= 44.7$; range $= 11.3$; $s^2 \approx 18.33$; $s \approx 4.28$

    New York City: $\bar{x} \approx 50.91$; median $= 50.6$; range $= 17.8$; $s^2 \approx 50.36$; $s \approx 7.10$

    (b) It appears from the data that the annual salaries in New York City are more variable than the annual salaries in Dallas. The annual salaries in Dallas have a lower mean and a lower median than the annual salaries in New York City.

27. (a) Males: $\bar{x} \approx 1643$; median $= 1679.5$; range $= 1087$; $s^2 \approx 116,477.4$; $s \approx 341.3$

    Females: $\bar{x} \approx 1709.1$; median $= 1686.5$; range $= 947$; $s^2 \approx 91,625.0$; $s \approx 302.7$

    (b) It appears from the data that the SAT scores for males are more variable than the SAT scores for females. The SAT scores for males have a lower mean and median than the SAT scores for females.

29. (a) Greatest sample standard deviation: (ii)

    Data set (ii) has more entries that are farther away from the mean.

    Least sample standard deviation: (iii)

    Data set (iii) has more entries that are close to the mean.

    (b) The three data sets have the same mean but have different standard deviations.

31. (a) Greatest sample standard deviation: (ii)

    Data set (ii) has more entries that are farther away from the mean.

    Least sample standard deviation: (iii)

    Data set (iii) has more entries that are close to the mean.

    (b) The three data sets have the same mean, median, and mode but have different standard deviations.

33. 68%   **35.** (a) 51   (b) 17

37. $2180, $1000, $2000, $950; $2180 is very unusual because it is more than 3 standard deviations from the mean.

39. 24   **41.** $\bar{x} \approx 2.1$, $s \approx 1.3$

43.

Class	Midpoint, x	f	xf
1–3	2	3	6
4–6	5	6	30
7–9	8	13	104
10–12	11	7	77
13–15	14	3	42
		$N = 32$	$\Sigma xf = 259$

$x - \mu$	$(x - \mu)^2$	$(x - \mu)^2 f$
-6.1	37.21	111.63
-3.1	9.61	57.66
-0.1	0.01	0.13
2.9	8.41	58.87
5.9	34.81	104.43
		$\Sigma (x - \mu)^2 f = 332.72$

$\mu \approx 8.1$

$\sigma \approx 3.2$

45.

Midpoint, x	f	xf	$x - \bar{x}$	$(x - \bar{x})^2$	$(x - \bar{x})^2 f$
70.5	1	70.5	−44	1936	1936
92.5	12	1110.0	−22	484	5808
114.5	25	2862.5	0	0	0
136.5	10	1365.0	22	484	4840
158.5	2	317.0	44	1936	3872
	$n = 50$	$\Sigma xf = 5725$			$\Sigma(x - \bar{x})^2 f = 16{,}456$

$\bar{x} = 114.5$

$s \approx 18.33$

47.

Class	Midpoint, x	f	xf
0–4	2.0	22.1	44.20
5–14	9.5	43.4	412.30
15–19	17.0	21.2	360.40
20–24	22.0	22.3	490.60
25–34	29.5	44.5	1312.75
35–44	39.5	41.3	1631.35
45–64	54.5	83.9	4572.55
65+	70.0	46.8	3276.00
		$n = 325.5$	$\Sigma xf = 12{,}100.15$

$x - \bar{x}$	$(x - \bar{x})^2$	$(x - \bar{x})^2 f$
−35.17	1236.93	27,336.15
−27.67	765.63	33,228.34
−20.17	406.83	8624.80
−15.17	230.13	5131.90
−7.67	58.83	2617.94
2.33	5.43	224.26
17.33	300.33	25,197.69
32.83	1077.81	50,441.51
		$\Sigma(x - \bar{x})^2 f = 152{,}802.59$

$\bar{x} \approx 37.17$

$s \approx 21.70$

49. Summary statistics:

Column	n	Mean	Variance
Amount (in dollars)	15	58.8	239.74286

Std. Dev.	Median	Range	Min	Max
15.483632	60	59	30	89

51. $CV_{\text{heights}} = \dfrac{3.29}{72.75} \cdot 100\% \approx 4.5\%$

$CV_{\text{weights}} = \dfrac{17.69}{187.83} \cdot 100\% \approx 9.4\%$

It appears that weight is more variable than height.

53. (a) $\bar{x} \approx 41.5$, $s \approx 5.3$

(b) $\bar{x} \approx 43.6$, $s \approx 5.6$

(c) $\bar{x} \approx 3.5$, $s \approx 0.4$

(d) When each entry is multiplied by a constant k, the new sample mean is $k \cdot \bar{x}$, and the new sample standard deviation is $k \cdot s$.

55. (a) Males: 249, Females: 245.4; The mean absolute deviation is less than the sample standard deviation.

(b) Team A: 0.0315, Team B: 0.0199; The mean absolute deviation is less than the sample standard deviation.

57. (a) $P \approx -2.61$

The data are skewed left.

(b) $P \approx 4.12$

The data are skewed right.

(c) $P = 0$

The data are symmetric.

(d) $P = 1$

The data are skewed right.

Section 2.4 Activity *(page 98)*

1. When a point with a value of 15 is added, the mean remains constant and the standard deviation decreases; When a point with a value of 20 is added, the mean is raised and the standard deviation increases. (Answers will vary.)

2. To get the largest standard deviation, plot four of the points at 30 and four of the points at 40; To get the smallest standard deviation, plot all of the points at the same number.

Section 2.5 *(page 107)*

1. The soccer team scored fewer points per game than 75% of the teams in the league.

3. The student scored higher than 78% of the students who took the actuarial exam.

5. The interquartile range of a data set can be used to identify outliers because data values that are greater than $Q_3 + 1.5(\text{IQR})$ or less than $Q_1 - 1.5(\text{IQR})$ are considered outliers.

7. False. The median of a data set is a fractile, but the mean may or may not be a fractile depending on the distribution of the data.

9. True

11. False. The 50th percentile is equivalent to Q_2.

13. False. A z-score of -2.5 is considered unusual.

15. (a) Min = 10, $Q_1 = 13$, $Q_2 = 15$, $Q_3 = 17$, Max = 20

(b) IQR = 4

17. (a) Min = 900, $Q_1 = 1250$, $Q_2 = 1500$, $Q_3 = 1950$, Max = 2100

(b) IQR = 700

19. (a) Min = -1.9, $Q_1 = -0.5$, $Q_2 = 0.1$, $Q_3 = 0.7$, Max = 2.1

(b) IQR = 1.2

21. (a) Min = 24, Q_1 = 28, Q_2 = 35, Q_3 = 41, Max = 60

(b)

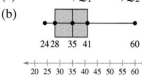

24 28 35 41   60

20 25 30 35 40 45 50 55 60

23. (a) Min = 1, Q_1 = 4.5, Q_2 = 6, Q_3 = 7.5, Max = 9

(b)

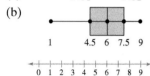

1   4.5 6 7.5 9

0 1 2 3 4 5 6 7 8 9

25. None. The data are not skewed or symmetric.

27. Skewed left. Most of the data lie to the right on the box plot.

29. Q_1 = B, Q_2 = A, Q_3 = C, because about one quarter of the data fall on or below 17, 18.5 is the median of the entire data set, and about three quarters of the data fall on or below 20.

31. (a) Q_1 = 2, Q_2 = 4, Q_3 = 5

(b) **Watching Television**

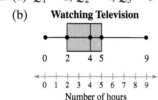

0   2   4 5   9

0 1 2 3 4 5 6 7 8 9
Number of hours

33. (a) Q_1 = 3, Q_2 = 3.85, Q_3 = 5.2

(b) **Airplane Distances**

1.8 3 3.85 5.2 6

0   1   2   3   4   5   6
Distance (in miles)

35. (a) 5   (b) 50%   (c) 25%

37. A → z = −1.43

B → z = 0

C → z = 2.14

A z-score of 2.14 would be unusual.

39. (a) Statistics: $z = \dfrac{75 - 63}{7} \approx 1.71$

Biology: $z = \dfrac{25 - 23}{3.9} \approx 0.51$

(b) The student did better on the statistics test.

41. (a) Statistics: $z = \dfrac{78 - 63}{7} \approx 2.14$

Biology: $z = \dfrac{29 - 23}{3.9} \approx 1.54$

(b) The student did better on the statistics test.

43. (a) $z_1 = \dfrac{34,000 - 35,000}{2250} \approx -0.44$

$z_2 = \dfrac{37,000 - 35,000}{2250} \approx 0.89$

$z_3 = \dfrac{30,000 - 35,000}{2250} \approx -2.22$

The tire with a life span of 30,000 miles has an unusually short life span.

(b) For 30,500, 2.5th percentile

For 37,250, 84th percentile

For 35,000, 50th percentile

45. 72 inches; 60% of the heights are below 72 inches.

47. $z_1 = \dfrac{74 - 69.9}{3.0} \approx 1.37$

$z_2 = \dfrac{62 - 69.9}{3.0} \approx -2.63$

$z_3 = \dfrac{80 - 69.9}{3.0} \approx 3.37$

The heights of 62 and 80 inches are unusual.

49. $z = \dfrac{71.1 - 69.9}{3.0} = 0.4$

About the 50th percentile

51. (a) Min = 27, Q_1 = 42, Q_2 = 49, Q_3 = 56, Max = 82

(b) **Ages of Executives**

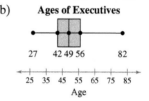

27   42 49 56   82

25   35   45   55   65   75   85
Age

(c) Half of the executives are between 42 and 56 years old.

(d) 49, because half of the executives are older and half are younger.

(e) The age groups 20–29, 70–79, and 80–89 would all be considered unusual because they lie more than two standard deviations from the mean.

53. 33.75   **55.** 19.8

57. **Credit Card Purchases**

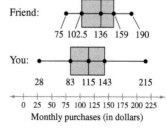

Friend:

75 102.5 136 159 190

You:

28   83 115 143   215

0 25 50 75 100 125 150 175 200 225
Monthly purchases (in dollars)

The shape of your bill is symmetric, and the shape of your friend's bill is uniform.

59. 40th percentile

61. (a) 62, 95

(b)
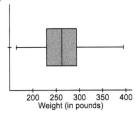

63. (a) **Summary statistics:**

Column	Min	Q1	Median	Q3	Max
Weight (in pounds)	165	230	262.5	294	395

(b) **Weights of Professional Football Players**

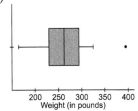

(c) **Weights of Professional Football Players**

200 250 300 350 400
Weight (in pounds)

Uses and Abuses for Chapter 2   *(page 113)*

1. Answers will vary.

2. No, it is not ethical because it misleads the consumer to believe that oatmeal is more effective at lowering cholesterol than it may actually be.

Review Answers for Chapter 2   *(page 115)*

1.

Class	Mid-point	Boundaries	Frequency, f	Relative frequency	Cumulative frequency
8–12	10	7.5–12.5	2	0.10	2
13–17	15	12.5–17.5	10	0.50	12
18–22	20	17.5–22.5	5	0.25	17
23–27	25	22.5–27.5	1	0.05	18
28–32	30	27.5–32.5	2	0.10	20
			$\Sigma f = 20$	$\Sigma \frac{f}{n} = 1$	

3. **Liquid Volume 12-oz Cans**

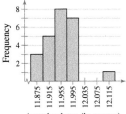

Actual volume (in ounces)

5.

Class	Midpoint	Frequency, f
79–93	86	9
94–108	101	12
109–123	116	5
124–138	131	3
139–153	146	2
154–168	161	1
		$\Sigma f = 32$

Rooms Reserved

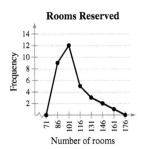

Number of rooms

7.

```
1 | 0 0                Key: 1|0 = 10
2 | 0 0 2 5 5
3 | 0 3 4 5 5 8
4 | 1 2 4 4 7 8
5 | 2 3 3 7 9
6 | 1 1 5
7 | 1 5
8 | 9
```

9. **Heights of Buildings**

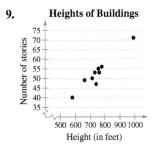

Height (in feet)

The number of stories appears to increase with height.

A61

11.

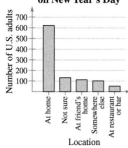

Location at Midnight on New Year's Day

13. $\bar{x} = 29.15$; median = 29.5; mode = 29.5

15. 17.8 **17.** 82.1 **19.** Skewed **21.** Skewed left

23. Median; When a distribution is skewed left, the mean is to the left of the median.

25. $2.80 **27.** $\mu \approx 6.9$, $\sigma \approx 4.6$

29. $\bar{x} = 2453.4$, $s \approx 306.1$

31. Between $41.50 and $56.50 **33.** 30 customers

35. $\bar{x} \approx 2.5$, $s \approx 1.2$

37. Min = 42, $Q_1 = 47.5$, $Q_2 = 53$, $Q_3 = 54$, Max = 60

39.

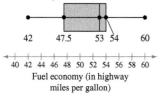

Motorcycle Fuel Economies

41. 4.5 **43.** 35% scored higher than 75.

45. Not unusual **47.** Unusual

Chapter Quiz for Chapter 2 *(page 119)*

1. (a)

Class	Midpoint	Class boundaries
101–112	106.5	100.5–112.5
113–124	118.5	112.5–124.5
125–136	130.5	124.5–136.5
137–148	142.5	136.5–148.5
149–160	154.5	148.5–160.5

Frequency, f	Relative frequency	Cumulative frequency
3	0.12	3
11	0.44	14
7	0.28	21
2	0.08	23
2	0.08	25

(b) Frequency histogram and polygon

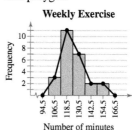

(c) Relative frequency histogram

(d) Skewed

(e)

10	1 8
11	1 4 6 7 8 9 9
12	0 0 3 3 4 7 7 8
13	1 1 2 5 9 9
14	
15	0 7

Key: $10|8 = 108$

(f)

Weekly Exercise

(g)

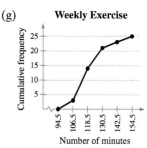

Weekly Exercise

2. 125.2, 13.0

3. (a)

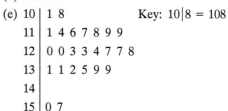

U.S. Sporting Goods

(b)

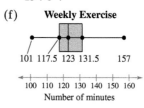

U.S. Sporting Goods

4. (a) $\bar{x} \approx 751.6$; median = 784.5; mode = none

The mean best describes a typical salary because there are no outliers.

(b) Range = 575; $s^2 \approx 48{,}135.1$; $s \approx 219.4$

5. Between $125,000 and $185,000

6. (a) $z = 3.0$, unusual (b) $z \approx -6.67$, very unusual

(c) $z \approx 1.33$ (d) $z = -2.2$, unusual

7. (a) Min = 59, $Q_1 = 74$, $Q_2 = 83.5$, $Q_3 = 88$, Max = 103

(b) 14

(c)

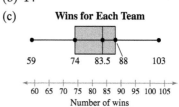

Wins for Each Team

Real Statistics–Real Decisions for Chapter 2 *(page 120)*

1. (a) Find the average cost of renting an apartment for each area and do a comparison.

(b) The mean would best represent the data sets for the four areas of the city.

(c) Area A: $\bar{x} = \$1005.50$

Area B: $\bar{x} = \$887.00$

Area C: $\bar{x} = \$881.00$

Area D: $\bar{x} = \$945.50$

2. (a) Construct a Pareto chart, because the data are quantitative and a Pareto chart positions data in order of decreasing height, with the tallest bar positioned at the left.

(b)

Cost of Monthly Rent per Area

(c) Yes. From the Pareto chart you can see that Area A has the highest average cost of monthly rent, followed by Area D, Area B, and Area C.

3. (a) You could use the range and sample standard deviation for each area.

(b)

Area A	*Area B*
$s \approx \$123.07$	$s \approx \$144.91$
range $= \$415.00$	range $= \$421.00$

Area C	*Area D*
$s \approx \$146.21$	$s \approx \$138.70$
range $= \$460.00$	range $= \$497.00$

(c) No. Area A has the lowest range and standard deviation, so the rents in Areas B–D are more spread out. There could be one or two inexpensive rents that lower the means for these areas. It is possible that the population means of Areas B–D are close to the populations mean of Area A.

4. (a) Answers will vary.

(b) Location, weather, population

Cumulative Review Answers for Chapters 1–2 *(page 124)*

1. Systematic sampling. A bias may enter this study if the machine makes a consistent error.

2. Random sampling. A bias of this type of study is that the researchers did not include people without telephones.

3.

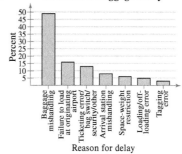

Reason for Baggage Delay

4. Parameter. All Major League Baseball players are included.

5. Statistic. The 19% is a numerical description of the 1000 voters surveyed in the United States.

6. (a) 95%   (b) 38

(c) $z_1 = \dfrac{90,500 - 83,500}{1500} \approx 4.67$

$z_2 = \dfrac{79,750 - 83,500}{1500} = -2.5$

$z_3 = \dfrac{82,600 - 83,500}{1500} = -0.6$

The salaries of $90,500 and $79,750 are unusual.

7. Population: Collection of the career interests of all college and university students

Sample: Collection of the career interests of the 195 college and university students whose career counselors were surveyed

8. Population: Collection of the life spans of all people

Sample: Collection of the life spans of the 232,606 people in the study

9. Census. There are only 100 members in the Senate.

10. Experiment. An experiment could compare a control group that has recess and a treatment group that has recess removed.

11. Quantitative. The data are at the ratio level.

12. Qualitative. The data are at the nominal level.

13. (a) Min $= 0$, $Q_1 = 2$, $Q_2 = 12.5$, $Q_3 = 39$, Max $= 136$

(b)

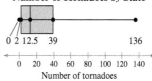

Number of Tornadoes by State

(c) The distribution of the number of tornadoes is skewed right.

14. 88.9

15. (a) $\bar{x} \approx 5.49$; median = 5.4; mode = none; Both the mean and the median accurately describe a typical American alligator tail length. (Answers will vary.)

(b) Range = 4.1; $s^2 = 2.34$; $s = 1.53$; The maximum difference in alligator tail lengths is about 4.1 feet, and about 68% of alligator tail lengths will fall between 3.96 feet and 7.02 feet.

16. (a) An inference drawn from the sample is that the number of deaths due to heart disease for women will continue to decrease.

(b) This inference may incorrectly imply that women will have less of a chance of dying of heart disease in the future.

17.

Class	Class boundaries	Midpoint
0–8	−0.5–8.5	4
9–17	8.5–17.5	13
18–26	17.5–26.5	22
27–35	26.5–35.5	31
36–44	35.5–44.5	40
45–53	44.5–53.5	49
54–62	53.5–62.5	58
63–71	62.5–71.5	67

Frequency, f	Relative frequency	Cumulative frequency
8	0.27	8
5	0.17	13
7	0.23	20
3	0.10	23
4	0.13	27
1	0.03	28
0	0.00	28
2	0.07	30
$\Sigma f = 30$	$\Sigma \dfrac{f}{n} = 1$	

18. The distribution is skewed right.

19.

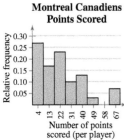

Montreal Canadiens Points Scored

Number of points scored (per player)

Class with greatest frequency: 0–8

Class with least frequency: 54–62

CHAPTER 3

Section 3.1 (page 138)

1. An outcome is the result of a single trial in a probability experiment, whereas an event is a set of one or more outcomes.

3. The probability of an event cannot exceed 100%.

5. The law of large numbers states that as an experiment is repeated over and over, the probabilities found in the experiment will approach the actual probabilities of the event. Examples will vary.

7. False. If you roll a six-sided die six times, the probability of rolling an even number at least once is approximately 0.984.

9. False. A probability of less than 0.05 indicates an unusual event.

11. b   **12.** d   **13.** c   **14.** a

15. {A, B, C, D, E, F, G, H, I, J, K, L, M, N, O, P, Q, R, S, T, U, V, W, X, Y, Z}; 26

17. {A♥, K♥, Q♥, J♥, 10♥, 9♥, 8♥, 7♥, 6♥, 5♥, 4♥, 3♥, 2♥, A♦, K♦, Q♦, J♦, 10♦, 9♦, 8♦, 7♦, 6♦, 5♦, 4♦, 3♦, 2♦, A♠, K♠, Q♠, J♠, 10♠, 9♠, 8♠, 7♠, 6♠, 5♠, 4♠, 3♠, 2♠, A♣, K♣, Q♣, J♣, 10♣, 9♣, 8♣, 7♣, 6♣, 5♣, 4♣, 3♣, 2♣}; 52

19.

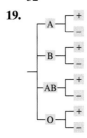

{(A, +), (A, −), (B, +), (B, −), (AB, +), (AB, −), (O, +), (O, −)}, where (A, +) represents positive Rh-factor with blood type A and (A, −) represents negative Rh-factor with blood type A; 8.

21. 1; Simple event because it is an event that consists of a single outcome.

23. 4; Not a simple event because it is an event that consists of more than a single outcome.

25. 204   **27.** 4500   **29.** 0.083   **31.** 0.667   **33.** 0.417

35. Empirical probability because company records were used to calculate the frequency of a washing machine breaking down.

37. 0.159   **39.** 0.000953   **41.** 0.042; Yes   **43.** 0.208; No

45. (a) 1000   (b) 0.001   (c) 0.999

47. {(SSS), (SSR), (SRS), (SRR), (RSS), (RSR), (RRS), (RRR)}

49. {(SSR), (SRS), (RSS)}

51. (a)

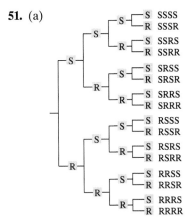

(b) {(SSSS), (SSSR), (SSRS), (SSRR), (SRSS), (SRSR), (SRRS), (SRRR), (RSSS), (RSSR), (RSRS), (RSRR), (RRSS), (RRSR), (RRRS), (RRRR)}

(c) {(SSSS), (SSRS), (SRSS), (RSSS)}

53. 0.399 **55.** 0.040 **57.** 0.936 **59.** 0.033 **61.** 0.275

63. Yes; The event in Exercise 55 can be considered unusual because its probability is 0.05 or less.

65. (a) 0.5 (b) 0.25 (c) 0.25

67. 0.795 **69.** 0.205

71. (a) 0.225 (b) 0.133

(c) 0.017; This event is unusual because its probability is 0.05 or less.

73. The probability of randomly choosing a tea drinker who does not have a college degree

75. (a)

Sum	Probability
2	0.028
3	0.056
4	0.083
5	0.111
6	0.139
7	0.167
8	0.139
9	0.111
10	0.083
11	0.056
12	0.028

(b) Answers will vary.

(c) Answers will vary.

77. The first game; The probability of winning the second game is $\frac{1}{11} \approx 0.091$, which is less than $\frac{1}{10}$.

79. $13:39 = 1:3$

81. p = number of successful outcomes

q = number of unsuccessful outcomes

$$P(A) = \frac{\text{number of successful outcomes}}{\text{total number of outcomes}} = \frac{p}{p+q}$$

Section 3.1 Activity *(page 144)*

1–2. Answers will vary.

Section 3.2 *(page 150)*

1. Two events are independent if the occurrence of one of the events does not affect the probability of the occurrence of the other event, whereas two events are dependent if the occurrence of one of the events does affect the probability of the occurrence of the other event.

3. The notation $P(B|A)$ means the probability of B, given A.

5. False. If two events are independent, then $P(A|B) = P(A)$.

7. Independent. The outcome of the first draw does not affect the outcome of the second draw.

9. Dependent. The outcome of a father having hazel eyes affects the outcome of a daughter having hazel eyes.

11. Dependent. The sum of the rolls depends on which numbers came up on the first and second rolls.

13. Events: moderate to severe sleep apnea, high blood pressure; Dependent. People with moderate to severe sleep apnea are more likely to have high blood pressure.

15. Events: exposure to aluminum, Alzheimer's disease; Independent. Exposure to everyday sources of aluminum does not cause Alzheimer's disease.

17. (a) 0.6 (b) 0.001

(c) Dependent.

$P(\text{developing breast cancer}|\text{gene})$

$\neq (\text{developing breast cancer})$

19. (a) 0.308 (b) 0.788 (c) 0.757 (d) 0.596

(e) Dependent.

$P(\text{taking a summer vacation}|\text{family owns a computer})$

$\neq P(\text{taking a summer vacation})$

21. (a) 0.093 (b) 0.75

(c) No, the probability is not unusual because it is not less than or equal to 0.05.

23. 0.745

25. (a) 0.017 (b) 0.757 (c) 0.243

(d) The event in part (a) is unusual because its probability is less than or equal to 0.05.

27. (a) 0.481 (b) 0.465 (c) 0.449

(d) Dependent.

$P(\text{having less than one month's income saved}|\text{being male})$

$\neq P(\text{having less than one month's income saved})$

29. (a) 0.00000590 (b) 0.624 (c) 0.376

31. (a) 0.25 (b) 0.063 (c) 0.000977

(d) 0.237 (e) 0.763

33. (a) 0.011 (b) 0.458 **35.** 0.444

37. 0.167 **39.** (a) 0.074 (b) 0.999 **41.** 0.954

Section 3.3 *(page 161)*

1. $P(A \text{ and } B) = 0$ because A and B cannot occur at the same time.

3. True

5. False. The probability that event A or event B will occur is $P(A \text{ or } B) = P(A) + P(B) - P(A \text{ and } B)$.

7. Not mutually exclusive. A student can be an athlete and on the Dean's list.

9. Not mutually exclusive. A public school teacher can be female and 25 years old.

11. Mutually exclusive. A student cannot have a birthday in both months.

13. (a) Not mutually exclusive. For five weeks the events overlapped.

(b) 0.423

15. (a) Not mutually exclusive. A carton can have a puncture and a smashed corner.

(b) 0.126

17. (a) 0.308   (b) 0.538   (c) 0.308

19. (a) 0.067   (b) 0.839   (c) 0.199

21. (a) 0.949   (b) 0.388

23. (a) 0.573   (b) 0.962   (c) 0.573

(d) Not mutually exclusive. A male can be a nursing major.

25. (a) 0.461   (b) 0.762   (c) 0.589   (d) 0.922

(e) Not mutually exclusive. A female can be frequently involved in charity work.

27. Answers will vary.   **29.** 0.55

Section 3.3 Activity   *(page 166)*

1. 0.333   **2.** Answers will vary.

3. The theoretical probability is 0.5, so the green line should be placed there.

Section 3.4   *(page 174)*

1. The number of ordered arrangements of n objects taken r at a time. An example of a permutation is the number of seating arrangements of you and three of your friends.

3. False. A permutation is an ordered arrangement of objects.

5. True   **7.** 15,120   **9.** 56   **11.** 203,490   **13.** 0.030

15. Permutation. The order of the eight cars in line matters.

17. Combination. The order does not matter because the position of one captain is the same as the other.

19. 5040   **21.** 720   **23.** 20,358,520   **25.** 320,089,770

27. 50,400   **29.** 6240   **31.** 86,296,950

33. (a) 720   (b) sample

(c) 0.0014; Yes, the event can be considered unusual because its probability is less than or equal to 0.05.

35. (a) 12   (b) tree

(c) 0.083; No, the event cannot be considered unusual because its probability is not less than or equal to 0.05.

37. (a) 907,200   (b) population

(c) 0.000001; Yes, the event can be considered unusual because its probability is less than or equal to 0.05.

39. 0.005   **41.** (a) 0.016   (b) 0.385

43. (a) 70   (b) 16   (c) 0.086

45. (a) 67,600,000   (b) 19,656,000   (c) 0.000000015

47. (a) 120   (b) 12   (c) 12   (d) 0.4

49. 0.000022   **51.** 6.00×10^{-20}

53. (a) 658,008   (b) 0.00000152

55. (a) 0.0002   (b) 0.0014   (c) 0.0211   (d) 0.0659

57. 1001; 1000

59.

Team (worst team first)	1	2	3	4	5
Probability	0.250	0.199	0.156	0.119	0.088

Team (worst team first)	6	7	8	9	10
Probability	0.063	0.043	0.028	0.017	0.011

Team (worst team first)	11	12	13	14
Probability	0.008	0.007	0.006	0.005

Events in which any of Teams 7–14 win the first pick would be considered unusual because the probabilities are all less than or equal to 0.05.

61. 0.314

Uses and Abuses for Chapter 3   *(page 179)*

1. (a) 0.000001   (b) 0.001   (c) 0.001

2. The probability that a randomly chosen person owns a pickup or an SUV can equal 0.55 if no one in the town owns both a pickup and an SUV. The probability cannot equal 0.60 because $0.60 > 0.25 + 0.30$. (Answers will vary.)

Review Exercises for Chapter 3   *(page 181)*

1. Sample space:

{HHHH, HHHT, HHTH, HHTT, HTHH, HTHT, HTTH, HTTT, THHH, THHT, THTH, THTT, TTHH, TTHT, TTTH, TTTT}; 4

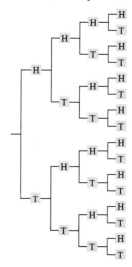

A66

3. Sample space:
   {January, February, March, April, May, June, July, August, September, October, November, December}; 3

5. 84

7. Empirical probability because it is based on observations obtained from probability experiments.

9. Subjective probability because it is based on opinion.

11. Classical probability because all of the outcomes in the event and the sample space can be counted.

13. 0.215   15. 1.25×10^{-7} 17. 0.92

19. Independent. The outcomes of the first four coin tosses do not affect the outcome of the fifth coin toss.

21. Dependent. The outcome of getting high grades affects the outcome of being awarded an academic scholarship.

23. 0.025; Yes, the event is unusual because its probability is less than or equal to 0.05.

25. Mutually exclusive. A jelly bean cannot be both completely red and completely yellow.

27. Mutually exclusive. A person cannot be registered to vote in more than one state.

29. 0.60   31. 0.538   33. 0.583   35. 0.291

37. 0.188   39. 0.703   41. 110   43. 35

45. 254,251,200   47. 2730   49. 2380

51. 0.00000923; unusual

53. (a) 0.955; not unusual   (b) 0.000000761; unusual
    (c) 0.045; unusual   (d) 0.999999239; not unusual

55. (a) 0.071; not unusual   (b) 0.005; unusual
    (c) 0.429; not unusual   (d) 0.114; not unusual

Chapter Quiz for Chapter 3   *(page 185)*

1. (a) 0.523   (b) 0.508   (c) 0.545   (d) 0.772
   (e) 0.025   (f) 0.673   (g) 0.094   (h) 0.574

2. The event in part (e) is unusual because its probability is less than or equal to 0.05.

3. Not mutually exclusive. A golfer can score the best round in a four-round tournament and still lose the tournament.

   Dependent. One event can affect the occurrence of the second event.

4. (a) 2,481,115   (b) 1   (c) 2,572,999

5. (a) 0.964   (b) 0.000000389   (c) 0.9999996

6. 450,000   7. 657,720

Real Statistics–Real Decisions for Chapter 3   *(page 186)*

1. (a) Answers will vary.
   (b) Use the Multiplication Rule, Fundamental Counting Principle, and combinations.

2. If you played only the red ball, the probability of matching it is $\frac{1}{39}$. However, because you must pick five white balls, you must get the white balls wrong. So, using the Multiplication Rule, you get

   P(matching only the red ball and not matching any of the five white balls) $= \frac{1}{39} \cdot \frac{54}{59} \cdot \frac{53}{58} \cdot \frac{52}{57} \cdot \frac{51}{56} \cdot \frac{50}{55}$

   ≈ 0.016

   $\approx \frac{1}{62}.$

3. The overall probability of winning a prize is determined by calculating the number of ways to win and dividing by the total number of outcomes.

   To calculate the number of ways to win something, you must use combinations.

CHAPTER 4

Section 4.1   *(page 197)*

1. A random variable represents a numerical value associated with each outcome of a probability experiment.
   Examples: Answers will vary.

3. No; Expected value may not be a possible value of x for one trial, but it represents the average value of x over a large number of trials.

5. False. In most applications, discrete random variables represent counted data, while continuous random variables represent measured data.

7. True

9. Discrete; Attendance is a random variable that is countable.

11. Continuous; Distance traveled is a random variable that must be measured.

13. Discrete; The number of books in a library is a random variable that is countable.

15. Continuous; The volume of blood drawn for a blood test is a random variable that must be measured.

17. Discrete; The number of messages posted each month on a social networking site is a random variable that is countable.

19. Continuous; The amount of snow that fell in Nome, Alaska last winter is a random variable that cannot be counted.

21. (a) 0.35   (b) 0.90   23. 0.22   25. Yes

27. (a)

x	$P(x)$
0	0.686
1	0.195
2	0.077
3	0.022
4	0.013
5	0.006
	$\sum P(x) \approx 1$

(b)
Dogs per Household

Skewed right

(c) 0.5, 0.8, 0.9

(d) The mean is 0.5, so the average number of dogs per household is about 0 or 1 dog. The standard deviation is 0.9, so most of the households differ from the mean by no more than about 1 dog.

29. (a)

x	P(x)
0	0.01
1	0.17
2	0.28
3	0.54

(b)

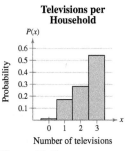

Televisions per Household

Skewed left

(c) 2.4, 0.6, 0.8

(d) The mean is 2.4, so the average household in the town has about 2 televisions. The standard deviation is 0.8, so most of the households differ from the mean by no more than about 1 television.

31. (a)

x	P(x)
0	0.031
1	0.063
2	0.151
3	0.297
4	0.219
5	0.156
6	0.083
	$\Sigma P(x) = 1$

(b)

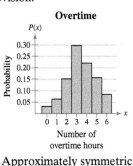

Overtime

Approximately symmetric

(c) 3.4, 2.1, 1.5

(d) The mean is 3.4, so the average employee worked 3.4 hours of overtime. The standard deviation is 1.5, so the overtime worked by most of the employees differed from the mean by no more than 1.5 hours.

33. An expected value of 0 means that the money gained is equal to the money spent, representing the break-even point.

35. (a) 5.3 (b) 3.3 (c) 1.8 (d) 5.3

(e) The expected value is 5.3, so an average student is expected to answer about 5 questions correctly. The standard deviation is 1.8, so most of the students' quiz results differ from the expected value by no more than about 2 questions.

37. (a) 2.0 (b) 1.0 (c) 1.0 (d) 2.0

(e) The expected value is 2.0, so an average hurricane that hits the U.S. mainland is expected to be a category 2 hurricane. The standard deviation is 1.0, so most of the hurricanes differ from the expected value by no more than 1 category level.

39. (a) 2.5 (b) 1.9 (c) 1.4 (d) 2.5

(e) The expected value is 2.5, so an average household is expected to have either 2 or 3 people. The standard deviation is 1.4, so most of the household sizes differ from the expected value by no more than 1 or 2 people.

41. (a) 0.881 (b) 0.314 (c) 0.294

43. A household with three dogs is unusual because the probability of this event is 0.022, which is less than 0.05.

45. −$0.05

47. (a)

x	P(x)
0	0.432
1	0.403
2	0.137
3	0.029
	$\Sigma P(x) \approx 1$

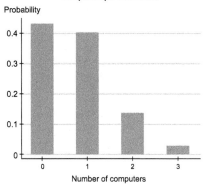

Computers per Household

(b) Skewed right

49. $38,800 **51.** 3020; 28

Section 4.2 *(page 211)*

1. Each trial is independent of the other trials if the outcome of one trial does not affect the outcome of any of the other trials.

3. (a) $p = 0.50$ (b) $p = 0.20$ (c) $p = 0.80$

5. (a) $n = 12$ (b) $n = 4$ (c) $n = 8$

As n increases, the distribution becomes more symmetric.

7. (a) $x = 0, 1, 2, 3, 4, 11, 12$

(b) $x = 0$

(c) $x = 0, 1, 2, 8$

9. Binomial experiment

Success: baby recovers

$n = 5, p = 0.80, q = 0.20, x = 0, 1, 2, 3, 4, 5$

11. Binomial experiment

Success: selecting an officer who is postponing or reducing the amount of vacation

$n = 20, p = 0.31, q = 0.69, x = 0, 1, 2, \ldots, 20$

13. 20, 12, 3.5 **15.** 32.2, 23.9, 4.9

17. (a) 0.088 (b) 0.104 (c) 0.896

19. (a) 0.111 (b) 0.152 (c) 0.848

21. (a) 0.257 (b) 0.220 (c) 0.780

23. (a) 0.187 (b) 0.605 (c) 0.084

25. (a) 0.255 (b) 0.562 (c) 0.783

27. (a) $n = 6$, $p = 0.63$ (b)

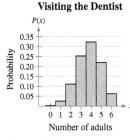

Visiting the Dentist

x	P(x)
0	0.003
1	0.026
2	0.112
3	0.253
4	0.323
5	0.220
6	0.063

Skewed left

(c) 3.8, 1.4, 1.2

(d) On average, 3.8 out of 6 adults are visiting the dentist less because of the economy. The standard deviation is 1.2, so most samples of 6 adults would differ from the mean by no more than 1.2 people. The values $x = 0$ and $x = 1$ would be unusual because their probabilities are less than 0.05.

29. (a) $n = 4$, $p = 0.05$ (b)

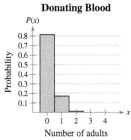

Donating Blood

x	P(x)
0	0.814506
1	0.171475
2	0.013538
3	0.000475
4	0.000006

Skewed right

(c) 0.2, 0.2, 0.4

(d) On average, 0.2 eligible adult out of every 4 gives blood. The standard deviation is 0.4, so most samples of four eligible adults would differ from the mean by at most 0.4 adult.

$x = 2$, 3, and 4 would be unusual because their probabilities are less than 0.05.

31. (a) $n = 6$, $p = 0.37$ (b) 0.323 (c) 0.029

x	P(x)
0	0.063
1	0.220
2	0.323
3	0.253
4	0.112
5	0.026
6	0.003

33. 2.2, 1.2

On average, 2.2 out of 6 travelers would name "crying kids" as the most annoying. The standard deviation is 1.2, so most samples of 6 travelers would differ from the mean by at most 1.2 travelers. The values $x = 5$ and $x = 6$ would be unusual because their probabilities are less than 0.05.

35. (a) 0.081 (b) 0.541

(c) 0.022; This event is unusual because its probability is less than 0.05.

37. 0.033

4.2 Activity *(page 216)*

1–3. Answers will vary.

Section 4.3 *(page 222)*

1. 0.080 **3.** 0.062 **5.** 0.175 **7.** 0.251

9. In a binomial distribution, the value of x represents the number of successes in n trials, and in a geometric distribution the value of x represents the first trial that results in a success.

11. Geometric. You are interested in counting the number of trials until the first success.

13. Binomial. You are interested in counting the number of successes out of n trials.

15. (a) 0.082 (b) 0.469 (c) 0.531

17. (a) 0.195 (b) 0.434 (*Tech:* 0.433)

(c) 0.566 (*Tech:* 0.567)

19. (a) 0.329 (b) 0.878 (c) 0.122

21. (a) 0.105 (b) 0.578 (c) 0.316

23. (a) 0.140

(b) 0.042; This event is unusual because its probability is less than 0.05.

(c) 0.064

25. (a) 0.1254235482

(b) 0.1254084986; The results are approximately the same.

27. (a) 1000, 999,000, 999.5

On average you would have to play 1000 times in order to win the lottery. The standard deviation is 999.5 times.

(b) 1000 times

Lose money. On average you would win $500 once in every 1000 times you play the lottery. So, the net gain would be −$500.

29. (a) 3.9, 2.0; The standard deviation is 2.0 strokes, so most of Phil's scores per hole differ from the mean by no more than 2.0 strokes.

(b) 0.385

Uses and Abuses for Chapter 4 *(page 225)*

1. 40, 0.081 **2.** 0.739; Answers will vary.

3. The probability of finding 36 adults out of 100 who prefer Brand A is 0.059. So, the manufacturer's claim is believable because 0.059 > 0.05.

4. The probability of finding 25 adults out of 100 who prefer Brand A is 0.000627. So, the manufacturer's claim is not believable.

Review Answers for Chapter 4 *(page 227)*

1. Continuous; The length of time spent sleeping is a random variable that cannot be counted.

3. Discrete   5. Continuous   7. No, $\sum P(x) \neq 1$.   9. Yes

11. (a)

x	f	$P(x)$
2	3	0.005
3	12	0.018
4	72	0.111
5	115	0.177
6	169	0.260
7	120	0.185
8	83	0.128
9	48	0.074
10	22	0.034
11	6	0.009
	$n = 650$	$\sum P(x) \approx 1$

(b)

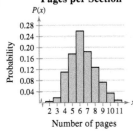

Pages per Section

Approximately symmetric

(c) 6.4, 2.9, 1.7

(d) The mean is 6.4, so the average number of pages per section is about 6 pages. The standard deviation is 1.7, so most of the sections differ from the mean by no more than about 2 pages.

13. (a)

x	f	$P(x)$
0	5	0.020
1	35	0.140
2	68	0.272
3	73	0.292
4	42	0.168
5	19	0.076
6	8	0.032
	$n = 250$	$\sum P(x) = 1$

(b)

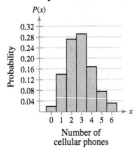

Cellular Phones per Household

Approximately symmetric

(c) 2.8, 1.7, 1.3

(d) The mean is 2.8, so the average number of cellular phones per household is about 3. The standard deviation is 1.3, so most of the households differ from the mean by no more than about 1 cellular phone.

15. 3.4

17. No; In a binomial experiment, there are only two possible outcomes: success or failure.

19. Yes; $n = 12$, $p = 0.24$, $q = 0.76$, $x = 0, 1, \ldots, 12$

21. (a) 0.208   (b) 0.322 (*Tech:* 0.321)   (c) 0.114

23. (a) 0.196   (b) 0.332   (c) 0.137

25. (a)

x	$P(x)$
0	0.125
1	0.323
2	0.332
3	0.171
4	0.044
5	0.005

(b)

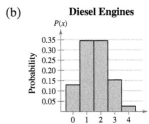

Help with Chores

Skewed right

(c) 1.7, 1.1, 1.1; The mean is 1.7, so an average of 1.7 out of 5 women have spouses who never help with household chores. The standard deviation is 1.1, so most samples of 5 women differ from the mean by no more than 1.1 women.

(d) The values $x = 4$ and $x = 5$ are unusual because their probabilities are less than 0.05.

27. (a)

x	$P(x)$
0	0.130
1	0.346
2	0.346
3	0.154
4	0.026

(b)

Diesel Engines

Skewed right

(c) 1.6, 1.0, 1.0; The mean is 1.6, so an average of 1.6 out of 4 trucks have diesel engines. The standard deviation is 1.0, so most samples of 4 trucks differ from the mean by no more than 1 truck.

(d) The value $x = 4$ is unusual because its probability is less than 0.05.

29. (a) 0.134   (b) 0.186   (c) 0.176

31. (a) 0.765   (b) 0.205   (c) 0.997   (d) 0.030; unusual

33. The probability increases as the rate increases, and decreases as the rate decreases.

Chapter Quiz for Chapter 4 *(page 231)*

1. (a) Discrete; The number of lightning strikes that occur in Wyoming during the month of June is a random variable that is countable.

(b) Continuous; The fuel (in gallons) used by the Space Shuttle during takeoff is a random variable that has an infinite number of possible outcomes and cannot be counted.

2. (a)

x	f	P(x)
1	114	0.400
2	74	0.260
3	76	0.267
4	18	0.063
5	3	0.011
	n = 285	$\sum P(x) \approx 1$

(b)

Hurricane Intensity

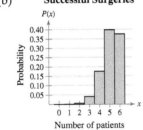

Skewed right

(c) 2.0, 1.0, 1.0

On average, the intensity of a hurricane will be 2.0. The standard deviation is 1.0, so most hurricane intensities will differ from the mean by no more than 1.0.

(d) 0.074

3. (a)

x	P(x)
0	0.00001
1	0.00039
2	0.00549
3	0.04145
4	0.17618
5	0.39933
6	0.37715

(b)

Successful Surgeries

Skewed left

(c) 5.1, 0.8, 0.9; The average number of successful surgeries is 5.1 out of 6. The standard deviation is 0.9, so most samples of 6 surgeries differ from the mean by no more than 0.9 surgery.

(d) 0.041; Yes, this event is unusual because 0.041 < 0.05.

(e) 0.047; Yes, this event is unusual because 0.047 < 0.05.

4. (a) 0.175   **(b)** 0.440   **(c)** 0.007

5. 0.038; Yes, this event is unusual because 0.038 < 0.05.

6. 0.335; No, this event is not unusual because 0.335 > 0.05.

Real Statistics–Real Decisions for Chapter 4 *(page 232)*

1. (a) Answers will vary. For instance, calculate the probability of obtaining 0 clinical pregnancies out of 10 randomly selected ART cycles.

(b) Binomial. The distribution is discrete because the number of clinical pregnancies is countable.

2. n = 10, p = 0.349, P(0) = 0.014

x	P(x)
0	0.01367
1	0.07329
2	0.17681
3	0.25277
4	0.23714
5	0.15256
6	0.06815
7	0.02088
8	0.00420
9	0.00050
10	0.00003

Answers will vary. *Sample answer:* Because P(0) = 0.014, this event is unusual but not impossible.

3. (a) Suspicious, because the probability is very small.

(b) Not suspicious, because the probability is not that small.

CHAPTER 5

Section 5.1 *(page 244)*

1. Answers will vary.

3. 1

5. Answers will vary.

Similarities: The two curves will have the same line of symmetry.

Differences: The curve with the larger standard deviation will be more spread out than the curve with the smaller standard deviation.

7. $\mu = 0, \sigma = 1$

9. "The" standard normal distribution is used to describe one specific normal distribution ($\mu = 0, \sigma = 1$). "A" normal distribution is used to describe a normal distribution with any mean and standard deviation.

11. No, the graph crosses the x-axis.

13. Yes, the graph fulfills the properties of the normal distribution.

15. No, the graph is skewed right.

17. It is normal because it is bell-shaped and nearly symmetric.

19. 0.0968   **21.** 0.0228   **23.** 0.4878   **25.** 0.5319

27. 0.005   **29.** 0.7422   **31.** 0.6387   **33.** 0.4979

35. 0.95   **37.** 0.2006 (*Tech:* 0.2005)

39. (a)

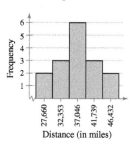

Life Spans of Tires

It is reasonable to assume that the life spans are normally distributed because the histogram is symmetric and bell-shaped.

(b) 37,234.7, 6259.2

(c) The sample mean of 37,234.7 hours is less than the claimed mean, so, on average, the tires in the sample lasted for a shorter time. The sample standard deviation of 6259.2 is greater than the claimed standard deviation, so the tires in the sample had a greater variation in life span than the manufacturer's claim.

41. (a) A = 105; B = 113; C = 121; D = 127

(b) −2.78; −0.56; 1.67; 3.33

(c) $x = 105$ is unusual because its corresponding z-score (−2.78) lies more than 2 standard deviations from the mean, and $x = 127$ is very unusual because its corresponding z-score (3.33) lies more than 3 standard deviations from the mean.

43. (a) A = 1241; B = 1392; C = 1924; D = 2202

(b) −0.86; −0.375; 1.33; 2.22

(c) $x = 2202$ is unusual because its corresponding z-score (2.22) lies more than 2 standard deviations from the mean.

45. 0.9750   **47.** 0.9775   **49.** 0.84   **51.** 0.9265

53. 0.0148   **55.** 0.3133   **57.** 0.901 (*Tech:* 0.9011)

59. 0.0098 (*Tech:* 0.0099)

61.

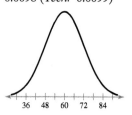

The normal distribution curve is centered at its mean (60) and has 2 points of inflection (48 and 72) representing $\mu \pm \sigma$.

63. (a) Area under curve = area of square = (1)(1) = 1

(b) 0.25   (c) 0.4

Section 5.2   *(page 252)*

1. 0.4207   **3.** 0.3446   **5.** 0.1787 (*Tech:* 0.1788)

7. 0.3442 (*Tech:* 0.3451)   **9.** 0.2747 (*Tech:* 0.2737)

11. 0.3387 (*Tech:* 0.3385)

13. (a) 0.0968   (b) 0.6612   (c) 0.2420

(d) No, none of the events are unusual because their probabilities are greater than 0.05.

15. (a) 0.1867 (*Tech:* 0.1870)   (b) 0.4171 (*Tech:* 0.4176)

(c) 0.0166 (*Tech:* 0.0167)

(d) Yes, the event in part (c) is unusual because its probability is less than 0.05.

17. (a) 0.0228   (b) 0.927   (c) 0.0013

19. (a) 0.0073   (b) 0.7215 (*Tech:* 0.7218)   (c) 0.0228

21. (a) 83.15% (*Tech:* 83.25%)

(b) 305 scores (*Tech:* 304 scores)

23. (a) 66.28% (*Tech:* 66.4%)   (b) 22 men

25. (a) 99.87%   (b) 1 adult

27. 1.5% (*Tech:* 1.51%); It is unusual for a battery to have a life span that is more than 2065 hours because the probability is less than 0.05.

29. (a) 0.3085   (b) 0.1499

(c) 0.0668; No, because 0.0668 > 0.05, this event is not unusual.

31. Out of control, because there is a point more than three standard deviations beyond the mean.

33. Out of control, because there are nine consecutive points below the mean, and two out of three consecutive points lie more than two standard deviations from the mean.

Section 5.3   *(page 262)*

1. −0.81   **3.** 2.39   **5.** −1.645   **7.** 1.555   **9.** −1.04

11. 1.175   **13.** −0.67   **15.** 0.67   **17.** −0.38

19. −0.58   **21.** −1.645, 1.645   **23.** −1.18   **25.** 1.18

27. −1.28, 1.28   **29.** −0.06, 0.06

31. (a) 68.58 inches   (b) 62.56 inches (*Tech:* 62.55 inches)

33. (a) 161.72 days (*Tech:* 161.73 days)

(b) 221.22 days (*Tech:* 221.33 days)

35. (a) 7.75 hours (*Tech:* 7.74 hours)

(b) 5.43 hours and 6.77 hours

37. 32.61 ounces

39. (a) 18.88 pounds (*Tech:* 18.90 pounds)

(b) 12.04 pounds (*Tech:* 12.05 pounds)

41. Tires that wear out by 26,800 miles (*Tech:* 26,796 miles) will be replaced free of charge.

Section 5.4   *(page 274)*

1. 150, 3.536   **3.** 150, 1.581

5. False. As the size of a sample increases, the mean of the distribution of sample means does not change.

7. False. A sampling distribution is normal if either $n \geq 30$ or the population is normal.

9. (c), because $\mu_{\bar{x}} = 16.5$, $\sigma_{\bar{x}} = 1.19$, and the graph approximates a normal curve.

11.

Sample	Mean	Sample	Mean	Sample	Mean
2, 2, 2	2	4, 4, 8	5.33	8, 16, 2	8.67
2, 2, 4	2.67	4, 4, 16	8	8, 16, 4	9.33
2, 2, 8	4	4, 8, 2	4.67	8, 16, 8	10.67
2, 2, 16	6.67	4, 8, 4	5.33	8, 16, 16	13.33
2, 4, 2	2.67	4, 8, 8	6.67	16, 2, 2	6.67
2, 4, 4	3.33	4, 8, 16	9.33	16, 2, 4	7.33
2, 4, 8	4.67	4, 16, 2	7.33	16, 2, 8	8.67
2, 4, 16	7.33	4, 16, 4	8	16, 2, 16	11.33
2, 8, 2	4	4, 16, 8	9.33	16, 4, 2	7.33
2, 8, 4	4.67	4, 16, 16	12	16, 4, 4	8
2, 8, 8	6	8, 2, 2	4	16, 4, 8	9.33
2, 8, 16	8.67	8, 2, 4	4.67	16, 4, 16	12
2, 16, 2	6.67	8, 2, 8	6	16, 8, 2	8.67
2, 16, 4	7.33	8, 2, 16	8.67	16, 8, 4	9.33
2, 16, 8	8.67	8, 4, 2	4.67	16, 8, 8	10.67
2, 16, 16	11.33	8, 4, 4	5.33	16, 8, 16	13.33
4, 2, 2	2.67	8, 4, 8	6.67	16, 16, 2	11.33
4, 2, 4	3.33	8, 4, 16	9.33	16, 16, 4	12
4, 2, 8	4.67	8, 8, 2	6	16, 16, 8	13.33
4, 2, 16	7.33	8, 8, 4	6.67	16, 16, 16	16
4, 4, 2	3.33	8, 8, 8	8		
4, 4, 4	4	8, 8, 16	10.67		

$\mu = 7.5, \quad \sigma \approx 5.36$

$\mu_{\bar{x}} = 7.5, \quad \sigma_{\bar{x}} \approx 3.09$

The means are equal but the standard deviation of the sampling distribution is smaller.

13. 0.9726; not unusual   **15.** 0.0351 (*Tech:* 0.0349); unusual

17. 7.6, 0.101   **19.** 235, 13.864

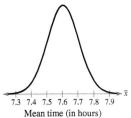
7.3 7.4 7.5 7.6 7.7 7.8 7.9 $\bar{x}$
Mean time (in hours)

207.3   235   262.7 $\bar{x}$
Mean price (in dollars)

21. 188.4, 10.9

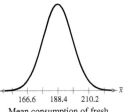
166.6   188.4   210.2 $\bar{x}$
Mean consumption of fresh vegetables (in pounds)

23. $n = 24$: 7.6, 0.07; $n = 36$: 7.6, 0.06

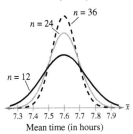
$n = 36$
$n = 24$
$n = 12$
7.3 7.4 7.5 7.6 7.7 7.8 7.9 $\bar{x}$
Mean time (in hours)

As the sample size increases, the standard error decreases, while the mean of the sample means remains constant.

25. 0.0003; Only 0.03% of samples of 35 specialists will have a mean salary less than $60,000. This is an extremely unusual event.

27. 0.9078 (*Tech:* 0.9083); About 91% of samples of 32 gas stations that week will have a mean price between $2.695 and $2.725.

29. ≈ 0 (*Tech:* 0.0000002); There is almost no chance that a random sample of 60 women will have a mean height greater than 66 inches. This event is almost impossible.

31. It is more likely to select a sample of 20 women with a mean height less than 70 inches because the sample of 20 has a higher probability.

33. Yes, it is very unlikely that you would have randomly sampled 40 cans with a mean equal to 127.9 ounces because it is more than 2 standard deviations from the mean of the sample means.

35. (a) 0.0008   (b) Claim is inaccurate.
(c) No, assuming the manufacturer's claim is true, because 96.25 is within 1 standard deviation of the mean for an individual board.

37. (a) 0.0002   (b) Claim is inaccurate.
(c) No, assuming the manufacturer's claim is true, because 49,721 is within 1 standard deviation of the mean for an individual tire.

39. No, because the z-score (0.88) is not unusual.

41. Yes, the finite correction factor should be used; 0.0003

43.

Sample	Number of boys from 3 births	Proportion of boys from 3 births
bbb	3	1
bbg	2	$\frac{2}{3}$
bgb	2	$\frac{2}{3}$
gbb	2	$\frac{2}{3}$
bgg	1	$\frac{1}{3}$
gbg	1	$\frac{1}{3}$
ggb	1	$\frac{1}{3}$
ggg	0	0

45.

Sample	Numerical representation	Sample mean
bbb	111	1
bbg	110	$\frac{2}{3}$
bgb	101	$\frac{2}{3}$
gbb	011	$\frac{2}{3}$
bgg	100	$\frac{1}{3}$
gbg	010	$\frac{1}{3}$
ggb	001	$\frac{1}{3}$
ggg	000	0

The sample means are equal to the proportions.

47. 0.0446 (*Tech:* 0.0441); About 4.5% (*Tech:* 4.4%) of samples of 105 female heart transplant patients will have a mean 3-year survival rate of less than 70%. Because the probability is less than 0.05, this is an unusual event.

Section 5.4 Activity (*page 280*)

1–2. Answers will vary.

Section 5.5 (*page 287*)

1. Properties of a binomial experiment:
  (1) The experiment is repeated for a fixed number of independent trials.
  (2) There are two possible outcomes: success or failure.
  (3) The probability of success is the same for each trial.
  (4) The random variable x counts the number of successful trials.

3. Cannot use normal distribution.

5. Cannot use normal distribution.

7. Cannot use normal distribution because $nq < 5$.

9. Can use normal distribution; $\mu = 27.5$, $\sigma \approx 3.52$

11. Cannot use normal distribution because $nq < 5$.

13. a   **14.** d   **15.** c   **16.** b

17. The probability of getting fewer than 25 successes; $P(x < 24.5)$

19. The probability of getting exactly 33 successes; $P(32.5 < x < 33.5)$

21. The probability of getting at most 150 successes; $P(x < 150.5)$

23. Can use normal distribution.
  (a) 0.0782 (*Tech:* 0.0785)   (b) 0.9147 (*Tech:* 0.9151)

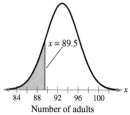

  (c) 0.0853 (*Tech:* 0.0849)

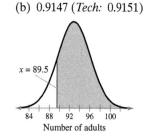

  (d) No, none of the probabilities are less than 0.05.

25. Can use normal distribution.
  (a) ≈ 1 (b) 0.9798 (*Tech:* 0.9801)

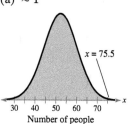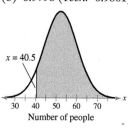

  (c) 0.6097 (*Tech:* 0.6109)

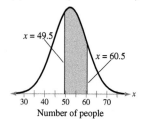

  (d) No, none of the probabilities are less than 0.05.

27. Can use normal distribution.
  (a) 0.0692 (*Tech:* 0.0691)   (b) 0.8770 (*Tech:* 0.8771)

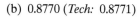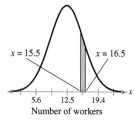

  (c) 0.8078 (*Tech:* 0.8080)   (d) 0.8212 (*Tech:* 0.8221)

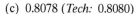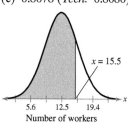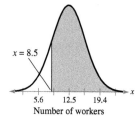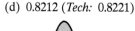

29. (a) Can use normal distribution.
    0.0069

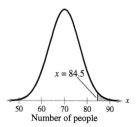

(b) Can use normal distribution.
0.3557 (*Tech:* 0.3545)

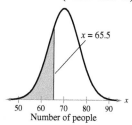

(c) Can use normal distribution.
0.0558 (*Tech:* 0.0595)

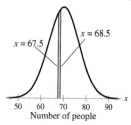

(d) Cannot use normal distribution because $np < 5$ and $nq < 5$; 0.002

31. Binomial: 0.549; Normal: 0.5463 (*Tech:* 0.5466); The results are about the same.

33. Highly unlikely. Answers will vary.   **35.** 0.1020

Uses and Abuses for Chapter 5   *(page 291)*

1. (a) Not unusual; A sample mean of 115 is less than 2 standard deviations from the population mean.

(b) Not unusual; A sample mean of 105 lies within 2 standard deviations of the population mean.

2. The ages of students at a high school may not be normally distributed.

3. Answers will vary.

Review Answers for Chapter 5   *(page 293)*

1. $\mu = 15, \sigma = 3$

3. Curve B has the greatest mean because its line of symmetry occurs the farthest to the right.

5. -2.25 ; 0.5; 2; 3.5   **7.** 0.6772   **9.** 0.6293   **11.** 0.7157

13. 0.00235 (*Tech:* 0.00236)   **15.** 0.4495

17. 0.4365 (*Tech:* 0.4364)   **19.** 0.1336

21. $A = 8; B = 17; C = 23; D = 29$ **23.** 0.8997

25. 0.9236 (*Tech:* 0.9237)   **27.** 0.0124   **29.** 0.8944

31. 0.2266   **33.** 0.2684 (*Tech:* 0.2685)

35. (a) 0.3156   (b) 0.3099   (c) 0.3446

37. No, none of the events are unusual because their probabilities are greater than 0.05.

39. -0.07 **41.** 1.13   **43.** 1.04   **45.** 0.51

47. 42.5 meters   **49.** 51.6 meters   **51.** 50.8 meters

53. $\mu = 145, \sigma = 45$

$\mu_{\bar{x}} = 145, \sigma_{\bar{x}} \approx 25.98$

The means are the same, but $\sigma_{\bar{x}}$ is less than σ.

55. 76, 3.465

57. (a) 0.0485 (*Tech:* 0.0482)

(b) 0.8180

(c) 0.0823 (*Tech:* 0.0829)

(a) and (c) are smaller, (b) is larger. This is to be expected because the standard error of the sample means is smaller.

59. (a) 0.1867 (*Tech:* 0.1855)   (b) ≈ 0

61. 0.0019 (*Tech:* 0.0018)

63. Cannot use normal distribution because $nq < 5$.

65. $P(x > 24.5)$ **67.** $P(44.5 < x < 45.5)$

69. Can use normal distribution.

≈ 0 (*Tech:* 0.0002)

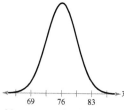

Chapter Quiz for Chapter 5   *(page 297)*

1. (a) 0.9945   (b) 0.9990

(c) 0.6212   (d) 0.83685 (*Tech:* 0.83692)

2. (a) 0.9198 (*Tech:* 0.9199)   (b) 0.1940 (*Tech:* 0.1938)

(c) 0.0456 (*Tech:* 0.0455)

3. 0.0475 (*Tech:* 0.0478); Yes, the event is unusual because its probability is less than 0.05.

4. 0.2586 (*Tech:* 0.2611); No, the event is not unusual because its probability is greater than 0.05.

5. 21.19%   **6.** 503 students (*Tech:* 505 students)

7. 125   **8.** 80

9. 0.0049; About 0.5% of samples of 60 students will have a mean IQ score greater than 105. This is a very unusual event.

10. More likely to select one student with an IQ score greater than 105 because the standard error of the mean is less than the standard deviation.

11. Can use normal distribution.

$\mu = 28.35, \sigma \approx 2.32$

12. 0.0004; This event is extremely unusual because its probability is much less than 0.05.

Real Statistics–Real Decisions for Chapter 5 *(page 298)*

1. (a) 0.0014 (b) 0.9495 (c) ≈ 0

 (d) There is a very high probability that at least 40 out of 60 employees will participate, and the probability that fewer than 20 will participate is almost 0.

2. (a) 0.2514 *(Tech: 0.2525)* (b) 0.4972 *(Tech: 0.4950)*

 (c) 0.2514 *(Tech: 0.2525)*

3. (a) 3; The line of symmetry occurs at $x = 3$.

 (b) Yes (c) Answers will vary.

Cumulative Review Answers for Chapters 3–5 *(page 300)*

1. (a) $np = 7.5 \geq 5, nq = 42.5 \geq 5$

 (b) 0.9973

 (c) Yes, because the probability is less than 0.05.

2. (a) 3.1 (b) 1.6 (c) 1.3 (d) 3.1

 (e) The size of a family household on average is about 3 persons. The standard deviation is 1.3, so most households differ from the mean by no more than about 1 person.

3. (a) 3.6 (b) 1.9 (c) 1.4 (d) 3.6

 (e) The number of fouls for a player in a game on average is about 4 fouls. The standard deviation is 1.4, so most of the player's games differ from the mean by no more than about 1 or 2 fouls.

4. (a) 0.476 (b) 0.78 (c) 0.659

5. (a) 43,680 (b) 0.0192

6. 0.7642 **7.** 0.0010 **8.** 0.7995

9. 0.4984 **10.** 0.2862 **11.** 0.5905

12. (a) 0.0462; unusual, because the probability is less than 0.05

 (b) 0.6029

 (c) 0.0139; unusual, because the probability is less than 0.05

13. (a) 0.0048 (b) 0.0149 (c) 0.9511

14. (a) 0.2777 (b) 0.8657

 (c) Dependent. $P(\text{being a public school teacher} \mid \text{having 20 years or more of full-time teaching experience}) \neq P(\text{being a public school teacher})$

 (d) 0.8881 (e) 0.4177

15. (a) 70, 0.1897 (b) 0.0006

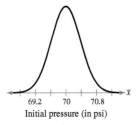

Initial pressure (in psi)

16. (a) 0.0548 (b) 0.6547 (c) 52.2 months
17. (a) 495 (b) 0.0020
18. (a) 0.0278; unusual, because the probability is less than 0.05

 (b) 0.2272 (c) 0.5982

CHAPTER 6

Section 6.1 *(page 311)*

1. You are more likely to be correct using an interval estimate because it is unlikely that a point estimate will exactly equal the population mean.

3. d; As the level of confidence increases, z_c increases, causing wider intervals.

5. 1.28 **7.** 1.15 **9.** -0.47 **11.** 1.76 **13.** 1.861

15. 0.192 **17.** c **18.** d **19.** b **20.** a

21. $(12.0, 12.6)$ **23.** $(9.7, 11.3)$ **25.** 1.4, 13.4

27. 0.17, 1.88 **29.** 126 **31.** 7 **33.** 1.95, 28.15

35. $(428.68, 476.92); (424.06, 481.54)$

 With 90% confidence, you can say that the population mean price is between $428.68 and $476.92; with 95% confidence, you can say that the population mean price is between $424.06 and $481.54. The 95% CI is wider.

37. $(87.0, 111.6); (84.7, 113.9)$

 With 90% confidence, you can say that the population mean is between 87.0 and 111.6 calories; with 95% confidence, you can say that the population mean is between 84.7 and 113.9 calories. The 95% CI is wider.

39. $(2532.20, 2767.80)$

 With 95% confidence, you can say that the population mean cost is between $2532.20 and $2767.80.

41. $(2556.87, 2743.13)$ *[Tech: $(2556.90, 2743.10)$]*

 The $n = 50$ CI is wider because a smaller sample is taken, giving less information about the population.

43. $(3.09, 3.15)$

 With 95% confidence, you can say that the population mean time is between 3.09 and 3.15 minutes.

45. $(3.10, 3.14)$

 The $s = 0.09$ CI is wider because of the increased variability within the sample.

47. (a) An increase in the level of confidence will widen the confidence interval.

 (b) An increase in the sample size will narrow the confidence interval.

 (c) An increase in the standard deviation will widen the confidence interval.

49. $(14.6, 15.6); (14.4, 15.8)$

 With 90% confidence, you can say that the population mean length of time is between 14.6 and 15.6 minutes; with 99% confidence, you can say that the population mean length of time is between 14.4 and 15.8 minutes. The 99% CI is wider.

51. 89

53. (a) 121 servings (b) 208 servings

(c) The 99% CI requires a larger sample because more information is needed from the population to be 99% confident.

55. (a) 32 cans (b) 87 cans

$E = 0.15$ requires a larger sample size. As the error size decreases, a larger sample must be taken to obtain enough information from the population to ensure the desired accuracy.

57. (a) 16 sheets (b) 62 sheets

$E = 0.0625$ requires a larger sample size. As the error size decreases, a larger sample must be taken to obtain enough information from the population to ensure the desired accuracy.

59. (a) 42 soccer balls (b) 60 soccer balls

$\sigma = 0.3$ requires a larger sample size. Due to the increased variability in the population, a larger sample size is needed to ensure the desired accuracy.

61. (a) An increase in the level of confidence will increase the minimum sample size required.

(b) An increase (larger E) in the error tolerance will decrease the minimum sample size required.

(c) An increase in the population standard deviation will increase the minimum sample size required.

63. (212.8, 221.4)

With 95% confidence, you can say that the population mean airfare price is between \$212.8 and \$221.4.

65. **80% confidence interval results:**

μ : population mean

Standard deviation = 344.9

Mean	n	Sample Mean	Std. Err.	L. Limit	U. Limit
μ	30	1042.7	62.969837	962.0009	1123.399

90% confidence interval results:

μ : population mean

Standard deviation = 344.9

Mean	n	Sample Mean	Std. Err.	L. Limit	U. Limit
μ	30	1042.7	62.969837	939.12384	1146.2761

95% confidence interval results:

μ : population mean

Standard deviation = 344.9

Mean	n	Sample Mean	Std. Err.	L. Limit	U. Limit
μ	30	1042.7	62.969837	919.2814	1166.1187

With 80% confidence, you can say that the population mean sodium content is between 962.0 and 1123.4 milligrams; with 90% confidence, you can say it is between 939.1 and 1146.3 milligrams; with 95% confidence, you can say it is between 919.3 and 1166.1 milligrams.

67. (a) 0.707 (b) 0.949 (c) 0.962 (d) 0.975

(e) The finite population correction factor approaches 1 as the sample size decreases and the population size remains the same.

69. *Sample answer:*

$$E = \frac{z_c \sigma}{\sqrt{n}} \qquad \text{Write original equation}$$

$$E\sqrt{n} = z_c \sigma \qquad \text{Multiply each side by } \sqrt{n}.$$

$$\sqrt{n} = \frac{z_c \sigma}{E} \qquad \text{Divide each side by } E.$$

$$n = \left(\frac{z_c \sigma}{E}\right)^2 \qquad \text{Square each side.}$$

Section 6.2 *(page 323)*

1. 1.833   **3.** 2.947   **5.** 2.7   **7.** 1.2

9. (a) (10.9, 14.1) (b) The t-CI is wider.

11. (a) (4.1, 4.5)

(b) When rounded to the nearest tenth, the normal CI and the t-CI have the same width.

13. 3.7, 18.4

15. 9.5, 74.1

17. 6.0; (29.5, 41.5); With 95% confidence, you can say that the population mean commute time is between 29.5 and 41.5 minutes.

19. 6.4; (29.1, 41.9); With 95% confidence, you can say that the population mean commute time is between 29.1 and 41.9 minutes. This confidence interval is slightly wider than the one found in Exercise 17.

21. (a) (3.80, 5.20)

(b) (4.41, 4.59); The t-CI in part (a) is wider.

23. (a) 90,182.9 (b) 3724.9 (c) (87,438.6, 92,927.2)

25. (a) 1767.7 (b) 252.2 (c) (1541.5, 1993.8)

27. Use a normal distribution because $n \geq 30$.

(26.0, 29.4); With 95% confidence, you can say that the population mean BMI is between 26.0 and 29.4.

29. Use a t-distribution because $n < 30$, the miles per gallon are normally distributed, and σ is unknown.

(20.5, 23.3) [*Tech:* (20.5, 23.4)]; With 95% confidence, you can say that the population mean is between 20.5 and 23.3 miles per gallon.

31. Cannot use normal or t-distribution because $n < 30$ and the times are not normally distributed.

33. 90% confidence interval results:

μ : mean of Variable

Variable	Sample Mean	Std. Err.
Time (in hours)	12.194445	0.4136141

DF	L. Limit	U. Limit
17	11.474918	12.91397

95% confidence interval results:

μ : mean of Variable

Variable	Sample Mean	Std. Err.
Time (in hours)	12.194445	0.4136141

DF	L. Limit	U. Limit
17	11.321795	13.067094

99% confidence interval results:

μ : mean of Variable

Variable	Sample Mean	Std. Err.
Time (in hours)	12.194445	0.4136141

DF	L. Limit	U. Limit
17	10.995695	13.393193

With 90% confidence, you can say that the population mean time spent on homework is between 11.5 and 12.9 hours; with 95% confidence, you can say it is between 11.3 and 13.1 hours; and with 99% confidence, you can say it is between 11.0 and 13.4 hours. As the level of confidence increases, the intervals get wider.

35. No; They are not making good tennis balls because the desired bounce height of 55.5 inches is not between 55.9 and 56.1 inches.

Activity 6.2 *(page 326)*

1–2. Answers will vary.

Section 6.3 *(page 332)*

1. False. To estimate the value of p, the population proportion of successes, use the point estimate $\hat{p} = x/n$.

3. 0.750, 0.250   **5.** 0.423, 0.577

7. $E = 0.014, \hat{p} = 0.919$ **9.** $E = 0.042, \hat{p} = 0.554$

11. (0.557, 0.619) [*Tech:* (0.556, 0.619)];

(0.551, 0.625) [*Tech:* (0.550, 0.625)];

With 90% confidence, you can say that the population proportion of U.S. males ages 18–64 who say they have gone to the dentist in the past year is between 55.7% (*Tech:* 55.6%) and 61.9%; with 95% confidence, you can say it is between 55.1% (*Tech:* 55.0%) and 62.5%. The 95% confidence interval is slightly wider.

A78

13. (0.438, 0.484);

With 99% confidence, you can say that the population proportion of U.S. adults who say they have started paying bills online in the last year is between 43.8% and 48.4%.

15. (0.622, 0.644)

17. (a) 601 adults   (b) 413 adults

(c) Having an estimate of the population proportion reduces the minimum sample size needed.

19. (a) 752 adults   (b) 483 adults

(c) Having an estimate of the population proportion reduces the minimum sample size needed.

21. (a) (0.234, 0.306)

(b) (0.450, 0.530)

(c) (0.275, 0.345)

23. (a) (0.274, 0.366)   (b) (0.511, 0.609)

25. No, it is unlikely that the two proportions are equal because the confidence intervals estimating the proportions do not overlap. The 99% confidence intervals are (0.260, 0.380) and (0.496, 0.624). Although these intervals are wider, they still do not overlap.

27. 90% confidence interval results:

p : proportion of successes for population

Method: Standard-Wald

Proportion	Count	Total	Sample Prop.
p	802	1025	0.78243905

Std. Err.	L. Limit	U. Limit
0.012887059	0.7612417	0.8036364

95% confidence interval results:

p : proportion of successes for population

Method: Standard-Wald

Proportion	Count	Total	Sample Prop.
p	802	1025	0.78243905

Std. Err.	L. Limit	U. Limit
0.012887059	0.75718087	0.8076972

99% confidence interval results:

p : proportion of successes for population

Method: Standard-Wald

Proportion	Count	Total	Sample Prop.
p	802	1025	0.78243905

Std. Err.	L. Limit	U. Limit
0.012887059	0.74924415	0.8156339

With 90% confidence, you can say that the population proportion of U.S. adults who disapprove of the job Congress is doing is between 76.1% and 80.4%; with 95% confidence, you can say it is between 75.7% and 80.8%; and with 99% confidence, you can say it is between 74.9% and 81.6%. As the level of confidence increases, the intervals get wider.

29. $(0.304, 0.324)$ is approximately a 97.6% CI.

31. If $n\hat{p} < 5$ or $n\hat{q} < 5$, the sampling distribution of $\hat{p}$ may not be normally distributed, so z_c cannot be used to calculate the confidence interval.

33.

$\hat{p}$	$\hat{q} = 1 - \hat{p}$	$\hat{p}\hat{q}$	$\hat{p}$	$\hat{q} = 1 - \hat{p}$	$\hat{p}\hat{q}$
0.0	1.0	0.00	0.45	0.55	0.2475
0.1	0.9	0.09	0.46	0.54	0.2484
0.2	0.8	0.16	0.47	0.53	0.2491
0.3	0.7	0.21	0.48	0.52	0.2496
0.4	0.6	0.24	0.49	0.51	0.2499
0.5	0.5	0.25	0.50	0.50	0.2500
0.6	0.4	0.24	0.51	0.49	0.2499
0.7	0.3	0.21	0.52	0.48	0.2496
0.8	0.2	0.16	0.53	0.47	0.2491
0.9	0.1	0.09	0.54	0.46	0.2484
1.0	0.0	0.00	0.55	0.45	0.2475

$\hat{p} = 0.5$ gives the maximum value of $\hat{p}\hat{q}$.

Activity 6.3 *(page 336)*

1–2. Answers will vary.

Section 6.4 *(page 341)*

1. Yes

3. 14.067, 2.167　　**5.** 32.852, 8.907　　**7.** 52.336, 13.121

9. (a) $(0.0000413, 0.000157)$　　(b) $(0.00643, 0.0125)$

With 90% confidence, you can say that the population variance is between 0.0000413 and 0.000157, and the population standard deviation is between 0.00643 and 0.0125 milligram.

11. (a) $(0.0305, 0.191)$　　(b) $(0.175, 0.438)$

With 99% confidence, you can say that the population variance is between 0.0305 and 0.191, and the population standard deviation is between 0.175 and 0.438 hour.

13. (a) $(6.63, 55.46)$　　(b) $(2.58, 7.45)$

With 99% confidence, you can say that the population variance is between 6.63 and 55.46, and the population standard deviation is between 2.58 and 7.45 dollars per year.

15. (a) $(380.0, 3942.6)$　　(b) $(19.5, 62.8)$

With 98% confidence, you can say that the population variance is between 380.0 and 3942.6, and the population standard deviation is between $19.5 and $62.8.

17. (a) $(22.5, 98.7)$　　(b) $(4.7, 9.9)$

With 95% confidence, you can say that the population variance is between 22.5 and 98.7, and the population standard deviation is between 4.7 and 9.9 beats per minute.

19. (a) $(128, 492)$　　(b) $(11, 22)$

With 95% confidence, you can say that the population variance is between 128 and 492, and the population standard deviation is between 11 and 22 grains per gallon.

21. (a) $(9,104,741, 25,615,326)$　　(b) $(3017, 5061)$

With 80% confidence, you can say that the population variance is between 9,104,741 and 25,615,326, and the population standard deviation is between $3017 and $5061.

23. (a) $(7.0, 30.6)$　　(b) $(2.6, 5.5)$

With 98% confidence, you can say that the population variance is between 7.0 and 30.6, and the population standard deviation is between 2.6 and 5.5 minutes.

25. **95% confidence interval results:**

σ^2 : variance of Variable

Variance	Sample Var.	DF	L. Limit	U. Limit
σ^2	11.56	29	7.332092	20.891039

$(2.71, 4.57)$

27. **90% confidence interval results:**

σ^2 : variance of Variable

Variance	Sample Var.	DF	L. Limit	U. Limit
σ^2	1225	17	754.8815	2401.4731

$(27, 49)$

29. Yes, because all of the values in the confidence interval are less than 0.015.

31. Answers will vary. *Sample answer:* Unlike a confidence interval for a population mean or proportion, a confidence interval for a population variance does not have a margin of error. The left and right endpoints must be calculated separately.

Uses and Abuses for Chapter 6　　*(page 344)*

1–2. Answers will vary.

Review Answers for Chapter 6　　*(page 346)*

1. (a) 103.5　(b) 9.0　　**3.** $(15.6, 16.0)$　　**5.** 1.675, 22.425

7. 47 people　　**9.** 49 people　　**11.** 1.383　　**13.** 2.624

15. $n = 20$　　**17.** 11.2　　**19.** 0.7　　**21.** $(60.9, 83.3)$

23. $(6.1, 7.5)$　　**25.** $(2050, 2386)$　　**27.** 0.81, 0.19

29. 0.540, 0.460　　**31.** 0.140, 0.860　　**33.** 0.490, 0.510

35. $(0.790, 0.830)$

With 95% confidence, you can say that the population proportion of U.S. adults who say they will participate in the 2010 Census is between 79.0% and 83.0%.

37. $(0.514, 0.566)$ [*Tech:* $(0.514, 0.565)$]

With 90% confidence, you can say that the population proportion of U.S. adults who say they have worked the night shift at some point in their lives is between 51.4% and 56.6% (*Tech:* 56.5%).

39. (0.112, 0.168)

With 99% confidence, you can say that the population proportion of U.S. adults who say that the cost of healthcare is the most important financial problem facing their family today is between 11.2% and 16.8%.

41. (0.466, 0.514)

With 80% confidence, you can say that the population proportion of parents with kids 4 to 8 years old who say they know their state booster seat law is between 46.6% and 51.4%.

43. (a) 385 adults (b) 359 adults

(c) Having an estimate of the population proportion reduces the minimum sample size needed.

45. 23.337, 4.404   **47.** 14.067, 2.167

49. (27.2, 113.5); (5.2, 10.7)   **51.** (0.80, 3.07); (0.89, 1.75)

Chapter Quiz for Chapter 6   *(page 349)*

1. (a) 6.85

(b) 0.65; You are 95% confident that the margin of error for the population mean is about 0.65 minute.

(c) (6.20, 7.50)

With 95% confidence, you can say that the population mean amount of time is between 6.20 and 7.50 minutes.

2. 39 college students

3. (a) 33.11; 2.38

(b) (31.73, 34.49)

With 90% confidence, you can say that the population mean time played in the season is between 31.73 and 34.49 minutes.

(c) (30.38, 35.84)

With 90% confidence, you can say that the population mean time played in the season is between 30.38 and 35.84 minutes. This confidence interval is wider than the one found in part (b).

4. (6510, 7138)

5. (a) 0.780   (b) (0.762, 0.798) [*Tech:* (0.762, 0.799)]

(c) 712 adults

6. (a) (2.10, 5.99)   (b) (1.45, 2.45)

Real Statistics–Real Decisions for Chapter 6   *(page 350)*

1. (a) Yes, there has been a change in the mean concentration level because the confidence interval for Year 1 does not overlap the confidence interval for Year 2.

(b) No, there has not been a change in the mean concentration level because the confidence interval for Year 2 overlaps the confidence interval for Year 3.

(c) Yes, there has been a change in the mean concentration level because the confidence interval for Year 1 does not overlap the confidence interval for Year 3.

2. The concentrations of cyanide in the drinking water have increased over the three-year period.

3. The width of the confidence interval for Year 2 may have been caused by greater variation in the levels of cyanide than in the other years, which may be the result of outliers.

4. (a) The sampling distribution of the sample means was used because the "mean concentration" was used. The sample mean is the most unbiased point estimate of the population mean.

(b) No, because typically σ is unknown. They could have used the sample standard deviation.

CHAPTER 7

Section 7.1   *(page 367)*

1. The two types of hypotheses used in a hypothesis test are the null hypothesis and the alternative hypothesis.

The alternative hypothesis is the complement of the null hypothesis.

3. You can reject the null hypothesis, or you can fail to reject the null hypothesis.

5. False. In a hypothesis test, you assume the null hypothesis is true.

7. True

9. False. A small P-value in a test will favor rejection of the null hypothesis.

11. H_0: $\mu \leq 645$ (claim); H_a: $\mu > 645$

13. H_0: $\sigma = 5$; H_a: $\sigma \neq 5$ (claim)

15. H_0: $p \geq 0.45$; H_a: $p < 0.45$ (claim)

17. c; H_0: $\mu \leq 3$ **18.** d; H_0: $\mu \geq 3$

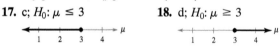

19. b; H_0: $\mu = 3$ **20.** a; H_0: $\mu \leq 2$

21. Right-tailed   **23.** Two-tailed

25. $\mu > 750$

H_0: $\mu \leq 750$; H_a: $\mu > 750$ (claim)

27. $\sigma \leq 320$

H_0: $\sigma \leq 320$ (claim); H_a: $\sigma > 320$

29. $\mu < 45$

H_0: $\mu \geq 45$; H_a: $\mu < 45$ (claim)

31. A type I error will occur if the actual proportion of new customers who return to buy their next piece of furniture is at least 0.60, but you reject H_0: $p \geq 0.60$.

A type II error will occur if the actual proportion of new customers who return to buy their next piece of furniture is less than 0.60, but you fail to reject H_0: $p \geq 0.60$.

33. A type I error will occur if the actual standard deviation of the length of time to play a game is less than or equal to 12 minutes, but you reject H_0: $\sigma \leq 12$.

A type II error will occur if the actual standard deviation of the length of time to play a game is greater than 12 minutes, but you fail to reject H_0: $\sigma \leq 12$.

35. A type I error will occur if the actual proportion of applicants who become police officers is at most 0.20, but you reject H_0: $p \leq 0.20$.

A type II error will occur if the actual proportion of applicants who become police officers is greater than 0.20, but you fail to reject H_0: $p \leq 0.20$.

37. H_0: The proportion of homeowners who have a home security alarm is greater than or equal to 14%.

H_a: The proportion of homeowners who have a home security alarm is less than 14%.

H_0: $p \geq 0.14$; H_a: $p < 0.14$

Left-tailed because the alternative hypothesis contains $<$.

39. H_0: The standard deviation of the 18-hole scores for a golfer is greater than or equal to 2.1 strokes.

H_a: The standard deviation of the 18-hole scores for a golfer is less than 2.1 strokes.

H_0: $\sigma \geq 2.1$; H_a: $\sigma < 2.1$

Left-tailed because the alternative hypothesis contains $<$.

41. H_0: The mean length of the baseball team's games is greater than or equal to 2.5 hours.

H_a: The mean length of the baseball team's games is less than 2.5 hours.

H_0: $\mu \geq 2.5$; H_a: $\mu < 2.5$

Left-tailed because the alternative hypothesis contains $<$.

43. (a) There is enough evidence to support the scientist's claim that the mean incubation period for swan eggs is less than 40 days.

(b) There is not enough evidence to support the scientist's claim that the mean incubation period for swan eggs is less than 40 days.

45. (a) There is enough evidence to support the U.S. Department of Labor's claim that the proportion of full-time workers earning over $450 per week is greater than 75%.

(b) There is not enough evidence to support the U.S. Department of Labor's claim that the proportion of full-time workers earning over $450 per week is greater than 75%.

47. (a) There is enough evidence to support the researcher's claim that the proportion of people who have had no health care visits in the past year is less than 17%.

(b) There is not enough evidence to support the researcher's claim that the proportion of people who have had no health care visits in the past year is less than 17%.

49. H_0: $\mu \geq 60$; H_a: $\mu < 60$

51. (a) H_0: $\mu \geq 15$; H_a: $\mu < 15$

(b) H_0: $\mu \leq 15$; H_a: $\mu > 15$

53. If you decrease α, you are decreasing the probability that you will reject H_0. Therefore, you are increasing the probability of failing to reject H_0. This could increase β, the probability of failing to reject H_0 when H_0 is false.

55. Yes; If the P-value is less than $\alpha = 0.05$, it is also less than $\alpha = 0.10$.

57. (a) Fail to reject H_0 because the confidence interval includes values greater than 70.

(b) Reject H_0 because the confidence interval is located entirely to the left of 70.

(c) Fail to reject H_0 because the confidence interval includes values greater than 70.

59. (a) Reject H_0 because the confidence interval is located entirely to the right of 0.20.

(b) Fail to reject H_0 because the confidence interval includes values less than 0.20.

(c) Fail to reject H_0 because the confidence interval includes values less than 0.20.

Section 7.2   *(page 381)*

1. In the z-test using rejection region(s), the test statistic is compared with critical values. The z-test using a P-value compares the P-value with the level of significance α.

3. $P = 0.0934$; Reject H_0.   **5.** $P = 0.0069$; Reject H_0.

7. $P = 0.0930$; Fail to reject H_0.

9. b   **10.** d   **11.** c   **12.** a

13. (a) Fail to reject H_0.

(b) Reject H_0.

15. Fail to reject H_0.

17. 1.645

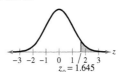

19. -1.88

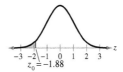

21. $-2.33, 2.33$

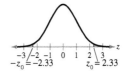

23. (a) Fail to reject H_0 because $z < 1.285$.

(b) Fail to reject H_0 because $z < 1.285$.

(c) Fail to reject H_0 because $z < 1.285$.

(d) Reject H_0 because $z > 1.285$.

25. Reject H_0. There is enough evidence at the 5% level of significance to reject the claim.

27. Reject H_0. There is enough evidence at the 2% level of significance to support the claim.

29. (a) $H_0: \mu \le 30$; $H_a: \mu > 30$ (claim)   (b) 2.83; 0.9977
(c) 0.0023   (d) Reject H_0.
(e) There is enough evidence at the 1% level of significance to support the student's claim that the mean raw score for the school's applicants is more than 30.

31. (a) $H_0: \mu = 28.5$ (claim); $H_a: \mu \ne 28.5$
(b) -1.71; 0.0436   (c) 0.0872 (*Tech:* 0.0878)
(d) Fail to reject H_0.
(e) There is not enough evidence at the 8% level of significance to reject the U.S. Department of Agriculture's claim that the mean consumption of bottled water by a person in the United States is 28.5 gallons per year.

33. (a) $H_0: \mu = 15$ (claim); $H_a: \mu \ne 15$
(b) -0.22; 0.4129 (*Tech:* 0.4135)
(c) 0.8258 (*Tech:* 0.8270)   (d) Fail to reject H_0.
(e) There is not enough evidence at the 5% level of significance to reject the claim that the mean time it takes smokers to quit smoking permanently is 15 years.

35. (a) $H_0: \mu = 40$ (claim); $H_a: \mu \ne 40$
(b) $-z_0 = -2.575$, $z_0 = 2.575$;
Rejection regions: $z < -2.575$, $z > 2.575$
(c) -0.584 (d) Fail to reject H_0.
(e) There is not enough evidence at the 1% level of significance to reject the company's claim that the mean caffeine content per 12-ounce bottle of cola is 40 milligrams.

37. (a) $H_0: \mu \ge 750$ (claim); $H_a: \mu < 750$
(b) $z_0 = -2.05$; Rejection region: $z < -2.05$
(c) -0.5 (d) Fail to reject H_0.
(e) There is not enough evidence at the 2% level of significance to reject the light bulb manufacturer's claim that the mean life of the bulb is at least 750 hours.

39. (a) $H_0: \mu \le 32$; $H_a: \mu > 32$ (claim)
(b) $z_0 = 1.555$; Rejection region: $z > 1.555$
(c) -1.478 (d) Fail to reject H_0.
(e) There is not enough evidence at the 6% level of significance to support the scientist's claim that the mean nitrogen dioxide level in Calgary is greater than 32 parts per billion.

41. (a) $H_0: \mu \ge 10$ (claim); $H_a: \mu < 10$
(b) $z_0 = -1.88$; Rejection region: $z < -1.88$
(c) -0.51 (d) Fail to reject H_0.
(e) There is not enough evidence at the 3% level of significance to reject the weight loss program's claim that the mean weight loss after one month is at least 10 pounds.

43. Hypothesis test results:
μ : population mean
$H_0: \mu = 58$
$H_A: \mu \ne 58$
Standard deviation = 2.35

Mean	n	Sample Mean	Std. Err.	Z-Stat	P-value
μ	80	57.6	0.262738	-1.5224292	0.1279

$P = 0.1279 > 0.10$, so fail to reject H_0. There is not enough evidence at the 10% level of significance to reject the claim.

45. Hypothesis test results:
μ : population mean
$H_0: \mu = 1210$
$H_A: \mu > 1210$
Standard deviation = 205.87

Mean	n	Sample Mean	Std. Err.	Z-Stat	P-value
μ	250	1234.21	13.020362	1.8593953	0.0315

$P = 0.0315 < 0.08$, so reject H_0. There is enough evidence at the 8% level of significance to reject the claim.

47. Fail to reject H_0 because the standardized test statistic $z = -1.86$ is not in the rejection region ($z < -2.33$).

49. b, d; If $\alpha = 0.05$, the rejection region is $z < -1.645$; because $z = -1.86$ is in the rejection region, you can reject H_0. If $n = 50$, the standardized test statistic is $z = -2.40$; because $z = -2.40$ is in the rejection region ($z < -2.33$), you can reject H_0.

Section 7.3   (*page 393*)

1. Identify the level of significance α and the degrees of freedom, d.f. $= n - 1$. Find the critical value(s) using the *t*-distribution table in the row with $n - 1$ d.f. If the hypothesis test is
(1) left-tailed, use the "One Tail, α" column with a negative sign.
(2) right-tailed, use the "One Tail, α" column with a positive sign.
(3) two-tailed, use the "Two Tails, α" column with a negative and a positive sign.

3. 1.717   **5.** -1.328 **7.** $-2.056, 2.056$

9. (a) Fail to reject H_0 because $t > -2.086$.
(b) Fail to reject H_0 because $t > -2.086$.
(c) Fail to reject H_0 because $t > -2.086$.
(d) Reject H_0 because $t < -2.086$.

11. (a) Fail to reject H_0 because $-2.602 < t < 2.602$.
(b) Fail to reject H_0 because $-2.602 < t < 2.602$.
(c) Reject H_0 because $t > 2.602$.
(d) Reject H_0 because $t < -2.602$.

13. Fail to reject H_0. There is not enough evidence at the 1% level of significance to reject the claim.

15. Reject H_0. There is enough evidence at the 1% level of significance to reject the claim.

17. (a) H_0: $\mu = 18{,}000$ (claim); H_a: $\mu \neq 18{,}000$

(b) $-t_0 = -2.145$, $t_0 = 2.145$;
Rejection regions: $t < -2.145$, $t > 2.145$

(c) 1.21   (d) Fail to reject H_0.

(e) There is not enough evidence at the 5% level of significance to reject the dealer's claim that the mean price of a 2008 Subaru Forester is \$18,000.

19. (a) H_0: $\mu \leq 60$; H_a: $\mu > 60$ (claim)

(b) $t_0 = 1.943$; Rejection region: $t > 1.943$

(c) 2.12   (d) Reject H_0.

(e) There is enough evidence at the 5% level of significance to support the board's claim that the mean number of hours worked per week by surgical faculty who teach at an academic institution is more than 60 hours.

21. (a) H_0: $\mu \leq 1$; H_a: $\mu > 1$ (claim)

(b) $t_0 = 1.356$; Rejection region: $t > 1.356$

(c) 6.44   (d) Reject H_0.

(e) There is enough evidence at the 10% level of significance to support the environmentalist's claim that the mean amount of waste recycled by adults in the United States is more than 1 pound per person per day.

23. (a) H_0: $\mu = \$26{,}000$ (claim); H_a: $\mu \neq \$26{,}000$

(b) $-t_0 = -2.262$, $t_0 = 2.262$;
Rejection regions: $t < -2.262$, $t > 2.262$

(c) -0.15 (d) Fail to reject H_0.

(e) There is not enough evidence at the 5% level of significance to reject the employment information service's claim that the mean salary for full-time male workers over age 25 without a high school diploma is \$26,000.

25. (a) H_0: $\mu \leq 45$; H_a: $\mu > 45$ (claim)

(b) 0.0052   (c) Reject H_0.

(d) There is enough evidence at the 10% level of significance to support the county's claim that the mean speed of the vehicles is greater than 45 miles per hour.

27. (a) H_0: $\mu = \$105$ (claim); H_a: $\mu \neq \$105$

(b) 0.0165   (c) Fail to reject H_0.

(d) There is not enough evidence at the 1% level of significance to reject the travel association's claim that the mean daily meal cost for two adults traveling together on vacation in San Francisco is \$105.

29. (a) H_0: $\mu \geq 32$; H_a: $\mu < 32$ (claim)

(b) 0.0344   (c) Reject H_0.

(d) There is enough evidence at the 5% level of significance to support the brochure's claim that the mean class size for full-time faculty is fewer than 32 students.

31. Hypothesis test results:

μ : population mean

$H_0 : \mu = 75$

$H_A : \mu > 75$

Mean	Sample Mean	Std. Err.	DF	T-Stat	P-value
μ	73.6	0.62757164	25	-2.2308211	0.9825

$P = 0.9825 > 0.05$, so fail to reject H_0. There is not enough evidence at the 5% level of significance to reject the claim.

33. Hypothesis test results:

μ : population mean

$H_0 : \mu = 188$

$H_A : \mu < 188$

Mean	Sample Mean	Std. Err.	DF	T-Stat	P-value
μ	186	4	8	-0.5	0.3153

$P = 0.3153 > 0.05$, so fail to reject H_0. There is not enough evidence at the 5% level of significance to support the claim.

35. Because the P-value $= 0.0748 > 0.05$, fail to reject H_0.

37. Use the t-distribution because the population is normal, $n < 30$, and σ is unknown.

Fail to reject H_0. There is not enough evidence at the 5% level of significance to reject the car company's claim that the mean gas mileage for the luxury sedan is at least 23 miles per gallon.

39. More likely; For degrees of freedom less than 30, the tails of a t-distribution curve are thicker than those of a standard normal distribution curve. So, if you incorrectly use a standard normal sampling distribution instead of a t-sampling distribution, the area under the curve at the tails will be smaller than what it would be for the t-test, meaning the critical value(s) will lie closer to the mean. This makes it more likely for the test statistic to be in the rejection region(s). This result is the same regardless of whether the test is left-tailed, right-tailed, or two-tailed; in each case, the tail thickness affects the location of the critical value(s).

Section 7.3 Activity   *(page 397)*

1–3. Answers will vary.

Section 7.4   *(page 401)*

1. If $np \geq 5$ and $nq \geq 5$, the normal distribution can be used.

3. Cannot use normal distribution.

5. Can use normal distribution.

Fail to reject H_0. There is not enough evidence at the 5% level of significance to support the claim.

7. Can use normal distribution.

Fail to reject H_0. There is not enough evidence at the 5% level of significance to reject the claim.

9. (a) H_0: $p \geq 0.25$; H_a: $p < 0.25$ (claim)

(b) $z_0 = -1.645$; Rejection region: $z < -1.645$

(c) -2.12 (d) Reject H_0.

(e) There is enough evidence at the 5% level of significance to support the researcher's claim that less than 25% of U.S. adults are smokers.

11. (a) H_0: $p \leq 0.50$ (claim); H_a: $p > 0.50$

(b) $z_0 = 2.33$; Rejection region: $z > 2.33$

(c) 1.96   (d) Fail to reject H_0.

(e) There is not enough evidence at the 1% level of significance to reject the research center's claim that at most 50% of people believe that drivers should be allowed to use cellular phones with hands-free devices while driving.

13. (a) H_0: $p \leq 0.75$; H_a: $p > 0.75$ (claim)

(b) $z_0 = 1.28$; Rejection region: $z > 1.28$

(c) 1.98   (d) Reject H_0.

(e) There is enough evidence at the 10% level of significance to support the research center's claim that more than 75% of females ages 20–29 are taller than 62 inches.

15. (a) H_0: $p \geq 0.35$; H_a: $p < 0.35$ (claim)

(b) $z_0 = -1.28$; Rejection region: $z < -1.28$

(c) 1.68   (d) Fail to reject H_0.

(e) There is not enough evidence at the 10% level of significance to support the humane society's claim that less than 35% of U.S. households own a dog.

17. Fail to reject H_0. There is not enough evidence at the 5% level of significance to reject the claim that at least 52% of adults are more likely to buy a product when there are free samples.

19. (a) H_0: $p \geq 0.35$; H_a: $p < 0.35$ (claim)

(b) $z_0 = -1.28$; Rejection region: $z < -1.28$

(c) 1.68   (d) Fail to reject H_0.

(e) There is not enough evidence at the 10% level of significance to support the humane society's claim that less than 35% of U.S. households own a dog.

The results are the same.

Section 7.4 Activity   *(page 403)*

1–2. Answers will vary.

Section 7.5   *(page 410)*

1. Specify the level of significance α. Determine the degrees of freedom. Determine the critical values using the χ^2- distribution. For a right-tailed test, use the value that corresponds to d.f. and α; for a left-tailed test, use the value that corresponds to d.f. and $1 - \alpha$; for a two-tailed test, use the values that correspond to d.f. and $\frac{1}{2}\alpha$, and d.f. and $1 - \frac{1}{2}\alpha$.

3. The requirement of a normal distribution is more important when testing a standard deviation than when testing a mean. If the population is not normal, the results of a χ^2-test can be misleading because the χ^2-test is not as robust as the tests for the population mean.

5. 38.885   **7.** 0.872   **9.** 60.391, 101.879

11. (a) Fail to reject H_0.   (b) Fail to reject H_0.

(c) Fail to reject H_0.   (d) Reject H_0.

13. (a) Fail to reject H_0.   (b) Reject H_0.

(c) Reject H_0.   (d) Fail to reject H_0.

15. Fail to reject H_0. There is not enough evidence at the 5% level of significance to reject the claim.

17. Reject H_0. There is enough evidence at the 10% level of significance to reject the claim.

19. (a) H_0: $\sigma^2 = 1.25$ (claim); H_a: $\sigma^2 \neq 1.25$

(b) $\chi_L^2 = 10.283$, $\chi_R^2 = 35.479$;

Rejection regions: $\chi^2 < 10.283$, $\chi^2 > 35.479$

(c) 22.68   (d) Fail to reject H_0.

(e) There is not enough evidence at the 5% level of significance to reject the manufacturer's claim that the variance of the number of grams of carbohydrates in servings of its tortilla chips is 1.25.

21. (a) H_0: $\sigma \geq 36$; H_a: $\sigma < 36$ (claim)

(b) $\chi_0^2 = 13.240$; Rejection region: $\chi^2 < 13.240$

(c) 18.076   (d) Fail to reject H_0.

(e) There is not enough evidence at the 10% level of significance to support the test administrator's claim that the standard deviation for eighth graders on the examination is less than 36 points.

23. (a) H_0: $\sigma \leq 25$ (claim); H_a: $\sigma > 25$

(b) $\chi_0^2 = 36.741$; Rejection region: $\chi^2 > 36.741$

(c) 41.515   (d) Reject H_0.

(e) There is enough evidence at the 10% level of significance to reject the weather service's claim that the standard deviation of the number of fatalities per year from tornadoes is no more than 25.

25. (a) H_0: $\sigma \geq \$3500$; H_a: $\sigma < \$3500$ (claim)

(b) $\chi_0^2 = 18.114$; Rejection region: $\chi^2 < 18.114$

(c) 37.051   (d) Fail to reject H_0.

(e) There is not enough evidence at the 10% level of significance to support the insurance agent's claim that the standard deviation of the total charges for patients involved in a crash in which the vehicle struck a construction barricade is less than $3500.

27. (a) H_0: $\sigma \leq 6100$; H_a: $\sigma > 6100$ (claim)

(b) $\chi_0^2 = 27.587$; Rejection region: $\chi^2 > 27.587$

(c) 27.897   (d) Reject H_0.

(e) There is enough evidence at the 5% level of significance to support the claim that the standard deviation of the annual salaries of environmental engineers is greater than $6100.

29. Hypothesis test results:

σ^2 : variance of Variable

$H_0 : \sigma^2 = 9$

$H_A : \sigma^2 < 9$

Variance	Sample Var.	DF	Chi-Square Stat	P-value
σ^2	2.03	9	2.03	0.009

Reject H_0. There is enough evidence at the 1% level of significance to reject the claim.

31. $\sigma^2 = 4.5^2 = 20.25$

$s^2 = 5.8^2 = 33.64$

Hypothesis test results:

σ^2 : variance of Variable

$H_0 : \sigma^2 = 20.25$

$H_A : \sigma^2 > 20.25$

Variance	Sample Var.	DF	Chi-Square Stat	P-value
σ^2	33.64	14	23.257284	0.0562

Fail to reject H_0. There is not enough evidence at the 5% level of significance to support the claim.

33. P-value $= 0.9059$ **35.** P-value $= 0.0462$

Fail to reject H_0.   Reject H_0.

Uses and Abuses for Chapter 7   *(page 413)*

1. Answers will vary.

2. H_0: $p = 0.73$; Answers will vary.

3. Answers will vary.

4. Answers will vary.

Review Answers for Chapter 7   *(page 417)*

1. H_0: $\mu \leq 375$ (claim); H_a: $\mu > 375$

3. H_0: $p \geq 0.205$; H_a: $p < 0.205$ (claim)

5. H_0: $\sigma \leq 1.9$; H_a: $\sigma > 1.9$ (claim)

7. (a) H_0: $p = 0.71$ (claim); H_a: $p \neq 0.71$

(b) A type I error will occur if the actual proportion of Americans who support plans to order deep cuts in executive compensation at companies that have received federal bailout funds is 71%, but you reject H_0: $p = 0.71$.

A type II error will occur if the actual proportion is not 71%, but you fail to reject H_0: $p = 0.71$.

(c) Two-tailed because the alternative hypothesis contains $\neq$.

(d) There is enough evidence to reject the news outlet's claim that the proportion of Americans who support plans to order deep cuts in executive compensation at companies that have received federal bailout funds is 71%.

(e) There is not enough evidence to reject the news outlet's claim that the proportion of Americans who support plans to order deep cuts in executive compensation at companies that have received federal bailout funds is 71%.

9. (a) H_0: $\sigma \leq 50$ (claim); H_a: $\sigma > 50$

(b) A type I error will occur if the actual standard deviation of the sodium content in one serving of a certain soup is no more than 50 milligrams, but you reject H_0: $\sigma \leq 50$.

A type II error will occur if the actual standard deviation of the sodium content in one serving of a certain soup is more than 50 milligrams, but you fail to reject H_0: $\sigma \leq 50$.

(c) Right-tailed because the alternative hypothesis contains $>$.

(d) There is enough evidence to reject the soup maker's claim that the standard deviation of the sodium content in one serving of a certain soup is no more than 50 milligrams.

(e) There is not enough evidence to reject the soup maker's claim that the standard deviation of the sodium content in one serving of a certain soup is no more than 50 milligrams.

11. 0.1736; Fail to reject H_0.

13. H_0: $\mu \leq 0.05$ (claim); H_a: $\mu > 0.05$

$z = 2.20$; P-value $= 0.0139$

$\alpha = 0.10 \Rightarrow$ Reject H_0.

$\alpha = 0.05 \Rightarrow$ Reject H_0.

$\alpha = 0.01 \Rightarrow$ Fail to reject H_0.

15. -2.05

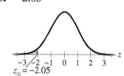

17. 1.96

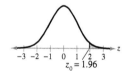

19. Fail to reject H_0 because $-1.645 < z < 1.645$.

21. Fail to reject H_0 because $-1.645 < z < 1.645$.

23. Reject H_0. There is enough evidence at the 5% level of significance to reject the claim.

25. Fail to reject H_0. There is not enough evidence at the 1% level of significance to support the claim.

27. Fail to reject H_0. There is not enough evidence at the 1% level of significance to reject the U.S. Department of Agriculture's claim that the mean cost of raising a child from birth to age 2 by husband-wife families in rural areas is $10,380.

29. $-2.093, 2.093$ **31.** -2.977

33. Fail to reject H_0. There is not enough evidence at the 5% level of significance to support the claim.

35. Fail to reject H_0. There is not enough evidence at the 10% level of significance to reject the claim.

37. Reject H_0. There is enough evidence at the 1% level of significance to reject the claim.

39. Fail to reject H_0. There is not enough evidence at the 10% level of significance to reject the advertisement's claim that the mean monthly cost of joining a health club is $25.

41. There is not enough evidence at the 1% level of significance to reject the education publication's claim that the mean expenditure per student in public elementary and secondary schools is at least $10,200.

43. Can use normal distribution.

Fail to reject H_0. There is not enough evidence at the 5% level of significance to reject the claim.

45. Can use normal distribution.

Fail to reject H_0. There is not enough evidence at the 8% level of significance to support the claim.

47. Cannot use normal distribution.

49. Can use normal distribution.

Fail to reject H_0. There is not enough evidence at the 2% level of significance to support the claim.

51. Reject H_0. There is enough evidence at the 2% level of significance to support the polling agency's claim that over 16% of U.S. adults are without health care coverage.

53. 30.144   **55.** 63.167

57. Reject H_0. There is enough evidence at the 10% level of significance to support the claim.

59. Fail to reject H_0. There is not enough evidence at the 5% level of significance to reject the claim.

61. Reject H_0. There is enough evidence at the 0.5% level of significance to reject the bolt manufacturer's claim that the variance is at most 0.01.

63. You can reject H_0 at the 5% level of significance because $\chi^2 = 43.94 > 41.923$.

Chapter Quiz for Chapter 7   (page 421)

1. (a) $H_0: \mu \geq 170$ (claim); $H_a: \mu < 170$

(b) One-tailed because the alternative hypothesis contains $<$; z-test because $n \geq 30$

(c) $z_0 = -1.88$; Rejection region: $z < -1.88$

(d) -2.59

(e) Reject H_0.

(f) There is enough evidence at the 3% level of significance to reject the service's claim that the mean consumption of vegetables and melons by people in the United States is at least 170 pounds per person.

2. (a) $H_0: \mu \geq 7.25$ (claim); $H_a: \mu < 7.25$

(b) One-tailed because the alternative hypothesis contains $<$; t-test because $n < 30$, σ is unknown, and the population is normally distributed

(c) $t_0 = -1.796$; Rejection region: $t < -1.796$

(d) -1.283

(e) Fail to reject H_0.

(f) There is not enough evidence at the 5% level of significance to reject the company's claim that the mean hat size for a male is at least 7.25.

3. (a) $H_0: p \leq 0.10$ (claim); $H_a: p > 0.10$

(b) One-tailed because the alternative hypothesis contains $>$; z-test because $np > 5$ and $nq > 5$

(c) $z_0 = 1.75$; Rejection region: $z > 1.75$

(d) 0.75

(e) Fail to reject H_0.

(f) There is not enough evidence at the 4% level of significance to reject the microwave oven maker's claim that no more than 10% of its microwaves need repair during the first 5 years of use.

4. (a) $H_0: \sigma = 112$ (claim); $H_a: \sigma \neq 112$

(b) Two-tailed because the alternative hypothesis contains $\neq$; χ^2-test because the test is for a standard deviation and the population is normally distributed

(c) $\chi_L^2 = 9.390$, $\chi_R^2 = 28.869$; Rejection regions: $\chi^2 < 9.390$, $\chi^2 > 28.869$

(d) 29.343

(e) Reject H_0.

(f) There is enough evidence at the 10% level of significance to reject the state school administrator's claim that the standard deviation of SAT critical reading test scores is 112.

5. (a) $H_0: \mu = \$62,569$ (claim); $H_a: \mu \neq \$62,569$

(b) Two-tailed because the alternative hypothesis contains $\neq$; t-test because $n < 30$, σ is unknown, and the population is normally distributed

(c) Not necessary   (d) -2.175; 0.0473

(e) Reject H_0.

(f) There is enough evidence at the 5% level of significance to reject the agency's claim that the mean income for full-time workers ages 25 to 34 with a master's degree is $62,569.

6. (a) $H_0: \mu = \$201$ (claim); $H_a: \mu \neq \$201$

(b) Two-tailed because the alternative hypothesis contains $\neq$; z-test because $n \geq 30$

(c) Not necessary

(d) 0.0030 (*Tech:* 0.0031)

(e) Reject H_0.

(f) There is enough evidence at the 5% level of significance to reject the tourist agency's claim that the mean daily cost of meals and lodging for a family of 4 traveling in the state of Kansas is $201.

1. (a)–(c) Answers will vary.

2. Fail to reject H_0. There is not enough evidence at the 5% level of significance to support PepsiCo's claim that more than 50% of cola drinkers prefer Pepsi® over Coca-Cola®.

3. Knowing the brand may influence participants' decisions.

4. (a)–(c) Answers will vary.

CHAPTER 8

Section 8.1 *(page 434)*

1. Two samples are dependent if each member of one sample corresponds to a member of the other sample. Example: The weights of 22 people before starting an exercise program and the weights of the same 22 people 6 weeks after starting the exercise program.

Two samples are independent if the sample selected from one population is not related to the sample selected from the other population. Example: The weights of 25 cats and the weights of 20 dogs.

3. Use P-values.

5. Independent because different students were sampled.

7. Dependent because the same football players were sampled.

9. Independent because different boats were sampled.

11. Dependent because the same tire sets were sampled.

13. (a) 2   (b) 2.95   (c) In the rejection region.

(d) Reject H_0. There is enough evidence at the 1% level of significance to reject the claim.

15. (a) 3   (b) 0.18   (c) Not in the rejection region.

(d) Fail to reject H_0. There is not enough evidence at the 5% level of significance to support the claim.

17. Fail to reject H_0. There is not enough evidence at the 1% level of significance to support the claim.

19. Reject H_0.

21. (a) The claim is "the mean braking distances are different for the two types of tires."

$H_0: \mu_1 = \mu_2$; $H_a: \mu_1 \neq \mu_2$ (claim)

(b) $-z_0 = -1.645$, $z_0 = 1.645$;

Rejection regions: $z < -1.645$, $z > 1.645$

(c) -2.786 (d) Reject H_0.

(e) There is enough evidence at the 10% level of significance to support the safety engineer's claim that the mean braking distances are different for the two types of tires.

23. (a) The claim is "Region A's average wind speed is greater than Region B's."

$H_0: \mu_1 \leq \mu_2$; $H_a: \mu_1 > \mu_2$ (claim)

(b) $z_0 = 1.645$; Rejection region: $z > 1.645$

(c) 1.53   (d) Fail to reject H_0.

(e) There is not enough evidence at the 5% level of significance to conclude that Region A's average wind speed is greater than Region B's.

25. (a) The claim is "male and female high school students have equal ACT scores."

$H_0: \mu_1 = \mu_2$ (claim); $H_a: \mu_1 \neq \mu_2$

(b) $-z_0 = -2.575$, $z_0 = 2.575$;

Rejection regions: $z < -2.575$, $z > 2.575$

(c) 0.202   (d) Fail to reject H_0.

(e) There is not enough evidence at the 1% level of significance to reject the claim that male and female high school students have equal ACT scores.

27. (a) The claim is "the average home sales price in Dallas, Texas is the same as in Austin, Texas."

$H_0: \mu_1 = \mu_2$ (claim); $H_a: \mu_1 \neq \mu_2$

(b) $-z_0 = -1.645$, $z_0 = 1.645$;

Rejection regions: $z < -1.645$, $z > 1.645$

(c) -1.30 (d) Fail to reject H_0.

(e) There is not enough evidence at the 10% level of significance to reject the real estate agency's claim that the average home sales price in Dallas, Texas is the same as in Austin, Texas.

29. (a) The claim is "the average home sales price in Dallas, Texas is the same as in Austin, Texas."

$H_0: \mu_1 = \mu_2$ (claim); $H_a: \mu_1 \neq \mu_2$

(b) $-z_0 = -1.645$, $z_0 = 1.645$;

Rejection regions: $z < -1.645$, $z > 1.645$

(c) 1.86   (d) Reject H_0.

(e) There is enough evidence at the 10% level of significance to reject the real estate agency's claim that the average home sales price in Dallas, Texas is the same as in Austin, Texas.

The new samples do lead to a different conclusion.

31. (a) The claim is "children ages 6–17 spent more time watching television in 1981 than children ages 6–17 do today."

$H_0: \mu_1 \leq \mu_2$; $H_a: \mu_1 > \mu_2$ (claim)

(b) $z_0 = 1.96$; Rejection region: $z > 1.96$

(c) 3.01   (d) Reject H_0.

(e) There is enough evidence at the 2.5% level of significance to support the sociologist's claim that children ages 6–17 spent more time watching television in 1981 than children ages 6–17 do today.

33. (a) The claim is "there is no difference in the mean washer diameter manufactured by two different methods."

$H_0: \mu_1 = \mu_2$ (claim); $H_a: \mu_1 \neq \mu_2$

(b) $-z_0 = -2.575$, $z_0 = 2.575$;

Rejection regions: $z < -2.575$, $z > 2.575$

(c) 64.978   (d) Reject H_0.

(e) There is enough evidence at the 1% level of significance to reject the production engineer's claim that there is no difference in the mean washer diameter manufactured by two different methods.

35. They are equivalent through algebraic manipulation of the equation.

$$\mu_1 = \mu_2 \Rightarrow \mu_1 - \mu_2 = 0$$

37. Hypothesis test results:

μ_1 : mean of population 1 (Std. Dev. = 5.4)

μ_2 : mean of population 2 (Std. Dev. = 7.5)

$\mu_1 - \mu_2$: mean difference

$H_0 : \mu_1 - \mu_2 = 0$

$H_A : \mu_1 - \mu_2 \neq 0$

Difference	n_1	n_2	Sample Mean
$\mu_1 - \mu_2$	50	45	4

Std. Err.	Z-Stat	P-value
1.3539572	2.9543033	0.0031

$P = 0.0031 < 0.01$, so reject H_0.

There is enough evidence at the 1% level of significance to support the claim.

39. Hypothesis test results:

μ_1 : mean of population 1 (Std. Dev. = 0.92)

μ_2 : mean of population 2 (Std. Dev. = 0.73)

$\mu_1 - \mu_2$: mean difference

$H_0 : \mu_1 - \mu_2 = 0$

$H_A : \mu_1 - \mu_2 < 0$

Difference	n_1	n_2	Sample Mean
$\mu_1 - \mu_2$	35	40	−0.32

Std. Err.	Z-Stat	P-value
0.193663	−1.6523548	0.0492

$P = 0.0492 < 0.05$, so reject H_0.

There is enough evidence at the 5% level of significance to reject the claim.

41. $H_0: \mu_1 - \mu_2 = -9$ (claim); $H_a: \mu_1 - \mu_2 \neq -9$

Fail to reject H_0.

There is not enough evidence at the 1% level of significance to reject the claim that children spend 9 hours a week more in day care or preschool today than in 1981.

43. $H_0: \mu_1 - \mu_2 \leq 10{,}000$; $H_a: \mu_1 - \mu_2 > 10{,}000$ (claim)

Reject H_0. There is enough evidence at the 5% level of significance to support the claim that the difference in mean annual salaries of microbiologists in Maryland and California is more than $10,000.

45. $-3.6 < \mu_1 - \mu_2 < -0.2$

47. $H_0: \mu_1 \geq \mu_2$; $H_a: \mu_1 < \mu_2$ (claim)

Reject H_0. There is enough evidence at the 5% level of significance to support the claim. You should recommend the DASH diet and exercise program over the traditional diet and exercise program because the mean systolic blood pressure was significantly lower in the DASH program.

49. The 95% CI for $\mu_1 - \mu_2$ in Exercise 45 contained only values less than 0 and, as found in Exercise 47, there was enough evidence at the 5% level of significance to support the claim.

If the CI for $\mu_1 - \mu_2$ contains only negative numbers, you reject H_0 because the null hypothesis states that $\mu_1 - \mu_2$ is greater than or equal to 0.

Section 8.2 *(page 446)*

1. (1) The samples must be randomly selected.

(2) The samples must be independent.

(3) Each population must have a normal distribution.

3. (a) $-t_0 = -1.714$, $t_0 = 1.714$

(b) $-t_0 = -1.812$, $t_0 = 1.812$

5. (a) $t_0 = -1.746$

(b) $t_0 = -1.943$

7. (a) $t_0 = 1.729$

(b) $t_0 = 1.895$

9. (a) -1.8 (b) -1.70

(c) Not in the rejection region.

(d) Fail to reject H_0.

11. (a) 105   (b) 2.05

(c) In the rejection region.

(d) Reject H_0.

13. (a) The claim is "the mean annual costs of routine veterinarian visits for dogs and cats are the same."

$H_0: \mu_1 = \mu_2$ (claim); $H_a: \mu_1 \neq \mu_2$

(b) $-t_0 = -1.943$, $t_0 = 1.943$;

Rejection regions: $t < -1.943$, $t > 1.943$

(c) 1.90   (d) Fail to reject H_0.

(e) There is not enough evidence at the 10% level of significance to reject the pet association's claim that the mean annual costs of routine veterinarian visits for dogs and cats are the same.

15. (a) The claim is "the mean bumper repair cost is less for mini cars than for midsize cars."

$H_0: \mu_1 \geq \mu_2$; $H_a: \mu_1 < \mu_2$ (claim)

(b) $t_0 = -1.325$; Rejection region: $t < -1.325$

(c) -0.93 (d) Fail to reject H_0.

(e) There is not enough evidence at the 10% level of significance to support the claim that the mean bumper repair cost is less for mini cars than for midsize cars.

17. (a) The claim is "the mean household income is greater in Allegheny County than it is in Erie County."

$H_0: \mu_1 \leq \mu_2$; $H_a: \mu_1 > \mu_2$ (claim)

(b) $t_0 = 1.761$; Rejection region: $t > 1.761$

(c) 1.99   (d) Reject H_0.

(e) There is enough evidence at the 5% level of significance to support the personnel director's claim that the mean household income is greater in Allegheny County than it is in Erie County.

19. (a) The claim is "the new treatment makes a difference in the tensile strength of steel bars."

$H_0: \mu_1 = \mu_2$; $H_a: \mu_1 \neq \mu_2$ (claim)

(b) $-t_0 = -2.831$, $t_0 = 2.831$;

Rejection regions: $t < -2.831$, $t > 2.831$

(c) -2.76 **(d)** Fail to reject H_0.

(e) There is not enough evidence at the 1% level of significance to support the claim that the new treatment makes a difference in the tensile strength of steel bars.

21. (a) The claim is "the new method of teaching reading produces higher reading test scores than the old method."

$H_0: \mu_1 \geq \mu_2$; $H_a: \mu_1 < \mu_2$ (claim)

(b) $t_0 = -1.282$; Rejection region: $t < -1.282$

(c) -4.295 **(d)** Reject H_0.

(e) There is enough evidence at the 10% level of significance to support the claim that the new method of teaching reading produces higher reading test scores than the old method and to recommend changing to the new method.

23. Hypothesis test results:

μ_1 : mean of population 1

μ_2 : mean of population 2

$\mu_1 - \mu_2$: mean difference

$H_0: \mu_1 - \mu_2 = 0$

$H_A: \mu_1 - \mu_2 > 0$

(with pooled variances)

Difference	Sample Mean	Std. Err.
$\mu_1 - \mu_2$	-8	16.985794

DF	T-Stat	P-value
22	-0.47098184	0.6789

$P = 0.6789 > 0.10$, so fail to reject H_0.

There is not enough evidence at the 10% level of significance to support the claim.

25. Hypothesis test results:

μ_1 : mean of population 1

μ_2 : mean of population 2

$\mu_1 - \mu_2$: mean difference

$H_0: \mu_1 - \mu_2 = 0$

$H_A: \mu_1 - \mu_2 < 0$

(without pooled variances)

Difference	Sample Mean	Std. Err.
$\mu_1 - \mu_2$	-43	28.12301

DF	T-Stat	P-value
18.990595	-1.5289971	0.0714

$P = 0.0714 > 0.05$, so fail to reject H_0.

There is not enough evidence at the 5% level of significance to reject the claim.

27. $45 < \mu_1 - \mu_2 < 307$ **29.** $11 < \mu_1 - \mu_2 < 35$

Section 8.3 *(page 456)*

1. (1) Each sample must be randomly selected.

(2) Each member of the first sample must be paired with a member of the second sample.

(3) Both populations must be normally distributed.

3. Left-tailed test; Fail to reject H_0.

5. Right-tailed test; Reject H_0.

7. Left-tailed test; Reject H_0.

9. (a) The claim is "a grammar seminar will help students reduce the number of grammatical errors."

$H_0: \mu_d \leq 0$; $H_a: \mu_d > 0$ (claim)

(b) $t_0 = 3.143$; Rejection region: $t > 3.143$

(c) $\bar{d} \approx 3.143$; $s_d \approx 2.035$

(d) 4.085 **(e)** Reject H_0.

(f) There is enough evidence at the 1% level of significance to support the teacher's claim that a grammar seminar will help students reduce the number of grammatical errors.

11. (a) The claim is "a particular exercise program will help participants lose weight after one month."

$H_0: \mu_d \leq 0$; $H_a: \mu_d > 0$ (claim)

(b) $t_0 = 1.363$; Rejection region: $t > 1.363$

(c) $\bar{d} = 3.75$; $s_d \approx 7.841$

(d) 1.657 **(e)** Reject H_0.

(f) There is enough evidence at the 10% level of significance to support the nutritionist's claim that the exercise program helps participants lose weight after one month.

13. (a) The claim is "soft tissue therapy and spinal manipulation help to reduce the length of time patients suffer from headaches."

$H_0: \mu_d \leq 0$; $H_a: \mu_d > 0$ (claim)

(b) $t_0 = 2.764$; Rejection region: $t > 2.764$

(c) $\bar{d} \approx 1.255$; $s_d \approx 0.441$

(d) 9.429 **(e)** Reject H_0.

(f) There is enough evidence at the 1% level of significance to support the physical therapist's claim that soft tissue therapy and spinal manipulation help reduce the length of time patients suffer from headaches.

15. (a) The claim is "the new drug reduces systolic blood pressure."

$H_0: \mu_d \leq 0$; $H_a: \mu_d > 0$ (claim)

(b) $t_0 = 1.895$; Rejection region: $t > 1.895$

(c) $\bar{d} = 14.75$; $s_d \approx 6.861$

(d) 6.081 **(e)** Reject H_0.

(f) There is enough evidence at the 5% level of significance to support the pharmaceutical company's claim that its new drug reduces systolic blood pressure.

17. (a) The claim is "the product ratings have changed from last year to this year."

$H_0: \mu_d = 0$; $H_a: \mu_d \neq 0$ (claim)

(b) $-t_0 = -2.365$, $t_0 = 2.365$

Rejection regions: $t < -2.365$, $t > 2.365$

(c) $\bar{d} = -1$; $s_d \approx 1.309$

(d) -2.160 (e) Fail to reject H_0.

(f) There is not enough evidence at the 5% level of significance to support the claim that the product ratings have changed from last year to this year.

19. Hypothesis test results:

$\mu_1 - \mu_2$: mean of the paired difference between Cholesterol (before) and Cholesterol (after)

$H_0 : \mu_1 - \mu_2 = 0$

$H_A : \mu_1 - \mu_2 > 0$

Difference	Sample Diff.
Cholesterol (before) − Cholesterol (after)	2.857143

Std. Err.	DF	T-Stat	P-value
1.6822401	6	1.6984155	0.0702

$P = 0.0702 > 0.05$, so fail to reject H_0.

There is not enough evidence at the 5% level of significance to support the claim that the new cereal lowers total blood cholesterol levels.

21. Yes; $P \approx 0.0003 < 0.05$, so you reject H_0.

23. $-1.76 < \mu_d < -1.29$

Section 8.4 *(page 465)*

1. (1) The samples must be randomly selected.

(2) The samples must be independent.

(3) $n_1\bar{p} \geq 5$, $n_1\bar{q} \geq 5$, $n_2\bar{p} \geq 5$, and $n_2\bar{q} \geq 5$

3. Can use normal sampling distribution; Fail to reject H_0.

5. Can use normal sampling distribution; Reject H_0.

7. Can use normal sampling distribution; Fail to reject H_0.

9. (a) The claim is "there is a difference in the proportion of subjects who feel all or mostly better after 4 weeks between subjects who used magnetic insoles and subjects who used nonmagnetic insoles."

$H_0: p_1 = p_2$; $H_a: p_1 \neq p_2$ (claim)

(b) $-z_0 = -2.575$, $z_0 = 2.575$;

Rejection regions: $z < -2.575$, $z > 2.575$

(c) -1.24 (d) Fail to reject H_0.

(e) There is not enough evidence at the 1% level of significance to support the claim that there is a difference in the proportion of subjects who feel all or mostly better after 4 weeks between subjects who used magnetic insoles and subjects who used nonmagnetic insoles.

11. (a) The claim is "the proportion of males who enrolled in college is less than the proportion of females who enrolled in college."

$H_0: p_1 \geq p_2$; $H_a: p_1 < p_2$ (claim)

(b) $z_0 = -1.645$; Rejection region: $z < -1.645$

(c) -4.22 (d) Reject H_0.

(e) There is enough evidence at the 5% level of significance to support the claim that the proportion of males who enrolled in college is less than the proportion of females who enrolled in college.

13. (a) The claim is "the proportion of subjects who are pain-free is the same for the two groups."

$H_0: p_1 = p_2$ (claim); $H_a: p_1 \neq p_2$

(b) $-z_0 = -1.96$, $z_0 = 1.96$;

Rejection regions: $z < -1.96$, $z > 1.96$

(c) 5.62 (*Tech:* 5.58)   (d) Reject H_0.

(e) There is enough evidence at the 5% level of significance to reject the claim that the proportion of subjects who are pain-free is the same for the two groups.

15. (a) The claim is "the proportion of motorcyclists who wear a helmet is now greater."

$H_0: p_1 \leq p_2$; $H_a: p_1 > p_2$ (claim)

(b) $z_0 = 1.645$; Rejection region: $z > 1.645$

(c) 1.37   (d) Fail to reject H_0.

(e) There is not enough evidence at the 5% level of significance to support the claim that the proportion of motorcyclists who wear a helmet is now greater.

17. (a) The claim is "the proportion of Internet users is the same for the two age groups."

$H_0: p_1 = p_2$ (claim); $H_a: p_1 \neq p_2$

(b) $-z_0 = -2.575$, $z_0 = 2.575$;

Rejection regions: $z < -2.575$, $z > 2.575$

(c) 5.31   (d) Reject H_0.

(e) There is enough evidence at the 1% level of significance to reject the claim that the proportion of Internet users is the same for the two age groups.

19. There is enough evidence at the 5% level of significance to reject the claim that the proportion of customers who wait 20 minutes or less is the same at the Fairfax North and Fairfax South offices.

21. There is enough evidence at the 10% level of significance to support the claim that the proportion of customers who wait 20 minutes or less at the Roanoke office is less than the proportion of customers who wait 20 minutes or less at the Staunton office.

23. No; When $\alpha = 0.01$, the rejection region becomes $z < -2.33$. Because $-2.02 > -2.33$, you fail to reject H_0. There is not enough evidence at the 1% level of significance to support the claim that the proportion of customers who wait 20 minutes or less at the Roanoke office is less than the proportion of customers who wait 20 minutes or less at the Staunton office.

25. Hypothesis test results:

p_1 : proportion of successes for population 1

p_2 : proportion of successes for population 2

$p_1 - p_2$: difference in proportions

$H_0 : p_1 - p_2 = 0$

$H_A : p_1 - p_2 > 0$

Difference	Count1	Total1	Count2	Total2
$p_1 - p_2$	7501	13300	8120	14500

Sample Diff.	Std. Err.	Z-Stat	P-value
0.0039849626	0.0059570055	0.66895396	0.2518

$P = 0.2518 > 0.05$, so fail to reject H_0.

There is not enough evidence at the 5% level of significance to support the claim that the proportion of men ages 18 to 24 living in their parents' homes was greater in 2000 than in 2009.

27. Hypothesis test results:

p_1 : proportion of successes for population 1

p_2 : proportion of successes for population 2

$p_1 - p_2$: difference in proportions

$H_0 : p_1 - p_2 = 0$

$H_A : p_1 - p_2 \neq 0$

Difference	Count1	Total1	Count2	Total2
$p_1 - p_2$	7501	13300	5610	13200

Sample Diff.	Std. Err.	Z-Stat	P-value
0.13898496	0.006142657	22.626196	<0.0001

$P < 0.0001 < 0.01$, so reject H_0.

There is enough evidence at the 1% level of significance to reject the claim that the proportion of 18- to 24-year-olds living in their parents' homes in 2000 was the same for men and women.

29. $-0.028 < p_1 - p_2 < -0.012$

Uses and Abuses for Chapter 8   *(page 469)*

1. Answers will vary.

2. Blind: The patients do not know which group (medicine or placebo) they belong to.

Double Blind: Both the researcher and patient do not know which group (medicine or placebo) that the patient belongs to.

Review Answers for Chapter 8   *(page 471)*

1. Dependent because the same cities were sampled.

3. Fail to reject H_0. There is not enough evidence at the 5% level of significance to reject the claim.

5. Reject H_0. There is enough evidence at the 10% level of significance to support the claim.

7. (a) The claim is "the Wendy's fish sandwich has less sodium than the Long John Silver's fish sandwich."

$H_0: \mu_1 \geq \mu_2$; $H_a: \mu_1 < \mu_2$ (claim)

(b) $-z_0 = -1.645$; Rejection region: $z < -1.645$

(c) -9.20 (d) Reject H_0.

(e) There is enough evidence at the 5% level of significance to support the claim that the Wendy's fish sandwich has less sodium than the Long John Silver's fish sandwich.

9. Yes; The new rejection region is $z < -2.33$, which contains $z = -9.20$, so you still reject H_0.

11. Reject H_0. There is enough evidence at the 5% level of significance to reject the claim.

13. Fail to reject H_0. There is not enough evidence at the 5% level of significance to reject the claim.

15. Reject H_0. There is enough evidence at the 1% level of significance to support the claim.

17. (a) The claim is "third graders taught with the directed reading activities scored higher than those taught without the activities."

$H_0: \mu_1 \leq \mu_2$; $H_a: \mu_1 > \mu_2$ (claim)

(b) $t_0 = 1.645$; Rejection region: $t > 1.645$

(c) 2.267   (d) Reject H_0.

(e) There is enough evidence at the 5% level of significance to support the claim that third graders taught with the directed reading activities scored higher than those taught without the activities.

19. Two-tailed test; Reject H_0.

21. Right-tailed test; Reject H_0.

23. (a) The claim is "the men's systolic blood pressure decreased."

$H_0: \mu_d \leq 0$; $H_a: \mu_d > 0$ (claim)

(b) $t_0 = 1.383$; Rejection region: $t > 1.383$

(c) $\bar{d} = 5$; $s_d \approx 8.743$ (d) 1.808   (e) Reject H_0.

(f) There is enough evidence at the 10% level of significance to support the claim that the men's systolic blood pressure decreased.

25. Can use normal sampling distribution; Fail to reject H_0.

27. Can use normal sampling distribution; Reject H_0.

29. (a) The claim is "the proportions of U.S. adults who considered the amount of federal income tax they had to pay to be too high were the same for the two years."

$H_0: p_1 = p_2$ (claim); $H_a: p_1 \neq p_2$

(b) $-z_0 = -2.575$, $z_0 = 2.575$;

Rejection regions: $z < -2.575$, $z > 2.575$

(c) 2.65   (d) Reject H_0.

(e) There is enough evidence at the 1% level of significance to reject the claim that the proportions of U.S. adults who considered the amount of federal income tax they had to pay to be too high were the same for the two years.

31. Yes; When $\alpha = 0.05$, the rejection regions become $z < -1.96$ and $z > 1.96$. Because $2.65 > 1.96$, you still reject H_0. There is enough evidence at the 5% level of significance to reject the claim that the proportions of U.S. adults who considered the amount of federal income tax they had to pay to be too high were the same for the two years.

Chapter Quiz for Chapter 8 *(page 475)*

1. (a) $H_0: \mu_1 \leq \mu_2$; $H_a: \mu_1 > \mu_2$ (claim)

(b) One-tailed because H_a contains $>$; z-test because n_1 and n_2 are each greater than 30.

(c) $z_0 = 1.645$; Rejection region: $z > 1.645$

(d) 0.585   (e) Fail to reject H_0.

(f) There is not enough evidence at the 5% level of significance to support the claim that the mean score on the science assessment for the male high school students was higher than for the female high school students.

2. (a) $H_0: \mu_1 = \mu_2$ (claim); $H_a: \mu_1 \neq \mu_2$

(b) Two-tailed because H_a contains $\neq$; t-test because n_1 and n_2 are less than 30, the samples are independent, and the populations are normally distributed.

(c) $-t_0 = -2.779$, $t_0 = 2.779$;
Rejection regions: $t < -2.779$, $t > 2.779$

(d) 0.341   (e) Fail to reject H_0.

(f) There is not enough evidence at the 1% level of significance to reject the teacher's claim that the mean scores on the science assessment test are the same for fourth grade boys and girls.

3. (a) $H_0: p_1 = p_2$ (claim); $H_a: p_1 \neq p_2$

(b) Two-tailed because H_a contains $\neq$; z-test because you are testing proportions and $n_1\overline{p}$, $n_1\overline{q}$, $n_2\overline{p}$, and $n_2\overline{q} \geq 5$.

(c) $-z_0 = 1.645$, $z_0 = 1.645$;
Rejection regions: $z < -1.645$, $z > 1.645$

(d) 1.32   (e) Fail to reject H_0.

(f) There is not enough evidence at the 10% level of significance to reject the claim that the proportion of U.S. adults who are worried that they or someone in their family will become a victim of terrorism has not changed.

4. (a) $H_0: \mu_d \geq 0$; $H_a: \mu_d < 0$ (claim)

(b) One-tailed because H_a contains $<$; t-test because both populations are normally distributed and the samples are dependent.

(c) $t_0 = -2.718$; Rejection region: $t < -2.718$

(d) -5.07 (e) Reject H_0.

(f) There is enough evidence at the 1% level of significance to support the claim that the seminar helps adults increase their credit scores.

Real Statistics–Real Decisions for Chapter 8   *(page 476)*

1. (a) Answers will vary. *Sample answer:* Divide the records into groups according to the inpatients' ages, and then randomly select records from each group.

(b) Answers will vary. *Sample answer:* Divide the records into groups according to geographic regions, and then randomly select records from each group.

(c) Answers will vary. *Sample answer:* Assign a different number to each record, randomly choose a starting number, and then select every 50th record.

(d) Answers will vary. *Sample answer:* Assign a different number to each record, and then use a table of random numbers to generate a sample of numbers.

2. (a) Answers will vary.   (b) Answers will vary.

3. Use a t-test; independent; yes, you need to know if the population distributions are normal or not; yes. you need to know if the population variances are equal or not.

4. There is not enough evidence at the 10% level of significance to support the claim that there is a difference in the mean length of hospital stays for inpatients.

This decision does not support the calim.

Cumulative Review Chapters 6–8   *(page 480)*

1. (a) $(0.109, 0.151)$

(b) There is enough evidence at the 5% level of significance to support the researcher's claim that more than 10% of people who attend community college are age 40 or older.

2. There is enough evidence at the 10% level of significance to support the claim that the fuel additive improved gas mileage.

3. $(25.94, 28.00)$; z-distribution

4. $(2.75, 4.17)$; t-distribution

5. $(10.7, 13.5)$; t-distribution

6. $(7.69, 8.73)$; t-distribution

7. There is enough evidence at the 10% level of significance to support the pediatrician's claim that the mean birth weight of a single-birth baby is greater than the mean birth weight of a baby that has a twin.

8. $H_0: \mu \geq 33$; $H_a: \mu < 33$ (claim)

9. $H_0: p \geq 0.19$ (claim); $H_a: p < 0.19$

10. $H_0: \sigma = 0.63$ (claim); $H_a: \sigma \neq 0.63$

11. $H_0: \mu = 2.28$; $H_a: \mu \neq 2.28$ (claim)

12. (a) $(5.1, 22.8)$ (b) $(2.3, 4.8)$

(c) There is not enough evidence at the 1% level of significance to support the pharmacist's claim that the standard deviation of the mean number of chronic medications taken by elderly adults in the community is less than 2.5 medications.

13. There is enough evidence at the 5% level of significance to support the organization's claim that the mean SAT scores for male athletes and male non-athletes at a college are different.

14. (a) (37,732.2, 40,060.7)

(b) There is not enough evidence at the 5% level of significance to reject the claim that the mean annual earnings for translators is $40,000.

15. There is not enough evidence at the 10% level of significance to reject the claim that the proportions of players sustaining head and neck injuries are the same for the two groups.

16. (a) (41.5, 42.5)

(b) There is enough evidence at the 5% level of significance to reject the zoologist's claim that the mean incubation period for ostriches is at least 45 days.

CHAPTER 9

Section 9.1 *(page 495)*

1. Increase

3. The range of values for the correlation coefficient is −1 to 1, inclusive.

5. Answers will vary. *Sample answer:*

Perfect positive linear correlation: price per gallon of gasoline and total cost of gasoline

Perfect negative linear correlation: distance from door and height of wheelchair ramp

7. r is the sample correlation coefficient, while ρ is the population correlation coefficient.

9. Negative linear correlation

11. Perfect negative linear correlation

13. Positive linear correlation

15. c; You would expect a positive linear correlation between age and income.

16. d; You would not expect age and height to be correlated.

17. b; You would expect a negative linear correlation between age and balance on student loans.

18. a; You would expect the relationship between age and body temperature to be fairly constant.

19. Explanatory variable: Amount of water consumed

Response variable: Weight loss

21. (a)

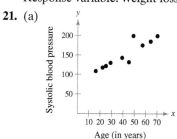

(b) 0.908

(c) Strong positive linear correlation

23. (a)

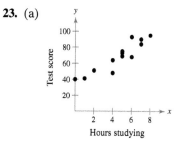

(b) 0.923   (c) Strong positive linear correlation

25. (a)

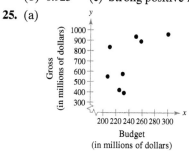

(b) 0.604   (c) Weak positive linear correlation

27. (a)

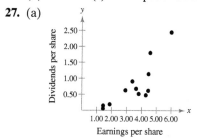

(b) 0.828   (c) Strong positive linear correlation

29. The correlation coefficient becomes $r \approx 0.621$. The new data entry is an outlier, so the linear correlation is weaker.

31. There is not enough evidence at the 1% level of significance to conclude that there is a significant linear correlation between vehicle weight and the variability in braking distance.

33. There is enough evidence at the 1% level of significance to conclude that there is a significant linear correlation between the number of hours spent studying for a test and the score received on the test.

35. There is enough evidence at the 1% level of significance to conclude that there is a significant linear correlation between earnings per share and dividends per share.

37. (a)

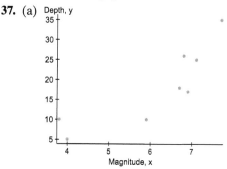

(b) 0.848

(c) Reject H_0. There is enough evidence at the 1% level of significance to conclude that there is a significant linear correlation between the magnitudes of earthquakes and their depths below the surface at the epicenter.

39. The correlation coefficient becomes $r \approx 0.085$. The new rejection regions are $t < -3.499$ and $t > 3.499$, and the new standardized test statistic is $t \approx 0.227$. So, you now fail to reject H_0.

41. 0.883; 0.883; The correlation coefficient remains unchanged when the x-values and y-values are switched.

43. Answers will vary.

Activity 9.1 *(page 500)*

1–4. Answers will vary.

Section 9.2 *(page 505)*

1. A residual is the difference between the observed y-value of a data point and the predicted y-value on the regression line for the x-coordinate of the data point. A residual is positive when the data point is above the line, negative when the point is below the line, and zero when the observed y-value equals the predicted y-value.

3. Substitute a value of x into the equation of a regression line and solve for y.

5. The correlation between variables must be significant.

7. b **8.** a **9.** e **10.** c **11.** f **12.** d

13. c **14.** b **15.** a **16.** d

17. $\hat{y} = 0.065x + 0.465$

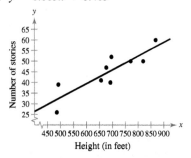

(a) 52 stories (b) 49 stories

(c) It is not meaningful to predict the value of y for $x = 400$ because $x = 400$ is outside the range of the original data.

(d) 41 stories

19. $\hat{y} = 7.350x + 34.617$

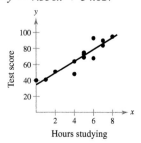

(a) 57 (b) 82

(c) It is not meaningful to predict the value of y for $x = 13$ because $x = 13$ is outside the range of the original data.

(d) 68

21. $\hat{y} = 2.472x + 80.813$

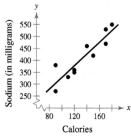

(a) 501.053 milligrams

(b) 328.013 milligrams

(c) 426.893 milligrams

(d) It is not meaningful to predict the value of y for $x = 210$ because $x = 210$ is outside the range of the original data.

23. $\hat{y} = 1.870x + 51.360$

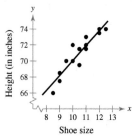

(a) 72.865 inches

(b) 66.32 inches

(c) It is not meaningful to predict the value of y for $x = 15.5$ because $x = 15.5$ is outside the range of the original data.

(d) 70.06 inches

25. Strong positive linear correlation; As the years of experience of the registered nurses increase, their salaries tend to increase.

27. No, it is not meaningful to predict a salary for a registered nurse with 28 years of experience because $x = 28$ is outside the range of the original data.

29. Answers will vary. *Sample answer:* Although it is likely that there is a cause-and-effect relationship between a registered nurse's years of experience and salary, you cannot use significant correlation to claim cause and effect. The relationship between the variables may also be influenced by other factors, such as work performance, level of education, or the number of years with an employer.

31. (a) $\hat{y} = -0.159x + 5.827$

(b) -0.852

(c)

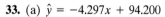
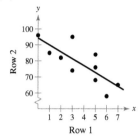

Fitted line plot

Earned run average, *y*

33. (a) $\hat{y} = -4.297x + 94.200$

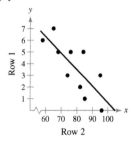

(b) $\hat{y} = -0.141x + 14.763$

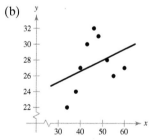

(c) The slope of the line keeps the same sign, but the values of *m* and *b* change.

35. (a) $\hat{y} = 0.139x + 21.024$

(b)

(c)

(d) The residual plot shows a pattern because the residuals do not fluctuate about 0. This implies that the regression line is not a good representation of the relationship between the two variables.

37. (a)

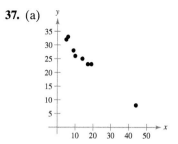

(b) The point $(44, 8)$ may be an outlier.

(c) The point $(44, 8)$ is not an influential point because the slopes and *y*-intercepts of the regression lines with the point included and without the point included are not significantly different.

39. $\hat{y} = 654.536x - 1214.857$

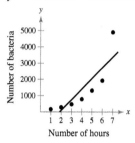

41. $y = 93.028(1.712)^x$

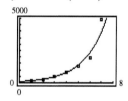

43. $\hat{y} = -78.929x + 576.179$

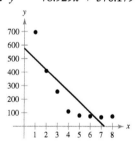

45. $y = 782.300x^{-1.251}$

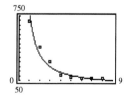

47. $y = 25.035 + 19.599 \ln x$

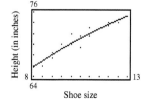

49. The logarithmic equation is a better model for the data. The graph of the logarithmic equation fits the data better than the regression line.

Activity 9.2 *(page 511)*

1–4. Answers will vary.

Section 9.3 *(page 519)*

1. The total variation is the sum of the squares of the differences between the y-values of each ordered pair and the mean of the y-values of the ordered pairs, or $\Sigma(y_i - \bar{y})^2$.

3. The unexplained variation is the sum of the squares of the differences between the observed y-values and the predicted y-values, or $\Sigma(y_i - \hat{y}_i)^2$.

5. Two variables that have perfect positive or perfect negative linear correlation have a correlation coefficient of 1 or -1, respectively. In either case, the coefficient of determination is 1, which means that 100% of the variation in the response variable is explained by the variation in the explanatory variable.

7. 0.216; About 21.6% of the variation is explained. About 78.4% of the variation is unexplained.

9. 0.916; About 91.6% of the variation is explained. About 8.4% of the variation is unexplained.

11. (a) 0.798; About 79.8% of the variation in proceeds can be explained by the variation in the number of issues, and about 20.2% of the variation is unexplained.

(b) 8064.633; The standard error of estimate of the proceeds for a specific number of issues is about $8,064,633,000.

13. (a) 0.981; About 98.1% of the variation in sales can be explained by the variation in the total square footage, and about 1.9% of the variation is unexplained.

(b) 30.576; The standard error of estimate of the sales for a specific total square footage is about $30,576,000,000.

15. (a) 0.963; About 96.3% of the variation in wages for federal government employees can be explained by the variation in wages for state government employees, and about 3.7% of the variation is unexplained.

(b) 20.090; The standard error of estimate of the average weekly wages for federal government employees for a specific average weekly wage for state government employees is about $20.09.

17. (a) 0.790; About 79.0% of the variation in the gross collections of corporate income taxes can be explained by the variation in the gross collections of individual income taxes, and about 21.0% of the variation is unexplained.

(b) 42.386; The standard error of estimate of the gross collections of corporate income taxes for a specific gross collection of individual income taxes is about $42,386,000,000.

19. $40{,}116.824 < y < 82{,}624.318$

You can be 95% confident that the proceeds will be between $40,116,824,000 and $82,624,318,000 when the number of initial offerings is 450 issues.

21. $1218.435 < y < 1336.829$

You can be 90% confident that the shopping center sales will be between $1,218,435,000 and $1,336,829,000 when the total square footage of shopping centers is 5,750,000,000.

23. $1007.82 < y < 1208.228$

You can be 99% confident that the average weekly wages of federal government employees will be between $1007.82 and $1208.23 when the average weekly wages of state government employees is $800.

25. $213.729 < y < 450.519$

You can be 95% confident that the corporate income taxes collected by the U.S. Internal Revenue Service for a given year will be between $213,729,000,000 and $450,519,000,000 when the U.S. Internal Revenue Service collects $1,250,000,000 in individual income taxes that year.

27.

29.

x_i	y_i	$\hat{y}_i$	$\hat{y}_i - \bar{y}$	$y_i - \hat{y}_i$	$y_i - \bar{y}$
9.4	7.6	7.1252	0.4372	0.4748	0.912
9.2	6.9	7.0736	0.3856	−0.1736	0.212
8.9	6.6	6.9962	0.3082	−0.3962	−0.088
8.4	6.8	6.8672	0.1792	−0.0672	0.112
8.3	6.9	6.8414	0.1534	0.0586	0.212
6.5	6.5	6.377	−0.311	0.123	−0.188
6	6.3	6.248	−0.44	0.052	−0.388
4.9	5.9	5.9642	−0.7238	−0.0642	−0.788

31. 0.746; About 74.6% of the variation in the median ages of trucks in use can be explained by the variation in the median ages of cars in use, and about 25.4% of the variation is unexplained.

33. $5.792 < y < 7.22$

You can be 95% confident that the median age of trucks in use will be between 5.792 and 7.22 years when the median age of cars in use is 7.0 years.

35. (a) 0.671   (b) 1.780   (c) $9.537 < y < 19.010$

37. Fail to reject H_0. There is not enough evidence at the 1% level of significance to support the claim that there is a linear relationship between weight and number of hours slept.

39. $-118.927 < B < 323.505$

$110.911 < M < 281.393$

A96

Section 9.4 *(page 527)*

1. (a) 39,103.5 pounds per acre
   (b) 39,939.1 pounds per acre
   (c) 38,063.5 pounds per acre
   (d) 39,052.4 pounds per acre

3. (a) 7.5 cubic feet   (b) 16.8 cubic feet
   (c) 51.9 cubic feet   (d) 62.1 cubic feet

5. $\hat{y} = -2518.364 + 126.822x_1 + 66.360x_2$

   (a) 28.489; The standard error of estimate of the predicted sales given specific total square footage and number of shopping centers is about $28.489 billion.

   (b) 0.985; The multiple regression model explains about 98.5% of the variation in y.

7. $\hat{y} = -2518.364 + 126.822x_1 + 66.360x_2$; The equation is the same.

9. 0.981; About 98.1% of the variation in y can be explained by the relationship between variables; $r_{adj}^2 < r^2$.

Uses and Abuses for Chapter 9   *(page 529)*

1. Answers will vary.   2. Answers will vary.

Review Answers for Chapter 9   *(page 531)*

1.

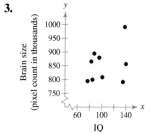

0.912; strong positive linear correlation; the number of passing yards increases as the number of pass attempts increases.

3.

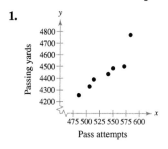

0.338; weak positive linear correlation; brain size increases as IQ increases.

5. There is not enough evidence at the 1% level of significance to conclude that there is a significant linear correlation.

7. There is enough evidence at the 5% level of significance to conclude that there is a significant linear correlation between a quarterback's pass attempts and passing yards.

9. There is not enough evidence at the 1% level of significance to conclude that there is a significant linear correlation between IQ and brain size.

11. $\hat{y} = 0.038x - 3.529$

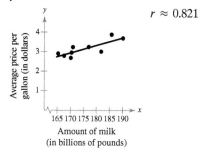

$r \approx 0.821$

13. $\hat{y} = -0.086x + 10.450$

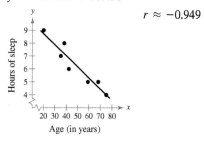

$r \approx -0.949$

15. (a) It is not meaningful to predict the value of y for $x = 160$ because $x = 160$ is outside the range of the original data.

   (b) $3.12   (c) $3.31

   (d) It is not meaningful to predict the value of y for $x = 200$ because $x = 200$ is outside the range of the original data.

17. (a) It is not meaningful to predict the value of y for $x = 18$ because $x = 18$ is outside the range of the original data.

   (b) 8.3 hours

   (c) It is not meaningful to predict the value of y for $x = 85$ because $x = 85$ is outside the range of the original data.

   (d) 6.15 hours

19. 0.203; About 20.3% of the variation is explained. About 79.7% of the variation is unexplained.

21. 0.412; About 41.2% of the variation is explained. About 58.8% of the variation is unexplained.

23. (a) 0.679; About 67.9% of the variation in the fuel efficiency of the compact sports sedans can be explained by the variation in their prices, and about 32.1% of the variation is unexplained.

   (b) 1.138; The standard error of estimate of the fuel efficiency of the compact sports sedans for a specific price of the compact sports sedans is about 1.138 miles per gallon.

25. $2.997 < y < 4.025$

   You can be 90% confident that the price per gallon of milk will be between $3.00 and $4.03 when 185 billion pounds of milk is produced.

27. $4.865 < y < 8.295$

You can be 95% confident that the hours slept will be between 4.865 and 8.295 hours for a person who is 45 years old.

29. $16.119 < y < 25.137$

You can be 99% confident that the fuel efficiency of the compact sports sedan that costs $39,900 will be between 16.119 and 25.137 miles per gallon.

31. $\hat{y} = 3.674 + 1.287x_1 - 7.531x_2$

33. (a) 21.705   (b) 25.21   (c) 30.1   (d) 25.86

Chapter Quiz for Chapter 9   *(page 535)*

1.

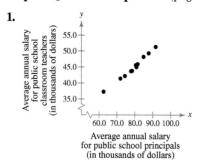

The data appear to have a positive linear correlation. As x increases, y tends to increase.

2. 0.993; strong positive linear correlation; public school classroom teachers' salaries increase as public school principals' salaries increase.

3. Reject H_0. There is enough evidence at the 5% level of significance to conclude that there is a significant linear correlation between public school principals' salaries and public school classroom teachers' salaries.

4. $\hat{y} = 0.491x + 5.977$

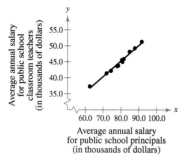

5. $50,412.50

6. 0.986; About 98.6% of the variation in the average annual salaries of public school classroom teachers can be explained by the variation in the average annual salaries of public school principals, and about 1.4% of the variation is unexplained.

7. 0.490; The standard error of estimate of the average annual salary of public school classroom teachers for a specific average annual salary of public school principals is about $490.

8. $46.887 < y < 49.273$

You can be 95% confident that the average annual salary of public school classroom teachers will be between $46,887 and $49,273 when the average annual salary of public school principals is $85,750.

9. (a) $59.30

    (b) $30.53

    (c) $45.67

    (d) $35.83

Real Statistics–Real Decisions for Chapter 9   *(page 536)*

1. (a)

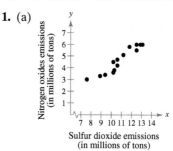

It appears that there is a positive linear correlation. As the sulfur dioxide emissions increase, the nitrogen oxides emissions increase.

(b) $r \approx 0.947$; There is a strong positive linear correlation.

(c) There is enough evidence at the 5% level of significance to conclude that there is a significant linear correlation between sulfur dioxide emissions and nitrogen oxides emissions.

(d) $\hat{y} = 0.652x - 2.438$

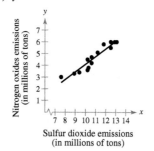

Yes, the line appears to be a good fit.

(e) Yes, for x-values that are within the range of the data set.

(f) $r^2 \approx 0.898$; About 89.8% of the variation in nitrogen oxides emissions can be explained by the variation in sulfur dioxide emissions, and about 10.2% of the variation is unexplained.

$s_e \approx 0.368$; The standard error of estimate of nitrogen oxides emissions for a specific sulfur dioxide emission is about 368,000 tons.

2. $1.358 < y < 3.286$

You can be 95% confident that the nitrogen oxide emissions will be between 1.358 and 3.286 million tons when the sulfur dioxide emissions are $17.3 - 10 = 7.3$ million tons.

CHAPTER 10

Section 10.1 (page 546)

1. A multinomial experiment is a probability experiment consisting of a fixed number of independent trials in which there are more than two possible outcomes for each trial. The probability of each outcome is fixed, and each outcome is classified into categories.

3. 45   5. 57.5

7. (a) H_0: The distribution of the ages of moviegoers is 26.7% ages 2–17, 19.8% ages 18–24, 19.7% ages 25–39, 14% ages 40–49, and 19.8% ages 50+. (claim)

   H_a: The distribution of ages differs from the claimed or expected distribution.

   (b) $\chi_0^2 = 7.779$; Rejection region: $\chi^2 > 7.779$

   (c) 7.256

   (d) Fail to reject H_0.

   (e) There is enough evidence at the 10% level of significance to conclude that the distribution of the ages of moviegoers and the claimed or expected distribution are the same.

9. (a) H_0: The distribution of the days people order food for delivery is 7% Sunday, 4% Monday, 6% Tuesday, 13% Wednesday, 10% Thursday, 36% Friday, and 24% Saturday.

   H_a: The distribution of days differs from the claimed or expected distribution. (claim)

   (b) $\chi_0^2 = 16.812$;

   Rejection region: $\chi^2 > 16.812$

   (c) 17.595

   (d) Reject H_0.

   (e) There is enough evidence at the 1% level of significance to conclude that there has been a change in the claimed or expected distribution.

11. (a) H_0: The distribution of the number of homicide crimes in California by season is uniform. (claim)

    H_a: The distribution of homicides by season is not uniform.

    (b) $\chi_0^2 = 7.815$; Rejection region: $\chi^2 > 7.815$

    (c) 0.727

    (d) Fail to reject H_0.

    (e) There is not enough evidence at the 5% level of significance to reject the claim that distribution of the number of homicide crimes in California by season is uniform.

13. (a) H_0: The distribution of the opinions of U.S. parents on whether a college education is worth the expense is 55% strongly agree, 30% somewhat agree, 5% neither agree nor disagree, 6% somewhat disagree, and 4% strongly disagree.

    H_a: The distribution of opinions differs from the claimed or expected distribution. (claim)

    (b) $\chi_0^2 = 9.488$; Rejection region: $\chi^2 > 9.488$

    (c) 65.236

    (d) Reject H_0.

    (e) There is enough evidence at the 5% level of significance to conclude that the distribution of the opinions of U.S. parents on whether a college education is worth the expense differs from the claimed or expected distribution.

15. (a) H_0: The distribution of prospective home buyers by the size they want their next house to be is uniform. (claim)

    H_a: The distribution of prospective home buyers by the size they want their next house to be is not uniform.

    (b) $\chi_0^2 = 5.991$; Rejection region: $\chi^2 > 5.991$ (c) 10.308

    (d) Reject H_0.

    (e) There is enough evidence at the 5% level of significance to reject the claim that the distribution of prospective home buyers by the size they want their next house to be is uniform.

17. **Chi-Square goodness-of-fit results:**

    Observed: Recent survey

    Expected: Previous survey

N	DF	Chi-Square	P-Value
400	9	18.637629	0.0285

    $P = 0.0285$, so reject H_0. There is enough evidence at the 10% level of significance to conclude that there has been a change in the claimed or expected distribution of U.S. adults' favorite sports.

19. (a) The expected frequencies are 17, 63, 79, 34, and 5.

    (b) $\chi_0^2 = 13.277$; Rejection region: $\chi^2 > 13.277$

    (c) 0.613   (d) Fail to reject H_0.

    (e) There is not enough evidence at the 1% level of significance to reject the claim that the test scores are normally distributed.

Section 10.2 (page 557)

1. Find the sum of the row and the sum of the column in which the cell is located. Find the product of these sums. Divide the product by the sample size.

3. Answers will vary. *Sample answer:* For both the chi-square test for independence and the chi-square goodness-of-fit test, you are testing a claim about data that are in categories. However, the chi-square goodness-of-fit test has only one data value per category, while the chi-square test for independence has multiple data values per category.

   Both tests compare observed and expected frequencies. However, the chi-square goodness-of-fit test simply compares the distributions, whereas the chi-square test for independence compares them and then draws a conclusion about the dependence or independence of the variables.

5. False. If the two variables of a chi-square test for independence are dependent, then you can expect a large difference between the observed frequencies and the expected frequencies.

7. (a)–(b)

	Athlete has		
Result	**Stretched**	**Not stretched**	**Total**
Injury	18 (20.82)	22 (19.18)	40
No injury	211 (208.18)	189 (191.82)	400
Total	229	211	440

9. (a)–(b)

	Preference			
Bank employee	**New procedure**	**Old procedure**	**No preference**	**Total**
Teller	92 (133.80)	351 (313.00)	50 (46.19)	493
Customer service	76 (34.20)	42 (80.00)	8 (11.81)	126
Total	168	393	58	619

11. (a)–(b)

	Type of car				
Gender	**Compact**	**Full-size**	**SUV**	**Truck/ van**	**Total**
Male	28 (28.6)	39 (39.05)	21 (22.55)	22 (19.8)	110
Female	24 (23.4)	32 (31.95)	20 (18.45)	14 (16.2)	90
Total	52	71	41	36	200

13. (a) H_0: Skill level in a subject is independent of location. (claim)

H_a: Skill level in a subject is dependent on location.

(b) d.f. = 2; $\chi_0^2 = 9.210$; Rejection region: $\chi^2 > 9.210$

(c) 0.297

(d) Fail to reject H_0. There is not enough evidence at the 1% level of significance to reject the claim that skill level in a subject is independent of location.

15. (a) H_0: The number of times former smokers tried to quit is independent of gender.

H_a: The number of times former smokers tried to quit is dependent on gender. (claim)

(b) d.f. = 2; $\chi_0^2 = 5.991$; Rejection region: $\chi^2 > 5.991$

(c) 0.002

(d) Fail to reject H_0. There is not enough evidence at the 5% level of significance to conclude that the number of times former smokers tried to quit is dependent on gender.

17. (a) H_0: Results are independent of the type of treatment.

H_a: Results are dependent on the type of treatment. (claim)

(b) d.f. = 1; $\chi_0^2 = 2.706$; Rejection region: $\chi^2 > 2.706$

(c) 5.106

(d) Reject H_0. There is enough evidence at the 10% level of significance to conclude that results are dependent on the type of treatment. Answers will vary.

19. (a) H_0: Reasons are independent of the type of worker.

H_a: Reasons are dependent on the type of worker. (claim)

(b) d.f. = 2; $\chi_0^2 = 9.210$; Rejection region: $\chi^2 > 9.210$

(c) 7.326

(d) Fail to reject H_0. There is not enough evidence at the 1% level of significance to conclude that reasons for continuing education are dependent on the type of worker. On the basis of these results, marketing strategies should not differ between technical and nontechnical audiences in regard to reasons for continuing education.

21. (a) H_0: Type of crash is independent of the type of vehicle.

H_a: Type of crash is dependent on the type of vehicle. (claim)

(b) d.f. = 2; $\chi_0^2 = 5.991$; Rejection region: $\chi^2 > 5.991$

(c) 144.099

(d) Reject H_0. There is enough evidence at the 5% level of significance to conclude that the type of crash is dependent on the type of vehicle.

23. (a)–(b) **Contingency table results:**

Rows: Expected income

Columns: None

Cell format
Count
Expected count

	More likely	**Less likely**	**Did not make a difference**
Less than $35,000	37 / 33.33	10 / 6.836	22 / 27.99
$35,000 to $50,000	28 / 25.17	12 / 5.164	15 / 21.14
$50,000 to $100,000	55 / 62.75	9 / 12.87	65 / 52.7
Greater than $100,000	36 / 34.75	1 / 7.127	29 / 29.18
Total	156	32	131

	Did not consider it	**Total**
Less than $35,000	25 / 25.85	94
$35,000 to $50,000	16 / 19.52	71
$50,000 to $100,000	48 / 48.68	177
Greater than $100,000	32 / 26.95	98
Total	121	440

Statistic	**DF**	**Value**	**P-value**
Chi-square	9	26.22966	0.0019

(c) Reject H_0. There is enough evidence at the 1% level of significance to conclude that the decision to borrow money is dependent on the child's expected income after graduation.

25. Fail to reject H_0. There is not enough evidence at the 5% level of significance to reject the claim that the proportions of motor vehicle crash deaths involving males or females are the same for each age group.

27. Right-tailed

29.

Status	Educational attainment			
	Not a high school graduate	High school graduate	Some college, no degree	Associate's, bachelor's, or advanced degree
Employed	0.055	0.183	0.114	0.290
Unemployed	0.006	0.011	0.005	0.007
Not in the labor force	0.073	0.118	0.053	0.085

31. Several of the expected frequencies are less than 5.

33. 45.2%

35.

Status	Educational attainment			
	Not a high school graduate	High school graduate	Some college, no degree	Associate's, bachelor's, or advanced degree
Employed	0.411	0.587	0.660	0.759
Unemployed	0.046	0.036	0.030	0.019
Not in the labor force	0.544	0.377	0.311	0.223

37. 4.6%

39. Answers will vary. *Sample answer:* As educational attainment increases, employment increases.

Section 10.3   *(page 571)*

1. Specify the level of significance α. Determine the degrees of freedom for the numerator and denominator. Use Table 7 in Appendix B to find the critical value F.

3. (1) The samples must be randomly selected, (2) the samples must be independent, and (3) each population must have a normal distribution.

5. 2.54   **7.** 2.06   **9.** 9.16   **11.** 1.80

13. Fail to reject H_0. There is not enough evidence at the 10% level of significance to support the claim.

15. Fail to reject H_0. There is not enough evidence at the 1% level of significance to reject the claim.

17. Reject H_0. There is enough evidence at the 1% level of significance to reject the claim.

19. (a) H_0: $\sigma_1^2 \leq \sigma_2^2$; H_a: $\sigma_1^2 > \sigma_2^2$ (claim)
   (b) $F_0 = 2.11$; Rejection region: $F > 2.11$
   (c) 1.08
   (d) Fail to reject H_0.
   (e) There is not enough evidence at the 5% level of significance to support Company A's claim that the variance of the life of its appliances is less than the variance of the life of Company B's appliances.

21. (a) H_0: $\sigma_1^2 = \sigma_2^2$; H_a: $\sigma_1^2 \neq \sigma_2^2$ (claim)
   (b) $F_0 = 6.23$; Rejection region: $F > 6.23$
   (c) 2.10
   (d) Fail to reject H_0.
   (e) There is not enough evidence at the 5% level of significance to conclude that the variances of the prices differ between the two companies.

23. (a) H_0: $\sigma_1^2 = \sigma_2^2$ (claim); H_a: $\sigma_1^2 \neq \sigma_2^2$
   (b) $F_0 = 2.635$; Rejection region: $F > 2.635$
   (c) 1.282
   (d) Fail to reject H_0.
   (e) There is not enough evidence at the 10% level of significance to reject the administrator's claim that the standard deviations of science assessment test scores for eighth grade students are the same in Districts 1 and 2.

25. (a) H_0: $\sigma_1^2 \leq \sigma_2^2$; H_a: $\sigma_1^2 > \sigma_2^2$ (claim)
   (b) $F_0 = 2.35$; Rejection region: $F > 2.35$
   (c) 2.41
   (d) Reject H_0.
   (e) There is enough evidence at the 5% level of significance to conclude that the standard deviation of the annual salaries for actuaries is greater in New York than in California.

27. Hypothesis test results:
   σ_1^2: variance of population 1
   σ_2^2: variance of population 2
   σ_1^2/σ_2^2: variance ratio
   H_0: $\sigma_1^2/\sigma_2^2 = 1$
   H_A: $\sigma_1^2/\sigma_2^2 \neq 1$

Ratio	n1	n2	Sample Ratio	F-Stat	P-value
σ_1^2/σ_2^2	15	18	0.5281571	0.5281571	0.2333

$P = 0.2333 > 0.10$, so fail to reject H_0. There is not enough evidence at the 10% level of significance to reject the claim.

29. Hypothesis test results:

σ_1^2: variance of population 1

σ_2^2: variance of population 2

σ_1^2/σ_2^2: variance ratio

$H_0: \sigma_1^2/\sigma_2^2 = 1$

$H_A: \sigma_1^2/\sigma_2^2 > 1$

Ratio	n1	n2	Sample Ratio	F-Stat	P-value
σ_1^2/σ_2^2	22	29	2.153926	2.153926	0.0293

$P = 0.0293 < 0.05$, so reject H_0. There is enough evidence at the 5% level of significance to reject the claim.

31. Right-tailed: 14.73   **33.** $(0.375, 3.774)$

Left-tailed: 0.15

Section 10.4 *(page 581)*

1. $H_0: \mu_1 = \mu_2 = \mu_3 = \ldots = \mu_k$

H_a: At least one of the means is different from the others.

3. The MS_B measures the differences related to the treatment given to each sample. The MS_W measures the differences related to entries within the same sample.

5. (a) $H_0: \mu_1 = \mu_2 = \mu_3$

H_a: At least one mean is different from the others. (claim)

(b) $F_0 = 3.37$;

Rejection region: $F > 3.37$

(c) 1.02

(d) Fail to reject H_0.

(e) There is not enough evidence at the 5% level of significance to conclude that the mean costs per ounce are different.

7. (a) $H_0: \mu_1 = \mu_2 = \mu_3$

H_a: At least one mean is different from the others. (claim)

(b) $F_0 = 5.49$; Rejection region: $F > 5.49$

(c) 21.99

(d) Reject H_0.

(e) There is enough evidence at the 1% level of significance to conclude that at least one mean salary is different.

9. (a) $H_0: \mu_1 = \mu_2 = \mu_3 = \mu_4 = \mu_5$

H_a: At least one mean is different from the others. (claim)

(b) $F_0 = 4.37$; Rejection region: $F > 4.37$

(c) 12.61

(d) Reject H_0.

(e) There is enough evidence at the 1% level of significance to conclude that at least one mean cost per mile is different.

11. (a) $H_0: \mu_1 = \mu_2 = \mu_3 = \mu_4$ (claim)

H_a: At least one mean is different from the others.

(b) $F_0 = 4.54$; Rejection region: $F > 4.54$

(c) 0.56

(d) Fail to reject H_0.

(e) There is not enough evidence at the 1% level of significance for the company to reject the claim that the mean number of days patients spend at the hospital is the same for all four regions.

13. (a) $H_0: \mu_1 = \mu_2 = \mu_3 = \mu_4$

H_a: At least one mean is different from the others. (claim)

(b) $F_0 = 2.255$; Rejection region: $F > 2.255$

(c) 3.107

(d) Reject H_0.

(e) There is enough evidence at the 10% level of significance to conclude that the mean energy consumption of at least one region is different from the others.

15. Analysis of Variance results:

Data stored in separate columns.

Column means

Column	n	Mean	Std. Error
Grade 9	8	84.375	9.531784
Grade 10	8	79.25	9.090321
Grade 11	8	76.625	7.648383
Grade 12	8	70.75	6.9224014

ANOVA table

Source	df	SS	MS
Treatments	3	771.25	257.08334
Error	28	15674.75	559.8125
Total	31	16446	

F-Stat	P-Value
0.45923114	0.7129

$P = 0.7129 > 0.01$, so fail to reject H_0. There is not enough evidence at the 1% level of significance to reject the claim that the mean numbers of female students who played on a sports team are equal for all grades.

17. Fail to reject all null hypotheses. The interaction between the advertising medium and the length of the ad has no effect on the rating and therefore there is no significant difference in the means of the ratings.

19. Fail to reject all null hypotheses. The interaction between age and gender has no effect on GPA and therefore there is no significant difference in the means of the GPAs.

21. $CV_{Scheffé} = 10.98$

$(1, 2) \rightarrow 22.396 \rightarrow$ Significant difference

$(1, 3) \rightarrow 40.837 \rightarrow$ Significant difference

$(2, 3) \rightarrow 2.749 \rightarrow$ No difference

23. $CV_{Scheffé} = 6.84$

$(1, 2) \rightarrow 0.032 \rightarrow$ No difference

$(1, 3) \rightarrow 1.490 \rightarrow$ No difference

$(1, 4) \rightarrow 1.345 \rightarrow$ No difference

$(2, 3) \rightarrow 2.374 \rightarrow$ No difference

$(2, 4) \rightarrow 1.122 \rightarrow$ No difference

$(3, 4) \rightarrow 7.499 \rightarrow$ Significant difference

Uses and Abuses for Chapter 10 *(page 587)*

1–2. Answers will vary.

Review Answers for Chapter 10 *(page 589)*

1. (a) H_0: The distribution of the allowance amounts is 29% less than \$10, 16% \$10 to \$20, 9% more than \$21, and 46% don't give one/other.

H_a: The distribution of amounts differs from the claimed or expected distribution. (claim)

(b) $\chi_0^2 = 6.251$; Rejection region: $\chi^2 > 6.251$

(c) 4.886　(d) Fail to reject H_0.

(e) There is not enough evidence at the 10% level of significance to conclude that there has been a change in the claimed or expected distribution.

3. (a) H_0: The distribution of responses from golf students about what they need the most help with is 22% approach and swing, 9% driver shots, 4% putting, and 65% short-game shots. (claim)

H_a: The distribution of responses differs from the claimed or expected distribution.

(b) $\chi_0^2 = 7.815$; Rejection region: $\chi^2 > 7.815$

(c) 0.503　(d) Fail to reject H_0.

(e) There is enough evidence at the 5% level of significance to conclude that the distribution of golf students' responses is the same as the claimed or expected distribution.

5. (a) $E_{1, 1} = 63$, $E_{1, 2} = 356.4$, $E_{1, 3} = 319.8$, $E_{1, 4} = 310.8$, $E_{2, 1} = 147$, $E_{2, 2} = 831.6$, $E_{2, 3} = 746.2$, $E_{2, 4} = 725.2$

(b) Reject H_0.

(c) There is enough evidence at the 1% level of significance to conclude that public school teachers' gender and years of full-time teaching experience are related.

7. (a) $E_{1, 1} \approx 54.86$, $E_{1, 2} \approx 40.38$, $E_{1, 3} \approx 22.10$, $E_{1, 4} = 16.00$, $E_{1, 5} \approx 26.67$, $E_{2, 1} \approx 17.14$, $E_{2, 2} \approx 12.62$, $E_{2, 3} \approx 6.90$, $E_{2, 4} = 5.00$, $E_{2, 5} \approx 8.33$

(b) Reject H_0.

(c) There is enough evidence at the 1% level of significance to conclude that a species' status (endangered or threatened) is dependent on vertebrate group.

9. 2.295　**11.** 2.39　**13.** 2.06　**15.** 2.08

17. Fail to reject H_0. There is not enough evidence at the 1% level of significance to reject the claim.

19. Fail to reject H_0. There is not enough evidence at the 10% level of significance to support the claim that the variation in wheat production is greater in Garfield County than in Kay County.

21. Fail to reject H_0. There is not enough evidence at the 1% level of significance to support the claim that the test score variance for females is different from that for males.

23. Reject H_0. There is enough evidence at the 10% level of significance to conclude that at least one of the mean costs is different from the others.

Chapter Quiz for Chapter 10 *(page 593)*

1. (a) H_0: $\sigma_1^2 = \sigma_2^2$; H_a: $\sigma_1^2 \neq \sigma_2^2$ (claim)

(b) 0.01　(c) $F_0 = 3.80$

(d) Rejection region: $F > 3.80$

(e) 2.12　(f) Fail to reject H_0.

(g) There is not enough evidence at the 1% level of significance to conclude that the variances in annual wages for San Francisco, CA and Baltimore, MD are different.

2. (a) H_0: $\mu_1 = \mu_2 = \mu_3$ (claim)

H_a: At least one mean is different from the others.

(b) 0.10　(c) $F_0 = 2.44$

(d) Rejection region: $F > 2.44$

(e) 27.48　(f) Reject H_0.

(g) There is enough evidence at the 10% level of significance to reject the claim that the mean annual wages are equal for all three cities.

3. (a) H_0: The distribution of educational achievement for people in the United States ages 35–44 is 13.4% not a high school graduate, 31.2% high school graduate, 17.2% some college, no degree; 8.8% associate's degree, 19.1% bachelor's degree, and 10.3% advanced degree.

H_a: The distribution of educational achievement for people in the United States ages 35–44 differs from the claimed distribution. (claim)

(b) 0.05　(c) $\chi_0^2 = 11.071$

(d) Rejection region: $\chi^2 > 11.071$

(e) 3.799　(f) Fail to reject H_0.

(g) There is not enough evidence at the 5% level of significance to conclude that the distribution for people in the United States ages 35–44 differs from the distribution for people ages 25 and older.

4. (a) H_0: The distribution of educational achievement for people in the United States ages 65–74 is 13.4% not a high school graduate, 31.2% high school graduate, 17.2% some college, no degree; 8.8% associate's degree, 19.1% bachelor's degree, and 10.3% advanced degree.

H_a: The distribution of educational achievement for people in the United States ages 65–74 differs from the claimed distribution. (claim)

(b) 0.01

(c) $\chi_0^2 = 15.086$

(d) Rejection region: $\chi^2 > 15.086$

(e) 26.175

(f) Reject H_0.

(g) There is enough evidence at the 1% level of significance to conclude that the distribution for people in the United States ages 65–74 differs from the distribution for people ages 25 and older.

Real Statistics–Real Decisions for Chapter 10 *(page 594)*

1. Reject H_0. There is enough evidence at the 1% level of significance to conclude that the distribution of responses differs from the claimed or expected distribution.

2. (a) $E_{1,1} = 15$, $E_{1,2} = 120$, $E_{1,3} = 165$, $E_{1,4} = 185$, $E_{1,5} = 135$, $E_{1,6} = 115$, $E_{1,7} = 155$, $E_{1,8} = 110$, $E_{2,1} = 15$, $E_{2,2} = 120$, $E_{2,3} = 165$, $E_{2,4} = 185$, $E_{2,5} = 135$, $E_{2,6} = 115$, $E_{2,7} = 155$, $E_{2,8} = 110$

(b) There is enough evidence at the 1% level of significance to conclude that the ages of the victims are related to the type of fraud.

CHAPTER 11

Section 11.1 *(page 604)*

1. A nonparametric test is a hypothesis test that does not require any specific conditions concerning the shapes of populations or the values of population parameters.

A nonparametric test is usually easier to perform than its corresponding parametric test, but the nonparametric test is usually less efficient.

3. When n is less than or equal to 25, the test statistic is equal to x (the smaller number of + or − signs).

When n is greater than 25, the test statistic is equal to

$$z = \frac{(x + 0.5) - 0.5n}{\frac{\sqrt{n}}{2}}.$$

5. Identify the claim and state H_0 and H_a. Identify the level of significance and sample size. Find the critical value using Table 8 (if $n \le 25$) or Table 4 ($n > 25$). Calculate the test statistic. Make a decision and interpret it in the context of the problem.

7. (a) H_0: median $\le$ \$300; H_a: median $>$ \$300 (claim)

(b) 1   (c) 5   (d) Fail to reject H_0.

(e) There is not enough evidence at the 1% level of significance for the accountant to conclude that the median amount of new credit card charges for the previous month was more than \$300.

9. (a) H_0: median $\le$ \$198,000 (claim)

H_a: median $>$ \$198,000

(b) 1   (c) 4   (d) Fail to reject H_0.

(e) There is not enough evidence at the 5% level of significance to reject the agent's claim that the median sales price of new privately owned one-family homes sold in the past year is \$198,000 or less.

11. (a) H_0: median $\ge$ \$3000 (claim); H_a: median $<$ \$3000

(b) −2.05   (c) −1.47   (d) Fail to reject H_0.

(e) There is not enough evidence at the 2% level of significance to reject the institution's claim that the median amount of credit card debt for families holding such debts is at least \$3000.

13. (a) H_0: median $\le$ 30; H_a: median $>$ 30 (claim)

(b) 4   (c) 10   (d) Fail to reject H_0.

(e) There is not enough evidence at the 1% level of significance to support the research group's claim that the median age of Twitter® users is greater than 30 years old.

15. (a) H_0: median $=$ 4 (claim); H_a: median $\ne$ 4

(b) −1.96   (c) −1.90   (d) Fail to reject H_0.

(e) There is not enough evidence at the 5% level of significance to reject the organization's claim that the median number of rooms in renter-occupied units is 4.

17. (a) H_0: median $=$ \$37.06 (claim); H_a: median $\ne$ \$37.06

(b) −2.575   (c) −0.91   (d) Fail to reject H_0.

(e) There is not enough evidence at the 1% level of significance to reject the labor organization's claim that the median hourly wage of computer systems analysts is \$37.06.

19. (a) H_0: The lower back pain intensity scores have not decreased.

H_a: The lower back pain intensity scores have decreased. (claim)

(b) 1   (c) 0   (d) Reject H_0.

(e) There is enough evidence at the 5% level of significance to conclude that the lower back pain intensity scores were lower after the acupuncture.

21. (a) H_0: The SAT scores have not improved.

H_a: The SAT scores have improved. (claim)

(b) 2   (c) 4   (d) Fail to reject H_0.

(e) There is not enough evidence at the 5% level of significance to conclude that the critical reading SAT scores improved.

23. (a) Reject H_0.

(b) There is enough evidence at the 5% level of significance to reject the claim that the proportion of adults who feel older than their real age is equal to the proportion of adults who feel younger than their real age.

25. Hypothesis test results:

Parameter : median of Variable

H_0 : Parameter $= 22.55$

H_A : Parameter $\neq 22.55$

Variable	n	n for test
Hourly wages (in dollars)	14	13

Sample Median	Below	Equal	Above	P-value
26.075	2	1	11	0.0225

$P = 0.0225 < 0.05$, so reject H_0. There is enough evidence at the 5% level of significance to reject the labor organization's claim that the median hourly wage of tool and die makers is $22.55.

27. (a) H_0: median $\leq$ $638 (claim); H_a: median $>$ $638

(b) 2.33   (c) 1.46   (d) Fail to reject H_0.

(e) There is not enough evidence at the 1% level of significance to reject the organization's claim that the median weekly earnings of female workers is less than or equal to $638.

29. (a) H_0: median $\leq$ 26 (claim); H_a: median $>$ 26

(b) 1.645   (c) 1.302   (d) Fail to reject H_0.

(e) There is not enough evidence at the 5% level of significance to reject the counselor's claim that the median age of brides at the time of their first marriage is less than or equal to 26 years.

Section 11.2 *(page 615)*

1. If the samples are dependent, use a Wilcoxon signed-rank test. If the samples are independent, use a Wilcoxon rank sum test.

3. (a) H_0: There is no reduction in diastolic blood pressure. (claim)

H_a: There is a reduction in diastolic blood pressure.

(b) Wilcoxon signed-rank test

(c) 10   (d) 17   (e) Fail to reject H_0.

(f) There is not enough evidence at the 1% level of significance to reject the claim that there was no reduction in diastolic blood pressure.

5. (a) H_0: The cost of prescription drugs is not lower in Canada than in the United States.

H_a: The cost of prescription drugs is lower in Canada than in the United States. (claim)

(b) Wilcoxon signed-rank test

(c) 4   (d) 6   (e) Fail to reject H_0.

(f) There is not enough evidence at the 5% level of significance for the researcher to conclude that the cost of prescription drugs is lower in Canada than in the United States.

7. (a) H_0: There is no difference in salaries.

H_a: There is a difference in salaries. (claim)

(b) Wilcoxon rank sum test

(c) ± 1.96 (d) -1.94 (e) Fail to reject H_0.

(f) There is not enough evidence at the 5% level of significance to support the representative's claim that there is a difference in the salaries earned by teachers in Wisconsin and Michigan.

9. Reject H_0. There is enough evidence at the 10% level of significance for the engineer to conclude that the gas mileage is improved.

Section 11.3 *(page 623)*

1. The conditions for using a Kruskal-Wallis test are that each sample must be randomly selected and the size of each sample must be at least 5.

3. (a) H_0: There is no difference in the premiums.

H_a: There is a difference in the premiums. (claim)

(b) 5.991   (c) 9.506   (d) Reject H_0.

(e) There is enough evidence at the 5% level of significance to conclude that the distributions of the annual premiums of the three states are different.

5. (a) H_0: There is no difference in the salaries.

H_a: There is a difference in the salaries. (claim)

(b) 6.251   (c) 1.202   (d) Fail to reject H_0.

(e) There is not enough evidence at the 10% level of significance to conclude that the distributions of the annual salaries in the four states are different.

7. Kruskal-Wallis results:

Data stored in separate columns.

Chi Square $= 8.0965185$ (adjusted for ties)

DF $= 2$

P-value $= 0.0175$

Column	n	Median	Ave. Rank
A	6	5	6.75
B	6	8.5	14.5
C	6	5	7.25

$P = 0.0175 > 0.01$, so fail to reject H_0. There is not enough evidence at the 1% level of significance to conclude that the distributions of the number of job offers at Colleges A, B, and C are different.

9. (a) Fail to reject H_0.   (b) Fail to reject H_0.

Both tests come to the same decision, which is that there is not enough evidence to support the claim that there is a difference in the number of days spent in the hospital.

Section 11.4 *(page 628)*

1. The Spearman rank correlation coefficient can be used to describe the relationship between linear or nonlinear data. Also, it can be used for data at the ordinal level and it is easier to calculate by hand than the Pearson correlation coefficient.

3. The ranks of the corresponding data are identical when r_s is equal to 1. The ranks are in "reverse" order when r_s is equal to -1. The ranks have no relationship when r_s is equal to 0.

5. (a) H_0: $\rho_s = 0$; H_a: $\rho_s \neq 0$ (claim)
   (b) 0.929
   (c) 0.857
   (d) Fail to reject H_0.
   (e) There is not enough evidence at the 1% level of significance to support the claim that there is a correlation between debt and income in the farming business.

7. (a) H_0: $\rho_s = 0$; H_a: $\rho_s \neq 0$ (claim)
   (b) 0.833
   (c) 0.950
   (d) Reject H_0.
   (e) There is enough evidence at the 1% level of significance to conclude that there is a correlation between the oat and wheat prices.

9. Fail to reject H_0. There is not enough evidence at the 5% level of significance to conclude that there is a correlation between science achievement scores and GNI.

11. Reject H_0. There is enough evidence at the 5% level of significance to conclude that there is a correlation between science and mathematics achievement scores.

13. Fail to reject H_0. There is not enough evidence at the 5% level of significance to conclude that there is a correlation between average hours worked and the number of on-the-job injuries.

Section 11.5 *(page 637)*

1. Answers will vary. *Sample answer:* It is called the runs test because it considers the number of runs of data in a sample to determine whether the sequence of data was randomly selected.

3. Number of runs: 8
   Run lengths: 1, 1, 1, 1, 3, 3, 1, 1

5. Number of runs: 9
   Run lengths: 1, 1, 1, 1, 6, 3, 2, 4

7. n_1 = number of T's = 6
   n_2 = number of F's = 6

9. n_1 = number of M's = 10
   n_2 = number of F's = 10

11. too high: 11; too low: 3

13. too high: 14; too low: 5

15. (a) H_0: The coin tosses were random.
    H_a: The coin tosses were not random. (claim)
    (b) lower critical value = 4
    upper critical value = 14
    (c) 9   (d) Fail to reject H_0.
    (e) There is not enough evidence at the 5% level of significance to support the claim that the coin tosses were not random.

17. (a) H_0: The sequence of leagues of winning teams is random.
    H_a: The sequence of leagues of winning teams is not random. (claim)
    (b) ± 1.96 (c) 1.79   (d) Fail to reject H_0.
    (e) There is not enough evidence at the 5% level of significance to conclude that the sequence of leagues of World Series winning teams is not random.

19. (a) H_0: The microchips are random by gender. (claim)
    H_a: The microchips are not random by gender.
    (b) lower critical value = 8
    upper critical value = 18
    (c) 12   (d) Fail to reject H_0.
    (e) There is not enough evidence at the 5% significance level to reject the claim that the microchips are random by gender.

21. Fail to reject H_0. There is not enough evidence at the 5% level of significance to support the claim that the daily high temperatures do not occur randomly.

23. Answers will vary.

Uses and Abuses for Chapter 11 *(page 639)*

1. Answers will vary.
2. Sign test $\rightarrow$ z- or t-test
   Paired-sample sign test $\rightarrow$ t-test
   Wilcoxon signed-rank test $\rightarrow$ t-test
   Wilcoxon rank sum test $\rightarrow$ z- or t-test
   Kruskal-Wallis test $\rightarrow$ one-way ANOVA
   Spearman rank correlation coefficient $\rightarrow$ Pearson correlation coefficient

Review Answers for Chapter 11 *(page 641)*

1. (a) H_0: median ≤ 650 (claim); H_a: median > 650
   (b) 2   (c) 7   (d) Fail to reject H_0.
   (e) There is not enough evidence at the 1% level of significance to reject the bank manager's claim that the median number of customers per day is no more than 650.

3. (a) H_0: median $= 2$ (claim); H_a: median $\neq 2$
   (b) -1.645 (c) -3.26 (d) Reject H_0.
   (e) There is enough evidence at the 10% level of significance to reject the agency's claim that the median sentence length for all federal prisoners is 2 years.

5. (a) H_0: There is no reduction in diastolic blood pressure. (claim)

H_a: There is a reduction in diastolic blood pressure.

(b) 2   (c) 3   (d) Fail to reject H_0.

(e) There is not enough evidence at the 5% level of significance to reject the claim that there was no reduction in diastolic blood pressure.

7. (a) Independent; Wilcoxon rank sum test

(b) H_0: There is no difference in the total times to earn a doctorate degree by female and male graduate students.

H_a: There is a difference in the total times to earn a doctorate degree by female and male graduate students. (claim)

(c) ± 2.575 (d) -1.357 (*or* 1.357)

(e) Fail to reject H_0.

(f) There is not enough evidence at the 1% level of significance to support the claim that there is a difference in the total times to earn a doctorate degree by female and male graduate students.

9. (a) H_0: There is no difference in the ages of doctorate recipients among the fields of study.

H_a: There is a difference in the ages of doctorate recipients among the fields of study. (claim)

(b) 9.210   (c) 6.741   (d) Fail to reject H_0.

(e) There is not enough evidence at the 1% level of significance to conclude that the distributions of ages of the doctorate recipients in these three fields of study are different.

11. (a) H_0: $\rho_s = 0$; H_a: $\rho_s \neq 0$ (claim)

(b) 0.786   (c) 0.830   (d) Reject H_0.

(e) There is enough evidence at the 5% level of significance to conclude that there is a correlation between overall score and price.

13. (a) H_0: The traffic stops were random by gender.

H_a: The traffic stops were not random by gender. (claim)

(b) lower critical value $= 8$

upper critical value $= 19$

(c) 14   (d) Fail to reject H_0.

(e) There is not enough evidence at the 5% level of significance to support the claim that the stops were not random by gender.

Chapter Quiz for Chapter 11   (*page 645*)

1. (a) H_0: There is no difference in the hourly earnings.

H_a: There is a difference in the hourly earnings. (claim)

(b) Wilcoxon rank sum test

(c) ± 1.645 (d) -3.326 (*or* 3.326)   (e) Reject H_0.

(f) There is enough evidence at the 10% level of significance to support the organization's claim that there is a difference in the hourly earnings of union and nonunion workers in state and local governments.

2. (a) H_0: median $= 52$ (claim); H_a: median $\neq 52$

(b) Sign test   (c) ± 1.96 (d) -2.75 (e) Reject H_0.

(f) There is enough evidence at the 5% level of significance to reject the organization's claim that the median number of annual volunteer hours is 52 hours.

3. (a) H_0: There is no difference in the sales prices among the regions.

H_a: There is a difference in the sales prices among the regions. (claim)

(b) Kruskal-Wallis test

(c) 11.345   (d) 25.957   (e) Reject H_0.

(f) There is enough evidence at the 1% level of significance to conclude that the distributions of the sales prices in these regions are different.

4. (a) H_0: The days with rain are random.

H_a: The days with rain are not random. (claim)

(b) Runs test

(c) lower critical value $= 10$

upper critical value $= 22$

(d) 16   (e) Fail to reject H_0.

(f) There is not enough evidence at the 5% level of significance for the meteorologist to conclude that days with rain are not random.

5. (a) H_0: $\rho_s = 0$; H_a: $\rho_s \neq 0$ (claim)

(b) Spearman rank correlation coefficient

(c) 0.829   (d) 0.886   (e) Reject H_0.

(f) There is enough evidence at the 10% level of significance to conclude that there is a correlation between the number of larceny-thefts and the number of motor vehicle thefts.

Real Statistics–Real Decisions for Chapter 11   (*page 646*)

1. (a) Answers will vary.

(b) Answers will vary.

(c) Answers will vary.

2. (a) Answers will vary.

(b) Sign test; You need to use the nonparametric test because nothing is known about the shape of the population.

(c) H_0: median ≥ 4.1; H_a: median < 4.1 (claim)

(d) Fail to reject H_0. There is not enough evidence at the 5% level of significance to support the claim that the median tenure for workers from the representative's district is less than 4.1 years.

3. (a) Wilcoxon rank sum test; You need to use the nonparametric test because nothing is known about the shape of the population.

(b) H_0: There is no difference between the median tenures for male workers and female workers.

H_a: There is a difference between the median tenures for male workers and female workers. (claim)

(c) Fail to reject H_0. There is not enough evidence at the 5% level of significance to support the claim that there is a difference between the median tenures for male workers and female workers.

Cumulative Review for Chapters 9–11 *(page 648)*

1. (a)

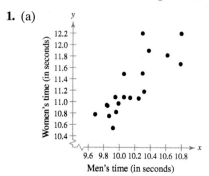

Men's time (in seconds)

$r \approx 0.815$; strong positive linear correlation

(b) Reject H_0. There is enough evidence at the 5% level of significance to conclude that there is a significant linear correlation between the men's and women's winning 100-meter times.

(c) $\hat{y} = 1.264x - 1.581$

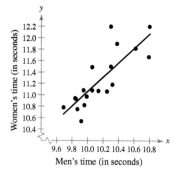

Men's time (in seconds)

(d) 10.93 seconds

2. There is enough evidence at the 5% level of significance to support the agency's claim that there is a difference in the weekly earnings of workers who are union members and workers who are not union members.

3. There is not enough evidence at the 1% level of significance to reject the company's claim that the median age of people with mutual funds is 50 years.

4. There is enough evidence at the 10% level of significance to reject the claim that the mean expenditures are equal for all four regions.

5. (a) 17,876.15 pounds per acre

(b) 20,148.12 pounds per acre

6. There is not enough evidence at the 10% level of significance to reject the administrator's claim that the standard deviations of reading test scores for eighth grade students are the same in Colorado and Utah.

7. There is enough evidence at the 1% level of significance for the representative to conclude that the distributions of annual household incomes in these regions are different.

8. There is not enough evidence at the 5% level of significance to conclude that the distribution of how much parents intend to contribute to their children's college costs differs from the claimed or expected distributions.

9. (a) 0.733; About 73.3% of the variation in height can be explained by the variation in metacarpal bone length; About 26.7% of the variation is unexplained.

(b) 4.255; The standard error of estimate of the height for a specific metacarpal bone length is about 4.255 centimeters.

(c) $168.026 < y < 190.83$; You can be 95% confident that the height will be between 168.026 centimeters and 190.83 centimeters when the metacarpal bone length is 50 centimeters.

10. There is enough evidence at the 10% level of significance to conclude that there is a correlation between the overall score and the price.

Selected Answers

CHAPTER 1

Section 1.1

10. False. A sample statistic can differ from sample to sample.

12. Population, because it is a collection of the energy collected from all the wind turbines on the wind farm.

14. Population, because it is a collection of the salaries of all the pharmacists at the pharmacy.

22. Population: The students who donate at a blood drive

Sample: The students who donate that have type O^+ blood

24. Population: Incomes of all home owners in Texas

Sample: Incomes of home owners in Texas with mortgages

26. Population: Collection of the heart rhythms of all infants in Italy

Sample: Collection of the heart rhythms of the 33,043 infants in the study

28. Population: Collection of the operating system choices of all people

Sample: Collection of the operating system choices of 1600 people surveyed

30. Population: Collection of the travel plans of all students at a college

Sample: Collection of the travel plans of 496 students surveyed

32. Population: Collection of the vacation budgets of all vacationers from the United States

Sample: Collection of the vacation budgets of the 791 U.S. vacationers surveyed

46. (a) The sample is the responses of the volunteers in the study.

(b) The population is the collection of the responses of all individuals who completed the math test.

(c) The statement "three times more likely to answer questions correctly" is an example of descriptive statistics.

(d) An inference drawn from the sample is that individuals who are not sleep deprived will be more likely to answer math questions correctly than individuals who are sleep deprived.

Section 1.2   *(page 13)*

4. False. For data at the interval level, you can calculate meaningful differences between data entries. You cannot calculate meaningful differences at the nominal or ordinal level.

6. False. Data at the ratio level can be placed in a meaningful order.

8. Quantitative, because balloon heights are numerical measurements.

10. Qualitative, because eye colors are attributes.

12. Quantitative, because carrying capacities are numerical measurements.

14. Qualitative, because student ID numbers are labels.

16. Qualitative, because tree species are labels.

18. Quantitative, because wait times are numerical measurements.

20. Qualitative. Nominal. No mathematical computations can be made, and data are categorized using names.

22. Quantitative. Ratio. A ratio of two data values can be formed, so one data value can be expressed as a multiple of another.

Section 1.3   *(page 23)*

2. A census includes the entire population; a sample includes only a portion of the population.

4. Replication is the repetition of an experiment under the same or similar conditions. It is important because it enhances the validity of the results.

6. False. A double-blind experiment is used to decrease the placebo effect.

8. False. A census is a count of an entire population.

12. Perform an observational study because you want to observe and record motorcycle helmet usage.

14. Use a survey to collect these data because it would be nearly impossible to ask all consumers whether they would still buy a product with a warning label.

16. Perform an observational study because you want to observe and record how often people wash their hands in public restrooms.

18. (a) The experimental units are the 80 people with early signs of arthritis. The treatment is the experimental sneaker.

(b) A problem with the design is that the sample size is small. The experiment could be replicated to increase validity.

(c) In a placebo-controlled, double-blind experiment, neither the subject nor the experimenter knows whether the subject is receiving a treatment or a placebo. The experimenter is informed after all the data have been collected.

(d) The groups could be randomly split into 20 males and 20 females in each treatment group.

26. Simple random sampling is used because each telephone number has an equal chance of being dialed, and all samples of 1012 phone numbers have an equal chance of being selected. The possible source of bias is that telephone sampling samples only those individuals who have telephones, who are available, and who are willing to respond.

32. Sampling, because the population of subscribers is too large for their most popular movie types to be easily recorded. Random sampling would be advised because it would be easy to select subscribers randomly and then record their most popular movie types.

40. Open Question

Advantage: Allows respondent to express some depth and shades of meaning in the answer. Allows for new solutions to be introduced.

Disadvantage: Not easily quantified and difficult to compare surveys.

Closed Question

Advantage: Easy to analyze results.

Disadvantage: May not provide appropriate alternatives and may influence the opinion of the respondent.

44. The placebo and Hawthorne effects are similar in that they both affect experimental results. These effects are different because the Hawthorne effect occurs when a subject changes behavior because the subject is in an experiment, whereas the placebo effect occurs when a subject reacts favorably to a placebo the subject has been given.

Review Answers for Chapter 1 *(page 29)*

2. Population: Collection of the opinions on managed health care of all nurses in the San Francisco area

Sample: Collection of the opinions on managed health care of 38 nurses in the San Francisco area that were sampled

10. The statement that 60% of physicians surveyed had considered leaving the practice of medicine because they were discouraged over the state of U.S. health care is an example of descriptive statistics.

An inference drawn from the sample is that 60% of all physicians had considered leaving the practice of medicine because of the state of U.S. health care.

20. Ratio. The data are quantitative, and it makes sense to say one amount is twice as large as another amount.

24. Take a survey because asking every college professor about teaching classes online would be nearly impossible.

26. Answers will vary. *Sample answer:* Number the volunteers and then use a random number generator to assign subjects randomly to one of the treatment groups or the control group.

28. Sampling. Take a survey, because asking all the students at the university about their favorite spring break destinations would be nearly impossible.

34. Convenience sampling is used because of the convenience of asking the people waiting for their baggage.

CHAPTER 2

Section 2.1 *(page 47)*

2. If there are too few or too many classes, it may be difficult to detect patterns because the data are too condensed or too spread out.

4. Relative frequency of a class is the portion or percentage of the data that falls in that class.

Cumulative frequency of a class is the sum of the frequencies of that class and all previous classes.

6. A frequency polygon displays relative frequencies whereas an ogive displays cumulative frequencies.

12. Class width = 13; Lower class limits: 12, 25, 38, 51, 64, 77; Upper class limits: 24, 37, 50, 63, 76, 89

14. Class width = 20; Lower class limits: 54, 74, 94, 114, 134, 154, 174, 194, 214, 234; Upper class limits: 73, 93, 113, 133, 153, 173, 193, 213, 233, 253

16. (a) Class width = 10

(b) and (c)

Class	Midpoint	Class boundaries
0–9	4.5	−0.5–9.5
10–19	14.5	9.5–19.5
20–29	24.5	19.5–29.5
30–39	34.5	29.5–39.5
40–49	44.5	39.5–49.5
50–59	54.5	49.5–59.5
60–69	64.5	59.5–69.5

18.

Class	Frequency, f	Midpoint	Relative frequency	Cumulative frequency
0–9	188	4.5	0.15	188
10–19	372	14.5	0.30	560
20–29	264	24.5	0.22	824
30–39	205	34.5	0.17	1029
40–49	83	44.5	0.07	1112
50–59	76	54.5	0.06	1188
60–69	32	64.5	0.03	1220
	$\Sigma f = 1220$		$\Sigma \dfrac{f}{n} = 1$	

30.

Class	Frequency, f	Midpoint	Relative frequency	Cumulative frequency
30–113	5	71.5	0.1724	5
114–197	7	155.5	0.2414	12
198–281	8	239.5	0.2759	20
282–365	2	323.5	0.0690	22
366–449	3	407.5	0.1034	25
450–533	4	491.5	0.1379	29
	$\Sigma f = 29$		$\Sigma \dfrac{f}{n} = 1$	

Class with greatest frequency: 198–281

Class with least frequency: 282–365

32.

Class	Frequency, f	Midpoint	Relative frequency	Cumulative frequency
32–35	3	33.5	0.1250	3
36–39	9	37.5	0.3750	12
40–43	8	41.5	0.3333	20
44–47	3	45.5	0.1250	23
48–51	1	49.5	0.0417	24
	$\Sigma f = 24$		$\Sigma \dfrac{f}{n} = 1$	

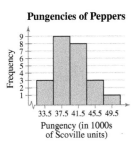

Pungencies of Peppers

The graph shows that most of the pungencies of the peppers were between 36 and 43 Scoville units. (Answers will vary.)

34.

Class	Frequency, f	Midpoint	Relative frequency	Cumulative frequency
2456–2542	7	2499	0.28	7
2543–2629	3	2586	0.12	10
2630–2716	2	2673	0.08	12
2717–2803	4	2760	0.16	16
2804–2890	9	2847	0.36	25
	$\Sigma f = 25$		$\Sigma \dfrac{f}{n} = 1$	

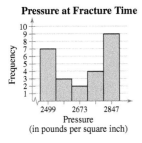

Pressure at Fracture Time

The graph shows that the most common pressures at fracture time were between 2804 and 2890 pounds per square inch. (Answers will vary.)

36.

Class	Frequency, f	Midpoint	Relative frequency	Cumulative frequency
10–24	11	17	0.3438	11
25–39	9	32	0.2813	20
40–54	6	47	0.1875	26
55–69	2	62	0.0625	28
70–84	4	77	0.1250	32
	$\Sigma f = 32$		$\Sigma \dfrac{f}{n} \approx 1$	

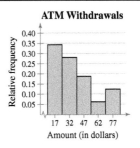

ATM Withdrawals

Class with greatest relative frequency: 10–24
Class with least relative frequency: 55–69

38.

Class	Frequency, f	Midpoint	Relative frequency	Cumulative frequency
6–7	3	6.5	0.12	3
8–9	10	8.5	0.38	13
10–11	6	10.5	0.23	19
12–13	6	12.5	0.23	25
14–15	1	14.5	0.04	26
	$\Sigma f = 26$		$\Sigma \dfrac{f}{n} = 1$	

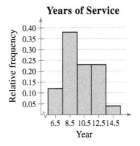

Years of Service

Class with greatest relative frequency: 8–9
Class with least relative frequency: 14–15

40.

Class	Frequency, f	Relative frequency	Cumulative frequency
16–22	2	0.10	2
23–29	3	0.15	5
30–36	8	0.40	13
37–43	5	0.25	18
44–50	0	0.00	18
51–57	2	0.10	20
	$\sum f = 20$	$\sum \dfrac{f}{n} = 1$	

Daily Saturated Fat Intake

Location of the greatest increase in frequency: 30–36

42.

Class	Frequency, f	Mid-point	Relative frequency	Cumulative frequency
0–2	17	1	0.3953	17
3–5	17	4	0.3953	34
6–8	7	7	0.1628	41
9–11	1	10	0.0233	42
12–14	0	13	0.0000	42
15–17	1	16	0.0233	43
	$\sum f = 43$		$\sum \dfrac{f}{n} \approx 1$	

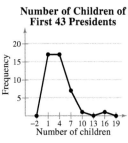

Number of Children of First 43 Presidents

The graph shows that most of the first 43 presidents had fewer than 6 children. (Answers will vary.)

44. (a)

Class	Frequency, f	Midpoint	Relative frequency	Cumulative frequency
7–53	22	30	0.44	22
54–100	16	77	0.32	38
101–147	6	124	0.12	44
148–194	2	171	0.04	46
195–241	2	218	0.04	48
242–288	0	265	0	48
289–335	0	312	0	48
336–382	2	359	0.04	50
	$\sum f = 50$		$\sum \dfrac{f}{n} = 1$	

(b)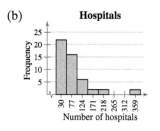

Hospitals

(c)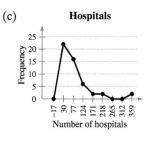

Hospitals

(d)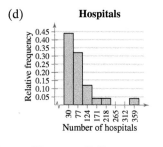

Hospitals

(e)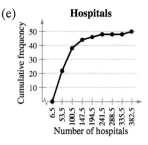

Hospitals

46.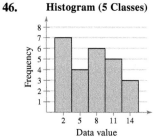

Histogram (5 Classes)   Histogram (10 Classes)

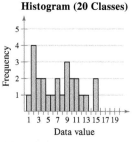

Histogram (20 Classes)

48. (a)

Class	Frequency, f	Relative frequency
976–1103	1	0.02
1104–1231	1	0.02
1232–1359	4	0.08
1360–1487	6	0.12
1488–1615	7	0.14
1616–1743	9	0.18
1744–1871	6	0.12
1872–1999	8	0.16
2000–2127	5	0.10
2128–2255	3	0.06
	$\sum f = 50$	$\sum \dfrac{f}{n} = 1$

Section 2.2   *(page 60)*

12. 214, 214, 214, 216, 216, 217, 218, 218, 220, 221, 223, 224, 225, 225, 227, 228, 228, 228, 228, 230, 230, 231, 235, 237, 239

Max: 239; Min: 214

22.

Housefly Life Spans

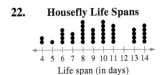

24.

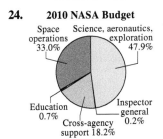

It appears that most of NASA's budget was spent on science, aeronautics, and exploration. (Answers will vary.)

26.

Ultraviolet Index

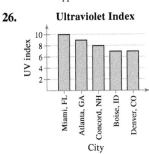

It appears that Boise, ID and Denver, CO have the same UV index. (Answers will vary.)

28.

Teachers' Salaries

It appears that there is no relation between a teacher's average salary and the number of students per teacher. (Answers will vary.)

30.

Manufacturing (percent of GDP)

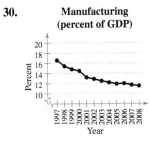

It appears that the largest decrease in manufacturing as a percent of GDP was from 2000 to 2001. (Answers will vary.)

32.

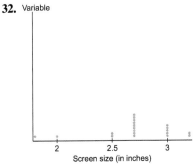

34. (a)

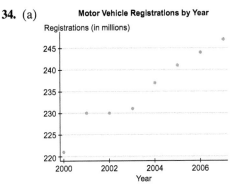

It appears that the number of registrations is increasing over time. (Answers will vary.)

(b)

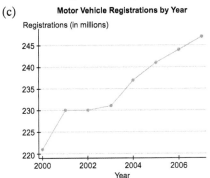

It appears that the number of crashes is decreasing over time. (Answers will vary.)

(c)

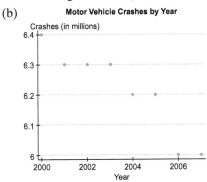

It appears that the number of registrations is increasing over time. (Answers will vary.)

(d)

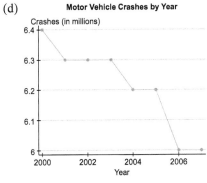

It appears that the number of crashes is decreasing over time. (Answers will vary.)

A113

36. (a) The graph is misleading because the vertical axis has no break. The percent of middle schoolers that responded "yes" appears three times larger than either of the others when the difference is only 10%. (Answers will vary.)

(b)

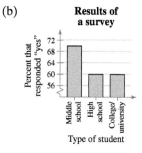

Results of a survey

38. (b)

U.S. Crude Oil Imports by Country of Origin 2008

40. (a)

```
      3:00 P.M. Class        8:00 P.M. Class
                      1 | 8 8 8 8 8 9 9 9 9 9
                      2 | 0 0 0 2 3 4 4 5 5 8 9 9
            8 5       3 | 1 1 9
              0       4 | 3 4 4
        9 7 5 3 1     5 | 6
    9 8 8 8 8 4 2 0   6 |           Key: 5|3|1 = 35-year-old
    7 7 6 5 5 5 3 3   7 | 1         in 3:00 P.M. class and
              5 4     8 |           31-year-old in 8:00 P.M. class
```

(b) In the 3:00 P.M. class, the lowest age is 35 years old and the highest age is 85 years old.

In the 8:00 P.M. class, the lowest age is 18 years old and the highest age is 71 years old.

(c) There are 26 participants in the 3:00 P.M. class and there are 30 participants in the 8:00 P.M. class.

(d) The participants in each class are clustered at one of the ends of their distribution range. The 3:00 P.M. class mostly has participants over 50 years old and the 8:00 P.M. class mostly has participants under 50 years old. (Answers will vary.)

Section 2.3 *(page 72)*

16. (10), because the distribution is rather symmetric due to the tendency of the weights of seventh grade boys.

24. $\bar{x}$ = not possible; median = not possible; mode = "Money needed"; The mean and median cannot be found because the data are at the nominal level of measurement.

26. $\bar{x}$ = not possible; median = not possible; mode = "Mashed"; The mean and median cannot be found because the data are at the nominal level of measurement.

A114

56.

Class	Frequency, f
1	6
2	5
3	4
4	6
5	4
6	5
	$\Sigma f = 30$

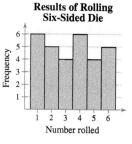

Results of Rolling Six-Sided Die

Uniform

60. Summary statistics:

Column	n	Mean	Median	Min	Max
Price (in dollars)	30	47.26433	43.585	15.9	132.39

Section 2.4 *(page 90)*

26. (a) Boston: $\bar{x} \approx 74.16$; median = 71.6; range = 29.8; $s^2 \approx 93.30$; $s \approx 9.66$

Chicago: $\bar{x} \approx 66.61$; median = 68.5; range = 11.6; $s^2 \approx 19.84$; $s \approx 4.45$

(b) It appears from the data that the annual salaries in Boston are more variable than the annual salaries in Chicago. The annual salaries in Boston have a higher mean and median than the annual salaries in Chicago.

28. (a) Team A: $\bar{x} \approx 0.2993$; median = 0.297; range = 0.149; $s^2 \approx 0.0019$; $s \approx 0.0435$

Team B: $\bar{x} \approx 0.2989$; median = 0.292; range = 0.067; $s^2 \approx 0.0006$; $s \approx 0.0242$

(b) It appears from the data that the batting averages for Team A are more variable than the batting averages for Team B. The batting averages for Team A have a lower mean and a higher median than those for Team B.

44.

Class	Midpoint, x	f	xf
145–164	154.5	8	1236.0
165–184	174.5	7	1221.5
185–204	194.5	3	583.5
205–224	214.5	1	214.5
225–244	234.5	1	234.5
		$N = 20$	$\Sigma xf = 3490.0$

$x - \mu$	$(x - \mu)^2$	$(x - \mu)^2 f$
−20	400	3200
0	0	0
20	400	1200
40	1600	1600
60	3600	3600
		$\Sigma (x - \mu)^2 f = 9600$

$\mu = 174.5$

$\sigma \approx 21.9$

46.

x	f	xf	$x - \bar{x}$	$(x - \bar{x})^2$	$(x - \bar{x})^2 f$
0	1	0	−1.9	3.61	3.61
1	9	9	−0.9	0.81	7.29
2	13	26	0.1	0.01	0.13
3	5	15	1.1	1.21	6.05
4	2	8	2.1	4.41	8.82
	$n = 30$	$\Sigma xf = 58$			$\Sigma (x - \bar{x})^2 f = 25.9$

$\bar{x} \approx 1.9$

$s \approx 0.9$

48.

Midpoint, x	f	xf
4.5	35.4	159.30
14.5	35.3	511.85
24.5	33.5	820.75
34.5	33.6	1159.20
44.5	28.8	1281.60
54.5	21.9	1193.55
64.5	13.9	896.55
74.5	7.2	536.40
84.5	2.6	219.70
94.5	0.3	28.35
	$n = 212.5$	$\Sigma xf = 6807.25$

$x - \bar{x}$	$(x - \bar{x})^2$	$(x - \bar{x})^2 f$
−27.53	757.90	26,829.66
−17.53	307.30	10,847.69
−7.53	56.70	1899.45
2.47	6.10	204.96
12.47	155.50	4478.40
22.47	504.90	11,057.31
32.47	1054.30	14,654.77
42.47	1803.70	12,986.64
52.47	2753.10	7158.06
62.47	3902.50	1170.75
		$\Sigma (x - \bar{x})^2 f = 91,301.02$

$\bar{x} \approx 32.03$

$s \approx 20.78$

50. Summary statistics:

Column	n	Mean	Variance
Price (in dollars)	12	216.65666	14,442.424

Std. Dev.	Median	Range	Min	Max
120.176636	189.99	410	89.99	499.99

Section 2.5 *(page 107)*

2. The salesperson sold more hardware equipment than 80% of the other salespeople.

4. The child has a higher IQ than 93% of the children in the same age group.

6. Quartiles are special cases of percentiles. Q_1 is the 25th percentile, Q_2 is the 50th percentile, and Q_3 is the 75th percentile.

10. False. The five numbers you need to graph a box-and-whisker plot are the minimum, the maximum, Q_1, Q_3, and the median (Q_2).

16. (a) Min = 100, $Q_1 = 130$, $Q_2 = 205$, $Q_3 = 270$, Max = 320

(b) IQR = 140

18. (a) Min = 25, $Q_1 = 50$, $Q_2 = 65$, $Q_3 = 70$, Max = 85

(b) IQR = 20

20. (a) Min = −1.3, $Q_1 = -0.3$, $Q_2 = 0.2$, $Q_3 = 0.4$, Max = 2.1

(b) IQR = 0.7

40. (a) Statistics: $z = \dfrac{60 - 63}{7} \approx -0.43$

Biology: $z = \dfrac{22 - 23}{3.9} \approx -0.26$

(b) The student did better on the biology test.

56. (a) The distribution of Concert 1 is symmetric. The distribution of Concert 2 is skewed right; Concert 1 has less variation.

(b) Concert 2 is more likely to have outliers because it has more variation.

(c) Concert 1, because 68% of the data should be between ±16.3 of the mean.

(d) No, you do not know the number of songs played at either concert or the actual lengths of the songs.

62. (a) **Summary statistics:**

Column	Min	Q1	Median	Q3	Max
Speed (in miles per hour)	52	65	70	72	88

(b)

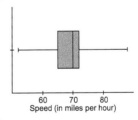

Vehicle Speeds

Speed (in miles per hour)

(c)

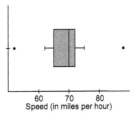

Vehicle Speeds

Speed (in miles per hour)

2. **Students Per Faculty Member**

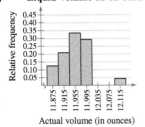

Class 13–17 has the greatest relative frequency, and class 23–27 has the least relative frequency.

4. **Liquid Volume 12-oz Cans**

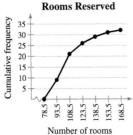

6. **Rooms Reserved**

8. **Air Quality of U.S. Cities**

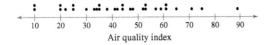

CHAPTER 3

Section 3.1 *(page 138)*

2. (a) Could represent the probability of an event. The probability of an event occurring must be contained in the interval $[0, 1]$ or $[0\%, 100\%]$.

(b) Could not represent the probability of an event. The probability of an event occurring cannot be less than 0.

(c) Could represent the probability of an event. The probability of an event occurring must be contained in the interval $[0, 1]$ or $[0\%, 100\%]$.

(d) Could represent the probability of an event. The probability of an event occurring must be contained in the interval $[0, 1]$ or $[0\%, 100\%]$.

(e) Could represent the probability of an event. The probability of an event occurring must be contained in the interval $[0, 1]$ or $[0\%, 100\%]$.

(f) Could not represent the probability of an event. The probability of an event occurring cannot be greater than 1.

18. {HHH, HHT, HTH, HTT, THH, THT, TTH, TTT}; 8

20. {(1, 1), (1, 2), (1, 3), (1, 4), (1, 5), (1, 6), (2, 1), (2, 2), (2, 3), (2, 4), (2, 5), (2, 6), (3, 1), (3, 2), (3, 3), (3, 4), (3, 5), (3, 6), (4, 1), (4, 2), (4, 3), (4, 4), (4, 5), (4, 6), (5, 1), (5, 2), (5, 3), (5, 4), (5, 5), (5, 6), (6, 1), (6, 2), (6, 3), (6, 4), (6, 5), (6, 6)}; 36

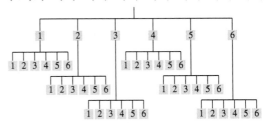

Section 3.2 *(page 150)*

18. (c) Dependent.

P(driving a pickup|driving a Ford)
$\neq P$(driving a pickup)

20. (e) Dependent.

P(being a nursing major|being a male)
$\neq P$(being a nursing major)

28. (d) Dependent.

P(owner spent \$100 or more|dog was a mixed breed)
$\neq P$(owner spent \$100 or more)

Review Exercises for Chapter 3 (page 181)

2. Sample space:

{(1, 1), (1, 2), (1, 3), (1, 4), (1, 5), (1, 6), (2, 1), (2, 2), (2, 3), (2, 4), (2, 5), (2, 6), (3, 1), (3, 2), (3, 3), (3, 4), (3, 5), (3, 6), (4, 1), (4, 2), (4, 3), (4, 4), (4, 5), (4, 6), (5, 1), (5, 2), (5, 3), (5, 4), (5, 5), (5, 6), (6, 1), (6, 2), (6, 3), (6, 4), (6, 5), (6, 6)}; 7

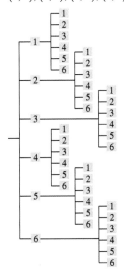

4. Sample space:

{GGG, GGB, GBG, GBB, BGG, BGB, BBG, BBB}; 3

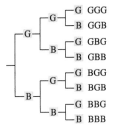

CHAPTER 4

Section 4.1 (page 197)

2. A discrete probability distribution lists each possible value a random variable can assume, together with its probability.

Condition 1: $0 \leq P(x) \leq 1$

Condition 2: $\sum P(x) = 1$

4. The mean of a probability distribution represents the "theoretical average" of a probability experiment.

8. False. The expected value of a discrete random variable is equal to the mean of the random variable.

10. Continuous; The length of time is a random variable that has an infinite number of possible outcomes and cannot be counted.

12. Discrete; The number of fatalities is a random variable that is countable.

14. Continuous; The length of time it takes to get to work is a random variable that cannot be counted.

16. Discrete; The number of tornadoes in the month of June in Oklahoma is a random variable that is countable.

18. Continuous; The tension at which a randomly selected guitar's strings have been strung is a random variable that cannot be counted.

20. Discrete; The total number of die rolls required for an individual to roll a five is a random variable that is countable.

28. (a)

x	$P(x)$
4	0.190
5	0.219
6	0.219
7	0.343
8	0.029

(b)

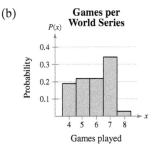

Skewed left

30. (a)

x	$P(x)$
0	0.250
1	0.297
2	0.229
3	0.168
4	0.034
5	0.021

(b)

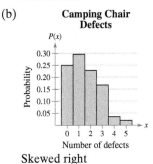

Skewed right

(c) 1.5, 1.5, 1.2

(d) The mean is 1.5, so the average batch of camping chairs has 1 or 2 defects. The standard deviation is 1.2, so most of the batches differ from the mean by no more than about 1 defect.

32. (a)

x	$P(x)$
0	0.059
1	0.122
2	0.163
3	0.178
4	0.213
5	0.128
6	0.084
7	0.053

(b)

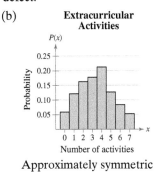

Approximately symmetric

(c) 3.3, 3.4, 1.8

(d) The mean is 3.3, so the average student is involved in about 3 extracurricular activities. The standard deviation is 1.8, so most of the students differ from the mean by no more than about 2 activities.

36. (a) 3.0 (b) 1.9 (c) 1.4 (d) 3.0

(e) The expected value is 3.0, so in an average hour the expected number of calls is 3. The standard deviation is 1.4, so the number of calls for most of the hours should differ from the expected value by no more than about 1 or 2 calls.

48. (a)

x	$P(x)$
1	0.128
2	0.124
3	0.124
4	0.122
5	0.123
6	0.125
7	0.127
8	0.128
	$\sum P(x) \approx 1$

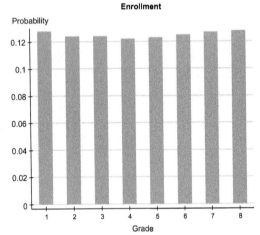

Section 4.2 *(page 211)*

28. (b)

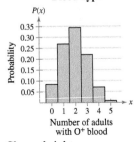

Skewed right

30. (a) $n = 5$, $p = 0.39$ (b)

x	$P(x)$
0	0.084
1	0.270
2	0.345
3	0.221
4	0.071
5	0.009

Skewed right

Review Answers for Chapter 4 *(page 227)*

12. (a)

x	f	$P(x)$
0	29	0.207
1	62	0.443
2	33	0.236
3	12	0.086
4	3	0.021
5	1	0.007
	$n = 140$	$\sum P(x) = 1$

(b)

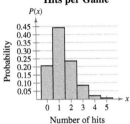

Skewed right

14. (a)

x	f	$P(x)$
15	76	0.134
30	445	0.786
60	30	0.053
90	3	0.005
120	12	0.021
	$n = 566$	$\sum P(x) \approx 1$

(b)

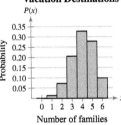

Skewed right

26. (a)

x	$P(x)$
0	0.001
1	0.014
2	0.073
3	0.206
4	0.328
5	0.279
6	0.099

(b) **Vacation Destinations**

Skewed left

A118

28. (a)

x	$P(x)$
0	0.007
1	0.059
2	0.201
3	0.342
4	0.291
5	0.099

(b)

Fast Food

Skewed left

46.

Proportion of boys from 4 births	Probability
0	$\frac{1}{16}$
$\frac{1}{4}$	$\frac{1}{4}$
$\frac{1}{2}$	$\frac{3}{8}$
$\frac{3}{4}$	$\frac{1}{4}$
1	$\frac{1}{16}$

CHAPTER 5

Section 5.4 *(page 274)*

18.

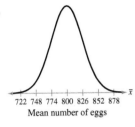

20.

22.

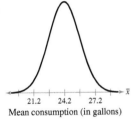

24.

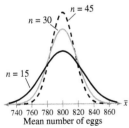

44.

Proportion of boys from 3 births	Probability
0	$\frac{1}{8}$
$\frac{1}{3}$	$\frac{3}{8}$
$\frac{2}{3}$	$\frac{3}{8}$
1	$\frac{1}{8}$

Section 5.5 *(page 287)*

24. Can use normal distribution.

(a) 0.0005   (b) 0.0342 (*Tech:* 0.0334)

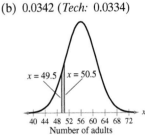

(c) 0.1357 (*Tech:* 0.1362)

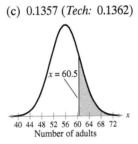

(d) Yes, the events in parts (a) and (b) are unusual because their probabilities are less than 0.05.

26. Can use normal distribution.

(a) 0.0396 (*Tech:* 0.0393)   (b) 0.0262

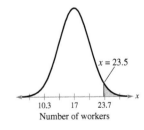

(c) 0.6736 (*Tech:* 0.6728)   (d) 0.0071

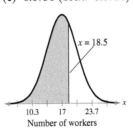
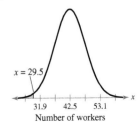

CHAPTER 6

Section 6.1 *(page 311)*

4. No; The 95% confidence interval means that with 95% confidence you can say that the population mean is in this interval. If a large number of samples is collected and a confidence interval created for each, approximately 95% of these intervals will contain the population mean.

56. (a) 35 bottles   (b) 9 bottles

$E = 1$ requires a larger sample size. As the error size decreases, a larger sample must be taken to obtain enough information from the population to ensure the desired accuracy.

60. (a) 12 soccer balls   (b) 3 soccer balls

$\sigma = 0.20$ requires a larger sample size. Due to the increased variability in the population, a larger sample size is needed to ensure the desired accuracy.

66. **80% confidence interval results:**

μ : mean of Variable
Std. Dev. not specified

Variable	n	Sample Mean
Carbohydrates (grams)	30	41.966667

Std. Err.	L. Limit	U. Limit
2.1493027	39.212223	44.721107

90% confidence interval results:

μ : mean of Variable
Std. Dev. not specified

Variable	n	Sample Mean
Carbohydrates (grams)	30	41.966667

Std. Err.	L. Limit	U. Limit
2.1493027	38.431377	45.501953

95% confidence interval results:

μ : mean of Variable
Std. Dev. not specified

Variable	n	Sample Mean
Carbohydrates (grams)	30	41.966667

Std. Err.	L. Limit	U. Limit
2.1493027	37.754112	46.179222

With 80% confidence, you can say that the population mean carbohydrate content is between 39.2 and 44.7 grams; with 90% confidence, you can say it is between 38.4 and 45.5 grams; with 95% confidence, you can say it is between 37.8 and 46.2 grams.

Section 6.2 *(page 323)*

18. 7.2; (15.0, 29.4); With 95% confidence, you can say that the population mean driving distance is between 15.0 and 29.4 miles.

20. 4.6; (17.6, 26.8); With 95% confidence, you can say that the population mean driving distance is between 17.6 and 26.8 miles. This confidence interval is narrower than the one found in Exercise 18.

34.

Mean	Sample Mean	Std. Err.	DF	L. Limit	U. Limit
μ	7.2	0.57287157	10	6.1616926	8.238307

Mean	Sample Mean	Std. Err.	DF	L. Limit	U. Limit
μ	7.2	0.57287157	10	5.9235625	8.476438

Mean	Sample Mean	Std. Err.	DF	L. Limit	U. Limit
μ	7.2	0.57287157	10	5.3844137	9.015586

Section 6.3 *(page 332)*

28. **90% confidence interval results:**

p : proportion of successes for population
Method: Standard-Wald

Proportion	Count	Total	Sample Prop.
p	734	2303	0.3187147

Std. Err.	L. Limit	U. Limit
0.009709986	0.30274323	0.33468622

95% confidence interval results:

p : proportion of successes for population
Method: Standard-Wald

Proportion	Count	Total	Sample Prop.
p	734	2303	0.3187147

Std. Err.	L. Limit	U. Limit
0.009709986	0.2996835	0.33774593

99% confidence interval results:

p : proportion of successes for population

Method: Standard-Wald

Proportion	Count	Total	Sample Prop.
p	734	2303	0.3187147

Std. Err.	L. Limit	U. Limit
0.009709986	0.29370347	0.34372598

With 90% confidence, you can say that the population proportion of U.S. adults who believe in UFOs is between 30.3% and 33.5%; with 95% confidence, you can say it is between 30.0% and 33.8%; with 99% confidence, you can say it is between 29.4% and 34.4%. As the level of confidence increases, the intervals get wider.

32. Answers will vary. *Sample answer:*

$$E = z_c \sqrt{\frac{\hat{p}\hat{q}}{n}} \qquad \text{Write original equation.}$$

$$\frac{E}{z_c} = \sqrt{\frac{\hat{p}\hat{q}}{n}} \qquad \text{Divide each side by } z_c.$$

$$\left(\frac{E}{z_c}\right)^2 = \frac{\hat{p}\hat{q}}{n} \qquad \text{Square each side.}$$

$$\hat{p}\hat{q}\left(\frac{z_c}{E}\right)^2 = n \qquad \text{Solve for } n.$$

Section 6.4 *(page 341)*

12. (a) (0.00467, 0.0195)

(b) (0.0683, 0.140)

With 95% confidence, you can say that the population variance is between 0.00467 and 0.0195, and the population standard deviation is between 0.0683 and 0.140 inch.

26. 99% confidence interval results:

σ^2 : variance of Variable

Variance	Sample Var.	DF	L. Limit	U. Limit
σ^2	0.64	6	0.20703505	5.6827703

(0.46, 2.38)

28. 97% confidence interval results:

σ^2 : variance of Variable

Variance	Sample Var.	DF	L. Limit	U. Limit
σ^2	77339.6	44	50970.645	130062.46

(225.8, 360.6)

Review Answers for Chapter 6 *(page 346)*

52. (0.94, 2.42); (0.97, 1.56)

With 90% confidence, you can say that the population variance is between 0.94 and 2.42 and the population standard deviation is between 0.97 and 1.56 seconds. The confidence intervals in Exercise 52 are narrower than those in Exercise 51.

CHAPTER 7

Section 7.1 *(page 367)*

2. A type I error occurs if the null hypothesis is rejected when it is true.

A type II error occurs if the null hypothesis is not rejected when it is false.

4. No; Failing to reject the null hypothesis means that there is not enough evidence to reject it.

10. False. If you want to support a claim, write it as the alternative hypothesis.

34. A type I error will occur if the actual proportion of U.S. adults who own a video game system is 0.26, but you reject H_0: $p = 0.26$.

A type II error will occur if the actual proportion of U.S. adults who own a video game system is not 0.26, but you fail to reject H_0: $p = 0.26$.

36. A type I error will occur if the actual mean cost of removing a virus infection is greater than or equal to $100, but you reject H_0: $\mu \geq 100$.

A type II error will occur if the actual mean cost of removing a virus infection is less than $100, but you fail to reject H_0: $\mu \geq 100$.

38. H_0: The mean time that the manufacturer's clocks lose is less than or equal to 0.02 second per day.

H_a: The mean time that the manufacturer's clocks lose is greater than 0.02 second per day.

H_0: $\mu \leq 0.02$; H_a: $\mu > 0.02$

Right-tailed because the alternative hypothesis contains $>$.

40. H_0: The proportion of lung cancer cases that are due to smoking is 87%.

H_a: The proportion of lung cancer cases that are due to smoking is not 87%.

H_0: $p = 0.87$; H_a: $p \neq 0.87$

Two-tailed because the alternative hypothesis contains $\neq$.

42. H_0: The mean tuition of the state's universities is less than or equal to $25,000 per year.

H_a: The mean tuition of the state's universities is greater than $25,000 per year.

H_0: $\mu \leq 25,000$; H_a: $\mu > 25,000$

Right-tailed because the alternative hypothesis contains $>$.

44. (a) There is enough evidence to reject the claim that the standard deviation of the life of the lawn mower is at most 2.8 years.

(b) There is not enough evidence to reject the claim that the standard deviation of the life of the lawn mower is at most 2.8 years.

46. (a) There is enough evidence to reject the automotive manufacturer's claim that the standard deviation for the gas mileage of its models is 3.9 miles per gallon.

(b) There is not enough evidence to reject the automotive manufacturer's claim that the standard deviation for the gas mileage of its models is 3.9 miles per gallon.

48. (a) There is enough evidence to reject the sports drink maker's claim that the mean calorie content of its beverages is 72 calories per serving.

(b) There is not enough evidence to reject the sports drink maker's claim that the mean calorie content of its beverages is 72 calories per serving.

Section 7.2 *(page 381)*

18.

20.

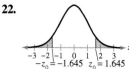

22.

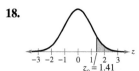

34. (e) There is not enough evidence at the 9% level of significance to support the analyst's claim that the mean annual salary for advertising account executives in Denver, Colorado is more than the national mean, $66,200.

36. (a) $H_0: \mu = 874$ (claim); $H_a: \mu \neq 874$

(b) $-z_0 = -1.96$, $z_0 = 1.96$; Rejection regions: $z < -1.96$, $z > 1.96$

(c) 1.98   (d) Reject H_0.

(e) There is enough evidence at the 5% level of significance to reject the association's claim that the mean monthly residential electricity consumption in your town is 874 kilowatt-hours.

40. (a) $H_0: \mu \geq 10{,}000$ (claim); $H_a: \mu < 10{,}000$

(b) $z_0 = -1.34$; Rejection region: $z < -1.34$

(c) -1.38 (d) Reject H_0.

(e) There is enough evidence at the 9% level of significance to reject the lamp manufacturer's claim that the mean life of fluorescent lamps is at least 10,000 hours.

42. (a) $H_0: \mu \geq 60$; $H_a: \mu < 60$ (claim)

(b) $z_0 = -2.33$; Rejection region: $z < -2.33$

(c) -3.62 (d) Reject H_0.

(e) There is enough evidence at the 1% level of significance to support the engineering company's claim that the mean time it takes an employee to evacuate a building during a fire drill is less than 60 seconds.

44. **Hypothesis test results:**

μ : population mean

$H_0: \mu = 495$

$H_A: \mu > 495$

Standard deviation = 17.8

Mean	n	Sample Mean	Std. Err.	Z-Stat	P-value
μ	65	498.4	2.2078183	1.5399818	0.0618

$P = 0.0618 > 0.05$, so fail to reject H_0. There is not enough evidence at the 5% level of significance to support the claim.

46. **Hypothesis test results:**

μ : population mean

$H_0: \mu = 28750$

$H_A: \mu \neq 28750$

Standard deviation = 3200

Mean	n	Sample Mean	Std. Err.	Z-Stat	P-value
μ	600	29130	130.63945	2.9087691	0.0036

$P = 0.0036 < 0.01$, so reject H_0. There is enough evidence at the 1% level of significance to support the claim.

50. b, d; If $\alpha = 0.07$, the rejection region is $z > 1.48$; because $z = 1.548$ is in the rejection region, you can reject H_0. If $n = 80$, the standardized test statistic is $z = 2.31$; because $z = 2.31$ is in the rejection region ($z > 1.645$), you can reject H_0.

Section 7.3 *(page 393)*

2. Identify the claim. State H_0 and H_a. Specify the level of significance. Identify the degrees of freedom. Determine the critical value(s) and rejection region(s). Find the standardized test statistic. Make a decision and interpret it in the context of the original claim.

The population must be normal or nearly normal.

10. (a) Fail to reject H_0 because $-1.372 < t < 1.372$.

(b) Reject H_0 because $t < -1.372$.

(c) Reject H_0 because $t > 1.372$.

(d) Fail to reject H_0 because $-1.372 < t < 1.372$.

12. (a) Fail to reject H_0 because $-1.725 < t < 1.725$.

(b) Reject H_0 because $t < -1.725$.

(c) Fail to reject H_0 because $-1.725 < t < 1.725$.

(d) Reject H_0 because $t > 1.725$.

14. Reject H_0. There is enough evidence at the 5% level of significance to support the claim.

16. Fail to reject H_0. There is not enough evidence at the 10% level of significance to support the claim.

18. (a) $H_0: \mu \leq 7$ (claim); $H_a: \mu > 7$

(b) $t_0 = 1.372$; Rejection region: $t > 1.372$

(c) 2.09

(d) Reject H_0.

(e) There is enough evidence at the 10% level of significance to reject the Internal Revenue Service's claim that the mean wait time for callers during a recent tax filing season was at most 7 minutes.

20. (a) $H_0: \mu \geq 30$ (claim); $H_a: \mu < 30$

(b) $t_0 = -2.567$; Rejection region: $t < -2.567$

(c) -3.74

(d) Reject H_0.

(e) There is enough evidence at the 1% level of significance to reject the company's claim that the mean battery life of their MP3 player is at least 30 hours.

22. (a) $H_0: \mu \leq 4$; $H_a: \mu > 4$ (claim)

(b) $t_0 = 2.518$; Rejection region: $t > 2.518$

(c) 1.94

(d) Fail to reject H_0.

(e) There is not enough evidence at the 1% level of significance to support the claim that the mean amount of waste generated by adults in the United States is more than 4 pounds per person per day.

28. (a) $H_0: \mu \geq \$240$ (claim); $H_a: \mu < \$240$

(b) 0.0058

(c) Reject H_0.

(d) There is enough evidence at the 10% level of significance to reject the travel association's claim that the mean daily lodging cost for two adults traveling together on vacation in San Francisco is at least $240.

30. (a) $H_0: \mu = 11.0$ (claim); $H_a: \mu \neq 11.0$

(b) 0.3155   (c) Fail to reject H_0.

(d) There is not enough evidence at the 1% level of significance to reject the dean's claim that the mean number of classroom hours per week for full-time faculty is 11.0.

32. **Hypothesis test results:**

μ : population mean

$H_0 : \mu = 27$

$H_A : \mu \neq 27$

Mean	Sample Mean	Std. Err.	DF	T-Stat	P-value
μ	31.5	1.3567731	11	3.316693	0.0069

$P = 0.0069 < 0.01$, so reject H_0. There is enough evidence at the 1% level of significance to support the claim.

34. **Hypothesis test results:**

μ : population mean

$H_0 : \mu = 2118$

$H_A : \mu < 2118$

Mean	Sample Mean	Std. Err.	DF	T-Stat	P-value
μ	1787	93.13368	16	-3.5540311	0.0013

$P = 0.0013 < 0.10$, so reject H_0. There is enough evidence at the 10% level of significance to reject the claim.

Section 7.4   *(page 401)*

2. Verify that $np \geq 5$ and $nq \geq 5$. State H_0 and H_a. Specify the level of significance α. Determine the critical value(s) and rejection region(s). Find the standardized test statistic. Make a decision and interpret it in the context of the original claim.

6. Can use normal distribution.

Fail to reject H_0. There is not enough evidence at the 4% level of significance to support the claim.

10. (a) $H_0: p \geq 0.40$ (claim); $H_a: p < 0.40$

(b) $z_0 = -2.05$; Rejection region: $z < -2.05$

(c) -2.5

(d) Reject H_0.

(e) There is enough evidence at the 2% level of significance to reject the research center's claim that at least 40% of U.S. adults think the census count is accurate.

12. (a) $H_0: p = 0.05$ (claim); $H_a: p \neq 0.05$

(b) $-z_0 = -1.75$, $z_0 = 1.75$; Rejection regions: $z < -1.75$, $z > 1.75$

(c) 3.34

(d) Reject H_0.

(e) There is enough evidence at the 8% level of significance to reject the researcher's claim that 5% of children under 18 years of age have asthma.

16. (a) $H_0: p = 0.30$ (claim); $H_a: p \neq 0.30$

(b) $-z_0 = -1.96$, $z_0 = 1.96$; Rejection regions: $z < -1.96$, $z > 1.96$

(c) 1.85

(d) Fail to reject H_0.

(e) There is not enough evidence at the 5% level of significance to reject the humane society's claim that 30% of U.S. households own a cat.

Section 7.5   *(page 410)*

2. No; In a χ^2-distribution, all χ^2-values are greater than or equal to 0.

4. State H_0 and H_a and identify the claim. Specify the level of significance. Determine the degrees of freedom. Determine the critical value(s) and rejection region(s). Find the standardized test statistic. Make a decision and interpret it in the context of the original claim.

18. Fail to reject H_0. There is not enough evidence at the 1% level of significance to support the claim.

20. (a) H_0: $\sigma^2 = 1.0$ (claim); H_a: $\sigma^2 \neq 1.0$

   (b) $\chi_L^2 = 12.401$, $\chi_R^2 = 39.364$;
   Rejection regions: $\chi^2 < 12.401$, $\chi^2 > 39.364$

   (c) 39.6

   (d) Reject H_0.

   (e) There is enough evidence at the 5% level of significance to reject the manufacturer's claim that the variance of the gas mileages of its hybrid vehicles is 1.0.

22. (a) H_0: $\sigma \geq 30$; H_a: $\sigma < 30$ (claim)

   (b) $\chi_0^2 = 6.408$; Rejection region: $\chi^2 < 6.408$

   (c) 21.325   (d) Fail to reject H_0.

   (e) There is not enough evidence at the 1% level of significance to support the state school administrator's claim that the standard deviation of test scores for eighth grade students who took a U.S. history assessment test is less than 30 points.

24. (a) H_0: $\sigma = 6.14$ (claim); H_a: $\sigma \neq 6.14$

   (b) $\chi_L^2 = 8.907$, $\chi_R^2 = 32.852$;
   Rejection regions: $\chi^2 < 8.907$, $\chi^2 > 32.852$

   (c) 21.293   (d) Fail to reject H_0.

   (e) There is not enough evidence at the 5% level of significance to reject the doctor's claim that the standard deviation of the lengths of stay for patients involved in a crash in which the vehicle struck a tree is 6.14 days.

26. (a) H_0: $\sigma \leq \$30$ (claim); H_a: $\sigma > \$30$

   (b) $\chi_0^2 = 37.566$;
   Rejection region: $\chi^2 > 37.566$

   (c) 27.613   (d) Fail to reject H_0.

   (e) There is not enough evidence at the 1% level of significance to reject the travel agency's claim that the standard deviation of the room rates of hotels in the city is no more than $30.

28. (a) H_0: $\sigma \geq \$10,600$ (claim); H_a: $\sigma < \$10,600$

   (b) $\chi_0^2 = 11.651$; Rejection region: $\chi^2 < 11.651$

   (c) 11.607

   (d) Reject H_0.

   (e) There is enough evidence at the 10% level of significance to reject the organization's claim that the standard deviation of the annual salaries of commodity buyers is at least $10,600.

30. **Hypothesis test results:**

   σ^2 : variance of Variable

   $H_0 : \sigma^2 = 14.85$

   $H_A : \sigma^2 \neq 14.85$

Variance	Sample Var.	DF	Chi-Square Stat	P-value
σ^2	28.75	16	30.97643	0.0271

   $P = 0.0271 < 0.05$, so reject H_0. There is enough evidence at the 5% level of significance level to reject the claim.

32. $\sigma^2 = 418^2 = 174,724$
   $s^2 = 305^2 = 93,025$

   Hypothesis test results:

   σ^2 : variance of Variable

   $H_0 : \sigma^2 = 174,724$

   $H_A : \sigma^2 \neq 174,724$

Variance	Sample Var.	DF	Chi-Square Stat	P-value
σ^2	93,025	23	12.245456	0.067

   $P = 0.067 < 0.10$, so reject H_0. There is enough evidence at the 10% level of significance level to support the claim.

Review Answers for Chapter 7   (page 417)

8. (a) H_0: $\mu \geq 400$ (claim); H_a: $\mu < 400$

   (b) A type I error will occur if the actual mean shelf life of the dried fruit is at least 400 days, but you reject H_0: $\mu \geq 400$.

   A type II error will occur if the actual mean shelf life of the dried fruit is less than 400 days, but you fail to reject H_0: $\mu \geq 400$.

   (c) Left-tailed because the alternative hypothesis contains $<$.

   (d) There is enough evidence to reject the agricultural cooperative's claim that the mean shelf life of the dried fruit is at least 400 days.

   (e) There is not enough evidence to reject the agricultural cooperative's claim that the mean shelf life of the dried fruit is at least 400 days.

10. (a) H_0: $\mu \geq 25$; H_a: $\mu < 25$ (claim)

   (b) A type I error will occur if the actual mean number of grams of carbohydrates in one bar is at least 25, but you reject H_0: $\mu \geq 25$.

   A type II error will occur if the actual mean number of grams of carbohydrates in one bar is less than 25, but you fail to reject H_0: $\mu \geq 25$.

   (c) Left-tailed because the alternative hypothesis contains $<$.

   (d) There is enough evidence to support the energy bar maker's claim that the mean number of grams of carbohydrates in one bar is less than 25.

(e) There is not enough evidence to support the energy bar maker's claim that the mean number of grams of carbohydrates in one bar is less than 25.

14. $H_0: \mu = 230$; $H_a: \mu \neq 230$ (claim)

$z = -5.41$; P-value ≈ 0

$\alpha = 0.10 \Rightarrow$ Reject H_0.

$\alpha = 0.05 \Rightarrow$ Reject H_0.

$\alpha = 0.01 \Rightarrow$ Reject H_0.

16. $-2.81, 2.81$

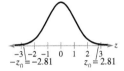

18. $-1.75, 1.75$

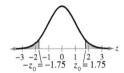

28. Fail to reject H_0. There is not enough evidence at the 5% level of significance to reject the tourist agency's claim that the mean daily cost of meals and lodging for a family of 4 traveling in Hawaii is at most $650.

42. Fail to reject H_0. There is not enough evidence at the 5% level of significance to reject the association's claim that the typical household spends a mean amount of $2698 per year on food away from home.

CHAPTER 8

Section 8.1 (page 434)

22. (a) The claim is "the mean braking distance for Type C is greater than the mean braking distance for Type D."

$H_0: \mu_1 \leq \mu_2$; $H_a: \mu_1 > \mu_2$ (claim)

(b) $z_0 = 1.28$;

Rejection region: $z > 1.28$

(c) 3.92   (d) Reject H_0.

(e) There is enough evidence at the 10% level of significance to support the safety engineer's claim that the mean braking distance for Type C is greater than the mean braking distance for Type D.

24. (a) The claim is "the wind speeds in Region C and Region D are equal."

$H_0: \mu_1 = \mu_2$ (claim); $H_a: \mu_1 \neq \mu_2$

(b) $-z_0 = -2.17$, $z_0 = 2.17$;

Rejection regions: $z < -2.17, z > 2.17$

(c) -2.21 (d) Reject H_0.

(e) There is enough evidence at the 3% level of significance to reject the researcher's claim that the wind speeds for Region C and Region D are equal.

26. (a) The claim is "high school students in a college preparation program have higher ACT scores than those in a general program."

$H_0: \mu_1 \leq \mu_2$; $H_a: \mu_1 > \mu_2$ (claim)

(b) $z_0 = 1.28$;

Rejection region: $z > 1.28$

(c) 2.07   (d) Reject H_0.

(e) There is enough evidence at the 10% level of significance to support the guidance counselor's claim that the ACT scores are higher for high school students in a college preparation program.

28. (a) The claim is "households in the United States headed by people under the age of 25 spend less on food away from home than do households headed by people ages 65–74."

$H_0: \mu_1 \geq \mu_2$; $H_a: \mu_1 < \mu_2$ (claim)

(b) $z_0 = -1.645$;

Rejection region: $z < -1.645$

(c) -0.08 (d) Fail to reject H_0.

(e) There is not enough evidence at the 5% level of significance to support the restaurant association's claim that households in the United States headed by people under the age of 25 spend less on food away from home than do households headed by people ages 65–74.

30. (a) The claim is "households in the United States headed by people under the age of 25 spend less on food away from home than do households headed by people ages 65–74."

$H_0: \mu_1 \geq \mu_2$; $H_a: \mu_1 < \mu_2$ (claim)

(b) $z_0 = -1.645$;

Rejection region: $z < -1.645$

(c) -3.19 (d) Reject H_0.

(e) There is enough evidence at the 5% level of significance to support the restaurant association's claim that households in the United States headed by people under the age of 25 spend less on food away from home than do households headed by people ages 65–74.

The new samples do lead to a different conclusion.

32. (a) The claim is "middle school boys spent less time studying in 1981 than middle school boys do today."

$H_0: \mu_1 \geq \mu_2$; $H_a: \mu_1 < \mu_2$ (claim)

(b) $z_0 = -1.88$;

Rejection region: $z < -1.88$

(c) -14.17 (d) Reject H_0.

(e) There is enough evidence at the 3% level of significance to support the sociologist's claim that middle school boys spent less time studying in 1981 than middle school boys do today.

34. (a) The claim is "there is no difference in the mean nut diameter manufactured by two different methods."

$H_0: \mu_1 = \mu_2$ (claim); $H_a: \mu_1 \neq \mu_2$

(b) $-z_0 = -2.05$, $z_0 = 2.05$;

Rejection regions: $z < -2.05$, $z > 2.05$

(c) -66.64 (d) Reject H_0.

(e) There is enough evidence at the 4% level of significance to reject the production engineer's claim that there is no difference in the mean nut diameter manufactured by two different methods.

36. They are equivalent through algebraic manipulation of the equation.

$\mu_1 \geq \mu_2 \Rightarrow \mu_1 - \mu_2 \geq 0$

38. Hypothesis test results:

μ_1 : mean of population 1 (Std. Dev. = 20.8)

μ_2 : mean of population 2 (Std. Dev. = 24.6)

$\mu_1 - \mu_2$: mean difference

$H_0 : \mu_1 - \mu_2 = 0$

$H_A : \mu_1 - \mu_2 > 0$

Difference	n_1	n_2	Sample Mean
$\mu_1 - \mu_2$	80	80	3.6

Std. Err.	Z-Stat	P-value
3.6017356	0.9995181	0.1588

$P = 0.1588 > 0.10$, so fail to reject H_0.

There is not enough evidence at the 10% level of significance to support the claim.

40.

Difference	n_1	n_2	Sample Mean
$\mu_1 - \mu_2$	100	100	44.58

Std. Err.	Z-Stat	P-value
23.294636	1.9137454	0.0278

44. $H_0: \mu_1 - \mu_2 \leq 15{,}000$; $H_a: \mu_1 - \mu_2 > 15{,}000$ (claim)

Fail to reject H_0. There is not enough evidence at the 10% level of significance to support the claim that the difference in mean annual salaries is greater than $15,000.

48. $H_0: \mu_1 \leq \mu_2$; $H_a: \mu_1 > \mu_2$ (claim)

Reject H_0. There is enough evidence at the 5% level of significance to support the claim. You should recommend Irinotecan over Fluorouracil because the average number of months with no reported cancer-related pain was significantly higher with Irinotecan.

50. The 95% CI for $\mu_1 - \mu_2$ in Exercise 46 contained only values greater than 0 and, as found in Exercise 48, there was enough evidence at the 5% level of significance to support the claim.

If the CI for $\mu_1 - \mu_2$ contains only positive numbers, you reject H_0 because the null hypothesis states that $\mu_1 - \mu_2$ is less than or equal to 0.

Section 8.2   *(page 446)*

2. State hypotheses and identify the claim. Specify the level of significance. Determine the degrees of freedom. Find the critical value(s) and identify the rejection region(s). Find the standardized test statistic. Make a decision and interpret it in the context of the original claim.

16. (a) The claim is "the mean footwell intrusions for small pickups and small SUVs are equal."

$H_0: \mu_1 = \mu_2$ (claim); $H_a: \mu_1 \neq \mu_2$

(b) $-t_0 = -2.878$, $t_0 = 2.878$;

Rejection regions: $t < -2.878$, $t > 2.878$

(c) 0.87   (d) Fail to reject H_0.

(e) There is not enough evidence at the 1% level of significance to reject the claim that the mean footwell intrusions for small pickups and small SUVs are equal.

20. (a) The claim is "the experimental method produces steel with greater mean tensile strength."

$H_0: \mu_1 \leq \mu_2$; $H_a: \mu_1 > \mu_2$ (claim)

(b) $t_0 = 1.350$;

Rejection region: $t > 1.350$

(c) 3.429   (d) Reject H_0.

(e) There is enough evidence at the 10% level of significance to support the claim that the experimental method produces steel with greater mean tensile strength and to recommend using the experimental method.

22. (a) The claim is "the mean science test score is lower for students taught using the traditional lab method than it is for students taught using the interactive simulation software."

$H_0: \mu_1 \geq \mu_2$; $H_a: \mu_1 < \mu_2$ (claim)

(b) $t_0 = -1.645$;

Rejection region: $t < -1.645$

(c) -1.721 (d) Reject H_0.

(e) There is enough evidence at the 5% level of significance to support the claim that the mean science test score is lower for students taught using the traditional lab method than it is for students taught using the interactive simulation software.

24. Hypothesis test results:

μ_1 : mean of population 1

μ_2 : mean of population 2

$\mu_1 - \mu_2$: mean difference

$H_0 : \mu_1 - \mu_2 = 0$

$H_A : \mu_1 - \mu_2 \neq 0$

(with pooled variances)

Difference	Sample Mean	Std. Err.
$\mu_1 - \mu_2$	-11	2.889967

DF	T-Stat	P-value
11	-3.8062718	0.0029

$P = 0.0029 < 0.01$, so reject H_0.

There is enough evidence at the 1% level of significance to support the claim.

26. Hypothesis test results:

μ_1: mean of population 1

μ_2: mean of population 2

$\mu_1 - \mu_2$: mean difference

$H_0: \mu_1 - \mu_2 = 0$

$H_A: \mu_1 - \mu_2 \neq 0$

(without pooled variances)

Difference	Sample Mean	Std. Err.
$\mu_1 - \mu_2$	22.4	8.114349

DF	T-Stat	P-value
6.2159977	2.7605417	0.0317

$P = 0.0317 < 0.10$, so reject H_0.

There is not enough evidence at the 10% level of significance to reject the claim.

Section 8.3 *(page 456)*

2. The symbol $\bar{d}$ represents the mean of the differences between the paired data entries in the dependent samples.

The symbol s_d represents the standard deviation of the differences between the paired data entries in the dependent samples.

10. (a) The claim is "an SAT preparation course improves the test scores of students."

$H_0: \mu_d \geq 0$; $H_a: \mu_d < 0$ (claim)

(b) $t_0 = -2.821$;

Rejection region: $t < -2.821$

(c) $\bar{d} = -59.9$; $s_d \approx 26.831$

(d) -7.060 (e) Reject H_0.

(f) There is enough evidence at the 1% level of significance to support the SAT preparation course's claim that its course improves the test scores of students.

20.

Difference	Sample Diff.
Time (beginning) − Time (end)	8.2375

Std. Err.	DF	T-Stat	P-value
1.8464191	7	4.461338	0.0029

Section 8.4 *(page 465)*

10. (a) The claim is "the proportion of subjects who are cancer-free after one year is greater for subjects who took the drug than for subjects who took a placebo."

$H_0: p_1 \leq p_2$; $H_a: p_1 > p_2$ (claim)

(b) $z_0 = 1.28$;

Rejection region: $z > 1.28$

(c) 5.72   (d) Reject H_0.

(e) There is enough evidence at the 10% level of significance to support the claim that the proportion of subjects who are cancer-free after one year is greater for subjects who took the drug than for subjects who took a placebo.

12. (a) The claim is "there is no difference in the proportion of females who have reduced the amount they spend on eating out and the proportion of males who have reduced the amount they spend on eating out."

$H_0: p_1 = p_2$ (claim); $H_a: p_1 \neq p_2$

(b) $-z_0 = -2.575$, $z_0 = 2.575$;

Rejection regions: $z < -2.575$, $z > 2.575$

(c) 2.36 (*Tech:* 2.38)   (d) Fail to reject H_0.

(e) There is not enough evidence at the 1% level of significance to reject the claim that there is no difference in the proportion of females who have reduced the amount they spend on eating out and the proportion of males who have reduced the amount they spend on eating out.

14. (a) The claim is "the proportion of subjects who are free of nausea is greater for subjects who took the drug than for subjects who took a placebo."

$H_0: p_1 \leq p_2$; $H_a: p_1 > p_2$ (claim)

(b) $z_0 = 1.28$; Rejection region: $z > 1.28$

(c) 3.46 (*Tech:* 3.48)   (d) Reject H_0.

(e) There is enough evidence at the 10% level of significance to support the claim that the proportion of subjects who are free of nausea is greater for subjects who took the drug than for subjects who took a placebo.

16. (a) The claim is "the proportion of motorcyclists who wear a helmet in the Northeast is less than the proportion of motorcyclists who wear a helmet in the Midwest."

$H_0: p_1 \geq p_2$; $H_a: p_1 < p_2$ (claim)

(b) $z_0 = -1.28$; Rejection region: $z < -1.28$

(c) -1.53 (d) Reject H_0.

(e) There is enough evidence at the 10% level of significance to support the claim that the proportion of motorcyclists who wear a helmet in the Northeast is less than the proportion of motorcyclists who wear a helmet in the Midwest.

18. (a) The claim is "the proportion of adults who use the Internet is greater for adults who live in an urban area than for adults who live in a rural area."

$H_0: p_1 \le p_2$; $H_a: p_1 > p_2$ (claim)

(b) $z_0 = 1.28$;

Rejection region: $z > 1.28$

(c) 1.20   (d) Fail to reject H_0.

(e) There is not enough evidence at the 10% level of significance to support the claim that the proportion of adults who use the Internet is greater for adults who live in an urban area than for adults who live in a rural area.

22. There is not enough evidence at the 5% level of significance to support the claim that there is a difference between the proportion of customers who wait 20 minutes or less at the Roanoke office and the proportion of customers who wait 20 minutes or less at the Fairfax North office.

24. Yes; When $\alpha = 0.10$, the rejection regions become $z < -1.645$ and $z > 1.645$. Because $-1.645 < -1.15 < 1.645$, you still fail to reject H_0. There is not enough evidence at the 10% level of significance to support the claim that there is a difference between the proportion of customers who wait 20 minutes or less at the Roanoke office and the proportion of customers who wait 20 minutes or less at the Fairfax North office.

26. Hypothesis test results:

p_1 : proportion of successes for population 1

p_2 : proportion of successes for population 2

$p_1 - p_2$: difference in proportions

$H_0 : p_1 - p_2 = 0$

$H_A : p_1 - p_2 > 0$

Difference	Count1	Total1	Count2	Total2
$p_1 - p_2$	5610	13200	6362	14200

Sample Diff.	Std. Err.	Z-Stat	P-value
−0.023028169	0.005996968	−3.8399684	0.9999

$P = 0.9999 > 0.05$, so fail to reject H_0.

There is not enough evidence at the 5% level of significance to support the claim that the proportion of women ages 18 to 24 living in their parents' homes was greater in 2000 than in 2009.

28. Hypothesis test results:

p_1 : proportion of successes for population 1

p_2 : proportion of successes for population 2

$p_1 - p_2$: difference in proportions

$H_0 : p_1 - p_2 = 0$

$H_A : p_1 - p_2 \ne 0$

Difference	Count1	Total1	Count2	Total2
$p_1 - p_2$	8120	14500	6362	14200

Sample Diff.	Std. Err.	Z-Stat	P-value
0.11197183	0.005902886	18.968998	<0.0001

$P < 0.0001 < 0.10$, so reject H_0.

There is enough evidence at the 10% level of significance to reject the claim that the proportion of 18- to 24-year-olds living in their parents' homes in 2009 was the same for men and women.

Review Answers for Chapter 8   *(page 471)*

8. (e) There is enough evidence at the 10% level of significance to reject the agency's claim that the mean annual salary of civilian federal employees in California is the same as those in Illinois.

30. (a) The claim is "the proportion of U.S. adults who believe it is likely that life exists on other planets is less now than in 2007."

$H_0: p_1 \le p_2$; $H_a: p_1 > p_2$ (claim)

(b) $z_0 = 1.645$;

Rejection region: $z > 1.645$

(c) 1.80   (d) Reject H_0.

(e) There is enough evidence at the 5% level of significance to support the claim that the proportion of U.S. adults who believe it is likely that life exists on other planets is less now than in 2007.

CHAPTER 9

Section 9.1   *(page 495)*

4. The sample correlation coefficient r measures the strength and direction of a linear relationship between two variables; $r = -0.932$ indicates a stronger correlation because $|-0.932| = 0.932$ is closer to 1 than $|0.918| = 0.918$.

28. (a)

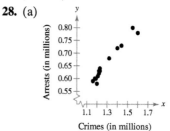

38. (a)

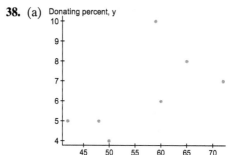

Section 9.2 (page 505)

18.

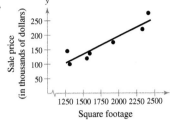

20.

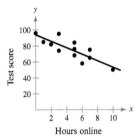

22.

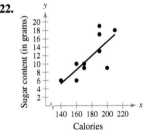

24.

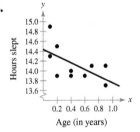

26.

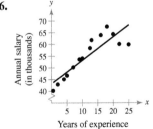

30. (c)

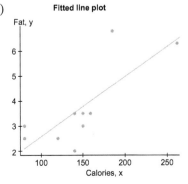

32. (a)

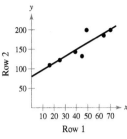

(b)

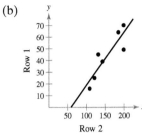

34. (b)

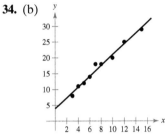

(c)

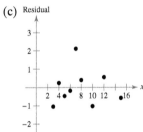

36. (a)

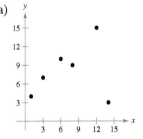

(b) The point $(14, 3)$ may be an outlier.

(c) The point $(14, 3)$ is an influential point because the slopes and y-intercepts of the regression lines with the point included and without the point included are significantly different.

38. The point $(201.4, 45,364)$ is an outlier because it is far removed from the other entries in the data set; The point is an influential point because the slopes and y-intercepts of the regression lines with the point included and without the point included are significantly different.

40. $\log y = 0.233x + 1.969$

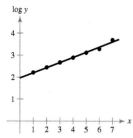

A linear model is more appropriate for the transformed data.

44. $\log y = -1.251 \log x + 2.893$

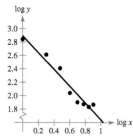

A linear model is more appropriate for the transformed data.

48. $y = 13.667 - 0.471 \ln x$

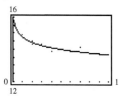

50. The logarithmic equation is a better model for the data. The graph of the logarithmic equation fits the data better than the regression line.

Section 9.3   *(page 519)*

2. The explained variation is the sum of the squares of the differences between the predicted y-values and the mean of the y-values of the ordered pairs, or $\sum(\hat{y}_i - \bar{y})^2$.

4. The coefficient of determination

$$r^2 = \frac{\sum(\hat{y}_i - \bar{y})^2}{\sum(y_i - \bar{y})^2}$$

is the ratio of the explained variation to the total variation and is the percent of variation of y that is explained by the relationship between x and y; $1 - r^2$ is the percent of the variation that is unexplained.

6. Two variables have a bivariate normal distribution when, for any fixed values of either variable, the corresponding values of the other variable are normally distributed.

16. (a) 0.919; About 91.9% of the variation in the turnout for federal elections can be explained by the variation in the voting age population, and about 8.1% of the variation is unexplained.

(b) 2.307; The standard error of estimate of the turnout in a federal election for a specific voting age population is about 2,307,000.

18. (a) 0.768 (*Tech:* 0.770); 76.8% (*Tech:* 77.0%) of the variation in assets in federal pension plans can be explained by the variation in IRAs, and 23.2% of the variation is unexplained.

(b) 77.710 (*Tech:* 77.701); The standard error of estimate of assets in federal pension plans for a specific total of IRA assets is about $77,710,000,000 (*Tech:* $77,701,000,000).

28. $\hat{y} = 0.258x + 4.700$

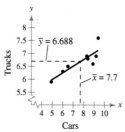

Review Answers for Chapter 9   *(page 531)*

2.

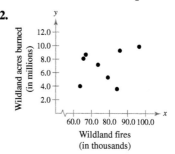

4.

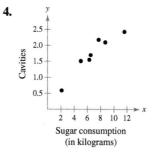

12. $\hat{y} = 1.076x + 0.299$

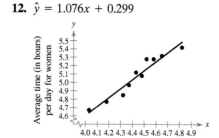

$r \approx 0.967$

A130

14. $\hat{y} = -0.090x + 44.675$

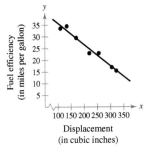

$r \approx -0.984$

18. (a) It is not meaningful to predict the value of y for $x = 86$ because $x = 86$ is outside the range of the original data.

(b) 26.86 miles per gallon

(c) 18.67 miles per gallon

(d) It is not meaningful to predict the value of y for $x = 407$ because $x = 407$ is outside the range of the original data.

26. $4.733 < y < 5.011$

You can be 90% confident that the average time women spend per day watching television will be between 4.733 and 5.011 hours when the average time men spend per day watching television is 4.25 hours.

28. $16.668 < y < 24.982$

You can be 95% confident that the fuel efficiency will be between 16.668 miles per gallon and 24.982 miles per gallon when the engine displacement is 265 cubic inches.

CHAPTER 10

Section 10.1 *(page 546)*

8. (e) There is not enough evidence at the 5% level of significance to conclude that the distribution of the amounts of coffee people drink and the claimed or expected distribution are the same.

10. (a) H_0: The distribution of responses is 41% limited advancement potential, 25% lack of recognition, 15% low salary/benefits, 10% unhappy with management, and 9% bored/don't know.

H_a: The distribution of responses differs from the claimed or expected distribution. (claim)

(b) $\chi_0^2 = 13.277$;

Rejection region: $\chi^2 > 13.277$

(c) 2.025

(d) Fail to reject H_0.

(e) There is not enough evidence at the 1% level of significance to conclude that the distribution of responses differs from the claimed or expected distribution.

14. (e) There is enough evidence at the 10% level of significance to conclude that the distribution of the opinions of U.S. female adults on which is more important to save for, your child's college education or your own retirement, differs from the distribution of the opinions of U.S. male adults.

18.

N	DF	Chi-Square	P-Value
250	7	9.523134	0.2172

Section 10.2 *(page 557)*

8. (a)–(b)

Result	Treatment		Total
	Drug	Placebo	
Nausea	36 (25.15)	13 (23.85)	49
No nausea	254 (264.85)	262 (251.15)	516
Total	290	275	565

10. (a)–(b)

Size of restaurant	Rating			Total
	Excellent	Fair	Poor	
Seats 100 or fewer	182 (165.92)	203 (235.58)	165 (148.5)	550
Seats over 100	180 (196.08)	311 (278.42)	159 (175.5)	650
Total	362	514	324	1200

12. (a)–(b)

Type of movie rented	Age		
	18–24	25–34	35–44
Comedy	38 (28.37)	30 (25.32)	24 (25.75)
Action	15 (15.99)	17 (14.27)	16 (14.52)
Drama	12 (20.63)	11 (18.41)	19 (18.73)
Total	65	58	59

Type of movie rented	Age		Total
	45–64	65 and older	
Comedy	10 (19.21)	8 (11.35)	110
Action	9 (10.83)	5 (6.40)	62
Drama	25 (13.97)	13 (8.25)	80
Total	44	26	252

24. (a)–(b) **Contingency table results:**

Rows: Gender

Columns: None

Cell format
Count
Expected count

	16–20	21–30	31–40	41–50
Male	45	170	90	72
	44.8	165.9	92.09	73.84
Female	9	30	21	17
	9.2	34.07	18.91	15.16
Total	54	200	111	89

	51–60	61 and older	Total
Male	45	26	448
	45.63	25.72	
Female	10	5	92
	9.37	5.281	
Total	55	31	540

Statistic	DF	Value	P-value
Chi-square	5	1.2078544	0.9441

32.

	Educational attainment			
Status	Not a high school graduate	High school graduate	Some college, no degree	Associate's, bachelor's, or advanced degree
Employed	0.086	0.285	0.177	0.452
Unemployed	0.207	0.379	0.172	0.241
Not in the labor force	0.221	0.358	0.163	0.259

38.

Section 10.3 *(page 571)*

2. (1) The F-distribution is a family of curves determined by two types of degrees of freedom, d.f.$_N$ and d.f.$_D$.

(2) F-distributions are positively skewed.

(3) The area under the F-distribution curve is equal to 1.

(4) F-values are always greater than or equal to 0.

(5) For all F-distributions, the mean value of F is approximately equal to 1.

4. Determine the sample whose variance is greater. Use the size of this sample n_1 to find d.f.$_N = n_1 - 1$. Use the size of the other sample n_2 to find d.f.$_D = n_2 - 1$.

14. Fail to reject H_0. There is not enough evidence at the 5% level of significance to reject the claim.

16. Reject H_0. There is enough evidence at the 5% level of significance to support the claim.

18. Fail to reject H_0. There is not enough evidence at the 5% level of significance to support the claim.

20. (a) $H_0: \sigma_1^2 \le \sigma_2^2$

$H_a: \sigma_1^2 > \sigma_2^2$ (claim)

(b) $F_0 = 3.08$; Rejection region: $F > 3.08$

(c) 3.21 (d) Reject H_0.

(e) There is enough evidence at the 1% level of significance to conclude that the variance of fuel consumption for the company's hybrid vehicles is less than that of the competitor's hybrid vehicles.

22. (a) $H_0: \sigma_1^2 = \sigma_2^2$; $H_a: \sigma_1^2 \ne \sigma_2^2$ (claim)

(b) $F_0 = 3.18$; Rejection region: $F > 3.18$

(c) 3.45

(d) Reject H_0.

(e) There is enough evidence at the 10% level of significance to conclude that the variances of the numbers of calories differ between the two brands.

26. (a) $H_0: \sigma_1^2 \le \sigma_2^2$

$H_a: \sigma_1^2 > \sigma_2^2$ (claim)

(b) $F_0 = 1.985$; Rejection region: $F > 1.985$

(c) 2.490 (d) Reject H_0.

(e) There is enough evidence at the 5% level of significance to support the service's claim that the standard deviation of the annual salaries for public relations managers is greater in Florida than in Louisiana.

28. Hypothesis test results:

σ_1^2 : variance of population 1

σ_2^2 : variance of population 2

σ_1^2/σ_2^2 : variance ratio

$H_0 : \sigma_1^2/\sigma_2^2 = 1$

$H_A : \sigma_1^2/\sigma_2^2 \ne 1$

Ratio	n1	n2	Sample Ratio	F-Stat	P-value
σ_1^2/σ_2^2	24	20	2.712803	2.712803	0.0308

$P = 0.0308 < 0.05$, so reject H_0. There is enough evidence at the 5% level of significance to support the claim.

30. Hypothesis test results:

σ_1^2 : variance of population 1

σ_2^2 : variance of population 2

σ_1^2/σ_2^2 : variance ratio

$H_0 : \sigma_1^2/\sigma_2^2 = 1$

$H_A : \sigma_1^2/\sigma_2^2 > 1$

Ratio	n1	n2	Sample Ratio	F-Stat	P-value
σ_1^2/σ_2^2	7	13	1.1879483	1.1879483	0.3751

$P = 0.3751 > 0.01$, so fail to reject H_0. There is not enough evidence at the 1% level of significance to support the claim.

Section 10.4 *(page 581)*

6. (a) $H_0: \mu_1 = \mu_2 = \mu_3$

H_a: At least one mean is different from the others. (claim)

(b) $F_0 = 3.74$;

Rejection region: $F > 3.74$

(c) 0.42

(d) Fail to reject H_0.

(e) There is not enough evidence at the 5% level of significance to conclude that at least one of the mean battery prices is different from the others.

12. (a) $H_0: \mu_1 = \mu_2 = \mu_3 = \mu_4 = \mu_5 = \mu_6$

16. Column means

Column	n	Mean	Std. Error
Gainesville	11	188.67273	12.685002
Orlando	10	209.56	11.292428
Tampa	10	173.96	14.59111

ANOVA table

Source	df	SS	MS
Treatments	2	6404.441	3202.2205
Error	28	48337.77	1726.3489
Total	30	54742.21	

F-Stat	P-Value
1.8549093	0.1752

Review Answers for Chapter 10 *(page 589)*

2. (a) H_0: The distribution of the lengths of office visits is 4% 1–5 minutes, 24% 6–10 minutes, 34% 11–15 minutes, 31% 16–30 minutes, 6% 31–60 minutes, and 1% 61 minutes and over.

H_a: The distribution of the lengths differs from the claimed or expected distribution. (claim)

(b) $\chi_0^2 = 15.086$; Rejection region: $\chi^2 > 15.086$

(c) 13.809

(d) Fail to reject H_0.

(e) There is not enough evidence at the 1% level of significance to conclude that there has been a change in the claimed or expected distribution.

CHAPTER 11

Section 11.1 *(page 604)*

4. Answers will vary. *Sample answer:* It is called the sign test because each value in a sample is compared to the hypothesized median and assigned a + or − sign, based on whether the difference is positive or negative. The numbers of + signs and − signs are used to determine whether the null hypothesis should be rejected.

6. A sample must be randomly selected from each population, and the samples must be dependent.

8. (a) H_0: median = 83 (claim); H_a: median ≠ 83

(b) 1　(c) 4　(d) Fail to reject H_0.

(e) There is not enough evidence at the 1% level of significance to reject the meteorologist's claim that the median daily high temperature for the month of July in Pittsburgh is 83° Fahrenheit.

10. (a) H_0: median = 66 (claim); H_a: median ≠ 66

(b) 2　(c) 0　(d) Reject H_0.

(e) There is enough evidence at the 1% level of significance to reject the meteorologist's claim that the median daily high temperature for the month of January in San Diego is 66° Fahrenheit.

12. (a) H_0: median ≥ \$65,000; H_a: median < \$65,000 (claim)

(b) −1.96　(c) −2.51　(d) Reject H_0.

(e) There is enough evidence at the 2.5% level of significance to support the accountant's estimate that the median amount of financial debt for families holding such debts is less than \$65,000.

14. (a) H_0: median ≥ 32; H_a: median < 32 (claim)

(b) 5　(c) 5　(d) Reject H_0.

(e) There is enough evidence at the 5% level of significance to support the research group's claim that the median age of Facebook® users is less than 32 years old.

22. (a) H_0: The SAT scores have not improved.

H_a: The SAT scores have improved. (claim)

(b) 1   (c) 3   (d) Fail to reject H_0.

(e) There is not enough evidence at the 1% level of significance to conclude that the critical reading SAT scores have improved.

26. Hypothesis test results:

Parameter : median of Variable

H_0 : Parameter $= 57$

H_A : Parameter > 57

Variable	n	n for test	Sample Median
Age	24	21	50

Below	Equal	Above	P-value
14	3	7	0.9608

$P = 0.9608 > 0.01$, so fail to reject H_0. There is not enough evidence at the 1% level of significance to support the television network's claim that the median age of viewers for the Masters Golf Tournament is greater than 57 years.

Section 11.3 *(page 623)*

4. (a) H_0: There is no difference in the hourly pay rates.

H_a: There is a difference in the hourly pay rates. (claim)

(b) 5.991   (c) 0.395   (d) Fail to reject H_0.

(e) There is not enough evidence at the 5% level of significance to conclude that the distributions of the registered nurses' hourly pay rates in the three states are different.

8. Kruskal-Wallis results:

Data stored in separate columns.

Chi Square $= 14.57108$ (adjusted for ties)

DF $= 3$

P-value $= 0.0022$

Column	n	Median	Ave. Rank
A	6	5	12.083333
B	6	8.5	20.5
C	6	5	12.333333
D	6	3	5.0833335

$P = 0.0022 < 0.01$, so reject H_0. There is enough evidence at the 1% level of significance to conclude that the distributions of the number of job offers at all four colleges are different.

10. (a) Reject H_0.   (b) Reject H_0.

Both tests come to the same decision, which is that the mean energy consumptions are different.

Section 11.5 *(page 637)*

2. When both n_1 and n_2 are less than or equal to 20, the test statistic is equal to the number of runs G. When either n_1 or n_2 is greater than 20, the test statistic is equal to

$$\frac{G - \mu_G}{\sigma_G}.$$

4. Number of runs: 9

Run lengths: 2, 2, 1, 1, 2, 2, 1, 1, 2

6. Number of runs: 10

Run lengths: 3, 3, 1, 2, 6, 1, 2, 1, 1, 2

8. $n_1 =$ number of U's $= 8$

$n_2 =$ number of D's $= 6$

10. $n_1 =$ number of A's $= 13$

$n_2 =$ number of B's $= 9$

16. (a) H_0: The sequence of majority parties is random.

H_a: The sequence of majority parties is not random. (claim)

(b) ± 1.96 (c) -3.12 (d) Reject H_0.

(e) There is enough evidence at the 5% level of significance to conclude that the sequence is not random.

20. (a) H_0: The sequence is random.

H_a: The sequence is not random. (claim)

(b) ± 1.96 (c) 2.61   (d) Reject H_0.

(e) There is enough evidence at the 5% level of significance to conclude that the sequence is not random.

INDEX

A

Addition Rule, 189
  for the probability of A
    and B, 157, 160
alternative formula
  for the standardized test
    statistic for a proportion,
    402
  for variance and standard
    deviation, 96
alternative hypothesis
  one-sample, 357
  two-sample, 430
analysis of variance
    (ANOVA) test
  one-way, 574, 575
  two-way, 580
approximating binomial
    probabilities, 284
area of a region
  under a probability curve, 239
  under a standard normal
    curve, 239

B

back-to-back stem-and-leaf
    plot, 64
Bayes' Theorem, 154
biased sample, 20
bimodal, 67
binomial distribution, 221
  mean of a, 210
  normal approximation to a,
    281
  population parameters of a,
    210
  standard deviation of a, 210
  variance of a, 210
binomial experiment, 202, 540
  notation for, 202
binomial probabilities, using
    the normal distribution to
    approximate, 284
binomial probability
    distribution, 205, 221
binomial probability formula, 204
bivariate normal distribution, 517
blinding, 18
blocks, 18
box-and-whisker plot, 102
  side-by-side, 111
boxplot, 102
  modified, 112

C

calculating a correlation
    coefficient, 488
categories, 540
c-confidence interval
  for the population mean, 307

for the population
    proportion, 328
cell, 551
census, 3, 20
center, 38
central angle, 56
Central Limit Theorem, 268
chart
  control, 256
  Pareto, 57
  pie, 56
  time series, 59
Chebychev's Theorem, 87
Chi-square
  distribution, 337
  goodness-of-fit test, 540, 542
  independence test, 553
  test
    finding critical values for, 404
    for independence, 553
    for standard deviation, 406,
      415
    test statistic for, 406
    for variance, 406, 415
class, 38
  boundaries, 42
  mark, 40
  width, 38
class limit
  lower, 38
  upper, 38
classical probability, 132, 160
closed question, 25
cluster sample, 21
clusters, 21
  of data, 71
coefficient
  correlation, 487
    t-test for, 492
  of determination, 514
  of variation, 96
combination
    of n objects taken r at a
      time, 170, 171
complement of event E, 136
complementary events, 160
completely randomized
    design, 18
conditional probability, 145
conditional relative frequency,
    563
confidence, level of, 305
confidence interval, 307
  for σ_1^2/σ_2^2, 573
  for the difference between
    means, 440, 450
  for the difference between
    two population
      proportions, 468
  for the mean of the
    differences of paired
      data, 460

for a population mean,
    finding a, 307
for a population proportion,
    constructing a, 328
for a population standard
    deviation, 339
for a population variance,
    339
for slope, 523
for y-intercept, 523
confounding variable, 18
constructing
  a confidence interval for the
    difference between
      means, 440, 450
  a confidence interval for the
    difference between two
      population
        proportions, 468
  a confidence interval for the
    mean of the differences
      of paired data, 460
  a confidence interval for the
    mean: t-distribution, 320
  a confidence interval for a
    population proportion,
    328
  a confidence interval for a
    population standard
      deviation, 339
  a confidence interval for a
    population variance, 339
  a discrete probability
    distribution, 192
  a frequency distribution
    from a data set, 38
  an ogive, 45
  a prediction interval for y
    for a specific value of x,
    517
contingency table, 551
contingency table cells, finding
    the expected frequency
    for, 551
continuity correction, 283
continuous probability
    distribution, 236
continuous random variable,
    190, 236
control
  chart, 256
  group, 16
convenience sample, 22
correction, continuity, 283
correction factor
  finite, 278
  finite population, 316
correlation, 484
correlation coefficient, 487
  Pearson product moment, 487
  Spearman rank, 625
  t-test for, 492

using a table for, 490
counting principle,
    fundamental, 130, 171
c-prediction interval, 517
critical region, 376
critical value, 305, 376
  in a normal distribution,
    finding, 376
  in a t-distribution, finding, 387
cumulative frequency, 40
  graph, 44
curve, normal, 236

D

data, 2
  qualitative, 9
  quantitative, 9
data sets
  center of, 38
  paired, 58
  shape of, 38
  variability of, 38
decision rule
  based on P-value, 364, 371
  based on rejection region, 378
degrees of freedom, 318
  corresponding to the
    variance in the
      denominator, 565
  corresponding to the
    variance in the
      numerator, 565
density function, probability, 236
dependent
  event, 146
  random variable, 201
  sample, 428
  variable, 484
descriptive statistics, 5
designing a statistical study, 16
determination, coefficient of, 514
deviation, 81
  explained, 513
  total, 513
  unexplained, 513
d.f.$_D$, 565
d.f.$_N$, 565
diagram, tree, 128
discrete probability
    distribution, 191
discrete random variable, 190
  expected value of a, 196
  mean of a, 194
  standard deviation of a, 195
  variance of a, 195
distinguishable permutation, 170
distribution
  binomial, 221
  binomial probability, 205, 221
  bivariate normal, 517
  chi-square, 337
  continuous probability, 236

discrete probability, 191
F-, 565
frequency, 38
  geometric, 218, 221
hypergeometric, 224
normal, 236
  finding critical values in, 376
  properties of a, 236
Poisson, 219, 221
sampling, 266
standard normal, 239, A1, A2
  properties of, 239, A2
t-, 318
  finding critical values in, 387
uniform, 248
dot plot, 55
double-blind experiment, 18
drawing a box-and-whisker
  plot, 103

E

e, 219
effect
  Hawthorne, 18
  interaction, 580
  main, 580
  placebo, 18
elements of well-designed
  experiment, 18
empirical probability, 133, 160
Empirical Rule (or 68-95-99.7
  Rule), 86
equation
  exponential, 510
  logarithmic, 510
  multiple regression, 524
  of a regression line, 502
  power, 510
error
  of estimate
    maximum, 306
    standard, 515
  margin of, 306
  of the mean
    standard, 266
  sampling, 20, 266, 306
  tolerance, 306
  type I, 359
  type II, 359
estimate
  interval, 305
  point, 304
  pooled, of the standard
    deviation, 442
  standard error of, 515
estimating p by minimum
  sample size, 331
estimator, unbiased, 304
event, 128
  complement of an, 136
  dependent, 146
  independent, 146, 160
  mutually exclusive, 156, 160
  simple, 129
expected frequency, 541

finding for contingency
  table cells, 551
expected value, 196
  of a discrete random
    variable, 196
experiment, 16
  binomial, 202, 540
  double-blind, 18
  multinomial, 215, 540
  probability, 128
  well-designed, elements of, 18
experimental design
  completely randomized, 18
  matched-pairs, 19
  randomized block, 18
experimental unit, 16
explained deviation, 513
explained variation, 513
explanatory variable, 484
exploratory data analysis
  (EDA), 53
exponential equation, 510

F

factorial, 168
false positive, 155
F-distribution, 565
finding areas under the
  standard normal curve,
  241, A4
finding a confidence interval
  for a population mean, 307
finding critical values
  for the chi-square test, 404
  for the F-distribution, 566
  in a normal distribution, 376
  in a t-distribution, 387
finding the expected frequency
  for contingency table
    cells, 551
finding the mean of a
  frequency distribution, 70
finding a minimum sample size
  to estimate μ, 310
  to estimate p, 331
finding the P-value for a
  hypothesis test, 371
finding the standard error of
  estimate, 515
finding the test statistic for the
  one-way ANOVA test, 575
finite correction factor, 278
finite population correction
  factor, 316
first quartile, 100
five-number summary, 103
formula, binomial probability,
  204
fractiles, 100
frequency, 38
  conditional relative, 563
  cumulative, 40
  expected, 541
  joint, 551
  marginal, 551
  observed, 541

relative, 40
frequency distribution, 38
  mean of, 70
  rectangular, 71
  skewed left (negatively
    skewed), 71
  skewed right (positively
    skewed), 71
  symmetric, 71
  uniform, 71
frequency histogram, 42
  relative, 44
frequency polygon, 43
F-test for variances,
  two-sample, 568
function, probability density, 236
Fundamental Counting
  Principle, 130, 171

G

Gallup poll, 351
gaps, 68
geometric distribution, 218, 221
  mean of a, 224
  variance of a, 224
geometric probability, 218
goodness-of-fit test, chi-square,
  540, 542
graph
  cumulative frequency, 44
  misleading, 64

H

Hawthorne effect, 18
histogram
  frequency, 42
  relative frequency, 44
history of statistics timeline, 33
homogeneity of proportions
  test, 561
hypergeometric distribution, 224
hypothesis
  alternative, 357, 430
  null, 357, 430
  statistical, 357
hypothesis test, 356
  finding the P-value for, 371
hypothesis testing
  for slope, 523
  steps for, 365
  summary of, 414, 415

I

independence test, chi-square,
  553
independent, 551
  event, 146, 160
  random variable, 201
  sample, 428
  variable, 484
inferential statistics, 5
inflection points, 236, 237
influential point, 509
inherent zero, 11
interaction effect, 580
interquartile range (IQR), 102

interval, c-prediction, 517
interval estimate, 305
interval level of measurement,
  11, 12
intervals, 38

J

joint frequency, 551

K

Kruskal-Wallis test, 619
  test statistic for, 619

L

law of large numbers, 134
leaf, 53
left, skewed, 71
left-tailed test, 362
  for a population correlation
    coefficient, 492
length of a run, 631
level of confidence, 305
level of significance, 361, 490
levels of measurement
  interval, 11, 12
  nominal, 10, 12
  ordinal, 10, 12
  ratio, 11, 12
limit
  lower class, 38
  upper class, 38
line
  of best fit, 501
  regression, 490, 501
linear transformation of a
  random variable, 201
logarithmic
  equation, 510
  transformation, 510
lower class limit, 38

M

main effect, 580
making an interval estimate, 305
margin of error, 306
marginal frequency, 551
matched samples, 428
matched-pairs design, 19
maximum error of estimate, 306
mean, 65
  of a binomial distribution, 210
  difference between
    two-sample t-test for, 442
    two-sample z-test for, 431
  of a discrete random
    variable, 194
  of a frequency distribution, 70
  of a geometric distribution,
    224
  standard error, 266
  trimmed, 78
  t-test for, 389
  weighted, 69
mean absolute deviation
  (MAD), 97
mean square

between, 574
within, 574
means, sampling distribution
    of sample, 266
measure of central tendency, 65
measurement
    interval level of, 11, 12
    nominal level of, 10, 12
    ordinal level of, 10, 12
    ratio level of, 11, 12
median, 66
midpoint, 40
midquartile, 111
midrange, 78
minimum sample size
    to estimate μ, 310
    to estimate p, 331
misleading graph, 64
mode, 67
modified boxplot, 112
multinomial experiment, 215, 540
multiple regression equation, 524
Multiplication Rule for the
    probability of A and B,
    147, 160
mutually exclusive, 156, 160

N

n factorial, 168
negative linear correlation, 484
negatively skewed, 71
no correlation, 484
nominal level of
    measurement, 10, 12
nonlinear correlation, 484
nonparametric test, 598
normal approximation to a
    binomial distribution, 281
normal curve, 236
normal distribution, 236
    bivariate, 517
    finding critical values in, 376
    properties of a, 236
    standard, 239, A1, A2
        finding areas under, 241, A4
        properties of, 239, A2
normal probability plot, A28
normal quantile plot, A28
notation for binomial
    experiment, 202
null hypothesis
    one-sample, 357
    two-sample, 430

O

observational study, 16
observed frequency, 541
odds, 143
    of losing, 143
    of winning, 143
ogive, 44
one-way analysis of variance, 574
    test, 574, 575
        finding the test statistic
            for, 575
open question, 25

ordered stem-and-leaf plot, 53
ordinal level of measurement,
    10, 12
outcome, 128
outlier, 54, 68

P

paired data sets, 58
paired samples, 428
    sign test, performing a, 602
parameter, 4
    population
        binomial distribution, 210
Pareto chart, 57
Pearson product moment
    correlation coefficient, 487
Pearson's index of skewness, 97
performing
    a chi-square goodness-of-fit
        test, 542
    a chi-square test for
        independence, 554
    a Kruskal-Wallis test, 620
    a one-way analysis
        of variance test, 576
    a paired-sample sign test, 602
    a runs test for randomness, 633
    a sign test for a
        population median, 599
    a Wilcoxon rank sum test, 612
    a Wilcoxon signed-rank test,
        609
permutation, 168, 171
    distinguishable, 170, 171
    of n objects taken r at a
        time, 168, 171
pie chart, 56
placebo, 16
    effect, 18
plot
    back-to-back stem-and-leaf, 64
    box-and-whisker, 102
    dot, 55
    normal probability, A28
    normal quantile, A28
    residual, 509
    scatter, 58, 484
    side-by-side box-and-whisker,
        111
    stem-and-leaf, 53
point, influential, 509
point estimate, 304
    for σ, 337
    for σ^2, 337
    for p, 327
Poisson distribution, 219, 221
    variance of a, 224
polygon, frequency, 43
pooled estimate of the
    standard deviation, 442
population, 3
    correlation coefficient
        using Table 11 for the, 490
        using the t-test for the, 492
    mean, finding a confidence
        interval for, 307

parameters
    of a binomial
        distribution, 210
    proportion, 327
        constructing a confidence
            interval for, 328
    standard deviation, 82
    variance, 81, 82
positive linear correlation, 484
positively skewed, 71
power equation, 510
power of the test, 361
principle, fundamental
    counting, 130, 171
probability
    Addition Rule for, 157, 160
    classical, 132, 160
    conditional, 145
    curve, area of a region
        under, 239
    density function, 236
    empirical, 133, 160
    experiment, 128
    formula, binomial, 204
    geometric, 218
    Multiplication Rule for, 147,
        160
    rule, range of, 135, 160
    statistical, 133
    subjective, 134
    that the first success will
        occur on trial number x,
        218, 221
    theoretical, 132
    value, 361
probability distribution
    binomial, 205, 221
    chi-square, 337
    continuous, 236
    discrete, 191
    geometric, 218, 221
    normal, properties of a, 236
    Poisson, 219, 221
    sampling, 266
    standard normal, 239
probability plot, normal, A28
properties
    of a normal distribution, 236
    of sampling distributions of
        sample means, 266
    of the standard normal
        distribution, 239, A2
proportion
    population, 327
        confidence interval for, 328
        z-test for, 398
    sample, 279
proportions, sampling
    distribution of sample, 279
proportions test, homogeneity
    of, 561
P-value, 361
    decision rule based on, 364,
        371
    for a hypothesis test, finding
        the, 371

Q

qualitative data, 9
quantile plot, normal, A28
quantitative data, 9
quartile, 100
    first, 100
    second, 100
    third, 100
question
    closed, 25
    open, 25

R

random sample, simple, 20
random sampling, 3
random variable, 190
    continuous, 190, 236
    dependent, 201
    discrete, 190
        expected value of a, 196
        mean of a, 194
        standard deviation of a, 195
        variance of a, 195
    independent, 201
    linear transformation of a, 201
randomization, 18
randomized block design, 18
randomness, runs test for, 632
range, 38, 80
    interquartile, 102
    of probabilities rule, 135, 160
rank correlation coefficient,
    Spearman, 625
rank sum test, Wilcoxon, 611
ratio level of measurement,
    11, 12
rectangular, frequency
    distribution, 71
region
    critical, 376
    rejection, 376
regression equation, multiple, 524
regression line, 490, 501
    deviation about, 513
    equation of, 502
    variation about, 513
rejection region, 376
    decision rule based on, 378
relative frequency, 40
    conditional, 563
    histogram, 44
replacement
    with, 21
    without, 21, 203
replication, 19
residual plot, 509
residuals, 501
response variable, 484
right, skewed, 71
right-tailed test, 362
    for a population correlation
        coefficient, 492
rule
    addition, 157, 160
    decision
        based on P-value, 364, 371

based on rejection region, 378
empirical, 86
multiplication, 147, 160
range of probabilities, 135, 160
run, 631
runs test for randomness, 632

S

sample, 3
biased, 20
cluster, 21
convenience, 22
dependent, 428
independent, 428
matched, 428
paired, 428
random, 20
simple, 20
stratified, 21
systematic, 22
sample means
sampling distribution for the difference of, 430
sampling distribution of, 266
sample proportion, 279
sample proportions, sampling distribution of, 279
sample size, 19
minimum to estimate μ, 310
minimum to estimate p, 331
sample space, 128
sample standard deviation, 83
for grouped data, 88
sample variance, 83
sampling, 20
sampling distribution, 266
for the difference of the sample means, 430
for the difference between the sample proportions, 461
for the mean of the differences of the paired data entries in dependent samples, 451
properties of, 266
of sample means, 266
of sample proportions, 279
sampling error, 20, 266, 306
sampling process
with replacement, 21
without replacement, 21
scatter plot, 58, 484
Scheffé Test, 586
score, standard, 105
second quartile, 100
shape, 38
side-by-side box-and-whisker plot, 111
sigma, 39
sign test, 598
performing a paired-sample, 602
test statistic for, 599
signed-rank test, Wilcoxon, 609

significance, level of, 361, 490
simple event, 129
simple random sample, 20
simulation, 17
skewed
left, 71
negatively, 71
positively, 71
right, 71
slope
confidence interval for, 523
hypothesis testing for, 523
Spearman rank correlation coefficient, 625
standard deviation
of a binomial distribution, 210
chi-square test for, 406, 415
confidence intervals for, 339
of a discrete random variable, 195
point estimate for, 337
pooled estimate of, 442
population, 82
sample, 83
standard error
of estimate, 515
of the mean, 266
standard normal curve, finding areas under, 241, A4
standard normal distribution, 239, A1, A2
properties of, 239, A2
standard score, 105
standardized test statistic,
for a chi-square test
for standard deviation, 406, 415
for variance, 406, 415
for the correlation coefficient
t-test, 492
for the difference between means
t-test, 452
z-test, 431
for the difference between proportions z-test, 462
for a t-test
for a mean 389, 415
two-sample, 442
for a z-test
for a mean, 373, 415
for a proportion, 398, 415
two-sample, 431
statistic, 4
statistical hypothesis, 357
statistical probability, 133
statistical process control (SPC), 256
statistical study, designing a, 16
statistics, 2
descriptive, 5
history of, timeline, 33
inferential, 5
status, 2
stem, 53
stem-and-leaf plot, 53

back-to-back, 64
ordered, 53
unordered, 53
steps for hypothesis testing, 365
strata, 21
stratified sample, 21
study
observational, 16
statistical, designing a, 16
subjective probability, 134
successes, population proportion of, 327
sum of squares, 81–83
sum test, Wilcoxon rank, 611
summary
of counting principles, 171
of discrete probability distributions, 221
five-number, 103
of four levels of measurement, 12
of hypothesis testing, 414, 415
of probability, 160
survey, 17
survey questions
closed question, 25
open question, 25
symmetric, frequency distribution, 71
systematic sample, 22

T

table, contingency, 551
t-distribution, 318
constructing a confidence interval for the mean, 320
finding critical values in, 387
test
chi-square
goodness-of-fit, 540, 542
independence, 553
homogeneity of proportions, 561
hypothesis, 356
Kruskal-Wallis, 619
left-tailed, 362
nonparametric, 598
one-way analysis of variance, 574, 575
paired-sample sign, 602
power of the, 361
for randomness, runs, 632
right-tailed, 362
Scheffé, 586
sign, 598
two-tailed, 362
two-way analysis of variance, 580
Wilcoxon rank sum, 611
Wilcoxon signed-rank, 609
test statistic, 361
for a chi-square test, 406, 415
for the correlation coefficient, 492
for the difference between means, 452

for the difference between proportions, 462
for the Kruskal-Wallis test, 619
for a mean
large sample, 373, 415
small sample, 389, 415
for a proportion, 398, 415
for the runs test, 633
for the sign test, 599
for a two-sample t-test, 442
for a two-sample z-test, 431
for the Wilcoxon rank sum test, 612
testing the significance of the Spearman rank correlation coefficient, 626
Theorem
Bayes', 154
Central Limit, 268
Chebychev's, 87
theoretical probability, 132
third quartile, 100
time series, 59
chart, 59
timeline, history of statistics, 33
total deviation, 513
total variation, 513
transformation, logarithmic, 510
transformations to achieve linearity, 510
transforming a z-score to an x-value, 259
treatment, 16
tree diagram, 128
trimmed mean, 78
t-test
for the correlation coefficient, 492
for the difference between means, 452
for a mean, 389, 415
two-sample
for the difference between means, 442
two-sample
F-test for variances, 568
t-test, 442
z-test
for the difference between means, 431
for the difference between proportions, 462
two-tailed test, 362
for a population correlation coefficient, 492
two-way analysis of variance test, 580
type I error, 359
type II error, 359

U

unbiased estimator, 304
unexplained deviation, 513
unexplained variation, 513

uniform, frequency
    distribution, 71
uniform distribution, 248
upper class limit, 38
using
    the chi-square test for a
        variance or standard
        deviation, 406
    the normal distribution
        to approximate binomial
        probabilities, 284
    P-values for a z-test for
        a mean, 373
    rejection regions for a
        z-test for a mean, 378
    Table 11 for the
        correlation coefficient,
        490
    the t-test
        for the correlation
            coefficient ρ, 492
        for the difference between
            means, 452
        for a mean, 389
    a two-sample F-test to
        compare σ_1^2 and σ_2^2, 568
    a two-sample t-test for
        the difference between
        means, 443

a two-sample z-test
    for the difference
        between means, 431
    for the difference
        between proportions,
        462
a z-test for a proportion,
    398

V

value
    critical, 305, 376
    expected, 196
    probability, 361
variable
    confounding, 18
    dependent, 484
    explanatory, 484
    independent, 484
    random, 190
        continuous, 190, 236
        discrete, 190
    response, 484
variability, 38
variance
    of a binomial distribution, 210
    chi-square test for, 406, 415
    confidence intervals for, 339

of a discrete random variable,
    195
of a geometric distribution,
    224
mean square
    between, 574
    within, 574
one-way analysis of, 574
point estimate for, 337
of a Poisson distribution, 224
population, 81, 82
sample, 83
two-sample F-test for, 568
two-way analysis of, 580
variation
    coefficient of, 96
    explained, 513
    total, 513
    unexplained, 513

W

weighted mean, 69
Wilcoxon rank sum test, 611
    test statistic for, 612
Wilcoxon signed-rank test, 609
with replacement, 21
without replacement, 21, 203

X

x, random variable, 190

Y

y-intercept, confidence interval
    for, 523

Z

zero, inherent, 11
z-score, 105
z-test
    for a mean, 373, 415
        test statistic for, 373, 415
        using P-values for, 373
        using rejection regions
            for, 378
    for a proportion, 398, 415
        test statistic for, 398, 415
    two-sample
        difference between
            means, 431
        difference between
            proportions, 462

Photo Credits

Table of Contents Ch. 1 Tetra Images/Alamy; **Ch. 2** AP Images; **Ch. 3** Monty Brinton/CBS/Getty Images; Valerie Macon/Getty Images; **Ch. 4** Monkey Business Images/Shutterstock; **Ch. 5** Dndavis/Dreamstime; **Ch. 6** Blend Images/Veers; **Ch. 7** Mark Sykes/Computer Screen Concepts/Alamy; **Ch. 8** Thinkstock; **Ch. 9** Henny Ray Abrams/AP Images; **Ch. 10** Rob Wilson/Shutterstock; **Ch. 11** Andresr/Shutterstock

Chapter 1 p. xx Tetra Images/Alamy; **p. 3** Collage Photography/Veer/Corbis; **p. 4** Blend Images Photography/Veer/Corbis; Globe: OJO Images Photography/Veer/Corbis; **p. 6** Ocean Photography/Veer/Corbis; **p. 19** Olinchuk/Shutterstock; **p. 20** Fancy Photography/Veers; **p. 33** John W. Tukey/Stanhope Hall/Princeton University; Karl Pearson/SPL/Photo Researchers, Inc; Sir Ronald Aylmer Fisher/A. Barrington Brown/Photo Researchers, Inc

Chapter 2 p. 36 AP Images; **p. 38** RubberBall Photography/Veer/Corbis; **p. 43** Collage Photography/Veer/Corbis; **p. 53** Andres Rodriguez/123RF; **p. 99** Katrina Brown/iStockphoto; **p. 100** Peter Nad/Shutterstock; **p. 106** Kevin Winter/Getty Images; **p. 113** Thumb/Shutterstock; **p. 120** Logo: Coutesy of the National Apartment Association; RF/Pakhnyushcha/Shutterstock; **p. 124** Veer/Corbis

Chapter 3 p. 126 Monty Brinton/CBS/Getty Images; Sign: Valerie Macon/Getty Images; **p. 148** iStockPhoto/Thinkstock; **p. 173** Andrew Brookes/Corbis; **p. 186** MUSL Multi-State Lottery Association

Chapter 4 p. 188 Monkey Business Images/Shutterstock

Chapter 5 p. 234 Dndavis/Dreamstime

Chapter 6 p. 302 Blend Images/Veers; **p. 317** Rob Wilson/Shutterstock; **p. 321** The Granger Collection

Chapter 7 p. 354 Mark Sykes/Computer Screen Concepts/Alamy; **p. 356** Michael Shake/Shutterstock; **p. 379** S06/Zuma Press/Newscom

Chapter 8 p. 426 Thinkstock; **p. 441** Blend Images/Alamy; **p. 443** Shutterstock; **p. 453** By permission of EastBay Shoes, Inc.

Chapter 9 p. 482 Henny Ray Abrams/AP Images; **p. 537** U. S. Food and Drug Administration

Chapter 10 p. 538 Rob Wilson/Shutterstock; **p. 587** Superstock; **p. 594** Courtesy of the National Consumer League's Fraud Center

Chapter 11 p. 596 Andresr/Shutterstock; **p. 639** Ersin Kurtdal/Shutterstock; **p. 646** Bureau of Labor Statistics

CHAPTER 2

$$\text{Class Width} = \frac{\text{Range of data}}{\text{Number of classes}}$$

(round up to next convenient number)

$$\text{Midpoint} = \frac{(\text{Lower class limit}) + (\text{Upper class limit})}{2}$$

$$\text{Relative Frequency} = \frac{\text{Class frequency}}{\text{Sample size}} = \frac{f}{n}$$

Population Mean: $\mu = \dfrac{\Sigma x}{N}$

Sample Mean: $\bar{x} = \dfrac{\Sigma x}{n}$

Weighted Mean: $\bar{x} = \dfrac{\Sigma(x \cdot w)}{\Sigma w}$

Mean of a Frequency Distribution: $\bar{x} = \dfrac{\Sigma(x \cdot f)}{n}$

Range = (Maximum entry) − (Minimum entry)

Population Variance: $\sigma^2 = \dfrac{\Sigma(x - \mu)^2}{N}$

Population Standard Deviation:

$$\sigma = \sqrt{\sigma^2} = \sqrt{\frac{\Sigma(x - \mu)^2}{N}}$$

Sample Variance: $s^2 = \dfrac{\Sigma(x - \bar{x})^2}{n - 1}$

Sample Standard Deviation: $s = \sqrt{s^2} = \sqrt{\dfrac{\Sigma(x - \bar{x})^2}{n - 1}}$

Empirical Rule (or 68-95-99.7 Rule) For data with a (symmetric) bell-shaped distribution:

1. About 68% of the data lies between $\mu - \sigma$ and $\mu + \sigma$.
2. About 95% of the data lies between $\mu - 2\sigma$ and $\mu + 2\sigma$.
3. About 99.7% of the data lies between $\mu - 3\sigma$ and $\mu + 3\sigma$.

Chebychev's Theorem The portion of any data set lying within k standard deviations $(k > 1)$ of the mean is at least $1 - \dfrac{1}{k^2}$.

Sample Standard Deviation of a Frequency Distribution:

$$s = \sqrt{\frac{\Sigma(x - \bar{x})^2 f}{n - 1}}$$

Standard Score: $z = \dfrac{\text{Value} - \text{Mean}}{\text{Standard deviation}} = \dfrac{x - \mu}{\sigma}$

CHAPTER 3

Classical (or Theoretical) Probability:

$$P(E) = \frac{\text{Number of outcomes in event } E}{\substack{\text{Total number of outcomes} \\ \text{in sample space}}}$$

Empirical (or Statistical) Probability:

$$P(E) = \frac{\text{Frequency of event } E}{\text{Total frequency}} = \frac{f}{n}$$

Probability of a Complement: $P(E') = 1 - P(E)$

Probability of occurrence of both events A and B:

$$P(A \text{ and } B) = P(A) \cdot P(B|A)$$

$$P(A \text{ and } B) = P(A) \cdot P(B) \text{ if } A \text{ and } B \text{ are independent}$$

Probability of occurrence of either A or B or both:

$$P(A \text{ or } B) = P(A) + P(B) - P(A \text{ and } B)$$

$$P(A \text{ or } B) = P(A) + P(B) \text{ if } A \text{ and } B \text{ are mutually exclusive}$$

Permutations of n objects taken r at a time:

$$_nP_r = \frac{n!}{(n - r)!}, \text{where } r \leq n$$

Distinguishable Permutations: n_1 alike, n_2 alike, $\ldots$, n_k alike:

$$\frac{n!}{n_1! \cdot n_2! \cdot n_2! \cdots n_k!},$$

where $n_1 + n_2 + n_3 + \cdots + n_k = n$

Combination of n objects taken r at a time:

$$_nC_r = \frac{n!}{(n - r)!r!}$$

Key Formulas

From Larson/*Farber Elementary Statistics: Picturing the World,* Fifth Edition
© 2012 Prentice Hall

CHAPTER 4

Mean of a Discrete Random Variable: $\mu = \Sigma x P(x)$

Variance of a Discrete Random Variable:

$$\sigma^2 = \Sigma(x - \mu)^2 P(x)$$

Standard Deviation of a Discrete Random Variable:

$$\sigma = \sqrt{\sigma^2} = \sqrt{\Sigma(x - \mu)^2 P(x)}$$

Expected Value: $E(x) = \mu = \Sigma x P(x)$

Binomial Probability of x successes in n trials:

$$P(x) = {}_nC_x p^x q^{n-x} = \frac{n!}{(n-x)!x!}p^x q^{n-x}$$

Population Parameters of a Binomial Distribution:

Mean: $\mu = np$ Variance: $\sigma^2 = npq$

Standard Deviation: $\sigma = \sqrt{npq}$

Geometric Distribution: The probability that the first success will occur on trial number x is $P(x) = p(q)^{x-1}$, where $q = 1 - p$.

Poisson Distribution: The probability of exactly x occurrences in an interval is $P(x) = \dfrac{\mu^x e^{-\mu}}{x!}$, where $e \approx 2.71828$ and μ is the mean number of occurences per interval unit.

CHAPTER 5

Standard Score, or z-Score:

$$z = \frac{\text{Value} - \text{Mean}}{\text{Standard deviation}} = \frac{x - \mu}{\sigma}$$

Transforming a z-Score to an x-Value: $x = \mu + z\sigma$

Central Limit Theorem ($n \geq 30$ or population is normally distributed):

Mean of the Sampling Distribution: $\mu_{\bar{x}} = \mu$

Variance of the Sampling Distribution: $\sigma^2_{\bar{x}} = \dfrac{\sigma}{n}$

Standard Deviation of the Sampling Distribution (Standard Error): $\sigma_{\bar{x}} = \dfrac{\sigma}{\sqrt{n}}$

$$z\text{-Score} = \frac{\text{Value} - \text{Mean}}{\text{Standard Error}} = \frac{\bar{x} - \mu_{\bar{x}}}{\sigma_{\bar{x}}} = \frac{\bar{x} - \mu}{\sigma/\sqrt{n}}$$

CHAPTER 6

c-Confidence Interval for μ: $\bar{x} - E < \mu < \bar{x} + E$,

where $E = z_c \dfrac{\sigma}{\sqrt{n}}$ if σ is known and the population is normally distributed or $n \geq 30$, or $E = t_c \dfrac{s}{\sqrt{n}}$ if the population is normally or approximately normally distributed, σ is unknown, and $n < 30$

Minimum Sample Size to Estimate μ: $n = \left(\dfrac{z_c \sigma}{E}\right)^2$

Point Estimate for p, the population proportion of successes: $\hat{p} = \dfrac{x}{n}$

c-Confidence Interval for Population Proportion p (when $np \geq 5$ and $nq \geq 5$): $\hat{p} - E < p < \hat{p} + E$, where

$$E = z_c \sqrt{\frac{\hat{p}\hat{q}}{n}}$$

Minimum Sample Size to Estimate p: $n = \hat{p}\hat{q}\left(\dfrac{z_c}{E}\right)^2$

c-Confidence Interval for Population Variance σ^2:

$$\frac{(n-1)s^2}{\chi^2_R} < \sigma^2 < \frac{(n-1)s^2}{\chi^2_L}$$

c-Confidence Interval for Population Standard Deviation σ:

$$\sqrt{\frac{(n-1)s^2}{\chi^2_R}} < \sigma < \sqrt{\frac{(n-1)s^2}{\chi^2_L}}$$

Key Formulas

From Larson/*Farber Elementary Statistics: Picturing the World,* Fifth Edition
© 2012 Prentice Hall

CHAPTER 7

z-Test for a Mean μ: $z = \dfrac{\bar{x} - \mu}{\sigma/\sqrt{n}}$, for σ known with a

normal population, or for $n \geq 30$

t-Test for a Mean μ: $t = \dfrac{\bar{x} - \mu}{s/\sqrt{n}}$, for σ unknown,

population is normal or nearly normal, and $n < 30$.
(d.f. $= n - 1$)

z-Test for a Proportion p (when $np \geq 5$ and $nq \geq 5$):

$$z = \frac{\hat{p} - \mu_{\hat{p}}}{\sigma_{\hat{p}}} = \frac{\hat{p} - p}{\sqrt{pq/n}}$$

Chi-Square Test for a Variance σ^2 or Standard Deviation σ:

$$\chi^2 = \frac{(n-1)s^2}{\sigma^2} \quad \text{(d.f. } = n - 1)$$

CHAPTER 8

Two-Sample z-Test for the Difference Between Means (Independent samples; n_1 and $n_2 \geq 30$ or normally distributed populations):

$$z = \frac{(\bar{x}_1 - \bar{x}_2) - (\mu_1 - \mu_2)}{\sigma_{\bar{x}_1 - \bar{x}_2}},$$

where $\sigma_{\bar{x}_1 - \bar{x}_2} = \sqrt{\dfrac{\sigma_1^2}{n_1} + \dfrac{\sigma_2^2}{n_2}}$

Two-Sample t-Test for the Difference Between Means (Independent samples from normally distributed populations, n_1 or $n_2 < 30$):

$$t = \frac{(\bar{x}_1 - \bar{x}_2) - (\mu_1 - \mu_2)}{\sigma_{\bar{x}_1 - \bar{x}_2}}$$

If population variances are equal, d.f. $= n_1 + n_2 - 2$ and

$$\sigma_{\bar{x}_1 - \bar{x}_2} = \sqrt{\frac{(n_1 - 1)s_1^2 + (n_2 - 1)s_2^2}{n_1 + n_2 - 2}} \cdot \sqrt{\frac{1}{n_1} + \frac{1}{n_2}}.$$

If population variances are not equal, d.f. is the

smaller of $n_1 - 1$ or $n_2 - 1$ and $\sigma_{\bar{x}_1 - \bar{x}_2} = \sqrt{\dfrac{s_1^2}{n_1} + \dfrac{s_2^2}{n_2}}$.

t-Test for the Difference Between Means (Dependent samples):

$$t = \frac{\bar{d} - \mu_d}{s_d/\sqrt{n}}, \text{ where } \bar{d} = \frac{\Sigma d}{n}, \ s_d = \sqrt{\frac{\Sigma(d - \bar{d})^2}{n - 1}}$$

and d.f. $= n - 1$

Two-Sample z-Test for the Difference Between Proportions ($n_1\bar{p}$, $n_1\bar{q}$, $n_2\bar{p}$, and $n_2\bar{q}$ must be at least 5):

$$z = \frac{(\hat{p}_1 - \hat{p}_2) - (p_1 - p_2)}{\sqrt{\bar{p}\bar{q}\left(\dfrac{1}{n_1} + \dfrac{1}{n_2}\right)}}, \text{ where } \bar{p} = \frac{x_1 + x_2}{n_1 + n_2}$$

and $\bar{q} = 1 - \bar{p}$.

CHAPTER 9

Correlation Coefficient:

$$r = \frac{n\Sigma xy - (\Sigma x)(\Sigma y)}{\sqrt{n\Sigma x^2 - (\Sigma x)^2}\sqrt{n\Sigma y^2 - (\Sigma y)^2}}$$

t-Test for the Correlation Coefficient:

$$t = \frac{r}{\sqrt{\dfrac{1 - r^2}{n - 2}}} \quad \text{(d.f. } = n - 2)$$

Equation of a Regression Line: $\hat{y} = mx + b$,

where $m = \dfrac{n\Sigma xy - (\Sigma x)(\Sigma y)}{n\Sigma x^2 - (\Sigma x)^2}$ and

$$b = \bar{y} - m\bar{x} = \frac{\Sigma y}{n} - m\frac{\Sigma x}{n}$$

Coefficient of Determination:

$$r^2 = \frac{\text{Explained variation}}{\text{Total variation}} = \frac{\Sigma(\hat{y}_i - \bar{y})^2}{\Sigma(y_i - \bar{y})^2}$$

Standard Error of Estimate: $s_e = \sqrt{\dfrac{\Sigma(y_i - \hat{y}_i)^2}{n - 2}}$

c-Prediction Interval for y: $\hat{y} - E < y < \hat{y} + E$, where

$$E = t_c s_e \sqrt{1 + \frac{1}{n} + \frac{n(x_0 - \bar{x})^2}{n\Sigma x^2 - (\Sigma x)^2}} \quad \text{(d.f. } = n - 2)$$

CHAPTER 10

Chi-Square: $\chi^2 = \Sigma \dfrac{(O - E)^2}{E}$

Goodness-of-Fit Test: d.f. $= k - 1$

Test of Independence:

d.f. $=$ (no. of rows $- 1$)(no. of columns $- 1$)

Two-Sample F-Test for Variances: $F = \dfrac{s_1^2}{s_2^2}$, where

$s_1^2 \geq s_2^2$, d.f.$_N = n_1 - 1$, and d.f.$_D = n_2 - 1$

One-Way Analysis of Variance Test:

$F = \dfrac{MS_B}{MS_W}$, where $MS_B = \dfrac{SS_B}{k - 1} = \dfrac{\Sigma n_i\left(\overline{x}_i - \overline{\overline{x}}\right)^2}{k - 1}$

and $MS_W = \dfrac{SS_W}{N - k} = \dfrac{\Sigma (n_i - 1)s_i^2}{N - k}$

(d.f.$_N = k - 1$, d.f.$_D = N - k$)

CHAPTER 11

Test Statistic for Sign Test:

When $n \leq 25$, the test statistic is the smaller number of $+$ or $-$ signs.

When $n > 25$, $z = \dfrac{(x + 0.5) - 0.5n}{\dfrac{\sqrt{n}}{2}}$, where x is the

smaller number of $+$ or $-$ signs and n is the total number of $+$ and $-$ signs.

Test Statistic for Wilcoxon Rank Sum Test:

$z = \dfrac{R - \mu_R}{\sigma_R}$, where $R =$ sum of the ranks for the

smaller sample, $\mu_R = \dfrac{n_1(n_1 + n_2 + 1)}{2}$,

$\sigma_R = \sqrt{\dfrac{n_1 n_2(n_1 + n_2 + 1)}{12}}$, and $n_1 \leq n_2$

Test Statistic for the Kruskal-Wallis Test:

Given three or more independent samples, the test statistic for the Kruskal-Wallis test is

$H = \dfrac{12}{N(N + 1)}\left(\dfrac{R_1^2}{n_1} + \dfrac{R_2^2}{n_2} + \cdots + \dfrac{R_k^2}{n_k}\right)$

$- 3(N + 1).$ (d.f. $= k - 1$)

Spearman Rank Correlation Coefficient:

$r_s = 1 - \dfrac{6\Sigma d^2}{n(n^2 - 1)}$

Test Statistic for the Runs Test:

When $n_1 \leq 20$ and $n_2 \leq 20$, the test statistic is G, the number of runs.

When $n_1 > 20$ or $n_2 > 20$, the test statistic is

$z = \dfrac{G - \mu_G}{\sigma_G}$, where $G =$ number of runs,

$\mu_G = \dfrac{2n_1 n_2}{n_1 + n_2} + 1$, and

$\sigma_G = \sqrt{\dfrac{2n_1 n_2(2n_1 n_2 - n_1 - n_2)}{(n_1 + n_2)^2(n_1 + n_2 - 1)}}.$

Table 4 — Standard Normal Distribution

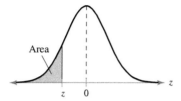

z	.09	.08	.07	.06	.05	.04	.03	.02	.01	.00
− 3.4	.0002	.0003	.0003	.0003	.0003	.0003	.0003	.0003	.0003	.0003
− 3.3	.0003	.0004	.0004	.0004	.0004	.0004	.0004	.0005	.0005	.0005
− 3.2	.0005	.0005	.0005	.0006	.0006	.0006	.0006	.0006	.0007	.0007
− 3.1	.0007	.0007	.0008	.0008	.0008	.0008	.0009	.0009	.0009	.0010
− 3.0	.0010	.0010	.0011	.0011	.0011	.0012	.0012	.0013	.0013	.0013
− 2.9	.0014	.0014	.0015	.0015	.0016	.0016	.0017	.0018	.0018	.0019
− 2.8	.0019	.0020	.0021	.0021	.0022	.0023	.0023	.0024	.0025	.0026
− 2.7	.0026	.0027	.0028	.0029	.0030	.0031	.0032	.0033	.0034	.0035
− 2.6	.0036	.0037	.0038	.0039	.0040	.0041	.0043	.0044	.0045	.0047
− 2.5	.0048	.0049	.0051	.0052	.0054	.0055	.0057	.0059	.0060	.0062
− 2.4	.0064	.0066	.0068	.0069	.0071	.0073	.0075	.0078	.0080	.0082
− 2.3	.0084	.0087	.0089	.0091	.0094	.0096	.0099	.0102	.0104	.0107
− 2.2	.0110	.0113	.0116	.0119	.0122	.0125	.0129	.0132	.0136	.0139
− 2.1	.0143	.0146	.0150	.0154	.0158	.0162	.0166	.0170	.0174	.0179
− 2.0	.0183	.0188	.0192	.0197	.0202	.0207	.0212	.0217	.0222	.0228
− 1.9	.0233	.0239	.0244	.0250	.0256	.0262	.0268	.0274	.0281	.0287
− 1.8	.0294	.0301	.0307	.0314	.0322	.0329	.0336	.0344	.0351	.0359
− 1.7	.0367	.0375	.0384	.0392	.0401	.0409	.0418	.0427	.0436	.0446
− 1.6	.0455	.0465	.0475	.0485	.0495	.0505	.0516	.0526	.0537	.0548
− 1.5	.0559	.0571	.0582	.0594	.0606	.0618	.0630	.0643	.0655	.0668
− 1.4	.0681	.0694	.0708	.0721	.0735	.0749	.0764	.0778	.0793	.0808
− 1.3	.0823	.0838	.0853	.0869	.0885	.0901	.0918	.0934	.0951	.0968
− 1.2	.0985	.1003	.1020	.1038	.1056	.1075	.1093	.1112	.1131	.1151
− 1.1	.1170	.1190	.1210	.1230	.1251	.1271	.1292	.1314	.1335	.1357
− 1.0	.1379	.1401	.1423	.1446	.1469	.1492	.1515	.1539	.1562	.1587
− 0.9	.1611	.1635	.1660	.1685	.1711	.1736	.1762	.1788	.1814	.1841
− 0.8	.1867	.1894	.1922	.1949	.1977	.2005	.2033	.2061	.2090	.2119
− 0.7	.2148	.2177	.2206	.2236	.2266	.2296	.2327	.2358	.2389	.2420
− 0.6	.2451	.2483	.2514	.2546	.2578	.2611	.2643	.2676	.2709	.2743
− 0.5	.2776	.2810	.2843	.2877	.2912	.2946	.2981	.3015	.3050	.3085
− 0.4	.3121	.3156	.3192	.3228	.3264	.3300	.3336	.3372	.3409	.3446
− 0.3	.3483	.3520	.3557	.3594	.3632	.3669	.3707	.3745	.3783	.3821
− 0.2	.3859	.3897	.3936	.3974	.4013	.4052	.4090	.4129	.4168	.4207
− 0.1	.4247	.4286	.4325	.4364	.4404	.4443	.4483	.4522	.4562	.4602
− 0.0	.4641	.4681	.4721	.4761	.4801	.4840	.4880	.4920	.4960	.5000

Critical Values

Level of Confidence c	z_c
0.80	1.28
0.90	1.645
0.95	1.96
0.99	2.575

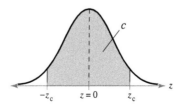

Table 4 — Standard Normal Distribution *(continued)*

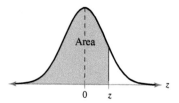

z	.00	.01	.02	.03	.04	.05	.06	.07	.08	.09
0.0	.5000	.5040	.5080	.5120	.5160	.5199	.5239	.5279	.5319	.5359
0.1	.5398	.5438	.5478	.5517	.5557	.5596	.5636	.5675	.5714	.5753
0.2	.5793	.5832	.5871	.5910	.5948	.5987	.6026	.6064	.6103	.6141
0.3	.6179	.6217	.6255	.6293	.6331	.6368	.6406	.6443	.6480	.6517
0.4	.6554	.6591	.6628	.6664	.6700	.6736	.6772	.6808	.6844	.6879
0.5	.6915	.6950	.6985	.7019	.7054	.7088	.7123	.7157	.7190	.7224
0.6	.7257	.7291	.7324	.7357	.7389	.7422	.7454	.7486	.7517	.7549
0.7	.7580	.7611	.7642	.7673	.7704	.7734	.7764	.7794	.7823	.7852
0.8	.7881	.7910	.7939	.7967	.7995	.8023	.8051	.8078	.8106	.8133
0.9	.8159	.8186	.8212	.8238	.8264	.8289	.8315	.8340	.8365	.8389
1.0	.8413	.8438	.8461	.8485	.8508	.8531	.8554	.8577	.8599	.8621
1.1	.8643	.8665	.8686	.8708	.8729	.8749	.8770	.8790	.8810	.8830
1.2	.8849	.8869	.8888	.8907	.8925	.8944	.8962	.8980	.8997	.9015
1.3	.9032	.9049	.9066	.9082	.9099	.9115	.9131	.9147	.9162	.9177
1.4	.9192	.9207	.9222	.9236	.9251	.9265	.9279	.9292	.9306	.9319
1.5	.9332	.9345	.9357	.9370	.9382	.9394	.9406	.9418	.9429	.9441
1.6	.9452	.9463	.9474	.9484	.9495	.9505	.9515	.9525	.9535	.9545
1.7	.9554	.9564	.9573	.9582	.9591	.9599	.9608	.9616	.9625	.9633
1.8	.9641	.9649	.9656	.9664	.9671	.9678	.9686	.9693	.9699	.9706
1.9	.9713	.9719	.9726	.9732	.9738	.9744	.9750	.9756	.9761	.9767
2.0	.9772	.9778	.9783	.9788	.9793	.9798	.9803	.9808	.9812	.9817
2.1	.9821	.9826	.9830	.9834	.9838	.9842	.9846	.9850	.9854	.9857
2.2	.9861	.9864	.9868	.9871	.9875	.9878	.9881	.9884	.9887	.9890
2.3	.9893	.9896	.9898	.9901	.9904	.9906	.9909	.9911	.9913	.9916
2.4	.9918	.9920	.9922	.9925	.9927	.9929	.9931	.9932	.9934	.9936
2.5	.9938	.9940	.9941	.9943	.9945	.9946	.9948	.9949	.9951	.9952
2.6	.9953	.9955	.9956	.9957	.9959	.9960	.9961	.9962	.9963	.9964
2.7	.9965	.9966	.9967	.9968	.9969	.9970	.9971	.9972	.9973	.9974
2.8	.9974	.9975	.9976	.9977	.9977	.9978	.9979	.9979	.9980	.9981
2.9	.9981	.9982	.9982	.9983	.9984	.9984	.9985	.9985	.9986	.9986
3.0	.9987	.9987	.9987	.9988	.9988	.9989	.9989	.9989	.9990	.9990
3.1	.9990	.9991	.9991	.9991	.9992	.9992	.9992	.9992	.9993	.9993
3.2	.9993	.9993	.9994	.9994	.9994	.9994	.9994	.9995	.9995	.9995
3.3	.9995	.9995	.9995	.9996	.9996	.9996	.9996	.9996	.9996	.9997
3.4	.9997	.9997	.9997	.9997	.9997	.9997	.9997	.9997	.9997	.9998

Table 5— *t*-Distribution

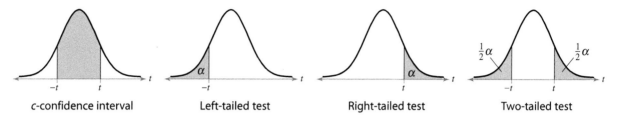

c-confidence interval	Left-tailed test	Right-tailed test	Two-tailed test			

d.f.	Level of confidence, *c*	0.50	0.80	0.90	0.95	0.98	0.99
	One tail, α	0.25	0.10	0.05	0.025	0.01	0.005
	Two tails, α	0.50	0.20	0.10	0.05	0.02	0.01
1		1.000	3.078	6.314	12.706	31.821	63.657
2		.816	1.886	2.920	4.303	6.965	9.925
3		.765	1.638	2.353	3.182	4.541	5.841
4		.741	1.533	2.132	2.776	3.747	4.604
5		.727	1.476	2.015	2.571	3.365	4.032
6		.718	1.440	1.943	2.447	3.143	3.707
7		.711	1.415	1.895	2.365	2.998	3.499
8		.706	1.397	1.860	2.306	2.896	3.355
9		.703	1.383	1.833	2.262	2.821	3.250
10		.700	1.372	1.812	2.228	2.764	3.169
11		.697	1.363	1.796	2.201	2.718	3.106
12		.695	1.356	1.782	2.179	2.681	3.055
13		.694	1.350	1.771	2.160	2.650	3.012
14		.692	1.345	1.761	2.145	2.624	2.977
15		.691	1.341	1.753	2.131	2.602	2.947
16		.690	1.337	1.746	2.120	2.583	2.921
17		.689	1.333	1.740	2.110	2.567	2.898
18		.688	1.330	1.734	2.101	2.552	2.878
19		.688	1.328	1.729	2.093	2.539	2.861
20		.687	1.325	1.725	2.086	2.528	2.845
21		.686	1.323	1.721	2.080	2.518	2.831
22		.686	1.321	1.717	2.074	2.508	2.819
23		.685	1.319	1.714	2.069	2.500	2.807
24		.685	1.318	1.711	2.064	2.492	2.797
25		.684	1.316	1.708	2.060	2.485	2.787
26		.684	1.315	1.706	2.056	2.479	2.779
27		.684	1.314	1.703	2.052	2.473	2.771
28		.683	1.313	1.701	2.048	2.467	2.763
29		.683	1.311	1.699	2.045	2.462	2.756
∞		.674	1.282	1.645	1.960	2.326	2.576

Table 6— Chi-Square Distribution

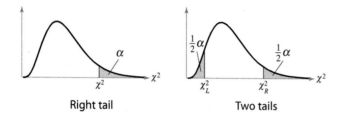

Right tail   Two tails

Degrees of freedom	α									
	0.995	**0.99**	**0.975**	**0.95**	**0.90**	**0.10**	**0.05**	**0.025**	**0.01**	**0.005**
1	—	—	0.001	0.004	0.016	2.706	3.841	5.024	6.635	7.879
2	0.010	0.020	0.051	0.103	0.211	4.605	5.991	7.378	9.210	10.597
3	0.072	0.115	0.216	0.352	0.584	6.251	7.815	9.348	11.345	12.838
4	0.207	0.297	0.484	0.711	1.064	7.779	9.488	11.143	13.277	14.860
5	0.412	0.554	0.831	1.145	1.610	9.236	11.071	12.833	15.086	16.750
6	0.676	0.872	1.237	1.635	2.204	10.645	12.592	14.449	16.812	18.548
7	0.989	1.239	1.690	2.167	2.833	12.017	14.067	16.013	18.475	20.278
8	1.344	1.646	2.180	2.733	3.490	13.362	15.507	17.535	20.090	21.955
9	1.735	2.088	2.700	3.325	4.168	14.684	16.919	19.023	21.666	23.589
10	2.156	2.558	3.247	3.940	4.865	15.987	18.307	20.483	23.209	25.188
11	2.603	3.053	3.816	4.575	5.578	17.275	19.675	21.920	24.725	26.757
12	3.074	3.571	4.404	5.226	6.304	18.549	21.026	23.337	26.217	28.299
13	3.565	4.107	5.009	5.892	7.042	19.812	22.362	24.736	27.688	29.819
14	4.075	4.660	5.629	6.571	7.790	21.064	23.685	26.119	29.141	31.319
15	4.601	5.229	6.262	7.261	8.547	22.307	24.996	27.488	30.578	32.801
16	5.142	5.812	6.908	7.962	9.312	23.542	26.296	28.845	32.000	34.267
17	5.697	6.408	7.564	8.672	10.085	24.769	27.587	30.191	33.409	35.718
18	6.265	7.015	8.231	9.390	10.865	25.989	28.869	31.526	34.805	37.156
19	6.844	7.633	8.907	10.117	11.651	27.204	30.144	32.852	36.191	38.582
20	7.434	8.260	9.591	10.851	12.443	28.412	31.410	34.170	37.566	39.997
21	8.034	8.897	10.283	11.591	13.240	29.615	32.671	35.479	38.932	41.401
22	8.643	9.542	10.982	12.338	14.042	30.813	33.924	36.781	40.289	42.796
23	9.260	10.196	11.689	13.091	14.848	32.007	35.172	38.076	41.638	44.181
24	9.886	10.856	12.401	13.848	15.659	33.196	36.415	39.364	42.980	45.559
25	10.520	11.524	13.120	14.611	16.473	34.382	37.652	40.646	44.314	46.928
26	11.160	12.198	13.844	15.379	17.292	35.563	38.885	41.923	45.642	48.290
27	11.808	12.879	14.573	16.151	18.114	36.741	40.113	43.194	46.963	49.645
28	12.461	13.565	15.308	16.928	18.939	37.916	41.337	44.461	48.278	50.993
29	13.121	14.257	16.047	17.708	19.768	39.087	42.557	45.722	49.588	52.336
30	13.787	14.954	16.791	18.493	20.599	40.256	43.773	46.979	50.892	53.672
40	20.707	22.164	24.433	26.509	29.051	51.805	55.758	59.342	63.691	66.766
50	27.991	29.707	32.357	34.764	37.689	63.167	67.505	71.420	76.154	79.490
60	35.534	37.485	40.482	43.188	46.459	74.397	79.082	83.298	88.379	91.952
70	43.275	45.442	48.758	51.739	55.329	85.527	90.531	95.023	100.425	104.215
80	51.172	53.540	57.153	60.391	64.278	96.578	101.879	106.629	112.329	116.321
90	59.196	61.754	65.647	69.126	73.291	107.565	113.145	118.136	124.116	128.299
100	67.328	70.065	74.222	77.929	82.358	118.498	124.342	129.561	135.807	140.169

Applet Correlation

Applet	Concept Illustrated	Descriptor	Applet Activity
Random numbers	This applet simulates selecting a random sample from a population by first assigning a unique integer to each experimental unit and then using the random numbers generated to determine the experimental units that will be included in the sample.	This applet generates random numbers from a range of integers specified by the user.	1.3
Mean versus median	The mean and the median of a data set respond differently to changes in the data. This applet investigates how skewedness and outliers affect measures of central tendency.	This applet allows the user to visualize the relationship between the mean and median of a data set. The user may easily add and delete data points. The applet automatically updates the mean and median for each change in the data.	2.3
Standard deviation	Standard deviation measures the spread of a data set. Use this applet to investigate how the shape and spread of a distribution affect the standard deviation.	This applet allows the user to visualize the relationship between the mean and standard deviation of a data set. The user may easily add and delete data points. The applet automatically updates the mean and standard deviation for each change in the data.	2.4
Simulating the stock market	Theoretical probabilities are long run experimental probabilities.	This applet simulates fluctuation in the stock market, where on any given day going up is equally likely as going down. The user specifies the number of days and the applet reports whether the stock market goes up or down each day and creates a bar graph for the outcomes. It also calculates and plots the proportion of days that the stock market goes up during the simulation.	3.1
Simulating the probability of rolling a 3 or 4	Theoretical probabilities are long run experimental probabilities. Use this applet to investigate the relationship between the theoretical and experimental probabilities of rolling a 3 or a 4 as the number of times the die is rolled increases.	This applet simulates rolling a fair die. The user specifies the number of rolls and the applet reports the outcome of each roll and creates a frequency histogram for the outcomes. It also calculates and plots the proportion of 3s and 4s rolled during the simulation.	3.3
Binomial distribution	As the number of samples increases, the estimated probability gets closer to the true value.	This applet simulates values from a binomial distribution. The user specifies the parameters for the binomial distribution (n and p) and the number of values to be simulated (N). The applet plots N values from the specified binomial distribution in a bar graph and reports the frequency of each outcome.	4.2
Sampling distributions	The mean and standard deviation of the distribution of sample means are unbiased estimators of the mean and standard deviation of the population distribution. This applet compares the means and standard deviations of the distributions and assesses the effect of sample size.	This applet simulates repeatedly choosing samples of a fixed size n from a population. The user specifies the size of the sample, the number of samples to be chosen, and the shape of the population distribution. The applet reports the means, medians, and standard deviations of both the sample means and the sample medians and creates plots for both.	5.4

(continued on next page)

Applet	Concept Illustrated	Descriptor	Applet Activity
Confidence intervals for a mean (the impact of not knowing the standard deviation)	Confidence intervals obtained using the sample standard deviation are different from those obtained using the population standard deviation. This applet investigates the effect of not knowing the population standard deviation.	This applet generates confidence intervals for a population mean. The user specifies the sample size, the shape of the distribution, the population mean, and the population standard deviation. The applet simulates selecting 100 random samples from the population and finds the 95% z-interval and 95% t-interval for each sample. The confidence intervals are plotted and the number and proportion containing the true mean are reported.	6.2
Confidence intervals for a proportion	Not all confidence intervals contain the population mean. This applet investigates the meaning of 95% and 99% confidence.	This applet generates confidence intervals for a population proportion. The user specifies the population proportion and the sample size. The applet simulates selecting 100 random samples from the population and finds the 95% and 99% confidence intervals for each sample. The confidence intervals are plotted and the number and proportion containing the true proportion are reported.	6.3
Hypothesis tests for a mean	Not all tests of hypotheses lead correctly to either rejecting or failing to reject the null hypothesis. This applet investigates the relationship between the level of confidence and the probabilities of making Type I and Type II errors.	This applet performs hypotheses tests for a population mean. The user specifies the shape of the population distribution, the population mean and standard deviation, the sample size, and the null and alternative hypotheses. The applet simulates selecting 100 random samples from the population and calculates and plots the t statistic and P-value for each sample. The applet reports the number and proportion of times the null hypothesis is rejected at both the 0.05 level and the 0.01 level.	7.2
Hypothesis tests for a proportion	Not all tests of hypotheses lead correctly to either rejecting or failing to reject the null hypothesis. This applet investigates the relationship between the level of confidence and the probabilities of making Type I and Type II errors.	This applet performs hypotheses tests for a population proportion. The user specifies the population proportion, the sample size, and the null and alternative hypotheses. The applet simulates selecting 100 random samples from the population and calculates and plots the z statistic and P-value for each sample. The applet reports the number and proportion of times the null hypothesis is rejected at both the 0.05 level and the 0.01 level.	7.4
Correlation by eye	The correlation coefficient measures the strength of a linear relationship between two variables. This applet teaches the user how to assess the strength of a linear relationship from a scatter plot.	This applet computes the correlation coefficient r for a set of bivariate data plotted on a scatter plot. The user can easily add or delete points and guess the value of r. The applet then compares the guess to its calculated value.	9.1
Regression by eye	The least squares regression line has a smaller SSE than any other line that might approximate a set of bivariate data. This applet teaches the user how to approximate the location of a regression line on a scatter plot.	This applet computes the least squares regression line for a set of bivariate data plotted on a scatter plot. The user can easily add or delete points and guess the location of the regression line by manipulating a line provided on the scatter plot. The applet will then plot the least squares line. It displays the equations and the SSEs for both lines.	9.2